ENVIRONMENTAL ENGINEERING AND SUSTAINABLE DESIGN

SECOND EDITION

Bradley A. Striebig

Professor of Engineering

James Madison University

Maria Papadakis

Professor of Integrated Science and Technology

James Madison University

Lauren G. Heine

Director of Science & Data Integrity

Chem*FORWARD*

Adebayo A. Ogundipe

Associate Professor of Engineering

James Madison University

Australia • Brazil • Canada • Mexico • Singapore • United Kingdom • United States

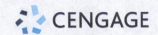

Environmental Engineering and Sustainable Design, **Second Edition**
Bradley A. Striebig, Maria Papadakis, Lauren G. Heine, and Adebayo A. Ogundipe

SVP, Higher Education Product Management: Erin Joyner

VP, Product Management, Learning Experiences: Thais Alencar

Product Director: Mark Santee

Senior Product Manager: Timothy L. Anderson

Product Assistant: Simeon Lloyd-Wingard

Learning Designer: MariCarmen Constable

Content Manager: Alexander Sham

Digital Delivery Quality Partner: Nikkita Kendrick

Director, Product Marketing: Jennifer Fink

Product Marketing Manager: Taylor Shenberger

IP Analyst: Deanna Ettinger

IP Project Manager: Nick Barrows

Production Service: RPK Editorial Services, Inc.

Compositor: MPS Limited

Manufacturing Planner: Ron Montgomery

Designer: Nadine Ballard

Cover Image: Bradley A. Striebig

For product information and technology assistance, contact us at **Cengage Customer & Sales Support, 1-800-354-9706** or **support.cengage.com**.

For permission to use material from this text or product, submit all requests online at **www.copyright.com**.

Library of Congress Control Number: 2021946436

Student Edition:
ISBN: 978-0-357-67585-4

Loose-leaf Edition:
ISBN: 978-0-357-67779-7

Cengage
200 Pier 4 Boulevard
Boston, MA 02210
USA

Cengage is a leading provider of customized learning solutions with employees residing in nearly 40 different countries and sales in more than 125 countries around the world. Find your local representative at **www.cengage.com**.

To learn more about Cengage platforms and services, register or access your online learning solution, or purchase materials for your course, visit **www.cengage.com**.

Printed in Mexico
Print Number: 01   Print Year: 2022

For Echo and Zachary

—Bradley A. Striebig

In memory of Pete Papadakis

—Maria Papadakis

In memory of P. Aarne Vesilind

—Lauren G. Heine

To all my Teachers

—Adebayo A. Ogundipe

Contents

Part 2 ENGINEERING ENVIRONMENTAL AND SUSTAINABLE PROCESSES

Part 3   DESIGNING RESILIENT AND SUSTAINABLE SYSTEMS

CHAPTER 12   Designing for Sustainability   818

Preface

Environmental Engineering and Sustainable Design, Second Edition is an invaluable resource for today's engineering and applied environmental science students. As engineering curriculum becomes more crowded, challenges arise in addressing the new paradigm of engineering in a resource-limited environment and adapting design to a new climactic condition. The authors have developed a comprehensive text that provides foundational knowledge and traditional engineering skills while also integrating our present understanding of resource consumption and climate issues into this new edition. This curriculum is focused upon applying engineering principles to real-world design and problem analysis. It includes specific step-by-step examples and case studies for solving complex conceptual and design problems related to sustainable design and engineering. This textbook also applies the principles of sustainable design to issues in both developed and developing countries. Instructors will benefit from having this updated best seller to bring sustainability science, environmental impact analysis, and models of sustainability to the undergraduate and graduate level.

Sustainability is important in manufacturing, construction, planning, and design. Allenby et al. state that: "Sustainable engineering is a conceptual and practical challenge to all engineering disciplines." The teaching of sustainability has sometimes been pigeonholed into graduate level courses in Industrial Ecology or Green Engineering. Environmental engineering and chemical engineering textbooks may cover some basic concepts of sustainability, but the extent and breadth of knowledge is insufficient to meet the multifaceted demand required to engineer sustainable processes and products.

Dr. John Crittenden, 2002, suggests that sustainable solutions include the following important elements/steps: (a) translating and understanding societal needs into engineering solutions such as infrastructures, products, practices, and processes; (b) explaining to society the long-term consequences of these engineering solutions; and (c) educating the next generation of scientists and engineers to acquire both the depth and breadth of skills necessary to address the important physical and behavioral science elements of environmental problems and to develop and use integrative analysis methods to identify and design sustainable products and systems.

New to the Second Edition

The Second Edition has been expanded to appeal to traditional foundational environmental engineering courses.

The content has been organized into three key sections:

- Part I: Environmental and Sustainability Science Principles
- Part II: Engineering Environmental and Sustainable Processes
- Part III: Designing Resilient and Sustainable Systems

Significant content from this textbook is adapted from *Introduction to Environmental Engineering*, Third Edition by P. Aarne Vesilind, Susan M. Morgan, and Lauren G. Heine. This content expands the use of the textbook to traditionally taught environmental engineering courses. This text is also used in courses focused on sustainable design and engineering, and this update provides content that is suitable to teaching a course on climate adaptation and resilience, as illustrated in Table P.1.

Topics new or significantly expanded in this edition include:

- Chapter 4: Material Flow and Processes in Engineering
- Chapter 5: Natural Resources, Materials, and Sustainability
- Chapter 6: Hazardous Substances and Risk Assessment
- Chapter 8: Wastewater Treatment
- Chapter 11: Energy Conservation, Development, and Decarbonization
- Chapter 12: Designing for Sustainability
- Chapter 15: Assessing Alternatives

New homework problems have been added and integrated into this textbook. Each chapter includes both qualitative and quantitative problems that cover a range of difficulty and complexity. Additional Active Learning Exercises have been added, with a focus on peer-to-peer learning activities to stimulate discussion, including the incorporation of climate and energy simulations for group role playing activities.

Organization and Potential Syllabus Topics

Sustainability is most often covered in existing environmental engineering courses; however, these courses are typically limited to civil and environmental engineering majors. Introductory environmental engineering courses often have objectives focused more upon historical perspectives in remediation and large-scale treatment systems than upon forward-looking sustainability concepts. Students will benefit from having methods for quantifying sustainability through environmental impacts, case studies, Life Cycle Analysis (LCA) models, and best practices. Case studies and active learning exercises make the learning experience real-world and hands-on. This title is the first to bring sustainability science, environmental impact analysis, and models of sustainability to the undergraduate level. Prerequisites for such a course are the foundational courses in calculus, chemistry, and physics.

Environmental Engineering and Sustainable Design, Second Edition is clearly arranged in three parts. *Part I: Environmental and Sustainability Science Principles* includes foundational content in the physical and social sciences that describe sustainability. *Part II: Engineering Environmental and Sustainable Processes* describes processes that relate to understanding and creating more sustainable systems for water development, air quality, climate adaptation, and energy development. *Part III: Designing Resilient and Sustainable Systems* addresses new tools and models that can be used in the design of products and infrastructure to create systems adapted to living in a resource-limited world that requires more sustainable approaches to the lifestyle of the developed world's nations. Suggested topics for courses are shown in Table P.1.

TABLE P.1 Suggested topics for courses in environmental engineering, sustainable design and engineering, and climate adaptation and resilience. New or reorganized chapters are bolded

CHAPTER	ENVIRONMENTAL ENGINEERING	SUSTAINABLE DESIGN AND ENGINEERING	CLIMATE ADAPTATION AND RESILIENCE
PART I: ENVIRONMENTAL AND SUSTAINABILITY SCIENCE PRINCIPLES			
Ch. 1 Sustainability, Engineering, and Design	X	X	X
Ch. 2 Analyzing Sustainability Using Engineering Science	X		X
Ch. 3 Biogeochemical Cycles	X		X
Ch. 4 Material Flow and Processes in Engineering	X		
Ch. 5 Natural Resources, Materials, and Sustainability	X	X	X
Ch. 6 Hazardous Substances and Risk Assessment	X		
PART II: ENGINEERING ENVIRONMENTAL AND SUSTAINABLE PROCESSES			
Ch. 7 Water Quality Impacts	X		
Ch. 8 Wastewater Treatment	X		
Ch. 9 Impacts on Air Quality	X		
Ch. 10 The Carbon Cycle and Energy Balances	X	X	X
Ch. 11 Energy Conservation, Development, and Decarbonization		X	X
PART III: DESIGNING RESILIENT AND SUSTAINABLE SYSTEMS			
Ch. 12 Designing for Sustainability		X	X
Ch. 13 Industrial Ecology		X	
Ch. 14 Life Cycle Analysis		X	
Ch. 15 Assessing Alternatives		X	X
Ch. 16 Sustainability and the Built Environment		X	X
Ch. 17 Challenges and Opportunities for Sustainability in Practice		X	X

Supplements

Additional instructor resources for this product are available online. Instructor assets include a Solution Answer Guide, Image Library, and PowerPoint® slides. Sign up or sign in at www.cengage.com to search for and access this product and its online resources.

Acknowledgments

The authors are grateful to their colleagues and students for their contributions to the development of this textbook. Although there are too many contributors to name, a few deserve special mention. Professor P. Aarne Vesilind was an inspiration for much of the content of this textbook. Professor Vesilind's devotion to the ethical uses of engineering skills to improve the human condition were foundational for many of his students, colleagues, and peers, and his lifelong works and words are influential in this edition of *Environmental Engineering and Sustainable Design*.

Dr. Striebig would like to thank two mentors, Dr. Raymond Regan and Dr. Robert J. Heinsohn, for encouragement throughout his career and with the development of this curriculum and textbook. He would like to acknowledge the support of his parents Janet and Ronald for making education a priority. Dr. Striebig is also indebted to his children, Echo and Zachary, who are a constant inspiration for the hope and potential of future generations.

Dr. Papadakis owes her love of technology to her father, Pete Papadakis, a career-long experimental stress and design engineer. Tom Walls, Robert Zetzl, Gladys Good, and Donald Clodfelter had a bigger impact than they could possibly know, and innately understood the needs of K-12 girls in STEM decades before this issue came to national prominence. She would like to thank her husband Fred Copithorn for his steady presence and love of the Earth, and her mother Iris Joseph continues to be a profound role model.

Dr. Heine was a graduate student of P. Aarne Vesilind and is profoundly grateful for his mentorship and love of teaching. Lauren is grateful to her parents Arthur and Sherry Glass for their support and for showing her how environmental engineering is integral to our daily lives. She would like to thank her husband Carl, and Hani and Cooper for their love and support; as well as dear friend and chemist Laura Gray for helping to ensure that the problems and active learning exercises are engaging and challenging. She would also like to thank Chem*FORWARD* friend and co-founder, Stacy Glass, for her vision and for valuing educational endeavors.

Dr. Ogundipe will be forever indebted to Dr. Washington Braida; a terrific teacher and friend who inspired him and pointed him in the right direction. He would also like to thank Tola, Safiyyah, Haneef, and Sumayyah for being the reason.

Last but certainly not least, the authors wish to thank the Global Engineering team at Cengage Learning for their dedication to this new edition: Timothy Anderson, Senior Product Manager; MariCarmen Constable, Learning Designer; Alexander Sham, Content Manager; and Simeon Lloyd-Wingard, Product Assistant. Thanks are also due to Rose P. Kernan of RPK Editorial Services.

Bradley A. Striebig
Maria Papadakis
Lauren G. Heine
Adebayo A. Ogundipe

About the Authors

BRADLEY A. STRIEBIG

Professor of Engineering, James Madison University, Harrisonburg, Virginia

Professor Striebig earned his Ph.D. from Pennsylvania State University. He has served as editor on major journals in his subject area. He has led major, funded, award-winning research activities focused on working with developing communities and natural treatment systems. He has written over 100 publications, including several book chapters, numerous peer-reviewed journal articles, and many peer-reviewed conference presentations and proceedings.

MARIA PAPADAKIS

Professor of Integrated Science and Technology, James Madison University, Harrisonburg, Virginia

Professor Papadakis is a political economist with expertise in energy management and sustainable manufacturing. Her research has been published in specialized reports of the National Science Foundation and in journals such as *Evaluation and Program Planning, Journal of Technology Transfer, The Scientist*, and the *International Journal of Technology Management*.

LAUREN G. HEINE

Director of Science & Data Integrity, ChemFORWARD

Lauren earned her doctorate in Civil and Environmental Engineering from Duke University. Dr. Heine applies green chemistry, green engineering, alternatives assessment, and multi-stakeholder collaboration to develop tools that result in safer and more sustainable chemical products and processes. Her work with Chem*FORWARD* builds on prior experience developing *GreenScreen®* for Safer *Chemicals*, a pioneering method for chemical hazard assessment to enable informed substitution; and *CleanGredients™*, a web-based information platform for identifying greener chemicals for use in cleaning products; both tools are designed to scale access to information needed to develop materials and products that are safe and circular. She served on the California Green Ribbon Science Panel and co-chairs the Apple Green Chemistry Advisory Board, tasked with helping to integrate green chemistry into Apple's products and supply chain. She began her career as a Fellow with the American Association for the Advancement of Science (AAAS) in the Green Chemistry Program at the U.S. Environmental Protection Agency.

ADEBAYO A. OGUNDIPE

Associate Professor of Engineering, James Madison University, Harrisonburg, Virginia

Professor Ogundipe's academic background is in Chemical and Environmental Engineering. His current areas of specialization and scholarship include Life Cycle Analysis, Industrial Ecology, and developing methods for assessing sustainability. His ongoing cross disciplinary work involves international collaborations aimed at developing appropriate educational modules to help engineering students develop global cultural competencies.

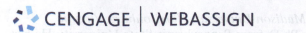

New Digital Solution for Your Engineering Classroom

WebAssign is a powerful digital solution designed by educators to enrich the engineering teaching and learning experience. With a robust computational engine at its core, *WebAssign* provides extensive content, instant assessment, and superior support.

WebAssign's powerful question editor allows engineering instructors to create their own questions or modify existing questions. Each question can use any combination of text, mathematical equations and formulas, sound, pictures, video, and interactive HTML elements. Numbers, words, phrases, graphics, and sound or video files can be randomized so that each student receives a different version of the same question.

In addition to common question types such as multiple choice, fill-in-the-blank, essay, and numerical, you can also incorporate robust answer entry palettes (mathPad, chemPad, calcPad, physPad, Graphing Tool) to input and grade symbolic expressions, equations, matrices, and chemical structures using powerful computer algebra systems.

WebAssign Offers Engineering Instructors the Following

- The ability to create and edit algorithmic and numerical exercises.
- The opportunity to generate randomized iterations of algorithmic and numerical exercises. When instructors assign numerical *WebAssign* homework exercises (engineering math exercises), the *WebAssign* program offers them the ability to generate and assign their students differing versions of the same engineering math exercise. The computational engine extends beyond and provides the luxury of solving for correct solutions/answers.
- The ability to create and customize numerical questions, allowing students to enter units, use a specific number of significant digits, use a specific number of decimal places, respond with a computed answer, or answer within a different tolerance value than the default.

Visit www.webassign.com/instructors/features/ to learn more. To create an account, instructors can go directly to the signup page at www.webassign.net/signup.html.

WebAssign Features for Students

- **Review Concepts at Point of Use**
  Within *WebAssign*, a "Read It" button at the bottom of each question links students to corresponding sections of the textbook, enabling access to the MindTap Reader at the precise moment of learning. A "Watch It" button allows a short video to play. These videos help students understand and review the problem they need to complete, enabling support at the precise moment of learning.

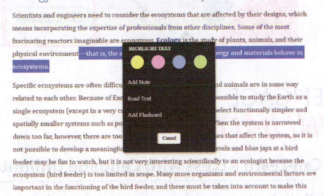

- **My Class Insights**

  WebAssign's built-in study feature shows performance across course topics so that students can quickly identify which concepts they have mastered and which areas they may need to spend more time on.

- **Ask Your Teacher**

  This powerful feature enables students to contact their instructor with questions about a specific assignment or problem they are working on.

MindTap Reader

Available via *WebAssign* and our digital subscription service, Cengage Unlimited, **MindTap Reader** is Cengage's next-generation eTextbook for engineering students.

The MindTap Reader provides more than just text learning for the student. It offers a variety of tools to help our future engineers learn chapter concepts in a way that resonates with their workflow and learning styles.

- **Personalize their experience**

  Within the MindTap Reader, students can highlight key concepts, add notes, and bookmark pages. These are collected in My Notes, ensuring they will have their own study guide when it comes time to study for exams.

- **Flexibility at their fingertips**

  With access to the book's internal glossary, students can personalize their study experience by creating and collating their own custom flashcards. The ReadSpeaker feature reads text aloud to students, so they can learn on the go—wherever they are.

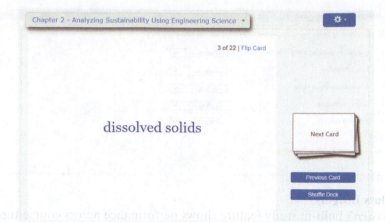

The Cengage Mobile App

Available on iOS and Android smartphones, the Cengage Mobile App provides convenience. Students can access their entire textbook anyplace and anytime. They can take notes, highlight important passages, and have their text read aloud whether they are online or off.

To learn more and download the mobile app, visit www.cengage.com/mobile-app/.

All-You-Can-Learn Access with Cengage Unlimited

Cengage Unlimited is the cost-saving student plan that includes access to our entire library of eTextbooks, online platforms and more—in one place, for one price. For just $119.99 for four months, a student gets online and offline access to Cengage course materials across disciplines, plus hundreds of student success and career readiness skill-building activities. To learn more, visit www.cengage.com/unlimited.

## PART 1

Environmental and Sustainability Science Principles

CONTENTS

## CHAPTER 1

Sustainability, Engineering, and Design

FIGURE 1.1 A high-resolution photo of our planet showing various ecosystems and weather patterns. Many believe the first images of Earth taken from space had a profound effect on how people in general perceived the interconnectedness between people, the planet, and future prosperity.

Source: NASA Goddard Space Flight Center Image by Reto Stöckli (land surface, shallow water, clouds). Enhancements by Robert Simmon (ocean color, compositing, 3D globes, animation). Data and technical support: MODIS Land Group; MODIS Science Data Support Team; MODIS Atmosphere Group; MODIS Ocean Group. Additional data: USGS EROS Data Center (topography); USGS Terrestrial Remote Sensing Flagstaff Field Center (Antarctica); Defense Meteorological Satellite Program (city lights).

It is known that there are an infinite number of worlds, simply because there is an infinite amount of space for them to be in. However, not every one of them is inhabited. Any finite number divided by infinity is as near nothing as makes no odds, so the average population of all the planets in the Universe can be said to be zero. From this it follows that the population of the whole Universe is also zero, and that any people you may meet from time to time are merely products of a deranged imagination.

—DOUGLAS ADAMS, FROM *THE RESTAURANT AT THE END OF THE UNIVERSE* (1980, p. 142)

GOALS

THE EDUCATIONAL GOALS OF THIS CHAPTER are to define sustainability and understand how social norms influence discussions about sustainability. We also examine how population changes and resource consumption have created the need for engineers, economists, scientists, and policymakers to consider sustainability in the design of products, infrastructure, and systems. The key concepts that are used to quantitatively consider sustainable design include the human development index, population growth models, and the ecological footprints analysis. This chapter also provides a greater context for the social and economic factors that shape successful design. In this chapter, we explore the ethical basis of human-centered design as a way of meeting the essential needs of people, which is an explicit element of sustainable development. In addition, we explain the dynamics of the adoption and diffusion of innovations, which is a critical prerequisite to the widespread social impact of more sustainable practices, products, and processes. Finally, we address the economic concepts that help us understand why achieving greater environmental sustainability can be a challenge and the role of governmental policymaking in surmounting those obstacles.

OBJECTIVES

At the conclusion of this chapter, you should be able to:

1.1 Calculate and relate the Human Development Index to indices for lifespan, education, and income.

1.2 Discuss ethical frameworks and engineering ethics in relation to sustainability.

1.3 Explain the different ethical principles that inform sustainable development, and discuss how these affect engineering design.

1.4 Give examples of successful and unsuccessful technologies appropriate for meeting the essential needs of people, and explain the reasons for their success or failure.

1.5 Define and discuss different definitions of sustainability, sustainable design, and sustainable development.

1.6 Evaluate global trends in population and describe how those trends challenge engineers to develop sustainable products, infrastructure, and systems.

1.7 Define and evaluate the carrying capacity of systems of various scales.

1.8 Define and discuss quantitatively the indicators of sustainable design, including the ecological footprint and the impact, population, affluence, and technology (IPAT) equation.

1.9 For a given innovation, summarize and analyze the social, cultural, technical, and economic factors that affect its potential impacts.

Introduction

Genetically modern humans appeared on Earth about 200,000 years ago, and biologically and behaviorally modern humans appeared about 70,000 years ago. The number of people and their effects on the planet were negligible for most of the history of the planet (Figure 1.2).

The number of humans on the planet remained very small until a few hundred years ago when advances in farming, energy, and mechanization took place, allowing the human population to increase exponentially (see Figure 1.3). Rapid changes in technology allowed humans to live longer; the decreasing death rates contributed to the high rate of human population growth over the past thousand years. Some time shortly after the year 1800, the world population reached 1 billion people for the first time (UN, 1999).

Demographers, people who study trends in population, say we are likely heading toward a world population of 9.5 to 12.5 billion over the next century (UN, 2019). While the human population on the planet is growing, natural resources that we have relied on for food, energy, and water are shrinking owing to the increasing human consumption of those resources. The human species has had a profound environmental impact on the planet, threatening the Earth's biodiversity, climate, energy resources, and water supply.

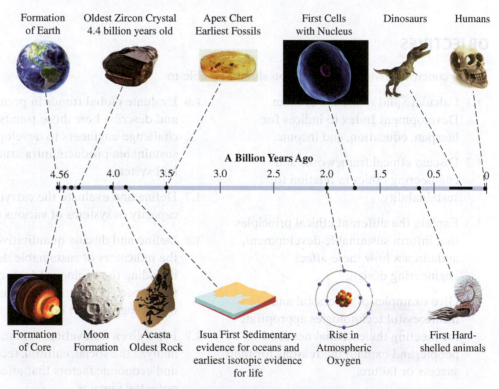

FIGURE 1.2 A timeline of planetary history showing the relatively short time humans have existed on Earth compared to the entirety of the history of Earth.

Source: Based on www.geology.wisc.edu/zircon/Earliest%20Piece/Images/28.jpg; leonello calvetti/Shutterstock.com; Johan Swanepoel/Shutterstock.com; Ortodox/Shutterstock.com; Imfoto/Shutterstock.com; falk/Shutterstock.com; oorka/Shutterstock.com; Sebastian Kaulitzki/Shutterstock.com; Number001/Shutterstock.com; DM7/Shutterstock.com; Empiric7/Shutterstock.com; SciePro/Shutterstock.com.

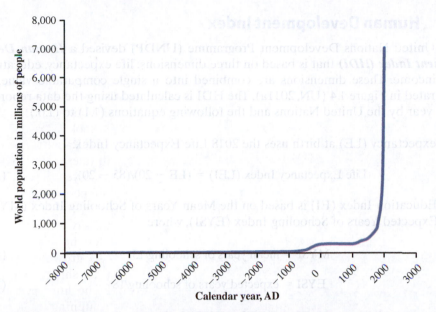

FIGURE 1.3 Historic estimates of human population from pre-history until 2011.

Source: Based on Kremer, M. (1993). "Population Growth and Technological Change: One Million B.C. to 1990." *The Quarterly Journal of Economics* 108(3): 681–716. AD 0–1990: United Nations Population Division Report, *The World at Six Billion. AD 1995–2012*: U.S. Census Bureau Data: *The World Population Clock.*

In the industrialized world, many people move faster, eat more, know more, and live in larger homes than even royalty could have dreamed of only a few centuries ago. Yet despite the great advances in science, technology, government, economics, education, and medicine over the past hundred years, these resources are not distributed equally on the planet. Economic, scientific, and technological advances have increased the lifespan and improved access to many marvelous things in the industrialized world, but this overall increase in the standard of living has failed to raise many people out of poverty. The standard of living relates income, comfort, and material goods to the socioeconomic classification of people. Scientists and engineers have played a key role in increasing both the average human life span and standard of living through applications of energy development and distribution, water treatment, sanitation, and other technological advances. As we will see later in this chapter, those who have not benefited from modern science, technology, and industrialization may not be able to meet their basic needs for food, clothing, shelter, water, and sanitation.

ACTIVE LEARNING EXERCISE 1.1   Preconceptions about Sustainability

Define "sustainability" in your own words to the best of your ability. Sketch a visualization of your definition using a cartoon or mind map. Show the linkages to things you perceive are related to sustainability on your sketch. Share your sketch with peers, and listen to how your peers think your sketch illustrates concepts of sustainability.

1.1   Human Development Index

The United Nations Development Programme (UNDP) devised a ***Human Development Index (HDI)*** that is based on three dimensions: life expectancy, education, and income. These dimensions are combined into a single comparable value, as illustrated in Figure 1.4 (UN, 2011a). The HDI is calculated using the data reported each year by the United Nations and the following equations (1.1) to (1.6.)

Life expectancy (LE) at birth uses the 2018 Life Expectancy Index:

$$\text{Life Expectancy Index (LEI)} = (LE - 20)/(85 - 20) \tag{1.1}$$

The Education Index (EI) is based on the Mean Years of Schooling Index (MYSI) and Expected Years of Schooling Index (EYSI), where

$$MYSI = \text{mean years of schooling}/15 \tag{1.2}$$

$$EYSI = \text{expected years of schooling}/18 \tag{1.3}$$

$$EI = (MYSI + EYSI)/2 \tag{1.4}$$

The Income Index (II) is based on the gross national income (GNI_{pc}) at purchasing power parity (PPP) per capita, which is an estimate and standardization of each individual's income in a country:

$$II = \{\ln(GNI_{pc}) - \ln(100)\}/\{\ln(75{,}000) - \ln(100)\} \tag{1.5}$$

The Human Development Index is determined from the geometric mean of the Life Expectancy Index, the Education Index, and the Income Index:

$$HDI = (LEI \times EI \times II)^{1/3} \tag{1.6}$$

Based on this index, the United Nations categorizes countries as Very High Human Development (HDI $\geq$ 0.800), High Human Development (0.800 > HDI $\geq$ 0.700), Medium Human Development (0.700 > HDI $\geq$ 0.550), and Low Human Development (HDI < 0.550).

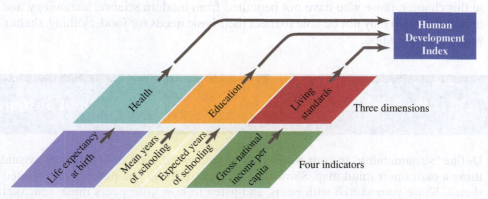

FIGURE 1.4 Components of the Human Development Index.

Source: Based on *Human Development Report 2011. Sustainability and Equity: A Better Future for All.* United Nations Development Programme.

EXAMPLE 1.1   Calculating the Human Development Index

Calculate the Human Development Index for the selected countries from 2018 data.

TABLE 1.1 Component values of the Human Development Index for selected countries

COUNTRY	LIFE EXPECTANCY AT BIRTH (YEARS)	EXPECTED YEARS OF SCHOOLING (YEARS)	MEAN YEARS OF SCHOOLING (YEARS)	GROSS NATIONAL INCOME (GNI) PER CAPITA (2011 PPI $)
Benin	60.2	12.6	3.6	2,061
Costa Rica	80.0	15.4	8.8	14,636
India	68.8	12.3	6.4	6,353
Jordan	74.5	13.1	10.4	8,288
Norway	82.3	17.9	12.6	68,012
United States	79.5	16.5	13.4	54,941

Source: Based on the *Human Development Report 2019*. Beyond income, beyond averages, beyond today: Inequalities in human development in the 21st century. New York. ISBN: 978-92-1-126439-5.

For Benin, the Life Expectancy Index can be calculated using Equation (1.1) from the life expectancy at birth:

$$(LEI) = (LE - 20)/(85 - 20)$$

$$= (60.2 - 20)/65$$

$$= 0.618$$

In order to calculate the Education Index, we first need to calculate the Mean Years of Schooling Index (MYSI) and the Expected Years of Schooling Index (EYSI) from Equations (1.2) and (1.3), respectively:

$$MYSI = \text{mean years of schooling}/15$$

$$= 12.6/15 = 0.840$$

$$EYSI = \text{expected years of schooling}/18$$

$$= 3.6/18 = 0.20$$

Substituting into the equation for the education index yields

$$EI = (MYSI + EYSI)/2$$

$$= (0.840 + 0.20)/2 = 0.520$$

We can calculate the Income Index (II) from the gross national income (GNI_{pc}) at purchasing power parity per capita using Equation (1.5):

$$II = \{\ln(GNI_{pc}) - \ln(100)\}/\{\ln(75,000) - \ln(100)\}$$

$$= \{\ln(2,061) - \ln(100)\}/\{\ln(75,000) - \ln(100)\} = 0.457$$

We can then use the Life Expectancy Index, Education Index, and Income Index to calculate the HDI using Equation (1.6):

$$HDI = (LEI \times EI \times II)^{1/3}$$

$$= (0.618 \times 0.520 \times 0.457)^{1/3} = 0.528$$

The HDI of 0.528 is much less than 1. Using the 2018 data from the United Nations for the countries given yields the HDI values for selected countries shown in Table 1.2.

TABLE 1.2 Calculated values for the subparts of the Human Development Index

COUNTRY	LEI	EI	II	HDI	2018 HDI RANKING OF COUNTRIES
Benin	0.618	0.520	0.457	0.528	163
Costa Rica	0.923	0.758	0.753	0.808	63
India	0.751	0.588	0.627	0.652	130
Jordan	0.838	0.726	0.667	0.740	95
Norway	0.958	0.947	0.985	0.963	1
United States	0.915	0.922	0.953	0.953	13

From Table 1.2, we can see significant gaps in resources associated with life expectancy, education, and income. Norway had the highest HDI score in 2018, followed by the United States. Both Norway and the United States are listed in the United Nations' Very High Human Development category. People living in Costa Rica and Jordan have a similar life expectancy as those living in the United States and Norway, but they would have lower education and income expectations. The United Nations classifies Jordan as a High Human Development nation. India has a lower life expectancy, educational index, and significantly lower income index, and is listed in the United Nations' Medium Human Development category. Benin has a much lower index score in each category than all the previous countries we mentioned, as is the case of many sub-Saharan African nations. This discrepancy in development is illustrated in Figure 1.5. Benin and other countries with little infrastructure, challenged educational systems, low life expectancy, and low expected income values are categorized as Low Human Development countries by the United Nations.

Figure 1.5 illustrates the uneven distribution and ranking of HDIs. By most definitions, in the year 2012, a total of 2.8 billion people lived in poverty or had income levels of less than 2 U.S. dollars per day. Nearly 1.4 billion lived in extreme poverty, earning less than 1.25 U.S. dollars per day (UN, 2012a). Over 850 million people were undernourished and lacked access to food. Approximately 2.5 billion people lacked access to either clean water or sanitation (UN, 2012a). These numbers illustrate the need for a large percentage of the world's population to improve their standard of living. Population numbers alone do not tell the whole story of resource consumption, the uneven distribution of scarce resources, and the desire of many people living in poverty to improve their access to food, water, energy, education, and economic development.

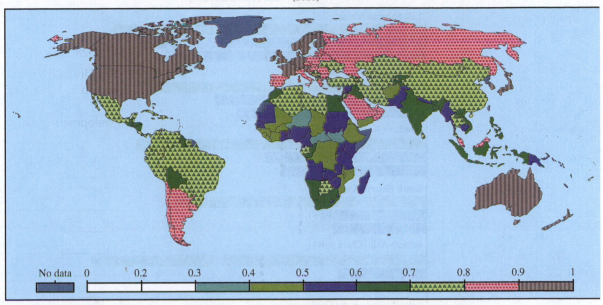

FIGURE 1.5 A map of country rankings based on the United Nations' Human Development Index.

Source: Based on Roser, M. (2019). *"Human Development Index (HDI)."* Published online at OurWorldInData.org.

Mahbub ul Haq (1934–1998), the founder of the *Human Development Report*, said that the purpose of development is:

> *to enlarge people's choices. In principle, these choices can be infinite and can change over time. People often value achievements that do not show up at all, or not immediately, in income or growth figures: greater access to knowledge, better nutrition and health services, more secure livelihoods, security against crime and physical violence, satisfying leisure hours, political and cultural freedoms and sense of participation in community activities. The objective of development is to create an enabling environment for people to enjoy long, healthy and creative lives.*

Sustainable development in one sense is the desire to improve the worldwide standard of living while considering the effects of economic development on natural resources. Since 1990, significant strides have been taken to decrease the percentage of the world's population living in poverty (Figure 1.6). The most significant gains have come from the industrialization of large population centers in Asia.

As economic centers and industrial centers continue to develop and transform our landscape, more and more people are looking to these centers as a means to improve their standard of living. As a result, current trends show that populations are migrating toward more centralized cities and urban areas (Figure 1.7). This rural-to-urban migration places significant strain on the regions surrounding these cities and mega-cities (cities with more than 10 million people) (UN, 2008). Many countries that are becoming more industrialized are struggling to develop the infrastructure required to provide food, water, sanitation, and shelter for the rural migrants. Peri-urban areas are substantially increasing. ***Peri-urban areas***, characterized by very high population densities, lack the infrastructure to distribute energy, water, and sanitation services. These areas severely strain natural resources, especially water and energy.

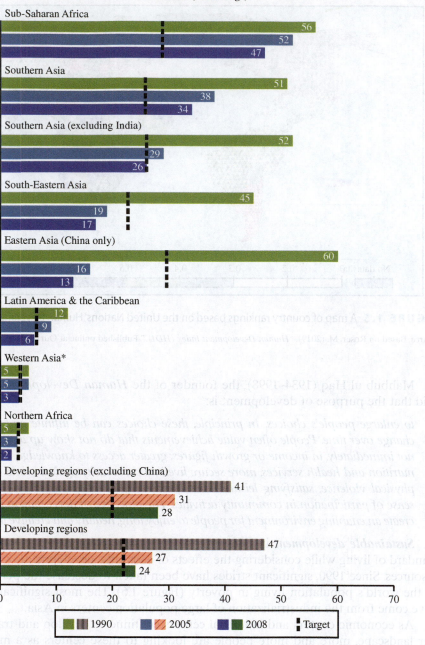

Proportion of people living on less than $1.25 a day in 1990, 2005, and 2008 (Percentage)

Sub-Saharan Africa — 56 / 52 / 47

Southern Asia — 51 / 38 / 34

Southern Asia (excluding India) — 52 / 29 / 26

South-Eastern Asia — 45 / 19 / 17

Eastern Asia (China only) — 60 / 16 / 13

Latin America & the Caribbean — 12 / 9 / 6

Western Asia* — 5 / 5 / 3

Northern Africa — 5 / 3 / 2

Developing regions (excluding China) — 41 / 31 / 28

Developing regions — 47 / 27 / 24

1990 2005 2008 Target

* The aggregate value is based on 5 of 13 countries in the region.

Note: No sufficient country data are available to calculate the aggregate values for Oceania.

FIGURE 1.6 Since 1990, there has been a significant decrease in the number of people living in economic poverty, defined as subsistence on less than one dollar per day.

Source: Based on *The Millennium Development Goals Report 2012*. United Nations.

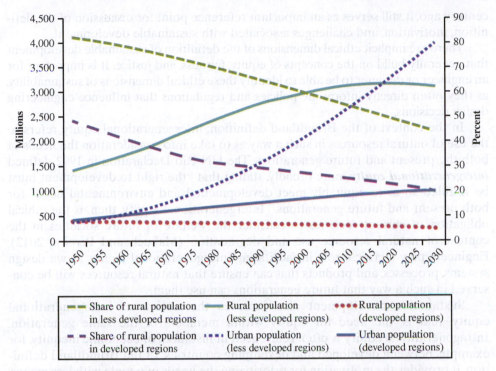

FIGURE 1.7 Trends in population migration into urban areas compared to decreasing population trends in rural areas for developed and less developed countries.

Source: Based on UN (2008). *Trends in Sustainable Development: Agriculture, rural development, land desertification and drought*. United Nations Department of Economic and Social Affairs Division for Sustainable Development.

1.2   Sustainable Development and Social Ethics

In many ways, the narrative that informs our contemporary understanding of sustainable development began over 50 years ago. Scientists, environmentalists, and economists identified a number of environmental and economic challenges associated with the unprecedented increase in global population growth and overconsumption described in the previous section. Connecting these threads, the United Nations requested that the World Commission on Environment and Development formulate "a global agenda for change." The commission articulated the concept of sustainable development in its holistic report *Our Common Future* (WCED, 1987 p. 41):

> *Sustainable development is development that meets the needs of the present without compromising the ability of future generations to meet their own needs. It contains within it two key concepts: (1) the concept of "needs," in particular the essential needs of the world's poor, to which overriding priority should be given; (2) the idea of limitations imposed by the state of technology and social organization on the environment's ability to meet present and future needs.*

This definition of sustainable development is often referred to as the ***Brundtland definition***, named after the chairperson of the UN commission that produced the report. Importantly, the report and the definition of sustainable development linked the three key tenets (known as the three pillars) of sustainability: the environment, society, and the economy. Although the report was published over a quarter of a century ago, it still serves as an important reference point for discussion of the definition, motivation, and challenges associated with sustainable development.

There are implicit ethical dimensions of the definition of sustainable development that generally build on the concepts of equity, fairness, and justice. It is important for an engineer or designer to be able to identify these ethical dimensions of sustainability, as they often directly inform the policies and regulations that influence engineering design decisions.

In the context of the Brundtland definition, intergenerational equity refers to the use of natural resources in such a way as to take into consideration the needs of both the present and future generations. The UN Rio Declaration in 1992 defined ***intergenerational equity*** more broadly, stating that "the right to development must be fulfilled so as to equitably meet developmental and environmental needs for both present and future generations." Intergenerational equity, then, is the ethical obligation of current societies to consider the welfare of future societies in the context of natural resource use and degradation (Makuch and Pereira, 2012). Engineers play a critical role in facilitating intergenerational equity, as we design systems, processes, and products that can ensure that natural resources will be conserved in such a way that future generations can use them.

Sustainable development also includes the notion of intragenerational equity, that is, the need for equity within members of the same generation. Intragenerational equity is often referenced in instances of economic inequity, for example, between developed and developing countries. In the Brundtland definition, it provides the motivation for prioritizing the needs of people with less means. In international treaties such as the Kyoto Protocol, intragenerational equity is the principle used to define common but differing responsibilities between countries at various stages of economic development. In the context of the Kyoto Protocol, developed and developing countries do not have identical obligations to mitigate

BOX 1.1 Formative Works Related to Sustainable Development

In 1956 a geoscientist who worked at Shell research lab named Dr. M. King Hubbert suggested that oil production in the United States would peak and followed a bell-shaped curve (Hubbert and American Petroleum, 1956). His work became a cornerstone of the notion of peak oil (i.e., the point in time where the rate of total petroleum extraction begins to decline permanently) and underlined the point that oil is a finite and depleting resource used as a fuel for economic development.

A few years later (Figure 1.8), the marine biologist Rachel Carson published *Silent Spring* (1962), which chronicled the negative impacts associated with pesticides and facilitated the ban of the pesticide DDT a decade later. This book is often cited as helping to begin the environmental movement in America.

In "The Tragedy of the Commons" (1968), ecologist Dr. Garret Hardin explored some of the moral and social challenges related to population growth and the management and use of natural resources.

The Limits to Growth (1972), by Donella Meadows, Dennis Meadows, Jørgen Randers, and William W. Behrens III, provided the first effort to holistically model the global system with respect to five key indicators of sustainability: world population, industrialization, pollution, food production, and resource depletion.

In *Small Is Beautiful* (1973), E. F. Schumacher critiqued modern economists, for example, for their treatment of natural resources such as oil as nondepleting. His work has since been applied to engineering design through appropriate technology.

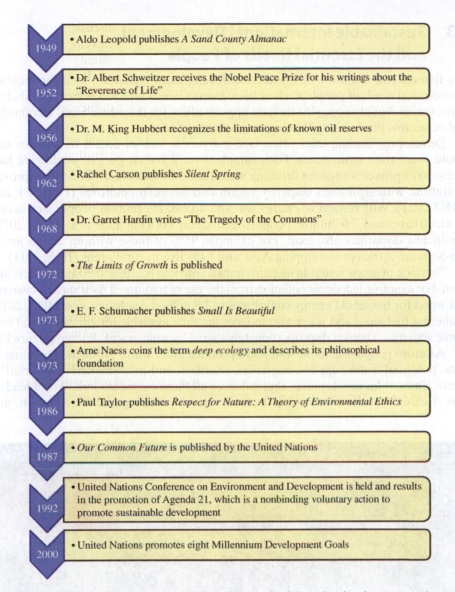

FIGURE 1.8 Formative works in developing the current philosophical underpinnings that support sustainable development.

climate change. Rather, their requirements for action under the treaty are equitably scaled to reflect differences in their economic and infrastructure capacities (Makuch and Pereira, 2012). *Social justice* is similar to intragenerational equity, but it is concerned more specifically with the fair distribution (or sharing) of the advantages and disadvantages (or benefits and burdens) that exist within society.

Given the evolution of the concept of sustainable development, it is not surprising that arguments for sustainability hinge upon the concept of social justice. In the context of sustainable development, environmental justice deals with the fair distribution of environmental benefits and burdens. The lines between environmental and social justice are occasionally blurred. However, because their objectives may at times differ, it is useful to keep them distinct.

1.3   Sustainable International Development and the Essential Needs of People

The Brundtland definition of sustainable development includes a focus on meeting the essential needs of people, such as basic human needs for food, water, and shelter. However, essential needs also include opportunities for the educational advantages and economic productivity that improve quality of life.

Developing communities often face a systemic lack of access to services that would meet their basic needs. For example, over 2 billion people worldwide lack access to improved sources of drinking water, and 3.6 billion lack access to improved sanitation, with significant disparity within and between countries (UNICEF and WHO, 2021). With respect to energy use and access, 759 million people lack access to electricity, and 2.6 billion people need clean cooking facilities (IEA, 2021). Significant disparities also exist. For example, 95% of those without access are in sub-Saharan Africa or developing Asia, and 84% live in rural areas (IEA, 2011).

This lack of access results in negative impacts on both health and economic development. For example, indoor air pollution from the use of traditional fuels such as charcoal and wood for household energy contributes to 3.8 million deaths per year (IEA, 2021). Gathering fuel wood and water also reduces the time available for education and economic activity—a burden disproportionately shared by women and children (Figure 1.9).

As more people use more water and produce more waste products that contaminate potential water supplies, engineers, scientists, and policymakers have tried to create a model for development that balances all these considerations. World leaders have focused on the relationship between development, population growth, and

FIGURE 1.9 Gathering fuel wood and water in Benin, West Africa. The physical labor of gathering fuel wood for cooking and collecting water is often the work of women and children in low-income countries. This represents a high social cost, as it takes them away from opportunities for schooling or engaging in business endeavors.

Source: Bradley Striebig.

natural resource management for many years. One of the most profound statements about these interrelationships is the United Nations Agenda 21 from the ***Conference on Environment and Development***, held in Rio de Janeiro, Brazil, in 1992:

> *Humanity stands at a defining moment in history. We are confronted with a perpetuation of disparities between and within nations, a worsening of poverty, hunger, ill health and illiteracy, and the continuing deterioration of the ecosystems on which we depend for our well-being.*

The statement continues:

> *Human beings are at the center of concerns for sustainable development. They are entitled to a healthy and productive life in harmony with nature. The right to development must . . . meet developmental and environmental needs of present and future generations. All States and all people shall cooperate in the essential task of eradicating poverty as an indispensable requirement for sustainable development. . . . To achieve sustainable development and a higher quality of life for all people, States should reduce and eliminate unsustainable patterns of production and consumption and promote appropriate demographic policies.*

In the year 2000, the United Nations specified eight ***Millennium Development Goals*** to address prior to 2015. Overall the Millennium Development Goals have potentially saved over 21 million extra lives due to accelerated progress, as illustrated in Figure 1.10. "The MDGs helped to lift more than one billion people out of extreme poverty, to make inroads against hunger, to enable more girls to attend school than ever before and to protect our planet," the UN Secretary General Ban Ki-moon explained. But he did not finish there. "Yet for all the remarkable gains, I am keenly aware that inequalities persist and that progress has been uneven." The progress on each individual goal and target varied and results also varied across different geographic regions, as shown in Figure 1.11.

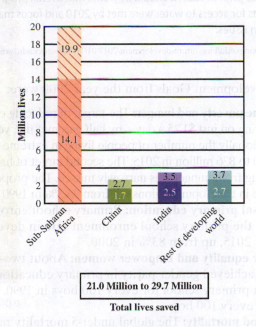

FIGURE 1.10 Indicators that the UN Millennium Development Goals successfully reduced mortality.

Source: Data from www.brookings.edu/blog/future-development/2017/01/11/how-successful-were-the-millennium
-development-goals/.

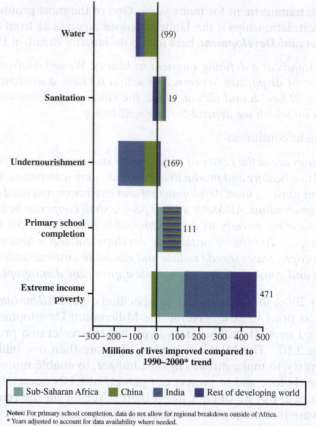

FIGURE 1.11 Indicators that the UN Millennium Development Goals had mixed results compared to trends in the prior decade. It should be noted that overall progress was still important; for example, MGD targets for access to water were met by 2010 and focus may have appropriately shifted toward sanitation issues.

Source: Data from www.brookings.edu/blog/future-development/2017/01/11/how-successful-were-the-millennium-development-goals/.

The Millennium Development Goals from the year 2000 were:

1. **Eradicate extreme poverty and hunger:** The target of reducing extreme poverty rates—people living on just $1.25 a day—by half was met five years ahead of the 2015 deadline. Globally the number of people living in extreme poverty fell from 1.9 billion in 1990 to 836 million in 2015. The second target of halving the proportion of people suffering from hunger was narrowly missed. The proportion of undernourished people in developing regions fell from 23.3% in 1990 to 12.9% in 2014.

2. **Achieve universal primary education:** Primary school enrollment figures rose substantially, as the primary school enrollment rate in developing regions reached 91% in 2015, up from 83% in 2000.

3. **Promote gender equality and empower women:** About two-thirds of developing countries achieved gender parity in primary education. Only 74 girls were enrolled in primary school for every 100 boys in 1990. Today, 103 girls are enrolled for every 100 boys.

4. **Reduce childhood mortality:** The global under-5 mortality rate declined by more than half since 1990—dropping from 90 to 43 deaths per 1,000 live births. This falls short of the targeted drop of two-thirds. In practical terms, this means 16,000 children under 5 continue to die every day from preventable causes.

5. **Improve maternal health:** Since 1990, the maternal mortality rate has been cut nearly in half, albeit falling short of the two-thirds reduction goal. There were an estimated 289,000 maternal deaths in 2013.

6. **Combat HIV/AIDS, malaria, and other diseases:** Although the number of new HIV infections fell by 40% between 2000 and 2013, the target of halting and beginning to reverse the spread of HIV/AIDS has not been met. According to the UN, over 6.2 million malaria deaths were averted between 2000 and 2015, primarily those of children under 5 years old in sub-Saharan Africa. The global malaria incidence rate fell by an estimated 37% and the mortality rate by 58%.

7. **Ensure environmental sustainability:** The target of halving the proportion of people without access to safe water was achieved. Between 1990 and 2015, 2.6 billion people gained access to improved drinking water. However, 663 million people across the world still do not have access to improved drinking water, and proper sanitation remains a challenge for low development areas of the world.

8. **Develop a global partnership for development:** Official development assistance from wealthy countries to developing countries increased by 66% between 2000 and 2014.

The Millennium Development Goals ended in 2015, after successfully achieving many of the desired goals in some parts of the planet. The MDG's successor—the 17 Sustainable Development Goals (SDGs) shown in Figure 1.12—were adopted by world leaders at a summit in New York in September 2015. Countries are tasked to mobilize efforts to end all forms of poverty, fight inequalities, and tackle climate change, while ensuring that no one is left behind.

FIGURE 1.12 The 17 United Nations Sustainable Development Goals.

Source: MintBlak/Shutterstock.com.

The SDGs build on the success of the Millennium Development Goals and aim to go further to end all forms of poverty. The new goals are unique in that they call for action by all countries to promote prosperity while protecting the planet. They recognize that ending poverty must go hand-in-hand with strategies that build economic growth and address a range of social needs, including education, health, social protection, and job opportunities, while also tackling climate change and environmental protection.

How can sustainable development be achieved if there are more people using more resources? In order to answer this complex question, we must first examine the definition of sustainability and understand the context associated with the term. The term *sustainable* or *sustainability* is widely used in a variety of applications, contexts, and marketing materials. Most of the uses of the word "sustainable" infer a qualitative comparison to something "other"; for example, "*our new green Excellon automobile is the sustainable solution to yesterday's sports utility vehicle (SUV)."* There is little or no quantifiable way to compare and contrast the advantages and disadvantages of the two products. Nor is there an attempt to explain how a product can be sustainably produced, sold, and disposed of for any significant period of time. The marketing of "sustainable" goods and products also typically infers that the "sustainable" product is morally superior to the less sustainable product.

Do you believe that sustainable products are morally superior to less sustainable products? If so, what does this belief imply about the developed world's largely consumer-based economic system of retail merchandise? How are technology and moral convictions woven into the fabric of our definitions of sustainable design?

The importance of defining quantitative measures of sustainability for design will be addressed in this chapter and throughout subsequent chapters. In order to determine if a product or process is sustainable, we must examine the consumption of raw materials, design life, cost, and ultimate disposal of a product. The context in which the product and process are developed should also be considered together with the societal implications of worldwide population growth and consumption patterns as well as with an evaluation of how an individual's environmental footprint may be affected by the product or process.

1.4   Engineering and Developing Communities

Engineers and designers play a key role in developing solutions for meeting the essential needs of people. However, to do this well, we must pay attention to the technical, cultural, economic, and environmental contexts that can affect project outcomes. Countless engineering and design efforts in developing communities have failed to meet expectations for sustained (and positive) societal impact. For example, in sub-Saharan Africa, 35% of rural water systems are nonfunctioning, with some countries experiencing an operational failure rate of 30% to 60% (Harvey and Reed, 2007). These high failure rates generally do not occur because of technical design flaws, but because of the failure to incorporate other salient factors through the design process, including social, economic, and environmental influences. Point-of-use (POU) water treatment technologies, like those described in Table 1.3, may be better suited to significantly reduce the risk of exposure to pathogenic organisms in drinking water.

Engineers and designers are now better equipped to incorporate these key social, economic, and environmental factors into their design and implementation decisions than they were several decades ago. Generally, two dimensions of design require a shift in thinking for sustainable engineering in developing countries. The first dimension involves a shift in thinking with regard to *product*, and the second, with regard to *process*.

TABLE 1.3 Point-of-use (POU) appropriate water treatment technologies considered for implementation in the model home near Kigali, Rwanda

TECHNOLOGY	DESCRIPTION	ADVANTAGE	DISADVANTAGE	REFERENCES
Biosand™	Sand filtration	• High removal efficiency for microorganisms	• Needs continual use and regular maintenance • Cost	Duke et al. (2006); Stauber et al. (2006)
Filtron™	Ceramic filter	• High removal efficiency for microorganisms • Sized for individual homes • Relatively inexpensive	• Requires fuel for construction • Limited lifetime • Requires regular cleaning	Bielefeldt et al. (2010); Brown and Sobsey (2010); Clasen et al. (2004); Striebig et al. (2007); van Halem et al. (2009)
SODIS™	Solar water disinfection	• Highly effective • Inexpensive • Can reuse a waste product (PET bottles)	• Long treatment time (6 to 48 hours) • Does not remove other potential pollutants • Requires warm climate and sunlight	Conroy et al. (2001); Kehoe et al. (2001); Mania et al. (2006); Meierhofer and Wegelin (2002); Sommer et al. (1997)

Source: Based on Striebig, B., Atwood, S., Johnson, B., Lemkau, B., Shamrell, J., Spuler, P., Stanek, K., Vernon, A., and Young., J. (2007). "Activated carbon amended ceramic drinking water filters for Benin." *Journal of Engineering for Sustainable Development* 2(1):1–12.

In terms of products, engineers and designers have transitioned to the concept of ***appropriate technology***. While this term has many differing meanings, it is used here to refer generally to engineering design that takes into consideration the key local social, economic, environmental, and technical factors that influence the success or failure of a design solution. That is, a technology (or design) is appropriate "when it is compatible with local, cultural, and economic conditions (i.e., the human, material and cultural resources of the economy), and utilizes locally available materials and energy resources, with tools and processes maintained and operationally controlled by the local population" (Conteh, 2003, p. 3). Appropriate technologies commonly considered when building homes or community buildings are briefly described in Table 1.4.

What does appropriate technology look like in practice? Consider, for example, the design of a point-of-use water supply and purification system for a rural household. In a high-income country, engineers would likely base their design decision on the amount of water needed by the household, size the system components for filtration, disinfection, and pumping accordingly, and then balance component quality and selection against the budgetary constraints of the household. In the case of a low-income country, engineers must still make these technical design decisions about system size and affordability, but they must also consider:

- Are parts readily available, either locally or nationally, if a component were to fail?
- Are there individuals who have the necessary skill or technical training to repair the component or system if it were to fail?
- Would members of the household readily understand how to use this system?
- What is the local availability of required infrastructures, such as electric power?

There is a significant body of literature on how to develop and design appropriate technology solutions for developing communities. These solutions tend to be

TABLE 1.4 Applications and uses of appropriate technology for medium- and low-income indexed countries

APPLICATION	TECHNOLOGY
Building design	Right-sized homes that maximize storage, comfort, social interactions, and use while minimizing the use of materials and energy.
	Natural ventilation can be integrated into a design by incorporating porches, central courtyards, other outside features, and strategically placed windows.
	Passive solar design maximizes exposure to the sun and takes advantage of the natural energy characteristics of building materials and air that are exposed to the energy of the sun.
	Overhangs take advantage of the thermal properties of the sun during the winter months while minimizing the sun's impact during the warmer summer months.
Power generation	Biogas power generation or microbial fuel cell technology can be integrated with waste management.
	Simple wind turbines may be made out of containers that would otherwise be disposed of as solid waste.
Material use in building construction	Appropriate, local, nontoxic, and reusable materials such as lime-stabilized rammed earth blocks, adobe, and straw bales can be promoted for building construction.
Stormwater management	Green roofs built on top of residential, commercial, and industrial structures not only effectively manage stormwater but also have benefits of reducing a building's energy consumption and regional urban heat island effect.
Water supply	Rainwater harvesting can assist groundwater recharge and provide all or a portion of domestic, commercial, and agricultural needs. It can also be incorporated into a building's cooling system.
Water treatment	*Moringa oleifera* tree seeds can be used to reduce turbidity.
	Other point-of-use treatment technologies described in Table 1.3 may be useful, especially in areas where power interruptions are frequent and those power interruptions frequently result in contamination of centralized piped drinking water supplies.

Source: Based on Hazeltine, B., and Bull, C. (2003). *Field Guide to Appropriate Technology*.

low-cost, culturally sensitive, and community-focused. They are also usually made and sourced from local supply chains. An added benefit of appropriate technology is that it can often be used as a tool for building human resource capacities in key areas, such as in electrical and mechanical skills and training. A wealth of information is now available on emerging appropriate technologies and addresses a wide range of essential human needs—for water, sanitation, energy, shelter, and so on—in developing communities. Global knowledge sharing about appropriate technology has greatly increased with the development of several online communities, such as *Engineering for Change* and *Appropedia* (Box 1.2).

Sustainable engineering focuses on both the design process itself and the implementation plan, and both are equally important to the technical design. The sustainable design process involves working with local community members to co-define the problem and its possible solutions, as well as to develop sound implementation plans. Community involvement is a hallmark of successful engineering design projects (Figure 1.13). It is through this process that appropriate technology can be used as a means of capacity building. In Bangladesh, an effort to provide solar energy and household lighting resulted in microlending opportunities that built local banking and financing capabilities (Box 1.3).

BOX 1.2   Online Resources for Appropriate Technology Solutions

Working on engineering solutions to meet the challenges in a developing community? Here are great resources to begin your exploration of possible strategies and ideas.

Engineering for Change (E4C): Founded by the American Society of Mechanical Engineers (ASME), Institute of Electrical and Electronics Engineers (IEEE), and Engineers Without Borders—USA (EWB-USA), E4C is a community of engineers, technologists, social scientists, non-governmental organizations (NGOs), local governments, and community advocates who work to develop locally appropriate and sustainable solutions for pressing humanitarian challenges. They maintain a "Solutions Library," which is a catalogue of appropriate technology solutions and case studies.

Website: www.engineeringforchange.org

Appropedia: This is a wiki that enables users to catalogue and collaborate on sustainability, appropriate technology, and poverty-reduction solutions. As with all crowd-sourced and community-managed wikis, the information on the site is best used for idea generation, as technical details are generally not independently validated.

Website: www.appropedia.org

Solar Cookers World Network: Another wiki-based site, the Solar Cookers World Network enables knowledge sharing on the design and construction of solar cookers.

Website: www.solarcooking.org

FIGURE 1.13 A Kenyan *Jiko*. This Kenyan cookstove is made of ceramic and tin and is an improved design that conserves scarce and costly charcoal and wood fuel. The final design resulted from collaboration between those who use them and the Massachusetts Institute of Technology engineers working on the project.

Source: Bradley Striebig.

BOX 1.3 Grameen Shakti (*Grameen Energy*)

Founded in 1996, Grameen Shakti is one of the world's largest suppliers of solar technologies, having installed nearly 100,000 solar photovoltaic systems in homes at a current rate of approximately 3,500 per month. Grameen Shakti is part of the Grameen Bank family, a microfinance enterprise for which its founder, Muhammad Yunus, won the Nobel Peace Prize.

Microfinance goes by several names, including microcredit and microlending. It operates by making extremely small loans to the poor without collateral or contract, and it relies on community peer pressure to assure loan repayment. Loans are given to individuals as investments that will allow borrowers an opportunity to surmount their poverty. An example is lending a woman money to purchase a sewing machine with which she can become a seamstress and earn an income. Indeed, the vast majority of Grameen microcredit is given to women.

With respect to Grameen Shakti, customers purchase their energy systems from the organization on a payment plan, usually a monthly installment over two or three years. Staff visit monthly to collect fees and perform maintenance on the systems. A typical photovoltaic system is used to power four light bulbs for four hours at night, enabling children to study and do their homework. Grameen Shakti also works with other renewable energy technologies, such as biogas applications and cook stoves.

Grameen Shakti is a unique social business in that it offers a complete package of technology integration: financing options to purchase the system, operational support of the technology, maintenance if the system breaks down, and technical advice on how to convert the product into a money-making tool. It is also unique in that it is location specific, providing remote, rural areas of Bangladesh with access to renewable energy technologies. This allows Grameen Shakti to make the design of the social aspects of its business sensitive to sociocultural factors that contribute to its success.

EXAMPLE 1.2   EWB House in Rwanda

Rwanda is a small country that lies in the heart of Central Africa. The capital of Rwanda is Kigali, a city that has grown very rapidly after the civil unrest that racked the country in the mid-1990s. As of 2002, Kigali City had 131,106 households, a total population of 604,966 (approximately 56% of which is age 20 years or under), and an annual growth rate of 10% (i.e., population will double every seven years) (Rwanda Ministry of Infrastructure, 2006).

The average Rwandan citizen consumed only 15.2 liters of water per day in 2002. The water cost was about a nickel (U.S.) each day. This may not sound like much, but that cost was more than 10% of the average daily wage! Water supply is still limited in Rwanda. Children are usually responsible for collecting the water. In Kimisange, a peri-urban area surrounding Kigali, the children carry the water over 0.25 kilometers from the nearest public water source, called a tapstand. However, this tapstand often runs dry, forcing the children to collect standing surface water in locations such as the one shown in Figure 1.14. The children may carry the water nearly 2 kilometers uphill back to their homes. Cooking fuel is also in short supply in Rwanda due to deforestation. Cooking fires with poor fuel continue to result in 10 to 11 deaths each day in many low-development countries such as Rwanda. Sewage in Rwanda is often discharged directly into the streets, which can contaminate the local water supply. The lack of access to clean

FIGURE 1.14 Photo showing typical housing, agriculture, and water sources in the peri-urban area near Kigali, Rwanda.

Source: Bradley Striebig.

water, energy, and sanitation has had a profound negative effect on human health. The average life expectancy in Rwanda in 2002 was only 47.3 years. The infant mortality rate was 89.61 infant deaths per 1,000 live births (WHO/UNICEF, 2005). Waterborne disease is suspected in killing one out of every five children born before the age of 5.

In 2004, an international group consisting of architects, engineers, and scientists led a project to demonstrate a low-income model home with sustainable on-site water, sanitation, and renewable sources of cooking fuel and fertilizer.

The local population of Kigali does not have the financial resources to invest in sanitation systems because they suffer from extreme poverty. Traditional centralized collection and treatment systems in the United States cost the average consumer 0.62 U.S. dollars per day (CIA, 2007). This would be nearly an entire day's wages or more for most people living in Kigali, so it is not a practical solution for Rwanda or most of the developing world. Existing centralized treatment facilities are expensive and require large amounts of energy for treatment and pumping. Ultimately, the traditional approach would not yield economically or environmentally sustainable solutions for sanitation or resource recovery in Kigali. Decentralized technologies are subject to failure due to poor maintenance and the costly disposal of residual wastes.

A group from Engineers Without Borders—USA and the Kigali Institute of Science and Technology worked with a community in Kigali to develop and optimize a scalable low-cost sustainable home. The home is made from reinforced low-cost bamboo and lime-stabilized earthen blocks, as shown in Figures 1.15 and 1.16. These blocks were produced at the site of the model home and will provide good insulation, keeping the interior warm during cool evenings and cool during warm days in Rwanda's very mild climate. Recovery of biogas will be used for cooking fuel and will help reduce deforestation due to the need to harvest wood for cooking. Rainwater will

FIGURE 1.15 Interlocked earth-block construction methods use inexpensive soil-based construction blocks that are made primarily from materials available near the construction site and use locally harvested bamboo reinforcement for construction.

Source: Photo used by permission of Chris Rollins.

FIGURE 1.16 Completed model home in Rwanda made from earth-block construction with a rainwater harvesting system shown in the foreground that is sitting on top of the visible portion of an underground biogas digester.

Source: Photo used by permission of Chris Rollins.

be collected from rooftops. This water can be used for drinking water, cooking, and sanitation. Drinking water is treated in model homes using a point-of-use treatment technology. Three treatment technologies (shown in Table 1.3) were considered for drinking water purification.

Construction and implementation of the model home occurred in 2007. Community-focused projects, such as this one, directly address eight of the Sustainable Development Goals set by the United Nations.

Sustainable Development Goals 1 and 2: No Poverty and Zero Hunger

Problems of poverty are inextricably linked to the availability and quality of water. Improving access to sanitation and the quality of water in the watershed can make a major contribution to eradicating poverty. Reliable shelter and biogas cooking fuel has the potential to reduce poverty and hunger.

Sustainable Development Goal 3: Good Health and Well-being
Sustainable Development Goal 4: Quality Education

Implementing safe sanitation practices and improving the water quality may reduce absenteeism and help students concentrate more fully on their education. Students and staff may benefit from the significant reduction in illness and absence from school due to the reduced exposure to pathogenic organisms. Furthermore, less time may be spent finding fuel for cooking since wood can be replaced by biogas.

Sustainable Development Goal 6: Clean Water and Sanitation

Increasing access to water can decrease the amount of time spent by young women retrieving water for use in the home, thereby increasing the likelihood those young women will do well in school and stay in school. Millions of children under the age of 5 die each year from preventable water-related diseases. The UN *Water World Development Report* states that, of all the people who died of diarrhea infections in 2001, 70% (or 1.4 million) were children. People weakened by HIV/AIDS are likely to suffer the most from the lack of a safe water supply and sanitation, especially from diarrhea and skin diseases. Access to pathogen-free drinking water may reduce childhood mortality and also reduce the suffering of people weakened by HIV/AIDS.

Waterborne diseases spread pathogens by the ingestion of urine- or feces-contaminated water. Typhoid fever, amoebic dysentery, schistosomiasis, and cholera are just a few of the diseases spread by contaminated water. Maternal mortality rates were estimated by the World Health Organization (WHO) to be 1 maternal death per 10 births. It is estimated that 203 of every 1,000 children die before the age of 5 in Rwanda. The model-building design increases access to water for drinking and hand washing. Promoting proper sanitation and providing the technology to implement point-source water treatment in the community will likely decrease childhood and maternal mortality rates in Kigali, Rwanda.

Sustainable Development Goal 11: Sustainable Cities and Communities

The earth-block model home addressed many of the indicators of sustainable development by meeting the current needs for wastewater treatment in the urban fringe of Kigali while conserving resources for future generations. The strategy reduced the spread of waterborne disease, while effectively treating wastewater so as not to exceed the assimilative capacity of native wetlands and also reduce the dependency on imported materials such as concrete and cement.

Sustainable Development Goal 17: Partnerships for the Goals

This project involved partnerships with the following organizations: the German Development Corporation (DED), Gonzaga University, Engineers Without Borders—USA, Tetra-Tech, Inc., the city of Kigali, the Kigali Institute for Science and Technology, and the National University of Rwanda.

ACTIVE LEARNING EXERCISE 1.2 Expectations about Common Resources

The next time you drink from a water fountain or buy a bottle of water, what are your expectations about the safety of the water? Who, if anyone, makes a profit from the sale of tap water? Who, exactly, would be responsible for fulfilling these expectations?

Imagine you have purchased bottled water from a vending machine, and answer the following questions:

- Where did the water originate?
- Where did the plastic materials originate?
- How much did the bottled water cost compared to tap water? Who makes a profit, if anyone, from the sale of bottled water?
- Does drinking bottled water present less risk than drinking tap water? Explain the factors you have considered.
- What are the limitations of providing bottled water to meet the need for drinking water in Rwanda? How do cost, waste production, and social justice factor into this equation?

1.5 Definitions of Sustainability

The *Merriam-Webster Dictionary* defines **"sustainable"** as "capable of being sustained." This first dictionary definition does not shed much light on our discussion. The second definition listed begins to illuminate our topic: "of, relating to, or being a method of harvesting or using a resource so that the resource is not depleted or permanently damaged." Within this definition, we begin to see some key topics, including resource depletion and the term *damage* associated with nonsustainable practices.

The United States Environmental Protection Agency (EPA) provides a more useful working definition of sustainability:

> *Sustainability is based on a simple principle: Everything that we need for our survival and well-being depends, either directly or indirectly, on our natural environment. Sustainability creates and maintains the conditions, under which humans and nature can exist in productive harmony, that permit fulfilling the social, economic and other requirements of present and future generations. Sustainability is important to making sure that we have and will continue to have, the water, materials, and resources to protect human health and our environment. (Federal Register, 2009)*

Within the EPA definition, we see the words "harmony" and "protect" being applied to a relationship between humans and nature, which infers a moral virtue associated with sustainability. When we begin to think about sustainability in moralistic terms, we venture into the world of ethics and conflicting or sometimes contradictory moral quandaries.

For our simple analysis, we can describe *morals* as the values people adopt to guide the way they ought to treat each other. When we have a conflict between morals, we can use ethics to guide us toward the best outcome based on our ethical reasoning. *Ethics*, therefore, provides a framework for making difficult choices when we face a problem involving moral conflict. These working definitions are much easier to understand when we evaluate a few of the following examples of the applications of an ethical code.

Most ethical thinking over the past 2,500 years has been a search for the appropriate ethical theory to guide our behavior in human–human relationships. Some of the most influential theories in Western ethical thinking, theories that are most defensible, are based on consequences or on acts. In the former, moral dilemmas are resolved on the basis of what the consequences are. If it is desired to maximize good, then the alternative that creates the greatest good is correct (moral). In the latter, moral dilemmas are resolved on the basis of whether the alternative (act) is considered good or bad; consequences are not considered.

The most influential consequentialist ethical theory is *utilitarianism*, described by Jeremy Bentham (1748–1832) and John Stuart Mill (1806–1873). In utilitarianism, the pain and pleasure of all actions are calculated and the worth of all actions is judged on the basis of the total happiness achieved, where happiness is defined as the highest pleasure/pain ratio. The so-called utilitarian calculus allows for the calculation of happiness for all alternatives being considered. To act ethically, then, is to choose the alternative that produces the highest level of pleasure and the lowest level of pain. Benefit/cost analysis can be considered to be utilitarian in its origins because money is presumed to equate with happiness.

A second group of ethical theories is based on the notion that human conduct should be governed by the morality of acts and that certain rules (such as "do not lie") should always be followed. These theories, often called *deontological* theories, emphasize the goodness of the act and not its consequence. Supporters of these theories hold that acts must be judged as good or bad, right or wrong, *in themselves*, irrespective of the consequences of these acts. An early system of deontological rules is the Ten Commandments, as these rules were meant to be followed *regardless of consequences*.

Possibly the best-known deontological system is that of Immanuel Kant (1724–1804), who suggested the idea of the ***categorical imperative***—the concept that one develops a set of rules for making value-laden decisions such that one would wish that all people obeyed the rules. Once these rules are established, one must always follow them; only then can that person be acting ethically because it is the act that matters. A cornerstone of Kantian ethics is the principle of *universalizability*, a simple test for the rationality of a moral principle. In short, this principle holds that, if an act is acceptable for one person, then *it must be equally acceptable for others*. Kant thus proposed that an act is either ethical or unethical if, when it is universalized, it makes for a better world. An example can be found in the *Code of Ethics for Engineers* shown in Box 1.4, which states "the engineer shall hold paramount the health, safety, and welfare of the public" (NSPE, 2007).

There are, of course, many more systems of ethics that could be discussed and that have relevance to the engineering and science professions, but it should be clear that traditional ethical thinking represents a valuable source of insight in one's search for a personal and professional lifestyle.

The concepts presented in both the EPA and Merriam-Webster definitions of sustainability suggest that conditions for the planet and humans inhabiting the Earth would be better if sustainable practices were adopted. However, the *Code of Ethics for Engineers* takes a human-centered view of a system, holding the "public" good in the highest regard. This might be thought of as an anthropocentric ethical framework in which nature is considered and valued based solely on the goods that nature can provide to humans or the "public."

Long before Kant, Aristotle appeared to support this anthropocentric view in his statement that "plants exist to give food to animals, and animals, to give food to men. . . . Since nature makes nothing purposeless or in vain, all animals must have

BOX 1.4 The National Society of Professional Engineers Code of Ethics for Engineers

Fundamental Canons: Engineers, in the fulfillment of their professional duties, shall:

1. Hold paramount the safety, health, and welfare of the public.
2. Perform services only in areas of their competence.
3. Issue public statements only in an objective and truthful manner.
4. Act for each employer or client as faithful agents or trustees.
5. Avoid deceptive acts.
6. Conduct themselves honorably, responsibly, ethically, and lawfully so as to enhance the honor, reputation, and usefulness of the profession.

Source: National Society of Professional Engineers (NSPE) (2007) Code of Ethics for Engineers, Publication #1102. Alexandria, VA.

been made by nature for the sake of man" (Vesilind et al., 2010, p. 71). While the anthropocentric view is simple and concise, it does not seem to adequately reflect the feeling toward animals and nature held in most modern societies. For example, it is illegal in the United States and many other countries to encourage or promote animal fights for one's own pleasure.

This change in ethics was taking place in the Western world during the 1800s and was spearheaded by prominent writers and philosophers of the time. The growth of ethical arguments to include animals and nature has been termed *existentialist ethical thinking*, which extends the moral community to include creatures other than humans. Aldo Leopold articulated this viewpoint in *A Sand County Almanac* (1949). Leopold's work led to the development of a new ethical framework called the **land ethic**, which encourages people to extend their thinking about communities to which we should behave ethically to include soil, water, plants and animals, or collectively, the land. Paul Taylor (1981, p. 207) describes a **biocentric outlook** that values all living things in Earth's community, so each organism is "a center of life pursuing its own good in its own way," and all organisms are interconnected. Arne Naess (1989, p. 166) took the biocentric outlook one step further when he wrote: "The right of all the forms to live is a universal right which cannot be quantified. No single species of living being has more of this particular right to live than any other species." The ethical framework in which humans have no greater importance than any other component of our world is sometimes referred to as deep ecology. The **deep ecology** ethic lies at the opposite side of the spectrum from an anthropocentric ethical code in evaluating the relationships between humankind and nature. However, most societies do not hold to the deep ecology worldview or paradigm. If modern societies did apply the deep ecology ethic, then modern medicine would not try to kill pneumonia bacteria with antibiotics while saving a human life.

So are there any ethical systems that modern societies have adopted? Certainly there are, and in the engineering profession, we are expected to uphold to specific ethical canons. Aldo Leopold suggests that "a thing is right when it tends to preserve the integrity, stability, and beauty of the biotic community. It is wrong when it tends otherwise." Certainly Leopold considered the human species to be part of this biotic community. Vesilind and Gunn (1998, p. 466) suggest another approach, which they describe as the **environmental ethic**—"recognizing that we are, at least at the present time, unable to explain rationally our attitude toward the environment and that these attitudes are deeply felt, unlike the feeling of spirituality." Furthermore, this approach embodies a sense of obligation to future generations of our species. When we consider the future conditions of our planet and species, we are using an intergenerational ethical model. If we consider future generations, we are perhaps making a moral choice to preserve and protect the things we value but have difficulty explaining.

Stewart Collis takes an agnostic approach to ethics:

Both polytheism and monotheism have done their work. The images are broken; the idols are all overthrown. This is now regarded as a very irreligious age. But perhaps it only means that the mind is moving from one state to another. The next state is not belief in many gods. It is not a belief in one god. It is not a belief at all—not a conception of the intellect. It is an extension of consciousness so that we may feel God. (Collis, 1954, p. 72)

In his 2002 published letter, Pope John Paul II suggested an intergenerational and environmental ethical framework that blends advances in science and technology:

We therefore invite all men and women of good will to ponder the importance of the following ethical goals: To think of the world's children when we reflect on and evaluate our options for action. To be open to study the true values based on the natural law that sustain every human culture. To use science and technology in a full and constructive way, while recognizing that the findings of science have always to be evaluated in the light of the centrality of the human person, of the common good and of the inner purpose of creation. Science may help us to correct the mistakes of the past, in order to enhance the spiritual and material well-being of the present and future generations. It is love for our children that will show us the path we must follow in the future.

Throughout the ages we have blended our ethical, moral, and spiritual beliefs in an effort to apply definitions to the actions we take both individually and as a species. The most commonly referenced definition of sustainability is derived from the Brundtland Commission's report on practices for sustainable development and approaches to reduce the number of people living in poverty. The report of this commission, called *Our Common Future*, defines sustainable development as "development that meets the needs of the present without compromising the ability of future generations to meet their own needs" (WCED, 1987).

Within this definition of sustainability, we see that human needs are placed somewhat above the needs of other animals and plants. However, this definition recognizes the inherent value of the natural world and the role of the natural world in meeting the basic needs of humanity. Furthermore, this definition relies heavily on intergenerational equity to protect the natural world so that future generations will not live in an impoverished planet. Engineers, scientists, technicians, and policymakers are charged with the role of identifying technologies that can meet these needs and improve the standard of living on the planet, in spite of significant resource constraints.

Sustainable design is the design of products, processes, or systems that balance our beliefs in the sanctity of human life and promote an enabling environment for people to enjoy long, healthy, and creative lives, while protecting and preserving natural resources for both their intrinsic value and the natural world's value to humankind. Engineers who practice sustainable design must have a grasp of the social, economic, and environmental consequences of their design decision and a thorough

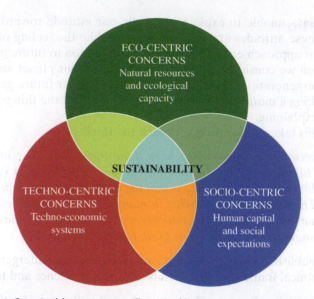

FIGURE 1.17 Sustainable systems are illustrated by those systems that balance eco-centric, techno-centric, and socio-centric concerns.

Source: Based on Elkington, J., "Towards the Sustainable Corporation: Win-Win-Win Business Strategies for Sustainable Development," *California Management Review* 36, no. 2 (1994): 90–100.

understanding of the scientific principles of the technology available, as illustrated in Figure 1.17. The EPA defined sustainability as "the continued protection of human health and the environment while fostering economic prosperity and societal well-being" (Fiksel et al., 2012, p. 5). The only way we can actually achieve sustainable development is for scientists, engineers, technicians, and policymakers to develop and apply more efficient technologies that improve the standard of living and are adaptable to the global marketplace. Furthermore, we need to identify and relate the limitations of our natural resources to our desire for continual development.

ACTIVE LEARNING EXERCISE 1.3 Sustainable Policy

Write down as many sustainability policies or regulations as you can. These might be laws related to the environment, health, or resources—such as limitations on fish catch, and so on. How would we engineer a future for more people, more stuff, and more energy that is all available for a longer time (into the future)? Describe how you might develop a sustainability policy for one aspect of something important in your life, and explain how that policy might be applied locally, nationally, and internationally.

In 1948, an air pollution event in Donora, Pennsylvania, resulted in the deaths of 20 people and thousands of pets as people were ordered to evacuate their homes and leave their pets behind. The fact that pets suffered greatly in the Donora events has been almost completely ignored by all accounts. Why do you think this information is largely ignored? Is this an ethically acceptable behavior? Why are we mostly concerned about only our own species?

1.6   Populations and Consumption

The relationship between consumption of a natural resource, waste production, and regeneration of that resource is described mathematically as the carrying capacity. Dr. Paul Bishop (1999) defines the **carrying capacity** as "the maximum rate of resource consumption and waste discharge that can be sustained indefinitely in a given region without progressively impairing the functional integrity and productivity of the relevant ecosystem." A sustainable economic system operating within the Earth's carrying capacity demands the following.

The usage of renewable resources is not greater than the rates at which they are regenerated.

The rates of use of nonrenewable resources do not exceed the rates at which renewable substitutes are developed.

The rates of pollution or waste production do not exceed the capacity of the environment to assimilate these materials.

In order to calculate the carrying capacity of the Earth and its resources, we must be able to make some predictions about human population and consumption patterns. The world's steadily increasing population has put stress on available natural resources, including food, water, energy, phosphorus, fossil fuels, and precious metals. Both human population and consumption of resources are *increasing* at an *increasing rate*. Thus, linear mathematical models do not accurately estimate population trends; instead, we must apply an exponential model to estimate population growth and resource consumption patterns.

Exponential growth occurs when the rate of change, dA/dt, is proportional to the instantaneous value of A at some time t:

$$\frac{dA}{dt} = kA \tag{1.7}$$

We can integrate this expression with respect to time, which yields

$$A_{(t)} = A_o \exp(k(t - t_o)) \tag{1.8}$$

where A_o is the value of A at our initial time, t_o. The variable k is the exponential growth rate constant and typically has units of [1/time].

EXAMPLE 1.3   Historical World Population Growth

The world's population was approximately 370 million in AD 1350. By the year 1804, the world's population had reached 1 billion people. Find the exponential growth rate during that time period.

The term k in Equation (1.8) represents the population growth constant. We can rearrange Equation (1.8) to solve for the rate constant:

$$A_{(t)} = A_o \exp(k(t - t_o))$$

$$\ln\left(\frac{A_{(t)}}{A_o}\right) = k(t - t_o)$$

$$k = \left(\ln\left(\frac{A_{(t)}}{A_o}\right)\right)\left(\frac{1}{(t - t_o)}\right)$$

In this example, the variables are

- Initial population = A_o = 370 × 10⁶ people
- Initial time = t_o = 1350
- Final population = $A_{(t)}$ = 1 × 10⁹ people
- Final time = t = 1804

Rearranging Equation (1.8) and substituting the appropriate values into our exponential equation yield

$$k = \left[\ln\left(\frac{1 \times 10^9}{370 \times 10^6} \right) \right] \left(\frac{1}{1804 - 1350} \right) = 0.0022 \, \frac{1}{\text{year}} \text{ or } 0.22\%$$

EXAMPLE 1.4   Current World Population Growth

The world's population was estimated to be 6 billion in 1999, but by 2012 the world's population had already grown to 7 billion people. What was the world's population growth rate? When should we expect the population to reach 9 billion people if the growth rate between 1999 and 2012 remains constant?

In this example, the variables are

- Initial population = A_o = 6 × 10⁹ people
- Initial time = t_o = 1999
- Final population = $A_{(t)}$ = 7 × 10⁹ people
- Final time = t = 2012

From Example 1.3 the exponential growth rate constant, k, is found from

$$k = \left(\ln\left(\frac{A_{(t)}}{A_o} \right) \right) \left(\frac{1}{(t - t_o)} \right)$$

$$k = \left[\ln\left(\frac{7 \times 10^9}{6 \times 10^9} \right) \right] \left(\frac{1}{2012 - 1999} \right) = 0.0119 \, \frac{1}{\text{year}} \text{ or } 1.19\%$$

Then, using the current rate constant and Equation (1.8), we can solve for the time when we expect the population to reach 9 billion as

$$\ln\left(\frac{A_{(t)}}{A_o} \right) = k(t - t_o)$$

$$t = \left[\ln\left(\frac{A_{(t)}}{A_o} \right) \right] \left(\frac{1}{k} \right) + t_o = \left[\ln\left(\frac{9 \times 10^9}{7 \times 10^9} \right) \right] \left(\frac{1}{0.0119} \right) + 2012 = 21.2 + 2012 = 2033$$

So, the population may reach 9 billion in 2033.

One of the most famous demographers of all time, Thomas Robert Malthus (1766–1834), wrote the following in his book *An Essay on the Principle of Population* (1798):

The power of population is so superior to the power of the earth to produce subsistence for man, that premature death must in some shape or other visit the human race.

Thomas Malthus made this argument because he believed the human population would very soon outpace the production of food and the regeneration of natural resources. He believed this would create a catastrophe that would surely reduce the population to a number that would be more sustainable. What Malthus did not foresee was a change in technology that would allow for the use of mechanical power and industrial fertilizer. These technological changes allowed for greater production of food than Malthus could conceive based upon the technology that was available during his lifetime.

The Malthusian Catastrophe, as Malthus's hypothesis was called, seemed to be reaching fulfillment during Malthus's life since the signs of his times were quite bleak. London was becoming a huge city—one of the largest on the planet at that time. The famed Thames River had become an open sewer and a source for miasma, which people at the time associated with sickness from ill winds and rotten smells. As the population of London grew, the waste from this population accumulated in the Thames. In the summer of 1848, a deadly cholera outbreak in London killed at least 14,600 people, according to published records. Shortly after that outbreak, in 1854, yet another outbreak of cholera occurred. At this time, Dr. John Snow began to trace the cause of the disease by mapping the cases and neighborhoods where the cholera was occurring. Dr. Snow was able to trace the cases to a community well, known as the Broad Street Well. He was able to convince people to remove the handle to the well pump, and shortly thereafter, the cholera cases in that community subsided. Thus disease was linked for the first time with water contaminated by human waste. In spite of Dr. Snow's efforts, 10,675 deaths from cholera were reported in 1854. The link between the contaminated water and cholera soon led to acceptance of the germ theory and to important advances in public health, epidemiology, engineering, and water treatment, which eventually had a profound influence on the standard of living and lifespan throughout the rapidly industrializing nations.

Cholera still plagues those nations with inadequate water and sanitation systems. In Zimbabwe in December of 2008, for example, a massive cholera outbreak occurred that is believed to have infected more than 57,000 people and resulted in more than 3,000 deaths. The continuing closing of several local hospitals and the scarcity of basic medical commodities such as medicine and health personnel are believed to have been a major contributor to the spread. The state media reported that most of the capital city of Harare had been left without water after the city ran out of chemicals for its treatment plant. Chlorination has been shown to effectively prevent cholera, as shown in Figure 1.18, which illustrates how implementing chlorination in the United States, starting in the 1920s, greatly reduced the incidence of waterborne disease. Yet Zimbabwe, Haiti, and many other low- and medium-income countries still lack the infrastructure to adequately provide clean water. Despite readily available and effective methods of treating drinking water, major outbreaks of cholera still occur regularly in West Africa and Haiti (Figure 1.19).

When Malthus published his dissertation in 1798, the world's population was just approaching 1 billion people. In 2012, when the world's population

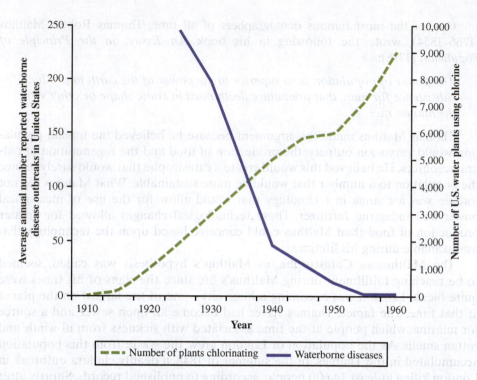

FIGURE 1.18 Chlorination of water supplies is an effective method of preventing cholera.

Source: Based on Culp, G.L., and Culp, R.L. (1974). *New Concepts in Water Purification*. New York: Litton Educational.

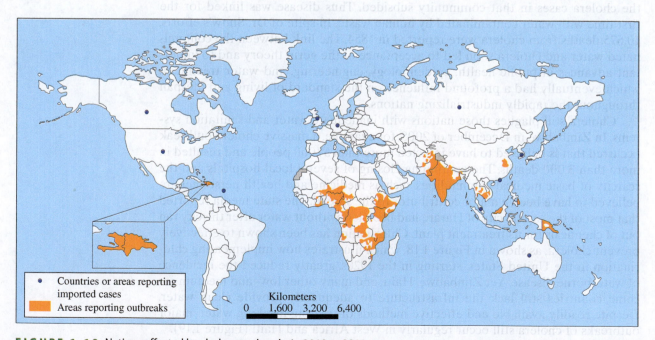

FIGURE 1.19 Nations affected by cholera outbreaks in 2010 to 2011.

Source: Based on World Health Organization (2012). reliefweb.int/map/world/cholera-areas-reporting-outbreaks-2010-2011.

surpassed 7 billion people, many economists, environmentalists, and scientists were once again gravely concerned that we would not be able to retain and improve the standard of living as the human population put greater strain on available natural resources.

In 1968, Garrett Hardin published an essay called "The Tragedy of the Commons" that discussed the problems with Thomas Malthus's original arguments and dates, but Hardin generally supported the notion that either population or the standard of living must be limited by the available natural resources. Hardin believed the first reason that there would be a great tragedy related to the environment is that it is not mathematically possible to maximize for two (or more) variables at the same time.

The second reason for Hardin's pessimism springs directly from biological facts. To live, any organism must have a source of energy (for example, food). This energy is utilized for two purposes: maintenance and work. For humans, life requires about 1,600 kilocalories a day just for "maintenance calories." Anything that they do over and above merely staying alive will be defined as work and is supported by "work calories" that they take in. Work calories are used not only for what we call work in common speech; they are also required for all forms of enjoyment—from swimming and automobile racing to playing music and writing poetry. If our goal is to maximize population, it is obvious what we must do: We must make the work calories per person approach as close to zero as possible. No gourmet meals, no vacations, no sports, no music, no literature, and no art (Hardin, 1968, p. 1).

Both Malthus and Hardin present a formidable argument: increasing populations lead to more difficulty in accomplishing the United Nations goals associated with sustainable development. The National Academy of Engineers has recognized that the tasks associated with sustainable development represent the grandest challenges to engineers in this century. These challenges for the next generation of engineers and scientists are illustrated in Figure 1.20.

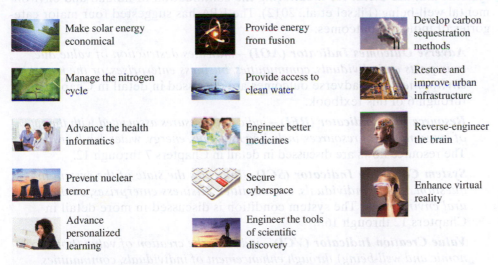

FIGURE 1.20 Grand Challenges in engineering for the next century.

Source: Based on National Academy of Engineering. www.engineeringchallenges.org/; c12/Shutterstock.com; Roman Sigaev/Shutterstock.com; Andriano/Shutterstock.com; Brian A Jackson/Shutterstock.com; Sl_photo/Shutterstock.com; hxdbzxy/Shutterstock.com; wavebreakmedia/Shutterstock.com; Pressmaster/Shutterstock.com; ollyy/Shutterstock .com; hxdyl/Shutterstock.com; Foxaon1987/Shutterstock.com; Syda Productions/Shutterstock.com; Stockfour/ Shutterstock.com; Sergey Nivens/Shutterstock.com.

ACTIVE LEARNING EXERCISE 1.4 Tragedy of the Commons

Read and analyze Garrett Hardin's essay "The Tragedy of the Commons." Do you agree with his premises and conclusions? What are your thoughts on the implications for freedom, the role of governmental authorities, and the "public good?" Be prepared to discuss your conclusions.

1.7 Technical Approaches to Quantifying Sustainability

It is vitally important that we develop quantitative methods of determining which technologies of the future truly represent progress toward more sustainable development. In this book, we will explore several methods of developing and evaluating sustainably designed products and processes. In order to effectively develop quantitative relationships, we must understand the fundamental units of measure that will help us compare and contrast the sustainable technologies of today and tomorrow.

Various terms are used to describe measures of sustainability and their interrelationships to one another. More specifically, a *measure* of sustainability is a value that is quantifiable against a standard at a point in time. A sustainability *metric* is a standardized set of measurements or data related to one or more sustainability indicators. A sustainability *indicator* is a measurement or metric based on verifiable data that can be used to communicate important information to decision makers and the public about processes related to sustainable design or development (Biodiversity Indicators Partnership, 2010). Sustainability indicators are those measurable aspects of economic, environmental, or societal systems that are useful for monitoring the continuation of human and environmental well-being (Fiksel et al., 2012). The EPA has suggested four major categories of indicator outcomes:

> *Adverse Outcomes Indicator (AOI)—indicates destruction of value due to impacts on individuals, communities, business enterprises, or the natural environment.* The adverse outcomes are discussed in detail in Chapters 2 through 6 of this textbook.
>
> *Resource Flow Indicator (RFI)—indicates pressures associated with the rate of consumption of resources, including materials, energy, water, land, or biota.* The resource flows are discussed in detail in Chapters 7 through 12.
>
> *System Condition Indicator (SCI)—indicates the state of the system in question, that is, individuals, communities, business enterprises, or the natural environment.* The system condition is discussed in more detail in Chapters 12 through 16.
>
> *Value Creation Indicator (VCI)—indicates the creation of value (both economic and well-being) through enhancement of individuals, communities, business enterprises, or the natural environment.* Value Creation typically lies outside the realm of the traditional engineering sciences and interfaces closely with the social and economic sciences. Value creation is discussed in detail in Chapters 13 through 17.

The relationships between these four indicator categories are illustrated in Figure 1.21. The resource flows have value, and the value is distributed by human actions and system relationships among natural capital, economic capital, and human capital. We may attempt to measure various indicators within the complex system. Examples of specific indicators of each indicator category are shown in Table 1.5. These individual indicators may be combined to form more complex tools that make up a sustainability index. A *sustainability index* is a numerical-based scale used to compare alternative designs or processes with one another. Examples of sustainability indices include the IPAT equation and the environmental footprint.

1.7.1 The IPAT Equation

In the early 1970s, environmentalist Paul Ehrlich suggested that environmental impact from human activities was the result of three contributing factors, and he proposed a conceptualized mathematical formula to represent this concept. This formula has become known as the **IPAT equation:**

$$I = P \times A \times T \tag{1.9}$$

where

I = environmental impact

P = population

A = affluence

T = technology

The IPAT equation is often used as a starting point to study the relationships between population, economics, and technological development and how they

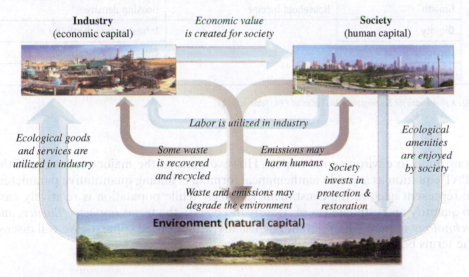

FIGURE 1.21 The resource flows, system dependence, and values among economic, human, and natural capital. Value, or capital, may be moved from one area of sustainability to another; maintaining a balance of capital resources for future generations is an intrinsic property of sustainability.

Source: Fiksel, J., et al. (2012). *A Framework for Sustainability Indicators at EPA*, Figure 3, p. 10.

TABLE 1.5 The major categories of sustainability indicators and examples of measurable properties

INDICATOR CATEGORY	INDICATOR TYPES	NATIONAL SCALE EXAMPLES	COMMUNITY SCALE EXAMPLES
Resource Flow Indicators	Volume	Greenhouse gas emissions	Greenhouse gas emissions
	Intensity	Material flow volume	Material flow analysis
	Recovery	Resource depletion rate	Water treatment efficacy
	Impact		Recycling rate
	Quantity		Land use
Value Creation Indicators	Profitability	Cost (reduction)	Cost (reduction)
	Economic output	Fuel efficiency (gain)	Fuel efficiency (gain)
	Income	Energy efficiency (gain)	Energy efficiency (gain)
	Capital investment		Vehicle use (miles per capita)
	Human development		
Adverse Outcome Indicators	Exposure	Health impacts of air pollution	Health impacts of air pollution
	Risk	Public safety	Public safety
	Incidence	Life cycle footprint of energy use	Sewer overflow frequency
	Impact		
	Loss		
	Impairment		
System Condition Indicators	Health	Air quality	Air and water quality
	Wealth	Water quality	Local employment
	Satisfaction	Employment	Local household income
	Growth	Household income	Housing density
	Dignity		Infrastructure durability
	Capacity		Community educational equity
	Quality of life		

Source: Based on Fiksel, J., et al. (2012). *A Framework for Sustainability Indicators at EPA*, Table 1, p. 11.

contribute to environmental impact. However, one of the major challenges of the IPAT equation as a usable mathematical formula is finding quantitative parameters to represent each of the constituent variables. While population is relatively easy to quantify and estimate (as previously discussed), the terms *impact*, *affluence*, and *technology* require better definitions, usually within varied contexts. We will discuss the terms before returning to the IPAT equation.

1.7.2 Impact

Environmental impact can be defined in terms of the carrying capacity of the planet. Anything that contributes to diminishing the sustainable rate of resource consumption and waste discharge can be classified as an *impact*. Although the Earth has natural mechanisms that replenish most resources and manage most waste discharges, these processes typically occur at very slow rates compared to the rates at which current impacts occur. Current rates of most environmental impacts can be directly linked to human-made developmental activities that will be discussed in Part II of the text.

Impacts can be classified based on the degree of severity and the scale of the impact. Critical impacts include global climate change, loss of water quality and quantity, depletion of fossil fuel resources, loss of biodiversity, land degradation, and stratospheric ozone depletion. Critical impacts can potentially lead to irreversible consequences. Significant but noncritical impacts include depletion of nonfossil fuel resources, acid precipitation, smog, and aesthetic degradation. Less significant impacts include radionuclide contamination, depletion of landfill capacity, thermal pollution, oil spills, and odors.

Impacts may also be classified based on the scale of the area directly affected. Hence, global-scale concerns—like global climate change, ozone layer depletion, and loss of habitat and biodiversity—ultimately affect the entire planet regardless of the source of the impacts.

Regional-scale concerns have more limited impacts; these include soil degradation, acid precipitation, and changes in surface water chemistry leading to consequences such as acidification and eutrophication. The impacts associated with the use of herbicides and pesticides may also be included in this category.

The impacts of local-scale concerns tend to be localized to the areas of origin of the impacts. These include groundwater pollution, oil spills, hazardous waste sites, and photochemical smog.

Most of these impacts are the direct or indirect results of technological developments and advancements in society. Understanding these classifications and consequently applying them to preventive efforts are central to industrial ecology studies.

1.7.3 Population

Population refers to the number of individual organisms present in a particular ecological system. Since organisms consume resources and generate waste as part of their metabolic activities, the population of an ecological system has a direct impact on the sustainability of that system. Population growth follows an exponential mathematical relationship described by Equation (1.8).

For a given area or region of study, the effective population growth rate, r, is equivalent to the generic exponential growth rate, k, in Equation (1.8). The population growth rate of a defined region is calculated from four factors:

- Rate of birth, r_b
- Rate of death, r_d
- Rate of immigration into the area, r_i
- Rate of emigration from the area, r_e

$$r = r_b + r_i - r_d - r_e \tag{1.10}$$

If we take the entire planet as our area of interest, we can conclude that the global population growth rate is not affected by immigration and emigration from Earth (not yet anyway). Therefore, the birth rate, r_b, and the death rate, r_d, are the contributing factors to population growth.

To reduce the rate of population growth or even reverse the growth trend, the rate of birth has to decrease or the rate of death has to increase, or a combination of both. Advances in health care, public health education, and policies, as well as modern methods of resolving social and political conflicts, all contribute to increases in the average global life expectancy. In general, people are living longer today than in previous generations. As shown in Table 1.6, data from the Population Division of the United Nations, Department of Economic and Social Affairs indicate that the global death rate, measured in deaths per 1,000 persons per year, is currently at a historical low, with projections that the death rate may start to rise in the next decade. These estimates were calculated from information collected for "every country in the world, including estimates and projections of 60 demographic indicators such as infant mortality rates and life expectancy." These values are for medium variant projections that balance between the upper and lower bound projection values.

Even with reductions in the growth rate in recent decades, in the foreseeable future it is projected that the overall global population growth rate, r, will remain positive, leading to an overall increase in global population over time.

1.7.4 Affluence

Affluence is a bit more difficult to define and quantify. In general, *affluence* refers to a measure of the quality of life of individual members of the society. It represents the complex link between the economic well-being of the society as a whole and the consumption patterns of the average member of that society. One indicator of affluence

TABLE 1.6 World historical and predicted crude death rates. Medium variant projection.

YEARS	CRUDE DEATH RATE (per 1,000 per year)	YEARS	CRUDE DEATH RATE (per 1,000 per year)
1950–1955	18.713	2030–2035	8.728
1955–1960	17.135	2035–2040	9.112
1960–1965	16.122	2040–2045	9.514
1965–1970	12.964	2045–2050	9.896
1970–1975	11.751	2050–2055	10.239
1975–1980	10.636	2055–2060	10.553
1980–1985	10.097	2060–2065	10.834
1985–1990	9.554	2065–2070	11.066
1990–1995	9.245	2070–2075	11.256
1995–2000	8.974	2075–2080	11.393
2000–2005	8.689	2080–2085	11.463
2005–2010	8.390	2085–2090	11.479
2010–2015	8.194	2090–2095	11.492
2015–2020	8.159		
2020–2025	8.239		
2025–2030	8.426		

Source: Based on UN data, data.un.org/

is the gross domestic product (GDP), which is an estimate of the market value of all goods and services produced. The GDP can be measured on a per capita basis, as illustrated in Figure 1.22. Figure 1.22 illustrates the disproportion of economic means allocated on a per capita basis, but GDP alone is insufficient to adequately describe the broader term *affluence*. The easiest way to think about affluence is to consider the access to and consumption of resources by members of the society. Figure 1.23 illustrates the correlation between GDP, population, and consumption. Consumption refers to the selection, use, reuse, maintenance, repair, and disposal of goods and services (Leslie, 2009). It has also been defined as the "human transformations of materials and energy" (Myers, 1997). Consumption becomes a problem when it makes materials and energy less available for future use and threatens human health and welfare.

Professor Thomas Princen of the University of Michigan has identified three layers of consumption:

- **Background consumption**: This term refers to the normal biological functioning of all organisms to meet physical and/or psychological needs in order to survive and reproduce. The total impact of background consumption is a function of the aggregate consumption of the total population. In this case, population is the significant driver of impact.

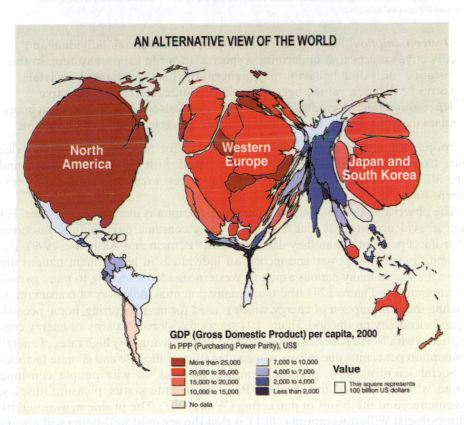

FIGURE 1.22 A carotid illustration of the distribution of wealth sizes of the countries according to their relative financial status based on GDP per capita.

Source: GRID-Arendal, World economy cartogram, www.grida.no/graphicslib/detail/world-economy-cartogram_1551.

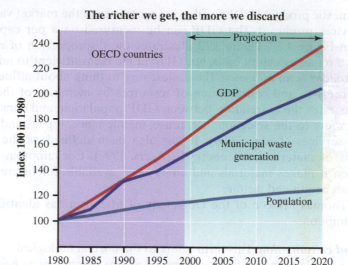

FIGURE 1.23 The correlation between GDP, population, and consumption shows that according to various scenarios, the GDP will most likely continue to increase for the next decades—but at a slower pace for those countries that can afford advanced waste management strategies.

Source: Based on GRID-Arendal, The richer we get, the more we discard—human consumption, waste and living standards, 2005, www.grida.no/graphicslib/detail/the-richer-we-get-the-more-we-discard-human-consumption -waste-and-living-standards_5bcc.

- *Overconsumption*: This is a level of consumption based on individual and collective choices that undermine a species' own life support system. In this instance, individual behavior may seem rational and conform to societal norms or dictates, yet the aggregate effect is injurious to the collective.
- *Misconsumption*: Misconsumption is when individuals consume in a way that undermines their own well-being even if there are no aggregate effects on the collective.

People sometimes mistake overconsumption for misconsumption and vice versa. Consumption trends are often correlated with economic metrics, and more affluent societies tend to consume more resources per capita than less affluent ones.

The observable trend is that consumption per capita is increasing over time. For example, evidence suggests that Americans today consume more natural resources and artifacts per capita than they did in the past (Putnam and Allshouse, 1999). The per capita purchasing power and income of individuals in less affluent nations are also on the rise as many nations aspire to Western standards, leading to more overall consumption (see Figure 1.24). There are many potential indicators of consumption, including the consumption of energy, which is used for manufacturing, home necessities, and electricity production. Figure 1.24 shows the global disparity of energy consumption—with North America and northern Europe having very high rates of energy consumption per capita, due in part to a combination of lifestyle and climate factors.

Social scientists have proposed various theories on why people consume beyond what they need for survival. Reasons include status, pleasure, display, convenience, and the result of marketing (Wilk, 2002). The problem, as noted by anthropologist Willett Kempton (2001), is that the sense of well-being will always be relative, generating an unending spiral of increasing consumption. So the implication of population growth for the carrying capacity of Earth is magnified by the fact that the average person today is consuming significantly more than the

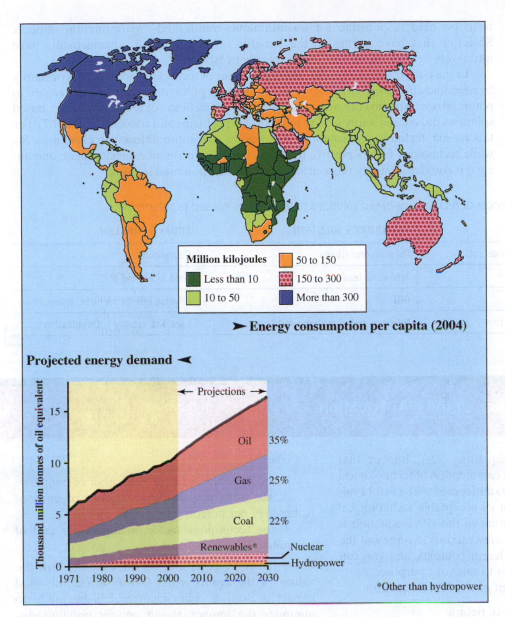

FIGURE 1.24 Worldwide energy consumption per capita.

Source: Based on GRID-Arendal, Energy consumption per capita (2004), www.grida.no/graphicslib/detail
/energy-consumption-per-capita-2004_5dca.

average person of a few decades ago. Moreover, the overconsumption of goods
leads to increased waste production.

1.7.5 Technology

Technology refers to all the artifacts of human development as well as the
processes and systems that contribute to these artifacts. It can be said that observed
trends in consumption are driven directly or indirectly by achievements made
in technological development. The role of technology in facilitating social and
economic development is not as contentious as the question of whether technology

is to be held responsible for unsustainable trends and environmental impacts. Societies that have harnessed the benefits of technological advancements have ultimately become more socioeconomically prosperous.

In the past, the development and deployment of technology were driven primarily by need and effectiveness in dealing with immediate problems, with little or no regard to potential downstream consequences and implications. Some of those technology-based solutions did consequently lead to modern-day problems and concerns (Table 1.7). For this reason, technological solutions may lead to only more unforeseen challenges. So, while technology fosters positive trends in social and economic development, unmanaged it could also nullify those achievements by its potential impacts.

TABLE 1.7 Consequences of technology-based solutions to technology-caused problems

YESTERDAY'S NEED	YESTERDAY'S SOLUTION	TODAY'S PROBLEM
Nontoxic, nonflammable refrigerants	Chlorofluorocarbons	Ozone hole
Automobile engine knock	Tetraethyl lead	Lead in air and soil
Locusts, malaria	DDT	Adverse effects on birds, mammals
Fertilizer to aid food production	Nitrogen and phosphorus fertilizer	Lake and estuary eutrophication

BOX 1.5 Application of the IPAT Equation

The IPAT equation [Equation (1.9)] implies that impacts occur owing to a combination of technological and socioeconomic factors and provides a basis for evaluating the parameters of an acceptable technological solution. One useful feature of the IPAT equation is that it can be written in many variants to represent the same relationship in different contexts, and you can develop your own variant to suit your purpose.

Consider one variant of the IPAT equation:

Total impact = number of people

$$\times \frac{\text{number of units of technology}}{\text{person}}$$

$$\times \frac{\text{impact}}{\text{unit of technology}} \quad (1.11)$$

The number of people represents the population factor, P

The number of units of technology per person represents the affluence factor, A

And the impact per unit of technology represents the technology factor, T

where

P, the population factor = number of people

A, the affluence factor = number of units of technology per person

T, the technology factor = impact per unit of technology

Using the IPAT equation, we can estimate how the values of the contributing factors affect the total impact as well as what changes can be made to minimize the impact. Based on our previous discussions on the rising trends in population (P) and consumption (A), to keep the current total impact value unchanged requires a significant reduction in the value of the technology factor (T). And in order to reverse the trend and bring the impact values to naturally manageable equilibrium values, even more significant reductions must be achieved.

For example, consider the impact associated with the combustion of gasoline in automobiles. The overall impact is related to the technology of the internal combustion engine, which burns gasoline and emits carbon dioxide, contributing to global climate change, but it is also a function of how many people use this technology—that is, how many cars are driven and how much gasoline is consumed per car, or the average distance traveled.

This relationship may be represented by an equation:

Total CO_2 emitted = no. of automobiles in use
$\times$ average distance traveled per automobile
$\times$ CO_2 emitted per distance traveled   (1.12)

To reduce the total CO_2 emitted from gasoline-powered automobiles, one could

1. Reduce the total number of automobiles in use
2. Reduce the average distance traveled per automobile
3. Reduce the CO_2 emitted per distance traveled

The U.S. Department of Transportation study of travel trends in the United States shows that since 1970, the increase in vehicle miles traveled has far outpaced population growth (Figure 1.25). This trend aligns closely with economic trends as seen in the GDP. The average annual driving distance in the United States increased by nearly 300% in the three decades from 1970 to 2000. A similar study by

the Department of Energy shows that the number of vehicles in the United States is growing faster than the population and that the percentage of households with three or more vehicles has increased significantly over the past 50 years (Table 1.8).

For example, the total CO_2 emitted represents the impact, I; the number of automobiles in use represents the population, P; the average distance traveled per automobile represents a measure of affluence, A; and the CO_2 emitted per distance traveled represents the technology, T. In this case, it gives an indication of the efficiency of the engine. Based on the observed trends (Figures 1.26 and 1.27) and the fact that options 1 and 2 are related to socioeconomic factors previously discussed, a technological approach to reducing total CO_2 emissions would be to reduce the CO_2 emitted per distance traveled. More efficient engines will emit less CO_2 per distance traveled by requiring less gasoline to travel.

The IPAT equation allows us to calculate what rate of reduction of CO_2 emissions per distance traveled we would need to achieve to balance the effect of the increasing number of automobiles in use and the average distance traveled per automobile.

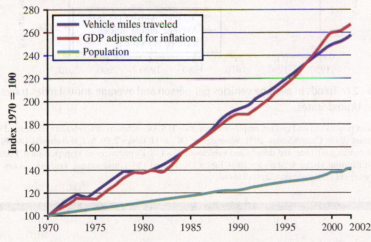

FIGURE 1.25 Vehicle miles traveled as a function of population and GDP in the United States.

Source: U.S. Department of Transportation, Transportation Air Quality Facts and Figures January 2006, Travel Trends, www.fhwa.dot.gov/environment/air_quality/publications/fact_book/page06.cfm.

(Continued)

BOX 1.5   Application of the IPAT Equation *(Continued)*

TABLE 1.8 Household vehicle ownership

YEAR	NO VEHICLES	ONE VEHICLE	TWO VEHICLES	THREE OR MORE VEHICLES
1960	21.53%	56.94%	19.00%	2.53%
1970	17.47%	47.71%	29.32%	5.51%
1980	12.92%	35.53%	34.02%	17.52%
1990	11.53%	33.74%	37.35%	17.33%
2000	9.35%	33.79%	38.55%	18.31%
2009	8.90%	33.69%	37.56%	19.85%

Source: Based on 1960–2009 Census. U.S. Department of Transportation, Volpe National Transportation Systems Center, Journey-to-Work Trends in the United States and its Major Metropolitan Areas, 1960–1990, Cambridge, MA, 1994, p. 2–2. 2000 data – U.S. Bureau of the Census, American Fact Finder, factfinder.census.gov, Table QT-04, August 2001. (Additional resources: www.census.gov) 2009 data – U.S. Bureau of the Census, American Community Survey, Table DP-4, 2009. cta.ornl.gov/data/chapter8.shtml.

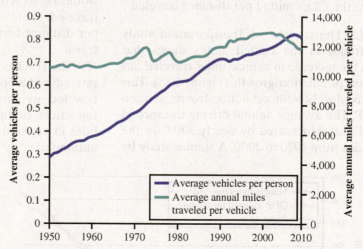

FIGURE 1.26 Trends in average vehicles per person and average annual miles traveled per vehicle in the United States.

Sources: Resident population and civilian employed persons – U.S. Department of Commerce, Bureau of the Census, Statistical Abstract of the United States–2011, Washington, DC, 2011, tables 2, 59, 601, and annual. (Additional resources: www.census.gov). Licensed drivers and vehicle-miles – U.S. Department of Transportation, Federal Highway Administration, Highway Statistics 2009, Tables DL-1C and VM-1, and annual. (Additional resources: www.fhwa.dot.gov) cta.ornl.gov/data/chapter8.shtml.

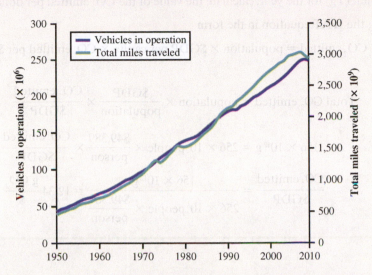

FIGURE 1.27 Trends in vehicles in operation and total miles traveled for the United States.

Source: U.S. Department of Transportation, Federal Highway Administration, Highway Statistics 2009, Tables DL-1C and VM-1.

Another variant of the IPAT equation is

$$\text{Total impact} = \text{population} \times \$\text{GDP per capita} \times \text{impact per } \$\text{GDP} \quad (1.13)$$

The gross domestic product, GDP, is a measure of the monetary value of all goods and services produced by a country over a specific time period. The GDP per capita of a country is the total GDP divided by the population of the country. It is a measure of the affluence and standard of living of the country. While some countries may have high GDP values, their GDP per capita may be comparatively low because of high population values.

The impact per GDP refers to the impact of all activities associated with the generation of one unit of GDP. This may include manufacturing, transportation, packaging, sales, and so on. The value of the impact per GDP of a country is often seen as a measure of that country's environmental efficiency. More efficient countries have lower impact per GDP values.

EXAMPLE 1.5 Using the IPAT Equation

The population of a country is 256 million, and the annual GDP per capita is estimated at \$49,389 per person. If the total CO_2 emission of the country is estimated at 156 teragram (Tg) for the year, calculate the value of the CO_2 emitted per dollar of GDP.

Using the IPAT equation in the form

Total CO_2 emitted = population × \$GDP per capita × CO_2 emitted per \$GDP

or

$$\text{Total } CO_2 \text{ emitted} = \text{population} \times \frac{\text{\$GDP}}{\text{population}} \times \frac{CO_2 \text{ emitted}}{\text{\$GDP}}$$

$$156 \times 10^{12}\,\text{g} = 256 \times 10^6\,\text{people} \times \frac{\text{\$49,389}}{\text{person}} \times \frac{CO_2 \text{ emitted}}{\text{\$GDP}}$$

$$\frac{CO_2 \text{ emitted}}{\text{\$GDP}} = \frac{156 \times 10^{12}\,\text{g}}{256 \times 10^6\,\text{people} \times \dfrac{\text{\$49,389}}{\text{person}}} = 12.34\,\frac{\text{g } CO_2}{\text{\$}}$$

EXAMPLE 1.6 Using the IPAT Equation to Incentivize Emissions Reduction

If the population of a country increased by 8% in the same time that the GDP per capita increased by 5%, what percentage change in the emissions per dollar of GDP would be required to maintain the total emissions at the original values?

Using the IPAT equation

$$I = P \times A \times T$$

$$\text{Impact} = \text{population} \times \text{affluence} \times \text{technology}$$

$$\text{Total emissions} = \text{population} \times \frac{\text{\$GDP}}{\text{population}} \times \frac{\text{emissions}}{\text{\$GDP}}$$

Let I_i, P_i, A_i, and T_i represent the initial impact, population, affluence, and technology factors, respectively, and I_f, P_f, A_f, and T_f represent the final impact, population, affluence, and technology factors, respectively.

If the final impacts are not to be greater than the initial impacts, then

$$I_i = I_f$$

then substituting from Equation (1.9) yields

$$P_i \times A_i \times T_i = P_f \times A_f \times T_f$$

From the information given with a population increase of 8%,

$$P_f = P_i + \left(\frac{8}{100} \times P_i\right) = 1.08 P_i$$

GDP per capita, the measure of affluence, increased by 5% is

$$A_f = A_i + \left(\frac{5}{100} \times A_i \right) = 1.05 A_i$$

Substituting the relationships between final and initial population and affluence into the previous equation yields

$$P_i \times A_i \times T_i = 1.08 P_i \times 1.05 A_i \times T_f$$

Solving for the technology terms, T:

$$T_i = 1.134 T_f$$

$$\frac{T_f}{T_i} = \frac{1}{1.134} = 0.88 \quad or \quad T_f = 0.88 T_i$$

The technology factor defined by the emissions per dollar of GDP needs to be reduced by 12%.

ACTIVE LEARNING EXERCISE 1.5 IPAT: Predicting the Future We Will Create

Pick one of the three cases listed below and define each of the four terms in the IPAT equation for the chosen scenario:

- Gasoline used in automobiles
- Fertilizers used to grow corn
- Greenhouse gas (GHG) emissions from coal-burning plants

Which term has the greatest influence on the *impact*? What term is the most variable or uncertain?

1.8 Productivity, Consumption, and the Ecological Footprint

Prior to the Industrial Revolution, the number of humans and their impact on the planet were relatively small. As human population growth and resource consumption accelerated, the impact on natural resources also accelerated, to the point where many scientists believe we have already passed the sustainable carrying capacity of the planet. Scientists estimate that sometime in the mid-1970s, demand on natural resources was outpaced by what resources the planet could produce. In order to understand these potential limitations on the Earth's capacity, one must have methods to measure the production and consumption of resources.

1.8.1 Biocapacity

Biocapacity may be thought of as the ecosystem's capacity to produce biological materials used by people and to absorb waste material generated by humans, under current management schemes and extraction technologies. Biocapacity is dependent upon the state of the natural environment and environmental factors like climate; thus the measured biocapacity in the year 2020 is vastly different than the biocapacity in the year 1600. Humans consume large amounts of biological material as food and fuel, and thus humans are dependent upon the productivity of natural ecosystems. The bioproductivity of land and water (both marine and inland waters) areas can be estimated from the area's photosynthetic activity and the accumulation of biomass used by humans and varies across the surface of the planet, as shown in Figure 1.28. This productivity includes many of the ecosystem services shown in

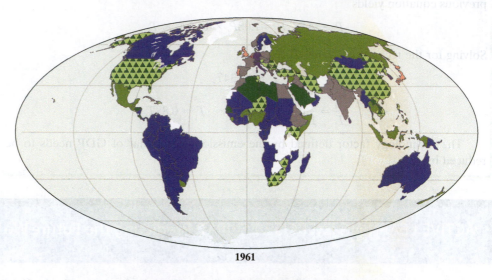

1961

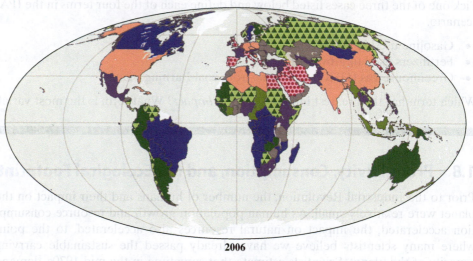

2006

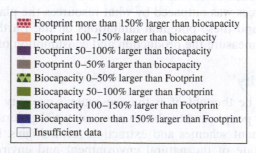

Footprint more than 150% larger than biocapacity
Footprint 100–150% larger than biocapacity
Footprint 50–100% larger than biocapacity
Footprint 0–50% larger than biocapacity
Biocapacity 0–50% larger than Footprint
Biocapacity 50–100% larger than Footprint
Biocapacity 100–150% larger than Footprint
Biocapacity more than 150% larger than Footprint
Insufficient data

FIGURE 1.28 Areas that exceed and have excess capacity in the rate of consumption of resources.

Source: Based on *The Ecological Wealth of Nations: Earth's biocapacity as a new framework for international cooperation 2010*. Figure 9, p. 21. Oakland, CA: Global Footprint Network.

FIGURE 1.29 Ecosystem services are the products obtained from ecosystems, regulating services are the benefits obtained from the regulation of ecosystem processes, cultural services are the nonmaterial benefits people obtain from ecosystems, and supporting services are those services that are necessary for the production of all other ecosystem services.

Source: *WWF (2018) Living Planet Report 2018. Aiming Higher*. M. Grooten and R.E.A. Almond (eds.). WWF, Gland, Switzerland.

Figure 1.29. Nonproductive areas with patchy vegetation as well as biomass that is not of use to humans are excluded from measures of bioproductivity. The total biologically productive area on land and water in 2019 was approximately 12.2 billion hectares.

The natural resources provide ecosystem services, such as providing clean air, clean water, and food for human society. The biosphere's productivity is variable and affected by human consumption of biological production and also humans' increased waste production that may damage biological productivity. Economic growth is thus generally predicated on the choices made by human society and the productivity of the biosphere; this in an inherent motivation for many of the SDGs, as illustrated in Figure 1.30.

The biocapacity of the Earth can be compared to the consumption of resources by human society. Consumption rates that are less than bioproductivity rates are generally sustainable over time, while consumption that outpaces bioproductivity will erode the natural capital of ecosystems and likely reduce future bioproductivity over time, and would therefore be unsustainable. The total of the consumptive activities of human society is called the *ecological footprint*. When the ecological footprint exceeds the biocapacity of the planet, the condition is called ecological overshoot.

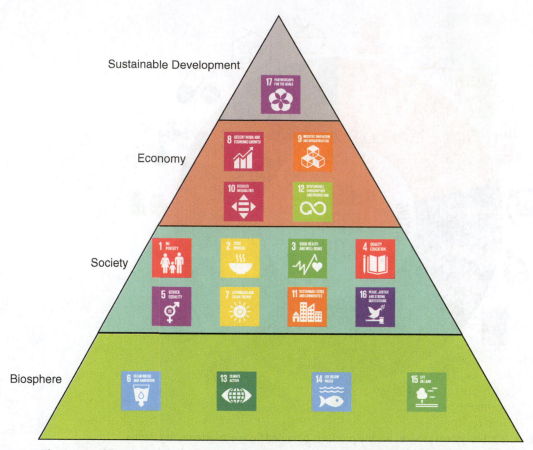

FIGURE 1.30 The Sustainable Development Goals are directly and indirectly related to bioproductivity of the biosphere.

Source: Based on *WWF (2018) Living Planet Report 2018. Aiming Higher*. M. Grooten and R.E.A. Almond (eds.). WWF, Gland, Switzerland; Artvictory/ Shutterstock.com.

This is illustrated in Figure 1.31, which shows the biocapacity, ecological footprint, and amount of overshoot in 2010.

1.8.2 Footprint Indicators of Sustainability

Footprint analysis relies on the mass balance and accounting methods developed through a process called a material flow analysis. For example, the water footprint estimates all of the water used by a person, product, or region; the water footprint tries to account for all of the water embodied in an individual's lifestyle, in a product's life cycle, or in a region's overall economic and social systems. Similarly, the carbon footprint uses mass balance and thermodynamics to calculate greenhouse gas emissions from direct or embodied energy use.

Several types of footprint analyses are becoming popular, especially for illustrating the consequences of particular lifestyle choices, predominant technologies, or production systems (Table 1.9). The ecological footprint is the oldest of the indicators and was developed to convey an understanding of overall natural resource sustainability. Designed by William Rees and Mathis Wackernagel in the 1990s, the ecological footprint is usually expressed as the amount of land (hectares) required to support particular patterns of lifestyle and consumption. The carbon footprint was developed during approximately the same timeframe, and it is widely used to calculate the reduction in carbon emissions from specific CO_2 mitigation strategies. The carbon footprint is typically expressed in CO_2 equivalents.

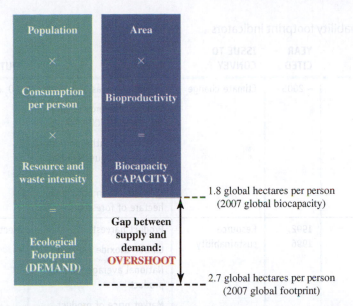

FIGURE 1.31 The relationship between the Earth's biocapacity, human society's ecological footprint, and the overshoot for the year 2010.

Source: Based on Ewing, B., Moore, D., Goldfinger, S., Oursler, A., Reed, A., and Wackernagel, M. (2010). *The Ecological Footprint Atlas 2010*. Oakland, CA: Global Footprint Network.

The water footprint was introduced in 2002 to help communicate information about water consumption and quality issues at multiple levels, from personal to national. Water footprints are usually expressed as a water volume. The nitrogen footprint was most recently developed (Leach et al., 2011) as a means of helping people understand their impact on the nitrogen cycle. The nitrogen footprint is measured in kilograms of nitrogen per person per year, and it is used to inform consumers how their food consumption impacts reactive nitrogen cycling.

1.8.3 The Ecological Footprint

People rely on the production of goods, services, infrastructure, and waste absorption provided by natural resources. The *ecological footprint* is a metric based on the mass balance of human consumption; it also takes into account the impacts of different prevailing technologies and resource management strategies on the environment (Box 1.6). The ecological footprint relies on the concept of biocapacity—the biologically productive land and sea area available to provide the ecosystem services that humanity consumes (Wackernagel et al., 2002). Ecological footprint calculations are based on international data sets and literature cited in the *Ecological Footprint Atlas* (Ewing et al., 2010).

Several simplifying assumptions must be made to calculate the ecological footprint. The following assumptions were used in calculating the ecological footprint of nations in 2010 (Ewing et al., 2010, p. 3):

- *The majority of the resources people consume and the wastes they generate can be quantified and tracked.*
- *An important subset of these resources and waste flows can be measured in terms of the biologically productive area necessary to maintain flows. Resource and waste flows that cannot be measured are excluded from the assessment, leading to a systematic underestimation of humanity's true ecological footprint.*

TABLE 1.9 A comparison of sustainability footprint indicators

TYPE OF FOOTPRINT INDICATOR	AUTHORS CITED	YEAR CITED	ISSUE TO CONVEY	INPUTS	OUTPUT
Carbon		~ 2005	Climate change	Greenhouse gas emissions	CO_2 equivalents
				Embodied energy of products from UN COMTRADE	
				Fraction of anthropogenic emissions sequestered by the ocean	
				Rate of carbon uptake per hectare of forestland	
Ecological	Rees (1992); Wackernagel and Rees (1996)	1992, 1996	Resource sustainability	Product harvest	Hectares of land
				Carbon dioxide emitted	
				National average yield of product	
				Market price of product	
				Carbon uptake capacity	
				Area available for given land use	
Nitrogen	Leach, Galloway, Bleeker, Erisman, Kohn, and Kitzes	2011	Food impacts and sustainability	Electricity use	kg of reactive nitrogen per capita per year
				Food consumption	
				Food production	
				Sanitation system	
				Heating system	
				Transportation	
Water	Hoekstra	2008	Water scarcity	Evapotranspiration	Green, blue, or gray water volume
				Effective precipitation	
				Environmental flow requirements	
				Crop water requirements	
				Crop yield	
				Water stress coefficient	
				Irrigation schedule	
				Anthropogenic pollutant concentration	
				Natural background concentration	
				Natural assimilation capacity	

- *By weighting each area in proportion to its bioproductivity, different types of areas can be converted into the common unit of global hectares, hectares with world average bioproductivity.*
- *Because a single global hectare represents a single use, and each global hectare in any given year represents the same amount of bioproductivity, they can be added up to obtain an aggregate indicator of ecological footprint or biocapacity.*
- *Human demand, expressed as the ecological footprint, can be directly compared to nature's supply, biocapacity, when both are expressed in global hectares.*
- *Area demanded can exceed area supplied if demand on an ecosystem exceeds the ecosystem's regenerative capacity.*

BOX 1.6 Calculating the Ecological Footprint

The ecological footprint of production that represents the demand for biocapacity is shown in Equation (1.14):

$$EF_{\hat{P}} = \frac{\hat{P}}{Y_N} \times YF \times EQF \qquad (1.14)$$

where P is either the amount harvested or the carbon dioxide emitted, Y_N is the average yield for the product or carbon dioxide uptake, and YF and EQF are the yield factor and equivalence factor for the type of land use as defined by Ewing et al. (2010a), respectively.

Yield factors, YF, are used in the ecological footprint calculations to differentiate between the productivity associated with different types of land use. The yield factor for a particular nation is the ratio of the national average to the world average yields, as illustrated in Table 1.10.

Equivalent area factors, EQF, are used to convert from actual hectares of different types of land use into equivalent *global hectares*, at global average bioproductivity across all land-use types. The equivalence factors are calculated based on the ratio of the world average suitability index for a given land-use

type to the average suitability index for all land-use types. For example, the world average productivity for cropland, as the land-use type, was more than 2.5 times as productive as the average productivity of all land on a global basis. Equivalence factors vary by land-use type and year, as shown in Table 1.11. Land use globally is divided into the following five categories based on crop productivity (Ewing et al., 2010b):

- Very suitable (VS)—0.9
- Suitable (S)—0.7
- Moderately suitable (MS)—0.5
- Marginally suitable (mS)—0.3
- Not suitable (NS)—0.1

The ecological footprint associated with consumption is the sum of the production footprint, $EF_{\hat{P}}$ and the footprint for imported commodities, EF_I, minus the footprint for exported commodities, EF_E, as shown in Equation (1.15):

$$EF_C = EF_{\hat{P}} + EF_I - EF_E \qquad (1.15)$$

A country's biocapacity, BC, for any land type is calculated from

$$BC = A \times YF \times EQF \qquad (1.16)$$

(Continued)

BOX 1.6 Calculating the Ecological Footprint *(Continued)*

TABLE 1.10 Example yield factors for selected countries based on 2007 data

YIELD	CROPLAND	FOREST	GRAZING LAND	FISHING GROUNDS
World Average	1.0	1.0	1.0	1.0
Algeria	0.3	0.4	0.7	0.9
Germany	2.2	4.1	2.2	3.0
Hungary	1.1	2.6	1.9	0.0
Japan	1.3	1.4	2.2	0.8
Jordan	1.1	1.5	0.4	0.7
New Zealand	0.7	2.0	2.5	1.0
Zambia	0.2	0.2	1.5	0.0

Source: Based on Ewing, B., Reed, A., Galli, A., Kitzes, J., and Wackernagel, M. (2010b). *Calculation Methodology for the National Footprint Accounts, 2010 Edition*. Oakland, CA: Global Footprint Network.

TABLE 1.11 Equivalence factors for land-use types based on 2007 data

LAND-USE TYPE	EQUIVALENCE FACTOR (GLOBAL HECTARES PER ACTUAL HECTARE OF LAND USE TYPE)
Cropland	2.51
Forest	1.26
Grazing land	0.46
Marine and inland water	0.37
Built-up land	2.51

Source: Based on Ewing, B., Reed, A., Galli, A., Kitzes, J., and Wackernagel, M. (2010b). *Calculation Methodology for the National Footprint Accounts, 2010 Edition*. Oakland, CA: Global Footprint Network.

where A is the actual area in hectares available for a specified land-use type, and YF and EQF are the associated yield factors and equivalence factors for that country's land-use type.

Land-use types are associated with either cropland, grazing land, forestland, fishing grounds, or built-up land. The carbon use is also accounted for, but in the case of the ecological footprint the carbon is accounted for based on the associated area of land required for carbon uptake.

Cropland in the ecological footprint consists of the area required to grow all crop products, including livestock feeds, fish meals, oil crops, and rubber (Ewing et al., 2010b). The footprint associated with each type of crop is calculated as the area of cropland required to produce the reported harvest amount of the crop based on worldwide average yield for that crop.

Grazing land accounts for the land area required for grazing and the additional area required for growing feed crops to support livestock. The total demand for pasture grass, P_{GR}, can be calculated from Equation (1.17):

$$P_{GR} = TFR - F_{Mkt} - F_{Crop} - F_{Res} \qquad (1.17)$$

where TFR is the total feed requirement, F_{Mkt} is the feed from marketed crops, F_{Crop} is the feed from crops grown specifically for fodder, and F_{Res} is feed from crop residues.

Similarly, the footprint associated with fishing grounds is based on the primary production requirement, PPR, or the ratio of the mass of fish harvest to the mass needed to sustain the species (Pauly and Christensen, 1995):

$$PPR = CC \times DR \times \left(\frac{1}{TE}\right)^{(TL-1)} \qquad (1.18)$$

where CC is the carbon content of wet-weight fish biomass, DR is 1.27 (the global average discard rate for bycatch), TE is 0.1 (that assumes the transfer efficiency of biomass between trophic levels is 10%), and TL is the trophic level of the particular fish species. The total sustainably harvestable primary production requirement, PP_S, may be calculated from

$$PP_S = \Sigma(Q_{S,i} \times PPR_i) \qquad (1.19)$$

where $Q_{S,i}$ is the estimated sustainable catch for a species i. The worldwide average yield, Y_M, can be calculated from the total harvest PP_S per area of fishing along the continental shelf, A_{CS}:

$$Y_M = \frac{PP_S}{A_{CS}} \qquad (1.20)$$

The forestland footprint is a measure of the annual harvest of fuelwood and timber for all forest supply products. The world average yield was reported by Ewing et al. (2010b) to be 1.81 m³ of harvestable wood per hectare per year. The built-up land footprint is calculated based on the area of land covered by human infrastructure, including transportation, housing, industrial structures, and reservoirs for hydroelectric power generation. In 2007, the world's total estimated built-up land area was estimated to be 169.59 million hectares (Ewing et al., 2010b).

The carbon footprint is typically the largest contributing component of the ecological footprint for people living in higher-income countries. The carbon footprint component of the ecological footprint is calculated based on the total land area required to remove carbon dioxide in the atmosphere associated with human activities (such as fossil fuel combustion and changes in land use), as well as natural carbon dioxide emissions in an average year from forest fires, volcanoes, and respiration from animals and microorganisms. Carbon dioxide uptake in ecological footprint calculations consider only forest uptake of carbon dioxide and partitioning of carbon to the oceans as shown in

$$EF_C = \frac{P_C(1 - S_{Ocean})}{Y_C} \times EQF \qquad (1.21)$$

where P_C is the annual emissions of carbon dioxide, S_{Ocean} is the fraction of anthropogenic emissions sequestered by oceans in a given year, and Y_C is the annual rate of carbon uptake per hectare of forestland at world average yield.

ACTIVE LEARNING EXERCISE 1.6   How Big Is Your Footprint?

Source: hddigital/Shutterstock.com

The ecological footprint assesses how our wealth, lifestyle, production technologies, and resource management affect our consumption of the Earth's ecological services and natural resources. The ecological footprint metric estimates the amount of biologically productive area (land and oceanic) that is required to sustain a particular lifestyle. Although there are criticisms of the ecological footprint methodology as described in the text, it is a useful way to understand how individual consumption affects ecosystems and natural resources.

In this activity, you will use an online calculator to estimate your own ecological footprint. Go to the ecological footprint quiz administered by the Center for Sustainable Economy at *footprintnetwork.org*. Read the webpage "About the Quiz," then take the quiz. Answer the following questions:

1. What exactly does this ecological footprint calculator measure? What is the global average footprint?
2. If everyone in the world **had your lifestyle**, how many "planet Earths" would be required to sustain the global population?
3. How does your footprint for carbon, food, housing, and goods and services compare to the national average? Why do you think your footprint for these aspects of your lifestyle is above or below average?
4. For which type of biome (ecosystem) do you have the largest footprint? Pasture, forestland, fisheries, or cropland?
5. Based on your footprint analysis, what can you do to *reduce* your ecological footprint?

The ecological footprint essentially reports how many hectares of biocapacity are required to support human consumption for a given lifestyle, at a particular level of economic development, and using a prevailing suite of production technologies and natural resource management strategies. The ecological footprint of a person in Japan is therefore likely to be different from that of an individual in the United States, even though both countries are at similar levels of economic development and enjoy comparable standards of living. Japanese lifestyles (diet, consumerism, housing needs, energy conservation, and so on) are not those of an American, the technological efficiencies of their economies are not the same, nor are their farming and forestry practices identical.

The ecological footprint provides a widely accepted method for assessing the mass balance between human consumption and biocapacity. However, there are several limitations and biases associated with the simplifications and calculations used for this footprint. Human demand on resources is *underestimated* because it does not consider all possible resource impacts; for example, the ecological footprint excludes water consumption, water and air pollution, soil erosion, and the impacts of greenhouse gases (GHGs) other than carbon dioxide. In contrast, biocapacity is likely to be *overestimated* by the method because neither land degradation nor the long-term sustainability of resource extraction are considered in footprint calculations.

In sum, footprint indicators provide a measure of mass accounting between the natural world and human consumption. Intrinsic to the footprint concept is the dynamic of material flow, and many footprints directly or indirectly capture critical biogeochemical cycles. As we shall see over and over again, mass balance and the virtually closed-loop biogeochemical cycles of Earth are a defining feature of sustainable engineering as both a design constraint and as an ideal to be imitated.

The ecological footprint calculations show that the land area required to sustain our current average lifestyle has more than doubled since 1961. In 2021, it was estimated that humanity used 1.7 planet's worth of resources (www.footprintnetwork.org)! Or, put another way, by July 29, 2021, humanity would have exhausted the Earth's available resources, so from August through December 2021, humanity started depleting the Earth's resources. The rate of resource consumption continues to increase.

In 2012, more than 80% of the world's population lived in countries that exceed the supply of resources within their borders each year. The desire to achieve higher levels of economic development will likely lead to higher levels of resource consumption, as illustrated in Figure 1.32, which shows that the ecological footprint is higher for countries with a higher HDI value. Generally, this implies that countries that have greater levels of development also use a disproportionate amount of the world's available resources. As the worldwide population continues to migrate

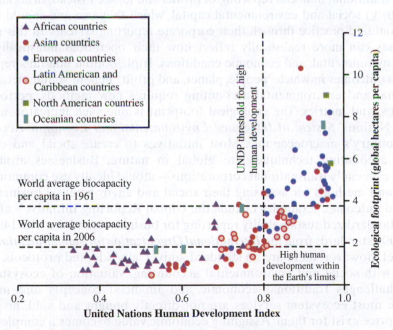

FIGURE 1.32 The relationship between the HDI and ecological footprint. Generally, a country's ecological footprint increases as the HDI value increases.

Source: *The Ecological Wealth of Nations: Earth's biocapacity as a new framework for international cooperation* (2010). Figure 7, p. 15. Oakland, CA: Global Footprint Network.

toward urban areas, the ecological footprint can be used as a tool to identify cities and regions that create stress on natural resources and biodiversity. The ecological footprint calculations can be used as one tool to measure and illustrate the sustainability of the current and future infrastructure and design decisions. The relationship between the HDI and the ecological footprint also demonstrates the great challenge facing engineers and scientists: to reduce poverty and achieve higher levels of development while simultaneously trying to reduce the environmental footprint historically associated with greater levels of development.

1.9   The Difficulty of Environmental Valuation

Environmental sustainability challenges traditional economic approaches in yet another way, which is how we monetarily value and account for the impacts of our actions on natural resources and a clean environment. We see this with the emergence of the ecosystem services concept as well as sustainable development's "three pillars" of society–environment–economy.

Several important changes in economic culture are taking place throughout the world today in order to accommodate the growing emphasis on *social and environmental accounting* by business enterprises. Generally, this term refers to a formal assessment of how business and economic activities are benefiting people and the environment. In business, two terms are used somewhat synonymously to reflect this notion: *corporate social responsibility* and *triple bottom line*. The British visionary John Elkington has been an advocate of corporate social responsibility for nearly three decades; he also coined the term *triple bottom line*. Both notions stress the idea that business accountability to investors and society involves more than just the traditional financial reporting of profits and losses. Instead, firms also build (or destroy) social and environmental capital, which they can and should account for as standard practice through their corporate reporting systems. In this manner, businesses can more realistically reflect how their operations holistically affect social, environmental, and economic conditions. Triple bottom line thus represents business outcomes in which "people, planet, and profit" benefit simultaneously.

Social and environmental accounting requires new methods, performance measures, and metrics. The ecological footprint is one such indicator, as is the United Nations' *System of Integrated Environmental and Economic Accounting* for a country's macroeconomy. Most initiatives to create social and environmental accounting techniques are global in nature. Businesses around the world—especially transnational corporations—should ideally use common terminology and methods for assessing their social and environmental business practices and outcomes. Examples include the Global Reporting Initiative's efforts to foster standardized sustainability reporting for businesses, and the ISO 14000 and ISO 26000 standards from the *International Organization for Standardization (ISO)*. As of yet, however, there are no standard, universally adopted protocols.

As with social and environmental accounting, valuation of ecosystem services challenges traditional economic and financial concepts and methods. Because most ecosystem services are not directly bought and sold, no market and no price exist for them. Assigning economic value becomes a complex effort and is based on methods of *nonmarket valuation* (assigning monetary value that is not based on the price of actually buying and selling the good or service in question). With individual services, such as water purification, valuation is relatively straightforward and is estimated from the cost of operating water treatment and supply facilities and/or the supply of potable water. The process becomes more difficult once we allow for the complexity of ecological systems, such as the role of healthy forests in preventing soil erosion or water contamination. Monetary valuation of ecosystems is not a venal process of "putting a price tag on nature," but rather a way of illustrating the opportunity costs associated with exploiting nature in a manner that jeopardizes more easily measurable, conventional, and long-term profits.

1.10   Summary

Only in the recent history of the Earth have there been enough humans present to significantly affect natural ecosystems. With the advances in modern energy use and technology, both the population and the impacts associated with human population have increased. Based on the global environmental footprint, humans have exceeded the biocapacity of the Earth, and the Earth's natural resources may be significantly reduced for future generations.

One of the core motivations for sustainable development is the principle that humanity in the current generation should be able to pursue progress and development in a manner that does not jeopardize the ability of future generations to do the same. Entrenched in this notion is the principle that has come to be known as intergenerational equity. Along with this principle is the supposition that achieving true progress and development would require equitable distribution of available natural resources among people in every generation. Social and economic justice have already been identified as integral to human development. Communities advancing toward sustainable development must develop and implement principles for the planning and equitable management of natural resources.

If renewable resources are not consumed at a rate greater than their rate of regeneration and if waste discharges are maintained at or below the capacity of the natural system, the viability of the first two items of natural capital can, in theory, be maintained over generations.

Sustainability often takes on different meanings in developed and developing countries. While developed countries focus on economic development while simultaneously preserving the natural environment, developing countries tend to concentrate on increasing the number of people who have access to adequate shelter, potable water, effective waste disposal, employment, and food. It is therefore not strange that developing countries will often mimic the path of developed countries to achieve this end. The notion exists that developed countries have reached their levels of economic prosperity by uneven distribution and utilization of global natural resources. Models show that if the living standards of the average citizen of developed nations are applied to the entire earthly population, the resources from a multiple number of Earths would be required. For example, it is estimated that if everyone in the world consumed resources at the rate the average American does, we would need four Earths to meet the needs of everyone (source: *www.footprintnetwork.org*). However, if everyone adopted the current resource consumption pattern of the average Nigerian, 0.8 Earths would be sufficient. Intragenerational equity implies an equitable distribution of global resources for the population within a generation.

Resource allocation is a method of allocating available resources in a sustainable way that has been proposed to advance intergenerational and intragenerational equity in sustainable development and planning. It starts with identifying the total stock of a particular resource that is available to be used sustainably over a specific time period and distributing this resource equally among the population. The resulting value indicates the allocation to individual members of the population and puts a limit on how much of that resource individuals can sustainably use over specific time periods. For example, if we estimate the maximum sustainable amount of greenhouse gases that can be emitted into the atmosphere without causing impacts on climate and other natural processes, then we can divide this number by the global population to establish the allowable emissions that each person in the world is allocated. This emissions per capita value would be the total from various activities such as driving, traveling by air, buying manufactured goods, consuming food (especially meats), and burning fuel to heat houses using electricity. A focus on defining scientifically measurable sustainable product design, process design, and development will be necessary if future generations are to achieve the same standards of living as people currently living in countries with a High Human Development ranking.

One potential application is in estimating the amount of a resource allocated in the sustainable design of a new product to be used over a specified time period. This could also be applied to environmental quality by determining the allocation of a sustainable rate of waste discharge per capita and incorporating this limit in processes that generate environmental emissions in new products.

The science of sustainability is complex and requires analysis of systems with several dimensions. Science and engineering use mathematical tools to measure various economic, environmental, and social indicators and metrics. Because sustainability is multidimensional—consisting of economic, environmental, and social components—various indices have been created to help the public and decision makers compare and contrast the complex issues related to sustainable design and development. In order to properly understand how to develop, use, and interpret the sustainability indices, it will be helpful to have a good understanding of the foundational principles and measures used to develop these indices.

In the next chapter, we will begin by developing the scientific foundational tools to describe individual sustainability indicators and add layers of sustainability metrics so that proper sustainability indices may be applied and interpreted with the help of the tools described throughout the subsequent chapters.

References

Bielefeldt, A. R., Kowalski, K., Schilling, C., Schreier, S., Kohler, A., and Summers, R. S. (2010). "Removal of virus to protozoan sized particles in point-of-use ceramic water filters." *Water Research* 44:1482–1488.

Biodiversity Indicators Partnership. (2010). *Guidance for national biodiversity indicator development and use.* UNEP World Conservation Monitoring Centre, Cambridge, UK.

Bishop, P. (1999). *Pollution Prevention: Fundamentals and Practice.* New York: McGraw-Hill.

Brown, J., and Sobsey, M. D. (2010). "Microbiological effectiveness of locally produced ceramic filters for drinking water treatment in Cambodia." *Journal of Water and Health* 8(1):1–10.

Carson, R. (1962). *Silent Spring.* New York: Houghton Mifflin.

CIA World FactBook. (2006–2007). Retrieved October 29, 2006, March 24, 2007, at www .cia.gov/cia/publications/factbook/geos/rw.html.

Clasen, T., Brown, J., Suntura, O., and Collin, S. (2004). "Safe household water treatment and storage using ceramic drip filters: A randomized controlled trial in Bolivia." *Water Science and Technology* 50(1):111–115.

Collis, J. S. (1954). *The Triumph of the Tree*. New York: Viking.

Conroy, R. M., Meegan, M. E., Joyce, T. M., McGuigan, K. G., and Barnes, J. (2001). "Use of solar disinfection protects children under 6 years from cholera." *Archives of Diseases in Childhood* 85:293–295.

Conteh, A. (2003). *Culture and the Transfer of Technology: Field Guide to Appropriate Technology*. B. Hazeltine and C. Bull (eds.). New York: Academic Press.

Culp, G. L., and Culp, R. L. (1974). *New Concepts in Water Purification*. New York: Litton Educational.

Duke, W., Nordin, R., Baker, D., and Mazumder, A. (2006). "The use and performance of BioSand filters in the Artibonit Valley of Haiti: A field study of 107 households." *Rural Remote Health* 6(3):570.

Ehrlich, P. R., and Holdren, J. P. (2009, October). "Impact of population growth." *Science*. 171(3977):1212–1217.

Ewing, B., Reed, A., Galli, A., Kitzes, J., and Wackernagel, M. (2010). *Calculation Methodology for the National Footprint Accounts*, 2010 ed. Oakland, CA: Global Footprint Network.

Federal Register. (2009). Executive Order 13514. Federal Leadership in Environmental, Energy, and Economic Performance onto the Agency's Green Purchasing Plan. Washington, DC, 74(194):52117–52127.

Fiksel, J., Eason, T., and Frederickson, H. (2012). A Framework for Sustainability Indicators at EPA. National Risk Management Research Laboratory. Office of Research and Development, U.S. Environmental Protection Agency. EPA/600/R/12/687.

Goldfinger, S., and Poblete, P. (eds.). (2010). *The Ecological Wealth of Nations*. Oakland, CA: Global Footprint Network. p. 38.

Hardin, G. J. (1968). "The tragedy of the commons: The population problem has no technical solution; it requires a fundamental extension in morality." *Science* 162(859):8.

Harvey, P. A., and Reed, R. A. (2007). "Community-managed water supplies in Africa: Sustainable or dispensable?" *Community Development Journal* 42(3):365–378.

Hazeltine, B., and Bull, C. *Field Guide to Appropriate Technology*, Elsevier Science. San Diego, CA.

Hoekstra, A. Y., Chapagain, A. K., Aldaya, M. M., and Mekonnen, M. M. (2011). *The Water Footprint Assessment Manual: Setting the Global Standard*. Washington, DC: Earthscan.

Hubbert, M. K. (1956). Nuclear energy and the fossil fuels. American Petroleum Institute, Shell Development Co., San Antonio, TX.

IEA. (2011). Energy for all: financing access for the poor. Special early excerpt of the World Energy Outlook 2011. International Energy Agency (IEA) and OECD.

IEA, IRENA, UNSD, World Bank, WHO. (2021). Tracking SDG 7: The Energy Progress Report. World Bank, Washington DC. © World Bank. License: Creative Commons Attribution—NonCommercial 3.0 IGO (CC BYNC 3.0 IGO).

John Paul II. (2002)."Common Declaration of John Paul II and the Ecumenical Patriarch His Holiness Bartholomew I". *Rome - Venice, 10 June 2002*: Libreria Editrice Vaticana. Retrieved from www.vatican.va/holy_father/john_paul_ii/speeches/2002 /june/documents/hf_jp-ii_ spe_20020610_venice-declaration_en.html, 2002.

Kehoe, S. C., Joyce, T. M., Ibrahim, P., Gillespie, J. B., Shahar, R. A., and McGuigan, K. G. (2001). "Effect of agitation, turbidity, aluminum foil reflectors and volume on inactivation efficiency of batch-process solar disinfectors." *Water Research* 35(4): 1061–1065.

Kempton, W. (2001). "Cognitive anthropology and the environment." In *New Directions in Anthropology and Environment*. C. L. Crumley (Ed.), pp. 49–71. Walnut Creek, CA: AltaMira Press.

Kremer, M. (1993). "Population growth and technological change: One million B.C. to 1990." *The Quarterly Journal of Economics* 108(3):681–716.

Leslie, D. (2009). "Consumption." *International Encyclopedia of Human Geography*. Ed. Rob Kitchen and Nigel Thrift. Elsevier. Volume 2: 268–274.

Leopold, A. (1949). *A Sand County Almanac and Sketches Here and There*. New York: Oxford University Press.

Makuch, K. E., and Pereira, R. (2012). *Environmental and Energy Law*. Chichester, UK: John Wiley & Sons, Inc.

Malthus, T. (1789). *An Essay on the Principle of Population*. London: J. Johnson.

Mania, S. K., Kanjura, R., Singha, I. S. B., and Reed, R. H. (2006). "Comparative effectiveness of solar disinfection using small-scale batch reactors with reflective, absorptive and transmissive rear surfaces." *Water Research* 40(4):721–727.

Meadows D. H., et al. (1972). *The Limits to Growth*, first edition. Universe Books.

Meierhofer, R., and Wegelin, M. (2002, October). "Solar water disinfection—A guide for the application of SODIS." Swiss Federal Institute of Environmental Science and Technology (EAWAG) Department of Water and Sanitation in Developing Countries (SANDEC).

Myers, N. (1997). "Consumption in relation to population, environment and development." *Environmentalist* 17: 33–44.

Naess, A. (1989). *Ecology, Community and Lifestyle: Outline of an Ecosphere*. Cambridge University Press, UK, p. 166.

National Society of Professional Engineers (NSPE). (2007). *Code of Ethics for Engineers*, Publication #1102. Alexandria, VA.

Pauly, D., and Christensen, V. "Primary production required to sustain global fisheries." *Nature*, 374, 255–257 (1995).

Princen, T., Maniates, M., and Conca, K. (Eds.) (2002). *Confronting Consumption*. MIT Press, Cambridge, MA, p. 392.

Putnam, J. J., and Allshouse, J. (1999). "Food consumption, prices and expenditures, 1970–1997." U.S. Department of Agriculture, *Statistical Bulletin* No. 965.

Rees, W. E. (1992). Ecological footprints and appropriated carrying capacity. *Environment and Urbanization*. 4(2): 120–130.

Rwanda Ministry of Infrastructure. (2006). Kigali Conceptual Master Plan. Existing Conditions Analysis by the Conceptual Master Plan Team: OZ Architecture, EDAW, Tetra Tech, Sypher, ERA, Geopmaps, Engineers Without Borders—USA and Water for People.

Schumacher, E. F. (1973). *Small Is Beautiful: Economics as if People Mattered*. London: Blond & Briggs.

Snow, J. (1855). *On the Mode of Communication of Cholera*. London: John Churchill.

Sommer, B., Mariño, A., Solarte, Y., Salas, M. L., Dierolf, C., Valiente, C., Mora, D., Rechsteiner, R., Setter, P., Wirojanagud, W., Ajarmeh, H., Al-Hassan, A., and Wegelin M. (1997). "SODIS—an emerging water treatment process." *Journal of Water Supply Research and Technology, Aqua* 46(3):127–137.

Stauber, C., Elliott, M., Koksal, F., Ortiz, G., DiGiano, F., and Sobsey, M. (2006). "Characterization of the biosand filter for E. coli reductions from household drinking water under controlled laboratory and field use conditions." *Water Science and Technology* 54(3):17.

Striebig, B., Atwood, S., Johnson, B., Lemkau, B., Shamrell, J., Spuler, P., Stanek, K., Vernon, A., and Young, J. (2007). "Activated carbon amended ceramic drinking water filters for Benin." *Journal of Engineering for Sustainable Development* 2(1):1–12.

Taylor, P. (1981). "The ethics of respect for nature." *Environmental Ethics* 3(3):197–218.

UN. (1992). Agenda 21. United Nations Conference on Environment and Development. Rio de Janeiro, Brazil, June 3–14, 1992.

UN. (1999). *The World at Six Billion*. Population Division. Department of Economic and Social Affairs. United Nations Secretariat. ESA/WP.154.

UN. (2000). Millennium Declaration. UN A/Res/55/2.

UN. (2008). Trends in Sustainable Development: Agriculture, rural development, land desertification and drought. United Nations Department of Economic and Social Affairs Division for Sustainable Development.

UN. (2011). Report of the Special Rapporteur on the promotion and protection of the right to freedom of opinion and expression, Frank La Rue. Human Rights Council, Seventeenth session, Agenda item 3, Promotion and protection of all human rights, civil, political, economic, social and cultural rights, including the right to development, United Nations General Assembly.

UN. (2012a). *The Millennium Development Goals Report* 2012.

UN. (2012b). Glossary of climate change acronyms. Retrieved October 31, 2012, at unfccc.int/essential_background/glossary/items/3666.php.

UN. (2020). *The Sustainable Development Goals Report* 2020.

UN Department of Economic and Social Affairs, Population Division. (2019). World Population Prospects 2019: Highlights (ST/ESA/SER.A/423).

UNDP. (2009). Human Development Report 2009: Overcoming barriers—human mobility and development.

UNDP. (2019). Human Development Report 2019. Beyond income, beyond averages, beyond today: Inequalities in human development in the 21st century. New York. ISBN: 978-92-1-126439-5.

UNICEF and WHO. (2021). "Progress on household drinking water, sanitation and hygiene 2000–2020: five years into the SDGs." Geneva: World Health Organization (WHO) and the United Nations Children's Fund (UNICEF), 2021.

U.S. Census Bureau Data: The World Population Clock. www.census.gov/popclock/.

U.S. Energy Information Administration. (2012). Annual Energy Review 2011. DOE/EIA-0384(2011), September 2012. www.eia.gov/aer.

van Halem, D., van der Laan, H., Heijman, S. G. J., van Dijk, J. C., and Amy, G. L. (2009). "Assessing the sustainability of the silver-impregnated ceramic pot filter for low-cost household drinking water treatment." *Physics and Chemistry of the Earth*, Parts A/B/C, 34(1–2):36–42.

Vesilind, P. A., and Gunn, A. S. (1998). *Engineering, Ethics, and the Environment*. New York: Cambridge University Press.

Vesilind, P. A., et al. (2010). *Introduction to Environmental Engineering*, third edition, p. 71. Stamford, CT: Cengage Learning.

Wackernagel, M., and Rees, W. (1996). *Our Ecological Footprint: Reducing Human Impact on the Earth*. New Society Publishers.

Wackernagel, M., et al. (2002). "Tracking the ecological overshoot of the human economy." *PNAS*. 99(14): 9266–9271.

WCED. (1987). *Our Common Future*. World Commission on Environment and Development. New York: Oxford University Press.

Wilk, R. (2002). "Consumption, human needs, and global environmental change." *Global Environmental Change* 12:5–13.

World Health Organization/UNICEF Joint Monitoring Programme for Water Supply and Sanitation. (2005). *Water for Life: Making It Happen*.

World Wildlife Fund. (2016). Living Planet Report 2016. "Risk and resilience in a new era." *Living Planet Report*, WWF International, Gland, Switzerland.

Key Concepts

Demographers

Human Development Index (HDI)

Sustainable development

Peri-urban areas

Brundtland definition

Intergenerational equity

Social justice

Conference on Environment
  and Development

Millennium Development Goals

Appropriate technology

Sustainability

Morals

Ethics

Categorical imperative

Land ethic

Biocentric outlook

Deep ecology

Environmental ethic

Sustainable design

Carrying capacity

Exponential growth

Measure

Metric

Indicator

Impact

Population

Affluence

Background consumption

Overconsumption

Misconsumption

Technology

Biocapacity

Ecological footprint

Social and environmental accounting

Corporate social responsibility

Triple bottom line

International Organization for
  Standardization (ISO)

Nonmarket valuation

Problems

1-1   Genetically modern humans appeared on Earth about 200,000 years ago, and biologically and behaviorally modern humans appeared about 70,000 years ago. The number of people and their effects on the planet were negligible, or as Douglas Adams says, "as near nothing as makes no odds," for most of the history of the planet. When did the planet's population reach 1 billion people? Assuming that the population has grown exponentially since that time, what was the time interval required to increase by 1 billion people—for up to 7 billion people, which was the approximate global population in 2012?

1-2   List the three dimensions and four categories used to calculate the Human Development Index (HDI) for a country.

1-3   Calculate the HDI for Australia given the following information: life expectancy at birth = 83.1, mean years of schooling = 12.9, expected years of schooling = 22.9, GNI per capita in PPP terms (constant 2011 international $) = 43,560.

1-4   Calculate the HDI for Japan given the following information: life expectancy at birth = 83.9, mean years of schooling = 12.8, expected years of schooling= 15.2, GNI per capita in PPP terms (constant 2011 international $) = 38,986.

1-5   Calculate the HDI for Turkey given the following information: life expectancy at birth = 76.0, mean years of schooling = 8.0, expected years of schooling = 15.2, GNI per capita in PPP terms (constant 2011 international $) = 24,804.

1-6   Calculate the HDI for Vietnam given the following information: life expectancy at birth = 76.5, mean years of schooling = 8.2, expected years of schooling = 12.7, GNI per capita in PPP terms (constant 2005 international $) = 5,859.

1-7   Calculate the HDI for Argentina given the following information: life expectancy at birth = 76.7, mean years of schooling = 9.9, expected years of schooling = 17.4, GNI per capita in PPP terms (constant 2005 international $) = 18,461.

1-8   Calculate the HDI for Vanuatu given the following information: life expectancy at birth = 72.3, mean years of schooling = 6.8, expected years of schooling = 10.9, GNI per capita in PPP terms (constant 2005 international $) = 2,995.

1-9   Calculate the HDI for Niger given the following information: life expectancy at birth = 60.4, mean years of schooling = 2.0, expected years of schooling = 5.4, GNI per capita in PPP terms (constant 2005 international $) = 906.

1-10   For each country listed in the accompanying table, calculate
    a. Life Expectancy Index
    b. Educational Index
    c. Income Index
    d. Human Development Index

COUNTRY AND 2011 DATA	LIFE EXPECTANCY AT BIRTH (YEARS)	EXPECTED YEARS OF SCHOOLING	MEAN YEARS OF SCHOOLING	GNI PER CAPITA IN PPP TERMS (CONSTANT 2005 INTERNATIONAL $)
Australia	81.9	18.0	12.0	34,431
China	73.5	11.6	7.5	7,476
Ireland	80.6	18.0	11.6	29,322
Kenya	57.1	11.0	7.0	1,492
South Africa	52.8	13.1	8.5	9,469

Source: Based on UN (2011a). *Human Development Report 2011, Sustainability and Equity: A Better Future for All*. United Nations Development Programme.

1-11   For each country listed in the accompanying table, calculate
    a. Life Expectancy Index
    b. Educational Index
    c. Income Index
    d. Human Development Index
    e. United Nations' development category

COUNTRY AND 2011 DATA	LIFE EXPECTANCY AT BIRTH (YEARS)	EXPECTED YEARS OF SCHOOLING	MEAN YEARS OF SCHOOLING	GNI PER CAPITA IN PPP TERMS (CONSTANT 2005 INTERNATIONAL $)
Canada	81.0	16.0	12.1	35,166
Japan	83.4	15.1	11.6	32,295
Mexico	77.0	13.9	8.5	13,245
Nigeria	51.9	8.9	5.0	2,069
United Kingdom	80.2	16.1	9.3	33,296

Source: Based on UN (2011a). *Human Development Report 2011, Sustainability and Equity: A Better Future for All*. United Nations Development Programme.

1-12 For each country listed in the accompanying table, calculate
 a. Life Expectancy Index
 b. Educational Index
 c. Income Index
 d. Human Development Index

COUNTRY AND 2011 DATA	LIFE EXPECTANCY AT BIRTH (YEARS)	EXPECTED YEARS OF SCHOOLING	MEAN YEARS OF SCHOOLING	GNI PER CAPITA IN PPP TERMS (CONSTANT 2005 INTERNATIONAL $)
Benin	56.1	3.3	9.2	1,364
Costa Rica	79.3	11.7	8.3	10,497
India	65.4	10.3	4.4	3,468
Malta	79.6	14.4	9.9	21,460
New Zealand	80.7	18.0	12.5	23,737
Rwanda	55.4	11.1	3.3	1,133

Source: Based on UN (2011a). *Human Development Report 2011, Sustainability and Equity: A Better Future for All*. United Nations Development Programme.

1-13 What are the HDI categories defined by the United Nations? For each of the four categories, describe what you think people may drink, eat, and wear, the type of homes they may live in, the types of school they are likely to attend, and the type of transportation they are most likely to use.

1-14 Describe, using your own words, the purpose of human development.

1-15 Define the following terms:
 a. Urban
 b. Suburban
 c. Peri-urban
 d. Rural

1-16 What is the "Brundtland definition" of sustainable development?

1-17 How would you describe sustainability to a 12-year-old student at your local school?

1-18 Create a graph with an x-axis as a linear sustainability scale. Sort the 30 companies that make up the Dow-Jones Industrial Average from "least sustainable" to "most sustainable" and place them along this axis. Describe the characteristic units of measure on the scale of the sustainability axis you created.

1-19 Look up and describe one of the formative written works related to sustainable development. Research this work more and summarize its main premise in a short 500-word essay.

1-20 Create a sustainability indicator (similar to the HDI or the ecological footprint). What actions, processes, or goods would you measure for your indicator? How would you collect and find the data for your indicator? What are the advantages and disadvantages of your proposed indicator? Note: Useful resources include the United Nations Development Program (UNDP) web page, the UNICEF web page, the World Bank web page, and the U.S. Environmental Protection Agency web page.

1-21 What characteristics define unsustainable development? Make a table of characteristics that might negatively affect development. Mark which of these characteristics are important in the following:
   a. Very High Human Development countries
   b. Low Human Development countries
   c. Both Very High and Low Human Development countries

1-22 Create a schematic or cartoon that communicates how the following concepts are related or the trends toward change in:
   a. Human population
   b. Resource consumption
   c. Educational resources
   d. Economic resources

1-23 List and describe in your own words the United Nations Sustainable Development Goals. Specifically, describe the economic, environmental, social, and technical challenges associated with meeting each of the goals within the next five years.

1-24 Do you believe that sustainable products are morally superior to nonsustainable products? If so, what does this belief imply about the developed world's largely consumer-based economic system of retail merchandise? How are technology and moral convictions woven into the fabric of our definitions of sustainable design?

1-25 If you had to live on $2 per day, how would you meet your basic needs for food, shelter, water, sanitation, and other requirements?
   a. Determine from recent utility bills how much you spend per day on
      i)   Water
      ii) Sanitation (sewer or wastewater company bill)
      iii) Garbage collection services
      iv) Energy
      v)   Heating/cooling
      vi) Communications (phone, cell phone, Internet, etc.)
      vii) Food
      viii) Shelter (based on rent or mortgage payment)
      ix) Entertainment
   b. Determine your total daily expenditure.
   c. If you were to pay 25% of your income on taxes, how much would your income need to be each year to pay for your daily expenses?
   d. With what level of the Human Development Index would this income be associated?

1-26 Imagine you are part of a company designing a school for a low-income country (based on the country's HDI). Use online resources to help address the following questions for the design of the proposed school:
   a. Are parts readily available, either locally or nationally, if a component were to fail?
   b. Are there individuals who have the necessary skill or technical training to repair the component or system if it were to fail?
   c. Would members of the household readily understand how to use this system?
   d. What is the local availability of required infrastructures, such as electric power?

1-27 Compare and contrast the definitions of sustainability given by the *Merriam-Webster Dictionary*, the U. S. Environmental Protection Agency, and the Brundtland Commission's *Our Common Future*.

1-28 Describe how concepts of sustainability might be applied to the fundamental canons of the National Society of Professional Engineers.

1-29 It took about 12 years, between 2000 and 2012, for the world population to increase by 1 billion people. In contrast, the world's population was estimated to be 300 million people in the year AD 0. By the year 1500, the world's population was estimated to be 500 million.

    a. Assuming exponential growth, what was the percentage of the world's population growth rate (in percent) between 2000 and 2012?

    b. Assuming exponential growth, what was the percentage of the world's population growth rate (in percent) between AD 0 and 1500?

    c. How many times greater was the population growth rate in the 20th century than the rate between AD 0 and 1500?

1-30 The world population in 1850 has been estimated at about 1 billion. The world population reached 4 billion in 1975. What was the percentage of the exponential growth rate during this time?

1-31 Tuition at a university rose from $1,500/year in 1962 to $25,000/year in 2010.

    a. What was the exponential growth rate during that period of time?

    b. If that rate of growth were to continue until 2050 (when your children might be paying tuition), what would the tuition be?

1-32 In 1999, tuition at a university was $1,963 per semester. In 2009, RSU tuition was $3,622 per semester. This increase is represented by an exponential growth rate of 6.1%. If tuition rates increase exponentially, what value is closest to the in-state semester tuition cost predicted in 2035?

1-33 In 2007, the world's population was estimated to be 6.7 billion. The UN forecasts the population will begin to level off at 9.2 billion in 2050. What will be the population growth rate (in percent) over this time period?

1-34 It has been estimated that 139.2×10^6 m^2 of rainforest is destroyed each day. Assume that the initial area of tropical rainforest is 20×10^{12} m^2.

    a. What is the exponential rate of rainforest destruction in units of 1/days?

    b. If there were 24.5×10^{12} m^2 of tropical rainforest on Earth in 1975, how much tropical rainforest would be left on Earth in 2015 if the exponential rate of destruction determined in part (a) stayed constant over this time interval?

    c. If tropical rainforests remove 0.83 kg (of C)/m^2-year from the atmosphere, how much less carbon [kg (of C)] would be removed in 2025 compared to that removed in 1975?

1-35 The world's population 10,000 years ago has been estimated at about 5 million. What exponential growth rate would have resulted in the population in 1800, which is estimated at 1 billion? Had that rate continued, what would have been the world's population in 2010?

1-36 In 2007, the population of the world's 50 least-developed countries was estimated to be 0.8 billion. The UN expects the population in these countries to grow exponentially at 1.75% until 2050. What is the predicted population of the least-developed countries in 2050?

1-37 A manure storage facility collapsed and caused manure to spill into a local pond. The nutrients from the manure caused the algae in the lake to grow exponentially. What is the constant value (in day^{-1}) that describes the exponential rate of growth of algae in the pond?

a. Prior to the spill there were 12 mg of algae per liter of pond water. After 5 days there were 470 mg/L of algae.

b. Prior to the spill there were 5 mg of algae per liter of pond water. After 7 days there were 500 mg/L of algae.

c. Prior to the spill there were 10 mg of algae per liter of pond water. After 3 days there were 1,000 mg/L of algae.

d. Prior to the spill there were 20 mg of algae per liter of pond water. After 2 days there were 800 mg/L of algae.

1-38 What must engineers hold paramount in their designs according to most professional ethics codes?

1-39 Describe quantitatively (use numeric values) the differences between access to improved drinking water supplies in the United States and access in the countries in Africa or the Caribbean.

1-40 List the UN Millennium Development Goals, and describe briefly how they might relate to access to drinking water.

1-41 What mathematical expression defines the term *biocapacity*?

1-42 What mathematical expression defines the global ecological footprint (when comparing it to biocapacity)?

1-43 Using a mathematical expression, describe how overshoot (related to global resource consumption) is related to biocapacity and the global ecological footprint.

1-44 Describe how affluence contributes to environmental impact.

1-45 If the population of a country in 2010 was 72 million and the projected *exponential rate* of increase is 6.3 per 1,000, what total percentage reduction of environmental impact per GDP will be required by 2050 to keep the environmental impact at the 2010 levels if the GDP per capita is predicted to increase at a rate of 5% per year between 2010 and 2050?

1-46 In 2012, the population of a country was 65 million and the projected *exponential rate* of increase is 5.2 per 1,000. If the environmental impact per GDP is reduced by 15% by 2040, how will this affect the GDP per capita assuming the total environmental impact in 2040 is maintained at the 2010 levels?

1-47 The annual exponential population growth rate is estimated to be 11.9 per 1,000 persons from 2012 to 2030 and 12.1 per 1,000 persons from 2030 to 2060. If the GDP per capita is estimated to increase at a yearly rate of 2.5% from 2012 to 2030 and 1.5% from 2030 to 2060, calculate the annual rate of emissions reductions from 2012 to 2060 needed to keep impact levels in 2060 at the 2012 values.

1-48 Energy derived from nuclear power has grown since 1970 according to the data from the U.S. Energy Information Administration that is summarized in the table below.

a. Plot the energy production from nuclear power between 1970 and 1990.

b. Find the best-fit curve for the plot (use a linear fit, polynomial fit, or power function). What is the equation for this best-fit curve?

c. Take the mathematical or graphical derivative of the function from the plot and graph the rate of change (first derivative) of energy derived from nuclear power between 1970 and 1990 in Excel or a similar spreadsheet program.

YEAR	1958	1960	1962	1964	1966	1968	1970	1972	1974	1976	1978	1980	1982	1984
Billions	0.2	0.5	2.3	3.3	5.5	12.5	21.8	54.1	114.0	191.1	276.4	251.1	282.8	327.6
Kilowatt-hours														

YEAR	1986	1988	1990	1992	1994	1996	1998	2000	2002	2004	2006	2008	2010
Billions	414.0	527.0	576.9	618.8	640.4	674.7	673.7	753.9	780.1	788.5	787.2	806.2	807.0
Kilowatt-hours													

Source: Based on U.S. Energy Information Administration (2012). *Annual Energy Review 2011*. DOE/EIA-0384(2011), September 2012. www.eia.gov/aer.

1-49 Thomas Malthus described a situation in which the population could overcome the available supply of natural resources near the year 1800. Over 200 years later, scientists, policymakers, and demographers fear the same situation may be occurring—we may exceed the biocapacity of the planet. Malthus's original arguments have been reworked in modern writings such as "The Tragedy of the Commons" and *The Population Bomb* theorized by Ehrlich (2009). What role do scientists and engineers play in the debate about the likelihood that humans, based on our current lifestyle, will exceed the planet's biocapacity? Base your essay on economic, environmental, social, and technical parameters.

1-50 Describe which of the grand challenges of engineering most interests you. Frame the problems that must be overcome associated with the challenge you have selected in terms of the variables in the IPAT equation.

1-51 Sustainable development is extremely difficult, since the environmental footprint of a nation generally increases with increasing development. Use the IPAT equation and determine whether each variable is likely to increase, decrease, or remain unchanged if the HDI of a country increases. What must the response of each variable in the IPAT equation be (increase, decrease, or no change) if development is to be truly sustainable?

Analyzing Sustainability Using Engineering Science

FIGURE 2.1 The Earth's surface undergoes constant change. Carbon dioxide concentrations were quite high until plant life developed and transformed the carbon dioxide in the atmosphere into oxygen as part of the photosynthetic process. Later on, animals would adapt to the increased oxygen concentration in the atmosphere and use the oxygen for energy and growth during respiration. The physical and chemical transformations in the Earth's surface are imperative for life as we know it.

Source: From the thermodynamic evolution of life: atmosphere, 129-162.

I consider nature a vast chemical laboratory in which all kinds of composition and decomposition are formed.

—Antoine Lavoisier

Analyzing Sustainability Using Engineering Science

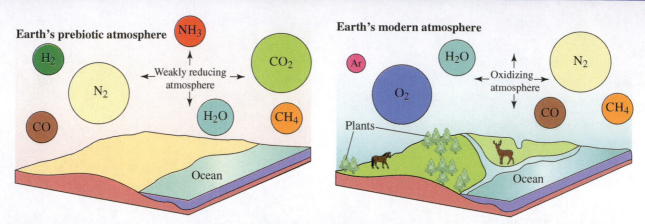

FIGURE 2.1 The Earth's surface undergoes constant change. Carbon dioxide concentrations were quite high until plant life developed and transformed the carbon dioxide in the atmosphere into oxygen as part of the photosynthetic process. Later on, animals would adapt to the increased oxygen concentration in the atmosphere and use the oxygen for energy and growth during respiration. The physical and chemical transformations in the Earth's surface are imperative for life as we know it.

Source: www.britannica.com/topic/evolution-of-the-atmosphere-1703862.

I consider nature a vast chemical laboratory in which all kinds of composition and decomposition are formed.

— ANTOINE LAVOISIER

GOALS

THE EDUCATIONAL GOALS OF THIS CHAPTER are to develop an understanding of how the basic sciences, biology, chemistry, mathematics, and physics may be applied to develop quantitative indicators of sustainable design and development described in Chapter 1. Fundamental units of measure are defined and applied to solve engineering problems. This chapter examines the characteristics and units of measure used to define environmental quality. Principles of environmental chemistry are applied to understand the fate of acids and bases in the natural environment. Chemical properties that influence the fate and transport of chemicals in the environment are also examined. Inorganic and organic chemical compounds and their benefits and risks to ecosystems, human health, and societal uses are investigated. The ethical responsibilities of engineers and scientists to manage and minimize the risk factors associated with the use of modern chemicals are also explored. These fundamental principles are used repeatedly in Chapters 4 through 11 to predict environmental impacts and also in Chapters 12 through 17 to compare and evaluate the sustainability of green design processes.

OBJECTIVES

At the conclusion of this chapter, you should be able to:

2.1 Define, convert, and determine appropriate uses for conventional units of measure in environmental systems.

2.2 Model the chemical partitioning between the gas phase and aqueous phase of dilute substances in the environment.

2.3 Define the relationship between pollutants in the gas phase and the atmospheric concentration of those pollutants.

2.4 Model the behavior of dissolved ionic species in aqueous solutions.

2.5 Model the relationship between pH and dissolved acids and bases in aqueous solutions in equilibrium.

2.6 Calculate chemical partitioning of solids and liquid phase substances in the environment.

Introduction

Chemical spill on local highway—get the full story on this late-breaking news!

Have you seen this headline before? If so, what is being inferred about "chemicals"? It is true that there have been many historical examples of chemicals that have had devastating environmental impacts. These include mercury pollution in Japan, arsenic contamination in Bangladesh, toxic waste contamination in Love Canal, New York, and the Chernobyl nuclear disaster in the former Soviet Union. Chemicals are also used to provide clean energy, disinfect and clean drinking water, purify food, manufacture medicines, and provide many other benefits associated with today's highly developed lifestyle. Chemicals support and comprise the molecules of life and the environment in which we live.

In order to get a quantitative understanding of what is happening in a system, we must also have knowledge of the units associated with the values provided. The topic of sustainability covers a very large breadth of information and spans multiple traditional fields, including chemistry, chemical engineering, energy engineering, environmental engineering and science, industrial engineering, and policy analysis. All of these fields have slightly different conventions in units and symbols, so if you are working in the field of sustainable engineering, you must become fluent in all these areas of study and also familiarize yourself with the units and terminology associated with the field. Both English standard units and the International System of Units (SI) are widely used in this field.

In analyzing various situations and responses, for example, we will discuss global energy scales and energy use that deal with numbers on the order of terawatts of energy. We will also investigate the effects of substances in the environment at concentrations less than 1 microgram per liter of fluid. You will need to become conversant in the standard prefixes that describe the standard scales of measure in our subsequent analysis.

Chemicals in the environment affect biodiversity, human health, and the Earth's climate. The fundamental sciences of biology, chemistry, mathematics, and physics provide a basis for analyzing how we achieve sustainable designs and development. Mass and energy balances are helpful in tracking chemicals as they move from one repository to another. However, we must study the applications and principles of chemistry more closely to appreciate how these chemicals are transformed in the environment and also transform the Earth's environment.

For example, carbon dioxide levels in the atmosphere have increased by more than 30% since the start of the Industrial Revolution. This increase has changed our atmosphere and the energy balance that affects how our climate responds to the sun's radiation. The carbon dioxide levels have also had a profound influence on ocean chemistry. The oceans are becoming much more acidic due to the increased carbon dioxide in the atmosphere. The increasing acidity inhibits the production of corals and calcifying phytoplankton and zooplankton, which in turn negatively affects the marine food web. Quantitative evaluations of chemical reactions in the environment, or compositions and decompositions in nature as Monsieur Lavoisier stated, are required to understand how carbon dioxide added to the atmosphere will lead to changes in ocean chemistry and subsequent marine life.

2.1   Elemental Analysis

The periodic table shown in Figure 2.2 lists all the elements found on Earth. While this list is imposing, the periodic table groups the elements in a way that greatly helps us understand how these chemicals react and are stored in the environment.

Legend (example cell):

- Atomic number
- Electron configuration
- Symbol
- Melting point [°C]
- Boiling point [°C]
- Atomic mass (mean relative)
- Oxidation states
- Name
- Density [g/cm³], for gases [g/l] (0° C, 1013 mbar)
- Electronegativity
- Radioactive

Example:

43 (98.91)
[Kr] $4d^5 5s^2$
Tc
Technetium
2140   11.5
5030 * 1.9

Group 1 (1A)

#	Mass	Config	Ox.	Symbol	Name	mp	density	bp	EN
1	1.01	$1s$		H	Hydrogen	-259	0.09	-253	2.1
3	6.94	[He]$2s$	1	Li	Lithium	181	0.53	1330	1.0
11	22.99	[Ne]$3s$	1	Na	Sodium	98	0.97	892	0.9
19	39.10	[Ar]$4s$	1	K	Potassium	64	0.86	760	0.8
37	85.47	[Kr]$5s$	1	Rb	Rubidium	39	1.53	688	0.8
55	132.91	[Xe]$6s$	1	Cs	Cesium	29	1.90	690	0.7
87	(223.0)	[Rn]$7s$	1	Fr	Francium	(27)	*	677	0.7

Group 2 (2A)

#	Mass	Config	Ox.	Symbol	Name	mp	density	bp	EN
4	9.01	[He]$2s^2$	2	Be	Beryllium	1277	1.85	2970	1.5
12	24.31	[Ne]$3s^2$	2	Mg	Magnesium	650	1.74	1107	1.2
20	40.08	[Ar]$4s^2$	2	Ca	Calcium	838	1.55	1440	1.0
38	87.62	[Kr]$5s^2$	2	Sr	Strontium	768	2.6	1380	1.0
56	137.33	[Xe]$6s^2$	2	Ba	Barium	714	3.76	1640	0.9
88	226.03	[Rn]$7s^2$	2	Ra	Radium	700	5.0	1140	0.9

Group 3 (3B/3A)

#	Mass	Config	Ox.	Symbol	Name	mp	density	bp	EN
21	44.96	[Ar]$3d4s^2$	3	Sc	Scandium	1539	3.0	2730	1.3
39	88.91	[Kr]$4d5s^2$	3	Y	Yttrium	1509	4.47	2927	1.2
71	174.97	[Xe]$4f^{14}5d6s^2$	3	Lu	Lutetium	1652	9.84	3327	1.2
103	(262.1)	[Rn]$5f^{14}6d7s^2$	3	Lr	Lawrencium	*	*	*	*

Group 4

#	Mass	Config	Ox.	Symbol	Name	mp	density	bp	EN
22	47.88	[Ar]$3d^24s^2$	4,3	Ti	Titanium	1668	4.54	3260	1.5
40	91.22	[Kr]$4d^25s^2$	4	Zr	Zirconium	1852	6.49	3580	1.4
72	178.49	[Xe]$4f^{14}5d^26s^2$	4	Hf	Hafnium	2222	13.31	5400	1.3
104	(261.1)	[Rn]$5f^{14}6d^27s^2$		Rf	Rutherfordium	*	*	*	*

Group 5

#	Mass	Config	Ox.	Symbol	Name	mp	density	bp	EN
23	50.94	[Ar]$3d^34s^2$	5,4,3,2	V	Vanadium	1900	6.11	3450	1.6
41	92.91	[Kr]$4d^45s$	5,3	Nb	Niobium	2468	8.57	4927	1.6
73	180.95	[Xe]$4f^{14}5d^36s^2$	5	Ta	Tantalum	2996	16.5	5425	1.5
105	(262.1)	[Rn]$5f^{14}6d^37s^2$		Db	Dubnium	*	*	*	*

Group 6

#	Mass	Config	Ox.	Symbol	Name	mp	density	bp	EN
24	52.00	[Ar]$3d^54s$	6,3,2	Cr	Chromium	1875	7.19	2200	1.6
42	95.94	[Kr]$4d^55s$	6,5,4,3,2,0	Mo	Molybdenum	2610	10.2	5560	1.8
74	183.85	[Xe]$4f^{14}5d^46s^2$	6,5,4,3,2,0	W	Tungsten	3410	19.3	5930	1.7
106	(263.1)	[Rn]$5f^{14}6d^47s^2$		Sg	Seaborgium	*	*	*	*

Group 7 (7A/7B)

#	Mass	Config	Ox.	Symbol	Name	mp	density	bp	EN
25	54.94	[Ar]$3d^54s^2$	7,6,4,3,2,0,-1	Mn	Manganese	1245	7.43	2097	1.5
43	(98.91)	[Kr]$4d^55s^2$	7	Tc	Technetium	2140	11.5	5030	1.9
75	186.21	[Xe]$4f^{14}5d^56s^2$	7,6,4,2,-1	Re	Rhenium	3180	21.0	5900	1.9
107	(264.1)	[Rn]$5f^{14}6d^57s^2$		Bh	Bohrium	*	*	*	*

Group 8

#	Mass	Config	Ox.	Symbol	Name	mp	density	bp	EN
26	55.85	[Ar]$3d^64s^2$	6,3,2,0,-2	Fe	Iron	1536	7.86	3000	1.8
44	101.07	[Kr]$4d^75s$	8,6,4,3,2,1,0,-2	Ru	Ruthenium	2500	12.4	3900	2.2
76	190.20	[Xe]$4f^{14}5d^66s^2$	8,6,4,3,2,0	Os	Osmium	3050	22.6	5500	2.2
108	(265.1)	[Rn]$5f^{14}6d^67s^2$		Hs	Hassium	*	*	*	*

Group 9

#	Mass	Config	Ox.	Symbol	Name	mp	density	bp	EN
27	58.93	[Ar]$3d^74s^2$	3,2,0,-1	Co	Cobalt	1495	8.9	2900	1.9
45	102.91	[Kr]$4d^85s$	5,4,3,1,2,0	Rh	Rhodium	1966	12.4	3730	2.2
77	192.22	[Xe]$4f^{14}5d^76s^2$	6,4,3,2,1,0,-1	Ir	Iridium	2454	22.7	4500	2.2
109	(268)	[Rn]$5f^{14}6d^77s^2$		Mt	Meitnerium	*	*	*	*

Group 10

#	Mass	Config	Ox.	Symbol	Name	mp	density	bp	EN
28	58.70	[Ar]$3d^84s^2$	3,2,0,-1	Ni	Nickel	1453	8.9	2730	1.9
46	106.42	[Kr]$4d^{10}$	4,2,0	Pd	Palladium	1552	12.0	3140	2.2
78	195.08	[Xe]$4f^{14}5d^96s$	4,2,0	Pt	Platinum	1769	21.4	3830	2.2
110	(269)	[Rn]$5f^{14}6d^87s^2$		Uun	Uun	*	*	*	*

Group 11 (1B)

#	Mass	Config	Ox.	Symbol	Name	mp	density	bp	EN
29	63.55	[Ar]$3d^{10}4s$	2,1	Cu	Copper	1083	8.9	2595	1.9
47	107.87	[Kr]$4d^{10}5s$	1	Ag	Silver	961	10.5	2210	1.9
79	196.97	[Xe]$4f^{14}5d^{10}6s$	3,1	Au	Gold	1063	19.3	2970	2.4
111	(272)	[Rn]$5f^{14}6d^97s^2$		Uuu	Uuu	*	*	*	*

Group 12 (2B)

#	Mass	Config	Ox.	Symbol	Name	mp	density	bp	EN
30	65.38	[Ar]$3d^{10}4s^2$	2	Zn	Zinc	420	7.13	906	1.6
48	112.41	[Kr]$4d^{10}5s^2$	2	Cd	Cadmium	321	8.65	765	1.7
80	200.59	[Xe]$4f^{14}5d^{10}6s^2$	2,1	Hg	Mercury	-38	13.6	357	1.9
112	(277)			Uub	Uub	*	*	*	*

Group 13 (3A/3B)

#	Mass	Config	Ox.	Symbol	Name	mp	density	bp	EN
5	10.81	[He]$2s^22p$	3	B	Boron	(2030)	2.35	2550	2.0
13	26.98	[Ne]$3s^23p$	3	Al	Aluminum	660	2.70	2680	1.5
31	69.72	[Ar]$3d^{10}4s^24p$	3	Ga	Gallium	30	5.91	2237	1.6
49	114.82	[Kr]$4d^{10}5s^25p$	3	In	Indium	156	7.31	2080	1.7
81	204.38	[Xe]$4f^{14}5d^{10}6s^26p$	3,1	Tl	Thallium	303	11.85	1457	1.8

Group 14 (4A/4B)

#	Mass	Config	Ox.	Symbol	Name	mp	density	bp	EN
6	12.01	[He]$2s^22p^2$	4,2,-4	C	Carbon	(3550)	2.2	4830	2.5
14	28.09	[Ne]$3s^23p^2$	4,-4	Si	Silicon	1410	2.33	2680	1.8
32	72.59	[Ar]$3d^{10}4s^24p^2$	4,2	Ge	Germanium	937	5.32	2830	1.8
50	118.69	[Kr]$4d^{10}5s^25p^2$	4,2	Sn	Tin	232	7.30	2270	1.8
82	207.20	[Xe]$4f^{14}5d^{10}6s^26p^2$	4,2	Pb	Lead	327	11.4	1725	1.9
114	(289)			Uuq	Uuq	*	*	*	*

Group 15 (5A/5B)

#	Mass	Config	Ox.	Symbol	Name	mp	density	bp	EN
7	14.01	[He]$2s^22p^3$	5,4,3,2,-3	N	Nitrogen	-210	1.25	-196	3.0
15	30.97	[Ne]$3s^23p^3$	5,3,-3	P	Phosphorus	44	1.82	280	2.1
33	74.91	[Ar]$3d^{10}4s^24p^3$	5,3,-3	As	Arsenic	Subl.	5.72	685	2.0
51	121.75	[Kr]$4d^{10}5s^25p^3$	5,3,-3	Sb	Antimony	631	6.69	1380	1.9
83	208.98	[Xe]$4f^{14}5d^{10}6s^26p^3$	5,3	Bi	Bismuth	271	9.8	1560	1.9

Group 16 (6A/6B)

#	Mass	Config	Ox.	Symbol	Name	mp	density	bp	EN
8	16.00	[He]$2s^22p^4$	-2,-1	O	Oxygen	-219	1.43	-183	3.5
16	32.06	[Ne]$3s^23p^4$	6,4,-2	S	Sulfur	119	2.07	445	2.5
34	78.96	[Ar]$3d^{10}4s^24p^4$	6,4,-2	Se	Selenium	217	4.79	685	2.4
52	127.60	[Kr]$4d^{10}5s^25p^4$	6,4,-2	Te	Tellurium	450	6.24	990	2.1
84	(209)	[Xe]$4f^{14}5d^{10}6s^26p^4$	6,4,2	Po	Polonium	254	9.3	962	2.0
116	(289)			Uuh	Uuh	*	*	*	*

Group 17 (7A/7B)

#	Mass	Config	Ox.	Symbol	Name	mp	density	bp	EN
9	19.00	[He]$2s^22p^5$	-1	F	Fluorine	-220	1.7	-188	4.0
17	35.45	[Ne]$3s^23p^5$	7,5,3,1,-1	Cl	Chlorine	-101	3.2	-35	3.0
35	79.90	[Ar]$3d^{10}4s^24p^5$	7,5,3,1,-1	Br	Bromine	-7	3.12	58	2.8
53	126.90	[Kr]$4d^{10}5s^25p^5$	7,5,1,-1	I	Iodine	114	4.94	183	2.5
85	(210)	[Xe]$4f^{14}5d^{10}6s^26p^5$	7,5,3,1,-1	At	Astatine	302	—	337	2.2

Group 18 (0)

#	Mass	Config	Symbol	Name	mp	density	bp	EN
2	4.00	$1s^2$	He	Helium	—	0.18	-269	—
10	20.18	[He]$2s^22p^6$	Ne	Neon	-249	0.9	-246	—
18	39.95	[Ne]$3s^23p^6$	Ar	Argon	-189	1.78	-183	—
36	83.80	[Ar]$3d^{10}4s^24p^6$	Kr	Krypton	-157	3.7	-152	—
54	131.29	[Kr]$4d^{10}5s^25p^6$	Xe	Xenon	-112	5.89	-108	—
86	(222)	[Xe]$4f^{14}5d^{10}6s^26p^6$	Rn	Radon	(-71)	9.73	-62	—
118	(293)		Uuo	Uuo	—	—	—	—

---

6 Lanthanoids

#	Mass	Config	Ox.	Symbol	Name	mp	density	bp	EN
57	138.91	[Xe]$5d6s^2$	3	La	Lanthanum	920	6.17	3470	1.1
58	140.12	[Xe]$4f5d6s^2$	4,3	Ce	Cerium	795	6.67	3468	1.1
59	140.91	[Xe]$4f^36s^2$	4,3	Pr	Praseodymium	935	6.77	3027	1.1
60	144.24	[Xe]$4f^46s^2$	3	Nd	Neodymium	1024	7.00	3027	1.2
61	(145)	[Xe]$4f^56s^2$	3	Pm	Promethium	(1027)	7.22	2460	*
62	150.36	[Xe]$4f^66s^2$	3,2	Sm	Samarium	1072	7.54	1790	1.1
63	151.96	[Xe]$4f^76s^2$	3,2	Eu	Europium	828	5.26	1439	*
64	157.25	[Xe]$4f^75d6s^2$	3	Gd	Gadolinium	1312	7.89	3000	1.1
65	158.93	[Xe]$4f^96s^2$	4,3	Tb	Terbium	1356	8.27	2800	1.2
66	162.50	[Xe]$4f^{10}6s^2$	3	Dy	Dysprosium	1407	8.54	2600	1.2
67	164.93	[Xe]$4f^{11}6s^2$	3	Ho	Holmium	1461	8.80	2600	1.2
68	167.26	[Xe]$4f^{12}6s^2$	3	Er	Erbium	1497	9.05	2900	1.2
69	168.93	[Xe]$4f^{13}6s^2$	3,2	Tm	Thulium	1545	9.33	1727	1.2
70	173.04	[Xe]$4f^{14}6s^2$	3,2	Yb	Ytterbium	824	6.98	1196	1.1

7 Actinoids

#	Mass	Config	Ox.	Symbol	Name	mp	density	bp	EN
89	(227)	[Rn]$6d7s^2$	3	Ac	Actinium	1050	10.1	3470	1.1
90	232.04	[Rn]$6d^27s^2$	4	Th	Thorium	1750	11.7	3850	1.3
91	231.04	[Rn]$5f^26d7s^2$	5,4	Pa	Protactinium	(1230)	15.4	—	1.4
92	238.03	[Rn]$5f^36d7s^2$	6,5,4,3	U	Uranium	1132	19.07	3818	1.4
93	(237.05)	[Rn]$5f^46d7s^2$	6,5,4,3	Np	Neptunium	637	19.5	3900	1.3
94	(244)	[Rn]$5f^67s^2$	6,5,4,3	Pu	Plutonium	640	19.81	3235	1.3
95	(243)	[Rn]$5f^77s^2$	6,5,4,3	Am	Americium	994	13.7	2607	1.3
96	(247)	[Rn]$5f^76d7s^2$	3	Cm	Curium	(1340)	13.51	3100	*
97	(247)	[Rn]$5f^97s^2$	4,3	Bk	Berkelium	986	*	*	*
98	(251)	[Rn]$5f^{10}7s^2$	3	Cf	Californium	*	*	*	*
99	(254)	[Rn]$5f^{11}7s^2$		Es	Einsteinium	*	*	*	*
100	(257)	[Rn]$5f^{12}7s^2$		Fm	Fermium	*	*	*	*
101	(258)	[Rn]$5f^{13}7s^2$		Md	Mendelevium	*	*	*	*
102	(259)	[Rn]$5f^{14}7s^2$	3,2	No	Nobelium	*	*	*	*

FIGURE 2.2 Periodic table of elements (also in front cover).

Furthermore, of the 118 elements on the periodic table, there are a few elements that you will get to know very well. All chemicals are important, but only a few of them are regularly encountered.

Most life forms on the planet are composed primarily of five elements: carbon, hydrogen, oxygen, nitrogen, and phosphorus. Carbon is the most basic building block of life. It is the primary component of cell mass and also of fossil fuels. Carbon bonds strongly with hydrogen (and other elements) by sharing electrons to form relatively "small" molecules like methane (CH_4). Carbon is also a primary component of very large molecules that include oxygen, nitrogen, phosphorus, and chains of carbon molecules, like our DNA. Carbon also bonds strongly with oxygen, forming carbon dioxide (CO_2) in the gas phase. Carbon and oxygen compounds, called carbonates, are an important part of natural waters (which we will discuss in greater detail in Chapters 7 through 10).

When we examine the elements in the Earth's atmosphere, we see in Table 2.1 that the atmosphere is composed of a mixture of gases. Nitrogen, in the form of N_2, and oxygen, in the form of O_2, are the two most prominent gases in our atmosphere. Argon, helium, neon, and krypton, with their place in the last column of the periodic chart, are "noble gases" and relatively inert. Although these noble gases are present in the atmosphere in relatively large amounts, they do not participate in environmentally or industrially common reactions.

TABLE 2.1 Composition of the dry standard atmosphere

GAS	FORMULA	VOLUME FRACTION (BY NUMBER OF MOLES)
Nitrogen	N_2	0.7808
Oxygen	O_2	0.2095
Argon	Ar	0.0093

GAS	FORMULA	VOLUME FRACTION (IN PARTS PER MILLION)
Carbon dioxide*	CO_2	419
Neon	Ne	18
Helium	He	5.2
Methane	CH_4	1.5
Krypton	Kr	1.1
Hydrogen	H_2	0.5

Source: Based on Hart, J. (1988). *Consider a Spherical Cow: A Course in Environmental Problem Solving*. Sausalito, CA: University Books.
*NOAA, June 7, 2021.

While we are all aware of the importance of oxygen in the atmosphere, most of the mass of oxygen is present in the oceans, in water, in biomass, and in the Earth's crust, as shown in Table 2.2. Silicon, aluminum, iron, calcium, magnesium, sodium, and potassium are all significant components of the Earth's crust and important to biological life, food production, and industry. Most of the Earth is blanketed in water, which contains more of the Earth's hydrogen, but we cannot overlook the importance of biomass as a chemical source of hydrogen as well.

TABLE 2.2 Major elements on the Earth's surface

ELEMENT	EARTH'S CRUST (PPM BY WEIGHT)	SEAWATER (PPM)	BIOMASS (PPM)
Oxygen, O	456,000	857,000	630,000
Silicon, Si	273,000	3	15,000
Aluminum, Al	83,600	0.01	500
Iron, Fe	62,200	0.01	1,000
Calcium, Ca	46,600	400	40,000
Magnesium, Mg	27,640	1,350	4,000
Sodium, Na	22,700	10,500	2,000
Potassium, K	18,400	380	20,000
Titanium, Ti	6,320	0.001	100
Hydrogen, H	1,520	108,000	80,000

Source: Based on Hart, J. (1988). *Consider a Spherical Cow: A Course in Environmental Problem Solving*. Sausalito, CA: University Books.

Combinations of elements form molecules. The form that the molecules take in the natural environment usually requires the least energy for which the elements can exist. In the gas phase, less energy is required for two molecules of oxygen to share electrons than for an individual molecule of oxygen to exist alone. Most of the oxygen in the gas phase exists as O_2, although other forms, like O_3 or ozone, may also exist depending on surrounding environmental conditions. Chemical reactions, when they do occur, are balanced based on the number of atoms it takes to convert from one molecule to another. For example, when we add energy and ignite methane (CH_4) in the presence of oxygen (O_2), the resulting products are carbon dioxide, water, and more energy:

$$\text{Energy for ignition} + CH_4 + O_2 \rightarrow CO_2 + H_2O + \text{more energy for heat or work} \quad (2.1)$$

Equation (2.1) is not balanced. The same number of carbon atoms exist on each side of the equation, so the carbon atoms are "balanced" in this chemical equation. However, there are more hydrogen atoms on the left side of the equation than on the right, so hydrogen atoms must be added to the right-hand side of the equation. Similarly, the numbers of oxygen atoms also must be balanced. Two hydrogen atoms are added to the right-hand side of the equation, in the form of H_2O, so that there are four molecules of hydrogen on both sides of the equation:

$$\text{Energy for ignition} + CH_4 + O_2 \rightarrow CO_2 + 2\,H_2O + \text{more energy for heat or work} \quad (2.2)$$

The oxygen atoms must also be balanced. There are a total of four oxygen atoms on the right-hand side of the equation: two used to form the carbon dioxide molecule (CO_2) and two molecules of water (H_2O)—each of which contains one atom of oxygen. Thus, we need to add one more molecule of diatomic oxygen gas (O_2) to the left-hand side of the equation to balance all the elements:

$$CH_4 + 2\,O_2 \rightarrow CO_2 + 2\,H_2O \quad (2.3)$$

Atoms are incredibly tiny forms of matter. In practice as engineers, we don't really try to balance the exact number of atoms required to burn gasoline in an engine. We need to think on scales that are much larger than the atom. If we only needed an order of magnitude greater in scale than the atom for a meaningful mass of a chemical, we could use a dozen atoms. Even a dozen atoms is an unimaginably small amount of a substance. Instead of individual atoms or a dozen atoms, we talk in terms of moles, where one mole of atoms is equivalent to 6.02×10^{23} atoms. The mole then is just a chemical unit of measure that relates a significant mass of atoms for chemists, engineers, and scientists to use in balancing equations and performing calculations in chemistry:

$$1 \text{ mole} = 6.02 \times 10^{23} \text{ atoms} \qquad (2.4)$$

We must use the molar mass of an element to find the mass of a given molecule associated with 1 mole of material:

$$\text{Mass [g]} = \text{number of moles} \times \text{molar mass [g/mol]} \qquad (2.5)$$

EXAMPLE 2.1   Atoms, Molecules, and Mass

Hydrogen (H_2) molecules and oxygen (O_2) molecules may combine to form water. If there were 24×10^{23} atoms of hydrogen (H) available to react with oxygen, how many moles of water could we produce, and how many moles of oxygen would we need to utilize all the hydrogen? How many grams of water would be produced?

The unbalanced equation looks like

$$H_2 + O_2 \rightarrow H_2O$$

Balancing the equation yields

$$2\,H_2 + O_2 \rightarrow 2H_2O$$

Four hydrogen atoms, or two hydrogen molecules (H_2), are needed to react with one molecule of oxygen gas (O_2). The reaction yields two molecules of H_2O.

The problem states that there are 24×10^{23} atoms of hydrogen as H.

$$\frac{24 \times 10^{23}\,\text{H atoms}}{2\,\text{H atoms per molecule of }H_2} = 12 \times 10^{23}\,\text{molecules of }H_2$$

$$\frac{12 \times 10^{23}\,\text{molecules of }H_2}{6.02 \times 10^{23}\,\dfrac{\text{molecules}}{\text{mole}}} = 2.0 \text{ moles of }H_2$$

From the balanced equation, 1.0 mole of O_2 is required to react with 2.0 moles of H_2, and this will yield 2.0 moles of H_2O.

$$2.0 \text{ mol } H_2 \times 18\,\frac{\text{grams }H_2O}{\text{mol }H_2O} = 36 \text{ grams }H_2O$$

EXAMPLE 2.2   Mass, Density, and Concentration

If you have 55.5 moles of distilled water in a 1-liter Nalgene® bottle at standard temperature and pressure, what is the density of the fluid in g/liter, kg/m³, and lb/ft³?

$$55.5\,\frac{\text{mol }H_2O}{\text{liter}} \times 18\,\frac{\text{grams}}{\text{mol}} = 1{,}000\,\frac{\text{g}}{\text{L}}$$

$$1{,}000 \, \frac{g}{L} \times \frac{1 \text{ kg}}{1{,}000 \text{ g}} \times \frac{1{,}000 \text{ L}}{m^3} = 1{,}000 \, \frac{kg}{m^3}$$

$$1{,}000 \, \frac{kg}{m^3} \times 2.205 \, \frac{lb}{kg} \times \frac{1 \text{ m}^3}{35.31 \text{ ft}^3} = 62.4 \, \frac{lb}{ft^3}$$

Most compounds found naturally in the environment are mixtures of elements and molecules. The air surrounding us is a mixture of nitrogen, oxygen, argon, carbon dioxide, and other gases, as shown in Table 2.2. Seawater contains high levels of dissolved salts, including sodium, chlorine, calcium, and magnesium. Even freshwater contains many of these same dissolved salts, but usually at lower concentrations than seawater. Both seawater and freshwater normally contain various concentrations of dissolved gases, including dissolved oxygen, carbon dioxide, and several other dissolved gases (see Box 2.1). So, most of the water in water is water,

BOX 2.1   How Much Oxygen Do Fish Need to Breathe?

Mossy Creek, in Virginia, is the home to a trout fishing tourist industry. The trout species of fish (Figure 2.3) requires very clean, cool water as its habitat. Trout generally need about 5×10^{-4} pounds of dissolved oxygen gas (O_2) per cubic foot of water to remain healthy. How many milligrams of dissolved oxygen per liter of water are needed for the trout in Mossy Creek to stay healthy?

$$5 \times 10^{-4} \, \frac{lb \, O_2}{ft^3} \times 35.31 \, \frac{ft^3}{m^3} \times \frac{1 \text{ m}^3}{1{,}000 \text{ L}}$$

$$\times \frac{1 \text{ kg}}{2.205 \text{ lb}} \times 10^6 \, \frac{mg}{kg} = 8 \, \frac{mg}{L}$$

About 8 milligrams of dissolved oxygen are required in each liter of water for a healthy trout habitat. Trout populations become stressed at values below 8 mg/L.

While fish can breathe the small amount of oxygen in the water, we, as humans, cannot. The air consists of about 21% oxygen, while water only contains on the order of 10×10^{-6} parts mass of oxygen compared to the total mass of the solution. Since only a small amount of oxygen dissolves into the water, we say that oxygen, as O_2, is only slightly soluble in water. The term *solubility* is used to describe the amount of a substance that will dissolve

in water. Substances like table salt (NaCl) and sugar ($\sim C_6H_{12}O_6$) are very soluble in water.

FIGURE 2.3 Brook trout require a high level of dissolved oxygen in the water for growth and reproduction. The dissolved oxygen concentration is sensitive to water temperature and the degree of mixing, so that cold, fast-moving water provides an ideal habitat for the brook trout species native to Virginia, and ponds and slower moving bodies of water do not have a high enough dissolved oxygen concentration in the water to allow this species to survive. Increasing surface temperatures and water pollution threaten to further decrease the trout habitat in temperate regions in North America and Europe.

Source: Bradley Striebig.

but not all of it! Or to be more precise, most of the liquid fluid we commonly call water is made up mostly of H_2O, but the fluid also contains other molecules. When we want to examine human-related (anthropogenic) impacts on the environment, we are concerned about very small quantities of mass that occur within the naturally occurring mixtures we call air, water, and soil. So, while most of the fluid in water is H_2O, it is the other chemical components that are of concern when we are trying to identify and evaluate the impacts of anthropogenic activities on a region's environment, society, and economy.

The mass of pollutants or compounds of concern in our environment may be very, very small—so small in fact, that these compounds may be much, much less than 1% of the total mass of the fluid in water. In order to avoid writing mass fractions of a pollutant in water that are very small decimal fractions, we define a unit fraction called a *part per million on a mass basis* (ppm_m) for compounds in water:

$$ppm_m = \frac{m_i}{m_{total}} \times 10^6 \qquad (2.6)$$

Notice that under conditions of standard pressure and temperature, water has a specific gravity of 1.00 or a density of 1.00 g/mL. When water comprises most of the fluid of concern, the specific gravity of the fluid is very nearly equal to 1.0, so

$$1\ ppm_m = 1\ mg/L \text{ for a fluid with a specific gravity} = 1.0\ g/mL \qquad (2.7)$$

EXAMPLE 2.3   Calcium in Seawater

Calcium from dissolved limestone and other minerals is a common component of freshwater. If 1 liter of a water solution contains 0.100 grams of calcium, what is the concentration of calcium in the following units?

(a) Grams per liter:

$$\frac{0.100 \text{ grams}}{1 \text{ liter}} = 0.100\ \frac{g}{L}$$

(b) Mass fraction:

From Example 2.2, we know that 1 liter of water weighs 1,000 grams; thus the mass fraction of calcium in water is

$$\frac{0.100 \text{ grams}}{1,000 \text{ grams}} = 0.000100\ \frac{\text{grams Ca}^{2+}}{\text{grams water solution}}$$

(c) Mass percentage:

$$0.000100\ \frac{\text{grams Ca}^{2+}}{\text{grams water solution}} \times 100\% = 0.01\% \text{ of the mass of water is Ca}^{2+}$$

(d) Parts per million on a mass basis:

$$0.000100\ \frac{\text{grams Ca}^{2+}}{\text{grams water solution}} \times 10^6 = 100\ ppm_m$$

(e) Milligrams per liter:

$$\frac{0.100 \text{ grams}}{1 \text{ liter}} \times \frac{1{,}000 \text{ milligrams}}{1 \text{ gram}} = 100 \frac{\text{mg}}{\text{L}}$$

2.2 Solubility and Henry's Law Constant

Most gases are only slightly soluble in water. In dilute aqueous (or water) solutions, the concentration of a substance in the gas phase is linearly related to the concentration of that substance in the aqueous phase. The ***Henry's law constant*** (H_i) is the name given to the constant that describes the slope of this linear relationship:

$$C_i = H_i P_i \tag{2.8}$$

where

P_i = partial pressure of the substance i in the gas. Typical units for the gas pressure are atmospheres [atm]

C_i = molar fraction of the substance i in the aqueous solution [moles of i per liter of fluid]

H_i = Henry's law constant

TABLE 2.3 Henry's law constant, H_i, for oxygen and carbon dioxide in water in units of mol/L-atm

TEMPERATURE (°C)	O_2	CO_2
0	0.0021812	0.076425
5	0.0019126	0.063532
10	0.0016963	0.053270
15	0.0015236	0.045463
20	0.0013840	0.039172
25	0.0012630	0.033363

Source: Based on Masters, G.M., and Ela, W.P. (2008). *Introduction to Environmental Engineering and Science*, 3rd ed. Upper Saddle River, NJ: Prentice Hall.

EXAMPLE 2.4 Investing in a Trout Farm for Aquaculture

A local farmer wants to diversify their business by creating a pay-per-fish trout farm near Mossy Creek and raise trophy trout for tourists and local restaurants. The water temperature of the pond in July and August averages 77°F. Find the dissolved oxygen (DO) level in the pond water in order to determine if this is a good investment. (Assume that the total atmospheric pressure is 1 atm.)

To convert from degrees Fahrenheit to degrees Celsius, use

$$°C = \frac{5}{9}(°F - 32) = \frac{5}{9}(77 - 32) = 25°C$$

The Henry's law values for oxygen and carbon dioxide are given in Table 2.3. The partial pressure of oxygen is given in Table 2.1:

$$P_{O_2} = 0.2095 \text{ atm}$$

Then the concentration of dissolved oxygen, C_{DO}, in the liquid phase can be determined from the Henry's law equation:

$$C_{DO} = P_{O_2} \times H_{O_2} = (0.2095 \text{ atm}) \left(0.0012630 \frac{\text{mol}}{\text{atm-L}} \right) = 0.0002646 \frac{\text{mol}}{\text{L}}$$

$$C_{DO} = 0.0002646 \frac{\text{mol}}{\text{L}} \times \frac{1,000 \text{ mmol}}{\text{mol}} \times \frac{32 \text{ mg}}{\text{mmol}} = 8.467 \frac{\text{mg}}{\text{L}} \text{ dissolved oxygen}$$

The Henry's law constant does not have standard units. In this example, H_i had units of [mol$_i$/L-atm]. However, the Henry's law constant units vary greatly from one academic field to another. Care must be taken to ensure that the units reported are correctly used in any application of this equation. The Henry's law constant varies with temperature, and these correlations can be estimated from data reported in the scientific literature. If the concentration of the gas compound dissolved in the fluid is very high (if it can be expressed as a mass percentage), Henry's law values become nonlinear and can no longer be used to model these highly concentrated gas-liquid systems.

2.3   The Ideal Gas Law

The units of concentration used in the gas phase tend to vary slightly, but significantly, from those used in the liquid phase. It is essential to understand these differences. Fundamentally, the causes of these differences are (1) a gas is a compressible fluid and subject to deformation with small changes in pressure, and (2) the density of air is much, much lower than the density of water. The ideal gas law governs the relationship between the pressure, volume, mass, and temperature of a substance in the gas phase:

$$PV = nRT \tag{2.9}$$

where

    P = absolute pressure (atm)

    V = volume (L)

    n = number of moles (mol)

    T = absolute temperature (K)

    R = universal gas constant (0.08206 L-atm/mol-K, or other units depending on the referenced source)

The numerical value of R is related to the units we use to define P, V, n, and T. R has a different numerical value if different units are used; for instance, another common value is

$$R = 8.314 \text{ (J/mol-K)} \tag{2.10}$$

As discussed previously, our atmosphere is composed of a mixture of various gases shown in Table 2.1. Each of these gases contributes to the total pressure the atmosphere exerts on the Earth. The International Union of Pure and Applied Chemistry (IUPAC) has defined the Standard Ambient Temperature and Pressure (SATP) as 298.15 K (25°C, 77°F) and an absolute pressure of exactly 105 Pa (1 bar, 0.986923 atm). Recall that the pressure of 105 pascal is the force in newtons (N) applied over an area of 1 m². Thus 1 bar (105 Pa or 0.98623 atm) is the approximate force exerted by the average weight of the atmosphere on the Earth. Each component of the gas contributes to the total pressure or weight of the atmosphere. The pressure of each part of a gas mixture exerted on its surrounding is called the *partial pressure* of that gas. When a system is at equilibrium, the partial pressure exerted by that gas is equal to the vapor pressures of the gas. When the partial pressures of each component of the gas are summed, the resultant pressure is the total pressure of the gas. The relationship between total pressure and the pressure of each component of the gas mixture is called Dalton's law of partial pressure:

$$P_t = \Sigma \, P_i \tag{2.11}$$

EXAMPLE 2.5   Atmospheric Pressure

The volume fractions of each of the major gases in the atmosphere are listed in Table 2.1. From this table, determine the partial pressure of nitrogen, oxygen, and argon at standard temperature and pressure (STP) in units of bar. What percentage of the total pressure of the dry atmosphere (neglecting water in the atmosphere) is made up of these three gases?

Nitrogen: 1 bar × 0.7808 = 0.7808 bar
Oxygen: 1 bar × 0.2095 = 0.2095 bar
Argon: 1 bar × 0.0093 = 0.0093 bar

Sum of the partial pressures of N, O, and Ar is 0.7808 bar + 0.2095 bar + 0.0093 bar = 0.9996 bar.

$$\frac{0.9996 \text{ bar}}{1.0000 \text{ bar}} \times 100\% = 99.96\% \text{ of the dry atmosphere is composed of these three gases.}$$

For compounds in the gas phase, the pressure and volume that the substance occupies are related to the number of moles or mass of the compound. A volumetric basis is used for comparing the concentration of anthropogenic emissions and low-level, naturally occurring compounds in the atmosphere. The mole fraction of an individual substance, i, divided by the total moles of all compounds in the gas mixture is directly proportional to the volumetric fraction if pressure and temperature remain constant:

$$P_{total} V_{total} = n_{total} RT \tag{2.12}$$

$$P_i V_i = n_i RT \tag{2.13}$$

so

$$V_i / V_{total} = n_i / n_{total} = y_i \tag{2.14}$$

where y is the molar fraction of the individual component i in the gas phase. The molar gas fraction is commonly used in chemical engineering analysis where the chemical engineer is concerned with the change of a compound from one form to another, and those changes are several percentages of the total gas. However, as in aqueous solutions, the environmental sciences are usually concerned with far smaller concentrations. We will again use the concept of parts per million, but on a *volume* basis (ppm$_v$) to describe the atmospheric concentration of pollutants. It is extremely important to note that in the gas phase, parts per million is a volumetric fraction, not a mass fraction, and thus the conversion to a mass density is quite different.

EXAMPLE 2.6   Historical Values of Carbon Dioxide

Prior to the Industrial Revolution, the atmosphere contained 0.00028 mole of carbon dioxide (CO_2) and 0.99972 mole of air (mainly from N_2, O_2, and Ar). What was the concentration of CO_2 expressed in ppm$_v$?

$$\frac{0.0028 \text{ mol } CO_2}{0.99972 \text{ mol air} + 0.0028 \text{ mol } CO_2} = 0.0028 \text{ mole fraction } CO_2 \text{ in air}$$

$$0.0028 \text{ mole fraction } CO_2 \times 10^6 = 280 \text{ ppm}_v CO_2$$

In order to determine the mass concentration in a given volume of air, we must apply the ideal gas equation. In order to use the ideal gas law, we must make the following substitutions:

$$n_i = \frac{m_i}{MW_i} \tag{2.15}$$

where

m_i = mass of component i

MW_i = molar mass of component i

Also from the law of partial pressure:

$$P_i = y_i \times P_{total} \tag{2.16}$$

Substituting the above expressions into the ideal gas law yields

$$(y_i \times P_{total})V = \left(\frac{m_i}{MW_i}\right)RT \tag{2.17}$$

Rearranging yields

$$C_i = \frac{m_i}{V} = \left(\frac{P_i}{T}\right)\left(\frac{MW_i}{R}\right) = y_i\left(\frac{P_t}{T}\right) \tag{2.18}$$

Substituting the values (P_{total} = 1 atm, T = 289 K, R = 0.08206 L-atm/mol-K) for standard temperature and pressure and converting to units of mg/m^3 yield

$$C_i\left[\frac{mg}{m^3}\right] = \frac{C_i[\text{ppm}_v]MW_i}{24.5} \tag{2.19}$$

and for nonstandard conditions

$$C_i\left[\frac{\mu g}{m^3}\right] = \text{ppb}_v \times MW_i \times \frac{1,000P}{RT} \tag{2.20}$$

Note how very different the conversion is between ppm$_v$ and mg/m^3 in the gas phase and ppm$_m$ and mg/L in an aqueous solution.

The human species has adapted over history to be able to react to and detect extremely small quantities of pollutants associated with spoiled food that can cause disease. Historically, doctors practicing medicine would even make diagnoses based on their ability to smell conditions in a patient that indicated pathogenic bacteria were likely present. We are able to smell some compounds in air at concentrations on the order of a few parts per billion (ppb$_v$), or three orders of magnitude smaller than a ppm$_v$:

$$1 \text{ ppm}_v = 1{,}000 \text{ ppb}_v \qquad (2.21)$$

EXAMPLE 2.7   Conversion of Gas Concentration Between Volumetric Fractions and Mass Concentration

Table 2.4 shows the highest reported concentrations of (a) carbon monoxide and (b) nitrogen dioxide, criteria air pollutants, in 2008 in Fairfax County, a suburban area near Washington, DC. Convert these concentrations from ppm$_v$ and ppb$_v$ to mg/m^3 and μg/m^3, respectively.

TABLE 2.4 Highest concentrations of carbon monoxide and nitrogen dioxide pollutant levels in the air in Fairfax, Virginia

CRITERIA POLLUTANT	VALUE	UNIT
CO	3.7	ppm$_v$
NO$_2$	13	ppb$_v$

Source: Based on *Virginia Ambient Air Monitoring 2010 Data Report* (2011). Office of Air Quality Monitoring, Virginia Department of Environmental Quality.

From the periodic table, find the molar mass, *MW*, of each element:

$MW_C = 12$ g/mol
$MW_N = 14$ g/mol
$MW_O = 16$ g/mol

(a) Find the concentration of carbon monoxide in the units specified.

Find the molar mass of one molecule of carbon monoxide:

$$12 \text{ g/mol}(1) + 16 \text{ g/mol}(1) = 28 \text{ g/mol}$$

Substitute the molar mass for carbon monoxide and the concentration of carbon monoxide given in Table 2.4 into Equation (2.18):

$$C_{CO}\left[\frac{\text{mg}}{\text{m}^3}\right] = \frac{C_{CO}[\text{ppm}_v]\,MW_{CO}}{24.5} = \frac{(3.7 \text{ ppm}_v)\left(28\dfrac{\text{g}}{\text{mol}}\right)}{24.5} = 4.2 \frac{\text{mg}}{\text{m}^3}$$

(b) Find the concentration of nitrogen dioxide in the units specified.

Find the molar mass of nitrogen dioxide:

$$14 \text{ g/mol}(1) + 16 \text{ g/mol}(2) = 46$$

Convert from ppb_v to ppm_v of nitrogen dioxide:

$$13\ ppb_v \times \frac{1\ ppm_v}{1{,}000\ ppb_v} = 0.013\ ppm_v$$

Substitute the molar mass for carbon monoxide and the concentration of carbon monoxide given in Table 2.4 into Equation (2.20):

$$C_{NO_2}\left[\frac{\mu g}{m^3}\right] = C_{NO_2}[ppm_v]\,MW_{NO_2}\left[\frac{g}{mol}\right]\frac{1{,}000(P)}{(RT)} = (0.013\ ppm_v)\left(46\frac{g}{mol}\right)\frac{1{,}000}{24.5} = 24\ \frac{\mu g}{m^3}$$

Notice that we could convert to mg/m^3 and then convert from mg to μg and achieve the same results as using Equation (2.20).

ACTIVE LEARNING EXERCISE 2.1 Invasion of the Toxic Marbles

This exercise is designed to help visualize and reinforce the concepts of ppm_m and ppm_v. All that is needed are a few marbles (preferably two different colors) and a drinking container. Any container will do, as long as the volume is known or can be assumed. If each of the marbles weighs 1 mg and six marbles of one color have $MW = 20$ g/mol while four marbles of another color have $MW = 40$ g/mol:

- What is the concentration of marbles in the container if it were filled with water in the given units?
  - mg/L
  - ppm_m
- What is the concentration of marbles in the 0.5-liter jar filled with air at STP in the given units?
  - mg/m^3
  - ppm_v

The conversions between units are an extremely important part of understanding quantitatively how we can assess measures of sustainability. Understanding these units and how to convert from one to another is a prerequisite to understanding environmental impacts, associating societal consequences of actions (or the lack of actions), and forecasting economic impacts. A deep understanding of unit analysis is also vital to performing Life Cycle Analysis based on material usage and transformation of those materials into different pathways. If that is not enough of a reason to understand and practice unit conversions, the bottom line is that if you work in the field of environmental impact analysis, environmental policy, or sustainable engineering, converting from one unit basis to another is a vital skill you will need to be successfully employed. Furthermore, people outside this field of study will likely lack this skill and will need to pay you to perform these calculations!

ACTIVE LEARNING EXERCISE 2.2 Conversion of Gas Concentration Between Volumetric Fractions and Mass Concentration

The United States Environmental Protection Agency (U.S. EPA) reported the concentrations of criteria air pollutants in 2008 in Fairfax County, a suburban area near Washington, DC. Convert these concentrations from ppm_v and ppb_v to mg/m^3 and $\mu g/m^3$, respectively.

Criteria pollutants observed in Fairfax, Virginia

CRITERIA POLLUTANT	VALUE	UNIT
O_3	107	ppb_v
SO_2	16	ppb_v

Source: Based on *Virginia Ambient Air Monitoring 2010 Data Report* (2011). Office of Air Quality Monitoring, Virginia Department of Environmental Quality.

2.4 Chemistry of Natural Systems

Carbon dioxide dissolves in water (see Figure 2.4) in accordance with Henry's law, as discussed earlier in the chapter. The dissolved carbon dioxide takes several forms in solution: dissolved carbon dioxide gas, carbonic acid (H_2CO_3), bicarbonate ion (HCO_3^-), and the carbon ion (CO_3^{2-}). Collectively, these make up the ***dissolved inorganic carbon*** (DIC) in water, which is sometimes designated with the symbol ($H_2CO_3^*$). The dissolved inorganic carbon compounds play a major role in regulating the acidity and basicity of natural and marine water.

$$CO_{2(g)} + H_2O \leftrightarrow H_2CO_3^* \tag{2.22}$$

$$[H_2CO_3^*] = [CO_{2(g)}] + [H_2CO_3] + [HCO_3^-] + [CO_3^{2-}] \tag{2.23}$$

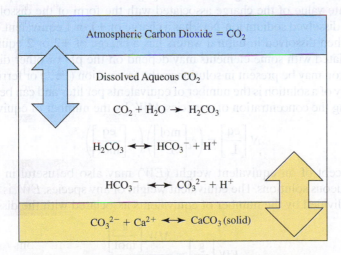

FIGURE 2.4 Predominant phases of inorganic carbon in natural systems.

EXAMPLE 2.8   Carbon Dioxide Dissolution in Natural Waters

Use Henry's law to calculate the concentration of dissolved inorganic carbonates at standard temperature and pressure (STP) in solution if the atmospheric concentration of carbon dioxide is 419 ppm_v CO_2 and the temperature is 15°C.

The Henry's law constant, from Table 2.3, for carbon dioxide at 15°C is

$$H_{CO_2} = 0.045463 \frac{mol}{L\text{-}atm}$$

The average partial pressure of carbon dioxide at STP is

$$P_{CO_2} = 419 \ ppm_v = 419 \times 10^{-6} \ atm$$

Using Henry's law yields

$$C_{H_2CO_3^*} \left[\frac{mol}{L} \right] = [H_2CO_3^*] = (H_{CO_2})(P_{CO_2}) = \left(0.045463 \frac{mol}{L\text{-}atm} \right) (419 \times 10^{-6} \ atm)$$

$$= 1.90 \times 10^{-5} \frac{mol}{L}$$

2.4.1 Law of Electroneutrality

The dissolved bicarbonate and carbonate ions are the major components of a natural water's alkalinity, or the abilities of water to neutralize acids. These ions also typically make up a large percentage of the negative ions in a freshwater solution. The amounts of positive and negative dissolved ions in a solution must equal one another. The *law of electroneutrality* states that the sum of all the positive ions (cations) in a solution must equal the sum of all the negative ions (anions) in a solution, so that the net charge of all natural waters is equal to zero:

$$\sum cations - \sum anions = 0 \tag{2.24}$$

Normality and equivalents represent a useful method of associating the charge of ions in solution. The number of charge equivalents (eq) associated with a compound is equal to the absolute value of the charge associated with the form of the dissolved ion. For example, the dissolved sodium ion, Na^+, has a charge of $+1$ or 1 equivalent. The calcium ion (Ca^{2+}) when dissolved in natural waters has a charge of $+2$ or 2 equivalents. The charge associated with some elements may depend on the pH or other dissolved ions in solution. Iron may be present in solution as the ferrous ion (Fe^{2+}) or ferric ion (Fe^{3+}). The normality of a solution is the number of equivalents per liter and can be determined by multiplying the concentration of a species, MW_i, by the number of equivalents, z_i:

$$N_i \left[\frac{eq}{L} \right] = \left(c_i \left[\frac{mol}{L} \right] \right) \left(z_i \left[\frac{eq}{mol} \right] \right) \tag{2.25}$$

The concept of an equivalent weight (EW) may also be useful in calculations involving aqueous solutions. The equivalent weight of any species, EW_i, is equal to the molar mass divided by the number of equivalents associated with the dissolved ion:

$$EW_i \left[\frac{g}{eq} \right] = \frac{MW_i \left[\frac{g}{mol} \right]}{z_i \left[\frac{eq}{mol} \right]} \tag{2.26}$$

2.4.2 Ionic Strength

The *ionic strength*, I, of a solution is the estimate of the overall concentration of dissolved ions in solution. The strength of the ionic interactions is strongly correlated with the square of the ionic charge of the individual ion:

$$I = \frac{1}{2} \sum_{\text{all ions}} c_i z_i^2 \tag{2.27}$$

where

I = ionic strength [mol/L]

c_i = concentration of each individual ion [mol/L]

z_i = charge associated with each ion species, i

| EXAMPLE 2.9 | Estimating the Ionic Strength from Dissolved Ion Concentrations |

Estimate the ionic strength of the Ganges River and the Dead Sea, given the data in Table 2.5.

TABLE 2.5 Comparison of major dissolved ion concentrations in the Dead Sea and the Ganges River

WATER BODY	CONCENTRATION OF IONS [meq/L]						
	Cations				Anions		
	[Na$^+$]	[K$^+$]	[Ca^{2+}]	[Mg^{2+}]	[SO$_4^{2-}$]	[Cl$^-$]	[HCO$_3^-$]
Dead Sea	1,519	193	788	3,453	11	5,859	4
Ganges River	0.28	0.06	1.10	0.40	0.06	0.16	1.70

Source: Based on Meybeck, M., Chapman, D., and Helmer, R. (eds.) (1989). *Global Freshwater Quality: A First Assessment.* London: Blackwell.

Note that the concentrations are given in meq/L. To covert to mol/L, divide each ion by its associated charge and multiply by the appropriate unit conversion:

$$c_{Na^+} = 0.28 \frac{\text{meq}}{\text{L}} \times \frac{\text{eq}}{1{,}000 \text{ meq}} \times \frac{1 \text{ mol}}{1 \text{ eq}} = 2.8 \times 10^{-4} \frac{\text{mol}}{\text{L}}$$

The same procedure can be used to determine the concentration of potassium ion K^+ in the solutions. For the diprotic ions, such as the calcium ion:

$$c_{Ca^{+2}} = 1.10 \frac{\text{meq}}{\text{L}} \times \frac{\text{eq}}{1{,}000 \text{ meq}} \times \frac{1 \text{ mol}}{2 \text{ eq}} = 5.5 \times 10^{-4} \frac{\text{mol}}{\text{L}}$$

The same procedure can be used to determine the concentration of magnesium ion Mg^{2+} in the solutions. Notice that the absolute value of the charge is used for the negative ions in solution:

$$c_{Cl^-} = 0.16 \frac{\text{meq}}{\text{L}} \times \frac{\text{eq}}{1{,}000 \text{ meq}} \times \frac{1 \text{ mol}}{1 \text{ eq}} = 1.6 \times 10^{-4} \frac{\text{mol}}{\text{L}}$$

The same procedure can be used to determine the concentration of other anions in the solutions. The charge associated with each ion and the sum of the terms, $c_i z_i^2$, are shown in Table 2.6.

TABLE 2.6 Tabular calculation of major dissolved ion concentrations and ionic strength components in the Dead Sea and the Ganges River

WATER BODY	CATIONS				ANIONS		
	$[Na^+]$	$[K^+]$	$[Ca^{2+}]$	$[Mg^{2+}]$	$[SO_4^{2-}]$	$[Cl^-]$	$[HCO_3^-]$
z_i	1	1	2	2	2	1	1
	Concentration, C_i (mol/L)						
Dead Sea	1.519	0.193	0.394	1.726	5.5×10^{-3}	5.859	4×10^{-3}
Ganges River	2.8×10^{-4}	0.6×10^{-4}	5.5×10^{-4}	2.0×10^{-4}	0.3×10^{-4}	1.6×10^{-4}	1.7×10^{-4}
	$c_i z_i^2$						
Dead Sea	1.519	0.193	1.576	6.906	22×10^{-3}	5.859	4×10^{-3}
Ganges River	2.8×10^{-4}	0.6×10^{-4}	22×10^{-4}	8.0×10^{-4}	1.2×10^{-4}	1.6×10^{-4}	1.7×10^{-4}
	Summation						
Dead Sea	10.2				5.88		
Ganges River	3.34×10^{-3}				1.98×10^{-3}		
	Ionic Strength (mol/L)						
Dead Sea	8.04						
Ganges River	2.66×10^{-3}						

The ionic strength for the solution can be determined using Equation (2.27):

$$I_{\text{Dead Sea}} = \frac{1}{2} \sum_{\text{all ions}} c_i z_i^2 = \frac{1}{2}(10.2 + 5.88) = 8.04 \frac{\text{mol}}{\text{L}}$$

$$I_{\text{Ganges River}} = \frac{1}{2} \sum_{\text{all ions}} c_i z_i^2 = \frac{1}{2}(3.34 + 1.98) \times 10^{-3} = 2.66 \times 10^{-3} \frac{\text{mol}}{\text{L}}$$

The Dead Sea is so called because the ionic strength of its waters is greater than that of ocean water due to its high concentration of salts; few life forms can survive in water with such large dissolved ion content. The Dead Sea's ionic strength is nearly four orders of magnitude greater than that of the Ganges River.

2.4.3 Solids and Turbidity

Because it is often difficult to account for and measure every dissolved ion in solution, several other methods for estimating ionic strength from simple laboratory or field measurements have been developed, as shown in Tables 2.7 and 2.8. Several of these methods are based on laboratory procedures (see Figure 2.5) to measure the total dissolved solids (TDS). There are two types of materials in water that are described by the term *solids*: suspended solids and dissolved solids. **Suspended solids** are the materials that are floating or suspended in the water. These solids may consist of silt, sand, and soil or organic

TABLE 2.7 Summary of procedures for calculating the solids content of a water sample

PARAMETER	ABBREVIATION	EQUATION
Total Solids	TS	$\dfrac{\text{(weight of dried unfiltered sample and flask − weight of empty sample flask)}}{\text{volume of water sample}}$ or TSS + TDS
Total Suspended Solids	TSS	$\dfrac{\text{(weight of dried sample on the filter − weight of filter only)}}{\text{volume of water sample}}$
Fixed Suspended Solids	FSS	$\dfrac{\text{(weight of dried sample and filter − weight of solids after heating to 550°C)}}{\text{volume of water sample}}$
Volatile Suspended Solids	VSS	TSS − FSS
Total Dissolved Solids	TDS	$\dfrac{\text{(weight of dried filtered sample and flask − weight of empty sample flask)}}{\text{volume of water sample}}$
Fixed Dissolved Solids	FDS	$\dfrac{\text{(weight of dried sample and flask − weight of flask after heating to 550°C)}}{\text{volume of water sample}}$
Volatile Dissolved Solids	VDS	TDS − FDS

TABLE 2.8 Empirical relationships for estimating ionic strength from laboratory or field measurements

MEASURED PARAMETERS	EQUATION	RATIONALE	REFERENCE
Individual species concentration, C_i, Individual species valence, z_i, and unaccounted for TDS, R	$I = \frac{1}{2}\Sigma C_i z_i^2 - 2.50 \times 10^{-5}\,R$ where $R = \text{TDS}_{measured} - \text{TDS}_{known\ species\ concentrations}$	Accounts for individual species concentrations	(Butler, 1982)
Total dissolved solids (TDS) [mg/L]	$I = 2.04 \times 10^{-5}\,(\text{TDS})$	Assumes the ionic strength to TDS ratio is equal to that of seawater	(Butler, 1982)
Total dissolved solids (TDS) in [mg/L]	$I = 2.50 \times 10^{-5}\,(\text{TDS})$	Assumes the ionic strength to TDS ratio is equal to that of a 40 g/mol monovalent salt	(Langelier, 1936)
Conductivity, κ in [μmho/cm = μS/cm; where S = siemens]	$I = 1.4\ \text{to}\ 1.6 \times 10^{-5}\,\kappa$	From data surveys	(Snoeyink and Jenkins, 1980)

Source: Based on Benjamin, M.M. (2001). *Water Chemistry*. New York: McGraw-Hill.

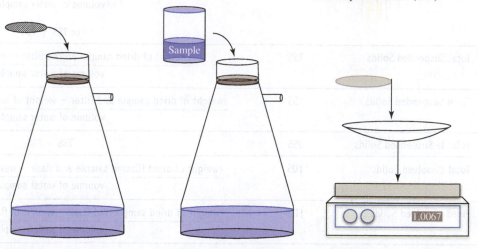

Step 1: Place a fiber and filter disc in filter holder. Record the dry weight of the filter.

Step 2: Filter a known volume of sample water by applying a vacuum to the bottom of the filtering flask.

Step 3: Remove the filter and place watch glass to dry in oven at 103–105°C. Re-weigh filter using an analytical balance. Calculate Total Suspended Solids (TSS).

FIGURE 2.5 EPA-approved method for measuring total dissolved solids in water and wastewater.

particles such as decaying leaves, algae, and bacteria. As you may expect, these suspended solids make the water cloudy or turbid. The clarity or turbidity of the water can be measured optically as an indicator of the level of suspended solids in the water. Alternatively, these solid particles can be collected on a filter and then weighed to obtain the mass of suspended solids per volume of water that passes through the filter (see Figure 2.5). The total suspended solids (TSS) are defined by the procedure as those particles greater than about 0.45 μm that are removed when the water passes through a standardized paper filter and the filter is dried at 103°C. TSS are divided into the particles that are volatile, called *volatile suspended solids* (VSS), and fixed solids. The volatile solids are determined by the weight of any particles that evaporate after the filter is heated to 550°C. The TSS concentration of China's Yellow River is shown in Figure 2.6. The Yellow River contains a high amount of eroded soil particles that make the water very turbid compared to more pristine, unpolluted waters.

Unlike the suspended solids, dissolved solids cannot be seen by the naked eye. The *dissolved solids* consist of salts and minerals that have been dissolved through natural weathering of soils or through the anthropogenic process. Since these solids are dissolved, they are not removed or measured by filtering the water. The salts and minerals that were dissolved in the water sample will remain in the container, which can be weighed again to determine the mass of the dissolved solids in the water sample following the calculations outlined in Table 2.7. TDS is calculated by determining the mass of dissolved solids that remain in the flask after the water has passed through the filter, and the water from the flask has been evaporated by placing the flask in a furnace or oven at 103°C.

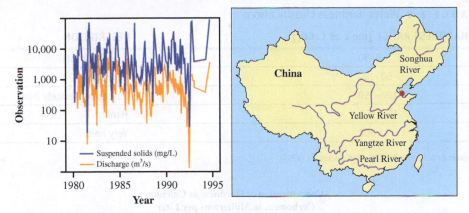

FIGURE 2.6 The suspended solids concentration reported in mg/L and instantaneous flow rate measured in m/s in the Yellow River at Lijin, China, approximately 80 km from the mouth of the river. The eroded soil particles and turbid brown- and green-colored water can be seen in satellite images of the discharge of the Yellow River.

Source: Based on UNEP GEMS (2006). *Water Quality for Ecosystem and Human Health.* United Nations Environment Programme Global Environment Monitoring System/Water Programme.

2.4.4 Water Hardness

Many concepts involving water quality, such as the definitions of TDS and TSS, have been derived from empirical laboratory procedures that have been used for several decades to indicate water quality and potential uses for water of that quality. *Water hardness* is one such historically defined water quality parameter. As the name implies, it is related to the part of the water mixture, or more precisely, the aqueous solution, that will become a solid (and therefore "hard") when the water is heated. Typically, water hardness is a result of high levels of dissolved calcium and carbonate ions that are found in groundwater in limestone-rich geologic strata. Hardness may also be caused by magnesium, strontium, manganese, and iron. The dissolved calcium and carbonate ions have the unusual property of becoming less soluble in solution with increasing temperature when both ions are found in groundwater at high concentrations. As water is heated in home water heaters or industrial boilers, these ions combine and form a solid precipitate or "scale" that decreases the overall efficiency of these appliances and may eventually cause the water heaters and boilers to cease working. Hardness also reduces the effectiveness of soaps.

The specific definition of hardness in water is the sum of the concentration of the divalent cations (species with a charge of 2+) in water. For most waters, the most important species that contribute to hardness are calcium (Ca^{2+}) and magnesium (Mg^{2+}). Most public water in the United States ranges in hardness from about 25 to 150 mg/L of $CaCO_3$. Hardness is most often expressed in mg/L as $CaCO_3$, since it is typically assumed that most of the hardness in the water is associated with the formation of calcium carbonate solids. Water with a hardness greater than 150 mg/L of $CaCO_3$ is usually considered hard, as shown in Table 2.9, and may undergo a water "softening" process where calcium ions are replaced with single-valent cations such as sodium or potassium ions. Hardness values are shown for surface water in the United States and various other sampling locations in Figures 2.7 and 2.8.

TABLE 2.9 Water hardness classifications

HARDNESS RANGE [mg/L as CaCO₃]	DESCRIPTION
0–50	Extremely soft
50–100	Very soft
100–150	Soft to moderately hard
150–300	Hard
> 300	Very hard

Source: Based on Dufor, C.N., and Becker, E. (1964).

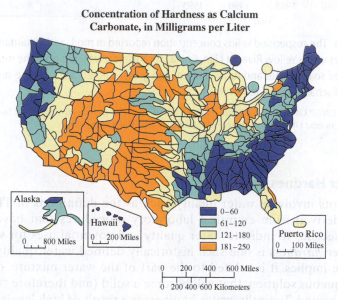

Concentration of Hardness as Calcium Carbonate, in Milligrams per Liter

FIGURE 2.7 Mean water hardness values as calcium carbonate at water monitoring sites in 1975 in the United States (adapted from Briggs and Ficke, 1977).

Source: Based on USGS, *USGS Water-Quality Information, Water Hardness*, water.usgs.gov/owq/hardness-alkalinity.html.

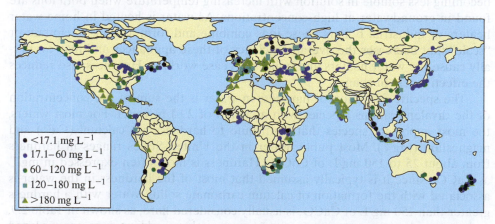

FIGURE 2.8 Water hardness values in mg/L reported at surface water monitoring stations on different continents.

Source: Based on UNEP GEMS (2006). *Water Quality for Ecosystem and Human Health*. United Nations Environment Programme Global Environment Monitoring System/Water Programme.

Hardness is divided into three components: total hardness, carbonate hardness, and noncarbonated hardness. Total hardness is determined by the sum of the diprotic cations. **Carbonate hardness** represents the portion of the diprotic ions that can combine with carbonates to form scaling. It is equal to the smaller value between the total hardness and the total bicarbonate concentration expressed as mg/L $CaCO_3$. **Noncarbonate hardness** is present only if the concentration of the bicarbonate ion is less than the total hardness. Noncarbonated hardness is the difference between total hardness and carbonate hardness, as shown in Table 2.10.

TABLE 2.10 Summary of hardness calculations

PARAMETER	RELEVANT EQUATIONS	UNITS	COMMENT
Total hardness	$= \sum$ concentration of diprotic cations	mg/L as $CaCO_3$	
Carbonate hardness	$= [HCO_3^-]$ or $=$ total hardness	mg/L as $CaCO_3$	Whichever value is smaller
Noncarbonated hardness	$=$ total hardness $-$ carbonate hardness	mg/L as $CaCO_3$	Or zero, if $[HCO_3^-]$ > total hardness

EXAMPLE 2.10   Determining Water Hardness

Spring water in Cumberland County, Pennsylvania, was analyzed, and the data shown in Table 2.11 were reported. Determine the water's total hardness, carbonate hardness, and noncarbonated hardness (if any) in units of mg/L as $CaCO_3$.

TABLE 2.11 Water analysis in groundwater from Cumberland County, Pennsylvania

WATER SOURCE	CONCENTRATION OF IONS [mg/L]						
	Cations			Anions			
	$[Na^+]$	$[Ca^{2+}]$	$[Mg^{2+}]$	$[NO_3^-]$	$[SO_4^{2-}]$	$[Cl^-]$	$[HCO_3^-]$
Spring Cu 29	13	95	13	24	34	12	300

Source: Based on Flippo, H.N. (1974).

Calculate the amount of calcium in solution as if it were to form calcium carbonate:

$$95 \frac{\text{mg } Ca^{2+}}{L} \times \frac{1 \text{ mmol } Ca^{2+}}{40 \text{ mg}} \times \frac{1 \text{ mmol } CaCO_3}{1 \text{ mmol } Ca^{2+}} \times \frac{100 \text{ mg } CaCO_3}{1 \text{ mmol } CaCO_3} = 237.5 \frac{\text{mg}}{L} \text{ as } CaCO_3$$

Similarly, calculate the amount of magnesium in solution as if it were to form calcium carbonate:

$$13 \frac{\text{mg } Mg^{2+}}{L} \times \frac{1 \text{ mmol } Mg^{2+}}{24.4 \text{ mg } Mg^{2+}} \times \frac{1 \text{ mmol as if } CaCO_3}{1 \text{ mmol } Mg^{2+}} \times \frac{100 \text{ mg } CaCO_3}{1 \text{ mmol } CaCO_3} = 53.3 \frac{\text{mg}}{L} \text{ as } CaCO_3$$

Total hardness [mg/L as $CaCO_3$] = $[Ca^{2+}] + [Mg^{2+}]$ as mg/L $CaCO_3$
                = 237.5 + 53.3 = 290.8

Therefore, this particular groundwater is hard.

Calculate the amount of bicarbonate in solution as if it were to form calcium carbonate:

$$300 \ \frac{mg \ HCO_3^-}{L} \times \frac{1 \ mmol \ HCO_3^-}{61 \ mg} \times \frac{1 \ mmol \ CaCO_3}{1 \ mmol \ HCO_3^-} \times \frac{100 \ mg \ CaCO_3}{1 \ mmol \ CaCO_3} = 491 \ \frac{mg}{L} \ as \ CaCO_3$$

Since $[HCO_3^-] >$ total hardness, carbonate hardness = total hardness = 290.8 mg/L as $CaCO_3$.

Noncarbonated hardness = 0, since CH > TH.

2.4.5 Chemical Reactivity, Activity, and the Activity Coefficient

Carbon in the dissolved carbonates is an important negative dissolved ion in natural water chemistry. The other dissolved species in the solution may also influence the form and chemical behavior of the dissolved carbon in aqueous solutions. The ionic strength is used as an indicator of the influence of the dissolved ions in chemical reactions. This influence is determined by the calculation of the chemical activity of the dissolved ions in a solution.

The effect of dissolved ions on chemical reaction rates must be accounted for to accurately model and predict how chemicals will react in freshwater (with very low concentrations of dissolved ions) compared to saltwater solutions (with very high concentrations of dissolved ions). In order for one chemical to react with another in a solution, the two chemicals must first come in contact with one another. The *chemical reactivity* refers to the chemical's overall tendency to participate in a reaction.

In freshwater solutions, the likelihood of one atom, A, coming into contact with another, B, is directly related to the concentration of each compound $[A]$ and $[B]$ or to how many atoms of a given chemical are found within a given volume of water. The situation of freshwater is very similar to what chemists would define as an ideal solution or a solution where the dissolved ions behave independently of each other. The reaction rate of formation of AB, r_{AB}, is proportional to the concentration, as illustrated in Equations (2.28) and (2.29). The reaction rate constant, k_{AB}, relates the change in concentration to the rate of the reaction. The greater the number of each atom, the more likely they are to contact one another and create a chemical reaction. The pure water case is analogous to two people trying to leave a football stadium through a single gate when the stadium is empty: The two people may meet at the gate (come into contact with one another) and pass through the door together (create a reaction) without anyone else interfering. In pure freshwater in the natural environment, the concentration of other chemical species is so low that other chemicals are unlikely to inhibit chemical reactions in most instances.

$$A + B \xrightarrow{\text{pure freshwater}} AB \tag{2.28}$$

$$r_{AB} = k_{AB}[A][B] \tag{2.29}$$

In solutions with significant concentrations of other dissolved ions in solution, or nonideal solutions, the dissolved ions may interfere or shield one reactant, C, from coming into contact with another reactant, D. The case with increased dissolved solids is analogous to two people on opposite sides of a stadium trying to leave a football stadium through a single gate when the football game is over and the stadium is full: The two people will find it very difficult to meet at the gate (come into contact with one another) and pass through the door at the same time (create a reaction) because there are so many other people (dissolved ions) interfering with their progress. In salt water in the natural environment, the concentration of other dissolved ions is so high that other chemicals inhibit chemical reactions. Even if, in these two cases, the concentrations of all the species are identical:

$$[A] = [C] \text{ and } [B] = [D]$$

The rate of the reaction between compounds C and D will be much slower than the reaction between A and B because of the interference of the other chemicals in the solution.

$$r_{AB} > r_{CD}$$

In order to account for the effects of other species in a solution, the concept of chemical activity has been developed. The **chemical activity** is a standardized measure of chemical reactivity within a defined system. The activity of a compound A is denoted by $\{A\}$. Standard state conditions are arbitrarily defined in environmental science as 25°C and atmospheric pressure equal to 1 bar (or 1 atm). (Other fields, such as oceanography, may define the standard state differently.) If the concentration of A in water is equal to 1 mol/L, then

$\{A\} = 1$ when under standard state conditions

If $\{A\} > 1$, then the system is not at standard state and has a greater chemical reactivity.

If $\{A\} < 1$, then the system is not at standard state and has a lower chemical reactivity.

At standard state, other conditions are also defined. The activity of the liquid solvent, in our case water, is also defined as 1, as indicated in Table 2.12:

$\{\text{pure liquid}\} \approx \{H_2O\} \approx 1$ at standard state for an ideal solution

TABLE 2.12 Chemical activity definitions of standard state conditions

STATE	CONCENTRATION	TEMPERATURE	PRESSURE
Solid	Pure		
Liquid	Pure	25°C	1 bar
Gas	Pure		
Solute	1.0 molar		

Since the standard conditions and activity have been defined, a relationship between activity and reactivity can be developed. The **activity coefficient**, γ_A, is used to relate the standard chemical activity and the conditional chemical reactivity. The activity coefficient is defined as the ratio of the reactivity per mole of A in a real system compared to the reactivity of A in the standard reference state:

$$\gamma_A = \frac{\text{real reactivity per mole of } A}{\text{standard activity per mole of } A} = \frac{\{A\}}{[A]} \tag{2.30}$$

Since both the activity of A, $\{A\}$, and the concentration of A, $[A]$, have units of mol/L, the activity coefficient is unitless.

Several empirical relationships have been developed to estimate the value of the activity coefficient based on the ionic strength of the aqueous solution, as shown in Table 2.13.

TABLE 2.13 Common approximations for individual ion activity coefficients in water at 25°C

ACTIVITY CORRELATION EQUATION	NAME	COMMENTS
$\gamma_i = 10^{-0.51z_i^2 I^{0.5}}$	Debye–Hückel limiting law	$I \leq 5 \times 10^{-3}$ M
$\gamma_i = 10^{\frac{-(0.51z_i^2 I^{0.5})}{(1+0.33aI^{0.5})}}$	Extended Debye–Hückel limiting law[a]	$I \leq 0.1$ M
$\gamma_i = 10^{\frac{-(0.51z_i^2 I^{0.5})}{(1+I^{0.5})}}$	Güntelberg	Mixtures with $I \leq 0.1$ M
$\gamma_i = 10^{\frac{-(0.51z_i^2 I^{0.5})}{(1+I^{0.5})-0.2I}}$	Davies[b]—for common ions such as Cl⁻ and OH⁻	$I \leq 0.5$ M

a. The variable a is an adjustable parameter based on the size of the ion. $a = 9$ for H^+, Al^{3+}, Fe^{3+}, La^{3+}, Ce^{3+}. $a = 8$ for Mg^{2+} and Be^{2+}. $a = 6$ for Ca^{2+}, Zn^{2+}, Cu^{2+}, Sn^{2+}, Mn^{2+}, Fe^{2+}. $a = 5$ for Ba^{2+}, Sr^{2+}, Pb^{2+}, CO_3^{2-}. $a = 4$ for Na^+, HCO_3^-, $H_2PO_4^-$, $CHCOO^-$, SO_4^{2-}, HPO_4^{2-}, PO_4^{3-}. $a = 3$ for K^+, Ag^+, NH_4^+, OH^-, Cl^-, ClO_4^-, NO_3^-, I^-, HS^-.
b. Davies has proposed using 0.3 instead of 0.2 for the coefficient in the last term of the equation.

Source: Adapted from Benjamin, M.M. (2001). *Water Chemistry*. New York: McGraw-Hill.

EXAMPLE 2.11 | Determination of the Activity Coefficient for the Ganges River

Estimate the activity coefficient associated with the sodium ion, calcium ion, and bicarbonate ion concentrations given in Table 2.5 in the Ganges River.

From Example 2.9, the ionic strength of the Ganges River water was determined to be $I = 2.66 \times 10^{-3}$ mol/L.

As shown in Table 2.13, the Debye–Hückel limiting law is valid for solutions with this approximate ionic strength. Substituting the values for the sodium ion into the Debye–Hückel limiting law yields

$$\log(\gamma_{Na^+}) = -0.5z_{Na^+}^2 I^{1/2} = -0.5(1)^2(2.66 \times 10^{-3})^{1/2} = -0.0258$$

$$\gamma_{Na^+} = 0.942$$

Similarly, for the calcium ion:

$$\log(\gamma_{Ca^+}) = -0.5z_{Ca^+}^2 I^{1/2} = -0.5(2)^2(2.66 \times 10^{-3})^{1/2} = -0.103$$

$$\gamma_{Ca^{+2}} = 0.788$$

And for the bicarbonate ion:

$$\log(\gamma_{HCO_3^-}) = -0.5z_{HCO_3^-}^2 I^{1/2} = -0.5(-1)^2(2.66 \times 10^{-3})^{1/2} = -0.0258$$

$$\gamma_{HCO_3^-} = 0.942$$

Notice that since the charge is squared, the activity coefficient for compounds with the same net charge, regardless of whether the charge is positive or negative, will have the same activity coefficient. The electrostatic forces that create the shielding are proportional to the square of the ionic strength, so the activity coefficient will change significantly for compounds with a different ionic charge. That is to say, the activity effects associated with the calcium ions are much greater than the activity effects we would expect to observe for the sodium or bicarbonate ions.

ACTIVE LEARNING EXERCISE 2.3 Activity and Concentration Relationships

The concepts of activity and reactivity are not difficult, but unfortunately, the language used to describe them can be intimidating. An example will help demonstrate the differences between activity and reactivity. Consider a slow reaction process, where compound X reacts slowly with compound Y to form XY:

$$X + Y \rightarrow XY$$

In the lab you are given a stop watch and a lab instrument that measures the mass of X and the mass of XY. You are asked to conduct an experiment to determine the activity coefficient for X in a sample of each of the following:

a) Pure freshwater
b) Estuary water, approximately a mixture of freshwater and salt water
c) Salt water

How would you set up your experiment? Under what conditions would you conduct the experiment in the laboratory environment? What information would each piece of equipment provide?

2.5 Equilibrium Models for Estimating Environmental Impacts

Many chemical processes reach equilibrium in the natural environment very quickly, particularly acid–base reactions. Equilibrium-based calculations and models are very useful for estimating the potential environmental impacts of many chemicals in the environment. Equilibrium models are also useful for designing systems to treat industrial and municipal gas, liquid, and solid waste streams. A basic understanding of equilibrium chemistry will help engineers design systems to reduce pollutant emissions as well as understand the fate and transport of chemicals that are emitted into the environment. Engineers can make more informed choices about sustainable systems and sustainable design by applying the basic concepts of chemical equilibrium.

It is also important to recognize the limitations of equilibrium-based models. Environmental systems are rarely static; instead, these systems are dynamic, and equilibrium-based models cannot be expected to adequately represent each and every element of complex natural systems. The engineer or scientist must be aware of the limitations of the mathematical assumptions and simplifications. There are also those reactions that simply occur too slowly within the system of concern, particularly some oxidation–reduction reactions and many biogeochemical reactions. For these relatively slow reactions, the assumptions required to develop an equilibrium-based model are simply not valid. Finally, the dynamic environmental systems differ from the ideal conditions from which much of our basic data is derived, yielding a significant level of uncertainty in calculations. It is not unusual for the reported "standard" values of key parameters to differ by 0.5% to 10% under laboratory conditions, and extrapolating these uncertainties to field conditions is a difficult task. Nonetheless, the equilibrium-based models provide powerful tools to estimate the bounds of environmental conditions that might be expected from relatively fast reactions and the extent to which reactions may proceed. They may also indicate if a particular chemical transformation is possible.

The fundamental parameter for equilibrium-based models is the equilibrium constant, k. The equilibrium constant describes the proportionality between reactants and products that will occur when the reaction has reached its minimum energy state and is complete. For a given chemical reaction, the equilibrium constant is defined as

$$A + B \leftrightarrow C + D$$

$$k_{equilibrium} = \frac{\{products\}}{\{reactants\}} = \frac{\{C\}\{D\}}{\{A\}\{B\}} \tag{2.31}$$

2.5.1 Acid and Base Definitions

Acids and bases influence water quality by controlling the pH of the aqueous solution, which in turn affects the dissolution and precipitation of compounds, the solubility of gases, and even the interactions between chemicals and living organisms. Equilibrium conditions are used to model the effects of pH in the natural environment, including the effects of pH on the carbonate system, and vice versa. The pH of natural waters is important to biodiversity since most species are only tolerant of natural waters in the pH range from 6.5 to 8.5, as shown in Figure 2.9. However, the pH of rainfall is largely dependent on the concentration of pollutants in the atmosphere and the solubility of those pollutants, as discussed in the following examples. Airborne sulfates emitted from coal-burning power plants have the greatest impact on the pH of rainfall; consequently, the biodiversity of natural bodies of water may be threatened, as shown in Figure 2.10. Acidified rainfall is also discussed in more detail in Chapter 9.

An *acid* may be defined as any substance that can donate a hydrogen ion, H^+ (or proton). A *base* may be defined as any substance that can accept an H^+ ion (or proton). The acid that donates the hydrogen ion and the base that accepts the ion are collectively known as an acid conjugate base pair, as

$$\text{Acid} \rightarrow \text{proton} + \text{conjugate base pair} \tag{2.32}$$

$$HA \rightarrow H^+ + A^- \tag{2.33}$$

or

$$\text{Base} + \text{proton} \rightarrow \text{conjugate acid pair} \tag{2.34}$$

$$A^- + H^+ \rightarrow HA \tag{2.35}$$

Either of the previous reactions may take place simultaneously, so that the net reaction becomes

$$HA \leftrightarrow H^+ + A^- \tag{2.36}$$

When written in this fashion, with the acid as a reactant and the base as a product, the acid equilibrium constant for the acid–base equilibrium is

$$k_a = \frac{\{H^+\}\{A^-\}}{\{HA\}} = \frac{\gamma_{H^+}[H^+]\gamma_{A^-}[A^-]}{\gamma_{HA}[HA]} \tag{2.37}$$

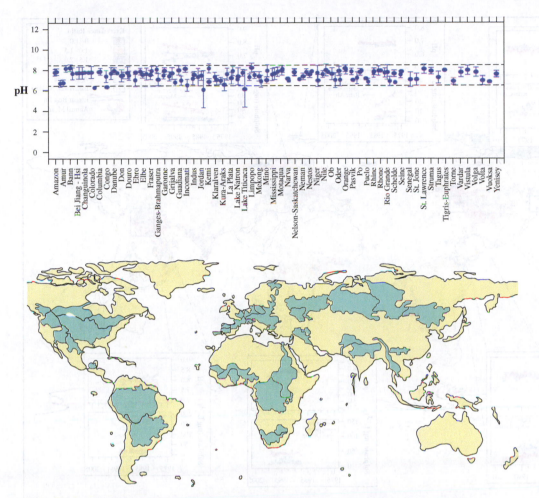

FIGURE 2.9 The mean pH (±1 standard deviation) of major drainage basins in the world. The dashed lines indicate the approximate pH range suitable for protecting biodiversity in natural waterways. The shaded areas on the map show the areas that have been sampled as part of the Global Environment Monitoring System.

Source: Based on UNEP GEMS (2006). *Water Quality for Ecosystem and Human Health*. United Nations Environment Programme Global Environment Monitoring System/Water Programme.

Acid equilibrium constants can be found online and in introductory chemisty textbooks. If the dissolved ion concentration in the solution is low, then we may assume that the solution is approximately ideal, $\gamma \cong 1$.

For an ideal solution, Equation (2.37) may be simplified to

$$k_a = \frac{\{H^+\}\{A^-\}}{\{HA\}} = \frac{[H^+][A^-]}{[HA]} \tag{2.38}$$

The water molecule may act as either an acid or a base (see Box 2.2). Substances that can either donate or receive a proton are called **ampholytes** (or they are described as amphoteric compounds). Equation (2.39) is directly analogous to

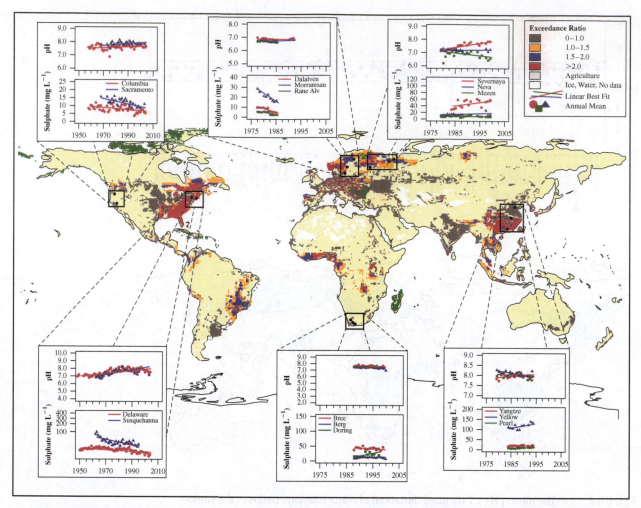

FIGURE 2.10 Trends in pH and sulphate concentrations in rivers around the world are overlaid onto a map showing the sensitivity of an area to acidification. Sensitivity is measured as an exceedance ratio. Exceedance ratios greater than 1 indicate areas that are sensitive to acid deposition (Bouwman et al., 2002). Note the general decreases in in-stream sulphate concentrations in the United States and Sweden, compared to increases in China and Russia. There was very little change in pH and sulphate concentrations in South Africa, which is considered at relatively low risk of acidification based on exceedance ratios. Changes in pH do not always parallel changes in sulphate concentrations, likely due to the effects of other acidifying chemicals such as nitrates.

Source: Based on UNEP GEMS (2006). *Water Quality for Ecosystem and Human Health.* United Nations Environment Programme Global Environment Monitoring System/Water Programme.

Equation (2.36). However, because water is the solvent in natural aqueous solutions, its acid equilibrium constant is given a special subscript, k_w. Recall that the arbitrary standard state chosen for environmental science defines the activity of water, $\gamma_{HOH} = 1$. Values for k_w are well documented and vary slightly with temperature, as shown in Table 2.14.

$$H_2O \leftrightarrow H^+ + OH^- \qquad (2.39)$$

$$k_w = \frac{\{H^+\}\{OH^-\}}{\{HOH\}} = \{H^+\}\{OH^-\} = 10^{-14} \qquad (2.40)$$

BOX 2.2 Acid and Base Properties of Pure Water

A review of the definitions of k_a and k_b will show many similarities. These equilibrium constants are related mathematically to the water equilibrium constant, k_w.

It will be helpful, in this proof of concept, to note that the actual form an acid takes in water is more complex than simply H^+. The acid becomes linked to several water molecules, as illustrated in Figure 2.11, in what is appropriately called a chemical complex. The chemical water complex is a molecular group that is more accurately represented by the formula, $H_9O_4^+$. Rather than write this "complex" formula, acid in water may be abbreviated in the shorthand notation as H_3O^+ or H^+. It will be helpful to use the slightly more formal H_3O^+ and include water molecules in this example. If you proceed with the correct algebra, it will become apparent why we've neglected this nuance in nomenclature in earlier examples.

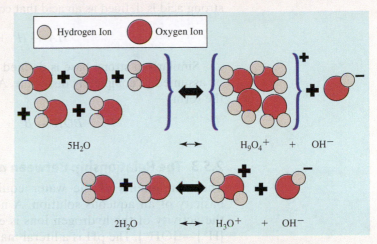

FIGURE 2.11 Visualization of how water partially dissociates to form very small amounts of acid molecules designated with the symbol H^+ and the hydroxide ion base OH^-, even in pure water solutions.

TABLE 2.14 Variation in k_w with variations in temperature

TEMPERATURE °C	LOG k_w
0	−14.93
10	−14.53
20	−14.17
30	−13.83
50	−13.26

Source: Based on Stumm, W., and Morgan, J.J. (1996). *Aquatic Chemistry: Chemical Equilibria and Rates in Natural Waters*, 3rd ed. New York: John Wiley.

A base–equilibrium constant may also be written for the following general equilibrium equation:

$$B^- + H^+ \leftrightarrow BH \tag{2.41}$$

$$k_b = \frac{BH}{\{B^-\}\{H^+\}} = \frac{\gamma_{BH}[BH]}{\gamma_{B^-}[B^-]\,\gamma_{H^+}[H^+]} \tag{2.42}$$

2.5.2 Strong Acids and Strong Bases

A very large value for the equilibrium means that the concentration of the products is much greater than the concentration of the reactants. If $\{H^+\}\{A^-\} \gg \{HA\}$, then the reactions proceed in one direction only, and the acid completely dissociates. A strong acid is defined as an acid that completely dissociates when added to water:

$$HA \rightarrow H^+ + A^- \text{ and } \{HA\} \approx 0 \tag{2.43}$$

Similarly, a strong base is defined as one that has a large value of k_b and may be assumed to completely dissociate. A typical strong base equation may appear in the form

$$BOH \rightarrow B^+ + OH^- \text{ and } \{BOH\} \approx 0 \tag{2.44}$$

2.5.3 The Relationship Between pH and pOH

The components of the water equilibrium equation define the acidity and basicity of an aqueous solution. A neutral solution is defined as one in which the activity of the hydrogen ions is equal to the activity of the hydroxide ions: $\{H^+\} = \{OH^-\}$. The pH is a literal mathematical definition that is defined as the negative logarithm of the activity of the hydrogen ion in the solution. (Note that the symbol p in pX represents the mathematical function of the negative logarithm, $pX = -\log\{X\}$.)

$$pH = -\log\{H^+\} \tag{2.45}$$

Similarly,

$$pOH = -\log\{OH^-\} \tag{2.46}$$

The pH and pOH of a solution are related by the equilibrium expression for water, k_w, at standard conditions by Equation (2.40), so

$$k_w = \{H^+\}\{OH^-\} = 10^{-14} \tag{2.47}$$

Applying the negative logarithm function to both sides of Equation (2.47) yields

$$-\log\{k_w\} = -\log\{H^+\} + -\log\{OH^-\} = -\log\{10^{-14}\} \tag{2.48}$$

$$pk_w = pH + pOH = 14 \tag{2.49}$$

Several steps are involved in modeling the effects of acids or bases in the natural environment:

Step 1: Define the system boundaries.

Step 2: Identify all the chemical species of interest.

Step 3: Write the constraining chemical equations for the system, including equilibrium equations, the electroneutrality equation, and mass balance equations.

Step 4: Make any simplifying assumptions that are possible.

Step 5: Algebraically solve the remaining independent equations and check the assumptions made to aid in solving the equations.

EXAMPLE 2.12 What Is the pH of Pure Water?

Determine the pH and pOH of pure water at 25°C and 1 bar.

Step 1: Define the system boundaries.

The container of the solution provides the system boundaries.

Step 2: Identify all the chemical species of interest.

$$H_2O, H^+, \text{ and } OH^-$$

Step 3: Write the constraining chemical equations for the system, including equilibrium equations, the electroneutrality equation, and mass balance equations.

The relevant equations are (2.40), (2.45), (2.46), and (2.49).

Step 4: Make any simplifying assumptions that are possible.

No simplifying assumptions are required to solve this problem.

Step 5: Algebraically solve the remaining independent equations and check the assumptions made to aid in solving the equations.

Substitute $\{H^+\} = \{OH^-\}$ into Equation (2.40):

$$k_w = \{H^+\}\{OH^-\} = \{H^+\}\{H^+\} = \{H^+\}^2 = 10^{-14}$$

Therefore, in a pure, neutral water solution,

$$\{H^+\} = \{OH^-\} = 10^{-7} \text{ mol/L}$$

The pH of the solution is determined by substituting the activity of the hydrogen ion into Equation (2.45):

$$pH = -\log\{H^+\} = -\log\{10^{-7}\} = 7$$

The pOH is found by using either Equation (2.46), since the hydroxide ion activity is known, or Equation (2.49):

$$pOH = -\log\{OH^-\} = -\log\{10^{-7}\} = 7$$

or

$$pk_w = pH + pOH = 14$$

$$pOH = 14 - pH = 14 - 7 = 7$$

EXAMPLE 2.13 Determining the pH of a Solution to Which a Strong Acid Has Been Added

Determine the pH for a 1-liter solution to which the strong acid, hydrochloric acid (HCl), has been added to produce a total acid concentration in the solution of 10^{-3} mol/L.

Step 1: Define the system boundaries.

The container of the solution provides the system boundaries.

Step 2: Identify all the chemical species of interest.

$$H_2O, H^+, Cl^-, \text{ and } OH^-$$

Step 3: Write the constraining chemical equations for the system, including equilibrium equations, the electroneutrality equation, and mass balance equations.

The relevant equations are (2.43) and (2.45).

The total concentration of the acid is given as: $H^+ = 10^{-3}$ mol/L.

Step 4: Make any simplifying assumptions that are possible.

It is assumed that HCl is a strong acid and completely dissociates. Since HCl is a strong acid,

$$HCl \rightarrow H^+ + Cl^-$$

Step 5: Algebraically solve the remaining independent equations and check the assumptions made to aid in solving the equations.

The pH of the solution is determined by substituting the activity of the hydrogen ion into Equation (2.45):

$$pH = -\log\{H^+\} = -\log\{10^{-3}\} = 3$$

EXAMPLE 2.14   Approximating the pH of Nitrogen Acidified Rain

Nitrous oxide (N_2O) is emitted to the atmosphere through naturally occurring biological transformations. The atmospheric concentration of nitrous oxide has increased from the anthropogenic use of nitrogen-based fertilizers, as shown in Figure 2.12. Although the concentration of nitrous oxides in the atmosphere is

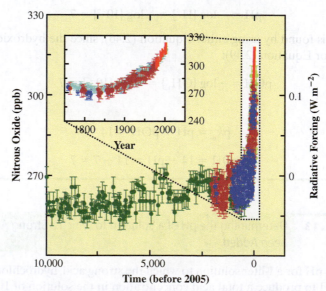

FIGURE 2.12 Atmospheric abundance of N_2O as determined from ice cores (various colors) samples. The inset contains changes over the last 10,000 years and since 1750.

Source: IPCC (2007). *Climate Change 2007: Working Group I: The Physical Science Basis*, Figure SPM.1.

almost 1,000 times less than the concentration of carbon dioxide, it is an influential greenhouse gas. Nitrous oxide has a very long residence time in the atmosphere, and it has a relatively large energy absorption capacity.

Nitrogen oxides can be removed by rainfall when the nitrogen oxides are absorbed by raindrops. The nitrogen oxides go through several reactions and form a strong acid, nitric acid (HNO_3), in the raindrop. Given today's average concentration of approximately 320 ppb N_2O in the atmosphere, what would be the pH of a raindrop, and the concentration of other dissolved ions in equilibrium with N_2O in the atmosphere?

Step 1: Define the system boundaries.

The system boundaries are defined by the raindrop in a standard atmosphere that contains 320 ppb N_2O.

Step 2: Identify all the chemical species of interest.

$$N_2O_{(g)}, H_2O, H^+, NO_3^-, HNO_3, \text{ and } OH^-$$

Step 3: Write the constraining chemical equations for the system, including equilibrium equations, the electroneutrality equation, and mass balance equations.

The relevant equations are (2.43) and (2.45).

The total concentration of the acid can be calculated from Henry's law:

$$k_H = 0.025 \frac{mol}{kg\text{-}bar}$$

$$c_{HNO_3(aq)} \cong c_{N_2O(aq)} = k_H P_{N_2O}$$

Step 4: Make any simplifying assumptions that are possible.

It is assumed that HNO_3 is a strong acid and completely dissociates. Since HNO_3 is a strong acid,

$$HNO_3 \rightarrow H^+ + NO_3^-$$

Step 5: Algebraically solve the remaining independent equations and check the assumptions made to aid in solving the equations.

Determine the aqueous concentration of the nitric acid:

$$c_{HNO_3(aq)} \cong c_{N_2O(aq)} = k_H P_{N_2O} = 0.025 \frac{mol}{L\text{-}bar} \times \frac{1.013\,bar}{atm} \times (320 \times 10^{-9}\,atm) = 8 \times 10^{-9} \frac{mol}{L}$$

The activity of the hydrogen ion generated from the nitrous oxide in the air is approximately

$$c_{HNO_3(aq)} \cong [H^+] = 8 \times 10^{-9} \frac{mol}{L}$$

Notice that this is substantially less than the hydrogen ion concentration in pure water:

$$[H^+]_{\text{pure water}} = 10^{-7} > 8 \times 10^{-9}\ \frac{\text{mol}}{\text{L}}$$

The pH of the solution is determined by substituting the activity of the greater of the hydrogen ion activities into Equation (2.45):

$$pH = -\log\{H^+\} \cong -\log\{10^{-7}\} = 7$$

In this example, even though nitric acid is a strong acid, not enough is added from average N_2O levels in the atmosphere to significantly change the pH of a pure water raindrop. Rainfall does acidify due to air pollutants in the atmosphere, but these occur at a more local or regional level, and the acidity largely comes from sulfur oxides emitted to the atmosphere. Atmospheric components that form a weaker acid may change the pH of rainfall, as will be illustrated in the next section.

2.5.4 Modeling Natural Waters That Contain a Weak Acid

An acid that only partially dissociates in an aqueous solution is called a weak acid. A weak acid has a small value for its equilibrium constant, k_a. The degree to which the acid dissociates depends on the amount of acid added to the solution and the overall pH of the solution if other dissolved ions are present. It should be noted that adding a weak acid to a solution may still produce a very acidic solution, with a very low pH, if enough of the acid is added to the solution.

A base that does not completely dissociate and has an associated small k_b value for its equilibrium constant is called a weak base and is analogous to a weak acid.

The pH of an aqueous solution controls the dissolved form of many weak acids and bases that are important to environmental systems, industrial systems, and human health. The pH can be thought of as a master variable that can be used in computer programs and graphical analyses to quickly identify the species of greatest concentration in aqueous systems. The same steps to solve acid–base equilibrium problems will be used, but more equations must be considered.

EXAMPLE 2.15 An Algebraic Solution for Determining the pH of a Weak Acid

Vinegar is formed biologically from the decomposition of sugars in water. Vinegar is a mixture of several organic compounds; the largest by concentration is acetic acid (CH_3COOH). The acid dissociates in solution to form hydrogen ions and acetate ions (CH_3COO^-). The acetate ion is commonly abbreviated as Ac^-. If acetic acid were added to a 1-liter flask of distilled water so that the total concentration of all acetic acid species was 10^{-4} mol/L, what would the pH of the aqueous solution be? Acetic acid has a $pK_a = 4.7$.

Step 1: Define the system boundaries.

The system boundaries are the walls of the 1-liter flask of water.

The total amount of the acetic acid species in the water = 10^{-4} mol/L.

Step 2: Identify all the chemical species of interest.

$$H_2O, HAc, H^+, Ac^-, \text{ and } OH^-$$

Step 3: Write the constraining chemical equations for the system, including equilibrium equations, the electroneutrality equation, and mass balance equations.

Relevant equilibrium equations are

$$k_w = \{H^+\}\{OH^-\} = 10^{-14}$$

$$k_a = \frac{\{H^+\}\{Ac^-\}}{\{HAc\}} = 10^{-4.7}$$

Relevant mass balance equation:

$$C_{HAc_{total}} = 10^{-4} = \{HAc\} + \{Ac^-\}$$

Electroneutrality equation:

$$\{H^+\} = \{AC^-\} + \{OH^-\}$$

Step 4: Make any simplifying assumptions that are possible.

If we add an acid to water, the hydrogen ion concentration increases, and the hydroxide ion concentration must decrease proportionally. Therefore, we will make the assumption that the hydroxide ion is small compared to the other possible variable, in order to simplify the algebraic analysis:

$$\{H^+\} \gg \{OH^-\}$$

The electroneutrality equation simplifies to:

$$\{H^+\} \cong \{AC^-\}$$

This example and others that occur in pure water are nearly ideal solutions; therefore, we may assume that the activity of the compound is equal to the concentration of the same compound, $\{A\} = [A]$.

Step 5: Algebraically solve the remaining independent equations and check the assumptions made to aid in solving the equations.

Substituting the above identity into the equilibrium expression yields:

$$k_a = \frac{\{H^+\}\{Ac^-\}}{\{HAc\}} = \frac{\{Ac^+\}\{Ac^-\}}{\{HAc\}} = 10^{-4.7}$$

Rearranging and solving for {HAc} yields

$$\{HAc\} = \frac{\{Ac^-\}^2}{10^{-4.7}}$$

Substituting the expression for {HAc} into the mass balance equation yields a second-order polynomial equation that can be solved using the standard quadratic equation solution:

$$c_{HAc_{total}} = 10^{-4} = \{HAc\} + \{Ac^-\} = \frac{\{Ac^-\}^2}{10^{-4.7}} + \{Ac^-\}$$

$$10^{4.7}\{Ac^-\}^2 + \{Ac^-\} - 10^{-4} = 0$$

$$\{Ac^-\} = \{H^+\} = \frac{-1 \pm \sqrt{1^2 - 4(10^{4.7})(-10^{-4})}}{2(10^{4.7})} = 3.6 \times 10^{-5} \frac{mol}{L}$$

The activity of the undissociated acetic acid in solution can be found from the mass balance equation:

$$\{HAc\} = 10^{-4} - \{Ac^-\} = 6.4 \times 10^{-5} \frac{mol}{L}$$

Check the assumption made and the concentration of the hydroxide ion:

$$\{OH^-\} = \frac{k_w}{\{H^+\}} = \frac{10^{-14}}{3.6 \times 10^{-5}} = 2.8 \times 10^{-10} \frac{mol}{L} << 3.6 \times 10^{-5} \frac{mol}{L}$$

The pH of the solution is determined by substituting the activity of the hydrogen ion activities into Equation (2.45):

$$pH = -\log\{H^+\} \cong -\log\{3.6 \times 10^{-5}\} = 4.4$$

ACTIVE LEARNING EXERCISE 2.4 Carbonate Species in Bottled Soft Drinks

Bottled soft drinks, such as Coca-Cola® or Pepsi®, contain a large amount of dissolved carbon dioxide. The carbon dioxide escaping the liquid phase and entering the gas phase produces the bubbles or "fizziness" of these drinks.

- Using pH paper or a pH probe, measure the pH of your favorite soft drink.
- Which of the following assumptions do you hypothesize is true?

  - $[H^+] = [HCO_3^-]$

    or

  - $[H^+] = [H_2CO_3]$

- Draw a pC versus pH diagram for the soft drink that shows the concentrations of $[H^+]$, $[OH^-]$, $[H_2CO_3]$, $[HCO_3^-]$, and $[CO_3^{2-}]$.

  Show how these equilibrium constants are related.

BOX 2.3 Acid Mine Drainage and Restoration

The Dents Run watershed is located in the northern part of Pennsylvania, as shown in Figure 2.13. This area is home to historic mining and forestry industries, as well as cold-water trout streams and one of the few elk herds east of the Mississippi River. Coal mining began in Dents Run in the late 1800s. These mining operations expanded to include underground "room-and-pillar" mines in the early 1900s. Larger mechanized "strip" mines began operating in the 1940s, and similar mines are still in operation in the region.

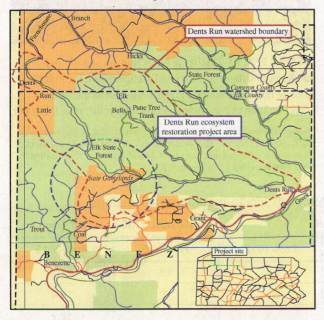

FIGURE 2.13 A map showing the Elk State Forest and the Dents Run watershed in northern Pennsylvania.

Source: Based on Cavazza, E.E., Malesky, T., and Beam, R. (2012). *The Dents Run AML/AMD Ecosystem Restoration Project*. Pennsylvania Department of Environmental Protection, Bureau of Abandoned Mine Reclamation.

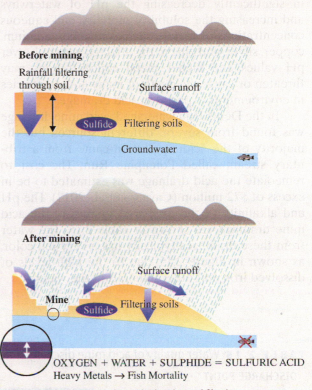

OXYGEN + WATER + SULPHIDE = SULFURIC ACID
Heavy Metals → Fish Mortality

Extraction decreases groundwater depth and natural filtration, and increases the groundwater contamination.

FIGURE 2.14 Mining may expose sulfide-bearing minerals in the earth to natural weathering processes. The exposed sulfides react with oxygen and water to form sulfuric acid, resulting in acid mine drainage (AMD).

Source: Based on Corcoran, E., Nellemann, C., Baker, E., Bos, R., Osborn, D., and Savelli, H. (eds.) (2010). *Sick Water? The central role of wastewater management in sustainable development. A Rapid Response Assessment.* United Nations Environment Programme, UN-HABITAT, GRID-Arendal. www.grida.no.

The waste products from mining operations consist of a variety of waste rocks and minerals that become exposed to natural weathering processes (see Figure 2.14). Acid rock drainage (ARD) or acid mine drainage (AMD) refers to the acidic drainage that comes from mine waste rock, tailings, and remaining mining structures. The formation of ARD/AMD from existing and past mining operations represents a significant environmental and health threat in many areas throughout the world. For example, by some estimates, more than 7,000 kilometers of waterways east of the Mississippi River and 8,000 to 16,000 kilometers of waterways west of the Mississippi have been negatively impacted by acid drainage in the United States (Kim et al., 1982; U.S. Forest Service, 1993). The acid drainage results

(Continued)

BOX 2.3   Acid Mine Drainage and Restoration (Continued)

in significantly decreasing the pH of waterways and increasing the solubility, mobility, and aqueous concentrations of metals, such as arsenic, cadmium, copper, silver, and zinc in the affected waters. Lower pH values and higher metal concentrations may threaten or cause the death of many aquatic species in environments exposed to acid drainage.

In the Dents Run watershed, acid mine drainage was found from several different sources, but the majority of the acidic pollution came from a tributary stream called Porcupine Run. The cost to remediate the acid drainage was estimated to be in excess of $7.2 million (Cavazza et al., 2012). The pH and alkalinity of the discharge sources of the acid mine drainage are shown in Table 2.15. The water from these drainages typically has a very red color, as shown in Figure 2.15, due to the high levels of dissolved iron and other minerals in the water.

FIGURE 2.15 Discharge water from an abandoned underground mine flowing into Porcupine Run. The reddish/orange hue of the water is due in large part to the high concentration of iron and other dissolved minerals in the water.

Source: Cavazza, E.E., Malesky, T., and Beam, R. (2012). *The Dents Run AML/AMD Ecosystem Restoration Project*. Pennsylvania Department of Environmental Protection, Bureau of Abandoned Mine Reclamation. P.O. Box 8476 Harrisburg, PA. Figure 6, p. 6.

TABLE 2.15 Water quality of acid mine discharge into Dents Run prior to and after remedial efforts

DISCHARGE POINT	pH		NEUTRALIZING CAPACITY (mg/L)	
	Pre-2010	Post-2012	Pre-2010	Post-2012
Dents Run upstream of Porcupine Run	6.3	6.6	9.6	12.4
Dents Run downstream of Porcupine Run	3.6	5.6	0	10.4
Porcupine Run at mouth	3.4	4.8	0	12.2
Tributary to Porcupine Run	3.9	5.85	0	16.3
Porcupine Run midstream	3.3	4.75	0	12.7
Porcupine Run headwaters	2.9	6.55	0	37.4
Dents Run at mouth	4.7	6.19	7.2	9.3

Source: Based on Cavazza, E.E., Malesky, T., and Beam, R. (2012). *The Dents Run AML/AMD Ecosystem Restoration Project*. Pennsylvania Department of Environmental Protection, Bureau of Abandoned Mine Reclamation.

The amount of acid from mine waste drainage was determined by analyzing Porcupine Run. Limestone ($CaCO_3$) forms the basic carbonate ion, CO_3^{2-}, when dissolved in water, which may react with the H^+ acid to neutralize the acidic drainage. However, because limestone is only partially soluble, an excess amount of limestone must be added to the acid waste in order to bring the water to a pH that is acceptable to native freshwater species. Approximately 500,000 metric tons of limestone from nearby mines were added to neutralize 1.8 million cubic meters of mine waste from 14 AMD sites. The pH of the resulting drainage was increased to levels that are acceptable for aquatic life in Dents Run as a result of the chemical neutralization.

The total cost of the stream restoration process, the results of which are shown in Figures 2.16 and 2.17, was in excess of $14 million. The restoration effort and cost were shared among the project partners that included federal, state, and local government agencies, private foundations, the coal industry, and local grassroots organizations (Cavazza et al., 2012). The restored watershed has resulted in improved habitat for trout, elk (see Figure 2.17), and other species in the region. It also has created jobs in environmental restoration and maintained tourism in the area, where each year 75,000 people visit the region to see the elk herd.

FIGURE 2.16 Unvegetated mine waste shown in the left-hand-side photo, compared to the same area after undergoing the neutralization and restoration process near Porcupine Run.

Source: Cavazza, E.E., Malesky, T., and Beam, R. (2012). *The Dents Run AML/AMD Ecosystem Restoration Project*. Pennsylvania Department of Environmental Protection, Bureau of Abandoned Mine Reclamation. P.O. Box 8476 Harrisburg, PA. Figure 11, p. 16.

FIGURE 2.17 Elk cooling off during the summer in a pond used to neutralize acid mine drainage in the Dents Run watershed.

Source: Cavazza, E.E., Malesky, T., and Beam, R. (2012). *The Dents Run AML/AMD Ecosystem Restoration Project*. Pennsylvania Department of Environmental Protection, Bureau of Abandoned Mine Reclamation. P.O. Box 8476 Harrisburg, PA. Figure 15, p. 20.

2.6 Environmental Fate and Partitioning of Chemicals

Chemicals may move between three phases: gas, liquid, and solid (Figure 2.18). In the natural environment, chemicals usually exist in a mixture. For example, nitrogen, oxygen, argon, water vapor, and other gases make up the mixture we know as air. When water in the air moves from the gaseous state to the liquid state, through the process called condensation, the water falls out of the air in the form of precipitation. Soils consist of a complex mixture of silicon, calcium, carbon, nitrogen, phosphorus, and many other elements that are distributed unevenly in soils and sediments. These elements may dissolve and become a part of an aqueous mixture in runoff and groundwater. Dissolved minerals may also precipitate and fall out of an aqueous solution if the chemistry, pH, or temperature of the solution changes.

The movement of chemicals from one phase or mixture to another is governed by a variety of environmental chemistry principles. Two general types of reactions may occur within a system: *Reduction–oxidation processes* occur when the oxidation states of participating atoms change. *Ionic reactions* are those reactions, like acid–base reactions, where there is a change in ion–ion interactions and relationships. Ionic reactions also frequently occur when a metal ion reacts with a base in precipitation and dissolution reactions, as in

$$Fe^{2+}_{(aq)} + 2OH^- \leftrightarrow Fe(OH)_{2(s)} \tag{2.50}$$

The dissolution of minerals is a large factor in determining the chemical composition of much natural water. Typically, precipitation and dissolution reactions occur much more slowly than acid–base reactions, so that both equilibrium considerations and the rate of the reaction are important in determining the extent and amount of chemical change that may occur within a system.

The extent of a compound's solubility is strongly influenced by ionic factors. Compounds that have a positive or negative charge tend to bond more strongly to water and may be more soluble (Figure 2.19). Compounds that bond strongly to water and tend to dissolve or stay in solution are called hydrophilic (water loving) compounds. Hydrophobic compounds are more strongly bonded to compounds with chemistries similar to those of their own and tend not to dissolve as much in water.

Salts are ionic compounds that dissociate completely in solution. An example is sodium chloride, otherwise known as table salt:

$$NaCl \rightarrow Na^+ + Cl^-$$

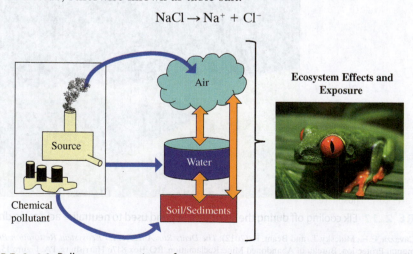

FIGURE 2.18 Pollutants may migrate from an anthropogenic or natural source through the air, water, or soil where they may come into contact with plants or animals.

Source: Ryan M. Bolton/Shutterstock.com.

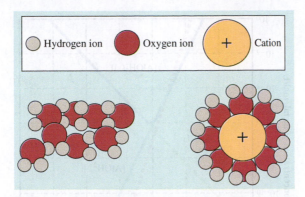

FIGURE 2.19 Compounds with an ionic charge as shown on the left tend to align and interact with water molecules (shown in red and gray), making ionic compounds more likely to dissolve into a solution.

Salts are very soluble compounds. The term *solubility* only has meaning in comparing one compound to another. In environmental chemistry, the solubility of carbonate compounds in water is often used as a reference point in comparing the solubility of different species and dissolved compounds. Some general guidelines for estimating chemical solubility are shown in Table 2.16.

Some metals may dissolve in water to form compounds that behave like acids or bases in solutions by abstracting a hydrogen or hydroxide ion from water, as in the case of the reaction of aluminum in solution shown in the following equations.

In an acidic solution, aluminum is present in the form Al^{+3}:

$$Al(OH)_{3(aq)} + 6H_2O^+_{(aq)} \leftrightarrow Al^{+3}_{(aq)} + 6H_2O \qquad (2.51)$$

In a basic solution, aluminum reacts with water to form a complex aluminum hydroxide compound:

$$Al(OH)_{3(aq)} + OH^-_{(aq)} \leftrightarrow Al(OH)^-_{4(aq)} \qquad (2.52)$$

The solubility of a dissolved compound is influenced by the concentration of the given compound, the pH of the solution, and other dissolved components in solution. Phase diagrams for pure solutions illustrate the relationship between pH and

TABLE 2.16 Solubility guidelines for common inorganic compounds

ION	CHARACTERISTIC SOLUBILITY
Nitrate, NO^{3-}	Soluble
Chloride, Cl^-	Soluble, except AgCl, PbCl, and HgCl
Sulfate, SO_4^{2-}	Soluble, except $BaSO_4$ and $PbSO_4$; $AgSO_4$, $CaSO_4$, and $HgSO_4$ are only slightly soluble
Carbonate, CO^{3-}; Phosphate, PO_4^{3-}; Silicate, SiO_4^{4-}	Insoluble, except those of Na, K, and NH_4^+
Hydroxide, OH^-	Insoluble, exceptions include LiOH, NaOH, KOH, NH_4OH, $Ba(OH)_2$, $Ca(OH)_2$, and $Sr(OH)_2$
Sulfide, S^{2-}	Insoluble, except for alkali metal sulfides, $(NH_4)_2S$, MgS, CaS, and BaS
Sodium, Na^+; Potassium, K^+; Ammonium, NH_4^+	Soluble, except for iron (Fe) and compounds that contain these ions with a heavy metal

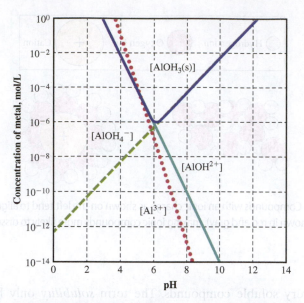

FIGURE 2.20 Soluble forms of aluminum in pure water between pH 0 and 14, and the range in which aluminum precipitation may occur.

Source: Based on Tchobanoglous, G., Burton, F.L., and Stensel, H.D. (2003). *Wastewater Engineering: Treatment and Reuse, 4th ed*. New York: McGraw-Hill.

solubility in Figure 2.20, where the lines approximate the total concentration of the stable residual aluminum concentration after precipitation of any insoluble components. Hydroxide ions or sulfur ions may be added to water to create less soluble forms of many metals in solution. Similar hydroxide and sulfur species solubility curves are shown for other common metals in Figures 2.21 through 2.23.

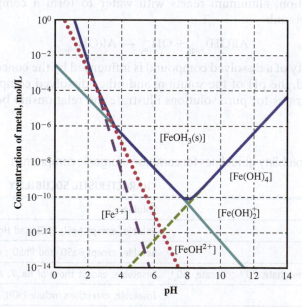

FIGURE 2.21 Soluble forms of ferric iron in pure water between pH 0 and 14, and the range in which iron precipitation may occur.

Source: Based on Tchobanoglous, G., Burton, F.L., and Stensel, H.D. (2003). *Wastewater Engineering: Treatment and Reuse, 4th ed*. New York: McGraw-Hill.

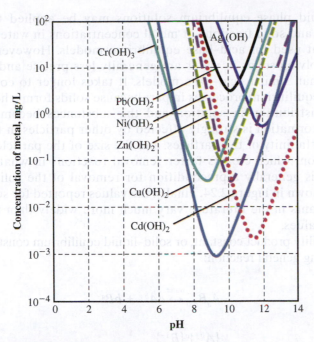

FIGURE 2.22 Hydroxide metal solubility curves.

Source: Based on Tchobanoglous, G., Burton, F.L., and Stensel, H.D. (2003). *Wastewater Engineering: Treatment and Reuse, 4th ed.* New York: McGraw-Hill.

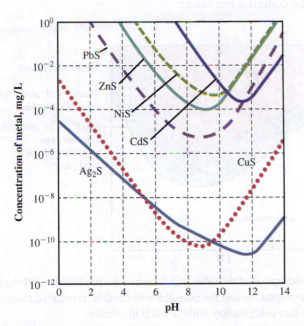

FIGURE 2.23 Sulfide metal solubility curves.

Source: Based on Tchobanoglous, G., Burton, F.L., and Stensel, H.D. (2003). *Wastewater Engineering: Treatment and Reuse, 4th ed.* New York: McGraw-Hill.

Solid–liquid phase equilibrium solutions may be applied to calculating the dissolved and solid fractions of metal concentrations in water in a method similar to that used for acid–base equilibrium models. However, equilibrium reactions involving solids in water are generally less precise and have greater uncertainty than similar acid–base models. It takes longer to complete solid–liquid phase equilibrium processes in part because solids form a heterogeneous, or uneven, distribution in solution and the substances become less evenly mixed. Solid formation is strongly affected by other particles in the water, the degree of crystallinity of the particles, and the size of the particles in solution. Supersaturation, which creates faster removal reactions in what is called the labile phase, is generally a precondition for removal of the solid phase from solution, as shown in Figure 2.24. Finally, the values reported for solubility equilibrium constants in the literature vary much more widely than the acid–base equilibrium values.

The solubility product constant, or solid–liquid equilibrium constant, is defined by the following general reaction:

$$A_a B_{b(s)} \overset{water}{\longleftrightarrow} aA^{b+} + bB^{a-} \tag{2.53}$$

$$k_{sp} = \frac{\{A^{b+}\}^a \{B^{a-}\}^b}{\{AB_{(s)}\}} = \{A^{b+}\}^a \{B^{a-}\}^b \tag{2.54}$$

Recall that the solid phase activity $\{AB_{(s)}\}$ is equal to 1 as defined by our standard state conditions. The equilibrium expressions and mass balance equations may be used to predict the solubility of compounds. Generally, the solubility of a given compound is reduced if the same ions are already present in solution; this effect is referred to as the common ion effect.

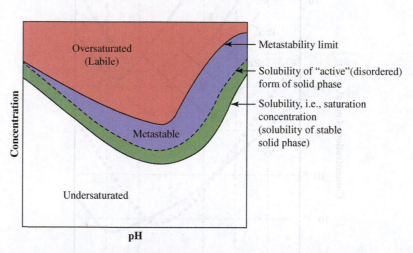

FIGURE 2.24 Illustration of the difference between the predicted equilibrium conditions to remove metals by precipitation and the oversaturated conditions required to achieve conditions that in practice produce precipitation of metal solids in solution.

Source: Based on Stumm, W., and Morgan, J.J. (1996). *Aquatic Chemistry: Chemical Equilibria and Rates in Natural Waters, 3rd ed.* New York: John Wiley.

EXAMPLE 2.16   Estimating the Solubility of Calcium Hydroxide

Calcium hydroxide, $Ca(OH)_2$, is widely used in industry as well as in some water treatment applications to modify the pH of solutions. Estimate the solubility of $Ca(OH)_2$ in distilled water at 25°C. Assume the solution acts as an ideal fluid. The solubility product constant is $k_{sp} = 10^{-5.19}$.

$$Ca(OH)_2 \leftrightarrow Ca^{2+} + 2(OH^-)$$

$$k_{sp} = [Ca^{2+}][OH^-]^2 = 10^{-5.19}$$

Inspecting the balanced chemical equation shows that two moles of hydroxide ion are formed for each mole of calcium ion formed. Using this relationship, let

$$[Ca^{2+}] = S \text{ and then } [OH^-] = 2S$$

Substituting the above into the equation for the solubility product constant yields

$$k_{sp} = [Ca^{2+}][OH^-]^2 = S(2S)^2 = 4S^3 = 10^{-5.19}$$

Solving for S yields

$$S = [Ca^{2+}] = 0.012 \text{ mol/L}$$

$$[OH^-] = 2S = 0.024 \text{ mol/L}$$

EXAMPLE 2.17   The Common Ion Effect for Fluoride in Water

Fluoride is commonly added to drinking water to help prevent tooth decay. Calculate the solubility of calcium fluoride, CaF_2, at 25°C if the water already contains 16 mg/L of the fluoride ion. Assume that the solution acts as an ideal fluid. The solubility product constant is $k_{sp} = 10^{-10.50}$.

$$CaF_2 \leftrightarrow Ca^{2+} + 2F^-$$

$$k_{sp} = [Ca^{2+}][F^-]^2 = 10^{-10.50}$$

The molecular mass of fluoride is 19 g/mol. The pre-existing concentration of fluoride in the water is

$$[F^-_{(initial)}] = 16 \frac{mg}{L} \times \frac{mmol}{19 \text{ mg}} \times \frac{mol}{1,000 \text{ mmol}} = 10^{-3.1}$$

Inspecting the balanced chemical equation shows that two moles of fluoride ion are formed for each mole of calcium ion formed. Using this relationship, let

$$[Ca^{2+}] = S \text{ and}$$

$$[F^-] = 2S + 10^{-3.1}$$

Substituting the above into the equation for the solubility product constant yields

$$k_{sp} = [Ca^{2+}][F^-]^2 = S(2S + 10^{-3.1})^2 = 10^{-10.50}$$

To simplify the algebraic solution, assume $2S \ll 10^{-3.1}$; then

$$k_{sp} = S(10^{-3.1})^2 = 10^{-10.50}$$

Solving for S yields

$$S = [Ca^{2+}] = 4.45 \times 10^{-5}$$

$2S = 8.90 \times 10^{-5} < 10^{-3.91}$ so the above assumption is close to this estimate, but $2S$ is not completely negligible. A more precise solution can be determined by iterating based on the previous solution. It will be assumed that

$$S = [Ca^{2+}] = 3.75 \times 10^{-5} = 10^{-4.43}$$

Substituting into the equilibrium expression:

$$S(2S + 10^{-3.1})^2 = 3.75 \times 10^{-5}(2(3.75 \times 10^{-5}) + 10^{-3.1})^2 = 10^{-10.50}$$

Then after adding CaF_2:

$$[F^-] = 2(3.75 \times 10^{-5}) + 10^{-3.1} = 0.00092 = 10^{-3.04} \text{ mol/L} = 17.4 \text{ mg/L}$$

Metals may have several forms in solution by reacting with other inorganic compounds, other organic compounds, and other particles in solution, as shown in Figure 2.25. In complex environmental systems, these complicating factors should be considered when evaluating the impacts of metals in the environment.

Other equilibrium-based coefficients may also be used to estimate the concentration, form, and fate of chemical compounds in the environment. In the most general form, the equilibrium concentrations of any compound, A, between two phases, x and y, can be compared by

$$A_{(x - \text{phase})} \leftrightarrow A_{(y - \text{phase})} \tag{2.55}$$

Then

$$k_{A_{x-y}} = \frac{[A]_{x - \text{phase}}}{[A]_{y - \text{phase}}} \tag{2.56}$$

The fraction of the total amount of A at equilibrium in the generic x-phase is

$$f_{Ax} = \frac{\text{mass of } A \text{ in phase } x}{\text{total mass of } A} = \frac{(C_{Ax})(V_x)}{(C_{Ax})(V_x) + (C_{Ay})(V_y)} = \frac{1}{1 + \dfrac{C_{Ay} V_y}{C_{Ax} V_x}} \tag{2.57}$$

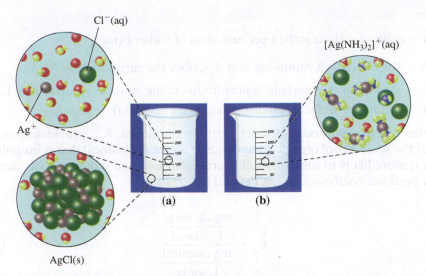

FIGURE 2.25 Silver chloride, AgCl(s), in distilled water is largely insoluble as shown in part (a). However, when the ammonium ion is added to water, the silver and ammonium ions form a complex molecule $[Ag(NH_3)_2]^+$ that dissolves into the aqueous solution in part (b).

Source: Based on Petrucci, R., Madura, J., Herring, F., and Bissonnette, C. (2011). *General Chemistry: Principles and Modern Applications, 10th ed.* Toronto: Pearson Prentice Hall.

The term V_y/V_x is simply the ratio of the two volumes, which can be defined as R_{xy}. Substituting the volume ratio and Equation (2.30) into Equation (2.31) yields

$$f_{Ax} = \frac{1}{1 + \dfrac{1}{(k_{A_{xy}})(R_{xy})}} = \frac{(k_{A_{xy}})(R_{xy})}{(k_{A_{xy}})(R_{xy}) + 1} \qquad (2.58)$$

Because our environment is not a homogeneous mixture of air and water, but a complex mixture that involves heterogeneous solid mixtures as well, various equilibrium approaches have been used to describe the gas–solid and liquid–solid phase equilibrium. Adsorption describes the process of a chemical adhering to a solid *surface*, whereas absorption describes the process of a chemical becoming incorporated into the bulk phase of a given volume of a liquid or solid. Because it is often difficult to determine precisely if a material is adsorbed on a surface or absorbed into a given material, the two processes are often described collectively as *sorption*. The **sorbate** is the substance that is transferred from one phase to another. The **sorbent** is the material into or onto which the sorbate is transferred. Sorption occurs due to the electrostatic or van der Waals force, chemical bonding or surface coordination reactions, or partitioning of the compound into the bulk phase of the sorbent.

Activated carbon is used in Brita-style drinking water filters, water treatment plants, and air pollution control facilities to remove chemical pollutants from either the gas or liquid phase. The Freundlich isotherm is used to estimate the portioning of pollutants from a more mobile phase in gas or water to the solid-activated carbon material. The Freundlich equation has the form

$$q = KC^{1/n} \qquad (2.59)$$

where

q = mass of sorbate sorbed per unit mass of sorbent (mg/g)

K = the Freundlich parameter that describes the partitioning $\left(\frac{mg/g}{(mg/L)^{\frac{1}{n}}}\right)$

C = the equilibrium sorbate concentration in the aqueous solution (mg/L)

n = Freundlich isotherm intensity parameter (unitless)

The dimensionless octanol/water partition coefficient, K_{ow}, provides an indication of the solubility of organic compounds in water. A chemical that is insoluble in water is more likely to adhere to solid particles in soils and sediments. The octanol/water partition coefficient, K_{ow}, is defined by

$$K_{ow} = \frac{\dfrac{mg\ chemical}{L\ octanol}}{\dfrac{mg\ chemical}{L\ water}} \qquad (2.60)$$

Octanol ($C_7H_{15}CH_2OH$) is an organic chemical with moderate polar and nonpolar properties, so that a wide variety of organic compounds are soluble in octanol solutions. The octanol/water partition coefficient has been used in the pharmaceutical industry because the partitioning between octanol and water simulates the partitioning of drugs between water and fat; as such, it is also a useful factor in estimating the bioaccumulation of organic compounds in animal fat. Smaller, more polar compounds tend to be more soluble in water, yielding a lower value for K_{ow}. Larger and fewer polar molecules tend to have a greater solubility in octanol and hence also tend to attach more readily to soil particles.

The organic carbon normalized partition coefficient, K_{oc}, is a useful parameter for predicting the partitioning of chemicals between soil and water and can be estimated from the octanol/water partition coefficient. The organic carbon normalized partition coefficient is defined as

$$K_{oc} = \frac{\dfrac{mg\ chemical}{kg\ organic\ carbon}}{\dfrac{mg\ chemical}{L\ water}} \qquad (2.61)$$

Empirical correlations have been developed for estimating the K_{oc} value based on a chemical's K_{ow} value:

Karickhoff et al. correlation: $\log(K_{oc}) = 1.00 \log(K_{ow}) - 0.21$ \qquad (2.62)

Schwarzenbach and Westall correlation: $\log(K_{oc}) = 0.72 \log(K_{ow}) + 0.49$ \quad (2.63)

If the decimal fraction of the organic carbon present in the soil f_{oc} is known, then the sediment/water partition coefficient can be calculated from

$$K_p = f_{oc}K_{oc} = \frac{\dfrac{mg\ chemical}{kg\ solids}}{\dfrac{mg\ chemical}{L\ water}} \qquad (2.64)$$

EXAMPLE 2.18   Chemical Partitioning in the Environment

Vinyl chloride is a colorless, odorless gas used in the production of polyvinyl chloride (PVC) plastics and in numerous industries. Vinyl chloride is also a known human carcinogen, and the EPA's maximum contaminant level (MCL) in drinking water is 0.002 mg/L, or 2 ppb. The vinyl chloride concentration in the soil outside of a plastic factory is 0.007 mg/kg, and the soil has a 15% organic content. Vinyl chloride (chloroethene) has a reported octanol–water coefficient equal to 4. The organic carbon normalized partition coefficient, K_{oc}, can be estimated using the procedures based on Equations (2.62) and (2.63) and the sediment/water partition coefficient. The sediment/water partition coefficient can be used to predict the possible equilibrium concentration of vinyl chloride in the groundwater.

Karickhoof et al. correlation: $\log(K_{oc}) = 1.00 \log(K_{ow}) - 0.21$

$$\log(K_{oc}) = 1.00 \log(4) - 0.21 = 0.4$$

$$K_{oc} = 2.5$$

Schwarzenbach and Westall correlation: $\log(K_{oc}) = 0.72 \log(K_{ow}) + 0.49$

$$\log(K_{oc}) = 0.72 \log(4) + 0.49 = 0.9$$

$$K_{oc} = 8.4$$

The sediment/water partition coefficient can be determined from the following empirical equations.

Using the Karickhoof approximation yields

$$K_p = f_{oc} K_{oc} = 0.15 \times 2.5 = 0.37$$

Using the Schwarzenbach and Westall approximation yields

$$K_p = f_{oc} K_{oc} = 0.15 \times 8.4 = 1.26$$

The partition coefficient in Equation (2.64) can be rearranged to yield

$$C_{water}\left[\frac{mg}{L}\right] = \frac{C_{soil}\left[\frac{mg_{pollutant}}{kg_{soil}}\right]}{K_p} = \frac{0.007\left[\frac{mg_{pollutant}}{kg_{soil}}\right]}{0.37} = 0.019\,\frac{mg}{L}$$

Using the values from the Schwarzenbach and Westall approximation yields $C_{water} = 0.006$ mg/L.

The expected range based in the groundwater would be 19 ppb $> C_{water} > 6$ ppb; in both cases this concentration is greater than the MCL of 2 ppb.

BOX 2.4 Metal Mobility in the Environment

For decades, lead has been used in making bullets and other ammunition because it has been readily available and is easy to form. However, exposure to lead in the environment has been linked to learning disabilities and other environmental and health concerns. The EPA estimates that roughly 7.26×10^7 kg of lead shot and bullets are released into the environment each year in the United States just from nonmilitary outdoor ranges. Concerns about the toxicity of lead in the environment have led to the development of a "green bullet" that replaces lead with tungsten munitions. Tungsten is a heavy metal with several unique physical and mechanical properties that make it appealing as a replacement for lead.

In shooting ranges, where lead munitions have traditionally been fired, fragments of lead fall on the soil. Once lead enters the soil, various factors will determine the extent of the actual hazard it might pose. The rate at which lead may move through the soil and groundwater depends on various factors, including soil type, soil pH, and rainfall volume. At a near-neutral pH, lead is relatively insoluble. However, if soils become more acidic (decreasing pH), lead solubility tends to increase, making it easier for the pollutant to spread over a larger area. The more easily the lead travels through the soil, the greater the impact it will have. However, the solubility and transport of lead in the soil are generally limited by the reaction of lead with natural carbonate in the soil that forms lead carbonate (Cerussite), which is minimally soluble. In fact, the standard treatment recommended to limit lead transport in natural environments is to add pulverized limestone ($CaCO_3$) to raise the soil pH and increase the total soil carbonate concentration. The increased soil pH reduces the corrosion rate of Pb by lowering the concentration of available H^+ ions and reducing the solubility of the Cerussite that is formed.

Now that "green" bullets are being used in the same outdoor firing ranges, tungsten can be released into soils that already contain "immobilized lead." However, fate and transport research on tungsten and tungsten alloys has shown that tungsten corrodes through a process that leads to dissolved oxygen depletion and pH reduction in water and certain soils. The corrosion of tungsten can be represented by two reactions involving the oxidation of metallic tungsten and the reduction of dissolved molecular oxygen:

$$\frac{3}{2}O_2 + 5H_2O + 6e^- \rightarrow 8OH^- + 2H^+ \qquad (2.65)$$

$$W + 8OH^- \rightarrow WO_4^{2-} + 4H_2O + 6e^- \qquad (2.66)$$

The rate of dissolution, and by extension, the rate of acidification of the soil system, depend on several factors, including the chemical properties of the soil, level of oxygen in the soil, and the ionic capacity of the soil.

Results of experiments performed by some researchers suggest that when tungsten is released in some soils that contain immobilized legacy lead, the corrosion of tungsten can lead to a reduction in soil pH (acidification), leading to increased solubility and transport of lead through the soil and subsequent contamination and impact. This is an example of how trying to solve one environmental problem can lead to another one if proper analysis and modeling are not employed. In 2002, the National Center for Environmental Health of the Centers for Disease Control and Prevention recommended toxicological and carcinogenesis studies. This recommendation was largely driven by the discovery of elevated tungsten concentrations in biological and drinking water samples in three different communities with leukemia clusters (CDC, 2003).

2.7 Summary

Chemicals in the environment move between the gas, liquid, and solid phases based on their chemical properties and other properties of the system. Chemicals in the gas phase move with local air movements and patterns and may be carried long distances and broadly dispersed. Chemicals that dissolve in water may move from the gas phase into the liquid phase and vice versa. Chemicals that have a preference for the liquid phase may move with local water movements through precipitation, runoff, infiltration, and groundwater movements. Chemicals that have an affinity for the solid phase or solid surfaces tend not to be very mobile, but may often persist in the environment with little change in concentration for extremely long time spans. The ability to assess and analyze risk associated with remediation or mitigation of potentially harmful chemicals in the environment must be considered, along with the transport and fate of those chemicals in the environment.

References

Benjamin, M. M. (2001). *Water Chemistry*. New York: McGraw-Hill.

Bouwman, A. F., Van Vuuren, D. P., Derwent, R. G., and Posch, M. (2002). "A global analysis of acidification and eutrophication of terrestrial ecosystems." *Water, Air and Soil Pollution* 141:349–382.

Briggs, J. C., and Ficke, J. F. (1977). *Quality of Rivers of the United States, 1975 Water Year—Based on the National Stream Quality Accounting Network (NASQAN)*: U.S. Geological Survey Open-File Report 78–200.

Butler, J. N. (1982). *Carbon Dioxide Equilibria and Their Applications*. Reading, MA: Addison-Wesley.

Cavazza, E. E., Malesky, T., and Beam, R. (2012). The Dents Run AML/AMD Ecosystem Restoration Project. Pennsylvania Department of Environmental Protection, Bureau of Abandoned Mine Reclamation. P.O. Box 8476, Harrisburg, PA.

Centers for Disease Control and Prevention. National Center for Environmental Health. Cancer Clusters. Churchill County (Fallon), Nevada Exposure Assessment. [online] 2003.

Clark, M. L., Sadler, W. J., and Ney, S. E. (2004). "Water-Quality Characteristics of the Snake River and Five Tributaries in the Upper Snake River Basin, Grand Teton National Park, Wyoming, 1998–2002." Scientific Investigations Report 2004-5017. U.S. Geological Survey, Reston, VA.

Corcoran, et al. (2010). *Sick Water? The Central Role of Wastewater Management in Sustainable Development. A Rapid Response Assessment*. UNEP, UN-HABITAT, GRID-Arendal.

Dufor, C. N., and Becker, E. (1964). "Public Water Supplies of the 100 Largest Cities in the United States, 1962." U.S. Geological Survey, Water Supply Paper 1812.

Flippo, H. N., Jr. (1974). Springs of Pennsylvania. U.S. Department of the Interior, Geological Survey, Harrisburg, PA.

Hart, J. (1988). *Consider a Spherical Cow: A Course in Environmental Problem Solving*. Sausalito, CA: University Science Books.

IPCC. (2007). *Climate Change* 2007: "The Physical Science Basis." Contribution of Working Group I to the Fourth Assessment Report of the Intergovernmental Panel on Climate Change. [Solomon, S., D. Qin, M. Manning, Z. Chen, M. Marquis,

K. B. Averyt, M. Tignor, and H. L. Miller (Eds.)] Cambridge, UK: Cambridge University Press.

Kim, A. G., Heisey, B., Kleinmann, R., and Duel, M. (1982). *Acid Mine Drainage: Control and Abatement Research.* U.S. DOI, Bureau of Mines IC 8905, p. 22.

Langelier, W. F. (1936). "The analytical control of anti-corrosion water treatment." *Journal of American Water Works Association* 28(1936):1500–1521.

Masters, G. M., and Ela, W. P. (2008). *Introduction to Environmental Engineering and Science,* 3rd ed. Upper Saddle River, NJ: Prentice-Hall.

Meybeck, M., Chapman, D., and Helmer, R. (Eds.) (1989). *Global Freshwater Quality: A First Assessment.* London: Blackwell.

Nahle, Nasif. (2007). "Cycles of global climate change." Biology Cabinet Journal Online. Article no 295.

Snoeyink, V. L., and Jenkins, D. (1980). *Water Chemistry.* New York: Wiley.

Stumm, W., and Morgan, J. J. (1996). *Aquatic Chemistry: Chemical Equilibria and Rates in Natural Waters*, 3rd ed. New York: John Wiley.

Tchobanoglous, G., Burton, F. L., and Stensel, H. D. (2003). *Wastewater Engineering: Treatment and Reuse*, 4th ed. New York: McGraw-Hill.

UNEP GEMS. (2006). *Water Quality for Ecosystem and Human Health.* United Nations Environment Programme, Global Environment Monitoring System/Water Programme.

USDA Forest Service. (1993). *Acid Mine Drainage from Mines on the National Forests, A Management Challenge.* Program Aid 1505, p.12.

U.S. EPA. (2021). "Criteria air pollutants | US EPA." www.epa.gov/criteria-air-pollutants /naaqs-table.

Virginia Ambient Air Monitoring 2010 Data Report. (2011). Office of Air Quality Monitoring, Virginia Department of Environmental Quality.

Visser, S. A., and Villeneuve, J. P. (1975). "Similarities and differences in the chemical composition of waters from West, Central, and East Africa." *Verh int. Verein. Theor. Angew. Limnology* 19:1416–1425.

Key Concepts

Henry's law constant	Chemical activity
Partial pressure	Activity coefficient
Dissolved inorganic carbon	Acid
Law of electroneutrality	Base
Ionic strength	Ampholyte
Suspended solids	Solubility
Volatile suspended solids	Precipitate
Dissolved solids	Reduction–oxidation process
Water hardness	Ionic reactions
Carbonate hardness	Dissolution of salts
Noncarbonate hardness	Sorbate
Chemical reactivity	Sorbent

Problems

2-1 Define the following terms or concepts:
    a. Law of electroneutrality
    b. Normality
    c. Equivalents
    d. Total dissolved solids (TDS)
    e. Total suspended solids (TSS)
    f. Volatile suspended solids (VSS)
    g. Fixed solids
    h. Total hardness
    i. Carbonate hardness
    j. Noncarbonate hardness
    k. Chemical reactivity
    l. Chemical activity
    m. Standard state activity
    n. Activity coefficient
    o. Solubility
    p. Precipitate
    q. Reduction–oxidation process
    r. Ionic reaction
    s. Dissolution of salts
    t. Sorbent
    u. Sorbate

2-2 Fill in the blank: Mass and energy balances are helpful to track chemicals as they move from one _____ to another; however, we must study the applications and principles of chemistry more closely to appreciate how these chemicals are transformed in the environment and also how they transform the Earth's environment.

2-3 How much have carbon dioxide levels in the atmosphere increased since the start of the Industrial Revolution?

2-4 What five elements are found in life on the Earth?

2-5 Write the balanced equation for the combustion of methane (CH_4) in the presence of oxygen.

2-6 How many atoms are in a mole?

2-7 What is the numerical value of R in [J/mol-K] in the ideal gas law?

2-8 Define the partial pressure of a gas.

2-9 What is the relationship between the concentration of a gas in air in mg/m^3 and ppm_v at standard temperature and pressure?

2-10 What is the relationship between the concentration of a gas in air in mg/m^3 and ppm_v at nonstandard temperature and pressure?

2-11 Relate parts per billion (ppb_v) to ppm_v.

2-12 Use balanced chemical equations to describe the forms of dissolved carbon dioxide in aqueous solution. There should be a diprotic acid, a monovalent anion, and a divalent conjugate base.

2-13 The ionic strength, I, of a solution is the estimate of the overall concentration of dissolved ions in solution and is defined by what?

2-14 Fill in the blank: The pH of natural waters is important to biodiversity, since most species are only tolerant of natural waters in the pH range from _____.

2-15 Phosphoric acid has the form H_3PO_4. List all the acids that may donate a hydrogen ion and the conjugate bases that may accept the proton for this acid compound.

2-16 What is the primary characteristic of an amphoteric compound?

2-17 Define the equilibrium constant for pure water.

2-18 What are the primary characteristics of a strong acid?

2-19 What are the primary characteristics of a strong base?

2-20 Write the equation that defines pH.

2-21 Write the equation that defines pOH.

2-22 Write the equation that relates k_w, pH, and pOH.

2-23 What are the steps involved in modeling the effects of acids or bases in the natural environment?

2-24 Define the primary characteristic of a weak acid that partially dissociates in an aqueous solution. *Hint*: It may help to use the acid equilibrium constant in your description.

2-25 Define the primary characteristic of a weak base that partially dissociates in an aqueous solution. *Hint*: It may help to use the base equilibrium constant in your description.

2-26 The solubility product constant, or solid–liquid equilibrium constant, is defined by what general reaction?

2-27 The Freundlich isotherm is used to estimate the portioning of pollutants from a more mobile phase in gas or water to the solid activated carbon material. What is the form of the Freundlich equation?

2-28 Describe in a few sentences how to set up an experiment in a laboratory to measure the octanol water partition coefficient K_{ow}.

2-29 Describe in a few sentences how to set up an experiment in a laboratory to measure the organic carbon normalized partition coefficient K_{oc}.

2-30 A 12-ounce can of soda contains about 40 grams of sugar. What is the concentration of sugar in a can of soda in mg/L?

2-31 A 2-ounce serving of espresso contains 100 mg of caffeine. Professor Coffy often has a 16-ounce iced latte with three shots of espresso before his 8 a.m. class.
a. What is the concentration of caffeine in mg/L in a single shot of espresso?
b. What is the concentration of caffeine in ppm in a single shot of espresso?
c. What is the concentration of caffeine in his coffee drink
   i) in mg/L?
   ii) in ppm?

2-32 There are about 5 mg of caffeine in each shot of decaf espresso. A barista is being paid under the table by Mrs. Coffy to change the espresso shots to decaffeinated espresso in Professor Coffy's 16-ounce latte drink cup from Problem 2-31.
a. How much caffeine in mg does Professor Coffy consume if he drinks decaffeinated lattes before class?
b. How many 16-ounce decaf lattes must he drink before he consumes the amount of caffeine equivalent to his old three shots of caffeinated espresso (from Problem 2-31)?

2-33 A 12 fluid ounce serving of soda pop contains mostly water and about 85 mg of sodium. What is the concentration of sodium in the aqueous liquid in units of parts per million on a mass basis (ppm_m)? There are 33.814 fluid ounces per liter.

2-34 A 12 fluid ounce serving of milk contains mostly water and other vitamins and minerals. What is the concentration of each compound in the aqueous liquid in units of parts per million on a mass basis (ppm_m)?
  a. 184 mg of sodium
  b. 413 mg of calcium
  c. 43 mg of magnesium
  d. 351 mg of chloride

2-35 The average concentration of dissolved oxygen (DO) in the Shenandoah River was reported as 9.7 ppm_m in 2006. What is the concentration in mg/L in the river water?

2-36 Table 2.17 includes typical constituents in water in mg/L. Complete the table by converting the concentrations to units given in the table, and find the mass in 1,000 liters of water—the amount of water you would typically ingest over the course of a year.

TABLE 2.17 Typical constituents in water

CONSTITUENT	mg/L	ppm	MOLECULAR WEIGHT	mmol/L	kg/YEAR
Bicarbonate (HCO_3)	75				
Carbonate (CO_3)	5				
Chloride (Cl)	35				
Sulfate (SO_3)	27				
Calcium (Ca)	11				
Magnesium (Mg)	7				
Potassium (K)	11				
Sodium (Na)	55				
Aluminum (Al)	0.2				
Fluoride (F)	0.3				

Source: Based on Tchobanoglous, G., Asano, T., Burton, F., Leverenz, H., Tsuchihashi, R. (2007). *Water Reuse: Issues, Technologies, and Applications*. McGraw-Hill.

2-37 The total volume of the oceans on Earth is 1.35×10^{18} m³. What are the masses of the following elements in the ocean in units of kg?
  a. Oxygen (O), for which the concentration in seawater is 857,000 ppm
  b. Hydrogen (H), for which the concentration in seawater is 108,000 ppm
  c. Sodium (Na), for which the concentration in seawater is 10,500 ppm

2-38 Air has a molecular weight of 28.967 g/mol. What is the density of air in units of g/m³ at 1 atm and 200°C?

2-39 The reported value of carbon dioxide in the atmosphere in 2010 was approximately 385 ppm_v. What was the concentration of carbon dioxide in the atmosphere in mg/m³?

2-40 The mass of the oceans is 1.4×10^{21} kg. (Assume the density of ocean water on average is 1.03 g/mL.) The concentration of potassium (K) in the oceans is 380 ppm_m. What is the total mass of the potassium stored in the oceans?

2-41 The mass of the Earth's troposphere, the lower part of the atmosphere, is approximately 4.4×10^{18} kg. What would the total volume of the Earth's troposphere be in cubic meters (m³) if it were all under the constraints of standard temperature and pressure? Use the ideal gas law to calculate the average conditions for a standard state ($P = 1$ atm, $T = 298$ K, $R = 0.0821$ atm-L/mol-K). *Note*: The average molecular weight of air is 28.96 g/mol.

a. The average mass of water in the troposphere is 1.3×10^{13} kg. Using your answer, calculate the concentration of water vapor in the troposphere in mg/m^3.

b. Using your answer from part (a), determine the concentration of water vapor in the troposphere in ppm_v.

2-42 Table 2.18 shows the National Ambient Air Quality Standards (NAAQS).

a. Express the standards in $\mu g/m^3$ at 1 atm pressure and 25°C.

b. At the elevation of Denver, the pressure is about 0.82 atm. Express the standards in $\mu g/m^3$ at that pressure and a temperature of 5°C.

TABLE 2.18 National Ambient Air Quality Standards

POLLUTANT	AVERAGING TIME	LEVEL	STANDARD CONDITIONS ($\mu g/m^3$)	DENVER IN WINTER ($\mu g/m^3$)
Carbon monoxide (CO)	8-hour	9 ppm		
	1-hour	35 ppm		
Nitrogen dioxide (NO_2)	1-hour	100 ppb		
	Annual	53 ppb		
Ozone (O_3)	8-hour	0.0070 ppm		
Sulfur dioxide (SO_2)	1-hour	75 ppb		
	3-hour	0.5 ppm		

Source: Based on U.S. EPA (2021).

2-43 Use Henry's law to calculate the concentration of dissolved inorganic carbonates at standard temperature and pressure (STP) in a raindrop if the atmospheric concentration of carbon dioxide in the Ordovician Epoch was 2,240 ppm_v CO_2, as illustrated in Figure 2.26.

2-44 The concentration of CO_2 was approximately 2,300 ppm_v in a 19°C atmosphere in the Cambrian period. What would the concentration of dissolved carbon dioxide in water at equilibrium with the atmosphere be based on Henry's law? Assume the atmospheric pressure was 1 atm and express your answer in units of mg/L.

2-45 The concentration of CO_2 was approximately 180 ppm_v in an 11°C atmosphere in the Quaternary glaciation period. What would the concentration of dissolved carbon dioxide in water at equilibrium with the atmosphere be based on Henry's law? Assume the atmospheric pressure was 1 atm and express your answer in units of mg/L.

2-46 The concentration of CO_2 was approximately 210 ppm_v in a 19°C atmosphere in the Triassic period. What would the concentration of dissolved carbon dioxide in water at equilibrium with the atmosphere be based on Henry's law? Assume the atmospheric pressure was 1 atm and express your answer in units of mg/L.

2-47 The concentration of CO_2 was approximately 340 ppm_v in a 19°C atmosphere in the Cretaceous period. What would the concentration of dissolved carbon dioxide in water at equilibrium with the atmosphere be

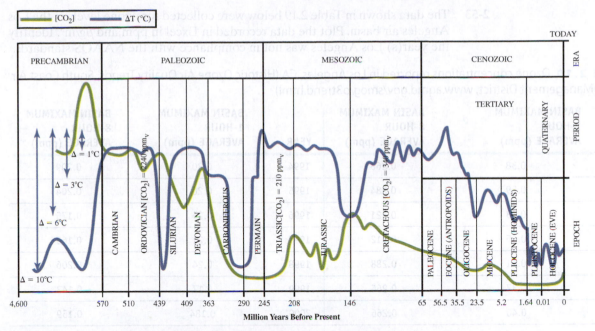

FIGURE 2.26 Fluctuation of carbon dioxide and temperature over geological timescales.

Source: Based on Nahle, N. (2007). Cycles of Global Climate Change. *Biology Cabinet Journal Online*. Article no 295.

based on Henry's law? Assume the atmospheric pressure was 1 atm and express your answer in units of mg/L.

2-48 The concentration of CO_2 was approaching 500 ppm_v in a 16°C atmosphere in the Anthropocene period. What would the concentration of dissolved carbon dioxide in water at equilibrium with the atmosphere be based on Henry's law? Assume the atmospheric pressure was 1 atm and express your answer in units of mg/L.

2-49 The concentration of benzene in ambient air in several U.S. locations was measured and found to be 0.48 mg/m³. The molecular formula for benzene is C_6H_6. Assume the air is nearly at standard temperature and pressure. What is the concentration expressed in units of parts per billion by volume, ppb_v?

2-50 The concentration of carbon tetrachloride in ambient air in several U.S. locations was measured and found to be 0.88 mg/m³. The molecular formula for carbon tetrachloride is CCl_4. Assume the air is nearly at standard temperature and pressure. What is the concentration expressed in units of parts per billion by volume, ppb_v?

2-51 The concentration of formaldehyde in ambient air in several U.S. locations was measured and found to be 0.25 mg/m³. The molecular formula for formaldehyde is CH_2O. Assume the air is nearly at standard temperature and pressure. What is the concentration expressed in units of parts per billion by volume, ppb_v?

2-52 The concentration of chloromethane in ambient air in several U.S. locations was measured and found to be 0.88 mg/m³. The molecular formula for chloromethane is CH_3Cl. Assume the air is nearly at standard temperature and pressure. What is the concentration expressed in units of parts per billion by volume, ppb_v?

2-53   The data shown in Table 2.19 below were collected for ozone levels in the Los Angeles air basin. Plot the data recorded in Excel in ppm and $\mu g/m^3$. Identify the year(s) Los Angeles was not in compliance with the NAAQS standards.

TABLE 2.19 Ozone concentrations reported in Los Angeles, CA (Historic Ozone Air Quality Trends, South Coast Air Quality Management District, www.aqmd.gov/smog/o3trend.html)

YEAR	BASIN MAXIMUM 1-HOUR AVERAGE (ppm)	BASIN MAXIMUM 8-HOUR AVERAGE (ppm)	YEAR	BASIN MAXIMUM 1-HOUR AVERAGE (ppm)	BASIN MAXIMUM 8-HOUR AVERAGE (ppm)
1976	0.38	0.268	1994	0.30	0.209
1977	0.39	0.284	1995	0.26	0.204
1978	0.43	0.321	1996	0.25	0.175
1979	0.45	0.312	1997	0.21	0.165
1980	0.41	0.288	1998	0.24	0.206
1981	0.37	0.266	1999	0.17	0.143
1982	0.40	0.266	2000	0.184	0.159
1983	0.39	0.245	2001	0.190	0.146
1984	0.34	0.249	2002	0.167	0.148
1985	0.39	0.288	2003	0.216	0.200
1986	0.35	0.251	2004	0.163	0.148
1987	0.33	0.210	2005	0.182	0.145
1988	0.35	0.258	2006	0.175	0.142
1989	0.34	0.253	2007	0.171	0.137
1990	0.33	0.194	2008	0.176	0.131
1991	0.32	0.204	2009	0.176	0.128
1992	0.30	0.219	2010	0.143	0.123
1993	0.28	0.195			

Source: Based on *Historic Ozone Air Quality Trends*, South Coast Air Quality Management District.

2-54   Butane and oxygen combine in the combustion process. Balance the following chemical equation that describes the combustion of butane:

$$C_4H_{10} + O_2 \rightarrow CO_2 + H_2O$$

a. How many moles of oxygen are required to burn 1 mol of butane?
b. How many grams of oxygen are required to burn 1 kg of butane?
c. At standard temperature and pressure, what volume of oxygen would be required to burn 100 g of butane?
d. What volume of air at STP is required to burn 100 g of butane?

2-55   What is the molarity of 25 g of glucose ($C_6H_{12}O_6$) dissolved in 1 L of water?

2-56   Wine contains about 15% ethyl alcohol (CH_3CH_2OH) by volume. If the density of ethyl alcohol is 0.79 kg/L, what is its molarity in wine? (Wine is an aqueous solution; that is, most of the rest of the volume of wine consists of water, and the density of wine is 0.98 g/mL.)

2-57 Table 2.20 shows ions measured in the Columbia River, with their values expressed in units of mol/L. Determine the ionic strength of the water in this reach of the Columbia River, and express your answer in mol/L.

TABLE 2.20 Data for analysis of water in Problem 2-57

CATION	K^+	Na^+	Ca^{2+}	MG^{2+}	H^+
mol/L	1.75×10^{-5}	1.00×10^{-4}	3.74×10^{-4}	1.48×10^{-4}	$10^{-7.1}$
ANION	HCO_3^-	NO_3^-	Cl^-	SO_4^{2-}	OH^-
mol/L	9.18×10^{-4}	8.55×10^{-6}	3.10×10^{-5}	7.08×10^{-6}	$10^{-6.9}$

2-58 Table 2.21 shows ions measured in the Colorado River, with their values expressed in units of mol/L. Determine the ionic strength of the water in this reach of the Colorado River, and express your answer in mol/L.

TABLE 2.21 Data for analysis of water in Problem 2-58

CATION	K^+	Na^+	Ca^{2+}	MG^{2+}	H^+
mol/L	3.32×10^{-5}	5.22×10^{-4}	6.74×10^{-4}	1.97×10^{-4}	$10^{-7.8}$
ANION	HCO_3^-	NO_3^-	Cl^-	SO_4^{2-}	OH^-
mol/L	1.23×10^{-3}		4.23×10^{-4}	2.60×10^{-4}	$10^{-6.2}$

2-59 A waste stream of 20,000 gal/min contains 270 mg/L of cyanide as NaCN.
   a. What is the ionic strength of the solution in mmol/L?
   b. What is the appropriate activity coefficient for dissolved cyanide (CN^-) in this solution if sodium cyanide is the only dissolved species? (Use the Güntelberg approximation.)

2-60 Table 2.22 represents a "total analysis" of wastewater that has been reported. Note that the pH is not given.

TABLE 2.22 Data for analysis of wastewater in Problem 2-60

CATION	CONCENTRATION (mg/L)	MOLAR MASS	ANION	CONCENTRATION (mg/L)	MOLAR MASS
NH_3	0.08	as N	F^-	21.2	as F
Na^+	18.3	as Na	Cl^-	24.1	as Cl
K^+	18.3	as K	HCO_3^-	15	as C
Ca^{+2}	1.7	as $CaCO_3$	SO_4^{-2}	20	as SO_4
			NO_3^-	2.0	as N
			NO_2^-	0.008	as N

   a. Complete a charge balance analysis on the data to see if such a solution would be electrically neutral.
   b. If H^+ and OH^- are the only ions missing from the analysis, what must their concentration be? That is, what are the pH and pOH?

2-61 Calculate the activity coefficient and activity of each ion in a solution containing 300 mg/L $NaNO_3$ and 150 mg/L $CaSO_4$.

2-62 The major dissolved species in the Colorado River near Phoenix are given in mg/L in Table 2.23.

TABLE 2.23 Approximate concentration of dissolved ions in the Colorado River near Phoenix, AZ

CATION	CONCENTRATION (mg/L)	MOLAR MASS	ANION	CONCENTRATION (mg/L)	MOLAR MASS
Ca^{2+}	83	as Ca	SO_4^{-2}	250	as SO_4
K^+	5.1	as K	Cl^-	88	as Cl
H^+	0.000065	as H	HCO_3^-	135	as $CaCO_3$

a. Express the concentration of Ca^{+2} in the Colorado River in moles of Ca^{+2} per liter and in milligrams per liter as $CaCO_3$.

b. The concentration of Na^+ is not given in the table. Assuming that Na^+ is the only significant species missing from the analysis, compute its value based on the law of electroneutrality. *Note*: The HCO_3^- concentration is expressed in terms of $CaCO_3$ hardness.

c. Calculate the total hardness, carbonate hardness, and noncarbonate hardness for the Colorado River using the data in the table.

2-63 From the water quality data shown in Table 2.24, find the ionic strength and the activity coefficient for the following:

a. The Snake River
b. The Mississippi River
c. The Lower Congo River
d. The Dead Sea

TABLE 2.24 Concentration of major dissolved ions in various water bodies

ION	SNAKE RIVER	MISSISSIPPI RIVER	LOWER CONGO, KINSHASA	GANGES RIVER	DEAD SEA
Cation			Concentration (mg/L)		
Ca^{2+}	16.1	162	10.8	88	63,000
Mg^{2+}	2.91	45	3.9	19	168,000
K^+	4.14	3		2.3	7,527
Na^+	28.4	21	14.2	6.4	35,000
Fe^{2+}	12				
Anions					
Cl^-	15.4	54	6.1	5.7	207,000
F^-	2.1				
HCO_3^-	73	124		104	244
SO_4^{2-}	25.2	100	7.8	371	2,110
Reference:	Clark et al., 2004	Meybeck et al., 1989	Visser and Villeneuve, 1975	Meybeck et al., 1989	Meybeck et al., 1989

2-64 Calculate the total hardness, carbonate hardness, and noncarbonate hardness for the Ganges River using the data in Table 2.24.

2-65 Calculate the total hardness, carbonate hardness, and noncarbonate hardness for the Mississippi River using the data in Table 2.24.

2-66 Calculate the total hardness, carbonate hardness, and noncarbonate hardness for the Dead Sea using the data in Table 2.24.

2-67 The laboratory data shown in Table 2.25 have been recorded for a 100-mL sample of water. Calculate the concentrations of total solids, total dissolved solids, and total volatile solids in mg/L.

TABLE 2.25 Representative data for calculating solids concentration in the water sample for Problem 2-67

DESCRIPTION	UNIT	RECORDED VALUE
Sample size	mL	100
Mass of evaporating dish	g	34.9364
Mass of evaporating dish plus residue after evaporating at 105°C	g	34.9634
Mass of evaporating dish plus residue after evaporating at 550°C	g	34.9606
Mass of Whatman paper filter	g	1.6722
Mass of Whatman paper filter after evaporating at 105°C	g	1.6834
Mass of Whatman paper filter after evaporating at 550°C	g	1.6773

2-68 The laboratory data shown in Table 2.26 have been recorded for a 1-L sample of water. Calculate the concentrations of total solids, total dissolved solids, and total volatile solids in mg/L.

TABLE 2.26 Representative data for calculating solids concentration in the water sample for Problem 2-68

DESCRIPTION	UNIT	RECORDED VALUE
Sample size	mL	1,000
Mass of evaporating dish	g	62.1740
Mass of evaporating dish plus residue after evaporating at 105°C	g	62.9712
Mass of evaporating dish plus residue after evaporating at 550°C	g	62.8912
Mass of Whatman paper filter	g	1.4671
Mass of Whatman paper filter after evaporating at 105°C	g	1.7935
Mass of Whatman paper filter after evaporating at 550°C	g	1.6188

2-69 Find the hardness of the following groundwater for a water sample from a well in Pennsylvania.

WATER SOURCE	CONCENTRATION OF IONS (mg/L)						
	Cations			Anions			
	[Na$^+$]	[Ca^{2+}]	[Mg^{2+}]	[NO$_3^-$]	[SO$_4^{2-}$]	[Cl$^-$]	[HCO$_3^-$]
Spring Mf 1	16	51	15	25	24	12	197

2-70 Find the hardness of the following groundwater for a water sample from a well in Pennsylvania.

WATER SOURCE	CONCENTRATION OF IONS (mg/L)						
	Cations			Anions			
	[Na$^+$]	[Ca^{2+}]	[Mg^{2+}]	[NO$_3^-$]	[SO$_4^{2-}$]	[Cl$^-$]	[HCO$_3^-$]
Spring Ln 12	4.7	48	7.3	25	26	9.2	114

2-71 What is the pH of a solution that contains hydroxide ion {OH$^-$} in the following concentrations?
   a. 10^{-8} mol/L
   b. 10^{-5} mol/L
   c. 10^{-3} mol/L
   d. 10^{-10} mol/L

2-72 Find the hydrogen concentration and the hydroxide concentration in tomato juice having a pH of 5.

2-73 What is [H$^+$] of water (in mol/L) in an estuary at pH 8.2 with an ionic strength of $I = 0.05$ mol/L?

2-74 Calculate the normality of the following solutions:
   a. 36.5 g/L hydrochloric acid [HCl]
   b. 80 g/L sodium hydroxide [NaOH]
   c. 9.8 g/L sulfuric acid [H$_2$SO$_4$]
   d. 9.0 g/L acetic acid [CH$_3$COOH]

2-75 Find the pH of a solution containing 10^{-3} mol/L of hydrogen sulfide (H$_2$S), where pK_a = 7.1.

2-76 The Henry's law constant for H$_2$S is 0.1 mol/L-atm, and

$$H_2S(aq) \leftrightarrow HS^- + H^+$$

The acid equilibrium constant for this reaction is $k_a = 10^{-7}$. If you bubble pure H$_2$S gas into a beaker of water, what is the concentration of HS$^-$ if the pH is controlled and remains at 5?
   a. in mol/L
   b. in mg/L
   c. in ppm$_m$

2-77 What is the pH of a solution containing 1×10^{-6} mg/L of OH^- (25°C)?

2-78 Find the hydrogen ion concentration and the hydroxide ion concentration in a baking soda solution with a pH of 8.5.

2-79 If 40 g of the strong base, NaOH, are added to 1 L of distilled water, what would the pH of the solution be, if
a. Activity effects are neglected?
b. Activity effects are estimated using the Debye–Hückel approximation?

2-80 Write a complete mathematical model for a closed system consisting of water only at equilibrium. Be sure to include the starting materials, a species list, equilibria expressions, mass balances, a charge balance, and other constraints. Assuming activity effects are negligible, how do $[H^+]$ and $[OH^-]$ compare?

2-81 If 35 milligrams (mg) of hydrochloric acid (HCl), a strong acid, are added to distilled water, what would the pH of the resulting solution be?

2-82 If 8.1 milligrams (mg) of hydrobromic (HBr) acid, a strong acid, are added to distilled water, what would the pH of the resulting solution be?

2-83 If 630 milligrams (mg) of nitric acid (HNO_3), a strong acid, are added to distilled water, what would the pH of the resulting solution be?

2-84 Estimate the pH of a solution containing 10^{-2} M of HCl.

2-85 Estimate the pH of a solution containing 10^{-4} M of HNO_3.

2-86 Calculate the pH of a 1-L aqueous solution that has 10^{-3} mol of formic acid ($HCHO_2$) added to distilled water. Note the formic acid is a weak acid with $k_a = 1.8 \times 10^{-4}$.

2-87 Calculate the pH of a 1-L aqueous solution that has 10^{-5} mol of lactic acid ($CH_3CH(OH)CO_2H$) added to distilled water. Note the lactic acid is a weak acid with $k_a = 1.38 \times 10^{-4}$.

2-88 Calculate the pH of a 1-L aqueous solution that has 10^{-2} mol of hydrocyanic acid (HCN) added to distilled water. Note the hydrocyanic acid is a weak acid with $k_a = 6.2 \times 10^{-10}$.

2-89 Calculate the pH of a 1-L aqueous solution that has 10^{-3} mol of benzoic acid ($C_6H_5CO_2H$) added to distilled water. Note the benzoic acid is a weak acid with $k_a = 6.4 \times 10^{-5}$.

2-90 Use a graphical approach to estimate the pH of pristine rainfall during the Ordovician Epoch, when atmospheric CO_2 concentrations were approximately 2,240 ppm_v.

2-91 Acid rainfall due to sulfur emission led to the addition of 10^{-4} M of H_2SO_4 in rainfall over the northeastern United States during the 1970s and 1980s. Carbon dioxide levels at that time were approximately 375 ppm_v.
a. Use the graphical approach to show the pH of rainfall over the northeastern United States during that time period.
b. Estimate the concentrations of carbonic acid and bicarbonate ion in the rainfall.

2-92 Calculate the solubility of magnesium carbonate ($MgCO_3$) at 25°C in distilled water. Assume the solution acts as an ideal fluid. The solubility product constant is $k_{sp} = 4 \times 10^{-5}$.

2-93 Calculate the solubility of calcium phosphate ($Ca_3(PO_4)_2$) at 25°C in distilled water. Assume the solution acts as an ideal fluid. The solubility product constant is $k_{sp} = 2.07 \times 10^{-33}$.

2-94 Calculate the solubility of silver chloride (AgCl) at 25°C if the water already contains 10 mg/L of the chloride ion. Assume the solution acts as an ideal fluid. The solubility product constant is $k_{sp} = 3 \times 10^{-10}$.

2-95 Calculate the solubility of aluminum phosphate ($AlPO_4$) at 25°C in distilled water. Assume the solution acts as an ideal fluid. The solubility product constant is $k_{sp} = 9.84 \times 10^{-21}$.

2-96 Calculate the solubility of iron(II) sulfide (FeS) at 25°C in distilled water. Assume the solution acts as an ideal fluid. The solubility product constant is $k_{sp} = 8 \times 10^{-19}$.

2-97 Calculate the solubility of lead(II) chloride ($PbCl_2$) at 25°C in distilled water. Assume the solution acts as an ideal fluid. The solubility product constant is $k_{sp} = 9.84 \times 10^{-21}$.

2-98 Calculate the solubility of mercury(I) sulfate ($HgSO_4$) at 25°C in distilled water. Assume the solution acts as an ideal fluid. The solubility product constant is $k_{sp} = 6.5 \times 10^{-7}$.

2-99 1,4-Dioxane is typically found at solvent release sites and PET manufacturing facilities. The 1,4-dioxane concentration in the soil outside of a factory is 2.0 mg/kg, and the soil has an 8.0% organic content. 1,4-Dioxane has a reported octanol–water coefficient equal to 0.53, and the organic carbon normalized partition coefficient, K_{oc}, is given as 17.0. Use the sediment/water partition coefficient to estimate the equilibrium concentration of 1,4-Dioxane in the groundwater.

2-100 The benzene concentration in the soil outside of a factory is 0.032 mg/kg, and the soil has a 2% organic content. Benzene has a reported octanol–water coefficient equal to 200, and the organic carbon normalized partition coefficient, K_{oc}, is given as 70.8. Use the sediment/water partition coefficient to determine if the equilibrium concentration of benzene in the nearby groundwater will exceed the U.S. EPA maximum contaminant level (MCL) of 0.0005 mg/L.

2-101 The polychlorinated biphenol (PCB) concentration in the soil on a riverbank is 0.5 mg/kg, and the soil has a 23% organic content. The reported octanol–water coefficient for a mixture of PCBs is approximately equal to $10^{5.58}$, and the organic carbon normalized partition coefficient, K_{oc}, is estimated as $10^{5.49}$. Use the sediment/water partition coefficient to determine if the equilibrium concentration of benzene in the nearby groundwater will exceed the U.S. EPA maximum contaminant level (MCL) of 0.0005 mg/L.

2-102 Vinyl chloride is a colorless, odorless gas used in the production of polyvinyl chloride (PVC) plastics and in numerous industries. Vinyl chloride is also a known human carcinogen, and the U.S. EPA's maximum contaminant level (MCL) in drinking water is 0.002 mg/L or 2 ppb. The vinyl chloride concentration in the soil outside of a plastics factory is 0.007 mg/kg, and the soil has a 15% organic content. Vinyl chloride (chloroethene) has a reported octanol–water coefficient equal to 4. Estimate the organic carbon normalized partition coefficient, K_{oc}, using the estimation procedures based on Equations (2.62) and (2.63) and the sediment/water partition coefficient. Use the sediment/water partition coefficient to predict the possible equilibrium concentration of vinyl chloride in the groundwater.

2-103 The EPA's maximum contaminant level (MCL) in drinking water is 0.005 mg/L or 5 ppb for carbon tetrachloride. The carbon tetrachloride concentration in the soil outside of a factory is 0.01 mg/kg, and the soil has a 1% organic content. Carbon tetrachloride has a reported octanol–water coefficient equal to 400. Estimate the organic carbon normalized partition coefficient, K_{oc}, using the estimation procedures based on Equations (2.62) and (2.63) and the sediment/water partition coefficient. Use the sediment/water partition coefficient to predict the possible equilibrium concentration of carbon tetrachloride in the groundwater.

2-104 The EPA's maximum contaminant level (MCL) in drinking water is 0.1 mg/L or 100 ppb for chlorobenzene. The chlorobenzene concentration in the soil outside of a factory is 0.01 mg/kg, and the soil has a 6% organic content. Chlorobenzene has a reported octanol–water coefficient equal to 700. Estimate the organic carbon normalized partition coefficient, K_{oc}, using the estimation procedures based on Equations (2.62) and (2.63) and the sediment/water partition coefficient. Use the sediment/water partition coefficient to predict the possible equilibrium concentration of chlorobenzene in the groundwater.

2-105 The EPA's MCL in drinking water is 0.002 mg/L or 2 ppb for aroclor. The Aroclor 1221 concentration in the soil outside of a factory is 0.028 mg/kg, and the soil has a 12% organic content. Aroclor 1221 has a reported octanol–water coefficient equal to 800. Estimate the organic carbon normalized partition coefficient, K_{oc}, using the estimation procedures based on Equations (2.62) and (2.63) and the sediment/water partition coefficient. Use the sediment/water partition coefficient to predict the possible equilibrium concentration of Aroclor 1221 in the groundwater.

## CHAPTER 3

Biogeochemical Cycles

FIGURE 3.1 This is a famous view of the lower falls of the Yellowstone River in Yellowstone National Park in Wyoming. In this photo, we see the obvious liquid water cascading over the waterfall, along with the mist of fine water droplets and gas-phase water particles; close inspection also shows ice and snow, which are the solid phase of water along the canyon walls. The river has formed a canyon through the softer underlying stone. Not only does the water flow over the waterfall and continue on downstream, but you can also see the ongoing erosion along the side of the canyon that will carry soil and sediments downstream.

Source: Bradley Striebig.

Everything flows and nothing stands still.

—PLATO

GOALS

MODERN SCIENTISTS RECOGNIZE THAT THERE ARE more than the four elements—earth, fire, water, and wind—as defined by the ancient philosophers. Yet the words of these early philosophers of science still ring very true in many ways. Plato is credited with observing that "everything flows and nothing stands still." In this chapter, you will examine the flow of gas and liquids on the surface of the Earth. The chemical and physical properties of substances describe how those substances move from one part of the Earth to another. This chapter will examine the movement of water through biogeochemical cycles and how this movement transforms our everyday world.

The educational goals of this chapter are to introduce the concepts of biogeochemical cycles and their interrelationship with society. We begin with the water cycle to provide a tangible system with which students are somewhat familiar. The concept of a mass balance and how it may be used to identify sustainable and unsustainable uses of natural resources is also introduced via the water budget equation. Students will use basic principles of chemistry, physics, and math to define and solve engineering problems—in particular, mass and energy balance problems.

OBJECTIVES

At the conclusion of this chapter, you should be able to:

3.1 Describe energy flow through an ecosystem.

3.2 Define and discuss the importance of biogeochemical cycles.

3.3 Understand where global freshwater supplies are in relationship to the distribution of people.

3.4 Describe the boundaries, transformations, and reservoirs that make up the water cycle.

3.5 Describe the rationale for sustainable water consumption and how unsustainable water usage may lead to problems in communities.

3.6 Describe how anthropogenic nutrient use has altered the rate of nutrient transformation in biogeochemical cycles.

3.7 Describe the importance of biogeochemical cycles in determining the impacts of anthropogenic activities on the natural environment and ecosystems.

Introduction

Humans are an integral part of the Earth's environment. We are, however, the only species that can alter their surroundings in such a comprehensive fashion that the environment in today's cities does not remotely resemble the environment at the same location a few hundred years ago. In the 4.5-billion-year history of the planet, the environmental changes made by humans since the Industrial Revolution in the mid-1800s have occurred in a geological blink of an eye. These radical changes to the environment are in part a consequence of many humans living longer and more comfortable lives.

The environment includes the world and all that surrounds us. However, for engineers and scientists, a more precise definition is required. Peavy, Rowe, and Tchobanglous (1985) define the environment as a system that "may take on global dimensions, may refer to a very localized area in which a specific problem must be addressed, or may, in the case of contained environments, refer to a small volume of liquid, gaseous, or solid materials within a . . . reactor." Engineering decisions and designs influence the environment in which we live. These design choices have significantly influenced our lifestyle, expectations, and the environment. We have altered the atmosphere (gas phase) and local watersheds (liquid phase), and we have created mountains of solid waste (solid phase). These lifestyle choices have led to more leisure time, longer lifespans, and less biodiversity. Engineers and scientists may use the science and math-based tools presented in this text to help forecast the impact of our engineering choices on the environment in which we live. By examining the relationship between engineering choices and their impacts on the environment, society, and the economy, engineers can make more informed decisions and create more sustainable infrastructure and manufacturing processes.

The anthropogenic (human-made) changes to the environment can be quantitatively described. The increase in the consumption of raw materials since 1900 is shown in Figure 3.2. This graph illustrates the increasing rate of consumption of raw

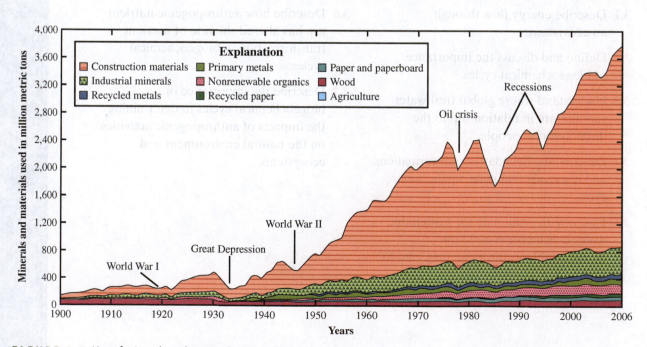

FIGURE 3.2 Use of minerals and materials in the United States.

Source: U.S. Geological Survey (2009, April). *Use of Minerals and Materials in the United States from 1900 Through 2006*, USGS Fact Sheet 2009-3008.

materials as the world has industrialized. When the rate of consumption of natural resources exceeds the rate at which these resources can be regenerated, an unsustainable process results. The mass rate of material consumed can be compared to the mass rate of material being generated. If the mass rate of consumption is less than or equal to the mass rate of generation of that resource, then the process can be sustained indefinitely, or at least until the system changes. The mass of materials moving from one portion of the Earth to another can be measured, and the rates of production can be compared to the rate of consumption. Biogeochemical cycles describe the amount of material stored, produced, or consumed within a repository, as well as the conversion of material from one repository to another.

3.1   Energy and Material Flows in Ecosystems

Scientists and engineers need to consider the ecosystems that are affected by their designs, which means incorporating the expertise of professionals from other disciplines. Some of the most fascinating reactors imaginable are *ecosystems*. **Ecology** is the study of plants, animals, and their physical environment—that is, the study of ecosystems and how energy and materials behave in ecosystems.

Specific ecosystems are often difficult to define because all plants and animals are in some way related to each other. Because of Earth's sheer complexity, it is not possible to study the Earth as a single ecosystem (except in a very crude way), so it is necessary to select functionally simpler and spatially smaller systems such as ponds, forests, or even gardens. When the system is narrowed down too far, however, there are too many ongoing external processes that affect the system, so it is not possible to develop a meaningful model. The interaction of squirrels and blue jays at a bird feeder may be fun to watch, but it is not very interesting scientifically to an ecologist because the ecosystem (bird feeder) is too limited in scope. Many more organisms and environmental factors are important in the functioning of the bird feeder, and these must be taken into account to make this ecosystem meaningful. The problem is deciding where to stop. Is the backyard large enough to study, or must the entire neighborhood be included in the ecosystem? If this is still too limited, where are the boundaries? There are none, of course, and everything truly is connected to everything else.

Both energy and materials flow inside ecosystems, but with a fundamental difference: energy flow is in only one direction, while material flow is cyclical.

All energy on Earth originates from the sun as light energy. Plants trap this energy through a process called **photosynthesis** and, using nutrients and carbon dioxide, convert the light energy to chemical energy by building high-energy molecules of starch, sugar, proteins, fats, and vitamins. In a crude way, photosynthesis can be pictured as

$$[\text{Nutrients}] + CO_2 \xrightarrow{\text{Sunlight}} O_2 + [\text{High-energy molecules}] \qquad (3.1)$$

All other organisms must use this energy for nourishment and growth through a process called **respiration**:

$$[\text{High-energy molecules}] + O_2 \rightarrow CO_2 + [\text{Nutrients}] \qquad (3.2)$$

This conversion process is highly inefficient, with only about 1.6% of the total energy available converted into carbohydrates through photosynthesis.

There are three main groups of organisms in an ecosystem. Plants, because they manufacture the high-energy molecules, are called **producers**, and the animals that use these molecules as a source of energy are called **consumers**. Both plants and animals produce wastes and eventually die. This material forms a pool of dead organic matter known as **detritus**, which still contains a considerable amount of energy. The organisms that use this detritus are known as **decomposers**.

This one-way flow is illustrated in Figure 3.3. The rate at which this energy is extracted slows considerably as the energy level decreases. Because energy flow is one way (from the sun to the plants, to be used by the consumers and decomposers for making new cellular material and for maintenance), energy is not recycled within an ecosystem, as illustrated by the following argument.

Suppose a plant receives 1,000 J of energy from the sun. Of that amount, 760 J are rejected (not absorbed) and only 240 J are absorbed. Most of this energy is released as heat, and only 12 J are used for production, 7 J of which must be used for respiration (maintenance) and the remaining 5 J for building new tissue. If a consumer eats the plant, 90% of the 5 J goes toward the animal's maintenance and only 10% (or 0.5 J) to new tissue. If this animal is, in turn, eaten, then again only 10% (or 0.05 J) is used for new tissue, and the remaining energy is used for maintenance.

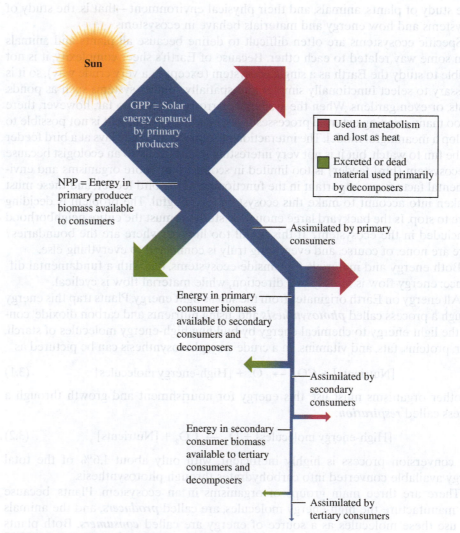

Sun

GPP = Solar energy captured by primary producers

NPP = Energy in primary producer biomass available to consumers

Used in metabolism and lost as heat

Excreted or dead material used primarily by decomposers

Assimilated by primary consumers

Energy in primary consumer biomass available to secondary consumers and decomposers

Assimilated by secondary consumers

Energy in secondary consumer biomass available to tertiary consumers and decomposers

Assimilated by tertiary consumers

FIGURE 3.3 Flow of energy through ecosystems.

If the second animal is a human being, then of the 1,000 J coming from the sun, only 0.05 J or 0.005% is used for tissue building—a highly inefficient system.

Although energy flow is in one direction only, nutrient flow through an ecosystem is cyclical, as shown in Figure 3.4. Starting with the dead organics, or the detritus, the initial decomposition by microorganisms produces compounds such as ammonia (NH_3), carbon dioxide (CO_2), and hydrogen sulfide (H_2S) for nitrogenous, carbonaceous, and sulfurous matter, respectively. These products are, in turn, decomposed further until the final stabilized, or fully oxidized, forms are nitrates (NO_3^-), carbon dioxide, sulfates (SO_4^{2-}), and phosphates (PO_4^{3-}). The carbon dioxide is, of course, used by the plants as a source of carbon, while the nitrates, phosphates, and sulfates are used as nutrients, or the building blocks for the formation of new plant tissue. The plants die or are used by consumers, which eventually die, returning us to decomposition.

There are three types of microbial decomposers: aerobic, anaerobic, and facultative. Microbes are classified according to whether or not they require molecular oxygen in their metabolic activity—that is, whether or not the microorganisms have the ability to use dissolved oxygen (O_2) as the *electron acceptor* in the decomposition reaction. The general equations for aerobic and anaerobic decomposers are

$$\textit{Aerobic}: \quad [\text{Detritus}] + O_2 \rightarrow CO_2 + H_2O + [\text{Nutrients}]$$
$$\textit{Anaerobic}: \quad [\text{Detritus}] \rightarrow CO_2 + CH_4 + H_2S + NH_3 + \cdots + [\text{Nutrients}] \tag{3.3}$$

Obligate aerobes are microorganisms that must dissolve oxygen to survive because they use oxygen as the electron acceptor. The hydrogen from the organic compounds ends up combining with the reduced oxygen to form water, as in the aerobic part of Equation (3.3). For **obligate anaerobes**, dissolved oxygen is, in fact, toxic, so they must use anaerobic decomposition processes. In anaerobic processes, the electron acceptors are inorganic oxygen-containing compounds, such as nitrates and sulfates. The nitrates

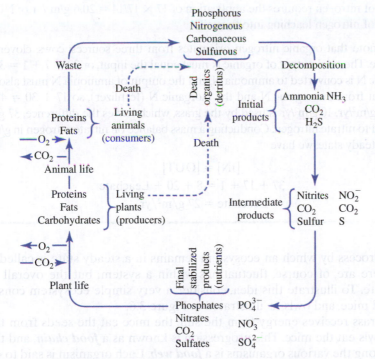

FIGURE 3.4 Aerobic cycle for phosphorus, nitrogen, carbon, and sulfur.

Source: McGaughy, P.H. (1968). *Engineering Management of Water Quality*. New York: McGraw-Hill.

are converted to nitrogen or ammonia (NH_3), and the sulfates to hydrogen sulfide (H_2S), as shown in the anaerobic part of Equation (3.3). The microorganisms find it easier to use nitrates, so this process occurs more often. *Facultative microorganisms* use oxygen when it is available; they also use anaerobic reactions if it is not available.

The decomposition carried out by the aerobic organisms is much more complete because some of the end products of anaerobic decomposition (e.g., ammonia nitrogen) are not in their final fully oxidized state. For example, aerobic decomposition is necessary to oxidize ammonia nitrogen to the fully oxidized nitrate nitrogen (see Figure 3.5). All three types of microorganisms are used in wastewater treatment, as discussed in Chapter 8.

Because nutrient flow in ecosystems is cyclical, it is possible to analyze these flows by using the techniques for material flow analysis:

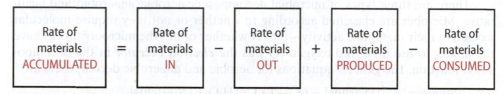

$$\boxed{\begin{array}{c}\text{Rate of}\\\text{materials}\\\text{ACCUMULATED}\end{array}} = \boxed{\begin{array}{c}\text{Rate of}\\\text{materials}\\\text{IN}\end{array}} - \boxed{\begin{array}{c}\text{Rate of}\\\text{materials}\\\text{OUT}\end{array}} + \boxed{\begin{array}{c}\text{Rate of}\\\text{materials}\\\text{PRODUCED}\end{array}} - \boxed{\begin{array}{c}\text{Rate of}\\\text{materials}\\\text{CONSUMED}\end{array}}$$

EXAMPLE 3.1　Nitrogen Balance

A major concern with the wide use of fertilizers is the leaching of nitrates into the groundwater. Such leaching is difficult to measure unless it is possible to construct a nitrogen balance for a given ecosystem. Consider the diagram in Figure 3.5, which illustrates nitrogen transfer in a meadow fertilized with 34 g/m²/yr of ammonia + nitrate nitrogen (17 + 17), both expressed as nitrogen. (Recall that the atomic weight of N is 14 and H is 1.0, so that 17 g/m²/yr of nitrogen requires the application of $17 \times 17/14 = 20.6$ g/m²/yr of NH_3.) What is the rate of nitrogen leaching into the soil?

First, note that organic nitrogen originates from three sources: cows, clover, and the atmosphere. The total output of organic N must equal the input, or $22 + 7 + 1 = 30$ g/m²/yr. The organic N is converted to ammonia N, and the output of ammonia N must also be equal to the input from the organic N and the inorganic N (fertilizer), so $17 + 30 = 47$ g/m²/yr. Of this 47 g/m²/yr, 10 g/m²/yr is used by the grass, which leaves the difference, 37 g/m²/yr, to be oxidized to nitrate nitrogen. Conducting a mass balance on nitrate nitrogen in g/m²/yr and assuming steady state, we have

$$[\text{IN}] = [\text{OUT}]$$
$$37 + 17 + 1 = 8 + 20 + \text{Leachate}$$
$$\text{Leachate} = 27 \text{ g/m}^2/\text{yr}$$

The process by which an ecosystem remains in a steady state is called *homeostasis*. There are, of course, fluctuations within a system, but the overall effect is steady state. To illustrate this idea, consider a very simple ecosystem consisting of grass, field mice, and owls, as illustrated in Figure 3.6.

The grass receives energy from the sun, the mice eat the seeds from the grass, and the owls eat the mice. This progression is known as a *food chain*, and the interaction among the various organisms is a *food web*. Each organism is said to occupy a *trophic level*, depending on its proximity to the producers. Because the mice eat the plants, they are at trophic level 1; the owls are at trophic level 2. It is also possible

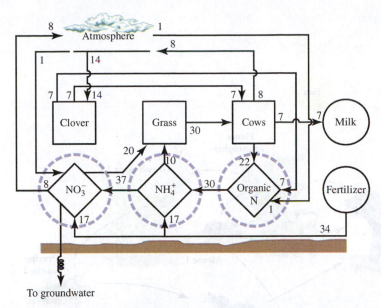

FIGURE 3.5 A nitrogen balance for an ecosystem.

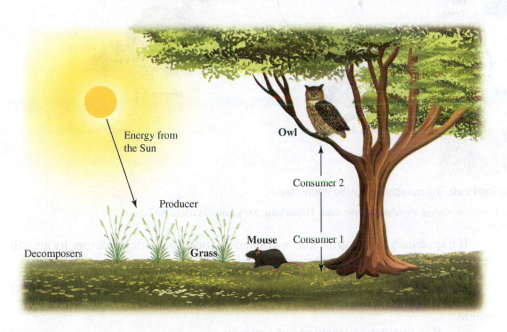

FIGURE 3.6 A simple ecosystem. The numbers indicate the trophic levels.

that a grasshopper eats the grass (trophic level 1), a praying mantis eats the grasshopper (trophic level 2), and a shrew eats the praying mantis (trophic level 3). If an owl then gobbles up the shrew, it is at trophic level 4. Figure 3.7 illustrates this land-based food web. The arrows show how energy is received from the sun and flows through the system.

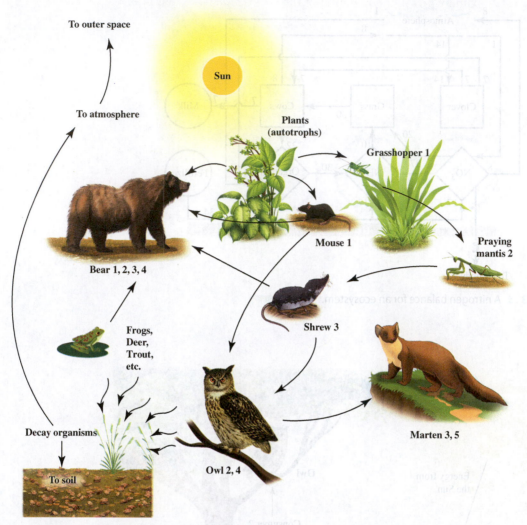

FIGURE 3.7 A terrestrial food web. The numbers show the trophic levels.

Source: Based on Turk. A. (1975). *Ecosystem, Energy, Population, First Edition,* Philadelphia, PA: Saunders College Publishing.

If a species is free to grow unconstrained by food, space, or predators, its growth is described as a first-order reaction:

$$\frac{dN}{dt} = kN \tag{3.4}$$

where

N = number of organisms of a species
k = rate constant
t = time

Fortunately, populations (except for human populations!) within an ecosystem are constrained by food availability, space, and predators. If we consider the first two constraints, the maximum population that can exist can be described in mathematical terms as

$$\frac{dN}{dt} = kN - \frac{k}{K}N^2 \tag{3.5}$$

where K is the maximum possible population in the ecosystem. Note that at steady state, $dN/dt = 0$, so

$$0 = kN - \frac{k}{K}N^2 \tag{3.6}$$

and

$$K = N \tag{3.7}$$

or the population is the maximum population possible in that system. If, for whatever reason, N is reduced to less than K, the population increases to eventually again attain the level K. Various species can, of course, affect the population level of any other species. Suppose M organisms of another species are in competition with the species of N organisms. Then the growth rate of the original species is expressed as

$$\frac{dN}{dt} = kN - \frac{k}{K}N^2 - sMN \tag{3.8}$$

where

$\quad s$ = growth rate constant for the competitive species
$\quad M$ = population of the competitive species

In terms of a rate balance expression:

$$
\boxed{\text{Growth Rate}} \;=\; \boxed{\begin{array}{c}\text{Unlimited}\\\text{Growth}\\\text{Rate}\end{array}} \;-\; \boxed{\begin{array}{c}\text{Self-}\\\text{crowding}\\\text{Effects}\end{array}} \;-\; \boxed{\begin{array}{c}\text{Competitive}\\\text{Effects}\end{array}}
$$

If s is small, then both species should be able to exist. If the first species of population N is not influenced by overcrowding, then the competition from the competitive species will be able to keep the population in check.

Note again that at steady state ($dN/dt = 0$):

$$N = K\left[1 - \frac{Ms}{k}\right] \tag{3.9}$$

so that, if $M = 0$, $N = K$.

Competition in an ecosystem occurs in niches. A *niche* is an animal or plant's best accommodation with its environment, where it can best exist in the food web. Returning to the simple grass/mouse/owl example, we see that each of the participants occupies a niche in the food web. If there is more than one kind of grass, each may occupy a very similar niche, but there are always extremely important differences. Two kinds of clover, for example, may seem to occupy the same niche until it is recognized that one species blossoms early and the other late in the summer; thus, they do not compete directly.

The greater the number of organisms available to occupy various niches within the food web, the more stable the system will be. If, in the above example, the single species of grass dies due to a drought or disease, the mice would have no food, and they as well as the owls would die of starvation. If there were *two* kinds of grasses, however, each of which the mouse could use as a food source (i.e., they both fill almost the same niche relative to the mouse's needs), the death of one grass would not result in the collapse of the system. This system is, therefore, more stable because it can withstand perturbations without collapsing. Examples of very stable ecosystems are tropical forests and estuaries, while unstable ecosystems include the northern tundra and the deep oceans.

Some of the perturbations to ecosystems are natural (witness the destruction caused by the eruption of Mount St. Helens in Washington State), while many more are caused by human activities. Humans have, of course, adversely affected ecosystems on both a small scale (streams and lakes) and a large scale (global).

BOX 3.1   The Kelp/Urchin/Sea Otter Ecological Homeostasis

A simplified example of a complex natural interaction is the restoration of the giant kelp forests along the Pacific coasts of Canada, the United States, and Mexico. This valuable seaweed, *Macrocystes*, forms 200-ft-long streamers that are fastened on the ocean floor and rise to the surface. Twice each year the kelp dies and regrows, being one of the fastest growing plants. In addition to the wealth of marine life for which the kelp forests provide habitat, kelp is a source of algin, a chemical used in foods, paints, cosmetics, and pharmaceuticals. Commercial harvesting is controlled to ensure the reproduction of the forests.

Some years ago, however, the destruction of these beds seemed inevitable. They were rapidly disappearing, leaving behind a barren ocean floor with no habitat for the marine life that had been so prevalent. The reason for the disappearance was disputed. Was it pollution? Overharvesting? Or some other more subtle cause?

The mystery was solved when it was discovered that sea urchins, which feed on the bottom, were eating the lower parts of the kelp plants, thus weakening their hold on the ocean bottom and allowing them to float away. The sea urchins in turn are the main source of food for sea otters and the otters keep the urchins from extirpating the kelp (see Figure 3.8). However, the hunting of sea otters had sufficiently depleted their numbers to bring about an explosion of the sea urchin population, which in turn resulted in the depletion of the kelp forests. The sea otter is considered a keystone predator, a top predator that has a major influence on a community's structure. Kelp, on the other hand, is a dominator species because there is so much of it. Its very abundance helps keep things in balance. In each case, however, the species provides a platform on which a complicated food web is built. With the protection of the sea otters, the balance of life was restored, and the kelp forests are again growing in the Pacific.

Source: Reprinted from Vesilind, P.A., and DiStefano, T.D. (2006). *Controlling Environmental Pollution*. Lancaster, PA: DEStech Publications, Inc.

(a)   (b)

FIGURE 3.8 (a) Sea otter feeding on a sea urchin. (b) Sea urchins feeding on kelp.

Source: (a) Noel Hendrickson/Photodisc/Getty Images; (b) Brent Durand/Moment/Getty Images.

3.2 Biogeochemical Cycles

Elements such as carbon, oxygen, hydrogen, nitrogen, and phosphorus are prerequisites for life. However, an overabundance of any of these elements in different parts of the ecosystem can cause an imbalance in the system and produce unwanted changes. Biogeochemical cycles help us understand and interpret how chemicals interact in the environment. *Biogeochemical cycles* are defined as "the biological and chemical reservoirs, agents of change and pathways of flow from one reservoir of a chemical on earth to another reservoir" (Mays, 2007, p. 57). These biogeochemical cycles help us understand what happens to important elements and compounds that influence our lifestyle. Perturbations, or changes to the water and carbon cycles, have long-range and important consequences for our environment and future generations.

Biogeochemical cycles describe the transport, storage, and conversion of compounds. The chemical form and process may determine how much of a compound is found in the atmosphere, oceans, or soil. Scientists and engineers use mathematical models, or descriptions, of these processes to predict how long compounds will reside in a certain reservoir and how they are converted to different reservoirs or different forms. The reservoir or repository is the place where these compounds may be found on Earth. For instance, most of the water on our planet is found in the oceans, as shown in Table 3.1. Water is also found in the atmosphere, but cycles through the atmosphere rapidly, spending, on average, less than 10 days in the atmosphere before falling out as precipitation. Oceans are also an important chemical repository for water and carbon. Water and carbon cycle through the oceans very slowly.

The atmosphere is one of the Earth's major repositories for water. We can estimate the average length of time that water remains in the atmosphere from the flow to and from the atmosphere shown in Table 3.2. The length of time that a compound remains in a defined system, in this case the atmosphere, is referred to as the *residence time*, t_r. We can determine the residence time in any system by summing the mass of the water in the atmosphere, M, and dividing by the mass flow rate, F, of water from the system:

$$t_r = \frac{M}{F}$$

(3.10)

TABLE 3.1 Major repositories for carbon, nitrogen, and water

	MAGNITUDE × 10^{12} kg		
REPOSITORY	Carbon (C)	Water (H₂O)	Nitrogen (N)
Atmosphere	735	13,000	3.9×10^6
Geologic (soils, ice, groundwater, etc.)	10^7	37,600	150
Ocean seawater	40,000	1.4×10^9	350
Dead organic mater	3,540	Negligible	400
Living organisms	562	3	7.8

Source: Based on Harte, J. (1988). *Consider a Spherical Cow*, pp. 238–262. Sausalito, CA: University Sciences Books.

TABLE 3.2 Mean annual flows of water on Earth

FLOW	MAGNITUDE (10^{12} m³/yr)
Precipitation on land	108
Precipitation on the sea	410
Evaporation from the land	62
Evaporation from the sea	456
Runoff	46

Source: Based on Harte, J. (1988). *Consider a Spherical Cow*, pp. 238–262. Sausalito, CA: University Sciences Books.

EXAMPLE 3.2   How Long Does Water Remain in the Atmosphere?

In this case, the water that flows from the atmosphere is equal to the amount of precipitation on land and on the sea (see Table 3.2):

$$P_{\text{total}} = P_{\text{precipitation land}} + P_{\text{precipitation in the sea}} = 108 \times 10^{12} \frac{m^3}{yr} + 410 \times 10^{12} \frac{m^3}{yr} = 518 \times 10^{12} \frac{m^3}{yr}$$

The units, m³/yr, represent the volumetric flow rate of water from the atmosphere, where m³ is the volume per unit time—in this case one year. In order to convert the volumetric flow rate, we must multiply by the density of the water in the precipitation. We will assume that the mean or average density, ρ, of water in a raindrop is equal to 1,000 kg/m³, which is the density of liquid water at 25°C and 1 atmosphere of pressure. Now we can calculate the mass flow rate:

$$F = P_{\text{total}} \times \rho = 518 \times 10^{12} \frac{m^3}{yr} \times 1,000 \frac{kg}{m^3} = 518 \times 10^{15} \frac{kg}{yr}$$

As shown in Table 3.1, the mass of water in the atmosphere is $13,000 \times 10^{12}$ kg. Substituting the mass of water and mass flow rate into the equation for the residence time yields

$$t_r = \frac{M}{F} = \frac{13,000 \times 10^{12}}{518 \times 10^{15}} = 0.025 \text{ years}$$

Since it is difficult to relate 2.5×10^{-2} years to time spans discussed in everyday conversation, we will convert the time to more useful units. From our calculations, we can see that water spends, on average, about 9.1 days in the atmosphere.

$$t_r = 0.025 \text{ years} \times \frac{365 \text{ days}}{\text{year}} = 9.1 \text{ days}$$

These cycles have changed little since the last ice age. Major shifts in biogeochemical cycles occurred over the Earth's history, as discussed in Box 3.2. The change to a cooler climate influenced the mass extinction of the age of the large dinosaur reptiles and ushered in a new age dominated by a newer species—mammals. A significant climate change in the past illustrates how easily a dominant animal, like *Tyrannosaurus rex*, can become a victim of an environment that is changing. Species incapable of adapting to changes in their environment face extinction. The dinosaurs had relatively small brains and lacked the capacity and technology to radically change and adapt to their surrounding environment. Scientists are observing a climate now that is changing at a rate unequaled in human history. How species adapt to these changes and interact with the biogeochemical cycles will influence the extent of the extinction period that is currently happening. Biologists are documenting the greatest extinction since the age of the dinosaurs due to the changing climate and human interactions with the environment. Human activities will determine the fate of many species over the next few centuries and may even influence our own fate as a species.

BOX 3.2 Chaco Canyon Based on *Collapse: How Societies Choose to Fail or Succeed* by Professor Jared Diamond (2005)

A beautiful high desert plateau runs along the border of New Mexico and Colorado. Amazing cliff dwellings from the Anasazi civilization such as those shown in Figure 3.9 can still be seen in this area at the Chaco Canyon National Historic Park and Mesa Verde National Park. The Anasazi civilization that ranged across the North American Southwest, as shown in Figure 3.10, constructed the tallest building in North America until skyscrapers were built in Chicago from steel girders in the 1880s.

Chaco Canyon's Anasazi society flourished from approximately AD 600 until sometime between AD 1150 and 1200. The narrow canyon at this time created environmental advantages for a fledgling

FIGURE 3.9 Anasazi ruins are still visible in parts of the southwestern United States. The complex ruins show a high level of technical sophistication and skill. The multistory buildings were some of the highest buildings constructed in North America.

Source: kravka/Shutterstock.com.

civilization. In this arid region of the United States, precipitation is a relatively rare, life-sustaining event. Runoff from surrounding areas was channeled through the narrow canyon, creating a reliable water supply in this otherwise arid region. The Anasazi relied on the runoff they collected to irrigate fields near their settlements. However, by about AD 900, deep channels called arroyos were formed that prohibited irrigation of crops unless the channels were completely filled with water. The nearby ecosystems were also degraded by deforestation. Deforestation occurred because the rate of logging was greater than the speed at which the slow-growing pinion and juniper pine trees grew.

Once the trees had been removed, the Anasazi could no longer use pine nuts as a source of protein in their diet or use pine timber for their buildings. The Anasazi population continued to grow in spite of the degraded environment. The population within Chaco Canyon received supplies of food and timber from satellite communities surrounding the canyon. Archaeologists determined that construction persisted in Chaco Canyon until approximately AD 1110. After this time, it has been theorized that the timber supplies were depleted in the surrounding area and the Chaco Canyon settlements began to decline and were eventually abandoned altogether.

Professor Jared Diamond, in his book *Collapse*, describes the conditions that led to the decline of the Anasazi civilization in and around Chaco Canyon after AD 1250. Diamond believes that population growth outstripped the rate of replenishment of the area's natural resources. Archaeological evidence shows that food, timber, and other natural resources were getting increasingly scarce as the

(Continued)

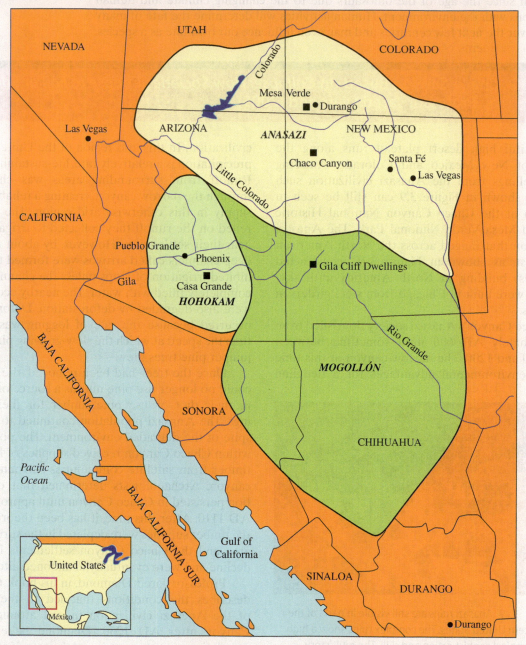

FIGURE 3.10 Map of the location and extent of the Anasazi civilization from AD 600 to approximately AD 1200 in the southwestern part of North America.

Source: Based on Angelakis, A.N., Mays, L.W., Koutsoyiannis, D., and Mamassis, N. (2012). *Evolution of Water Supply Through the Millennia*. IWA Publishing Alliance House, 12 Caxton Street, London, SW1H 0QS, UK.

Chaco Canyon population declined. Finally, water scarcity due to a drought that began in AD 1130 forced the residents of Chaco Canyon to abandon this region between AD 1150 and 1200.

According to Diamond, four significant factors led to the collapse of the Anasazi civilization in Chaco Canyon (and generally to other civilizations as well):

- Human-made environmental degradation occurred, caused by deforestation and arroyo cutting.
- Climate change associated with rainfall and temperature exacerbated the anthropogenic environmental factors.

- The Chaco Canyon civilization and population became unsustainable because of the fragile ecosystem in which the civilization developed.
- Societal and cultural factors apparently helped promote unsustainable practices within the community.

Diamond sees these characteristics as indicators of the potential collapse of a community. Sustainable societies must consider the needs of future generations and not just those of the current generation. The infrastructure and economic models of past and present societies that avoid the characteristics described by Professor Diamond are more sustainable.

Engineers and scientists continue to explore and research the biogeochemical cycles. Mathematical models that describe the chemical reactions, transformations, and pathways of water and carbon are being investigated to help populations adapt to our changing environments and avoid a fate similar to that of the residents of Chaco Canyon. Satellite technology and more accurate analyses of our atmosphere, oceans, and soil allow engineers and scientists to gain a better understanding of how biogeochemical cycles work during the present time. Analyzing biogeochemical cycles also reveals something about how these cycles worked thousands of years ago. Mathematical models of these systems based on data from the past and present allow scientists and engineers to predict how human behavior might impact these cycles and vice versa: specifically, how changes in biogeochemical cycles may impact human lifestyles in the future.

ACTIVE LEARNING EXERCISE 3.1 Your Water Use

Many of our day-to-day activities depend on a constant and inexpensive supply of water. Just as residents of the Chaco Canyon civilization relied on water and imported resources to support their way of life, so too do we today in the industrialized world.

In order to gain a better understanding of water consumption and its impacts, examine the consumptive use of water in your home or living space. List the uses of water in your home. For each use, determine what happens to the water after it is used. What part of the used water goes directly to the environment—for example, through watering the grass or washing the car? What fraction of the used water goes into the sanitary sewer? Is your wastewater treatment plant in the same region, or is the wastewater diverted to another location far away? What effects might diverting water from your local area have on local streams, wetlands, or groundwater?

3.3   The Hydrologic Cycle

The *atmosphere* is a mixture of gases extending from the surface of the Earth toward space; the *lithosphere* is the soil crust that lies on the surface of the planet where we live; and the *hydrosphere* is the portion of the Earth that accounts for most of the water storage and consists of oceans, lakes, streams, and shallow groundwater bodies. *Hydrology* is "the science that treats the waters of the Earth, their occurrence, circulation, and distribution, their chemical and physical properties, and their reaction with the environment, including the relations to living things" (Mays, 2007, p. 191). The *hydrologic cycle*, or water cycle, describes the movement of water from one biogeochemical cycle to another (Figure 3.11). The hydrologic cycle is one example of a biogeochemical cycle.

3.3.1   Water Repositories

Oceans contain 97% of the water on the planet (Figures 3.12 and 3.13). They provide many definable separate ecosystems based on the chemistry of the water. Ocean water contains approximately 35 grams of dissolved minerals (or salts) per liter. Dissolved mineral concentrations above 2 g/liter make water unsuitable for human consumption. However, for ocean-dwelling animals and plants, including mammals such as dolphins and whales, ocean water is consumed for hydration.

The largest repository of freshwater on the planet lies in the Arctic and Antarctic ice caps (Table 3.3). Marine species of microorganisms, fish, and large marine mammals such as whales and leopard seals have adapted to life at or near the Arctic. Microbial blooms in these cold waters feed fish, and the fish draw larger predatory fish which in turn attract even larger predators. The water chemistry and temperature control the movement of water through large circulation patterns in

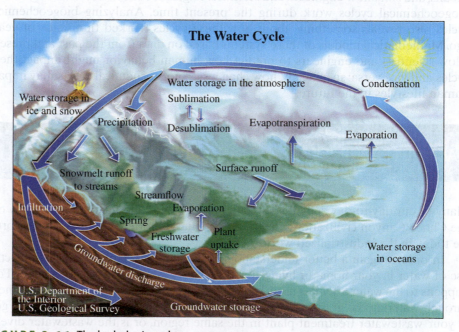

FIGURE 3.11 The hydrologic cycle.

Source: U.S. Department of the Interior, *U.S. Geological Survey*, John Evans, Howars Perlman, USGS, water .usgs.gov/edu/watercycleprint.html.

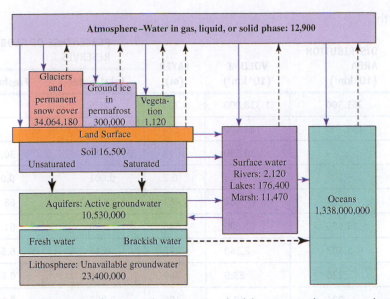

FIGURE 3.12 Water distribution with reserves in cubic kilometers and movement between biogeochemical repositories.

Source: Based on Murakami, M. (1995). *Managing Water for Peace in the Middle East: Alternative Strategies.* United Nations University (1976).

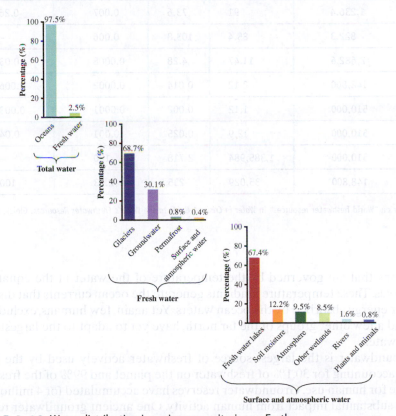

FIGURE 3.13 Water distribution in various repositories on Earth.

Source: Based on Arthurton, R., Barker, S., Rast, W., Huber, M., Alder. J., Chilton, J., Gaddis, E., Pietersen, K., and Zockler, C. (2007). *Water. State and Trends of the Environment: 1987–2007,* Chapter 4. Global Environment Outlook (GEO4) Environment for Development. United Nations Environment Programme. DEW/1001/NA.

TABLE 3.3 Water reserves on Earth

RESERVE	DISTRIBUTION AREA (10^3 km^2)	VOLUME (10^3 km^3)	LAYER (m)	PERCENTAGE OF GLOBAL RESERVES	
				Total water	Freshwater
World ocean	361,300	1,338,000	3,700	96.5	–
Groundwater	134,800	23,400	174	1.7	–
Freshwater	–	10,530	78	0.76	30.1
Soil moisture	–	16.5	0.2	0.001	0.05
Glaciers and permanent snow cover	16,227	24,064	1,463	1.74	68.7
Antarctic	13,980	21,600	1,546	1.56	61.7
Greenland	1,802	2,340	1,298	0.17	6.68
Arctic islands	226	83.5	369	0.006	0.24
Mountainous regions	224	40.6	181	0.003	0.12
Ground ice/permafrost	21,000	300	14	0.022	0.86
Water reserves in lakes	2,058.7	176.4	85.7	0.013	
Fresh	1,236.4	91	73.6	0.007	0.26
Saline	822.3	85.4	103.8	0.006	–
Swamp water	2,682.6	11.47	4.28	0.0008	0.03
River flows	148,800	2.12	0.014	0.0002	0.006
Biological water	510,000	1.12	0.002	0.0001	0.003
Atmospheric water	510,000	12.9	0.025	0.001	0.04
Total water reserves	510,000	1,385,984	2,718	100	–
Total freshwater reserves	148,800	35,029	235	2.53	100

Source: Based on Shiklomanov, I. (1993). Chapter on "World freshwater resources" in *Water in Crisis: A Guide to the World's Freshwater Resources*. Gleick, P.H. (ed.) Oxford: Oxford University Press.

the oceans that are governed by the temperature of the water at the equator and at the poles. These temperature gradients generate the ocean currents that distribute the solar energy absorbed by the ocean waters. Yet again, few humans, excluding the Inuit and a few other groups of the far north, have yet to adapt to the largest source of freshwater.

Groundwater is the largest source of freshwater actively used by the human species, accounting for 30.1% of freshwater on the planet and 99% of the freshwater available for human use. Groundwater reserves have accumulated for 4 million years without substantial impact from human activity. One ancient groundwater resource is the Ogallala Aquifer, which humans have used to transform the midwestern part of North America, as shown in Figure 3.14. Water has been flowing into the Ogallala

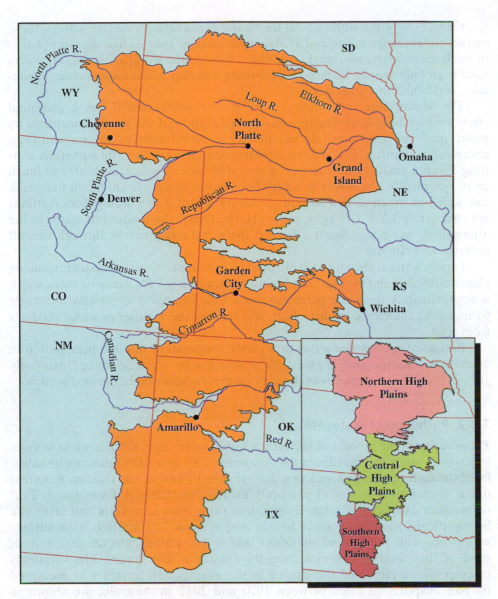

FIGURE 3.14 Extent of the Ogallala Aquifer.

Source: Based on USGS National Water-Quality Assessment (NAWQA) Program—High Plains Regional Groundwater (HPGW) Study, High Plains Aquifer System.

Aquifer near the base of the Rocky Mountains for 2 to 6 million years. It serves as a source of drinking water and irrigation water throughout the farms and cities of the Midwest. Today, however, more water is being withdrawn from the aquifer than is entering. In 2005, the United States Geological Service (USGS) estimated that the total water stored in the aquifer was 2.925 million acre-feet (3,608 km^3). While this is a vast amount of water, the USGS estimates that it is 9% (or 253 million acre-feet) less than predevelopment levels due to the rate of water use for irrigation of agricultural fields in the midwestern United States.

Only 0.3% of the world's total freshwater is available in surface waters. These surface waters range in size and availability from large sources like the Great Lakes to small seasonal streams, creeks, and rivers. Most of the world's population centers have been built around surface water sources, and they affect today's cities, economies, and health to a large degree.

The atmosphere is another repository for water, and it has a profound effect on our daily lives. The humidity, or saturation of the local atmosphere by water, plays an important role in climate, the biodiversity of species, and even energy consumption. The amount of water contained in the atmosphere is startling. Even a small cloud consists of over 400 tons of water, or nearly as much as 100 elephants. The water in a very large storm cloud system like a hurricane can weigh as much as 100 million elephants! In total, the atmosphere contains approximately 1.3×10^{16} kg of water (Harte, 1988). The amount of water in the atmosphere has a significant effect on the world's population through rainfall patterns and climate.

Finally, water is also stored in biological components of our ecosystem, including plants, animals—and us! Our own brains contain nearly 70% water by mass, which is approximately the same percentage as a tree. Biological organisms use water to transport chemicals, regulate heat, and remove the buildup of waste products from cells.

Biological species use water to regulate temperature through perspiration in animals and transpiration in plants. Plant transpiration accounts for approximately 10% of the water that moves to the atmosphere, as discussed in the next section.

3.3.2 Pathways of Water Flow

Evaporation is the process of converting liquid water from surface water sources to gaseous water that resides in the atmosphere. Evaporation rates can be calculated through observation and vary dramatically by region and climate. Scientists use a 48-inch-diameter steel pan aptly named the "Class A Evaporation Pan" to measure evaporation rates. Evaporation rates from reservoirs and lakes can be significant sources of water loss in arid regions. Evaporation from surface waters accounts for 90% of the water that is transported into the atmosphere; most of that water evaporates from the surface of the oceans. The variations in evaporation rates and the change in the average soil moisture as measured by pan evaporation rates between 1970 and 2012 in Australia are shown in Figures 3.15 and 3.16.

Transpiration occurs when water is conveyed from living plant tissue, especially leaves, to the atmosphere. In wetter climates, transpiration remains fairly constant and is closely related to the growing season. In arid regions, transpiration is much more closely related to the depth of the roots for the plants growing in that region. Shallow-rooted plants may die from lack of moisture availability, while deep-rooted plants may continue to transpire water by utilizing deeper groundwater supplies for their growth and maintenance. During a growing season, a leaf will transpire many times more water than its own weight. An acre of corn gives off about 3,000–4,000 gallons (11,400–15,100 L) of water each day, and a large oak tree can transpire 40,000 gallons (151,000 L) per year.

Evaporation and transpiration are often grouped into one term— evapotranspiration—that represents the overall pathways of water moving into the

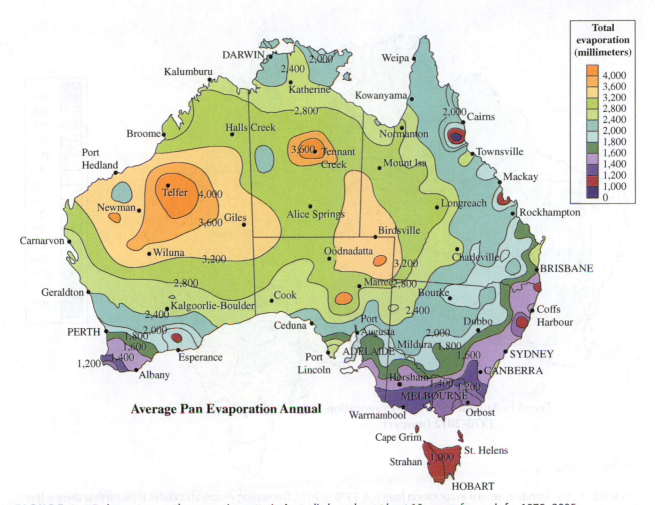

Average Pan Evaporation Annual

FIGURE 3.15 Average annual evaporation rates in Australia based on at least 10 years of records for 1975–2005.

Source: Based on Australian Government Bureau of Meteorology.

atmosphere. Evapotranspiration includes all evaporation and transpiration from plants. It varies with temperature, wind, and ambient moisture conditions and can be calculated using a lysimeter (a buried soil-filled tank). Global evaporation rate estimates for Africa, Australia, India, and the United States are shown in Figures 3.17 to 3.21 (see pages 167 to 169). The change in annual average evapotranspiration rates across the United States is shown in Figure 3.22 (see page 170). Evapotranspiration (ET) estimates for individual nations can be found on the Environmental Data Explorer website of the United Nations Environmental Programme (geodata.grid .unep.ch/).

Condensation converts water in the gas phase to liquid water by cooling the water molecules. This process may result in the formation of clouds or fog, which is composed of very fine, lightweight water droplets. These droplets may collide and combine to form larger droplets that may eventually grow large enough to produce precipitation.

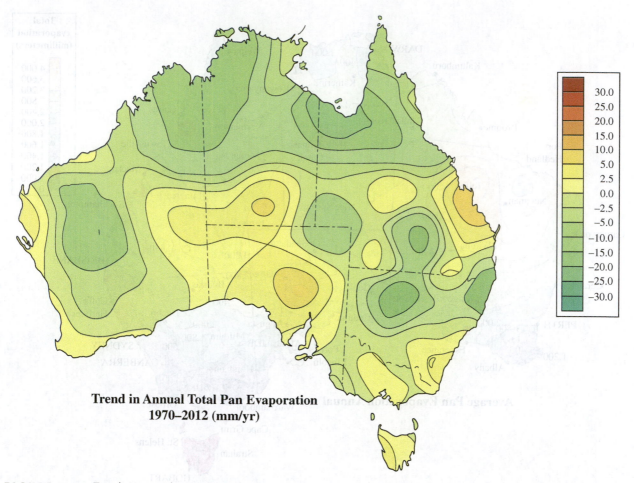

**Trend in Annual Total Pan Evaporation
1970–2012 (mm/yr)**

FIGURE 3.16 Trends in annual evaporation from soil, 1970 to 2012. Decreasing evaporation rates indicate that there is less moisture retained and available in the soils.

Source: Based on Australian Government Bureau of Meteorology.

3.3.3 Precipitation

Water moves from the atmosphere to the surface of the planet through **precipitation**. Precipitation may occur when the atmosphere becomes completely saturated with water (100% humidity) and the droplets have enough mass to fall from the atmosphere. Evaporation from the oceans and eventual cooling of the water vapor from the oceans account for approximately 90% of the Earth's precipitation. Locations near the oceans generally receive greater rainfall than those in the interior of the continents, as illustrated in Table 3.4 (see page 173). Variation in precipitation patterns is shown in Figures 3.23 to 3.27 (see pages 170 to 174). However, some areas, such as the desert highland of Chile, are the most arid in the world, even though they are not far removed from the Pacific Ocean. Ocean currents, weather patterns, latitude, and geography are important influences on precipitation patterns. Detailed precipitation data for individual nations can be found on the AQUASTAT website of the Food and Agricultural Organization of the United Nations (www.fao.org /aquastat/en/).

(a) Annual E (1984–1998)

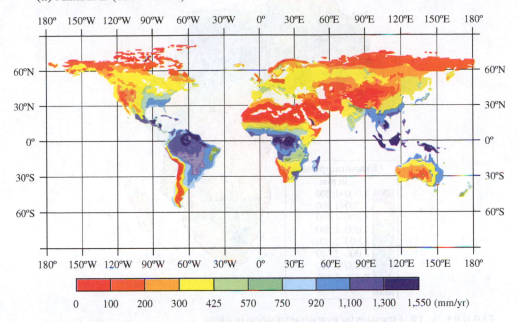

(b) Ratio of annual E to annual precipitation (1984–1998)

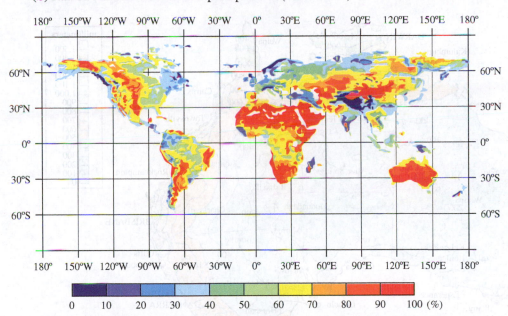

FIGURE 3.17 Global estimates of evapotranspiration in mm/yr compared to the ratio of annual precipitation. Red areas show regions where water loss is greater than precipitation, and blue areas show regions where excess water is available.

Source: Based on Yan, H., Wang, S.Q., Billesbach, D., Oechel, W., Zhang, J.H., Meyers, T., Martin, T.A., Matamala, R., Baldocchi, D., Bohrer, G., Dragoni, D., and Scott, R. (2012). "Global estimation of evapotranspiration using a leaf area index-based surface energy and water balance model." *Remote Sensing of Environment*, 124:581–595. dx.doi.org /10.1016/j.rse.2012.06.004.

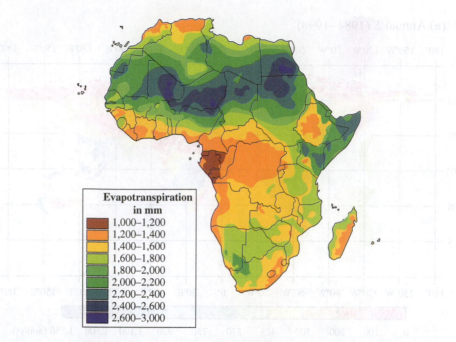

FIGURE 3.18 Estimates for evapotranspiration in Africa.

Source: Based on FAO.

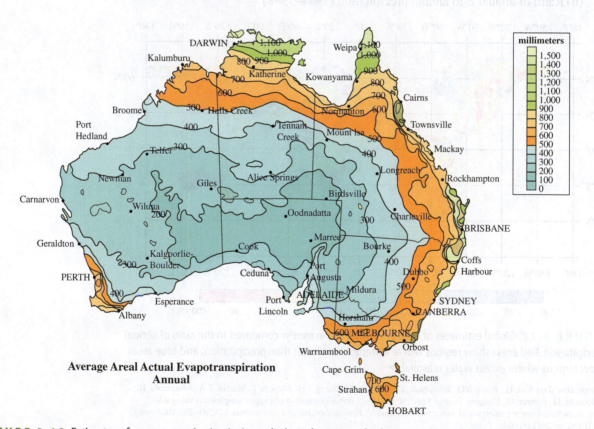

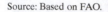

**Average Areal Actual Evapotranspiration
Annual**

FIGURE 3.19 Estimates of evapotranspiration in Australia based on a standard 30-year climatology (1961–1990).

Source: Based on Australian Government Bureau of Meteorology.

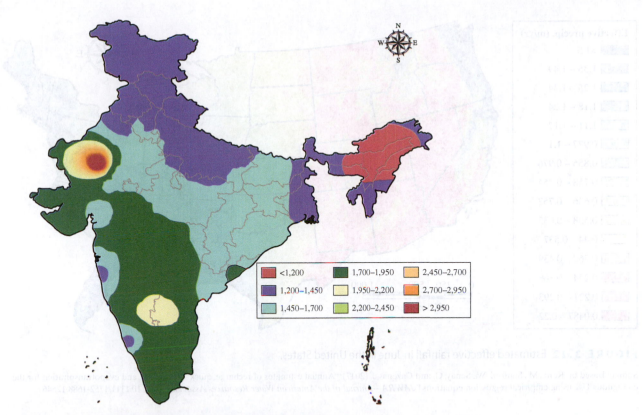

FIGURE 3.20 Estimated evapotranspiration rates and variations (in mm) for Indian conditions by the Penman–Montieth method.

Source: Based on Rao, B., Sandeep, V.M., Rao, V.U.M., and Venkateswarlu, B. (2012). *Potential evapotranspiration estimation for Indian conditions: Improving accuracy through calibration coefficients.* Tech. Bull. No 1/2012. All India Co-ordinated Research Project on Agrometeorology, Central Research Institute for Dryland Agriculture, Hyderabad.

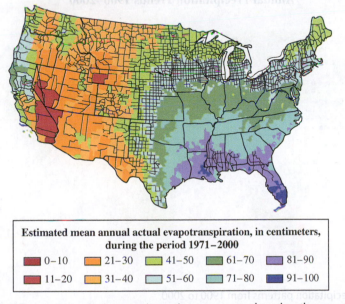

FIGURE 3.21 U.S. evapotranspiration rates in centimeters per year based on the years 1971–2000.

Source: Based on UC Santa Barbara Geography.

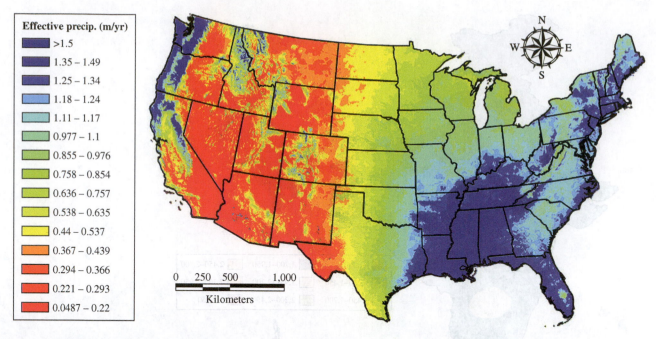

FIGURE 3.22 Estimated effective rainfall in June in the United States.

Source: Based on Reitz, M., Sanford, W., Senay, G. and Cazenas, J. (2017). "Annual estimates of recharge, quick-flow runoff, and evapotranspiration for the contiguous U.S. using empirical regression equations." *JAWRA—Journal of the American Water Resources Association*. 53, 10.1111/1752-1688.12546.

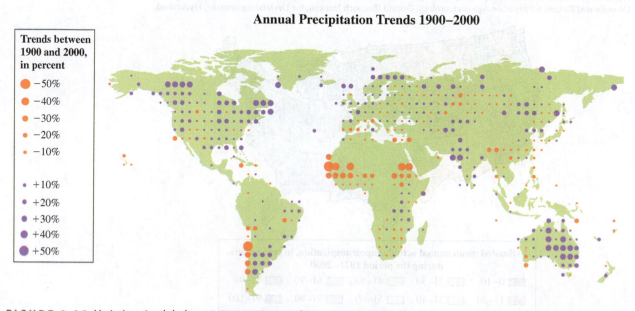

FIGURE 3.23 Variations in global precipitation patterns from 1900 to 2000.

Source: Based on Arthurton, R., Barker, S., Rast, W., Huber, M., Alder. J., Chilton, J., Gaddis, E., Pietersen, K., and Zockler, C. (2007). Chapter 4: *Water. State and Trends of the Environment: 1987–2007.* Global Environment Outlook (GEO4) Environment for Development. United Nations Environment Programme. DEW/1001/NA.

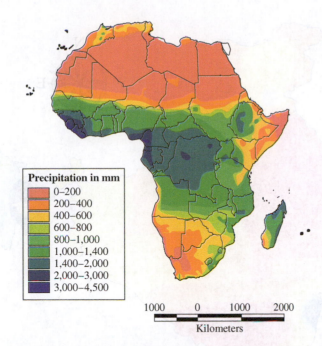

FIGURE 3.24 Precipitation patterns in Africa.

Source: Based on FAO.

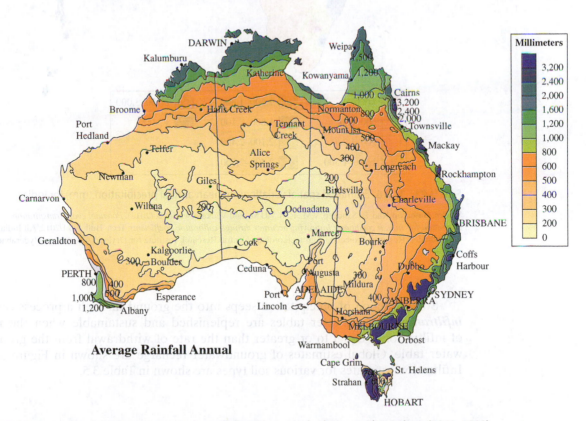

Average Rainfall Annual

FIGURE 3.25 Precipitation patterns in Australia based on standard 30-year climatology (1961–1990).

Source: Based on Australian Government Bureau of Meteorology.

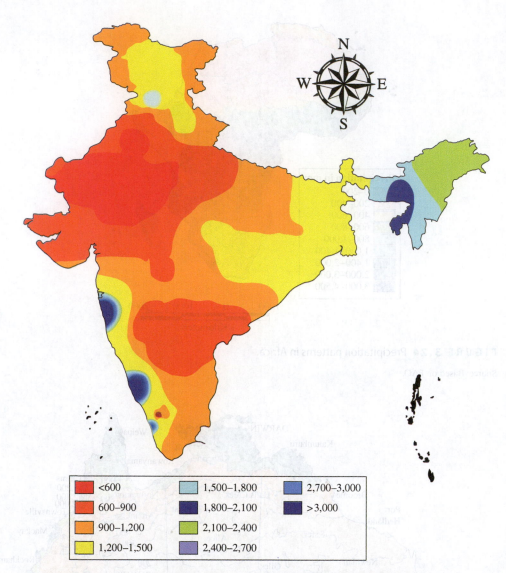

FIGURE 3.26 Estimated spatial variations in mean annual precipitation (mm) for India.

<600	1,500–1,800	2,700–3,000
600–900	1,800–2,100	>3,000
900–1,200	2,100–2,400	
1,200–1,500	2,400–2,700	

Source: Based on Rao, B., Sandeep, V.M., Rao, V.U.M., and Venkateswarlu, B. (2012). *Potential evapotranspiration estimation for Indian conditions: Improving accuracy through calibration coefficients.* Tech. Bull. No 1/2012. All India Co-ordinated Research Project on Agrometeorology, Central Research Institute for Dryland Agriculture, Hyderabad.

Some fraction of precipitation seeps into the ground through a process called *infiltration*. Groundwater tables are replenished and sustainable when the rate of infiltration is equal to or greater than the rate of withdrawal from the groundwater table. Global estimates of groundwater recharge are shown in Figure 3.28. Infiltration estimates for various soil types are shown in Table 3.5.

TABLE 3.4 Average annual precipitation for selected U.S. cities from the National Climate Data Center

CITY	AMOUNT Inches	Millimeters
Phoenix, Arizona	7.6	193
El Paso, Texas	8.6	218
Los Angeles, California	11.9	302
Denver, Colorado	15.4	391
Salt Lake City, Utah	15.6	296
Fargo, North Dakota	19.6	498
Dallas, Texas	35.0	889
Chicago, Illinois	35.8	909
Portland, Oregon	36.3	922
Columbus, Ohio	37.8	960
Seattle, Washington	38.1	968
Kansas City, Missouri	38.0	970
Bangor, Maine	40.5	1,029
New York, New York	41.5	1,054
Atlanta, Georgia	49.8	1,265
Memphis, Tennessee	52.7	1,339
Miami, Florida	59.0	1,499

Source: Based on National Climate Data Center, National Oceanic and Atmospheric Administration, U.S. Department of Commerce, Ashville, NC.

TABLE 3.5 Infiltration rate estimates for various soils

SOIL TYPE	BASIC INFILTRATION RATE (mm/hr)
Sand	less than 30
Sandy loam	20–30
Loam	10–20
Clay loam	5–10
Clay	1–5

Source: Based on FAO (1988). *Irrigation Water Management: Irrigation Methods.*

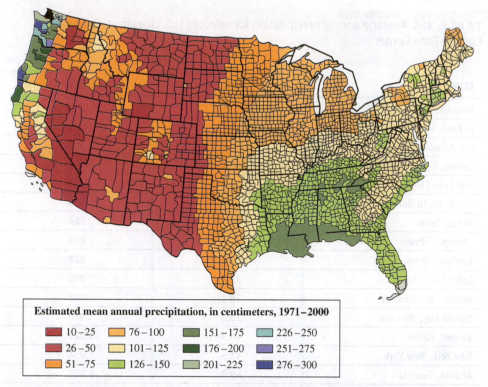

Estimated mean annual precipitation, in centimeters, 1971–2000

10–25	76–100	151–175	226–250
26–50	101–125	176–200	251–275
51–75	126–150	201–225	276–300

FIGURE 3.27 Estimated mean annual precipitation (cm) patterns in the United States, 1971–2000.

Source: USGS. Accessed June, 2020. www.usgs.gov/media/images/map-annual-average-precipitation-us-1981–2010.

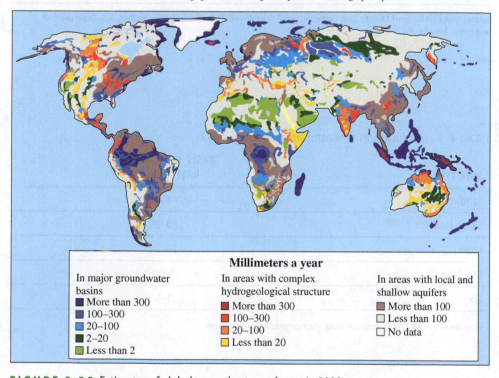

Millimeters a year

In major groundwater basins
- More than 300
- 100–300
- 20–100
- 2–20
- Less than 2

In areas with complex hydrogeological structure
- More than 300
- 100–300
- 20–100
- Less than 20

In areas with local and shallow aquifers
- More than 100
- Less than 100
- No data

FIGURE 3.28 Estimates of global groundwater recharge in 2008.

Source: Based on World Water Assessment Programme (2009). *The United Nations World Water Development Report 3: Water in a Changing World*. Paris: UNESCO, and London: Earthscan.

ACTIVE LEARNING EXERCISE 3.2 How Does a Single Large Storm Affect Runoff (Or Flooding) and Overall Water Availability for a Region?

Stormwater runoff and stream flow are measured by the United States Geological Survey (USGS). Find a stream or river near where you live. Create and download a graph of the recent stream flow rate and long-term average flow for a stream/river from the USGS website (waterdata.usgs.gov/nwis/sw) for a one-week period and for a one-year period. On each graph, identify the largest single storm event, the period of most likely flooding, and the period of most likely drought. Using triangles and rectangles to approximate the area under the graph, estimate:

a. The total volume of water flowing past the point of measurement for the week
b. The total volume of water flowing past the point of measurement for the year
c. The long term yearly average volume of water

3.4 Watersheds and Runoff

The *watershed* is the region that collects rainfall. When the height of the ground-water table is equal to the surface, a stream is formed. The precipitation may also flow over saturated land through *runoff* in small rivulets that may be collected into intermittent streams. The intermittent streams flow into groundwater-based streams or river systems. The river systems transport the water back to the oceans. Mays (2011) defines drainage basins, catchments, and watersheds as "three synonymous terms that refer to the topographic area that collects and discharges surface stream flow through one outlet or mouth." All the runoff within a watershed exits a single point downhill or downstream from all other points in the watershed. Figure 3.29 is a simple visualization of a watershed.

Watersheds vary in scale based on the analysis needed. They may be very small in scale if we are concerned with a small parcel of land—for instance, residential developments. Alternatively, they may be large enough to contain many states or regions when we analyze resources on a regional and global scale. The Amazon River collects water from the largest watershed in the world, which is 2.38 million square miles (Cech, 2010). Table 3.6 lists the 10 largest watersheds in the world and shows the area extent of the watershed and the *volumetric flow rate*, or volume of water exiting the watershed each day. Runoff formed from precipitation that is not absorbed by the soils is the major source of drinking water for many communities around the globe. Average long-term global runoff trends in a region along with human population in the region are illustrated in Figure 3.30 (see page 178). As shown in Figure 3.31 (see page 179), surface water supplies from runoff are the major source of water for all human uses including agriculture, drinking water, and energy production. Generally, regions of high

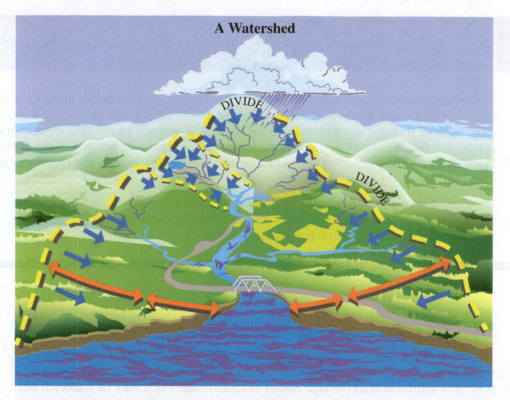

FIGURE 3.29 Illustration of a watershed, showing the downward flow of water (in blue) and the watershed boundaries (in yellow).

Source: Based on Recycle Works.

human population have high water demands, but water demand is also related to affluence, agricultural practices, and industrial water consumption, as discussed in Box 3.3 (see pages 180 to 182) and the following sections.

3.5 Water Budget

The water budget balances the flows of water into and out of a watershed or system. The watershed boundaries determine the boundaries for the system of concern. In the case of the funnel system, the edges of the funnel, or the highest point, represent the boundaries of the watershed. Likewise, in natural watersheds, the highest points in those watersheds create the watershed boundaries. The boundaries can be determined by outlining the highest points on a topographic map. By connecting the dots between these highest points to the lowest point occurring in the stream that exits the watershed, the watershed boundaries can be determined.

We can sum all these terms to create a **water budget** for the watershed, where the sum of all the terms balance the inputs and outputs into a watershed:

$$\text{Inputs} = \text{outputs} + \text{storage} \tag{3.11}$$

$$P = R + I + E + T + S \tag{3.12}$$

TABLE 3.6 World's largest watersheds

RIVER	WATER RUNOFF VOLUME (km³/yr)	WATERSHED POPULATION (MILLION PEOPLE)	WATERSHED LAND AREA (MILLION km²)		
			Minimum	Maximum	Average
Amazon	6.92	14.3	6,920	8,510	5,790
Ganges	1.75	439	1,389	1,690	1,220
Congo	3.50	48.3	1,300	1,775	1,050
Orinoco	1.00	22.4	1,010	1,380	706
Yangtze	1.81	346	1,003	1,410	700
La Plata	3.10	98.4	811	1,860	453
Yenisei	2.58	4.77	618	729	531
Mississippi	3.21	72.5	573	880	280
Lena	2.49	1.87	539	880	280
Ob	2.99	22.5	404	567	270
Mekong	0.79	75.0	505	610	376
Mackenzie	1.75	0.35	333	420	281
Amur	1.86	4.46	328	483	187
Niger	2.09	131	303	482	163
Volga	1.38	43.3	255	390	161
Danube	0.82	85.1	225	231	137
Indus	0.96	150	220	359	126
Nile	2.87	89.0	161	248	94.8
Amu Dana	0.31	15.5	77.1	118	56.7
Yellow	0.75	82.0	66.1	97	22.1
Dneiper	0.50	36.6	53.3	95	21.7
Syr Danya	0.22	13.4	38.3	75	26.2
Don	0.42	17.5	26.9	52	11.9
Murray	1.07	2.1	24	129	1.16

Source: Based on Revenga, C., Nackoney, J., Hoshino, E., Kura, Y., and Maidens, J. (2003). *Watersheds of the World*. IUCN, IWMI, Rasmar Convention Bureau and WRI. Washington, DC: World Resources Institute.

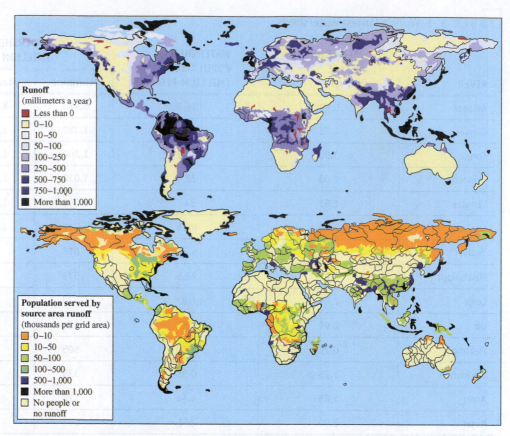

FIGURE 3.30 The runoff available in selected global watersheds compared to the population that uses the water. The consumptive use of water for agriculture, drinking water, industry, and other uses is called the water demand; areas of high population generally use a larger portion of the available runoff. The top map shows runoff-producing areas in absolute terms, with darker blue indicating areas that generate intense local runoff. This is the traditional view of the global distribution of the renewable water resource base. The bottom map shows the importance of all the world's runoff-producing areas as measured by the human population served. Thus, runoff produced across a relatively unpopulated region, like Amazonia, while a globally significant source of water to the world's oceans, is much less critical to the global water resources base than runoff across a region like south Asia.

Source: Based on World Water Assessment Programme (2009). *The United Nations World Water Development Report 3: Water in a Changing World*. Paris: UNESCO, and London: Earthscan.

Precipitation (P) is the input into the watershed on the left-hand side of the water budget equation. Water is removed from the watershed by evaporation (E), transpiration (T) [sometimes these terms are combined into the term evapotranspiration (ET)], runoff (R), and infiltration (I) into the soil. Water stored (S) in the system is also accounted for on the right-hand side of the equation. Civilization has tended to cluster and grow in or around watersheds where there are natural storage and plentiful water. Examples include the Egyptian civilization's growth around the Nile River, the location of present-day London around the Thames River, and almost any other city of significant size.

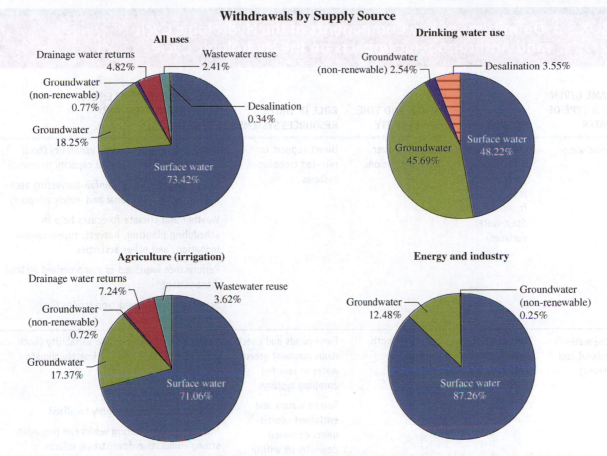

FIGURE 3.31 Global source and uses of drinking water for major water use sectors.

Source: Based on World Water Assessment Programme (2009). *The United Nations World Water Development Report 3: Water in a Changing World.* Paris: UNESCO, and London: Earthscan.

Humans have created aquifers to increase storage of water within a watershed in order to use this water for agriculture, industry, energy, or flood management, as shown in Tables 3.7 to 3.10 (see pages 183 to 184). Anthropogenic uses of water can significantly alter water-flow patterns within a watershed, even in watersheds of significant size. Consumptive human water use refers to water extracted from natural storage and water diverted or lost from the watershed.

Consumptive water use is defined as "water removed from available supplies without return to a water resources system (e.g., water used in manufacturing, agriculture, and food preparation that is not returned to a stream, river, or water treatment plant)" (Womach, 2005). Viessman and Hammer (1985) define two additional terms that help describe consumptive use:

Withdrawal use is "the use of water for any purpose which requires that it be physically removed from the source."

Nonwithdrawal use is "the use of water for any purpose which does not require that it be removed from the original source, such as water used for navigation."

BOX 3.3 Definitions of Key Components of the Hydrologic Cycle and Anthropogenic Impacts on the Hydrologic Cycle

NAME GIVEN TO A TYPE OF WATER	DEFINITION	SPACE AND TIME VARIABILITY	ROLE IN WATER RESOURCES SYSTEMS	MANAGEMENT CHALLENGES, VULNERABILITIES, AND OPPORTUNITIES
Green water	Soil moisture (nonproductive green water is evaporated from soil and open-water surfaces)	Very high over both dimensions	Direct support to rain-fed cropping systems	Highly sensitive to climate variability (both drought and flood); limited capacity to control Can be augmented by rainfall-harvesting techniques (many traditional and widely adopted) Weather and climate forecasts help in scheduling planting, harvest, supplemental irrigation, and other activities Performance improved or compromised by land management Selection of improved crop strains for climate-proofing
Blue water (natural and altered)	Net or local groundwater recharge and surface runoff, streamflow	High over both dimensions	Farm ponds and check dams augment green water in rain-fed cropping systems Source waters and entrained constituents delivered downstream within watershed	Highly sensitive to climate variability (both drought and flood) and ultimately climate change Some capacity to control Habitat management highly localized Many small engineering works can propagate strong cumulative downstream effects Poor land management heightens possibilities of flash flooding followed by dry streambeds
	Inland water systems (lakes, rivers, wetlands)	Decreased variability with increased size	Key resources over district, national, and multinational domains Important role in transport, waste management, and domestic, industrial, and agricultural sectors	Water losses through net evaporation occur naturally and through human use Legacy of upstream management survives downstream (e.g., irrigation losses, pollution) Multiple-sector management objectives may be difficult to attain simultaneously Potential upstream–downstream conflicts (human to human; human to nature), including international
	Groundwater (shallow)	Moderate over both dimensions; links to streams	Locally distributed shallow well systems serving drinking water and irrigation needs	Intimate connection to weather and climate means water yields subject to precipitation extremes Easily polluted Easily overused, resulting in temporary depletion; some loss of regional importance to oceans

NAME GIVEN TO A TYPE OF WATER	DEFINITION	SPACE AND TIME VARIABILITY	ROLE IN WATER RESOURCES SYSTEMS	MANAGEMENT CHALLENGES, VULNERABILITIES, AND OPPORTUNITIES
	Fossil ground-water (deep)	Extremely stable	Critical (and often sole) source of water in arid and semi-arid regions	Large repositories of water but with limited recharge potential
				Use typically nonsustainable, leading to declining water levels and pressure, increasing extraction costs
				Low-replenishment rates mean pollution often effectively becomes permanent
Blue water (engineered)	Diversions, including reservoirs and interbasin transfers	Stable to very stable	Critical (and often sole) source of water in arid and semi-arid regions	Large quantities of water with high recharge potential
				Modified flow regime, with positive and negative impacts on humans and ecosystems
	Reuse waters		Altered blue water balance as flows stabilized or redirected from water-rich times and places	Can destroy river fish habitat while creating lake fisheries by fragmenting habitat
				Natural ecosystem "cues" for breeding and migration removed
			Multiple uses: hydropower, irrigation, domestic, industrial, recreational, flood control	Water supplies stabilized for use when needed most by society
				Sediment trapping, leading to downstream inland waterway, coastal zone problems
				Potential for introduction of exotic species
			Secondary reuse as effluents in irrigation	Greenhouse gas emission from stagnant water
				Health problems (e.g., schistosomiasis) from stagnant water
				Social instability due to forced resettlement
Grey water	Recycled, reus-able wastewater from residential, commercial, or industrial bathroom sinks, bathtubs, shower drains, and clothes washing	Stable	May offset consumptive use for nonpotable water such as agriculture and land-scaping or some industrial process waters	May be used for groundwater recharge
				May augment surface water reservoirs
				May be used to prevent saltwater intrusion in coastal areas
				Technical feasibility and risk still to be fully determined

(Continued)

BOX 3.3 Definitions of Key Components of the Hydrologic Cycle and Anthropogenic Impacts on the Hydrologic Cycle (Continued)

NAME GIVEN TO A TYPE OF WATER	DEFINITION	SPACE AND TIME VARIABILITY	ROLE IN WATER RESOURCES SYSTEMS	MANAGEMENT CHALLENGES, VULNERABILITIES, AND OPPORTUNITIES
Virtual water	Not an additional water system element	Stable, but linked to fluctuations in global economy	Water embodied in production of goods and services, typically with crops traded on the international market Not explicitly recognized as a water resources management tool until recently	Can implicitly offload water use requirements from more water-poor to more water-rich locations Particularly important where rain-fed agriculture is restricted and irrigation relies on rapidly depleting fossil groundwater sources
Desalination		Stable	Augmentation in water-scarce areas	Costly, special-use water supply, technologies rapidly developing for cost effectiveness

Source: Adapted from World Water Assessment Programme (2009). *The United Nations World Water Development Report 3: Water in a Changing World*. Paris: UNESCO, and London: Earthscan, and U.S. EPA.

Global water consumption rates and forecasted consumption rates are shown in Figure 3.32 (see page 185). Trends of water use for public supply, agriculture, electricity production, and other industries in the United States between 1950 and 2000 are shown in Figure 3.33 (see page 186). Water demand in the United States peaked in the mid-1970s, at about the time water conservation, water recycling, and water regulations were established. In contrast, world water demand is expected to continue to climb into the foreseeable future. A portion of the worldwide demand is due to the related demand for more food to feed more people, for the globally increasing energy demand, and for other industrial needs. Water required for various energy generation scenarios is illustrated in Figure 3.34 (see page 186).

We can add the consumption term, C, to the output side of the water balance equation:

$$P = R + I + E + T + S + C \tag{3.13}$$

In some cases, water can be extracted from other watersheds and rerouted to a different watershed via piping or aqueducts. For instance, the Romans used this method to increase the availability of water in their cities. In these cases, consumption from one watershed can lead to a net input of water into another watershed. Human interactions with the hydrologic cycle change the ecosystems within the affected watersheds. For instance, consumptive withdrawals from the Colorado River have been so great that in some years no flow from the Colorado River has reached the Pacific Ocean.

Engineers are currently examining water availability and have identified many water-scarce areas in both the developed and developing worlds where there are

TABLE 3.7 Typical water use, calorie return, and economic return on selected agricultural products

PRODUCT	KILOGRAM PER CUBIC METER	$ PER KILOGRAM	$ PER CUBIC METER	PROTEIN GRAMS PER CUBIC METER	KILOCALORIES PER CUBIC METER
Cereal					
Wheat	0.2–1.2	0.2	0.04–0.24	50–150	660–4,000
Rice	0.15–1.6	0.31	0.05–0.18	12–50	500–2,000
Maize	0.30–2.00	0.11	0.03–0.22	30–200	1,000–7,000
Legumes					
Lentils	0.3–1.0	0.3	0.09–0.30	90–150	1,060–3,500
Fava beans	0.3–0.8	0.3	0.09–0.24	100–150	1,260–3,360
Groundnut	0.1–0.4	0.8	0.08–0.32	30–120	800–3,200
Vegetables					
Potatoes	3.0–7.0	0.1	0.3–0.7	50–120	3,000–7,000
Tomatoes	5.0–20.0	0.15	0.75–3.0	50–200	1,000–4,000
Onions	3.0–10.0	0.1	0.3–1.0	20–67	1,200–4,000
Fruits					
Apples	1.0–5.0	0.8	0.8–4.0	Negligible	520–2,600
Olives	1.0–3.0	1.0	1.0–3.0	10–30	1,150–3,450
Dates	0.4–0.8	2.0	0.8–1.6	8–16	1,120–2,240
Other					
Beef	0.03–0.1	3.0	0.09–0.3	10–30	60–210
Fish (aquaculture)	0.05–1.0		0.07–1.35	17–340	85–1,750

Source: Comprehensive Assessment of Water Management in Agriculture (2007). *Water for Food, Water for Life: A Comprehensive Assessment of Water Management in Agriculture*. London: Earthscan, and Colombo: International Water Management Institute.

TABLE 3.8 Water consumption per metric ton of product produced

PRODUCT	WATER USE (m³/tonne)
Paper	80–2,000
Sugar	3–400
Steel	2–350
Petrol	0.1–40
Soap	1–35
Beer	8–25

Source: Based on Margat., J., and Andreassian, V. (2008). *L'Eau, les Idees Receus*. Paris, Editions le Cavalier Bleu.

TABLE 3.9 Typical household water use in the United States before and after installation of low flow appurtenances and water conservation measures

USE	GALLONS PER CAPITA: BEFORE	GALLONS PER CAPITA: AFTER	SAVINGS: GALLON(S) DAILY WATER USE
Showers	11.6	8.8	2.8
Clothes washer	15.0	10.0	5.0
Toilets	18.5	8.2	10.3
Dishwasher	1.0	0.7	0.3
Baths	1.2	1.2	0.0
Leaks	9.5	4.0	5.5
Faucets	10.9	10.8	0.1
Other domestic uses	1.6	1.6	0.0
Total	69.3	45.3	24.0

Source: Mays, L. (2011). *Water Resources Engineering: Second Edition*. John Wiley & Sons, Inc. Hoboken, NJ. p. 347.

TABLE 3.10 Water use for energy production

POWER PROVIDER	GALLONS EVAPORATED PER kWh AT THERMOELECTRIC PLANTS	GALLONS EVAPORATED PER kWh AT HYDROELECTRIC PLANTS	WEIGHTED GALLONS EVAPORATED PER kWh OF SITE ENERGY
Western Interconnect	0.38 (1.4 L)	12.4 (47.0 L)	4.42 (16.7 L)
Eastern Interconnect	0.49 (1.9 L)	55.1 (208.5 L)	2.33 (8.8 L)
Texas Interconnect	0.44 (1.7 L)	0.0 (0.0 L)	0.43 (1.6 L)
U.S. Aggregate	0.47 (1.8 L)	18.0 (68.0 L)	2.00 (7.6 L)

Source: Mays, L. (2011). *Water Resources Engineering: Second Edition*. John Wiley & Sons, Inc. Hoboken, NJ. p. 347.

existing or expected water shortages. A sustainable level of consumption is one where the net storage term is greater than zero for a watershed. When the consumption terms increase from human demand to the point where the storage term becomes negative in value, then there is nonsustainable water use in the watershed. Archaeologists believe that water shortages have been a significant factor in the collapse of many civilizations. Today in regions where consumptive withdrawals are greater than inputs into the watershed, water must be acquired from beyond the watershed boundaries, or water-rationing measures can be adopted to prevent the degradation and disruption of ecosystems and human systems within the watershed of concern.

The human uses of water may be part of a growing concern about global water scarcity, or the lack of available water supplies to meet human demands, as well as maintaining balanced ecosystem requirements. Human population growth, economic development, and advancing societal needs may outstrip the water available

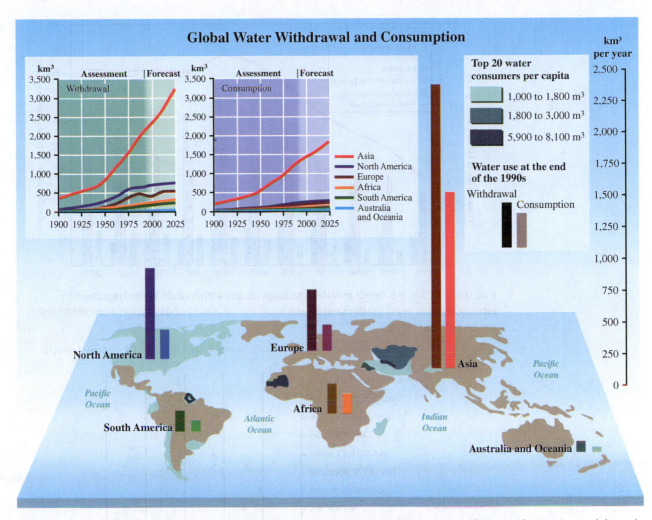

FIGURE 3.32 Global water use and consumption trends from 1900 to 2000 and expected forecast of water demand through 2025. The Asian continent has the highest rates of both water withdrawal and consumptive use.

Source: Based on UNEP/GRID-Arendal, *Vital Water Graphics, Water withdrawal and consumption*, www.grida.no/graphicslib/collection/vital-water-graphics. Data from Igor A. Shiklomanov, SHI (State Hydrological Institute, St. Petersburg), and UNESCO, Paris, 1999; *World Resources 2000–2001: People and Ecosystems: the Fraying Web of Life*, WRI, Washington DC, 2000; Paul Harrison and Fred Pearce, *AAAS Atlas of Population 2001*, AAAS, University of California Press, Berkeley.

from current surface water and groundwater resources. "The rapid global rise in living standards combined with population growth presents the major threat to sustainability of water resources and environmental services." (World Water Assessment Programme, 2009). Technological advances in water development and energy development have the potential to help meet the growing water demand. At the same time, technological and industrial development may also increase the demand for water associated with agriculture, energy demand, and industrialization, as illustrated in Tables 3.7 through 3.10. Engineers, scientists, and policymakers must work across regional and national boundaries to meet the future demand for water in many water-scarce areas, such as the American Southwest and all along the Mediterranean Sea, as shown in Figures 3.35 through 3.37 (see pages 187 to 188) and discussed in Box 3.4 (see pages 188 to 190).

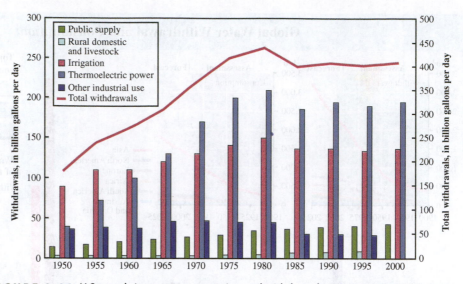

FIGURE 3.33 U.S. trends in water consumption and withdrawals. Water consumption increased steadily with the demand for electricity demand and public supplies from 1950 until peaking around 1980.

Source: From Hendrix, M. and Thompson, G.R. 2015. *Earth 2, Second Edition.* Stamford, CT: Cengage Learning.

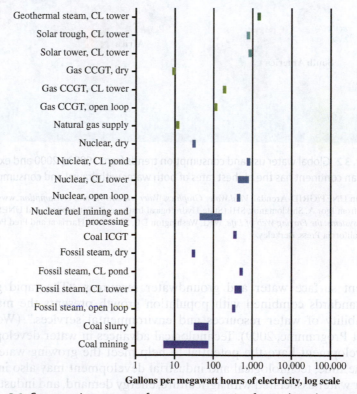

FIGURE 3.34 Consumptive water use for energy generation from selected generation scenarios in the United States in 2006. As shown in Figure 3.33, electricity supply is the largest use of water in the United States. This graphic shows the demand for closed-loop cooling (CL), combined cycle gas turbine (CCGT), and integrated gasification combined cycle (ICGT) generation of electricity.

Source: Based on World Water Assessment Programme (2009). *The United Nations World Water Development Report 3: Water in a Changing World.* Paris: UNESCO, and London: Earthscan.

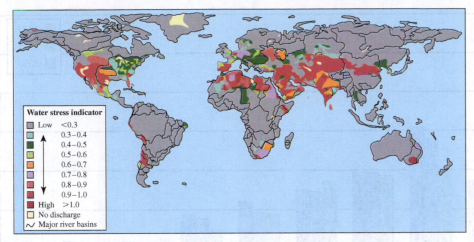

FIGURE 3.35 Water stress levels of major watersheds, worldwide in 2002.

Source: Based on World Water Assessment Programme (2009). *The United Nations World Water Development Report 3: Water in a Changing World*. Paris: UNESCO, and London: Earthscan.

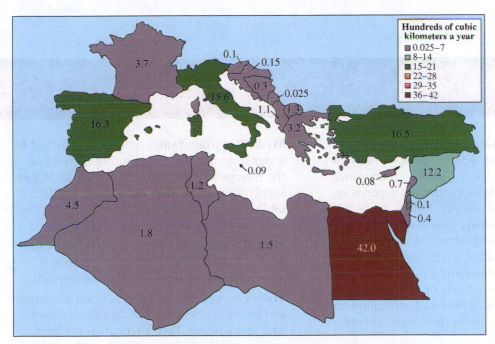

FIGURE 3.36 The values represent the overconsumption of available water resources in countries bordering the Mediterranean Sea for all uses of water from 2000 to 2005.

Source: Based on World Water Assessment Programme (2009). *The United Nations World Water Development Report 3: Water in a Changing World*. Paris: UNESCO, and London: Earthscan.

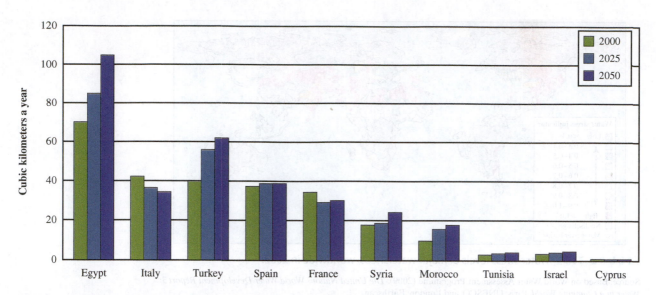

FIGURE 3.37 Water demand and consumption are likely to continue to increase for most Mediterranean countries in the foreseeable future.

Source: Based on World Water Assessment Programme (2009). *The United Nations World Water Development Report 3: Water in a Changing World.* Paris: UNESCO, and London: Earthscan.

BOX 3.4 Malta's Water Worries

Malta is a small island nation in the Mediterranean Sea. It is home to about 400,000 people, most of whom receive freshwater from a limited groundwater supply. Every sector of Malta's economy depends on this limited groundwater supply. Water needs for agriculture, home use, industry, and tourism threaten to overwhelm the dwindling supply of water for this island nation. Malta's limited water resources make the entire population particularly susceptible to climate change. Nonetheless, much of the population remains unaware of these issues and still clings to traditional cultural myths about their water supply.

Maltese water myths:

Myth: The natural occurring limestone removes salt from the groundwater, and thus as sea levels increase, there will be more freshwater. **Fact:** The limestone does not remove salt from the seawater. As sea level increases, the amount of available groundwater will decrease.

Myth: The groundwater table is replenished by water from the Nile River or other rivers on the European continent, and the groundwater supply is inexhaustible. **Fact:** The groundwater table is not linked to water from the Nile. The groundwater table is already being depleted.

Myth: Deeper wells will provide better water quality. **Fact:** The failure rate of all wells is increasing due to deteriorating water quality.

Myth: Crops grown with reused water are inferior and have a shorter shelf life than those grown with groundwater. **Fact:** There are some legitimate concerns or risks to using recycled water for agricultural use; however, there are also risks with using very saline water from inadequate wells.

The groundwater on Malta is stored in limestone caves, tunnels, and reservoirs about 97 meters beneath the surface (see Figure 3.38). These ancient water repositories are called the *Ta' Kandja Galleries*. Freshwater from runoff and infiltration lies only 10 meters above the denser saltwater that has infiltrated these galleries. Due to climate change and rising sea levels, the level of saltwater is rising, decreasing the available supply of freshwater that is suitable for drinking and irrigation.

Scheme showing a Ghyben-Herzberg (floating) groundwater body in an island

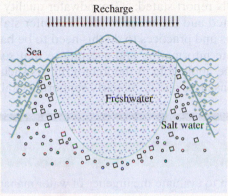

FIGURE 3.38 Illustration of the conceptual freshwater cone underlying Malta.

Source: Based on FAO (2006). *Malta Water Resources Review*. Food and Agriculture Organization of the United Nations, Rome.

The Maltese government has been working with the United Nations and scientists to determine the extent of and possible solutions to Malta's water worries. In April 2007, Malta's water worries made worldwide headlines when Malta water services engineer Paul Micallef told a reporter for the BBC that the rising sea would make the galleries very difficult to operate in the future. "According to recent studies, the water will rise about 96 cm by the year 2100. This will affect the availability of groundwater as the interface between seawater and freshwater will actually rise by about 1 m and the high salinity levels will be close to our extraction sources."

Decreasing water supplies threaten Malta's food supply. Urbanization, which has increased runoff, has led the government to create groundwater protection areas (Figure 3.39) to prevent development from encroaching upon groundwater recharge (Figure 3.40). Maltese farmers primarily use unregulated wells (called boreholes), which tap

Groundwater Protected Areas in the Maltese Islands

FIGURE 3.39 Area of Malta that is protected by the government for groundwater infiltration.

Source: Based on FAO (2006). *Malta Water Resources Review*. Food and Agriculture Organization of the United Nations, Rome.

Map Showing the Extent of Urbanization, 2000

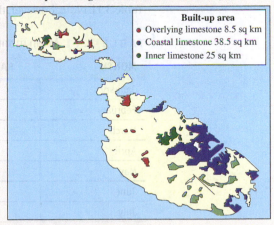

Built-up area
- Overlying limestone 8.5 sq km
- Coastal limestone 38.5 sq km
- Inner limestone 25 sq km

FIGURE 3.40 Areas of Malta that are already urbanized and lead to low rates of infiltration.

Source: Based on FAO (2006). *Malta Water Resources Review*. Food and Agriculture Organization of the United Nations, Rome.

(Continued)

BOX 3.4 Malta's Water Worries (*Continued*)

into the groundwater reservoirs, to irrigate their crops. Salty water prevents plants from obtaining the nutrients they need from the soil for growth. Plants and trees will die if they are irrigated with water that contains too much salt. David and Mary Mallia, organic farmers interviewed for the 2007 BBC report, observed significant changes in their orchard productivity in recent years. "Since the rainfall has become less, the salinity is becoming higher and higher." In the past, normal salt concentration in the irrigation water would be 200 microsiemens. (A microseimen is a measure of electrical conductivity that is directly proportional to the amount of salt in the water.) "This year (2007), with the lack of rain, it went up to 4,000. It's not good for irrigation. If you water your trees with this water, it will kill them."

A panel of engineers and scientists with the Food and Agriculture Organization (FAO) of the United Nations (UN) reviewed the state of Malta's water resources. These engineers and scientists are working on a plan to develop a policy for water management that will be socially and politically acceptable and geared to tackling the complex water-related challenges. This review showed that, although the demand for groundwater is outstripping supply, water conservation measures could be effective. The UN report stated: "Groundwater quality can be protected and the mean sea-level aquifer stabilized. Policies and practices to do this need to be based on accurate information and acceptance that solutions must be applicable in the long-term. As such, political consensus and cross-party support is vital."

EXAMPLE 3.3   Malta Water Budget

We can apply the water budget equation to evaluate the urgency of water management measures for the island nation of Malta.

We first need to determine the average annual input, or precipitation, of water. We can estimate this by calculating the average rainfall that occurs over a year on the islands listed in Table 3.11. Note, however, that reported rainfall varies over the island, as shown in Table 3.12.

TABLE 3.11 Mean monthly values of rainfall and temperature in Malta

MONTH	RAINFALL (mm)	MAX TEMP °C	MIN TEMP °C
January	86.4	14.9	10.0
February	57.7	15.2	10.0
March	41.8	16.6	10.7
April	23.2	18.5	12.5
May	10.4	22.7	15.6
June	2.0	27.0	19.2
July	1.8	29.9	21.9
August	4.8	30.1	22.5
September	29.5	27.7	20.9
October	87.8	23.9	17.7
November	91.4	20.0	14.4
December	104.3	16.7	11.4

Source: Based on FAO (2006). *Malta Water Resources Review*. Food and Agriculture Organization of the United Nations, Rome.

TABLE 3.12 Reported rates of rainfall, evapotranspiration, and effective rainfall (mm) for Malta, 1956–1991

AUTHOR	RAINFALL	EVAPOTRANSPIRATION	RUNOFF	EFFECTIVE RAINFALL
Morris (1952)/ Edelmann (1968)	522	392		130
ATIGA (Martin)	587	475		95
ATIGA (Verho-even/Gessel)	536	439		97
ATIGA (WWD Data)	551	431		120
FAO (2006)	587	437		150
Spiteri Staines (1987)	508	356	30	122
BRGM (1991)	551	348		203

Source: Based on FAO (2006). *Malta Water Resources Review*. Food and Agriculture Organization of the United Nations, Rome.

$$\text{Precipitation} = \frac{(86.4 + 57.7 + 41.8 + 23.2 + 10.4 + 2.0 + 1.8 + 4.8 + 29.5 + 87.8 + 91.4 + 104.3)}{12}$$

$$= \frac{45 \text{ mm}}{\text{month}}$$

We can convert the average monthly precipitation values expressed in mm/month to a yearly value expressed in m/year:

$$\frac{45 \text{ mm}}{\text{month}} \times \frac{12 \text{ months}}{\text{year}} = \frac{540 \text{ mm}}{\text{year}} \times \frac{1 \text{ m}}{1{,}000 \text{ mm}} = \frac{0.54 \text{ m}}{\text{year}}$$

We will assume this rainfall occurs equally over all the area of the islands, which is a reasonable assumption since the total land area is relatively small compared to much larger nations:

$$\text{Area} = 316 \text{ km}^2 = 316 \times \left(\frac{10^3 \text{ m}}{\text{km}}\right)^2 = 316 \times 10^6 \text{ m}^2$$

$$\text{Average yearly rainfall over Malta} = \frac{0.54 \text{ m}}{\text{year}} \times 316 \times 10^6 \text{ m}^2$$

$$= 171 \times 10^6 \frac{\text{m}^3}{\text{year}}$$

We might first ask how much water would be available per person if all the rainfall could be collected and used. Water consumption is normally expressed in liters per day, so we will convert our answer to liters per day:

$$\text{Population} = 398{,}000 + 2{,}400 \text{ per year}$$

The maximum consumption if all the rainfall were to be used is

$$\frac{\text{Average rainfall}}{\text{Population}} = \frac{\left(171 \times 10^6 \frac{\text{m}^3}{\text{year}}\right)}{398{,}000 \text{ people}} = \frac{430 \text{ m}^3}{\text{year}}$$

Converting to liters per day:

$$\frac{430\text{ m}^3}{\text{year}} \times \frac{1{,}000\text{ L}}{\text{m}^3} \times \frac{1\text{ year}}{365\text{ days}} = \frac{1{,}200\text{ L}}{\text{day}}$$

This is more than enough water to meet demand, but it does not account for natural losses in the water budget equation. We will now examine those losses.

Losses include runoff and evapotranspiration, listed in Table 3.13 in hm³/yr. One cubic hectometer = 1×10^6 m³.

$$\text{Yearly surface runoff to the sea} = 24\text{ hm}^3 = 24 \times 10^6\text{ m}^3$$

Yearly actual evapotranspiration (assumed to be 68% of the total surface water)

$$= 105\text{ hm}^3 = 105 \times 10^6\text{ m}^3$$

Not all the groundwater remains available for use. The groundwater mixes with seawater and also flows into the sea below the surface. The infiltration, I, and removal of water from the watershed by subsurface groundwater discharge to the sea are shown in Table 3.13. The water balance of each individual aquifer in Malta is shown in Table 3.14.

TABLE 3.13 Balance of flow into and from the groundwater aquifers in Malta

INFLOW	DESCRIPTION	hm³/YEAR	COMMENTS
A	Precipitation	174	Based on an average annual rainfall of 550 mm
B	Surface runoff to the sea	24	Based on a variable catchment area runoff coefficient (excluding coastal built-up areas)
C	Actual evapotranspiration	105	Assumed as 68% of the total surface water
D	Natural aquifer recharge	45	B and C deducted from A
E	Artificial recharge from leaks	12	Estimated inflow from potable water and sewage network leakages
F	Total groundwater inflow	57	Sum of variables D and E
Outflow			
G	Water Services Corporation (WSC) groundwater abstraction	16	Official WSC extraction for hydrological year 2002/2003
H	Private groundwater outflow	15	Estimate based on water demand of various sectors (industry and agriculture)
I	Subsurface discharge to the sea	23	Estimate based on groundwater modeling
J	Total groundwater outflow	54	Sum of variables G, H, and I
Balance			
K	Total groundwater inflow	57	Equal to variable F
L	Total groundwater outflow	54	Equal to variable J
M	Balance	3	Inflow (K) less outflow (L)

Source: Based on FAO (2006). *Malta Water Resources Review*. Food and Agriculture Organization of the United Nations, Rome.

TABLE 3.14 Water balance of flow into and from individual aquifers in Malta

GROUNDWATER CODE	GROUNDWATER BODY NAME	SIZE (km²)	INFLOW	OUTFLOW (hm²)	BALANCE	MAJOR EXTRACTION
MT001	Malta main mean sea level	216.6	34.27	36.65	−2.38	Abstraction for potable and agricultural purposes
MT002	Rabat-Dingli perched	22.6	4.64	4.62	0.02	Abstraction for agricultural purposes
MT003	Mgarr-Wardija perched	13.7	2.86	3.46	−0.59	Abstraction for potable and agricultural purposes
MT005	Pwales coastal	2.8	0.69	0.69	0.00	Abstraction for agricultural purposes
MT006	Mizieb mean sea level	5.2	1.11	0.96	0.15	Abstraction for potable and agricultural purposes
MT008	Mellieha perched	4.5	0.75	0.53	0.22	Abstraction for agricultural purposes
MT009	Mellieha coastal	2.9	0.69	0.38	0.31	Abstraction for agricultural purposes
MT010	Marfa coastal	5.5	0.89	0.62	0.27	Abstraction for agricultural purposes
MT011	Mqabba-Zurrieq perched	3.4	0.50	n/a	n/a	Abstraction for agricultural purposes
MT012	Comino mean sea level	2.7	0.52	0.30	0.22	Abstraction for agricultural purposes
MT013	Gozo mean sea level	65.8	8.66	9.78	−1.12	Abstraction for potable and agricultural purposes
MT014	Ghajnielem perched	2.7	0.73	0.34	0.39	Abstraction for agricultural purposes
MT015	Nadur perched	5.0	1.15	0.58	0.57	Abstraction for agricultural purposes
MT016	Xaghra perched	3.0	0.71	0.33	0.38	Abstraction for agricultural purposes
MT017	Zebbug perched	0.4	0.10	0.03	0.07	Abstraction for domestic purposes
MT018	Victoria-Kercem perched	1.5	0.39	0.14	0.25	Abstraction for domestic purposes

Source: Based on FAO (2006). *Malta Water Resources Review*. Food and Agriculture Organization of the United Nations, Rome.

Rearranging and grouping evaporation and transpiration into one terms yield

$$S = P - R - ET - I = 171 \times 10^6 \frac{m^3}{year} - 24 \times 10^6 \frac{m^3}{year} - 105 \times 10^6 \frac{m^3}{year} - 23 \times 10^6 \frac{m^3}{year}$$

$$S = 19 \times 10^6 \frac{m^3}{year}$$

The remaining term, S, in the water budget equation accounts for all water available in *both* groundwater and surface water storage.

Once again we can estimate the available water supply if all the stored water were used by each person in Malta:

$$\text{Annual total storage/population} = \frac{\left(19 \times 10^6 \frac{m^3}{year}\right)}{398,000 \text{ people}} = \frac{47 \text{ m}^3}{year} \times \frac{1,000 \text{ L}}{m^3} \times \frac{1 \text{ year}}{365 \text{ days}}$$

$$= 131 \text{ L per day}$$

TABLE 3.15 Water use in Malta by sector and source (10^3 m^3)

USE	WATER SERVICES CORPORATION				PRIVATE		TOTAL
	BILLED	UNBILLED	GROUNDWATER	RO	TREATED EFFLUENT	RUNOFF HARVESTING	
			REVERSE OSMOSIS DESALINIZATION				
Domestic	12,620	3,686	1,000			2,000	19,306
Tourism	1,134	331	500	1,000			2,965
Farms	1,336	390	500				2,226
Agriculture			14,500		1,500	2,000	18,000
Commercial	1,247	364					1,611
Industrial	941	275	1,000		500		2,716
Government	818	239					1,057
Others	869	254					1,123
Total consumption	18,965	5,540	17,500	1,000	2,000	4,000	49,005
Real losses		9,636					9,636
Total + losses	18,965	15,176	17,500	1,000	2,000	4,000	58,641
WSC							
Total apparent losses	5,540	16%					
Total loss	15,176	44%					

Source: Based on FAO (2006). *Malta Water Resources Review.* Food and Agriculture Organization of the United Nations, Rome.

Therefore, 131 liters per day is the maximum consumptive use of water that is sustainable for Maltese society, without additional sources of water or water reuse.

In Table 3.15, we see that the total water use for Malta in 2003 was 58.641×10^6 m^3/yr. This equates to a per capita consumption of 400 L per capita day.

We can compute the percent over consumption of the available water resources in Malta:

$$\frac{\text{Actual consumption}}{\text{Maximum sustainable consumption}} \times 100\%$$

$$= \frac{58.641 \times 10^6 \text{ m}^3/\text{yr}}{19 \times 10^6 \text{ m}^3/\text{yr}} \times 100\% = 309\%$$

Therefore, actual consumption is 309% greater than the sustainable withdrawals expected from rainfall.

We can compare the available sustainable water supply to actual water demand in Malta. In Example 3.3, it is shown that the water demand has outpaced the rate of replenishment of water to the groundwater supply on the island. Agricultural, industrial, and residential water demand puts a heavy burden on the limited water supply. Increasing water demands (see Figure 3.41) and a slowly changing climate that is

greenhouses are due to the annual temperature variations. The changing climate that in part may increase Malta's water demand. This is known as a monitoring spiral of causes and carry and will be exploited in general testing in research chapters.

What is expected which is slowly being adopted in Malta is adapting to learn infrastructure and technology assistance. Water resources controlled treatment of restwaters. Because water is valuable to Malta, as much water is used as cheap and relative recycled domestic flows within less but consume water must be.

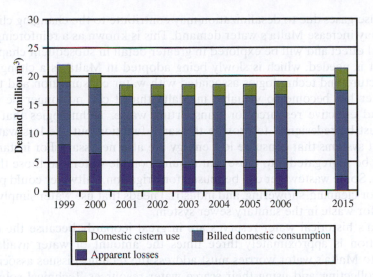

FIGURE 3.41 Maltese water consumption variations by year and expected demand.

Source: Based on FAO (2006). *Malta Water Resources Review*. Food and Agriculture Organization of the United Nations, Rome.

becoming dryer have caused Malta to recognize the close connection between water supply, energy consumption, and resource use and the importance of the water–energy nexus. (A nexus is the point of interaction of one system with another.)

Malta has begun building and using desalination plants to supplement its natural water resources (see Figure 3.42). Desalinization is an energy-intensive process that removes enough salt from seawater to make it suitable for drinking. Malta has had to increase the importation of oil from Middle Eastern countries; a significant portion of this energy must be used for desalinization. Increasing the use of energy has the potential to increase Malta's greenhouse gas emissions associated with fossil fuels. The greenhouse gases are regulated in the European Union and make the process of desalinization even more expensive. Furthermore, the emission of

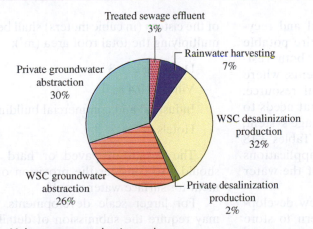

FIGURE 3.42 Maltese water production and sources.

Source: Based on FAO (2006). *Malta Water Resources Review*. Food and Agriculture Organization of the United Nations, Rome.

greenhouse gases due to desalinization may contribute to the changing climate that in part may increase Malta's water demand. This is known as a reinforcing spiral of cause and effect and will be explored in greater detail in subsequent chapters.

What is needed, which is slowly being adopted in Malta, is a change in basic infrastructure and technologies associated with water consumption and treatment. Freshwater has become so valuable in Malta that it can no longer be used as a cheap and effective resource for transporting waste. Technologies that consume water must be redesigned to provide the same function but use less water. Water treatment systems that consume less energy are also necessary. For instance, water uses may be segregated into those that require potable water and those that do not require it. Some wastewater can be reused for irrigation. Saltwater could potentially be used for cleaning, showering, and industrial needs that use water simply as a conveyance for waste in the sanitary sewer system.

Malta's historical water reserves are in great jeopardy because the estimated consumption is approximately three times the amount of water available. Any solution to Malta's water worries must address the technical issues associated with storing, collecting, and using their scarce water resources. Technical solutions will not be sufficient to prevent a water disaster in Malta. Social issues, such as rejecting long-held myths about their water supply, providing educational resources about the urgency of the problem, and ensuring government oversight and monitoring of the water resources must also be addressed. Private boreholes and wells are difficult to manage and regulate effectively. The FAO recommends significant investment in household rainwater collection systems, but private implementation is difficult to enforce (see Box 3.5). Neither technical nor social issues are likely to be addressed if the economic questions and concerns are not taken into account. Malta's economy relies on industrial productivity and a tourist industry, both of which create significant demands on Malta's limited water supply (Tables 3.16 and 3.17). However, if the

BOX 3.5 FAO Design Guidance and Recommendations for Malta

Rainwater runoff should be collected and recycled (for those uses that do not require potable water). This requirement applies to both residential and nonresidential developments, where the collected runoff may be a useful resource. Collection also reduces the amount that needs to be dealt with by the stormwater drainage system, and so may have wider benefits (see Tables 3.18 through 3.20). Plans submitted with applications should show the proposed location of the water cistern.

The FAO (2006) proposed that new developments be provided with a water cistern to store rainwater runoff from the built-up area. The volume of the cistern (in cubic meters) shall be calculated by multiplying the total roof area (m^2):

Dwelling × 0.3

Villas × 0.45

Industrial and commercial buildings × 0.45

Hotels × 0.6

The design of paved or hard surface areas should also consider the provision of water catchment for surface water runoff.

For larger scale developments, the authority may require the submission of details on how the water collection is to be used.

TABLE 3.16 Agricultural demand of water for animal production

ANIMAL CATEGORY	NUMBER OF ANIMALS	DAILY WATER DEMAND (liters/day)	ANNUAL DEMAND (m³)
Pigs	73,067	Summer: 20.25 / Winter: 13.5	450,000
Cattle			
Calves 1 year	4,909	Summer: 139.5 / Winter: 94.5	209,500
Cattle 1–2 years	4,983	Summer: 182.25 / Winter: 137.25	290,500
Cattle 2 years	8,093	Summer: 195.75 / Winter: 150.75	511,800
Total	17,985		1,011,800
Sheep	14,861	Summer: 18.0 / Winter: 13.5	85,500
Goats	5,374	Summer: 18.0 / Winter: 13.5	31,000
Rabbits	55,254	Summer: 3.0 / Winter: 1.5	45,000
Poultry			
Layers (others)	756,288	Summer: 0.36 / Winter: 0.32	94,000
Broilers	1,184,157	Summer: 0.22 / Winter: 0.28	108,000
Total			202,000
Equine	853	Summer: 58.5 / Winter: 49.5	16,800
Total			1,842,100

Source: Based on FAO (2006). *Malta Water Resources Review*. Food and Agriculture Organization of the United Nations, Rome.

TABLE 3.17 Water demand in Malta by the tourism industry

	1997/1998 (m³)	1998/1999 (m³)	1999/2000 (m³)
Total annual water production	40,772,926	37,963,808	36,604,128
Production to satisfy tourist demand	3,669,563	3,416,742	3,294,371
Per capita demand per day	0.324	0.293	0.321

Source: Based on FAO (2006). *Malta Water Resources Review*. Food and Agriculture Organization of the United Nations, Rome.

environmental issues associated with the rising sea level due to climate change and anthropogenic depletion of the groundwater supplies are not addressed, the entire Maltese economy and society could be jeopardized in the near future. Sustainable use of Malta's water resources will require improved governance, improved awareness, staged and adaptive implementation, demand management, supply augmentation, the ensuring of equity and justice, and targeted interventions.

The FAO report suggests that significant progress must be made immediately to protect Malta's fragile water resources. The report describes four demand scenarios (based on current levels of demand and the subsequent projections that have emerged from discussions with various stakeholders). It also discusses different strategies for meeting future demand and achieving the vision shown in Tables 3.18 and 3.19.

The water crisis on the small island of Malta brings up issues other nations and communities must address in the near future. Large cities in arid regions of the United States, such as Los Angeles, Las Vegas, and Phoenix, are already facing several water shortages despite importing huge quantities of water from adjacent watersheds. Even Atlanta in the southeastern United States, a city located in a relatively wet climate, faces severe water shortages. Atlanta's water shortages have been created by an unusual geology. A large granite monolith beneath Atlanta prevents the city from easily tapping into groundwater supplies. Surrounding cities

TABLE 3.18 Simplified plan for groundwater development in Malta

	STAGE 1 (YEARS PRIOR TO THE EARLY 1980s)	STAGE 2 (EARLY 1980s TO MID-2000s)	STAGE 3 (MID-2000s TO LATE 2000s)	STAGE 4 (LATE 2000s ONWARD)
Stages	Low-level irrigation meetings, local demand for vegetables, olives, grapes, wine, etc.	Agrarian boom initially by availability of drilling technology and subsequently sustained by EU-accession, land-based subsides. Large increase in area under olive cultivation and vineyards.	Symptoms of groundwater overexploitation increasingly apparent and starting to affect yields and crop quality.	Decline in overall groundwater use. Increased use of unconventional sources for irrigation.
Characteristics	Total irrigated area less than 500 ha. Use of spring water for irrigation and gradual development of the perched aquifer using shallow wells.	Total irrigated area increased to more than 250 ha. Large increase in use of supplemental irrigation. Deep boreholes used to exploit sea-level aquifers.	Peak reached in irrigated area and agricultural water use as a result of declining groundwater quality and introduction of regulations and tariffs.	Low-value crops no longer profitable. Area under irrigation declines. Agri-environment planning becomes the norm.
Impacts and sustainability	High-level sustainability. Some decline in water levels and spring flows from perched aquifer. Onset of a nitrate pollution problem.	Impacts of groundwater regulation, awareness campaigns, and improved planning starting to have an impact on sustainability.	Impacts of groundwater regulation awareness campaigns and improved planning starting to have an impact on sustainability.	A balance achieved between groundwater inflow and outflow. A strategic reverse created. Water quality improving steadily.
Interventions	Limited government support. No groundwater regulation.	Local market not protected. Groundwater regulation (including tariffs) and catchment planning introduced.	Local market not protected. Groundwater regulation (including tariffs) and catchment planning introduced.	Adaptive management of groundwater becomes the norm. Good environmental awareness established at all levels.

Source: Based on FAO (2006). *Malta Water Resources Review.* Food and Agriculture Organization of the United Nations, Rome.

TABLE 3.19 Expected demand and plans for Malta's water resources

WATER DEMAND SCENARIOS	POTENTIAL DEMAND SCENARIOS BASED ON CURRENT TRENDS	WATER SUPPLY STRATEGIES	STRATEGIES FOR MEETING CURRENT AND FUTURE DEMANDS (i.e., 2015)	ECONOMIC/SOCIAL IMPLICATIONS
I	Municipal demand remains fairly constant at current values or increases at about 1 to 2%/year, while agricultural demand increase reaches a maximum not exceeding 21 hm³ as projected.	I	Agriculture is given priority over the use of groundwater, and, consequently, the urban supply is increasingly sourced from RO plants.	The WSC has to source its supply increasingly from reverse osmosis desalination plants, with resulting price increases in the urban sectors. The quality of the domestic supply will increase. However, the country will be fully dependent on RO plants for the production of potable water, leaving the country vulnerable to fluctuations in the industrial and tourism sectors, which would be expected to reduce its economic competitiveness.
		II	No action is taken, and groundwater abstraction remains unregulated.	Over-abstraction will result in an increase in the salinity of groundwater abstracted from the sea-level aquifers and in the drying up of perched aquifers in the summer period. Degeneration in quality will make groundwater unsuitable for direct utilization, and extra treatment costs will be incurred by all sectors in the long term as groundwater becomes progressively unusable for all sectors.
		III	A reduction in groundwater abstraction is implemented in order to achieve a sustainable abstraction strategy allowing the setting up of a strategic groundwater reserve. The available abstractable groundwater quota is then allocated on a 50/50 basis between the WSC and all other users. Option involving artificial recharge of groundwater and improved rainwater harvesting will also have to be implemented in order to augment groundwater availability. Agri-environment schemes and smart irrigation techniques are used to encourage low-water-using farming systems.	The WSC will have to reduce the proportion of abstracted groundwater, while agriculture and industry will have to substantially increase the amount of recycled water used. Tariffs for the domestic/commercial sectors will be utilized to manage the sectoral demand. Agriculture will have to absorb a proportion of the cost of treating sewage effluent—possibly additional costs of desalination on the effluent treated to tertiary level. Groundwater will be viewed as a national strategic resource and will have a potential negative impact on the livelihood of some agricultural users. These drawbacks will be outweighed by the potential benefits to the economy as a whole. In the long term, better groundwater quality will result in decreasing treatment costs for all sectors.
II	Municipal demand remains fairly stable at current values or increases at about 1 to 2%/year, while agricultural demand decreases to pre-EU accession levels (of 10 hm³/year) driven mainly by market forces.	I	Agriculture is given priority over the use of groundwater, with the potable supply being increasingly sourced from RO plants. Effluent from wastewater treatment plants viewed primarily as an option to supplement water supply to the agricultural/industrial sector, thus further reducing the pressures on groundwater.	The cost of the WSC supply will increase in proportion to the increased dependence on RO water that will be required in order to maintain potable quality standards.

Source: Based on FAO (2006). *Malta Water Resources Review.* Food and Agriculture Organization of the United Nations, Rome.

TABLE 3.20 Timeline for implementing Malta's water resources protection plan

YEAR	REQUIREMENT
2003	Directive transposed into national legislation.
	Identification of river-basin districts and of the competent authorities that will be empowered to implement the directive.
2004	Completion of the first characterization process and the first assessment of impacts on the river-basin districts.
	Completion of the first economic analysis of water use.
	Establishment of a register of protected areas for the river-basin districts.
2006	Environmental monitoring programs established and operational.
	Work program for the production of the first River Basin Management Plan established.
2007	Public consultation process on significant water management issues in the river-basin districts initiated.
2008	Publication of the first River Basin Management Plan for public consultation.
2009	First River Basin Management Plan finalized and published.
	Program of measures required in order to meet the environmental objectives of the directive finalized.
2012	Program of measures to be fully operational.
	Work program for the production of the second River Basin Management Plan published.
2013	Review of the characterization and impact assessment of the river-basin districts.
	Review of the economic analysis of water use.
	Interim overview of significant water management issues published.
2014	Publication of the second River Basin Management Plan for public consultation.
2015	Achievement of the environmental objectives specified in the first River Basin Management Plan.
	Second River Basin Management Plan finalized and published with revised program of measures.
2021	Achievement of the environmental objectives specified in the second River Basin Management Plan.
	Third River Basin Management Plan to be published
2027	Achievement of the environmental objectives specified in the third River Basin Management Plan.
	Fourth River Basin Management Plan to be published.

Source: Based on FAO (2006). *Malta Water Resources Review*. Food and Agriculture Organization of the United Nations, Rome.

both upstream and downstream of Atlanta are concerned about the environmental health of their watersheds due to the amount of water the city of Atlanta demands from those surrounding watersheds.

Increasing populations and changing lifestyles that demand more water, together with changing climates, have profound effects on water supplies throughout the world.

3.6 Nutrient Cycles

Nitrogen and phosphorus are elements that are vital **nutrients** for plant growth. They are considered to be macronutrients because they are required in significant amounts for agricultural development (see Box 3.6). Plants require at least 17 essential elements for maximum growth potential. They obtain 14 of these elements

ACTIVE LEARNING EXERCISE 3.3 Engineering Solutions to Stormwater Runoff

When engineers design roads, buildings, and other infrastructure, they must consider how they will manage excess runoff from the proposed infrastructure. Typically, excess stormwater is managed by a series of engineered systems that you see around the place you live, but you may have never noticed them or the purpose they serve. The systems designed to control excess runoff are based on *best management practices (BMPs)*, which are used to reduce the amount of water that enters storm sewer systems. During the past few decades, engineers have designed the stormwater systems in a community to be separate from sanitary sewer systems that convey wastewater from indoor appurtenances like sinks, showers, and toilets. A *municipal separate storm sewer system (MS4)* is a conveyance or system of conveyances (including but not limited to streets, ditches, catch basins, curbs, gutters, and storm drains) that is used to direct stormwater away from community infrastructure. There are numerous stormwater control measures you likely pass during the week on your way home or to campus. Take a photo or make a sketch of a nearby engineered stormwater control structure. What is the purpose of the structure, and how does the structure function so that it fulfills this purpose? How does the structure reduce or control the flow of stormwater? Are there any features of the structure that are also designed to prevent soil erosion? If so, what do the erosion control features look like?

through the soil, and the rest are from air and water. Nitrogen, phosphorus, and potassium are the three elements most commonly used in fertilizers. Other elements, such as calcium, zinc, and many other compounds, are considered micronutrients because they are required in much smaller quantities.

A mass balance approach is used to determine how nitrogen and phosphorus flow from one repository to another. The largest natural nitrogen flux is between the atmosphere and oceans, as illustrated in Figure 3.43 (see page 203). Nitrogen fixation in the oceans removes nearly 140 Tg of nitrogen (as N) per year. Industrial synthetic fertilizer production nearly equals the largest natural flux. Synthetic fertilizer accounts for 120 Tg of N per year, and agricultural crop nitrogen fixation accounts for an additional 60 Tg per year. Thus, anthropogenic nitrogen transfer to soils is far greater than any natural nitrogen addition process to soils. It is estimated that 95 Tg of N per year (over 50% of the nitrogen added to soils) is leached, or moved, from the soils into freshwater, where this excess nitrogen significantly degrades water quality. Wastewater (treated and untreated) from domestic and industrial use also adds nitrogen to surface waters, as discussed in more detail in Chapters 7 and 8.

Natural minerals and phosphate deposits contribute most of the yearly flux of phosphorus to the environment, as illustrated in Figure 3.44 (see page 204). Much less phosphorus, on the order of 4–10 Tg of P, are leached from the soils into freshwaters, and because of this the effects of excess nutrients on freshwaters can often be minimized by reducing phosphorus loading to freshwater bodies. Thus, phosphorus tends to be the limiting nutrient for unwanted algae and other plant growth.

Unfortunately, nutrient inputs are not evenly distributed across the globe. In many areas, excessive use of nitrogen-based fertilizer, shown in Figure 3.45 (see

BOX 3.6 Definitions of Nutrients and Nutrient Sources

Anthropogenic sources of nitrogen and phosphorus come from products and materials used to increase plant growth and crop yield. The definitions presented here are those used by the Global Partnership on Nutrient Management (Sutton et al., 2013).

Nutrient refers here to an element needed to support plant and animal growth and development. The major nutrients, or "macronutrients," that are needed in greatest supply by plants are nitrogen (N), phosphorus (P), and potassium (K). Nutrients that are needed in only small amounts are called "micronutrients"; these include copper, zinc, molybdenum, and many others. Elements like sulphur (S) are needed in larger supply than the micronutrients and are sometimes referred to as major nutrients.

Fertilize is taken here to mean "making fertile" or more productive, in the sense of providing nutrients to allow more plants and animals to grow and develop. Land may be made more fertile by adding materials like manufactured fertilizers, animal manures, green manures, and other organic inputs.

Fertilizer means "a substance that makes fertile" because of its nutrient content. The word can cause confusion because both manures and manufactured compounds containing nutrients are used to make land more fertile. For many people, all anthropogenic nutrient products are simply called *fertilizers*. The concepts of "mining nutrients" and

"mineral fertilizers" apply to both P and N. In the case of reactive nitrogen (Nr) production, humans are "mining the atmosphere."

Where humans depend on actually digging up Nr and P nutrient sources from the ground (such as the minerals apatite, saltpetre, coal, and guano), we can refer to "fossil nitrogen" and even a "fossil nutrient economy."

Manure refers here to an organic (carbon-containing) nutrient source that can be applied to the land to grow crops and build soil quality. Specifically, we refer to recycled organic byproducts, especially animal excreta (mix of feces, urine, and bedding materials) and human wastes (including sewage sludge or biosolids), either collected locally (e.g., "night soil") or produced at centralized sewage treatment facilities.

Agronomists often talk about increasing the "fertilizer equivalence value" of organic manures, aiming to achieve the same efficiency of nutrient use as can be obtained from good practice when using mineral fertilizers. "Green manure" refers to an organic input entirely based on plant material.

Reactive nitrogen (Nr) includes all forms of N except di-nitrogen gas (N_2). The Earth's atmosphere is made up of 78% N_2; this gas is so stable that it is not usable by most organisms, which instead depend on small amounts of Nr entering the environment.

page 204), results in low fertilizer efficiency and large amounts of pollutants. Other areas, like western Kenya, may not have sufficient nitrogen to meet agricultural demands and basic nutrition needs, as illustrated in Figure 3.45. Areas of excessive phosphorus addition and areas where there is a deficiency of phosphorus are shown in Figure 3.46 (see page 205).

Prior to the Industrial Revolution, nitrogen and phosphorus were available for agriculture primarily through natural soil amendments, which limited the productivity of agricultural development. This led Thomas Malthus and others to predict the dire consequences and human suffering that would surely ensue as population growth outpaced food production, as described in

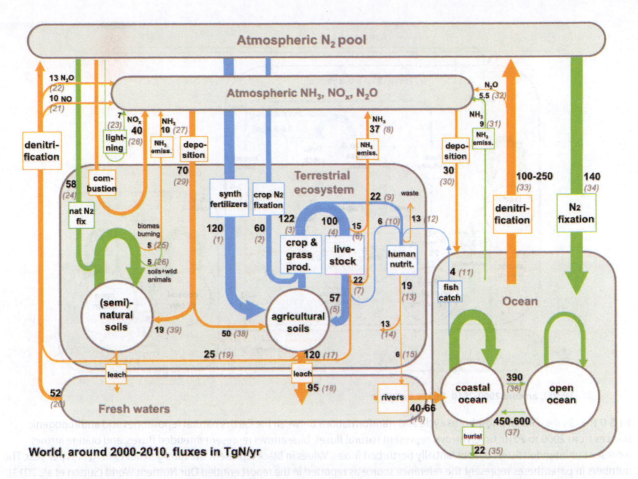

World, around 2000-2010, fluxes in TgN/yr

FIGURE 3.43 Global nitrogen cycle and transformations between the Earth's natural repositories and anthropogenic sources from 2000 to 2010. Green arrows represent natural fluxes, blue arrows represent intended fluxes, and orange arrows represent unintended fluxes or substantially perturbed fluxes. Values in black represent the nitrogen fluxes in Tg of N per year. The numbers in parentheses represent the reference source as reported in the report entitled Our Nutrient World (Sutton et al., 2013).

Source: Sutton, M.A., et al. (2013). "Our nutrient world: The challenge to produce more food and energy with less pollution." *Global Overview of Nutrient Management.* Centre for Ecology and Hydrology, Edinburgh, on behalf of the Global Partnership on Nutrient Management and the International Nitrogen Initiative, Figure 3.1.

Chapter 1. These early forecasters could not predict that low-cost fossil fuels would make it possible to produce effective reactive nitrogen fertilizers that would greatly increase agricultural productivity. In 1908, Fritz Haber and Carl Bosch developed a chemical process and full-scale production system to combine N_2 and H_2 gas to form reactive ammonia, NH_3 (Smil, 2001; Haber, 1920). This low-cost fertilizer process has since doubled the increase in the global flow of reactive nitrogen compounds illustrated in Figure 3.47 (Fowler et al., 2013) (see page 206). Both nitrogen and phosphorus uses continue to increase and are expected to increase into the foreseeable future in order to provide the nutritional and food needs associated with increasing human population.

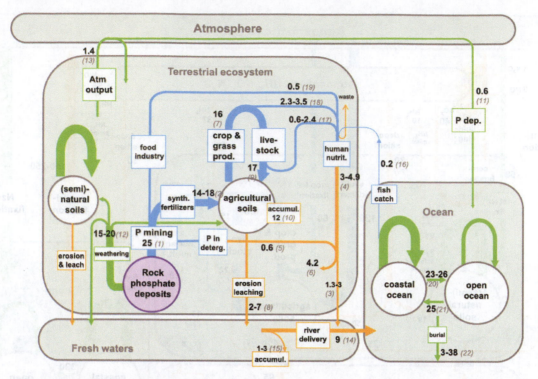

World, around 2000-2010, fluxes in TgP/yr

FIGURE 3.44 Global phosphorus cycle and transformations between the Earth's natural repositories and anthropogenic sources from 2000 to 2010. Green arrows represent natural fluxes, blue arrows represent intended fluxes, and orange arrows represent unintended fluxes or substantially perturbed fluxes. Values in black represent the nitrogen fluxes in Tg of N per year. The numbers in parentheses represent the reference source as reported in the report entitled *Our Nutrient World* (Sutton et al., 2013).

Source: Sutton, M.A., et al. (2013). "Our nutrient world: The challenge to produce more food and energy with less pollution." *Global Overview of Nutrient Management.* Centre for Ecology and Hydrology, Edinburgh, on behalf of the Global Partnership on Nutrient Management and the International Nitrogen Initiative, Figure 3.2.

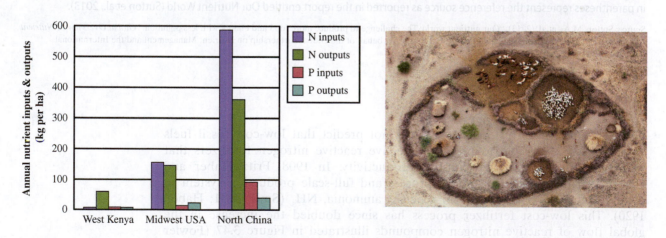

FIGURE 3.45 Nutrient distribution is uneven across the globe. China and the United States use high levels of nutrient fertilizers and inputs are in excess of crop consumption. In western Kenya, crop cultivation has created nitrogen-deficient soils illustrated here by the barren soil around the home. This in turn leads to decreasing crop yield and biodiversity.

Source: Based on Sutton, M.A., et al. (2013). "Our nutrient world: The challenge to produce more food and energy with less pollution." *Global Overview of Nutrient Management.* Centre for Ecology and Hydrology, Edinburgh, on behalf of the Global Partnership on Nutrient Management and the International Nitrogen Initiative, Figure 4.11. Photo: Martin Harvey/The Image Bank/Getty Images.

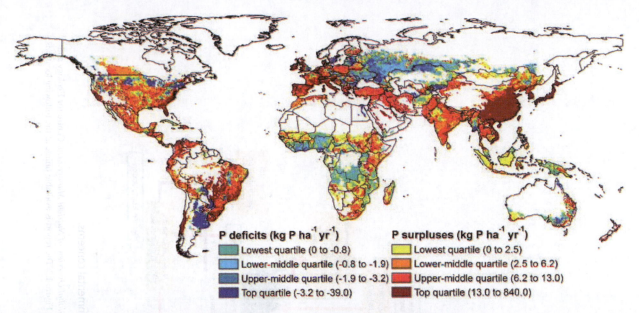

FIGURE 3.46 Estimated areas of phosphorus excess and areas of phosphorus deficiency.

Source: Sutton, M.A., et al. (2013). "Our nutrient world: The challenge to produce more food and energy with less pollution." *Global Overview of Nutrient Management.* Centre for Ecology and Hydrology, Edinburgh, on behalf of the Global Partnership on Nutrient Management and the International Nitrogen Initiative, Figure 3.7.

Excessive amounts of nutrients in soil may cause soil acidification and reduce the organic content of soils, which in turn harms native plant species and may increase the mobilization or leaching of pollutants. Leaching (movement from soil to water) of excess nitrogen and phosphorus has serious consequences on environmental water quality and access to water. Excess nutrients in water lead to eutrophication, which in turn may result in algal blooms and removal of dissolved oxygen.

Consider the model of a lake ecosystem shown in Figure 3.48 (see page 207). Note that the producers (algae) receive energy from the sun and through the process of photosynthesis produce biomass and oxygen. Because the producers (the algae) receive energy from the sun, they obviously must be restricted to the surface waters in the lake. Fish and other animals also exist mostly in the surface water because much of the food is there, but some scavengers are on the bottom. The decomposers mostly inhabit the bottom waters because this is the source of their food supply (the detritus).

Through photosynthesis, the algae use nutrients and carbon dioxide to produce high-energy molecules and oxygen. The consumers, including fish, plankton, and many other organisms, all use the oxygen, produce CO_2, and transfer the nutrients to the decomposers in the form of dead organic matter. The decomposers, including scavengers such as worms and various microorganisms, reduce the energy level further by the process of respiration, using oxygen and producing carbon dioxide. The nutrients, nitrogen and phosphorus (as well as other nutrients often called micronutrients), are then again used by the producers. Some types of algae are able to fix nitrogen from the atmosphere, while some decomposers produce ammonia nitrogen that bubbles out. (This discussion is considerably simplified from what actually occurs in an aquatic ecosystem. For a more accurate representation of such systems, see any modern text on aquatic ecology.)

The only element of major importance that does not enter the system from the atmosphere is phosphorus. A given quantity of phosphorus is recycled from the decomposers back to the producers. The fact that only a certain amount of phosphorus is available to the ecosystem limits the rate of metabolic activity. Were it

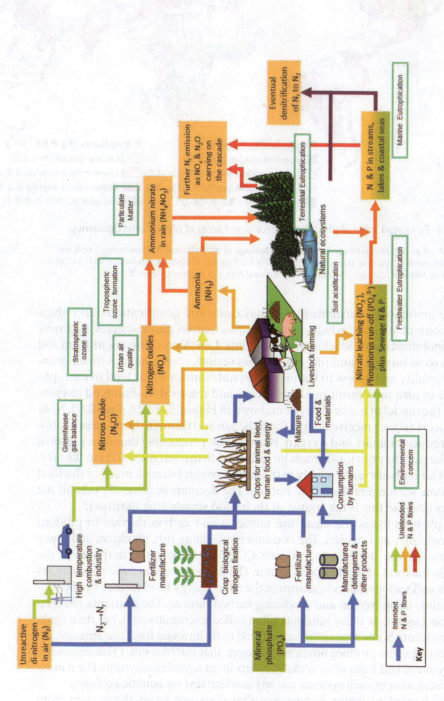

FIGURE 3.47 Schematic of nitrogen and phosphorus flows from modern processes and related environmental concerns.

Source: Sutton, M.A., et al. (2013). "Our nutrient world: The challenge to produce more food and energy with less pollution." *Global Overview of Nutrient Management*. Centre for Ecology and Hydrology, Edinburgh, on behalf of the Global Partnership on Nutrient Management and the International Nitrogen Initiative, Figure 1.2. The report is available online at the following locations: www.unep.org, www.gpa.unep.org/gpnm.html, www.initrogen.org, and www.igbp.net/publications.

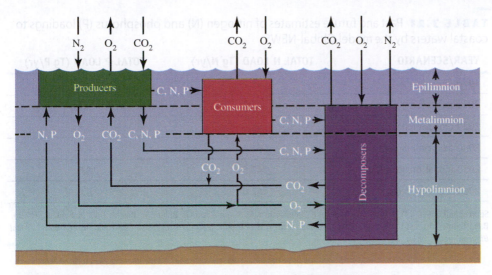

FIGURE 3.48 Movements of carbon, oxygen, and nutrients in a lake.

not for the limited quantity of phosphorus, the ecosystem's metabolic activity could accelerate and eventually self-destruct because all other chemicals (and energy) are in plentiful supply. For the system to remain at homeostasis (steady state), some key component—in this case, phosphorus—must limit the rate of metabolic activity by acting as a brake in the process.

Consider now what would occur if an external source of phosphorus, such as from farm runoff or wastewater treatment effluent, were introduced. The brake on the system would be released, and the algae would begin to reproduce at a faster rate, resulting in greater production of food for the consumers, which in turn would grow at a faster rate. All of this activity would produce ever-increasing quantities of dead organic matter for the decomposers, which would greatly multiply.

Unfortunately, the dead organic matter is distributed throughout the body of water while the algae, which produce the oxygen needed for the decomposition to take place, live only near the surface of the lake, where there is sunlight. The other supply of oxygen for the decomposers is from the atmosphere, also at the water surface. The oxygen, therefore, must travel through the lake to the bottom to supply the aerobic decomposers. If the water is well mixed, this would not provide much difficulty, but unfortunately, most lakes are thermally stratified and not mixed throughout the full depth. This means that not enough oxygen can be transported quickly enough to the lake bottom to avoid deoxygenation of the water. In stratified lakes with an increased food supply, oxygen demand by the decomposers increases and finally outstrips the supply. The aerobic decomposers die and are replaced by anaerobic forms, which produce large quantities of biomass and incomplete decomposition. Eventually, with increasing supplies of phosphorus, the entire lake becomes anaerobic, most fish die, and the algae is concentrated on the very surface of the water, forming slimy algae mats called *algal blooms*. Over time the lake fills with dead and decaying organic matter, and the result is a peat bog.

This process is called ***eutrophication***. It is actually a naturally occurring phenomenon, as all lakes receive some additional nutrients from the air and overland flow. Natural eutrophication is usually very slow, however, measured in thousands of years before major changes occur. What people have accomplished, of course, is to speed up this process with the introduction of large quantities of nutrients, resulting in ***accelerated eutrophication***. Thus, an aquatic ecosystem that might have been in

TABLE 3.21 Past and future estimates of nitrogen (N) and phosphorus (P) loadings to coastal waters by the model Global-NEWS.

YEAR/SCENARIO	TOTAL N LOAD (Tg N/yr)	TOTAL P LOAD (Tg P/yr)
1970	36.7	7.6
2000 reference	43.2	8.6
2030 global orchestration	45.5	8.6
2030 adapting mosaic	41.4	8.4
2050 global orchestration	47.5	8.5
2050 adapting mosaic	42.0	8.6

Source: Seitzinger, S.P., Mayorga, E., Kroeze, C., Bouwman, A.F., Beusen, A.H.W., Billen, G., Drecht, G.V., Dumont, E., Fekete, B.M., Garnier, J., and Harrison, J. (2010). "Global nutrient river export: a scenario analysis of past and future trends." *Global Biogeochemical Cycles*, 24.

essentially steady state (homeostasis) has been disturbed, and an undesirable condition results.

Incidentally, the process of eutrophication can be reversed by reducing nutrient flow into a lake and flushing or dredging as much of the phosphorus out as possible, but this is an expensive proposition.

The global nutrient loading of excess nutrients to water has been estimated in several studies, with the results shown in Table 3.21. Excess plant growth and low dissolved oxygen concentrations in water reduce biodiversity and negatively affect native aquatic species.

Nutrient pollutants may also threaten air quality, the greenhouse gas balance, biodiversity, and soil quality. Reactive nitrogen in the forms of NO_x and NH_3 can react in the atmosphere, as illustrated in Figure 3.49, to form photochemical smog, which negatively affects ecosystems and human health (as discussed in Chapter 9). Atmospheric nitrogen in the form of N_2O is now considered the main cause of stratospheric ozone depletion.

As discussed in Chapter 1, the economic, environmental, and societal effects of pollutants are intertwined and difficult to separate. It may also be difficult to assign an economic value to environmental damage or threats. The cost of removing nutrients from wastewater can be estimated based on current treatment effectiveness and cost. The costs associated with environmental damages due to excessive nutrients can be estimated from the value of lost productivity and human health costs associated with a particular pollutant. Jenkins et al. (2010) calls the environmental costs "shadow prices" and compares the costs of water treatment to the costs associated with the undesirable effects of specific pollutants in Tables 3.22 and 3.23. The expected environmental costs associated with nutrient pollution are much higher than the costs associated with suspended solids (SS) and organic pollutants. Currently, most wastewater treatment plants in operation are designed to remove SS and organic pollutants, but they have not been designed to remove the majority of nutrients in the wastewater. The minority of treatment plants that are removing substantial percentages of nitrogen and phosphorus are expected to accrue greater return on investment from treating the wastewater, since 90% of the expected environmental value of wastewater treatment is estimated to be due to nutrient removal. The expected return on investment is estimated to be $1,248/ha/yr of forested wetland nutrient removal in the Mississippi Alluvial Valley in the United States (Jenkins et al., 2010).

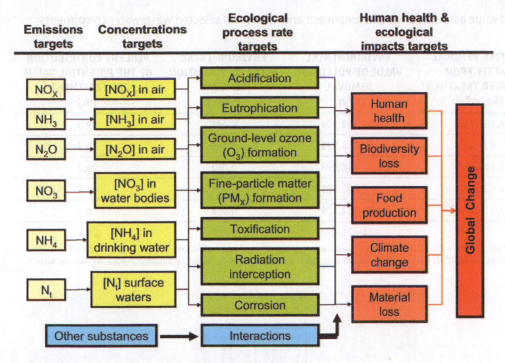

FIGURE 3.49 Nitrogen-based emissions to biogeochemical repositories and potential ecological and human health effects.

Source: Sutton, M.A., et al. (2013). "Our nutrient world: The challenge to produce more food and energy with less pollution." *Global Overview of Nutrient Management*. Centre for Ecology and Hydrology, Edinburgh, on behalf of the Global Partnership on Nutrient Management and the International Nitrogen Initiative, Figure 7.2.

TABLE 3.22 The price of water treatment compared to the shadow prices of the adverse effects from particular pollutants (Jenkins et al., 2010)

DESTINATION	WATER PRICE [€/m³]	SHADOW PRICE FOR UNDESIRABLE OUTPUTS [€/kg]				
		N	P	SS	BOD	COD
River	0.7	−16.353	−30.944	−0.005	−0.033	−0.098
Sea	0.1	−4.612	−7.533	−0.001	−0.005	−0.010
Wetlands	0.9	−65.209	−103.424	−0.010	−0.117	−0.12
Reuse	1.5	−26.182	−79.268	−0.010	−0.058	−0.140

Source: Corcoran, E., Nellemann, C., Baker, E., Bos, R., Osborn, D., and Savelli, H. (eds.). (2010). "Sick water? The central role of waste-water management in sustainable development. A Rapid Response Assessment." United Nations Environment Programme, UN-HABITAT, GRID-Arendal. www.grida.no. ISBN: 978-82-7701-075-5. Printed by Birkeland Trykkeri AS, Norway. Table 2.

The benefits of better management of nutrients in the environment can be determined by estimating the impacts and values associated with the efficiency of nitrogen and phosphorus use, the transport of nutrients in the environment, and the expected impacts of excess nutrient pollution. The "nutrient nexus," illustrated in Figure 3.50, shows the relationship between the economic, environmental, and societal impacts of nutrient consumption. When various factors are selected and tracked in the nutrient nexus, a nitrogen-based footprint can be calculated

TABLE 3.23 Estimated value associated with the treatment and removal of selected wastewater constituents (Jenkins et al., 2010)

POLLUTANT	POTENTIAL REMOVAL CAPACITY FROM WASTEWATER TREATMENT [kg/yr]	ENVIRONMENTAL VALUE OF POLLUTION REMOVAL [€/yr]	ENVIRONMENTAL VALUE OF POLLUTION REMOVAL [€/m³]	PERCENT CONTRIBUTION OF THE POTENTIAL VALUE OF TREATMENT [%]
N	4,287,717	98,133,996	0.481	59.6
P	917,895	50,034,733	0.245	30.4
SS	60,444,987	448,098	0.002	0.3
BOD	59,635,275	2,690,421	0.013	1.6
COD	113,510,321	13,364,429	0.066	8.1
Total		164,671,677	0.807	100.0

Source: Corcoran, E., et al. (eds.). (2010). "Sick water? The central role of waste-water management in sustainable development. A Rapid Response Assessment." United Nations Environment Programme, UN-HABITAT, GRID-Arendal. www.grida.no. ISBN: 978-82-7701-075-5. Printed by Birkeland Trykkeri AS, Norway. Table 3.

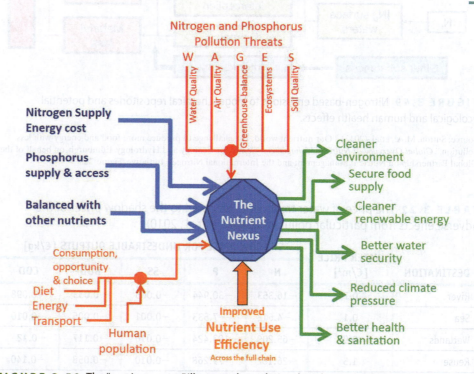

FIGURE 3.50 The "nutrient nexus" illustrates key relationships between the economic, environmental, and societal impacts related to nutrient consumption, use, and waste treatment.

Source: Sutton, M.A., et al. (2013). "Our nutrient world: The challenge to produce more food and energy with less pollution." *Global Overview of Nutrient Management*. Centre for Ecology and Hydrology, Edinburgh, on behalf of the Global Partnership on Nutrient Management and the International Nitrogen Initiative, Figure 7.7.

that relates lifestyle choices, such as food choices and energy consumption, to the expected rate of nitrogen pollution generation. Leach et al. (2012) compared the nitrogen footprint of the United States to that of the Netherlands based on consumption rates of food and energy. The results (Figure 3.51) show that the nitrogen footprint and thus the expected nutrient-related impacts in the United States are significantly greater than those in the Netherlands.

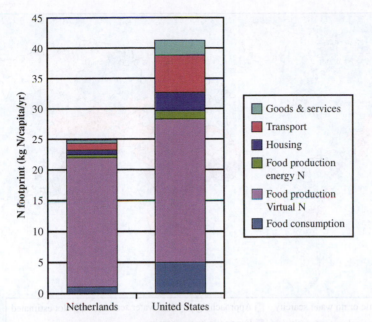

FIGURE 3.51 A comparison of food and energy lifestyle choices and their impacts on the mass of nitrogen consumed per capita year in the United States and the Netherlands.

Source: Based on Leach, A.M., Galloway, J.N., Bleeker, A., Erisman, J.W., Kohn, R., and Kitzes, J. (2012). "A nitrogen footprint model to help consumers understand their role in nitrogen losses to the environment." *Environmental Development* 1:40–66.

3.7 Summary

Our Common Future, a report of the Brundtland Commission for the United Nations (1987), defined sustainable development as "development that meets the needs of the present without compromising the ability of future generations to meet their own needs." Water is a primary resource and a need for all generations. The World Health Organization (WHO) reports that 2 billion people lack access to improved water supply, or a source of water protected from sewage contamination. Nonimproved water sources are often contaminated, and 3.6 billion people lack access to proper sanitation. The lack of adequate sanitation results in pollution of local and downstream watersheds and degradation of the local ecosystem, resulting in poor human health in those localities. WHO estimates that 25% of all preventable illnesses worldwide occur as a result of poor environmental quality.

Water will become scarcer as the planet's population continues to grow. Water scarcity occurs when there is insufficient water to meet the water demands for drinking water, washing, and cooking (see Figure 3.52). The WHO estimates that 20 liters per day of water located within 1 kilometer of the user's dwelling is the minimum amount of water necessary in most cultures to have reasonable access to water. Engineers are recognizing that the level of consumption of water is important for determining the requirements of the infrastructure we are developing for future generations. When consumption is greater than storage in a system, the level of water withdrawal is nonsustainable. Future generations will therefore be forced to ration water or find a technical solution to reuse water within the watershed. Otherwise, they will face similar circumstances to those faced by Chaco Canyon's residents 1,000 years ago.

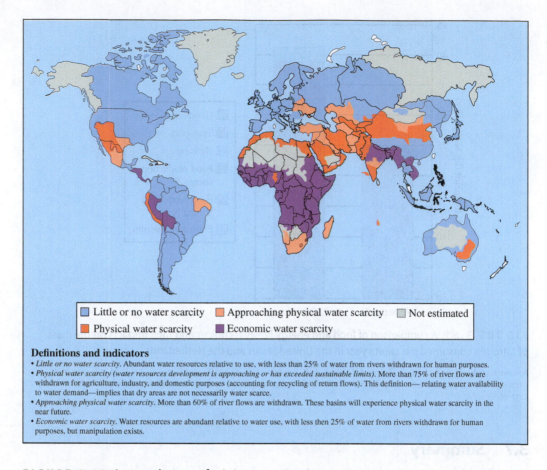

Definitions and indicators

- *Little or no water scarcity*. Abundant water resources relative to use, with less than 25% of water from rivers withdrawn for human purposes.
- *Physical water scarcity (water resources development is approaching or has exceeded sustainable limits)*. More than 75% of river flows are withdrawn for agriculture, industry, and domestic purposes (accounting for recycling of return flows). This definition— relating water availability to water demand—implies that dry areas are not necessarily water scarce.
- *Approaching physical water scarcity*. More than 60% of river flows are withdrawn. These basins will experience physical water scarcity in the near future.
- *Economic water scarcity*. Water resources are abundant relative to water use, with less then 25% of water from rivers withdrawn for human purposes, but manipulation exists.

FIGURE 3.52 Areas and causes of existing water scarcity.

Source: Based on UNEP (2008). GRID-Arendal, Maps & Graphics Library, maps.grida.no/go/graphic/areas-of-physical-and-economic-water-scarcity.

Anthropogenic processes are changing rapidly, in turn changing the Earth's environment. These changes will make our current lifestyles more difficult to maintain with existing technology and projected population growth. Engineers have made choices in the past that did not consider the sustainability of systems as part of the design process. Today's engineers and scientists must utilize new technical changes that are currently underway, such as green building codes and incentives that value water conservation, to create a more sustainable infrastructure. Only then can they better meet the needs of today's generations and generations to come.

References

Angelakis, A. N., Mays, L. W., Koutsoyiannis, D., and Mamassis, N. (2012). *Evolution of Water Supply Through the Millennia*. IWA Publishing Alliance House, 12 Caxton Street, London, SW1H 0QS, UK.

Arthurton, R., Barker, S., Rast, W., Huber, M., Alder, J., Chilton, J., Gaddis, E., Pietersen, K., and Zockler, C. (2007). *State and Trends of the Environment*: 1987–2007. Chapter 4: Water. Global Environment Outlook (GEO4) Environment for Development. United Nations Environment Programme. DEW/1001/NA. Australian Government

Bureau of Meteorology: www.bom.gov.au/jsp/ncc/climate_averages/evaporation/index.jsp. Product Code: IDCJCM006.

ATIGA. (1972). "Wastes disposal and water supply project in Malta." UNDP.

Bureau de Recherche Géologique et Minière (BRGM). (1991). "Study of the fresh-water resources of Malta." Government of Malta.

Cech, T. V. (2010). *Principles of Water Resources: History, Development, Management and Policy*, 3rd ed., p. 74. Hoboken, NJ: John Wiley.

Corcoran, E., Nellemann, C., Baker, E., Bos, R., Osborn, D., and Savelli, H. (Eds.). (2010). Sick Water? The central role of waste-water management in sustainable development. A Rapid Response Assessment. United Nations Environment Programme, UN-HABITAT, GRID-Arendal. www.grida.no. Printed by Birkeland Trykkeri AS, Norway.

Diamond, J. (2005). *Collapse: How Societies Choose to Fail or Succeed*. New York: Penguin Books.

Evan, J., and Perlman, H. U.S. Department of the Interior, U.S. Geological Survey, USGS, www.usgs.gov/media/images/natural-water-cycle-0.

FAO. (1988). *Irrigation Water Management: Irrigation Methods*. Food and Agriculture Organization of the United Nations.

FAO. (2006). Malta water resources review. Food and Agriculture Organization of the United Nations, Rome, 2006. Viale delle Terme di Caracalla, 00100 Rome, Italy.

Fowler, D., Coyle, M., Skiba, U., Sutton, M. A., Cape, J. N., Reis, S., Sheppard, L. J., Jenkins, A., Galloway, J. N., Vitousek, P., Leech, A., Bouwman, A. F., Butterbach-Bahl, K., Dentener, F., Stevenson, D., Amann, M., and Voss, M. (2013). "The global nitrogen cycle in the 21st century." *Phil. Trans. Roy. Soc., Ser. B.*

Haber, F. (1920). The synthesis of ammonia from its elements. Nobel Lecture (1920), available at www.nobelprize.org/no-bel_prizes/chemistry/laureates/1918/haber-lecture.pdf.

Harte, J. (1988). *Consider a Spherical Cow*. Sausalito, CA: University Sciences Books.

Hendrix, M., and Thompson, G. R. (2015). *Earth 2, Second Edition*. Stamford, CT: Cengage Learning.

Jenkins, W. A., Murray, B. C., Kramer, R. A., and Faulkner, S. P. (2010). "Valuing ecosystem services from wetlands restoration in the Mississippi Alluvial Valley." *Ecological Economics* 69:1051–1061.

Leach, A. M., Galloway, J. N., Bleeker, A., Erisman, J. W., Kohn, R., and Kitzes, J. (2012). "A nitrogen footprint model to help consumers understand their role in nitrogen losses to the environment." *Environmental Development* 1:40–66.

Margat, J., and Andreassian, V. (2008). *L'Eau, les Idees Receus*. Paris, Editions le Cavalier Bleu.

Mays, L. (2007). "Water sustainability of ancient civilizations in Mesoamerica and the American Southwest." *Water Science and Technology: Water Supply* 7(1):229–236.

Mays, L. (2011). *Water Resources Engineering*, 2nd ed., p. 4. Hoboken, NJ: John Wiley.

McGaughy, P. H. (1968). *Engineering Management of Water Quality*. New York: McGraw-Hill.

Molden, D. (Ed.). Comprehensive Assessment of Water Management in Agriculture (2007). *Water for Food, Water for Life: A Comprehensive Assessment of Water Management in Agriculture*. London: Earthscan, and Colombo: International Water Management Institute.

Morris, T. O. (1952). "The water supply resources of Malta." Government of Malta.

Murakami, M. (1995). *Managing Water for Peace in the Middle East: Alternative Strategies*. United Nations University, Tokyo, Japan.

Peavy, H. S., Rowe, D. R., and Tchobanoglous, G. (1985). *Environmental Engineering*, p. 1. New York: McGraw Hill.

Rao, B., Sandeep, V. M., Rao, V. U. M., and Venkateswarlu, B. (2012). Potential evapotranspiration estimation for Indian conditions: Improving accuracy through calibration coefficients. *Tech. Bull.* No 1/2012. All India Co-ordinated Research Project on Agrometeorology, Central Research Institute for Dryland Agriculture, Hyderabad.

Reitz, M., Sanford, W., Senay, G., and Cazenas, J. (2017). "Annual estimates of recharge, quick-flow runoff, and evapotranspiration for the contiguous U.S. using empirical regression equations." *JAWRA—Journal of the American Water Resources Association.* 53, 10.1111/1752-1688.12546.

Revenga, C., Nackoney, J., Hoshino, E., Kura, Y., and Maidens, J. (2003). Watersheds of the World. IUCN, IWMI, Rasmar Convention Bureau and World Resources Institute, Washington, DC. 2002 © 2003 World Resources Institute.

Seitzinger, S. P., Mayorga, E., Kroeze, C., Bouwman, A. F., Beusen, A. H. W., Billen, G., Drecht, G. V., Dumont, E., Fekete, B. M., Garnier, J., and Harrison, J. (2010). "Global nutrient river export: a scenario analysis of past and future trends." *Global Biogeochemical Cycles,* 24.

Shiklomanov, I. (1993). Chapter on "World freshwater resources" in *Water in Crisis: A Guide to the World's Freshwater Resources.* Gleick, P. H. (ed.) Oxford: Oxford University Press.

Smil, V. (2001). *Enriching the Earth: Fritz Haber, Carl Bosch, and the Transformation of World Food Production.* Cambridge, MA: MIT Press.

Spiteri Staines, E. (1987). Aspects of water problems in the Maltese Islands. Malta

Sutton, M. A., et al. (2013). Our Nutrient World: The Challenge to Produce More Food and Energy with Less Pollution. Global Overview of Nutrient Management. Centre for Ecology and Hydrology, Edinburgh UK on behalf of the Global Partnership on Nutrient Management and the International Nitrogen Initiative.

Turk. A. (1975). *Ecosystem, Energy, Population*, First Edition. Philadelphia, PA: Saunders College Publishing.

UNEP (United Nations Environment Programme). (2008). maps.grida.no/go /graphic/areas-of-physical-and-economic-water-scarcity.

United Nations Food and Agriculture Organization (UN FAO). (2013). www.fao.org/nr /water/aquastat/watresafrica/index3.stm.

U.S. Geological Survey. (2009, April). Use of Minerals and Materials in the United States from 1900 Through 2006. USGS Fact Sheet 2009-3008.

Vesilind, P. A., et al. (2010). *Introduction to Environmental Engineering*, Third Edition, P.71. Stamford, CT: Cengage Learning.

Vesilind, P. A., and DiStefano, T. D. (2006). *Controlling Environmental Pollution.* Lancaster, PA: DEStech Publications, Inc.

Viessman, W., Jr., and Hammer, M. J. (1985). *Water Supply and Pollution Control*, 4th ed., p. 70. New York: Harper & Row.

WCED. (1987). *Our Common Future.* World Commission on Environment and Development. New York: Oxford University Press.

Womach, J. (2005). Agriculture: A Glossary of Terms, Programs, and Laws, 2005 Edition. Congressional Research Service, The Library of Congress, Washington DC p. 62.

World Water Assessment Programme. (2009). *The United Nations World Water Development Report* 3: "Water in a Changing World." Paris: UNESCO, and London: Earthscan. ISBN: 978-9-23104-095-5. Map 10.5, p. 174.

Yan, H., Wang, S. Q., Billesbach, D., Oechel, W., Zhang, J. H., Meyers, T., Martin, T. A., Matamala, R., Baldocchi, D., Bohrer, G., Dragoni, D., and Scott, R. (2012, September). Global estimation of evapotranspiration using a leaf area index-based surface energy and water balance model. *Remote Sensing of Environment* 124:581–595. dx.doi.org/10.1016/j.rse.2012.06.004.

Key Concepts

Ecology
Photosynthesis
Respiration
Producer
Consumer
Detritus
Decomposer
Electron acceptor
Obligate aerobe
Obligate anaerobe
Facultative microorganism
Food chain
Food web
Trophic level
Biogeochemical cycle
Residence time
Atmosphere
Lithosphere
Hydrosphere

Hydrology
Hydrologic cycle
Evaporation
Transpiration
Condensation
Precipitation
Infiltration
Watershed
Runoff
Volumetric flow rate
Water budget
Nutrient
Fertilize
Fertilizer
Manure
Reactive nitrogen
Eutrophication
Accelerated eutrophication

Problems

3-1 What percentage of materials used for physical goods in the United States came from recycled materials in the year 2000?

3-2 Define and describe a biogeochemical cycle.

3-3 List and describe important biogeochemical repositories.

3-4 Define the following terms:
   a. Atmosphere
   b. Condensation
   c. Evaporation
   d. Hydrology
   e. Hydrosphere
   f. Hydrologic cycle
   g. Infiltration
   h. Lithosphere
   i. Precipitation
   j. Residence time
   k. Runoff
   l. Transpiration
   m. Watershed

3-5 Describe the mathematical equation and variables used to define the residence time of a system.

3-6 Sketch the nitrogen cycle.

3-7 Sketch the phosphorus cycle.

3-8 Sketch the carbon cycle.

3-9 How are potable water and nonpotable water distributed?

3-10 What four factors does Jared Diamond conclude caused the collapse of the Chaco Canyon civilization?

3-11   Define the hydrologic cycle. Sketch the hydrologic cycle showing repositories in your watershed.

3-12   Sketch and label the repositories and transformation processes in the oxygen cycle.

3-13   What are the chemical equations that describe the biological transformation of water by plants and animals?

3-14   What are the chemical equations that describe the biological transformation of oxygen by plants and animals?

3-15   What is the annual average precipitation in your region in inches per year and millimeters per year?

3-16   Write the water budget equation and define each variable.

3-17   Describe the differences between consumptive and nonconsumptive withdrawal, and provide examples of each.

3-18   Why would an environmental engineer possibly want to converse with a wildlife biologist during a project?

3-19   Besides playing important roles in ecosystems, microorganisms play important roles in engineering. What are some of these roles?

3-20   Estimate from the tables provided in the chapter how much water you consume each day through direct water use. How much water is this each year?

3-21   Estimate from your electric bill, or a family member's electric bill, how much water is used to generate the electricity you consume each day. How much water is this each year?

3-22   Estimate the residence time of water in the oceans.

3-23   Estimate the residence time of nitrogen in the atmosphere if the magnitude of flow of nitrogen from the atmosphere is 5×10^{12} kg(N)/yr.

3-24   Estimate the residence time of carbon in the atmosphere assuming that the CO_2 flow rate to the atmosphere is approximately equal to the flow due to decomposition and combustion of terrestrial organic matter and that from animal respiration is 50×10^{12} kg(C)/yr.

3-25   Estimate evaporation rates from Figure 3.21 or online resources in
   a. Eastern North Carolina
   b. Seattle, Washington
   c. Southwest Arizona

3-26   Estimate the evapotranspiration rates from Figures 3.17 through 3.21 or online resources such as the Environmental Data Explorer website of the United Nations Environmental Programme (geodata.grid.unep.ch/) in
   a. Benin, West Africa
   b. Central Tanzania, East Africa
   c. Rwanda, East Africa
   d. Sydney, Australia

3-27   Estimate precipitation from Figures 3.24 through 3.27 or online resources such as the AQUASTAT website of the Food and Agricultural Organization of the United Nations (www.fao.org/aquastat/en/)
   a. Eastern North Carolina
   b. Seattle, Washington
   c. Southwest Arizona
   d. Benin, West Africa
   e. Central Tanzania, East Africa
   f. Rwanda, East Africa
   g. Sydney, Australia

3-28 Assuming runoff and infiltration are equal to 40% of the amount of precipitation, calculate the available water for withdrawal based on your answers to Problems 3-25 through 3-27 in
   a. Eastern North Carolina
   b. Seattle, Washington; assume runoff and infiltration are equal to 70%
   c. Southwest Arizona; assume runoff and infiltration are equal to 20%
   d. Benin, West Africa
   e. Central Tanzania, East Africa
   f. Rwanda, East Africa
   g. Sydney, Australia

3-29 Describe the boundaries of your watershed and name the stream or river that provides the outflow from your immediate watershed. Use the World Resources Institute's website (www.wri.org) to find the average runoff and population served in your major drainage basin.

3-30 Describe how water scarcity affected the civilization living in and around Chaco Canyon in AD 1100, and describe what lessons from that civilization might be applicable to Malta in modern times.

3-31 What remaining sources of water can be tapped into for meeting the needs of water-scarce regions of the planet?

3-32 Examine the map of physical and economic water scarcity. Does this map align closely with maps of precipitation? What areas are similar, and what areas are different? What, if anything, do you think is responsible for the differences between the water-scarcity maps and precipitation maps?

3-33 The average yearly precipitation in the Shenandoah Valley watershed is 36 inches. The average rate of evapotranspiration is 28 inches per year. Typical runoff through the Shenandoah River is 163 ft^3/s (cfs). Reservoirs within the watershed store 28 million gallons of water. The area of the watershed is 7.5×10^{10} ft^2. Currently, consumptive use of water is 7 million gallons per day (MGD). There are 7.48 gallons per cubic foot.
   a. Determine the yearly infiltration rate to the groundwater table in MGD, based on the given information.
   b. If the 7 MGD accounted for all withdrawals from the groundwater table, what is the average net gain or loss from the groundwater table in MGD?

3-34 A river catchment has an area of 73,230 km^2 and an average annual runoff of 11,300,000 megaliters (ML). The climate is tropical and monsoonal, with a mean annual rainfall of approximately 800 mm; most precipitation occurs from November to April. Annual evapotranspiration rates in the region are approximately 2,000 mm per year. There are several dams within the watershed, with a total storage in all the dams of 425,779 ML. About 7,500 people live within the watershed. The average water withdrawal from groundwater in the watershed is 55,229 ML. A water management plan is currently being developed for the catchment. To assist the watershed management association, you have been asked to do the following:
   a. Write the water budget equation.
   b. Show the volume of water available each year for each term in the water budget equation in units of m^3.
   c. Compare the infiltration rate to the average withdrawal of water in the watershed and determine if this use is sustainable.

3-35 The yearly precipitation in a 23,000 m^2 watershed is 128 cm. The average rate of evapotranspiration is given as 53 cm/yr. Typical infiltration through the watershed is 19 cm/yr. Reservoirs within the watershed store 2,400 m³ of water each year.

    a. Determine the yearly runoff to the river exiting the watershed in m³/minute, based on the above information.

    b. What would happen to the flow rate of water downstream from the watershed if a proposed reservoir was built that would store 8,500 m³ of water per year to be used for irrigation?

    c. What might happen to the ecosystem around the river exiting the watershed if the reservoir was built?

3-36 Describe the four Maltese water myths and the actual facts about water related to the myths.

3-37 What geologic feature contains most of Malta's groundwater supply?

3-38 Based on Figures 3.39 and 3.40, estimate approximately what percentage of Malta's surface area is protected for groundwater recharge. Also estimate what percentage of Malta was already urbanized in 2007.

3-39 Apply the water budget equation to evaluate the urgency of water management measures for the island nation of Malta.

    a. Determine the average annual input, or precipitation, of water. Estimate this total by calculating the average rainfall that occurs during a year on the islands from Table 3.11.

    b. Convert the average monthly precipitation values expressed in mm/month to a yearly value expressed in m/year. Assume this rainfall occurs equally over all the area of the islands, which is a reasonable assumption since the total land area is relatively small compared to that of much larger nations.

    c. Calculate the volume of rainwater in liters per day for each of Malta's 398,000 people if all the rainfall over the entire area of the islands could be collected and used.

    d. Not all the precipitation flows into the ground for storage and later use. The water budget equation terms are shown in Table 3.13. Calculate the water budget equation parameters in liters per day from Table 3.13 listed below.

        i) Precipitation

        ii) Runoff

        iii) Evapotranspiration

        iv) Infiltration to the groundwater

        v) Consumption (private and public water use plus subsurface discharge to the sea)

        vi) Water storage, S, in the ground

    e. Estimate the liters of water available per day (lpcd) from the groundwater storage for each of the 398,000 people who live in Malta if 100% of the water stored in the ground was used by the population.

    f. In Table 3.15, the total water use for Malta in 2003 was given as 58.641 × 10⁶ m³/year. How many liters of water did each of the 398,000 people in Malta use per day? This is the water consumption expressed in liters per capita day (lpcd).

    g. Compare the available sustainable water supply (part e) to actual water demand in Malta (part f). The FAO report suggests that significant progress must be made to protect Malta's fragile water resources. The report describes four demand scenarios (based on current levels of demand and the subsequent projections that have emerged from discussions with various stakeholders). It also discusses different strategies for meeting future demand and achieving the vision as shown in Tables 3.16 and 3.17. Does the proposed strategy seem like it will be sufficient to ensure that future water resources are available in Malta? Discuss why or why not using the information calculated above.

3-40 Find the topographic map section in your library and locate the USGS topographic quadrangle map for the region in which you live.

a. Determine your watershed boundaries.

b. Estimate the area of your watershed from the map.

3-41 Obtain a copy of the USGS National Resources Conservation Service soil map either online or in your library, and determine the dominant soil type in your region. Estimate the infiltration rate based on the average soil type in your watershed.

3-42 Determine the annual average volumetric flow rate of runoff of the nearest stream to your watershed on the USGS webpage for streamflow gauges.

3-43 For your watershed:

a. Estimate the annual precipitation.

b. Estimate the annual evapotranspiration.

c. From your local water company, determine if there is any net storage in reservoirs within your watershed.

d. Solve the water budget equation.

e. What capacity of water can be sustainably withdrawn for your community?

f. Determine from your water company's website how much water is used within your community.

g. Does your community have a sustainable water supply? Why or why not?

3-44 It has been suggested that the limiting concentration of phosphorus for accelerated eutrophication is between 0.1 and 0.01 mg/L of P. Typical river water might contain 0.2 mg/L of P, 50% of which comes from farm and urban runoff, and 50% from domestic and industrial wastes. Synthetic detergents contribute 50% of the P in municipal and industrial waste.

a. If all phosphorus-based detergents are banned, what level of P would you expect in a typical river?

b. If this river flows into a lake, would you expect the phosphate detergent ban to have much effect on the eutrophication potential of the lakewater? Why or why not?

3-45 If algae contain P:N:C in the proportion 1:16:100, which of the three elements would limit algal growth if the concentration in the water were

- 0.20 mg/L of P
- 0.32 mg/L of N
- 1.00 mg/L of C

Show your calculations.

## CHAPTER 4

Material Flow and Processes in Engineering

FIGURE 4.1 A thermal vent under the ocean. In this location, warm sulfur-containing gases are vented from the ocean floor into the deep ocean, creating unique chemical conditions and reactions between the sulfur compounds and other salts and minerals found in ocean waters.

Source: www.whoi.edu/know-your-ocean/ocean-topics/seafloor-below/hydrothermal-vents/.

Life is a series of natural and spontaneous changes. Don't resist them—that only creates sorrow. Let reality be reality. Let things flow naturally forward in whatever way they like.

—LAO TZU

GOALS

MANY CHEMICAL PROCESSES OCCUR NATURALLY; there are natural products in foods, smoke from cooking fires, pesticides from plants, and fertilizing chemicals in wastewater. Industrialization has significantly changed the type and the amount of both natural and synthetic chemicals present in the environment. Understanding chemical reaction processes is critical in managing the risk that chemicals in the environment pose to ecosystems and humans. The second Global Chemicals Outlook, presented during the UN Environment Assembly in Nairobi in 2019, found that the current chemical production capacity of 2.3 billion tonnes is projected to double by 2030. Despite commitments to maximize the benefits and minimize the impacts of this industry, hazardous chemicals continue to be released into the environment in large quantities. Common environmental chemical pollutants include pesticides and herbicides; volatile organic compounds (VOCs) such as benzene, toluene, and chloroform; heavy metals such as lead, mercury, and arsenic; air contaminants such as carbon monoxide, ozone, particulate matter (PM), and secondhand smoke; and persistent organic pollutants, such as the dioxins, PCBs, and DDT. As scientists and engineers, we want to understand both the chemical transformations that are possible and how the concentrations of these chemicals change in natural and human-made systems. The following questions might be considered:

- Are the groundwater and earth beneath industrial and municipal dump sites contaminated with pollutants? If so, how can these sites be remediated?
- What happens to the household chemicals in the cleaners that run down your drain?
- What is the impact of factory emissions on our air quality and climate?

Except for processes involving nuclear reactions, with which we are not concerned in this text, a pound of any material, such as lead, in the beginning of any process will yield a pound of that material in the end, although perhaps in a different form. This concept of conservation of mass leads to a powerful engineering tool: the material balance. In this chapter, the material balance around a black box unit operation is introduced. We will begin to investigate methods used to understand and model chemical reactions and reaction processes that describe chemical transformations in the natural environment and human-made systems that harness the beneficial aspects of chemistry for human society. Students will use basic principles of chemistry, physics, and math to define and solve engineering problems—in particular, mass and energy balance problems.

OBJECTIVES

At the conclusion of this chapter, you should be able to:

4.1 Define the laws of conservation for mass and energy.

4.2 Determine system boundaries for solving mass and energy flow problems.

4.3 Solve mass balance problems involving steady state conservative substances.

4.4 Apply zero-, first-, and second-order reaction models to reactor design.

4.5 Determine the half-life of processes.

4.6 Describe the inputs and outputs of consecutive reactions with mathematical models.

4.7 Recognize and describe the characteristics of a batch reactor, plug-flow reactor, and constantly mixed-flow reactor.

4.8 Formulate and solve material and energy balances for ideal mixed-batch reactors, plug-flow reactors, completely mixed-flow reactors, and reactors in series.

Introduction

Many processes and reactions are predictable in nature. Engineers and scientists strive to understand these processes and create mathematical models to describe their behaviors. Such models may be simple or complex. If a pipeline under pressure maintains a specific flow of material, it is possible to predict the new flow if the pressure is doubled. Global climate models are intricate and complex. Despite their complexity, global climate models are built based on the same fundamental principles as simpler material flow and transformation models described in this chapter. The law of conservation of mass—that matter is neither created nor destroyed—leads to a powerful engineering tool, the ***material balance***. Except for processes involving nuclear reactions (with which we are not concerned in this text), a gram of any material, such as a gram of lead, in the beginning of any process will yield a gram of material at the end of the process. It may, however, be in a very different form. In most cases, simple rate equations and material balances can accurately predict a change in quantity (mass or volume) with time. These mathematical models provide engineers and scientists with the tools to understand and shape the future.

4.1   Material Balances with a Single Reaction

The ***law of conservation of mass*** states that mass cannot be created or destroyed. Balancing the mass flow into and out of a system allows engineers and scientists to quantitatively analyze the behavior of anthropogenic emissions in the environment.

Imagine a box with material flowing through it as shown in Figure 4.2. The flows into the box are called ***influents***, represented here by X. If the flow is described as a mass per unit time, X_0 is the mass per unit time flowing into the box.

Definition of the control volume

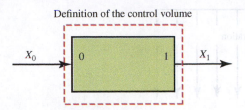

FIGURE 4.2 A black box process with one inflow and one outflow.

Similarly, X_1 is the outflow or ***effluent***. In this case, no material is being created or destroyed within the box and the flow remains constant with time—that is, at a steady state flow. The material balance using the box as the control volume of our analysis yields

$$\boxed{\begin{array}{c}\text{Mass per unit time} \\ \text{of X} \\ \text{IN}\end{array}} = \boxed{\begin{array}{c}\text{Mass per unit time} \\ \text{of X} \\ \text{OUT}\end{array}}$$

or
$$[X_0] = [X_1] \tag{4.1}$$

The volume of flow into and out of a system is also defined by the boundaries of the system under examination. The volume within these boundaries is the ***control volume***. The control volume for our water budget equation is defined by the boundaries of the watershed. A control volume in the laboratory is defined by the boundaries of the reactor's walls.

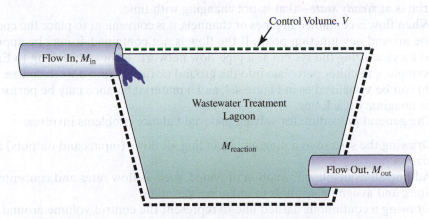

FIGURE 4.3 Schematic of a mass balance for a wastewater treatment process.

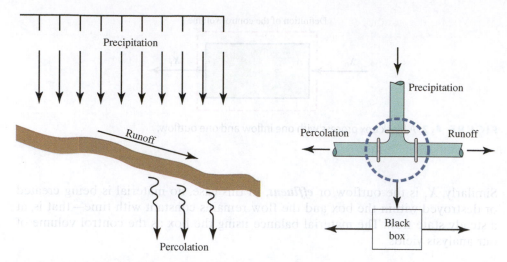

FIGURE 4.4 Precipitation, runoff, and percolation can be visualized and modeled as a black box.

For a given system, such as the wastewater treatment lagoon shown in Figure 4.3, we have

M_{in} = mass flow of any substance into the lagoon

M_{out} = mass flow of any substance out of the lagoon

$M_{reaction}$ = production or degradation of the mass flow of any substance within the lagoon

dM/dt = accumulation or removal of a mass of any substance within the lagoon

The black box can be used to establish a volume balance and a mass balance if the density does not change in the process. Since the definition of density is mass per unit volume, a mass can be converted to a volume balance by dividing each term by the (constant) density. Engineers typically use a volume balance for liquids and a mass balance for solids. In Sections 4.1 through 4.3, we will assume the flow of material is at *steady state*—that is, not changing with time.

When flow is contained in pipes or channels, it is convenient to place the control volume around any junction point. If the flow is not contained, it may be approximated by visualizing the system as a pipe flow network. Rainwater falling to Earth, for example, can either percolate into the ground or run off into a watercourse. This system can be visualized as in Figure 4.4, and a material balance may be performed on the imaginary black box.

The general procedure for solving material balance problems involves:

1. Drawing the system as a diagram, including all flows (inputs and outputs) as arrows.
2. Adding all available information provided, such as flow rates and concentrations, and assigning symbols to unknown variables.
3. Drawing a continuous dashed line to represent the control volume around the components or components that are to be balanced. This may include a unit operation, a junction, or a combination of these processes.
4. Deciding what material is to be balanced. This may be a volumetric or mass flow rate.

5. Writing the general material balance equation:

Mass or Volume Rate ACCUMULATED	$=$	Mass or Volume Rate IN	$-$	Mass or Volume Rate OUT	$+$	Mass or Volume Rate PRODUCED	$-$	Mass or Volume Rate CONSUMED

6. Solving for the unknown variables using algebraic techniques.

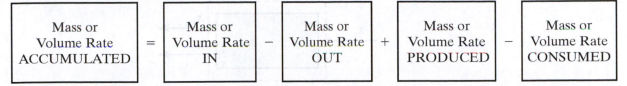

ACTIVE LEARNING EXERCISE 4.1   Black Box Modeling of Water in a Home

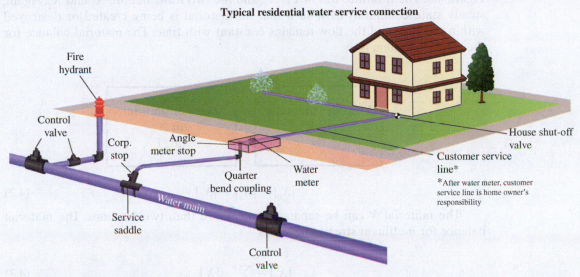

FIGURE 4.5 Water connection to a home.

Most homes in highly developed countries are connected to pressurized piped water from either a publicly owned water distribution system or a ground-water pump.

1. For your home or a home you are familiar with, draw a black box model that represents the home, the inputs of water to the home, and the outputs from the pressurized system. Be sure to include:

   - The control volume boundaries
   - The input to the home
   - All outputs from the pressurized lines

2. According to some estimates, over half the world's population does not have running water in their homes. If the water input to the home does not come from a pressurized pipe:

   - Where is the input source for the water to the home?
   - How does the input potentially affect the number and types of "outputs" from the pressurized pipes you have identified above?

Definition of the control volume

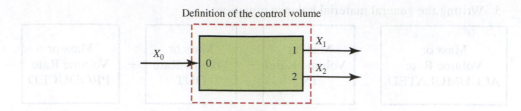

FIGURE 4.6 A separator with one influent and two effluents.

4.1.1 Splitting Single-Material Flow Streams

A flow into a box may be separated into two or more effluent streams, as shown in Figure 4.6. The flow into this box is X_0, and the two flows out are X_1 and X_2. Again, steady state conditions exist such that no material is being created or destroyed within the box, and the flow remains constant with time. The material balance for the splitting system is

Mass per unit time of X IN	=	Mass per unit time of X OUT

or
$$[X_0] = [X_1] + [X_2] \tag{4.2}$$

The material X can be separated into more than two fractions. The material balance for n effluent streams becomes

$$[X_0] = \sum_{i=1}^{n} [X_i] \tag{4.3}$$

4.1.2 Combining Single-Material Flow Streams

A black box can also receive numerous inflows and mix the flow streams together to discharge one effluent, as shown in Figure 4.7. The flows into this box are $X_1, X_2, \ldots, X_m$. The material balance for the mixing system yields the reverse of the splitting system:

$$\sum_{i=1}^{m} [X_i] = [X_e] \tag{4.4}$$

Control volume

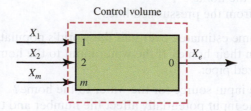

FIGURE 4.7 A mixer with several inflows (m) and one effluent flow (e).

EXAMPLE 4.1   Refuse Management

A city generates 102 tons/day of refuse, all of which goes to a transfer station. At the transfer station, the refuse is split into four flow streams headed for three incinerators and one landfill. If the capacities of the incinerators are 20, 50, and 22 tons/day, how much refuse must go to the landfill?

Consider the situation illustrated in Figure 4.8. The four output streams are the known capabilities of the three incinerators and one landfill. The input stream is the solid waste delivered to the transfer station.

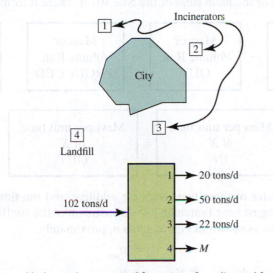

FIGURE 4.8 Material balance diagram used for a city refuse disposal facility.

Using the diagram, we set up the mass balance in terms of tons per day:

Mass or Volume Rate ACCUMULATED	=	Mass or Volume Rate IN	−	Mass or Volume Rate OUT	+	Mass or Volume Rate PRODUCED	−	Mass or Volume Rate CONSUMED

Mass per unit time of X IN	=	Mass per unit time of X OUT

[Mass per unit of refuse IN] = [mass per unit of refuse OUT]

102 tons/day = 20 tons/day + 50 tons/day + 22 tons/day + M

where M = mass of refuse to the landfill.

Solving for the unknown yields: M = 10 tons/day.

EXAMPLE 4.2   Flow in the Blue and White Nile Rivers

A common example of the application of a material flow analysis occurs when two rivers join together at a point called the confluence and these rivers form one larger river that continues to flow toward an ocean.

The United Nations Food and Agriculture Organization (FAO) reported that the peak flow from the Blue Nile in August at Khartoum near the confluence of the two rivers is approximately 15 km³ per month. The peak flow from the White Nile in August at Mogren near the confluence of the two rivers is approximately 2 km³ per month.

What is the flow rate of the main stem of the Nile River where it forms near Tamaniat?

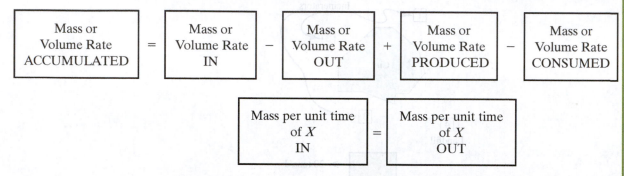

The mass flow rates of these two rivers are additive, and the flow rate of the two combined rivers in August near Tamaniat, just downstream of the confluence of the Blue and White Nile Rivers, as shown in Figure 4.9, is approximately

$$M_{\text{in}} = M_{\text{out}}$$

$$M_{\text{White Nile}} + M_{\text{Blue Nile}} = M_{\text{Combined Nile Rivers}}$$

In units of km³/month:

15 km³ per month + 2 km³ per month = 17 km³ per month

In units of liters per second for the Blue Nile River:

$$15\,\frac{\text{km}^3}{\text{month}} \times \left(\frac{1{,}000\,\text{m}}{\text{km}}\right)^3 \times \frac{1{,}000\,\text{L}}{\text{m}^3} \times \frac{1\,\text{month}}{30\,\text{days}} \times \frac{1\,\text{day}}{24\,\text{hours}} \times \frac{1\,\text{hour}}{60\,\text{minutes}} \times \frac{1\,\text{minute}}{60\,\text{seconds}} = 5.8 \times 10^6\,\frac{\text{L}}{\text{s}}$$

In units of liters per second for the White Nile River:

$$2\,\frac{\text{km}^3}{\text{month}} \times \left(\frac{1{,}000\,\text{m}}{\text{km}}\right)^3 \times \frac{1{,}000\,\text{L}}{\text{m}^3} \times \frac{1\,\text{month}}{30\,\text{days}} \times \frac{1\,\text{day}}{24\,\text{hours}} \times \frac{1\,\text{hour}}{60\,\text{minutes}} \times \frac{1\,\text{minute}}{60\,\text{seconds}} = 0.8 \times 10^6\,\frac{\text{L}}{\text{s}}$$

To convert liters per second to kg/s for the Blue Nile River, we must multiply by the density of the fluid. For freshwater under standard conditions, the density of water is 1 kg per liter:

$$5.8 \times 10^6\,\frac{\text{L}}{\text{s}} \times \frac{1\,\text{kg}}{1\,\text{L}} = 5.8 \times 10^6\,\frac{\text{kg}}{\text{s}}$$

To convert liters per second to kg/s for the White Nile River:

$$0.8 \times 10^6\,\frac{\text{L}}{\text{s}} \times \frac{1\,\text{kg}}{1\,\text{L}} = 0.8 \times 10^6\,\frac{\text{kg}}{\text{s}}$$

Then the combined mass flow rate of water in the Nile River is

$$5.8 \times 10^6 \frac{\text{kg}}{\text{s}} + 0.8 \times 10^6 \frac{\text{kg}}{\text{s}} = 6.6 \times 10^6 \frac{\text{kg}}{\text{s}}$$

Over six and a half million kilograms of water flow each second through the Nile River channel during the peak flow season!

The volumetric flow rate (km³ per month) must be multiplied by the density of the water in order to calculate the mass flow rate (kg/s).

FIGURE 4.9 Map of the Blue and White Nile River basins.

Source: lapi/Shutterstock.com with data added by author.

EXAMPLE 4.3 Collecting Wastewater Flow

A trunk sewer shown in Figure 4.10 has a flow capacity of 4.0 m³/s. If the flow to the sewer is exceeded, it will not be able to transmit all the sewage through the pipe, and backups will occur. Currently, three neighborhoods contribute to the sewer, and their maximum (peak) flows are 1.0, 0.5, and 2.7 m³/s. A builder wants to construct a development that will contribute a maximum flow of 0.7 m³/s to the trunk line (the main pipe that conveys wastewater to the treatment plant). Would this cause the sewer to exceed the capacity of the trunk line?

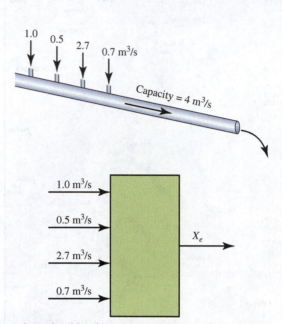

FIGURE 4.10 Sewer branches blending into one trunk sewer.

We can set up the material balance in terms of m³/s:

Mass or Volume Rate ACCUMULATED	=	Mass or Volume Rate IN	−	Mass or Volume Rate OUT	+	Mass or Volume Rate PRODUCED	−	Mass or Volume Rate CONSUMED

Mass per unit time of X IN	=	Mass per unit time of X OUT

[Volume/unit time of sewage IN] = [volume/unit time of sewage OUT]

$$[1.0 \text{ m}^3/\text{s} + 0.5 \text{ m}^3/\text{s} + 2.7 \text{ m}^3/\text{s} + 0.7 \text{ m}^3/\text{s}] = X_e$$

where X_e is the flow in the trunk line. Solving algebraically yields $X_e = 4.9$ m³/s, which is greater than the capacity of the trunk sewer pipe, so the sewer would be overloaded if the new development is allowed to attach to the trunk line. Even now the system is overloaded during the peak flow of 4.2 m³/s, and the only reason disaster has been avoided so far is that not all the neighborhoods produce the maximum flow at the same time of day.

4.1.3 Complex Processes with a Single Material

Previously, it was assumed that the flows were in steady state and the form of the material was neither destroyed (consumed) nor created (produced). The general form of the material balance equation without making any assumptions includes the possibility of material change over time, material production, and material consumption:

| Material per unit time ACCUMULATED | = | Material per unit time IN | − | Material per unit time OUT | + | Material per unit time PRODUCED | − | Material per unit time CONSUMED |

For a specific material labeled A, the mass balance becomes

| Mass of A per unit time ACCUMULATED | = | Mass of A per unit time IN | − | Mass of A per unit time OUT | + | Mass of A per unit time PRODUCED | − | Mass of A per unit time CONSUMED |

Or, if the density of the bulk material remains constant, then the equation can be written in volumetric terms as

| Volume of A per unit time ACCUMULATED | = | Volume of A per unit time IN | − | Volume of A per unit time OUT | + | Volume of A per unit time PRODUCED | − | Volume of A per unit time CONSUMED |

The flow of mass or volume per unit time is the **rate**. Thus, the material balance equation for either mass or volume is

| Rate of A ACCUMULATED | = | Rate of A IN | − | Rate of A OUT | + | Rate of A PRODUCED | − | Rate of A CONSUMED |

Many systems are designed to operate continuously and not change with time; the flows at one moment are exactly like the flows at a later time. This means that no material accumulates in the black box, either positively (material builds up in the box) or negatively (material is flushed out of the box). The general form of the material balance equation for a steady state system when there is no accumulation of material is

$$0 = \begin{array}{|c|} \text{Rate of A} \\ \text{IN} \end{array} - \begin{array}{|c|} \text{Rate of A} \\ \text{OUT} \end{array} + \begin{array}{|c|} \text{Rate of A} \\ \text{PRODUCED} \end{array} - \begin{array}{|c|} \text{Rate of A} \\ \text{CONSUMED} \end{array}$$

In many instances, the substances of concern are not produced or degraded over time. In this steady state condition, $dM/dt = 0$. If there is no decay or generation of the substance within the control volume, then the mass balance equation simplifies to

$$0 = \boxed{\begin{array}{c}\text{Rate of A}\\\text{IN}\end{array}} - \boxed{\begin{array}{c}\text{Rate of A}\\\text{OUT}\end{array}} + 0 - 0$$

or, more simply,

$$\boxed{\begin{array}{c}\text{Rate of A}\\\text{IN}\end{array}} = \boxed{\begin{array}{c}\text{Rate of A}\\\text{OUT}\end{array}}$$

EXAMPLE 4.4   Combining Flows in a Black Box

A sewer carrying stormwater to manhole 1 (Figure 4.11) has a constant flow of 2,000 L/min (Q_A). At manhole 1, it receives a constant lateral flow of 100 L/min (Q_B). What is the flow to manhole 2 (Q_C)?

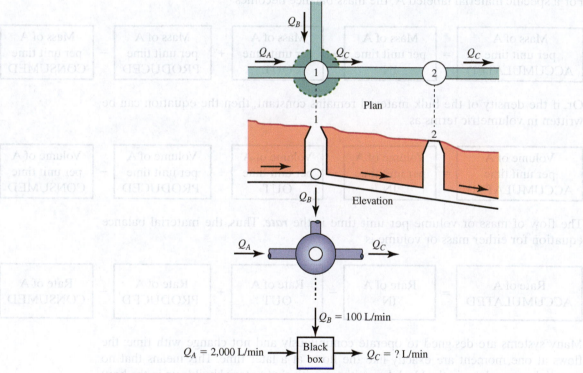

FIGURE 4.11 A sewer with two flows entering manhole 1, mixing and moving toward manhole 2.

Think of manhole 1 as a black box, as shown in Figure 4.11. We can write the material balance for water:

$$\boxed{\begin{array}{c}\text{Rate of A}\\\text{ACCUMULATED}\end{array}} = \boxed{\begin{array}{c}\text{Rate of A}\\\text{IN}\end{array}} - \boxed{\begin{array}{c}\text{Rate of A}\\\text{OUT}\end{array}} + \boxed{\begin{array}{c}\text{Rate of A}\\\text{PRODUCED}\end{array}} - \boxed{\begin{array}{c}\text{Rate of A}\\\text{CONSUMED}\end{array}}$$

Since no water accumulates in the black box, the system is defined as being in steady state and the first term is equal to zero. No water is produced or consumed, so the last two terms are also equal to zero:

$$0 = \left[\begin{array}{c} \text{Rate of A} \\ \text{IN} \end{array}\right] - \left[\begin{array}{c} \text{Rate of A} \\ \text{OUT} \end{array}\right] + 0 - 0$$

$$0 = (Q_A + Q_B) - (Q_C) + 0$$

Substituting into the above equation the given flow rates in liters per minute yields

$$0 = (2{,}000 + 100) - Q_C$$

and solving for Q_C yields

$$Q_C = 2{,}100 \text{ L/min}$$

A system may contain any number of processes or flow junctions, all of which could be treated as black boxes. For example, in the hydrologic cycle, precipitation falls and is added to the watershed into which it falls. At the Earth's surface, some of the rainfall flows out of the watershed while some of it percolates into the groundwater. If water in the area is used for irrigation, the water is removed from the groundwater reservoir through wells. The irrigation water percolates into the ground, is incorporated into vegetation, or flows back into the atmosphere through evaporation or transportation (water released to the atmosphere through plants). Both evaporation and transpiration are commonly combined in one term called *evapotranspiration*. This system can be visualized and modeled as a series of black boxes, as shown in Figure 4.12.

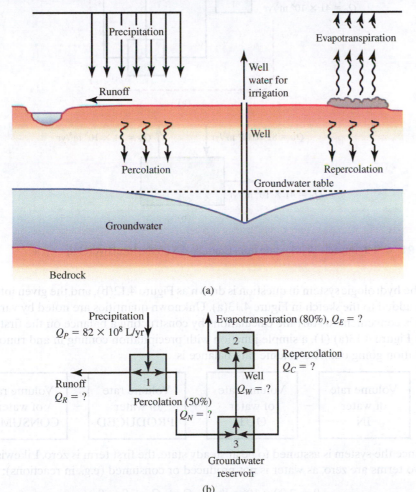

FIGURE 4.12 Visualization and modeling of the hydrologic cycle as black box processes.

EXAMPLE 4.5 Sustainable Water Withdrawal for Irrigation

Suppose the rainfall is 145 cm/yr, of which 50% percolates into the ground. A farmer irrigates crops using well water. Of the extracted well water, 80% is lost by evapotranspiration; the remainder percolates back into the ground. How much groundwater could a farmer on a 20,000-hectare farm extract from the ground per year without depleting the groundwater reservoir volume?

Recognizing this as a material balance problem, we first convert the rainfall to a rate: 145 cm/yr over 20,000 hectares is

$$145 \frac{cm}{yr}\left(\frac{1\ m}{100\ cm}\right) \times 20,000\ \text{hectares}\left(\frac{10,000\ m^2}{\text{hectare}}\right) = 2.90 \times 10^8 \frac{m^3}{yr}$$

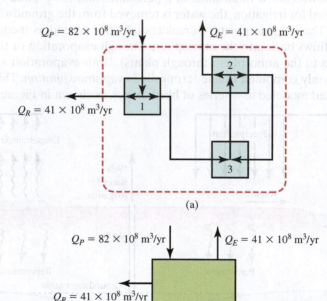

(a)

(b)

FIGURE 4.13 Impacts of irrigation from Example 4.5 on the local hydrologic cycle.

The hydrologic system in question is drawn as Figure 4.12(b), and the given information is added to the sketch in Figure 4.13(a). Unknown quantities are noted by variables.

It is convenient to start the calculations by constructing a balance on the first black box in Figure 4.13(a) (1), a simple junction with precipitation coming in and runoff and percolation going out. The volume rate balance is

Volume rate of water ACCUMULATED		Volume rate of water IN		Volume rate of water OUT		Volume rate of water PRODUCED		Volume rate of water CONSUMED
	=		−		+		−	

Since the system is assumed to be at steady state, the first term is zero. Likewise, the last two terms are zero, as water is not produced or consumed (e.g., in reactions):

$$0 = 2.90 \times 10^8\ m^3/yr - Q_R + Q_N + 0 - 0$$

As stated in the problem, half of the water percolates into the ground; the other half is runoff:

$$Q_R = 0.5Q_P = Q_N$$

Plugging this information into the material balance yields

$$0 = 2.90 \times 10^8 - 2Q_R$$

Solving for Q_R and Q_N, we find

$$Q_R = 1.45 \times 10^8 \frac{m^3}{yr} = Q_N$$

A balance on the second black box (2) yields

$$0 = [Q_W] - [Q_E + Q_C] + 0 - 0$$

As stated in the problem, 80% of the irrigation water is lost by evapotranspiration; therefore, 20% of the irrigation water percolates back into the ground:

$$0 = Q_W - 0.8Q_W - Q_C$$

$$Q_C = 0.2(Q_W)$$

Finally, a material balance on the groundwater reservoir (3) can be written if the quantity of groundwater in the reservoir is assumed not to change:

$$0 = [Q_N + Q_C] - [Q_W] + 0 - 0$$

From the first material balance, $Q_N = 1.45 \times 10^8$ m³/yr, and from the second, $Q_C = 0.2Q_W$. We can plug this information into the third material balance and solve:

$$0 = 1.45 \times 10^8 + 0.2Q_W - Q_W$$

$$Q_W = 1.81 \times 10^8 \text{ m}^3/\text{yr}$$

This is the maximum safe yield of well water for the farmer.

As a check, consider the entire system as a black box. This is illustrated in Figure 4.13(a). There is only one way water can get into this box (precipitation) and two ways out (runoff and evapotranspiration). By representing this black box as Figure 4.13(b), it is possible to write the material balance in cubic meters per year of water:

$$0 = (2.90 \times 10^8) - (1.45 \times 10^8) - (1.45 \times 10^8)$$

The balance of the overall system checks the calculations.

When a series of black boxes are combined to model processes within a larger system, it is possible to draw the control volume boundaries around multiple processes within that system. That larger system of processes can also be modeled as a material balance system. This may be used to check the solution to large complex systems, such as the hydrologic model illustrated in Example 4.5. If the calculations for all the flows are correct within each individual black box, then the flows in the larger system should also balance (as they do in Example 4.5).

4.2 Material Balances with Multiple Materials

Mass and volume balances can be developed with multiple materials flowing in a single system. Chemicals or physical substances in water are often of interest due in part to their potential negative (or sometimes positive) environmental impact. A mass balance on the mass flow of any substance, $M_{substance}$, in a given flow of water, Q_{fluid}, can be related to the concentration of the substance, $c_{substance}$, where

$$M \left[\frac{mass}{time} \right] = Q \left[\frac{volume}{time} \right] c \left[\frac{mass}{volume} \right] \tag{4.5}$$

$$M_{substance} \, [mg/s] = c_{substance} \, [mg/L] \, Q_{fluid} \, [L/s] \tag{4.6}$$

Substituting the relationship between volume, V, and concentration, c, for the mass flow rate yields the general mass balance equation in terms of the substance concentration:

Mass Rate ACCUMULATED	=	Mass Rate IN	−	Mass Rate OUT	+	Mass Rate PRODUCED	−	Mass Rate CONSUMED

$$V(dc/dt) = cQ_{in} - cQ_{out} \pm (dc/dt)V \tag{4.7}$$

4.2.1 Mixing Multiple-Material Flow Streams

Because the mass balance and volume balance equations are not independent equations, it is not possible to develop more than one material balance equation for a black box unless more than one material is involved in the flow. The following examples contain multiple streams, a water stream, and a silt (or solids) stream, so that two material balance equations may be written for each of the systems.

EXAMPLE 4.6   Silt Inputs to the Ohio River

The Allegheny and Monongahela Rivers meet at Pittsburgh to form the mighty Ohio. The Allegheny, flowing south through forests and small towns, runs at an average flow rate of 340 cfs (cubic feet per second) and has a low silt load, 250 mg/L. The Monongahela, on the other hand, flows north at a rate of 460 cfs through old steel towns and farm country, carrying a silt load of 1,500 mg/L.

a. What is the average flow in the Ohio River?
b. What is the silt concentration in the Ohio?

Follow the general rules.

Step 1: Draw the system. Figure 4.14 shows the confluence of the rivers and identifies the flows.

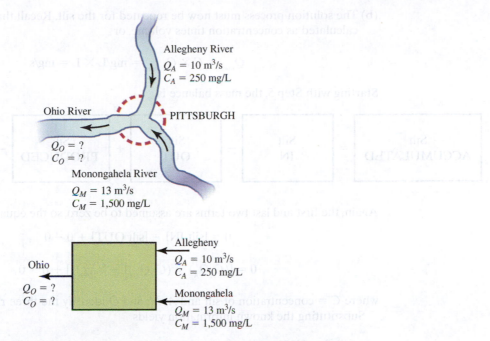

FIGURE 4.14 Confluence of the Allegheny and Monongahela to form the Ohio River.

Step 2: All the available information is added to the sketch, including the known and unknown variables.

Step 3: The confluence of the rivers is the black box, as shown by the dashed line.

Step 4: Water flow is to be balanced first.

Step 5: Write the balance equation:

Rate of water ACCUMULATED	=	Rate of water IN	−	Rate of water OUT	+	Rate of water PRODUCED	−	Rate of water CONSUMED

(a) Because this system is assumed to be in steady state, the first term is zero. Also, because water is neither produced nor consumed, the last two terms are zero. Thus, the material balance becomes

$$0 = [\text{water IN}] - [\text{water OUT}] + 0 - 0$$

Two rivers flow in and one flows out, so the equation in cfs reads

$$0 = [340 + 460] - [Q_o] + 0 - 0$$

where Q_o = flow in the Ohio.

Step 6: Solve for the unknown:

$$Q_o = 800 \text{ cfs}$$

(b) The solution process must now be repeated for the silt. Recall that mass flow is calculated as concentration times volume, or

$$Q_{mass} = C \times Q_{volume} = \text{mg/L} \times \text{L} = \text{mg/s}$$

Starting with Step 5, the mass balance is

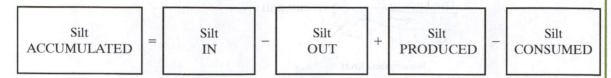

| Silt ACCUMULATED | = | Silt IN | − | Silt OUT | + | Silt PRODUCED | − | Silt CONSUMED |

Again, the first and last two terms are assumed to be zero, so the equation becomes

$$0 = [\text{silt IN}] - [\text{silt OUT}] + 0 - 0$$

$$0 = [(C_A Q_A) + (C_M Q_M)] - [C_O Q_O] + 0 - 0$$

where C = concentration of silt and A, M, and O identify the three rivers. Substituting the known information yields

$$0 = [(250 \text{ mg/L} \times 340 \text{ cfs}) + (1{,}500 \text{ mg/L} \times 460 \text{ cfs})] - [C_O \times 800 \text{ cfs}]$$

Note that the flow rate of the Ohio is 800 cfs as calculated from the volume balance (which is why the water balance was done first).

Note also that there is no need to convert the flow rate from ft³/s to L/s because the conversion factor would be a constant that would appear in every term of the equation and would simply cancel.

Solve the equation:

$$C_O = 969 \text{ mg/L} \approx 970 \text{ mg/L}$$

EXAMPLE 4.7 | Total Suspended Solids in the Nile River at Steady State with No Reaction

The Nile River carries a tremendous amount of eroded soil particles in its waters during the wet season. Volumetric flow rates in the rivers and pollutant levels vary with the time of year and wet or dry season, as shown in Figure 4.15. The solid particles in the water are filtered and measured, and these particles are referred to as the total suspended solids (TSS). The Nile Basin Initiative Transboundary Environmental Action Project reported that the TSS concentration in the Blue Nile during the wet season just upstream of the confluence of the two rivers is approximately 7,000 mg/L. The TSS concentration in the White Nile during the wet season just upstream of the confluence of the two rivers is approximately 70 mg/L. Determine the concentration of TSS just downstream of the confluence of the Blue Nile and White Nile. For this example we will assume that there is no change in concentration with time (i.e., $dM/dt = 0$) and the total suspended solids do not react with anything in the river at the confluence of the two rivers (i.e., $M_{reaction} = 0$).

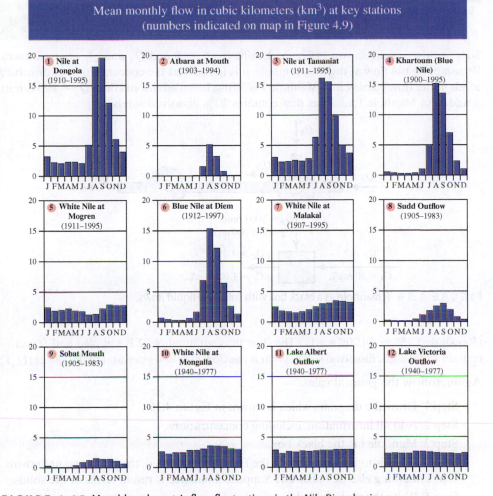

FIGURE 4.15 Monthly volumetric flow fluctuations in the Nile River basin.

Source: Based on FAO (2011). *Synthesis Report: FAO – Nile Basin Project. GCP/INT/945/ITA 2004 to 2009.* Food and Agriculture Organization of the United Nations, Rome, Italy.

$$M_{TSSin} = M_{TSSout}$$

$$M_{TSS\ White\ Nile} + M_{TSS\ Blue\ Nile} = M_{TSS\ Nile\ River}$$

Substituting the volumetric flow rates and concentrations for each tributary of the Nile yields

$$Q_{White\ Nile}\, c_{TSS\ White\ Nile} + Q_{Blue\ Nile}\, c_{TSS\ Blue\ Nile} = Q_{Nile\ River}\, c_{TSS\ Nile\ River}$$

Rearranging:

$$c_{TSS\ Nile} = \frac{Q_{White\ Nile}\, c_{TSS\ White\ Nile} + Q_{Blue\ Nile}\, c_{TSSBlue\ Nile}}{Q_{Nile}}$$

$$c_{TSS\ Nile} = \frac{0.8 \times 10^6\ \frac{L}{s} \times 70\ \frac{mg}{L} + 5.8 \times 10^6\ \frac{L}{s} \times 7{,}000\ \frac{mg}{L}}{6.6 \times 10^6\ \frac{L}{s}} \cong 6{,}000\ \frac{mg}{L} \text{ of TSS in the Nile River}$$

EXAMPLE 4.8   Solving a Material Balance Solution with Two Flows of the Same Type That Are Unknown

Suppose the sewers shown in Figure 4.16 have $Q_B = 0$ and Q_A and Q_C are unknowns. By sampling the flow at the first manhole, it is found that the concentration of dissolved solids in the flow coming into Manhole 1 is 50 mg/L. An additional flow, $Q_B = 100$ L/min, is added to Manhole 1, and this flow contains 20% dissolved solids.

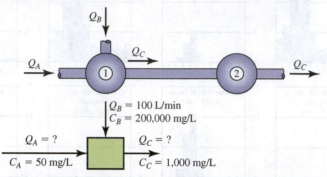

FIGURE 4.16 A manhole as a black box with solid and liquid flows.

(Recall that 1% = 10,000 mg/L.) The flow through manhole 2 is sampled and found to contain 1,000 mg/L dissolved solids. What is the flow rate of wastewater in the sewer (Q_A)?

Again, follow the general rules:

Step 1: Draw the diagram, which is shown in Figure 4.16.

Step 2: Add all information, including concentrations.

Step 3: Manhole 1 is the black box.

Step 4: What is to be balanced? If the flows are balanced, there are two unknowns. Can something else be balanced? Suppose a balance is run in terms of the solids.

Step 5: Write the material balance for solids:

$$\boxed{\begin{array}{c}\text{Solids}\\\text{ACCUMULATED}\end{array}} = \boxed{\begin{array}{c}\text{Solids}\\\text{IN}\end{array}} - \boxed{\begin{array}{c}\text{Solids}\\\text{OUT}\end{array}} + \boxed{\begin{array}{c}\text{Solids}\\\text{PRODUCED}\end{array}} - \boxed{\begin{array}{c}\text{Solids}\\\text{CONSUMED}\end{array}}$$

A steady state assumption allows the first term to be zero, and because no solids are produced or consumed in the system, the last two terms are zero, so the equation reduces to

$$0 = [\text{solids flow in}] - [\text{solids flow out}] + 0 - 0$$

$$0 = [Q_A C_A + Q_B C_B] - [Q_C C_C] + 0 - 0$$

$$0 = \left[Q_A \left(50 \, \frac{\text{mg}}{\text{L}} \right) + \left(100 \, \frac{\text{L}}{\text{min}} \right) \left(200{,}000 \, \frac{\text{mg}}{\text{L}} \right) \right] - \left[Q_C \left(1{,}000 \, \frac{\text{mg}}{\text{L}} \right) \right]$$

Note that (L/min) × (mg/L) = (mg solids/min). This results in an equation with two unknowns, so it is necessary to skip Step 6 and proceed to Step 7.

Step 7: If more than one unknown variable results from the calculation, establish another balance. A balance in terms of the volume flow rate is

Volume ACCUMULATED	=	Volume IN	−	Volume OUT	+	Volume PRODUCED	−	Volume CONSUMED

Again, the first term and the last two terms are assumed to be zero, so

$$0 = [Q_A + Q_B] - [Q_C]$$

and

$$0 = Q_A + 100 - Q_C$$

Now we have two equations with two unknowns. We can substitute $Q_A = (Q_C - 100)$ into the first equation:

$$50(Q_C - 100) + 200,000(100) = 1,000Q_C$$

and solve:

$$Q_C = 21,047 \text{ L/min} \approx 21,000 \text{ L/min and } Q_A = 20,947 \text{ L/min} \approx 20,900 \text{ L/min}$$

BOX 4.1   Raw Material Combinations Used to Manufacture Nitrogen-Based Fertilizer

Industrial production of low-cost fertilizer is critical to the food needs of Earth's ever-growing human population. Ammonia is a nitrogen-based fertilizer that can be produced from nitrogen in air and a hydrocarbon source—most commonly natural gas. Industrial production of fertilizer was made possible in the early 1900s by German chemists Fritz Haber and Carl Bosch, who developed a method to reliably convert atmospheric nitrogen to ammonia; it was subsequently called the Haber-Bosch process.

The Haber-Bosch process converts nitrogen gas (N_2) and hydrogen gas (H_2) to ammonia (NH_3) by using an iron metal catalyst in a reaction process at high temperatures and pressures:

$$N_2 + 3H_2 \rightarrow 2NH_3$$

The simplified ammonia production process can be illustrated by a series of material black boxes, as in Figure 4.17. The raw materials for ammonia production often come from air, water, and natural gas, which are refined to create methane (CH_4). The water is heated to high temperature and pressurized steam in the presence of a nickel catalyst to separate the carbon and hydrogen atoms, forming hydrogen gas and carbon monoxide. In the Haber-Bosch process, air is then added to supply the nitrogen. The oxygen is removed through combustion with excess methane, which produces carbon dioxide CO_2 as a byproduct that is separated and removed from the process as a waster gas. The remaining nitrogen gas and hydrogen gas are converted to ammonia in the presence of an iron catalyst and by adding energy through an electric current.

The simplified black box model shown in Figure 4.17 can be used to track and analyze material flow in the production of ammonia fertilizer. In actuality, the process is more complex when temperature changes and pressure changes are included. In manufacturing

(Continued)

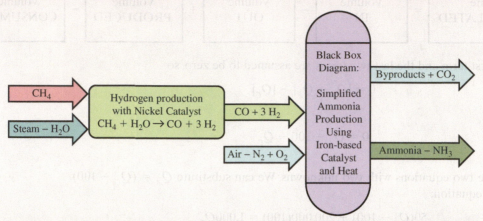

FIGURE 4.17 Black box diagram illustrating the production of ammonia, a nitrogen-based fertilizer.

facilities, the temperature and pressure changes necessary to produce the ammonia are created through electromechanical processes such as compressors, heat exchangers, separators, and condensers. In addition to these steps, a singular unit for ammonia production is often only about 15% efficient; thus, in most applications the product streams are recycled to increase the overall process efficiency. The manufacturing diagram for ammonia production is shown in Figure 4.18.

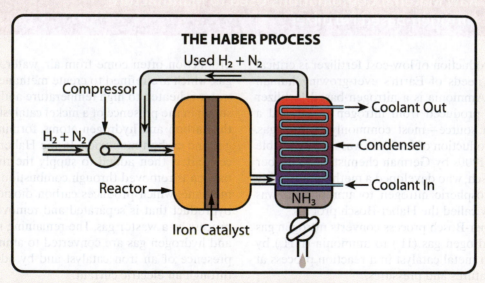

FIGURE 4.18 Illustration of the common steps in ammonia manufacturing.

Source: Gstraub/Shutterstock.com.

Ammonia can be used as a liquid fertilizer; however, it is difficult to store and handle, so it is often converted to other forms that are more easily handled and applied. Typically, this involves first converting ammonia to nitric acid (HNO_3) through an oxidation process (called the Ostwald process) that produces nitric oxide as an intermediary chemical that requires more energy and then adding water to produce nitric acid. The nitric acid (NO_3^-, an anion) and ammonium (NH_4^+, a cation) may then be mixed together in a tank to form ammonium nitrate (NH_4NO_3). The ammonium nitrate can be formed into granular pellets and stored and transported more easily than ammonia.

Black box diagrams can be used to understand the flow and conversion of raw materials through processes to product formation. Black box diagrams can also be used to identify byproducts and waste in manufacturing processes. More detailed flow diagrams are needed to fully account for energy flow throughout the processes, but the black box diagrams used for material flow analysis are a prerequisite for designing manufacturing processes and determining material conversion and flow through a manufacturing system.

ACTIVE LEARNING EXERCISE 4.2 Tracking Nutrient Mass Application and Utilization

Black box models are useful for understanding manufacturing processes, identifying waste streams, and understanding how various materials affect the life cycles of manufactured goods and foods. Fertilizers are used to produce agricultural products for food, biofuels, and manufactured goods, like some plant-based oils and waxes. Create a black box model for the materials required to grow a soybean plant. What are the materials that flow into the plant, and what compounds and materials are outputs of the soybean plant?

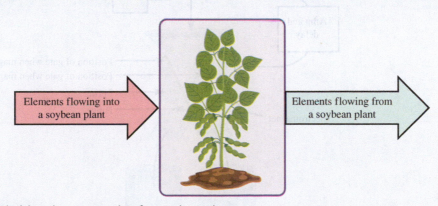

Elements flowing into a soybean plant

Elements flowing from a soybean plant

FIGURE 4.19 Black box diagram template for a soybean plant.

4.2.2 Separating Multiple-Material Flow Streams

The objective of a material separator is to split a mixed-feed material into its individual components by exploiting some difference in material properties. Consider a glass recycling facility that needs to separate transparent crushed glass from opaque (amber, brown, or green) glass into two different streams. The first step is to determine the material properties that determine the separate streams. This becomes the *code*, or signal, that will be used to tell the machine how to divide the individual particles in the stream. In the example of the glass recycling plant, this would be the transparency of the material, which is the property used in the design of the separator system illustrated in Figure 4.20. In this machine, the glass pieces drop off a conveyor belt and pass through a light beam. The amount of light transmitted is read by a photocell. If the light beam is interrupted by an opaque piece of glass, the photocell receives less light, which activates an electromagnet that pulls the gate left. A transparent piece of glass does not interrupt the light beam and the gate does not move. The gate, activated by the signal from the photocell, is the *switch*, separating the material according to the code. This simple device illustrates the nature of coding and switching, reading a property difference and then using that signal to achieve separation.

Material separation devices are not infallible; they make mistakes. It cannot be assumed that the light will always correctly identify the clear glass. Even for the most sophisticated and carefully designed devices, mistakes will occur, and an opaque piece of glass may end up in the container with the transparent glass, and vice versa. The measure of how well separation works uses two parameters: recovery and purity.

Two components, *x* and *y*, enter a black box separator, as shown in Figure 4.21. The two components are to be separated so that *x* goes to product stream 1 and *y* goes to product stream 2. Unfortunately, some of the *y* material is carried into

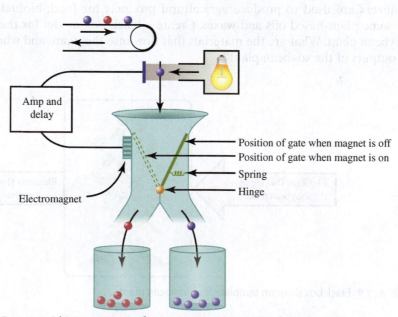

FIGURE 4.20 A binary separator for separating colored glass from clear glass.

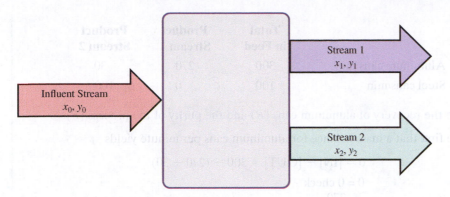

FIGURE 4.21 A binary separator.

stream 1, and likewise a small amount of the x material comes out in stream 2. The recovery of component x in product stream 1 is defined as

$$R_{x_1} = \frac{x_1}{x_0} \times 100 \qquad (4.8)$$

where x_1 is the material flow in the effluent stream 1 and x_0 is the material flow in the influent stream.

Similarly, the recovery of material y in product stream 2 can be expressed as

$$R_{y_2} = \frac{y_2}{y_0} \times 100 \qquad (4.9)$$

where y_2 is the material flow in effluent stream 2 and y_0 is the material flow in the influent stream.

If the materials are perfectly separated, then the recovery of both x and y is 100%.

If we want to maximize the recovery of x in product stream 1, then x_2 should be reduced to the smallest possible value. This can be accomplished by shutting off product stream 2 and diverting the entire stream to stream 1, which would achieve $R_x = 100\%$! Obviously, this is not achieving the desired goal of separation—we are achieving 100% recovery, but we aren't achieving anything useful. The separation purity is the second parameter in addition to recovery that describes a useful separation process. The purity of x in exit stream 1 is defined as

$$P_{x_1} = \frac{x_1}{x_1 + y_1} \times 100 \qquad (4.10)$$

A combination of recovery and purity is used to establish the performance criteria for a separator. The same principle can be applied to more complex environmental processes, such as gravitational thickness.

<table>
<tr><td>**EXAMPLE 4.9**</td><td>Separating Aluminum and Steel Cans in a Binary Separator</td></tr>
</table>

Assume that an aluminum can separator in a local recycling plant processes 400 cans/min. The two product streams consist of the following:

	Total in Feed	Product Stream 1	Product Stream 2
Aluminum cans/min	300	270	30
Steel cans/min	100	0	100

Calculate the recovery of aluminum cans (R) and the purity of the process (P).

Note first that a mass balance for aluminum cans per minute yields

$$0 = [\text{IN}] - [\text{OUT}] = 300 - (270 + 30)$$
$$0 = 0 \text{ check}$$
$$R_{\text{Al cans}_1} = \frac{270}{300} \times 100 = 90\%$$
$$P_{\text{Al cans}_1} = \frac{270}{270 + 0} \times 100 = 100\%$$

This separator has a high recovery and a very high purity. In fact, according to these data, there is no contamination (i.e., steel cans) in the final aluminum can stream (product stream 1).

EXAMPLE 4.10 Separating Liquids and Solids in a Gravitational Thickener

Figure 4.22(a) shows a gravitational thickener used in numerous water and wastewater treatment plants. This device separates suspended solids from the liquid (usually water) by taking advantage of the fact that the solids have a higher density than the water. The code for this separator is density, and the thickener is the switch, allowing the denser solids to settle to the bottom of a tank from which they are removed. The flow into the thickener is called the influent, or feed; the low solids exit stream is the overflow, and the heavy, or concentrated, solids exit stream is the underflow. Suppose a thickener in a metal plating plant receives a feed of 40 m³/hr of precipitated metal plating waste with a suspended solids concentration of 5,000 mg/L. If the thickener is operated in a steady state mode so that 30 m³/hr of flow exits as the overflow, and this overflow has a solids concentration of 25 mg/L, what is the underflow solids concentration and what is the recovery of the solids in the underflow?

Once again, we can consider the thickener as a black box, as in Figure 4.22(b), and proceed stepwise to balance first the volume flow and then the solids flow. Assuming steady state, the volume balance in cubic meters per hour is

Volume ACCUMULATED	=	Volume IN	−	Volume OUT	+	Volume PRODUCED	−	Volume CONSUMED

$$0 = 40 - (30 + Q_u) + 0 - 0$$
$$Q_u = 10 \text{ m}^3/\text{hr}$$

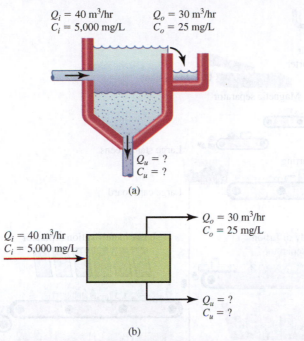

FIGURE 4.22 A gravity-based reactor to separate solid particles in wastewater from the liquid.

For the solids, the mass balance is

$$0 = (C_iQ_i) - [(C_uQ_u) + (C_oQ_o)] + 0 - 0$$
$$0 = (5,000 \text{ mg/L})(40 \text{ m}^3/\text{hr}) - [C_u(10 \text{ m}^3/\text{hr}) + (25 \text{ mg/L})(30 \text{ m}^3/\text{hr})]$$
$$C_u = 19,900 \text{ mg/L}$$

The recovery of solids is

$$R_u = \frac{C_uQ_u}{C_iQ_i} \times 100$$

$$R_u = \frac{[(19,900 \text{ mg/L})(10 \text{ m}^3/\text{hr}) \times 100]}{[(5,000 \text{ mg/L})(40 \text{ m}^3/\text{hr})]} = 99.5\%$$

It is possible to envision a polynary separator, illustrated in Figure 4.23 and modeled mathematically in Figure 4.24, that divides a mixed material into three or more components. The performance of a polynary separator can also be described by its recovery, purity, and efficiency. The recovery of component x_1 in effluent stream 1 is

$$R_{x_{11}} = \frac{x_{11}}{x_{10}} \times 100 \tag{4.11}$$

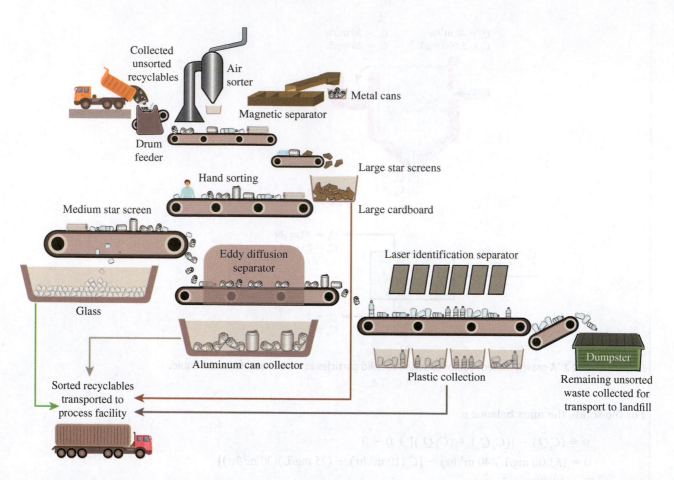

FIGURE 4.23 Sorting of solid waste for recycling and material recovery.

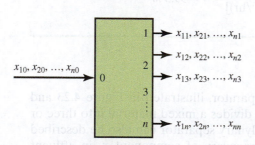

FIGURE 4.24 An example of a polynary separator.

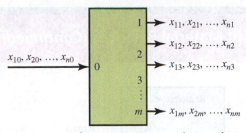

FIGURE 4.25 A polynary separator for *n* components and *m* product streams.

where x_{11} is the component of x_1 in effluent stream 1 and x_{10} is the component x_1 in the influent.

The purity of product stream 1 with respect to component x_1 is

$$P_{x_{11}} = \frac{x_{11}}{x_{11} + x_{21} + x_{31} + \cdots + x_{n1}} \times 100 \qquad (4.12)$$

In this polynary separator, it is assumed that the feed has *n* components and the separation process has *n* product streams. A more general condition is presented in Figure 4.25, where there are *m* product streams for a feed with *n* components. The equations for recovery and purity can be constructed in the manner presented.

BOX 4.2 Large-Scale Applications of Material Balances

Material balances, also called mass balances, can be applied at any scale. In the United Kingdom, mass balances were analyzed to minimize waste throughout the entire country. The material balance project examined resource flows from the point at which they were extracted or imported, manufactured into products, used, and disposed of or recycled, as illustrated in Figure 4.26. The mass of inputs—whether to a process, an industry, or a region—must balance the mass of outputs as products, emissions, and wastes, and it must account for any changes in the accumulated mass (referred to as stock, such as building infrastructure). When applied in a systematic manner, this concept can help industry, government, and others make more

informed decisions and prioritize both problems and opportunities.

Waste management in the United Kingdom prior to 2000 dealt mainly with the removal of waste and not with the minimization or management of waste products (Quested, Ingel, and Parry, 2013). A more sustainable approach to waste management was sought that considered energy, agricultural, and waste resource markets that were subject to distinct, often unconnected supply- and demand-side influences on costs and prices that resulted in increased greenhouse gas pollution. Several real threats to existing resources and disposal processes were identified that were related to waste production:

- Probable price increases in fossil carbon resources

(Continued)

BOX 4.2 Large-Scale Applications of Material Balances *(Continued)*

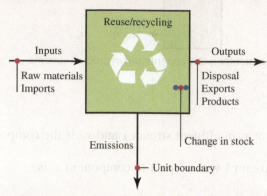

FIGURE 4.26 General material balance used to evaluate waste minimization in the United Kingdom.

- Taxation or pollution permits applied to carbon emissions
- Continued escalation in the UK landfill tax to £80 per tonne by 2012
- Closure of obsolete coal-fired and nuclear electrical generation capacity (equivalent to 33% of total UK electrical supply in 2008–2015)
- International treaties committing the UK to supplying 15% of total energy from renewables by 2020 (Jones, 2009)

Resource flow or material flow analysis (MFA) was used to compare consumption in human societies to other biological metabolic processes. A society was depicted as a living organism that continuously withdraws resources from nature, digests them in the transformation processes of production and consumption, and finally releases them back to nature as wastes/residuals. This mass balance approach has been further developed to create eight areas of specialized analyses:

- Economy-wide material flow analysis (EMFA)
- Bulk material flow or material systems analysis (BMFA/MSA)
- Physical input-output analysis (PIOA)
- Environmental input-output analysis (EIOA)
- Environmental accounts (EA)
- Life cycle analysis/inventories (LCA or LCI)
- Substance flow analysis (SFA)

- National accounting matrix including environmental accounts (NAMEA)

These material balance methodologies can contribute valuable evidence to inform policymakers and governments as they develop strategic material flow approaches. This requires linking MFA methodologies as closely as possible with consideration of environmental impacts, human (economic) activities, and development of materials with less harmful physical flows in an integrated approach. The interrelationship between material and economic analysis methods is illustrated in Figure 4.27.

Both waste and greenhouse gas emissions associated with waste were reduced in the UK as a result of the Courtauld Commitment, an agreement to develop waste reduction strategies. Key indicators of a successful waste minimization strategy included the following (European Commission, 2021; Municipal Waste Generation, 2008):

- Over £100 million business savings delivered by reducing food waste
- Product and packaging waste reduced by 3%
- Recovery and recycling rate increase from 95% in 2012 to 99% in 2015
- 7% reduction in carbon impact of food and drink packaging, as shown in Figure 4.28
- A notable increase in surplus food and drink redistributed for human consumption

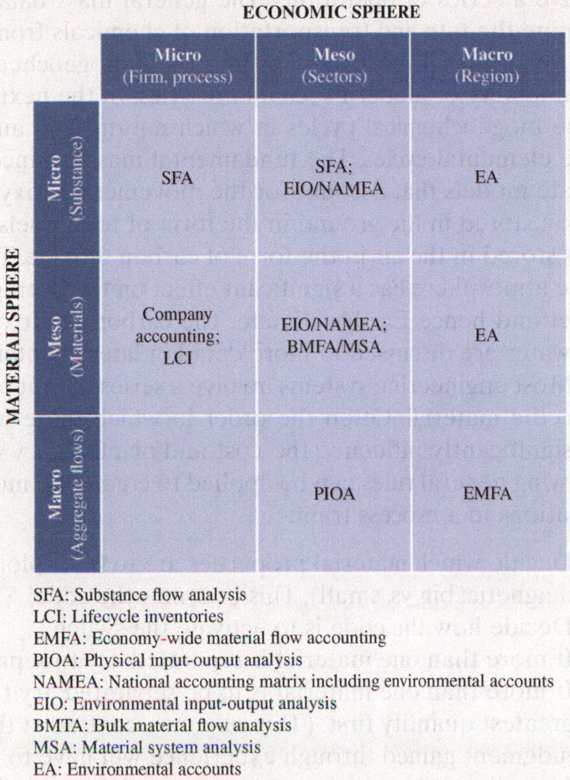

ECONOMIC SPHERE

		Micro (Firm, process)	Meso (Sectors)	Macro (Region)
MATERIAL SPHERE	**Micro (Substance)**	SFA	SFA; EIO/NAMEA	EA
	Meso (Materials)	Company accounting; LCI	EIO/NAMEA; BMFA/MSA	EA
	Macro (Aggregate flows)		PIOA	EMFA

SFA: Substance flow analysis
LCI: Lifecycle inventories
EMFA: Economy-wide material flow accounting
PIOA: Physical input-output analysis
NAMEA: National accounting matrix including environmental accounts
EIO: Environmental input-output analysis
BMTA: Bulk material flow analysis
MSA: Material system analysis
EA: Environmental accounts

FIGURE 4.27 The interrelationship between material balance models and economic models that may be used by policymakers.

Source: Wiedmann, T., Minx, J., Barrett, J., Vanner, R., and Ekins, P. (2006). *Sustainable Consumption and Production - Development of an Evidence Base: Project Ref.: SCP001 Resource Flows—Final Project Report.* Heslington, York, UK: University of York - Stockholm Environment Institute.

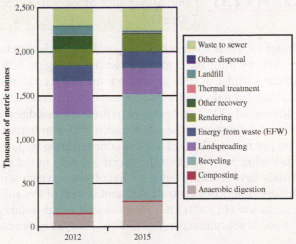

FIGURE 4.28 Reduction in carbon impact of food and drink packaging from material management policies in the UK.
Source: www.wrap.org.uk/content/courtauld-commitment-3-delivering-action-waste.

4.2.3 Complex Processes with Multiple Materials

The preceding principles of mixing and separation using black box models can be applied to a complex system or network with multiple materials by analyzing the system as a combination of several black boxes. Most environmental processes involve a series of operations. The general mass balance equation can be used to examine the fate and transportation of chemicals from one biogeochemical repository to another. There are many important biogeochemical cycles that influence the sustainability of modern societal lifestyles. In the next sections, we will look closely at the biogeochemical cycles in which natural and anthropogenic forces influence these elemental cycles. The fundamental mass balance equation is the basis of the climate models that account for the movement of oxygen and carbon dioxide from carbon stored in the ground in the form of fossil fuels to carbon and oxygen molecules stored in the air in the form of carbon dioxide. The amount of carbon dioxide in the atmosphere has a significant effect on the average surface temperature of the planet and hence Earth's climate. The carbon and oxygen cycles in the Earth's air and water are discussed in more detail in later chapters.

Most engineering systems involve a series of unit processes intended to separate a specific material. Often the order in which these unit operations are employed can significantly influence the cost and/or efficiency of the overall operation. The following general rules can be applied to create the most efficient placement of unit operations in a process train:

- Decide which material properties are to be exploited (e.g., magnetic vs. non-magnetic; big vs. small). This becomes the code.
- Decide how the code is to activate the switch.
- If more than one material is separated, try to separate the easiest one first.
- If more than one material is to be separated, try to separate the one in greatest quantity first. (This rule may contradict the prior rule; engineering judgment gained through experience will have to be exercised.)
- If at all possible, do not add any materials to facilitate separation because this often involves the use of another separation step to recover the material.

EXAMPLE 4.11   Using a Centrifuge to Separate Liquids and Solids in Wastewater Treatment

A sludge of solids concentration $C_0 = 4\%$ is to be thickened to a solids concentration $C_E = 10\%$ by using the centrifuge pictured in Figure 4.29. Unfortunately, the centrifuge produces a sludge at 20% solids from the 4% feed sludge. In other words, it works too well. It has been decided to bypass some of the feed sludge flow and blend it later with the dewatered (20%) sludge so as to produce a sludge with exactly 10% solids concentration. The question is, then, how much sludge to bypass. The influent flow rate (Q_0) is 4 liters per minute (lpm) at a solids concentration (C_0) of 4%. It is assumed here, and in the following problems, that the specific gravity of the sludge solids is 1.0 g/cm³; that is, the solids have a density equal to that of water, which is usually a good assumption. The centrifuge produces a centrate (effluent stream of low solids concentration) with a solids concentration (C_C) of 0.1% and a cake (the high solids concentrated effluent stream) with a solids concentration (C_K) of 20%. Find the required flow rates in liters per minute.

Step 1: Consider the centrifuge as a black box separator, shown in Figure 4.29(b). A volume balance yields

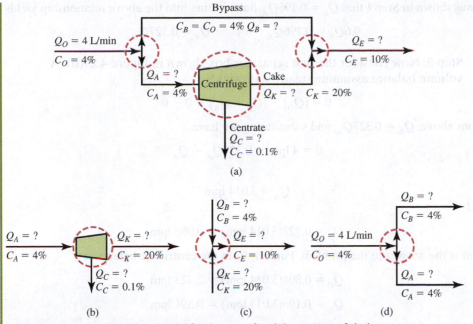

FIGURE 4.29 A black box model for the centrifugal dewatering of sludge.

Volume of sludge ACCUMULATED	=	Volume of sludge IN	−	Volume of sludge OUT	+	Volume of sludge PRODUCED	−	Volume of sludge CONSUMED

Assuming steady state,

$$0 = Q_A - [Q_C + Q_K] + 0 - 0$$

Note that the volume includes the volume of the sludge solids and the volume of the surrounding liquid. Assuming a steady state, a solids balance on the centrifuge gives

$$0 = [Q_A C_A] - [Q_K C_K + Q_C C_C] + 0 - 0$$

$$0 = Q_A (4\%) - Q_K (20\%) - Q_C (0.1\%)$$

Obviously, there are only two equations and three unknowns. For future use, we can solve both of these equations in terms of Q_A, as $Q_C = 0.804 Q_A$ and $Q_K = 0.196 Q_A$.

Step 2: Consider the second junction, in which two streams are blended; see Figure 4.29(c). A volume balance assuming steady state yields

$$0 = [Q_B + Q_K] - [Q_E] + 0 - 0$$

and a solids mass balance is

$$0 = [Q_B C_B + Q_K C_K] - [Q_E C_E] + 0 - 0$$

$$0 = Q_B (4\%) + Q_K (20\%) - Q_E (10\%)$$

Substituting and solving for Q_K yields

$$Q_K = 0.6 Q_B$$

It was shown in Step 1 that $Q_K = 0.196Q_A$. Substituting into the above relationship yields

$$0.6Q_B = 0.196Q_A \quad \text{or} \quad Q_B = 0.327Q_A$$

Step 3: Now consider the first separator box shown in Figure 4.29(d). A volume balance assuming steady state yields

$$0 = [Q_O] - [Q_B + Q_A] + 0 - 0$$

From above, $Q_B = 0.327Q_A$, and substituting, we have

$$0 = 4 \text{ lpm} - 0.327Q_A - Q_A$$

or

$$Q_A = 3.014 \text{ lpm}$$

and

$$Q_B = 0.327(3.014 \text{ lpm}) = 0.986 \text{ lpm}$$

That is the answer to the problem. Further, from the centrifuge balance:

$$Q_C = 0.804(3.014 \text{ lpm}) = 2.423 \text{ lpm}$$

$$Q_K = 0.196(3.014 \text{ lpm}) = 0.591 \text{ lpm}$$

And from the blender box the balance is

$$0 = Q_K + Q_B - Q_E$$

or

$$Q_E = Q_B + Q_K = 1.576 \text{ lpm}$$

Step 4: A check is also available. Using the entire system and assuming steady state, a volume balance in lpm gives

$$0 = Q_O - Q_C - Q_E$$
$$0 = 4.0 - 2.423 - 1.576 = 0.001 \quad \text{Check}$$

ACTIVE LEARNING EXERCISE 4.3 Separation of Plastics in a Recycling Plant

Suppose 7 tons of shredded waste plastics are classified into individual types of plastic. The four different plastics in the mix have the following quantities and densities:

PLASTIC	SYMBOL	RECYCLING CODE	QUANTITY (tons)	DENSITY, ρ (g/cm³)
Polyvinylchloride	PVC	3	4	1.313
Polystyrene	PS	4	1	1.055
Low density polyethylene	LDPE	5	1	0.916
Polypropylene	PP	6	1	0.901

Source: Berthourex, B.M., and Rudd, D.F. (1977). *Strategy for Pollution Control*. New York: John Wiley & Sons.

How would one determine the *best* method for separation?

- What is the code?
- What is the switch?

Write two possible black box models that use different sequential steps to illustrate how the separation process should be designed. The proposed separation process uses three different density fluids to float or sink each plastic type. Each of the three fluids has a different density (ρ): Water has a density of 1 g/cm³, palm oil has a density of 0.908 g/cm³, and a sodium chloride solution has a density of 1.20 g/cm³.

Although the differences are small, it still might be possible to use density, ρ, as the code, and a float/sink apparatus can be used as a switch by judiciously choosing a proper fluid density.

In this case, the easiest separation is the removal of the two heavy plastics (PVC and PS) from the two lighter ones (LDPE and PP) using water, which has a density of 1.0 ($\rho = 1.0$). Following this step, the two streams must be separated further. The most difficult separation is then the splitting of LDPE and PP, which must be done with great precision and is made somewhat easier if PVA and PS are removed first, as illustrated in Figure 4.30.

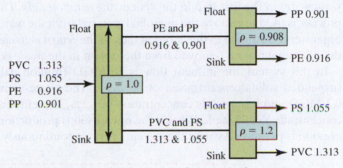

FIGURE 4.30 A process train for the separation of four types of plastics by float/sink.

In the first process train, the first separator receives a flow of 7 tons. The following two units receive 5 and 2 tons each, for a total of 14 tons handled in the process. Is the order of separation shown in Figure 4.30 best? Since the PVC represents the greatest bulk mass of material, it might be better to remove the PVC first, as shown in Figure 4.31. In this alternative process, the first separation device must again handle 7 tons of material, the next one 3 tons (4 have already been removed), and the last one 2 tons, for a total of 12 tons handled in the process. Since the latter process handles less mass, the units can likely be smaller in size and energy usage. Therefore, the alternative process is preferred in terms of the total quantity of material handled.

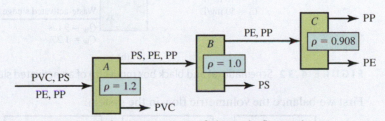

FIGURE 4.31 An alternative process train for the separation of plastics by float/sink.

4.3   Material Balances with Reactors

In the preceding examples, we have assumed that the system is in steady state and there is no production or destruction of the material of interest. ***Reactors*** are the unit processes used to change the form (destruction) of one material and produce another material form. We will continue to assume that the system does not change (steady state) with time, so that if the flows are sampled at any given moment, the result will always be the same.

EXAMPLE 4.12   Conversion of Dissolved Organic Matter in Wastewater to Microbial Cell Mass in a Sludge System

Dissolved organic waste material in wastewater is often converted to cellular mass by microorganisms in a wastewater treatment process. This cellular mass is removed from the flow stream by gravitational settling. This *activated sludge process* (illustrated in Figure 4.32) is one of the most common systems used to remove organic material from wastewater in the first step in the system, the *aeration tank*. The *settling tank*, the second process, is used to remove the microbial material from the water. A portion of the microorganisms is returned to the aeration tank in the *return activated sludge*, and a portion of the microbial mass is removed from the system in the *waste activated sludge*.

In this system, the influent flow is 10 MGD (million gallons per day) and has a suspended solids concentration of 50 mg/L. The waste-activated sludge flow rate is 0.2 MGD and has a solids concentration of 1.2%. The effluent (discharge) has a solids concentration of 20 mg/L. What is the rate of solids production, called the yield, in units of pounds per day? Assume the system operates continuously at steady state.

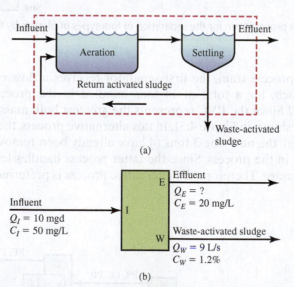

FIGURE 4.32 Schematic (a) and black box model (b) of an activated sludge process.

First we balance the volumetric flow in the system:

$$
0 = \boxed{\begin{array}{c} \text{Volumetric} \\ \text{flow rate} \\ \text{IN} \end{array}} - \boxed{\begin{array}{c} \text{Volumetric} \\ \text{flow rate} \\ \text{OUT} \end{array}} + \boxed{\begin{array}{c} \text{Volumetric} \\ \text{flow rate} \\ \text{PRODUCED} \end{array}} - \boxed{\begin{array}{c} \text{Volumetric} \\ \text{flow rate} \\ \text{CONSUMED} \end{array}}
$$

Flow in = flow out

$$0 = [10 \text{ MGD}] - [0.2 \text{ MGD} + Q_E] + 0 - 0$$

Solving for Q_E yields

$$Q_E = 9.8 \text{ MGD}$$

The second material balance is on the suspended solids:

$$0 = \begin{array}{|c|}\hline \text{Solids} \\ \text{IN} \\\hline\end{array} - \begin{array}{|c|}\hline \text{Solids} \\ \text{OUT} \\\hline\end{array} + \begin{array}{|c|}\hline \text{Solids} \\ \text{PRODUCED} \\\hline\end{array} - \begin{array}{|c|}\hline \text{Solids} \\ \text{CONSUMED} \\\hline\end{array}$$

$$0 = [Q_1 C_1] - [Q_E C_E + Q_W C_W] + X - 0$$

Solving for X yields

$$X = [Q_E C_E + Q_W C_W] - [Q_1 C_1]$$

where X is the rate at which solids are produced within the aeration tank. Because all the known terms are in units of MGD × mg/L, they must be converted to lb/day:

$$X = \left[\left\{\left(10 \frac{10^6 \text{ gallons}}{\text{day}}\right) \times \left(50 \frac{\text{mg}}{\text{L}}\right)\right\} + \left\{\left(9.8 \frac{10^6 \text{ gallons}}{\text{day}}\right) \times \left(20 \frac{\text{mg}}{\text{L}}\right)\right\}\right]$$

$$\times \left\{\left(\frac{3.78 \text{ L}}{\text{gallon}}\right) \times \left(\frac{2.2 \text{ lb}}{10^6 \text{ mg}}\right)\right\}$$

$$- \left[\left(0.2 \frac{10^6 \text{ gallons}}{\text{day}}\right) \times \left(12{,}000 \frac{\text{mg}}{\text{L}}\right)\right]\left[\left(\frac{3.78 \text{ L}}{\text{gallon}}\right) \times \left(\frac{2.2 \text{ lb}}{10^6 \text{ mg}}\right)\right]$$

We have $X = 17{,}438 \text{ lb/day} \approx 17{,}000 \text{ lb/day}$.

4.4 Defining the Order of Reactions

Chemical reactions, whether occurring in nature or human-made systems, are predictable. Engineers and scientists use mathematical models to describe the behavior of chemical reactions in a system. As in material balance analysis, we can use simplifying assumptions to create useful estimates of chemical reactions. However, some reaction systems, like ozone depletion and photochemical smog formation, are quite complex and oversimplifying these systems can lead to errors in the predicted behavior. Because these processes are complex, global models are intricate and highly interconnected.

Regardless of their complexity, all models start with simple processes that are built upon and modified as required for accurate analysis. Each model begins with the assumption that some quantity (mass or volume) changes with time and that the quantity of a component can be predicted using rate equations and material balances.

In earlier sections, material flow was analyzed as a steady state operation. Time was not a variable. In Sections 4.4.1 through 4.8, we will consider the case in which the material changes with time. For a completely mixed-batch system, a general mathematical expression describing a rate at which the mass or volume of some material, A, is changing with time, t, can be represented by

$$\frac{dA}{dt} = r \tag{4.13}$$

where r is the reaction rate.

4.4.1 Zero-Order Reactions

Many changes in nature occur at a constant rate. Consider the simple example of a bucket being filled from a garden hose. The volume of water in the bucket is changing with time, and this change is constant (assuming no one is opening or closing the faucet). If at time 0 the bucket has 2 liters of water in it, at 2 seconds it has 3 liters, at 4 seconds it has 4 liters, and so on, the change in the volume of water in the bucket is constant, at a constant rate of

$$1 \text{ liter/2 seconds} = 0.5 \text{ L/s}$$

For the condition where the reaction rate is constant in a system, the reaction is classified as a **zero-order reaction** and the **reaction rate constant**, k, governs the process, where

$$\frac{dA}{dt} = r = k \tag{4.14}$$

Integrating the expression for $A = A_0$ at time $t = t_0$ yields

$$\int_{A_0}^{A} dA = k \int_{t_0}^{t} dt$$

$$A - A_0 = -kt$$

Note that when the constituent of interest, A, is given in terms of mass, the unit for the reaction rate constant k is mass/time, such as kg/s. For a constituent with a concentration, C, the rate constant has units of mass/volume/time or mg/L/s if C is in milligrams per liter and t is in seconds.

The integrated form of the zero-order reaction when the concentration is increasing is

$$C = C_0 + kt \tag{4.15}$$

and if the concentration is decreasing, the equation is

$$C = C_0 - kt \tag{4.16}$$

This equation can be plotted as shown in Figure 4.33 if the material is being destroyed or consumed so that the concentration is decreasing, and the slope has a negative value. If, on the other hand, C is being produced and is increasing, the slope would be positive.

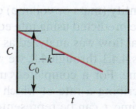

FIGURE 4.33 Plot of a zero-order reaction in which the concentration is decreasing.

EXAMPLE 4.13   Anteater Buffet

An anteater finds an anthill and starts eating. The ants are so plentiful that all it has to do is flick out its tongue and gobble them up at a rate of 200/min. How long will it take to have a concentration of 1,000 ants in the anteater?

We define

C = concentration of ants in the anteater at any time t, ants/anteater

C_0 = initial concentration of ants at time $t = 0$, ants/anteater

k = reaction rate, the number of ants consumed per minute

= 200 ants/anteater/minute (*Note*: k is positive because the concentration is increasing.)

According to Equation (4.15),

$$C = C_0 + kt$$
$$1,000 = 0 + 200(t)$$
$$t = 5 \text{ min}$$

EXAMPLE 4.14   Ozone Generation

Ozone can be generated through electrical discharge or through oxygen contact with ultraviolet (UV) light. Ozone generation is desirable for the removal of microorganisms and some organic compounds in wastewater treatment. In this case, a UV-ozone generator was constructed that produced 20 ppb of ozone in 36 seconds. What was the ozone generation rate constant, k, in ppb/s?

We define

C = concentration of ozone in ppb in the reactor at 36 seconds

C_0 = initial concentration of ozone in ppb entering the reactor at time zero, which is assumed to be zero

t = time, seconds

Using Equation (4.15), we have

$$C = C_0 + kt$$
$$20 = 0 + k(36)$$
$$k = 0.56 \text{ ppb/s}$$

4.4.2 First-Order Reactions

The ***first-order reaction*** of a material being consumed or destroyed can be expressed as

$$\frac{dA}{dt} = r = -kA \qquad (4.17)$$

The equation can be integrated between A_0 and A and between $t = 0$ and t:

$$\int_{A_0}^{A} \frac{dA}{A} = -k \int_{0}^{t} dt$$

$$\ln \frac{A}{A_0} = -kt$$

$$\text{or } \frac{A}{A_0} = e^{-kt}$$

$$\text{or } \ln A - \ln A_0 = -kt \tag{4.18}$$

Note that this integrated form yields the natural logarithm, or base e logarithm. You may recall that the conversion from natural (base e) to common (base 10) logarithms can be accomplished with:

$$\frac{\log A}{\ln A} = \frac{k't}{kt} = 0.434 = \frac{k'}{k}$$

For example, suppose $A = 10$; then

$$\frac{\log 10}{\ln 10} = \frac{1}{2.302} = 0.434 = \frac{k'}{k}$$

Thus,

$$k' = (0.434)k$$

$$k_{\text{base } 10} = (0.434)k_{\text{base } e} \tag{4.19}$$

Figure 4.34 shows that plotting the logarithm of the concentration ($\log C$) versus time (t) yields a straight line with slope $0.434k$. The y-intercept is equal to $\log C_0$. Note that in this plot the slope of the y-term is a logarithmic term. Also, the mass, A, and the concentration, C, are interchangeable in the equations if the volume is constant.

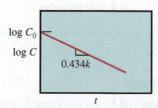

FIGURE 4.34 Plot of a first-order reaction in which the concentration is decreasing.

EXAMPLE 4.15   Owl Buffet

An owl eats frogs as a delicacy, and its intake of frogs is directly dependent on how many frogs are available. There are 200 frogs in the pond, and the rate constant is 0.1 days^{-1}. How many frogs are left at the end of 10 days?

Because the rate is a function of the concentration, this can be described as a first-order reaction. We define

C = number of frogs in the pond at any time t

C_0 = initial frog population, 200 frogs per pond

k = rate constant, 0.1 days^{-1}

Using Equation (4.18), we have

$$\ln \frac{C}{C_0} = -kt$$

$$\ln C - \ln (200) = -0.1(10)$$

$$\ln C = 4.3$$

$$C = 73 \text{ frogs}$$

4.4.3 Pseudo-First Order Reactions

There are many instances in the natural environment where the reaction may be second order or higher in a complex system. However, the kinetics of the system may be controlled by a single parameter, which can be represented by the term A in Equations (4.17) and (4.18). That is to say, the rate of the reaction is dependent on only one parameter in the system. The main difference between a true first-order reaction and a pseudo first-order reaction is the true fundamental nature of the basic chemical reaction. A ***pseudo first-order reaction*** is second order or higher by nature, but the system has been altered to make it simulate a first-order reaction. The other difference is that in a first-order reaction, the rate of reaction depends on all the reactants, whereas in a pseudo first-order reaction, the rate of reaction depends on only the isolated reactant, since a difference in the concentration of the reactant in excess will not affect the reaction. Pseudo first-order reactions can be modeled mathematically as first-order reactions, thus potentially greatly simplifying the analysis of what may otherwise be a complex system.

EXAMPLE 4.16   Ozone Reaction with Organic Compounds in Air

Ozone may react with organic air pollutants. The results of laboratory tests shown in Figure 4.35 illustrate the pseudo first-order reaction rate for the reaction of ethylbenzene in an ozonation chamber. In this test, the concentration of ethylbenzene was lower than the concentration of ozone and the kinetic model was developed based on first-order kinetics. Using the information in Figure 4.34, determine the reaction rate constant and how long it would take to remove 50% (half) of the ethylbenzene for the system modeled in Figure 4.35.

The best-fit slope of the line through the data set shown in Figure 4.35 is shown as $y = 3.2254x$—in this case, $y = \ln(C/C_0)$ and $x = t$—which follows the form of

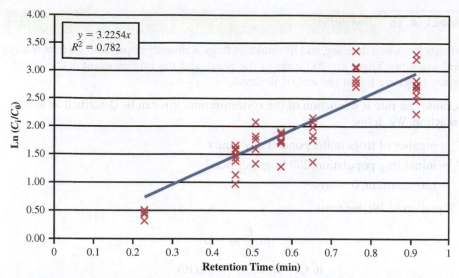

FIGURE 4.35 Pseudo first-order reaction rate for ethylbenzene in an ozonation chamber with excess ozone present.

Source: Striebig, B.A., and Showman, J.P. (1999). "Practicality of photo-oxidative treatment of ethylbenzene." The Fifth International Conference on Advanced Oxidation Technologies for Water and Air Remediation, Albuquerque, NM, May 24–28.

$$\ln\left(\frac{C}{C_0}\right) = -kt$$

Thus, $k = -3.2254$ min^{-1}.

To determine the time to remove 50% of the ethylbenzene, the ratio $C/C_0 = 0.5$. Substituting into the above equation yields

$$\ln(0.5) = -3.2254(t)$$
$$t = 0.215 \text{ min} = 13 \text{ s}$$

4.4.4 Second-Order and Noninteger-Order Reactions

The *second-order reaction* is defined as

$$\frac{dA}{dt} = r = -kA^2 \tag{4.20}$$

Integrated, it is

$$\int_{A_0}^{A} \frac{dA}{A^2} = -k \int_{0}^{t} dt$$

$$\frac{1}{A} \Big|_{A_0}^{A} = kt$$

$$\frac{1}{A} - \frac{1}{A_0} = kt$$

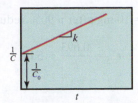

FIGURE 4.36 Plot of a second-order reaction in which the concentration is decreasing.

which plots as a straight line, as shown in Figure 4.36.

The **noninteger-order** (any number) **reaction** is defined as

$$\frac{dA}{dt} = r = -kA^n$$

where n is any number. Integrated, it is

$$\left(\frac{A}{A_0}\right)^{1-n} - 1 = \frac{(n-1)\,kt}{A_0^{(1-n)}} \tag{4.21}$$

These reactions are not as common in environmental engineering.

4.5 Half-Life and Doubling Time

The **half-life** is defined as the time required to convert one-half of a component. At $t = t_{1/2}$, the amount of A is 50% of A_0. The ratio of A at the half-life is known, according to the definition, as $[A]/[A_0] = 0.5$, which can be substituted into the equations for the various orders as shown:

First order: $$t_{1/2} = \frac{\ln 2}{k} = \frac{0.693}{k} \tag{4.22}$$

Second order: $$t_{1/2} = \frac{1}{kA_0} \tag{4.23}$$

Noninteger order: $$t_{1/2} = \frac{[(1/2)^{1-n} - 1] \times A_0^{1-n}}{(n-1)k} \tag{4.24}$$

Similarly, the doubling time is the amount of time required to double the amount of a component. At $t = t_2$, A is twice as much as A_0. By substituting $[A]/[A_0] = 2$ into the expressions, we can calculate the doubling times for various reaction orders.

EXAMPLE 4.17 How Long Before Bedtime Can You Drink Coffee?

The metabolism of caffeinated drinks in the human body can be modeled using first-order kinetics. If Professor Coffy drinks an espresso that contains 100 mg of caffeine, how long will it take him to metabolize 90% of the caffeine if the average half-life for the metabolism of caffeine is 5 hours?

Let $A_0 = 100$ mg. Then $A = 10$ mg after a 90% reduction. Now we have

$$t_{1/2} = \frac{0.693}{k} = 5 \text{ hr}$$

Rearranging yields

$$k = 0.1386 \text{ hr}^{-1}$$

Then

$$\ln\left(\frac{10}{100}\right) = -0.1386(t)$$

$$t = \frac{-2.302}{-0.1386} = 16.6 \text{ hr}$$

Caffeine acts quickly; many people notice the effects within minutes. The duration of the effects of caffeine depends on many factors, and each person may feel the effects differently. Due to the long-term effects of caffeine, the American Academy of Sleep Medicine recommends that you don't consume it for at least 6 hours before bedtime.

4.6 Consecutive Reactions

Some reactions occur consecutively such that

$$A \rightarrow B \rightarrow C \rightarrow \cdots$$

If the first reaction is first order, then

$$\frac{dA}{dt} = -k_1 A$$

Likewise, if the second reaction is first order with respect to B, the overall reaction is

$$\frac{dB}{dt} = -k_1 A - k_2 B$$

where k_2 is the rate constant for the reaction $B \rightarrow C$. Note that some B is being made at a rate of $k_1 A$ while some is being destroyed at a rate of $-k_2 B$. Integrating yields

$$B = \frac{k_1 A_0}{k_2 - k_1}(e^{-k_1 t} - e^{-k_2 t}) + B_0 e^{-k_2 t} \tag{4.25}$$

This equation is reintroduced in Chapter 7 when we analyze the oxygen level in a stream. The oxygen deficit is B, being increased as the result of the oxygen consumption by microorganisms and decreased by the oxygen diffusing into the water from the atmosphere.

Box 4.3 Using a Combination of Chemical Kinetics to Develop Treatment Processes

Typical painting operations involve painting or priming 2 to 4 days during the week. The concentrations of organic compounds in the air range from less than 100 ppm to a maximum of approximately 500 ppm during the painting sessions. The VOC emissions typically consist of less than 20% mixed alcohols and greater than 80% mixed aromatic benzene, ethylbenzene, toluene, and xylene (BTEX) compounds with other trace constituents.

An ozonation chamber was modeled based on the parameters determined for the degradation of ethylbenzene, a BTEX compound. Ethylbenzene degradation in an ozone reactor resulted in a mathematical model useful for predicting the scale-up requirements for oxidation, as shown in Figure 4.37. The model and empirical results predict two distinct operating regimes for the UV system: an oxidant-limited regime and a reactant-limited regime.

The concentration of pollutants entering the reactor in the oxidant-limited regime is greater than the oxidant generation capacity of the reactor. Therefore, the maximum destruction rate of the pollutants will occur at the maximum oxidant generation rate for the system. Maximum reactor efficiency will be achieved in the oxidant-limited regime when operating at the optimum residence time, T_{max}.

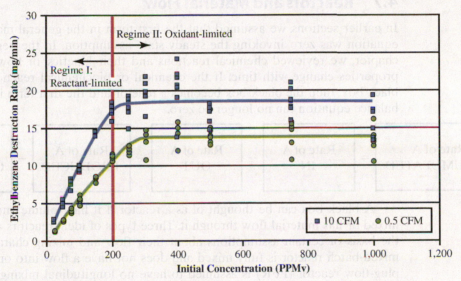

FIGURE 4.37 Regimes of operation in a UV/ozone reactor.

Source: Striebig, B.A., and Showman, J.P. (1999). "Practicality of photo-oxidative treatment of ethylbenzene." The Fifth International Conference on Advanced Oxidation Technologies for Water and Air Remediation, Albuquerque, NM, May 24–28.

In the reactant-limited regime, the production rate of oxidants is greater than or equal to the influent flow rate of the pollutants. In this regime, the reaction rates can be approximated by assuming the oxidant concentration is steady state. The destruction rate of the pollutants can then be modeled by using the first-order reaction rate as illustrated in Box 4.2.

For the ozone-ethylbenzene system, the calculations and design are based on the concentration data shown in Table 4.1. The concentrations are used for the design criteria of the system, as this portion of the system must be capable of achieving the desired removal rate at any moment of operation.

(Continued)

Box 4.3 Using a Combination of Chemical Kinetics to Develop Treatment Processes (Continued)

TABLE 4.1 Modeling and scale-up requirements for an ozone-ethylbenzene reactor

PARAMETER	UNITS	SCENARIO	
		LOW LOADING RATE	HIGH LOADING RATE
Flow rate	scfm	20,000	40,000
VOC concentration	ppm	100	400
VOC loading rate	lb/day	92	2,000
Operating time	hr	3	8
Number of stages in the reactor	#	1	2
Total reactor volume	m³	530	1,500

4.7 Reactors and Material Flow

In earlier sections, we assumed that the first term in the general material balance equation was zero, invoking the steady state assumption. In the beginning of this chapter, we reviewed chemical reactions and their kinetics, or how the chemical properties change with time. If the chemical or biochemical reactions occur in a black box, then the black box becomes a reactor and the first term in the material balance equation can no longer be zero.

$$\boxed{\begin{array}{c}\text{Rate of A}\\\text{ACCUMULATED}\end{array}} = \boxed{\begin{array}{c}\text{Rate of A}\\\text{IN}\end{array}} - \boxed{\begin{array}{c}\text{Rate of A}\\\text{OUT}\end{array}} + \boxed{\begin{array}{c}\text{Rate of A}\\\text{PRODUCED}\end{array}} - \boxed{\begin{array}{c}\text{Rate of A}\\\text{CONSUMED}\end{array}}$$

A black box can be thought of as a reactor if it has volume and if it either is mixed or has material flow through it. Three types of ideal reactors are defined on the basis of certain assumptions about their flow and mixing characteristics. The mixed-batch reactor is fully mixed and does not have a flow into or out of it. The plug-flow reactor (PFR) is assumed to have no longitudinal mixing but complete mixing across its cross-sectional area (latitudinal mixing). The completely mixed-flow reactor (also known as a continuously or completely stirred tank reactor, CSTR) has perfect mixing throughout its entire volume. Many natural systems, as well as engineered systems, can be analyzed using ideal reactor models.

When no reactions occur inside the reactor (nothing is consumed or produced), the mixing model of a reactor is adequate for analysis. The reaction model is used when reactions do occur within a system.

A *tracer* is a chemical, dye (color), or radioactive element (with a long half-life) that is used to trace the flow of materials in a system or reactor. To evaluate the mixing model, a device known as a *conservative and instantaneous signal* is used. A signal is simply a tracer placed into the flow as the flow enters the reactor. The signal allows for the characterization of the reactor by measuring the signal (typically the tracer concentration) with time. The term *conservative* means the signal does not react. For example, a colored dye introduced into water does not react chemically and neither loses nor gains color. *Instantaneous* means that the signal (like a colored dye)

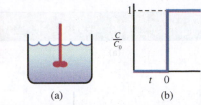

FIGURE 4.38 (a) A mixed-batch reactor and (b) its C-distribution curve. The propeller symbol means that it is perfectly mixed, with no concentration gradients.

is introduced to the reactor all at once—that is, not over time. In real life, this can be visualized as dropping a cup of dye into a bucket of water. Conservative and instantaneous signals can be applied to all three basic types of reactors: mixed-batch, plug-flow, and completely mixed-flow reactors.

4.7.1 Mixed-Batch Reactors

The *mixed-batch reactor* illustrated in Figure 4.38(a) has no flow in or out. If a conservative substance is introduced, it is assumed the signal is mixed instantaneously (zero mixing time). The mixing assumption can be illustrated by plotting the tracer concentration versus time, as shown in Figure 4.38(b). The ordinate indicates the concentration at any time, t, divided by the initial concentration of the tracer. Before $t = 0$, there is no signal (tracer) in the reactor. When the signal is introduced, the signal is immediately and evenly distributed in the reactor vessel (perfect mixing). The concentration immediately "jumps" to C_0, the concentration at $t = 0$ after the signal is introduced to the system. The concentration does not change after that time because there is no flow into or out of the reactor and because the dye is not destroyed (it is conservative).

The plot in Figure 4.38(b) is commonly called a *C-distribution* and is a useful way of graphically representing the behavior of reactors. A C-distribution curve can be plotted as simply C vs. t, but it is more often normalized and plotted as C/C_0 vs. t. For an ideal mixed-batch reactor at $t = 0$, after the signal has been introduced, $C = C_0$ and $C/C_0 = 1$.

Although mixed-batch reactors are useful in a number of industrial and pollution-control applications, a far more common reactor is one in which the flow into and out of the reactor is continuous. Such reactors can be described by considering two ideal reactors: the plug-flow and the completely mixed-flow reactor.

4.7.2 Plug-Flow Reactors

Figure 4.39(a) illustrates the characteristics of a *plug-flow reactor* (PFR), which may be visualized as a very long tube (such as a garden hose) into which a continuous flow is introduced. We assume that while the fluid is in the tube, it experiences no longitudinal mixing. If a conservative signal, such as a dye tracer, is instantaneously introduced into the reactor at the influent end, any two elements within that signal that enter the reactor together will always exit the effluent end at the same time. All signal elements have equal *retention times*, defined as the time between entering and exiting the reactor and calculated as

$$\bar{t} = \frac{V}{Q} \qquad (4.26)$$

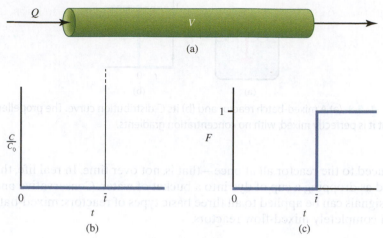

FIGURE 4.39 (a) A plug-flow reactor, (b) its C-distribution curve, and (c) the F-distribution curve.

where

$\bar{t}$ = hydraulic retention time, s
V = volume of the reactor, m³
Q = flow rate to the reactor, m³/s

Note that the retention time in an ideal PFR is the time *any* fluid particle spends in the reactor. Another way of thinking of retention time is the time necessary to *fill* a reactor. If it is plug flow, it is like turning on an empty garden hose and waiting for water to come out the other end. The same definition of retention time holds for *any* type of reactor.

If a conservative instantaneous signal is now introduced into the reactor, the signal moves as a plug through the reactor (which may be visualized as a solid disc moving through the hose), exiting at time *t*. Before time *t*, no portion of the signal (or plug) exits the reactor, and the concentration of the signal in the flow is zero ($C = 0$). Immediately after *t*, all of the signal has exited, and again $C = 0$ in an ideal reactor. The C-distribution curve is thus one instantaneous peak, shown in Figure 4.39(b).

Another convenient means of describing a PFR is to use the **F-distribution**, shown in Figure 4.39(c). F is defined as the fraction of the signal that has left the reactor at any time *t*:

$$F = \frac{A_0 - A_r}{A_0}$$ (4.27)

where

A_0 = amount (usually mass) of a tracer added to the reactor
A_R = amount of tracer remaining in the reactor

As shown in Figure 4.39(c), at $t = \bar{t}$, all the signal exits at the same time in an ideal reactor, so $F = 1$.

To recap, perfect plug flow occurs when there is no longitudinal mixing in the reactor. In practice, reactors are not ideal and perfect plug flow does not exist. The ratio of the C-distribution and the F-distribution curve can be used to create design factors that modify the ideal reactor equations. A full description of accounting methods for nonideal reactors is beyond the scope of this text, but these methods are readily available in the literature and through creating C-distribution and

F-distribution curves for tracers in existing reactors or systems. The impacts of non-ideal flow are often small for large-volume reactors, assuming ideal conditions is often useful in design and analysis.

4.7.3 Completely Mixed-Flow Reactors

In a *completely mixed-flow* (CMF) *reactor*, perfect mixing is assumed. There are no concentration gradients at any time, and a signal is mixed perfectly and instantaneously. Figure 4.40(a) illustrates a CMF reactor.

A conservative instantaneous signal can be introduced into the CMF reactor feed, and the mass balance equation can be written as

Rate of Signal ACCUMULATED	=	Rate of Signal IN	−	Rate of Signal OUT	+	Rate of Signal PRODUCED	−	Rate of Signal CONSUMED

Because the signal is assumed to be instantaneous, the *rate* at which the signal is introduced is zero. Likewise, the signal is conservative and, therefore, the rate produced and the rate consumed are both zero. Thus,

$$\boxed{\text{Rate of Signal ACCUMULATED}} = 0 - \boxed{\text{Rate of Signal OUT}} + 0 - 0$$

If the amount of the signal in the reactor at any time *t* is *A*, and A_0 is the amount of the signal at $t = 0$, then the concentration of the signal in the reactor at $t = 0$ is

$$C_0 = \frac{A_0}{V} \tag{4.28}$$

where
  C_0 = concentration of signal in the reactor at time zero
  A_0 = amount (e.g., mass) of signal in the reactor at time zero
  V = volume of the reactor

After the signal has been instantaneously introduced, the clear liquid continues to flow into the reactor, so the signal in the reactor is progressively diluted. At any time *t*, the concentration of the signal is

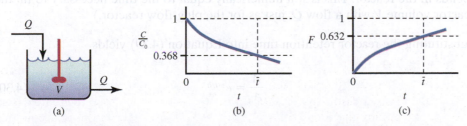

(a)   (b)   (c)

FIGURE 4.40 (a) A completely mixed-flow reactor, (b) its *C*-distribution curve, and (c) its *F*-distribution curve.

$$C = \frac{A}{V}$$

where

C = concentration of signal at any time t
A = amount of signal at any time t
V = volume of the reactor

Because the reactor is perfectly mixed, the concentration of the signal in the flow withdrawn from the reactor must also be C.

The *rate* at which the signal is withdrawn from the reactor is equal to the concentration times the flow rate. Substituting this into the mass balance equation yields

$$\boxed{\begin{array}{c}\text{Rate of Signal}\\ \text{ACCUMULATED}\end{array}} = -CQ = -\left(\frac{A}{V}\right)Q$$

where Q is the flow rate into and out of the reactor. The volume of the reactor, V, is assumed to be constant. The rate that signal A is accumulated is dA/dt, so

$$\frac{dA}{dt} = -\left(\frac{A}{V}\right)Q$$

This equation can be integrated:

$$\int_{A_0}^{A} \frac{dA}{A} = -\int_{0}^{t} \frac{Q}{V}\, dt$$

or

$$\ln A - \ln A_0 = -\frac{Q}{V}t$$

$$\frac{A}{A_0} = e^{-(Qt/V)} \tag{4.29}$$

Recall that the retention time is $\bar{t} = V/Q$.

(*Note*: Now the retention time is defined as the average time a particle of water spends in the reactor. This is still numerically equal to the time necessary to fill the reactor volume V with a flow Q, just as for the plug-flow reactor.)

Substituting the reactor retention time into Equation (4.29) yields

$$\frac{A}{A_0} = e^{-(t/\bar{t})} \tag{4.30}$$

Both A and A_0 may be divided by V to obtain an expression in terms of the concentrations C and C_0, so that

$$\frac{C}{C_0} = e^{-(t/\bar{t})} \tag{4.31}$$

The equation can now be plotted as the C-distribution shown in Figure 4.40(b). At the retention time, $t = \bar{t}$,

$$\frac{C}{C_0} = e^{-1} = 0.368 \tag{4.32}$$

which means that at the retention time, 36.8% of the signal is still in the reactor, and 63.2% of the signal has exited. This is best illustrated again with the F-distribution, such as that shown in Figure 4.40(c).

4.7.4 Completely Mixed-Flow Reactors in Series

The practicality of the mixing model of reactors can be greatly enhanced by considering one further reactor configuration—a series of CMF reactors, as shown in Figure 4.41. Each of the n reactors has a volume of V_0, so that $V = n \times V_0$.

Performing a mass balance on the first reactor and using an instantaneous, conservative signal as before results in

$$\frac{dA_1}{dt} = -\left(\frac{Q}{V_0}\right) A_1$$

where
  A_1 = amount of the signal in the first reactor at any time t, in units such as kg
  V_0 = volume of reactor 1, m³
  Q = flow rate, m³/s

Integrating and rearranging as before yields

$$A_1 = A_0 e^{-(Qt/V_0)} \tag{4.33}$$

where A_0 is the amount of the signal in the reactor at $t = 0$. For each reactor, the retention time is $\bar{t}_0 = V_0/Q$, so that

$$\frac{A_1}{A_0} = e^{-(t/\bar{t}_0)} \tag{4.34}$$

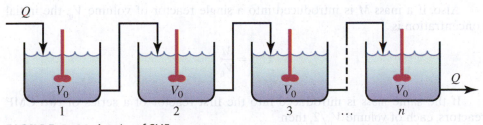

FIGURE 4.41 A series of CMF reactors.

For reactor 2, the mass balance reads as before, and the last two terms are again zero. But now the reactor is *receiving* a signal over time as well as *discharging* it. In addition to the signal accumulation, there are also an inflow and an outflow. In differential form,

$$\frac{dA_2}{dt} = \left(\frac{Q}{V_0}\right)A_1 - \left(\frac{Q}{V_0}\right)A_2$$

Substituting for A_1 from Equation (4.33), we have

$$\frac{dA_2}{dt} = \left(\frac{Q}{V_0}\right)A_0 e^{-(Qt/V_0)} - \left(\frac{Q}{V_0}\right)A_2$$

and integrating:

$$A_2 = \left(\frac{Qt}{V_0}\right)A_0 e^{-(Qt/V_0)}$$

For three reactors,

$$\frac{A_3}{A_0} = \frac{t}{\bar{t}_0}\left(\frac{e^{-(t/\bar{t}_0)}}{2!}\right)$$

Generally, for i reactors,

$$\frac{A_i}{A_0} = \left(\frac{t}{\bar{t}_0}\right)^{i-1}\left(\frac{e^{-(t/\bar{t}_0)}}{(i-1)!}\right) \tag{4.35}$$

For a series of n reactors in terms of concentration (obtained by dividing both A_i and A_0 by the reactor volume V_0), we have

$$\frac{C_n}{C_0} = \left(\frac{t}{\bar{t}_0}\right)^{n-1} e^{-(t/\bar{t}_0)}\left(\frac{1}{(n-1)!}\right) \tag{4.36}$$

Equation (4.36) describes the amount of a conservative, instantaneous signal in any one of a series of n reactors at time t and can be used again to plot the C-distribution. Also, recall that $nV_0 = V$, so that the *total volume* of the entire reactor never changes. The big reactor volume is simply divided into n equal smaller volumes, or

$$n\bar{t}_0 = \bar{t} = \frac{V}{Q}$$

Also, if a mass M is introduced into a single reactor of volume V_0, the initial concentration is

$$C_{0(1)} = \frac{M}{V_0}$$

If the same mass is introduced into the first reactor of a series of two CMF reactors, each of volume $V_0/2$, then

$$C_{0(2)} = \frac{M}{V_0/2} = \frac{2M}{V_0} = 2C_{0(1)}$$

For n reactors,

$$C_{0(n)} = \frac{nM}{V_0} = nC_{0(1)} \tag{4.37}$$

The F-distribution is a good descriptor of tracer behavior, as a CMF reactor is divided into smaller individual CMF volumes:

$$F = \frac{M_0 - M_R}{M_0}$$

The mass of tracer remaining in each of the reactors is

$$M_r = M_1 + M_2 + M_3 + \cdots + M_n = V_0(C_1 + C_2 + C_3 + \cdots + C_n)$$

$$M_0 = V_0 C_0$$

$$F = \frac{C_0 - (C_1 + C_2 + \cdots + C_n)}{C_0} = 1 - \left(\frac{C_1}{C_0} + \frac{C_2}{C_0} + \cdots + \frac{C_n}{C_0}\right)$$

Consider now what the concentration would be in the first reactor as the signal is applied. Because the volume of this first reactor of a series of n reactors is $V_0 = V/n$, the concentration at any time must be n times that of only one large reactor (the same amount of signal diluted by only one-nth of the volume). If it is then necessary to calculate the concentration of the signal in any subsequent reactor, the equation must be

$$\frac{C_n}{C_0} = n\left(\frac{t}{t_0}\right)^{(n-1)}\left(\frac{e^{-(t/t_0)}}{(n-1)!}\right) = n\left(\frac{nt}{\bar{t}}\right)^{(n-1)}\left(\frac{e^{-(nt/\bar{t})}}{(n-1)!}\right) \tag{4.38}$$

where C_0 is the concentration of the signal in the first reactor.

A reactor total volume divided into n volumes must be compared on the basis of the total retention time in the system, which is $\bar{t} = V/Q$.

Substituting the retention time and concentration terms from Equation (4.38) into the equation to develop the F-distribution for several reactors in series yields the curve shown in Figure 4.42 from the following equation:

$$F = 1 - \left[e^{-nt/\bar{t}} + \left(\frac{nt}{\bar{t}}\right)e^{-nt/\bar{t}} + \left(\frac{nt}{\bar{t}}\right)^2\left(\frac{1}{2!}\right)e^{-nt/\bar{t}} + \cdots + \left(\frac{nt}{\bar{t}}\right)^{n-1}\left(\frac{1}{(n-1)!}\right)e^{-nt/\bar{t}}\right]$$

$$= 1 - e^{-nt/\bar{t}}\left[1 + \left(\frac{nt}{\bar{t}}\right) + \left(\frac{nt}{\bar{t}}\right)^2\left(\frac{1}{2!}\right) + \cdots + \left(\frac{nt}{\bar{t}}\right)^{n-1}\left(\frac{1}{(n-1)!}\right)\right]$$

$$= 1 - e^{-nt/\bar{t}}\left[1 + \sum_{i=1}^{n}\left(\frac{nt}{\bar{t}}\right)^{i-1}\frac{1}{(i-1)!}\right] \tag{4.39}$$

As the number of reactors increases, the F-distribution curve becomes more and more like an S-shaped curve. At $n = $ infinity, it becomes exactly like the F-distribution curve for a plug-flow reactor, shown in Figure 4.39(c). This can be readily visualized as a great many reactors in series so that a plug moves rapidly from one very small reactor to another. Since each little reactor has a very short

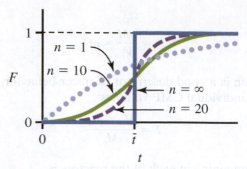

FIGURE 4.42 *F*-distribution curve for a series of CMF reactors.

residence time, as soon as it enters one reactor, it gets flushed out into the next one. This, of course, is exactly how the plug moves through the plug-flow reactor.

Note again, as the *C* and *F* curves clearly show, that as the number of small reactors (*n*) increases, the series of completely mixed reactors begin to behave increasingly like an ideal plug-flow reactor.

EXAMPLE 4.18   Analysis of an Aeration Pond

It is decided to estimate the effect of dividing a large, completely mixed aeration pond (used for wastewater treatment) into 2, 5, 10, and 20 sections so that the flow enters each section in series. Draw the *C*- and *F*-distributions for an instantaneous conservative signal for the single pond and the divided pond.

For a single pond,

$$\frac{C}{C_0} = e^{-(t/\bar{t})}$$

where

C = concentration of the signal in the effluent at time t

C_0 = concentration at time t_0

t = retention time

Substituting various values of t, we have

$$
\begin{array}{lll}
t = 0.25\bar{t} & C/C_0 = e^{-(0.25)} & = 0.779 \\
t = 0.5\bar{t} & C/C_0 = e^{-(0.50)} & = 0.607 \\
t = 0.75\bar{t} & C/C_0 = e^{-(0.75)} & = 0.472 \\
t = \bar{t} & C/C_0 = e^{-1} & = 0.368 \\
t = 2\bar{t} & C/C_0 = e^{-2} & = 0.135
\end{array}
$$

These results are plotted in Figure 4.43 and describe a single reactor with $n = 1$.

Similar calculations can be performed for a series of reactors. For example, for $n = 10$, recall that $10\,\bar{t}_0 = \bar{t}$, and using Equation (4.38):

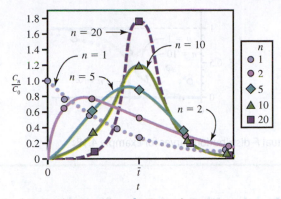

FIGURE 4.43 Actual C-distribution curves for Example 4.18.

$$t = \bar{t} \qquad \frac{C_{10}}{C_0} = (10)(10)^9\, e^{-10}\, \frac{1}{9!} = 1.25$$

$$t = 0.5\bar{t} \qquad \frac{C_{10}}{C_0} = (10)(5)^9\, e^{-5}\, \frac{1}{9!} = 0.36$$

These data are also plotted in Figure 4.43.

Now for the F-distribution: With one reactor, we have

$$t = 0.25\bar{t} \quad F = 1 - e^{-(0.25)} \quad = 0.221$$
$$t = 0.5\bar{t} \quad F = 1 - e^{-(0.50)} \quad = 0.393$$
$$t = \bar{t} \quad F = 1 - e^{-1} \quad = 0.632$$
$$t = 2\bar{t} \quad F = 1 - e^{-2} \quad = 0.865$$

and for the 10 reactors in series:

$$F = 1 - e^{-(nt/\bar{t})}\left[1 + \frac{nt}{\bar{t}} + \left(\frac{nt}{\bar{t}}\right)^2\left(\frac{1}{2!}\right) + \cdots + \left(\frac{nt}{\bar{t}}\right)^{n-1}\frac{1}{(n-1)!}\right]$$

At $t = 0.5\bar{t}$:

$$F = 1 - e^{-5}[1 + 5 + (5)^2\,(1/2!) + (5)^3\,(1/3!) + \cdots + (5)^9\,(1/9!)]$$
$$= 1 - 0.00674[1 + 5 + 12.5 + 12.5 + 20.83 + 26.04 + 26.04$$
$$+ 21.70 + 15.50 + 9.68 + 5.48]$$
$$= 1 - 0.969 = 0.031$$

Similarly, at $t/\bar{t} = 1$:

$$F = 1 - e^{-10}\left[1 + 10 + (10)^2\left(\frac{1}{2!}\right) + \cdots + (10)^9\left(\frac{1}{9!}\right)\right]$$

The plots of the above results are shown in Figure 4.44.

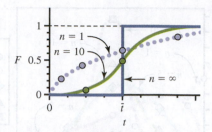

FIGURE 4.44 Actual *F*-distribution curves for Example 4.18.

4.7.5 Mixing Models with Continuous Signals

Thus far the signals used have been instantaneous and conservative. If the instantaneous constraint is removed, the signal can be considered continuous. A continuous conservative signal applied to an ideal PFR simply produces a *C*-distribution curve with a discontinuity going from $C = 0$ to $C = C_0$ at time t. For a CMF reactor, the *C*-distribution curve for when a signal is cut off is exactly like the curve for an instantaneous signal. The reactor is simply being flushed out with clear water, and the concentration of the dye decreases exponentially as before.

If, on the other hand, a continuous signal is introduced to a reactor at $t = 0$ and continued, what does the *C* curve look like? We start by writing a mass balance equation as before. Remember that the flow of dye *in* is fixed while again the rates of production and consumption are zero. The equation describing a CMF reactor with a continuous (from $t = 0$) conservative signal thus is written in differential form as

$$\frac{dC}{dt} = QC_0 - QC$$

and after integration,

$$C = C_0(1 - e^{Qt})$$

4.7.6 Arbitrary-Flow Reactors

Nothing in this world is ideal, including reactors. In plug-flow reactors, there obviously is *some* longitudinal mixing, producing a *C*-distribution more like that shown in Figure 4.45(a) than that in Figure 4.39(b). Likewise, a completely mixed-flow reactor cannot be ideally mixed, so its *C*-distribution curve behaves more like that shown in Figure 4.45(b) than that in Figure 4.40(b). These nonideal reactors are commonly called *arbitrary-flow reactors*.

It should be apparent that the actual *C*-distribution curve in Figure 4.45(b) for the nonideal CMF (arbitrary-flow) reactor looks a lot like the *C* curve for two CMFs in series, as shown in Figure 4.43. In fact, looking at Figure 4.43, we see that as the number of CMFs in series increases, the *C*-distribution curve approaches the *C* curve for the perfect (ideal) plug-flow reactor. Thus, it seems reasonable to expect that all real-life (arbitrary-flow) reactors really operate in the mixing mode as a series of CMF reactors. This observation allows for a quantitative description of reactor flow properties in terms of *n* CMF reactors in series, where *n* represents the number of CMFs in series and defines the type of reactor.

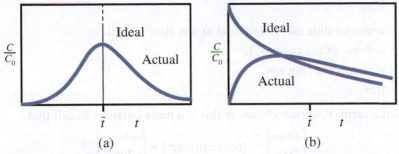

FIGURE 4.45 C-distribution curves for ideal and actual reactors: (a) a plug-flow reactor and (b) a completely mixed-flow reactor.

Further analysis of mixing models is beyond the scope of this brief discussion, and the student is directed to any modern text on reactor design theory for more advanced study.

4.8 Reactor Models

As noted earlier, reactors can be described in two ways: (1) in terms of their mixing properties only, with no reactions taking place; or (2) as true reactors, in which a reaction occurs. In the previous section, the mixing concept was introduced, and now it is time to introduce reactions into these reactors. Stated another way, the constraint that the signal is conservative is now removed.

Again, consider three different ideal reactors: the mixed-batch reactor, the plug-flow reactor, and the completely mixed-flow reactor.

4.8.1 Mixed-Batch Reactors

The same assumptions hold here as in the previous section—namely, perfect mixing (there are no concentration gradients). The mass balance in terms of the material undergoing some reaction is

$$\boxed{\text{Rate ACCUMULATED}} = \boxed{\text{Rate IN}} - \boxed{\text{Rate OUT}} + \boxed{\text{Rate PRODUCED}} - \boxed{\text{Rate CONSUMED}}$$

Because it is a batch reactor, there is no inflow or outflow, and thus

$$\boxed{\text{Rate ACCUMULATED}} = \boxed{\text{Rate PRODUCED}} - \boxed{\text{Rate CONSUMED}}$$

If the material is being produced and there is no consumption, this equation can be written as

$$\frac{dC}{dt} V = rV$$

where

C = concentration of the material at any time t, mg/L

V = volume of the reactor, L

r = reaction rate, mg/L/s

t = time, s

The volume term, V, appears because this is a mass balance. Recall that

$$\begin{bmatrix} \text{Mass} \\ \text{flow} \end{bmatrix} = [\text{concentration}] \times \begin{bmatrix} \text{volume} \\ \text{flow} \end{bmatrix}$$

The left (accumulation) term units are

$$\frac{(\text{mg/L})}{\text{s}} \times (\text{L}) = \frac{\text{mg}}{\text{s}}$$

and the right-side units are

$$(\text{mg/L/s}) \times (\text{L}) = \frac{\text{mg}}{\text{s}}$$

The volume term can be canceled, so that

$$\frac{dC}{dt} = r$$

and integrated

$$\int_{C_0}^{C} dC = r \int_{0}^{t} dt$$

For a zero-order reaction, $r = k$, where k is the reaction rate constant, and thus

$$C - C_0 = kt \tag{4.40}$$

Equation (4.40) holds when the material in question is being produced. In the situation where the reactor destroys the component, the reaction rate is negative, so

$$C - C_0 = -kt \tag{4.41}$$

In both cases,

C = concentration of the material at any time t

C_0 = concentration of the material at $t = 0$

k = reaction rate constant

so

If the reaction is first-order, $r = kC$,

$$\frac{dC}{dt} = kC$$

with the rate constant k having units of time^{-1}. Integration gives

$$\int_{C_0}^{C} \frac{dC}{C} = k \int_{0}^{t} dt$$

$$\ln \frac{C}{C_0} = kt$$

$$C = C_0 e^{kt} \tag{4.42}$$

If the material is being consumed, the reaction rate is negative, so

$$\ln \frac{C}{C_0} = -kt$$

$$C = C_0 e^{-kt} \tag{4.43}$$

EXAMPLE 4.19   Removal of Color by Activated Carbon

An industrial wastewater treatment process uses activated carbon to remove color from the water. The color is reduced as a first-order reaction in a batch adsorption system. If the rate constant, k, is 0.35 day^{-1}, how long will it take to remove 90% of the color?

Let C_0 = initial concentration of the color, and C = concentration of the color at any time t. It is necessary to reach $0.1C_0$. Using Equation (4.43), we find

$$\ln \left(\frac{C_0}{C} \right) = kt$$

$$\ln \left(\frac{C_0}{0.1C_0} \right) = 0.35t$$

$$\ln \left(\frac{1}{0.1} \right) = 0.35t$$

$$t = \frac{2.30}{0.35} = 6.6 \text{ days}$$

4.8.2 Plug-Flow Reactors

The equations for mixed-batch reactors apply equally well to plug-flow reactors because it is assumed that in perfect plug-flow reactors, a plug of reacting materials flows through the reactor and that this plug is itself like a miniature batch reactor. Thus, for a zero-order reaction occurring in a plug-flow reactor in which the material is produced, we have

$$C = C_0 + k\bar{t} \tag{4.44}$$

where

C = concentration of the effluent

C_0 = concentration of the influent

$\bar{t}$ = retention time of the reactor = V/Q

V = volume of the reactor

Q = flow rate through the reactor

If the material is consumed according to the kinetics of a zero-order reaction,

$$C = C_0 - k\bar{t} \tag{4.45}$$

If the material is being produced according to first-order reaction kinetics, then

$$\ln \frac{C}{C_0} = k\bar{t}$$

$$\ln \frac{C}{C_0} = k\left(\frac{V}{Q}\right)$$

$$V = \left(\frac{Q}{k}\right) \ln \frac{C}{C_0} \tag{4.46}$$

If the material is being consumed according to first-order reaction kinetics,

$$\ln \frac{C}{C_0} = -k\bar{t}$$

$$V = \left(-\frac{Q}{k}\right) \ln \frac{C}{C_0}$$

$$V = \frac{Q}{k} \ln \frac{C_0}{C} \tag{4.47}$$

EXAMPLE 4.20 Plug-Flow Reactor for Odor Abatement

An industry wants to use a long drainage ditch to remove odor from its waste. Assume that the ditch acts as a plug-flow reactor. The odor reduction behaves as a first-order reaction, with the rate constant $k = 0.35$ day^{-1}. How long must the ditch be if the velocity of the flow is 0.5 m/s and 90% odor reduction is desired?

Using Equation (4.47), we have

$$\ln \frac{C}{C_0} = -k\bar{t}$$

$$\ln \frac{0.1C_0}{C_0} = -(0.35)\bar{t}$$

$$\ln(0.1) = -(0.35)\bar{t}$$

$$\bar{t} = 6.58 \text{ days}$$

Length of ditch $= 0.5$ m/s $\times$ 6.58 days $\times$ 86,400 s/day

$$= 2.8 \times 10^5 \text{ m}(!)$$

4.8.3 Completely Mixed-Flow Reactors

The mass balance for the completely mixed-flow reactor can be written in terms of the material in question as

Rate ACCUMULATED	=	Rate IN	−	Rate OUT	+	Rate PRODUCED	−	Rate CONSUMED

$$\frac{dC}{dt}V = QC_0 - QC + r_1V - r_2V$$

where

C_0 = concentration of the material in the influent, mg/L

C = concentration of the material in the effluent, and at any place and time in the reactor, mg/L

r = reaction rate

V = volume of the reactor, L

It is now necessary to again assume a steady state operation. Although a reaction is taking place within the reactor, the effluent concentration (and hence the concentration in the reactor) is not changing with time; that is,

$$\frac{dC}{dt}V = 0$$

If the reaction is zero-order and the material is being produced, $r = k$, the consumption term is zero and

$$0 = QC_0 - QC + kV$$

$$C = C_0 + k\frac{V}{Q}$$

$$C = C_0 + k\bar{t} \tag{4.48}$$

Or, if the material is being consumed by a zero-order reaction,

$$C = C_0 - k\bar{t} \tag{4.49}$$

If the reaction is first-order and the material is produced, $r = kC$, and

$$0 = QC_0 - QC + kCV$$

$$\frac{C_0}{C} = 1 - k\frac{V}{Q}$$

$$\frac{C_0}{C} = 1 - k\bar{t}$$

$$C = \frac{QC_0}{Q - kV} \tag{4.50}$$

If the reaction is first-order and the material is being destroyed,

$$\frac{C_0}{C} = 1 + k\left(\frac{V}{Q}\right)$$

$$\frac{C_0}{C} = 1 + k\bar{t}$$

$$C = \frac{QC_0}{Q + kV}$$

$$V = \frac{Q}{k}\left[\frac{C_0}{C} - 1\right] \tag{4.51}$$

Suppose it is necessary to maximize the performance of a reactor that destroys a component—that is, it is desired to *increase* C_0/C or make C small. This can be accomplished in three ways:

1. Increasing the volume of the reactor, V
2. Decreasing the flow rate to the reactor, Q
3. Increasing the rate constant, k

The rate constant k is dependent on numerous variables, such as temperature and the intensity of mixing. The variability of k with temperature is commonly expressed in exponential form as

$$k_T = k_0 e^{\Phi(T - T_0)} \tag{4.52}$$

where

Φ = constant
k_0 = rate constant at temperature T_0
k_T = rate constant at temperature T

EXAMPLE 4.21   CMF Reactor for Disinfecting Water

A new disinfection process destroys coliform (*coli*) organisms in water by using a completely mixed-flow reactor. The reaction is first-order with $k = 1.0$ day^{-1}. The influent concentration is 100 *coli*/mL. The reactor volume is 400 L, and the flow rate is 1,600 L/day. What is the effluent concentration of coliforms?

The general material balance equation is

Rate ACCUMULATED	=	Rate IN	−	Rate OUT	+	Rate PRODUCED	−	Rate CONSUMED

$$0 = QC_0 - QC + 0 - rV$$

where $r = kC$. We have

$$0 = (1{,}600 \text{ L/day})(100 \text{ } coli/\text{mL}) - (1{,}600 \text{ L/day})C - (1.0 \text{ day}^{-1})(400 \text{ L})C$$

$$C = 80 \text{ } coli/\text{mL}$$

4.8.4 Completely Mixed-Flow Reactors in Series

For two CMF reactors in series, the effluent from the first is C_1, and assuming a first-order reaction in which the material is being destroyed, we have

$$\frac{C_0}{C_1} = 1 + k\left(\frac{V_0}{Q}\right)$$

$$\frac{C_1}{C_2} = 1 + k\left(\frac{V_0}{Q}\right)$$

where V_0 is the volume of the individual reactor. Similarly, for the second reactor, the influent is C_1 and the effluent is C_2. For the two reactors,

$$\frac{C_0}{C_1} \times \frac{C_1}{C_2} = \frac{C_0}{C_2} = \left[1 + k\left(\frac{V_0}{Q}\right)\right]^2 \tag{4.53}$$

For any number of reactors in series,

$$\frac{C_0}{C_n} = \left[1 + k\left(\frac{V_0}{Q}\right)\right]^n$$

$$\left(\frac{C_0}{C_n}\right)^{1/n} = 1 + k\frac{V}{nQ} \tag{4.54}$$

where

V = volume of all the reactors, equal to nV_0

n = number of reactors

V_0 = volume of each reactor

The equations for completely mixed-flow reactors are summarized in Table 4.2. Table 4.3 shows the distinction between the equations for completely mixed-flow reactors and plug-flow reactors for different order reactions.

TABLE 4.2 Performance Characteristics of Completely Mixed-Flow Reactors

Zero-order reaction, material produced	$C = C_0 + k\bar{t}$
Zero-order reaction, material destroyed	$C = C_0 - k\bar{t}$
First-order reaction, material produced	$\frac{C_0}{C} = 1 - k\bar{t}$
First-order reaction, material destroyed	$\frac{C_0}{C} = 1 + k\bar{t}$
Second-order reaction, material produced	$C = \frac{-1 + [1 + 4k\bar{t}C_0]^{1/2}}{2k\bar{t}}$
Series of n CMF reactors, zero-order reaction, material destroyed	$\frac{C_0}{C_n} = \left(1 + \frac{k\bar{t}}{C_0}\right)^n$
Series of n CMF reactors, First-order reaction, material destroyed	$\frac{C_0}{C_n} = (1 + k\bar{t}_0)^n$

Note: C = effluent concentration; C_0 = influent concentration; $\bar{t}$ = retention time = V/Q; k = rate constant; C_n = concentration of the nth reactor; $\bar{t}_0$ = retention time in each of n reactors

TABLE 4.3 Summary of Ideal Reactor Performance

| REACTION ORDER | CMF | | PLUG-FLOW REACTOR |
	SINGLE REACTOR	n REACTORS	
Zero	$V = \dfrac{Q}{k}(C_0 - C)$	$V = \dfrac{Q}{k}(C_0 - C_n)$	$V = \dfrac{Q}{k}(C_0 - C)$
First	$V = \dfrac{Q}{k}\left(\dfrac{C_0}{C} - 1\right)$	$V = \dfrac{Qn}{k}\left[\left(\dfrac{C_0}{C}\right)^{1/n} - 1\right]$	$V = \dfrac{Q}{k}\ln\dfrac{C_0}{C}$
Second	$V = \dfrac{Q}{k}\left(\dfrac{C_0}{C} - 1\right)\dfrac{1}{C}$	Complex	$V = \dfrac{Q}{k}\left(\dfrac{1}{C} - \dfrac{1}{C_0}\right)$

Note: Material destroyed; V = reactor volume; Q = flow rate; k = reaction constant; C_0 = influent concentration; C = effluent concentration; n = number of CMF reactors in series.

EXAMPLE 4.22   Comparison of CMF Reactor and PFR

Consider a first-order reaction that requires a 50% reduction in the concentration. Would a plug-flow or a CMF reactor require the least volume?

From Table 4.3,

$$\frac{V_{\text{CMF}}}{V_{\text{PF}}} = \frac{\frac{Q}{k}\left(\frac{C_0}{C} - 1\right)}{\frac{Q}{k}\left(\ln\frac{C_0}{C}\right)}$$

For 50% conversion,

$$\frac{C_0}{C} = 2$$

$$\frac{V_{\text{CMF}}}{V_{\text{PF}}} = \frac{(2 - 1)}{\ln 2} = 1.44$$

A CMF reactor would require 44% more volume than a PFR.

4.9 Summary

Material balances are used ubiquitously in all engineering fields. The same mathematical modeling process applied to measuring solids concentrations in wastewater can also be applied to producing pharmaceutical products in a bioreactor or the product production rate on an industrial assembly line. These are the basic steps in modeling a black box process with a material balance:

1. Drawing the system as a diagram, including all flows (inputs and outputs) as arrows.

2. Adding all available information provided, such as flow rates and concentrations. Assigning symbols to unknown variables.
3. Drawing a continuous dashed line to represent the control volume around the component or components that are to be balanced. This may include a unit operation, a junction, or a combination of these processes.
4. Deciding what material is to be balanced. This may be a volumetric or mass flow rate.
5. Writing the general material balance equation:

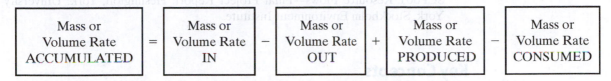

$$\begin{array}{c}\text{Mass or}\\\text{Volume Rate}\\\text{ACCUMULATED}\end{array} = \begin{array}{c}\text{Mass or}\\\text{Volume Rate}\\\text{IN}\end{array} - \begin{array}{c}\text{Mass or}\\\text{Volume Rate}\\\text{OUT}\end{array} + \begin{array}{c}\text{Mass or}\\\text{Volume Rate}\\\text{PRODUCED}\end{array} - \begin{array}{c}\text{Mass or}\\\text{Volume Rate}\\\text{CONSUMED}\end{array}$$

6. Solving for the unknown variables using algebraic techniques.

The general mass balance equation can be used to examine the fate and transportation of chemicals from one biogeochemical repository to another. Many important biogeochemical cycles influence the sustainability of modern societal lifestyles. In the following chapters, we will look closely at the nitrogen and phosphorus cycles, which are influenced by natural and anthropogenic forces. The fundamental mass balance equation is the basis of the climate models that account for the movement of oxygen and carbon dioxide from carbon stored in the ground in the form of fossil fuels to carbon and oxygen molecules stored in the air in the form of carbon dioxide. The amount of carbon dioxide in the atmosphere has a significant effect on the average surface temperature of the planet and, hence, the Earth's climate. The carbon and oxygen cycles in the Earth's air and water are discussed in more detail in Chapters 5 through 10.

The efficiency of ideal reactors can be compared by first solving all the descriptive equations in terms of reactor volume, as shown in Table 4.3 and Example 4.22. The conclusion reached in Example 4.22 is a very important concept used in many environmental engineering systems. Stated generally, for reaction orders greater than or equal to 1, the ideal plug-flow reactor will always outperform the ideal completely mixed-flow reactor. This fact is a powerful tool in the design and operation of treatment systems.

References

Berthourex, B. M., and Rudd, E. F. (1977). *Strategy for Pollution Control.* New York: John Wiley & Sons.

European Commission. Guidelines on Waste Prevention Programmes. ec.europa .eu/environment/waste/prevention/pdf/Waste%20Prevention_Handbook.pdf. Accessed 2021.

Jones, P. T. (2009). "Material flow and mass balance analysis in the United Kingdom." *Journal of Industrial Ecology* 13(6): 843–846.

Municipal Waste Generation, European Environment Agency assessment, 2008: themes.eea.europa.eu/IMS/IMS/ISpecs/ISpecification20041007131809 /IAssessment1183020255530/view_content.

Quested, T., Ingle, R., and Parry, A. (2013). Household Food and Drink Waste in the United Kingdom 2010: Final Report. Project Code: CFP102. ISBN: 978-1-84405-458-9.

Striebig, B. A., and Showman, J. P. (1999). "Practicality of photo-oxidative treatment of ethylbenzene." The Fifth International Conference on Advanced Oxidation Technologies for Water and Air Remediation, Albuquerque, NM, May 24–28.

Wiedmann, T., Minx, J., Barrett, J., Vanner, R., and Ekins, P. (2006). Sustainable Consumption and Production—Development of an Evidence Base: Project Ref.: SCP001 Resource Flows—Final Project Report. Heslington, York: University of York, Stockholm Environment Institute.

Key Concepts

Material balance

Law of conservation of mass

Influent

Effluent

Control volume

Steady state

Rate

Code

Switch

Reactor

Zero-order reaction

Reaction rate constant

First-order reaction

Pseudo first-order reaction

Second-order reaction

Noninteger-order reaction

Half-life

Tracer

Conservative and instantaneous signal

Mixed-batch reactor

Plug-flow reactor

Retention times

C-distribution

F-distribution

Completely mixed-flow reactor

Arbitrary flow reactor

Problems

4-1   A wastewater treatment plant receives 10 mg/L of flow. The wastewater has a solids concentration of 192 mg/L. How many kilograms of solids enter the plant every day?

4-2   A stream flowing at 225 L/min carries a sediment load of 2,000 mg/L. What is the sediment load in kilograms/day?

4-3   A power plant emits 55 kg of fly ash per hour up the smokestack. The flow rate of the hot gases in the stack is 2.5 m³/s. What is the concentration of the fly ash in micrograms per cubic meter?

4-4   A pickle-packing plant produces and discharges a waste brine solution with a salinity of 13,000 mg/L NaCl at a rate of 100 gal/min into a stream with a flow rate greater than the discharge of 1.2 million gal/day and a salinity of 20 mg/L. Below the discharge point is a prime sport fishing spot, and the fish are intolerant of salt concentrations higher than 200 mg/L.

   a. What must the level of salt in the effluent be to reduce the level in the stream to 200 mg/L?

   b. If the pickle-packing plant has to spend $8 million to reduce its salt concentration to this level, it cannot stay in business. Over 200 people will lose their jobs. All this because of a few people who like to fish in the stream? Write a letter to the editor expressing your views, pro or con, concerning the action of the state in requiring the pickle plant to clean up its effluent. Use your imagination.

c. Suppose now that the pickle plant is 2 miles upstream from a brackish estuary and asks permission to pipe its salty wastewater into the estuary where the salinity is such that the waste would actually *dilute* the estuary (the waste has a lower salt concentration than the estuary water). As the plant manager, write a letter to the director of the State Environmental Management Division requesting a relaxation of the salinity effluent standard for the pickle plant. Think about how you are going to frame your arguments.

4-5 Raw primary sludge with a solids concentration of 4% is mixed with waste-activated sludge with a solids concentration of 0.5%. The flows are 20 and 24 gal/min, respectively. What is the resulting solids concentration? This mixture is next thickened to a solids concentration of 8%. What are the quantities (in gallons per minute) of the thickened sludge and thickener overflow (water) produced? Assume 100% of the sludge solids are captured in the thickener.

4-6 A 10-MGD wastewater treatment plant's influent and effluent concentrations of several metals are shown here.

	CONCENTRATION OF THE METAL	
METAL	INFLUENT (mg/L)	EFFLUENT (mg/L)
Cd	0.012	0.003
Cr	0.32	0.27
Hg	0.070	0.065
Pb	2.42	1.26

The plant produces a dewatered sludge cake at a solids concentration of 22% (by weight). Plant records show that 45,000 kg of sludge (wet) are disposed of on a pasture per day. The state has restricted sludge disposal on land to only sludges that have metal concentrations less than the following:

METAL	MAXIMUM ALLOWABLE METAL CONCENTRATION (mg metal/kg dry sludge solids)
Cd	15
Cr	1,000
Hg	10
Pb	1,000

Does the sludge meet the state standards?

a. Assume that the flow rate (MGD of sludge) is negligible compared to the flow rate of the influent; that is, assume $Q_{influent} = Q_{effluent}$, where Q = flow rate in MGD. Check to see whether this is a valid assumption.

b. Assume instead that the density of the sludge solids is about that of water, so that 1 kg of sludge represents a volume of 1 L. Check to see whether this is a valid assumption.

4-7 Two flasks contain 40% and 70% by volume of formaldehyde, respectively. If 200 g from the first flask and 150 g from the second flask are mixed, what is the concentration (expressed as percent formaldehyde by volume) of the formaldehyde in the final mixture?

4-8 A textile plant discharges a waste that contains 20% dye from vats. The color intensity is too great, and the state has told the plant manager to reduce the color in the discharge. The plant chemist tells the manager that color would not be a problem if they could have no more than 8% dye in the wastewater. The plant manager decides that the least expensive way of doing this is to dilute the 20% waste stream with drinking water so as to produce an 8% dye waste. The 20% dye wastewater flow is 900 gal/min.

a. How much drinking water is necessary for the dilution?

b. What do you think of this method of pollution control?

c. Suppose the plant manager dilutes the waste and the state regulatory personnel assume that the plant is actually removing the dye before discharge. The plant manager, because the company has run a profitable operation, is promoted to corporate headquarters. One day they are asked to prepare a presentation for the corporate board on how they were able to save so much money on wastewater treatment. Develop a short play for this meeting, starting with the presentation by the former plant manager. They will, of course, try to convince the board that they did the right thing. What will be the board's reaction? Include in your script the company president, the treasurer, the legal counsel, and any other characters you want to invent.

4-9 A capillary sludge drying system operates at a feed rate of 200 kg/hr and accepts a sludge with a solids content of 45% (55% water). The dried sludge is 95% solids (5% water), and the liquid stream contains 8% solids. What is the quantity of dried sludge, and what are the liquid and solids flows in the liquid stream?

4-10 A solid waste processing plant has two classifiers that produce a refuse-derived fuel (RFD) from a mixture of organic (A) and inorganic (B) refuse. A portion of the plant schematic and the known flow rates are shown in Figure 4.46.

a. What is the flow of A and B from classifier I to classifier II $[Q_{A2}$ and $Q_{B2}]$?

b. What is the composition of the classifier II exit stream (Q_{A3} and Q_{B3})?

c. What are the purity of the RFD and the recovery of component A?

4-11 A separator accepts waste oil at 70% oil and 30% water by weight. The top product stream is pure oil, while the bottom underflow contains 10% oil. If a flow of 20 gal/min is fed to the tank, how much oil is recovered?

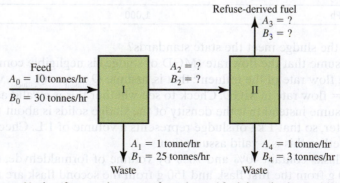

FIGURE 4.46 Air classifiers producing a refuse-derived fuel described in Problem 4-10.

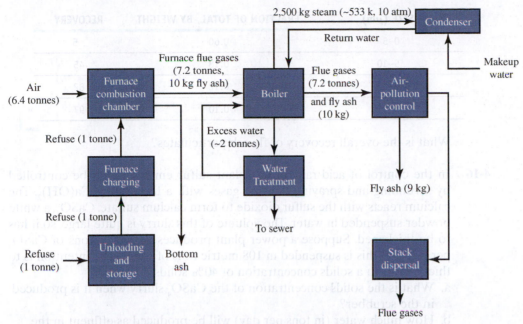

FIGURE 4.47 Material flow in the solid waste incinerator of Problem 4-13.

4-12 The Mother Goose Jam Factory makes jam by combining black currants and sugar at a weight ratio of 45:55. This mixture is heated to evaporate the water until the final jam contains one-third water. If the black currants originally contain 80% water, how many kilograms of berries are needed to make 1 kg of jam?

4-13 The flow diagram in Figure 4.47 shows the material flow in a heat-recovery incinerator.
    a. How much fly ash will be emitted out of the stack (flue) per tonne of refuse burned?
    b. What is the concentration of the particulates (fly ash) in the stack, expressed as $\mu g/m^3$ (μg of fly ash per m^3 of flue gases emitted from the stack)? Assume a density of 1 tonne/500 m^3.

4-14 An electrostatic precipitator for a coal-fired power plant has an overall particulate recovery of 95%. (It removes 95% of all the flue gas particulates coming to it.) The company engineer decides that this is too good, believing it is not necessary to be quite this efficient. The company proposed that part of the flue gas bypass the electrostatic precipitator so that the recovery of fly ash (particulates) from the flue gas would be only 85%.
    a. What fraction of the flue gas stream would need to be bypassed?
    b. What do you think the engineer means by "too good?" Explain the thought processes. Can any pollution control device ever be "too good?" If so, under what circumstances? Write a one-page essay entitled "Can Pollution Control Ever Be Too Effective?" Use examples to argue your case.

4-15 A cyclone used for removing fly ash from a stack in a coal-fired power plant has the following recoveries of the various size particulates:

SIZE (μm)	FRACTION OF TOTAL, BY WEIGHT	RECOVERY
0–5	0.60	5
5–10	0.18	45
10–50	0.12	82
50–100	0.10	97

What is the overall recovery of fly ash particulates?

4-16 In the control of acid rain, power plant sulfur emissions can be controlled by scrubbing and spraying the flue gases with a lime slurry, $Ca(OH)_2$. The calcium reacts with the sulfur dioxide to form calcium sulfate, $CaSO_4$, a white powder suspended in water. The volume of this slurry is quite large, so it has to be thickened. Suppose a power plant produces 27 metric tons of $CaSO_4$ per day and this is suspended in 108 metric tons of water. The intention is to thicken this to a solids concentration of 40% solids.

a. What is the solids concentration of the $CaSO_4$ slurry when it is produced in the scrubber?

b. How much water (in tons per day) will be produced as effluent in the thickening operation?

c. Acid rain problems can be reduced by lowering SO_2 emissions into the atmosphere. This can occur by controlling pollution, such as lime slurry scrubbing, or by reducing the amount of electricity produced and used. What responsibility do individuals have, if any, not to waste electricity? Argue both sides of the question in a two-page paper.

4-17 A stream (see Figure 4.48) flowing at 3 MGD with 20 mg/L suspended solids receives wastewater from three separate sources, A, B, and C:

SOURCE	QUANTITY (MGD)	SOLIDS CONCENTRATION (mg/L)
A	2	200
B	6	50
C	1	200

What are the flow and suspended solids concentration downstream at the sampling point?

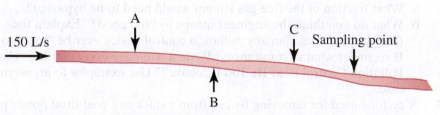

FIGURE 4.48 Stream receiving the discharge from sources A, B, and C as described in Problem 4-17.

4-18 An industrial flow of 12 L/min has a solids concentration of 80 mg/L. A solids removal process extracts 960 mg/min of the solids from the flow without affecting the liquid flow. What is the recovery (in percent)?

4-19 A manufacturer of beef sticks produces a wastewater flow of 2,000 m³/day containing 120,000 mg/L of salt. It is discharging into a river with a flow rate of 34,000 m³/day and a salt concentration of 50 mg/L. The regulatory agency has told the company to maintain a salt level no higher than 250 mg/L downstream of the discharge.

    a. What recovery of salt must be accomplished in the plant before the waste is discharged?

    b. What could be done with the recovered salt? Suggest alternatives.

4-20 A community has a persistent air pollution problem called an inversion (discussed in Chapter 9) that creates a mixing depth of only about 500 m above the community. The town's area is about 9 km by 12 km. A constant emission of particulates of 70 kg/m²/day enters the atmosphere.

    a. If there is no wind during one day (24 hr) and if the pollutants are not removed in any way, what will be the *lowest* concentration level of particles in the air above the town at the end of that period?

    b. If one of the major sources of particulates is wood-burning stoves and fireplaces, should the government have the power to prohibit the use of wood stoves and fireplaces? Why or why not?

    c. How much should the government restrict our lifestyle because of pollution controls? Why does the government seem to be doing more of this now than just a few years ago? Where do you think it's all headed, and why is it happening? Is there anything that we can do about it? Respond to these questions in a one-page paper.

4-21 The Wrightsville Beach, North Carolina, water treatment plant produces distilled water of zero salinity. To reduce the cost of water treatment and supply, the city mixes this desalinated water with untreated groundwater that has a salinity of 500 mg/L. If the desired drinking water is to have a salinity not exceeding 200 mg/L, how much of each water—distilled water and untreated groundwater—is needed to produce a finished water flow of 100 MGD?

4-22 A 1,000-tonne/day ore-processing plant processes an ore that contains 20% mineral. The plant recovers 100 tonne/day of the mineral. This is not a very effective operation, and some of the mineral is still in the waste ore (the tailings). These tailings are sent to a secondary processing plant that is able to recover 50% of the mineral in the tailings. The recovered mineral from the secondary recovery plant is sent to the main plant, where it is combined with the other recovered mineral, and then all of it is shipped to the user.

    a. How much mineral is shipped to the user?

    b. How much total tailings are to be wasted?

    c. How much mineral is wasted in the tailings?

4-23 An aluminum foundry emits particulates that are to be controlled by particulate control devices: first a cyclone, followed by a bag filter. For now, consider these as black boxes that remove particulates (solid particles suspended in air). These particulates can be described by their size, based upon the diameter in micrometers (μm). The effectiveness of the cyclone and the bag filter is as follows:

PARTICLE SIZE, μm	0–10	10–20	20–40	>40
Percentage by weight of each size	15	30	40	15
Percentage recovery of cyclone for each size of particles	20	50	85	95
Percentage recovery of bag filter for each size of particles	90	92	95	100

a. What is the overall percentage recovery for the cyclone? (*Hint*: Assume 100 tonne/day are treated by the cyclone.)
b. What would be the overall percentage recovery for the bag filter if this was the only treatment device?
c. What is the overall percentage recovery by this system? Remember that the cyclone is in front of the bag filter.
d. The system has the cyclone in front of the bag filter. Why wouldn't you have placed the bag filter in front of the cyclone?
e. Assuming all the particles were perfect spheres, would a cubic meter of 5 μm spherical particulates weigh more or less than a cubic meter of perfectly spherical 40 μm particles? Show the calculations on which you base your answer.

4-24 The die-off of coliform organisms below a wastewater discharge point can be described as a first-order reaction. It has been found that 30% of the coliforms die off in 8 hours and 55% die in 16 hours. About how long would it take to have 90% die off? Use semi log paper to solve this problem.

4-25 A radioactive nuclide is reduced by 90% in 12 min. What is its half-life?

4-26 In a first-order process, a blue dye reacts to form a purple dye. The amount of blue at the end of 1 hr is 480 g and at the end of 3 hr is 120 g. Graphically estimate the amount of blue dye present initially.

4-27 A reaction of great social significance is the fermentation of sugar with yeast. This is a zero-order (in sugar) reaction, where the yeast is a catalyst (it does not enter the reaction itself). If a ½-L bottle contains 4 g of sugar and it takes 30 min to convert 50% of the sugar, what is the rate constant?

4-28 A batch reactor is designed to remove gobbledygook by adsorption. The data are as follows:

TIME (min)	CONCENTRATION OF GOBBLEDYGOOK (mg/L)
0	170
5	160
10	98
20	62
30	40
40	27

What order of reaction does this appear to be? Graphically estimate the rate constant.

4-29 An oil storage area, abandoned 19 years ago, had spilled oil on the ground that saturated the soil at a concentration of 400 mg/kg of soil. A fast-food chain wants to build a restaurant there and samples the soil for contaminants only to discover that the soil still contains oil residues at a concentration of

20 mg/kg. The local engineer concludes that, since the oil must have been destroyed by the soil microorganisms at a rate of 20 mg/kg each year, in one more year the site will be free of all contamination.

a. Is this a good assumption? Why or why not? (*Hint*: Consider kinetics.)

b. How long do *you* think it will take for the soil to reach the acceptable contamination of 1 mg/kg?

4-30 Radioactive waste from a clinical laboratory contains 0.2 μCis (microcuries) of calcium-45 (^{45}Ca) per liter. The half-life of ^{45}Ca is 152 days.

a. How long must this waste be stored before the level of radioactivity falls below the required maximum of 0.01 μCi/L?

b. Radioactive waste can be stored in many ways, including deep well injection and above-ground storage. Deep well injection involves pumping the waste thousands of feet below the Earth's surface. Above-ground storage is in buildings in isolated areas, which are then guarded to prevent people from getting near the waste. What risk factors are associated with each of these storage methods? What could go wrong? Which system of storage do you think is superior for the waste?

c. How would a deep ecologist view the use of radioactive substances in the health field, considering especially the problem of storage and disposal of radioactive waste? Try to think like a deep ecologist when responding. Do you agree with the conclusion you have drawn?

4-31 A dye mill has a highly colored wastewater with a flow rate of 8 MGD. One suggestion has been to use biological means to treat this wastewater and remove the coloration, and a pilot study is performed. Using a mixed-batch reactor, the following data result:

TIME DYE (hr)	CONCENTRATION (mg/L)
0	900
10	720
20	570
40	360
80	230

A completely mixed aerated lagoon (batch reactor) is to be used. How large must the lagoon be to achieve an effluent of 50 mg/L?

4-32 A settling tank has an influent rate of 0.6 MGD. It is 12 ft deep and has a surface area of 8,000 ft^2. What is the hydraulic retention time?

4-33 An activated sludge tank, $30 \times 30 \times 200$ ft, is designed as a plug-flow reactor, with an influent Biochemical Oxygen Demand (BOD) of 200 mg/L and a flow rate of 1 MGD.

a. If BOD removal is a first-order reaction and the rate constant is 2.5 days^{-1}, what is the effluent BOD concentration?

b. If the same system operates as a completely mixed-flow reactor, what must its volume be (for the same BOD reaction)? How much bigger is this, as a percent of the plug-flow volume?

c. If the plug-flow system was constructed and found to have an effluent concentration of 27.6 mg/L, the system could be characterized as a series of completely mixed-flow reactors. How many reactors in series, n, would be required?

4-34 Some plant managers have a decision to make. They need to reduce the concentration of salt in an 8,000-gal tank from 30,000 mg/L to 1,000 mg/L. They can do it in one of two ways: They can start flushing it out by keeping the tank well mixed while running in a hose with clean water (zero salt) at a flow rate of 60 gal/min (with an effluent of 60 gal/min).

Or they can empty out some of the saline water and fill it up again with enough clean water to get 1,000 mg/L. The maximum rate at which the tank will empty is 60 gpm, and the maximum flow of clean water is 100 gpm.

a. If they intend to do this job in the shortest time possible, which alternative will they choose?

b. Jeremy Rifkin points out that the most pervasive concept of modern times is *efficiency*. Everything has to be done so as to expend the least energy, effort, and especially time. He notes that we are losing our perspective on time, especially if we think of time in a digital way (digital numbers on a watch) instead of in an analogue way (hands on a conventional watch). With the digits we cannot see where we have been, and we cannot see where we are going, and we lose all perspective of time. Why, indeed, would the plant managers want to empty out the tank in the shortest time? Why are they so hung up with time? Has the issue of having time (and not wasting it) become a pervasive value in our lives, sometimes overwhelming our other values? Write a one-page paper on how you value time in your life and how this value influences your other values.

4-35 A first-order reaction is employed in the destruction of a certain kind of microorganism. Ozone is used as the disinfectant, and the reaction is found to be

$$\frac{dC}{dt} = -kC$$

where

C = concentration of microorganisms, microbes/mL

k = rate constant, 0.1 min^{-1}

t = time, min

The present system uses a completely mixed-flow reactor, and a system of baffles may be added to create a series of CMF reactors.

a. If the objective is to increase the percentage microorganism destruction from 80% to 95%, how many CMF reactors are needed in series?

b. We routinely kill microorganisms and think nothing of it. But do microorganisms have the same right to exist as larger organisms such as whales, for example, or as people? Should we afford moral protection to microorganisms? Can a microorganism ever become an endangered species? Write a letter to the editor of your school newspaper on behalf of Microbe Coliform, a typical microorganism who is fed up with not being given equal protection and consideration in the human society, and who is demanding microbe rights. What philosophical arguments can be mounted to argue for microbe rights? Do not make this a silly letter. Consider the question seriously because it reflects on the entire problem of environmental ethics.

4-36 A completely mixed-flow reactor, with volume, V, and flow rate, Q, has a zero-order reaction, $dA/dt = k$, where A is the material being produced and k is the rate constant. Derive an equation that would allow for the direct calculation of the required reactor volume.

4-37 A completely mixed-flow continuous bioreactor used for growing penicillin operates as a zero-order system. The input, glucose, is converted to various organic yeasts. The flow rate to this system is 20 L/min, and the conversion rate constant is 4 mg/(min-L). The influent glucose concentration is 800 mg/L, and the effluent must be less than 100 mg/L. What is the smallest reactor capable of producing this conversion?

4-38 Suppose you are to design a chlorination tank for killing microorganisms in the effluent from a wastewater treatment plant. It is necessary to achieve 99.99% kill in a wastewater flow of 100 m³/hr. Assume the disinfection is a first-order reaction with a rate constant of 0.2 min⁻¹.

a. Calculate the tank volume if the contact tank is a CMFR.
b. Calculate the tank volume if the contact tank is a PFR.
c. What is the retention time of both reactors?

4-39 A completely mixed-flow reactor, operating at steady state, has an inflow of 4 L/min and an inflow "gloop" concentration of 400 mg/L. The volume is 60 L; the reaction is zero-order. The gloop concentration in the reactor is 100 mg/L.

a. What is the reaction rate constant?
b. What is the hydraulic retention time?
c. What is the outflow (effluent) gloop concentration?

## CHAPTER 5

Natural Resources, Materials, and Sustainability

FIGURE 5.1 A drive-through tree in Humboldt County, California. Sustainability requires that people rethink their relationship with the natural world. This giant redwood tree in California was cut to allow a road to pass through it for our recreational entertainment, not out of any real necessity.

Source: Image Source/Getty Images.

There is hope—I've seen it—but it does not come from the governments or corporations, it comes from the people. The people who have been unaware are now starting to wake up, and once we become aware, we change.

—GRETA THUNBERG, CLIMATE ACTIVIST

GOALS

THE PURPOSES OF THIS CHAPTER are to review the concept of sustainable development, to introduce ways of thinking about natural resources, and to link natural resources to modern engineered materials and the environmental issues associated with them. We will present the waste management hierarchy and material life cycle model as engineering strategies for a more sustainable use of Earth's natural resources and for reducing the environmental impacts of products throughout their life cycle.

OBJECTIVES

At the conclusion of this chapter, you should be able to:

5.1 Compare and contrast the role of natural resources in the constrained growth and the resource maintenance models of sustainable development.

5.2 Explain and give examples of the different features of nonrenewable and renewable resources. Compare and contrast these traditional concepts with ecosystem services as approaches to the conservation of natural resources.

5.3 List, explain, and give examples of the five basic categories of engineered materials and their potential for recyclability.

5.4 Discuss the challenges presented by each stage of the linear materials economy for sustainable development.

5.5 Describe how the waste management hierarchy and life cycle approaches to materials flow can act as models for more sustainable engineering in practice.

Introduction

Activist Greta Thunberg is known for her unsparing criticism of inaction on the climate crisis. In her 2019 address to the United Nations Climate Change Conference, she took the world's leaders to task for jeopardizing the future of the next generation. In that same speech, however, she also acknowledged the momentum of hope. While Thunberg speaks of climate change specifically, her insights about the reason for hope—*once we become aware, we change*—apply to sustainable development broadly. Sustainability reflects a new awareness about our relationship to the natural world and the changes that it inspires.

Earth is our *life support system*. Over its 4.5-billion-year history, it has created environments in which single-celled organisms, tiny sea creatures, carbon-rich fern forests, dinosaurs, and humankind have evolved and thrived. At all times and in all places on our planet, the health and survival of living things depend on complex webs of interdependence with other living systems and Earth's physical biogeochemical cycles. The emergence of sustainability as a concept and a way of thinking requires us to acknowledge that human actions have profound consequences for the natural world and can disrupt the ability of the planet's living systems and physical cycles to support human and other life.

In this chapter, we introduce you to ways of thinking that are changing how we engineer our world. We review concepts and models that you can use to develop more sustainable materials, products, and processes as well as think more holistically about engineering design and problem solving. Topics include the nature of natural resources, engineered materials and their environmental issues, the concepts of waste in ecological and human systems, and the basis of ecological design.

5.1 Sustainability and Natural Resources

Sustainability is a term that carries a wide variety of meanings depending on context. "Sustainability" to a biologist studying a small wetlands ecosystem is different than "sustainability" for a commercial forester, which is yet again different than "sustainability" to a businessperson. So what are the implications of multiple contexts and meanings for engineers?

The *social* (as opposed to environmental or ecological) concept of sustainability came into widespread use after a 1987 United Nations report entitled *Our Common Future*. Referred to as the Brundtland Report, this study explored the interdependent relationship between the environment and economic development and addressed the consequences of environmental degradation for long-term economic growth and social well-being. The concept that the Brundtland Report put forward was **sustainable development**, which the report defined as "development that meets the needs of the present without compromising the ability of future generations to meet their own needs." What is critical about this definition is *time*, the sense that in providing for human societies today, we should not jeopardize the ability of our descendants to provide adequately for themselves. Because human survival and material quality of life depend on the natural world, sustainable development requires us to consider the long-term consequences of our impact on the environment for others. This sense of obligation to future generations is referred to as **intergenerational ethics**.

There is no uniform, standardized approach to sustainable development in practice. Nonetheless, all discussions explore the relationships among the *three pillars of sustainable development*—the environment, society, and the economy—with an understanding that:

- The environment and biosphere should be protected as a *dynamic system.*
- Socioeconomic development is necessary for all societies and should be equitable; no one group should bear an unfair burden of environmental harm.
- Economic considerations reflect legitimate concerns about the cost and affordability of solutions to environmental problems, but economic profitability may also be achieved by more environmentally sustainable actions.

What continues to be debated is the relationship between the environment, society, and the economy. Figure 5.2 illustrates two of the most common ways of representing these relationships. In the Venn diagram on the left, sustainable solutions are those that acceptably integrate environmental, societal, and economic goals. This model is known as the *constrained growth* model. This image suggests that there are a broad range of activities in the three pillars that have little bearing on one another, and the model is based on the need for economic growth and expansion. In the diagram on the right, the nested concentric circles suggest that environmental dynamics represent a defining limit on economic and social activity; this approach is referred to as the *resource maintenance* model. From this perspective, human action, quality of life, and survival are ultimately dependent on and bounded by processes of the natural world.

The value of sustainability concepts is that they encourage us *to think holistically about systems.* In a world in which nature, human well-being, and economic production are interdependent, it is critical that we understand how change in one aspect of these relationships has consequences for the others. In popular terminology, this is referred to as a "butterfly effect," but in systems dynamics language, it represents feedback and system disturbance.

As you will see throughout this chapter, sustainability concepts *do* generate a variety of ideals for practice. These are to minimize our use of raw materials and finite natural resources, to strive for zero waste, to protect ecological processes, and to avoid creating toxic and hazardous substances.

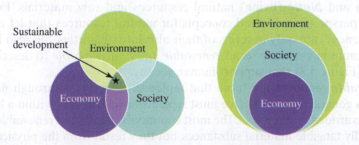

FIGURE 5.2 Proposed relationships among the three pillars of sustainable development. It is widely accepted that sustainable development integrates considerations of the environment, societal needs, and economic constraints. What is still debated is the relationship among these three pillars. The diagrams in this figure illustrate the two most commonly discussed models: *constrained growth* (left) and *resource maintenance* (right).

5.2   The Nature of Natural Resources

The first step toward developing new insights into sustainability is understanding the nature of natural resources. From a scientific perspective, Earth is a biophysical system constructed of matter (the elements) and energy flows. From a practical perspective, Earth provides the natural resources on which our survival and material well-being depend. The simple and obvious substances are air and water. But people do not survive like other living things, where existence is based on constant, direct interaction with nature. Instead, human survival is mediated by a technological and material culture: We exploit plant and animal fibers for clothing (cotton, wool, and flax, for example), we construct buildings, we manufacture synthetic substances that are not found naturally in the physical world, we cultivate animals and scientifically hybridize crops for food, we burn fossil fuels, we design complex transportation systems, we have toilets and sanitation, we fabricate metals, we invent chemical pharmaceuticals, and we manufacture a nearly incomprehensible array of consumer goods and products. In an industrialized society, virtually nothing about human existence—except breathing—reflects a "natural" interaction with our environment. Even in hunter-gatherer communities, material culture, technology, and dwellings leave a small ecological footprint. To develop models for sustainable engineering, we need to understand both the finite nature of our material resources and the ecological processes on which our lives depend.

5.2.1   Traditional Concepts of Natural Resources

Laws of physics limit and govern our material world of natural resources. First and foremost, all tangible things that we drink, eat, use, touch, or consume are made of matter. The periodic table identifies the elemental matter on Earth, and all material substances are composed of combinations of elements. When a naturally occurring substance or living organism is exploited by human beings, we refer to it as a natural resource. **Raw materials** are those naturally occurring resources that we extract from the environment and then process into things useful for people. As you saw in Chapter 3, the Earth's biogeochemical cycles are responsible for the environmental cycling and transformation of the elements and essential resources such as water.

As a reminder, matter is finite. Mass on our planet is neither created nor destroyed, but material substances are formed from the elements and transformed by the Earth's systems and by human action. We usually distinguish between **abiotic** (nonliving) and **biotic** (living) natural resources and raw materials. From a sustainability perspective, we need concepts for natural resources that let us evaluate their "finiteness" not only in terms of their absolute quantities, but also in terms of *time*. We commonly use the terms *renewable* and *nonrenewable* to describe natural resources (Figure 5.3) that support human society.

Renewable resources are those that replenish themselves through natural processes. As a general rule, a resource must replenish itself roughly within a human lifetime to be considered renewable. The most commonly mentioned renewable resources are not really tangible material substances, but they result from the physical forces of our planetary system, such as solar energy, wind and tidal power, and the Earth's atmosphere. What makes these resources unique is that they are virtually infinite—their presence is the result of the mechanics of our solar system and planet, and human use does not diminish their availability. In contrast, many of our other renewable natural resources are biotic, such as animals, fish, plants, and forests. As growing, living things, they have variable rates of regeneration through biological reproduction.

Nonrenewable Resources				Renewable Resources	
Nonregenerative	Theoretically recoverable	Recyclable	Regenerative	Regenerative	Infinite
Coal Natural gas Oil Fossil water	All elemental minerals and metals Water	Metals Water	Soil Surface water Groundwater	Animals Fish Vegetation/ forests Surface water Groundwater	Air Solar energy Wind energy Geothermal energy Tidal/wave energy Water

FIGURE 5.3 A typology of natural resources. This schematic organizes critical natural resources according to their renewability, recyclability, and regenerative properties.

Therefore, these resources are renewable *only to the extent that we do not over-harvest them.* If we consistently harvest biotic resources at a rate faster than they regenerate, one of two things will generally happen. The most extreme is that the species is driven to extinction. Alternatively, the species population collapses to low numbers and can no longer serve as a natural resource. Classic examples of overharvest and collapse include many fisheries and forests worldwide.

Nonrenewable resources are those that regenerate themselves extremely slowly, if at all. Metals, for example—iron, copper, tin, gold, silver, aluminum, and so forth—are elemental substances that are absolutely finite in the conventional physical and chemical sense. Soil is an amalgam of organic life and minerals and can be regenerated, but depending on climate and location, it requires 100 to 500 years to create an inch of topsoil. It took millions of years to form fossil fuels, and as a consequence these natural resources must be considered finite and nonregenerative. The rare earths are intriguing nonrenewable resources because they are not particularly rare geologically; instead, geopolitical "scarcity" is one of the forces driving technological innovation to recycle and recover these elements (see Box 5.1).

Because of the finiteness of elemental matter and some raw materials, as well as the extremely slow regenerative times of others, the recoverability of nonrenewable resources is of interest. Some raw materials are entirely nonrecoverable because the process of using them prevents us from physically capturing the substance and reusing it in its original form. Fossil fuels are the most extreme example of nonrecoverability because combustion effectively consumes them. All elemental minerals and metals are theoretically recoverable because they are elemental, but we may lack the practical know-how to do so or the required technologies may be prohibitively expensive. Notably, elemental metals are considered to be infinitely recyclable because they can be continuously recovered and reused.

BOX 5.1 How Rare Are Rare Earths?

The rare earth elements are 17 natural elements with exotic names such as yttrium, neodymium, and samarium. They aren't particularly rare, and until recently we weren't especially concerned about their relative scarcity. What is unusual about most rare earths is that they are not found in concentrated deposits, but are instead diffused throughout the Earth's crust. They were labeled rare not because they lacked abundance, but because it is difficult to locate them in economically viable concentrations that can be readily mined.

So what do we use these elements for (Figure 5.4)? There are a number of rare earth powerhouses, all of which promote a high-tech world. Lanthanum is a critical element in modern battery technology on which electric vehicles like the Toyota Prius depend. Europium was used to create the bright red color in old cathode-ray TV picture tubes, but it is finding new life in white light-emitting diode (LED) light bulbs. Erbium, when added in minute quantities, is used for lasers and fiber optics. Samarium and neodymium are used to create tiny electromagnets so that we can have miniature consumer electronics. Cerium promotes cleaner air as a component of catalytic converters. The list goes on, and only trace amounts of a rare earth are usually needed to significantly change the properties of a material or its performance.

Rare earth elements are critical for sustainability because of their role in clean energy production and energy efficiency. For example, they play an essential role in neodymium-iron-boron permanent magnets, which are high-efficiency magnets used in wind power and electric vehicles. Because of the ongoing demand for rare earths in electronics and their new role in low-carbon energy, world demand for these elements is expected to double by 2030.

Geopolitical attention to rare earth resources intensified in 2010 when China significantly reduced

FIGURE 5.4 The use of permanent magnets in wind turbines, electric vehicles, and other energy applications is expected to account for 40% of all consumption of rare earth elements by 2030. Neodymium-iron-boron magnets create the strongest magnetic field per unit of volume of all magnets (Williams, 2021).

Source: W. Scott McGill/Shutterstock.com.

its exports and world prices increased steeply. Global supply chains are gradually changing in reaction to China's previous monopoly on these raw materials. China still dominates rare earth markets, however, accounting for 85% of all refined rare earth products and 70% of global rare earth element consumption (Williams, 2021).

As industries become concerned over the cost, relative scarcity, and reliability of rare earth supplies, recycling has become more important. Innovations in recycling technology have made recovery of some rare earth elements technically feasible and cost effective in light of rising global prices. Key product classes for recycling and rare earth recovery are fluorescent light bulbs, batteries, magnets, consumer electronics, and computer components.

As you can see in Figure 5.3, water appears in several categories and is a challenging natural resource to classify. Traditionally, water has been regarded as an infinitely renewable natural resource because of the biogeochemical water cycle, but this perspective has changed. Because of the Earth's hydrological cycle, the amount of water in the biosphere stays relatively constant, and human use does not diminish the absolute quantity of water molecules. At the planetary scale, it is appropriate to regard water as an infinite natural resource. At the local scale, however, freshwater in aquifers, groundwater, lakes, rivers, and streams can be either a regenerative renewable or nonrenewable resource depending on climatic and geological conditions.

Freshwater regenerates locally in surface and groundwater as a consequence of regional weather, climate, and hydrologic cycles. As a result, the rate of regeneration can fluctuate and there is a risk of overabstraction of water—we withdraw it at rates faster than it can replenish. In the case of aquifer *fossil water*, overabstraction will result in a permanent loss of groundwater. This is what is happening to the Ogallala Aquifer in the U.S. Great Plains.

In the case of rivers and streams, overabstraction can reduce flow to a trickle and make water unavailable to people and other living things downstream, as we see with the Colorado River. Overabstraction of river water for agricultural irrigation has also contributed to the virtual disappearance of the Aral Sea, one of the four largest lakes in the world until it was reduced to only one-tenth of its original size by 2010 (Figure 5.5). Sometimes freshwater can be recovered and recycled, but it depends on how it is being used. It is also a costly process.

FIGURE 5.5 These satellite images show the Aral Sea in 1989 and 2010. Located in Eurasia, the Aral Sea was previously one of the four largest lakes in the world, but it is now only one-tenth of its original size by volume. Water for irrigation was withdrawn from rivers that fed the lake, to the point that the lake could not replenish itself. Local economies have collapsed, and the region is plagued by environmental degradation.

Source: NASA.

In sum, a useful way of understanding natural resources with regard to more sustainable engineering is in the context of stocks and flows. ***Natural resources management*** is the field of study that scientifically explores how to manage natural resources in a sustainable manner, in which human rates of use and extraction are in balance with the biogeochemical regeneration rates for abiotic resources and within the reproduction and population limits of biotic systems. As engineers, we can reflect on the absolute amount of resources (their finiteness at any given point in time), the degree and speed with which they can regenerate, and whether or not the resources are recoverable or recyclable in principle. These considerations can, for example, help us decide on materials selection, reevaluate the water intensity of industrial processes, or develop systems for recovering diminishing and costly resources.

5.2.2 Ecosystem Services and Natural Capital

The Earth is a biosphere. Traditional ways of thinking about natural resources as stocks and flows of discrete materials and substances are a useful but incomplete way of understanding the complexity of Earth as a life support system. For example, many fish that we eat are themselves supported by a complex marine ecology in which the fish reproduce only in coastal estuaries, and those estuaries are in turn the product of habitats provided by mangrove trees, diurnal variations in water levels, microscopic bacteria, and so on. For many living species, the problem is not that we are overharvesting them, but that we are disturbing, degrading, or even destroying the habitats through which their own life cycles are maintained in delicate balance. Similarly, once deforested, many tropical regions cannot regenerate vegetation. Because of the unique mineral composition of soils in some parts of the tropics, direct sunlight and the loss of organic matter from decomposing vegetation rapidly harden the soil to the point that it cannot naturally regenerate or support the biodiversity of tropical forests.

Human threats to natural resources are therefore not just through the overconsumption of individual species, substances, or materials. By disrupting the ecological balances in which living things directly and indirectly survive, we affect not only the number and diversity of species, but also biogeochemical processes. As a consequence, we must take into account the physical integrity of ecosystems themselves. Ecosystems are communities of living organisms, where ecosystem dynamics represent the flow of nutrients and energy between living entities and their abiotic physical environment (water, soil minerals, air, climate). In an ecosystem, the living organisms and the abiotic environment are mutually interdependent. For example, forests require rainfall, but forests themselves create the microclimates in which rainfall can occur. Biological processes are significant contributors to elemental matter cycling on Earth, particularly for carbon and nitrogen.

Two concepts have emerged over the last decade or so to help us better model the usefulness of ecosystem dynamics. These two concepts are ecosystem services and natural capital. The term ***ecosystem services*** represents the idea that a wide variety of ecological dynamics support humankind in a way that a simple consideration of natural resource use does not. Traditionally, we have thought of natural resources in terms of their material products, such as oil, metals, food, and potable water. In the ecosystem services model, ecosystems generate tangible and intangible provisioning, regulating, and cultural services for people (Figure 5.6). Provisioning services are analogous to the production of material natural resources such as the food, freshwater, fuels, fibers, minerals, and metals discussed in the previous section.

Cultural Services	Regulating Services	Provisioning Services
Nonmaterial benefits obtained from ecosystems	*Benefits obtained from regulation of ecosystem processes*	*Products obtained from ecosystems*
● Spiritual and religious	● Climate regulation	● Food
● Recreation and ecotourism	● Disease regulation	● Freshwater
● Aesthetic	● Water regulation	● Fuelwood
● Inspirational	● Water purification	● Fiber
● Educational	● Pollination	● Biochemicals
● Sense of place		● Genetic resources
● Cultural heritage		

Supporting Services

Services necessary for the production of all other ecosystem services

● Soil formation   ● Nutrient cycling   ● Primary production

FIGURE 5.6 The scope of ecosystem services. Ecosystems provide a wealth of services to humankind, ranging from traditional raw materials to water purification to our spiritual values. Not all of these services have direct monetary value, and as a consequence, they are hard to protect through market economic systems.

Source: Based on Board of the Millennium Ecosystem Assessment (2003). *Millennium Ecosystem Assessment: Ecosystems and Human Well-Being—A Framework for Assessment.* Island Press.

Regulating services are ecosystem processes that moderate harmful impacts on human communities, such as controlling disease, purifying water, and mitigating floods. Regulating services also include processes that are essential for the provisioning of natural resources, such as pollination of food crops and climate dynamics that regulate temperature and rainfall (Figure 5.7). Finally, ecosystems provide services to humans in the form of culture and values, such as an aesthetic appreciation of the natural world, spiritual and religious beliefs, recreational opportunities, and our sense of place in the world. The concept and model of ecosystem services provide us with a more holistic portrait of the natural world, of its functions, and of its benefits to humanity.

Natural capital is a concept developed by Paul Hawken and Amory and Hunter Lovins in their book *Natural Capitalism: Creating the Next Industrial Revolution* (1999). We will provide a simplified version of this concept here, because a background in business or economics is necessary to understand its finer points. Imagine that natural resources and ecosystem services are the Earth's "money in the bank" and that humankind lives off of this money. Ideally, we would protect this investment and its ability to generate interest. That is, we would conserve natural resources by not overconsuming them (driving down the principal) or degrading them (lowering their "interest rate"). Both overconsumption and degradation reduce the wealth that Earth's investment can generate.

To give a concrete example, groundwater in an aquifer offers us the "interest income" of freshwater. If that water is degraded through pollution, we may be forced to buy water elsewhere or to treat the water to make it drinkable. Either way, it costs us money, and the return on investment is reduced. Similarly, if we continually extract the water faster than it can recharge, then eventually the aquifer will go dry. At that point, we have spent all of our principal and gone bankrupt. The threats

FIGURE 5.7 Ecosystem pollination services. We can buy honey, but the real value of honeybees and other insects is in their ecosystem service as pollinators. Almost all major human food crops depend on pollinators to produce the grains, fruits, nuts, and vegetables of our food supply. From an economic perspective, these pollination services are worth billions of dollars, even though we do not have to pay for them most of the time. Because of shrinking populations of wild pollinators from habitat loss, farmers must now hire commercial pollination services or risk losing their crops. The almond industry in California is particularly dependent on commercial honeybee services.

Source: Photo by Bradley Striebig.

to an extraordinary number of ecosystem services, as well as all regenerative and finite natural resources, can be understood with this analogy.

The dilemma is that conventional economics gives us no good way of valuing natural capital, of accounting for it in the prices of the goods and services that we buy and sell, or of incentivizing its stewardship. Natural capitalism as a concept helps us understand why ecosystem services are valuable in economic terms and why they should be protected as investments for the future. As a consequence, it can inspire people and businesses to consider their environmental impacts in a different way and to be willing to conserve resources. In addition, natural capitalism requires new techniques that will allow us to better represent and incorporate the monetary value of natural capital in our economic systems.

Ecosystem services and natural capital are not simple concepts. They require that we understand the relationship between complex ecological processes and human systems, and that we integrate environmental values and costs into our economic decision making. Later in this chapter, we will explore models for applying ideas about sustainable development and resource use to engineering design, thinking, and problem solving. To get there, we need to understand how natural resources are processed into engineered materials. We also need to address the environmental issues associated with our largely linear materials economy.

5.3 From Natural Resources to Engineered Materials

All of the manufactured products used and consumed by humankind and the global economy come from Earth's natural resources. The extent to which a natural resource is recoverable or recyclable depends on what kind of material it is engineered into; the complexity of the product that it is embodied in; and whether an infrastructure and market are in place for reclaiming, recycling, and reprocessing the product and the material. An understanding of engineered materials can help us understand the potential for more sustainable resource use.

We cannot use Earth's substances directly in industry; they are too raw. Natural resources must be cleaned, purified, and processed into physically and chemically uniform feedstocks before they can be used as manufacturing inputs. As raw substances, natural resources undergo several transformations to become engineered materials and then finished products. Every stage along the way has environmental impacts that depend, in part, on the type of material that is being processed and used.

5.3.1 Traditional Engineered Solid Materials

Manufactured goods and construction materials are made from engineered feedstocks. These material feedstocks have been intentionally made with desired physical and chemical properties; these properties in turn enable products to achieve their desired performance, safety, quality, and manufacturing specifications. Most products in our daily lives are made from one or more of the four traditional solids: (1) metals, (2) ceramics, (3) polymers, and (4) composites. "Advanced" or "high-technology" materials are another class of engineered material, but they are not considered traditional solids.

Materials are classified as metals, ceramics, or polymers according to their atomic structure and chemical composition. A composite material is created when two or more of the other traditional solids are combined (like metals with ceramics). There is no consistent definition of an advanced or high-tech material, but they are broadly understood to have extremely enhanced material properties that exceed the performance of traditional solids.

Metals

Metals are so significant in our material culture that technological eras have been named after them. The Copper, Bronze, and Iron Ages represent notable advances in our ability to shape and strengthen metals into ever more useful products (and weapons). You are familiar with many common metals like copper, iron, aluminum, steel, and silver because of their widespread use. Metals are literally elemental substances—they are the metallic elements of the periodic table.

Metals have crystalline structures and some properties that make them notably different from ceramics and polymers. Metals conduct heat and electricity; they are dense, stiff, and strong; and they are highly malleable, meaning they can be shaped into very thin sheets without fracturing. An *alloy* is a metal that has been combined with another metal or element. Steel, bronze, and brass are all alloys. Metals are also commonly grouped as either ferrous (iron-based) or nonferrous. An unfortunate material property of many metals is that they corrode easily, so they often need to have protective coatings or to be converted into alloys that resist the chemical reactivity that causes corrosion (like oxidation). This is why aluminum is such a significant metal; it has many of the desirable strength and stiffness properties of ferrous metals while also being highly resistant to corrosion.

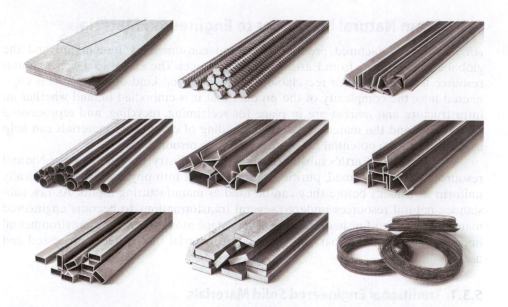

FIGURE 5.8 Engineered material feedstocks. Earth's natural resources are too raw to be used directly in manufacturing. They must first be extracted, purified, and formed into uniform bulk materials, then further processed and shaped into the final material feedstocks that will be used as manufacturing inputs. This image shows a variety of milled steel products that are ready for manufacturing, including sheets, rods, tubes, bars, and wire. The transformation processes for getting to this stage are extremely energy intensive.

Source: iStockPhoto.com/urfinguss..

The processes for converting most commercial metals into bulk uniform materials suitable for manufacturing are often highly energy intensive (Figure 5.8). A critical processing stage in converting iron, copper, and aluminum into an intermediate feedstock is smelting using heat-intensive furnaces. Magnesium processing is typically done through electrolysis. Therefore, a distinct advantage of metal recycling is not just conserving a finite resource, but also reducing the energy consumption needed for processing the raw material. Recycled aluminum, for example, uses 90% less energy than that required for manufacturing aluminum from raw resources.

In theory, all elemental metals are infinitely recyclable—they can be recovered and reprocessed repeatedly without any change in their fundamental material properties. This is not the case with alloys, however. Each time alloys are recycled, their quality and material integrity degrade. Metals are not biodegradable outdoors in the natural environment, and not all metals are harmless. For example, mercury and lead are toxic and can create serious health hazards if they contaminate food or water supplies.

Ceramics

Ceramic materials are compounds of metallic and nonmetallic elements. Common ceramic compounds are oxides (oxygen is the nonmetallic element), nitrides (nitrogen-metal compounds), and carbides (carbon-metal compounds). We all use ceramic products every day. Pottery, brick, porcelain, and tile are made from clay minerals; most glass is made from silica; and cement, an essential ingredient of concrete, is made from calcium and silicon.

Like metals, ceramic materials and products are very energy and heat intensive. Items made from clay must be fired in a kiln. Glass manufacturing requires melting a blend of sand, limestone, and soda ash at temperatures above 850°C. Cement manufacturing, widely considered to be the most energy-intensive industry in the world, involves making "clinker" by burning limestone, chalk, and clay at temperatures above 1,300°C, a heat intensity that can be efficiently obtained only by burning coal or coke.

Ceramics have desirable properties that make them useful for many purposes. They are good insulators because they do not transfer heat or electricity readily. They can also withstand many harsh or corrosive environments, so they resist the degradation that affects metals and polymers. Ceramic materials are known for being hard, inelastic, and strong, with material properties often comparable to those of metals. Ceramics are quite brittle, however, and they easily fracture or shatter. Modern ceramics are engineered to be far more resistant to fracture and stress, and they can be found in applications as wide-ranging as kitchen knives and aerospace components. However, they are usually considered advanced materials.

Traditional ceramics do not biodegrade and can persist for ages in the environment. These materials last from a few years or decades for clay-fired items to thousands of years for glass. Modern glass is largely inert and nontoxic outdoors, however. Ceramics are also somewhat limited in their recyclability. Once clay is fired, it cannot be reconstituted as a clay material feedstock; the same holds true for cement. Cement (embodied in concrete) can be crushed into aggregate and used as a substitute for gravel in many applications, but it can never be reprocessed back into cement. Glass food and beverage containers can be recycled indefinitely back into a glass feedstock without any real degradation of physical integrity or purity, similar to many metals. However, most other types of glass items (such as windows) cannot be recycled easily—if at all—because of the complexity of their chemical composition and molecular bonding.

Polymers

Polymers are a confusing material because we often think of them exclusively as synthetic plastics. However, polymers can be both natural and synthetic. What defines them as a class of materials is their chemical structure. Polymers are macromolecules composed of chains of thousands of repeating, smaller molecular units called monomers (Figure 5.9). All monomers in a polymer do not have to be identical molecules; what matters is that they are connected in a pattern of identical repeating building blocks.

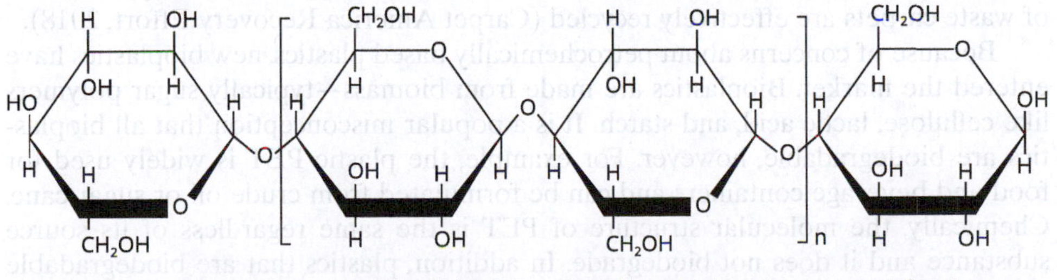

FIGURE 5.9 Chemical structure of cellulose. As a class of materials, polymers are defined as macromolecules composed of long chains of repeating blocks of monomers. Polymers can be natural or synthetic. Here we see the structure for cellulose, a natural polymer that is the material substance of plant fibers and cell walls. Wood and cotton are examples of natural cellulosic polymers.

Source: petrroudny43/Shutterstock.com.

Dozens of natural polymers have been used as materials since ancient times. Cellulose is a polymer and the primary substance of plant cell walls. Wood and plant fibers like cotton and linen are cellulose and hence natural polymers. Proteins are natural polymers as well, so animal hair (like wool), fur, leather, and silk are polymers. Because natural polymers are organic compounds, they are biodegradable in the environment. In some circumstances, natural polymers can be recycled and reprocessed into a feedstock, but the reconstituted material is generally of lesser quality and integrity. Wool, cotton, and linen can all be shredded and re-spun, for example, but the recycled fibers do not have the same advantages as those from first-use resources. Used paper can be re-pulped, but it can typically only be reused to make the same or poorer quality paper.

Synthetic plastics and rubbers are the most significant artificial polymeric materials. There are many material advantages to synthetic polymers: they are relatively inert in many environmental conditions, are nonmagnetic, and do not conduct electricity. Most important, they have very low densities per unit of mass and are consequently lightweight. Many types of plastic are surprisingly strong and nearly equivalent to metals and ceramics per unit of mass. Because they can also be easily shaped into complex forms, plastics have become a ubiquitous, low-cost, versatile material and have replaced metals and ceramics in a wide variety of applications.

Crude oil and natural gas are the raw materials for synthetic polymers. These fossil fuels are refined and cracked before being further processed to become resins. Resins are then polymerized by heat and pressure into one of many different types of plastic. As with metals and ceramics, the industrial processes required to purify, synthesize, and form the engineered plastic feedstock are profoundly heat and energy intensive.

There are only two broad types of finished plastics: thermosets and thermoplastics. These two types have very different material properties, especially when exposed to heat. In fact, it is the thermal property of thermosets and thermoplastics that fundamentally determines their recyclability. ***Thermoset*** plastics are irreversibly hardened in the polymerization process. Because their molecular structure is quite rigid, thermosets will burn and not melt when exposed to heat (Figure 5.10). The advantage of thermoset plastics is that they can withstand high heat and some harsh chemical environments while maintaining their structural integrity. Thermoset plastics cannot be recycled because they do not melt. In contrast, ***thermoplastics*** have a chemical bonding that enables them to melt when heat and pressure are applied. This makes them somewhat inherently recyclable, although they are not necessarily recycled (Figure 5.11). For example, nylon carpets are thermoplastics and recyclable. However, millions of pounds of carpet are discarded each year. Nylon carpets are only collected and reprocessed through specialized industry efforts. Only about 5% of waste carpets are effectively recycled (Carpet America Recovery Effort, 2018).

Because of concerns about petrochemically based plastics, new bioplastics have entered the market. Bioplastics are made from biomass—typically sugar polymers like cellulose, lactic acid, and starch. It is a popular misconception that all bioplastics are biodegradable, however. For example, the plastic PET is widely used for food and beverage containers and can be formulated from crude oil or sugar cane. Chemically, the molecular structure of PET is the same regardless of its source substance and it does not biodegrade. In addition, plastics that are biodegradable typically require an industrial composting process or specialized recycling facility to break them down, and they cannot disintegrate in the natural outdoor environment. Polylactic acid (PLA) bioplastic is an example. To decompose it requires sustained temperatures over 130°F, a condition not naturally found on Earth.

No one knows how long plastics persist in the outdoor environment before they are reduced to a state of matter that can be safely processed by Earth's natural

FIGURE 5.10 Thermoset and thermoplastic analogy. The critical distinction between thermosets and thermoplastics is how they behave when heated. When thermosets are formed by polymerization, they become hard like a fried egg. When a cooled fried egg is heated, it does not melt—it only burns. In contrast, thermoplastics are like butter. When heated, they melt. Thermoplastics are therefore potentially recyclable, while thermosets are not.

Source: iStockPhoto.com/Pineapple Studio; iStockPhoto.com/Maxsol7.

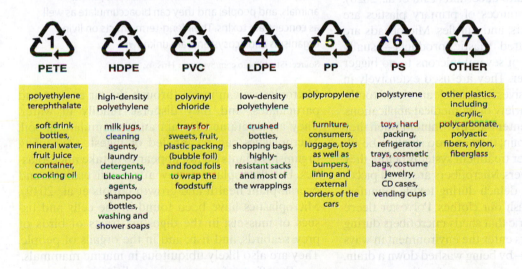

♻1 PETE	♻2 HDPE	♻3 PVC	♻4 LDPE	♻5 PP	♻6 PS	♻7 OTHER
polyethylene terephthalate	high-density polyethylene	polyvinyl chloride	low-density polyethylene	polypropylene	polystyrene	other plastics, including acrylic, polycarbonate, polyactic fibers, nylon, fiberglass
soft drink bottles, mineral water, fruit juice container, cooking oil	milk jugs, cleaning agents, laundry detergents, bleaching agents, shampoo bottles, washing and shower soaps	trays for sweets, fruit, plastic packing (bubble foil) and food foils to wrap the foodstuff	crushed bottles, shopping bags, highly-resistant sacks and most of the wrappings	furniture, consumers, luggage, toys as well as bumpers, lining and external borders of the cars	toys, hard packing, refrigerator trays, cosmetic bags, costume jewelry, CD cases, vending cups	

FIGURE 5.11 Resin identification codes. The numeric symbols on plastics are not recycling codes, but identification codes for the resin used for a particular plastic. These codes do not mean that a particular plastic is recycled. Many types of plastic are not reclaimed for complex economic and technical reasons. Only plastics #1 and #2 are widely collected and recycled in the United States.

Source: Norberthos/Shutterstock.com.

biogeochemical cycles. Common estimates range from 50 to more than 400 years, depending on the type of plastic and the environment it is in (for example, on land or in the ocean). The inability of plastic to decompose in the outdoor environment has created a new kind of pollutant—microplastics. Microplastics are a growing concern because they seem to be present everywhere on the planet, they are absorbed by living things, they persist in living tissue, and they appear to be transmissible through food chains (see Box 5.2).

BOX 5.2 Microplastics

Microplastics are a newly recognized type of environmental contaminant. Microplastics are plastic particles 5 mm or less in length, or slightly shorter than 0.25 inch (Figure 5.12). Some plastics are already this tiny when they enter the environment and are classified as *primary* microplastics. *Secondary* microplastics are created when larger pieces of plastic trash break down outdoors. Plastic does not biodegrade in the natural environment, but it does physically and chemically degrade into smaller and smaller fragments through weathering, exposure to ultraviolet light, and friction. Research suggests that microplastics in the ocean may photodegrade from sunlight and eventually dissolve. The length of time this takes, and the consequences of plastic's solubility for marine life and ecosystems, are still quite uncertain (Zhu et al., 2020).

The three basic sources of primary plastics are microbeads, microfibers, and nurdles. Microbeads are purposely manufactured plastic products—usually spherical—ranging from several microns to no bigger than 1 mm in diameter. They are used extensively in cosmetics, as fine abrasives in facial cleaners and toothpaste, and in a wide variety of biomedical applications. Microbeads typically enter the environment when they are washed down a drain; they are too small to be filtered out by wastewater treatment systems before they are discharged into rivers. Microfibers are small pieces of synthetic fiber that detach during textile manufacturing or when we wash our clothes. Polyester fleece is an example of a fabric that sheds microfibers during laundering. Microfibers enter the environment in ways similar to microbeads—by being washed down a drain.

Nurdles is the common name for the manufactured resin pellets that are used as feedstocks for thermoplastic products. Nurdles can enter the environment at the factory if they are spilled and enter the wastewater system. More often, nurdles contaminate environments when they are spilled or mishandled during shipping or when there are accidents. For example, when the X-Press Pearl container ship caught fire and sank off the coast of Sri Lanka in May-June 2021, tons of nurdles washed ashore.

A wide range and large number of studies indicate that microplastics are found everywhere in the world—from the Arctic to the Antarctic and from Mount Everest to ocean floor sediments.

FIGURE 5.12 Microplastics are pieces of plastic that range from several microns to 5 mm or less in length. They are concerning because they have been found virtually everywhere on Earth; they are detectable in plants, animals, and people; and they can bioaccumulate as well as concentrate toxins. Their long-term effects on living organisms and ecosystems are unknown.

Source: iStockPhoto.com/Svetlozar Hristov.

Microplastics can move through the air as dust and particulates, and they disperse readily in water. They are contained in rain and accumulate in soil. Microplastics can be inhaled or ingested by living organisms. Plants do not appear to uptake microplastics, but microplastics may affect soil ecology and overall plant health and growth (Boots et al., 2019). Microplastics have been found in the cells and tissues of mussels; in the digestive tracts of birds of prey, seabirds, and fish; and in the organs of people. They are also likely ubiquitous in marine mammals.

The effects of microplastics on living organisms and ecosystems are not clear, and this is a rapidly growing field of study. Early research shows that microplastics have the potential to bind to environmental toxins that can then be ingested, and a range of studies indicate that microplastics can disrupt immune systems and contribute to inflammations. Other concerns include the unknown effects of microplastics on the microbes responsible for Earth's biogeochemical cycling processes. And not all microplastics have the same effects: much depends on their resins as well as their size and shape (Thompson, 2018).

Over time, it is likely that microplastics will become a universal, standard pollutant that must be analyzed for risk, prevention, treatment, and control.

Composites

Composites are not defined as a class based on their physical, chemical, or molecular properties. Rather, composites have been fabricated from two or more other solids. Composites are desirable because they combine the advantages of the individual substances. The material properties of the resulting composite are truly greater than the sum of their parts. Composites are often designed to be lightweight, durable, and resistant to corrosion.

Composites are usually made of just two constituent substances, and they are often formed by wrapping, weaving, or adhesive bonding. Composites are not created by dissolving or molecular blending. The two most widely used composite materials in the world are concrete and fiberglass, and both are examples of adhesive bonding techniques. Concrete is made by binding gravel and stone in a matrix of cement, while fiberglass is made by binding glass fibers in a resin (polymer) epoxy. Fiberglass is widely used for consumer products such as bathtubs, automobile bodies, swimming pools, and sporting goods; it also has extensive uses in industry.

Printed circuit boards are another example of a composite (Figure 5.13). The classic green substrate of these essential electronic components is an amalgam of paper, glass fiber, and epoxy. This composite material does not conduct electricity and is stable in the presence of heat, which makes it ideal for the delicate electrical circuitry embedded in these devices.

Composites are challenging materials to recycle because their substances must be isolated from one another for recovery and reuse, and they often have been integrated at very small scales. Separating the constituent substances can

FIGURE 5.13 A printed circuit board. Printed circuit boards are the backbone of modern electronics. Electrical circuitry is embedded in the green substrate, which is a composite material of glass fiber and polymers.

Source: iStockPhoto.com/LongHa2006.

involve multiple processes such as grinding or shredding, heating, and chemical separation. It is therefore difficult to generalize about the sustainability of these materials. Fiberglass recycling is technically possible but very expensive, so there is little demand for it; innovation is coming from the automobile and boat-building industries. Concrete is now more extensively reclaimed and reused as crushed aggregate. It is now cheaper to reuse concrete in many places than to pay the landfill disposal costs.

5.3.2 Advanced Materials

Advanced materials are frequently described as novel, high-value-added substances with enhanced properties. They are often created with highly specialized, very sophisticated manufacturing techniques. Advanced (or high-tech) materials are not really a "class" of material because there are no characteristics or properties that all these materials share. Nothing unites them from a chemical, physical, or molecular perspective. They can be metals, ceramics, polymers, composites, or something else entirely.

Nanotechnology, quantum technologies, and complex semiconductors are all considered advanced materials. Biomaterials used in medicine as artificial tissues are advanced materials, as are ceramics used in applications such as turbines and jet engines. Smart materials that change their properties in reaction to a stimulus are advanced materials. An example is switchable glass, which transforms from clear to opaque when exposed to heat or an electric current.

Graphene (Figure 5.14) is a good example of an advanced material that is expected to play a revolutionary role in composite materials over the coming decades.

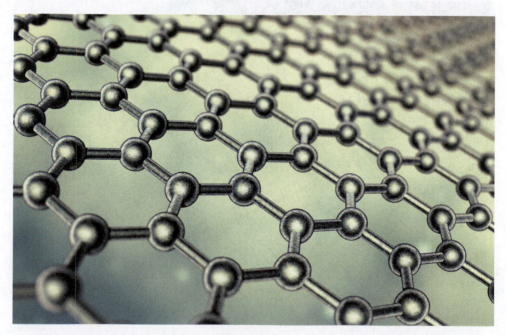

FIGURE 5.14 Graphene is carbon arranged in a honeycomb lattice one atom thick. It is an advanced material expected to revolutionize key technologies needed for a more sustainable future, including electronics, batteries, and water filtration.

Source: iStockPhoto.com/iLexx.

Graphene was first isolated in 2004 and is a honeycomb lattice of carbon. Graphene is basically graphite—the stuff of common pencil lead—*but it is only one atomic layer thick*. It is therefore a flat, two-dimensional nanomaterial. Carbon in this form has extraordinary material properties. It is the strongest known material, about 200 times stronger than steel but lighter than paper. It has about 1,000 times the current-carrying capacity of copper and has an electron mobility up to 200 times faster than silicon (Olson, 2018; Berger, undated).

Graphene conducts and dissipates heat at superior levels, enabling it to function as a cooling material as well as being extraordinarily efficient at electrical conductivity and electron mobility. Graphene is in commercial use today in a variety of applications, but its great promise is expected to be in electronics, batteries, and water filtration systems (because of its high surface-to-volume ratio and porosity). Graphene is manufactured through specialized fabrication, and there are currently no mass production techniques for this material.

We cannot generalize about the sustainability or potential recyclability of advanced materials because, as a group, they are too diverse and too new. Questions about many of these materials remain because of their inherent chemical and physical complexity. How to assess their potential toxicity is one set of questions; another relates to how they behave in the natural environment over time because their material properties are not yet fully understood.

To summarize, the potential sustainability of materials is inherently established by their physical and chemical properties. The common perception that almost everything can be reclaimed is not true. Not all plastics can be recycled, composites and advanced materials are challenging to separate into their constituent elements and compounds, and not all "biodegradable" material will actually breakdown naturally in the outdoor environment. Even relatively benign ceramic materials like glass and clay can persist in the environment for hundreds of years. In addition, there are always trade-offs in our material choices. Such dilemmas are illustrated by building insulation products, which are designed to reduce energy use but have other types of sustainability impacts (see Active Learning Exercise 5.1). Nonetheless, many materials can be cost effectively recycled or can be recovered and reused in applications other than those for which they were originally used. Approaches to more sustainable product design and materials selection are explored in Chapter 12.

ACTIVE LEARNING EXERCISE 5.1 Which Insulation Is Most Sustainable?

Good product engineering requires that we design with our end users in mind. In this activity, you will integrate your knowledge of materials with market analysis to evaluate which type of building insulation is most sustainable for a start-up firm. This activity also illustrates the powerful role that economics plays in getting (or not getting) sustainable products into the marketplace.

Assume that you are the founder of a start-up building materials company committed to green engineering and making sustainable resources readily available to the do-it-yourself home owner market. You conduct some market research and find that the U.S. building insulation market accounts for more

(Continued)

than $8 billion in sales each year, with annual increases in sales of 5%. The strongest growth will be in the residential sector, as existing home owners focus on greater energy conservation. Because of this market data, you decide to expand your product line to include insulation products.

Conduct an Internet search in which you obtain information about fiberglass batt, mineral wool, cellulose, polyurethane spray foam, and soy spray foam insulation. As you search, look for the following types of information:

- Materials content, including any use of toxic or hazardous substances in the manufacturing process. What category of traditional materials does each of these types of insulation belong to?
- Safety and health issues associated with do-it-yourself installation of such products.
- Ease of installation, and flexibility of the product (for example, it can be used in a wide variety of climates or building conditions).
- The cost of the amount of insulation required to achieve roughly R-12 to R-15 for 1,000 square feet of surface.

The **material safety data sheet** (SDS or MSDS) for these products will give you a wide variety of information, as will various websites that provide insulation cost calculators. Safety data sheets are standardized summaries of the material content, risks, and safety precautions for handling a product or material that contains a hazardous substance. See Chapter 6 for a review of SDS.

Use the three pillars of sustainable development to evaluate which of these materials are the most sustainable. Which one do you select to develop for your company? What would have to change (if anything) for your product to be more competitive in the insulation market?

5.4 Sustainability and the Linear Materials Economy

The flow of natural resources into engineered materials and manufactured products is part of a larger social, economic, and consumption dynamic that is described as the *linear materials economy* (Figure 5.15). In this model, natural resources are extracted, processed, and engineered into uniform material feedstocks from which manufactured goods are made. After the goods are bought and used, they eventually reach the end of their commercial or physical life. At this point, they become post-consumer solid waste that is disposed of in a municipal landfill.

The linear materials economy is how our industrial world has operated over the past century, and with limited exceptions, it is still the predominant way end-of-life products are handled. When products are used up, worn out, fail, or are no longer wanted, they are "trashed" with no reclaiming or reusing of any of their material content. This linearity of product life cycles and materials is problematic in several important ways from a sustainability standpoint. First and perhaps most obviously, there is the fossil fuel use and pollution that result from transporting materials and products between each stage of their transformation and use.

Second, every stage of the material transformation process is associated with environmental degradation in some manner. The mining and extraction of many metal and mineral resources are notoriously harmful to local ecosystems and have

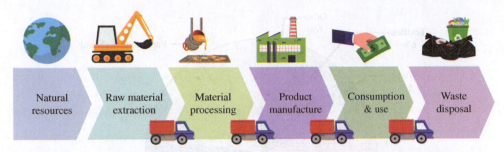

FIGURE 5.15 The linear materials economy. The linear materials economy is a model of the sequential flow of Earth's natural resources as they are extracted and converted into consumer and industrial products. Raw materials are embodied in physical products that are ultimately thrown away as waste at their end of life. Despite awareness and attention given to material recovery, reuse, and recycling, this is still the dominant way that economies and material flows occur throughout the world.

been associated with horrific labor practices in some instances. Landscapes and habitats can be bulldozed, blasted, or scraped into barren ecosystems. Tailings, the residues from mining after valuable ores have been extracted, are often dumped into surface waters or holding ponds; some residues can contain heavy metals or radioactive substances. Water degradation can also be created by the caustic and acidic chemicals used for many types of mineral and metal resource extraction methods.

Mining and extraction can be unsustainable in other ways. Many natural resources have been in the news because of the exploitive and inhumane labor conditions associated with them. "Blood" or "conflict" diamonds are a well-known example. Conflict metals and minerals are those for which mining is controlled by armed militants, who physically force children and adults to work and may capture or enslave them. Other conflict materials include tantalum, tin, tungsten, and gold. Mica is widely known to involve child labor, and several ethical mica advocacy movements exist.

Processing metals, ceramics, and polymers into engineered intermediate and final feedstocks is very energy intensive and can likewise result in significant air and water pollution. The production of finished manufactured items always requires energy, many industries are water intensive, and some sectors generate hazardous wastes as part of their industrial processes.

Third, without reclamation, recycling, or reuse, there is the problem of waste itself. Garbage must be *safely* disposed of *somewhere*. Americans generated about 5 pounds of municipal solid waste per person *per day* in 2018 (Figure 5.16). Sanitary landfills are required to safely dispose of this waste to protect surrounding areas from water, soil, and air contamination, and to protect public health; these are also costly to build and maintain. In addition, land itself is a scarce and expensive resource in many parts of the world, which drives up the cost of waste management and disposal.

Fourth, the linear materials economy depletes finite natural resources faster than what would occur with reclamation, recycling, or reuse. We cannot assume there will always be an abundance of raw materials. Indeed, one of the most surprising resource shortages in the world today is sand (Figure 5.17).

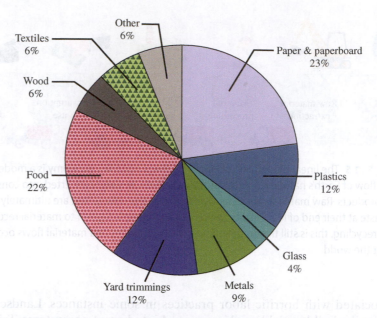

FIGURE 5.16 U.S. municipal solid waste by material weight. Food, paper, and yard waste accounted for 57% of the municipal solid waste generated in 2018. Some of this material was diverted from landfill disposal because of recycling or composting programs.

Source: Data from U.S. Environmental Protection Agency (2020).

FIGURE 5.17 Coarse sand. There is an astonishing and growing global scarcity of sand, a material that is critical as an input for construction and computer chips. Sand seems to be abundant, but desert or common beach sand is not useful because it is too smooth and rounded. Coarse, angular sand is needed for the construction industry and commercial purposes. Organized and violent crime, sand theft, sand shortages, and true physical sand scarcity are cropping up throughout the world.

Source: iStockPhoto.com/Waithaya Palee.

By reclaiming, reusing, and recycling materials, we can make our linear economy more circular. This clearly reduces the demand on Earth's natural resources and enables finite substances to be more "regenerative" from a technical perspective. When we think about material flows critically and try to close loops, we enhance sustainability in other critical ways as well. We avoid some of the environmental damage from mining, extraction, industrial pollution, and transportation that comes from processing nature's raw materials. We can also make more ethical choices about the sources of our materials.

5.5 Waste Management and Material Life Cycles

Traditional concepts about pollution, toxicity, natural resource abundance, scarcity, recoverability, and recyclability are all integrated by engineering models dealing with waste management and material life cycles. As a consumer, you are already familiar with waste management principles if you know the slogan "Reduce, Reuse, Recycle." If you have ever heard the expressions "cradle-to-grave" or "cradle-to-cradle" in reference to products and manufacturing, then you are also familiar with life cycle concepts. In this section, we will summarize how waste management and product life cycle models represent a more sustainable engineering approach to natural resource use and environmental protection.

5.5.1 The Waste Management Hierarchy

Waste management evolved as a field of study and profession from sanitary engineering, which dealt with the public health threats created by urban sewage, garbage, and human excrement. Over time, other types of waste required management as well. Today, *waste management* broadly refers to the collection, transport, processing, disposal, and monitoring of waste materials. Such wastes may be liquid, solid, or gaseous, and they may be hazardous or nonhazardous. Pollution is a type of waste, even though we don't usually think of it this way. The purpose of waste management is to minimize the impacts of waste on human health and the environment.

Waste management is guided by a ranking of most preferred to least preferred strategies known as the *waste hierarchy*, or waste pyramid (Figure 5.18). The most preferred action is to prevent waste by using only those natural resources that are absolutely necessary, by reducing packaging, by optimizing production processes so that scrap waste is minimized, and by avoiding materials and processes that will create toxic or hazardous waste and pollutants. In the waste management hierarchy, *source reduction* represents the prevention of waste in the early stages of a product's life cycle. Source reduction conserves resources and energy and can reduce pollution and toxicity as well. Source reduction and waste prevention result from many engineering actions, including:

- Redesigning products to reduce their materials content, a process known as *dematerialization*
- Selecting better materials and manufacturing processes to minimize scrap waste, pollution, and toxicity
- Designing more durable consumer products that will last longer and that can be repaired or refurbished more easily for continued use
- Reusing products that have not reached the end of their useful life
- *Remanufacturing* products by recovering modules and components for reassembly or reuse

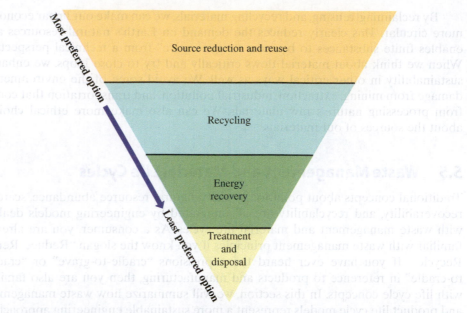

Most preferred option

Source reduction and reuse

Recycling

Energy recovery

Treatment and disposal

Least preferred option

FIGURE 5.18 The waste management hierarchy. Waste management is guided by a ranking of most to least preferred actions, where the most preferred is to prevent waste from being created and the least preferred is to arrange for its permanent disposal. "Reduce, Reuse, Recycle" is a slogan modeled on the waste pyramid and encourages consumers and businesses to reduce their waste stream.

After source reduction ("reduce and reuse" in Reduce, Reuse, Recycle), waste management becomes a process of diverting waste streams for productive purposes. *Recycling* refers to the activities involved in collecting waste that has been disposed of, separating out usable materials, and processing these materials back into usable feedstocks for manufacturing. Ideally, recycled substances would be used again as a source material for the product from which they came, such as aluminum for beverage cans, glass for containers, and so on. In practice, many recycled wastes cannot be readily or cost effectively reprocessed in this way. When a recycled material or product is used in an application that is of lower quality or has more limited functionality, it is known as *downcycling*. Examples include asphalt roofing shingles that are recycled into roadway materials and recycled office paper that is used for lower quality paperboard.

When material waste cannot be recycled, recovering its embodied energy is the next most desirable step in the waste management hierarchy. This process generally involves "burning trash" to recover embodied energy as steam or electricity. The least desirable strategy in the pyramid is permanent waste disposal. This typically takes the form of burying trash in a landfill, also known as *end-of-life disposal*.

Incentives to practice aggressive waste management are based largely on government policies and economic costs. Government regulations that control pollution and limit the use of toxic substances put pressure on manufacturers for source reduction. When landfill space is scarce, municipalities may charge large disposal fees or ban certain kinds of solid waste altogether to avoid making costly new landfills. This forces large producers of waste to reduce at the source, creatively reuse their waste, or recycle.

High or rising costs of raw materials and energy also facilitate materials reuse, recovery, and recycling. For example, it is far more economical to recycle aluminum than to create it from raw materials. Successful recycling markets do, however, depend on the demand for the recycled material. Without demand, there would be no one to buy recovered materials and products. Weak markets at the regional scale are often one of the largest barriers to effective recycling programs. Recycling relies heavily on a cost-effective business case for the recycled materials or on government policies that incentivize recycling markets.

Today, the waste management field is moving toward a philosophy of ***zero waste***. This goal is based on the principle that there is no waste in natural ecological systems and that human communities should have similarly closed-loop waste systems. The zero waste concept not only encourages a looping waste stream in which waste is processed and reused as an input, but it also places an emphasis on nontoxic materials that can biodegrade naturally if they reach end-of-life disposal. Biodegradable plastics (polymers) made from corn starch are an example of materials selection that would foster zero waste sustainability (see Box 5.3). Recycling markets and technology, as well as closed-loop engineering principles, are explored in more detail in Chapter 12.

BOX 5.3 Source Reduction and Sustainability for LEGO

The Danish company LEGO manufactures some of the most famous toys in the world and has set out ambitious sustainability goals for itself (Figure 5.19). The company is striving to make all of its packaging material sustainable by 2025, replace its petroleum-based toy bricks with biodegradable or recycled feedstock by 2030, send no solid manufacturing waste to landfills by 2025, and be a carbon-neutral factory by 2022. These goals are remarkable because LEGO must also compete in a fast-moving toy market that favors electronics and technological gadgets over hands-on play. The company still has to balance on the three pillars of sustainable development— while reducing its ecological footprint, it also has to be a profitable business and offer affordable, safe, engaging toys to children around the world.

With respect to packaging, LEGO plans to eliminate the single-use plastic bags in boxes that contain the bricks and replace them with paper bags. Although replacing plastic with paper sounds simple,

FIGURE 5.19 LEGO toy bricks are instantly recognizable worldwide. The company has set impressive environmental goals for itself, but it must also invest millions of dollars to develop sustainable materials that meet stringent product requirements.

Source: 3d_kot/Shutterstock.com.

(Continued)

BOX 5.3 Source Reduction and Sustainability for LEGO *(Continued)*

it is more challenging than it seems and requires innovation. The paper must be lightweight to keep transportation damage and costs to a minimum but be durable as packaging material and a potential play item for children. LEGO also plans to use only paper that is certified by the Forest Stewardship Council, an organization that offers independent certification of sustainably sourced forest products.

LEGO is committed to replacing the petroleum-based plastic of its toy bricks with feedstock that is either plant-based or from recycled materials. The colorful, shiny, and painful-to-step-on bricks are made of ABS plastic, a somewhat challenging material from a sustainability perspective. ABS is a thermoplastic derived from petroleum and is not biodegradable. It uses a nonrenewable fossil fuel as its base raw material and doesn't break down naturally through Earth's biogeochemical cycles. As a thermoplastic, ABS is 100% recyclable back into ABS, which is a sustainability mark in its favor. But in the United States and many other countries, ABS is grouped into recycling code 7 for "other" plastics, a category that goes straight to the landfill because of the recycling challenges of mixed materials.

Sugar cane is the natural resource that LEGO is experimenting with in addition to recycled plastics.

Inventing a new sustainable material is challenging, and success has been elusive to date. LEGO bricks have stringent product requirements that demand very specific material properties for the polymers from which they are made. The bricks must be in bold primary colors, resist breaking and fracturing, and allegedly be able to withstand just over 4,200 newtons of force (about 950 pounds) before they begin to deform. They must be shiny with smooth surfaces. And, of course, they have to click precisely together and hold snugly after hundreds (if not thousands) of uses!

LEGO is investing heavily to invent a sustainable plastic that meets its product design standards. More than 200 materials have been tested and about 100 people are working on the effort. LEGO originally budgeted about 1 billion kroner (approximately $162 million) for its plastics research and development (Reed, 2018). In the fall of 2020, the company announced it was expecting to spend up to $400 million between 2020 and 2023 to accelerate all its various corporate sustainability and social responsibility initiatives. These efforts include sustainable materials, zero waste, carbon neutrality, and LEGO Replay, which is a free take-back program for used bricks so they can be shared and redistributed to other children.

5.5.2 Life Cycle Approaches

Waste management principles are reflected in several different types of engineering models. *Industrial ecology* is a systems approach to industrial processes that models the material and energy flows of ecological systems. Its goal is to close the loop in production systems by optimizing processes and finding opportunities in which wastes can be reused as inputs by other processes. Industrial ecology derives its analytical concepts and design principles from ecosystem dynamics and biogeochemical cycles; it applies these to manufacturing and fabrication processes. Industrial ecology as an engineering model and method is the basis of Chapter 13 and is explored more thoroughly there.

Many engineering practices based on waste management principles are variations on what we refer to as *life cycle models*. Life cycle models are those that analyze a product or material from its raw materials (the "cradle"), through its fabrication and use, to its ultimate end-of-life disposal (the "grave"). Unlike the linear materials economy, life cycle models strive to "close the loop" of material flows and avoid waste. Figure 5.20 illustrates the extraction and processing of raw materials

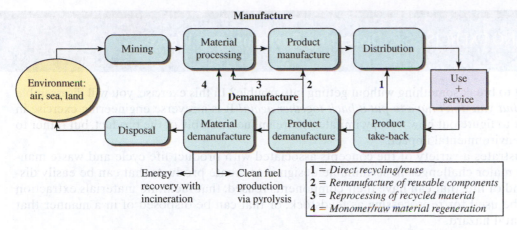

FIGURE 5.20 Life cycle pathways for manufactured goods. There are multiple pathways that the material stream for manufactured goods can follow over their life cycle. In this diagram, materials recovery, as well as product take-back and remanufacture, illustrates the principles of cradle-to-cradle design and best practices in waste management.

Source: Based on Bras, B. (1997). "Incorporating Environmental Issues in Product Design and Realization," *Industry and Environment* 20 (1–2).

for manufactured goods and the multiple pathways that these products can follow for reuse, recycling, remanufacture, and disposal. The architect William McDonough popularized life cycle models with the term cradle-to-cradle design in his book *Cradle-to-Cradle: Remaking the Way We Make Things* (2002). The cradle-to-cradle model also derives from the absence of "waste" in natural ecological systems and idealizes closed-loop raw material, product, and waste flows.

In engineering, innovations based on waste management principles and life cycle concepts have already taken place to facilitate waste minimization and to move us toward closed-loop manufacturing systems. These innovations broadly fall under the term **design for environment (DfE)** and are grounded in the idea that environmental protection and waste avoidance should be *designed into* products and processes rather than managed after the fact. Major design for environment practices include:

- Dematerialization, the reduction in the amount of materials required for a product without changing its functionality
- Design for recyclability, to facilitate materials recovery and reuse
- Design for disassembly, in which products are designed to readily come apart into their constituent components for reuse, remanufacture, or recycling
- Remanufacturing, the process of recovering product modules and components, repairing and refurbishing them, and then reusing them again in new production or sales
- Minimized use of energy, toxic materials, and toxic production processes to limit the environmental release or disposal of contaminants

Life cycle analysis has emerged as a standard method for evaluating material, energy, and waste flows in product design and manufacturing, and it is an essential engineering tool for creating more sustainable products, processes, and systems. You will have the opportunity to perform simple life cycle analysis in Chapter 14.

ACTIVE LEARNING EXERCISE 5.2 Dissecting Postconsumer Waste

How often do you get to break something without getting into trouble? In this exercise, you will disassemble a consumer product, *but you don't have to put it back together again!* It is a reverse engineering exercise in which your goal is not to figure out how to copy, imitate, or streamline assembly of the product, but rather to redesign it for lesser environmental impact.

This exercise illustrates a variety of the concepts associated with product life cycle and waste management. One of the major challenges in sustainable design is to create products that can be easily disassembled at their end of life, that can have their components reused, that facilitate materials extraction for recycling or can be used for reprocessing as feedstock, or that can be disposed of in a manner that mitigates environmental hazard.

Begin by obtaining a small consumer product that you don't mind breaking or that may already not work. Examples include kitchen appliances (toasters, coffee pots, blenders), consumer electronics (cell phones, calculators, game consoles, TV sets, radios, stereo equipment, fax machines, alarm clocks, electric staplers), personal hygiene products (electric razors, blow dryers, curling irons), power tools, and toys. You might have something lying around at home that you can use; thrift stores and charity shops are good sources for very cheap consumer products. You'll also need a tool set with an assortment of screwdrivers and pliers.

Clear and set up a workspace, ideally covered with newspaper or other material to protect your work surface and corral small parts. Have a notebook and pencil ready; a smart phone or a digital camera is also useful. Begin by sketching and photographing the product in its fully assembled form, and make notes about how it appears to be put together. Start disassembling the product. As you take it apart, draw a series of sketches and/or take photographs of the units you remove and disassemble. Keep track of the *sequence* in which the item's components and modules come apart, and make sketches of how components and modules fit together. Also keep a record of how the product is physically held together in terms of screws, fasteners, clips, adhesives, soldering, and so on.

Break down your product into the smallest units for which you can safely do so. Keep your materials organized into their dissected groupings to make it easier to identify your product's substructure. Once your item is completely taken apart, reflect on the following:

- Count and describe the discrete modules and components you dissected. At what point could you no longer easily reduce your product into smaller units? How complicated was the disassembly? What are the implications of your product's complexity for waste management or Design for Environment?
- Are there any component units that could be refurbished and used for remanufacture? Which ones? What cannot be easily refurbished? Why?
- Can you readily identify the materials used in your product? Are any of these recyclable? Can these materials be easily physically extracted?
- What is the potential for "zero waste" disposal of your product once it is at the end of its useful life?
- Are there ways in which the design and fabrication of this product could be simplified for easier disassembly and component or materials recovery? What do you suggest, and why?

5.6 Summary

In this chapter, we introduced a number of concepts and models that will help you develop analytical skills for creating more sustainable materials, products, and processes. Natural resources and raw materials are limited substances, and the ecological dynamics of our biosphere support life through a variety of ecosystem services. By both emulating and protecting the resources, biodiversity, and biogeochemical cycles of the Earth, we have a foundation for sustainable engineering and design. Waste management and ecological principles provide us with industrial ecology and life cycle models of zero waste and closed-loop waste systems.

References

Berger, M. (Undated). "Graphene Description." Nanowerks.com. www.nanowerk.com /what_is_graphene.php.

Boots, B., et al. (2019). "Effects of microplastics in soil ecosystems: Above and below ground." *Environmental Science & Technology* 53(19):11496–11506. DOI: 10.1021 /acs.est.9b03304.

Carpet America Recovery Effort. (2018). CARE Annual Report 2017. carpetrecovery .org/wp-content/uploads/2018/05/CARE-Annual-Report-2017-FINAL.pdf.

Hawken, P., Lovins, A. B., and Lovins, L. H. (1999). *Natural Capitalism: Creating the Next Industrial Revolution.* Boston: Little, Brown and Co.

McDonough, W., and Braungart, M. (2002). *Cradle-to-Cradle: Remaking the Way We Make Things.* New York: North Point Press.

Olson, D. (2018). "Mineral resource of the month: Graphite." Earth Magazine online, 12 June. www.earthmagazine.org/article/mineral-resource-month-graphite.

Parker, L. (2020). "Microplastics have invaded virtually every crevice on Earth." National Geographic online, 10 August. www.nationalgeographic.co.uk/environment -and-conservation/2020/08/microplastics-have-moved-into-virtually-every -crevice-on-earth.

Reed, S. (2018). "LEGO wants to completely remake its toy bricks (without anyone noticing)." *New York Times*, 31 August. www.nytimes.com/2018/08/31/business/energy -environment/lego-plastic-denmark-environment-toys.html.

Thompson, A. (2018). "From fish to humans, a microplastic invasion may be taking a toll." Scientific American online, 4 September. www.scientificamerican.com/article /from-fish-to-humans-a-microplastic-invasion-may-be-taking-a-toll/.

U.S. Environmental Protection Agency. (2020). National Overview: Facts and Figures on Materials, Wastes and Recycling. www.epa.gov/facts-and-figures-about-materials -waste-and-recycling/national-overview-facts-and-figures-materials.

Williams, C. A. (2021) "China continues dominance of rare earths markets to 2030, says Roskill." Mining.com, 26 February. www.mining.com/china-continues-dominance-of -rare-earths-markets-to-2030-says-roskill/.

Zhu, L., et al. (2020). "Photochemical dissolution of buoyant microplastics to dissolved organic carbon: Rates and microbial impacts." *Journal of Hazardous Materials.* doi .org/10.1016/j.jhazmat.2019.121065.

Key Concepts

Sustainable development

Intergenerational ethics

Three pillars of sustainable development

Raw materials
Abiotic
Biotic
Renewable resource
Nonrenewable resource
Fossil water
Natural resources management
Ecosystem services
Natural capital
Metals
Alloy
Ceramics
Polymers
Thermosets
Thermoplastics
Microplastics

Composites
Advanced materials
Linear materials economy
Waste management
Waste hierarchy
Source reduction
Dematerialization
Remanufacturing
Recycling
Downcycling
End-of-life disposal
Zero waste
Industrial ecology
Life cycle models
Design for environment

Problems

5-1 Can all of Earth's abiotic substances and biotic species be considered natural resources? Explain why or why not.

5-2 What is the difference between a regenerative renewable resource and a regenerative nonrenewable resource?

5-3 "All renewable resources are the same." Do you agree with this statement? Explain why or why not.

5-4 Classify the following resources as either renewable or nonrenewable:
a. Biogas
b. Tin
c. Fossil water
d. Tuna
e. Geothermal energy
f. Diamonds

5-5 Classify the following as either an infinitely available renewable resource or a regenerative renewable resource:
a. River water
b. Solar energy
c. Wheat
d. Algae
e. Wind energy
f. Bagasse

5-6 Why are all elemental minerals and metals theoretically recoverable?

5-7 What economic and technical factors affect whether a substance is recoverable and/or recyclable?

5-8 Consider the biogeochemical cycle for water (see Chapter 3). Why *isn't* water always an infinitely renewable resource?

5-9 Use the second law of thermodynamics to explain why fossil fuels and their energy are not recoverable once combusted.

5-10 Classify the following by the type of ecosystem service(s) they represent:
a. Vegetation growing along a streambed
b. Sisal production

    c. A scenic mountain vista

    d. Insect pollination

    e. A sacred flower

5-11 A forest provides multiple ecosystem services. Identify and describe five of these services. (*Hint*: Some can be found in this chapter, but you may need to do a quick Internet search.)

5-12 How do the concepts of ecosystem services and natural capital help us better understand the importance of environmental sustainability?

5-13 List, explain, and give examples of the five basic categories of engineered materials.

5-14 What challenges and opportunities for recycling are presented by each of the five basic categories of engineered materials?

5-15 What are the stages of the linear materials economy, and what are the environmental issues associated with each?

5-16 Identify and briefly explain each of the four steps of the waste management hierarchy. Why is source reduction the most preferred waste management strategy?

5-17 Think about the slogan "Reduce, Reuse, Recycle." Why do you think so much emphasis is placed on recycling in our popular culture, and not on reduce or reuse?

5-18 Give examples of five different engineering methods and practices that reduce and minimize the waste stream.

5-19 Which is better from a sustainability perspective—recycling or downcycling? Why?

5-20 Is it possible to have a truly zero waste production system? Why or why not?

5-21 What is design for environment (DfE)? List and briefly describe five different engineering practices that reflect DfE.

Team-Based Problem Solving

5-22 A household in the city of San Antonio, Texas, wants to design a rainwater harvesting system that will let them water their garden without using the municipal water supply, which comes from the increasingly stressed Edwards Aquifer. The San Antonio climate is mild, so it has two growing seasons per year: spring (February 1 through July 31) and fall (September 1 through December 31). These home owners would like to have a total supply of 1,000 gallons of water for spring and summer and 500 gallons for fall. Design a rainwater harvest system that meets these goals. Assume that your materials budget is limited to $750. The roof area of the house is 1,200 square feet.

5-23 Coordinate a "garbage analysis" event with your school's facilities management or sustainability coordinator (if it has one). Select a classroom building of manageable size, gather up all of the waste in the building that has been thrown away in trash cans, and take it outdoors. Sort and separate the trash to see what has been thrown away (make sure to wear gloves when you do this). How much of the trash could have been recycled at your school or in your town? What do the results suggest about recycling awareness and efforts on your campus? (This is a good event for your school newspaper to cover, too. Photos and videos can help document your activity.)

## CHAPTER 6

Hazardous Substances and Risk Assessment

FIGURE 6.1 A chemical spill of benzene-containing products that occurred in Deer Park, Texas, in 2019. How would you determine the extent of this problem? What is the right response to protect the health of those who live nearby, the health of those responding to the chemical spill, and the health of the surrounding ecosystem?

Source: Brett Coomer/Houston Chronicle/AP Images.

It is essential that we take steps to prevent chemical substances from becoming environmental hazards. Unless we develop better methods to assure adequate testing of chemicals, we will be inviting the environmental crisis of the future.

—RICHARD M. NIXON

GOALS

THE EDUCATIONAL GOALS OF THIS CHAPTER are to develop an understanding of hazards and risk and how they can be assessed using science and mathematics. In Chapters 2 through 5, we studied how chemicals and materials move through the natural environment and the factors that affect this movement. Some materials are particularly bad for human health and the environment. In this chapter, we quantify what "bad" means and discuss how to prioritize and manage substances that have adverse effects on human health and the environment. The major legal frameworks that regulate hazardous materials are explained. The methodologies used to determine human and environmental risk are illustrated. We present several case studies that illustrate historic failures in managing hazardous materials and the severe environmental, economic, and social consequences associated with those failures. Strategies exist to minimize and possibly prevent the negative impacts of hazardous materials, and those strategies are explored in the last two sections of this chapter.

OBJECTIVES

At the conclusion of this chapter, you should be able to:

6.1 Describe the relationship between a hazard and its risk.

6.2 Identify the laws and frameworks that relate to hazardous chemicals used in commerce and those that address the management of hazardous wastes.

6.3 Explain the process of risk assessment and how risk is calculated, including hazard identification, dose–response assessment, exposure assessment, and risk characterization.

6.4 Identify different types of hazardous wastes and how their properties are characterized.

6.5 Describe how radioactive waste is managed.

Introduction

An estimated 60,000 chemicals and millions of mixtures are in commercial use in the United States, with more than 1,000 new chemicals synthesized each year, many of which are considered hazardous. The situation is similar in Europe. Managing the impacts from making, using, and disposing of chemicals after use is daunting. Some hazardous chemicals end up in products, while others end up as process wastes. Different wastes are controlled under different laws and regulations depending on their potential for exposure and whether or not they are hazardous.

The inadequacy of data on the nature of chemical hazards and on their level of exposure to humans and the environment from products and processes makes the situation even more challenging. It is the goal of scientists, engineers, policymakers, and regulatory agencies to work in concert to detect, mitigate, and reduce exposure to hazardous chemicals. The design of products and processes that reduce, if not eliminate, the use and production of hazardous chemicals is increasingly seen as the key to a more sustainable future.

6.1 Understanding Hazard and Risk

A *hazard* is anything that has the potential to produce conditions that endanger safety and health. *Toxicity* is a type of hazard that refers to the quality of being toxic or poisonous. The word hazard (or toxicity) by itself does not express the likelihood that a dangerous incident will occur. *Risk* is a measure of probability. It is the likelihood that harm will result from a hazard, as illustrated in Figure 6.2. When it comes to chemicals, risk is a function of both hazard and exposure. The engineer or scientist who ignores the risks and hazards associated with a product or process creates a professional risk, with the possibility of being liable for any damages associated with their work that are perceived to have occurred. Liability is a legal concept associated with a wrongful or injurious act for which a civil court action occurs. There are two types of liability: negligence and strict liability. Negligent liability may occur if an engineer or scientist is careless or does not adhere to professional standards, with the result that some harm or damages occur. Engineers and scientists are assumed to have certain responsibilities associated with their position; in a strict liability lawsuit, plaintiffs need only to show that those responsibilities were not upheld.

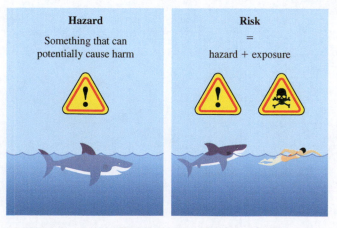

FIGURE 6.2 The difference between hazard and risk.

The level of risk that the public is willing to accept is closely related to societal norms and whether the risk is voluntary or involuntary. Consider the risk you believe is reasonable of having disease-causing (possibly deadly) microorganisms present in your drinking water compared to the risk you would accept of going hang gliding or skydiving on semester break. You may choose to go hang gliding or skydiving and even pay to do so despite the elevated level of voluntary risk. A *voluntary risk* is a risk individuals take of their own free will. An *involuntary risk* is one imposed on individuals because of circumstances beyond their control. The level of pathogens in public drinking water or even the risk associated with certain high-speed highways is an example of involuntary risk. Individuals are typically willing to accept a higher level of risk when undertaking voluntary activities (especially in high-income developed nations) but expect a lower level of risk in involuntary activities. As shown in Table 6.1, the risk of death due to hang gliding is much

TABLE 6.1 Annual mortality risk associated with giving birth in selected countries, disease, physical activities, and environmental exposures

ACTIVITY/EXPOSURE	ANNUAL RISK (DEATHS PER 100,000 PERSONS AT RISK)
Maternal mortality, Somalia	1,000
Maternal mortality, Benin	350
Smoking, all causes	284
Maternal mortality, India	200
Cancer	186
Heart disease	180
Hang gliding	179
Parachuting	175
Pedestrian transportation	142
Motorcyclist	131
Boxing	45
Coal mining	29
Agriculture	29
Maternal mortality, China	37
Maternal mortality, United States	21
Motor vehicle accidents	12
Death due to firearms	10
Unintentional fall deaths	8
Maternal mortality, Australia	7
Mountain hiking	6
Cataclysmic storm	3

(Continued)

TABLE 6.1 *(Continued)*

ACTIVITY/EXPOSURE	ANNUAL RISK (DEATHS PER 100,000 PERSONS AT RISK)
Scuba diving	3
American football	2
Contact with hornets, wasps, and bees	1
Earthquakes and other earth movements	1
Chlorinated drinking water (chloroform)	0.8
4 tbsp peanut butter per day (aflatoxin)	0.8
Lightning strike	0.7
3 oz charcoal-broiled steak per pay (PAHs)	0.5
Flooding	0.2
Skiing	0.1

Source: CDC, (2009); Heran, M. (2012); Rose, G. (1985); Windsor et al. 2009.

higher than the risk of death related to having residual chloroform in drinking water; however, the public is generally less concerned with hang gliding regulations than it is with drinking water regulations. Voluntary risks are typically undertaken because there is a perceived benefit. Assuming you survive hang gliding, your rewards are the thrill and associated memories. Involuntary risks are sometimes insidious, and the people who create the risk are not necessarily those who receive the benefit. For example, a facility that emits wastewater containing a carcinogen into a waterway may do so because it is cheaper than treating the water to remove it. The facility benefits financially by avoiding costs associated with proper hazardous waste management while increasing the public health risk. The "Right to Know" is a legal principle underlying some workplace and environmental laws. It ensures that workers and the public know the chemicals to which they may be exposed at work or in their daily living and that the information is communicated in a way that helps them understand the associated risks.

It is the responsibility of engineers and scientists to design chemicals, products, and processes that are not intentionally harmful or defective. Engineers, scientists, and industry may be liable for exposure that results in adverse impacts to the environment or human health. Hazard assessment and hazard communication are needed to understand a chemical's inherent hazardous properties and how to handle it safely. The physical and chemical properties of a substance determine its likely fate and method of transport in the environment and subsequent environmental exposure. For example, physical and chemical properties determine whether or not a chemical is soluble and mobile in water, or volatile and mobile in air. The concentration, duration, and types of exposure are also important factors in understanding relative and quantitative risks. Risk assessment tools are useful for estimating or predicting the environmental and human health risks associated with various use patterns. These assessments contain significant uncertainties that should be communicated clearly to public policymakers and product designers.

6.2   Legal Frameworks for Managing Hazardous Substances

6.2.1 The Toxic Substances Control Act (TSCA)

The Toxic Substances Control Act (TSCA) was created in 1976 as the primary law for managing chemicals in the United States. The purpose of the TSCA is to empower the U.S. Environmental Protection Agency (EPA) to evaluate risks from new and existing chemicals and to address any unreasonable risks that chemicals may pose for human health and the environment. The EPA maintains the TSCA Chemical Substances Control Inventory (TSCA Inventory) of chemicals that are manufactured or processed, including imports. The TSCA Inventory now lists about 86,000 substances, but it is unknown exactly how many of these substances are really found in commerce; a reasonable estimate is about 60,000 chemicals. Manufacturers of new substances must provide the EPA with pre-manufacture notification prior to allowing those substances to enter commerce. The EPA evaluates chemical substances for risks and exposures of concern based on their intended use and may require testing, leading to restrictions such as mandatory exposure controls. Once approved, substances are added to the TSCA Inventory for the uses specified. If a significant new use of a chemical could result in risk from exposures or releases, then the EPA may issue Significant New Use Rules (SNURs) to provide additional oversight of the chemical for the new uses.

The TSCA was updated in 2016 as the Frank R. Lautenberg Chemical Safety for the 21st Century Act. The amendment contained some needed improvements, including mandatory requirements for the EPA to evaluate existing chemicals with enforceable deadlines. The TSCA was created primarily to regulate new substances and significant new uses of existing substances. That means that approximately 60,000 chemicals in commerce prior to the creation of the TSCA were "grandfathered in" and not subject to EPA review. The updated TSCA requires the EPA to perform risk-based chemical assessments and to increase public transparency for chemical information. It also provides a source of funding for the EPA to carry out these responsibilities under the new law.

6.2.2 Resource Conservation and Recovery Act

In the United States, both solid and hazardous wastes are controlled by the EPA under the authority of the **Resource Conservation and Recovery Act (RCRA)** (42 U.S.C. §6901 et seq. (1976)). Nonhazardous solid waste is regulated under Subtitle D of the RCRA, while hazardous waste is regulated under Subtitle C.

Ideally, the TSCA is designed to prevent the creation of materials that may eventually prove damaging or difficult to dispose of safely, whereas the RCRA addresses the disposal of hazardous wastes by establishing standards for hazardous waste generators, transporters, secure landfills, and treatment processes. This includes permitting requirements, enforcement, corrective action, and clean-up, as well as the ability to authorize states to implement key provisions in lieu of the federal government.

The RCRA has been amended and strengthened by Congress multiple times, including in November 1984 with the passage of the federal Hazardous and Solid Waste Amendments (HSWA). The HSWA required phasing out the disposal of hazardous waste on land and required facilities that treat, store, or dispose of hazardous waste to take responsibility for clean-up or "corrective action" of any releases. The HSWA also puts forward waste minimization as a strategy for creating less hazardous waste by the use of source reduction and/or environmentally sound recycling methods prior to treating or disposing of hazardous wastes.

The Comprehensive Environmental Response, Compensation, and Liability Act (CERCLA) is directed at correcting the mistakes of the past by cleaning up old hazardous waste sites. It is commonly referred to as Superfund because it set up a large trust fund paid for by the chemical and petroleum industries for the purpose of providing resources to clean up abandoned sites that could endanger public health or the environment. The EPA uses these funds to address acute problems created by improper hazardous waste disposal, accidental discharges, and the clean-up of old sites, as discussed in Sections 6.4 and 6.5.

6.2.3 The Globally Harmonized System (GHA) of Classification and Labelling of Chemicals

The *Globally Harmonized System of Classification and Labelling of Chemicals (GHS)* is a worldwide initiative to promote guidelines for ensuring the safe production, transport, handling, use, and disposal of hazardous chemicals. The GHS is not a regulation, but it has been written into regulations in many countries. The GHS provides standardized criteria for *hazard classification* and guidance on how to communicate the health, physical, and environmental hazards of chemicals through labels. Hazard classification indicates the type of chemical hazard and its severity. The GHS was proposed at the United Nations 1992 Rio Conference on Environment and Development to reduce international disagreements over what materials were classified as hazardous. It was launched in 2002 and is updated every two years. The GHS uses pictograms, hazard statements, hazard statement codes, and signal words such as "Danger" and "Warning" to communicate hazard information on product labels and safety data sheets (SDSs), formerly called material safety data sheets (MSDSs) (see Table 6.2). The GHS "Purple Book" defines how to classify hazards based on scientific information and standardized toxicological tests. It addresses hazards associated with individual chemicals as well as with chemical mixtures. As of 2021, the GHS has been adopted fully or in part by about 70 countries.

TABLE 6.2 Examples of hazard communication standard pictograms

HAZARD	SYMBOL
Acute Toxicity (fatal or toxic) through any exposure route	Skull and crossbones
	Signal Word: DANGER
Carcinogens Mutagenicity Reproductive Toxicity Respiratory Sensitizer Target Organ Toxicity Aspiration Toxicity	Health Hazard Signal Word: DANGER

HAZARD	SYMBOL	
Irritant (skin and eye)	Exclamation Mark	
Skin Sensitizer		
Acute Toxicity (harmful)	Signal Word:	
Narcotic Effects	WARNING	
Respiratory Tract Irritant		
Hazardous to Ozone Layer		
Skin Corrosion/Burns	Corrosion	
Eye Damage	Signal Word:	
Corrosive to Metals	DANGER	
Aquatic Toxicity	Environment	
	Signal Word:	
	WARNING	
Flammables	Flame	
Pyrophorics		
Self-Heating	Signal Word:	
Emits Flammable Gas	DANGER	
Self-Reactives		
Organic Peroxides		
Explosives	Exploding Bomb	
Self-Reactives		
Organic Peroxides	Signal Word:	
	DANGER	

The primary goal of the GHS is to protect human health and the environment worldwide by providing chemical users and handlers with clear and consistent information on chemical hazards. Sound management of chemicals requires that chemical manufacturers identify the relevant hazards and then communicate them. The GHS categorizes individual hazard endpoints (i.e., carcinogenicity, skin irritation, etc.) as Category 1 (most hazardous) to Categories 2–5 (less hazardous) depending on the endpoint. Each category has an associated Hazard Statement Code. For example, a Category 1 Carcinogen has the H statement code H350, while a Category 2 Carcinogen has the H Statement Code H351. This system takes a little getting used to, but it is clear and consistent. Substances that don't qualify for a classification are "Not Classified" for that hazard endpoint. For example, substances that do not cause skin reactions when applied to the skin are Not Classified for Skin Sensitization. More detail on hazard assessment and GHS classification is provided later in this chapter.

The GHS helps ensure more consistency in the hazard classification process, thereby improving and simplifying hazard communication. This improved communication system informs users of the presence of a hazard and the need to minimize risk whether by minimizing exposure or by finding safer alternatives. The end result is safer transportation, handling, and use of chemicals.

In addition, the GHS reduces the inefficiencies caused by the need to comply with multiple classification and labeling systems. Historically, companies that did business in many countries had to comply with the hazard label requirements specific to each country. At times, product labels could not contain all of the information needed to properly label a product for sale in multiple countries. When each country requires different systems of hazard communication and labeling, the product may not be big enough to contain the label needed to sell it! The GHS provides a harmonized basis for the communication of hazards. For countries without well-developed regulatory systems, the GHS is particularly useful because those countries can simply adopt it and automatically have a system equivalent to the rest of the world.

The GHS also provides a standardized approach to Safety Data Sheets (SDSs). SDSs accompany products when sold. (*Note*: In the United States, SDSs are mandatory only for industrial products. Products purchased for the home do not require them, which makes it difficult for consumers to know about the chemical composition and hazards associated with the products they buy. The GHS does not change that.)

The GHS also has its limitations, in large part due to limitations in the state of the science of toxicology and in regulatory toxicology. There are existing and emerging chemical hazards that are not included in the GHS. For example, there are no criteria for endocrine-disrupting substances in the GHS. Endocrine disruptors are substances that mimic or interfere with the body's hormones and that can have negative impacts at very low concentrations. When people or other organisms susceptible to endocrine-disrupting substances are exposed to them, particularly at key points of development, there can be adverse effects. Other hazards not currently included in the GHS that relate to chemicals and materials include impacts from microplastics and substances that are persistent, mobile, and toxic, such as some of the per- and polyfluorinated substances (PFAS) used to impart water and grease repellency in textiles, electronics, food packaging, firefighting foam, and more.

6.2.4 Registration, Evaluation, Authorisation and Restriction of Chemicals (REACH)

In the European Union (EU), legislation on the Registration, Evaluation, Authorisation and Restriction of Chemicals (REACH) was enacted in 2006 and implemented on June 1, 2007. The aim of REACH was to improve the protection of human health and the environment by generating information on the intrinsic properties of

BOX 6.1 Tools for Finding Safer Consumer Products

Although Safety Data Sheets (SDSs) may not be readily available for consumer products such as cleaning products and cosmetics, a number of phone apps have been developed by mostly nonprofit organizations to provide information on ingredients and hazardous chemicals for those who are concerned about exposure to toxic chemicals in consumer products. Table 6.3 provides a few examples.

TABLE 6.3 Phone applications to help consumers find information on safer cosmetics

NAME OF APP	PURPOSE	URL
Think Dirty	Users learn about potential toxins in primarily personal care and beauty products by scanning the product barcode.	www.thinkdirtyapp.com
Cosmethics	With this European app, users scan a product barcode and analyze cosmetic products in seconds; it also suggests alternatives when products are deemed to contain toxic ingredients.	www.cosmethics.com
EWG's Healthy Living	Users find ratings for more than 120,000 food and personal care products. They can scan a product's barcode, review its rating, and make their choice (i.e., EWG Verified).	www.ewg.org/apps/#

FIGURE 6.3 Clear and simple visuals can quickly communicate information on products to people who have limited time and chemistry knowledge.

Source: CosmEthics.com.

chemicals as a prerequisite for market access (i.e., "no data, no market"). REACH also set out to enhance innovation and the competitiveness of the EU chemicals industry, a potential benefit from driving substitution with safer alternatives. REACH requires industry to manage the risks from chemicals and to provide safety information on the substances. Manufacturers and importers are required to gather information on the properties of their chemical substances and to register the information in a central database maintained by the European Chemicals Agency in Helsinki, Finland (ECHA 2020). REACH integrates with the GHS by using the hazard classification categories and hazard statement codes defined in the GHS system. The rapid growth in chemical hazard data and integration with the GHS have made ECHA and REACH powerful global resources for information and data on chemicals.

REACH requires the identification of *Substances of Very High Concern* (SVHCs). These are chemicals determined to have hazardous properties that cause cancer, infertility, birth defects, gene mutations, and equivalent concerns, along with chemicals that are persistent and toxic and/or can bioaccumulate in humans and other organisms. Under REACH, companies must provide information to consumers when SVHCs are present and work to substitute them with less dangerous chemicals. It is possible to receive authorization to continue using SVHCs, but authorization is a challenging process. While authorization may allow for ongoing use of an SVHC, it will be for limited applications and for limited time periods. REACH is beneficial to both consumers and producers because providing information on a chemical's intrinsic properties, including its human health and environmental properties, can enhance understanding and ensure safe handling. It also makes markets more competitive by leveling the playing field with respect to the availability of data and provides a service to consumers by shifting the onus of responsibility for chemical substance property information and safety toward producers.

ACTIVE LEARNING EXERCISE 6.1 Tracking Substances of Concern in Products

Recently, the European Chemicals Agency (ECHA) created a database to track **S**ubstances of **C**oncern **I**n articles as such or in complex **P**roducts—the **SCIP** Database (echa.europa.eu/scip). Companies that supply articles containing substances of very high concern (SVHCs) in a concentration higher than 0.1% by weight on the European market were required to submit information on these articles starting on January 5, 2021. The SCIP database ensures that information on SVHCs is available throughout the entire life cycle of products and materials, including at the waste stage. The information in the database is made available to waste operators and consumers. It is intended to encourage manufacturers to substitute SVHCs in products with safer alternatives and to assist waste operators in making informed decisions about recycling products containing toxic substances that lower the value of recycled materials. The SCIP is designed to foster a shift to a "safe and circular" economy.

Imagine that you are part of a company that manufactures vinyl flooring. Your current vinyl product contains the plasticizer diethyl hexyl phthalate (DEHP), which has been identified as an SVHC. Assign roles to individuals in your group consistent with roles that would exist in a company. These roles might include individuals from the executive team, legal, sales and marketing, environmental health and safety, regulatory compliance, and product design and development. Decide as a team how you will respond to the legal requirements of the SCIP database. What are the pros and cons of reporting your products with SVHCs? How might you approach finding alternative chemicals or materials to DEHP?

6.3 Risk Assessment

6.3.1 Risk Assessment, Risk Perception, and Risk Management

In the United States, many risk-based measures of public protection, such as regulations on environmental pollutants and workforce exposure to chemicals, are designed to reduce the level of risk associated with a given workplace or environmental exposure below a one-in-a-million lifetime risk. This risk level is comparable to some of the activities shown to increase the mortality risk by one in a million illustrated in Table 6.4. Calculating this risk is subject to many interpretations in toxicity and exposure data. Some measures, such as hazard quotients (discussed later in this chapter), are not technically based on risk because they are not probabilities. But they have become associated with risk assessment because they provide a threshold (i.e., one) above which risk is assumed to be significant. Certain assumptions must be made to determine risk. For example, an environmentally related risk calculation might assume that a person lives in the same location for 70 years and is exposed to a given environmental substance present in their home or local environment 24 hours per day, 365 days per year for 70 years. Exposure in the workforce might be calculated based on a 30-year career, with exposure 8 hours per day for 250 days per year. Risk factors need to be estimated, and these estimates may be used to make decisions about publicly acceptable levels of risk compared to publicly

TABLE 6.4 Activities that increase mortality risk by one-in-a-million

ACTIVITY	TYPE OF RISK
Smoking 1.4 cigarettes	Cancer, heart disease
Drinking 0.5 liters of wine	Cirrhosis of the liver
Spending 1 hour in a coal mine	Black lung disease
Spending 3 hours in a coal mine	Accident
Living 2 days in New York or Boston	Air pollution
Drinking Miami water for 1 year	Cancer from chloroform exposure
Traveling 300 miles by car	Accident
Flying 1,000 miles by jet	Accident
Flying 6,000 miles by jet	Accident
Traveling 10 miles by bicycle	Accident
Traveling 6 miles by canoe	Accident
Living 2 summers in Denver (vs. sea level)	Cancer from cosmic radiation
Living 2 months with a cigarette smoker	Cancer, heart disease
Eating 40 tablespoons of peanut butter	Cancer from aflatoxin
Eating 100 charcoal-broiled steaks	Cancer from benzopyrene
Living 50 years within 5 miles of a nuclear reactor	Accident releasing radiation

Source: Based on Wilson, R. (1979, February). "Analyzing the Daily Risks of Life." *Technology Review*, 41–44.

accepted levels of regulation. Often risk assumptions are conservative in order to account for uncertainty in the data and variability between individuals, due in part to age, gender, existing health conditions, and other factors.

The public's perception of risk may be very different from that of someone trained in risk assessment. Figure 6.4 illustrates a variety of types of risks that are perceived as high or low risks and the degree of understanding of those risks. If risk is a function of both hazard and exposure, then we as individuals may overestimate or underestimate either the severity of the hazard or the likelihood of an event occurring. Mathematical understanding of the risk due to exposure to chemical hazards will be discussed later in this section. The risk of common voluntary choices, like riding a bike or bouncing on a trampoline, are not always well understood by the public in terms of their likelihood to cause harm. However, the perception of risk may be quite low due to the perception of choice in choosing to drive a car or bounce on a trampoline. Education about the risks associated with hazardous

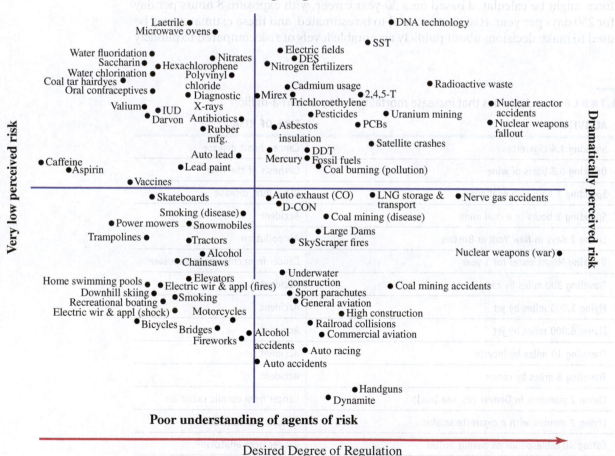

FIGURE 6.4 The public's perception of risk may differ dramatically from that of someone trained in risk assessment. Public perception may be related to the degree of the perceived risk and the understanding of the perceived risk.

Source: Reprinted with permission from Stern, P.C., and Fineberg, H.V. (eds.) (1996). *Understanding Risk: Informing Decisions in a Democratic Society.* Committee on Risk Characterization, Commission on Behavioral and Social Sciences and Education, Figure 2.4, page 62, National Research Council, 1996, by the National Academy of Sciences, Courtesy of the National Academies Press, Washington, DC.

substances is often a prerequisite for the implementation of regulations through the political process in the United States. The public is most amenable to the regulation of hazardous substances when they apply to well-understood health risks and when exposure to them is not voluntary.

Government agencies in many countries are responsible for creating regulations that minimize the risks, particularly involuntary risks, associated with the use of hazardous chemicals, transportation systems, workplace exposure, air quality, and water quality. ***Risk assessment*** is the process of estimating the spectrum and frequency of negative impacts to "receptors" (i.e., us or other humans or organisms in the environment) using numerical values to indicate the hazard and the probability of exposure for certain activities. Risk assessment of chemicals requires adequate knowledge of the inherent hazards of the chemical of interest, the potency of those hazards, the physical form in which the contaminant is present, the methods by which the chemical may be transported to receptors that will result in an exposure pathway, and exposure estimates to humans and other organisms in the environment. Typically, a human health risk assessment includes four steps:

- Hazard identification
- Dose–response assessment
- Exposure assessment
- Risk characterization

Risk management is the process of identifying, selecting, and implementing appropriate actions, controls, or regulations to reduce risk for a group of people. Scientists and engineers develop the scientific comparisons risk managers use to relate the public acceptance of risk to the acceptance of government regulations and restrictions. The overall risk assessment process is illustrated in Figure 6.5.

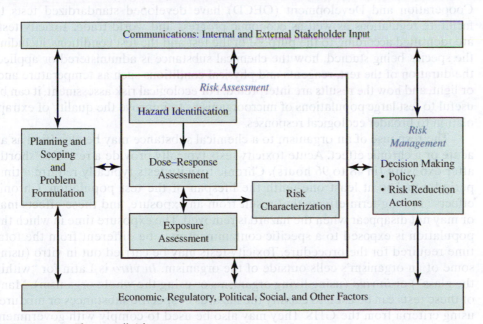

FIGURE 6.5 The overall risk assessment process.

Source: www.epa.gov/sites/production/files/2020-01/documents/guidelines_for_human_exposure_assessment_final2019.pdf.

Risk assessment is an example of applied science used at the nexus of science and policymaking. Figure 6.5 illustrates the steps in a risk assessment and where discussions between the risk assessor and risk manager take place. Risk assessors and risk managers are not typically the same people. Risk assessors are trained in various aspects of risk assessment, including human and environmental toxicology, chemistry, environmental fate and transport, uncertainty analysis, and exposure assessment. Risk managers must be able to interpret the science and the results of the risk assessment; understand the level of uncertainty; gauge economic, political, and social factors; and determine appropriate and acceptable risk control or policy interventions. Whether or not risk control or policy interventions are acceptable depends in part on cultural mores, whether the risks are voluntary or involuntary, and perceived options. It may be possible to avoid the risk altogether by choosing other options.

6.3.2 Hazard Identification

Toxicology is the science that studies the adverse effects of chemicals, substances, or situations on biological systems. Regulatory toxicology is the application of toxicology to inform decision making concerning health hazards and risks involved with the use, production, and disposal of such agents as pesticides, drugs, cosmetics, food, and household goods. Risk assessment is an important practice in regulatory toxicology as applied in the academic, industrial, or governmental sectors. Broadly speaking, regulatory toxicology tends to be more concerned with the use of standardized tests and protocols such as risk assessment than basic research. Decisions need to be made with the best available information, and regulatory toxicology helps to ensure consistency and best practices.

Toxicity tests have been developed to enhance the understanding of how organisms and ecosystems respond to chemical substances. Government agencies such as the U.S. EPA and global cooperative efforts such as the Organization for Economic Cooperation and Development (OECD) have developed standardized tests to facilitate regulations as well as economic progress and world trade. Toxicity tests are identified according to the purpose of the test and the test conditions, including the species being studied, how the chemical substance is administered or applied, the duration of the test, reagents and physical conditions such as temperature and/ or light, and how the results are interpreted. For ecological risk assessment, it can be useful to test large populations of microorganisms to improve the quality of extrapolation to broader ecological responses.

The response of an organism to a chemical substance may be classified as an acute or a chronic effect. Acute toxicity tests typically provide a response shortly after exposure (in 48 to 96 hours). Chronic toxicity tests typically require a time period equal to at least one-tenth the lifespan of the test population. Chronic effects are long-term effects that result from an exposure, and these effects may or may not disappear when the hazard is removed. The exposure time in which the population is exposed to a specific contaminant may be different from the total time required for the procedure. Toxicity tests may be carried out in vitro (using some of an organism's cells outside of the organism. *In vitro* is Latin for "within the glass") or *in vivo* (using living organisms or using the whole organism). Many of these tests can be used to classify the hazards of specific substances or mixtures using criteria from the GHS. They may also be used to comply with government water quality and other standards that are not based on the GHS.

Some types of chemical substances that cause long-term chronic effects include:

- **Carcinogens**, which may cause cancer
- **Teratogens**, which may cause physical or functional defects in an embryo
- **Neurotoxins**, which may cause impaired brain or nervous system function
- **Mutagens**, which may cause genetic mutations
- **Specific target organ toxicants**, which may cause disorders of the pulmonary, cardiovascular, skeletal, and/or immune systems
- **Sensitizers**, which may cause hypersensitivity of the skin and/or respiratory tract

Toxicity is a form of hazard. The term **hazard endpoint** is used to define the type of adverse outcome that occurs from exposure to chemicals with certain inherent hazardous properties. For example, skin irritation is a hazard endpoint. If a chemical causes skin irritation, then the chemical is classified as a skin irritant. Other environmental hazards cause toxicity to aquatic organisms and physical hazards such as flammability or explosiveness. Physical hazards may be dangerous, but technically they are not toxicants. Table 6.5 presents the hazards included in the GHS.

TABLE 6.5 Chemical hazards included in the GHS

PHYSICAL HAZARDS	HEALTH HAZARDS	ENVIRONMENTAL HAZARDS
Explosives	Acute toxicity (oral, dermal, and inhalation)	Hazardous to the aquatic environment, short-term (acute)
Flammable gases	Skin corrosion/irritation	Hazardous to the aquatic environment, long-term (chronic)
Aerosols	Serious eye damage/eye irritation	Hazard to the ozone layer
Oxidizing gases	Respiratory sensitizer	
Gases under pressure	Skin sensitizer	
Flammable liquids	Germ cell mutagenicity	
Self-reactive substances	Carcinogenicity	
Pyrophoric liquids	Toxic to reproduction	
Pyrophoric solids	Specific target organ toxicity following single exposure	
Self-heating substances	Specific target organ toxicity following repeated exposure	
Substances that, in contact with water, emit flammable gases	Aspiration hazard	
Oxidizing liquids		
Oxidizing solids		
Organic peroxides		
Corrosive to metals		
Desensitized explosives		

Toxicity tests may involve animal and non-animal tests. Historically, animals have been used as the "gold standard" test model for assessing human effects. But there is a global movement toward 21st-century toxicology and the development and use of *New Approach Methods (NAMs)*, sometimes called Non-Animal Methods, that can provide validated predictive power for human toxicity without using animals. A well-known aphorism ascribed to statistician George Box is that "all models are wrong but some are useful." Both animal and non-animal tests are models for predicting human toxicity, and some non-animal tests are better predictors of human toxicity than their animal counterparts.

NAMs may include the following:

- In vitro tests
- In silico tests (using a computer model; Latin for "in silicon")
- Genomic analyses that look at how chemical substances affect gene or protein expression
- "Read-across," which is a structured way of extrapolating toxicology assessment from a chemical with test data to one that has limited test data but a very similar chemical structure

A growing number of NAMs can be used as part of regulatory toxicology. Validating and approving the use of NAMs for regulatory toxicology has been a relatively slow process. However, adopting them is being aided by laws in Europe and California that forbid the use of animal testing for cosmetics, especially when there are proven methods for testing hazards such as eye irritation or sensitization without the use of animals. Results from toxicological tests are increasingly seen as threads of evidence that can be used together to paint a portrait of the inherent hazards of a chemical substance or mixture.

Not all regulatory toxicology tests focus on toxicity. Some of them measure parameters such as persistence and bioaccumulation potential, which are indicators of environmental fate. For example, some chemicals biodegrade in water, others photodegrade in air, and others do not degrade very much in any environmental medium (air, water, soil, or sediment) while also being soluble in water and highly mobile. Environmental fate and mobility are important parameters because they are proxies for estimating the exposure that is necessary for risk assessment. The longer a toxic substance sticks around in the environment, the more likely there will be exposure to it.

Bioaccumulation potential is the intake and retention of a substance by an organism using all possible means, including air, water, or ingestion. Bioaccumulation potential reflects the ability of a material to be retained in animal tissue to the extent that organisms higher up the trophic level will have increasingly higher concentrations of this chemical. Many pesticides, for example, reside in the fatty tissues of animals and do not break down very quickly. As the smaller creatures containing the pesticide are eaten by larger ones, the concentration in the fatty tissues of the larger organisms can increase and eventually reach toxic levels for them. Of most concern are aquatic animals and birds that feed on fish, such as seals and pelicans, as well as other carnivorous birds such as eagles, falcons, and condors. *Bioconcentration* refers to the uptake and accumulation of an aquatic substance from water. It can be estimated by measuring the octanol-water coefficient of the substance.

In both human health assessment and toxicity evaluations for other species, bioaccumulation of substances in the environment is an important consideration in risk evaluation. A *bioconcentration factor (BCF)* with units of L/kg is used to estimate the equilibrium concentration in fish (in mg/kg) from a given concentration of a contaminant in water (in mg/L):

$$C_{fish} \frac{mg}{kg} = C_w \frac{mg}{L} \times BCF \frac{L}{kg} \quad (6.1)$$

The BCF characterizes the tendency of some chemicals to partition to the fatty tissue found in fish. Some examples of bioaccumulation factors are listed in Table 6.6.

Persistence reflects the ability of a substance to break down in various environmental media. Biodegradability is the ability of a substance to be broken down by microbes. Not all degradation pathways are biological, however; they can also be chemical, such as hydrolysis, or stimulated by light, such as photodegradation. In tests of persistence, it is important to specify the conditions under which biodegradation is likely to occur and how the degradation is measured. The OECD has tests that simulate biodegradability in aerobic freshwater, anaerobic freshwater, wastewater, sediment, marine water, and more. They test for the percentage of degradation over time to allow users to calculate the half-life of a substance in different media. Substances that are very difficult to break down are called recalcitrant and are typically defined as having a half-life in water longer than 180 days. Substances that have a half-life of 16 days in water are considered readily biodegradable.

Tests must be carefully selected to make sure that they fit the expected fate of a substance. For example, tests for the compostability of a material measure how well it biodegrades in the industrial compost environment, which has controlled conditions of time, moisture, and heat. Materials such as polylactic acid (PLA), a commercial plastic predominantly derived from corn, will degrade in a commercial compost facility. But PLA does not biodegrade well in a backyard compost bin where ideal compost conditions are not maintained. To determine the persistence

TABLE 6.6 Bioaccumulation factors for selected compounds

COMPOUND	BIOCONCENTRATION FACTOR (L/kg)
4,4—Methylene dianiline	11.1
Arsenic	4
Cadmium	366
Chromium	2
Dioxins and furans	19,000
Hexachlorocyclohexanes	456
Hexochlorobenzene	13,130
Lead	155
Mercury (inorganic)	5,000
PAH as benzo[a]pyrene	583
Polychlorinated biphenyls	99,667
Diethylhexylphthalate	483.1

Source: Based on Marty, M.A., and Blaisdell, R.J. (Eds.) (2000). *Air Toxics Hot Spots Program Risk Assessment Guidelines: Part IV Technical Support Document for Exposure Assessment and Stochastic Analysis.* California Office of Environmental Health and Hazard Assessment.

of a compound, it is important to understand whether the test is measuring the disappearance of the parent compound into smaller constituents or measuring the complete degradation (mineralization) of the substance into carbon dioxide and water. It is also important to remember that just because something is biobased does not mean that it is biodegradable.

Mobility is a parameter of increasing regulatory importance, especially when combined with persistence. It reflects the likelihood that a persistent substance will migrate in water. If a chemical is recalcitrant, toxic, and mobile, then it has the potential for ongoing exposure. Persistent, toxic, and mobile substances, such as the per- and polyfluoroalkyl substances (PFAS), are sometimes referred to as "forever chemicals" because of their persistence. For example, Mark Ruffalo portrayed the corporate attorney Rob Bilott in the movie *Dark Waters*, where Bilott brought a case against the chemical manufacturing corporation DuPont after DuPont surreptitiously contaminated a West Virginia town's water with persistent, toxic, and mobile perfluorooctanoic acid (PFOA) from the production of Teflon.

6.3.3 Dose–Response Assessment

The *dose* of a contaminant is defined as the amount of a chemical received by an individual that can interact with an individual's biological functions and receptors. The dose is typically measured in mg of contaminant per kg of the person's body mass (mg/kg). The dose may be any one of the following:

- The amount of a contaminant that is administered to the individual
- The amount administered to a specific location of an individual (for example, the individual's kidneys)
- The amount available for interaction after the contaminant crosses a barrier such as the skin or stomach wall

Toxicity test results may be reported as an inhibiting or effective concentration. The *effective concentration (EC)* is the concentration of the contaminant that produces a specified response (such as lack of cell reproduction) in a specific time period. The concentration of a contaminant that produces a 50% reduction in cell reproduction over a 96-hour period is called a 96-h EC_{50}. The *lethal concentration (LC)* is the concentration of the contaminant that results in the death of a specific percentage of organisms in a specific time period (e.g., 96-h LC_{50}), as shown in Figure 6.6. The animals in a toxicity study are fed progressively higher doses of the chemical until half of them die, and this dose is known as the median *lethal dose* (50%), or LD_{50}. The smaller the amount of the toxin used to kill 50% of the specimens, the higher the toxic value of the chemical. Some chemicals, such as dioxin and PCBs, show incredibly low LC_{50} values, which suggests that they are extremely dangerous to test species and presumably humans as well.

The lowest measured value that produces a result statistically different from the control population is called the *lowest observable effect concentration (LOEC)*. The dose at which there is no measurable response at all is the *no observed adverse effect level (NOAEL)*. The allowable concentration that does not cause any adverse effects to aquatic life is called the *maximum acceptable toxicant concentration (MATC)*. Some typical toxicity test conditions for aquatic species are shown in Tables 6.7 and 6.8. There are, however, some serious uncertainties with such tests. First, the tests are conducted on laboratory specimens and then the effects are extrapolated to other organisms, including humans. Second, there can be a lot of variability within species, including humans, with respect to how they respond to the same hazards. Third, the tests are

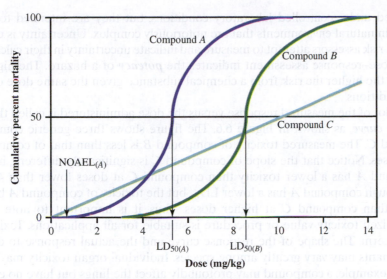

FIGURE 6.6 Theoretical dose–response acute toxicity curve for three generic compounds, *A*, *B*, and *C*.

Source: Based on Masters, G.M., and Ela, W.P. (2008). *Introduction to Environmental Engineering, 3rd ed*, Fig. 4.11. Upper Saddle River, NJ: Pearson Education, Inc.

TABLE 6.7 Typical toxicity test conditions for freshwater species

SPECIES/COMMON NAME	DURATION	ENDPOINTS
Cladoceran (*Ceriodaphnia dubia*)	Approx. 7 days	Survival, reproduction
Flathead minnow (*Pimephales promelas*)	7 days	Larval growth
	9 days	Embryo–larval survival, percent hatch, percent abnormality
Freshwater algae (*Selastrum capriocomutum*)	4 days	Growth

TABLE 6.8 Typical toxicity test conditions for marine and estuarine species

SPECIES/COMMON NAME	DURATION	ENDPOINTS
Sea urchin (*Arbacia puntulata*)	1.5 hours	Fertilization
Red macroalgae (*Champia parvula*)	7–9 days	Cystocarp production (fertilization)
Mysid (*Mysidopsis bahia*)	7 days	Growth survival, fecundity
Sheephead minnow (*Caprinodon variegatus*)	7 days	Larval growth
	9 days	Embryo–larval survival, percent hatch, percent abnormality
Inland silverside (*Menidia beryijina*)	7 days	Larval growth, survival

conducted under controlled laboratory conditions, but they are intended to predict impacts in natural environments that are notoriously complex. Uncertainty is unavoidable, and risk assessors attempt to measure and indicate uncertainty in their calculations.

A dose–response assessment indicates the *potency* of a hazard. The higher the potency, the higher the risk from a chemical substance given the same dose or exposure conditions.

A plot of the measured response versus the dose administered is called the *dose–response curve*, as shown in Figure 6.6. The figure shows three generic compounds: *A*, *B*, and *C*. The measured toxicity of compound *B* is less than that of compound *A* at all doses. Notice that the slope of compound *C* is significantly different, however. Compound *A* has a lower toxicity than compound *C* at doses lower than 4 mg/kg, even though compound *A* has a lower LD_{50}, but the toxicity of compound *A* becomes greater than compound *C* at higher doses. Thus, it is important to note that no standardized toxicity value or procedure is suitable for all applications. Toxicity is a relative term. The shape of the response curve and the actual response to different contaminants may vary greatly among species. Individual organ toxicity may vary as well; for example, a compound may profoundly affect the lungs but have no effect on the nervous system. As a rough guideline, a waste is considered toxic to mammals if it is found to have an LD_{50} lower than 50 mg/kg body weight or if the LC_{50} is lower than 2 mg/kg.

EXAMPLE 6.1 Calculating the Lethal Dose from Experimental Data

A toxicity study on the resistance of mice to a new pesticide has been conducted, with the results shown in the table below. What is the LD_{50} of this pesticide for a mouse that weighs 20 g? What is the LD_{50} for a person who weighs 70 kg?

AMOUNT INGESTED (mg)	FRACTION THAT DIED AFTER 4 HOURS
0 (control)	0
0.01	0
0.02	0.1
0.03	0.1
0.04	0.3
0.05	0.7
0.06	1.0
0.07	1.0

The data are plotted in Figure 6.7 (known as a dose–response relationship), and the point at which 50% of the mice die is identified. The mouse LD_{50} is, therefore, about 0.043 mg. The human lethal dose is estimated as

$$\frac{70,000 \text{ g}}{20 \text{ g}} (0.043 \text{ mg}) = 150 \text{ mg}$$

This is a shaky conclusion, however. First, the physiology of a person is quite different from that of a mouse, so a person may be able to ingest relatively either more or less of a toxin before showing an adverse effect. Second, there is a lot of variability among people in age, gender, general health conditions, and other characteristics. Third, the effect measured is an *acute* effect, not a long-term (*chronic*) effect. Thus, the toxicity of chemicals that affect the body slowly, over years, is not measured, since the mouse experiments are done in hours. Finally, the chemical being investigated may act synergistically with other toxins, and this technique assumes that there is only one adverse effect at a time. As an example of such problems, consider the case of dioxin, which has been shown to be extremely toxic to small laboratory animals. All available epidemiological data, however, show that humans appear to be considerably more resilient to acute effects from dioxins than the test data suggest.

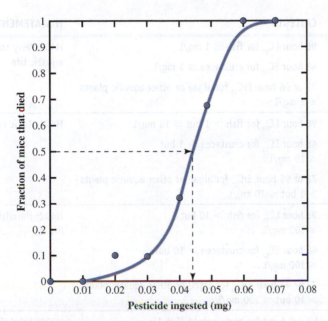

FIGURE 6.7 Calculation of LD$_{50}$ from mouse data.

The GHS classifies chemical hazards into categories for many hazard endpoints, with Category 1 being the most severe. Some of the endpoints can be classified based on potency, as shown in Table 6.9, where LD$_{50}$ or LC$_{50}$ effects are seen with very low concentrations of Category 1 substances and at higher concentrations of higher-category substances, indicating potency. Other endpoints are classified based on the strength of evidence. For example, a Category 1 Carcinogen is one that is "*known*" to have carcinogenic potential based on human or animal data. A Category 2 Carcinogen is "*suspected*" of being a carcinogen based on human or animal studies, but the evidence is not sufficiently convincing to place the substance in Category 1.

Substances that are "*Not Classified*" may still have some level of toxicity, but it is below thresholds of concern. Technically, any substance, even water, can be toxic if a person ingests, absorbs, or inhales enough if it. The GHS provides reasonable thresholds for identifying substances that are hazardous enough to require hazard communication and ostensibly risk assessment. Substances that show no toxicity after the highest thresholds are not considered toxic. For example, substances that have a 96-hour LC$_{50}$ for fish that is higher than 100 mg/L are *classified as* Not Classified. It is considered wasteful and inhumane to test at very high doses that are unrealistic. Non-lethal tests for environmental hazards are based upon the growth of organisms in aquatic environments. The median effective concentration is the concentration of a substance that reduces the mass of crustacea growth by 50 percent (EC_{50}) or reduces the growth rate (ErC_{50}) for algae or Daphnia (U.S. EPA, 1998).

TABLE 6.9 Environmental hazard classifications and criteria

HAZARD ENDPOINT	CATEGORY LEVEL	CRITERIA	H STATEMENT CODE
Hazardous to the aquatic environment, acute	1	96 hour LC_{50} for fish $\leq$ 1 mg/L 48 hour EC_{50} for crustacea $\leq$ 1 mg/L 72 or 96 hour ErC_{50} for algae or other aquatic plants $\leq$ 1 mg/L	H400: Very toxic to aquatic life
Hazardous to the aquatic environment, acute	2	96 hour LC_{50} for fish > 1 but $\leq$ 10 mg/L 48 hour EC_{50} for crustacea > 1 but $\leq$ 10 mg/L 72 or 96 hour ErC_{50} for algae or other aquatic plants > 1 but $\leq$ 10 mg/L	H401: Toxic to aquatic life
Hazardous to the aquatic environment, acute	3	96 hour LC_{50} for fish > 10 but $\leq$ 100 mg/L 48 hour EC_{50} for crustacea > 10 but $\leq$ 100 mg/L 72 or 96 hour ErC_{50} for algae or other aquatic plants > 10 but $\leq$ 100 mg/L	H402: Harmful to aquatic life
Acute mammalian toxicity, oral	1, 2	$LD_{50} \leq$ 5 mg/kg body weight (Cat 1); $\leq$ 50 mg/kg body weight (Cat 2)	H300: Fatal if swallowed
Acute mammalian toxicity, oral (mg/kg)	3	50 mg/kg < $LD_{50} \leq$ 300 mg/kg body weight	H301: Toxic if swallowed
Acute mammalian toxicity, oral (mg/kg)	4	300 mg/kg < $LD_{50} \leq$ 2,000 mg/kg body weight	H302: Harmful if swallowed
Acute mammalian toxicity, dermal	1, 2	$LD_{50} \leq$ 50 mg/kg body weight (Cat 1), $\leq$ 200 mg/kg body weight (Cat 2)	310: Fatal in contact with skin
Acute mammalian toxicity, dermal	3	200 mg/kg < $LD_{50} \leq$ 1,000 mg/kg body weight	311: Toxic in contact with skin
Acute mammalian toxicity, dermal	4	1,000 mg/kg < $LD_{50} \leq$ 2,000 mg/kg body weight	312: Harmful in contact with skin
Acute mammalian toxicity, inhalation (vapor/gas)	1, 2	$LC_{50} \leq$ 0.5 mg/L (Cat 1); $\leq$ 2 mg/L (Cat 2)	330: Fatal if inhaled
Acute mammalian toxicity, inhalation (vapor/gas)	3	2 < $LC_{50} \leq$ 10 mg/L	331: Toxic if inhaled
Acute mammalian toxicity, inhalation (vapor/gas)	4	10 < $LC_{50} \leq$ 20 mg/L	332: Harmful if inhaled

ACTIVE LEARNING EXERCISE 6.2 Evaluating Substance Classification in Product Design

Polyvinyl chloride (PVC) is a versatile polymer used in rigid applications such as pipes and siding, and flexible applications such as flooring and cable jackets. It is an inherently rigid polymer that is made flexible by the addition of plasticizers to increase its flow and thermoplasticity. Plasticizers can slowly migrate out of plastics, resulting in low levels of exposure to home occupants over time. Imagine that you are a member of a product development team working for a company that manufactures flexible PVC for use in flooring. Because your products are used indoors and because children and pets typically crawl around on floors, you want to use the safest possible plasticizers in your flooring product. You also want your product to be successful in the marketplace.

The options available in your market are listed in the table below. You know the CAS number and product trade name for each, and you have a Safety Data Sheet that lists Hazard Statement Codes associated with each plasticizer. Look up each plasticizer to determine its full chemical name. Then look up the hazard endpoints and GHS Category Levels that apply to each Hazard Statement Code, and the associated pictogram if required. Assign roles to individuals in your group consistent with roles that would exist in a company. These roles might include individuals from the executive team, legal, sales and marketing, environmental health and safety, regulatory compliance, and product design and development. Which plasticizer would you choose to use to make your PVC flooring and why?

CHEMICAL/CAS NUMBER	CHEMICAL NAME	HAZARD STATEMENT CODE	PICTOGRAM	HAZARD ENDPOINT	CATEGORY LEVEL
DEHP/117-81-7		H351, H360			
DINCH/166412-78-8		H316			
TXIB/6846-50-0		H361, H401, H412			
ATBC/77-90-7		H401, H411			

6.3.4 Exposure Assessment

The process of determining how, and to what extent, human or other "receptors" are exposed to a chemical is called *exposure assessment*. Important parameters for exposure include magnitude, frequency, and duration. The size, nature, and types of human or other populations exposed to chemicals and the uncertainties in exposure should be thoroughly documented as part of the exposure assessment. The total exposure to a study population may be estimated from measured concentrations in the environment, consideration of models of chemical transport and fate, and estimates of human intake over time resulting in estimates of dose. The exposure assessment must also consider both the pathway of exposure (how a contaminant moves though the environment to contact the study population) and the route of exposure (how the contaminant enters an individual's body—typically via ingestion, dermal contact, or inhalation). A total exposure assessment considers the cumulative risk of all exposure pathways, as illustrated in Figure 6.8 and Table 6.10.

Risk is a function of both hazard and exposure. Extrapolating toxicity data from animal or other models to humans is quite difficult. Risk assessment involves predicting the nature and severity of adverse human or other species' responses to chemicals based on exposure. In order to use animal-based toxicity data to predict human responses, a scaling factor or uncertainty factor is used in risk assessment

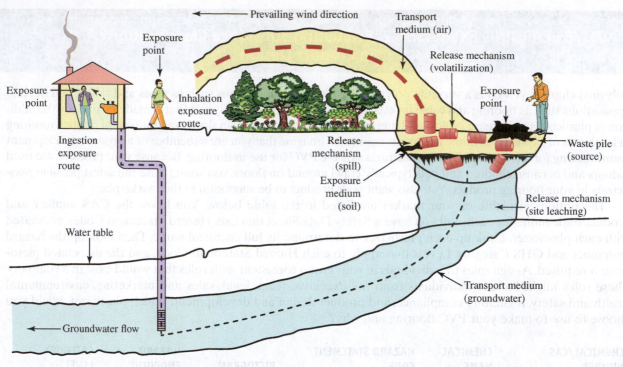

FIGURE 6.8 Illustration of exposure pathways.

Source: Based on U.S. EPA (1989). *Risk Assessment Guidance for Superfund: Volume I Human Health Evaluation Manual (Part A)*. Office of Emergency and Remedial Response. U.S. Environmental Protection Agency, Washington, DC.

TABLE 6.10 Routes of exposure from environmental contaminants

ENVIRONMENTAL SOURCE	ROUTES OF EXPOSURE
Groundwater	Ingestion, dermal contact, inhalation during showering
Surface water	Ingestion, dermal contact, inhalation during showering
Sediments	Ingestion
Air	Inhalation of gases and particles
Soil and dust	Incidental ingestion (especially children), dermal contact
Food	Ingestion

Source: Davis, M., and Masten, S. 2004. Principles of Environmental Engineering and Science. New York: McGraw-Hill. 704p.

models. Risk assessment procedures typically compare the body mass of the test animal to the average body mass of a human adult (typically considered to be 70 kg) to predict human risk from toxicity data from animal tests.

Risk assessment can be complemented with the science of epidemiology. *Epidemiology* is the study of the determinants and distributions of diseases and adverse responses in specified populations. Epidemiological work may be based on historical data and toxicity data, and insights can be derived statistically. Direct harm to an individual exposed to low levels of a toxicant may be very different from exposure to a population. The epidemiologist Geoffrey Rose observed that a large number of people at low risk gives rise to more cases of disease than a small number of people at high risk. A strength of epidemiology is that with good data on populations and exposure, researchers can gain insight into complex interactions between relatively low risks. For example, in *The Ecology of Breast Cancer* (2013), Ted Schettler clearly details a number of nongenetic risk factors that can lead to breast cancer and explains that it is the interaction between them and their cumulative impact that cause breast cancer. The risk factors include (1) diet, nutrition, and the food environment, including vitamin D; (2) physical activity and exercise; (3) environmental chemicals and contaminants; (4) regular night work (light at night); and (5) ionizing radiation. Not one of these risks is likely to lead to breast cancer alone. There is no single "smoking gun" cause, and there is no single "silver bullet" cure. Reducing multiple small risks can lead to large population benefits.

Epidemiological studies can be hard to control for and are often complicated by a lack of information about the duration and exposure to a toxicant, the change in exposure as individuals and populations change locations, long lag times, and multiple pathways of exposure to multiple toxicants. The lack of data and data

BOX 6.2 Undark and the Radium Girls

From 1917 to 1926, the element radium (Ra) was extracted from uranium-rich carnotite ore and used to make paints that spontaneously emitted light. These luminous paints were marketed as "Undark" by the Radium Dial Corporation (see Figure 6.9). Radium Dial and U.S. Radium Corporation (USRC) were two U.S. corporations engaged in the extraction and purification of radium and the painting of dials on clocks and watches. An estimated 4,000 workers were employed by corporations in North America to paint watch faces with radium. Primarily female, the painters mixed their paint in small crucibles and then used camel hair brushes to apply the paint by hand onto dials. Painters were paid by the piece, so they sought to paint as efficiently as possible. USRC supervisors famously encouraged workers to keep the brushes pointed by shaping them with their lips ("lip, dip, paint").

Managers at USRC told the workers that radium was safe, that the concentration of radium in the paint was negligible, and that it would not harm them. When faced with an independent study confirming otherwise, the president of USRC was outraged and paid for studies that came to the opposite conclusion. Because they were misinformed, employees painted their nails, teeth, and faces for fun. Employees at Radium Dial began to show signs of radium poisoning in 1926 and 1927. The first problems were dental, including dental pain, loose teeth, lesions, ulcers, and tooth extractions that

would not heal. Many of the women later began to suffer from anemia and "radium jaw," which is necrosis of the jaw. Some of the women also became sterile. The cancers that developed were gruesome (Moore, 2017). In 1923, 24-year-old Amelia Maggia became the first dial painter to die. By 1924, 50 painters had become ill and a dozen had died. In 1928, the inventor of radium dial paint, Dr. Sabin Von Sochocky, died as the 16th known victim of poisoning by radium dial paint.

Workers asked their employers to help with medical expenses. In response, the companies attempted to discredit the women and avoid responsibility for their illnesses, including by engaging medical professionals to suggest that the symptoms were due to sexually transmitted diseases like syphilis. One worker, Grace Fryer, decided to sue. It took her two years to even find a lawyer willing to take on USRC. She was joined by Edna Hussman, Katherine Schaub, and sisters Quinta McDonald and Albina Larice. Together the five women became known as the "Radium Girls." The litigation process moved slowly, and at their first appearance in court in 1928, two of the women were bedridden. The lawsuit went before the Illinois Industrial Commission (IIC), which ruled in favor of the women. Radium Dial appealed over and over again, taking the case all the way to the Supreme Court. But in October 1939, the Court decided not to take the case and the lower

(Continued)

BOX 6.2 Undark and the Radium Girls *(Continued)*

ruling was upheld. In the end, the case was won eight times before Radium Dial was finally forced to pay a $10,000 settlement to each of the Radium Girls (equivalent to $149,000 in 2019).

From the historical and legal perspective, the case of the Radium Girls helped to establish occupational disease labor law and the right of individual workers to sue for damages from corporations. It also triggered labor safety standards to improve the work environment. What we take for granted today was achieved by the suffering, courage, and tenacity of the Radium Girls. This story also serves as a reminder of why it is so important to have the means to test and to provide transparency about the hazards associated with chemicals used in manufacturing and in products, and why it is so important to continue to innovate with chemicals and products that are safe across their full life cycle.

The Power of Radium at Your Disposal

Twenty-three years ago radium was unknown. Today, thanks to constant laboratory work, the power of this most unusual of elements is at your disposal. Through the medium of Undark, radium serves you safely and surely.

Does Undark really contain radium? Most assuredly. It is radium, combined in exactly the proper manner with zinc sulphide, which gives Undark its ability to shine *continuously* in the dark.

Manufacturers have been quick to recognize the value of Undark. They apply it to the dials of watches and clocks, to electric push buttons, to the buckles of bed room slippers, to house numbers, flashlights, compasses, gasoline gauges, autometers and many other articles which you frequently wish to see in the dark.

The next time you fumble for a lighting switch, bark your shins on furniture, wonder vainly what time it is *because of the dark*—remember Undark. *It shines in the dark.* Dealers can supply you with Undarked articles.

For interesting little folder telling of the production of radium and the uses of Undark address

RADIUM LUMINOUS MATERIAL CORPORATION
50 PINE STREET NEW YORK CITY
Factories: Orange, N. J. Mines: Colorado and Utah

To Manufacturers

The number of manufactured articles to which Undark will add increased usefulness is manifold. From a sales standpoint, it has many obvious advantages. We gladly answer inquiries from manufacturers and, when it seems advisable, will carry on experimental work for them. Undark may be applied either at your plant, or at our own.

The application of Undark is simple. It is furnished as a powder, which is mixed with an adhesive. The paste thus formed is painted on with a brush. It adheres firmly to any surface.

UNDARK
Radium Luminous Material
Shines in the Dark

FIGURE 6.9 Radium serving you "safely and surely," in an advertisement for "Undark."

Source: Undark, Radium Luminous Material Corporation.

variability can create uncertainty in the results of epidemiological studies. However, historical examples of a few specific environmental exposure cases (see Box 6.2) have led to models that are useful to estimate human risk based on exposure data and toxicological data. While engineers and scientists don't need to be experts in risk assessment and epidemiology, they need to understand the strengths and weaknesses of the fields and how to work with risk assessors and epidemiologists on complex challenges.

A significant confounding factor in risk assessment, which is open to debate among toxicologists and risk assessors, is how to model risk from low concentrations of compounds. Most often, environmental exposure to toxicants occurs at very low doses. The data available on human response to very low doses of chemicals or toxicants are very limited. Figure 6.6 illustrates two models of potential response in the low dose region. For compounds *A* and *B*, no response would be expected below the NOAEL dose. However, toxicity data normally are available only for higher dose values. Therefore, the linear dose model, as illustrated by compound *C*, is used. The linear dose model predicts a higher response and higher risk for compounds at low exposure concentrations. Potentially carcinogenic compounds with limited data are usually modeled using the linear model illustrated by compound *C*, which results in higher estimates of risk at low doses. However, nonlinear methods may be used where there is sufficient evidence supporting a nonlinear mode of action. An example is Proposition 65 in California, a regulation that allows for the use of nonzero, no significant risk levels, as a "Safe Harbor" in lieu of a linear model that goes to the origin of a plot.

The U.S. EPA uses a nonlinear model for noncarcinogenic risk assessment. This method is based on the assumption that there is a NOAEL value for which no toxic effects of exposure exist. The ***reference dose (RfD)*** is defined as an estimate (with uncertainty spanning perhaps an order of magnitude) of a daily oral exposure by a human population (including sensitive groups, such as asthmatics, children, or seniors) that is likely to have no appreciable risk of adverse effects during a lifetime. ***Uncertainty factors (UF)*** take into account variability and uncertainty that are reflected in differences between animal and human toxicity data and variations in human populations. The RfD is typically reported in units of mg/kg-day:

$$RfD = \frac{NOAEL}{UF} \tag{6.2}$$

The inhalation risk is characterized by a similar term, the ***reference concentration (RfC)***, which is typically expressed in units of mg/m³.

The EPA uses the ***hazard quotient (HQ)*** to describe the probability of a noncarcinogenic toxicity risk:

$$HQ = \frac{intake}{RfD} \tag{6.3}$$

The HQ is not a strict probability. Instead, a risk is considered significant if the HQ is greater than one, and in general, the higher the HQ value, the greater the risk. As indicated by the GHS, there are multiple noncarcinogenic hazard endpoints and HQ values should be grouped by hazard endpoints or related toxicological mechanisms.

When there is suspected exposure to more than one contaminant, the total risk is estimated to be the ***hazard index (HI)***:

$$HI = \sum HQ_{ij} \tag{6.4}$$

The EPA has created an online ***integrated risk information system (IRIS)*** that provides a database for human health assessment data (see IRIS assessments at *www.epa.gov/iris*).

Through the IRIS Program, the EPA provides science-based human health assessment data to support the agency's regulatory activities. The IRIS database is web accessible and contains NOAEL levels, *RfD*s, slope factors, and carcinogenic health risk information on more than 550 chemical substances.

The EPA relates the low-dose response for carcinogenic compounds to human response in risk assessment models with the slope factor (also sometime called the potency factor). The EPA method is referred to as the ***reasonable maximum exposure (RME)*** technique. The slope factor assumes a linear dose–response relationship and typically is in units of risk per unit dose (kg-day/mg):

$$SF = \frac{\text{incremental lifetime cancer risk}}{\text{chronic daily intake}\left(\dfrac{\text{mg}}{\text{kg-day}}\right)} \quad (6.5)$$

The total cancer risk is determined by the sum of the risks associated with each potential intake pathway (i.e., ingestion and dermal exposure). The single-exposure pathway cancer risk is determined by multiplying the ***chronic daily intake (CDI)*** by the slope factor (SF):

$$\text{Cancer risk} = CDI \times SF \quad (6.6)$$

The *CDI* is a composite estimate of various exposure scenarios and assumptions used to estimate contaminant intake, as shown in Table 6.11. The *CDI* has units of mg/kg-day.

The EPA recommends a modified approach to estimating the risk of cancer from inhaled chemicals (U.S. EPA, 2009). An inhalation unit risk (*IUR*) factor was created to better estimate the cancer risk from inhaled chemicals based on the physiological response to inhaled chemicals. *IUR* factors are also found on the IRIS website. The EPA recommends that risk assessors use the concentration of the chemical in air (C_A) for exposure rather than the ingestion rate (*IR*) and body weight (*BW*). The exposure concentration (*EC*) given in Equation (6.12) replaces the *CDI* in the general risk equation given in Equation (6.6). The *IUR* replaces the slope factor in Equation (6.6). The cancer risk for air inhalation then becomes:

$$\text{Cancer risk} = EC \times IUR. \quad (6.7)$$

TABLE 6.11 Intake due to exposure routes for various pathways

ROUTE	INTAKE EQUATION	VARIABLES	UNITS	EQ. #
Ingestion in drinking water	$CDI = \dfrac{(C_W)(IR)(EF)(ED)}{(BW)(AT)}$	C_W = concentration in water	mg/L	(6.7)
		IR = ingestion rate	L/day	
		EF = exposure frequency	Days, years, or events	
		ED = exposure duration	Years	
		BW = body weight	kg	
		AT = averaging time	Days	
Ingestion while swimming	$CDI = \dfrac{(C_W)(CR)(ET)(EF)(ED)}{(BW)(AT)}$	CR = contact rate	L/hr	(6.8)
		ET = exposure time	hr/event	

TABLE 6.11 *(Continued)*

ROUTE	INTAKE EQUATION	VARIABLES	UNITS	EQ. #
Dermal contact with water	$AD = \dfrac{(C_w)(SA)(P_c)(ET)(EF)(ED)(CF_w)}{(BW)(AT)}$	AD = absorbed dose	$\dfrac{mg}{kg\text{-}day}$	(6.9)
		SA = skin surface area for contact	cm^2	
		P_c = chemical specific dermal permeability constant	cm/hr	
		CF_w = conversion factor for water	$10^{-3}\ L/cm^3$	
Ingestion of chemicals in soil	$CDI = \dfrac{(C_s)(IR)(CF_s)(FI)(EF)(ED)}{(BW)(AT)}$	C_s = concentration in soil	mg/kg	(6.10)
		CF_s = conversion factor for soil	$10^{-6}\ kg/mg$	
		FI = fraction ingested		
Dermal contact with soil	$AD = \dfrac{(C_s)(SA)(AF)(AB_s)(EF)(ED)(CF_s)}{(BW)(AT)}$	AF = soil-to-skin adherence factor	mg/cm^2	(6.11)
		AB_s = absorption factor for soil contaminants		
Inhalation of airborne (vapor phase) chemicals	$EC = \dfrac{(C_A)(ET)(EF)(ED)}{(AT)}$	C_A = concentration in air	mg/m^3	(6.12)
		EC = exposure concentration	$\mu g/m^3$	
Ingestion of contaminated foods	$CDI = \dfrac{(CF_s)(IR_F)(FI)(EF)(ED)}{(BW)(AT)}$	IR_F = ingestion rate of food	$kg/meal$	(6.13)

Source: Based on U.S. EPA (1989). *Risk Assessment Guidance for Superfund: Volume I Human Health Evaluation Manual (Part A)*. Office of Emergency and Remedial Response. U.S. Environmental Protection Agency, Washington, DC.

EXAMPLE 6.2   Reasonable Maximum Exposure (RME) Risk of Benzo[a]pyrene in Drinking Water

Suppose an individual has lived in the same home for 30 years, their entire lifetime. The water in the home comes from a well, and the well has been contaminated with 0.2 ppb of benzo[a]pyrene from an industrial leak and plume that has migrated into the groundwater table. Estimate the cancer risk to the person from ingesting 0.2 ppb of benzo[a]pyrene in the well water using the EPA procedure to calculate the RME.

We can use Equation (6.7) in Table 6.11 to determine the ingestion from drinking water, where C_w is the concentration of benzo[a]pyrene in the water in units of mg/L. (1 ppm$_m$ = 1,000 ppb$_m$ and 1 ppm$_m$ = 1 mg/L, where ppm$_m$ is parts per million by mass.) Thus:

$$C_w = 0.2\ \text{ppb}_m\left(\frac{1\ \text{ppm}_m}{1,000\ \text{ppb}_m}\right)\left(\frac{1\ \frac{mg}{L}}{1\ \text{ppm}_m}\right) = 0.0002\ \frac{mg}{L}$$

We use the U.S. EPA Exposure Factors for Risk Assessment from Appendix D. From Table D.1, the per capita ingestion rate (IR) for drinking water is broken down into the age groups shown in Table 6.12. We can set up a spreadsheet to calculate the CDI for each age group that will be used to calculate the risk for the individual as they live through each age group. The age groups are chosen in large part because the risk is proportional to body mass. In other words, the mass of benzo[a]pyrene to which an infant is exposed—in terms of mass of contaminant (mg) to body mass (kg)—is typically much greater than for an adult exposed to the same concentration of the benzo[a]pyrene in the drinking water because the larger body mass of an adult effectively dilutes the amount of contaminant distributed over the body. Each row in the spreadsheet will be used to enter the information for each appropriate risk for that particular age. For example, the mean ingestion rate of water for infants from birth to 1 month is 184 mL/day, expressed in liters per day:

$$IR = 184 \, \frac{mL}{day} \times \frac{L}{1{,}000 \, mL} = 0.184 \, \frac{L}{day}$$

The body weight for each category is also provided in Appendix Table D.9 and can be entered into the calculations represented in Table 6.11. Notice that the body weight for age groups 16–18 years and 18–21 years is identical.

The reasonable maximum exposure factor (EF) is 365 days per year at a residence because the data used in estimating the water ingestion rate are assumed to represent average daily intake over the long term (i.e., over a year). This is an estimate of a reasonable worst-case scenario. In actuality, most people spend far fewer hours per day and days per year in their home, but as the year 2020 taught us, sometimes we may be unexpectedly homebound. If we were modeling a worker's exposure in a factory, we would consider the amount of time the worker was likely to spend in that particular area.

The exposure duration (ED) is the average residency time in the home. In the scenario provided, it was specified that this individual lived in the home for 30 years. If we were considering an average risk, we would choose 12 years from the mean residential occupancy period provided in Table D.18 for Population Mobility, but this value is given in the described scenario. Notice, however, that the duration is the specified duration *in each age bracket*. For example, for a newborn to 1 month of age, the exposure duration *in that age bracket* is 1 month, or 1/12 of a year. The individual would be exposed to the benzo[a]pyrene from ages 21 to 30 in the last age bracket, a period of 9 years that would be entered into the spreadsheet.

The averaging time (AT) is equivalent to the lifetime of the individual being evaluated. For the purposes of this example, the average lifetime for men and women is used because the exposures are assumed to reflect the general population and are not gender or age specific. This value is found in Table D.19 for the total life expectancy, which is 78 years. Notice that in order to keep the units consistent, the averaging time is entered in days:

$$AT = 78 \text{ years} \left(\frac{365 \text{ days}}{\text{year}} \right) = 28{,}470 \text{ days}$$

The chronic daily intake (CDI) can then be calculated for each row of data using Equation (6.7). For example, the CDI for the first month of exposure is

$$CDI = \frac{(C_w)(IR)(EF)(ED)}{(BW)(AT)}$$

$$= \frac{\left(0.0002 \, \frac{mg}{L} \right)\left(0.184 \, \frac{L}{day} \right)\left(365 \, \frac{days}{year} \right)(0.08 \text{ year})}{(4.8 \text{ kg})(28{,}470 \text{ days})}$$

$$= 8.19 \times 10^{-9} \, \frac{mg}{(kg \cdot day)}$$

The slope factor (SF) is found from the EPA IRIS website (*www.epa.gov/iris*), and searching for benzo[a]pyrene from the full list of IRIS chemicals yields the oral slope factor for benzo[a]pyrene of 1 (mg/kg-day)$^{-1}$. The slope factor can be entered into the table after obtaining this value from the IRIS website or other referenced sources.

Finally, we use the age dependent adjustment factor ($ADAF$) to account for the greater uncertainty in risk analysis extrapolation for the age groups as specified by the EPA Exposure Factor Handbook. The $ADAF$ values from Table D.21 are applied to the oral slope factors, drinking water unit risks, and inhalation unit risks for chemicals with a mutagenic mode of action. According to the EPA Toxicological Review, benzo[a]pyrene has been determined to be carcinogenic by a mutagenic mode of action, and $ADAF$ values are recommended to be applied in estimating cancer risk.

The cancer risk for each age group is found by multiplying the CDI by the slope factor from Equation (6.6). In this case, since benzo[a]pyrene is mutagenic, multiplying by $ADAF$ also yields the cancer risk for the specific age bracket:

$$\text{Cancer risk}_{0-1\,\text{mo}} = (CDI)(SF)(ADAF_{\text{mutagens}})$$

$$= \left(8.19 \times 10^{-9}\frac{\text{mg}}{\text{kg-day}}\right)\left(1\left(\frac{\text{mg}}{\text{kg-day}}\right)^{-1}\right)10 = 8.19 \times 10^{-8}$$

The cancer risk for the first 30 years of the individual's life is found by summing the risk in each age bracket:

$$\text{Cancer risk}_{\text{Birth}-30\,\text{years}} = \sum_{\text{Birth}}^{30\,\text{years}} \text{cancer risk} = 3.52 \times 10^{-6}$$

TABLE 6.12 Example spreadsheet for calculating the cumulative cancer risk for benzo[a]pyrene in drinking water

AGE BRACKET	CHRONIC DAILY INTAKE (CDI) FACTORS								
	C	IR	BW	EF	ED	AT	SF$_{\text{oral}}$		
	mg/L	L/day	kg	days/year	years	days	[mg/(kg-day)]$^{-1}$	ADAF	Risk
0–1 mo	0.0002	0.184	4.8	365	0.08	28,470	1	10	8.19×10^{-8}
1–3 mo	0.0002	0.227	5.9	365	0.17	28,470	1	10	1.64×10^{-7}
3–6 mo	0.0002	0.362	7.4	365	0.25	28,470	1	10	3.14×10^{-7}
6–12 mo	0.0002	0.360	9.2	365	0.5	28,470	1	10	5.02×10^{-7}
1–2 yr	0.0002	0.271	11.4	365	1	28,470	1	10	6.10×10^{-7}
2–3 yr	0.0002	0.317	13.8	365	1	28,470	1	3	1.76×10^{-7}
3–6 yr	0.0002	0.327	18.6	365	3	28,470	1	3	4.06×10^{-7}
6–11 yr	0.0002	0.414	31.8	365	5	28,470	1	3	5.01×10^{-7}
11–16 yr	0.0002	0.520	56.8	365	5	28,470	1	3	3.52×10^{-7}
16–18 yr	0.0002	0.573	71.6	365	2	28,470	1	1	4.10×10^{-8}
18–21 yr	0.0002	0.681	71.6	365	3	28,470	1	1	7.32×10^{-8}
21–65 yr	0.0002	1.043	80.0	365	9	28,470	1	1	3.01×10^{-7}
>65 yr	0.0002	1.046	80.0	365	0	28,470	1	1	0
Cumulative risk from birth to 30 years for a permanent resident of the same home									3.52×10^{-6}

The concentration of benzo[a]pyrene in this drinking water is very, very low—only 0.2 part per *billion*. The cumulative cancer risk may also seem like a very small number at 3.52×10^{-6}. However, this would be expected to result in three to four cancers per million people exposed to the drinking water. If this level were present in all drinking water across the United States, this very low level of benzo[a]pyrene would be expected to cause more than 1,000 cancer cases. Is this an acceptable risk in drinking water in general? What about in your drinking water?

6.3.5 Risk Characterization

Risk assessment is a complex process in which a risk assessor estimates hazards and exposure to calculate risk and also considers the level of uncertainty as part of the analysis. In risk characterization, the risk assessor conveys a judgment about the nature and severity of the risk, the type and levels of uncertainties, and where policy choices need to be made.

Data for risk assessments should be collaboratively gathered and reviewed in order to make quantitative decisions about risk. Detailed documentation of the risk assessment is essential; each component of the risk assessment (e.g., hazard assessment, dose–response assessment, exposure assessment) should be included along with the key findings, assumptions, limitations, and uncertainties. Judgments from risk characterization are used to inform the risk manager and others about the rationale and approach for conducting the risk assessment. The risk characterization should be transparent, clear, and consistent, and it should show that reasonable assumptions and judgment have been used in the assessment.

The EPA's risk characterization policy calls for conducting risk characterizations in a manner consistent with the following principles:

- *Transparency:* The characterization should fully and explicitly disclose the risk assessment methods, default assumptions, logic, rationale, extrapolations, uncertainties, and overall strength of each step in the assessment.
- *Clarity:* The products from the risk assessment should be readily understood by readers inside and outside of the risk assessment process. Documents should be concise and free of jargon and should include understandable tables, graphs, and equations as needed.
- *Consistency:* The risk assessment should be conducted and presented in a manner that is consistent with EPA policy and with other risk characterizations of similar scope prepared across programs within the EPA.
- *Reasonableness:* The risk assessment should be based on sound judgment based on methods and assumptions consistent with the current state of the science and conveyed in a manner that is complete, balanced, and informative.

These four principles—transparency, clarity, consistency, and reasonableness—are referred to collectively as TCCR.

6.4 Hazardous Waste

6.4.1 Characterizing Hazardous Waste

What is the difference between hazardous chemicals and hazardous waste? Perhaps the operative word is "waste" because a waste is not a waste until it is declared a waste!

Scientists and engineers design systems to reduce and properly manage waste, including hazardous waste. It is amazing, in retrospect, how nonchalantly the American people tolerated the improper disposal of highly toxic wastes ever since industry first started producing them. Open dumps, huge waste lagoons, blatant dumping into waterways—all existed for decades. Only since the early 1980s has the public become conscious of what such indiscriminate dumping can do. Since then, we have made heroic efforts to clean up the most odious examples of environmental insults and health dangers and to regulate industry so as to prevent future problems.

A *hazardous waste* is defined in the United States by the EPA under the RCRA as any waste that is dangerous or potentially harmful to our health or the environment. Hazardous wastes can be liquids, solids, gases, or sludges. They can be discarded commercial products, like cleaning fluids or pesticides, or the byproducts of manufacturing processes. Under the RCRA, the EPA differentiates between hazardous and nonhazardous solid wastes. It is important to note that the regulatory definition of solid waste does not refer only to wastes that are physically solid; many solid wastes are liquid, semi-solid, or gaseous materials.

Under the RCRA, the EPA takes two complementary approaches to managing hazardous wastes. The first approach is to specifically name chemicals and materials that are hazardous wastes. This approach is precise because chemicals and materials are "listed." However, substances that are hazardous but not listed could fall through the cracks. Therefore, the EPA's second approach is to define hazardous wastes based on broad chemical characteristics, including ignitability, corrosivity, reactivity, and toxicity.

There are two additional classes of hazardous wastes. Universal wastes include commonly used products such as batteries, pesticides, thermostats, and lamps containing mercury. Because these materials are moderately hazardous but very common, they have the potential to cause harm if not managed properly. *Mixed wastes* are those that contain both radioactive and hazardous waste components. Wastes are assigned waste codes, as presented in Table 6.13.

Under the RCRA, every state must, at a minimum, comply with the federal requirements for managing hazardous waste. States are allowed to impose more stringent regulations if they choose, however. For example, Washington State defines hazardous waste based on persistence and additional requirements for toxicity. The additional toxicity requirements can be applied by looking up LD_{50} values for known hazardous chemicals. This approach works well for relatively pure substances that have data. Another option is to perform an acute toxicity bioassay test on salmonids. If the waste is determined to be toxic to fish, then it is hazardous; this is a useful approach for mixtures. In contrast, Idaho is a "RCRA only" state, meaning that it meets the minimum federal requirements but does not go beyond them.

Listed Wastes

Under the RCRA, *listed wastes* are those that are included on the F-list and the K-list. Over 50,000 chemicals are currently identified. The F-list (40 CFR §261.31) contains nonspecific source wastes from common manufacturing and industrial processes, such as solvents that have been used for cleaning and degreasing. The K-list (40 CFR §261.32) contains source-specific wastes such as sludges or wastewater from industries such as petroleum refining or pesticide manufacturing. There are also P- and U-lists that contain wastes from specific commercial chemical products in an unused form. Some pesticides and some pharmaceutical products become P- or U-listed hazardous wastes when discarded.

TABLE 6.13 Examples of RCRA Hazardous Waste Codes (40 CFR Part 261)

CONTAMINANT	RCRA HAZARDOUS WASTE CODE
Characteristic waste	
Ignitable waste	D001
Corrosive waste	D002
Reactive waste	D003
Toxic waste	
Arsenic	D005
Mercury	D009
Benzene	D018
Waste from nonspecific sources	
Wastewater treatment sludges from electroplating	F006
Quenching bath sludge from metal heat treating operations	F010
Waste from specific sources	
Wastewater treatment sludges from the production of chlordane	K032
Ammonia still lime sludge from coking production	K060
Off-specification and discarded chemicals, spill and container residues	
Arsenic trioxide	P012
Tetraethyl lead	P110
Creosote	U051
Mercury	U151

Source: www.epa.gov/hw/defining-hazardous-waste-listed-characteristic-and-mixed-radiological-wastes.

Characteristic Wastes

Characteristic wastes are those that do not qualify as "listed" but exhibit any of the following properties as defined (40 CFR Part 261 Subpart C):

Ignitability: Ignitable wastes have RCRA Hazardous Waste Code D001. They can create fires under certain conditions, are spontaneously combustible, or have a flash point below 60°C (140°F). Such wastes can be tested using standard ignitability tests and typically apply to waste oils and used solvents.

Corrosivity: Corrosive wastes have RCRA Hazardous Waste Code D002. They are acids or bases (pH ≤ 2 or ≥ 12.5) that are capable of corroding steel containers, such as storage tanks, drums, and barrels. Battery acid is an example.

Reactivity: Reactive wastes have RCRA Hazardous Waste Code D003. They are unstable under "normal" conditions. They can cause explosions, toxic fumes, gases, or vapors when heated, compressed, or mixed with water. Examples are lithium-sulfur batteries and explosives.

Toxicity Toxic wastes have RCRA Hazardous Waste Code D004-D043. They are harmful or fatal when ingested or absorbed (e.g., containing mercury, lead).

One concern in hazardous waste disposal is the speed with which the chemical can be released to produce toxic effects in plants or animals. For example, one commonly used method of hazardous waste disposal is to mix the waste with a slurry consisting of cement, lime, and other materials (a process known as stabilization/solidification). When the mixture is allowed to harden, the toxic material is safely buried inside the block of concrete, from which it cannot escape and cause trouble.

Or can it? This question is also relevant in cases such as the disposal of incinerator ash. Many toxic materials, such as heavy metals, are not destroyed during incineration and escape with the ash. If these metals are safely tied to the ash and cannot leach into the groundwater, then there seems to be no problem. If, however, they leach into the water when the ash is placed in a landfill, then the ash must be treated as a hazardous waste and disposed of accordingly.

How then is it possible to measure the rate at which such potential toxins can escape the material in which they are embedded? As a crude approximation of such potential leaching, the EPA uses an extraction procedure in which the solidified waste is crushed, mixed with weak acetic acid, and shaken for a number of hours. This process is known as the *Toxicity Characteristic Leaching Procedure (TCLP)* (Method 1311). The TCLP helps identify wastes that are likely to leach concentrations of contaminants that may be harmful to human health or the environment.

Once the leaching has taken place, the leachate is analyzed for possible hazardous materials. The EPA has developed a list of contaminants that constitute possible acute harm and determined that the level of these contaminants must not be exceeded in the leachate. Table 6.14 includes a few such chemicals. Many of the numbers on this list are EPA Drinking Water Standards multiplied by 100 to obtain the leaching standard.

Critics of the test point out that calculating toxicity on the basis of drinking water standards compounds a series of potential errors. The drinking water standards are, after all, based on scarce data and often set on the basis of expediency. How is it possible then to multiply these spurious numbers by 100 and simply say that something is or is not toxic?

TABLE 6.14 A few examples of EPA's maximum allowable concentrations in leachates from the TCLP

CONTAMINANT	ALLOWABLE LEVEL (mg/L)
Arsenic	5.0
Benzene	0.5
Cadmium	1.0
Chromium	5.0
Chloroform	6.0
2,4-D	10.0
Heptachlor	0.008
Lead	5.0
Pentachlorophenol	100.0
Trichloroethylene	0.5
Vinyl chloride	0.2

The TCLP cannot accommodate every possible scenario. Conditions such as water, pH, temperature, the turbulence of the mixing, and the condition of the solid placed into the mixer may all be important. The financial implications to an industry can be staggering if one of their leaching tests shows contaminant levels that exceed the allowable concentrations. The test results that mean so much to industrial firms often have a limited epidemiological base. Unfortunately, until we can come up with something better, this technique, while conservative, gives us an estimate on the possible detrimental effects of hazardous wastes.

In summary, a waste is a waste once it is declared a waste. And it is a hazardous waste if it fails any of the tests that would keep it from being classified so (and is not specifically exempted in the regulations). Using this criterion, the EPA has developed a list of chemicals it considers hazardous. The list is long and growing as new chemicals and waste streams are identified. Once a chemical or process stream is listed, it needs to be treated as a hazardous waste and is subject to all the applicable regulations. Getting delisted is a difficult and expensive process, and the burden of proof is on the petitioner. This is, in a way, a situation in which the chemical is considered guilty until proven innocent, a protective approach for chemicals deemed hazardous.

6.4.2 Disposal of Hazardous Waste

If a waste is listed, its disposal becomes more difficult because only a limited number of licensed hazardous waste disposal facilities are available. Usually, long and costly transportation is necessary. Some companies have attempted to avoid the trouble by disposing of hazardous waste surreptitiously (and illegally). Two hundred miles of roadway in North Carolina are contaminated by PCBs from transformer waste. Drivers opened the drain valves and drained their trucks instead of making the long trips to the disposal facilities. Others have used skill, creativity, and capital to address the root cause by redesigning their plant so as not to create waste. More than one industry has had to close when viable options were unavailable.

The disposal of hazardous waste is similar in many ways to the disposal of non-hazardous solid waste. Because disposal in the oceans is prohibited and outer space disposal is still far too expensive, the final resting place must be on land.

Deep-well injection has been used in the past and is still the method of choice for land disposal in the petrochemical industry. The idea is to inject the waste so deep into the Earth that it may not reappear and cause damage. This is, of course, the problem. Once the waste is deep in the ground, it is impossible to tell where its final destination will be and whether it will eventually contaminate groundwater.

Another method of land disposal is to spread the waste on land and allow the soil microorganisms to metabolize the organics. This technique was widely used in oil refineries and seemed to work well. The EPA, however, has restricted the practice because there is little control over the chemical once it is on the ground.

The method most widely used for the disposal of hazardous waste is the secure landfill designated for hazardous waste. Hazardous waste landfills and sanitary landfills are similar. It is important to note here that today's modern landfills differ from "dumps" used prior to the regulation of the disposal of solid and hazardous waste, as shown in Figure 6.10. Instead of one impervious liner, hazardous waste landfills now have multiple liners. Liquid waste is banned; all waste must be stabilized or in containers. As in sanitary landfills, leachate is collected, and a cap is placed on the landfill once it is complete. Continued care is also required, although the EPA presently requires only 30 years of monitoring.

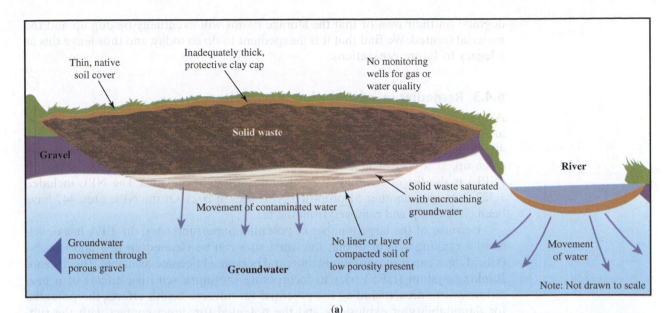

Thin, native soil cover

Inadequately thick, protective clay cap

No monitoring wells for gas or water quality

Solid waste

Gravel

River

Solid waste saturated with encroaching groundwater

Movement of contaminated water

Movement of water

Groundwater movement through porous gravel

Groundwater

No liner or layer of compacted soil of low porosity present

Note: Not drawn to scale

(a)

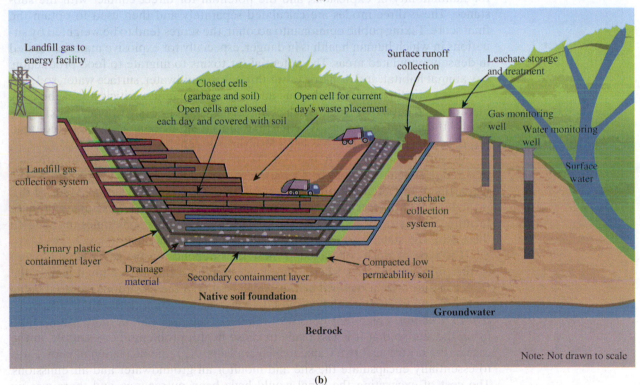

Landfill gas to energy facility

Closed cells (garbage and soil) Open cells are closed each day and covered with soil

Open cell for current day's waste placement

Surface runoff collection

Leachate storage and treatment

Gas monitoring well

Water monitoring well

Landfill gas collection system

Surface water

Leachate collection system

Primary plastic containment layer

Drainage material

Secondary containment layer

Compacted low permeability soil

Native soil foundation

Groundwater

Bedrock

Note: Not drawn to scale

(b)

FIGURE 6.10 (a) A pre-regulatory waste "dump" compared to (b) a modern engineered landfill that complies with RCRA regulatory requirements for monitoring and leachate control.

Most hazardous waste engineers agree that there is no such thing as a secure landfill and that eventually all the material will find its way into the water or air. What we are betting on, therefore, is that the wastes in these storage basins will degrade on their own or that the storage basins will eventually be dug up and the material treated. We find that it is inexpedient to do so today, and thus leave this as a legacy to future generations.

6.4.3 Remediation of Hazardous Waste Sites

As part of the Superfund program, the EPA prioritizes hazardous waste sites that are in need of clean-up and that are eligible for funding under the CERCLA. About 40,000 federal Superfund sites in the United States are in need of some degree of clean-up, at a cost of millions to billions of dollars! The National Priorities List (NPL) is the list of the highest priority hazardous waste sites. The NPL included 1,322 sites, with another 51 proposed NPL sites in 2021. Of the NPL sites, 447 have been cleaned up and removed from the list.

Because of the large number of potential Superfund sites, the EPA has developed a ranking system so that the worst sites can be cleaned up first and the less critical ones can wait until funds, time, and personnel become available. This Hazard Ranking System (HRS) tries to incorporate the more sensitive effects of a hazardous waste area, including the potential for adverse health effects, the potential for flammability or explosions, and the potential for direct contact with the substance. These three modes are calculated separately and then used to obtain the final score. Taking public opinion into account, the scores tend to be weighted by situations in which human health is in danger, especially for explosive materials stored in densely populated areas. The potential for toxins to migrate to food production; key animal habitat; and drinking water through groundwater, surface water, soil, and air is also prioritized. This technique addresses the common criticism that the EPA has not done enough to clean up the Superfund sites. They have, in fact, tackled the most difficult and most sensitive sites first and should be commended for the rapid response to a problem that has been festering for generations.

In addition to the EPA list, the Department of Energy has about 107 sites that require clean-up, and the Department of Defense has thousands more. The Department of Defense itself has spent *billions* of dollars on clean-up. Much of this is the legacy of nuclear weapons testing and manufacturing because many of these facilities were strictly secret, and the contracting operators sometimes mishandled the waste. The states also have lists of sites to be remediated.

The type of work conducted at these sites depends on the severity and extent of the problem. In cases where there is an imminent threat to human health, the EPA can authorize a ***removal action***, which results in the hazardous material being removed and treated or safely disposed. In less acute instances, the EPA authorizes ***remedial action***, which may consist of removing the material or, more often, stabilizing the site so that it is less likely to cause health problems. For example, in the case of the notorious Love Canal in Niagara Falls, the remedial action of choice was to essentially encapsulate the site and monitor all groundwater and air emissions. The cost of excavating the canal would have been outrageous, and there was no guarantee that the eventual disposition of the hazardous materials would have ultimately been safer than leaving them in place. In addition, the actual act of removal and transport might have resulted in significant human health problems. Remedial action, therefore, simply implies that action has been taken at the site to minimize the risk of having the hazardous material present there.

In the most common situation, hazardous substances have already contaminated the groundwater, so remediation is necessary. For example, a dry cleaner may accidentally (or purposefully) discharge its waste cleaning fluid, and this may find its way into a drinking water well for a nearby residence. Once the problem is detected, the first question is whether or not the contamination is life-threatening or poses a significant threat to the environment. Then a series of tests are run using monitoring wells or soil samples to determine the geology of the area and the size and shape of the plume or range of the contaminated area.

Depending on the seriousness of the situation, several options for remedial action are available. If there is no threat to life and if it can be expected that the chemical will eventually metabolize into harmless end products, then one solution is to do nothing except let the material degrade and continue monitoring—a method known as ***natural attenuation***. In most cases, this is not a viable option, and direct intervention is necessary.

Containment is the option used if there is no need to remove the offending material and/or if the cost of removal is prohibitive, as was the case at Love Canal. Containment usually involves the installation of slurry walls, which are deep trenches filled with bentonite clay or some other highly nonpermeable material, and continuous monitoring for leakage out of the containment. With time, the offending material might slowly biodegrade or chemically change to a nontoxic form, or new treatment methods may become available for detoxifying the waste.

Extraction and treatment is the pumping of contaminated groundwater to the surface for either disposal or treatment, or the excavation of contaminated soil for disposal or treatment. Sometimes air is blown into the ground and the contaminated air is collected.

The physical characteristics of a hazardous chemical in part determine its location underground and, therefore, dictate the remediation process. If the chemical is immiscible with water and is lighter than water, it most likely will float on top of the aquifer. Pumping this chemical out of the ground is relatively simple. If the chemical readily dissolves in water, however, then it can be expected to be mixed in a contaminated plume. In that case, it is necessary to contain the plume either by installing barrier walls and pumping the waste to the surface for treatment or by sinking a discharge well to reverse the flow of water. A third option is that the chemical is denser than water and is immiscible, in which case it is expected to lie on an impervious layer somewhere under the aquifer.

Wells are commonly drilled around the contaminated site so that the groundwater flow can be reversed or the groundwater can be contained in the area. Once the contaminated water is extracted, it must be treated, and the choice of treatment obviously depends on the nature of the problem. If the contamination is from hydrocarbons, such as trichloroethylene, they may be removed with activated carbon. Depending on the contaminant, biodegradation in reactors or piles may be used. If the contamination is from metals, then a precipitation or redox process may be used.

Some soils may be so badly contaminated that the only option is to excavate the site and treat the soil ***ex-situ***. This is usually the case with PCB contamination because no other method seems to work well. The soil is dug out and usually incinerated to remove the PCBs, and then returned to the site or landfilled. (Excavation and landfilling are also commonly used to meet construction and real estate transaction schedules.)

Treatment of the contaminated soil ***in-situ*** involves injecting either bacteria or chemicals that destroy the offending material. If heavy metals are of concern, these can be tied up chemically, or fixed, so that they will not leach into the groundwater. Organic solvents and other chemicals can be degraded by injecting freeze-dried bacteria or by making conditions suitable for indigenous bacteria to degrade the waste (e.g., by injecting air and nutrients). Microorganisms have been found that are able to decompose materials that were previously thought to be refractory or even toxic.

BOX 6.3 Love Canal: A Project Ahead of Its Time

It was a grand dream. William T. Love wanted to build a canal between the two great lakes, Erie and Ontario, to allow ships to pass around Niagara Falls. The project started in the 1890s but soon floundered due to inadequate financing. Love walked away from a big hole in the ground, and in 1942 Hooker Chemical found this to be the perfect place to dump its industrial waste. This was wartime in the United States, and there was little concern for possible environmental consequences. Hooker Chemical (which became Occidental Chemical Corporation) disposed of over 21,000 tons of chemical wastes, including halogenated pesticide and chlorobenzenes, into the old Love Canal. The disposal continued until 1952, at which time the company covered the site with soil and deeded it to the City of Niagara Falls, which wanted to use it for a public park. In the transfer of the deed, Hooker specifically stated that the site was used for the burial of hazardous materials and warned the city that this fact should govern future decisions on the use of the land. Everything Hooker Chemical did during those years was legal and above board.

About this time the Niagara Falls Board of Education was looking around for a place to construct a new elementary school, and the old Love Canal seemed like a perfect spot. This area was a growing suburb, with densely packed single-family residences on streets paralleling the old canal. A school was placed by this site, as shown in Figure 6.11.

In the 1960s the first complaints began, and they intensified during the early 1970s. The groundwater table rose during those years and brought to the surface some of the buried chemicals. Children in the school playground were seen playing with 55-gallon drums that popped out of the ground. The contaminated liquids started to ooze into the basements of the nearby residents, causing odor and health problems. More importantly, the contaminated liquid was found to have entered the storm sewers and was being discharged upstream of the water intake for the Niagara Falls water treatment plant. The situation reached a crisis point and President Carter declared an environmental emergency in 1978,

resulting in the evacuation of 950 families in an area of 10 square blocks around the canal from homes like the one shown in Figure 6.12.

The solution presented a difficult engineering problem. Excavating the waste would have been dangerous work and would probably have caused the death of some of the workers. Digging up the waste would also have exposed it to the atmosphere, resulting in uncontrolled toxic air emissions. Finally, there was the question as to what would be done with the waste. Because it was all mixed up, no single solution such as incineration would have been appropriate. The U.S. EPA finally decided that the only thing to do with this dump was to isolate it and continue to monitor and treat the groundwater. The contaminated soil on the school site was excavated, detoxified, and stabilized, and the building itself was razed (Figure 6.13). All of the sewers were cleaned, removing 62,000 tons of sediment that had to be treated and transported to a remote site. The groundwater was pumped and treated to prevent further contamination. The cost, of course, was staggering, paid for by Occidental Chemical, the Federal Emergency Management Agency, and the U.S. Army (which was found to have contributed waste to the canal).

The Love Canal story had the effect of galvanizing the American public into understanding the problems of hazardous waste and was the impetus for the passage of several significant pieces of legislation, including the Resource Conservation and Recovery Act, the Comprehensive Environmental Response, Compensation, and Liability Act, and the Toxic Substances Control Act.

Environmental scientists continue to uncover the long-term health effects of chemical exposure. Endocrine-disrupting chemicals (including dioxin), which were virtually unknown in 1978, are currently one of the hottest topics in environmental health science. Researchers have found that, in some cases, these chemicals can cause reproductive and developmental effects that carry forward for multiple generations. The follow-up health study of Love Canal finds a disturbing trend that echoes

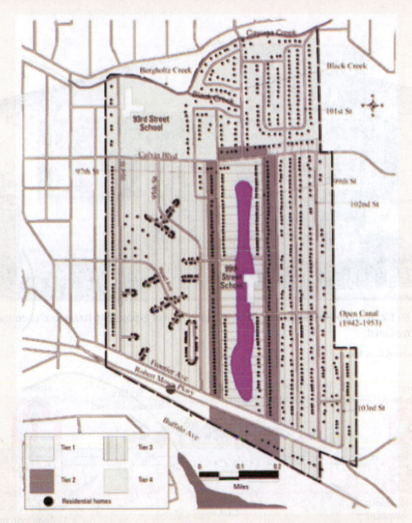

FIGURE 6.11 A map of the Love Canal disposal site, showing the location of the school over the chemical disposal site.

Source: Environmental Health Perspectives, Mortality among Former Love Canal Residents, Vol. 117, No. 2.

that pattern: children born to mothers who lived on the canal during pregnancy have increased risk of adverse pregnancy outcomes themselves later in life, including low birth weight, preterm birth, and babies born small for their gestational age.

The New York State Department of Health found that the birth defect rate for children born to parents who had lived near the canal was higher than for Niagara County and the rest of the state. However, cancer rates of the Love Canal residents were no higher than the rates throughout the region. An epidemiological study on Love Canal residents published in 2009 that evaluated causes of death from 1979 to 1996 did not observe increased

(Continued)

BOX 6.3 Love Canal: A Project Ahead of Its Time (Continued)

FIGURE 6.12 A home evacuated due to the high potential of exposure to hazardous chemicals buried in Love Canal.

Source: Bettmann/Getty Images.

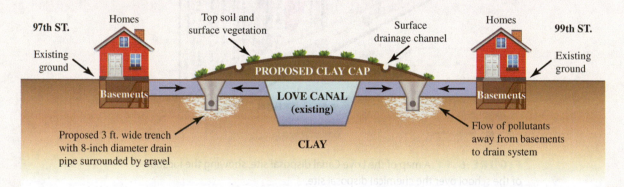

FIGURE 6.13 Proposed remediation of the Love Canal site for collection and treatment of the leachate from the former canal.

Source: Based on ACS, pubsapp.acs.org/cen/government/86/8646gov2.html.

mortality for residents of Love Canal. However, people who lived in tiers 1 and 2 adjacent to the site during the closed period (1954–1978) had a higher risk of death from acute myocardial infarction (heart attacks). The role of exposure to the landfill contaminants is limited by the data and population size available. The study could not rule out direct cardiotoxic and neurotoxic effects from exposure to landfill chemicals or indirect effects, but neither could those impacts be identified from the data available (Gensburg et al., 2009).

By August 2020, forty-two years had passed since President Carter declared Love Canal a federal health emergency. The area between 95th Street and 101st Street is still blocked off by a chain link fence designating the Superfund site. The areas immediately adjacent to the Love Canal site were zoned for commercial/light industrial use. It is estimated that over 21,000 tons of toxic wastes remain buried below 18 inches of soil and a thick plastic liner. Over 5 million gallons per year of leachate were collected and treated onsite prior to being discharged to the sanitary sewer system. Additionally, over 150 monitoring wells are in use to monitor contaminants in the canal and surrounding soils. In 2017, the data from all but one of the monitoring well samples revealed very limited to no exceedances above detection levels for volatile organic compounds (VOCs), semi-VOCs (SVOCs), polychlorinated bi-phenyls (PCBs), and pesticides.

Ongoing operation and maintenance of the Superfund site require frequent visual inspections of the monitoring wells, the barrier drain collection system, and the landfill cap. The barrier drain system contains and continues to collect and treat the leachate so that the direction of groundwater flow is toward the water wells within the collection system. A scientific review of the data and operation of the facility is required every five years, with the previous review completed in 2018. Each five-year review concluded that (1) the remedies implemented at the Site have alleviated any risks to the Love Canal area community, as originally identified early in the initial investigations, and continue to be protective of human health and the environment; and (2) the ongoing operation and monitoring activities were continuing to ensure that there are no exposures of site-related hazardous materials to human and environmental receptors.

Based on Vesilind, P.A., and DiStefano, T.D. (2006). *Controlling Environmental Pollution*. Lancaster, PA: DEStech Publications, Inc. Reproduced with permission of DEStech Publications, Inc.

6.4.4 Treatment of Hazardous Wastes

A number of treatment technologies can be considered depending on the nature of the hazardous wastes.

Chemical treatment is commonly used, especially for inorganic wastes. In some cases, a simple neutralization of the hazardous material renders the chemical harmless. In other cases, oxidation is used, such as for the destruction of cyanide. Ozone is often the oxidizing agent. When heavy metals must be removed, precipitation is the method of choice. Most metals become extremely insoluble at high pH ranges, so the treatment consists of the addition of a base, such as lime or caustic, and the settling of the precipitate (similar to the lime-soda softening process in water treatment). Other physical/chemical methods employed in industry include reverse osmosis, electrodialysis, solvent extraction, and ion exchange.

If the hazardous material is organic and readily biodegradable, most often the least expensive and most dependable strategy is *biological treatment*. The situation becomes interesting, however, when the hazardous material is an anthropogenic compound (created by people). Because these combinations of carbon, hydrogen, and oxygen are new to nature, there may be no microorganisms that can readily use them as an energy source. In some cases, it is still possible to find a microorganism that will use this chemical as an energy source, and treatment would then consist of a biological contact tank in which the pure culture is maintained. Alternatively, it is now increasingly possible to design specific microorganisms by gene manipulation to attack certain especially difficult-to-treat organic wastes.

Finding the specific organism can be a difficult and arduous task. There are millions to choose from, so how would we know that *Corynebacterium pyrogens* just happens to like toxaphene, a particularly refractory organic pesticide? The pathway is often convoluted, and a single microorganism can only break down the chemical to another refractory compound, which would then be attacked by a different organism. For example, DDT is metabolized by *Hydrogenomonas* to *p*-chlorophenylacetic acid, which is then attacked by various *Arthrobacter* species. Tests in which only a single culture is used to study metabolism would fail to note the need for a sequence of species.

One interesting development has been the use of cometabolism to treat organic chemicals that were once considered biologically nonbiodegradable. With this technique, the hazardous material is mixed with a nonhazardous and at least partially biodegradable material, and the mixture is treated in a suspended- or fixed-film bioreactor. The microorganisms apparently are so busy making the necessary enzymes for the degradation of the biodegradable material (the metabolite) that they forget they cannot treat the toxic material (the cometabolite). The enzymes produced, however, will biodegrade both the chemicals. The addition of phenol to the soil, for example, tricks some types of microorganisms, such as fungus, into decomposing chlorophenol, even though chlorophenol is usually toxic to the fungus.

One of the most widely used treatment techniques for organic wastes is incineration. Ideally, hazardous waste incinerators produce carbon dioxide, water vapor, and an inert ash. In fact, no incinerator achieves complete combustion of the organics. It discharges some chemicals in the emissions, concentrates some in the bottom ash and some in the fly ash, and produces various compounds called products of incomplete combustion (PIC). For example, polychlorinated biphenyls (PCBs) are thought to decompose within the incinerator to highly toxic chlorinated dibenzo furans (CDBF), which, although organic, do not oxidize at normal incinerator temperatures. Despite these problems, hazardous waste incinerators must achieve high levels of removal efficiencies, often at least 99.99%, which is commonly referred to as "four nines." In some cases, removal efficiencies require five or even six nines.

Incineration often reduces the original mass of solid waste by 80% to 85% and the volume by 95% to 96%. Although it does not eliminate the need for landfilling, it does reduce the volume of waste. The ash itself is a hazardous material that may sometimes require further treatment. Separation technologies such as magnetic separation can pull out ferrous metals or extract heavy metals. There are also thermal technologies that are intended to stabilize the residue in a way that will prevent toxins from leaching out, by melting residues at high temperatures with glass-forming additives. This process is energy intensive, but it reduces the ability of toxic metals to leach from the vitrified remains.

> **EXAMPLE 6.3**   Hazardous Waste Destruction Removal Efficiency

A hazardous waste incinerator is to burn 100 kg/hr of a PCB waste that is 22% PCBs and 78% organic solvents. In a trial run, the concentration of PCBs in the emission is measured as 0.02 g/m³, and the stack gas flow rate is 40 m³/min. No PCBs are detected in the ash. What destruction removal efficiency (destruction of PCBs) does the incinerator achieve?

A black box is once again useful: What comes in, must go out. The rate in is

$$(100 \text{ kg/hr})(0.22)(1,000 \text{ g/kg}) = 22,000 \text{ g/hr}$$

The rate out of the stack is

$$QC = (40 \text{ m}^3/\text{min})(0.02 \text{ g/m}^3)(60 \text{ min/hr}) = 48 \text{ g/hr}$$

Because there are no PCBs in the ash, the amount of PCBs combusted must be the difference between what enters and what leaves:

$$22,000 - 48 = 21,952 \text{ g/hr}$$

so the combustion efficiency is

$$\text{Efficiency} = \frac{21,952 \text{ g/hr}}{22,000 \text{ g/hr}} \times 100 = 99.78\%$$

There are only two nines, so this test does not meet the four nines criterion.

6.5 Radioactive Waste Management

A final criterion for being hazardous is whether or not the material is **_radioactive_**. Radioactive wastes are, however, handled separately and are governed by separate rules and regulations. Environmental engineers do not usually get involved in radiation safety, which is a specialized field, but they nevertheless should know about both the risk and the disposal technology of radioactive materials. In addition to the EPA, the Nuclear Regulatory Commission and the Department of Energy have authority over the management of radioactive waste. The EPA is particularly involved when radioactive waste is mixed with RCRA hazardous waste (substances known as mixed wastes).

6.5.1 Ionizing Radiation

Radiation is a form of energy caused by the decay of isotopes. Compared to a standard element, the **_isotope_** of an element has the same atomic number (number of protons) but a different mass number (number of neutrons and protons). (In other words, it has a different number of neutrons.) Remember that an element is defined by its atomic number; for example, the atomic number of uranium is 92. Uranium-235 (U-235), therefore, is an isotope of uranium (U-238).

To regain equilibrium, isotopes decay by emitting protons, neutrons, or electromagnetic radiation to carry off energy. This natural spontaneous process is **_radioactivity_**. The isotopes that decay in this manner are called **_radioisotopes_**. The energy emitted by this decay that is strong enough to strip electrons and sever chemical bonds is called **_ionizing radiation_**.

There are four kinds of ionizing radiation: alpha particles, beta particles, gamma (or photon) rays, and X-rays. Alpha particles consist of two protons and two neutrons, so they are the equivalent of the nucleus of a helium atom being ejected. Alpha particles are quite large and do not penetrate material readily; therefore, they can be stopped by skin and paper and are not a major health concern unless they are taken internally, in which case they can cause a great deal of damage. The decay changes the parent element into a different element, known as a daughter product.

Beta radiation results from an instability in the nucleus between the protons and neutrons. When there are too many neutrons, some of them decay into a proton and an electron. The proton stays in the nucleus to reestablish the neutron/proton balance, while the electron is ejected as a beta particle. The beta particle is much smaller than the alpha particle and can penetrate living tissue. However, about 1 cm of material can shield against beta particles. Beta decay, as with alpha decay, changes the parent into a new element.

Both alpha and beta decay are accompanied by gamma radiation, which is a release of energy from the change of a nucleus in an excited state to a more stable state. The element thus remains the same, and energy is emitted as gamma radiation. Related to gamma radiation are X-rays, which result from an energy release when electrons transfer from a higher to a lower energy state. Both types of radiation have more energy and penetrating power than the alpha and beta particles, so a dense material, such as lead or concrete, is required to stop them.

All radioactive isotopes decay and eventually reach stable energy levels. The decay of radioactive material is first order, meaning that the change in the activity during the decay process is directly proportional to the original activity present, or

$$\frac{dA}{dt} = -kA \tag{6.14}$$

where

A = activity
t = time
k = radioactive decay constant, inverse time units

As before, integration yields

$$A = A_0 e^{(-kt)} \tag{6.15}$$

where A_0 = the activity at time zero.

Of particular interest is the half-life of the isotope, which is the time required for half of the nuclei to decay. Inserting $A = A_0/2$ into Equation (6.15) and solving, we can calculate the half-life as

$$t_{1/2} = \frac{\ln 2}{k} = \frac{0.693}{k} \tag{6.16}$$

Half-lives of radioisotopes can range from the almost instantaneous (e.g., polonium-212 with a half-life of 3.03×10^{-7} seconds!) to very long (e.g., carbon-14 with a half-life of 5,730 years) (see Table 6.15). The half-life is characteristic of an isotope. Therefore, if you know the isotope, you know the half-life, and vice versa.

TABLE 6.15 Examples of half-lives

RADIOISOTOPE	HALF-LIFE
Carbon	
C-14	5,730 yr
Gold	
Au-95	183 days
Au-98	2.696 days
Hydrogen	
H-3	12.28 yr
Iron	
Fe-55	2.7 days
Fe-59	44.63 days
Lead	
Pb-210	22.26 yr
Pb-212	10.643 hr
Mercury	
Hg-203	46.6 days
Polonium	
Po-210	138.38 days
Po-212	3.03×10^{-7} s
Po-214	1.64×10^{-4} s
Po-218	3.05 min
Radon	
Rn-220	55.61 s
Rn-222	3.824 days
Uranium	
U-234	2.45×10^5 yr
U-235	7.04×10^8 yr
U-238	4.47×10^9 yr

EXAMPLE 6.4 Radioactive Decay

What will be the activity after 5 days of a 1.0-curie (Ci) Rn-222 source? (A curie, Ci, and a becquerel, Bq, are measures of the emission or decay rate.)

To determine the activity, we need to use

$$A = A_0 e^{(-kt)}$$

We are given A_0 and t, and we can calculate the constant k because we know the element, Rn-222. Using Table 6.15, we see that Rn-222 has a half-life of 3.824 days. Then

$$t_{1/2} = \frac{\ln 2}{k}$$

gives $k = 0.181/\text{day}$. So, the activity will be 0.40 Ci in 5 days.

6.5.2 Risks Associated with Ionizing Radiation

Soon after Wilhelm Roentgen (1845–1923) discovered X-rays, the detrimental effect of ionizing radiation on humans became known, and the emergent field of health physics set about creating safety standards with the assumption that it would be possible to identify levels of exposure that are safe. Researchers figured that there had to be a threshold, an exposure below which no effects would be found. Over about 30 years of studying both acute and chronic cases, they found that the threshold kept getting lower and lower. While small amounts of radiation above background levels have been found to stimulate biological systems (a process known as radiation hormesis), researchers finally concluded that assuming there is no threshold in radiation is the safest course of action.

The exposure of human tissue to ionizing radiation is complicated by the fact that different tissues absorb radiation differently. Historically, the *roentgen* was defined as the exposure from gamma or X-ray radiation equal to a unit quantity of electrical charge produced in air. The roentgen is a purely physical quantity that measures the rate of ionization; it has nothing to do with the absorption or effect of the radiation.

Different types of radioactivity create different effects, and not all tissues react the same way to radiation. Thus, the *rem* (or roentgen equivalent man) was invented. The rem accounts for the biological effect of absorbed nuclear radiation, so it measures the extent of biological injury. The rem, in contrast to the roentgen, is a biological dose. When different sources of radiation are compared for possible damage to human health, rems are used as the unit of measurement.

Modern radiological hygiene has replaced the roentgen with a new unit, the *gray* (Gy), defined as the quantity of ionizing radiation that results in the absorption of 1 joule of energy per kg of absorbing material. But the same problem exists, as the absorption may be the same but the damage might be different. Hence, the *sievert* (Sv) was invented. A Sv is an absorbed radiation dose that does the same amount of biological damage to tissue as 1 Gy of gamma radiation or X-ray. One Sv is numerically equal to 100 rem.

The damage from radiation is chronic as well as acute. The bombings of Hiroshima and Nagasaki at the end of World War II demonstrated that radiation can kill within a few hours or days. Over time, lower levels of radiation exposure can lead to cancer and mutagenic effects, and these are the exposures that are of most concern to the public. Exposure to radioactivity can be classified as involuntary background radiation, voluntary radiation, involuntary incidental radiation, and involuntary radiation exposure due to accidents.

Background radiation is due mostly to cosmic radiation from space, the natural decay of radioactive materials in rocks (terrestrial), and radiation inside buildings (internal). *Radon* is a special kind of background radiation. Radon-222 is a natural isotope with a half-life of about 3.8 days; it is the product of uranium decay in Earth's surface. Radon is a gas, and because uranium is ubiquitous in soil and rock, there is a lot of radon present. With its relatively long half-life, it stays around long enough to build up high concentrations in buildings. Its decay products are known to be dangerous isotopes that, if breathed into the lungs, can produce cancer. The best technique for reducing the risk from radon is to first monitor to see if a problem exists and then, if necessary, ventilate the radon to the outside.

Voluntary radiation can occur from sources such as diagnostic X-rays. One dental X-ray can produce hundreds of times the background radiation. The damaging effects of X-rays (called Roentgen rays everywhere in the world except the United States) were discovered after the early deaths of many friends of Wilhelm Roentgen. Another source of voluntary exposure is high-altitude flights in commercial airlines. Earth's atmosphere is a good filter for cosmic radiation, but there is little filtering at high altitudes.

Involuntary incidental radiation stems from such sources as nuclear power plants, weapons facilities, and industrial sources. While the Nuclear Regulatory Commission allows fairly high exposure limits for workers in these facilities, it is fastidious about not allowing radiation off premises. In all probability, the amount of radioactivity produced in such facilities would be less than the likely background exposure.

The fourth type of radiation exposure is from accidents, and this is a different matter. The most publicized accidental exposure to radiation has come from accidents or near-accidents at nuclear power plants. The 1979 accident near Harrisburg, Pennsylvania, at the Three Mile Island nuclear power facility was serious enough to bring a halt to the nuclear power industry in the United States, which was already struggling from lack of political acceptance. Following a series of operating errors, a large amount of fission products was released into the containment structure. The amount of fission products released into the atmosphere was minor, so at no time were the acceptable radiation levels around the plant exceeded. The release of radiation during this accident was so small, in fact, that it is estimated it would result in one additional cancer death over the 50-mile radius surrounding the Three Mile Island facility. However, the elevated risk from this exposure is impossible to detect because, within that population, over 500,000 cancers are expected to occur during this population's lifetime. After the accident, the damaged reactor was permanently closed and entombed in concrete. Fifteen tons of radioactive fuel waste were eventually moved to a nuclear waste facility in Idaho.

A much more far-reaching accident occurred in Chernobyl in the Ukraine. This actually was not an accident in the strict sense; the disaster was caused by several engineers who wanted to conduct some unauthorized tests on the reactor. The warning signals went off as they proceeded with their experiment, and they systematically turned off all the alarms until the reactor experienced a meltdown. The core of the reactor was destroyed; the reactor caught on fire, and the flimsy containment structure was blown off. Massive amounts of radiation escaped into the atmosphere, and while much of it landed within the first few kilometers of the facility, some reached as far as Estonia, Sweden, and Finland. In fact, the accident was first discovered when radiation safety personnel in Sweden detected unusually high levels of radioactivity. Only then did the officials of the former Soviet Union acknowledge that there had been an accident. In all, 31 people died of acute radiation poisoning, many by heroically going into the extremely highly radioactive facility to try to extinguish the fires. The best estimates are that the accident will produce up to 9,000 excess cancer deaths in the Ukraine,

BOX 6.4 Radiation Poisoning in Goiânia, Brazil

In the early 1980s, a small cancer clinic was opened in Goiânia (see Figure 6.14), but it closed five years later due to lack of business. A radiation therapy machine was left in the abandoned building, along with some canisters containing waste radioactive material: cesium-137, which has a half-life of 30 years. In 1987, the container of cesium-137 was discovered by local residents and was opened, revealing a luminous blue powder. The material was a local curiosity, and children even used it to paint their bodies, which caused them to sparkle. One of the little girls went home for lunch and ate a sandwich without first washing her hands. Six days later, she was diagnosed with radiation illness after having received a radiation dose estimated to be five to six times the lethal exposure for adults. The

ensuing investigation identified the true contents of the curious barrel. In all, over 200 people had been contaminated, and 54 developed illnesses serious enough to be hospitalized, with four people dying from the exposure (including the aforementioned little girl). By the time the government mobilized to respond to the disaster, the damage was done. A large fraction of the population had received excessive radiation, and the export of produce from Goiânia dropped to zero, creating a severe economic crisis. The disaster is now recognized as the second worst radiation accident in the world, second only to the explosion of the nuclear power plant in Chernobyl.

Source: Adapted from Vallero, D.A., and Vesilind, P.A. (2007). *Socially Responsible Engineering*. New York: John Wiley & Sons, Inc. Used by permission.

FIGURE 6.14 Location of Goiânia, Brazil.

Source: Based on Charobnica/Shutterstock.com.

Byelorussia, Lithuania, and the Scandinavian countries. This number is dwarfed by the expected total cancer deaths of 10 million during the lifetime of the people in these areas.

The total radiation we receive is, of course, related to how we live our lives, where we choose to work, if we choose to smoke cigarettes, whether we have radon in our basements, and other environmental factors. Natural background sources of radiation come from cosmic radiation, terrestrial radiation, and internal radiation. Cosmic radiation comes from the sun and stars and can vary based on differences in elevation, atmospheric conditions, and variations in the Earth's magnetic field. Terrestrial radiation is emitted by naturally occurring radioactive materials, such as uranium, thorium, and radon in the Earth. Radon, a product of radium breakdown, is responsible for most of the doses that individuals receive in the United States, primarily from the air. All organic matter contains radioactive carbon and potassium. All people have internal radiation primarily from potassium-40 and carbon-14, which is in their bodies from birth. Anthropogenic sources of radiation include medical sources and consumer products. Medical sources are the most significant anthropogenic sources; they include diagnostic X-rays and nuclear medicine procedures such as CT scans. Consumer products emit radiation from sources such as building and road construction materials, X-ray security systems, televisions, luminous watches, tobacco (polonium-210), and some glass and ceramics, to name only a few.

It is possible, nevertheless, to estimate the annual radiation dose for an average person in the United States. Figure 6.15 illustrates the percentage of radiation exposure in the United States from different sources. Radon is the highest single source, as some people in the United States live in homes that have high levels of radon in their basements, where radon problems generally occur.

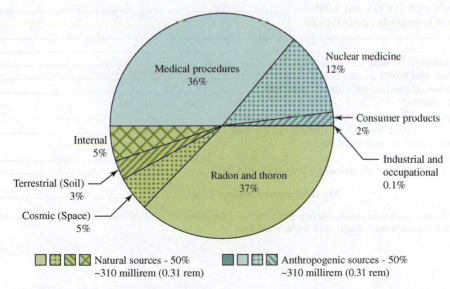

FIGURE 6.15 Sources of radiation exposure in the United States.

Source: Based on NCRP Report No. 160 (2009).

ACTIVE LEARNING EXERCISE 6.3 Calculate Your Average Annual Dose of Radiation

Use Figure 6.16 to calculate your average annual dose of radiation. An online calculator based on this worksheet is available at www.nrc.gov/about-nrc/radiation/around-us/calculator.html.

Where you live

1. Cosmic radiation at sea level (from outer space).. 26
2. Select the number of milirems for your elevation (in feet)
    up to 1,000 ft = **2**   1,000–2,000 ft = **5**
    2,000–3,000 ft = **9**   3,000–4,000 ft = **9**
    4,000–5,000 ft = **21**   5,000–6,000 ft = **29**
    6,000–7,000 ft = **40**   7,000–8,000 ft = **53**
    8,000–9,000 ft = **70**
    add this number: ....

    Elevation of some U. S. cities (in feet): Atlanta, 1,050; Chicago, 595; Dallas, 435; Denver, 5,280; Las Vegas, 2,000; Minneapolis, 815; Pittsburg, 1,200; Salt Lake City, 4,400; Spokane, 1,890; Washington, DC, 25.

3. Terrestrial (from the ground):
    If you live in states that border the Gulf or Atlantic Coast, **add 23** .... _____
    If you live in the Colorado Plateau area (around Denver), **add 90** .... _____
    If you live in middle America (rest of the U.S.), **add 46**... _____
4. House construction:
    If you live in a stone, brick, or concrete building, **add 7** .... _____

What you eat and drink

5. Internal radiation (in your body):*
    From food and water 40
    From air (radon).. 200

Other sources

6. Weapons test fallout (less than 1):** 1
7. Jet plane travel:
    For each 1,000 miles you travel, **add 1**... _____
8. If you have porcelain crowns or false teeth, **add 0.07**.. _____
9. If you use gas lantern mantles when camping, **add 0.003**.. _____
10. If you wear a luminous wristwatch (LCD), **add 0.006**.. _____
11. If you use luggage inspection at airports (using typical
    x-ray machine), **add 0.002**.. _____
12. If you watch TV**, **add 1**... _____
13. If you use a video display terminal**, **add 1**.. _____
14. If you have a smoke detector, **add 0.008** .... _____
15. If you wear a plutonium-powered cardiac pacemaker, **add 100**....................................... _____
16. If you have had medical exposures:*
    Diagnostic X-rays (e.g., upper and lower gastrointestinal, chest), **add 40** .... _____
    If you have had nuclear medical procedures (e.g., thyroid scans), **add 14** .... _____
17. If you live within 50 miles of a nuclear power plant
    (pressurized water reactor), **add 0.0009**.. _____
18. If you live within 50 miles of a coal-fired electrical utility plant, **add 0.03** .... _____

My total annual mrems dose: .... _____

Some of the radiation sources listed in this chart result in an exposure to only part of the body. For example, false teeth result in a radiation dose to the mouth. The annual dose numbers given here represent the "effective dose" to the whole body.

* These are yearly average dose.
**The value is actually less than 1.

FIGURE 6.16 Worksheet to calculate one's personal radiation dose.

Source: U.S. Nuclear Regulatory Commission Personal Radiation Dose Calculator, www.nrc.gov/reading-rm/basic-ref/students/for-educators/average-dose-worksheet.pdf.

6.5.3 Treatment and Disposal of Radioactive Waste

The most important distinction to be made in radioactive waste disposal involves the level of radioactivity emitted. While there appears to be an increasingly complex system for characterizing radioactive wastes, the broad classification is high-level or low-level waste. ***High-level wastes*** are the highly radioactive materials produced as byproducts from reactions that occur inside nuclear reactors. They include spent nuclear fuel and reprocessing waste. The majority of high-level wastes are left over from defense reprocessing plants such as those at Hanford, Washington, and Savannah River, South Carolina. ***Low-level radioactive wastes*** include items that have become contaminated with radioactive material or have become radioactive due to neutron radiation.

In nuclear power plants, nuclear fission occurs when fissionable material, such as uranium-235, is bombarded with neutrons and a chain reaction occurs. The fissionable material then splits (hence the term "fission") to release a huge quantity of heat, which is used to produce steam and then electricity. As the U-235 decays, it produces a series of daughter isotopes that themselves decay until the rate and hence the heat output are reduced. What is left over, commonly known as the fission fragments, represents the high-level radioactive material that requires cooling down, long-term storage, and eventual disposal. Safe long-term storage of high-level radioactive waste has been debated for decades. Over protests, a site in Nevada at Yucca Flats was selected, but that plan is currently on hold.

Low-level radioactive wastes can be less of a disposal problem. Because the activity levels of these wastes are low enough to handle by direct contact, it would seem that, with judicious volume reduction such as incineration, any secure landfill would be adequate. The U.S. Congress passed the Low-Level Waste Policy Act in 1980, which stipulates that states form compacts so that each state will take its turn in providing a disposal facility for a certain time. Unfortunately, the very mention of the word "radioactivity" is enough to arouse public concern and prevent the siting of these disposal facilities.

Mixed wastes are hazardous wastes that also contain radioactive materials; they are regulated under both the RCRA and the Atomic Energy Act. The EPA's Mixed Waste Rule allows the Nuclear Regulatory Commission (NRC) to manage mixed wastes and requires that they be stored and treated in a tank or a container. Generators of mixed wastes may store them as long as they are permitted to do so under their NRC or state license.

BOX 6.5 The Hanford Vitrification Site

The Hanford site is a decommissioned nuclear production complex operated by the U.S. government on the Columbia River in Washington State. It was used to produce plutonium for over 40 years to support U.S. efforts in World War II and during the Cold War. Plutonium manufactured at the site was used in the atomic bomb that was detonated over Nagasaki, Japan. Hanford is currently the most contaminated nuclear site in the United States and is the focus of the nation's largest environmental clean-up. Besides the clean-up project, Hanford is the site of the commercial nuclear power plant, the Columbia Generating Station, and various centers for scientific research and

(Continued)

BOX 6.5   The Hanford Vitrification Site *(Continued)*

development, such as the Pacific Northwest National Laboratory.

The 50-mile stretch of the Columbia River that snakes along the Hanford site is the last free-flowing stretch of the Columbia River. It is a natural wonder and designated as a National Monument. Today, 56 million gallons of waste are stored in 177 aging underground tanks in a tank "farm." Of these, more than 60 have leaked, contaminating the subsurface and threatening the nearby Columbia River.

According to the Vit Plant website:

- The tanks contain the most complex heterogeneous radioactive waste at any U.S. clean-up site.
- Waste is found in the form of sludge, salts, and liquids.
- Every tank has a different combination of waste.
- There are 1,800 different chemicals in the tank waste.

In 1989, the Washington Department of Ecology (Ecology), the U.S. EPA, and the U.S. Department of Energy (DOE) entered into the Tri-Party Agreement to set targets and milestones for clean-up. The EPA and Ecology share regulatory oversight through the CERCLA (Superfund) and the RCRA. Today, an enormous treatment operation is under construction. The Hanford Waste Treatment and Immobilization Plant, known as the Vit Plant, will use vitrification to immobilize most of Hanford's waste. *Vitrification* involves turning the waste into a solid glass form that is stable and ideally impervious to the environment. In this form, its stability will allow the radioactivity to dissipate over hundreds to thousands of years (see Figure 6.17).

The Vit Plant will cover 65 acres with four nuclear facilities—Pretreatment, High-Level Waste Vitrification, Low-Activity Waste (LAW) Vitrification, and an Analytical Laboratory—as well

FIGURE 6.17 Vitrified radioactive waste in canisters to make glass logs for long-term disposal.

Source: U.S. Department of Energy.

as operations and maintenance buildings and office space. The waste treatment process will begin in the Pretreatment Facility, where waste will be divided into high-level solids and low-activity liquids. From there, the low-activity waste will be transferred via underground pipes to the Low-Activity Waste (LAW) Vitrification Facility. This facility is approximately the size of one and a half football fields and is seven stories high (see Figure 6.18). The original plan was to treat high-level and low-activity wastes simultaneously. However, in order to begin treating wastes as soon as possible, the Department of Energy developed a sequenced approach that would start by treating low-activity waste with a direct-feed system. The Direct-Feed Low-Activity Waste (DFLAW) system requires the construction of an effluent management facility to handle liquid secondary waste from melters and offgas treatment. It is estimated that 90% of the radioactive

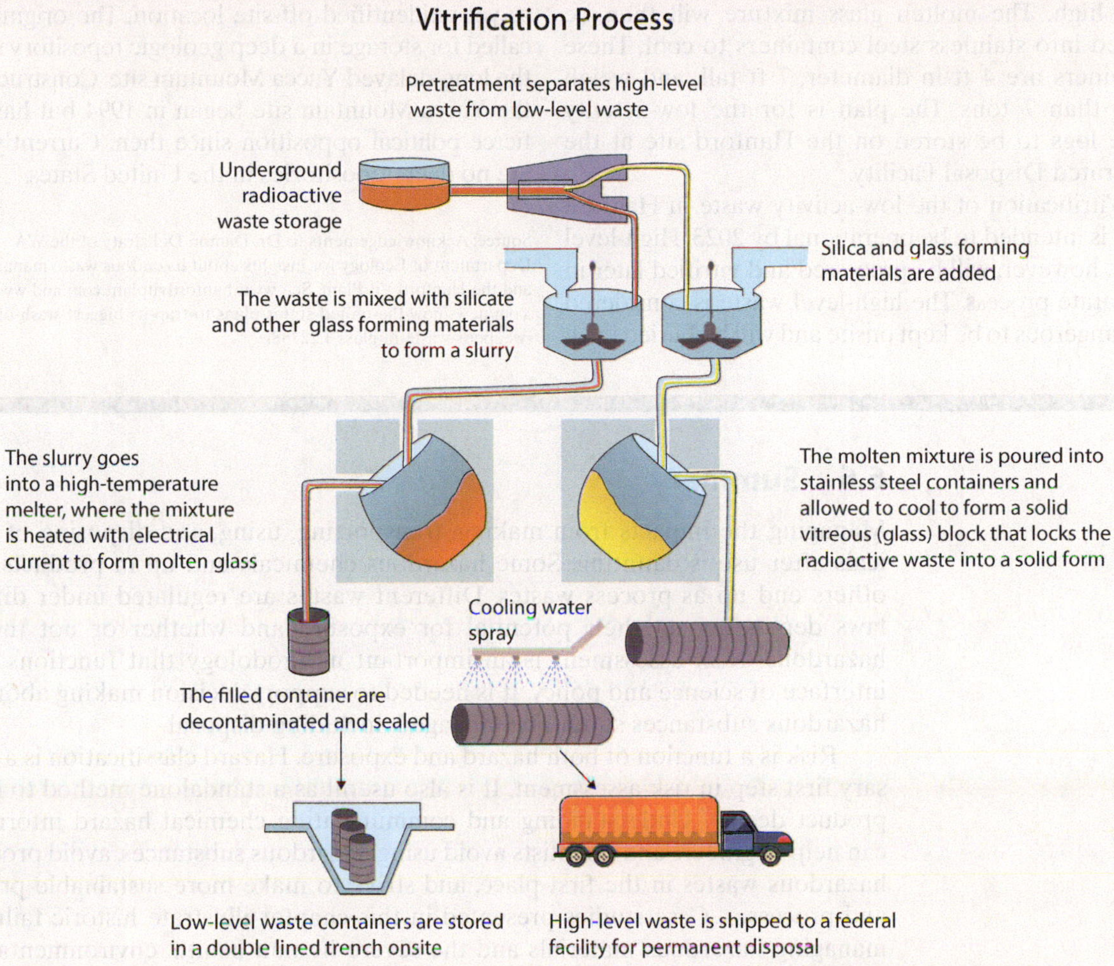

Vitrification Process

Pretreatment separates high-level waste from low-level waste

Underground radioactive waste storage

Silica and glass forming materials are added

The waste is mixed with silicate and other glass forming materials to form a slurry

The slurry goes into a high-temperature melter, where the mixture is heated with electrical current to form molten glass

The molten mixture is poured into stainless steel containers and allowed to cool to form a solid vitreous (glass) block that locks the radioactive waste into a solid form

Cooling water spray

The filled container are decontaminated and sealed

Low-level waste containers are stored in a double lined trench onsite

High-level waste is shipped to a federal facility for permanent disposal

FIGURE 6.18 The process for vitrification under development at the Hanford "Vit Plant."

waste stored in Hanford's underground tanks will be LAW. Given the number of leaking tanks in the tank farm, it is understandable that DOE wants to start vitrification as soon as possible by starting with the LAW.

In the LAW Facility, concentrated low-activity waste will be mixed with silica and other glass-forming materials. The mixture will be fed into the LAW's two melters and heated to 2,100°F. The 300-ton melters are approximately 20 ft × 30 ft and

BOX 6.5 The Hanford Vitrification Site (Continued)

16 ft high. The molten glass mixture will then be poured into stainless steel containers to cool. These containers are 4 ft in diameter, 7 ft tall, and weigh more than 7 tons. The plan is for the low-activity waste logs to be stored on the Hanford site at the Integrated Disposal Facility.

Vitrification of the low-activity waste in Hanford tanks is intended to be operational by 2023. High-level waste, however, will be processed and vitrified later in a separate process. The high-level waste is considered too dangerous to be kept onsite and will be buried in an as-yet-unidentified off-site location. The original plan called for storage in a deep geologic repository such as the long-delayed Yucca Mountain site. Construction of the Yucca Mountain site began in 1994 but has faced fierce political opposition since then. Currently there are no deep repositories in the United States.

Source: Acknowledgements to Dr. Damon Delistraty of the WA Department of Ecology for insights about hazardous waste management and the Hanford Vit Plant. See www.hanfordvitplant.com and www.nature .com/news/how-the-united-states-plans-to-trap-its-biggest-stash-of-nuclear -weapons-waste-in-glass-1.22788.

6.6 Summary

Managing the impacts from making, transporting, using, and disposing of chemicals after use is daunting. Some hazardous chemicals end up in products, while others end up as process wastes. Different wastes are regulated under different laws depending on their potential for exposure and whether or not they are hazardous. Risk assessment is an important methodology that functions at the interface of science and policy. It is needed to support decision making about how hazardous substances should be managed, including disposal.

Risk is a function of both hazard and exposure. Hazard classification is a necessary first step in risk assessment. It is also useful as a standalone method to inform product design. Understanding and communicating chemical hazard information can help engineers and scientists avoid using hazardous substances, avoid producing hazardous wastes in the first place, and strive to make more sustainable products and processes. Case studies presented in this chapter illustrate historic failures in managing hazardous materials and the severe human health, environmental, economic, and social consequences that have occurred because of those failures. We hope the lessons learned will not be forgotten by emerging scientists and engineers who understand the importance of transparency of information about chemical hazards and potential exposures to workers and the public.

References

CDC (Centers for Disease Control and Prevention). (2009). National Hospital Ambulatory Medical Care Survey: 2009 Emergency Department Summary Tables. U.S. Department of Health and Human Services, Washington, DC.

Davis, M. L., and Cornwell, D. A. (1991). Introduction to Environmental Engineering. New York: McGraw-Hill.

Gensburg, L. J., Pantea, C., Fitzgerald, E., Stark, A., Hwang, S. A., and Kim, N. (2009). "Mortality among former Love Canal residents." Environmental Health Perspectives 117(2):209–216. doi.org/10.1289/ehp.11350.

Heran, M. (2012). Deaths: Leading Causes for 2009. National Vital Statistics Report 60(7). U.S. National Vital Statistics Systems. National Center for Health Statistics, Centers for Disease Control and Prevention, U.S. Department of Health and Human Services, Washington, DC.

Marty, M. A., and Blaisdell, R. J. (Eds.). (2000). Air Toxics Hot Spots Program Risk Assessment Guidelines: Part IV. Technical Support Document for Exposure Assessment and Stochastic Analysis. California Office of Environmental Health and Hazard Assessment.

Moore, K. (2017). The Radium Girls: The Dark Story of America's Shining Women. Naperville, IL: SourceBooks. 496p.

Schettler, T. (2013). The Ecology of Breast Cancer: The Promise of Prevention and the Hope for Healing. Creative Commons. Accessed 2021 at www.healthandenvironment .org/docs/EcologyOfBreastCancer_Schettler.pdf.

Stern, P. C., and Fineberg, H. V. (Eds.). (1996). Understanding Risk: Informing Decisions in a Democratic Society. Committee on Risk Characterization, Commission on Behavioral and Social Sciences and Education. National Research Council. National Academies Press, Washington, DC.

United Nations. (2021). Globally Harmonized System of Classification and Labelling of Chemicals (GHS), Ninth Revised Edition. unece.org/sites/default/files/2021-09/GHS_Rev9E_0.pdf.

U.S. EPA. (1989). Risk Assessment Guidance for Superfund: Volume I, Human Health Evaluation Manual (Part A). Office of Emergency and Remedial Response, U.S. Environmental Protection Agency, Washington, DC.

U.S. EPA. (1998). Guidelines for Ecological Risk Assessment. Risk Assessment Forum, U.S. Environmental Protection Agency, Washington, DC.

U.S. EPA. (2000). Technical Support Document for Exposure Assessment and Stochastic Analysis. Risk Assessment Forum, U.S. Environmental Protection Agency, Washington, DC.

U.S. EPA. (2006). A Framework for Assessing Health Risks of Environmental Exposures to Children. National Center for Environmental Assessment, Office of Research and Development, U.S. Environmental Protection Agency, Washington, DC.

U.S. EPA. (2009). Risk Assessment Guidance for Superfund: Volume I, Human Health Evaluation Manual (Part F, Supplemental Guidance for Inhalation Risk Assessment) Final. Office of Superfund Remediation and Technology Innovation, U.S. Environmental Protection Agency, Washington, DC.

Wilson, R. (1979, February). "Analyzing the daily risks of life." *Technology Review*, 41–46.

Windsor, J. S., Firth, P. G., Grocott, M. P., Rodway, G. W., and Montgomery, H. E. (2009). "Mountain mortality: A review of deaths that occur during recreational activities in the mountains." *Postgraduate Medical Journal* 85:316–321.

Key Concepts

Hazard
Toxicity
Risk
Voluntary risk
Involuntary risk
Resource Conservation and Recovery Act (RCRA)

Globally Harmonized System of Classification and Labelling of Chemicals (GHS)
Hazard classification
Risk Assessment
Risk management
Toxicology

Carcinogens

Teratogens

Neurotoxins

Mutagens

Specific target organ toxicants

Sensitizers

Hazard endpoint

New Approach Methods (NAMs)

Bioaccumulation potential

Bioconcentration

Bioconcentration factor (BCF)

Persistence

Mobility

Dose

Effective concentration (EC)

Lethal concentration (LC)

Lethal dose (LD)

Lowest observable effect concentration (LOEC)

No observed adverse effect level (NOAEL)

Maximum allowable toxicant concentration (MATC)

Potency

Dose–response curve

Epidemiology

Reference dose

Uncertainty factor (UF)

Reference concentration (RfC)

Hazard quotient (HQ)

Hazard index (HI)

Integrated Risk Information System (IRIS)

Reasonable maximum exposure (RME)

Chronic daily intake (CDI)

Hazardous waste

Mixed wastes

Listed wastes

Characteristic wastes

Toxicity Characteristic Leaching Procedure (TCLP)

Removal action

Remedial action

Natural attenuation

Containment

Extraction and treatment

Ex-situ

In-situ

Chemical treatment

Biological treatment

Radioactive

Isotope

Radioactivity

Radioisotopes

Ionizing radiation

Roentgen

Rem

Gray

Sievert

Radon

High-level wastes

Low-level radioactive waste

Vitrification

Problems

6-1 The following paragraph appeared as part of a full-page ad in a professional journal:

Today's Laws
A Reason to Act Now

The Resource Conservation and Recovery Act provides for corporate fines up to $1,000,000 and jail sentences up to five years for officers and managers, for the improper handling of hazardous wastes. That's why identifying waste problems, and developing economical solutions is an absolute necessity. O'Brien & Gere can help you meet today's strict regulations, and today's economic realities, with practical, cost-effective solutions.

What do you think of this ad? Do you find anything unprofessional about it? It is, after all, factual. Suppose you are the president of O'Brien & Gere (one of the most reputable environmental engineering firms in the business) and you saw this in the journal. Write a one-page memo to your marketing director commenting on it.

6-2 Some old electrical transformers were stored in the basement of a university maintenance building and were "forgotten." One day a worker entered the basement and saw that some sticky, oily substance was oozing out of one of the transformers and into a floor drain. The worker notified the Director of Grounds, who immediately realized the severity of the problem. They called in hazardous waste consulting engineers, who first took out the transformers and eliminated the source of the polychlorinated biphenyls (PCBs) that were leaking into the storm drain. Then they traced the drain to a little stream and started taking water samples and soil samples. They discovered that the water was at 0.12 mg/L PCB, and the soil ranged from 32 mg PCB/kg (dry soil) to 0.5 mg/kg. The state environmental management requires that streams contaminated with PCBs be cleaned so that they are "free" of PCBs. Recall that PCBs are extremely toxic and stable in the environment, and they biodegrade very slowly. If nothing is done, the contaminant in the soil could remain for hundreds of years.

   a. If you were one of the engineers, how would you approach this problem? What would you do? Describe how you would clean this stream. Be as detailed as possible. Will your plan disturb the stream ecosystem? Is this of concern to you?

   b. Will it ever be possible for the stream to be free of PCBs? If not, what do you think the state means by this? How would you know that you have done all you can?

   c. There is no doubt that if we wait long enough, someone will eventually come up with a really slick way of cleaning PCB-contaminated streams, and this method will be substantially cheaper than anything we have available today. Why not just wait, then, and let some future generation take care of the problem? The stream does not threaten us directly, and the worst part of it can be cordoned off so people will not be able to get close to the stream. Then someday, when technology has improved, we can clean it up. What do you think of this approach? (It is, of course, illegal, but this is not the question. Is it ethical? Would you be prepared to promote this solution?)

6-3 In Problem 6-2, if the clean-up resulted in a PCB concentration in the water of 0.000073 mg/L, what is the percentage reduction in the contaminant? How many nines have been achieved?

6-4 A dry cleaning establishment buys 500 gal of carbon tetrachloride every month. As a result of the dry cleaning operation, most of this is lost to the atmosphere, and only 50 gal remain to be disposed of. What is the emission rate of carbon tetrachloride from this dry cleaner, in pounds per day? (The density of carbon tetrachloride is about 1.6 g/mL.)

6-5 Lead levels in Lake Roosevelt water were measured, and the average concentration was determined to be 3.9 micrograms per liter.

   a. What would you expect the concentration of lead to be in walleye, a top chain predatory fish, in units of mg-lead/kg-fish in Lake Roosevelt?

   b. Does this represent a hazard to fishermen consuming the fish?

   c. How would you determine the health risk to lead exposure from eating fish in Lake Roosevelt?

6-6 A toxicity study of minnows to a new herbicide has been conducted, with the results shown in the table below. What is the LD_{50} of this pesticide for a minnow that weighs 12 g? What is the LD_{50} for a young person who might weigh 25 kg?

AMOUNT INGESTED (mg)	FRACTION THAT DIED AFTER 4 HOURS
0 (control)	0
0.02	0
0.04	0.05
0.06	0.11
0.08	0.23
0.10	0.45
0.12	0.72
0.14	0.91
0.16	1

6-7 In the 1970s, many hazardous waste dumping sites were discovered in the United States. Some had been in operation for decades, and some were still active. Most of them, however, were abandoned, such as the notorious Valley of Drums in Tennessee. This was a rural valley where an almost unimaginable concoction of hazardous materials had been dumped. The owner had simply abandoned the site after collecting hefty fees for getting rid of the hazardous materials for large chemical companies. Why did we wait until the 1970s to discover the problem? Why weren't we aware of the problem of hazardous waste years ago? Why is the concern with hazardous waste relatively recent? Discuss your thoughts in a one-page paper.

6-8 Benzene (C_6H_6) is somewhat soluble in water (1.8 g/L) and is a known carcinogen. The maximum contaminant level (MCL) in drinking water is 5 ppb. Gasoline is restricted to 0.62% benzene by volume. How much water will 1 gal of gasoline contaminate to the MCL?

6-9 Determine the hazard quotient for an adult exposed to 2×10^{-4} mg/kg of acrylamide in the drinking water during 30-years in a home. Use the *Rfd* value of 2×10^{-3} mg/(kg-day) from the U.S. EPA IRIS website (iris.epa.gov /AtoZ).

6-10 Determine the hazard quotient for an adult exposed to 5×10^{-6} mg/kg of Aroclor 1016 in the drinking water during 10-years in a home. Use the *Rfd* value of 7×10^{-5} mg/(kg-day) from the U.S. EPA IRIS website (iris.epa.gov /AtoZ).

6-11 Determine the hazard quotient for an adult exposure to 0.1 mg/m³ of acrylamide in the air of a polymer production facility during a 30-year industrial career. Use the *Rfc* value of 6×10^{-3} mg/m³ from the U.S. EPA IRIS website (iris.epa.gov/AtoZ) as the *Rfd*.

6-12 Determine the hazard quotient for an adult exposure to 0.015 mg/m³ of benzene in the air at a refinery during a 30-year industrial career. Use the Rfc value of 3×10^{-2} mg/m³ from the U.S. EPA IRIS website (iris.epa.gov /AtoZ) as the *Rfd*.

6-13 A family living in a home is exposed to 0.008 mg/L of 1,4-dioxane in their drinking water. Calculate the lifetime risk for:
a. An adult male
b. An adult female
c. An 8-year-old child

6-14   A female mechanic working in an industrial facility is exposed to 18 mg/m³ of benzene in the air at the workplace. Calculate the lifetime cancer risk for this exposure.

6-15   How long will it take for the radiation from a spill of Po-210 and Po-214 to decay by 80%? See Table 6.15 for half-life values.

6-16   Among other contamination, a site was found to have 4.5 Ci of tritium (H-3) on January 15, 2000. Records indicate that the original release was likely in June 1963. The site is now owned by the city due to nonpayment of taxes.

a. How many curies of H-3 were originally released? See Table 6.15 for half-life values.

b. Who should be responsible for the clean-up of the site?

6-17   Radon-222 is released from the Earth and can become trapped in buildings. Suppose the concentration measured in a college classroom building is 1,500 pCi/m³, which is significantly higher than the recommended remedial action level of 4 to 8 pCi/m³.

a. If all leaks into the building are sealed (i.e., no more radon enters the building), how long will it take for the concentration of radon to reach the recommended level?

b. Do you recommend that some type of remedial action be taken instead (i.e., removal of the source)? Why or why not?

## PART 2

Engineering Environmental and Sustainable Processes

CONTENTS

## CHAPTER 7

Water Quality Impacts

FIGURE 7.1 Water quantity, quality, and availability are closely related to sustainable development. At the time of writing this book, approximately one-third of global population lacks access to clean water. The burden of finding water often falls on children and women in society.

Source: Photo by Bradley Striebig.

Water is life. A third of the nations on our shrinking planet, mostly in the developing world, suffer from water scarcity and the associated stress this scarcity places on societies. My home area of Kitale, Kenya, shares in this suffering.

—GILBERT NALELIA

GOALS

THE EDUCATIONAL GOALS OF THIS CHAPTER are to describe basic water quality parameters and water quality improvement processes. This chapter will describe the parameters and methods used to determine if water is of sufficient quality for use as drinking water. Water quality parameters that influence human and ecosystem health in the environment will be examined. Mathematical models will be used to demonstrate how high levels of BOD or nutrients may adversely affect ecosystems. Differences in water quality between high-income nations and low- to medium-income nations will also be demonstrated.

OBJECTIVES

At the conclusion of this chapter, you should be able to:

7.1 Describe economic, environmental, and societal implications of water quality in high-income nations and low- to medium-income nations.

7.2 Describe or model the major effects of pollutants on water quality and ecosystems.

7.3 Describe the parameters that influence dissolved oxygen, and model the levels of dissolved oxygen in surface water.

7.4 Describe basic centralized and decentralized drinking water treatment technologies.

Introduction

Every year more people die from unsafe water than from all forms of violence, including war. The most significant sources of water pollution are lack of adequate treatment of human wastes and inadequately managed and treated industrial and agricultural waste. Every day, 2 million tonnes of sewage and other effluents drain into the world's waters. The quality of water necessary for each human use varies, as do the criteria used to assess water quality.

Source: World Water Assessment Programme (2009).

7.1 The Water Crisis

A silent humanitarian crisis kills an estimated 3,900 children every day (Bartram et al., 2005). The root of this unrelenting catastrophe lies in these plain, grim facts: The United Nations has estimated that nearly half of the people in the world do not have access to improved sanitation and one quarter lack access to improved safe drinking water (Figure 7.2). Far more people endure the largely preventable effects of poor sanitation and water supply than are affected by war, terrorism, and weapons of mass destruction combined. Yet those other issues capture the public and political imagination and resources in a way that water and sanitation issues do not. Most people find it hard to imagine defecating daily in plastic bags, buckets, open pits, and public areas because there is no alternative; nor can they relate to the everyday life of the 1.1 billion people without access to a protected well or spring within reasonable walking distance of their homes (Bartram et al., 2005).

Water is fundamental for life and health. The human right to water is a prerequisite to the realization of all other human rights. Many governments recognize the human right to access water and sanitation, but lack of or poor use of funding and the scarcity of a trained workforce in water treatment and delivery have prevented all people from gaining access to water and sanitation. The poverty of a large amount of the world's population is both a symptom and a cause of the water crisis that affects low-income populations, ending in sickness, lost educational and employment opportunities, and for a staggeringly large number of people, early death (Gleick, 2002). Water-related diseases are among the most common causes of illness and death in developing countries, especially in sub-Saharan Africa (Figure 7.3) (Corcoran et al., 2010). The sad fact is that this disease burden is preventable. Furthermore, investing in water and sanitation services is expected to have positive economic benefits, with a 5 to 25 times return on each dollar invested (see Figure 7.4). In developing countries with poor water and sanitation systems, life expectancy is far lower than it is in industrialized countries. The causes of deaths are also quite different; infectious diseases account for more than 40% of deaths in developing countries, whereas in industrialized nations deaths are more often related to chronic disease and cancer (Kitawaki, 2002).

In addition to the Human Development Index, a ***Water Poverty Index (WPI)*** has also been developed; the WPI helps explain the relationship between poverty and five water-related criteria: resources, access, capacity, use, and environment (Lawrence et al., 2002; Sullivan, 2002). It is designed to allow a single measure to relate household welfare and access to water. The WPI value based on a ranking of 0 to 100 is given by the general equation

$$WPI = \frac{1}{3}\left[w_a A_s + w_s S + w_t(100 - T)\right] \qquad (7.1)$$

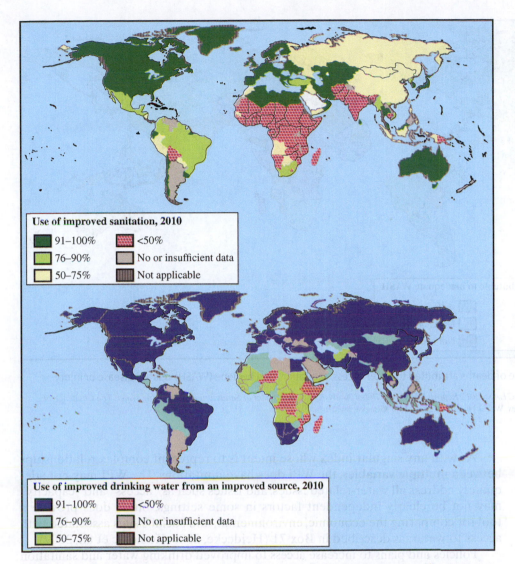

FIGURE 7.2 Percentage of population with access to improved water supply and sanitation.

Source: Based on WHO (2012). *GLASS 2012 Report: UN-Water Global Analysis and Assessment of Sanitation and Drinking Water: The Challenge of Extending and Sustaining Services.* World Health Organization. Geneva, Switzerland.

where

A_s = percent of adjusted water availability (AWA) for the population. This is calculated on the basis of ground and surface water availability related to ecological water requirements and a basic human requirement, plus all other domestic demands, as well as the demand for agriculture and industry. (The value of A should also recognize the seasonal variability of water supplies.)

S = percent of the population with access to safe water and sanitation.

T = an index between 0 and 100 that represents the time and effort required for water collection.

w_a, w_s, w_t = weights assigned to each of the components, A_s, S, and T, of the index.

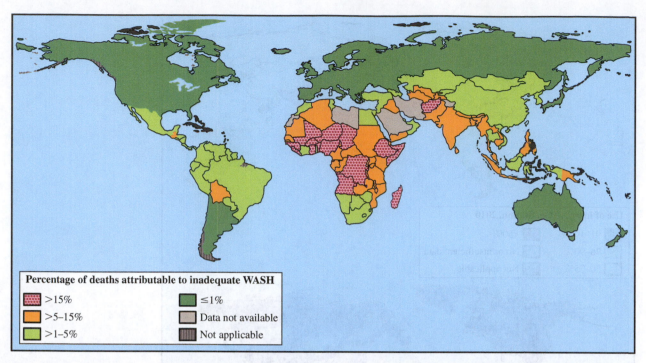

FIGURE 7.3 Percentage of deaths attributed to water access, sanitation, and hygiene (WASH)-related disease or injury.

Source: Based on WHO (2012). *GLASS 2012 Report: UN-Water Global Analysis and Assessment of Sanitation and Drinking Water: The Challenge of Extending and Sustaining Services.* World Health Organization. Geneva, Switzerland.

As with any singular index whose intent is to represent complex relationships between multiple variables, the WPI does have limitations. The WPI may not adequately address all water-related issues, and issues such as "access" and "capacity" may not be clearly independent factors in some settings, but it does provide a tool for comparing the economic, environmental, and social issues associated with access to water, as described in Box 7.1 (Heidecke, 2006; Komenic et al., 2009).

Policies and plans to increase access to improved drinking water and sanitation are a priority among many countries, including low- to medium-income countries (see Figure 7.8 on page 401).

Water and sanitation systems may be costly to construct. The operation and maintenance (O&M) expenses for water and sanitation systems are often overlooked and may cause failure due to the inability to collect revenue, which is especially problematic in low-income countries. Centralized water and sewer systems have proven to be very effective in improving and protecting water quality and reducing the spread of water-related disease. However, centralized systems require a large capital expense; sufficiently trained personnel to design, build, and operate the systems; and collection of revenue to pay for trained operators and O&M systems. Decentralized water and sanitation solutions exist and have also been shown to significantly improve water quality and sanitation (see Box 7.2 on page 404). Decentralized systems have a lower capital cost (they are generally designed for lower O&M costs), and as such are often implemented to meet rural community needs.

Growing urban centers challenge engineers, scientists, and policymakers to implement appropriate use of technology. Both the urban centers of Jakarta, Indonesia, and Sydney, Australia, generate approximately 1.2 to 1.3 million cubic meters of wastewater each day (see Figure 7.9 on page 402). Jakarta has a population of nearly

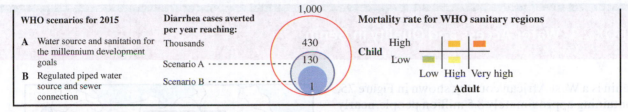

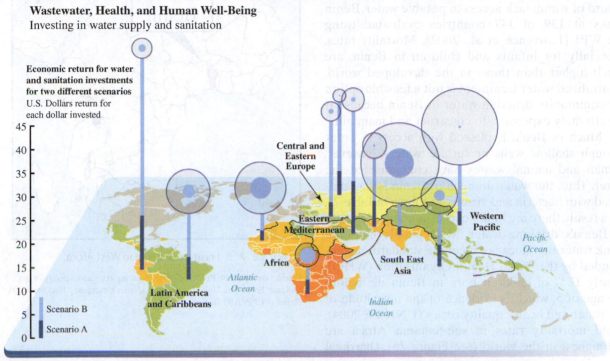

FIGURE 7.4 The relationship between mortality, access to improved water sources and sanitation, and expected economic return on water-related investments.

Source: Based on UNEP/GRID-Arendal, www.grida.no/graphicslib/detail/wastewater-health-and-human-well-being-investing-in-water
-supply-and-sanitation_120c, data from Hutton, G., et al. (2007). "Global cost-benefit analysis of water supply and sanitation interventions." *Journal of
Water and Health* www.who.int/water_sanitation_health/publications/2012/globalcosts.pdf.

9 million people, and approximately 60% of them get their water from private decentralized wells. Most of the wastewater in Jakarta flows into decentralized septic tanks that are improperly maintained. Existing sewage trenches, originally built to control flooding, have been partially clogged by silt and garbage. Increased runoff rates and lack of infrastructure create increasingly severe flooding events that often carry contaminants into the water supply. In addition, stagnant water remaining after the floods subside provides an ideal breeding ground for vectors, which is likely to account for the increasing incidence of dengue fever and other water-related disease. As a result, only about 3% of the wastewater in Jakarta is treated prior to discharge (see Figure 7.9 on page 402). In contrast, Sydney has a population of about 4 million people and possesses centralized water and sanitation systems that serve nearly the entire population. Almost 100% of Sydney's wastewater is treated prior to discharge.

Only 40% of low- to medium-income countries report fiscal resources and responsibility for decentralized services. Low- to medium-income countries also lack adequately trained staff to operate and manage facilities in urban and rural areas, as illustrated by the data reported in Figures 7.10 and 7.11 (see pages 402 and 403).

BOX 7.1 Water Access and Quality in Benin

Benin is a West African country, shown in Figure 7.5, containing approximately 8.5 million people, nearly a third of whom lack access to potable water. Benin ranks at 139 of 147 countries evaluated using the WPI (Lawrence et al., 2002). Mortality rates, especially for infants and children in Benin, are much higher than those in the developed world. Centralized water treatment is not a feasible option for community drinking water in Benin because it is extremely expensive to construct and maintain.

Much of Benin is blessed with access to water through shallow wells or surface waters; however, human and animal wastes have contaminated the water. Thus, the water in much of Benin is contaminated with bacteria and viruses, as shown in Table 7.1. As a result, there are major concerns about the safety of Benin's drinking water. Most people in Benin drink water that does not meet the standards recommended by the World Health Organization (WHO). Nearly 17% of children born in Benin die before the age of 5, which is evidence of the magnitude of the water and health quality issues (UNICEF, 2004). Child mortality rates in sub-Saharan Africa are the highest in the world (see Figure 7.6). Diarrheal

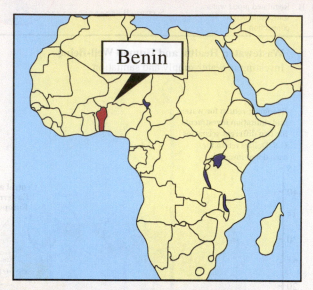

FIGURE 7.5 Location of Benin in West Africa.

Source: Based on PEPPFAR (2006). Building Human Capacity in the Defense Force. The United States President's Emergency Plan for AIDS Relief (PEPPFAR) Benin.

TABLE 7.1 Microbial and chemical contaminants measured in Benin water samples

CONTAMINANT	UNITS	CONCENTRATION IN BENIN WATER	WHO STANDARD	U.S. EPA STANDARD
Total coliforms	MPN/100 mL	>1,600	0	0
Fecal coliforms	MPN/100 mL	20	0	0
E. coli	MPN/100 mL	NA	0	0
Pathogens	MPN/100 mL	>8	0	0
Lead	μg/L Pb	4	10	15
Arsenic	μg/L As	ND	10	10
Nitrates	mg/L NO_3^--N	>30.0	50	10
Phosphate	mg/L PO_4^{3-}	0.19	NA	NA

Source: Based on Striebig, B., Atwood, S., Johnson, B., Lemkau, B., Shamrell, J., Spuler, P., Stanek, K., Vernon, A., and Young, J. (2007). "Activated carbon amended ceramic drinking water filters for Benin." *Journal of Engineering for Sustainable Development* 2(1):1–12.

disease is a significant cause of death among children under 5 and is closely linked to water and sanitation access (Figure 7.7) (Corcoran et al., 2010).

In rural Benin, the people obtain clean drinking water primarily by boiling water or purchasing imported bottled water. Boiling water requires wood and native vegetation, depleting local resources and emitting smoke into households and the atmosphere. While treated water is available in a few locations, less than 10% of the population has treated water piped into their homes. Bagged and bottled water can be purchased in the marketplace, but this water is expensive and unsustainable. Additionally, it is possible to tamper with this water, meaning it can become contaminated. The preferred treatment for the area would be a cost-effective point-of-use technology. However, not all of Benin's population understands the links between contaminated drinking water, sanitation, and disease, thereby creating a need for an educational program.

Most early deaths due to water-related disease are the result of diarrheal illnesses caused by ingesting water contaminated by fecal matter, as well as by inadequate sanitation and hygiene. For public health purposes, water-related disease is typically grouped within the classifications defined in Table 7.2.

The lack of sanitation increases the likelihood of water-related diseases by contaminating water supplies; this contaminated water causes skin disease

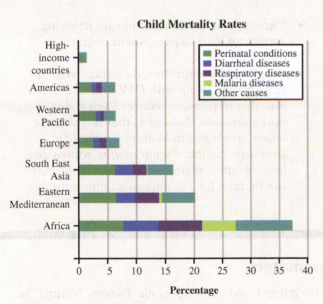

FIGURE 7.6 Childhood mortality rates among high-income and lower-income countries in different regions.

Source: Based on Corcoran, E., Nellemann, C., Baker, E., Bos, R., Osborn, D., and Savelli, H. (eds.) (2010). *Sick Water? The Central Role of Waste-water Management in Sustainable Development. A Rapid Response Assessment.* United Nations Environment Programme, UN-HABITAT, GRID-Arendal. Data source: WHO, 2008.

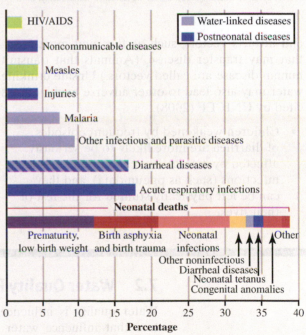

FIGURE 7.7 Water-related factors that contribute to childhood mortality rates.

Source: Based on Corcoran, E., Nellemann, C., Baker, E., Bos, R., Osborn, D., and Savelli, H. (eds.) (2010). *Sick Water? The Central Role of Waste-water Management in Sustainable Development. A Rapid Response Assessment.* United Nations Environment Programme, UN-HABITAT, GRID-Arendal. Data source: WHO, 2008.

(Continued)

BOX 7.1 Water Access and Quality in Benin *(Continued)*

TABLE 7.2 Bradley classification of diseases related to water and sanitation issues

CATEGORY	EXAMPLE	INTERVENTION
Waterborne	Diarrheal disease, cholera, dysentery, typhoid, infectious hepatitis	Improve quality of drinking water, prevent casual use of unprotected sources
Water-washed	Diarrheal disease, cholera, dysentery, trachoma, scabies, skin and eye infections, acute respiratory infections	Increase water quantity used, improve hygiene
Water-based	Schistosomiasis, guinea worm	Reduce need for contact with contaminated water, reduce surface water contamination
Water-related (insect vector)	Malaria, onchocerciasis, dengue fever, Gambian sleeping sickness	Improve surface water management, destroy insect breeding sites, use mosquito netting

Source: Based on UNICEF (2008). *UNICEF Handbook on Water Quality.* United Nations Children's Fund (UNICEF) New York, New York, USA.

and attracts vectors, such as flies and mosquitoes, that may transfer disease. (Animals that transmit human disease are called vectors.) Unsafe drinking water may also lead to other adverse health effects cited by UNICEF (2008).

- Children weakened by frequent episodes of diarrhea are more likely to be seriously affected by malnutrition and opportunistic infections (such as pneumonia), and they can be left physically stunted for the rest of their lives.

- Chronic consumption of unsafe drinking water can lead to permanent cognitive damage.
- People with compromised immune systems (e.g., people living with HIV and AIDS) are less able to resist or recover from water-borne diseases. Pathogens that might cause minor symptoms in healthy people (e.g., *Cryptosporidium*, *Pseudomonas*, rotaviruses, heterotrophic plate count microorganisms) can be fatal for the immunocompromised.

7.2 Water Quality Parameters

Water quality is influenced by natural and anthropogenic factors. Natural factors that influence water quality include weathering of bedrock and minerals, atmospheric deposition of dissolved gases and windblown particles, leaching of nutrients and organic matter from soils, biological processes, and the hydrologic factors discussed in Chapters 2 and 3. Declining water quality is a concern owing to the stresses imposed by increased human water consumption, agricultural and industrial discharges, and negative effects related to climate change. Poor water quality is a concern owing to the economic, environmental, and social impacts that occur if water supplies become too polluted for drinking, washing, fishing, or recreation.

Water quality is determined by measuring factors that are suitable for potential use. Drinking water should be pathogen free, but the microbial content of the water may not be a concern for some industrial uses. Broad categories of water quality include

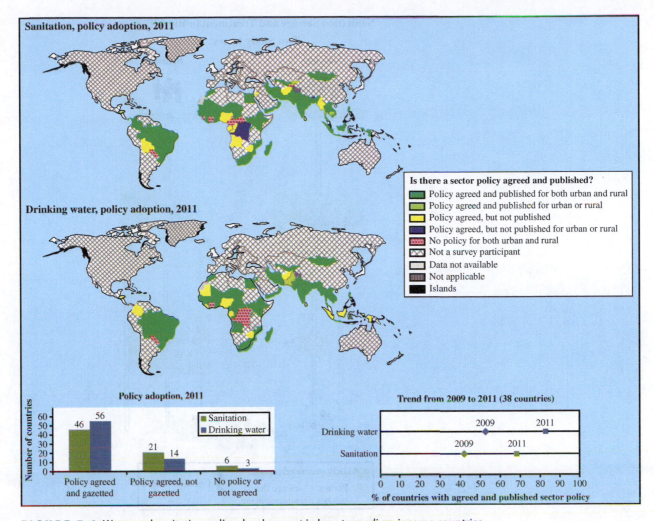

FIGURE 7.8 Water and sanitation policy development in low- to medium-income countries.

Source: Based on WHO (2012). *GLASS 2012 Report: UN-Water Global Analysis and Assessment of Sanitation and Drinking Water: The Challenge of Extending and Sustaining Services*. World Health Organization. Geneva, Switzerland.

microbial content, pH, solids content, ionic strength, dissolved oxygen, oxygen demand, nutrient levels, and toxic chemical concentrations. Water pollution may be any condition caused naturally or by human activity that adversely affects the quality of a stream, lake, ocean, or source of groundwater. Pollutants may also be defined as any harmful chemical or constituent present in the environment at concentrations above the naturally occurring background levels. Major water pollutants include microorganisms, nutrients, heavy metals, organic chemicals, oil, sediments, and heat. Water pollution issues affect every country in every economic category on Earth; however, the water issues vary with country and region. The types and origins of water pollutants likewise vary, but they all have negative economic, environmental, and societal impacts if the pollutants remain untreated, as illustrated in Figure 7.14 (see page 408). Pollutants discharged into the environment negatively affect water quality, as illustrated in Figure 7.15 (see page 409), but wastewater can be reused and treated to positively affect the economy, environment, and society with forethought and investment in basic infrastructure. Improving water access and sanitation directly impacts the UN Sustainable Development Goal number 6—sustainable management of water—and directly affects many of the other goals.

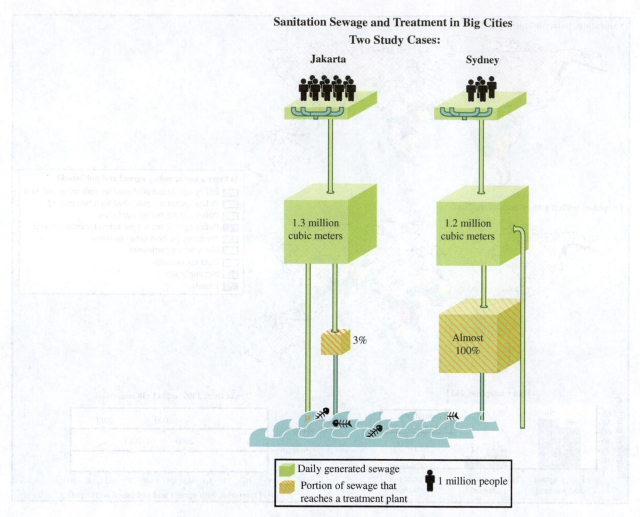

FIGURE 7.9 Comparison of the wastewater treatment between Jakarta, Indonesia, and Sydney, Australia.

Source: Based on Corcoran, E., Nellemann, C., Baker, E., Bos, R., Osborn, D., and Savelli, H. (eds.) (2010). *Sick Water? The Central Role of Waste-water Management in Sustainable Development. A Rapid Response Assessment.* United Nations Environment Programme, UN-HABITAT, GRID-Arendal.

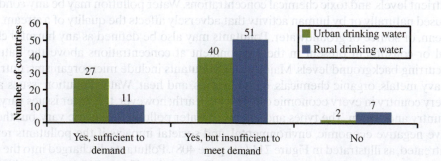

FIGURE 7.10 Availability of trained staff to meet water and sanitation needs reported by low- to medium-income countries.

Source: Based on WHO (2012). *GLASS 2012 Report: UN-Water Global Analysis and Assessment of Sanitation and Drinking Water: The Challenge of Extending and Sustaining Services.* World Health Organization. Geneva, Switzerland.

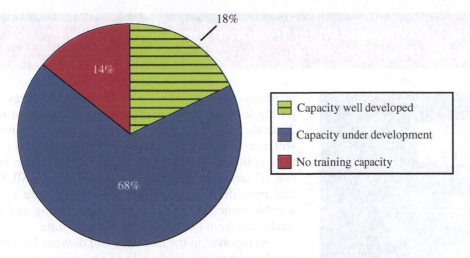

FIGURE 7.11 Training capacity for artisans and technicians to meet rural sanitation needs.

Source: Based on WHO (2012). *GLASS 2012 Report: UN-Water Global Analysis and Assessment of Sanitation and Drinking Water: The Challenge of Extending and Sustaining Services.* World Health Organization. Geneva, Switzerland.

BOX 7.2 Point-of-Use Decentralized Water Treatment Process for Porto-Novo, Benin

The Songhai Center (SC), a United Nations (UN) Center of Excellence for sustainable agricultural practices and technology, is headquartered in Porto-Novo, the capital of Benin. During a visit to Benin in 2004, Father Nzamujo Godfrey, director of the Songhai Center, identified the need for low-cost, sustainable point-of-use water treatment technology. Community-focused projects, such as this one, directly address the eight Millennium Development Goals set forth by the UN (2000).

The colloidal silver-impregnated ceramic water filter (CWF) technology was selected for potential implementation. Potters for Peace designed the chosen CWF technology to fit within a five-gallon plastic pail or clay container (see Figure 7.12). In addition to porous ceramic filtration, colloidal silver is used to inhibit bacterial growth (Fahlin, 2002; Clausen et al., 2004; van Halem et al., 2009). The UN has cited the Filtrón in its *Appropriate Technology Handbook*, which is used by both the Red Cross and Doctors Without Borders (Lantagne et al., 2010). Currently, several countries around the globe have employed this low-cost, appropriate technology filter

with good results. Most other water treatment technologies require more energy (ultraviolet disinfection systems) or chemical additives (chlorination or other chemical disinfectants) as reported in Table 7.3. Energy- and chemical-intensive disinfection systems may provide comparable or even better disinfection; however, the cost and availability of energy and chemical supplies are not sustainable within the community.

Ceramic water filters (CWFs) are manufactured using the process designed by Potters for Peace (Lantagne et al., 2010). This process uses a recipe for mixing clay and sawdust, forming the filters in a mold, and following a predetermined firing procedure in a high-temperature pottery kiln. The effectiveness of the colloidal silver-impregnated filters was evaluated using several methods. Potters for Peace recommends that the filters meet specific flow rate requirements in order to provide the required amount of water needed each day and as an indicator of the pore integrity in the CWF. The CWFs were evaluated in the laboratory using several established procedures as

(Continued)

BOX 7.2 Point-of-Use Decentralized Water Treatment Process for Porto-Novo, Benin (Continued)

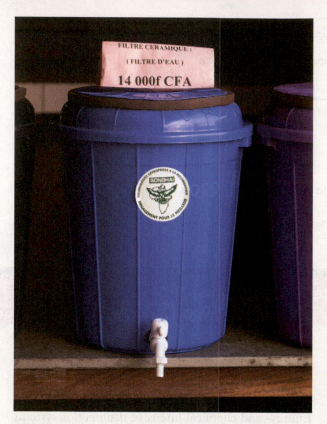

FIGURE 7.12 Ceramic point-of-use water filter, called a Filtrón, produced at the Songhai Center for Sustainable Development in Porto-Novo, Benin.

Source: Photo by Bradley Striebig.

indicators of drinking water quality, including the Hach LPT-MUG test for total coliform and *E. coli* and the Hach Pathoscreen test for hydrogen sulfide-producing organisms. The results of these tests are shown in Table 7.4. The Hach Pathoscreen test is widely accepted as a reasonable cost-effective indicator of potential contamination by mammalian wastes. If the water sample is contaminated, the water sample turns black, indicating the presence of hydrogen sulfide-producing bacteria and the possible presence of pathogenic bacteria. Because this test is inexpensive, does not require special

incubators (in the climate in Benin), and turns a jet black color, it is particularly useful as a tool to illustrate how debris-free, clear-colored water may still be contaminated. A negative LPT-MUG test for total coliforms indicates that there are likely to be few (if any) bacteria in the water. The LPT-MUG test procedure is commonly used to provide reasonable indication water is safe for drinking and is likely free from bacteria in the United States.

As reported in the literature and discussed in personal communications, the Filtrón performed very well in removing biological constituents (Lantagne et al., 2010; Clausen et al., 2004). The micropore structure of the ceramic filtration was able to prevent more than 99% of the total coliforms, fecal coliforms, *E. coli*, and pathogens from passing through the filter with the water in laboratory testing.

The Songhai Center sold approximately 60 filters in the first year of production. This number reflects start-up lessons, climatic considerations, and a conscious effort to build up an inventory for supply of the filters prior to developing a market plan. Also during this start-up phase, several manufacturing issues were identified and addressed. Fourteen years after the initial process development, the Songhai Center is still supplying water filters that have addressed the need for clean water for thousands of people in the community.

During follow-up visits in 2008 and 2009, U.S. university students and Songhai Center staff tested local water supplies and the effectiveness of the filters. Samples of local well water, the Songhai Center (unfiltered) tap water, Pur™ treated water from the local market, and well water were treated with the ceramic filters. Ten CWFs were randomly chosen from the Songhai Center's filter inventory for testing. Sample containers were prepared by cleaning with a dilute bleach solution, rinsing with filtered soapy water, and rinsing one final time with bottled water. All water samples were prepared using sterilized sample vials. The water samples were evaluated using presence/absence methods with Hach test packets and prepared vials for total coliform bacteria and

TABLE 7.3 A comparison of point-of-use appropriate water treatment technologies

TECHNOLOGY	ADVANTAGE	DISADVANTAGE	FILTER TIME
Biosand™ sand filtration	High removal efficiency for microorganisms	Needs continual use and regular maintenance Cost	1–2 hours
Filtrón™ ceramic filter	High removal efficiency for microorganisms Sized for households Relatively inexpensive	Requires fuel for construction Limited lifetime Requires regular cleaning	2–8 hours
SODIS™ solar water disinfection	Highly effective Inexpensive Can reuse a waste product (PET bottles)	Long treatment time (12 to 48 hours) Does not remove other pollutants Requires warm climate and sunlight	12–48 hours

TABLE 7.4 Some laboratory results showing the potential reduction in microbial contaminants with the Filtrón ceramic water filters

CONTAMINANT	UNITS	FILTERED WATER	AVERAGE REMOVAL
Fecal coliforms	MPN/100 mL	< 2 ± 0	> 99.92%
Total coliforms	MPN/100 mL	< 2 ± 0	> 99.97%
E. coli	MPN/100 mL	< 2 ± 0	> 99.0%
Pathogens (H_2S-producing bacteria)	MPN/100 mL	< 2 ± 0	> 99.7%
Streptococci	MPN/100 mL	< 2 ± 0	NA
Amoeba	MPN/100 mL	37,000 ± 115,000	99.5%

Source: Based on Striebig, B., Atwood, S., Johnson, B., Lemkau, B., Shamrell, J., Spuler, P., Stanek, K., Vernon, A., and Young, J. (2007). "Activated carbon amended ceramic drinking water filters for Benin." *Journal of Engineering for Sustainable Development* 2(1):1–12.

sulfide-producing bacteria (Pathoscreen™). None of the field samples collected in Benin appeared dirty or contaminated to the naked eye (see Figure 7.13). In spite of being clear of dirt and turbidity, however, untreated water samples (well water and tap water) indicated the presence of total coliforms and hydrogen sulfide-producing bacteria, which are indicative of pathogen organisms and likely contamination by human or animal wastes. The treated water samples were negative for all of the bacterial tests.

The data in Table 7.5, though limited in scope, are consistent with results from previous testing of water in Benin and of filtered water in both lab and field tests of new filters. The ceramic filters performed very well as point-of-use filters. The chemically treated Pur™ water was also free of bacteria. Although both methods of treating the water appeared successful, there were significant differences in the economic, cultural, and societal impacts of the two technologies.

(Continued)

BOX 7.2 Point-of-Use Decentralized Water Treatment Process for Porto-Novo, Benin (Continued)

FIGURE 7.13 The original water samples are both clear and free from debris. However, the analytical methods show that bacteria contaminate the untreated water, while no contamination was observed in the filtered water. These samples also provide a visual method for demonstrating contamination of water that otherwise appears "clean."

Source: Photo by Bradley Striebig.

Field tests of ceramic filters reported in the literature indicate that the average lifespan of a filter is approximately 2 years. If the filter is filled twice per day, it can easily supply about 20 liters per day. (Actual maximum capacity is approximately 45 to 75 liters per day.) The average household size in Benin is six family members who range widely in age. However, for comparing the costs of the methods, we have assumed only four members per household, which would provide a more conservative estimate of the daily water cost per household. The cost per liter of Pur-treated water in the local marketplace was approximately 0.10 USD. The economic savings for treated water for those able to invest in the technology are self-evident. At the current sales price, using conservative estimates for water usage and household size, the filter would pay for itself in 37 days, as illustrated in Table 7.6 (Striebig et al., 2010).

Development and implementation of the CWF process are a complex issue. Considerations of transportation, scheduling, climatic variations, material availability, and educational resources should all be considered when planning a CWF implementation. Monitoring and evaluating the program have provided the initiative to continue process improvements and verified that the technology is effective. Addressing health education and maintenance procedures in the community is important if the filters are to have the maximum impact in protecting drinking water quality. A plan for long-term monitoring and evaluation should be considered prior to initial implementation, as this step has proven critical in making the CWF manufacturing process a sustainable solution to improving the quality of drinking water in Benin.

TABLE 7.5 Presence of bacterial contamination in the untreated water samples compared to treated water samples

SOURCE	NO. OF SAMPLES	NO. INDICATING BACTERIAL PRESENCE (POSITIVE RESULT)		
		Total Coliform (LPT-MUG)	E. coli	Hydrogen Sulfide-Producing Bacteria (Pathoscreen™)
Well water	4	4	4	4
Tap water	10	6	1	6
Pur™ water	4	0	0	0
Filtered water	10	0	0	0

Source: Based on Striebig, B., Gieber, T., Cohen, B., Norwood, S., Godfrey, N., and Elliot, W. (2010). "Implementation of a Ceramic Water Filter Manufacturing Process at the Songhai Center: A Case Study in Benin." International Water Association's World Water Congress in Montreal 2010.

TABLE 7.6 Return on investment (ROI) in a ceramic water filter compared to drinking bottled (or bagged) water available in the marketplace

TREATED WATER TYPE	LITERS PER DAY PER PERSON	PERSONS PER HOUSEHOLD	PURCHASE COST (USD)	COST PER HOUSEHOLD PER DAY (USD)	COST PER LITER (USD)	ROI (DAYS)
Pur	2	4	$0.11	$0.89	$0.11	
Ceramic filter	5	4	$31.11	$0.04	$0.01	37

Source: Based on Striebig, B., Gieber, T., Cohen, B., Norwood, S., Godfrey, N., and Elliot, W. (2010). "Implementation of a Ceramic Water Filter Manufacturing Process at the Songhai Center: A Case Study in Benin." International Water Association's World Water Congress in Montreal 2010.

ACTIVE LEARNING EXERCISE 7.1 Analysis of Ceramic Water Filter

Describe what characteristics of the ceramic water filters that are produced by the Songhai Center in Benin are sustainable and why. What characteristics or aspects of the ceramic filter design are least sustainable? Why? Do this for each of the following aspects of sustainability:

a. Technological
b. Economic
c. Social
d. Environmental

7.2.1 Microorganisms in Water

Microorganisms are a natural part of the environment; in fact, they outnumber all other species combined on Earth. Microorganisms are the principal decomposers of natural and anthropogenic waste. Microorganisms convert organic waste in landfills to carbon dioxide, methane, and water. Microorganisms can also convert atmospheric nitrogen to ammonia through a process called nitrogen fixation, and they help fertilize soils. In their water treatment role, microorganisms are very helpful in remediating contaminated water, sediments, and soil in engineered systems. Microorganisms also sometimes consume oxygen in water to the exclusion of other species. A very small number, compared to all types of organisms, are pathogenic.

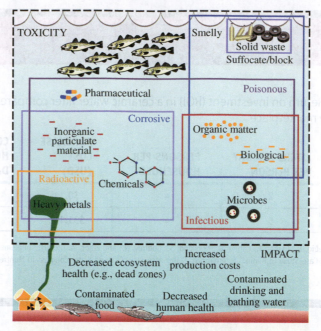

FIGURE 7.14 Sources, types, impacts, and management strategies for water pollutants.

Sources: Based on Corcoran, E., Nellemann, C., Baker, E., Bos, R., Osborn, D., and Savelli, H. (eds.) (2010). *Sick Water? The Central Role of Waste-water Management in Sustainable Development. A Rapid Response Assessment.* United Nations Environment Programme, UN-HABITAT, GRID-Arendal.

There are two broad classifications of microorganisms, prokaryotes and eukaryotes, the characteristics of which are summarized in Table 7.7. **Prokaryotes** have the simplest cell structure. Prokaryotes, illustrated in Figure 7.16, are classified by their lack of a nucleus membrane that contains cellular DNA. Bacteria, blue-green algae, and archaea are classified as types of prokaryotes. **Eukaryotes**, illustrated in Figure 7.17, are more complex organisms. Eukaryotes have a nucleus or nuclear

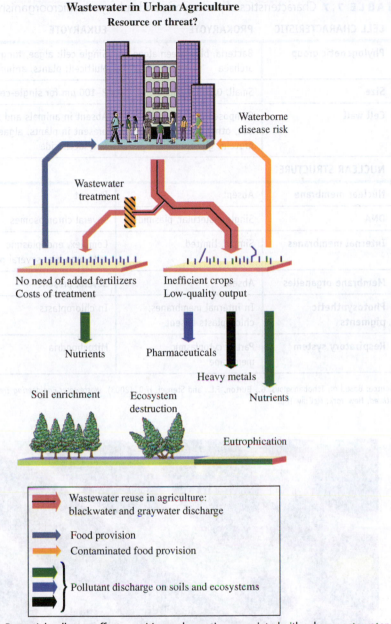

Wastewater in Urban Agriculture
Resource or threat?

Waterborne disease risk

Wastewater treatment

No need of added fertilizers
Costs of treatment

Inefficient crops
Low-quality output

Nutrients

Pharmaceuticals

Heavy metals

Nutrients

Soil enrichment

Ecosystem destruction

Eutrophication

Wastewater reuse in agriculture:
blackwater and graywater discharge

Food provision

Contaminated food provision

Pollutant discharge on soils and ecosystems

FIGURE 7.15 Potential pollutant effects, positive and negative, associated with urban wastewater.

Source: Based on Corcoran, E., Nellemann, C., Baker, E., Bos, R., Osborn, D., and Savelli, H. (eds.) (2010). *Sick Water? The Central Role of Waste-water Management in Sustainable Development. A Rapid Response Assessment.* United Nations Environment Programme, UN-HABITAT, GRID-Arendal.

envelope and endoplasmic reticulum, or interconnected organelles. Eukaryotes include protozoa, fungi, and green algae.

The most significant concern associated with drinking water quality is preventing pathogens from contaminating the water supply. Pathogenic organisms are typically excreted by human beings and other warm-blooded animals. Pathogens are usually classified as bacteria, protozoa, helminths, or viruses. Figure 7.18 illustrates the approximate concentration of microorganisms found in a typical liter of

TABLE 7.7 Characteristics of prokaryotic and eukaryotic microorganisms

CELL CHARACTERISTIC	PROKARYOTE	EUKARYOTE
Phylogenetic group	Bacteria, blue-green algae, archaea	Single cell: algae, fungi, protozoa Multicell: plants, animals
Size	Small, 0.2–3.0 μm	2–100 μm for single-cell organisms
Cell wall	Composed of peptidoglycan, other polysaccharides, protein, glycoprotein	Absent in animals and most protozoans; present in plants, algae, fungi; usually polysaccharide
NUCLEAR STRUCTURE		
Nuclear membrane	Absent	Present
DNA	Single molecular, plasmids	Several chromosomes
Internal membranes	Simple, limited	Complex, endoplasmic reticulum, golgi, mitochondria; several present
Membrane organelles	Absent	Several present
Photosynthetic pigments	In internal membranes; chloroplasts absent	In chloroplasts
Respiratory system	Part of cytoplasmic membrane	Mitochondria

Source: Based on Tchobanoglous, G., Burton, F.L., and Stensel, H.D. (2003). *Wastewater Engineering Treatment and Reuse, 4th ed.* New York: McGraw-Hill.

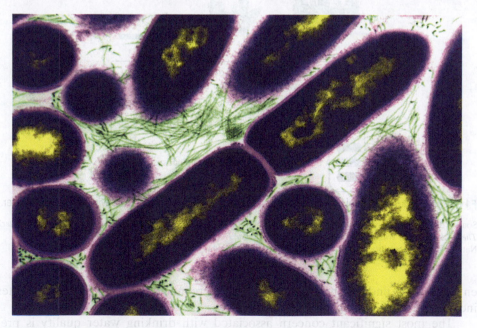

FIGURE 7.16 Color-enhanced transmission electron micrograph image of the prokaryote *Haemophilus ducreyi* bacteria associated with some types of food poisoning. The cellular DNA material is distributed and there is no distinct nucleus in the prokaryotic bacteria.

Source: David M. Phillips/Science Source.

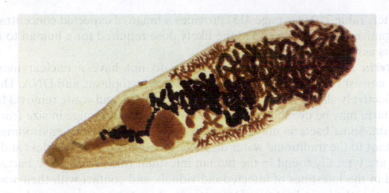

FIGURE 7.17 Stained light micrograph showing a small eukaryotic liver fluke (*Dicrocoelium*) that illustrates distinct nucleus, ribosomes, and mitochondria.

Source: M. I. Walker/Science Source.

A Look Inside

Concentrations of microorganisms excreted in one liter of wastewater

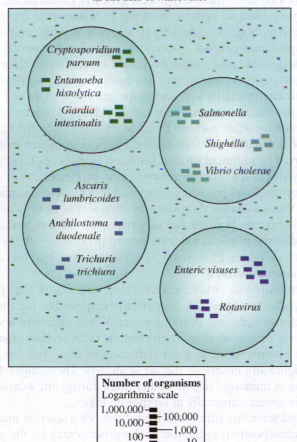

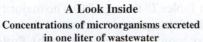

Number of organisms
Logarithmic scale

FIGURE 7.18 Logarithmic estimate of typical microorganism levels in 1 liter of wastewater.

Source: WHO (2006). *Guidelines for the Safe Use of Wastewater, Excreta and Greywater, Volume 2, Wastewater Use in Agriculture.*

wastewater. Table 7.10 (see page 415) provides a range of expected concentrations of microorganisms in wastewater and the likely dose required for a human to develop symptoms associated with the pathogen.

Bacteria are unicellular organisms that do not have a nuclear membrane. Bacteria consist only of a cell wall, cell membrane, cytoplasm, and DNA. The membrane selectively allows passage of nutrients for growth and waste removal from the cell. Bacteria may be cylindrical to spherical in shape and range in size from about 0.1 to 5 μm. Some bacteria may form endospores in unfavorable environments that are resistant to the traditional water disinfection process. Both harmless and helpful bacteria are typically found in the human intestinal tract. Pathogenic bacteria may be found in the intestines of infected individuals, and contact with their waste may spread the pathogenic bacteria. Domestic wastewater contains a wide variety of nonpathogenic and pathogenic bacteria (listed in Tables 7.8 through 7.10). Archaea are similar to bacteria in size and cellular components, but the cell wall, cell material, and RNA composition are different. Archaea are important in the anaerobic degradation of waste and are found in many environments not tolerated by bacteria.

Protozoa are motile, usually single-celled, microscopic eukaryotes, some examples of which are listed in Tables 7.8 through 7.10. The majority of protozoa are aerobic (use oxygen) heterotrophs. Some protozoa are aerotolerant anaerobes, and a few are anaerobic (i.e., they do not require oxygen for growth). Protozoa are often predatory as well and may be used to remove bacteria in wastewater treatment processes. Many protozoa are pathogenic, especially to individuals with compromised immune systems. *Cryptosporidium parvum*, cyclospora, and *Giardia lamblia* are particularly problematic pathogens because they are resistant to traditional chlorine disinfection.

Helminths are parasitic worms that are transmitted by wastewater and wastewater solids; their characteristic size and shape are listed in Tables 7.8 through 7.10. They are one of the leading causes of disease worldwide. Helminths are transmitted in wastewater when an adult parasite living in an infected individual lays eggs that are excreted into untreated wastewater. The eggs range in size from 10 to 100 μm and can be removed by sedimentation or filtration. Some eggs, however, are resistant to the common water and wastewater disinfection processes.

Viruses are intracellular parasites composed of a nucleic acid core with an outer protein coating. Viruses cannot reproduce by themselves, as they require a host cell to reproduce. The extremely small size of viruses, 0.2–0.8 μm, makes them difficult to remove in traditional physical water treatment processes. Other characteristics of viruses are shown in Tables 7.8 through 7.10. Viruses may also be resistant to some chemical disinfection processes.

Algae are chlorophyll-containing eukaryotic organisms that produce oxygen through photosynthesis. Algae may be partially responsible for the negative taste and odor characteristic of some drinking water. When excessive levels of nutrients enter a water body, excessive algae reproduction rates (called *algae blooms*) may occur. The algae blooms result in a net reduction of dissolved oxygen available in the water for higher organisms as the algae decays. These algae blooms are frequently the cause of increased fish mortality rates during hot weather when oxygen levels are at their lowest, especially in temperate lakes.

Engineers and scientists often find it useful to characterize microorganisms by their metabolic classification, as the metabolic process may be the greatest concern or utility related to the microorganisms. Metabolism refers to the sum of all processes required for the organisms to convert other materials to energy for internal use for growth and reproduction of new cellular mass. All known life forms require carbon for growth, and microorganisms may be classified by the carbon source that they use.

TABLE 7.8 Potential infectious agents found in wastewater

ORGANISM	DISEASE	SYMPTOMS
Bacteria		
Campylobacter jejuni	Gastroenteritis	Diarrhea
Escherichia coli	Gastroenteritis	Diarrhea
Legionella pneumophila	Legionnaires' disease	Fever, respiratory illness
Leptospira	Leptospirosis	Jaundice, fever
Salmonella	Salmonellosis	Food poisoning
Salmonella typhi	Typhoid fever	Fever, diarrhea, ulceration
Shigella	Shigellosis	Bacillary dysentery
Vibrio cholerae	Cholera	Diarrhea, dehydration
Yersinia enterocolitica	Yersiniosis	Diarrhea
Protozoa		
Balantidium coli	Balantidiasis	Diarrhea, dysentery
Cryptosporidium parvum	Cryptosporidosis	Diarrhea
Cyclospora cayetanesis	Cyclosporasis	Severe diarrhea and nausea
Entamoeba histolytica	Ambiasis (amoebic dysentery)	Severe diarrhea and abscesses
Giardia lamblia	Giadiasis	Diarrhea, nausea
Helminths		
Ascaris lumbrocoides	Ascariasis	Roundworm infestation
Enterobius vermicularis	Enterobiasis	Pinworm
Fasciola hepatica	Fascioliasis	Sheep liver fluke
Hymenolepis nana	Hymenolepiasis	Dwarf tapeworm
Tanenia saginata	Taeniasis	Beef tapeworm
T. solium	Taeniasis	Pork tapeworm
Trichuris triciura	Trichuriasis	Whipworm
Viruses		
Adenovirus	Respiratory disease	
Enteroviruses	Gastroenteritis, heart anomalies, meningitis	Polio, etc.
Hepatitis A virus	Infectious hepatitis	Jaundice, fever
Norwalk agent	Gastroenteritis	Vomiting
Parvovirus	Gastroenteritis	
Rotavirus	Gastroenteritis	

Source: Based on Tchobanoglous, G., Burton, F.L., and Stensel, H.D. (2003). *Wastewater Engineering Treatment and Reuse, 4th ed.* New York: McGraw-Hill.

TABLE 7.9 Characteristics of selected microorganisms commonly found in wastewater

MICROORGANISM	SHAPE	SIZE, μm	RESISTANT FORM
Protozoa			
Cryptosporidium			
Oocysts	Spherical	3–6	Oocysts
Sporozoite	Teardrop	1–3 W × 6–8 L	Oocysts
Entamoeba histolytica			
Cysts	Spherical	10–155	Cysts
Trophozoite	Semispherical	10–20	Cysts
Giardia lamblia			
Cysts	Ovid	6–8 W × 8–14 L	Cysts
Trophozoite	Pear or Kite	6–8 W × 12–16 L	Cysts
Helminths			
Ancylostoma duodenale (hookworm) eggs	Elliptical or egg	36–40 W × 55–70 L	Filariform larva
Ascaris lummbicoides (roundworm) eggs	Lemon or egg	35–50 W × 45–70 L	Embryonated egg
Trichuris trichiura (whipworm) eggs	Elliptical or egg	20–24 W × 50–55 L	Embryonated egg
Viruses			
MS2	Spherical	0.022–0.026	Viron
Enterovirus	Spherical	0.020–0.030	Viron
Norwalk	Spherical	0.020–0.035	Viron
Polio	Spherical	0.025–0.030	Viron
Rotavirus	Spherical	0.070–0.080	Viron

Source: Based on Tchobanoglous, G., Burton, F.L., and Stensel, H.D. (2003). *Wastewater Engineering Treatment and Reuse, 4th ed*. New York: McGraw-Hill.

Adenosine triphosphate (ATP) is the primary chemical used to store energy within a cell. Nicotinamide adenine dinucleotide (NAD^+) is the second most important chemical compound used to store energy. Heterotrophs obtain carbon from other organic matter or proteins. In some instances, autotrophs obtain carbon from inorganic carbon sources like carbonates. Chemotrophs produce energy within the cells through chemical oxidation of electron donors in their environment. Phototrophs obtain energy for reproduction through photon capture and photooxidation of a food or substrate.

Microorganisms may also be classified by the source of oxygen they use. **Aerobic organisms** require molecular oxygen (O_2) for respiration. Aerobic organisms typically utilize substrates more efficiently and more rapidly than nonaerobic organisms. Anaerobic organisms do not require molecular oxygen; they may obtain oxygen from inorganic ions, such as nitrates, sulfates, or proteins.

Microorganisms in water are difficult to enumerate owing to their small size and frequently uneven distribution in water sources. Pathogenic organisms, those we are most concerned about, are difficult to identify because they are so few in number relative to all other species. A variety of indicator tests have been developed as a

TABLE 7.10 Microorganism concentration and infectious dose in typical wastewater

ORGANISM	CONCENTRATION	
	MPN/100 mL	**Infectious Dose**
Bacteria		
Bacterioides	10^7–10^{10}	
Coliform, total	10^7–10^9	
Coliform, fecal	10^6–10^8	10^6–10^{10}
Clostridium perfringens	10^3–10^5	1–10^{10}
Enterococci	10^4–10^5	
Fecal streptococci	10^4–10^7	
Pseudomonas aeruginosa	10^3–10^6	
Shigella	10–10^3	10–20
Salmonella	10^2–10^4	10–10^8
Protozoa		
Cryptosporidium parvum oocysts	10–10^3	1–10
Entamoeba histolytica cysts	10^{-1}–10	10–20
Giardia lamblia cysts	10^3–10^4	<20
Helminths		
Ova	10–10^3	
Ascaris lumbricoides	10^{-2}–10	1–10
Viruses		
Enteric virus	10^3–10^4	1–10
Coliphage	10^3–10^4	

Source: Based on Tchobanoglous, G., Burton, F.L., and Stensel, H.D. (2003). *Wastewater Engineering Treatment and Reuse, 4th ed.* New York: McGraw-Hill.

surrogate measure of pathogen measurement in water samples. The indicator tests focus on measuring a broad category of microorganisms that are numerous in water samples and much easier to measure than particular pathogenic organisms.

The ideal *indicator organisms* would be present to indicate fecal contamination by warm-blooded mammals and would be numerous in the sample. Ideally, the organisms would be more resistant to typical disinfection processes than comparable pathogenic organisms, but they should not be a health threat to laboratory personnel preparing samples and conducting the tests. The ideal organism would be easy to isolate and quantify. Unfortunately, the currently available tests do not meet all these ideals, so the engineer or scientist must recognize the potential shortcomings of the standardized test procedures, summarized in Table 7.11, that are available for determining water quality.

The presence of indicator organisms in the existing standardized tests, listed in Table 7.12, does not necessarily mean enteric viruses or other pathogens are present in the water sample. The tests listed in the table are insufficient for some pathogens in water, especially those that are particularly resistant to

TABLE 7.11 Bacterial indicator organisms

INDICATOR ORGANISM	CHARACTERISTICS
Total coliform bacteria	Gram-negative rods that produce colonies in 24 to 48 hr at 35°C
	Includes *Eschericha, Citobacter, Enterobacter,* and *Klebsiella*
Fecal coliform bacteria	Able to produce colonies at elevated incubation temp. (44.5°C for 24 ± 2 hr)
Klebsiella	Included in fecal and total coliform tests. Cultured at 35°C for 24 ± 2 hr
E. coli	Most representative of fecal sources
Bacteroides	Anaerobic organism proposed as a human-specific indicator
Fecal streptococci	Used with fecal coliforms to determine the source of fecal contamination
	Several strains are ubiquitous
Enterococci	Two strains, *S. faecalis* and *S. faecium,* are most human-specific organisms
Clostridium perfringens	Spore-forming anaerobe
	Desirable indicator for disinfection, past pollution or for "old" samples
P. aeruginosa	Present in large numbers in domestic wastewater
A. hydrophila	Can be recovered in the absence of immediate sources of fecal pollution

Source: Based on Tchobanoglous, G., Burton, F.L., and Stensel, H.D. (2003). *Wastewater Engineering Treatment and Reuse, 4th ed.* New York: McGraw-Hill.

TABLE 7.12 Standardized indicator-based criteria tests

WATER USE	INDICATOR ORGANISMS
Drinking water	Total coliform
Fresh water recreation	Fecal coliform
	E. coli and *Enterococci*
Saltwater recreation	Fecal and total coliform
	Enterococci
Shellfish-growing areas	Fecal and total coliform
Agricultural reclamation	Total coliform
Wastewater effluent	Total coliform
Disinfection	Fecal coliform
	MS2 coliphage

Source: Based on Tchobanoglous, G., Burton, F.L., and Stensel, H.D. (2003). *Wastewater Engineering Treatment and Reuse, 4th ed.* New York: McGraw-Hill.

disinfection, like *Giardia lamblia*. As a result, water-disease outbreaks have occurred sporadically in water systems that have passed the prescribed standard microbial test procedures. The science of microbial testing is quickly evolving, and there is greater interest, especially in bacteriophage tests, in indicating the possible presence of enteric viruses in water. Thus, it is likely that the tests listed in Table 7.12 may change as science advances.

Several procedures have been standardized that provide estimates of the type and number of indicator organisms listed in Table 7.12. Individual bacteria enumeration methods include direct microscopic counting, pour and spread plate counts, membrane filtration, and multiple tube fermentation. Colonies of bacteria may be quantified using the heterotrophic plate count. Specific bacteria species can be identified by staining or fluorescence methods.

The pour plate and spread plate methods may be used to culture, identify, and enumerate bacteria (see Figure 7.19). The process uses a series of sample dilutions that are then mixed with the warm agar medium placed into the Petri dish. The dishes are incubated under controlled conditions, and the bacterial colonies are counted based on the known sample volume added to each Petri dish. Values are typically reported in colony-forming units (cfu) per volume (mL), and since the colonies must grow, they indicate only viable organisms and do not include nonviable organisms in the results.

The membrane filter technique, illustrated in Figure 7.20, is an effective technique for quickly estimating the number of indicator organisms in a sample. In this method, a known sample volume of water is passed through a filter, typically with a 0.45 μm pore size. Bacteria are retained on the filter, and the filter is placed into a Petri dish with an agar substrate and incubated for 24 hours. The viable colonies are counted after incubation. The membrane filter technique is commonly used for total coliform, fecal streptococcus, and *E. coli*.

A rapid, statistical method for estimating the number of microorganisms in water is based on a process of dilution until extinction and is called either a multiple tube fermentation test or the most probable number (MPN) test.

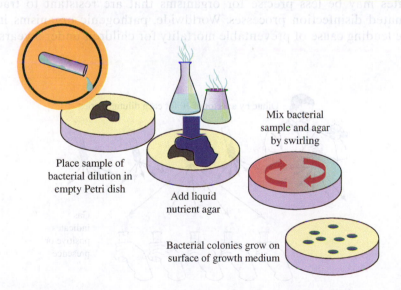

Place sample of bacterial dilution in empty Petri dish

Add liquid nutrient agar

Mix bacterial sample and agar by swirling

Bacterial colonies grow on surface of growth medium

FIGURE 7.19 Pour plate method for enumerating colony-forming units (cfu) in water samples.

disinfection, like *Giardia lamblia*. As a result, some outbreaks have occurred sporadically in water systems that have passed all prescribed standard microbial test procedures. The science of microbial testing is evolving quickly, and there is greater interest in evaluating bacteriophages as indicating the possible presence of enteric viruses in water. Thus, it is likely that the tests listed in Table 7.12 may change as better methods emerge.

Several procedures have been developed that provide estimates of the type and number of indicator organisms. As shown in Table 7.12, typical bacteria enumeration methods include direct *microscopic* counts, standard or spread plate counts, membrane filtration, and multiple tube fermentation. Groups of bacteria may be quantified using direct microscopic plate count. Specific organisms or species can be identified by staining or other methods.

The pour plate (Figure 7.19) method can be used to culture, identify, and enumerate bacteria (cfu). This method requires a series of sample dilutions that are then mixed with the warm agar medium placed into the Petri dish. The *colonies that form on the surface* of the agar medium, as in Figure 7.20, are counted based on the known sample volume added to each Petri dish. Values are typically reported in colony-forming units (cfu) per unit volume. Because the colonies must grow, they indicate only viable organisms and do not include nonviable organisms in the results.

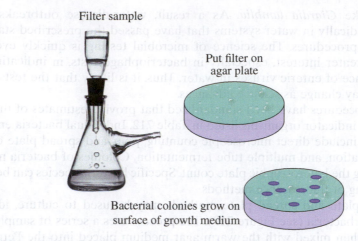

Filter sample

Put filter on
agar plate

Bacterial colonies grow on
surface of growth medium

FIGURE 7.20 Membrane filter technique for enumerating colony-forming units (cfu) in water samples.

Source: iStockPhoto.com/Chromatos.

The MPN is based on statistical equations, such as the Poisson equation, or on MPN tables. The results are typically reported as MPN/100 mL. In this process, illustrated in Figure 7.21, a water sample is diluted by an order of magnitude in each of five sample vials. The resulting test thus covers three to five orders of magnitude. The number of tubes with a viable sample are counted, and the MPN is determined from a statistical equation or MPN tables approved for the method.

Quantitative information about the number of microorganisms can be gained from indicator tests; typically, these tests can be completed in 24 to 48 hours. Keep in mind that indicator tests are not precise counts of specific pathogens, but instead estimates based on commonly occurring organisms. These estimates may be less precise for organisms that are resistant to traditional chlorinated disinfection processes. Worldwide, pathogenic organisms in water are the leading cause of preventable mortality for children under 5 years of age.

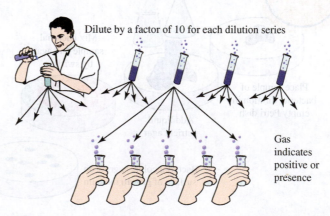

Dilute by a factor of 10 for each dilution series

Gas
indicates
positive or
presence

FIGURE 7.21 Illustration of the multiple tube fermentation tests for determining the MPN per 100 mL of a water sample.

Water quality controls and treatment methods that decrease pathogen levels in water are extremely important. Most microorganisms in our environment are nonpathogenic, and many are considered helpful. Microorganisms help degrade a wide variety of "waste" products and pollutants in the environment. Engineers have harnessed this process to use microorganisms in water treatments, wastewater treatments, environmental remediation of hazardous wastes, and pharmaceutical products.

7.2.2 Dissolved Oxygen

The amount of oxygen in the standard atmosphere is stable at 20.95%. However, the amount of *dissolved oxygen (DO)* in water changes dramatically with the temperature of the water; the consumption rate or deoxygenation rate in the water due to uptake from microorganisms and higher organisms; and the reaeration rate or how quickly oxygen is reabsorbed and mixed in the water body. The Henry's law constant value for oxygen was presented in Table 2.3 in Chapter 2. This information has been translated for standard conditions to a simple paper-based computer called a nomograph, shown in Figure 7.22. A straight line between any two points on the

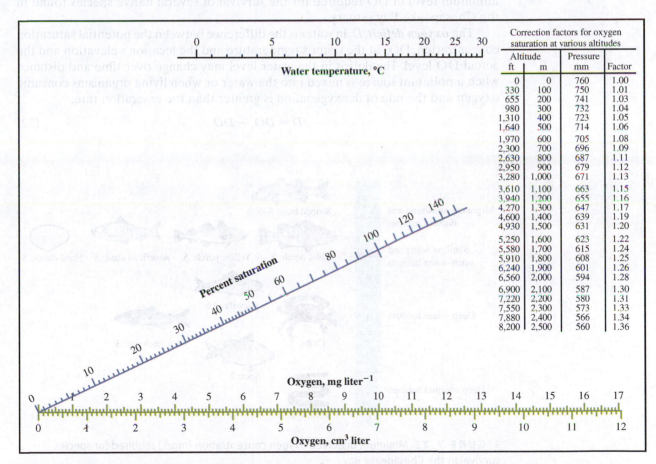

FIGURE 7.22 Nomograph that can be used to calculate the concentration of dissolved oxygen or percent saturation. Corrections are required to account for elevations above sea level.

Source: Based on Mortimer, C. H. (1981). "The oxygen content of air-saturated fresh waters over ranges of temperature and atmospheric pressure of limnological interest." *Mitt. int. Ver. Limnol.* pp. 22–23.

nomograph yields the value for the third point on the chart. The amount of dissolved oxygen in water is critical to animal life in aquatic systems. There is approximately 25 times more oxygen in air than in water, but due to the added weight of water, it requires about 800 times more energy to gain enough oxygen to support life in water than in air. Thus, the balance of dissolved oxygen in water is delicate, and small changes can and do result in large fish kills, especially in slow-moving rivers and lakes.

When oxygen concentrations are in equilibrium with the oxygen in the atmosphere, the aqueous solution is saturated with oxygen. The **saturated dissolved oxygen (DO_s)** concentration decreases with increasing water temperature and increases with increasing atmospheric pressure. Since atmospheric pressure generally decreases with elevation, the DO_s concentration in water tends to decrease as elevation increases. Thus, there are correction factors on the nomograph to account for variations in DO_s in locations with different elevations.

Different species of organisms require different levels of DO in water. Generally, smaller organisms and those that feed near the bottom of waterways require less oxygen. Larger predatory species that feed on smaller organisms higher in the water column require higher levels of DO. Figure 7.23 shows the minimum level of DO required for the survival of several native species found in the Chesapeake Bay estuary.

The **oxygen deficit**, *D*, in water is the difference between the potential saturation concentration, DO_s, at the water's temperature and the location's elevation and the actual DO level. The deficit in the water level may change over time and distance when a pollutant source is mixed into the water or when living organisms consume oxygen and the rate of deoxygenation is greater than the reaeration rate:

$$D = DO_s - DO \tag{7.2}$$

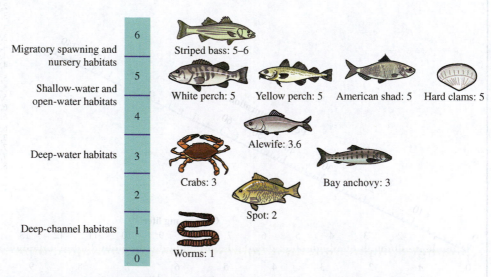

FIGURE 7.23 Minimum dissolved oxygen concentration (mg/L) required for species survival in the Chesapeake Bay.

Source: U.S. EPA (2003). *Ambient Water Quality Criteria for Dissolved Oxygen, Water Clarity and Chlorophyll a for the Chesapeake Bay and Its Tidal Tributaries*. U.S. Environmental Protection Agency. Region III. Chesapeake Bay Program Office. Annapolis, Maryland. EPA 903-R-03-002.

7.2.3 Biochemical Oxygen Demand

Microorganisms in the environment and in engineered systems remove oxygen in proportion to their initial number and rate of growth. The growth rate of microorganisms can be modeled, and from the growth rate, the oxygen uptake rate can be determined. In a closed system, microorganisms go through four stages of growth: a lag phase, an exponential growth phase, a stationary phase, and a death phase. During the lag phase, the amount of nutrients and substrate is much greater than the number of microorganisms. Microbial growth is limited in the lag phase, as the microorganisms first manufacture enzymes necessary to metabolize the substrate and the number of microorganisms present is small. Once the enzymes have been produced, growth becomes exponential as most of the microorganism's energy is used for growth and reproduction. In the exponential growth phase, the substrate and nutrients are still abundant and waste products from microbial growth do not yet inhibit growth. During exponential growth, the growth rate is proportional to the number of microorganisms. In the exponential phase, growth is governed by the exponential equations and is typically represented as

$$X = X_o e^{\mu t} \tag{7.3}$$

where

X = concentration of the microorganisms (g/m^3)

μ = specific growth rate, with units of inverse time (time^{-1})

The population tends to reach a maximum value and then plateaus, as shown in Figure 7.24. The population growth rate approaches zero as the substrate concentration diminishes or waste products build up to toxic levels in the closed system. As substrate levels continue to diminish and waste products increase, the number of viable organisms begins to decrease as the population enters a death phase in a closed system.

In many cases in the environment and engineered systems, growth is limited by the amount of substrate present in the system. A maximum specific growth rate can be estimated during substrate-limited growth by the Michaelis–Menton equation:

$$r_{su} = -\frac{kXS}{K_s + S} \tag{7.4}$$

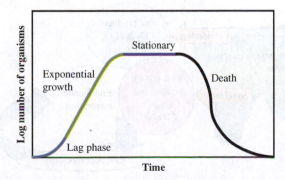

FIGURE 7.24 Stages of microbial growth in a closed system.

where

r_{su} = rate of substrate concentration change $\frac{g}{m^3 - day}$
k = the maximum specific substrate utilization rate $\frac{g\ of\ substrate}{g\ of\ microorganism - day}$
X = concentration of the microorganisms (g/m³)
S = growth-limiting substrate concentration (g/m³)
K_s = half-velocity degradation coefficient (g/m³)

The half-velocity degradation coefficient, also called the Monod constant, corresponds to the substrate concentration at which the specific growth rate is one-half the maximum specific growth rate (i.e., $\frac{1}{2}k$). The maximum specific *substrate* utilization rate is related to the maximum specific bacterial growth rate by

$$k = \frac{\mu_{max}}{Y} \qquad (7.5)$$

where

μ_{max} = the maximum specific bacterial growth rate
(g new cells/g cell · day)
Y = true substrate yield coefficient (g cell/g substrate used)

The **substrate utilization rate** provides information about energy conservation from the substrate to cell maintenance and growth (see Figure 7.25). The fraction of energy from the substrate that is used for cell growth is equal to the substrate yield, Y. The substrate utilization rate may then also be written as

$$r_{su} = \frac{dS}{dt} = -\frac{\mu_{max} XS}{Y(K_s + S)} \qquad (7.6)$$

Aerobic microorganisms (chemoheterotrophs) generate the energy for cell maintenance and biomass growth from the substrate from respiration according to the following generalized equation:

$$C - based\ substrate + O_2 \rightarrow biomass + CO_2 + H_2O \qquad (7.7)$$

The substrate utilization rate and change in biomass produce a **carbonaceous oxygen demand (COD)** in water. Similarly, chemoautotrophs use carbon dioxide as a carbon source and inorganic matter as an energy source. Chemoautotrophs also

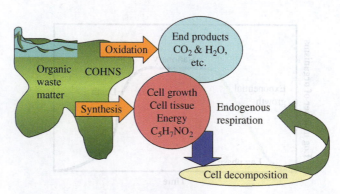

FIGURE 7.25 Conversion of organic waste matter to biomass and carbon dioxide through respiration.

require energy in this energy conversion process according to the general equation for *nitrification*:

$$NH_3 + O_2 \rightarrow biomass + NO_3^- + H_2O + H^+ \tag{7.8}$$

The oxygen consumed for the nitrification of ammonia (NH_3) for the production of biomass is called the *nitrogenous oxygen demand (NOD)*.

Microorganisms convert a portion of the organic waste (represented by COHNS, where the amount of each element is unknown) into carbon dioxide to gain energy for cell maintenance through oxidation of the organic waste:

$$COHNS + O_2 + bacteria \rightarrow CO_2 + H_2O + NH_3 \tag{7.9}$$

A portion of the organic waste is converted into new biomass through synthesis. Biomass is represented by an average equation based on the elemental ratios of the components of cell tissue ($C_5H_7NO_2$):

$$COHNS + O_2 + bacteria + energy \rightarrow C_5H_7NO_2 \tag{7.10}$$

Microorganisms begin to convert their own cell tissue into energy for cell maintenance through *endogenous respiration* when substrate levels become very low:

$$C_5H_7NO_2 + 5O_2 \rightarrow 5CO_2 + NH_3 + 2H_2O \tag{7.11}$$

If the particular substrate in a system is known, a *theoretical oxygen demand (ThOD)* can be calculated in a series of three steps from the stoichiometric relationship. The first step is to write the equation describing the reaction for oxidation of the carbon-based substrate. The second step is to balance the chemical equation. It is usually easiest to balance the number of carbon atoms on both sides of the equation first and then balance the hydrogen atoms. The stoichiometric value for the oxygen required can be found by adding the oxygen required by the right-hand side and subtracting any oxygen in the substrate. The third step is to use the stoichiometric relationship from the balanced chemical equation to convert to the units required for the carbonaceous ThOD, usually mg/L.

EXAMPLE 7.1   Theoretical Carbonaceous Oxygen Demand of a Can of Soda

Sucrose ($C_{12}H_{22}O_{11}$) is the disaccharide that makes up table sugar. One can of soda typically contains 40 grams equivalent of table sugar. (Look on your favorite soda label to see if it contains sucrose or fructose, $C_6H_{12}O_6$.) What is the ThOD in one can of soda?

Step 1: Write the unbalanced equation for the reaction:

$$C_{12}H_{22}O_{11} + ?O_2 \rightarrow ?CO_2 + ?H_2O$$

Step 2: Balance the chemical equation.

There are 12 atoms of carbon on the left-hand side of the equation, so 12 molecules of carbon dioxide will be needed to balance the carbon on the right-hand side of the equation.

There are 22 atoms of hydrogen on the left-hand side of the equation, so 11 molecules of water will be needed to balance the hydrogen on the right-hand side of the equation.

Adding the atoms of oxygen from carbon dioxide and water molecules on the right-hand side of the equation and subtracting the atoms of water contained in sucrose yield the atoms of oxygen needed:

$$12 \times 2 + 11 \times 1 - 11 = 24 \text{ additional atoms of oxygen are required}$$

The balanced chemical equation showing the molecular oxygen (O_2) required is

$$C_{12}H_{22}O_{11} + 12O_2 \rightarrow 12CO_2 + 11H_2O$$

Step 3: Use the balanced equation and unit conversions to find the carbonaceous ThOD in mg/L:

$$\frac{40 \text{ g sucrose}}{12 \text{ oz}} \times \frac{33.814 \text{ oz}}{L} \times \frac{\text{mole sucrose}}{342 \text{ g sucrose}} \times \frac{12 \text{ mole } O_2}{\text{mole sucrose}} \times \frac{32,000 \text{ mg } O_2}{\text{mole } O_2} = 1.27 \times 10^5 \frac{\text{mg}}{L}$$

Therefore, ThOD = 1.27×10^5 mg/L, so during the summer a truckload of soda that spills into a river or lake could have dramatic consequences due to the amount of oxygen that will need to be removed from the water as microorganisms in the water will have a carbonated soda party.

A variety of microorganisms (both chemoheterotrophs and chemoauto-trophs) are present in the natural environment, and both use oxygen in water as an electron acceptor to convert a substrate to biomass. The amount of oxygen required by microorganisms to convert a substrate to biomass is called the **biochemical oxygen demand (BOD)**. The BOD is the most widely used water quality parameter to quantify water quality related to organic pollutants from public sewage, agricultural runoff, and industrial discharges. A measure of BOD provides a useful estimate of the overall strength of pollutants in a water sample. BOD measurements also provide estimates for the amount of oxygen demanded if wastewater enters a river, lake, or engineered treatment system. A standardized procedure has been developed and widely used to measure the effectiveness of treatment processes and determine compliance with industrial and municipal discharge permits.

The standardized five-day BOD test is used to measure the oxygen con-sumed during the decomposition of organic waste in water. The test measures the total amount of oxygen required by microorganisms in five days of degra-dation in a sample containing an organic waste material. The sample bottles are either made of opaque glass or incubated in a way that prevents light from entering the bottle in order to prevent background algae from adding oxygen to the sample through photosynthesis. The standard BOD procedure consists of the following steps:

- Add a sample volume, ranging from 1 to 300 mL according to the type of water described in Table 7.13, to a 300-mL BOD bottle.
- Fill the remaining volume of the 300-mL bottle with dilution water that has been saturated with oxygen and contains a phosphate buffer and nutrients, as illustrated in Figure 7.26. The dilution water contains the phosphate buffer to keep changes in pH from the degradation of wastes from inhibiting sub-strate utilization. Nutrients are added in the dilution water to ensure that

TABLE 7.13 Approximate sample volumes required for standard five-day BOD test

SAMPLE TYPE	Est. BOD (mg/L)	Sample size (mL)	Estimated BOD (mg/L) Sea Level	Estimated BOD (mg/L) 305 m (1,000 ft)	Estimated BOD (mg/L) 1,524 m (5,000 ft)	At Sample Size (mL)
Strong trade waste	600	1	2,460	2,380	2,032	1
	300	2	1,230	1,189	1,016	2
	200	3	820	793	677	3
Raw and settled sewage	150	4	615	595	508	4
	120	5	492	476	406	5
	100	6	410	397	339	6
	75	8	304	294	251	8
	60	10	246	238	203	10
	50	12	205	198	169	12
	40	15	164	158	135	15
Oxidized effluents	30	20	123	119	101	20
	20	30	82	79	68	30
	10	60	41	40	34	60
Polluted river waters	6	100	25	24	21	100
	4	200	12	12	10	200
	2	300	8	8	7	300

Under MINIMUM SAMPLE SIZE: Est. BOD (mg/L) and Sample size (mL). Under MAXIMUM SAMPLE SIZE: Estimated BOD (mg/L) with columns Sea Level, 305 m (1,000 ft), 1,524 m (5,000 ft), and At Sample Size (mL).

Source: Based on U.S. EPA (2003). *Ambient Water Quality Criteria for Dissolved Oxygen, Water Clarity and Chlorophyll a for the Chesapeake Bay and Its Tidal Tributaries*. U.S. Environmental Protection Agency. Region III. Chesapeake Bay Program Office. Annapolis, Maryland.

all the required elements for cell growth are present in the sample bottle. Microorganisms may also be added via the dilution water (called *seeding* the sample) in order to ensure that the number of organisms is sufficient in the sample to achieve logarithmic growth.

- Measure the initial DO of the bottle, DO_i.
- Incubate the bottle at 20°C in the dark.
- Measure DO after five days (±4 hours) of incubation, DO_f.

Water samples that contain microorganisms, such as raw sewage, treated unchlorinated effluent from a wastewater treatment facility, or natural waters, do not require the addition of extra microorganisms, called a sample seed, to perform the BOD test. However, water samples that do not contain microorganisms, such as chlorinated water from a treatment facility or industrial-strength wastewater, require the addition of microorganisms through a seed. Tests that require a seed also require a sample blank that consists of a bottle containing only the sample seed and dilution water in addition to another sample that contains the water sample, the seed, and the dilution water. At least three blanks should be included with each set of BOD samples.

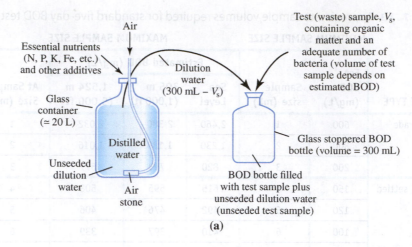

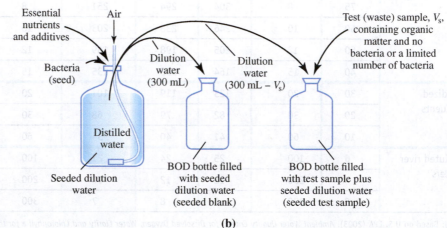

FIGURE 7.26 Standardized five-day BOD test procedure.

Source: Based on Tchobanoglous, G., Burton, F.L., and Stensel, H.D. (2003). *Wastewater Engineering Treatment and Reuse, 4th ed*. New York: McGraw-Hill.

The five-day BOD of an *unseeded sample* is determined from

$$BOD_5 = \frac{DO_i - DO_f}{P} \tag{7.12}$$

where

BOD_5 = standard five-day BOD (mg/L)

DO = measured dissolved oxygen in the sample bottle (mg/L)

$$P = \frac{\text{volume of sample (mL)}}{\text{volume (mL) of sample + dilution water}} \overset{\text{def}}{=} \frac{\text{volume of sample (mL)}}{300 \ (\text{mL})} \tag{7.13}$$

The five-day BOD of a *seeded sample* is determined from

$$BOD_5 = \frac{(DO_i - DO_f) - (1 - P)(B_i - B_f)}{P} \tag{7.14}$$

where B = measured dissolved oxygen in the blank bottle (mg/L).

A wastewater treatment plant must maintain its effluent BOD_5 concentration at less than 20 mg/L to remain in compliance with its discharge permit. The laboratory provided the data shown in Table 7.14 to the facility engineer. Is the plant in compliance or not?

TABLE 7.14 Example BOD data

	SAMPLE	BLANK
Initial dissolved oxygen (mg/L)	8.06	8.12
Final dissolved oxygen (mg/L)	6.48	8.01
Volume of the sample (mL)	30	30
Volume of the BOD bottle (mL)	300	300

Using Equation (7.13) for the dilution factor and Equation (7.14) for the seeded sample:

$$P = \frac{30\text{ mL}}{300\text{ mL}} = 0.1$$

$$BOD_5 = \frac{(8.06 - 6.48) - (1 - 0.10)(8.12 - 8.01)}{0.1} = 14.8\text{ mg/L}$$

There are several limitations associated with the BOD_5 test. The microorganisms must be readily able to degrade the waste. Chemicals resistant to biodegradation may not be properly accounted for in a standard BOD test. Only wastes that are biodegradable can be measured; nonbiodegradable organic wastes and nonorganic wastes are not accounted for in the BOD test. Since specific chemical formulas for the waste mixture and microorganism composition are not typically known, the results cannot be used in stoichiometric equations. The test requires a relatively long period to complete, which may not represent the state of the systems at the completion of the test. Nonetheless, the BOD test is extremely helpful for predicting potential environmental impacts of organic wastes.

7.2.4 Nutrients in Water

Excessive levels of nutrients in water have caused major deterioration of many of the world's watersheds resulting from higher than normal levels of nitrogen-containing and phosphorus-containing compounds in water. All living things need a variety of organisms to support and maintain healthy ecosystems, including nitrogen and phosphorus. The ecosystem balance may be upset when high levels of nitrogen and phosphorus are discharged into waterways. Excessive nitrogen and phosphorus loading is usually due in large part to use of excessive fertilizer in agricultural practices and the discharge of municipal wastewater. The distribution of areas and sources that contribute to nitrogen and phosphorus levels into the Chesapeake Bay in the eastern United States is shown in Figures 7.27 and 7.28.

Excessive levels of nutrients may have adverse effects on drinking water supply, recreational water use, aquatic life, and fisheries. Nitrate levels above

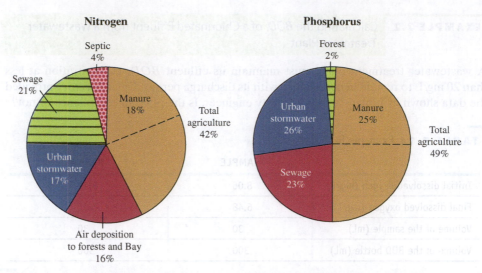

FIGURE 7.27 Sources of nitrogen and phosphorus in the Chesapeake Bay watershed and estuary.

Source: Based on Chesapeake Bay Foundation (2004). *Manure's Impact on Rivers, Streams and the Chesapeake Bay: Keeping Manure Out of the Water.*

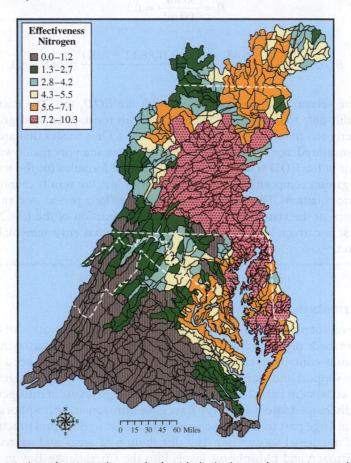

FIGURE 7.28 Area showing subwatersheds with the highest to lowest nitrogen-loading effect on the Chesapeake Bay.

Source: U.S. EPA (2010). *Chesapeake Bay Total Maximum Daily Load for Nitrogen, Phosphorus and Sediment.* U.S. Environmental Protection Agency. Region 3. Water Protection Division. Philadelphia, PA. Figure ES-2, page ES-6.

10 mg/L in drinking water have been linked to methemoglobinemia (blue baby syndrome), which may lead to premature death in infants. ***Eutrophication***, or the nutrient enrichment of aquatic ecosystems, is a natural process that can be accelerated by excessive anthropogenic emissions of nitrogen and phosphorus. Accelerated eutrophication increases the growth rate of aquatic plants, algae, and microorganisms.

Eutrophication often results in significantly increasing the oxygen deficit in bodies of water as the higher than normal levels of plant matter settle to the bottom of the water body. Microorganism concentrations increase as they use the dead plant matter as a substrate, and oxygen levels decrease as the plant matter decays. Eventually, oxygen levels may become completely depleted in portions of the water body. This occurred in the summer of 2005 throughout large sections of the Chesapeake Bay (shown in Figure 7.29). The pollution loading to the bay changed

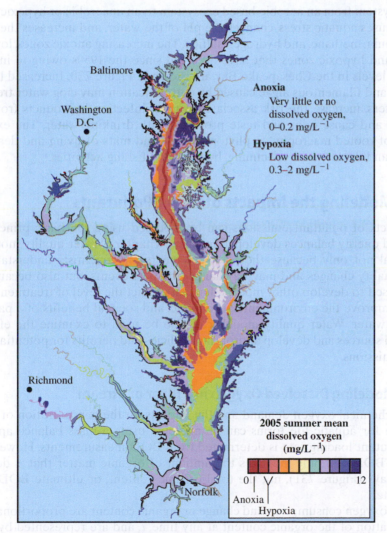

Anoxia
Very little or no dissolved oxygen, 0–0.2 mg/L^{-1}

Hypoxia
Low dissolved oxygen, 0.3–2 mg/L^{-1}

2005 summer mean dissolved oxygen (mg/L^{-1})

0 | | 12
Anoxia |
Hypoxia

FIGURE 7.29 Large areas of the Chesapeake Bay in the eastern United States suffer from low DO concentrations due in large part to excessive nutrient applications in the watershed.

Source: Based on Wicks, C., Jasin, D., and Longstaff, B. (2007). *Breath of Life: Dissolved Oxygen in Chesapeake Bay.* May 27, 2007 Newsletter. Chesapeake Bay Foundation. Annapolis, MD.

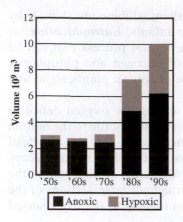

FIGURE 7.30 The increase in the volume of anoxic and hypoxic zones occurring in the Chesapeake Bay.

Source: Based on Chesapeake Bay Foundation (2004). *Manure's Impact on Rivers, Streams and the Chesapeake Bay: Keeping Manure Out of the Water.*

the ecosystem from an aerobic-based ecosystem to an anaerobic or hypoxic system, which causes aquatic stress, changes the pH of the water, and increases the release of ammonia, methane, and hydrogen sulfide. The increasing anoxic zones, low levels of DO, and hypoxic zones that have occurred since the 1950s owing to increased nutrient levels in the Chesapeake Bay are shown in Figure 7.30. Increased levels of diatoms and filamentous algae caused by eutrophication may clog water treatment plant filters, increase the risk associated with disinfection byproducts from chlorination, and cause odor and taste problems with drinking water. The extensive growth of rooted macrophytes, phytoplankton, and mats of living and dead plant matter can interfere with swimming, boating, and fishing activities.

7.3 Modeling the Impacts of Water Pollutants

The effects of pollutant emissions can be modeled using the basic principles of mass and energy balances developed in earlier chapters. Water quality models are beneficial not only because they help engineers and scientists understand how water quality changes and may be changed by past events, but also because they can be used to develop other models that can predict the level of treatment necessary to improve the environmental, economic, and societal benefits of a particular body of water. Water quality models can also be used to examine the effects of potential sources and develop appropriate policies and permits for potential wastewater emissions.

7.3.1 Modeling Dissolved Oxygen in a River or Stream

The biochemical oxygen demand and the impacts on the concentration of oxygen available for aquatic life forms can be modeled using a mass balance approach. The pollutant loading rate is determined from BOD measurements. However, the five-day BOD test only indicates the amount of organic matter that is degraded in five days (Figure 7.31), not the total organic content, or ultimate BOD, of the wastewater.

The oxygen consumption and change in organic content are proportional to the concentration of the organic content at any time, t, and are represented by a first-order rate constant, r_b:

$$-r_b = \frac{dL}{dt} = -kL \qquad (7.15)$$

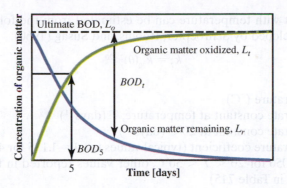

FIGURE 7.31 The total organic matter remaining in a water sample decreases with time and biological activity. The organic matter oxidized, which is measured by a change in oxygen concentration in the sample, is the reciprocal, or mirror image, of the organic matter remaining in the solution. The organic matter oxidized is equivalent to the BOD at the given time, and typically about two-thirds of the organic matter is oxidized by day five.

where

    L = the oxygen demand associated with the organic content (mg/L)
    k = reaction rate constant [days^{-1}]

Rearranging and integrating Equation (7.15) yields the exponential function that describes the change in organic content with time:

$$L_t = L_o e^{-kt} \tag{7.16}$$

where

    L_o = the oxygen demand associated with the organic content at the initial time (mg/L)
    L_t = the oxygen demand associated with the organic content at any time t (mg/L)

The BOD_t is equivalent to the difference between the initial organic content and the organic content at time t, as shown in Figure 7.31:

$$BOD_t = L_o - L_t \tag{7.17}$$

Substituting Equation (7.16) into Equation (7.17) yields

$$BOD_t = L_o - L_o e^{-kt} = L_o(1 - e^{-kt}) \tag{7.18}$$

The ultimate BOD is equivalent to the initial organic content of the wastewater, L_o, and is the maximum amount of oxygen that would be consumed by microbial degradation of the waste. The relationship between the measured BOD at five days, BOD_5, and the ultimate BOD, L_o, is

$$L_o = \frac{BOD_5}{(1 - e^{-5k})} \tag{7.19}$$

The value for the BOD rate constant depends on how rapidly the waste is degraded, or how biodegradable the waste is, and the temperature. The change in the rate constant with temperature can be estimated from the following empirical correlations developed by Schroepfer, Robins, and Susag (1960):

$$k_T = k_{20}(\theta)^{T-20} \tag{7.20}$$

where

T = temperature (°C)

k_T = BOD rate constant at temperature, T (days^{-1})

k_{20} = BOD rate constant at 20°C (days^{-1})

θ = temperature coefficient (typical values for θ = 1.135 for $4 < T < 20$°C and θ = 1.056 for $20 < T < 30$°C, other values reported in the literature are shown in Table 7.15)

A conservative mass balance approach is useful for modeling a point-source pollutant discharge or tributary input into a river or stream, as illustrated in Figure 7.32. For simplicity, it is assumed that the river water and inputs are instantaneously well

TABLE 7.15 Reported values for the temperature compensation coefficient used for carbonaceous BOD decay

θ, TEMPERATURE CORRECTION FACTOR	TEMPERATURE LIMITS (°C)	REPORTED REFERENCES IN U.S. EPA REPORT
1.047		Chen (1970)
		Harleman et al. (1977)
		Medina (1979)
		Genet et al. (1974)
		Bauer et al. (1979)
		JRD (1983)
		Bedford et al. (1983)
		Thomann and Fitzpatrick (1982)
		Velz (1984)
		Roesner et al. (1981)
1.05		Rich (1973)
1.03–1.06	(0–5)–(30–35)	Smith (1978)
1.075		Imhoff et al. (1981)
1.024		Metropolitan Washington Council of Governments (1982)
1.02–1.06		Baca and Arnett (1976)
		Baca et al. (1973)
1.04		Di Toro and Connolly (1980)
1.05–1.15	5–30	Fair et al. (1968)

Source: Based on Bowie, G.L, Mills, W.B., Porcella, D.B., Campbell, C.L., Pagenkopf, J.R., Rupp, G.L., Johnson, K.M., Chan, P.W.H., Gherini, S.A., and Chamberlin, C.E. (1985). *Rates, Constants, and Kinetics Formulations in Surface Water Quality Modeling, 2nd ed.* Environmental Research Laboratory. Office of Research and Development, U.S. Environmental Protection Agency. Athens, Georgia.

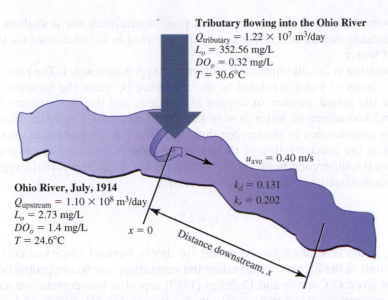

Tributary flowing into the Ohio River
$Q_{tributary} = 1.22 \times 10^7$ m^3/day
$L_o = 352.56$ mg/L
$DO_o = 0.32$ mg/L
$T = 30.6°C$

$u_{ave} = 0.40$ m/s

$k_d = 0.131$
$k_r = 0.202$

Ohio River, July, 1914
$Q_{upstream} = 1.10 \times 10^8$ m^3/day
$L_o = 2.73$ mg/L
$DO_o = 1.4$ mg/L
$T = 24.6°C$

$x = 0$

Distance downstream, x

FIGURE 7.32 A moving river system can be modeled with a mass balance approach. The input or tributary stream is considered to be instantaneously mixed with the upstream flow at the point of discharge, $x = 0$.

mixed at the junction point. The conservative mass balance equation may be used to find the ultimate BOD, dissolved oxygen content, temperature, and combined flow rate of the combined river and input flows:

$$Q_{\text{river upstream}} + Q_{\text{input}} = Q_{\text{river downstream}} \qquad (7.21)$$

and for any parameter C in the water:

$$Q_{\text{river upstream}} C_{\text{upstream}} + Q_{\text{input}} C_{\text{input}} = Q_{\text{river downstream}} C_{\text{downstream}} \qquad (7.22)$$

As the water containing an organic waste moves downstream, the waste will be biologically degraded and the microorganisms utilizing the organic waste as a substrate will consume oxygen. The rate of oxygen consumption, or deoxygenation, is proportional to the ultimate BOD at any time, t, multiplied by the deoxygenation rate constant for the moving water body:

$$r_d = k_d L_t \qquad (7.23)$$

where
r_d = rate of deoxygenation
L_t = ultimate BOD remaining at time t some distance downstream (mg/L)
k_d = the deoxygenation rate constant (day^{-1})

It is usually assumed that $kd = k$ in Equation (7.17). Then, substituting Equation (7.17) into Equation (7.23) yields the decay rate in terms of the ultimate BOD of the mixed water occurring at the point of mixing:

$$r_d = k_d L_o e^{-k_d t} \qquad (7.24)$$

Note that the rate constant k_d must still be corrected for temperature if the water is not at 20°C. The assumption that $k_d = k$ is generally appropriate for slow-moving bodies of water, but it is a poor assumption for a shallow, rapidly moving stream. Additional values for k_d are reported in the literature for different bodies of water.

Reaeration is the dissolution of molecular oxygen into water. The rate of reaeration of a body of water is related to the difference between the saturated oxygen level and the actual amount of oxygen in solution, and the temperature, velocity, depth, and turbulence of water flow in the stream channel. The reaeration rate, r_r, is directly proportional to the oxygen deficit, D. The gas diffusion rate into a liquid increases as the concentration of the gas in solution decreases. The oxygen deficit represents the difference between the saturation level and the actual oxygen level in solution as stated in Equation (7.2); therefore,

$$r_r = k_r D \tag{7.25}$$

where k_r is the reaeration rate constant (in day^{-1}). Several empirical correlations can be found in the literature that relate the reaeration rate to the properties of the stream or river. O'Connor and Dobbins (1958) reported one correlation indicating that the reaeration rate increases with increasing stream velocity, u, and decreases as the average depth, H, increases:

$$k_r = \frac{3.9u^{0.5}}{H^{1.5}} \tag{7.26}$$

where

k_r = reaeration rate constant (day^{-1})
u = average stream velocity (m/s)
H = average stream depth (m)

The equation developed by O'Connor and Dobbins for streams and rivers was limited in depth and velocity. Churchill et al. (1962) and Owens et al. (1964) extended this range, and Covar (1976) compiled these relationships into Figure 7.33, which allows for an estimation of the reaeration coefficient across a wide range of depths and velocities.

The change in the oxygen deficit, dD/dt, as the water moves downstream from the point of discharge is equal to the rate of deoxygenation minus the rate of reaeration:

$$\frac{dD}{dt} = k_d L_o e^{-k_d t} - k_r D \tag{7.27}$$

The integration of the two curves yields the shape of the curves shown in Figure 7.34. If the organic waste concentration in the stream is relatively high, then the rate of deoxygenation is greater than the rate of reaeration, and the dissolved oxygen concentration in the stream will decrease with time. However, eventually the organic waste concentration decreases, subsequently decreasing the rate of deoxygenation at the same time that the oxygen deficit has been increasing, which increases the rate of reaeration. The point at which the rate of deoxygenation is equal to the rate of reaeration is the critical point at which the lowest oxygen concentration is expected in the stream or river, as shown in Figure 7.35. After the critical point, the rate of reaeration begins to decrease as the oxygen deficit decreases, and the rate of deoxygenation continues to

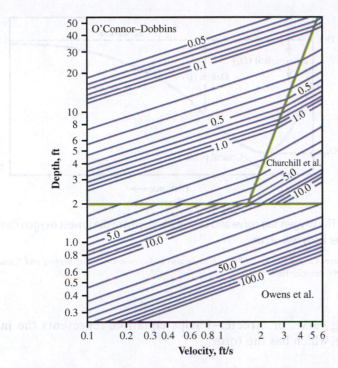

FIGURE 7.33 Reaeration correlations compiled by Covar (1976) for estimating the reaeration coefficient (day⁻¹) at a specific depth and velocity.

Source: EPA kinetic and constants report.

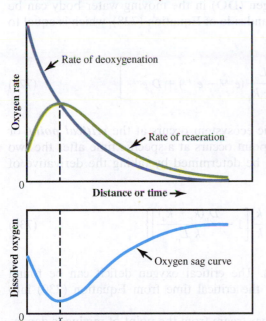

FIGURE 7.34 The individual curves representing the rate of deoxygenation and the rate of reaeration in a moving body of water into which an organic waste has been introduced. The combined curves form the oxygen sag curve.

Source: Based on Masters, G., and Ela, W. P. (2007). *Introduction to Environmental Engineering and Science, 3rd ed.* Upper Saddle River, NJ: Prentice Hall.

decrease along with decreasing levels of organic waste in the water. The plot of the combined deoxygenation and reaeration curves is called the oxygen sag curve. The curve is often also referred to as the Streeter–Phelps curve, named after the scientists who first described this process along the Ohio River in 1914.

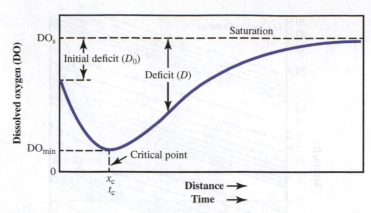

FIGURE 7.35 The oxygen sag curve and critical point, where the lowest oxygen concentration in the moving water body will occur.

Source: Based on Masters, G., and Ela, W. P. (2007). *Introduction to Environmental Engineering and Science, 3rd ed.* Upper Saddle River, NJ: Prentice Hall.

The oxygen sag curve, or Streeter–Phelps equation, represents the integral of Equation (7.27), which has the form

$$D = \frac{k_d L_o}{k_r - k_d}(e^{-k_d t} - e^{-k_r t}) + D_o e^{-k_r t} \tag{7.28}$$

The actual level of dissolved oxygen (DO) in the moving water body can be determined by substituting the right-hand side of Equation (7.28), which is equal to the deficit, into Equation (7.2):

$$DO = DO_s - \left[\frac{k_d L_o}{k_r - k_d}(e^{-k_d t} - e^{-k_r t}) + D_o e^{-k_r t} \right] \tag{7.29}$$

The greatest impact on the aquatic ecosystem occurs at the **critical point**, or maximum oxygen deficit. The critical point occurs at a specific time after the two streams have mixed together and can be determined by taking the derivative of Equation (7.29), which yields

$$t_c = \frac{1}{k_r - k_d} \ln\left\{ \frac{k_r}{k_d}\left[1 - \frac{D_o(k_r - k_d)}{k_d L_o} \right] \right\} \tag{7.30}$$

where t_c is the critical time (in days). The critical oxygen deficit can be found by substituting the resulting value for the critical time from Equation (7.30) into Equation (7.28).

The point, at some distance, x_c, downstream from the point of mixing is dependent on the average velocity of the stream or river:

$$x_c = u(t_c) \tag{7.31}$$

where u is the average stream velocity (m/day), and x_c (m) is the distance downstream from the mixing point where the critical deficit will occur.

EXAMPLE 7.3   A Mass Balance Approach to Modeling the Impact of Organic Waste on the Carbonaceous Oxygen Demand

In 1914, Streeter and Phelps (1925) studied the effects of organic waste loading on the oxygen content of the Ohio River. The researchers collected data over the course of a year over long stretches of the Ohio River. Their data have been modified slightly to suit this example, but are representative of the actual data they collected in 1914 (and sometimes their actual data are used). Streeter and Phelps published the completed report in Public Health Bulletin 146: A Study of the Pollution and Natural Purification of the Ohio River, by the United States Public Health Service in 1925. The long-term trends in the summer flow rate for the Ohio River are shown in Figure 7.36.

- Part I: Using the data set collected for July 1914 (illustrated in Figure 7.32), we will calculate the critical time and maximum oxygen deficit expected in the Ohio River. We will also plot the oxygen sag curve downstream from the point of mixing. Using the model, we will determine at what distance downstream the critical point occurs and where there might be a negative impact on the water quality in the Ohio River.
- Part II: Assuming that 90% of the BOD in the Ohio River and tributary has been removed as a result of passage of the Clean Water Act in 1972 and also assuming that present-day conditions in the Ohio River are as shown in Figure 7.32, we will calculate the critical time and maximum oxygen deficit expected in the Ohio River. The oxygen sag curve downstream from the point of mixing will also be shown graphically. The graphical analysis will show at what distance downstream the critical point occurs and where there might be a negative impact on the water quality in the Ohio River.

Part I:

The first step is to use conservative mass balance to determine the volumetric flow rate, organic waste loading rate, initial dissolved oxygen, and temperature that result from the two tributaries mixing together (see Figure 7.32).

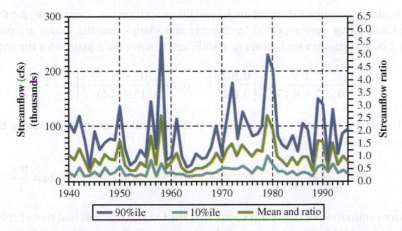

FIGURE 7.36 Long-term trends in mean, 10th, and 90th percentile statistics computed for summer (July–September) streamflow in the Ohio River at Louisville, Kentucky.

Source: U.S. EPA (2000). *Progress in Water Quality: An Evaluation of the National Investment in Municipal Wastewater Treatment.* U.S. Environmental Protection Agency. Office of Water. Washington, DC. EPA-832-R-00-008. Chap. 11.

From Equation (7.21),

$$Q_{downstream} = 1.10 \times 10^8 + 1.22 \times 10^7 = 1.22 \times 10^8 \ \frac{m^3}{day}$$

From Equation (7.22),

$$Q_{up}L_{o_{up}} + Q_{trib}L_{o_{trib}} = Q_{downstream}L_{o_{downstream}}$$

Rearranging yields

$$L_{o_{downstream}} = \frac{Q_{up}L_{o_{up}} + Q_{trib}L_{o_{trib}}}{Q_{downstream}} = \frac{(1.10 \times 10^8)(2.73) + (1.22 \times 10^7)(352.56)}{1.22 \times 10^8} = 41.6 \ \frac{mg}{L}$$

Similarly, for the initial dissolved oxygen:

$$DO_{o_{downstream}} = \frac{Q_{up}DO_{o_{up}} + Q_{trib}DO_{o_{trib}}}{Q_{downstream}} = \frac{(1.10 \times 10^8)(1.4) + (1.22 \times 10^7)(0.32)}{1.22 \times 10^8} = 1.28 \ \frac{mg}{L}$$

And the temperature of the Ohio River after mixing was

$$T_{o_{downstream}} = \frac{Q_{up}T_{o_{up}} + Q_{trib}T_{o_{trib}}}{Q_{downstream}} = \frac{(1.10 \times 10^8)(24.6) + (1.22 \times 10^7)(30.6)}{1.22 \times 10^8} = 25.8 \ \frac{mg}{L}$$

The saturated dissolved oxygen concentration for water at 25.8°C may be determined from either Henry's law value corrected for temperature or the nomograph in Figure 7.22. A straight line drawn from the temperature through the 100% saturation level indicates that the DO_s concentration would be 8.1 mg/L, as shown in Figure 7.37.

The initial deficit is determined from Equation (7.2):

$$D_o = 8.1 - 1.3 = 6.8 \ \frac{mg}{L}$$

The deoxygenation rate constant, k_d, and the reaeration rate constant, k_r, for this stretch of the Ohio River were reported by Streeter and Phelps, and the values are provided in Figure 7.38. The known variables are now sufficient to solve the equation for the critical time:

$$t_c = \frac{1}{0.202 - 0.131} \ln\left[\frac{0.202}{0.131}\left(1 - (6.8)\frac{0.202 - 0.131}{0.131(41.6)}\right)\right] = 4.8 \text{ days}$$

The maximum deficit can be found at the critical point by substituting the above results into Equation (7.28):

$$D_c = \frac{0.131(41.6)}{0.202 - 0.131}(e^{-0.131(4.8)} - e^{-0.202(4.8)}) + 6.8(e^{-0.202(4.8)}) = 14.4 \ \frac{mg}{L}$$

An examination of this result shows that the Ohio River was bad news for fish in the summer of 1914. Closer inspection of the maximum deficit and the predicted dissolved oxygen concentration reveals one aspect that the basic mathematical equation does not account for, as shown below:

$$DO_c = DO_s - D_c = 8.1 - 14.4 = -6.3 \ \frac{mg}{L}$$

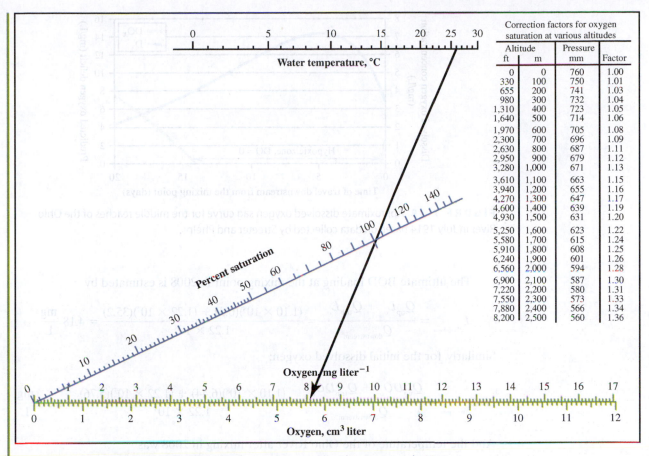

Correction factors for oxygen saturation at various altitudes

Altitude ft	Altitude m	Pressure mm	Factor
0	0	760	1.00
330	100	750	1.01
655	200	741	1.03
980	300	732	1.04
1,310	400	723	1.05
1,640	500	714	1.06
1,970	600	705	1.08
2,300	700	696	1.09
2,630	800	687	1.11
2,950	900	679	1.12
3,280	1,000	671	1.13
3,610	1,100	663	1.15
3,940	1,200	655	1.16
4,270	1,300	647	1.17
4,600	1,400	639	1.19
4,930	1,500	631	1.20
5,250	1,600	623	1.22
5,580	1,700	615	1.24
5,910	1,800	608	1.25
6,240	1,900	601	1.26
6,560	2,000	594	1.28
6,900	2,100	587	1.30
7,220	2,200	580	1.31
7,550	2,300	573	1.33
7,880	2,400	566	1.34
8,200	2,500	560	1.36

FIGURE 7.37 Determining the DO_s value for conditions in the Ohio River in July 1914.

Source: Based on Mortimer, C.H. (1981). "The oxygen content of air-saturated fresh waters over ranges of temperature and atmospheric pressure of limnological interest." *Mitt. int. Ver. Limnol.* pp. 22–23.

The mathematical theory predicts a negative dissolved oxygen concentration. However, it is not possible in practice to have a negative DO concentration; instead, the DO concentration would be zero. The initial DO concentration in the Ohio River was quite low to begin with, and shortly after mixing with the higher concentration waste-containing tributary water, the water turned anaerobic, as illustrated in the oxygen sag curve for this system shown in Figure 7.38. If these data are truly representative of the Ohio River in 1914, and all indications point to that, the Ohio River would have been devoid of any fish and the ecosystem would have been very unhealthy for both aquatic organisms and humans.

Part II:

What has happened in the Ohio River since 1914? Collection and treatment of sewage water and industrial wastewater began in earnest in the early 1900s in the United States. As a result, significant changes in water quality in the Ohio River have occurred. It is estimated that the BOD levels in the water have decreased by approximately 90% as a result of treating municipal and industrial wastewater as part of the Clean Water Act. In July 2008, upstream dissolved oxygen levels were measured and were found to be significantly higher than in 1914. The water temperature in 2008 was similar to that in 1914. For the example below, we will assume that the same average flow rates occur in the river and tributary and that the deoxygenation and reaeration constants also are the same as those Streeter and Phelps used in 1914 (as shown in Figure 7.32).

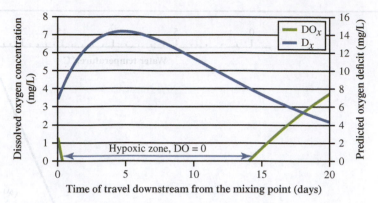

FIGURE 7.38 Approximate dissolved oxygen sag curve for the middle reaches of the Ohio River in July 1914 based on data collected by Streeter and Phelps.

The ultimate BOD loading at the mixing point in 2008 is estimated by

$$L_{o_{downstream}} = \frac{Q_{up}L_{o_{up}} + Q_{trib}L_{o_{trib}}}{Q_{downstream}} = \frac{(1.10 \times 10^8)(0.3) + (1.22 \times 10^7)(35.2)}{1.22 \times 10^8} = 4.18 \frac{mg}{L}$$

Similarly, for the initial dissolved oxygen:

$$DO_{o_{downstream}} = \frac{Q_{up}DO_{o_{up}} + Q_{trib}DO_{o_{trib}}}{Q_{downstream}} = \frac{(1.10 \times 10^8)(6.44) + (1.22 \times 10^7)(5.76)}{1.22 \times 10^8} = 6.36 \frac{mg}{L}$$

And the temperature of the Ohio River after mixing in 2008 was

$$T_{o_{downstream}} = \frac{Q_{up}T_{o_{up}} + Q_{trib}T_{o_{trib}}}{Q_{downstream}} = \frac{(1.10 \times 10^8)(25.0) + (1.22 \times 10^7)(26.1)}{1.22 \times 10^8} = 25.1 \frac{mg}{L}$$

The saturated dissolved oxygen can be determined from the nomograph and is 8.2 for the 2008 conditions. The initial deficit is determined from Equation (7.2):

$$D_o = 8.2 - 6.36 = 1.8 \frac{mg}{L}$$

The known variables are now sufficient to solve the equation for the critical time:

$$t_c = \frac{1}{0.202 - 0.131} \ln\left[\frac{0.202}{0.131}\left(1 - (1.8)\frac{0.202 - 0.131}{0.131(4.16)}\right)\right] = 2.2 \text{ days}$$

The maximum deficit can be found from Equation (7.28):

$$D_c = \frac{0.131(4.16)}{0.202 - 0.131}(e^{-0.131(2.2)} - e^{-0.202(2.2)}) + 1.8(e^{-0.202(2.2)}) = 2.0 \frac{mg}{L}$$

The minimum expected dissolved oxygen content is

$$DO_c = DO_s - D_c = 8.2 - 2.0 = 6.2 \frac{mg}{L}$$

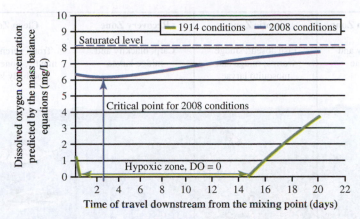

FIGURE 7.39 A comparison of expected dissolved oxygen levels based on reported conditions in 1914 and 2008 in the middle section of the Ohio River.

Figure 7.39 illustrates that the conditions in the Ohio River are much more conducive than in 1914 for both aquatic life and human use. Actual measurements of dissolved oxygen in 2008 show minimum oxygen levels of about 5.6 mg/L, which is very close to the levels predicted by the model Streeter and Phelps developed, since we have not accounted for additional BOD loading to the river downstream of this single tributary. The model also does not account for contributions to oxygen demand from nitrogen-containing compounds or materials found in the sediments in the river. As Figure 7.40 clearly shows, it is quite apparent that the Clean Water Act regulations and subsequent wastewater treatment processes realized in the watershed profoundly improved the water quality in the Ohio River over the past century. The overall aquatic effects of the DO curve are summarized in Figure 7.41.

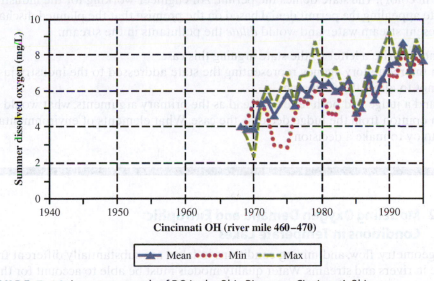

FIGURE 7.40 Long-term trends of DO in the Ohio River near Cincinnati, Ohio.

Source: U.S. EPA (2000). *Progress in Water Quality: An Evaluation of the National Investment in Municipal Wastewater Treatment.* U.S. Environmental Protection Agency. Office of Water. Washington, DC. EPA-832-R-00-008. Chap. 11.

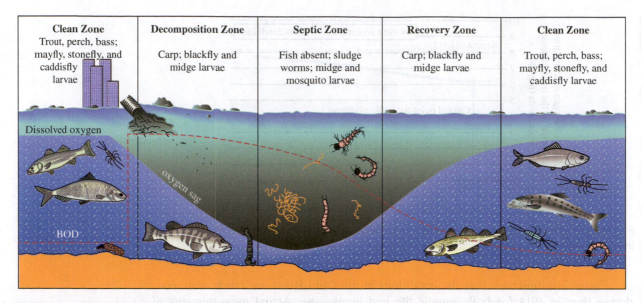

Clean Zone	Decomposition Zone	Septic Zone	Recovery Zone	Clean Zone
Trout, perch, bass; mayfly, stonefly, and caddisfly larvae	Carp; blackfly and midge larvae	Fish absent; sludge worms; midge and mosquito larvae	Carp; blackfly and midge larvae	Trout, perch, bass; mayfly, stonefly, and caddisfly larvae

Dissolved oxygen

oxygen sag

BOD

FIGURE 7.41 Aquatic and ecosystem impacts of oxygen demand, deoxygenation, and reaeration.

Source: Based on Davis, M.L., and Masten, S.J. (2009). *Principles of Environmental Engineering and Science, 2nd ed.* New York: McGraw-Hill.

ACTIVE LEARNING EXERCISE 7.2 Regulatory Ethics

An industry applies to the state for a discharge permit into a highly polluted stream (zero DO, a foul stench, oil slicks on the surface, black in color). The state denies the permit. An engineer working for the industry is told to write a letter to the state appealing the permit denial based on the premise that the planned discharge is actually *cleaner* than the present stream water and would *dilute* the pollutants in the stream.

a. Imagine you are the engineer. Write a letter to the state arguing this case.
b. Write a return letter from the regulatory agency representing the state addressed to the industry justifying the state's decision not to allow the discharge.
c. If the case went to court and a judge had both letters to read as the primary arguments, what would be the outcome? Write an opinion from the judge deciding the case. What elements of environmental ethics might the judge employ to make a decision?

7.3.2 Modeling Oxygen Demand and Eutrophic Conditions in Temperate Lakes

The geometry, flow, and mixing conditions in lakes are substantially different from those in rivers and streams. Water quality models must be able to account for those differences. The structure, or **morphology**, of a typical lake is shown in Figure 7.42. The important aspects of lake morphology related to water quality models are the regions where light penetrates the lake and where mixing with air occurs. Both light penetration and mixing are greater at the surface and decrease with depth. The light that penetrates the water also increases the water temperature and, as

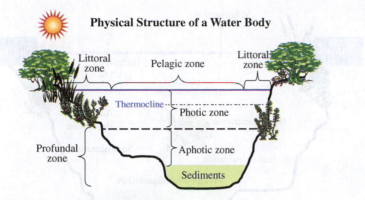

FIGURE 7.42 Physical structure of a typical lake.

the temperature of the water changes, so too does the density of the water, as illustrated in Figure 7.43. Water density and temperature differences create layers of water with different characteristics, or ***thermal stratification***, in a lake.

Thermal stratification affects the biological, chemical, and physical processes that occur in the water body. Mass balance, energy balance, and equilibrium models can be used to model the effects of the processes shown in Figure 7.44, as they occur in different strata in a lake or water body.

Oxygen distribution is correlated with light penetration and thermal differences in lakes. Oxygen is produced in the photic zone through photosynthesis. Oxygen at the

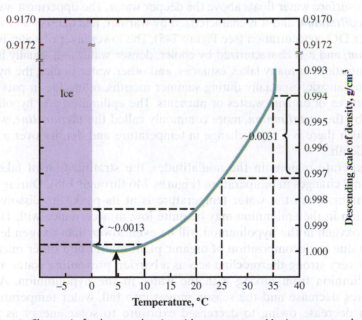

FIGURE 7.43 Change in freshwater density with temperature. Maximum water density occurs at about +4°C. Ice is much less dense than water and thus floats. Similarly, warmer water in the epilimnion may float over denser, cooler water in the hypolimnion. The rate of density change is also important to the thermal expansion of water due to global climate change—note the high rate of change in density of water between +15 and +25°C.

Source: Based on Goldman, C.R., and Horne, A. J. (1983). *Limnology*. New York: McGraw-Hill.

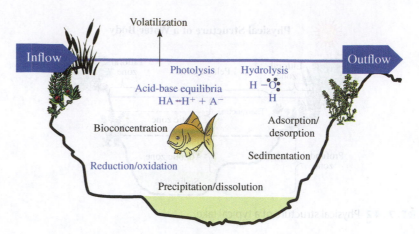

FIGURE 7.44 Biological, chemical, and physical processes that must be considered in modeling the water quality of a lake or water body.

surface of a water body is in equilibrium, or saturated, with oxygen concentrations in the atmosphere. The depth to which the oxygen level is saturated in the water body is dependent on the amount of mixing due to turbulence and the rate of deoxygenation due to microbial degradation of organic wastes. Oxygen is consumed in the aphotic zone through respiration and degradation of the organic wastes.

Light penetrating the water leads to plant growth, photosynthesis, and heating of the water. The warmer water in a lake is less dense than the cooler, deeper water, and the warmer surface water floats above the deeper water. The uppermost water layer is called the *epilimnion*, and it is characterized by warmer, less dense water and usually has a higher DO concentration (see Figure 7.45). The lower layer of water is called the *hypolimnion*, and it is characterized by cooler, denser water and usually has a lower DO concentration. In many lakes, estuaries, and other water bodies, the hypolimnion may become anoxic, especially during summer months, often due in part to anthropogenic sources of organic wastes or nutrients. The epilimnion and hypolimnion are separated by the *metalimnion*, more commonly called the *thermocline*, which is the layer in which there is a rapid change in temperature and density over a very small change in depth.

In temperate regions in the midlatitudes, the stratification of lakes changes with seasonal changes in temperature (Figures 7.46 through 7.48). During the height of the summer, when the water temperature is at its peak, the dissolved oxygen concentration in the epilimnion may be quite low, in accordance with Henry's law. Dissolved oxygen in the hypolimnion will be even lower than oxygen levels in the epilimnion due to decomposition of organic plant matter and other microbial processes. The very strong thermocline acts as a barrier, preventing water and oxygen in the epilimnion from mixing with the water in the hypolimnion. As summer temperatures decrease and the season moves into fall, water temperatures in the epilimnion decrease owing to decreased exposure to solar energy as days grow shorter (Figure 7.46). Lower water temperatures coincide with increasing oxygen levels in the epilimnion. However, the strong thermocline still prevents mixing between the epilimnion and hypolimnion, and the dissolved oxygen concentration in the hypolimnion becomes even lower. Fall is when hypolimnion DO levels are typically at their lowest, possibly forcing many fish species into shallower waters in the epilimnion to avoid oxygen deprivation.

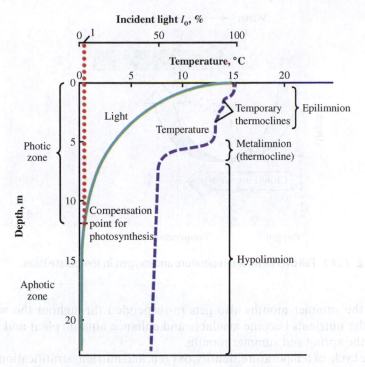

FIGURE 7.45 The relationship between light penetration, water temperature, and depth in a typical stratified lake or water body.

Source: Based on Davis, M.L., and Masten, S.J. (2009). *Principles of Environmental Engineering and Science, 2nd ed.* New York: McGraw-Hill.

As the temperature of the epilimnion water cools during the fall, the density difference between the epilimnion and hypolimnion becomes negligible, and the stratified layers disappear. The entire lake water may then "overturn," representing a well-mixed system for the entire depth of the water body (see Figure 7.47). Thus, the oxygen concentration increases in the deeper water and is equal throughout the water column. Some of the nutrients that settled into the deeper, denser water

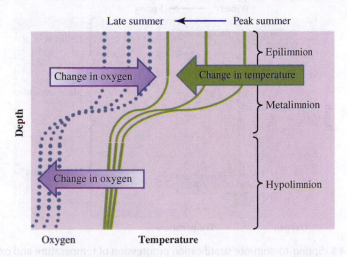

FIGURE 7.46 Summer-to-fall stratification of temperature and oxygen in temperate lakes.

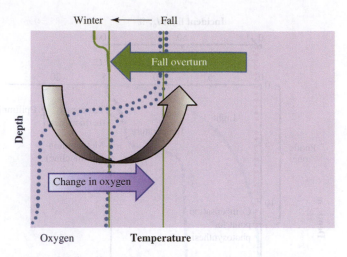

FIGURE 7.47 Fall overturns of temperature and oxygen in temperate lakes.

during the summer months also gets re-suspended throughout the water column, where the nutrients become available and enhance aquatic plant and algae growth during the spring and summer months.

The cycle of temperature, density, oxygen, and nutrient stratification begins again as the lake's water temperature increases in spring. As the thermocline develops, plants begin photosynthesis and oxygen levels remain quite high in the cool spring water in the epilimnion. As the thermocline is established and the epilimnion waters warm, dissolved oxygen levels begin to decrease in the hypolimnion (Figure 7.48). If excess nutrients are added to the water body, plant growth accelerates; microorganisms in the hypolimnion and sediments consume oxygen at a higher rate if more organic matter is present from the previous year's growing season.

The thermal stratification of lakes is dependent on climate systems. Climate changes are likely to affect the dates associated with the fall and spring turnover cycle. If the lake water does not cool enough, the lakes may remain permanently

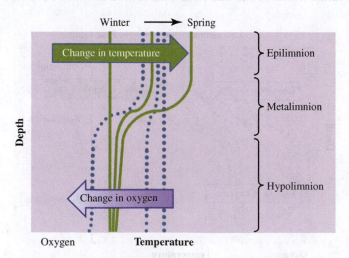

FIGURE 7.48 Spring-to-summer stratification progression of temperature and oxygen in temperate lakes.

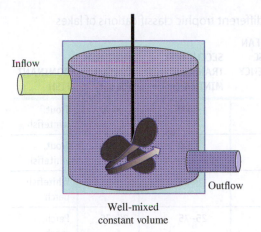

FIGURE 7.49 Constantly mixed flow reactor.

Inflow

Outflow

Well-mixed
constant volume

stratified, thereby significantly affecting oxygen and nutrient profiles through the water column in the lake.

The thermal stratification process of a lake or water body should be understood in order to properly represent the system with chemical and physical mathematical models. For instance, if the water in the lake is not thermally stratified during the fall-to-spring overturn, then the entire lake may be represented as a constantly mixed flow reactor (CMFR), as illustrated in Figure 7.49. (The constantly stirred tank reactor [CSTR] is another common name used for the type of reactor shown in Figure 7.49.) During the period that the lake is thermally stratified, the volume of the lake representative of the epilimnion may still be modeled as if it were a CMFR. However, the hypolimnion should be modeled as a constant-volume batch reactor without significant flow or mixing into the volume of the hypolimnion (Figure 7.50).

Lakes may be classified based on their trophic status, as determined by the organic matter (usually represented by chlorophyll) or the nutrient concentration in the water column (Table 7.16). Oligotrophic lakes are characterized by low primary productivity, low levels of biomass, low N and P concentrations, and cool, clear water high in dissolved oxygen concentrations. Eutrophic and hypereutrophic lakes contain a large amount of plant matter that has high nutrient concentrations in the water column, low visibility, and a low dissolved oxygen concentration. The nutrients in lakes may be derived from internal recycling of organic matter or from external

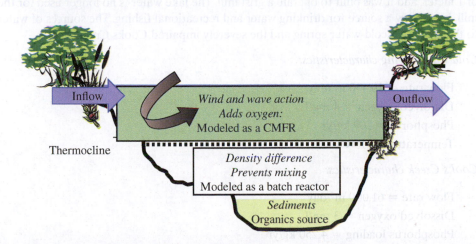

Inflow

Wind and wave action
Adds oxygen:
Modeled as a CMFR

Outflow

Thermocline

Density difference
Prevents mixing
Modeled as a batch reactor

Sediments
Organics source

FIGURE 7.50 Idealized model of a stratified lake.

TABLE 7.16 Typical nutrient levels, biomass, and productivity for different trophic classifications of lakes

TROPHIC CATEGORY	MEAN TOTAL PHOSPHORUS (mg/m³)	ANNUAL MEAN CHLOROPHYLL (mg/m³)	CHLOROPHYLL MAXIMA (mg/m³)	ANNUAL MEAN SECCHI DISC TRANSPARENCY (m)	SECCHI DISC TRANSPARENCY MINIMA (m)	MINIMUM OXYGEN (% sat)[a]	DOMINANT FISH
Ultra-oligotrophic	<4.0	1.0	2.5	12.0	6.0	<90	Trout, whitefish
Oligotrophic	4.0–10.0	2.5	8.0	8.0	8.0	8.0	Trout, whitefish
Mesotrophic	10–35	2.5–8	8–25	8–25	8–25	8–25	Whitefish, perch
Eutrophic	35–100	8–25	25–75	25–75	25–75	25–75	Perch, roach
Hypereutrophic	>100.0	25.0	75.0	75.0	75.0	75.0	Roach, bream

[a] % saturation in bottom waters depending on mean depth.

Source: Based on Chapman, D. (1996). *A Guide to Use of Biota, Sediments and Water in Environmental Monitoring, 2nd ed.*, Chap. 7, Lakes. UNESCO/WHO/UNEP.

inputs of organic matter, nitrogen-containing wastes, or phosphorus-containing wastes. Depending on nitrogen and phosphorus levels in the water, either nutrient could limit plant growth and eutrophication. Generally, in water with a nitrogen-to-phosphorus ratio (N/P) greater than 7, phosphorus is the limiting nutrient; if the ratio is less than 7, nitrogen is the limiting nutrient. Accelerated eutrophication may significantly degrade the water quality in a lake and limit the water uses for drinking, fishing, or recreation.

EXAMPLE 7.4 Phosphorus Loading and Classification of Greengots Lake

Greengots Lake in this example is based on a real 48,000-square-meter lake in a temperate region of the United States. The shallow lake is a human-made lake with an average depth of 1 meter, and it was built to operate a grist mill. The lake water is no longer used for the mill, but is now a source for drinking water and recreational fishing. The sources of water to the lake are a cold-water spring and the severely impaired Cooks Creek.

Cold water spring characteristics:

Flow rate = 10,600 m³/day
Dissolved oxygen = 4 mg/L
Phosphorus = 209 kg/yr
Temperature = 15°C

Cooks Creek characteristics:

Flow rate = 61,000 m³/day
Dissolved oxygen = 3 mg/L
Phosphorus loading = 4,250 kg/yr
Temperature = 23°C

Other phosphorus loads:

> Row crops from surrounding farmland = 7,410 kg/yr
> Pasture/hay = 700 kg/yr
> Undeveloped areas = 227 kg/yr
> Urban drainage = 1,020 kg/yr
> Septic drainage = 200 kg/yr

Calculate:

- Outflow from the lake
- Hydraulic residence time in the lake
- Steady state phosphorus concentration in the lake

Assume evaporation = precipitation and that the phosphorus settling rate can be estimated from an empirical correlation developed by Reckhow and Chapra (1983):

$$v_s \left[\frac{m}{yr}\right] = 11.6 + 0.2 \left(\frac{Q\left[\frac{m^3}{yr}\right]}{A\,[m^2]}\right) \tag{7.32}$$

$$Q_{out} = Q_{spring} + Q_{creek} + Q_{precipitation} - Q_{evaporation}$$

Since $Q_{precipitation} = Q_{evaporation}$, we have

$$Q_{out} = 10,600\,\frac{m^3}{day} + 61,000\,\frac{m^3}{day} = 71,600\,\frac{m^3}{day}$$

The average hydraulic retention time in the lake is

$$t_r = \frac{V}{Q} = \frac{48,000\,m^2 \times 1\,m}{71,000\,\frac{m^3}{day}} = 0.67\,day$$

The removal rate of the phosphorus settling rate constant is determined from

$$v_s\left[\frac{m}{yr}\right] = 11.6 + 0.2\left(\frac{Q\left[\frac{m^3}{yr}\right]}{A\,[m^2]}\right) = 11.6 + 0.2\left(\frac{71,600\left[\frac{m^3}{yr}\right]}{48,000\,[m^2]}\right) = 11.9\left[\frac{m}{yr}\right]$$

$$k_s = 11.9\,\frac{m}{yr} \times 1\,m = 11.9\,[yr^{-1}]$$

Assuming the lake is at steady state, we have:

$$\frac{dM}{dt} = 0 = (M_{in}) - (M_{out})$$

Rearranging and substituting the mass inputs on the right-hand side of the equation below, and the outputs on the left-hand side:

$$M_{spring} + M_{creek} + M_{crops} + M_{pasture} + M_{undeveloped} + M_{urban} + M_{septic} = M_{settling} + M_{outflow}$$

$$M_{spring} + M_{creek} + M_{crops} + M_{pasture} + M_{undeveloped} + M_{urban} + M_{septic} = k_s C_{lake} V + Q_{out} C_{lake}$$

$$209\ \frac{kg}{yr} + 4{,}250\ \frac{kg}{yr} + 7{,}410\ \frac{kg}{yr} + 700\ \frac{kg}{yr} + 227\ \frac{kg}{yr} + 1{,}020\ \frac{kg}{yr}$$

$$+ 200\ \frac{kg}{yr} = 11.9\ \frac{1}{yr} \times C_{lake} \times 48{,}000\ m^3 + 71{,}600\ \frac{m^3}{day} \left(365\ \frac{days}{yr}\right) C_{lake}$$

$$C_{lake} = \frac{14{,}016\ \dfrac{kg}{yr}}{26{,}705{,}000\ \dfrac{m^3}{yr}} \times 10^6\ \frac{mg}{kg} = 525\ \frac{mg}{m^3}$$

As Table 7.16 shows, the concentration in the lake is hypereutrophic. Indeed, the actual lake on which the example is based is listed as an impaired water body, and although in winter it is capable of supporting trout, in the summer any trout in the lake die and the only native species found in the lake during the summer months are a type of bream.

7.4 Water Treatment Technologies

Few diseases are spread by water in highly economically developed countries. In contrast, in less economically developed countries, a safe, reliable supply of water may be unavailable and a sanitation infrastructure is often absent. Environmental engineers apply the basic principles of science and engineering to design water treatment systems for drinking water and treating human and industrially contaminated wastewater. The engineer's goal is to design water treatment systems to protect public health while balancing environmental, economic, social, and political constraints. A proper water treatment design requires knowledge of the constituents of concern, the impact of these constituents, the transformation and fate of the constituents, treatment methods to remove or reduce the toxicity of the constituents, and methods to dispose of or recycle treatment byproducts.

Louis Pasteur explained the germ theory in the 1880s; however, sewer systems were used for centuries before the theory was scientifically understood to remove the miasma of odorous sewage containing water away from urban centers. Cholera and typhoid were endemic problems in urban areas, including England and the United States, throughout the 1800s. Cholera and typhoid outbreaks did not begin to decrease until chlorination for treating drinking water received widespread acceptance in 1908–1911. Mortality and water supply records kept by the Commonwealth of Massachusetts show how improved water filtration and chlorination led to a rapid decline in typhoid fever deaths (Figure 7.51). The United States Public Health Service adopted water quality-based standards in 1914 to improve water quality and reduce disease; this was the first of many policy measures the United States enacted over several decades to improve water quality (Figure 7.52).

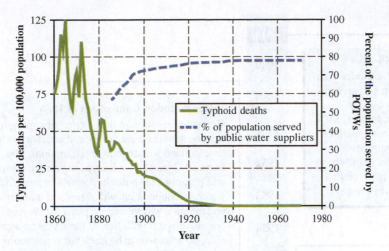

FIGURE 7.51 Comparison of death rates due to typhoid and the percentage of the population served by a treated public water supply in Massachusetts from 1860 to 1970.

Source: U.S. EPA (2000). Chapter 2 - An examination of BOD Loadings before and after the CWA. *Progress in Water Quality: An Evaluation of the National Investment in Municipal Wastewater Treatment.* U.S. Environmental Protection Agency. Office of Water. Washington, DC. EPA-832-R-00-008.

Water movements have been altered to benefit communities and urban areas. A schematic of the modified "urban water cycle" is shown in Figure 7.53. The public and industry consume water through withdrawals from groundwater or surface water. The water is treated at a treatment plant or at the point-of-use. Wastewater is generated in homes and in industrial processes. This water may be collected in a community sewer system, and the wastewater should be treated prior to discharge in a receiving body of water. Stormwater in urban areas may also be collected to minimize flooding in urban areas. In some cases, stormwater runoff is emptied directly into receiving waters, or the stormwater may pass through treatment processes dedicated to treating only the runoff; in many urban areas, stormwater and sewage systems are mixed together into *combined sewer overflow (CSO)* that may or may not be treated prior to discharge.

7.4.1 Water Treatment Technologies

Traditionally, engineers and hydrologists have used historical data and maximum possible demand (typically due to fire suppression flow rates) to make assumptions about the water supply. However, many developing communities and many large urban areas are located in regions of the world where water is scarce. Engineers now consider both the supply and demand of water resources. Thus, water engineers have begun to think of water as a limited resource and use water demand management strategies. *Water demand management* considers the total quantity of water abstracted from a source of supply using measures to control waste and excessive consumption. New approaches to water management consider efficiency, effectiveness, and demand management in each user sector, rather than simply providing for the maximum possible demand.

The highest quality water available should be used as the source of drinking water prior to treatment. Treatment technologies may be chemical or physical in nature and very simple or highly sophisticated, depending on the treatment needs and resources available. Groundwater from deep-well boreholes generally contains lower levels of microorganisms than surface waters (see Table 7.17) and may require less sophisticated treatment than surface water.

1948

1949

1950

Water Pollution Control Act of 1948

Water Pollution Control Act of 1948 authorized the U.S. Public Health Service to develop comprehensive basin plans for water pollution control and to encourage the adoption of uniform state laws. $100 million of loans were authorized to municipalities annually, but no appropriation for treatment facilities under this act was ever made. However, the act influenced the states to apply more control over the discharge of pollutants into their waters.

1951

1952

1953

1954

1955

1956

1957

1958

1959

1960

1961

1962

1963

1964

1965

1966

1967

1968

1969

1970

1971

Water Pollution Control Act of 1956

Grants for assisting in the construction of municipal treatment works were authorized and, for the first time, funded with federal appropriations. The Surgeon General was directed to prepare comprehensive programs for pollution control in interstate waters in cooperation with states and municipalities, and the state was to prepare plans for prevention and control of water pollution. If there was no approved plan, no grant was to be made for constructing treatment facilities. $50 million annually in grants was authorized. Grants were limited to 30% of the cost of construction, or $250,000, whichever was smaller. Legislation in the states increasingly required secondary treatment for polluted waters.

Water Pollution Control Act of 1961

Comprehensive programs and plans for water pollution abatement and control were still required. Grants were limited to 30% of the cost of construction or $600,000, whichever was less, or $2.4 million for multiple municipal plants. At least half of the appropriation was to go to cities of 125,000 or less. The Congress advocated 85% removal of pollutants in the hearings.

Water Quality Act of 1965

For the first time, each state was required to have water quality standards to receive grants, expressed as water quality criteria applicable to interstate waters. If the state did not develop standards, the Federal Water Pollution Control Administration (FWPCA) was required to do so. To comply with these standards and criteria, secondary treatment was increasingly necessary. Construction grants were raised to 30% of reasonable costs, and an additional 10% was allowed where the project conformed with a comprehensive plan for a metropolitan area. At least 50% of the first $100 million in appropriations had to go to municipalities of less than 125,000 population. Individual grants were limited to $1.2 million, with a limit of $4.8 million for multiple municipalities.

Clean Water Restoration Act of 1966

The requirements for state water quality standards were continued. Each state planning agency receiving a grant was to develop an effective, comprehensive pollution control plan for a basin. The FWPCA, in a guideline, attempted to require states to conform to a national uniform standard of secondary treatment or its equivalent. This action was challenged and the guideline was not enforced. Secretary Udall stated at House hearings that the states had agreed to the requirement for secondary treatment. Grants for POTWs are set at 30% with an increase to 40% if the state paid 30%. The maximum could be increased to 50% if the state agreed to pay 25%. A grant could be increased by 10% if it conformed to a comprehensive plan for the metropolitan area. The limit of $1.2 million and $4.8 million for grants was waived if the state matched equally all federal grants. At least 50% of the first $100 million in annual appropriations had to be directed to municipalities of <125,000 people.

Water Quality Improvement Act of 1970

The Water Quality Improvement Act of 1970 did not contain any new provisions regarding required standards. The requirements for state water quality standards were continued. However, in hearings for the act, the authority of EPA to require uniform treatment limitations for discharges, such as secondary treatment, was questioned.

FIGURE 7.52 Timeline of major water quality initiatives in the United States from 1948 to 1996.

Source: U.S. EPA (2000). Chapter 2 - An examination of BOD Loadings before and after the CWA. *Progress in Water Quality: An Evaluation of the National Investment in Municipal Wastewater Treatment*. U.S. Environmental Protection Agency. Office of Water. Washington, DC. EPA-832-R-00-008.

1972

1973

Federal Water Pollution Control Act Amendments of 1972

The Federal Water Pollution Control Act of 1972 (later to be renamed the Clean Water Act) contained the first statutory requirement for a minimum of secondary treatment by all Publicly owned treatment works (POTWs). The act also established the National Pollutant Discharge Elimination System (NPDES), under which every discharger of pollutants was required to obtain a permit. Under the permit, each POTW is to discharge only effluent that had received secondary treatment. EPA defined secondary treatment in a regulation as attaining an effluent quality of at least 30 mg/L BOD_5, 30 mg/L TSS, and 85% removal of these pollutants in a period of 30 consecutive days.

1974

1975

1976

1977

1978

1979

1980

1981

Clean Water Act Amendments of 1977

The Clean Water Act Amendments of 1977 created the 301(h) program, which waived the secondary treatment requirement for POTWs discharging to a marine environment if they could show that the receiving waters would not be adversely affected. Extensive requirements had to be met before such a waiver could be issued.

National Municipal Policy, January 30, 1984

The EPA National Municipal Policy was published on January 30, 1984. It was designed to ensure that all POTWs met the compliance deadlines for secondary or greater treatment of discharges. The key to the policy is that it provides for POTWs that had not complied by the July 1, 1988 deadline to be put on enforceable schedules. The policy has been outstandingly successful and has resulted in significant increases in compliance.

Clean Water Act Amendments of 1981, PL 97–117

The Clean Water Act Amendments of 1981 amended the Clean Water Act to the effect that "such biological treatment facilities as oxidation ponds, lagoons, ditches, and trickling filters shall be deemed the equivalent of secondary treatment." EPA is directed to provide guidance on design criteria for such facilities, taking into account pollutant removal efficiencies and assuring that water quality will not be adversely affected (Sec. 304(d)(4)). Regulations to this effect were published in final on September 20, 1984. Also, a notice was issued to solicit public comments on "problems related to meeting the percent removal requirements and on five options EPA was considering for amending the percent removal requirements."

1982

1983

1984

1985

1986

1987

1988

1989

1990

1991

1992

1993

1994

1995

1996

Secondary Treatment Regulations, June 3, 1985

The secondary treatment regulations published in final on June 3, 1985 revised the previous regulations published in Title 40, Part 133, of the Code of Federal Regulations. Specifically, on a 30-day average, the achievement of not less than 85% removal of BOD_5, $CBOD_5$, and suspended solids for conventional secondary treatment processes was required. However, for those treatment processes designated by the Congress as being equivalent to secondary treatment (such biological treatment facilities as oxidation ponds, lagoons, ditches, and trickling filters), at least 65% pollution removal was required, provided that water quality was not adversely affected. Waste stabilization ponds were given separate suspended solids limits. Special consideration was provided for various influent conditions and concentration limits.

Secondary Treatment Regulations, January 27, 1989

This secondary treatment regulation allows adjustments for dry weather periods for POTWs serving combined sewers.

FIGURE 7.52 *(Continued)*

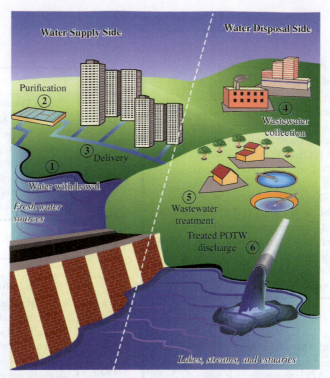

FIGURE 7.53 Simple schematic of the modified urban water cycle.

Source: U.S. EPA (2004). *Primer for Municipal Treatment Systems*.

TABLE 7.17 Fecal coliform levels in untreated domestic water sources in selected countries

COUNTRY	SOURCE	FECAL COLIFORMS/100 mL
Gambia	Open hand-dug wells, 15–18 m deep	Up to 100,000
Indonesia	Canals in central Jakarta	3,100–3,100,000
Lesotho	Streams	5,000
	Unprotected springs	900
	Water holes	860
	Protected springs	200
	Borehole	1
Uganda	Rivers	500–8,000
	Streams	2–1,000

Source: Based on UNICEF (2008). *UNICEF Handbook on Water Quality*. United Nations Children's Fund (UNICEF) New York, New York, USA.

7.4.2 Groundwater Resources

Groundwater is usually the preferred source for drinking water (also called **potable water**) because of its typically low concentration of microbial constituents. The water may be pumped from the groundwater aquifer (illustrated in Figure 7.54) into an elevated storage tank so that gravity can provide the force needed for fluid flow through the remaining treatment processes. A gravity flow-based system ensures that some water will be available even in the event of a power outage. The groundwater may be aerated in the storage tank if it contains elevated levels of dissolved gases, although this procedure is relatively uncommon. The water is typically passed through a granular filtration process, which can improve both the physical and microbial quality of the water. Numerous types of granular filters are available, including both slow and rapid sand filters.

Groundwater is derived from subsurface water, which may be either close to the surface or deep below the surface. The *vadose zone*, or *zone of aeration*, is near the surface of the Earth and the soil pore spaces contain both air and water. In some areas, such as swamplands, the depth to the water may approach zero, while in arid regions, the depth of the water may be several hundred feet. Moisture found in the zone of aeration is not a reliable supply of water because the water is held to the soil particles by capillary forces and is not easily released.

The area below the zone of aeration in which the void spaces are filled with water is called the *zone of saturation*, as illustrated in Figure 7.55. The surface of the zone of saturation is known as the *water table*, and the water found in this layer is called *groundwater*. A geologic stratum that contains a substantial amount of groundwater is an *aquifer*. Water that can flow from the surface into an aquifer that is underlain by an impervious stratum is an *unconfined aquifer*.

The term *confined aquifer* describes the groundwater trapped between two impervious layers, as shown in Figure 7.56. The water trapped in a confined aquifer can be pressurized by gravity in the aquifer, just like water in a pipe. If the pressurized water is tapped or comes to the surface, an *artesian well* results.

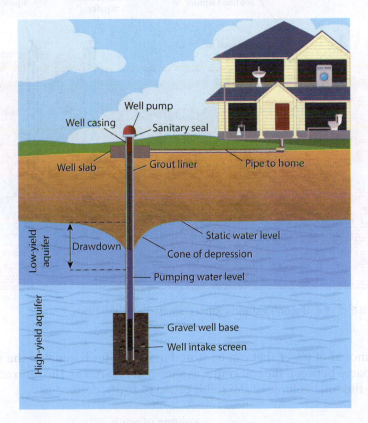

FIGURE 7.54 Groundwater sources and well apparatus.

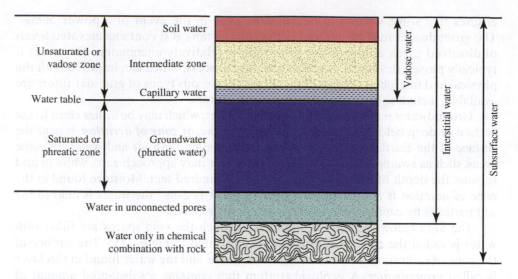

Soil water

Unsaturated or vadose zone — Intermediate zone

Water table — Capillary water

Saturated or phreatic zone — Groundwater (phreatic water)

Water in unconnected pores

Water only in chemical combination with rock

Vadose water

Interstitial water

Subsurface water

FIGURE 7.55 Schematic of groundwater layers under the ground surface.

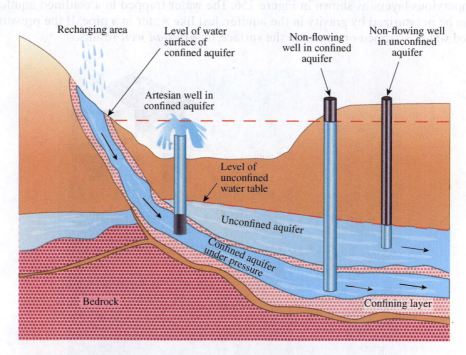

Recharging area

Level of water surface of confined aquifer

Non-flowing well in confined aquifer

Non-flowing well in unconfined aquifer

Artesian well in confined aquifer

Level of unconfined water table

Unconfined aquifer

Confined aquifer under pressure

Bedrock

Confining layer

FIGURE 7.56 Unconfined and confined aquifers and well types.

The amount of water that can be stored in an aquifer is equal to the volume of the void spaces between the soil grains, as shown in Figure 7.57. The ratio of the voids volume to the total volume of the soil is called *porosity*, defined as

$$\text{Porosity} = \frac{\text{volume of voids}}{\text{total volume}} \qquad (7.33)$$

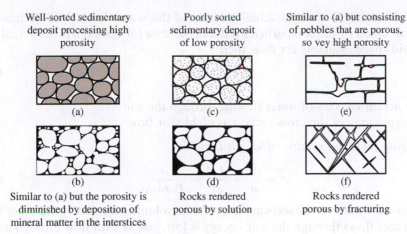

Well-sorted sedimentary deposit processing high porosity

(a)

Poorly sorted sedimentary deposit of low porosity

(c)

Similar to (a) but consisting of pebbles that are porous, so vey high porosity

(e)

(b)

(d)

(f)

Similar to (a) but the porosity is diminished by deposition of mineral matter in the interstices

Rocks rendered porous by solution

Rocks rendered porous by fracturing

FIGURE 7.57 Types of porosity in soil.

Source: Meinzer, O. E. (1923a). "The occurrence of ground water in the United States, with a discussion of principles." *U.S. Geol. Survey Water-Supply Paper 489*, p. 1–321, fig. 1–110, pl. 1–31.

A portion of this water is tightly bound to the soil particles. However, the amount of water that can be removed from the pore space is known as the *specific yield*:

$$\text{Specific yield} = \frac{\text{volume of water that will drain freely from a soil}}{\text{total volume of water in the soil}} \quad (7.34)$$

The flow of water out of a soil can be illustrated as in Figure 7.58. The flow rate must be proportional to the area through which the flow occurs multiplied by the velocity of the flow through the soil:

$$Q = v_s A \quad (7.35)$$

where

    Q = flow rate, m³/s

    A = total cross-sectional area of porous material, m²

    v_s = superficial velocity, m/s

Most of the cross-sectional area of soils is occupied by the soil particles; thus, only a portion of the total cross-sectional area is available for water flow. Likewise, the

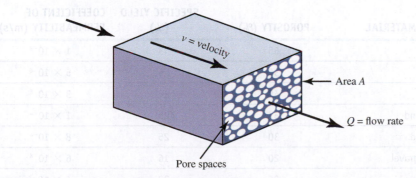

v = velocity

Area A

Q = flow rate

Pore spaces

FIGURE 7.58 Flow through a porous medium, such as soil.

superficial velocity is not the actual velocity of the water in the soil because the volume occupied by the soil particles reduces the area for flow. If a is the actual area of the void space available for flow, then

$$Q = Av = av'$$ (7.36)

where

v' = actual velocity of water flowing through the soil
a = void area of the cross section available for flow

Rearranging for a soil sample of length L yields

$$v' = \frac{Av}{a} = \frac{AvL}{aL} = \frac{v}{Porosity}$$ (7.37)

Since the total volume of the soil sample is AL, the volume occupied by the water is aL.

As water flows through the soil, energy is lost, just as with flow through a pipe. The energy loss per distance traveled is

$$\frac{dh}{dL}$$

where

h = energy, measured as the elevation of the water table in an unconfined aquifer or the pressure in a confined aquifer, m
L = horizontal distance in the direction of flow, m

In an unconfined aquifer, the drop in the elevation of the water table with distance is the slope of the water table, dh/dL, in the direction of flow. The elevation of the water surface is the potential energy of the water, and water flows from a higher energy state to a lower energy state. The Darcy equation relates the energy loss to the flow through a porous medium such as soil:

$$Q = KA\frac{dh}{dL}$$ (7.38)

where

K = coefficient of permeability, m/s
A = cross-sectional area, m^2

Different types of soils have different typical characteristics of groundwater flow, as shown in Table 7.18.

TABLE 7.18 Typical aquifer parameters

AQUIFER MATERIAL	POROSITY (%)	SPECIFIC YIELD (%)	COEFFICIENT OF PERMEABILITY (m/s)
Clay	55	3	1×10^{-6}
Loam	35	5	5×10^{-6}
Fine sand	45	10	3×10^{-5}
Medium sand	37	25	1×10^{-4}
Coarse sand	30	25	8×10^{-4}
Sand and gravel	20	16	6×10^{-4}
Gravel	25	22	6×10^{-3}

The Darcy equation illustrates that the greater the driving force (or energy available), the greater the flow. The coefficient of permeability, K, is an indirect measure of the ability of a soil to transmit water. It varies greatly for different soils—ranging from 0.05 m/day for clay to more than 5,000 m/day for gravel, as shown in Table 7.19.

When a well is drilled to gain access to an unconfined aquifer and water is pumped out, water in the aquifer will begin to flow toward the well, as illustrated in Figure 7.59. As the water approaches the well, the area through which it flows gets progressively smaller, thus producing a higher superficial (and actual) velocity. The removal of water at a higher velocity near the well results in a deficit of water near the well, which is called the *drawdown*. The lower level of water from the well to the water table is called the *cone of depression*. If the rate of water flowing toward the well is equal to the rate of withdraw, the system is at equilibrium and the drawdown remains constant. If the rate of withdraw is increased, a deeper drawdown and a deeper cone of depression result.

The Darcy equation may be applied to a cylindrical area with a radius r of withdraw that represents a well, as shown in Figure 7.60.

$$Q = KA \frac{dh}{dL} = K(2\pi rw) \frac{dh}{dr} \qquad (7.39)$$

TABLE 7.19 Range of values for hydraulic conductivity and permeability for selected soil types

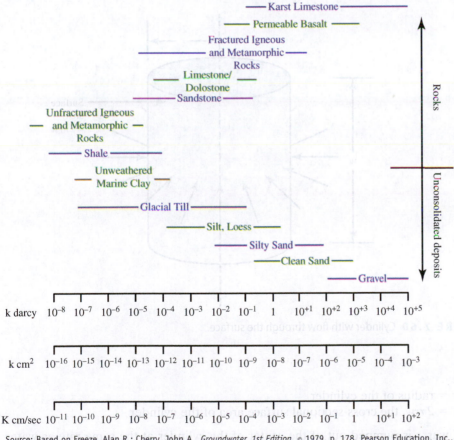

Source: Based on Freeze, Alan R.; Cherry, John A., *Groundwater, 1st Edition*, © 1979. p. 178. Pearson Education, Inc., Upper Saddle River, NJ.

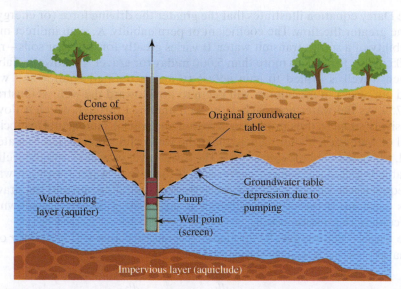

FIGURE 7.59 Representation of the drawdown curve in the water table due to pumping from a well.

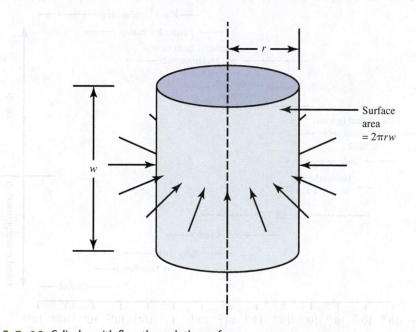

FIGURE 7.60 Cylinder with flow through the surface.

where

r = radius of the cylinder

$A = 2\pi r w$, the cross-sectional surface area of the cylinder

If water is pumped out of the center of the cylinder at the same rate as water is moving in through the cylinder's surface area, the depth of the cylinder through which the water flows into the well, w, can be replaced by the height of the water above the impermeable layer, h. The equation can be integrated to yield

$$\int_{r_2}^{r_1} Q \frac{dr}{r} = 2\pi K \int_{h_2}^{h_1} h\,dh$$

The integration between any two arbitrary values of r and h for steady state and uniform radial flow yields

$$Q = \frac{\pi k (h_1^2 - h_2^2)}{\ln \frac{r_1}{r_2}} \tag{7.40}$$

The equation describes the available water withdraw flow rate for a given drawdown any distance from a well in an unconfined aquifer by using water level measurements in two observation wells (as shown in Figure 7.61). This solution assumes that the aquifer is homogeneous and infinite and that the well penetrates

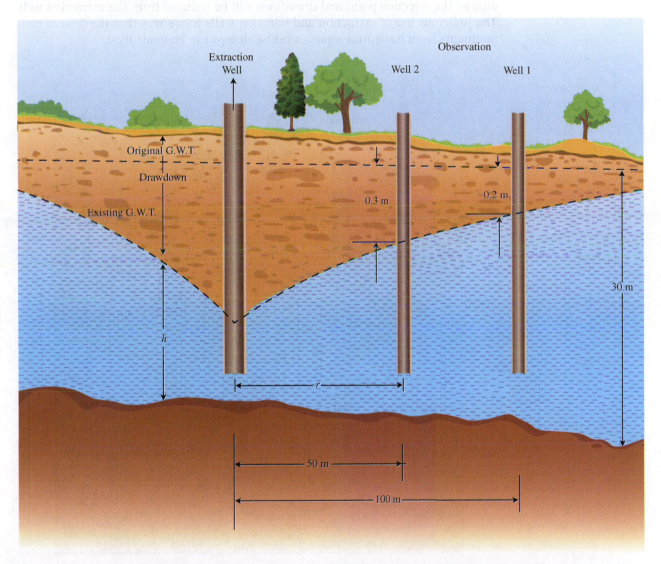

FIGURE 7.61 An example of multiple wells and the effect of extraction on the groundwater table (G.W.T.).

the entire aquifer and is open for the entire depth. For a well of a given diameter, we can estimate the critical point of the cone of depression and the drawdown of a well. If the drawdown is depressed all the way to the bottom of the aquifer, water will not be available and the well will go dry. Multiple wells in a single aquifer can have intersecting cones of depression; thus, one well can affect other nearby wells, as in Figure 7.62. If many wells are sunk into an aquifer, the combined effect of the wells could deplete the groundwater resources and potentially cause all the wells to go dry. This is a simple groundwater model based on several assumptions stated above. Modeling the flow of groundwater is complex and often requires sophisticated analysis for accurate predictions of the impact of wells on natural systems.

The reverse condition is also true for injection wells. Injection wells are sometimes used to return water to an aquifer for water recycling and are also associated with groundwater treatment of hazardous waste, advanced wastewater treatment plants, hydraulic fracturing of natural gas, and other uses. Suppose one of the wells shown in Figure 7.62 is used as an injection well; the injected water will build up around the injection point and drawdown will be reduced from the extraction well. The judicious use of extraction and injection wells is one way that the flow of contaminants from hazardous waste or refuse dumps can be controlled.

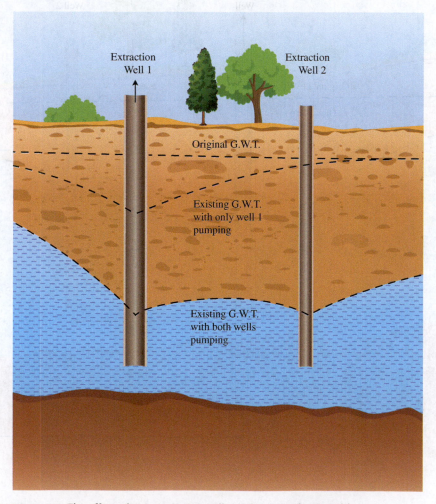

FIGURE 7.62 The effect of two extraction wells on the groundwater table.

EXAMPLE 7.5 Determining the Coefficient of Permeability

A soil sample is placed in a permeameter as shown in Figure 7.63. The length of the sample is 0.1 m, and it has a cross-sectional area of 0.05 m². The water pressure is 2.5 m on the upflow side and 0.5 m on the downstream side. A flow rate of 2.0 m³/day is observed. What is the coefficient of permeability?

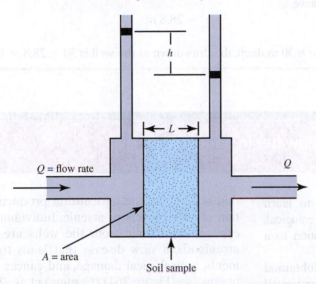

FIGURE 7.63 Permeameter used for measuring the coefficient of permeability according to the Darcy equation.

We will use Equation (7.38), the Darcy equation. The pressure drop is the difference between the upstream and downstream pressures, or $h = 2.5 - 0.5 = 2.0$ m. Solving for K, we have

$$K = \frac{Q}{A\dfrac{dh}{dL}} = \frac{2.0 \text{ m}^3/\text{day}}{0.05 \text{ m}^2 \times \dfrac{2 \text{ m}}{0.1 \text{ m}}} = 2 \text{ m/day} = 2 \times 10^{-5} \text{ m/s}$$

From Table 7.18, we see that the sample appears to contain fine sand.

EXAMPLE 7.6 Estimating Well Drawdown

A well is 0.2 m in diameter and pumps from an unconfined aquifer 30 m deep at an equilibrium (steady state) rate of 1,000 m³/day. Two observation wells are located at distances 50 m and 100 m from the well, and they have been drawn down by 0.3 m and 0.2 m, respectively. What are the coefficient of permeability and the estimated drawdown at the well? (See Figure 7.61.)

We can use Equation (7.40) with $h_1 = 30$ m $- 0.2$ m $= 29.8$ m and $h_2 = 30$ m $- 0.3$ m $= 29.7$ m:

$$K = \frac{Q \ln \dfrac{r_1}{r_2}}{\pi(h_1^2 - h_2^2)} = \frac{(1{,}000 \text{ m}^3/\text{day}) \ln \left(\dfrac{100 \text{ m}}{50 \text{ m}}\right)}{\pi[(29.8 \text{ m})^2 - (29.7 \text{ m})^2)]} = 37.1 \text{ m/day}$$

If the radius of the well is 0.2 m/2 = 0.1 m, this can be plugged into the same equation:

$$Q = \frac{\pi K(h_1^2 - h_2^2)}{\ln \frac{r_1}{r_2}} = \frac{\pi (37.1 \text{ m/day})[(29.7 \text{ m})^2 - h_2^2]}{\ln \frac{50 \text{ m}}{0.1 \text{ m}}} = 1{,}000 \text{ m}^3/\text{day}$$

Solving for h_2, we have

$$h_2 = 28.8 \text{ m}$$

Because the aquifer is 30 m deep, the drawdown at the well is 30 − 28.8 = 1.2 m.

BOX 7.3 Arsenic in the Groundwater in Bangladesh

As scientists and engineers, we continue to learn about the need to consider local cultural, geological, and logistical issues when designing a solution to a problem.

Until the 1960s, most Bangladeshis obtained their drinking and irrigation water from unimproved sources of bacteria-contaminated ponds, rivers, and shallow wells (Mukherjee and Bhattacharya, 2001). In an effort to reduce deaths from bacterial disease, the World Bank and UNICEF supported a project to dig 4.5 million tubewells, a technique that was effectively used in other countries, to provide drinking water to densely populated areas with no access to treated water (WHO, 2008). The well-drilling project succeeded in reducing diarrheal mortality from 250,000 in 1983 to 110,000 in 1996 (Mukherjee and Bhattacharya, 2001).

In Bangladesh, however, there are unusually high natural levels of arsenic in alluvial soils. A large number of diverse chemical and biological reactions, including oxidation, reduction, adsorption, precipitation, methylation, and volatilization, participate actively in the cycling of this toxic element in the groundwater table (Bhattacharya et al., 2017).

These reactions control the arsenic concentrations to which humans are exposed to arsenic. Disturbing the arsenic laden soils through agriculture and well drilling increases the rate at which arsenic partitions to the groundwater. The widespread use of well water in irrigation means that vegetables and agricultural products also contain elevated levels of arsenic. Individuals who are exposed to arsenic from the wells are at risk of arsenicosis, a slow disease that leads to disfigurements, neurological damage, and cancer in internal organs (see Figure 7.64) (Centeno et al., 2007).

The contamination of groundwater by arsenic in Bangladesh is a huge environmental disaster, likely exceeding the risk of the accidents at Bhopal, India, in 1984 and Chernobyl, Ukraine, in 1986. Between 35 million and 77 million people have a significant health risk associated with drinking contaminated water (Aggarwal et al., 2001). Figure 7.65 shows the distribution of arsenic concentrations in tubewell water samples.

Cancer caused by long-term low-dose arsenic exposure through the consumption of contaminated water is now an important concern in Bangladesh, as it is being increasingly reported from arsenic-exposed individuals (Ahmad et al., 2018). In 2018, a high proportion of Bangladeshis were still consuming arsenic-contaminated water because they lacked a sustainable water supply. To provide sustainable arsenic-safe water options, any option advocated should be cheap, easy to use, locally maintainable, and owned by the community. Solutions have included modifying existing tubewells; treating the water from the tubewells; and using environmental, social, and economic indicators to provide solutions to those most at risk. Social mapping is used as an important tool for optimizing sites to ensure easy access to new

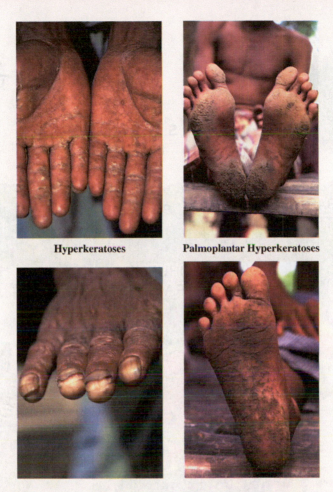

Hyperkeratoses   **Palmoplantar Hyperkeratoses**

FIGURE 7.64 Examples of the effects of arsenicosis.

Source: Adrian Page/Alamy Stock Photo.

safe well installations for a greater number of beneficiaries.

Various low-cost filters have been designed for arsenic removal, including the SONO filter developed by Dr. Abul Hussam in response to the Grainger Challenge Prize sponsored by the National Academy of Engineering in the United States (WHO, 2008). This point-of-use filter method can be used for two or three homes. Water first passes through a composite iron matrix that removes arsenic through surface complexation reactions. It then goes into a second bucket where it passes through coarse sand, gravel, and wood charcoal to remove organic compounds. A third container uses wet brick chips to remove fine particles. The result is a simple, low-cost method to achieve water that meets and exceeds the WHO standards for arsenic.

(Continued)

BOX 7.3   Arsenic in the Groundwater in Bangladesh *(Continued)*

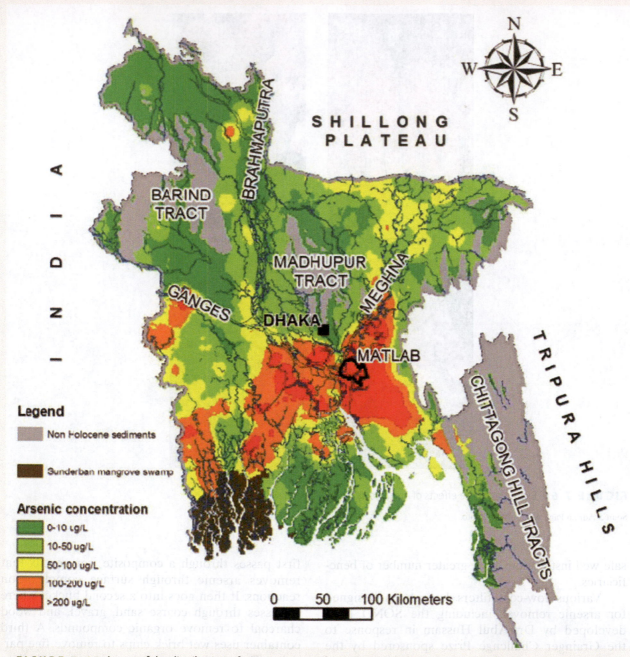

FIGURE 7.65 A map of the distribution of arsenic in groundwater in Bangladesh, showing the location of Matlab as a hotspot in the study area.

Source: Hossain, M., Rahman, S., Bhattacharya, P., Jacks, G., Saha, R., and Rahman, M. (2015). "Sustainability of arsenic mitigation interventions—an evaluation of different alternative safe drinking water options provided in Matlab, an arsenic hot spot in Bangladesh." *Frontiers in Environmental Science*, 3. 10.3389/fenvs.2015.00030.

FIGURE 7.66 A modified tubewell used throughout India for drinking water.

Source: Dinodia Photos/Dinodia Photos RM/Alamy Stock Photo.

7.4.3 Surface Water Supplies

Surface water supplies are dependent on a steady and predictable pattern of rainfall. Due in large part to increasing demand from growing populations in urban areas and greater variability in precipitation patterns that have resulted from a shifting climate, surface water supplies are less reliable than groundwater sources. Surface waters are also more susceptible to pollution from land runoff. The flows in rivers and streams may vary so much that demand may not be met during short dry periods; thus, *reservoirs* have been built to store water during wet periods for use during dry periods. The size of the reservoirs is based on geologic conditions and expected water demand, with the objective being to make the reservoir a sufficient size to provide a dependable water supply.

Reservoir capacity may be estimated using the mass curve method illustrated in Figure 7.67. In this method, the total flow in a stream at the point of a proposed reservoir is summed and plotted against time. The water demand is plotted on the same graph and the difference between the total mass of water flowing in and the demand is the quantity of water that must be stored. Streamflow data over multiple years must be statistically normalized based on the driest period of time for which water supply should be available. For example, it is typical for water reservoirs to be sized to supply water for 19 or 20 years. A *frequency analysis* may be used to estimate the streamflow and expected demand from the driest year in 20 years, which may be used to determine reservoir capacity. Note that in a dry year, water demand may significantly increase due in particular to increased irrigation.

Droughts may occur when natural or managed water systems do not provide enough water to meet established human and environmental uses due to shortfalls in precipitation or surface water flow. Engineers may use drought management

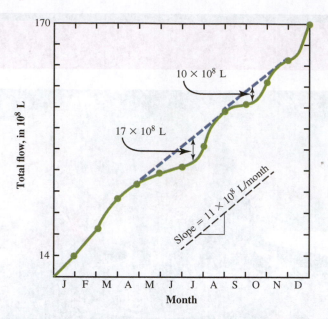

FIGURE 7.67 Mass curve showing required water storage volumes for service.

planning to help control the water supply during dry periods. Urban areas throughout the world have experienced significant water shortfalls and droughts recently, including extended droughts in California and India. Fifty percent of all water supply utilities asked their customers to reduce consumption during a drought in 1988 (Moreau, 1989). Drought management needs to consider the economic, environmental, and societal impacts of drought to prepare appropriate plans. The U.S. National Drought Management Center has promoted the following steps to prepare and mitigate the impacts of drought on a community.

A drought management team should be composed of community members—including engineers, policymakers, industry representatives, and the public—who understand the issues of appropriateness, urgency, equity, and cultural awareness in drought risk analysis. The advantage of publicly discussing questions and options is that the procedures used in making any decisions will be better understood, and it also demonstrates a commitment to participatory management. At a minimum, decisions and reasoning should be openly documented to build public trust and understanding.

The drought management working team should estimate a range of impacts related to the severity of the drought that may be based on past, current, and potential issues. These impacts highlight sectors, populations, or activities that are vulnerable to drought and, when evaluated with the probability of drought occurrence, identify varying levels of drought risk. The impacts should then be prioritized based on prevailing local conditions. Relevant questions might include:

- Which impacts are important to the affected individual's or group's way of life?
- If impacts are not distributed evenly, should hard-hit groups receive greater attention?
- Is there a trend of particular impacts becoming more of a problem than others?

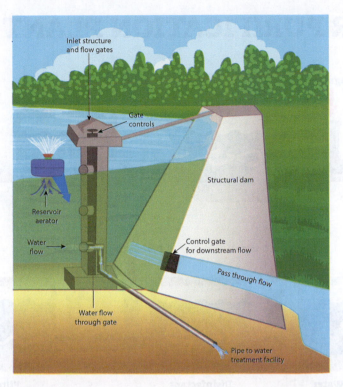

Inlet structure
and flow gates

Gate
controls

Structural dam

Reservoir
aerator

Water
flow

Control gate
for downstream flow

Pass through flow

Water flow
through gate

Pipe to water
treatment facility

FIGURE 7.68 Surface water inlet structures and appurtenances in storage reservoirs.

"Vulnerability assessment provides a framework for identifying the social, economic, and environmental causes of drought impacts. It bridges the gap between impact assessment and policy formulation by directing policy attention to underlying causes of vulnerability rather than to its result, the negative impacts, which follow triggering events such as drought" (Ribot et al., 1996, p. 4). The causes and effects of the potential impacts should be considered, and the hierarchy of vulnerabilities should be diagramed to determine the most vulnerable community resources.

The drought management team should then consider actions that may be taken to preserve the water supply or reduce water demand. This plan may be proposed in phases that are related to the degree of the drought. The National Drought Management Center provides access to the drought management plan for each state in the United States as well as many other resources to assist with developing action plans. The drought management plan is complete when the team decides on the list of tasks to accomplish prior to, during, and after a drought. Finally, the drought risk analysis should be reviewed periodically and revised as conditions and priorities change.

Whenever possible, water from impoundment reservoirs flows to the water treatment plant via a gravity-fed system. In low-lying areas, water must be pumped to the water supply infrastructure. A schematic of the water intake structure from a reservoir is shown in Figure 7.68.

Municipal water supplies for communities may receive various degrees of treatment based largely on the quality of the water from the source. A typical water

WATER PURIFICATION PLANT

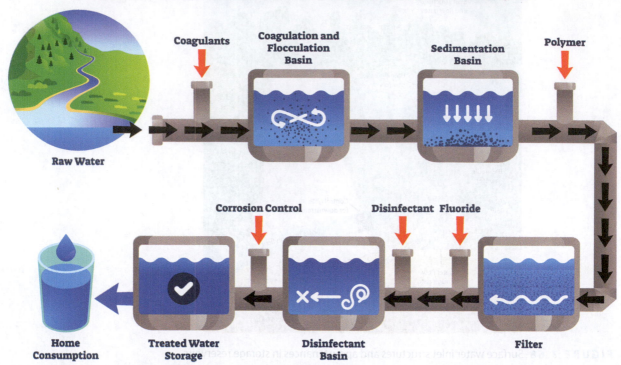

FIGURE 7.69 Processes in a conventional water treatment facility.

Source: VectorMine/Shutterstock.com.

treatment plant is made up of a series of reactors or unit processes, as illustrated in Figure 7.69. The order of treatment processes is important, and common water treatment methods are described in the following sections.

7.4.4 Water Softening

Some waters (both surface water and groundwater) are improved by removing hardness, which was described in Section 2.4.4. While hardness does not cause serious health problems, it does reduce the effectiveness of soaps and causes scale formation. Scale, which forms when calcium carbonate precipitates from heated water, reduces heat transfer efficiency by coating water heaters, boilers, heat exchangers, tea pots, and so on, and may eventually clog pipes. Excessive hardness may also result in objectionable tastes. Softening is the process of removing hardness.

Lime soda softening uses chemical precipitation to remove the ions that cause hardness; this process is illustrated in Figure 7.70. The pH of the water is increased, often through the addition of quicklime (CaO, unslaked lime), hydrated lime ($Ca(OH)_2$), slaked lime, or sodium hydroxide, as shown by the reactions in Figure 7.71. Calcium carbonate forms above a pH of approximately 10.3 and begins to precipitate. Magnesium hydroxide ($Mg(OH)_2$) begins to precipitate above a pH of approximately 11. Noncarbonate hardness is more expensive to precipitate

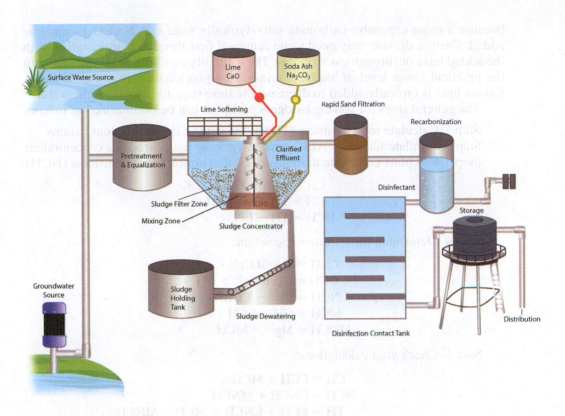

FIGURE 7.70 Example of a lime softening process.

Carbon dioxide
$$CO_2 + Ca(OH)_2 \rightleftharpoons CaCO_3 \text{ (s)} + H_2O$$

Calcium carbonate hardness (CCH)
$$Ca(HCO_3)_2 + Ca(OH)_2 \rightleftharpoons 2CaCO_3 \text{ (s)} + 2H_2O$$

Calcium noncarbonate hardness (CNCH)
$$CaSO_4 + Na_2CO_3 \rightleftharpoons CaCO_3 \text{ (s)} + Na_2SO_4$$

Magnesium carbonate hardness (MCH)
$$Mg(HCO_3)_2 + Ca(OH)_2 \rightleftharpoons CaCO_3 \text{ (s)} + MgCO_3 + 2H_2O$$
$$MgCO_3 + Ca(OH)_2 \rightleftharpoons Mg(OH)_2 \text{ (s)} + CaCO_3 \text{ (s)}$$

Magnesium noncarbonate hardness (MNCH)
$$MgSO_4 + Na_2CO_3 \rightleftharpoons MgCO_3 + Na_2SO_4$$
$$MgCO_3 + Ca(OH)_2 \rightleftharpoons Mg(OH)_2 \text{ (s)} + CaCO_3 \text{ (s)}$$

Species	Ratio of Chemical to Species (meq/meq)	
	Lime	Soda Ash
CO_2	1	0
CCH	1	0
CNCH	0	1
MCH	1	0
MNCH	1	1

FIGURE 7.71 Lime soda water softening reactions.

because a more expensive carbonate salt—typically, soda ash (Na_2CO_3)—must be added. Carbon dioxide may need to be removed first through neutralization with the added base or through gas stripping. The solubility equilibrium conditions limit the practical lower level of hardness removal by precipitation to about 40 mg/L. Excess lime is typically added to decrease the time required to soften the water.

The general steps for solving hardness problems can be summarized as follows:

Step 1. Calculate total hardness (TH) as the sum of the multivalent cations.
Step 2. Calculate alkalinity (ALK), which is typically the bicarbonate concentration.
Step 3. Calculate carbonate hardness (CH) and noncarbonate hardness (NCH):

$$CH = ALK \text{ if } TH > ALK$$
$$CH = TH \text{ if } TH < ALK$$
$$NCH = TH - CH$$

Step 4. Determine the hardness speciation:

$$CCH = Ca^{2+} \text{ if } Ca^{2+} < CH$$
$$CCH = CH \text{ if } Ca^{2+} > CH$$
$$CNCH = Ca^{2+} - CCH$$
$$MCH = CH - CCH$$
$$MNCH = Mg^{2+} - MCH$$

Step 5. Check your calculations:

$$CH = CCH + MCH$$
$$NCH = CNCH + MNCH$$
$$TH = CCH + CNCH + MCH + MNCH$$
$$Ca^{2+} = CCH + CNCH$$
$$Mg^{2+} = MCH + MNCH$$

The comparisons can be done either visually with bar charts or mathematically. Note that we do not need to know what the noncarbonate anions are; we simply need to know how much of the hardness is not associated with alkalinity.

EXAMPLE 7.7   Determining the Amount of Water Softening Chemicals Needed

Water from a well has a total hardness of 4.1 meq/L as shown in the bar chart in Figure 7.72. Determine the mass rate of chemicals required to soften the water to the practical solubility limit if the flow rate is 2 MGD and 93% pure quicklime (CaO) and 98% pure soda ash are used.

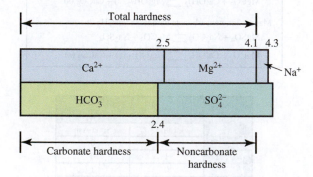

FIGURE 7.72 Hardness components in meq/L of the well water for Example 7.7.

Note that the equivalent weight of carbon dioxide is $(44 \text{ g/mol})/(2 \text{ eq/mol}) = 22 \text{ mg/meq}$. To determine the quantity of chemicals required, use the fact that each meq/L of hardness will require 1 meq/L of chemicals to remove it (see Figure 7.71). The meq/L values for lime and soda ash are shown in the following table. Note that the removal of carbon dioxide and CCH requires only lime, the removal of CNCH requires only soda ash, and the removal of MNCH requires both lime and soda ash. To soften to the practical solubility limit, 1.25 meq/L, excess CaO is needed. (For hydrated lime, $Ca(OH)_2$, 0.4 meq/L is needed.)

	CONCENTRATION (meq/L)		
COMPONENT	**COMPONENT**	**LIME**	**SODA ASH**
CO_2	0.27	0.27	0
CCH	2.4	2.4	0
CNCH	0.1	0	0.1
MCH	0	0	0
MNCH	1.6	1.6	1.6
Excess		1.25	
Total		5.52	1.7

The following equations can also be used to calculate the quantity of chemicals required:

$$\text{Lime} = (CO_2) + CH + MNCH + \text{excess}$$
$$\text{Soda ash} = NCH$$

To determine the mass rate of chemicals required, meq/L must be converted to mg/L by using the equivalent weights (EW) of the chemicals:

CHEMICAL	EW (mg/meq)
CaO	28
$Ca(OH)_2$	37
$Na_2(CO_3)$	53

From Chapter 4, concentration is converted to mass flow rate by using

$$Q_M = M = CQ_V$$

The purity of the chemicals used is less than 100%, so the mass rate must be divided by the purity. For this example, 2 MGD of water are being treated with 93% pure CaO and 98% pure soda ash, so the chemical amounts required are

$$M_{CaO} = \frac{(5.52 \text{ meq/L})(28 \text{ mg/meq})(2 \text{ MGD})\left(8.34 \frac{\text{lb}}{\text{(mil gal)(mg/L)}}\right)}{0.93}$$
$$= 2{,}770 \text{ lb/day} = 1.4 \text{ tons/day}$$

$$M_{Na_2CO_3} = \frac{(1.7 \text{ meq/L})(53 \text{ mg/meq})(2 \text{ MGD})\left(8.34 \frac{\text{lb}}{(\text{mil gal})(\text{mg/L})}\right)}{0.98}$$

$$= 1{,}530 \text{ lb/day} = 0.8 \text{ ton/day}$$

Many homeowners who derive their water from wells and those whose water supplies contain moderate amounts of hardness may opt to soften their water at home. Ion exchange and precipitation are the typical methods used, but reverse osmosis (RO) is also used.

Ion exchange, or zeolite, softening is most applicable for water that has high levels of noncarbonated hardness and less than 350 mg/L of $CaCO_3$ total hardness (Bowen, 1999). In an ion exchange-based unit (illustrated in Figure 7.73), the water passes through a column containing a resin material. The resin adsorbs the hardness ions, typically exchanging them for more soluble sodium ions. Once the resin no longer removes the amount of hardness desired, a concentrated salt solution (or brine) is used to regenerate the resin by removing the hardness ions so that the resin

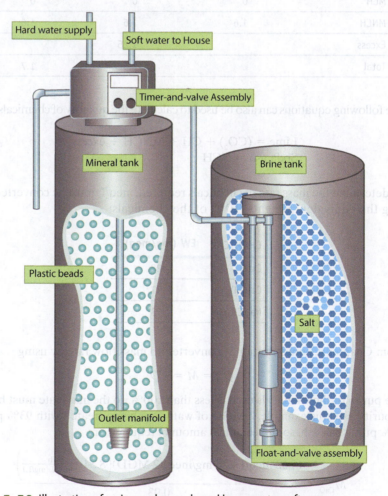

FIGURE 7.73 Illustration of an ion exchange-based home water softener.

can be reused. Usually, regeneration requires 2.1–3.5 kg salt/kg hardness removed, and regeneration rates are 2.2–4.4 L/s-m³ (1–2 gpm/ft³) of resin for 55 minutes followed by 6.6–11 L/s-m³ (3–5 gpm/ft³) of resin for 5 minutes for municipal facilities. A short backwash (backward flow through the system) is performed before regeneration to expand the bed and remove solid particles.

EXAMPLE 7.8 Capacity of an Ion Exchange-Based Water Softener

A water softener has 0.07 m³ of ion exchange resin with an exchange capacity of 46 kg/m³. The water use is 1,500 L/day. If the incoming water contains 245 mg/L of hardness as $CaCO_3$ and the softened water needs to have 100 mg/L as $CaCO_3$, how much water should bypass the softener? What is the time between regeneration cycles?

The amount of water that should bypass the softener is a function of the desired hardness and the initial hardness. A schematic of the process is shown in Figure 7.74. Writing a material balance equation around the mixing point (the circle in the diagram) provides the solution at a particular point in time:

$$Q_{IX}C_{IX} + Q_{BP}C_{BP} = Q_0C_f$$

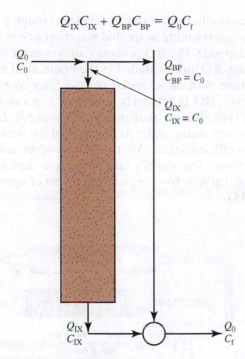

FIGURE 7.74 Schematic of a typical ion exchange-based water softener and the mass balance of flows in the system.

For the case of a fresh resin, the effluent hardness is zero, so

$$Q_{IX}(0) + Q_{BP}C_{BP} = Q_0C_f$$

$$\frac{Q_{BP}}{Q_0} = \frac{C_f}{C_{BP}}$$

or

$$\% \text{ Bypass} = \frac{\text{desired hardness}}{\text{initial hardness}}(100)$$

The amount that should be bypassed can be calculated as

$$\% \text{ Bypass} = \frac{100 \text{ mg/L as CaCO}_3}{245 \text{ mg/L as CaCO}_3}(100) \cong 41\%$$

$$\text{Amount to bypass} = 0.41(1{,}500 \text{ L/day}) \cong 610 \text{ L/day}$$

The length of the cycle, or the time to breakthrough, is a function of the exchange capacity of the resin. If we assume complete saturation of the resin before regenerating, then

$$\text{Breakthrough} = \frac{(\text{capacity})(V_{\text{resin}})}{Q_{\text{IX}}\, C_0} = \frac{(\text{capacity})(V_{\text{resin}})}{(1 - \text{bypass})\, Q_0 C_0}$$

$$\text{Breakthrough} = \frac{(46 \text{ kg/m}^3)(0.07 \text{ m}^3)(10^6 \text{ mg/kg})}{(1 - 0.41)(1{,}500 \text{ L/day})(245 \text{ mg/L as CaCO}_3)} \cong 15 \text{ day}$$

Therefore, salt will have to be added approximately every two weeks.

RO uses high pressure to push water molecules through a membrane that has very small pore openings, resulting in treated water on one side and concentrated wastewater on the other side. The RO process removes most of the dissolved minerals and some bacteria. RO units installed in the home, such as the one illustrated in Figure 7.75, also come with an activated carbon filter to remove chlorine and improve taste. However, RO is relatively slow and generates significant quantities of wastewater; a typical home unit produces about 5% to 15% of water for consumption and the remaining water is discharged as wastewater. Large-scale systems typically have efficiencies of 75% to 90%, however, with only 10% to 25% production of wastewater. The world's largest reverse osmosis facility in Sorek, Israel, produces 624,000 m³/day from seawater at a cost of approximately 0.58 US$ (2013)/m³ (Wang, 2015).

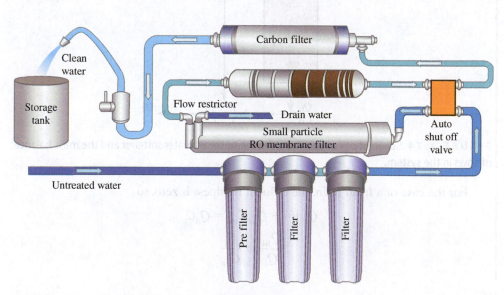

FIGURE 7.75 The components in a typical RO process.

7.4.5 Coagulation and Flocculation

Raw surface water entering a water treatment plant usually has significant turbidity caused by tiny (colloidal) clay and silt particles. These particles have a natural electrostatic charge that keeps them from colliding, sticking together, and forming larger particles. Chemicals called coagulants, such as alum (aluminum sulfate), and coagulant aids, such as lime and polymers, are added to the water to neutralize the surface charges on particles. The chemical coagulant alters the particles' surface chemistry and creates an electrostatic attraction between particles. The particles tend to clump together, increasing the effective size of particles in the water, which allows increased efficiencies in settling and filtration processes that remove the particles. *Coagulation* is a chemical process that alters the surface charge on particles suspended in the water to form larger particles.

The net effect of coagulation is to destabilize the colloidal particles so that they have the propensity to grow into larger particles. If the coagulant dose is too high, however, the particles' surface charge may switch and may not be neutralized. In this case, the particle size and density may not be sufficiently increased.

Flocculation gently mixes the wastewater and coagulants to increase the particle movement and attract the particles together, which increases the overall size of the particles. The particles and microorganisms clump together in a group called a "floc." A large, slow-speed paddle is commonly used in the treatment process to gently stir the water in this unit process. The coagulation process is illustrated in Figure 7.76. Once the coagulants and flocculation process increase the particle size, the particles are removed in a settling tank (also called a clarifier) or through a filtration process.

7.4.6 Settling

Gravimetric settling tanks allow the heavier-than-water floc particles to settle to the bottom of the tank. Settling tanks are designed to approximate a plug-flow reactor and minimize turbulence. The entrance and exit configurations of a reaction tank, such as the one illustrated in Figure 7.77, are critical for reducing turbulence and re-suspension as water enters and exits the tank. The particles that accumulate in a layer toward the bottom of the tank are called sludge, and the layers formed are called the sludge blanket. This sludge is composed of aluminum hydroxides, calcium carbonates, and clays, so it does not biodegrade or decompose at the bottom of the tank. Typically, the sludge is removed every few weeks through a mud valve on the tank bottom and is transferred to a sanitary sewer or sludge holding/drying pond.

The particle removal efficiency is a function of the particles' size, shape, and density, as well as the fluid's density and velocity. The downward settling velocity of the particles is determined by a force balance on the particle, as shown in Figure 7.78. Stokes' settling law says that the gravimetric force acting on the particle, F_g, must be equal and opposite to the sum of the buoyancy force, F_b, and the drag force, F_d:

$$F_g = F_d + F_b \tag{7.41}$$

Each force is defined as
Gravimetric force:

$$F_g = m_p g = \rho_p V_p g \tag{7.42}$$

Drag force:

$$F_d = \frac{1}{2} C_d A \rho_f v_s^2 \tag{7.43}$$

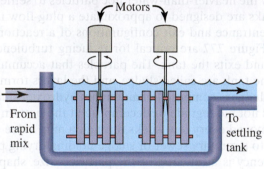

FIGURE 7.76 A typical flocculator used in water treatment.

Source: Photo Courtesy of P. Aarne Vesilind.

Buoyancy force:

$$F_b = V_p \rho_f g \tag{7.44}$$

where

m_p = mass of the particle
g = acceleration constant for gravity
ρ_p = density of the particle
V_p = volume of the particle (usually assumed that the particle is spherical)
C_d = particle drag coefficient
A = surface area of the particle (usually assumed that the particle is spherical)
ρ_f = density of the fluid
v_s = settling velocity of the particle

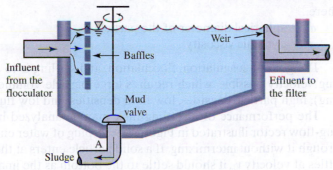

FIGURE 7.77 A typical settling tank used in water treatment plants.

Source: Photo Courtesy of P. Aarne Vesilind.

Substituting Equations (7.42) through (7.44) into Equation (7.41) yields

$$\rho_p V_p g = \frac{1}{2} C_d A \rho_f v_s^2 + V_p \rho_f g \qquad (7.45)$$

Solving Equation (7.45) for the settling velocity yields

$$v_s = \left[\frac{2(\rho_p - \rho_f) v_p g}{C_d A \rho_f} \right]^{\frac{1}{2}} \qquad (7.46)$$

The velocity of the water in a settling tank is quite slow, so if the particle is assumed to be spherical and the Reynolds number is low, then

$$C_d \approx \frac{24}{Re} \qquad (7.47)$$

FIGURE 7.78 Stokes' settling law for a particle of volume, V_p.

The Reynolds number is defined as

$$Re = \frac{d_p \rho_f v}{\mu}$$

(7.48)

Then

$$v_s = \left[\frac{2(\rho_p - \rho_f) v_p g}{C_d A \rho_f} \right]^{\frac{1}{2}} = \left[\frac{(\rho_p - \rho_f) d_p^2 g}{18\mu} \right]^{\frac{1}{2}}$$

(7.49)

where

d_p = spherical diameter of the particle
μ = dynamic viscosity

The goal of coagulation, flocculation, and settling is to achieve the highest settling velocity possible, which requires large particle volumes, compact shapes (low drag), high particle densities, low fluid densities, and low fluid viscosity.

The performance of a settling tank may be analyzed by modeling the perfect plug-flow rector illustrated in Figure 7.79. A plug of water enters the tank and moves through it without intermixing. If a solid particle enters at the top of the column and settles at velocity v_0, it should settle to the bottom as the imaginary column of water exits the tank, having moved across the tank with a horizontal velocity of v_h. The following assumptions are required in the analysis:

- Uniform flow occurs within the settling tank.
- All particles settling to the bottom are removed.

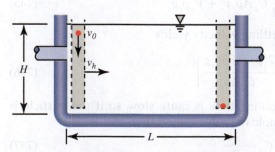

FIGURE 7.79 An ideal settling tank.

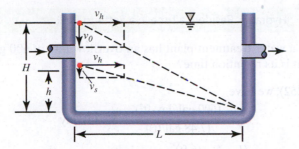

FIGURE 7.80 Particle trajectories in an ideal settling tank.

- Particles are evenly distributed in the flow as they enter the tank.
- All particles still suspended in the water when the column reaches the far end of the tank are not removed and escape the tank.

We can examine the expected motion of a particle entering the water surface at the left-hand side of the clarifying tank as shown in Figure 7.80. The particle has a settling velocity of v_0 and a horizontal velocity of v_h such that the resultant vector defines a trajectory from the top left to precisely the bottom right of the tank, as shown. Particles that have this velocity are called ***critical particles***: particles with lower settling velocities are not all removed and particles with higher settling velocities are all removed. Note that as a critical particle enters the tank at another height h, its trajectory carries it to the bottom of the tank.

The time that the water resides inside a tank or clarifier is called the hydraulic retention, or detention, time and is defined as

$$t_r = \frac{V}{Q} \tag{7.50}$$

The volume of a rectangular tank is calculated as $V = HLW$, where H is the height of the tank, W is the width of the tank, and L is the length of the settling zone (which in a clarifier is less than the length of the tank due to turbulence near the inlet and outlet of the tank). Using the mass continuity equation, we find that the flow rate is $Q = Av$, where A is the cross-sectional area, $H \times W$, through which flow occurs and v is the fluid velocity. Substituting these terms into the equation for the retention time yields

$$t_r = \frac{V}{Q} = \frac{HWL}{(HW)v} = \frac{L}{v} \tag{7.51}$$

In the vertical orientation, we can find the settling velocity of the particle over the depth of the tank using similar triangles:

$$v_0 = \frac{H}{t_r} = \frac{H}{A_S(H/Q)} = \frac{Q}{A_S} \tag{7.52}$$

where $A_s = WL$ and is the surface area of the settling tank or clarifier. This equation represents an important design parameter for settling tanks called the ***overflow rate***, which is the velocity of the critical particle. Note that the units for the overflow rate terms are Q in m^3/s and A in m^2, which leaves the overflow rate in m/s. The overflow rate has the same units as velocity. The overflow rate is commonly expressed in English units as gal/day-ft². Note that when any two of the following—overflow rate, retention time, or depth—are defined, the remaining parameter is also fixed.

EXAMPLE 7.9 Hydraulic Residence Time in a Settling Tank

A settling tank in a water treatment plant has an overflow rate of 600 gal/day-ft^2 and a depth of 6 ft. What is its retention time?

From Equation (7.52), we have

$$v_0 = \frac{(600 \text{ gal/day-ft}^2)}{(7.48 \text{ gal/ft}^3)} = 80.2 \text{ ft/day}$$

$$t_r = \frac{H}{v_0} = \frac{(6 \text{ ft})}{(80.2 \text{ ft/day})} = 0.0748 \text{ day} \cong 2 \text{ hr}$$

There is an important similarity between ideal settling tanks and the ideal world: neither one exists. A settling tank, for example, never has uniform flow. Wind, density, and temperature currents as well as inadequate baffling at the tank entrance can all cause nonuniform flow. Thus, settling tanks are designed with a large safety factor. Cost and hydraulic parameters are also important in the practical design of settling tanks, as is settling tank depth. Experience has shown that as particles settle, they can flocculate and stick together, creating higher settling velocities and enhancing particle removal. Settling tanks are typically built 3 m to 4 m deep to take advantage of the natural flocculation that occurs during settling.

EXAMPLE 7.10 Expected Particle Removal in a Settling Tank

A small water plant has a raw water inflow rate of 0.6 m^3/s. Laboratory studies have shown that the flocculated slurry can be expected to have a uniform particle size (only one size), and it has been found through experimentation that all the particles settle at a rate of v_s = 0.004 m/s. (This is unrealistic, of course.) A proposed rectangular settling tank has an effective settling zone of L = 20 m, H = 3 m, and W = 6 m. Can 100% removal be expected?

Remember that the overflow rate is actually the critical particle settling velocity. What is the critical particle settling velocity for the tank? We can use Equation (7.52):

$$v_0 = \frac{Q}{A_s} = \frac{0.6 \text{ m}^3/\text{s}}{(20 \text{ m})(6 \text{ m})} = 0.005 \text{ m/s}$$

The critical particle settling velocity is greater than the settling velocity of the particle to be settled; hence, not all the incoming particles will be removed. The same conclusion can be reached by using the particle trajectory. The velocity v through the tank is

$$v = \frac{Q}{HW} = \frac{0.6 \text{ m}^3/\text{s}}{(3 \text{ m})(6 \text{ m})} = 0.033 \text{ m/s}$$

Using similar triangles yields

$$\frac{v_s}{v} = \frac{H}{L'}$$

where L' is the horizontal distance the particle would need to travel to reach the bottom of the tank.

$$\frac{0.004 \text{ m/s}}{0.033 \text{ m/s}} = \frac{3}{L'}$$

$$L' = 25 \text{ m}$$

Hence, the particles would need 25 m to be totally removed, but only 20 m is available, so 100% removal cannot be expected.

EXAMPLE 7.11	Fraction of Particles Removed in a Settling Tank

In Example 7.10, what fraction of the particles will be removed?

We can assume that the particles entering the tank are uniformly distributed vertically. If the length of the tank is 20 m, the settling trajectory from the far bottom corner will intersect the front of the tank at height 4/5 (3 m), as shown in Figure 7.81. Particles that enter the tank below this point will be removed while particles entering the tank above this point will remain in the effluent water. The fraction of particles that will be removed is then 4/5, or 80%.

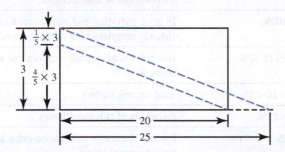

FIGURE 7.81 Schematic of an ideal settling tank described in Examples 7.10 and 7.11.

7.4.7 Filtration

Environmental engineers have observed that the movement of water through the ground removes many contaminants in the water. They have learned to apply this natural process to water treatment to develop sand filters.

Slow sand filters are characterized by a slow rate of filtration and the formation of an active layer called the schmutzdecke. A well-designed and properly maintained slow sand filter (SSF) effectively removes turbidity and pathogenic organisms through various biological, physical, and chemical processes in a single treatment step. Slow sand filtration systems are characterized by high reliability and rather low life cycle costs. Moreover, neither construction nor operation and maintenance require more than basic skills. Hence, slow sand filtration is a promising filtration method for small to medium-sized rural communities that have a fairly good quality of initial surface water.

Typically, SSFs are completely saturated with water. Large particles are removed in an upper coarse granular layer. In the lower finer granular layer, smaller particles are formed, and predatory microorganisms may attack and remove pathogens in the schmutzdecke layer. SSFs are usually operated in parallel, as the layers become clogged over time. They must be cleaned or drained, which temporarily upsets the active schmutzdecke layer, requiring some time for this layer to reform and achieve high pathogen removal efficiencies. SSFs may be very effective, as shown in Table 7.20, and are relatively inexpensive and simple to operate and maintain.

Rapid sand filtration, illustrated in Figure 7.82, achieves higher velocities through the filter. The filter process is more evenly distributed throughout the depth of the granular media. Rapid sand filters may remove 50% to 90% of larger pathogens, and the removal efficiency can be increased to 90% to 99% with the addition of a chemical coagulant. Rapid sand filters are periodically

TABLE 7.20 Typical removal efficiencies in slow sand filtration

WATER QUALITY PARAMETER	EFFLUENT OR REMOVAL EFFICIENCY	COMMENTS
Turbidity	<1 NTU	The level of turbidity and the nature and distribution of particles affect the treatment efficiency
Fecal bacteria	90 to 99.9%	Affected by temperature, filtration rate, size, uniformity and depth of sand bed, cleaning operation
Fecal viruses and *Giardia* cysts	99 to 99.99%	High removal efficiencies, even directly after cleaning (removal of the *schmutzdecke*)
Schistosomiasis *Cercaria*	100%	In good operation and maintenance conditions virtually complete removal is obtained
Color	25 to 30%	True color is associated with organic material and humic acids
Organic carbon	<15–25%	Total organic carbon
THM precursors	<25%	Precursors of trihalomethanes
Microcystins	85 to >95%	Cyanobacteria and their toxins extracted from a cyanobacterial bloom
Iron, manganese	30 to 90%	Iron levels above 1 mg/L reduce filter run length

Sources: Bellamy, W., et al. (1985); Grutzmacher, G., et al. (2002); IRC (2002).

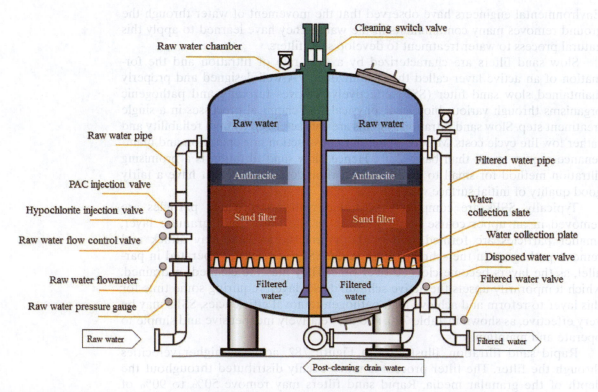

FIGURE 7.82 Rapid sand filtration process for drinking water.

Source: www.hitachizosen.co.jp/english/release/2015/12/001953.html.

backwashed—that is, the flow through the filters is reversed—to decrease clogging and pressure drop in the filters. The filters are operated in parallel to allow for frequent backwashing, and they may be used immediately after the backwashing procedure is completed. Rapid sand filtration is used commonly in developed countries for the treatment of large quantities of water where land is a strongly limiting factor and where material, skilled labor, and continuous energy supplies are available.

Filters are a very important process in meeting turbidity limits. A common design and operating parameter is the filtration rate (or filter loading), which is the rate of water applied to the surface area of the filter. The calculation and units, such as L/(s-m^2), are similar to those for the overflow rate. This rate can vary from about 1.4–7 L/s-m^2 but may be limited to 1.4-2 L/s-m^2 by regulation. Backwash rates generally range from 6.5 to 17 L/s-m^2. Large-scale filtration methods are very successful at reducing concentrations of pathogens in water. The filtration methods discussed above, however, may be disrupted by clogging and operational upset, and they do not remove viruses.

EXAMPLE 7.12   Filtration Rate

What is the filtration rate for a 10 m × 8 m filter if it receives water at a flow rate of 85 L/s?

$$\text{Filtration rate} = \frac{Q}{A_s} = \frac{85\,\frac{L}{s}}{(10\,\text{m})(8\,\text{m})} = 1.06\,\frac{L}{\text{s-m}^2} = 0.001\,\text{m/s}$$

EXAMPLE 7.13   Backwashing Requirements

How much backwash water is required to clean a 10 m × 8 m filter?

Assume that 15 L/s-m^2 will be used as the backwash rate and the filters will be cleaned for 15 min. Then

$$V = (\text{backwash rate})(A_s)(t) = \left(15\tfrac{1}{\text{s-m}^2}\right)(10\,\text{m})(8\,\text{m})(15\,\text{min})\left(60\,\tfrac{s}{\text{min}}\right)$$

$$= 1{,}080{,}000\,\text{L} = 1{,}080\,\text{m}^3 \text{ of backwash water needed}$$

BOX 7.4 The Biosand Filter

The BioSand Water Filter, illustrated in Figure 7.83, is an adaptation of slow sand filtration designed for use by families at the household level. This award-winning water filtration technology was developed by Dr. David Manz, a former University of Calgary professor. Biosand filters have been implemented throughout the world and demonstrated to successfully reduce water-related disease. Pure Water for the World, Inc. (PWW) is a nonprofit NGO headquartered in the United States and operating out of Honduras and Haiti. It is dedicated to improving lives by empowering people with access to life's most basic necessities: safe water and sanitation. PWW started implementing concrete biosand filters

(BSFs) in Honduras in the early 2000's and in Haiti in 2008. Later, it switched to Hydraid® plastic filter containers, as shown in Figure 7.84, to keep the price of the system low and make it easier to transport the filters. To date, PWW has installed BSFs in more than 16,975 homes, reaching 100,000+ people, and in more than 2,080 schools, reaching 500,000+ students. Through financial support from foundations, grants, and donations, PWW is able to subsidize the filter costs; however all families are required to contribute a small portion of the cost (<$10 USD) to ensure a strong sense of ownership.

PWW has invested in training community agents (CAs), members of the local communities who

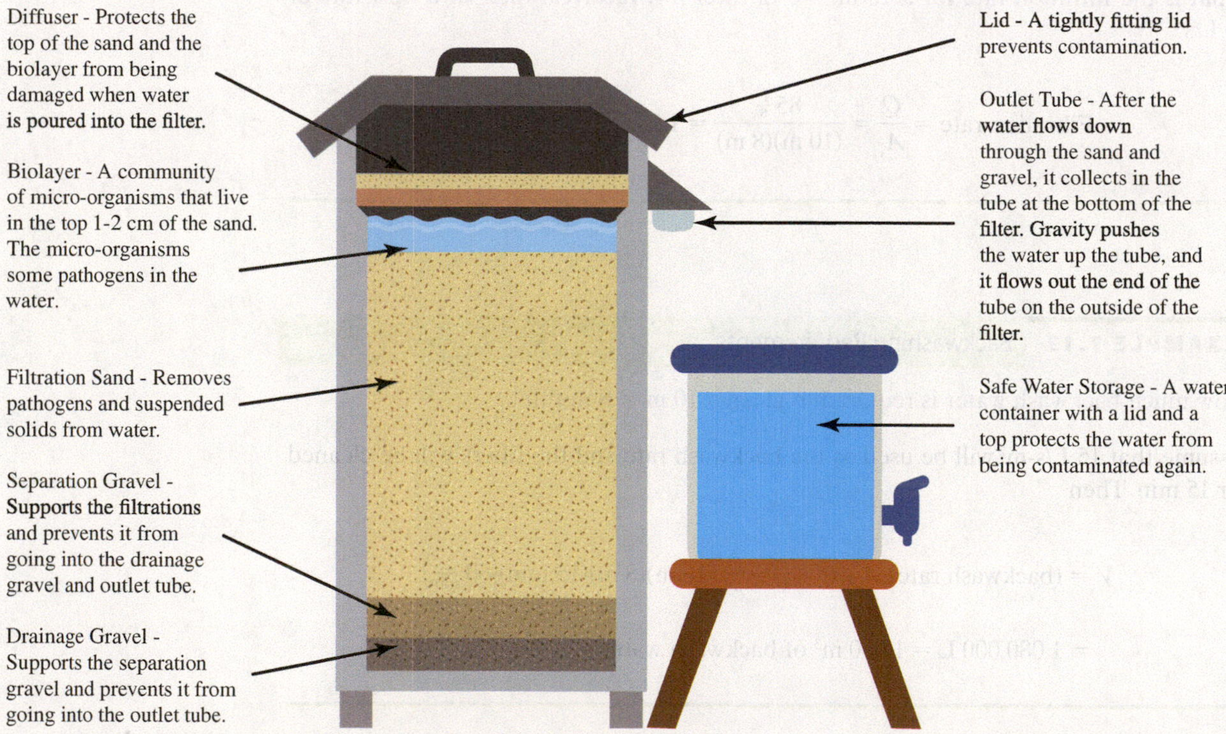

Diffuser - Protects the top of the sand and the biolayer from being damaged when water is poured into the filter.

Biolayer - A community of micro-organisms that live in the top 1-2 cm of the sand. The micro-organisms some pathogens in the water.

Filtration Sand - Removes pathogens and suspended solids from water.

Separation Gravel - **Supports the filtrations** and prevents it from going into the drainage gravel and outlet tube.

Drainage Gravel - Supports the separation gravel and prevents it from going into the outlet tube.

Lid - A tightly fitting lid prevents contamination.

Outlet Tube - After the water flows down through the sand and gravel, it collects in the tube at the bottom of the filter. Gravity pushes the water up the tube, and it flows out the end of the tube on the outside of the filter.

Safe Water Storage - A water container with a lid and a top protects the water from being contaminated again.

FIGURE 7.83 Layers of the biosand filter.

volunteer to serve as an extension of the PWW team, helping to install the filters and support families with the correct and consistent use of the filters and hygiene practices. CAs also provide feedback to PWW and do the follow-up visits scheduled for each household. Interested community members are eligible to become CAs, and women are encouraged to participate. The CAs monitor the biosand filters and go with a PWW staff health promoter to visit each filter. Together they check that the filter was installed

properly, reinforce the proper use and maintenance of the filter, and solve any question the users have.

PWW works to build and expand local capacity through comprehensive WASH (water, sanitation, and hygiene) programs, including the BSF technology and CA programs. PWW also presents workshops to raise awareness and build capacity among the communities about general WASH issues, including environmental hygiene, household hygiene, latrines, and personal hygiene.

FIGURE 7.84 A biosand filter used to remove pathogens at the point-of-use in the home.

Source: Pure Water for the World, Inc.

ACTIVE LEARNING EXERCISE 7.3 Sand Filter in a Bucket

The biosand filter is a small point-of-use modification of a slow sand filter. The point-of-use filter is designed to treat water in the home or at the point-of-use. The general design of the biosand filter is illustrated in Figure 7.85.

1. An even smaller version of the biosand filter can be used to demonstrate the principles of how the filter is designed, fabricated, operated, and maintained. The materials list to build your own sand filter is shown in Table 7.21. Based on Figure 7.85 and the materials list, design a sand filter on paper. Label each part and show how they will fit together. Create engineering drawings of your design. Your drawing should include the following elements:
   a. A profile view that shows:
      i) Dimensions and location of gravel
      ii) Dimensions and location of sand
      iii) Dimensions and location of PVC piping
      iv) Dimensions and location of the nozzle
      v) Call out any fittings required
   b. A top view
   c. A side view

TABLE 7.21 Materials list for a bucket sand filter

MATERIAL	AMOUNT	COST ($ U.S.)	AMOUNT PER FILTER	COST PER FILTER
Sand	0.5 cu ft	3.38		
Gravel	0.5 cu ft	3.48		
Bucket	1	2.54		
White spigot	1	1.76		
Black spigot with gaskets	1	2.81		
PVC pipe	10 ft	1.68		
PVC cement	1 pack	7.51		
Epoxy	1 pack	5.98		
Male adapter	1	0.23		
Elbow	1	0.26		
Threaded elbow	1	0.57		
End cap	1	0.32		
O rings	10 pack	1.97		
Other				

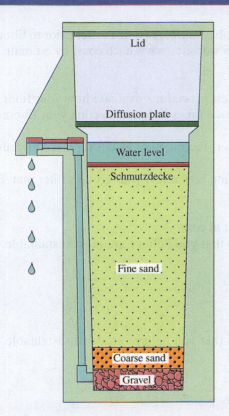

FIGURE 7.85 Schematic of a point-of-use biosand filter.

Source: Based on commons.wikimedia.org/wiki/File%3ABiosandFilter_Section.svg.

2. Meet in your predetermined groups. Compare and discuss your individual drawings. Note the differences. As a group, create another sketch with a similar level of detail. This drawing should include some approximate dimensions, but it need not be exactly to scale.

3. Construct your filter from the available supplies. Submit in hard copy an AS BUILT drawing of your filter. The drawing should comply with professional drafting rules and standards:
   a. Professional appearance (either hand drawn or drawn using computer software)
   b. Border
   c. Title block
   d. Since this is a relatively simple machine, you need only show a cross section of your design (side view)
   e. Use appropriate line weights and types (i.e., solid or hidden lines)
   f. Use callouts to label off-the-shelf parts; these are components that can be purchased and used unmodified, such as spigots, 5-gallon paint buckets, and PVC fittings
   g. Properly dimension all other components such as the height of the sand or gravel, height of the spigot, and PVC dimensions

(Continued)

ACTIVE LEARNING EXERCISE 7.3 Sand Filter in a Bucket *(Continued)*

4. Collect surface water from a local pond, puddle, or other available source. Test this water prior to filtering using the Hach MPN Pathoscreen methodology on the Hach website (www.hach.com) to estimate the number of organisms in 100 mL of the water.
   a. Report the MPN for the unfiltered pond water.
   b. Using the surface water and the Hach MPN Pathoscreen test procedure, evaluate how your filter works. You will need enough water to "rinse" your filter three times before collecting your filtered water sample.
   c. Compare your results to those of the rest of the class. Report the range of values reported in your section.
5. Using a spreadsheet and the materials list in Table 7.21, calculate how much your group's filter materials cost.
   a. What parts are most expensive?
   b. Where might you be able to achieve the greatest reduction in cost?
6. List the characteristics and qualities of the sand water filters that you believe are most sustainable. Do this for each of the following aspects of sustainability:
   a. Technological
   b. Economic
   c. Social
   d. Environmental
7. List the characteristics and qualities of the sand water filters that you believe are least sustainable. Do this for each of the following aspects of sustainability:
   a. Technological
   b. Economic
   c. Social
   d. Environmental
8. Describe and show how you would improve the filter design. The drawing should comply with professional drafting rules and standards:
   a. Professional appearance (either hand drawn or computerized)
   b. Border
   c. Title block
   d. Since this is a relatively simple machine, you need only show a cross section of your design (or side view)
   e. Use appropriate line weights and types (i.e., solid or hidden lines)
   f. Use callouts to label off-the-shelf parts; these are components that can be purchased and used unmodified, such as spigots, 5 gallon paint buckets, and PVC fittings
   g. Properly dimension all other components such as the height of the sand or gravel, height of the spigot, and PVC dimensions

7.4.8 Disinfection

Water treatment facilities add disinfectants to destroy microorganisms that can cause disease in humans. The Surface Water Treatment Rule requires public water systems to disinfect water obtained from surface water supplies or groundwater sources under the influence of surface water. Disinfection processes are capable of destroying bacteria, viruses, and amebic cysts, three types of pathogenic organisms that may be problematic in water. The chemical disinfection agents commonly used to treat water are listed in Table 7.22; they include free chlorine (OCl$^-$), combined chlorine (HOCl and OCl$^-$), ozone (O$_3$), chlorine dioxide (ClO$_2$), and ultraviolet irradiation. The effectiveness of the disinfection process is dependent on the pathogen's sensitivity to the particular disinfectant, the concentration of the disinfectant, the time of contact, and the possible presence of other substances in the water that could interfere with the disinfectant.

Chlorine is the water disinfectant most commonly used in the United States. Prechlorination may be done before filtration to help keep the filters free of growth and ensure adequate **contact time** with the disinfectant. Chlorine is available as compressed elemental gas, sodium hypochlorite solution (NaOCl), or solid calcium hypochlorite (Ca(OCl)$_2$). All forms of chlorine, when applied to water, form hypochlorous acid (HOCl). Gaseous chlorine acidifies the water and reduces the alkalinity, whereas the liquid and solid forms of chlorine increase the pH and the alkalinity at the application point. The pH of the water affects the dominant chlorine species such that HOCl dominates at lower pH and the hypochlorite ion (OCl$^-$) dominates at higher pH. Of the two species, HOCl is the stronger oxidant. Therefore, chlorine is more effective as an oxidant and a disinfectant at lower pH. Both forms, HOCl and OCl$^-$, are referred to as **free chlorine**. Adequate disinfection treatment processes must balance the concentration (C) of disinfectant added to the water and the contact time (T). The product of concentration and contact time (CT) is the most important operational parameter in disinfection and inactivation, as illustrated in Figure 7.86, which shows the relationship between inactivation for different species of microorganisms, the free chlorine concentration, and contact time.

Drinking water systems in the United States are required to have a *residual* level of active chlorine to guard against any microbial contamination in the distribution system, as shown in Table 7.23.

Several common minerals and other chemicals such as manganese, iron, nitrite, ammonia, and organics found in water may react with the chlorine between points 1 and 2 in Figure 7.87. There is no available chlorine residual, and no disinfection occurs between points 1 and 2. The additional chlorine added between points 2 and 3 in Figure 7.87 begins to react with organics and ammonia, forming chlorinated organics and chloroamines. The presence of all the various forms of

TABLE 7.22 Considerations and comparison of various chemical disinfectant processes

CONSIDERATION	CL$_2$	NaOCL	O$_3$	ClO$_2$	CHLORAMINE	UV
Residual persistence	Low	Low	None	Moderate	Low	None
pH dependence	Yes	Yes	Some	Yes	Some	None
Safety	Very high	Moderate	High	Very high	Very high	Moderate
Complex equipment	Yes	No	Yes	Yes	Yes	Yes
Equipment reliability	Good	Very good	Good	Good	Good	Moderate
Process control	Well developed	Well developed	Developing	Developing	Well developed	Developing
O&M requirements	Low	Low	High	High	Low	Low

Source: Davis, M. L. (2010). *Water and Wastewater Engineering Design Principles and Practices, 1st Ed.* New York: McGraw-Hill. pp. 10–21.

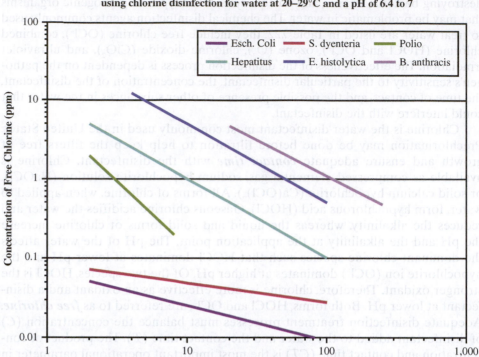

using chlorine disinfection for water at 20–29°C and a pH of 6.4 to 7

FIGURE 7.86 Variation of the impact of contact time and free chlorine dose upon various microorganisms.

Source: Based on Baumann, E.R., and Ludwig, D. D. 1962. *Free Available Chlorine Residuals for Small Nonpublic Water Supplies*. AWWA. 54(11):1379–1388.

TABLE 7.23 Chlorine residual levels for drinking water

DISINFECTANT	MINIMUM RESIDUAL* (mg/L)		MRDL** (mg/L)	MRDLG*** (mg/L)
	LEAVING PLANT	IN DISTRIBUTION SYSTEM		
Chloramines (as Cl_2)	2.0		4.0	4.0
Combined residual		0.5		
Chlorine (as Cl_2)	2.0		4.0	4.0
Chlorine dioxide (as ClO_2)	2.0		0.8	0.8
Free chlorine residual		0.2		

*From Illinois and Missouri regulations.
**MRDL = maximum residual disinfectant level.
***MRDLG = Maximum residual disinfectant level goal.

chlorine in the system is called the ***combined residual***. As additional chlorine is added between points 3 and 4, the chlororganics and chloroamines are partly destroyed. To obtain free chlorine in the reactor, chlorine must be added beyond the ***breakpoint dose***, point 4 in Figure 7.87. After the breakpoint, any added chlorine results in a residual-free chlorine level that is directly proportional to the amount of chlorine added.

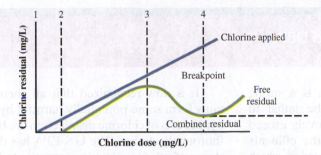

FIGURE 7.87 Chlorine breakpoint dosage curve.

EXAMPLE 7.14   Daily Chlorine Residual Dosage

A 12 m³/min water treatment plant uses 9.5 kg/day of chlorine for disinfection. If the daily chlorine demand is 0.5 mg/L, what is the daily chlorine residual?

The chlorine demand is the dose required to reach the desired residual level. Therefore, the residual is the difference between the chlorine applied and the chlorine demand.

Using the equation $M = QC$ yields

$$\text{Chlorine applied} = \frac{M}{Q} = \left[\frac{9.5 \text{ kg/day}}{12 \text{ m}^3/\text{min}}\right]\left(\frac{10^6 \text{ mg}}{\text{kg}}\right)\left(\frac{\text{m}^3}{1{,}000 \text{ L}}\right)\left(\frac{\text{day}}{24 \text{ hr} \times 60 \text{ min}}\right) = 0.55 \tfrac{\text{mg}}{\text{L}}$$

$$\text{Chlorine residual} = 0.55 \text{ mg/L} - 0.50 \text{ mg/L} = 0.05 \text{ mg/L}$$

This residual is below the minimum recommended free residual level of 0.2–0.5 mg/L.

Although increasing the dose increases the ability of chlorine to oxidize and disinfect, it may also lead to taste and odor issues and to the formation of disinfection byproducts (DBPs) by chlorine's reaction with natural organic matter (NOM). The dose is also affected by the application point, chlorine demand of the water, and desired residual concentration. Concerns about the halogenated disinfection byproducts (DBPs) of chlorination and the maximum contaminant level (MCL) of 0.10 mg of total trihalomethanes per liter set by the Environmental Protection Agency under the Safe Drinking Water Act (EPA 1979, 1980) have caused treatment facilities in several states to increase or switch to alternative chlorination methods. For example, chlorine dioxide does not react with organics and thus does not form trihalomethanes (THMs), so it is adopted more frequently. Ozonation and UV treatment are other alternatives to chlorine; however, neither of these disinfectants provides a residual level to protect the water while it is in the distribution system.

Ozonation for the treatment of drinking water has become increasingly popular, especially in European countries, as it reduces or eliminates the formation of THMs and HAAs in drinking water. The treatment of water with ozone has a wide range of applications, including efficient disinfection and the degradation of organic and inorganic pollutants. Ozone is a gas composed of three oxygen atoms (O_3) and is one of the most powerful oxidants. Ozone is produced with the use of energy by subjecting oxygen (O_2) to high-voltage electricity or to UV radiation, as illustrated in Figure 7.88. Its short half-life in water, approximately 10–30 minutes in practical treatment applications, requires that ozone be generated onsite for use as a disinfectant. Ozone is used as the primary disinfectant in many drinking water treatment plants, mostly in Europe and Canada.

BOX 7.5 Chlorine Toxicity and DBPs

Chlorine, particularly as hypochlorite, is a strong oxidizer. Chlorinated effluents may be lethal to aquatic life. When residual chlorine levels exceed 0.1 mg/L, laboratory tests show that the effluents are lethal to fish and invertebrate species, which may result in changes in aquatic ecosystems, such as reductions in diversity and shifts in the distribution of species. These effects, which may persist for up to 0.5 km downstream, can be caused by chlorine at levels as low as 0.02 mg/L.

When chlorine reacts with organic matter, it forms disinfection byproducts (DBPs). In recent years, regulators and the general public have focused more attention on potential health risks from DBPs. The predominant DBPs are trihalomethanes (THMs—chloroform, bromoform, bromodichloromethane, and dibro-mochloromethane) and haloacetic acids (HAAs—monochloro-, dichloro-, trichloro-, monobromo-, and dibromo-). THMs and HAAs form when chlorine reacts with organic material in source water (which comes from decomposing plant material, pesticides, etc.). The amount of THMs and HAAs in drinking water can change from day to day, depending on the season, water temperature, amount of chlorine added, amount of plant material in the water, and a variety of other factors. Concerned that these chemicals may be carcinogenic to humans, the U.S. EPA set the first regulatory limits for THMs in 1979. Since that time, a wealth of research has improved our understanding of how DBPs are formed, their potential health risks, and how they can be controlled.

It is now recognized that all chemical disinfectants form some potentially harmful byproducts. The byproducts of chlorine disinfection are by far the most thoroughly studied. The U.S. EPA has developed two stages of rules to minimize exposure to DBPs. Stage 1 Disinfectants and Disinfection Byproducts Rule (DBPR) reduces drinking water exposure to disinfection byproducts. The rule applies to community water systems and nontransient noncommunity systems, including those serving fewer than 10,000 people that add a disinfectant to the drinking water during any part of the treatment process.

The Stage 2 DBPR strengthens public health protection by tightening the requirements for compliance monitoring for THMs and HAAs. The rule targets public water systems (PWSs) with the greatest risk. Together, the Stage 1 and Stage 2 Disinfectants and Disinfection Byproducts Rules (DBPRs) improve drinking water quality and decrease health risks associated with certain cancers. Most water systems meet the DBPRs by controlling the amount of natural organic matter prior to disinfection.

Chlorine dioxide (ClO_2) is obviously a chlorine-based chemical, but it is not as strong an oxidizer as hypochlorite. It is often used as a preoxidant because, unlike chlorine, it does not chlorinate organic compounds and therefore does not react with organic matter in the water to form THMs. However, it does effectively oxidize reduced iron, manganese, sulfur compounds, and certain odor-causing organic substances in raw water.

Disinfection by ultraviolet irradiation at a wavelength of about 264 nm is also used for disinfecting water and wastewater. Economic factors that influence the cost effectiveness of UV treatment depend on the treatment plant capacity and requirements for storage and residual chlorine. The UV irradiation process does not produce significant quantities of known disinfection byproducts.

7.4.9 Finishing Steps and Distribution

In addition to the treatment processes already discussed, "finishing" processes may be added in some areas. Water must be stable, and it must not cause corrosion or

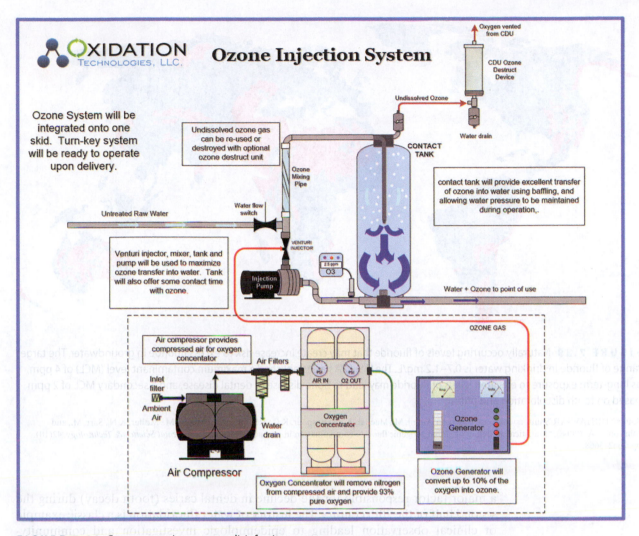

FIGURE 7.88 Components in an ozone disinfection system.

Source: www.oxidationtech.com/cws.html#1025=&1026=.

scaling prior to entering the distribution system. Two tests are used to determine the stability of water, the Marble Test and the Langelier Index, which indicate the calcium carbonate saturation level in the water. The water is considered stable if it is saturated with calcium carbonate. Unstable waters may be stabilized using recarbonation, acid addition, phosphate addition, alkali addition, or aeration.

Taste and odor (T&O) and color issues may also be addressed prior to distribution. These issues may arise in distribution pipes and home plumbing. T&O problems are prevented through source water management, proper plant operation and maintenance (O&M), and distribution system maintenance.

Most water has some fluoride, but usually not enough to prevent cavities. Community water systems can add the right amount of fluoride to the local drinking water to prevent cavities. Fluoride can occur naturally in surface waters from the deposition of particles in the atmosphere and the weathering of fluoride-containing rocks and soils, as shown in Figure 7.89. Fluoride in groundwater occurs from leaching from rock formations. Fluoridation of community drinking water is

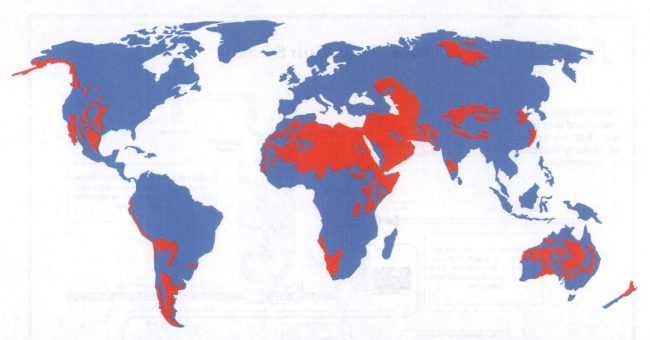

FIGURE 7.89 Naturally occurring levels of fluoride that may create increase risk to excess fluoride in groundwater. The target range of fluoride in drinking water is 0.7–1.2 mg/L. The U.S. EPA has established a maximum contaminant level (MCL) of 4 ppm, as long-term exposure to elevated levels of fluoride may lead to bone disease or dental disease, and a secondary MCL of 2 ppm based on tooth discoloration and pitting.

Source: DBHAVSAR/Shutterstock.com; Based on Amini, M., Mueller, K., Abbaspour, K.C., Rosenberg, T., Afyuni, M., Møller, K.N., Sarr, M., and Johnson, C.A. (2008). "Statistical modeling of global geogenic fluoride contamination in groundwaters." *Environmental Science & Technology*. 42(10), pp. 3662–3668.

a major factor responsible for the decline in dental caries (tooth decay) during the second half of the 20th century. The history of water fluoridation is a classic example of clinical observation leading to epidemiologic investigation and community-based public health intervention. Community water fluoridation is recommended by nearly all public health, medical, and dental organizations—most notably the American Dental Association, American Academy of Pediatrics, U.S. Public Health Service, and World Health Organization.

Water is typically stored in a clear well following treatment and prior to being pumped into the distribution system. The distribution system should supply water, without impairing its quality, in adequate quantities and at sufficient pressures to meet system requirements. The facilities that make up the distribution system include finished water storage; pumping, transmission, and distribution piping supply mains; and valves.

Drinking water is usually stored again, at a higher elevation, to ensure water availability in the event of a power outage, as illustrated in Figure 7.90. The water distribution system requires storage that is adequate for basic domestic purposes, commercial and industrial uses, and to accommodate the flows necessary for emergencies such as firefighting. Typically, a water tower is sized to hold approximately one day's worth of water for the community served by the tower. The pumps, fire

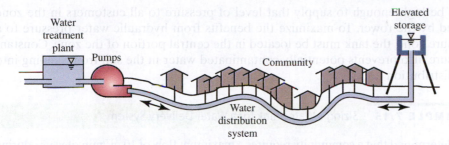

FIGURE 7.90 During periods of high water demand, the water flows from both the water plant and the elevated storage tanks. During low-demand periods, the pumps fill the elevated storage tanks.

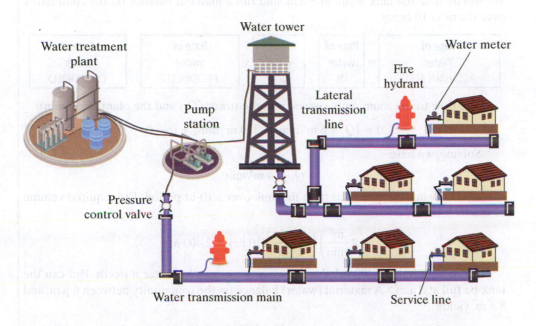

FIGURE 7.91 Appurtenances and parts of the water delivery system form the water source to the service line attached to a home or property.

hydrants, manholes, and other parts of the water delivery system are illustrated in Figure 7.91. In addition, the tower may play a major role during a fire and can affect the community's insurance rates. During a fire, the water demand increases significantly and may greatly exceed the capacity of the pumps at the community's water plant.

The elevation of the water tower determines the water pressure in the community. New systems are required to provide a minimum pressure of 138 kPa (20 psi) at ground level at all points in the distribution system under all flow conditions. Typically, normal working pressure in a distribution system is approximately 415–550 kPa (60–80 psi) and not less than 240 kPa (35 psi). The water level in the tower must be high enough to supply that level of pressure to all customers in the zone served by the tower. To maximize the benefits from hydraulic water pressure to a pressure zone, the tank must be located in the central portion of the zone. Constant pressure also prevents potentially contaminated water in the soil from seeping into the distribution lines.

EXAMPLE 7.15   Sizing a Water Tank for a Water Delivery System

It is determined that a community requires a maximum flow of 10 m³/min of water during 10 hours in a peak day, beginning at 8 a.m. and ending at 6 p.m. During the remaining 14 hours, it needs a flow of 2 m³/min. During the entire 24 hours, the water treatment plant is able to provide a constant flow of 6 m³/min, which is pumped into the distribution system. How large must the elevated storage tank be to meet this peak demand?

We will assume the tank is full at 8 a.m. and run a material balance on the community over the next 10 hours:

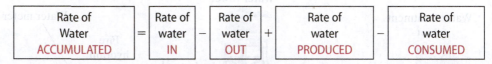

| Rate of Water ACCUMULATED | = | Rate of water IN | − | Rate of water OUT | + | Rate of water PRODUCED | − | Rate of water CONSUMED |

The flow to the community comes from the tower (Q_1) and the plant (6 m³/min):

$$0 = [Q_1 + 6 \text{ m}^3/\text{min}] - [10 \text{ m}^3/\text{min}] + 0 - 0$$

Solving, we have

$$Q_1 = 4 \text{ m}^3/\text{min}$$

This flow must be provided from the tank over a 10-hr period. The required volume of the tank is

$$\left(4 \frac{\text{m}^3}{\text{min}}\right)\left(60 \frac{\text{min}}{\text{hr}}\right)(10 \text{ hr}) = 2{,}400 \text{ m}^3$$

If the tank holds 2,400 m³, the community can get the water it needs. But can the tank be full at 8 a.m.? A material (water) balance on the community between 6 p.m. and 8 a.m. yields

$$0 = [6 \text{ m}^3/\text{min}] - [2 \text{ m}^3/\text{min} + Q_2] + 0 - 0$$

The flow coming into the community is 6 m³/min while the flow out is 2 m³/min (the water used) plus Q_2, which is water needed to fill the tank:

$$Q_2 = 4 \text{ m}^3/\text{min}$$

This is spread over 14 hours, so that

$$\left(4 \frac{\text{m}^3}{\text{min}}\right)\left(60 \frac{\text{min}}{\text{hr}}\right)(14 \text{ hr}) = 3{,}360 \text{ m}^3$$

will be supplied to the tower. There is no problem filling the tank.

7.5 Summary

Access to an appropriate quantity of water was discussed in detail in Chapter 3. In this chapter, we have investigated issues related to water quality. Water-related microorganisms in water mostly cause diseases, which are still endemic in large areas of the world. Microbial contamination of water is often caused by improper sanitation and lack of adequate wastewater treatment. Improper treatment of human and animal wastes also leads to environmental impacts from elevated levels of organic material and nutrients in water. Unnatural increases in organic material and nutrients decrease the available dissolved oxygen in water, negatively impacting many aquatic species in freshwater and marine ecosystems. Environmental engineers use mathematical models to estimate the causes, effects, and potential remediation strategies associated with water pollution. The quality of water required for drinking water differs substantially from the quality of water required for irrigation and some industrial uses. Water quality and water quantity are related in determining appropriate water reuse strategies that may be able to offset increasing water demand and greater global water scarcity. Methods to assess water reuse strategies are discussed in more detail in Chapters 8 and 16.

References

Aggarwal, P. K., Dargie, M., Groening, M., Kulkarni, K. M., and Gibson, J. J. (2001). "An inter-laboratory comparison of arsenic analysis in Bangladesh Draft report (INIS-XA-649)." International Atomic Energy Agency (IAEA).

Ahmad, S. A., Khan, M. H., and Haque, M. (2018). "Arsenic contamination in groundwater in Bangladesh: Implications and challenges for healthcare policy." *Risk Management and Healthcare Policy* 11:251–261. doi:10.2147/RMHP.S153188.

Amini, M., Mueller, K. Abbaspour, K. C., Rosenberg, T., Afyuni, M., Møller, K. N., Sarr, M., and Johnson, C. A. "Statistical modeling of global geogenic fluoride contamination in groundwaters." *Environmental Science & Technology*. 42(10), pp. 3662–3668.

Baca, R. G., and Arnett, R. C. (1976). A limnological model for eutrophic lakes and impoundments. Battelle, Inc., Pacific Northwest Laboratories, Richland, WA.

Baca, R. G., Waddel, W. W., Cole, C. R., Brandsletter, A., and Caerlock, D. B. (1973). EXPLORE-I: A river-basin water quality model. Battelle Laboratories, Richland, WA.

Bartram, J., Lewis, K., Lenton, R., and Wright, A. (2005). "Focusing on improved water and sanitation." *The Lancet* 365(9461):810–812.

Bauer, D. P., Rathbun, R. E., and Lowham, H.W. (1979). Travel time, unit-concentration, longitudinal dispersion, and reaeration characteristics of upstream reaches of the Yampa and Snake Rivers, Colorado and Wyoming. USGS Water Resources Investigation. 78–122.

Bedford, K. W., Sylees, R. M., and Libicki, C. (1983). "Dynamic advective water quality models for rivers." *ASCE Jour. Env. Engr.* 109:535–554.

Bellamy, W., Silverman, G., et al. (1985). "Removing *Giardia* cysts with slow sand filtration." *Journal of the American Water Works Association.* 77(2): pp. 52–60.

Bhattacharya, P., Polya, D. A., and Jovanovic, D. (2017). *Best Practice Guide on the Control of Arsenic in Drinking Water*. IWA Publishing. DOI: doi.org/10.2166/9781780404929. ISBN: 9781780404929.

Bowen, R. (1999). "Taste and odor control." *Water Treatment Plant Operation: A Field Study Training Program*. California Department of Health Services and U.S. EPA. Sacramento: California State University.

Bowie, G. L., Mills, W. B., Porcella, D. B., Campbell, C. L., Pagenkopf, J. R., Rupp, G. L., et al. (1985). *Rates, Constants, and Kinetics Formulations in Surface Water Quality Modeling*, 2nd ed. Environmental Research Laboratory. Office of Research and Development, U.S. Environmental Protection Agency, Athens, GA.

Centeno, J. A., et al. (2007). "Global impacts of geogenic arsenic: A medical geology research case." *Royal Swiss Academy of Sciences*. February, pp. 78–81.

Chapman, D. (1996). *A Guide to Use of Biota, Sediments and Water in Environmental Monitoring*, 2nd ed. Chapter 7—Lakes. UNESCO/WHO/UNEP.5.

Chen, C. W. (1970). "Concepts and utilities of ecological mode." *ASCE Journ. San. Eng. Div.* 96:SA5.

Chesapeake Bay Foundation. (2004). Manure's Impact on Rivers, Streams and the Chesapeake Bay: Keeping Manure Out of the Water. Annapolis, MD.

Churchill, M. A., Elmore, H. L., and Buckingham, R. A. (1962). "The prediction of stream reaeration rates." *ASCE, Journal Sanitary Engineering Division* 88(SA4):1–46.

Clausen, T., Brown, J., Suntura, O., and Collin, S. (2004). "Safe household water treatment and storage using ceramic drip filters: A randomized controlled trial in Bolivia." *Water Science and Technology* 50(1):111–115.

Corcoran, E., Nellemann, C., Baker, E., Bos, R., Osborn, D., and Savelli, H. (Eds.). (2010). *Sick Water? The Central Role of Wastewater Management in Sustainable Development*. A Rapid Response Assessment. United Nations Environment Programme, UN-HABITAT, GRID-Arendal. *www.grida.no*. Printed by Birkeland Trykkeri AS, Norway.

Covar, A. P. (1976). "Selecting the proper reaeration coefficient for use in water quality models." Presented at the U.S. EPA Conference on Environmental Simulation and Modeling. April 19–22, Cincinnati, OH.

Davis, M. L. (2010). *Water and Wastewater Engineering Design Principles and Practices*, 1st Ed. New York: McGraw-Hill. pp. 10–21.

Di Toro, D. M., and Connolly, J. P. (1980). Mathematical models of water quality in large lakes. Part 2: Lake Erie. U.S. Environmental Protection Agency, Duluth, MN. EPA-600/3-3-80-065.

Fahlin, C. (2002). Filtron Assessment for Drinking Water Treatment. Undergraduate Research Opportunities Program (UROP). University of Colorado—Engineers Without Borders.

Fair, G. M., Geyer, J. C., and Okun, D. A. (1968). *Water and Wastewater Engineering*. New York: John Wiley & Sons.

Genet, L. A., Smith, D. J., and Sonnen, M. B. (1974). Computer program documentation for the dynamic estuary model. U.S. Environmental Protection Agency, Systems Development Branch, Washington, DC.

Gleick, P. H. (2002). *Dirty Water: Estimated Deaths from Water-related Diseases 2000–2020*. Pacific Institute for Studies in Development, Environment and Security, Research Report. *www.pacinst.org*.

Goldman, C. R., and Horne, A. J. (1983). *Limnology*. New York: McGraw-Hill.

Grutzmacher, G., Bottcher, G., et al. (2002). "Removal of microcystins by slow sand filtration." *Environmental Toxicology* 17(4): pp. 386–394.

Harleman, D. R. F., Dailey, J. E., Thatcher, M. L., Najarian, T. O., Brocard, D. N., and Ferrara, R. A. (1977). User's manual for the M.I.T. transient water quality model network model—Including nitrogen-cycle dynamics for rivers and estuaries. R.M. Parsons Laboratory for Water Resources and Hydrodynamics, Massachusetts Institute of Technology, Cambridge, MA. For U.S. Environmental Protection Agency, Corvallis, OR. EPA-600/3-77-010.

Heidecke, C. (2006). Development and Evaluation of a Regional Water Poverty Index for Benin. International Food Policy Research Institute. Environment and Production Division, Washington, DC.

Imhoff, J. C., Kittle, J. L. Jr., Donigian, A. S. Jr., and Johanson, R. C. (1981). Users Manual for Hydrological Simulation Program—Fortran (HSPF). Contract 68-03-2895. U.S. Environmental Protection Agency, Athens, GA.

IRC. (2002). "Small community water supplies." Technical paper no. 40. The Hague: IRC International Water and Sanitation Centre.

Kitawaki, H. (2002). "Common problems in water supply and sanitation in developing countries." *International Review for Environmental Strategies* 3:264.

Komenic, V., Ahlers, R., and van der Zaag, P. (2009). "Assessing the usefulness of the water poverty index by applying it to a special case: Can one be water poor with high levels of access?" *Physics and Chemistry of the Earth*, Parts A/B/C. 34(4–5):219–224.

Kumar et al. (2015). "Ground water arsenic poisoning in "Tilak Rai Ka Hatta," village of Buxar District, Bihar, India, causing severe health hazards and hormonal imbalance." *J. Environ. Anal. Toxicol.* 5:4.

Lantagne, D., Klarman, M., Mayer, A., Preston, K., Napotnik, J., and Jellison, K. (2010). "Effect of production variables on microbial removal in locally-produced ceramic filters for household water treatment." *International Journal of Environmental Health Research* 20(3):171–187.

Lawrence, P., Meigh, J., and Sullivan, C. (2002). The Water Poverty Index: An International Comparison. Keel Economics Research Papers. Department of Economics. Keele University, Keele, Staffordshire, UK.

Masters, G., and Ela, W. P. (2007). *Introduction to Environmental Engineering and Science*, 3rd ed. Upper Saddle River, NJ: Prentice Hall.

Medina, M. A. Jr. (1979). Level II—Receiving water quality modeling for urban stormwater management. U.S. Environmental Protection Agency. Municipal Environmental Research Laboratory, Cincinnati, OH. EPA-600/2-79-100.

Metropolitan Washington Council of Governments. (1982). Application of HSPF to Seneca Creek watershed, Washington, DC.

Moreau, D., and Little, K. (1989). "Managing public water supplies during droughts; Experiences in the United States in 1986 and 1988." *Water Resources Research Institute Report*, No. 250; University of North Carolina.

Mukherjee, A. B., and Bhattacharya, P. (2001). "Arsenic in groundwater in the Bengal Delta Plain: Slow poisoning in Bangladesh." *Environmental Review*, pp. 189–220.

O'Conner, D. J., and Dobbins, W. E. (1958). "Mechanisms of reaeration in natural streams." *American Society of Civil Engineering Proceedings, Transactions* 123(2934):641–684.

Owens, M., Edwards, R. W., and Gibbs, J. W. (1964). "Some reaeration studies in streams." *Int. J. Air Wat. Poll.* 8:469–486.

Reckhow, H., and Chapra, S. C. (1983). *Engineering Approaches for Lake Management. Vol 1. Data Analysis & Empirical Modeling.* Butterworth-Heinemann. London, UK.

Ribbot, J. C., Najam, A., and Watson, G. (1996). *Climate Variability, Climate Change and Social Vulnerability in the in the Semi-arid Tropics.* Eds. Ribbot, J.C., Magalhaes, A. R., and Panagides, S. S. New York: University of Cambridge. pp. 13–54.

Rich, L. G. (1973) *Environmental Systems Engineering.* New York: McGraw-Hill.

Roesner, L. A., Giguere, P. A., and Evenson, D. E. (1981). User's Manual for Stream Quality Model (QUAL-II). U.S. Environmental Protection Agency, Environmental Research Laboratory, Athens, GA. EPA-600/a-81-015.

Schroepfer, G. J., Robins, M. L., and Susag, R. H. (1960). "A reappraisal of deoxygenation rates of raw sewage, effluents, and receiving waters." *Water Pollution Control Federation Jour.* 32(11):1212–1231.

Smith, D. J. (1978). WQRRS, Generalized Computer Program for River-Reservoir Systems. U.S. Army Corps of Engineers, Hydrologic Engineering Center (HEC), Davis, CA. User's Manual 401-100, 100A, 210 pp.

Streeter, H. W., and Phelps, E. B. (1925). A study of the pollution and natural purification of the Ohio River, III. Factors concerned in the phenomena of oxidation and reaeration. United States Public Health Service. Public Health Bulletin No. 146.

Striebig, B., Atwood, S., Johnson, B., Lemkau, B., Shamrell, J., Spuler, P., et al. (2007). "Activated carbon amended ceramic drinking water filters for Benin." *Journal of Engineering for Sustainable Development* 2(1):1–12.

Striebig, B., Gieber, T., Cohen, B., Norwood, S., Godfrey, N., and Elliot, W. (2010). "Implementation of a Ceramic Water Filter Manufacturing Process at the Songhai Center: A Case Study in Benin." *International Water Association's World Water Congress*, Montreal.

Sullivan, C. (2002). "Calculating a Water Poverty Index." *World Development* 30(7): 1195–1210.

Tchobanoglous, G., Burton, F. L., and Stensel, H. D. (2003). *Wastewater Engineering Treatment and Reuse*, 4th ed. New York: McGraw-Hill.

Thomann, R. V., and Fitzpatrick, J. J. (1982). Calibration and verification of a mathematical model of eutrophication of the Potomac Estuary. Department of Environmental Services, Government of the District of Columbia.

UN. (2000). Millennium Declaration, UN A/Res/55/2.

UNICEF. (2004). The Official Summary of the State of the World's Children. United Nations Children's Fund (UNICEF) New York.

UNICEF. (2008). UNICEF Handbook on Water Quality. United Nations Children's Fund (UNICEF) New York.

U.S. EPA. (2000). Progress in Water Quality: An Evaluation of the National Investment in Municipal Wastewater Treatment. U. S. Environmental Protection Agency, Office of Water, Washington, DC.

U.S. EPA. (2004). Primer for Municipal Treatment Systems. U.S. Environmental Protection Agency, Office of Wastewater Management, Office of Water, Washington, DC.

U.S. EPA. (2010). Chesapeake Bay Total Maximum Daily Load for Nitrogen, Phosphorus and Sediment. U.S. Environmental Protection Agency. Region 3. Water Protection Division. Philadelphia, PA.

van Halem, D., van der Lann, H., Heijman, S. G. J., van Dijk, J. C., and Amy, G. L. (2009). "Assessing the sustainability of the silver-impregnated ceramic pot filter for low-cost household drinking water treatment." *Physics and Chemistry of the Earth* 34(2009):36–42.

Velz, C. J. (1984). *Applied Stream Sanitation*. New York: John Wiley & Sons.

Wang, B. (2015). *Next Big Future: Israel scales up reverse osmosis desalination to slash costs with a fourth of the piping.* 19 February, 2015. The Next Big Future.com. Accessed 3/17/21. nextbigfuture.com/2015/02/isreal-scales-up-reverse-osmosis.html.

Wicks, C., Jasin, D., and Longstaff, B. (2007). "Breath of Life: Dissolved Oxygen in Chesapeake Bay." May 27, 2007 Newsletter. Chesapeake Bay Foundation, Annapolis, MD.

World Health Organization (WHO). (2008). "An interview with Mahmuder Rahman." *Bulletin of the World Health Organization*. January 2008, pp. 11–12.

World Health Organization (WHO). (2012). GLASS 2012 Report: UN-Water Global Analysis and Assessment of Sanitation and Drinking Water: The Challenge of Extending and Sustaining Services. World Health Organization. Geneva, Switzerland.

World Water Assessment Programme. (2009). *The United Nations World Water Development Report 3: Water in a Changing World*. Paris: UNESCO, and London: Earthscan.

Key Concepts

Water Poverty Index (WPI)
Prokaryotes
Eukaryotes
Bacteria
Protozoa
Helminths
Viruses
Algae
Algae blooms
Aerobic organisms
Indicator organisms
Dissolved oxygen (DO)
Saturated dissolved oxygen (DO_s)
Oxygen deficit
Substrate
Substrate utilization rate
Carbonaceous oxygen demand (COD)
Nitrification
Nitrogenous oxygen demand (NOD)
Endogenous respiration
Theoretical oxygen demand (ThOD)
Biochemical oxygen demand (BOD)
Eutrophication
Reaeration
Critical point
Morphology
Thermal stratification

Epilimnion
Hypolimnion
Metalimnion
Thermocline
Combined sewer overflow (CSO)
Water demand management
Potable water
Vadose zone
Zone of aeration
Zone of saturation
Water table
Groundwater
Aquifer
Unconfined aquifer
Confined aquifer
Artesian well
Reservoir
Frequency analysis
Droughts
Coagulation
Flocculation
Critical particles
Overflow rate
Contact time
Free chlorine
Combined residual
Breakpoint dose

Problems

7-1 Describe the human use, human health, or environmental concerns associated with the following water quality parameters:
a. Total suspended solids
b. Biochemical oxygen demand
c. Nutrients
d. Pathogens

7-2 Describe the differences between prokaryotes and eukaryotes.

7-3 A poorly designed latrine is located upstream of a surface drinking water source. Two cubic meters of stormwater contaminated by the latrine and containing 10^{10} MPN/100 mL of salmonella flow into a shallow pool for collecting drinking water. The total volume of the drinking water pool after the contaminated water flows into it is 5 m^3. If the typical infectious dose for salmonella in this community is 100 organisms, how much water would someone have to drink to become ill due to salmonella contamination?

7-4 A stream flowing through a rural agricultural region contains 10^{-1} MPN/100 mL of *Giardia lamblia* cysts. If a backpacking tourist drinks 5 L of water per day, how many days will pass before the hiker is likely to become ill due to salmonella contamination? Assume the typical infectious dose for *Giardia* in this community is 10 organisms.

7-5 What is the saturated dissolved oxygen concentration for water at sea level and 18°C?

7-6 What is the saturated dissolved oxygen concentration for water at 2,000 m and 15°C?

7-7 What is the percent saturation of water at sea level and 25°C containing 9.8 mg/L of dissolved oxygen? Would you expect the dissolved oxygen level to increase or decrease if the system moved toward equilibrium?

7-8 What is the dissolved oxygen deficit for water containing 5 mg/L of dissolved oxygen at sea level and 18°C?

7-9 What is the dissolved oxygen deficit for water containing 5 mg/L of dissolved oxygen at 2,000 m and 15°C?

7-10 What is the theoretical carbonaceous oxygen demand for water containing 25 mg/L of propyl alcohol, C_3H_7OH?

7-11 What is the theoretical carbonaceous oxygen demand for water containing 100 mg/L of acetic acid, CH_3COOH?

7-12 Bacterial cells are often represented by the chemical formula $C_5H_7NO_2$. Compute the theoretical carbonaceous oxygen demand (in mg/L) in a 1 molar solution of cells in water.

7-13 A 5-mL sample of municipal sewage is used for a BOD_5 analysis, and 4.7 mg/L of oxygen is consumed after the sample is incubated for five days in the dark at 20°C. What is the BOD_5 of the sewage?

7-14 A bottle of water has a BOD_5 of 10 mg/L. The initial DO in the BOD bottle is 8 mg/L and the dilution is 1:10. What is the final DO in the BOD bottle?

7-15 What is the BOD_5 of the domestic wastewater sample, given the following data collected from the laboratory?

WASTEWATER SAMPLE	VOLUME OF WASTEWATER (mL)	INITIAL DO OF WASTEWATER (mg/L)	VOLUME OF DILUTION WATER (mL)	INITIAL DO OF DILUTION WATER (mg/L)	FINAL DO (mg/L)
Sample A	2.0	3.0	298.0	10.5	6.2
Sample B	6.0	4.9	294.0	10.5	7.6
Sample C	10.0	3.8	290.0	10.5	8.2

7-16 What is the BOD_5 of the lake water sample, given the following data collected from the laboratory?

	VOLUME OF WASTEWATER (mL)	INITIAL DO OF WASTEWATER (mg/L)	VOLUME OF DILUTION WATER (mL)	INITIAL DO OF DILUTION WATER (mg/L)	FINAL DO (mg/L)
Sample A	100	6.0	200	9.2	5.4
Sample B	200	7.0	100	9.2	4.4
Sample C	300	8.1	0	9.2	6.1

7-17 Operators at the local wastewater treatment plant have reported the following data to determine how well the plant is operating.
a. What percent of the BOD does the treatment plant remove?
b. Is the plant in compliance with its permit if it is required to remove 85% of the BOD?

	VOLUME OF WASTEWATER (mL)	INITIAL DO OF WASTEWATER (mg/L)	VOLUME OF DILUTION WATER (mL)	INITIAL DO OF DILUTION WATER (mg/L)	FINAL DO (mg/L)
Influent to treatment facility	5	3.0	295	8.5	6.2
Effluent from treatment facility	25	8.2	275	8.5	5.9

7-18 Consider the following data from a BOD test:

DAY	DO (mg/L)	DAY	DO (mg/L)
0	9	5	6
1	9	6	6
2	9	7	4
3	8	8	3
4	7	9	3

If there is no dilution (i.e., the dilution factor is 1), what is the:
a. BOD_5?
b. ultimate carbonaceous BOD?
c. ultimate nitrogenous BOD?
d. Why do you think no oxygen is used until the third day?

7-19 Consider the following data from a BOD test. Assume there is no dilution. Plot the BOD values versus time. Calculate BOD_5.

DAY	DO (mg/L)	DAY	DO (mg/L)
0	9	4	5
1	8	5	4.5
2	7	6	4
3	6		

7-20 Some former scientists and opponents of dam removal in the Northwest have recently reversed their stance and now believe dam removal is the only way to help endangered salmon. They state that their position has changed because of growing climate change concerns.
a. If water temperatures increase from 20°C to 25°C in the Columbia River, what change in the saturated dissolved oxygen levels (in mg/L) would be expected in the river?
b. In what way does impounding (damming) a river reduce the dissolved oxygen? Please state the specific variable that might be affected.

7-21 A wastewater sample has $k_d = 0.20 \, day^{-1}$ and an ultimate BOD $L_o = 200 \, mg/L$. What is the final dissolved oxygen after five days in a BOD bottle in which the sample is diluted 1:20 and the initial DO is 10.2 mg/L?

7-22 Calculate the concentration of oxygen in milligrams per liter required to completely oxidize the following organic compounds. Also calculate the volume of air in liters per liter of solution treated. Use 500 mg/L as the concentration of each compound.

a. Methyl tertiary butyl ether (MTBE), $C_5H_{12}O$

b. Benzylmorphine, $C_{24}H_{25}NO_3$

c. Bepridil, $C_{24}H_{34}N_2O$

d. Toluene, $H_5C_6H_2C$-H

7-23 A student places two BOD bottles in an incubator, having measured the initial DO of both as 9.0 mg/L. Bottle A contains 100% sample, and bottle B has 50% sample and 50% unseeded dilution water. The final DO at the end of five days is 3 mg/L in bottle A and 4 mg/L in bottle B.

a. What was the BOD_5 of the sample as measured in each bottle?

b. What might have happened to make these values different?

c. Do you think the BOD measure included:

i)   only carbonaceous BOD?

ii)   only nitrogenous BOD?

iii)   both carbonaceous and nitrogenous BOD?

d. Why so you think so?

7-24 The BOD_5 of an industrial waste after pretreatment is 220 mg/L, and the ultimate BOD is 320 mg/L.

a. What is the deoxygenation constant k_d (base 10)?

b. What is the deoxygenation constant k_d (base e)?

7-25 The ultimate BOD of each of two wastes is 280 mg/L. For the first, the deoxygenation constant k_{d1} (base e) = 0.08 day^{-1}, and for the second, k_{d2} (base e) = 0.12 day^{-1}. What is the BOD_5 of each sample?

7-26 A mixture of river water and effluent from a municipal wastewater treatment plant is completely mixed near the point of discharge. The dissolved oxygen concentration in the river at the mixing point is 5.6 mg/L, and the ultimate BOD of the mixed water is 24 mg/L. The temperature of the river is 21°C. The deoxygenation constant for the river is k_d = 0.06/day.

a. Estimate the reoxygenation coefficient assuming the river speed is 0.25 m/s and the average depth is 3 m.

b. Find the critical time downstream at which the minimum DO occurs.

c. Find the minimum DO downstream.

7-27 The flow rate of the Madison River is about 500 cu ft per second (CFS). The Madison River has an ultimate BOD of 5.9 mg/L and DO of 6.2 mg/L. The Madison Wastewater Treatment Plant (WWTP) discharges 10 million gallons per day (MGD) of treated wastewater with an ultimate BOD of 20 mg/L and DO equal to 8.0 mg/L. (*Note:* 1 cu ft = 7.48 gal)

a. What is the BOD (mg/L) immediately downstream of the discharge?

b. What is the DO (mg/L) immediately downstream of the discharge?

c. If the temperature of the mixed water at the point of discharge is 20°C, what is the initial oxygen deficit in mg/L?

d. If the deoxygenation constant is 0.8 day^{-1} and the reoxygenation rate is 1.4 day^{-1}, what is the critical time in days?

e. What would the minimum dissolved oxygen concentration in the river be in mg/L?

7-28 The flow rate of the Spokane River is about 5,000 cu ft per second (CFS). The Spokane River has an ultimate BOD of 5 mg/L and DO of 6 mg/L. The

Spokane Wastewater Treatment Plant (WWTP) discharges 50 million gallons per day (MGD) of treated wastewater, with an ultimate BOD of 30 mg/L and DO equal to 8 mg/L.

a. What is the BOD (mg/L) immediately downstream of the discharge?

b. What is the DO (mg/L) immediately downstream of the discharge?

c. If the temperature of the mixed water at the point of discharge is 20°C, what is the initial oxygen deficit in mg/L?

d. If the deoxygenation constant is 0.4 day^{-1} and the reaeration rate is 0.8 day^{-1}, what is the critical time in days?

e. What would the minimum dissolved oxygen concentration in the Spokane River be in mg/L?

f. Plot the oxygen sag curve for the river.

7-29 A stream has a dissolved oxygen level of 9 mg/L, an ultimate oxygen demand of 12 mg/L, and an average flow of 0.2 m^3/s. Industrial waste at zero dissolved oxygen with an ultimate oxygen demand of 20,000 mg/L and a flow rate of 0.006 m^3/s is discharged into the stream. What are the ultimate oxygen demand and the dissolved oxygen in the stream immediately below the discharge?

7-30 A large stream with a velocity of 0.85 m/s, saturated with oxygen, has a reoxygenation constant of $k_r = 0.4\ day^{-1}$ and a temperature of $T = 12°C$, with an ultimate BOD = 13.6 mg/L and a flow rate $Q = 2.2\ m^3/s$. Wastewater with a dissolved oxygen concentration of 1.5 mg/L is discharged to this stream. The flow rate of the waste is 0.5 m^3/s, the temperature is $T = 26°C$, and the ultimate BOD is $L_o = 220$ mg/L. Downstream the deoxygenation constant is $k_d = 0.2\ day^{-1}$. What is the dissolved oxygen 48.3 km downstream?

7-31 Below a discharge from a wastewater treatment plant, an 8.6-km stream has a reoxygenation constant of 0.4 day^{-1}, a velocity of 0.15 m/s, a dissolved oxygen concentration of 6 mg/L, and an ultimate oxygen demand (L) of 25 mg/L. The stream is at 15°C. The deoxygenation constant is estimated at 0.25 day^{-1}.

a. Might there be a healthy population of fish in this stream?

b. Why should we care if there are fish in the stream? Do fish deserve moral consideration and protection?

7-32 A municipal wastewater treatment plant discharges into a stream that, during some points each year, has no other flow. The waste stream has a flow = 0.1 m^3/s, a dissolved oxygen concentration = 6 mg/L, a temperature = 18°C, a deoxygenation rate constant $k_d = 0.23\ day^{-1}$, and an ultimate BOD (L_o) = 280 mg/L. The velocity in the stream is 0.5 m/s, and the reoxygenation constant k_r is estimated to be 0.45 day^{-1}.

a. Will the stream maintain a minimum DO of 4 mg/L?

b. If the flow of the stream above the outfall has a temperature of 18°C, has no demand for oxygen, and is saturated with DO, how great must the streamflow be to ensure a minimum dissolved oxygen of 4 mg/L downstream of the discharge? (*Note:* Please do not try to answer this question by hand. Use a spreadsheet or similar mathematical software.)

7-33 Ellerbe Creek, a large stream at normal velocity, is the recipient of the wastewater from the 10 MGD Durham Northside Wastewater Treatment Plant. It has a mean summertime flow of 0.28 m^3/s, a temperature of 24°C, and a velocity of 0.25 m/s. Assume the dissolved oxygen is saturated and the ultimate BOD is 0 mg/L. The waste stream has a dissolved oxygen concentration = 2 mg/L, a temperature = 28°C, a deoxygenation rate constant $k_d = 0.23\ day^{-1}$, and an ultimate BOD (L_o) = 40 mg/L. The total stream length from the outfall to the river is 14 miles, at which point it empties into the Neuse River. The reoxygenation constant k_r is estimated to be 0.35 day^{-1} for the reach of the stream downstream of the discharge.

a. Should the state of North Carolina be concerned about the effect of this discharge on Ellerbe Creek?

b. Other than legal considerations, why *should* the state be concerned about the oxygen levels in Ellerbe Creek? It isn't much of a creek actually, and it empties into the Neuse River without being of much use to anyone. Yet the state has set dissolved oxygen levels of 4 mg/L for the 10-yr, 7-day low flow. Write a letter to the editor of a fictitious local newspaper decrying the spending of tax revenues for improvements to the Northside Treatment Plant just so the dissolved oxygen levels in Ellerbe Creek can be maintained above 4 mg/L.

c. Imagine you are a fish in Ellerbe Creek. You have read the letter to the editor in part b and it has upset you. Write a letter to the editor of the newspaper from the standpoint of the fish. The quality of your letter will be judged on the basis of the strength of your arguments.

7-34 Bald Eagle Creek flows into Bald Eagle reservoir, which acts as a completely mixed reactor. Bald Eagle Creek has a flow rate of 0.20 m^3/s and a dissolved oxygen concentration of 4.75 mg/L. Bald Eagle reservoir has a volume of 10^4 m^3. Oxygen is removed in the lake through biological processes, and the reaction rate, k_d, is 0.06 day^{-1}. Evaporation removes water from the surface of the lake at a rate of 0.002 m^3/s. Determine the effluent flow rate and the dissolved oxygen concentration for this completely mixed lake.

7-35 Sketch and briefly describe why thermal stratification occurs in a lake in Scotland over four seasons.

7-36 You have been assigned to conduct a site assessment for a lake in Montana. A new biofuels plant is operating and discharging into the stream that feeds the lake and provides water for agricultural production of cattle and switchgrass. You observed high levels of algae floating on the surface of the lake, and the water was highly turbid. You also measured the DO at 1 m to be 5 mg/L, and at a depth of 20 m the DO was 1.0 mg/L.

a. How would you describe the productivity level of the lake? Explain your reasoning.

b. Similar lakes have threatened Yellowstone cutthroat trout populations, but this lake does not. What changes would you suggest be made in managing the lake's water quality prior to reintroducing trout to the lake?

7-37 The Ganges River in India has an annual average flow rate of 12,105 m^3/s. The average temperature of the Ganges River is reported to be 24.6°C. The Ganges has become contaminated with raw sewage and industrial waste, so that the dissolved oxygen in the river upstream of sewage discharges is 4.6 mg/L. Suppose the raw sewage that is untreated flows into the Ganges at a rate of 460 m^3/s, the temperature is 25°C, and the dissolved oxygen in the sewage stream is completely depleted (i.e., DO = 0). Develop a plan for improving water quality in the Ganges River.

7-38 Refer to Problem 7-37 for the following questions:

a. Assuming the Ganges River is well mixed immediately at the point of discharge of the sewage stream, what is the DO in mg/L immediately downstream of the discharge?

b. What is the initial oxygen deficit in the mixed sewage and river water in mg/L?

c. Suppose the following data are collected to estimate the five-day BOD in the river and sewage water:

	VOLUME OF SAMPLE (mL)	INITIAL DO OF SAMPLE (mg/L)	VOLUME OF DILUTION WATER (mL)	INITIAL DO OF DILUTION WATER (mg/L)	FINAL DO (mg/L)
River water	25	4.6	275	7.8	4.7
Sewage	2	0	298	7.6	5.9

i) Calculate the BOD of the river water.
ii) Calculate the BOD of the sewage stream.
iii) Calculate the ultimate BOD of the river, assuming $k = 0.20$ day^{-1}.
iv) Calculate the ultimate BOD of the sewage stream, assuming $k = 0.50$ day^{-1}.

d. Calculate the ultimate BOD of the mixed river and sewage stream.

e. Estimate the reaeration constant assuming the mean depth of the river is 1.5 m and the mean velocity is 0.1 m/s.

f. Suppose the deoxygenation constant is 4.5(k_r) day^{-1}. At what time downstream will the critical point occur?

g. What would the minimum dissolved oxygen concentration in the Ganges River be in mg/L?

h. Plot the oxygen sag curve for the river.

i. Will there be negative effects on the aquatic ecosystem?

j. What effects might there be on human health and recreation? Note that the Ganges is a source of drinking water and water for washing, and it also has spiritual significance to the surrounding population.

k. If a dam were built to create a reservoir impounding part of the Ganges, what characteristic trophic state might you expect for the reservoir? Explain your answer.

l. Explain how you might design a series of water treatment facilities to treat the raw sewage. What processes would be required, and what positive impacts would treating the sewage have on water quality?

m. Suppose the upstream river water quality is improved so that the dissolved oxygen content of the river prior to receiving water from the sewage stream is equal to 90% of the saturated DO level. Suppose also that the sewage is treated in a series of wastewater treatment facilities.

    i) If the minimum allowable DO level was established to be 2.0 mg/L, what would the allowable ultimate BOD discharged from the wastewater treatment facilities be?

    ii) What percent removal efficiency would be needed?

7-39 A typical colloidal clay particle suspended in water has a diameter of 1.0 μm. If coagulation and flocculation with other particles increase its size to 100 times its initial diameter (at the same shape and density), how much shorter will be the settling time in 10 ft of water, such as in a settling tank?

7-40   An unconfined aquifer is 10 m thick and is being pumped so that one observation well placed at a distance of 76 m shows a drawdown of 0.5 m. On the opposite side of the extraction well is another observation well, 100 m from the extraction well, and this well shows a drawdown of 0.3 m. Assume the coefficient of permeability is 50 m/day.
   a. What is the discharge of the extraction well?
   b. Suppose the well at 100 m from the extraction well is now pumped. Show with a sketch what this will do to the drawdown.
   c. Suppose the aquifer sits on a geologic barrier that has a slope of 1/100. Show with a sketch how this would change the drawdown.

7-41   A settling tank in a water treatment plant has an inflow of 2 m³/min and a solids concentration of 2,100 mg/L. The effluent from this settling tank goes to sand filters. The concentration of sludge coming out of the bottom (the underflow) is 18,000 mg/L, and the flow to the filters is 1.8 m³/min.
   a. What is the underflow flow rate?
   b. What is the solids concentration in the effluent?
   c. How large must the sand filters be (in m²)?

7-42   A settling tank is 20 m long, 10 m deep, and 10 m wide. The flow rate to the tank is 10 m³/min. The particles to be removed all have a settling velocity of 0.1 m/min.
   a. What is the hydraulic retention time?
   b. Will all the particles be removed?

7-43   The settling basins for a 50-MGD wastewater treatment plant are operated in parallel with flow split evenly among 10 settling tanks, each of which is 3 m deep and 25 m wide, with a length of 32 m.
   a. What is the expected theoretical percentage removal for particles of 0.1 mm diameter that settle at 1×10^{-2} m/s?
   b. What theoretical percentage removal is expected for particles of 0.01 mm diameter that settle at 1×10^{-4} m/s?

7-44   A 0.1-m-diameter well fully penetrates an unconfined aquifer 20 m deep. The permeability is 2×10^{-3} m/s. How much can it pump for the drawdown at the well to reach 20 m and the well to start sucking air?

7-45   The settling velocity of a particle is 0.002 m/s, and the overflow rate of a settling tank is 0.008 m/s.
   a. What percent of the particles does the settling tank capture?
   b. If the particles are flocculated so that their settling rate is 0.05 m/s, what fraction of the particles is captured?
   c. If the particles are not changed and another settling tank is constructed to run in parallel with the original settling tank, will all of the particles be captured?

7-46   A water treatment plant is being designed for a flow of 1.6 m³/s.
   a. How many rapid sand filters, using only sand as the medium, are needed for this plant if each filter is 10 m × 20 m? What assumption do you have to make to answer this question?
   b. How can you reduce the number of filters?

7-47   A water treatment plant has a sedimentation basin receiving 2 MGD that has a diameter of 60 ft and an average water depth of 10 ft. What are the retention time and overflow rate in the basin?

7-48   A 4-MGD water treatment plant is being designed for an overflow rate of 0.5 gpm/ft². If two circular sedimentation basins will be used at all times, what should be the diameter of each basin?

7-49 Calculate the alkalinity, total hardness, carbonate hardness, and noncarbonate hardness for the following water in mg/L as $CaCO_3$:

CATION	CONCENTRATION (mg/L)	ANION	CONCENTRATION (mg/L)
Ca^{2+}	94	HCO_3^-	135
Mg^{2+}	28	SO_4^{2-}	134
Na^+	14	Cl^-	92
K^+	31	pH	7.8

7-50 Determine the quantity of quicklime and soda ash required to soften 1 MGD of the following water to the practical solubility limit. Also calculate the flow rate of sludge if it thickens to 7% solids.

COMPONENT	Ca^{2+}	Mg^{2+}	HCO_3^-	CO_2	pH
CONCENTRATION (mg/L)	53	12.1	285.0	7.2	8.1

7-51 Determine the quantity of quicklime and soda ash required to soften 500,000 gallons per day (gpd) of the following water to the practical solubility limit. Also calculate the flow rate of sludge if it thickens to 11% solids.

COMPONENT	Ca^{2+}	Mg^{2+}	HCO_3^-	CO_2	pH
CONCENTRATION (mg/L)	53	12.1	134.0	6.8	7.2

7-52 Determine the quantity of hydrated lime and soda ash required to soften 300,000 gpd of the following water to the practical solubility limit. Also calculate the flow rate of sludge if it thickens to 9% solids.

COMPONENT	Ca^{2+}	Mg^{2+}	HCO_3^-	CO_2	pH
CONCENTRATION (mg/L)	83	42	240.0	12	7.1

7-53 A water sample has the composition shown. It will be treated in a 3.5-MGD plant.

COMPONENT	Ca^{2+}	Mg^{2+}	HCO_3^-	CO_2
CONCENTRATION (meq/L)	4.1	1.8	3.8	0.23

a. If the sample will be treated with lime-soda softening to the minimum level using excess quicklime, how much lime and how much soda ash will be required in pounds per day?

b. If the sample will be softened using the selective calcium removal process, how much lime and how much soda ash will be required in pounds per day?

c. What will be the approximate final hardness of the softened water in part b?

7-54 A water treatment facility treats 225,000 gpd of effluent. Based on jar test results, 10 mg/L of alum is used to coagulate a sample containing 75 mg/L of suspended solids. Assume $Al(OH)_3$ is precipitated.

a. How many milligrams per liter as $CaCO_3$ of natural alkalinity are consumed?

b. What is the mass rate of sludge generated if suspended solids are reduced to 3 mg/L?

7-55 A water treatment facility treats 250,000 gpd of effluent. Based on jar test results, 10 mg/L of $FeCl_3$ is used to coagulate a sample containing 75 mg/L of suspended solids. Assume $Fe(OH)_3$ is precipitated.

a. How many milligrams per liter as $CaCO_3$ of natural alkalinity are consumed?

b. What is the mass rate of sludge generated if suspended solids are reduced to 3 mg/L?

7-56 Nearly all the water on Earth contains naturally occurring fluoride. Investigation of the tooth decay-preventing effects of naturally occurring fluoride in water led to the start of community water fluoridation in 1945. Fluorides are effective in dramatically reducing tooth decay, particularly in teenagers and young adults.

a. Are there negative effects of fluoridation? What else might the fluoride added to drinking water do?

b. Are there alternatives to adding fluoride that could achieve the same reduction in tooth decay?

Wastewater Treatment

FIGURE 8.1 This photo shows a water final clarifier at a modern urban wastewater treatment plant. Notice the clarity of the water flowing out of the clarifier on the left.

Source: M-Production/Shutterstock.com.

Sewage can lead to another disaster, which is disease.

—WALTER MAESTRI

GOALS

THE BUILDUP OF WASTE MATERIALS IN ecosystems decreases biological productivity and may increase the risk to human health. In today's technologically oriented society, water is used to convey waste from individual locations to community treatment systems. The modern wastewater treatment plant is a sophisticated, clean, and effective series of unit processes designed to prevent the waste generated by communities from negatively impacting the environment and health.

This chapter focuses on providing an understanding of the unit processes used in wastewater treatment systems. This includes describing the flows into and out of each unit process. The size and efficiency of these unit processes are based on the mass balance analysis.

Waste compounds in the water are concentrated and transformed whenever possible into benign compounds. However, the wastewater treatment process itself creates waste, or residuals. This waste material, which is made up of solid organic and inorganic compounds that settle from the flowing water stream, is generally called wastewater sludge, and this sludge must also ultimately be disposed of. The treatment and disposal of the water sludge are also discussed in this chapter.

OBJECTIVES

At the conclusion of this chapter, you should be able to:

8.1 Describe the unit processes—preliminary, primary, secondary, and tertiary—in a wastewater treatment system.

8.2 Describe the importance of preliminary and primary treatment to downstream processes in a wastewater treatment facility.

8.3 Formulate and solve material and energy balances for biological reactors to remove organic compounds and nutrients from wastewater.

8.4 Compare nutrient removal alternatives in wastewater treatment.

8.5 Describe alternative treatment technologies for treating nutrients and other secondary pollutants.

8.6 Compare disposal options for residual wastewater solids.

8.7 Describe the advantages and risks in recycling wastewater.

Introduction

Water has many uses, including drinking, commercial navigation, recreation, fish propagation, and waste disposal. It is easy to forget that a major use of water is simply as a vehicle for transporting wastes. In isolated areas where water is scarce, waste disposal becomes a luxury use for water, and other methods of waste carriage may be employed, such as pneumatic pipes or containers. In many industrialized countries, using water for waste transport is almost universal, and this results in large quantities of contaminated water.

Contamination by human waste, or sewage, increases the concentration of pathogens, organic waste, and nutrients in water. The negative effects of sewage were accepted long ago. Sewers were commonly used in ancient Roman cities. It was not until the outbreak of diseases in the 1800s, however, and the scientific association with disease in germ theory that the collection of wastewater became a public policy concern. The treatment of wastewater is a relatively new practice that resulted from fairly recent public policies regarding stormwater, wastewater, and amendments to the Clean Water Act.

8.1 Wastewater Treatment

Wastewater is discharged from homes, commercial establishments, and industrial plants by means of sanitary sewers. Sanitary sewers usually employ large pipes that are partially full (not under pressure). Sewage flows downhill by gravity drainage. The system of sewers has to be designed so that the collecting sewers, which collect the wastewater from homes and industries, all converge to a central point where the water flows by trunk sewers to the wastewater treatment plant. Sometimes it is impossible or impractical to install gravity sewers, so the waste has to be pumped by pumping stations through force mains or pressurized pipes.

The design and operation of sewers are complicated by the inflow of storm-water, which is supposed to flow off in separate storm sewers in newer communities; however, it often seeps into the wastewater through loose manhole covers and broken lines. Such an additional flow to the wastewater sewers is called *inflow*. (Older communities may have combined sewers designed to collect and transport both sanitary wastewater and stormwater. They frequently have problems with combined sewer overflows [CSOs] during rain events.) Furthermore, sewers often have to be installed below the groundwater table, so any breaks or cracks in the sewer (such as from tree roots seeking water) can result in water seeping into the sewers. This additional flow is called *infiltration*. Local communities often spend considerable time and expense in rehabilitating sewage systems to prevent such ***inflow and infiltration (I/I)*** because every gallon that enters the sewage system has to be treated at the wastewater treatment plant.

The wastewater diluted by I/I flows downhill and eventually to the edge of the community that the sewage system serves. In the past, this wastewater simply entered a convenient natural watercourse and was forgotten by the community. Today, however, the growth of our population and awareness of public health problems created by the raw sewage make such a discharge untenable and illegal; wastewater treatment is required.

Although many materials can pollute water, the most common contaminates found in domestic wastewater that can damage natural watercourses or create human health problems are discussed in Chapter 7. Municipal wastewater treatment plants are designed to remove the objectionable components from the influent sewage. A typical wastewater treatment process must be capable of operating 365 days a year, 24 hours a day. ***Publicly owned treatment works (POTWs)*** may be designed to remove or reduce the concentrations of organic matter, suspended solids, nutrients, and bacteria in municipal wastewater generated by homes and businesses. Industrial wastewater treatment facilities may be designed to remove only one or more targeted contaminants from the wastewater produced on the industrial site. Industrial wastewater typically may be discharged to the municipal sewer or receiving waters (a stream or river) if it meets the legal permit requirements.

Wastewater treatment processes are generally classified (see Table 8.1) into preliminary treatment, primary treatment processes, secondary treatment processes, tertiary treatment processes, and residuals (or solids) treatment, as illustrated in Figure 8.2.

Preliminary treatment removes large objects and large solid particles to keep them from clogging pipes and damaging pumps in the treatment works. A majority of the suspended solids in the incoming wastewater and about a third of the organic biochemical oxygen demand (BOD) are removed in primary treatment through settling. Biological degradation of the organic waste and removal of the microorganisms and most of the remaining solids occur in the secondary treatment stage. Secondary treatment often consists of a biological process followed by coagulation, flocculation, and settling of the biomass. Tertiary or advanced treatment processes may include nutrient removal, disinfection, or other specialized processes, depending on the constituents of the wastewater.

As shown in Table 8.2, there is a large range of typical concentrations between weak and strong sanitary wastewater and between sanitary wastewater and septage (which is the substance that remains in septic tanks after anaerobic treatment). These concentrations are typical in the United States but may not be applicable in other countries. For example, engineers in Thailand found that wastewater in Bangkok had a BOD between 50 and 70 mg/L and a suspended solids concentration between 90 and 110 mg/L (Klankrong et al., 2001). It is important for engineers to obtain current local information on wastewater composition and flow rates when designing a new treatment plant or upgrades to an existing plant.

TABLE 8.1 Classifications of municipal wastewater treatment processes

Preliminary	Removal of gross solids to prevent equipment damage
Primary	Removal of a portion of the suspended solids and organic matter
Advanced primary	Chemical addition or filtration for enhanced solids and organic matter removal
Secondary	Biological and chemical processes to remove biodegradable organic matter and suspended solids
Secondary with nutrient removal	Removal processes to reduce nitrogen and phosphorus
Tertiary/advanced	Removal of residual suspended solids usually by filtration or microscreens, usually includes disinfection and possible nutrient removal as well as removal of dissolved and suspended materials when required for water reuse
Residual (solids) treatment	Collection, stabilization, and subsequent disposal of solids removed from other processes

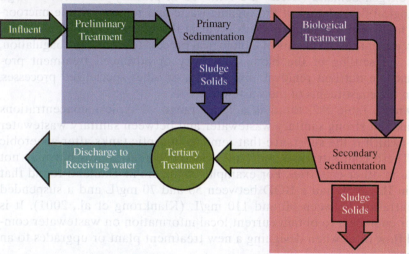

FIGURE 8.2 The photograph illustrates the scale of modern wastewater treatment facilities. The diagram illustrates the typical steps in wastewater treatment processes.

Source: Bim/E+/Getty Images.

TABLE 8.2 Typical concentrations of the components of U.S. municipal wastewater

COMPONENT	CONCENTRATION (mg/L)			
	WEAK SANITARY	**MEDIUM SANITARY**	**STRONG SANITARY**	**SEPTAGE**
TS	350	720	1,200	40,000
SS	100	220	350	15,000
BOD_5	110	220	400	6,000
N (as N)	20	40	85	700
P (as P)	4	8	15	250

Source: Metcalf and Eddy, Inc. (Revised by George Tchobanogous and Franklin L. Burton.) 1991. *Wastewater Engineering: Treatment, Disposal, and Reuse.* New York: McGraw-Hill.

8.2 Preliminary and Primary Treatment

8.2.1 Preliminary Treatment

The most objectionable aspect of discharging raw sewage into watercourses is the presence of floating material. It is logical, therefore, that screens were the first form of wastewater treatment used by communities, and even today screens are the first step in treatment plants. Typical screen systems (like those shown in Figure 8.3) consist of a series of steel bars that are typically spaced 2.5 cm (1 in.) apart. The purpose of screening in modern treatment plants is to remove large materials that might damage equipment or hinder further treatment. In some older treatment plants, screens are cleaned by hand, but mechanical cleaning equipment is used in almost all new plants. The cleaning rakes are automatically activated when the screens are clogged to a degree that raises the water level in front of the bars.

In many plants, the next step is a comminutor, a circular grinder designed to grind the solids coming through the screen into pieces about 0.3 cm (1/8 in.) or smaller. Many designs are in use; one common design is illustrated in Figure 8.4.

The third common preliminary treatment step involves the removal of grit or sand, as shown in Figure 8.5. Large grit and sand particles can wear out and damage equipment such as pumps and flow meters. The most common treatment method is the use of a grit chamber, which is simply a wide place in the channel where the flow is slowed sufficiently to allow the heavy grit to settle. Sand is about 2.5 times

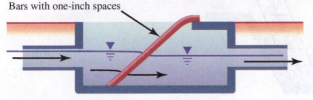

Bars with one-inch spaces

FIGURE 8.3 A typical bar screen.

Source: Sattawat Thuangchon/Shutterstock.com.

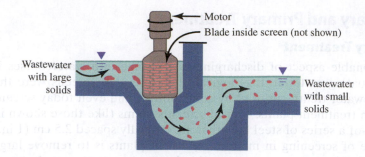

FIGURE 8.4 A typical comminutor.

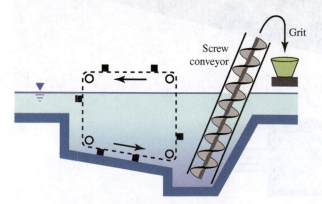

FIGURE 8.5 A typical grit chamber.

Source: iStockPhoto.com/Pedro Freithas.

heavier than most organic solids, and thus settles much faster than the lighter weight organic material. The collected grit can be dumped as fill or otherwise disposed of without significant odor or health concerns. One method of ensuring that the lightweight biological solids do not settle with the grit is to aerate the grit chamber, which allows the sand and other heavy particles to sink while the entrained air keeps the organics suspended or floating. Aeration has the additional advantage of driving some oxygen into the sewage, which may help reduce odors in the plant.

8.2.2 Primary Treatment

Primary wastewater treatment processes include flow equalization, coagulation, flocculation, and clarification, which are similar to the operations used in the treatment of drinking water. In most wastewater treatment plants, a settling tank follows the grit chamber to settle out as much of the solids material as possible. These tanks in principle are the same as the settling tanks used in water treatment and discussed in Chapter 7. The tanks are operated as a plug-flow reactor and turbulence is kept to a minimum. The solids settle to the bottom of the tank and are removed through a pipe, while the clarified liquid is removed over a V-notch weir. The V-notch weir consists of a notched steel plate that promotes equal distribution of the liquid discharge all the way around a tank. Settling tanks may be circular, with flow moving from the center outward (Figure 8.6), or rectangular (Figure 8.7).

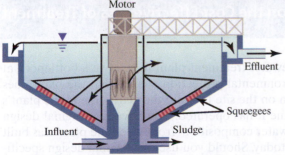

FIGURE 8.6 A typical circular settling tank (primary clarifier) used in wastewater treatment.

Source: Potus/Shutterstock.com.

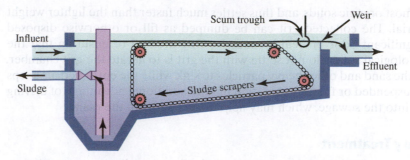

FIGURE 8.7 A typical rectangular settling tank (primary clarifier) used in wastewater treatment.

Settling tanks are also known as *clarifiers* or sedimentation tanks. The settling tank that follows preliminary treatment is called the primary clarifier. The solids that are removed from the bottom of the primary clarifier are removed as raw sludge.

Raw sludge is very odiferous, can contain pathogenic organisms, and is full of water—three characteristics that make its disposal difficult. It must be stabilized (further treated) to reduce its possible public health impact and to slow further decomposition. Sludge is also dewatered to reduce its weight and volume and to make disposal easier. In addition to the solids from the primary clarifier, solids from other processes must be similarly treated and disposed of. The treatment and disposal of wastewater solids (sludge) are an important and expensive part of wastewater treatment and are discussed further in Section 8.6.

Primary treatment typically removes nearly two-thirds of the solids in the influent stream, nearly one-third of the oxygen-demanding components of the water, and perhaps one-fifth of the phosphorus. If primary treatment is judged to be insufficient, then solids, BOD, and phosphorus removal can be enhanced by the addition of chemicals such as aluminum sulfate (alum) or calcium hydroxide (lime) to the primary clarifier's influent. The addition of these *chemical coagulants* may reduce the BOD to about 50 mg/L, which may meet required discharge standards for the effluent water. Adding chemicals during primary treatment is especially attractive in large coastal cities where discharge from the facility can be highly dispersed. However, most wastewater treatment facilities use primary treatment without chemical addition followed by secondary treatment designed specifically to remove the materials that consume oxygen in the wastewater.

ACTIVE LEARNING EXERCISE 8.1 Limitations on the Cost Effectiveness of Treatment

You are designing an additional sand filter for a local wastewater treatment plant to remove a bottleneck at the plant. However, you first must convince the state environmental protection agency that the plant does not need an additional secondary clarifier. There is no room on the site to build the clarifier, and the plant's effluent has had only minor violations during the entire time it has operated. You have the original design specifications, including the estimated flow rates and wastewater composition, from when the plant was built in the early 1970s. You must send your letter to the agency today. Should you use the original design specifications? Why or why not?

8.3   Secondary Treatment

Typically about one-half to two-thirds of the suspended organic matter is removed in the primary clarifier, but the components in the wastewater still have a high demand for oxygen due to the dissolved biodegradable organics. This biochemical oxygen demand (BOD) must be reduced prior to discharging the water to the environment. BOD is converted to biomass in *secondary treatment processes*. Secondary treatment processes are traditionally designed for the cost-effective removal of organic waste by utilizing microorganisms to degrade the organic waste and gravity to remove the microorganisms. Several methods may be employed to remove the biodegradable organic material; the basic differences among these alternatives relate to how the waste is brought into contact with the microorganisms.

8.3.1   Fixed Film Reactors

The *trickling filter* was the first modern method of secondary treatment of biodegradable organics broadly applied. The name *trickling filter* is a misnomer, however, as the organics are removed through biological processes and not through any actual filtering mechanism. A trickling filter consists of a bed of media (fist-sized rocks or engineered plastic media) over which the waste is trickled, as illustrated in Figure 8.8.

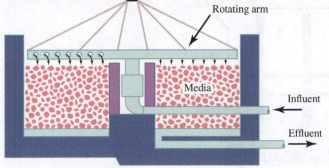

FIGURE 8.8 Cross section of a typical trickling filter.

Source: R Kawka/Alamy Stock Photo.

An active biological film forms on the surface of the media, and the biological organisms use the organic material in the wastewater dripping over the media as their food (or energy) source. Either air is forced through the media or, more commonly, air circulation is achieved naturally by temperature differences between the air in the trickling bed and the ambient temperature. In older trickling filters, the wastewater is typically sprayed onto rocks from fixed nozzles. Newer trickling filters typically use a rotating arm that is moved by the force of the water exiting the nozzles; this achieves a more even distribution of the wastewater over the entire bed. A portion of the effluent flow may be recirculated back over the bed if more treatment is required.

A ***rotating biological contactor (RBC)***, or rotating disc, is a modern modification of the traditional trickling filter design. The microbial film grows on media inside a rotating disc, as shown in Figure 8.9. The water still provides the energy for growth (food) for microorganisms in the biological film attached to the media in the disc. As the disc rotates into the open air, the microbes are able to obtain the necessary oxygen to maintain aerobic conditions at the surface of the biofilm. The rotation of the disc also helps to slough off some of the biofilm, which is removed from the bottom of the reactor as a secondary sludge waste.

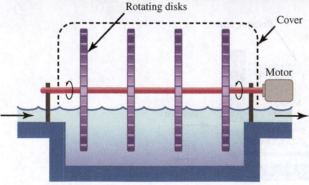

FIGURE 8.9 Cross section of a typical rotating biological contactor.

Source: Napier-Reid Ltd.

8.3.2 Suspended Growth Reactors

The *activated sludge system* was first demonstrated in 1914 as a method to create a biological reactor that did not require the space and weight that rocks occupy in a trickling filter. The key to the activated sludge system is the recycling of microorganisms from the effluent to maintain a sufficiently large "active" microbe population in the reactor (see Figure 8.10). The activated sludge processes use aerobic organisms to degrade the organic waste, since aerobic removal rates are higher than the rates for anaerobic processes. Aerobic processes also do not produce unwanted malodorous gases such as hydrogen sulfide and mercaptans, which are the byproducts of anaerobic degradation. Air is mixed into the reactor to maintain adequate concentrations of dissolved oxygen throughout the reactor. The microorganisms use the energy and

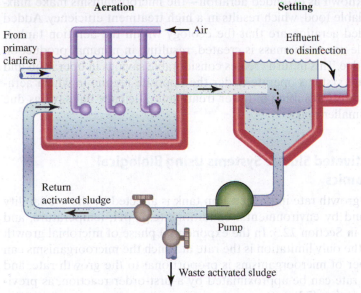

FIGURE 8.10 Cross section of a typical activated sludge process.

Source: Matthew Corley/Shutterstock.com.

carbon in the waste by oxidizing the waste materials to CO_2, H_2O, and some stable organic compounds, and in the process produce more microorganisms. The growth of new organisms is slow, and most of the volume of the aeration tank is required for biomass growth. The microorganisms are removed in the secondary clarifier. A portion of the microorganism-laden mixture is recycled into the biological reactor, and a portion of the microbes are removed with other suspended solids in what is called the secondary sludge.

The activated sludge process is a continuous operation. The excess microorganisms are heavier than water and turn the dissolved BOD material into biomass that can be removed by gravity in the clarifier. If the microorganisms were not removed, the concentration would eventually clog the process with settled solids. The separated microorganisms on the bottom of the final clarifier are starved for a food source and become "hungry." These microorganisms are said to be "activated"—hence, the term *activated sludge*. When the activated microorganisms are pumped to the head of the aeration tank, food is plentiful and the process cycle begins again. The portion of the sludge that is pumped from the bottom of the secondary clarifier to the aeration tank is known as the *return activated sludge*. The portion of the settling solids that are collected for disposal (and not recycled) is called the *waste activated sludge*. The disposal of the waste activated sludge is one of the most difficult challenges of wastewater treatment.

The activated sludge system is designed on the basis of *substrate loading*, or the amount of organic matter (food) added relative to the microorganisms available. The *food-to-microorganism ratio (F/M)* is a major design parameter. This ratio is approximated by measuring the incoming BOD as the food and the solids generation rate divided by the microorganism growth rate. The mixture of the liquid wastewater and the microorganisms undergoing aeration is called the *mixed liquor*, and the particles in the tank are called the *mixed liquor suspended solids (MLSS)*. The ratio of influent BOD to MLSS, the *F/M* ratio, is the loading on the system and is calculated as mass of BOD/day per mass of MLSS in the aeration tank.

If the *F/M* ratio is low and the aeration period (the detention time in the aeration tank) is long—known as extended aeration—the microorganisms make maximum use of the available food, which results in a high treatment efficiency. Added advantages of extended aeration are that the ecology within the aeration tank is quite diverse and little excess biomass is created, resulting in minimal production of waste activated sludge. This in turn creates considerable savings in operation and maintenance. High-rate systems operate under the opposite regime in which aeration periods are very short, resulting in lower treatment efficiency but savings due to the need for only smaller tanks.

8.3.3 Design of Activated Sludge Systems Using Biological Process Dynamics

The microorganisms' growth rate in the aeration tank is affected by the availability of food (substrate) and by environmental conditions (e.g., pH, temperature, and salinity), as discussed in Section 7.2.3. In the exponential phase of microbial growth shown in Figure 7.24, the only limitation is the rate at which the microorganisms can reproduce. The number of microorganisms is proportional to the growth rate, and therefore the growth rate can be approximated by a first-order reaction, as previously presented in Equation (7.3):

$$X = X_0 e^{\mu t} \tag{7.3}$$

The doubling (or generation) time can be determined by substituting $X = 2X_0$ into this equation:

$$t_D = \frac{\ln(2)}{\mu} \qquad (8.1)$$

The activated sludge process is a large-scale continuous growth reactor, or chemostat. Chemostats maintain their cell population in the exponential growth phase by controlling the dilution rate and the concentration of a limiting nutrient, such as carbon or nitrogen. Both the cell density (or population) and the growth rate can be controlled. Chemostat operation can be described by

$$\mu = \mu_{max} \frac{S}{K_S + S} \qquad (8.2)$$

where

μ_{max} = maximum specific growth rate (at nutrient saturation)
S = substrate or nutrient concentration
K_S = saturation, or half-velocity, constant

The saturation constant is the nutrient concentration at which the growth rate is half the maximum growth rate (see Figure 8.11). Estimates of μ_{max} and K_S are obtained by plotting $1/S$ versus $1/\mu$. The intercept is $1/\mu_{max}$ and the slope is K_S/μ_{max}.

The ideal reactor described above does not exist in practice. There are no pure cultures of a single type of microorganism in the activated sludge system. Exponential growth is limited by the availability of nutrients, competition among microorganisms, and predator-prey relationships. The actual maximum growth rate is typically much slower than the laboratory rate due to fluctuations in environmental conditions.

Microbial growth can be measured by counting or weighing cells. The direct microscopic count can be used to estimate the total cell count; however, one of the problems with this method is that both living and dead cells are counted. The colony count method described in Section 7.2.1 measures only living cells. An indirect measure of cell growth is determined from the cell mass either by centrifuging and weighing the cells or by measuring turbidity with a colorimeter or spectrophotometer. Turbidity is less sensitive than viable counting, but it is quick, easy, and does not change the sample.

Although treatment plant operators control the microbial concentration in the aeration basin, they are interested in doing so only to reduce the BOD. The

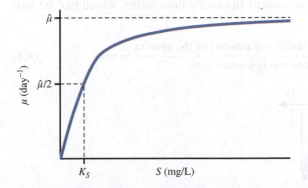

FIGURE 8.11 Specific growth rate of microbes.

microorganisms expressed as suspended solids biodegrade and use the BOD (substrate) at the rate r_{su}. As the substrate is consumed, new organisms are produced. The rate of production of new cell mass (microorganisms) as a result of the destruction of the substrate is

$$r_x = -Yr_{su}$$
(8.3)

where Y is the yield, or mass, of microorganisms produced per mass of substrate used, commonly expressed as kg SS produced per kg BOD used. Y is always less than 1 due to inefficiencies in energy conversion processes, as discussed in Section 7.2.3. Combining Equations (7.3), (8.2), and (8.3) yields Equation (7.6), the expression for substrate utilization:

$$r_{su} = \frac{-X}{Y}(\mu) = \frac{-X}{Y}\left(\frac{\mu_{max}S}{K_s + S}\right)$$
(7.6)

This expression, known as the Monod equation, is an empirical model based on experimental work with pure cultures. The two constants, K_s and μ_{max}, must be evaluated for each substrate and each microorganism culture. They remain constant for a given system, however, and S and X may be variable and are a function of the substrate and microbial mass.

Consider a simple, completely mixed continuous biological reactor of volume V with a flow rate of Q, as illustrated in Figure 8.12. Recall that this means the influent is dispersed within the tank immediately upon introduction; thus, there are no concentration gradients in the tank, and the quality of the effluent is exactly the same as that of the tank contents.

Material balances can be developed for the solids (microorganisms) and BOD (substrate) in the reactor. There are also two retention, or detention, times: one for the liquid and another for the solids in the reactor. The liquid, or hydraulic, retention time was introduced in Chapter 4 and is expressed as

$$t_r = \frac{V}{Q}$$
(4.26)

Recall that t_r can also be defined as the average time the liquid remains in the reactor. The **solids retention time** is analogous to the hydraulic retention time and represents the average time solids stay in the system (which is longer than the hydraulic retention time when the solids are recycled). The solids retention time is also known as the sludge age and the **mean cell residence (or detention) time (MCRT)**. All three names represent the same parameter, which can be calculated as

$$\theta_C = \frac{\text{mass of solids (microorganisms) in the system}}{\text{mass of solids wasted/time}} = \text{time}$$
(8.4)

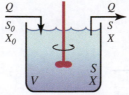

FIGURE 8.12 A suspended growth reactor with no recycle stream.

The numerator, the amount of solids in the simple reactor, is expressed as VX (volume $\times$ solids concentration), and the denominator, the rate of the solids wasted, is equal to QX (flow rate $\times$ solids concentration). Thus, the mean cell retention time is

$$\theta_C = \frac{VX}{QX} = \frac{V}{Q} \tag{8.5}$$

which, of course, is the same as the hydraulic retention time given in Equation (4.26) because the solids are not recycled.

The amount of solids wasted must also be equal to the rate at which they are produced, or dX/dt. Substituting $(dX/dt)(V)$ for QX and using Equation (8.3) yields

$$\theta_C = \frac{VX}{r_x V} = \frac{VX}{(-Yr_s)V} = \frac{-X}{Yr_s} \tag{8.6}$$

Note that the concentration terms have to be multiplied by volume to obtain mass. Remember that these two equations are for a reactor *without recycle*, as in Figure 8.12.

Using Figure 8.12, we can write a mass balance in terms of the microorganisms:

Rate of ACCUMULATION	=	Rate IN	−	Rate OUT	+	Rate of Microorganisms GROWTH	−	Rate of Microorganisms DEATH

If the growth and death rates are combined as *net* growth, this equation reads

$$V\frac{dX}{dt} = QX_0 - QX + r_x V$$

or

$$V\frac{dX}{dt} = QX_0 - QX - Yr_s V$$

In a steady state system, $dX/dt = 0$ and, assuming there are no cells in the inflow, $X_0 = 0$. Using this information and making use of the Monod substrate utilization model in Equation (7.6) gives us

$$r_s = \frac{-X}{Y}\left(\frac{Q}{V}\right) = \frac{-X}{Y}\left(\frac{1}{\theta_C}\right) = \frac{-X}{Y}\left(\frac{\hat{\mu} S}{K_S + S}\right)$$

Therefore,

$$\frac{1}{\theta_C} = \frac{\hat{\mu} S}{K_S + S} = \mu \tag{8.7}$$

or

$$S = \frac{K_S}{\hat{\mu}\theta_C - 1} \tag{8.8}$$

This is an important expression because we can infer from it that the substrate concentration, S, is a function of the kinetic constants (which are beyond our control for a given substrate) and the mean cell retention time. The value of S, which in real life would be the effluent BOD, is influenced by the mean cell retention time (or the sludge age as previously defined). If the mean cell retention time is increased, the effluent concentration should decrease.

EXAMPLE 8.1   Determining the Volume of a Biological Reactor

A biological reactor such as the one pictured in Figure 8.12 (with no solids recycle) must be operated so that an influent BOD of 600 mg/L is reduced to 10 mg/L. The kinetic constants have been found to be $K_S = 500$ mg/L and $\hat{\mu} = 4$ day^{-1}. If the flow is 3 m^3/day, how large should the reactor be?

Remember that the substrate concentration, S, in the effluent is the same as S in the reactor if the reactor is assumed to be perfectly mixed.

Using Equations (8.7) and (8.5), we find

$$\theta_C = \frac{K_S + S}{S\hat{\mu}} = \frac{500 \text{ mg/L} + 10 \text{ mg/L}}{(10 \text{ mg/L})(4 \text{ day}^{-1})} = 12.75 \text{ days}$$

$$V = \theta_C Q = (12.75 \text{ days})(3 \text{ m}^3/\text{day}) = 38.25 \text{ m}^3 \cong 38 \text{ m}^3$$

EXAMPLE 8.2   Substrate Removal in a Biological Reactor

Given the conditions in Example 8.1, suppose the only reactor available has a volume of 24 m^3. What is the percent reduction in the substrate (the substrate removal efficiency)?

Using Equations (8.5) and (8.8), we find

$$\theta_C = \frac{V}{Q} = \frac{24 \text{ m}^3}{3 \text{ m}^3/\text{day}} = 8 \text{ days}$$

$$S = \frac{K_S}{\hat{\mu}\theta_C - 1} = \frac{500 \text{ mg/L}}{(4 \text{ day}^{-1})(8 \text{ days}) - 1} = 16 \text{ mg/L}$$

$$\text{Recovery} = \frac{(600 - 16)}{600} \times 100 = 97\%$$

The activated sludge process differs from the simple biological reactor shown in Figure 8.12. In the activated sludge process, a clarifier removes the solids created in the biological reactor and returns a portion of the activated solids from the clarifier to the biological reactor. A high concentration of microorganisms in the reactor is maintained by the return of the microorganisms to the reactor, as illustrated in Figure 8.13. Similarly, in modeling an activated sludge process, two mass balance models must be developed: a model for the substrate (BOD) removal and a model for the solids (microorganism) production. There are also two retention or detention times: one for the liquid components and another for the solids.

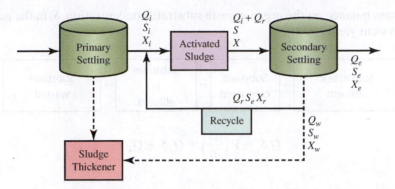

FIGURE 8.13 Schematic of wastewater flow, recycle flow, and effluent flow associated with a typical activated sludge process.

A mass balance on the biomass for the activated sludge process shown in Figure 8.12 is

Biomass in influent	+	Net biomass growth	=	Biomass in effluent	+	Biomass wasted

$$Q_i X_i + V\left(\frac{dX}{dt}\right) = Q_e X_e + Q_w X_w \tag{8.9}$$

where

X = concentration of cells in the activated sludge biological reactor
X_i = concentration of cells in the influent wastewater
X_e = concentration of cells in the effluent from the secondary settling tank
X_w = concentration of cells wasted to the sludge storage tank
V = volume of the activated sludge biological reactor
Q_i = volumetric flow of the influent wastewater from the primary settling tank
Q_e = volumetric flow of the effluent wastewater from the secondary settling tank
Q_w = volumetric flow of wastewater to the sludge storage tank

Recall that the growth of biomass may be described by the Monod equation, Equation (7.6). Rearranging Equation (8.9) and substituting yields the change in biomass concentration in the activated sludge biological reactor:

$$\frac{dX}{dt} = V\left(\frac{\mu_m SX}{K_S + S} - k_d X\right) \tag{8.10}$$

Then

$$Q_i X_i + V\left(\frac{\mu_m SX}{K_S + S} - k_{decay} X\right) = Q_e X_e + Q_w X_w \tag{8.11}$$

where k_{decay} is the decay coefficient and the other terms have been defined.

A mass balance on the organic waste substrate concentration, S, in the activated sludge system yields

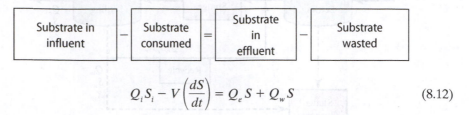

$$Q_i S_i - V\left(\frac{dS}{dt}\right) = Q_e S + Q_w S \tag{8.12}$$

where

$S = S_e$ = concentration of substrate in the activated sludge biological reactor, which is equal to the effluent concentration since the reactor is assumed to be well mixed

S_i = concentration of cells in the influent wastewater

S_w = concentration of cells wasted to the sludge storage tank

The substrate concentration is related to the biomass in the system by the yield in Equation (7.6), where

$$\frac{dS}{dt} = \frac{\mu_m SX}{Y(K_S + S)} \tag{8.13}$$

Then

$$Q_i S_i - V\left(\frac{\mu_m SX}{Y(K_S + S)}\right) = (Q_i - Q_w)S + Q_w S \tag{8.14}$$

The design equations that govern the activated sludge system require these assumptions:

- The influent (X_i) and effluent (X_e) biomass concentrations are very small compared to the concentration of biomass in the reactor (X).
- The biological reactor is well mixed, so that the concentration of the substrate in the reactor and the effluent from the reactor are equal (i.e., $S = S_e$).
- All measurable reactions occur in the biological reactor.

Equation (8.9) may then be simplified to

$$V\left(\frac{\mu_m SX}{K_S + S} - k_{decay} X\right) = Q_w X_w \tag{8.15}$$

Notice that both terms include the Monod equation components, so that from Equation (8.10), we have

$$\left(\frac{\mu_m S}{K_S + S}\right) = \frac{Q_w X_w}{VX} + k_{decay} \tag{8.16}$$

Rearranging Equation (8.9) gives

$$\left(\frac{\mu_m S}{K_S + S}\right) = \frac{Q_i Y}{VX}(S_i - S) \tag{8.17}$$

Then, combining Equations (8.11) and (8.12) yields

$$\left(\frac{Q_w X_w}{VX}\right) = \left(\frac{Q}{V}\right)\left(\frac{Y}{X}\right)(S_i - S) - k_{decay} \tag{8.18}$$

Equation (8.13) contains parameters that are fundamental to the design of activated sludge processes. The removal efficiency of the activated sludge process is strongly correlated with the average time that the microorganisms or biomass reside in the reactor, called the mean cell residence time or the solids retention time (SRT). The mean cell residence time, θ_C, is defined mathematically as the mass rate of cells in the system divided by the mass rate of cells leaving the system:

$$\theta_C = \frac{XV}{X_w Q_w + X_e Q_e} \approx \frac{XV}{X_w Q_w} \tag{8.19}$$

The hydraulic residence time, θ, is represented in the right-hand side of Equation (8.13):

$$\theta = \frac{V}{Q} \tag{8.20}$$

The mean cell residence time, θ_c, may be specified as a component of the reactor design and is controlled in part by the recycle ratio of the activated sludge system. If the mean cell residence time is specified, the substrate concentration in the reactor and in the effluent can be found by combining Equations (8.11) and (8.14):

$$S = \frac{K_s(1 + k_{decay}\theta_C)}{\theta_C(\mu_m - k_{decay}) - 1} \tag{8.21}$$

The concentration of biomass in the activated sludge reactor may be determined from the mean cell residence time and the hydraulic residence time:

$$X = \frac{\theta_C Y(S_i - S)}{\theta(1 + k_{decay}\theta_C)} \tag{8.22}$$

Typical values for the constants used in activated sludge wastewater treatment are listed in Table 8.3.

TABLE 8.3 Values of growth constants used for activated sludge wastewater treatment processes for municipal wastewater in temperate regions

PARAMETER	UNIT BASIS	VALUE RANGE	VALUE TYPICAL
K_s	mg/L BOD_5	25–100	60
k_{decay}	1/day	0–0.30	0.10
μ_m	1/day	1–8	3
Y	mg VSS/mg BOD_5	0.4–0.8	0.6

EXAMPLE 8.3   Solids Generation in an Activated Sludge System

An activated sludge wastewater treatment system uses a 2-million-gallon aeration basin. The mean cell retention time is 12 days. The MLSS is kept at 3,100 mg/L, and the recycled activated sludge (RAS) rate is 11,000 mg/L. What is the waste activated sludge (WAS) rate if the sludge is wasted from (a) the aeration basin and (b) the recycle line?

   a. When wasting occurs from the aeration basin, the wasting concentration is the same as the concentration in the aeration basin, which is the MLSS concentration. Assuming the effluent solids concentration (X_e) is negligible, the wasting rate from Equation (8.19) is

$$Q_w \approx \frac{VX}{\theta_c X} = \frac{V}{\theta_c} = \frac{2 \times 10^6 \text{ gal}}{12 \text{ days}} = 0.2 \text{ MGD}$$

   b. When wasting occurs from the recycle line, the wasting concentration is the same as the concentration in the recycle line, which is the RAS concentration. If we assume the effluent solids concentration (X_e) is negligible, the wasting rate from Equation (8.20) is

$$Q_w \approx \frac{VX}{\theta_c X_r} = \frac{(2 \times 10^6 \text{ gal})(3,100 \text{ mg/L})}{(12 \text{ days})(11,000 \text{ mg/L})} = 0.05 \text{ MGD}$$

In this case, wasting from the recycle line reduces the WAS rate by 75%. Therefore, less pumping is required.

The removal of substrate is often expressed in terms of the substrate removal velocity, q, defined as

$$q = \frac{\text{mass substrate removed/time}}{\text{mass microorganisms under aeration}}$$

or, using the previous notation,

$$q = \frac{\left(\dfrac{(S_0 - S)}{t_r}\right)V}{XV}$$

$$q = \frac{S_0 - S}{Xt_r} \tag{8.23}$$

The substrate removal velocity is a rational measure of the substrate removal activity, or the mass of BOD removed in a given time per mass of microorganisms doing the work. This is sometimes called the process loading factor, which may be a useful design parameter. Recall that the food-to-microorganism ratio (F/M) is the ratio of incoming BOD to MLSS:

$$\frac{F}{M} = \frac{QS_0}{VX} = \frac{S_0}{t_r X} \tag{8.24}$$

The substrate removal velocity can also be derived by conducting a mass balance in terms of the substrate on a continuous system with microorganism recycle (see Figure 8.13):

$$
\begin{array}{|c|}\hline \text{Rate of} \\ \text{ACCUMULATION} \\ \hline \end{array}
=
\begin{array}{|c|}\hline \text{Rate} \\ \text{IN} \\ \hline \end{array}
-
\begin{array}{|c|}\hline \text{Rate} \\ \text{OUT} \\ \hline \end{array}
+
\begin{array}{|c|}\hline \text{Rate of} \\ \text{Substrate} \\ \text{PRODUCTION} \\ \hline \end{array}
-
\begin{array}{|c|}\hline \text{Rate of} \\ \text{Substrate} \\ \text{CONSUMPTION} \\ \hline \end{array}
$$

The rate of substrate *production* is zero (no BOD is being produced). The rate of substrate *consumption* is equal to the substrate removal velocity multiplied by the solids concentration and the reactor volume:

$$\frac{dS}{dt} V = QS_0 - QS + 0 - qXV$$

Assuming steady state conditions, where $(dS/dt) = 0$, and solving for q yields

$$q = \frac{S_0 - S}{Xt_r} \tag{8.25}$$

The substrate removal velocity can also be expressed as

$$q = \frac{\mu}{Y} = \frac{\mu_{max} S}{Y(K_S + S)} = \frac{1}{\theta_c Y} \tag{8.26}$$

Equating these two expressions for substrate removal velocity and solving for $(S_0 - S)$ gives the substrate removal (reduction in BOD):

$$S_0 - S = \frac{\mu_{max} SXt_r}{Y(K_S + S)} \tag{8.27}$$

Solving for X in Equation (8.18) yields the concentration of microorganisms in the reactor (MLSS):

$$X = \frac{S_0 - S}{t_r q} \tag{8.28}$$

EXAMPLE 8.4 Designing an Activated Sludge System

An activated sludge system operates at a flow rate (Q) of 400 m³/day with an incoming BOD (S_0) of 300 mg/L. Through pilot plant work, the kinetic constants for this system are determined to be $Y = 0.5$ kg SS/kg BOD, $K_S = 200$ mg/L, and $\hat{u} = 2$ day^{-1}.

A solids concentration of 4,000 mg/L in the aeration tank is considered appropriate. A treatment system must be designed that will produce an effluent BOD of 30 mg/L (90% removal). Determine the following:

a. Volume of the aeration tank
b. Sludge age (or mean cell residence time, MCRT)
c. Quantity of sludge wasted daily
d. F/M ratio

The mixed liquor suspended solids concentration is usually limited by the ability to keep an aeration tank mixed and to transfer sufficient oxygen to the microorganisms.

A reasonable value for the solids under aeration is $X = 4,000$ mg/L, as stated in the problem.

a. The volume of a basin is typically calculated from the hydraulic retention time (HRT), which is currently unknown. Therefore, we need another equation to calculate HRT. Because we know the kinetic constants, we can use Equation (8.27):

$$S_0 - S = \frac{\hat{\mu} S X \bar{t}}{Y(K_S + S)}$$

Rearranged:

$$\bar{t} = \frac{Y(S_0 - S)(K_S + S)}{\hat{\mu} S X}$$

$$= \frac{(0.5 \text{ kg/kg})(300 \text{ mg/L} - 30 \text{ mg/L})(200 \text{ mg/L} + 30 \text{ mg/L})}{(2 \text{ day}^{-1})(30 \text{ mg/L})(4,000 \text{ mg/L})}$$

$$= 0.129 \text{ day} = 3.1 \text{ hr}$$

From Equation (8.20), the volume of the tank is then $V = tQ = (0.129 \text{ day})(400 \text{ m}^3/\text{day}) = 51.6 \text{ m}^3 \cong 52 \text{ m}^3$.

b. The sludge age is obtained from Equation (8.7):

$$\theta_C = \frac{1}{\mu} = \frac{K_S + S}{\hat{\mu} S} = \frac{200 \text{ mg/L} + 30 \text{ mg/L}}{(2 \text{ day}^{-1})(30 \text{ mg/L})} = 3.8 \text{ days}$$

Alternatively, the sludge age may be obtained from Equation (8.26):

$$q = \frac{\hat{\mu} S}{Y(K_S + S)} = \frac{\mu}{Y} = \frac{1}{\theta_C Y}$$

$$\theta_C = \frac{1}{qY}$$

First, we must calculate q. Using the kinetic constants in Equation (8.26) yields

$$q = \frac{\hat{\mu} S}{Y(K_S + S)} = \frac{(2 \text{ day}^{-1})(30 \text{ mg/L})}{(0.5 \text{ kg/kg})(200 \text{ mg/L} + 30 \text{ mg/L})} = 0.522 \text{ day}^{-1}$$

or, equivalently, using Equation (8.25) yields

$$q = \frac{S_0 - S}{X \bar{t}} = \frac{300 \text{ mg/L} - 30 \text{ mg/L}}{(4,000 \text{ mg/L})(0.129 \text{ day})}$$

$$= 0.523 \frac{\text{kg BOD removed/day}}{\text{kg SS in the reactor}} = 0.523 \text{ day}^{-1}$$

So

$$\theta_C = \frac{1}{qY} = \frac{1}{(0.522 \text{ day}^{-1})(0.5 \text{ kg/kg})} = 3.8 \text{ days}$$

c. Now that we know the MCRT, we can calculate the sludge wasting rate from Equation (8.19):

$$\theta_C = \frac{XV}{X_r Q_w - (Q - Q_w)X_C} \cong \frac{XV}{X_r Q_w}$$

$$X_rQ_w \cong \frac{XV}{\theta_C} = \frac{(4{,}000\ \text{mg/L})(51.6\ \text{m}^3)(10^3\ \text{L/m}^3)}{(3.8\ \text{days})(10^6\ \text{mg/kg})} \cong 54\ \text{kg/day}$$

d. Using Equation (8.24), we find

$$\frac{F}{M} = \frac{S_0}{\bar{t}X} = \frac{300\ \text{mg/L}}{(0.129\ \text{day})(4{,}000\ \text{mg/L})} = 0.58\ \frac{\text{kg BOD/day}}{\text{kg SS}}$$

EXAMPLE 8.5   Impact of Mixed Liquor Solids Concentration on the Efficiency of BOD Removal

Use the same data as in Example 8.4 to determine what mixed liquor solids concentration is necessary to attain a 95% BOD removal (i.e., S = 15 mg/L).

In this case, we can use the two equations for the substrate removal velocity: Equation (8.26)

$$q = \frac{\hat{\mu}S}{Y(K_s + S)} = \frac{(2\ \text{day}^{-1})(15\ \text{mg/L})}{(0.5\ \text{kg/kg})(200\ \text{mg/L} + 15\ \text{mg/L})} = 0.28\ \text{day}^{-1}$$

and Equation (8.28)

$$X = \frac{S_0 - S}{\bar{t}q} = \frac{300\ \text{mg/L} - 15\ \text{mg/L}}{(0.129\ \text{day})(0.28\ \text{day}^{-1})} = 7{,}890\ \text{mg/L}$$

Notice that the MLSS concentration must be doubled in order to reduce the effluent concentration by half. From Equation (8.26), the mean cell residence time would also now be almost twice as long:

$$\theta_C = \frac{1}{qY} = \frac{1}{(0.28\ \text{day}^{-1})(0.5\ \text{kg/kg})} = 7.1\ \text{days}$$

While more microorganisms are required in the aeration tank for higher removal efficiencies, the concentration of microorganisms depends on the settling efficiency in the final clarifier. If the sludge does not settle well, the return sludge solids concentration is low, and there is no way to increase the solids concentration in the aeration tank.

8.3.4 Gas Transfer

The activated sludge process is an aerobic process requiring that oxygen be readily available for the microorganisms in the system. The oxygen is typically added by either mechanically mixing the water or by bubbling compressed air through diffusers in the tank, as shown in Figure 8.14. In both cases the goal is gas transfer—the transfer of oxygen from the gas state to the dissolved liquid state, while also removing CO_2 gases from the liquid.

Gas transfer is the process of allowing any gas to dissolve in a fluid or the opposite, releasing dissolved gas from a fluid. Recall from Chapter 2 that most gases are only slightly soluble in water (H_2, O_2, N_2) in accordance with Henry's law (discussed in Section 2.2). Other gases may be much more soluble in water, including sulfur dioxide (SO_2), chlorine (Cl_2), and carbon dioxide (CO_2), and typically these gases

FIGURE 8.14 Mechanical and diffused aeration techniques.

then ionize in water. The transfer of oxygen takes place through the bubble gas/liquid interface illustrated in Figure 8.15.

Above the water surface in Figure 8.16 it is assumed that the air is well mixed so that there are no concentration gradients in the air. When the system reaches equilibrium, the concentration of dissolved oxygen in the water will eventually attain a saturation level S. Before equilibrium occurs, however, at some time t, the concentration of dissolved oxygen in the water is C, some value less than S. The difference between the saturation value S and the concentration C is the deficit D, so that

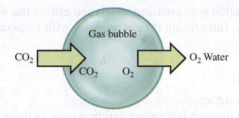

FIGURE 8.15 Gas transfer into and out of an air bubble in water.

$D = (S - C)$. As time passes, the value of C increases until it becomes S, which produces a saturated solution and reduces the deficit D to zero.

There is a diffusion layer, dC/dx, at the air/water interface shown in Figure 8.16. Oxygen from the air must pass through this diffusion zone before entering the "bulk" liquid. The concentration of the gas decreases through the interface until the concentration C is reached. If the water is well mixed, it may be assumed that the concentration of dissolved oxygen throughout the bulk water volume is C in units of mg/L. At the interface, the concentration increases at a rate of dC/dx, where x is the thickness of the diffusion layer. If dC/dx is large, then the rate at which oxygen is driven into the water is high, and vice versa, which may be expressed mathematically as

$$\frac{dC}{dx} \, \alpha (S - C)$$

As $(S - C)$ approaches zero, $dC/dx \rightarrow 0$, and when $(S - C)$ is large, dC/dx is large. It is assumed that the rate of change in concentration with time must be high when the slope (dC/dx) is large, and vice versa, or

$$\frac{dC}{dt} \, \alpha (S - C)$$

The gas transfer coefficient, $K_L a$, is defined as the proportionality constant such that

$$\frac{dC}{dt} = K_L a (S - C)$$

or, written in terms of the deficit,

$$\frac{dD}{dt} = -K_L a (D)$$

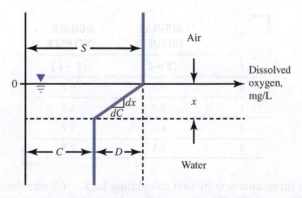

FIGURE 8.16 Visualization of the gas/liquid interface for describing gas transfer.

Note that since the deficit is decreasing as oxygen enters the water, K_La is negative in the equation above. Integrating that expression with respect to time yields

$$\ln \frac{D}{D_0} = -K_L at \qquad (8.29)$$

where D_0 is the initial deficit.

K_La is determined through laboratory aeration tests. In these tests, the water is first stripped of oxygen, usually by chemical means, so that C approaches zero. Air is then added and the dissolved oxygen concentration is measured with time. If, for example, different types of diffusers are to be tested, the K_La is calculated for all diffuser types using identical test conditions. A higher K_La implies that the diffuser is more effective in transferring oxygen into the water and thus presumably the least expensive to operate in the wastewater treatment plant. K_La is a function of, among other factors, the type of aerator, temperature, bubble size, volume of water, path taken by the bubbles, and presence of active surface agents. In Chapter 7, K_La was described as the aeration constant for streams used in the Streeter and Phelps dissolved oxygen model.

EXAMPLE 8.6   Selecting an Oxygen Diffuser for a Biological Reactor

Two diffusers are to be tested for their oxygen transfer capabilities. Tests were conducted at 20°C, using equipment like that shown in Figure 8.14 with the following results:

	DISSOLVED OXYGEN, C (mg/L)	
TIME (min)	AIR-MAX DIFFUSER	WONDER DIFFUSER
0	2.0	3.5
1	4.0	4.8
2	4.8	6.0
3	5.7	6.7

Note that the test does not need to start at $C = 0$ and $t = 0$. Plot the experimental gas transfer data and determine which diffuser increases the oxygen concentration more effectively.

With $S = 9.2$ mg/L (saturation at 20°C), we find

	AIR-MAX DIFFUSER	WONDER DIFFUSER
t	$(S - C)$	$(S - C)$
0	7.2	5.7
1	5.2	4.4
2	4.4	3.2
3	3.5	2.5

Now we can plot these numbers by first calculating $\ln(S - C)$ and then plotting against the time t (see Figure 8.17). The slope of the plot is the proportionality factor, or in this case, the gas transfer coefficient K_La. After the slopes are calculated, the K_La for the Air-Max diffuser is found to be 0.27 min^{-1}, while the Wonder diffuser has a K_La of 0.25 min^{-1}. The latter seems to be the better diffuser based on oxygen transfer capability.

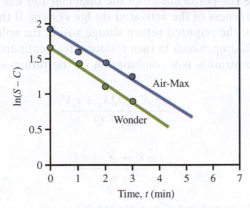

FIGURE 8.17 Experimental gas transfer data.

The aeration process typically consumes the most energy of any process at the treatment facility, as shown in Figure 8.18. Many treatment plants that have never optimized their aeration systems overaerate, which has a substantial energy and monetary cost associated with excess air feed to the system. Improving the control of aeration is often the best way to improve cost efficiency and potentially provide nitrogen reduction, as discussed later in the chapter.

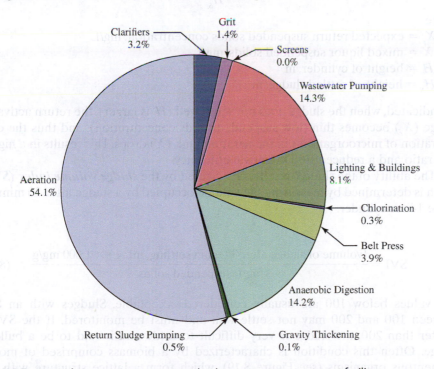

FIGURE 8.18 Average energy consumption in a wastewater treatment facility.

Source: Based on SAIC (Science Applications International Corporation) (2006). *Water and Wastewater Energy Best Practice Guidebook*. Madison, WI: Focus on Energy.

8.3.5 **Solids Separation**

The final clarifier is usually used to separate the solids from the liquid in an activated sludge process. The performance of the final clarifier has a significant impact on the overall effectiveness of the activated sludge system. If the final settling tank is not able to achieve the required return sludge solids, the solids concentration in the aeration tank will drop, which in turn reduces the treatment efficiency.

The MLSS concentration is a combination of the return solids diluted by the influent flow:

$$X = \frac{Q_r X_r + Q X_0 + r_x V}{Q_r + Q} \tag{8.30}$$

or

$$X = \frac{\alpha X_r + X_0 + r_x \bar{t}}{\alpha + 1} \tag{8.31}$$

where α is the recycle or recirculation ratio (Q_r/Q).

Again, if we assume no solids in the influent ($X_0 = 0$), we have

$$X = \frac{Q_r X_r + r_x V}{Q_r + Q} = \frac{\alpha X_r + r_x \bar{t}}{\alpha + 1} \tag{8.32}$$

Results from settling the sludge in a 1-liter cylinder can be used to estimate the return sludge concentration. After 30 minutes of settling, the solids in the cylinder are at an SS concentration that equals the expected return sludge solids, or

$$X_r = \frac{H}{H_S} (X) \tag{8.33}$$

where

X_r = expected return suspended solids concentration, mg/L
X = mixed liquor suspended solids, mg/L
H = height of cylinder, m
H_S = height of settled sludge, m

As indicated, when the sludge does not settle well (H_s is larger), the return activated sludge (X_r) becomes thin (low suspended solids concentration), and thus the concentration of microorganisms in the aeration tank (X) drops. This results in a higher *F/M* ratio and a reduced BOD removal efficiency.

The ability of the sludge to settle is indicated by the ***sludge volume index (SVI)***, which is determined by measuring the volume occupied by a sludge after 30 minutes in the 1-liter cylinder:

$$\text{SVI} = \frac{(\text{volume of sludge after } 30-\text{min settling, mL/L}) \times 1,000 \text{ mg/g}}{\text{mg/L suspended solids}} \tag{8.34}$$

SVI values below 100 are usually considered acceptable. Sludges with an SVI between 100 and 200 may not settle well and must be monitored. If the SVI is greater than 200, the sludge is very difficult to settle and is said to be a bulking sludge. Often this condition is characterized by a biomass comprised of mostly filamentous organisms (see Figure 8.19), which form a lattice structure with the filaments that do not settle. Operators of treatment plants must closely monitor settling characteristics, as a trend toward poor settling can indicate that the activated

FIGURE 8.19 Filamentous bacteria: (a) *Sphaerotilus natans* (false branching) and (b) *Nocardia* form (true branching).

Source: (a) Photo Researchers/Science History Images/Alamy Stock Photo; (b) Smith Collection/Gado /Archive Photos/Getty Images.

sludge process is becoming upset and may not be able to achieve the required BOD removal efficiency.

EXAMPLE 8.7   Sludge Volume

A sample of mixed liquor was found to have SS = 4,000 mg/L, and after settling for 30 min in a 1-L cylinder, it occupied 400 mL. Calculate the SVI.

Using Equation (8.34), we find

$$\text{SVI} = \frac{(400 \text{ mL/L})(1,000 \text{ mg/g})}{4,000 \text{ mg/L}} = 100$$

8.3.6 Secondary Effluent

Effluent from secondary treatment typically has a suspended solids concentration of about 20 mg/L and a BOD of about 15 mg/L. These SS and BOD levels are usually adequate for discharge from the plant, but the values are dependent on local conditions and the allowable total maximum daily load (TMDL) of the water into which the effluent is discharged.

Modern treatment plants are required to disinfect water prior to discharge to reduce the possibility of disease transmission. Chlorination is commonly used for disinfection because the process is well understood and relatively inexpensive. The chlorine contact tanks are designed as plug-flow reactors, as shown in Figure 8.20, with a residence time of at least 30 minutes. Prior to discharge, excess chlorine is removed through dechlorination. Sulfur dioxide gas is typically bubbled through the liquid near the end of the disinfection process to allow the SO_2 to be oxidized to sulfate and to chemically reduce the chlorine in the solution. Chlorination of wastewater is subject to the possible formation of disinfection byproducts (DBPs) that are chlorinated organic compounds (such as chloroform). As discussed in Chapter 7, these byproducts may have negative environmental impacts. Other disinfection processes may be used, such as ozonation or a UV treatment that does not produce DBPs.

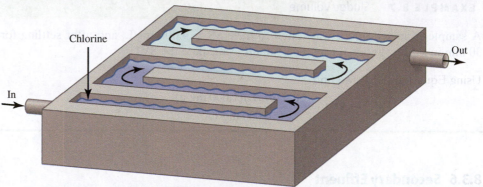

FIGURE 8.20 Plug-flow chlorine disinfection tank at a wastewater treatment facility.
Source: Siyanight/Shutterstock.com.

8.4 Nutrient Removal

Organic waste may be removed efficiently by using chemical and biological processes, as shown in Figure 8.21; however, nitrogen and phosphorus may be removed more cost effectively by using a combination of aerobic and anaerobic processes. Nutrient removal is one of the most widespread, costly, and challenging environmental problems. Typical nutrient concentrations found in a wastewater treatment plant are listed in Table 8.4; however, these values vary considerably among different system types, different geographical areas, and different sources of influent wastewater. Excessive concentrations of nitrogen and phosphorus can lead to a variety of problems, including eutrophication and harmful algal blooms, with impacts on drinking water, recreation, and aquatic life as discussed in Chapter 7. Various combinations of aerobic and anaerobic adaptation have been used in the traditional activated sludge process to improve the removal of nutrients; however,

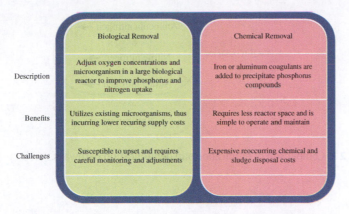

FIGURE 8.21 Benefits and challenges of chemical and biological nutrient removal.

Source: Based on www.pca.state.mn.us/sites/default/files/wq-wwtp8-21.pdf.

TABLE 8.4 Nutrient removal performance for wastewater treatment plants

TREATMENT SYSTEM	TOTAL NITROGEN (mg/L)	TOTAL PHOSPHORUS (mg/L)
Raw wastewater	40	7.0
Primary treatment	37	6.2
Activated sludge (with no nutrient removal)	25	5.6
Facultative lagoon	16	4.2
Trickling filter	25	5.8

Source: U.S. EPA. (2015). Case Studies on Implementing Low-Cost Modifications to Improve Nutrient Reduction at Wastewater Treatment Plants, Draft – Version 1.0. EPA-841-R-15-004. Washington DC.

these upgrades are not affordable or necessary for all facilities. Nutrient concentrations may be reduced in some plants by optimizing operation and maintenance practices without incurring large capital expenses.

8.4.1 Nitrogen Removal

Nitrogen is removed by treating the waste thoroughly enough in secondary treatment to oxidize all the nitrogen to nitrate. This usually involves longer detention times in secondary treatment, during which bacteria, such as *Nitrobacter* and *Nitrosomonas*, convert ammonia nitrogen to NO_3^-, a process called nitrification. ***Biological nitrogen removal (BNR)*** is the general term used to describe the two-step nitrification-denitrification process, which is the primary approach used to deliberately remove nitrogen during municipal wastewater treatment. The biological reactions for nitrification are

$$2NH_4^+ + 3O_2 \xrightarrow{Nitrosomonas} 2NO_2^- + 2H_2O + 4H^+$$
$$2NO_2^- + O_2 \xrightarrow{Nitrobacter} 2NO_3^-$$

Both reactions are slow and require sufficient oxygen and long detention times in the aeration tank. The rate of microorganism growth is also slow, resulting in low

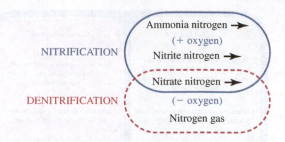

FIGURE 8.22 Comparison of nitrification and denitrification pathways.

net sludge production and making washout a constant danger. This process removes the oxygen demand caused by the nitrogen.

To remove the nutrient properties of nitrogen, the nitrate must be converted to nitrogen gas. Once the ammonia has been converted to nitrate, it can be reduced to nitrogen gas by a broad range of facultative and anaerobic bacteria, such as *Pseudomonas*. This reduction, called denitrification, requires a carbon source and methanol (CH_3OH) for that purpose. The nitrification and denitrification pathways are compared in Figure 8.22. The sludge containing NO_3^- is placed in an anoxic condition, in which the microorganisms use the nitrogen as the electron acceptor. Using methanol as the source of carbon, the facultative organisms can then convert the nitrate to nitrogen gas, N_2, which then bubbles out of the sludge into the atmosphere.

The three main types of denitrification processes shown in Figure 8.23 are: (a) pre-anoxic denitrification, (b) post-anoxic denitrification, and (c) single-reactor nitrification-denitrification. The first two processes involve the creation of dedicated unaerated or anoxic zones for denitrification. Single-reactor nitrification and denitrification provide nitrification and denitrification in the same space. This includes simultaneous nitrification-denitrification, which is promoted under low dissolved oxygen (DO) conditions, and cyclic processes where aeration is switched on and off, such as step-feed processes, slowly rotating contactors, and other techniques.

Pre-anoxic denitrification often utilizes carbon from BOD in the influent or the primary clarifier effluent to feed the denitrifying organisms that reduce nitrate. It must, therefore, be returned to the pre-anoxic zone in the return activated sludge (RAS) and/or internal recycle streams. In comparison, the post-anoxic zone follows the aerobic zone, and the carbon from endogenous decay of the cell mass (discussed in Chapter 7) is used for denitrification, which results in a much lower nitrate/nitrite reduction rate than in the pre-anoxic zone. Carbon from external sources such as methanol can be used to increase the denitrification rate.

Simultaneous and/or cyclic nitrification-denitrification, as shown in Figure 8.24, may be used in systems with long SRTs (20 days or longer) and long hydraulic retention times (HRT), such as oxidation ditches and lagoons. An oxidation ditch provides long retention times and has zones where there is no oxygen addition followed by zones with added aeration to create a two-step process within one reactor, as illustrated in Figure 8.25 (see page 549). Nitrification and denitrification rates are relatively slow, which is why longer SRTs are required to achieve complete nitrification.

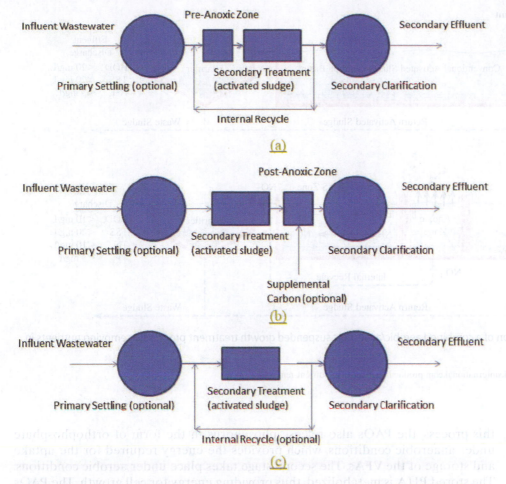

FIGURE 8.23 Schematic of common nitrification processes: (a) pre-anoxic denitrification, (b) post-anoxic denitrification, and (c) single-reactor nitrification/denitrification.

Source: U.S. EPA. (2015). Case Studies on Implementing Low-Cost Modifications to Improve Nutrient Reduction at Wastewater Treatment Plants, Draft – Version 1.0. EPA-841-R-15-004. Washington DC.

8.4.2 Phosphorus Removal

Phosphorus in water typically has the form of orthophosphate (PO_4^{3-}), polyphosphate (P_2O_7), and organically bound phosphorus, as shown in Table 8.5. Microorganisms use the poly- and organo-phosphates and produce the oxidized form of phosphorus, orthophosphate. The removal of phosphorus during wastewater treatment is typically the result of natural biological processes, including uptake and enhanced biological phosphorus removal (EBPR).

Specialized bacteria in activated sludge mixed liquors called *polyphosphate-accumulating organisms (PAOs)* can be used to biologically remove phosphorus from wastewater to levels that might meet water quality objectives. PAOs require two stages for phosphorus removal. The first stage is anaerobic, in which PAOs uptake *volatile fatty acids (VFAs)* from the organic carbon in the influent and store them as polyhydroxyalkanoate (PHA) for later oxidation in an aerobic zone. This first anaerobic stage is sometimes called an "anaerobic selector" because it preferentially selects for the proliferation of PAOs. During

Existing Activated Sludge Plant

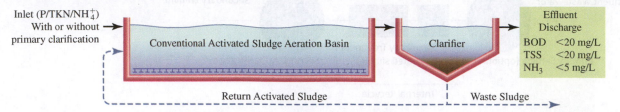

Converted to IFAS for BNR

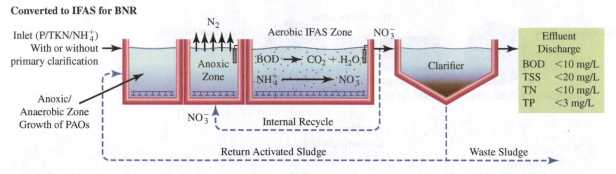

FIGURE 8.24 Illustration of a combined aerobic/anaerobic suspended growth treatment process for removing nutrients in municipal wastewater.

Source: Based on www.headworksinternational.com/product/bnr-biological-nutrient-removal/.

this process, the PAOs also release phosphorus in the form of orthophosphate under anaerobic conditions, which provides the energy required for the uptake and storage of the VFAs. The second stage takes place under aerobic conditions. The stored PHA is metabolized, thus providing energy for cell growth. The PAOs uptake and store more phosphorus under aerobic conditions than is released under anaerobic conditions, which results in net uptake and storage of phosphorus. This also provides the PAOs with a competitive advantage over other organisms, allowing them to thrive under these conditions. The stored phosphorus is then removed from the system with the waste sludge. Up to four times as much phosphorus can be removed biologically using EBPR than conventional activated sludge treatment. If secondary clarifiers are allowed to become anaerobic or the waste activated sludge (WAS) is treated in an anaerobic digester, the PAOs can release stored phosphorus back into the process stream. The functions of the various zones in biological nutrient removal processes are summarized in Table 8.6.

Chemical processes use metal salts to precipitate phosphorus to the solids (sludge) fraction. Chemical phosphorus removal requires that the phosphorus be fully oxidized to orthophosphate; hence, the most effective chemical removal occurs at the end of the secondary biological treatment system. The most common chemicals used for phosphorus removal are lime, $Ca(OH)_2$, and alum, $Al_2(SO_4)_3$. The amount of lime or alum required to remove a specified amount of phosphorus depends on the amount of phosphorus present as well as what other constituents are present in the water. The calcium ion at high pH combines with phosphate to form a white, insoluble precipitate called calcium hydroapatite that is settled and removed. Insoluble calcium carbonate is also formed and precipitates.

FIGURE 8.25 Nitrogen removal in an oxidation ditch.

Source: iStockPhoto.com/MarioGuti; iStockPhoto.com/Gyn9038.

EXAMPLE 8.8   Biological Phosphorus Removal

The BOD of influent wastewater is 275 mg/L and the influent phosphorus concentration is 12 mg/L. A BOD:phosphorus ratio greater than 20:1 is desired in order to meet the carbon requirements for the biological removal of the phosphorus. Is this wastewater amenable to biological phosphorus removal?

We can calculate the ratio of BOD to phosphorus:

$$\frac{\text{Influent BOD}}{\text{Influent P}} = \frac{275 \text{ mg/L}}{12 \text{ mg/L}} = 23:1$$

Since the ratio is greater than 20:1, the wastewater is amenable to biological phosphorus removal.

TABLE 8.5 Typical concentrations and forms of phosphorus in wastewater

PHOSPHATE FORM	TYPICAL CONCENTRATION (mg/L)
Orthophosphate, PO_4^{3-}-P	3.0–4.0
Polyphosphates	2.0–3.0
Organic phosphates	0.7–1.0
Total as P	5.7–8.0

Source: U.S. EPA 2015.

TABLE 8.6 Functions of the treatment zones in biological nutrient removal processes

ZONE	BIOCHEMICAL TRANSFORMATIONS	FUNCTIONS	ZONE REQUIRED FOR
Anaerobic	• Uptake and storage of VFAs by PAOs with associated phosphorus release • Fermentation of readily biodegradable organic matter by heterotrophic bacteria	• Selection of PAOs	Phosphorus removal
Anoxic	• Denitrification • Alkalinity production	• Conversion of NO_3-N to N_2 • Selection of denitrifying bacteria	Nitrogen removal
Aerobic	• Nitrification and associated alkalinity consumption • Metabolism of stored and exogenous substrate by PAOs • Metabolism of exogenous substrate by heterotrophic bacteria • Phosphorus uptake	• Conversion of NH_3-N to NO_3-N • Nitrogen removal through gas stripping • Formation of polyphosphate • Growth of nitrifiers • Growth of PAOs	Nitrogen removal Phosphorus removal

EXAMPLE 8.9   Solids Retention in Biological Phosphorus Removal

A biological phosphorus removal process is designed similarly to other biological reactors and requires a relatively long solids retention time. In one case, a biological phosphorus reactor is designed for a 0.02 MGD wastewater treatment plant. The mixed liquor suspended solids (MLSS) concentration in the reactors is 3,400 mg/L. The MLSS is maintained by returning a portion of the waste activated sludge (WAS) to the reactors. The concentration of solids in the WAS is 5,000 mg/L. An anaerobic selector with a volume of 50,000 gallons and an anoxic selector with a volume of 50,000 gallons are used for phosphorus removal followed by an aeration basin with a volume of 500,000 gallons. Determine the solids retention time in the biological phosphorus removal system.

We know from Equation (8.5) that the solids retention time is

$$\theta_C = \frac{\text{mass of solids in biological reactors}}{\text{mass rate of solids removed daily}} = \frac{VX}{QX}$$

$$\theta_C = \frac{(50{,}000 + 50{,}000 + 500{,}000) \text{ gallons} \times \dfrac{3{,}400 \text{ mg}}{L}}{20{,}000 \dfrac{\text{gallons}}{\text{day}} \times 5{,}000 \dfrac{\text{mg}}{L}} = 20.4 \text{ days}$$

Thus the solids retention time (SRT) is about 20 days.

Alum is typically added in the final clarifier to precipitate the phosphate. The aluminum ion from alum precipitates as poorly soluble aluminum phosphate, $AlPO_4$, and aluminum hydroxide, $Al(OH)_3$. The hydroxide forms attractive flocs that help to precipitate the phosphates.

EXAMPLE 8.10   Chemical Precipitation of Phosphorus

In the chemical reaction of phosphorus with a metal salt, a positively charged metal ion reacts with a negatively charged phosphate ion (PO_4^{3-}) ion to precipitate a solid salt. The positive and negative charges must balance. Determine the *weight ratio* on a pound-per-pound basis of ferrous iron ion, Fe^{2+}, to phosphorus required to remove 1 pound of phosphate ion.

The ferrous ion, a positively charged iron ion ($+2$), combines with a negatively charged phosphate ion (-3) to form the precipitate, iron phosphate:

$$3Fe^{2+} + 2PO_4^{3-} \rightarrow Fe_3(PO_4)_{2(s)}\downarrow$$

The problem asks for the weight ratio of iron and phosphorus, so we use the molar weights of those two compounds, where 1 mole of iron is 56 grams and 1 mole of phosphorus is 31 grams.

For the ferrous ion, 2 moles of molecular iron (in the form of the ferrous ion) are required to remove 2 moles of phosphorus (in the form of the phosphate ion). Thus, the weight ratio of Fe to P is

$$\frac{3 \times 56}{2 \times 31} = \frac{2.7}{1}$$

The Fe^{2+}:P ratio is 2.7:1, which means that it takes 2.7 pounds of ferrous ion to remove 1 pound of phosphorus.

Biological methods of removing phosphorus have the advantage of producing fewer solids for disposal. Most biological phosphorus removal systems rely on slow-growing microorganisms that are stressed by the controlled metabolic pathway. When oxygen is suddenly reintroduced into these processes, the microorganisms store much more phosphorus in their cells than would occur under their preferred growth conditions. This luxury uptake of phosphorus is followed by the removal of the cells that have stored the phosphorus as cellular material.

BOX 8.1 Alternatives to Phosphate Mining

Phosphorus (P), usually in the form of phosphate, is both a nutrient and a pollutant. It is considered a nutrient when it is found where it is needed—facilitating desired plant growth. It is a pollutant when it is found where it is not needed—facilitating unwanted plant growth that leads to eutrophication in lakes and other waterways.

Phosphate has negative impacts on the life cycle when it is mined. One of the unwanted byproducts of its mining is hydrogen fluoride. When the Rocky Mountain Phosphate Company in Garrison, Montana, began operations in the early 1960s, it had great difficulty with pollution control and emitted high concentrations of sulfur oxides and fluorides. By the summer of 1965, the vegetation in and around Garrison began to die. The land was barren of wildlife, with most animals dying off because of their inability to walk. Cattle on the farms around Garrison were unable to stand up and soon died. Excessive fluoride causes cattle (and people) to have fluorosis, a disease in which their joints swell and they are unable to walk. In effect, the fluoride etches the bones and makes them brittle. The same effect occurs when people drink water that has too much fluoride in it: their teeth become soft and mottled. Finally, in 1967, the U.S. government intervened and forced the company to cease operations until it installed emission equipment that removed 99.9% of the fluoride (Smith, 2003).

Additional negative consequences of mining phosphate rock are the release of cadmium and uranium. Phosphate rock may be contaminated with cadmium, a highly toxic metal. Uranium radionuclides may be released from mining, in the processing effluents, and from the use of phosphate rock products (such as fertilizer). Erosion of agricultural soils in areas with heavy fertilizer usage may input ^{238}U decay radionuclides into drinking water supplies (WEF, 2008).

Needless to say, getting phosphate to where it is wanted and keeping it from where it is not wanted is a challenge. One very promising technology that is consistent with sustainable materials management principles is the mining of phosphorus from human sewage. The city of Edmonton (pop. approx. 700,000) in Alberta, Canada, is home to the Gold Bar Wastewater Treatment Plant, which successfully operates the world's first industrial-sized nutrient treatment facility to remove phosphorus and other nutrients from municipal sludge and to recycle them into environmentally safe commercial fertilizer. Currently, the technology extracts more than 80% of the phosphorus on average and 10% to 15% of the ammonia from a flow of 500,000 L/day—approximately 20% of the plant's liquid sludge stream.

While many wastewater treatment plants remove phosphorus to reduce their nutrient loading on receiving waters, the nutrients end up in the sludge, which may be dewatered and either applied to the land or composted to take advantage of the nutrients. The liquid fraction from dewatering is also nutrient rich, however, and actually adds costs to a system by clogging pipes with a concrete-like scale called "struvite." Struvite is formed from phosphorus and ammonia (nitrogen) combined with magnesium. Rather than fight struvite production, innovative nutrient recovery processes promote it, processing the liquid fraction from sludge dewatering to recover phosphorus and other nutrients and then converting them into 99.9% pure struvite—a high-quality commercial fertilizer that can generate revenue for the municipality.

The fertilizer contains no heavy metals and has a commercially desirable formulation of nitrogen, phosphorus, and magnesium. Because it dissolves slowly over a 9-month period, it does not leach into water and cause eutrophication. It is used in turf (golf course) markets, container nurseries, agriculture, and other markets that value slow-release fertilizers. The reactor at Gold Bar produces approximately 500 kg/day. The product is sorted, dried, and bagged onsite and is immediately ready for commercial sale. In contrast to the negative impacts of mining phosphorus for fertilizer, the struvite recovery process recovers nutrients that would otherwise be released into the environment, helps sewage treatment plants reduce operating costs and meet environmental regulations, and provides municipalities with revenue from the sale of the valuable fertilizer.

The economics of phosphorus recovery was examined for the Metropolitan Water Reclamation District of Greater Chicago at Stickney, Illinois, which has a treatment capacity of 1,200 million gallons per day. The results show the promising potential of nutrient recovery from municipal wastewater and the thermodynamic benefits of struvite recovery. The chemical costs of phosphate recovery from the municipality are offset by the high energy required in mining (Theregowda et al., 2019). A direct financial benefit to the municipality can be created from the new revenue stream of fertilizer production (Shu et al., 2006; Kleemann, 2015). Furthermore, the slow-release nature of struvite implies that the amount required for the same level of crop production is less, leading to less polluted runoff (Talboys et al., 2016). Expanding the system boundary helps engineers understand all aspects of the entire nutrient cycle. The system efficiency is further revealed by comparing both fertilizers in terms of crop production, nutrient uptake, nutrient runoff or erosion, and eventual release in the waste stream.

In the policy arena, Germany and Sweden have set national targets to obtain phosphorus from internal recycling rather than from mining phosphate rock. On the global scale, if phosphorus mining from sewage were expanded to include manures from agriculture, it is postulated that phosphate mining could be completely displaced.

EXAMPLE 8.11 Comparing the Cost of Nutrient Removal Processes

Engineers are considering a biological phosphorus removal process that will potentially reduce the cost associated with a chemical phosphorus removal process recently added to a wastewater treatment plant. The plant produced and disposed of 25,000 pounds of biosolids per day prior to any phosphorus removal. The plant now regularly adds $500 per day of chemical coagulants to remove phosphorus. The chemical phosphorus removal process has increased biosolids production by 30%. The biological phosphorus removal process was expected to increase solids production by 25%. It costs the facility $40 to dispose of 1 ton of biosolids.

  a. How much does it cost in additional operating expenses per year for the chemical removal of phosphorus?
  b. What would the expected cost saving be in the first year of operation if the plant switched to a biological phosphorus removal process?

a. We first determine the added cost for the disposal of the solids from the chemical removal process:

$$\frac{25,000 \, \frac{\text{pounds}}{\text{day}} \times 0.3 \times \frac{\$40}{\text{ton}}}{2,000 \, \frac{\text{pounds}}{\text{ton}}} \left(365 \, \frac{\text{days}}{\text{year}} \right) = \$54,750/\text{year}$$

Next we add the cost of chemicals:

$$\$500/\text{day} \times 365 \, \text{days/year} = \$182,500/\text{year}$$

Thus, chemical phosphorus removal costs the facility a total of $237,250 per year.

b. Now we determine the cost for the biological removal of the solids:

$$\frac{25,000 \, \frac{\text{pounds}}{\text{day}} \times 0.25 \times \frac{\$40}{\text{ton}}}{2,000 \, \frac{\text{pounds}}{\text{ton}}} \left(365 \, \frac{\text{days}}{\text{year}} \right) = \$45,625/\text{year}$$

We can calculate the savings in operating costs per year if the plant switched to biological phosphorus treatment:

$$\$237,250 - \$45,625 = \$191,625$$

Thus, the plant would likely save $191,625 per year by switching to a biological treatment process, which sounds like a lot of money. This cost savings can be used in engineering economics calculations to determine the present worth of those savings over the design life of the process to determine whether these annual savings offset the cost of building a new biological reactor. Keep in mind that the biological reactor requires more space, and thus the cost of building a new reactor will be much higher in a region where the cost of land per acre is very high (cities) compared to areas where land is less expensive (rural areas).

8.5 Tertiary Treatment

Rapid sand filters similar to those in drinking water treatment plants can be used to remove additional residual suspended solids and further reduce other components. Finishing trickling filters are placed between the secondary clarifier and disinfection steps.

Oxidation ponds or lagoons are commonly used for BOD removal. An oxidation, or polishing, pond is essentially a hole in the ground—a large pond used to confine the plant effluent before it is discharged into a natural watercourse. Such ponds are designed to be aerobic, and because light penetration for algal growth is important, a large surface area is needed. The reactions that occur within an oxidation pond are depicted in Figure 8.26.

Activated carbon adsorption is another method of BOD removal, but this process has the added advantage that inorganics as well as organics are removed. The mechanism of adsorption onto activated carbon is both physical and chemical. Activated carbon is produced from a carbon source (like coconut shells) and is heated in the absence of oxygen to make a porous charcoal-like material. This material is then "activated" with pressure and very small amounts of oxygen to form microscopic pockets from the oxidized carbonaceous material. Organic materials chemically adhere to the surface of these pores, and small particles become trapped in the small openings of the activated carbon.

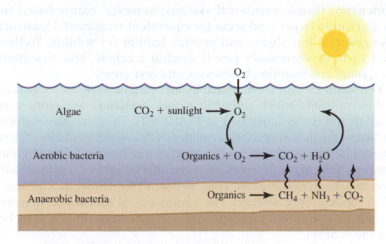

O_2

Algae CO_2 + sunlight $\longrightarrow$ O_2

Aerobic bacteria   Organics + O_2 $\longrightarrow$ CO_2 + H_2O

Anaerobic bacteria   Organics $\longrightarrow$ CH_4 + NH_3 + CO_2

FIGURE 8.26 Reactions in an oxidation (or stabilization) pond.

The activated carbon process is similar to and often a component of the rapid sand filters discussed in Chapter 7. Typically, dirty water enters the bottom of the tank and clear water exits the top. When the carbon's pore space becomes saturated with various materials, the carbon must be either regenerated or removed. The regeneration process usually involves heating the carbon in the absence of oxygen to drive off organic materials. Heating breaks down a small portion of the carbon, and not all the pore spaces can be completely cleaned, so that additional new carbon is added after the regeneration step.

The tertiary wastewater treatment processes described are not only complex but also commonly energy intensive and expensive. In order to conserve energy and reduce related carbon dioxide emissions, several less expensive modifications of natural processes are used to polish wastewater. These steps are appealing in small communities where land is available and relatively inexpensive.

Land treatment for polishing wastewater involves spraying secondary effluent on land and allowing the soil microorganisms to degrade the remaining organics. There are three types of land treatment: slow rate infiltration, rapid infiltration, and overland flow. Slow rate land treatment is the oldest and most widely used type. The nutrients and the water in partially treated wastewater may contribute to the maintenance of parks, pasture lands, and forests. Slow rate systems can produce a very high-quality percolate, but they also require the largest land area. On a worldwide basis, thousands of systems use wastewater for irrigation in variations of the slow rate process.

Proper soils and an adequate land area are critical for land application. The slow rate process is most suitable for soils with low to medium permeability. Other important considerations include the suitability of climate and soil conditions, length of the growing season, and public health regulations. Furthermore, the slope of the land should be less than 15% to promote infiltration rather than surface runoff.

Land application has been demonstrated to be effective for reducing BOD, TSS, nitrogen, phosphorus, metals, trace organics, and pathogens. Climatic variations may prohibit land application during periods of wet weather and freezing weather, thus creating a need for an adequate storage volume during these times.

Constructed wetlands are designed based on the assimilation of high levels of organic material, nitrogen, and phosphorus. Constructed wetlands rely on natural processes for the majority of the pollutant removal mechanisms. These natural treatment processes consume less energy and require less operation and maintenance than conventional systems; however, nature-based treatment methods do require larger land areas for equivalent treatment. Constructed wetlands also generate less sludge and provide habitat for wildlife. Wetlands have been used to treat stormwater runoff, landfill leachate, and wastewater from residences, small communities, businesses, and rest areas.

The two main categories of wetlands are surface flow and subsurface flow. Surface flow wetlands, which resemble natural wetlands, are more common in wastewater treatment, as illustrated in Figure 8.27. They are also known as free water surface wetlands and open water wetlands. A low-permeability material (such as clay, bentonite, or a synthetic liner) is used on the bottom to avoid groundwater contamination. Subsurface flow wetlands are also known as vegetated submerged bed, gravel bed, reed bed, and root zone wetlands (see Figure 8.27). These systems are used to replace septic systems. The advantages of subsurface systems include reduced odor and mosquito problems because the wastewater is kept below the surface of the medium.

Constructed wetlands are considered attached growth biological reactors. The major components of constructed wetland systems are plants, soils, and microorganisms. The plants serve as support media for microorganisms, provide shade (which reduces algal growth), insulate the water from heat loss, filter solids and pathogens, and provide dissolved oxygen (Crites and Tchobanoglous, 1998; Rittmann and McCarty, 2001). Cattails, reeds, rushes, bulrushes, arrowheads, and sedges are the plants most commonly used, and the choice of plant is dictated by the depth of water

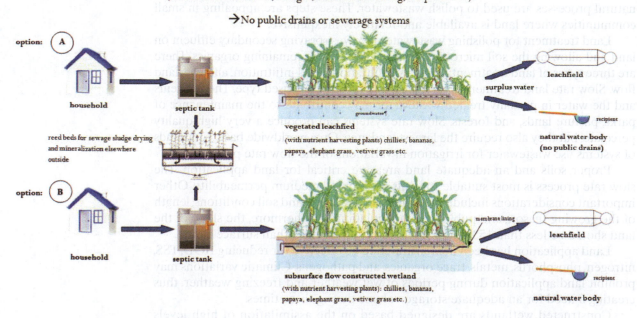

INDIVIDUAL ON-SITE TREATMENT AND DISPOSAL BY CONSTRUCTED WETLANDS AS A DECENTRALIZED TREATMENT STRATEGY

(also suitable for hotels, villages, nutrient harvesting, hospitals etc.)

→No public drains or sewerage systems

FIGURE 8.27 Surface and subsurface flow wetlands.

Source: blumberg-engineers.com/en/15/individual-sewage-treatment.

required for treatment and plant growth (Reed et al., 1998). These plants typically dominate the system due to the high nutrient levels in the water.

Typical design criteria include a detention time of 7 days (longer for cold regions) and a hydraulic loading of 200 m³/ha-day (Reed et al., 1998). BOD loadings up to 220 kg/ha-day, depth of 1.5 m, and length-to-width ratios of 3:1 have been successfully demonstrated (Rittmann and McCarty, 2001). Water depths in surface flow wetlands have been 4–18 in. (100–450 mm); bed depths of subsurface flow wetlands have been 1.5–3.3 ft (0.45–1 m) (Crites and Tchobanoglous, 1998). Wetland systems can achieve an effluent of 5–10 mg/L BOD and total nitrogen of 5–15 mg/L TSS (Reed et al., 1998).

8.6 Sludge Treatment and Disposal

Wastewater treatment facilities generate a tremendous amount of secondary waste, most of which is from the collection of suspended solids materials, collectively called wastewater residuals of sludge. Generally speaking, two types of sludges are produced in conventional wastewater treatment plants: raw primary sludge and biological or secondary sludge. The raw primary sludge comes from the bottom of the primary clarifier. The biological sludge is either solids that have grown on the fixed film reactor surfaces and sloughed off the media or waste activated sludge grown in the activated sludge system.

The quantity of sludge produced in a typical treatment plant can be determined by using the mass flow technique illustrated in Figure 8.28, where

S_0 = influent BOD, lb/day (kg/hr)
X_0 = influent suspended solids, lb/day (kg/hr)
h = fraction of BOD not removed in the primary clarifier
i = fraction of BOD not removed in the activated sludge system
X_e = plant effluent suspended solids, lb/day (kg/hr)
k = fraction of influent solids removed in the primary clarifier
ΔX = net solids produced by biological action, lb/day (kg/hr)
Y = yield, or the mass of biological solids produced in the aeration tank per mass of BOD destroyed, or $\Delta X/\Delta S$, where

$$\Delta S = hS_0 - ihS_0$$

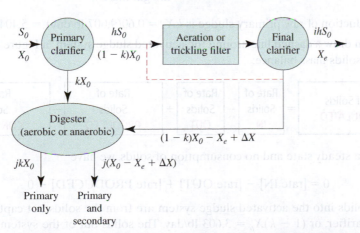

FIGURE 8.28 Typical wastewater treatment plant used for calculating sludge quantities.

EXAMPLE 8.12 Sludge Production

Wastewater enters a 6-MGD treatment plant with a BOD of 200 mg/L and suspended solids of 180 mg/L. The primary clarifier is expected to be 60% effective in removing the solids while it also removes 30% of the BOD. The activated sludge system removes 95% of the BOD that it receives, produces an effluent with a suspended solids concentration of 20 mg/L, and is expected to yield 0.5 lb of solids per pound of BOD destroyed. The plant is shown schematically in Figure 8.29. Find the quantities of raw primary sludge and waste activated sludge produced in this plant.

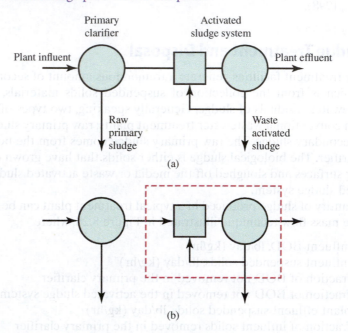

FIGURE 8.29 Sludge production from primary and secondary treatment.

The raw primary sludge from the primary clarifier is simply the fraction of solids removed, $k = 0.60$, times the influent solids. The influent solids flow is

$$X_0 = (180 \text{ mg/L})(6 \text{ MGD})\left(8.34 \frac{\text{lb}}{\text{mil gal-mg/L}}\right) = 9{,}007 \text{ lb/day}$$

so the production of raw primary sludge is $kX_0 = 0.60(9{,}007 \text{ lb/day}) = 5{,}404 \text{ lb/day}$.

We can draw a dashed line around the activated sludge system in Figure 8.29(b) and set up the solids mass balance:

Rate of Solids ACCUMULATED	=	Rate of Solids IN	−	Rate of Solids OUT	+	Rate of Solids PRODUCED	−	Rate of Solids CONSUMED

Assuming a steady state and no consumption of solids, we have

$$0 = [\text{rate IN}] - [\text{rate OUT}] + [\text{rate PRODUCED}] - 0$$

The solids into the activated sludge system are from the solids not captured in the primary clarifier, or $(1 - k)X_0 = 3{,}603 \text{ lb/day}$. The solids out of the system are of two kinds: effluent solids and waste activated sludge. The effluent solids are

$$X_e = (20\ \text{mg/L})(6\ \text{MGD})\left(8.34\ \frac{\text{lb}}{\text{mil gal-mg/L}}\right) = 1,001\ \text{lb/day}$$

The waste activated sludge is unknown.

The biological sludge is produced as the BOD is used. The amount of BOD entering the activated sludge system is

$$hS_0 = (0.7)(200\ \text{mg/L})(6\ \text{MGD})\left(8.34\ \frac{\text{lb}}{\text{mil gal-mg/L}}\right) = 7,006\ \text{lb/day}$$

The activated sludge system is 95% effective in removing this BOD, or $1 - i = 0.95$, so the amount of BOD destroyed within the system is $(1 - i) \times hS_0 = 6,655$ lb/day. The yield is assumed to be 0.5 lb of solids produced per lb of BOD destroyed, so the biological solids produced must be

$$Y \times [(1 - i) \times hS_0] = 0.5 \times 6,655 = 3,328\ \text{lb/day}$$

Plugging the known and unknown information into the mass balance in pounds per day yields

$$0 = [(1 - k)X_0] - [X_e + X_w] + [Y(1 - i)(hS_0)] - 0$$
$$0 = 3,603 - [1,001 + X_w] + 3,328 - 0$$

or

$$X_w = 5,930\ \text{lb/day}$$

or about 3 tons of dry solids per day!

8.6.1 Sludge Stabilization

Residual disposal and treatment generally are responsible for 30% to 40% of the capital costs and about 50% of the operating costs of a wastewater treatment plant. The residuals are typically treated in the wastewater treatment plant to remove water from the primary and secondary sludge, further reduce the organic materials in the residuals, reduce the odor, and reduce the pathogen content of the residuals. The primary means used are lime stabilization, aerobic digestion, and anaerobic digestion.

Lime stabilization is achieved by adding lime (either as hydrated lime, $Ca(OH)_2$, or as quicklime, CaO), which raises the sludge pH to about 11 or higher. This helps in the destruction of pathogens. The major disadvantages of lime stabilization are that it is temporary and lime is difficult to thoroughly mix with the material. In time (typically days), the pH drops and the sludge once again becomes putrescent.

Aerobic digestion is an extension of the activated sludge system where the residence time of the waste activated sludge is greatly extended in another aeration tank. In the aerobic digestion of sludge, the main carbon source comes from biological material as the concentrated solids are allowed to progress into the endogenous respiration phase. This results in the net reduction of total and volatile solids. Aerobic sludges are, however, more difficult to dewater than anaerobic sludges.

Anaerobic digestion is the third commonly employed method of sludge stabilization. The biochemistry of anaerobic decomposition is illustrated in Figure 8.30. Anaerobic digestion is a staged process where the organic material must first be partially dissolved by extracellular enzymes. Acid-forming microorganisms then produce organic acids from the organic material, and these organic materials can

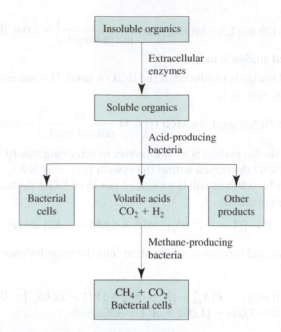

FIGURE 8.30 Anaerobic sludge digestion process.

be further decomposed by methane-forming microorganisms. The multistep process requires careful control and operation, and successful anaerobic digestion requires maintaining suitable conditions for the methane formers. The methane-forming organisms are obligate anaerobes, which are unable to function in the presence of oxygen and are very sensitive to environmental conditions such as temperature, pH, and the presence of toxins. If digestors go "sour" and smell heavily of putrid organic acids, the methane formers are inhibited in some way and the organic acid formers create a buildup of these undesirable organic acids. The gas created from anaerobic processes contains large amounts of methane gas as well as other combustible gases, collectively called *biogas*. The biogas may be recaptured and used to heat the anaerobic reactor.

Poor mixing is the cause of many issues in anaerobic digestion systems. Many anaerobic digestors used in the United States have primary and secondary anaerobic digestors, as illustrated in Figure 8.31. The primary digestor is typically covered, heated to approximately 35°C (95°F), and mixed to increase the reaction rate. The secondary digestors are not mixed or heated and are used for the storage of excess gas and for settling the sludge. As the sludge settles, the liquid supernatant is pumped back to the beginning of the plant for further treatment. The cover of the secondary digestor often floats up and down depending on the amount of gas stored.

A newer egg-shaped digester has been developed and is widely used in Europe. It provides improved mixing and thus typically better operation. In this design, shown in Figure 8.32, the digester gas is pumped into the bottom, which sets up a circulation pattern that improves the mixing.

8.6.2 Sludge Dewatering

Throughout this chapter, we have used the terms *solids* and *sludge*, which sound like materials that are composed of mostly solids; however, typical settled sludges are 97% to 99% water! The high water content allows the sludge to be moved

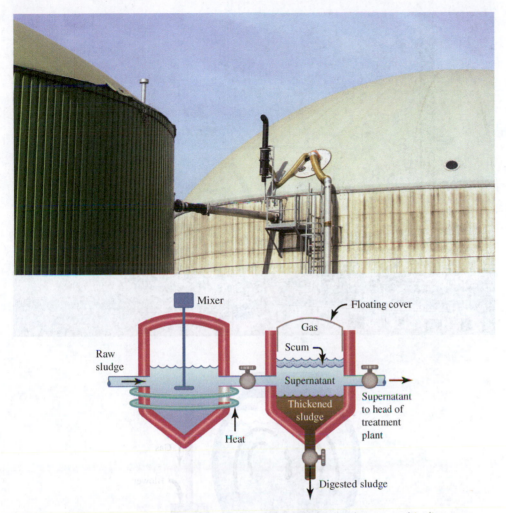

FIGURE 8.31 Two-stage anaerobic digestion: primary and secondary anaerobic digestors.

Source: iStockPhoto.com/Kontrast-fotodesign.

throughout the plant by pumping it through pipes. It is expensive to transport and ultimately dispose of the sludge, so the excess water is removed from the sludge prior to transporting it to its ultimate disposal site. In most wastewater treatment plants, dewatering is the final step to reduce the volume prior to ultimate disposal. In the United States, three dewatering techniques are most widely used: sand bed drying, belt filters, and centrifuges.

Sand beds have been in use for a long time and are still a cost-effective means of dewatering when large areas of land are available and labor costs are reasonable. As shown in Figure 8.34 (see page 565), sand beds consist of tile drains in gravel covered by about 0.25 m (10 in.) of sand. The sludge to be dewatered is poured on the beds to about 15 cm (6 in.) deep. Two mechanisms combine to separate the water from the solids: seepage and evaporation. Seepage into the sand and through the tile drains, although important in the total volume of water extracted, lasts for only a few days. As drainage into the sand ceases, evaporation takes over and this process is responsible for the conversion of liquid sludge to a solid form. In some northern climates, sand beds are enclosed under greenhouses to promote evaporation as well as to prevent rain from falling onto the beds.

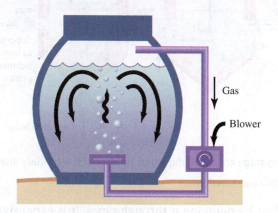

FIGURE 8.32 An egg-shaped anaerobic digester.

Source: Steve Liss/The Chronicle Collection/Getty Images.

For mixed digested sludge, the usual design is to allow 3 months of drying time. In some instances, the land area is allowed to "rest" for 1 month after the sludge has been removed, which appears to increase the drainage efficiency when sludge is reapplied. Raw primary sludge does not drain well on sand beds, usually has a noxious odor, and is therefore seldom dried on sand beds. Raw secondary sludges may either seep through the pore space or clog the pore space. When dewatering by sand beds is impractical, mechanical dewatering techniques are typically used.

The belt filter press uses both gravity filtration and pressurized filtration to remove water, as illustrated in Figure 8.35. Typically, the sludge is introduced on a moving belt, and the free water surrounding the sludge particles drips through the porous belt while the solids are retained on the belt. The belt then moves into a

BOX 8.2 Converting Biological Waste to Energy

Sewage treatment facilities process human waste 24 hours a day, 365 days a year. The volume varies per community, but the supply is constant. Managing this waste is an engineering marvel. Current municipal wastewater treatment results in large amounts of solids. Many wastewater treatment plants use anaerobic digestion to reduce the volume of these solids, resulting in the production of biogas.

Biogas (also known as digester gas) is actually a mixture of gases. Depending on the process, it is made up of methane (60% to 70%), carbon dioxide (20% to 30%), and small amounts of hydrogen sulfide, nitrogen, hydrogen, methylmercaptans, and oxygen. In many plants, the methane generated from anaerobic digestion is treated as a pollutant and is burned in flares to convert it to CO_2. It was not unusual to see a large gas flare looming high as if the plant were the next site for the Olympics.

Flaring reduces the negative climate impact of releasing methane because methane has approximately 21 times the climate change potential (i.e., potency) of CO_2 in the environment. But flaring also wastes the energy value inherent in the methane that gives the biogas an energy content of about 600 Btu/ft³. A more efficient alternative is to capture the energy of biogas using an engine or a fuel cell. For a fuel cell, the biogas, once treated, is converted to pure methane. Molten carbonate fuel cells operating at a high temperature (650°C) allow the methane to be converted to hydrogen within the fuel cell, eliminating the typical external reforming process and its associated costs and emissions (Figure 8.33).

As in all fuel cells, electricity is generated without combustion by using an electrolyte sandwiched between an anode that receives the fuel and a cathode over which oxygen passes, typically as plain air. The

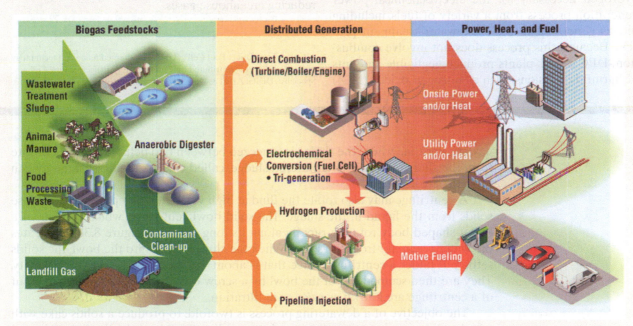

FIGURE 8.33 Biomass to electricity, heat, and hydrogen pathways.

Source: National Renewable Energy Laboratory (NREL) (2012). Biogas and Fuel Cells Workshop Summary Report Proceedings from the Biogas and Fuel Cells Workshop Golden, Colorado, June 11–13, 2012. Workshop Proceedings NREL/BK-5600-56523. Accessed March 18, 2021, at www.energy.gov/sites/prod/files/2014/03/f12/june2012_biogas_workshop_report.pdf.

(Continued)

BOX 8.2 Converting Biological Waste to Energy *(Continued)*

fuel is oxidized at the anode, releasing electrons that move to the cathode via an external circuit. These electrons meet the hydrogen and push charged ions across the electrolyte. The charged ions then move across the ion-conducting electrolyte membrane to complete the electric circuit. The waste products from this process include pure H_2O and nominal amounts of CO_2, as the CO_2 is cycled back into the process.

Reforming reaction: $CH_4 + 2H_2O \rightarrow CO_2 + 4H_2$ to H_2 create the

Reaction at the anode: $H_2 + CO_3^{2-} \rightarrow H_2O + CO_2 + 2e^-$

Reaction at the cathode: $0.5O_2 + CO_2 + 2e^- \rightarrow CO_3^{2-}$

The overall cell reaction: $H_2 + 0.5O_2 \rightarrow H_2O$

The heat generated is recovered to preheat waste sludge, optimizing the anaerobic digestion process. Direct fuel cell (DFC) power plants extract the hydrogen necessary for the electrochemical power generation process from a variety of fuels, including biogas generated in the wastewater treatment process. Because this process does not involve combustion, DFC power plants produce negligible amounts of harmful emissions, such as nitrogen oxides (NO_x)

and sulfur oxides (SO_x), as well as significantly reduced CO_2 compared to traditional fossil fuel power plants of equivalent size. A DFC plant sized for a municipal sewage treatment plant can generate 1 MW of power, enough for about 1,000 households.

Waste energy from the digestion process can be captured in several forms. The economics of biogas fuel cell projects depend on a number of factors, including the energy price for refined biogas used as natural gas and power produced from biogas in combined heat and power (CHP) applications that produce electricity (NREL, 2012). Waste heat from power generation can also be used to improve digester efficiency, especially in cold climates. Electricity, heat, and, in some cases, hydrogen are all needed during various steps of highly integrated biorefining processes. Fuel cells, powered by either biogas streams produced within the plant or externally sourced natural gas, offer the opportunity to provide these needs at high efficiency while potentially reducing greenhouse gases.

Adapted from: Direct Fuel Cell® (DFC®) is patented name and technology of FuelCell Energ. Contributed by Erin Kanoa, Hydrogen & Fuel Cell Specialist, Digital Artist & Designer, UrbanMarmot.

---

dewatering zone where the sludge is squeezed between two belts, forcing the filtrate from the sludge solids. The dewatered sludge, called cake, is then discharged when the belts are separated.

A centrifuge may also rotate the fluid at high velocities to separate more dense solids from the liquid in the sludge. The solid bowl centrifuge, which consists of a bullet-shaped body rotating rapidly along its long axis (see Figure 8.36), separates the water (*centrate*) from the solids. When the sludge is fed into the bowl, the solids settle out under a centrifugal force that is about 500 to 1,000 times that of gravity. They are then scraped out of the bowl by a screw conveyor. The solids coming out of a centrifuge are known as *cake*, as in filtration.

The objective of a dewatering process is twofold: to produce a solids cake with a high solids concentration and to make sure all the solids, and only the solids, end up in the cake. No process is 100% efficient; thus some of the solids still end up in the centrate and a substantial amount of water still ends up in the cake. The performance of centrifuges is measured by sampling the feed, centrate, and cake coming out of the machine. The centrifuge process makes it difficult to measure the flow rates of the centrate and cake; only the solids are easily sampled. Solids recovery and cake solids percent must be calculated using the mass balance technique.

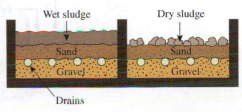

FIGURE 8.34 Sand drying bed for sludge dewatering.

Source: Courtesy of P. Aarne Vesilind.

The centrifuge is a two-material black box. A mass balance equation can be written assuming steady state operation and recognizing that no liquid or solids are produced or consumed in the centrifuge:

$$[\text{Rate of material IN}] = [\text{rate of material OUT}]$$

The volume balance in terms of the sludge flowing in and out is

$$Q_0 = Q_k + Q_c$$

and the mass balance in terms of the sludge solids is

$$Q_0 C_0 = Q_k C_k + Q_c C_c$$

where

Q_0 = flow of sludge as the feed, volume/unit time
Q_k = flow of sludge as the cake, volume/unit time
Q_c = flow of sludge as the centrate, volume/unit time

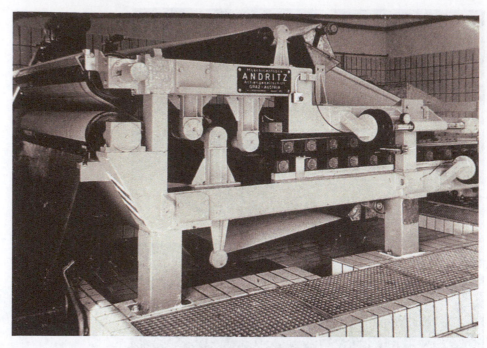

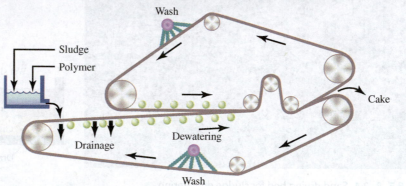

FIGURE 8.35 A filter belt used for sludge dewatering.

Source: Courtesy of P. Aarne Vesilind.

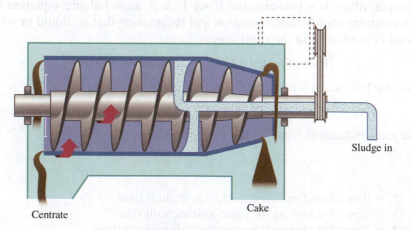

FIGURE 8.36 A solid bowl centrifuge used for solids dewatering.

For the recovery of solids, we have

$$\text{Solids recovery} = \frac{\text{mass of dry solids as cake}}{\text{mass of dry feed solids}} \times 100$$

$$\text{Solids recovery} = \frac{C_k Q_k}{C_0 Q_0} \times 100$$

Solving the first material balance for Q_c and substituting into the second material balance yield

$$Q_k = \frac{Q_0(C_0 - C_c)}{C_k - C_c}$$

Substituting this expression into the equation for solids recovery, we find

$$\text{Solids recovery} = \frac{C_k(C_0 - C_c)}{C_0(C_k - C_c)} \times 100$$

This expression allows for the calculation of solids recovery using only the concentration terms.

EXAMPLE 8.13   Solids Dewatering by a Centrifuge

A wastewater sludge centrifuge operates with a feed of 10 gpm and a feed solids concentration of 1.2%. The cake solids concentration is 22%, and the centrate solids concentration is 500 mg/L. What is the recovery of solids?

The solids concentrations must first be converted to the same units. Recall that if the density of the solids is almost 1 (a good assumption for wastewater solids), then 1% solids = 10,000 mg/L.

	SOLIDS CONCENTRATION (mg/L)
Feed solids	12,000
Cake solids	220,000
Centrate solids	500

$$\text{Solids recovery} = \frac{220,000(12,000 - 500)}{12,000(220,000 - 500)} \times 100 = 96\%$$

The centrifuge is an interesting solids separation device because it can operate at almost any cake solids and solids recovery, depending on the conditions of the sludge and the flow rate. As the flow rate increases, the solids recovery begins to decrease because the residence time in the centrifuge is shorter and only the larger and more dense particles are separated from the liquid. The cake solids concentration is higher as less water is entrained in these particles. The relationship between the flow rate, percent solids recovery, and cake solids concentration as shown in Figure 8.37 can be developed for any centrifugal sludge system and is known as the *operating curve*. At lower feed flows, long residence times allow for complete settling and solids removal, but the cake becomes wetter because all the soft and small particles entrain a lot of water when they are removed.

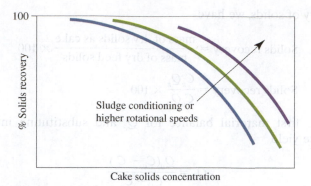

FIGURE 8.37 Centrifuge operating curve.

As mentioned above, there is a trade-off between solids recovery and cake solids concentration. There are two ways to improve both of these qualities: increase the centrifugal force imposed on the solids and improve sludge conditioning.

The centrifugal forces may be explained by Newton's law:

$$F = ma$$

where

    F = force, N
    m = mass, kg
    a = acceleration, m/s^2

If the acceleration is due to gravity, then $F = mg$, where g is the acceleration due to gravity, in m^2/s.

A spinning mass requires a force vector toward the center of rotation to keep it from flying off into space. The force is described as

$$F_c = m(w^2 r)$$

where

    F_c = centrifugal force, N
    m = mass, kg
    w = rotational speed, rad/s
    r = radius of rotation

The term $w^2 r$ is called the centrifugal acceleration and has units of meters per second squared (m/s^2). The gravities, G, produced by a centrifuge are

$$G = \frac{w^2 r}{g}$$

Increasing the rotational speed of the bowl therefore increases the centrifugal force and moves the entire operating curve to the right.

The second method of simultaneously increasing cake solids and solids recovery is to condition the sludge with chemicals, such as polyelectrolytes, prior to dewatering. These large polyelectrolyte molecules have molecular weights of several million and they have charged sites on them that appear to attach to sludge particles and bridge the smaller particles such that large flocs are formed. These larger flocs are able to expel water and compact to small solids, yielding both a cleaner centrate and a more compact cake. This, in effect, moves the operating curve to the right.

8.6.3 Ultimate Disposal

The EPA established the Standards for the Use or Disposal of Sewage Sludge, commonly referred to as Part 503.3, in 1993. The purpose of this regulation is to establish the maximum concentrations of several pollutants that can be present in land-applied sewage sludge and still protect a highly exposed person, animal, or plant. The regulations authorize the permit authority to be more restrictive on a case-by-case basis in order to address local environmental conditions.

Several categories relate to the regulation and disposal of wastewater sludge. The treated sludge may be categorized as either exceptional quality (EQ) or nonexceptional quality (non-EQ) sludge, as illustrated in Figure 8.38. *Exceptional quality sludge* (*EQ sludge*) must meet pollutant concentration limits, pathogen reduction standards, and vector attraction reduction. EQ sludge is largely free of regulatory requirements for land application. *Nonexceptional quality sludge* (*non-EQ sludge*) must also meet pollutant concentration limits, pathogen reduction standards, and vector attraction reduction, but the requirements are not as stringent as for EQ sludge. Non-EQ sludge is subject to considerable regulatory requirements for land application. Sewage sludge that does not meet the quality criteria for either EQ or non-EQ sludge may not be applied to land. Such material must be disposed of either by incineration or in a landfill.

It is expensive to incinerate sludge because of the high energy requirements to safely combust the material, although it may be the only option for communities with dense populations. Two types of incinerators are most commonly used for sludge incineration: multiple-hearth and fluidized bed incinerators. The multiple-hearth incinerator, illustrated in Figure 8.39, has several hearths stacked vertically with

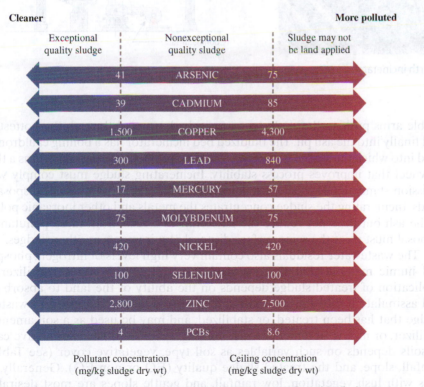

FIGURE 8.38 Criteria for exceptional quality and nonexceptional quality sludge.

Source: Based on extension.psu.edu/sewage-sludge-a-plain-english-tour-of-the-regulations.

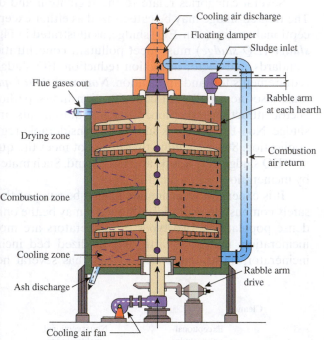

Labels on figure:
- Cooling air discharge
- Floating damper
- Sludge inlet
- Flue gases out
- Rabble arm at each hearth
- Drying zone
- Combustion air return
- Combustion zone
- Cooling zone
- Rabble arm drive
- Ash discharge
- Cooling air fan

FIGURE 8.39 A multiple-hearth incinerator.

Source: Courtesy of P. Aarne Vesilind.

rabble arms pushing the sludge progressively downward through the hottest layers and finally into the ash pit. The fluidized bed incinerator has a boiling cauldron of hot sand into which the sludge is pumped. The sand provides mixing and acts as a thermal flywheel that improves process stability. Incinerating sludge must comply with air emission standards. The ash must comply with all other sludge use or disposal standards. Incinerating the sludge concentrates the metals and other inorganic pollutants in the ash but does not reduce the environmental risks from these pollutants. Ash disposal must meet the same surface disposal requirements as other sludges.

The wastewater residuals also contain very high levels of nitrogen, phosphorus, and humic material that may be used as a soil conditioner or fertilizer. Land application of treated sludge depends on the ability of the land to absorb sludge and assimilate it into the soil matrix. *Biosolids* is the term applied to wastewater sludge that has been treated, or stabilized, and may be used as a soil amendment, fertilizer, or means to remediate acid mining waste sites. The assimilative capacity of soils depends on such variables as soil type, vegetative cover (see Table 8.7), rainfall, slope, and the treated sludge quality (EQ or non-EQ). Generally, sandy soils with lush vegetation, low rainfall, and gentle slopes are most desirable for sludge disposal.

TABLE 8.7 Variations in nitrogen assimilation for common crops

CROP	EXPECTED YIELD (bushel/acre/year)	NITROGEN REQUIREMENT (lb-N/acre/year)
Corn	100	100
Oats	90	60
Barley	70	60
Grass and hay	4 tons/acre	200
Sorghum	60	60
Peanuts	40	30
Wheat	70	105
Wheat	150	250
Soybeans	40	30
Cotton	1 bale/acre	50
Cotton	1.5 bales/acre	90

Note: Crop fertilization requirements vary greatly with soil type, expected yields, and climatic conditions. To get more information on crop fertilization needs specific to a locality, contact local agricultural agents.

Source: Domestic *Septage Regulatory Guidance: A Guide to the EPA 503 Rule* (U.S. EPA, 1983).

EXAMPLE 8.14 Land Application of Biosolids

Anaerobically digested biosolids with 5% nitrogen (dry weight N basis) were applied for 2 years on tree farmland. The application rate of the biosolids on the farmland in year 1 was 5,000 kg/ha. The application rate of the biosolids on the farmland in year 2 was 3,000 kg/ha. The mineralization rate of nitrogen in the field is based on the U.S. EPA factors shown in Table 8.8. Calculate the total mineralized organic nitrogen available from the application of the biosolids in the third year of the growing season.

TABLE 8.8 Mineralization rates for nitrogen applied from biosolids

TIME AFTER BIOSOLIDS APPLICATION (YR)	FRACTION OF ORG-N MINERALIZED FROM STABILIZED PRIMARY AND WASTE ACTIVATED SLUDGE	FRACTION OF ORG-N MINERALIZED FROM AEROBICALLY DIGESTED SLUDGE	FRACTION OF ORG-N MINERALIZED FROM ANAEROBICALLY DIGESTED SLUDGE	FRACTION OF ORG-N MINERALIZED FROM COMPOSTED SLUDGE
0–1	0.40	0.30	0.20	0.10
1–2	0.20	0.15	0.10	0.05
2–3	0.10	0.08	0.05	0.03

Note: The volatilization factors and mineralization rates are from the Process Design Manual for the Land Application for Sewage Sludge (U.S. EPA, 1983). Many localities have developed different values for the volatilization and mineralization rates which may yield more precise estimates of available nitrogen.

We can calculate the rate of nitrogen application in year 1:

$$\left(\frac{5{,}000 \text{ kg biosolids}}{\text{ha}}\right)\left(\frac{0.05 \text{ kg N}}{\text{kg biosolids applied}}\right) = 250 \frac{\text{kg}}{\text{ha}}$$

During year 1, 20% of the nitrogen is mineralized according to Table 8.8, so the mineralized organic nitrogen remaining in the soil at the beginning of year 2 from the first application is estimated to be

$$\left(\frac{250 \text{ kg total N}}{\text{ha}}\right)\left(\frac{0.2 \text{ kg mineralized org N}}{\text{kg N}}\right) = 50 \frac{\text{kg}}{\text{ha}}$$

The organic nitrogen remaining in the soil after year 1 and at the beginning of year 2 is

$$250 \text{ kg/ha} - 50 \text{ kg/ha} = 200 \text{ kg/ha}$$

Similar calculations done for year 2 are summarized in Table 8.9.

From Table 8.9, the total mineralized organic-N available for the growing season in year 3 is the sum of the mineralized org-N from rows c and e: $9 + 30 = 39$ kg org-N/ha.

TABLE 8.9 Summary table for the mineralization of nitrogen in Example 8.14

APPLICATION YEAR	STARTING ORG-N (kg/ha)	MINERALIZATION RATE (kg MINERALIZED-N/ ORG-N)	MINERALIZED ORG-N (kg/ha)	ORGANIC NITROGEN REMAINING (kg-ha)
First Application				
(a) 0–1	250	0.2	50	200
(b) 1–2	200	0.1	20	180
(c) 2–3	180	0.05	9	171
Second Application				
(d) 0–1	0	0	0	0
(e) 1–2	150	0.2	30	120
(f) 2–3	120	0.1	12	108

ACTIVE LEARNING EXERCISE 8.2 Reclamation Project Using Biosolids

Your city is contemplating placing its wastewater sludge on an abandoned strip mine to try to reclaim the land for productive use. As the city consulting engineer, you are asked for your opinion. You know that the city has been having difficulties with its anaerobic digesters, and although it can probably meet the Class B standards on pathogens most of the time, it probably will not be able to do so consistently. Reclaiming the strip mine is a positive step, as is the low cost of disposal. City engineers tell you that if they cannot use the strip mine disposal method, they must buy a very expensive incinerator to meet the new EPA regulation. The upgrade of the anaerobic digesters will also be prohibitively expensive for your city. What do you recommend? Analyze this problem in terms of the affected parties, possible options, and final recommendations for action.

TABLE 8.10 Cumulative pollutant loading limits for land application of sewage sludge

POLLUTANT	CUMULATIVE POLLUTANT LOADING RATE (lb/acre)	POLLUTANT	CUMULATIVE POLLUTANT LOADING RATE (lb/acre)
Arsenic	36	Mercury	15
Cadmium	34	Nickel	370
Copper	1,320	Selenium	88
Lead	264	Zinc	2,464

If bulk sewage sludge being applied to land is non-EQ due to pollutant levels, the applier must comply with cumulative pollutant loading rates (CPLRs). The amount of each pollutant applied to land sites (including previous applications) must be tracked for land application subject to CPLRs. The land applier must also calculate the amount of each pollutant applied to the land according to the procedure described in applicable regulations. The applier is required to maintain records of the amounts of sludge and component materials (typically nitrogen, phosphorus, and metals) applied to the site so that the allowable limits of pollutants shown in Table 8.10 are not exceeded. Part 503 requires that bulk non-EQ sewage sludge be applied to a site at a rate that is equal to or less than the agronomic rate for the site. The agronomic rate is the whole sewage sludge application rate (on a dry weight basis) that is designed to provide the amount of nitrogen needed by the crop or vegetation while also minimizing the amount of nitrogen that passes below the root zone of the crop or vegetation to the groundwater. A sewage sludge application rate that exceeds the agronomic rate can result in nitrate contamination of the groundwater. Sewage sludge should be applied as close to the time of maximum nutrient uptake by crops as feasible.

EXAMPLE 8.15 Estimating Pollutant Loading in Sludge

A sample of biosolids is analyzed in the lab and has a 20% solids content and a copper concentration of 250 mg/L on a wet basis. What is the concentration of copper on a dry basis?

We can assume the specific gravity of the sludge is approximately equal to that of water. If 76% of the weight (of the water) is removed, then

$$\frac{250 \dfrac{\text{mg copper}}{\text{L wet solids}}}{0.20 \dfrac{\text{kg dry solids}}{\text{L wet solids}}} = 1,250 \frac{\text{mg copper}}{\text{kg dry solids}}$$

8.7 Water Recycling and Reuse

Water reuse and recycling has been documented since the Bronze Age (ca. 3200–1100 BC), when many civilizations from China to Egypt used domestic wastewater for irrigation and aquaculture (Angelakis et al., 2018). Greek and Roman civilizations used wastewater for irrigation and fertilization, especially around more densely populated cities. Land application of wastewater for disposal and agricultural use is used today in European cities and in the United States.

There are many terms for reusing water, including water reuse, water recycling, and water reclamation. The steps in water reuse typically include:

1. Collecting water
2. Treating the water
3. Finding a beneficial reuse of the water, commonly for agriculture, irrigation, environmental restoration, industrial processes, and potable water supplies

Water reuse may be planned or unplanned. Unplanned water reuse is the use of water that has been previously used, such as when communities draw water from the downstream stretches of large river systems like the Colorado River or the Mississippi River. Both the Colorado and Mississippi Rivers receive treated wastewater from many communities, so that by the time the rivers reach their outlets (if they do, in the case of the Colorado River), a significant portion of the river water has been used at least once for human purposes.

The combination of population growth, urbanization, and changing climate has led more and more communities to consider planned water reuse strategies to combat current and future water scarcity. Planned water reuse refers to collection systems, wastewater treatment systems, and water supply systems that have been designed with the goal of beneficially reusing a recycled water supply. Agricultural and landscape irrigation, industrial process water, potable water supplies, and groundwater supply management are some examples of planned water reuse systems.

California has been recycling water for drinking water since the 1970s, when groundwater supplies began to suffer from sea water intrusion (much like in Malta) and the water supply from the Colorado River and Sierra Nevada Mountains was insufficient to meet water demand (Leslie, 2018). Orange County used highly treated wastewater and reinjected the water into the groundwater table to store the water in the local aquifer and to dilute it prior to withdrawing it for its potable water supply. Orange County was and remains the world's largest wastewater-to-drinking-water plant.

Orange County has recently made a multimillion-dollar investment to expand its water reuse capacity. The project takes treated wastewater effluent and further purifies it using a three-step process of microfiltration, reverse osmosis, and disinfection with ultraviolet light and hydrogen peroxide. The water that is injected into the ground meets or exceeds state and federal drinking water standards. Orange County generates 130 million gallons of drinking water a day, enough to serve about half of its 2.5 million customers (Burris, 2020). In 2018, Orange County produced four times more drinking water per day than the world's second largest sewage-to-drinking-water facility in Singapore (Leslie, 2018).

The shift toward reused and recycled drinking water is occurring in many water-scarce areas throughout the world. Although the wastewater treatment process is very intensive, the costs of recycling drinking water in Orange County at $850 per acre-foot are less than the cost of transporting water over long distances and storing water in reservoirs at $1,000 per acre-foot. The advantages to recycling water include:

- Lower costs
- Reliable water source
- Fewer impacts from excessive water withdraw to ecosystems
- Less costly infrastructure investments needed compared to building large reservoirs or conveyance systems

There are also some inherent risks to reusing wastewater. Depending on the level of treatment, they may include exposure to bacteria, heavy metals, or organic pollutants (including pharmaceuticals, personal care products, and pesticides).

Irrigation with wastewater can have both positive and negative effects on soil and plants, depending on the chemical composition of the wastewater, the types of soil, and the types of plants being irrigated. The National Academies of Science, Engineering, and Medicine evaluated the risks of water reuse under various scenarios (National Research Council, 2012). The analysis compared the risks of exposure to four pathogens (adenovirus, norovirus, *Salmonella*, and *Cryptosporidium*) and 24 chemical contaminants, including pharmaceuticals, personal care products, natural hormones, industrial chemicals, and byproducts from water disinfection processes. The Research Council concluded:

> *"the risk of exposure to certain microbial and chemical contaminants from drinking reclaimed water does not appear to be any higher than the risk experienced in at least some current drinking water treatment systems—and may, in fact, be orders of magnitude lower. The analysis revealed that carefully planned potable water reuse projects should be able to provide a level of protection from waterborne illness and chemical contaminants comparable to—and, in some cases, better than—the level of protection the public experiences in many drinking water supplies across the nation. However, the committee pointed out that the analysis was presented as an example and should not be used to endorse certain treatment schemes or to determine the risk at any particular site without site-specific analysis."*

Adopting technology to reuse water is critical to many of the fastest growing and most populous cities in the United States. The populations of cities such as Phoenix, Arizona, and Las Vegas, Nevada, located in the arid Southwest, continue to grow despite serious questions about available water supplies. Even in the Southeast, where there is much greater rainfall, population growth has led to water stresses in Atlanta, Georgia, due to slow recharge rates to the groundwater table and demand for water from the Chattahoochee River. There is vast potential for water reuse, since coastal areas in the United States discharge 12 billion gallons of wastewater into estuaries and oceans every day (equivalent to 6% of the country's total daily water use). Water reuse options illustrated in Figure 8.40 provide a more sustainable approach to meeting the demand for water in many growing communities.

Municipal wastewater treatment facilities receive varied amounts of wastewater with various concentrations of constituents. Wastewater treatment plants are designed to reduce the total suspended solids, organic wastes, nutrients, and pathogens in the water. Wastewater may be most efficiently collected and treated at publicly owned plants. In rural areas and some developing nations, however, the infrastructure does not exist to properly treat domestic wastewater. In this case, there are numerous decentralized processes that may be applicable. One effective method in small communities is to create small collective activated sludge or suspended growth treatment systems. Septic systems, sand filters, constructed wetlands, and composting toilets are other effective processes used to treat domestic wastewater.

A *septic tank*, illustrated in Figure 8.41, acts as a partially aerobic and partially anaerobic treatment system. It allows microbial treatment and requires a long hydraulic residence time. The wastewater solids build up in the septic tank over time, so the tanks must be emptied periodically to remove the wastewater solids and residuals. Septic tanks may be very effective in removing organic waste and suspended solids, but they must be properly maintained to prevent leaks from both the tank itself and the plumbing that connects the septic tank to the home.

Wastewater may be considered a resource that can supply communities with additional recycled water directly as well as energy and nutrients, as illustrated

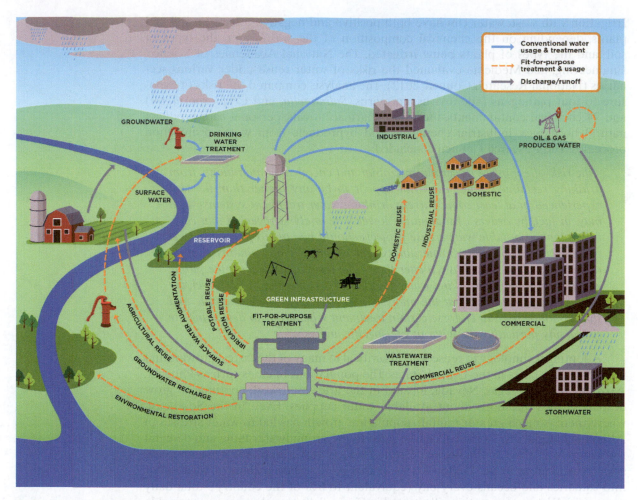

FIGURE 8.40 Opportunities for water reuse in communities, dependent on the level of wastewater treatment.

Source: U.S. EPA, www.epa.gov/waterreuse/basic-information-about-water-reuse.

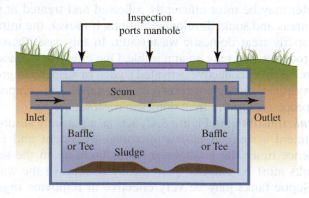

FIGURE 8.41 Example of a septic tank for decentralized treatment of domestic wastewater.

in Figure 8.42. The reuse of treated wastewater is increasingly being considered for urban community landscaping use, industrial use, agriculture, groundwater recharge, and augmentation of potable supplies. The water quality requirements for typical agricultural applications are presented in Table 8.11. Properly treated

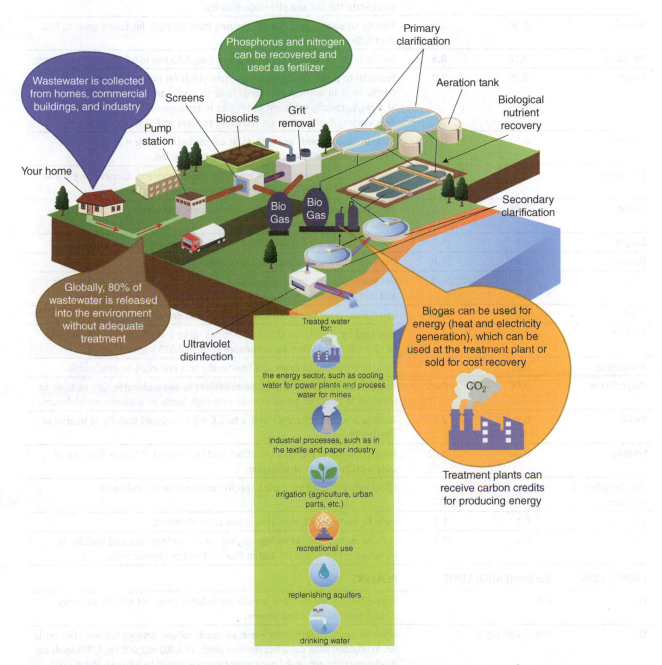

FIGURE 8.42 Opportunities to utilize wastewater components and characteristics as a resource for water, fertilizer, and energy.

TABLE 8.11 Maximum constituent concentration in reclaimed water for use in agricultural irrigation water

CONSTITUENT	LONG-TERM USE (mg/L)	SHORT-TERM USE (mg/L)	REMARKS
Aluminum	5.0	20	Can cause nonproductiveness in acid soils, but soils at pH 5.5 to 8.0 will precipitate the ion and eliminate toxicity.
Arsenic	0.10	2.0	Toxicity to plants varies widely, ranging from 12 mg/L for Sudan grass to less than 0.05 mg/L for rice.
Beryllium	0.10	0.5	Toxicity to plants varies widely, ranging from 5 mg/L for kale to 0.5 mg/L for bush beans.
Boron	0.75	2.0	Essential to plant growth, with optimum yields for many obtained at a few-tenths mg/L in nutrient solutions. Toxic to many sensitive plants (e.g., citrus) at 1 mg/L. Usually sufficient quantities in reclaimed water to correct soil deficiencies. Most grasses are relatively tolerant at 2.0 to 10 mg/L.
Cadmium	0.01	0.05	Toxic to beans, beets, and turnips at concentrations as low as 0.1 mg/L in nutrients solution. Conservative limits recommended.
Chromium	0.1	1.0	Not generally recognized as an essential growth element. Conservative limits recommended due to lack of knowledge on toxicity to plants.
Cobalt	0.05	5.0	Toxic to tomato plants at 0.1 mg/L in nutrient solution. Tends to be inactivated by neutral and alkaline soils.
Copper	0.2	5.0	Toxic to a number of plants at 0.1 to 1.0 mg/L in nutrient solution.
Fluoride	1.0	15.0	Inactivated by neutral and alkaline soils.
Iron	5.0	20.0	Not toxic to plants in aerated soils, but can contribute to soil acidification and loss of essential phosphorus and molybdenum.
Lead	5.0	10.0	Can inhibit plant cell growth at very high concentrations.
Lithium	2.5	2.5	Tolerated by most crops at concentrations up to 5 mg/L; mobile in soil. Toxic to citrus at low doses – recommended limit is 0.075 mg/L.
Manganese	0.2	10.0	Toxic to a number of crops at a few-tenths to a few mg/L in acidic soils.
Molybdenum	0.01	0.05	Nontoxic to plants at normal concentrations in soil and water. Can be toxic to livestock if forage is grown in soils with high levels of available molybdenum.
Nickel	0.2	2.0	Toxic to a number of plants at 0.5 to 1.0 mg/L; reduced toxicity at neutral or alkaline Ph.
Selenium	0.02	0.02	Toxic to plants at low concentrations and to livestock if forage is grown in soils with low levels of selenium.
Tin, Tungsten, & Titanium	—	—	Effectively excluded by plants; specific tolerance levels unknown.
Vanadium	0.1	1.0	Toxic to many plants at relatively low concentrations.
Zinc	2.0	10.0	Toxic to many plants at widely varying concentrations; reduced toxicity at increased pH (6 or above) and in fine-textured or organic soils.

CONSTITUENT	RECOMMENDED LIMIT	REMARKS
pH	6.0	Most effects of pH on plant growth are indirect (e.g., pH effects on heavy metals' toxicity described above).
TDS	500–2,000 mg/L	Below 500 mg/L, no detrimental effects are usually noticed. Between 500 and 1,000 mg/L, TDS in irrigation water can affect sensitive plants. At 1,000 to 2,000 mg/L, TDS levels can affect many crops and careful management practices shpuld be followed. Above 2,000 mg/L, water can be used regularly only for tolerant plants on permeable soils.
Free Chlorine Residual	<1 mg/L	Concentrations greater than 5 mg/L causes severe damage to most plants. Some sensitive plants may be damaged at levels as low as 0.05 mg/L.

Adapted from: U.S. EPA Water Reuse Guidelines (2004) and Rowe and Abdel-Magid, 1995.

water may supplement existing water supplies in a wide variety of industrial uses, landscaping uses, and agricultural uses. Several countries that already have water resources that do not meet current demand have adopted water reclamation and reuse strategies.

8.8 Summary

Who is to choose which of these treatment strategies will be used with any given wastewater and aquatic conditions? Such decisions are made by engineers and scientists who define what is to be achieved and then propose the treatment to accomplish the objective. Design engineers establish the objectives of treatment and then design a facility to meet these objectives.

In this role, engineers have considerable latitude and also considerable responsibility. Society asks them to design something that not only works but also works at the lowest possible cost, does not prove to be a nuisance to its neighbors, and looks nice. In this role, engineers become the repositories of the public trust. Because of this public and environmental responsibility, engineering is a profession, and as such, all engineers are expected to adhere to high professional standards. Not only are environmental engineers responsible for performing a job, but they also have another "client": the environment. Explaining the conflicts that arise in working with environmental concerns is a major role of the environmental engineer. The responsibilities far exceed those of an ordinary citizen.

References

Angelakis, A. N., Asano, T., Bahri, A., Jimenez, B. E., and Tchobanoglous, G. (2018). "Water reuse: From ancient to modern times and the future." *Front. Environ. Sci.* 6:26.

Burris, D. L. (2020). Groundwater Replenishment System 2019 Annual Report. Orange County Water District. Accessed March 18, 2021, at www.ocwd.com/gwrs/annual-reports/.

Centre Europeen d'Etudes des Polyphosphates. (2008). Accessed September 2008 at www.ceep-phosphates.org/.

Crites, R., and Tchobanoglous, G. (1998). *Small and Decentralized Wastewater Management Systems*. Boston: WCB McGraw-Hill.

Klankrong, T., and Worthley, T. S. (2001). "Rethinking Bangkok's wastewater strategy." *Civil Engineering* 71(no. 6):72–77.

Kleemann, R. (2015). Sustainable phosphorus recovery from waste. Ph.D. thesis. Surrey, UK: University of Surrey.

Leslie, J. (2018, May 1). Where Water Is Scarce, Communities Turn to Reusing Wastewater. Yale Environment 360. Accessed March 18, 2021, at e360.yale.edu/features/instead-of-more-dams-communities-turn-to-reusing-wastewater.

National Renewable Energy Laboratory (NREL). (2012). Biogas and Fuel Cells Workshop Summary Report Proceedings from the Biogas and Fuel Cells Workshop, Golden, Colorado, June 11–13, 2012. Workshop Proceedings NREL/BK-5600-56523.

National Research Council. (2012). *Understanding Water Reuse: Potential for Expanding the Nation's Water Supply Through Reuse of Municipal Wastewater*. Washington, DC: The National Academies Press. doi.org/10.17226/13514.

Reed, S. C., Middlebrooks, E. J., and Crites, R. W. (1988). *Natural Systems for Waste Management and Treatment*. New York: McGraw-Hill.

Rittmann, B. E., and McCarty, P. L. (2001). *Environmental Biotechnology: Principles and Applications*. Boston: McGraw-Hill.

Rodriguez, D. J., Serrano, H. A., Delgado, A., Nolasco, D., and Saltiel, G. (2020). "From Waste to Resource: Shifting paradigms for smarter wastewater interventions in Latin America and the Caribbean." World Bank, Washington, DC.

Shu, L., Schneider, P., Jegatheesan, V., and Johnson, J. (2006). "An economic evaluation of phosphorus recovery as struvite from digester supernatant." *Bioresour. Technol.* 91:2211

Smith, G. (2003). Why fluoride is an environmental issue. Earth Island Institute. *Earth Island Journal*.

Talboys, P. J., Heppell, J., Roose, T., Healey, J. R., Jones, D. L., and Withers, P. J. A. (2016). Struvite: A slow-release fertiliser for sustainable phosphorus management? *Plant Soil* 401, 109.

Theregowda, R. B., González-Mejía, A. M., Ma, X., and Garland, J. (2019, July 1). "Nutrient recovery from municipal wastewater for sustainable food production systems: An alternative to traditional fertilizers." *Environ. Eng. Sci.* 36(7):833–842.

U.S. EPA. (2015). Case Studies on Implementing Low-Cost Modifications to Improve Nutrient Reduction at Wastewater Treatment Plants, Draft–Version 1.0. EPA-841–R–15–004. Washington DC.

WEF. (2008). Recycling Nutrients into Environmentally Safe Commercial Fertilizer. Water Environment Federation 20(5). Accessed September 2008 at www.wef.org /ScienceTechnologyResources/Publications/WET/08/08May/08MayProblemSolver .htm.

Key Concepts

Inflow and infiltration (I/I)

Publicly owned treatment works (POTWs)

Primary wastewater treatment processes

Clarifiers

Chemical coagulants

Secondary treatment processes

Trickling filter

Rotating biological contactor (RBC)

Activated sludge system

Activated sludge

Return activated sludge (RAS)

Waste activated sludge (WAS)

Substrate loading

Food-to-microorganism ratio (F/M)

Mixed liquor

Mixed liquor suspended solids (MLSS)

Solids retention time (SRT)

Mean cell residence time (MCRT)

Sludge volume index (SVI)

Biological nitrogen removal (BNR)

Polyphosphate-accumulating organisms (PAOs)

Volatile fatty acids (VFAs)

Constructed wetlands

Biogas

Operating curve

Exceptional quality sludge

Nonexceptional quality sludge

Septic tank

Problems

8-1   The following data are from the operation of a wastewater treatment plant:

CONSTITUENT	INFLUENT (mg/L)	EFFLUENT (mg/L)
BOD_5	200	20
SS	220	15
P	10	0.5

a. What percent removal was experienced for each of these constituents?

b. What kind of treatment plant would produce such an effluent? Draw a block diagram showing one configuration of the treatment steps that would result in this plant performance.

8-2   Describe the condition of a primary clarifier one day after the raw sludge pumps broke down. What do you think would happen?

8-3   One operational problem with trickling filters is *ponding*, the excessive growth of slime on the rocks and subsequent clogging of the spaces so that the water no longer flows through the filter. Suggest some cures for the ponding problem.

8-4   One problem with sanitary sewers is illegal connections from roof drains. Suppose that for a family of four, living in a house with a roof area of 25 m × 15 m, the roof drain is connected to the sewer. Assume that 25 cm of rain falls in 1 day.

a. What percent increase will there be in the flow from the house over the dry weather flow (assumed at 190 gal/capita-day)?

b. What is wrong with connecting roof drains to the sanitary sewers? How would you explain this to someone who has just been fined for such illegal connections? (Don't just say, "It's the law." Explain *why* there is such a law.) Use ethical reasoning to fashion your argument.

8-5   What secondary and, if necessary, tertiary unit operations are needed to treat the following wastes to effluent levels of BOD_5 = 20 mg/L, SS = 20 mg/L, and P = 1 mg/L?

WASTE	BOD_5 (mg/L)	SS (mg/L)	P (mg/L)
a. Domestic	200	200	10
b. Chemical industry	40,000	0	0
c. Pickle cannery	0	300	1
d. Fertilizer mfg.	300	300	200

8-6   The success of an activated sludge system depends on the settling of the solids in the final settling tank. Suppose the sludge in a system starts to bulk (not settle very well) and the suspended solids concentration of the return activated sludge drops from 10,000 mg/L to 4,000 mg/L. (Do not answer quantitatively.)

a. What will this do to the mixed liquor suspended solids?

b. What will this do to the BOD removal? Why?

8-7 The MLSS in an aeration tank is 4,000 mg/L. The flow from the primary settling tank is 0.2 m³/s with an SS of 50 mg/L, and the return sludge flow is 0.1 m³/s with an SS of 6,000 mg/L. Do these two sources of solids make up the 4,000 mg/L SS in the aeration tank? If not, how is the 4,000-mg/L level attained? Where do the solids come from?

8-8 A 1-L cylinder is used to measure the ability to settle 0.5% by weight suspended solids sludge. After 30 minutes, the settled sludge solids occupy 600 mL. Calculate the SVI.

8-9 What measures of stability would you need if a sludge from a wastewater treatment plant was to be:

a. placed on the White House lawn?

b. dumped into a trout stream?

c. sprayed on the playground?

d. spread on a vegetable garden?

8-10 A sludge is thickened from 2,000 mg/L to 17,000 mg/L. What is the reduction in volume, in percent? (Use a black box and material balance.)

8-11 Sludge age (called mean cell residence time) is defined as the mass of sludge in the aeration tank divided by the mass of sludge wasted per day. Calculate the sludge age if the aeration tank has a hydraulic retention time of 2 hr, the suspended solids concentration is 2,000 mg/L, the flow rate of wastewater to the aeration basin is 1.5 m³/min, the return sludge solids concentration is 12,000 mg/L, and the flow rate of waste activated sludge is 0.02 m³/min.

8-12 The block diagram in the figure shows a secondary wastewater treatment plant.

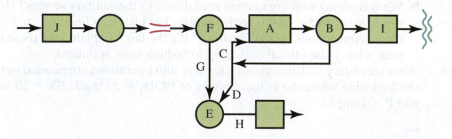

a. Identify the various unit operations and flows, and state their purpose or function. Why are they there, or what do they do?

b. Suppose you are a senior engineer in charge of wastewater treatment for a metropolitan region. You have hired a consulting engineering firm to design the plant shown. Four firms were considered for the job, and this firm was selected by the metropolitan authority board. However, you are concerned that the decision was influenced by large political contributions from the firm to the members of the board. Although this plant will probably work, it clearly will not be the caliber of plant you expected, and the municipality will probably need to spend a lot of money upgrading it in the near future. What ethical responsibilities and duty as an employee of the municipality do you have as a registered professional engineer? Analyze the problem, identify the people involved, state the options you have, and come to a conclusion.

8-13 Suppose a law requires that all wastewater discharges into a watercourse meet these standards:

BOD   Less than 20 mg/L
Suspended solids   Less than 20 mg/L
Phosphorus (total)   Less than 0.5 mg/L

Design treatment plants (block diagrams) for the following wastes:

WASTE	BOD (mg/L)	SS (mg/L)	P (mg/L)
a	250	250	10
b	750	30	0.6
c	30	450	20

8-14 Draw a block diagram for a treatment plant, showing the necessary treatment steps for achieving the desired effluent. Be sure to use only the treatment steps needed to achieve the desired effluent. Extraneous steps will be considered wrong.

	INFLUENT (mg/L)	EFFLUENT (mg/L)
BOD	1,000	10
Suspended solids	10	10
Phosphorus	50	5

8-15 A 36-in.-diameter centrifuge for dewatering sludge is rotated at 1,000 rotations per minute. How many gravities does this machine produce? (Remember that each rotation represents 2π radians.)

8-16 A gas transfer experiment results in the following data:

TIME (min)	DISSOLVED OXYGEN (mg/L)
0	2.2
5	4.2
10	5.0
15	5.5

The water temperature is 15°C. What is the gas transfer coefficient $K_L a$?

8-17 Derive the equation for calculating the recovery of centrifuge solids from concentration terms. That is, recovery is to be calculated only as a function of the solids concentrations of the three streams: feed, centrate, and cake. Show each step of the derivation, including the material balances.

8-18 A community with a wastewater flow of 10 MGD is required to meet effluent standards of 30 mg/L for both BOD_5 and SS. Pilot plant results with an influent of $BOD_5 = 250$ mg/L estimate the kinetic constants at $K_s = 100$ mg/L, $\hat{\mu} = 0.25$ day^{-1}, and $Y = 0.5$. It is decided to maintain the MLSS at 2,000 mg/L. What are the hydraulic retention time, the sludge age, and the required tank volume?

8-19 Wastewater from a peach packaging plant was tested in a pilot activated sludge plant. The kinetic constants were found to be $\hat{\mu} = 3$ day^{-1}, $Y = 0.6$, and $K_S = 450$ mg/L. The influent BOD is 1,200 mg/L, and a flow rate of 19,000 m^3/day is expected. The aerators to be used will limit the suspended solids in the aeration tank to 4,500 mg/L. The available aeration volume is 5,100 m^3.

a. What efficiency of BOD removal can be expected?

b. Suppose we find that the flow rate is actually much higher—say, 35,000 m^3/day—and the flow is more dilute, $S_0 = 600$ mg/L. What removal efficiency might we expect now?

c. At this flow rate and S_0, suppose we cannot maintain 4,500 mg/L solids in the aeration tank (why?). If the solids are only 2,000 mg/L, and if 90% BOD removal is required, how much extra aeration tank volume is needed?

8-20 An activated sludge system has a flow of 4,000 m^3/day with $X = 4,000$ mg/L and $S_0 = 300$ mg/L. From work in a pilot plant, the kinetic constants are $\hat{\mu} = 3$ day^{-1}, $Y = 0.5$, and $K_S = 200$ mg/L. We need to design an aeration system that will remove 90% of the BOD$_5$. Specifically, we need to know:

a. the volume of the aeration tank

b. the sludge age

c. the amount of waste activated sludge

What are their values?

8-21 A centrifuge manufacturer is trying to sell your city a new centrifuge that is supposed to dewater your sludge to 35% solids. You know, however, that their machines will most likely achieve only about 25% solids in the cake.

a. How much extra volume of sludge cake will you have to be able to handle and dispose of (35% versus 25%)?

b. You know that no competitors can provide a centrifuge that works any better. You like this machine and convince the city to buy it. A few weeks before the purchase is final, you receive a set of drinking glasses with the centrifuge company logo on them, along with a thank you note from the salesman. What do you do? Why? Use ethical reasoning to develop your answer.

8-22 What are the principal mechanisms used to remove and transform pollutants in surface water flow wetlands?

8-23 What are the principal mechanisms used to remove and transform pollutants in subsurface water flow wetlands?

8-24 How many gallons per day of sludge will be generated if a 2.5-MGD water treatment plant thickens 14,000 pounds/day of sludge to 9% solids?

8-25 A municipal water treatment plant uses two sedimentation tanks operating in parallel as part of a lime-soda softening process. Each clarifier has a diameter of 40 ft, a sidewater depth of 15 ft, and an outlet weir length of 270 ft. The design flow rate is 1.5 MGD. Do these two clarifiers meet the requirements of at least a 4-hr detention time and a surface overflow rate no greater than 700 gpd/ft^2?

8-26 A conventional wastewater treatment plant receives 2 MGD with an average BOD of 250 mg/L. The aeration basin is 100,000 ft^3. The MLSS is 2,800 mg/L, and the effluent SS is 25 mg/L. The waste activated sludge is 38,000 gpd from the recycle line. The SS of the recycle flow is 9,000 mg/L.

a. What is the F/M ratio?

b. What is the MCRT?

8-27 Estimate the sludge wasting rate and the required reactor volume of an activated sludge plant being designed. The flow to the plant is 25 MGD. The influent BOD_5 to the aeration basin is 200 mg/L, and the effluent BOD_5 from the plant is 8 mg/L. The design is for a MCRT of 10 days. The maximum yield coefficient is estimated to be 0.6 mg cells/mg substrate, and the endogenous decay coefficient is estimated to be 0.09/day. The aeration basin is expected to operate at an MLSS of 2,500 mg/L, and the return sludge and wasting concentrations are expected to be 9,000 mg/L.

8-28 What changes can occur over time in a community that will affect the waste-water composition and/or flow rate?

## CHAPTER 9

Impacts on Air Quality

FIGURE 9.1 Air pollution over the city of Kathmandu, Nepal, from photochemical oxidation of nitrogen oxides and volatile organic compounds.

Source: Jerome Lorieau Photography/Moment Open/Getty Images.

Our most basic common link is that we all inhabit this planet. We all breathe the same air. We all cherish our children's future. And we are all mortal.

—JOHN F. KENNEDY

You go into a community and they will vote 80 percent to 20 percent in favor of a tougher Clean Air Act, but if you ask them to devote 20 minutes a year to having their car emissions inspected, they will vote 80 to 20 against it. We are a long way in this country from taking individual responsibility for the environmental problem.

—WILLIAM D. RUCKELSHAUS, FORMER EPA ADMINISTRATOR, *NEW YORK TIMES*, NOVEMBER 30, 1988

GOALS

THE EDUCATIONAL GOALS OF THIS CHAPTER are to use basic principles of chemistry, physics, and math to define and solve engineering problems associated with air pollutant emissions, dispersion, and control. This chapter utilizes the fundamental physical and chemical models discussed in Chapters 2 through 4 to examine problems associated with air pollution emissions at local, regional, and global levels. Regulatory policies that limit air pollutants in the United States are discussed, as well as international regulations designed to protect the stratospheric ozone layer. This chapter describes the parameters and methods used to determine and predict air quality downwind from air pollution sources. Air quality constituents that influence human and ecosystem health in the environment are examined. Mathematical models are used to predict air pollutant emissions and transport, and to identify potential health risks. Many of the fundamental principles discussed in this chapter are also applicable to carbon dioxide emissions, which are discussed in more detail in Chapter 10. A basic understanding of air pollution provides a foundation for an understanding of energy issues, industrial ecology, life cycle analysis, and the rationale for more efficient building practices discussed in Parts II and III of the text.

OBJECTIVES

At the conclusion of this chapter, you should be able to:

9.1 Describe the economic, environmental, and societal implications of air quality in high-income nations and low- to medium-income nations.

9.2 Predict the health impacts from various types of air pollutants.

9.3 Estimate emissions of air pollutants using emission factors.

9.4 Estimate the dispersion of air pollutants downwind from a source.

9.5 Describe the processes that produce air pollution from combustion sources.

9.6 Select the appropriate air quality treatment process for various types of air pollutants.

9.7 Describe the global effects of air pollutants and international strategies to address the challenges of global air pollution.

Introduction

The average adult typically inhales 3 to 4 liters of air in each breath taken. The air quality in a region is dependent on local and regional factors. The proximity to power plants, highways, industry, and mountains significantly influences local air quality. Global air circulation patterns, latitude, and air temperature profoundly influence regional air quality. Scientists and engineers use the principles of chemistry, math, and physics to calculate the amount of pollutants emitted from a process, how those pollutants may be transported to expose a population downwind from the source, what health effects may be incurred by the exposed population, and what technologies may be used to reduce emissions and subsequent exposure.

The composition of clean air was given in Table 2.1. Clean air is made up primarily of nitrogen (78.08%), oxygen (20.95%), and argon (0.93%). The remaining components of the air that are of concern as environmental pollutants are present in quantities of less than 1% in the atmosphere. An air pollutant may be thought of as any chemical or substance that is present in the atmosphere in quantities that negatively impact human health, the environment, and the economy, or unreasonably interfere with the enjoyment of life, property, or recreation.

Many air pollutants are emitted as a result of the process involved in producing energy for heat, work, or electricity. Urbanization patterns, climate change, population growth, and increasing energy demand have decreased air quality in many locations. Premature death has been attributed to poor air quality in high- to low-income countries in rural and urban areas. Over 49,000 deaths per year are attributed to secondhand cigarette smoke indoors (CDC, 2008). The World Health Organization (WHO) estimates that 2 million people per year die prematurely from illness attributed to use of household solid fuels. WHO also estimates that 50% of pneumonia deaths among children under the age of 5 are related to particulate matter (PM) inhaled from indoor pollutants. However, improvements in air quality and decreased health effects have also been documented when regulations and investment in air treatment processes have been made to improve regional air quality.

9.1 Air Quality History and Regulations

One of the first laws against air pollution came in 1300 when King Edward I decreed the death penalty for burning of coal. At least one execution for that offense is recorded. But economics triumphed over health considerations, and air pollution became an appalling problem in England.

—GLENN T. SEABORG, ATOMIC ENERGY COMMISSION CHAIRMAN, SPEECH,

ARGONNE NATIONAL LABORATORY, 1969

Air pollution was documented as a problem in preindustrialized England, when burning "sea coal" or peat resulted in toxic indoor air conditions. Subsequently, King Edward I passed the first laws to prevent air pollution from burning this "dirty" fuel. Later, in 1661, pamphlets were printed and distributed suggesting ways to reduce air pollution. The industrial age increased the demand for burning fuels for energy, work, and electricity, which in turn increased air pollutant emissions from solid and fossil fuel combustion.

Air pollution continues to be a very significant human health threat. In 1948, industrial air pollutants that included sulfur dioxide, nitrogen dioxide, and fluorine caused approximately 20 human deaths and thousands of reported cases of respiratory distress due to atmospheric conditions that led to very high concentrations of a toxic smog in the town of Donora, Pennsylvania, in the United States. A few years later, in 1952, the "great smog" enveloped parts of London. The smog was formed when unusually still air conditions and cold weather combined to form a toxic mixture of gases. While acute effects were limited during the five-day event, more recent research estimates that this event caused more than 10,000 premature deaths.

Acid rainfall was first observed from plants damaged by the rain in the northeastern United States in the mid-19th century. Decades of acid rainfall in this region led to the acidification of many lakes and streams, damaging aquatic ecosystems and also causing architectural damage. These and other air pollution-related events resulted in the passage of numerous environmental laws designed to protect or improve air quality.

Regulations are laws used to attempt to protect the local or global environment. Regulations are, by political necessity, a balance between social and commercial interests. Laws may be enacted at multiple levels of government, as illustrated in the regulatory pyramid shown in Figure 9.2.

Regulation of air pollutants in the United States has evolved since 1955, as shown in Figure 9.3, when Public Law 159 authorized funding for the U.S. Public Health Service (the forerunner to the EPA) to initiate research into air pollution. The 1963 ***Clean Air Act*** was the first federal act in the United States to allow regulations for the control of air pollutants. The 1970 Clean Air Act (CAA) was the first comprehensive step by the federal government to limit emissions of air pollutants from both stationary and mobile sources. Facilities that did not comply with the CAA were subjected to fines for violating emissions limits. The 1970 CAA also established ***National Ambient Air Quality Standards (NAAQS)***. The NAAQS are goals set to achieve reasonable air quality in all regions of the United States. The 1977 amendments to the CAA were targeted at improving air quality in regions where concentrations of pollutants in the air were found to be higher than the target goals of the NAAQS. The additional amendments to the 1970 CAA passed in 1990 expanded programs to bring regions into compliance with the NAAQS, set more stringent standards for new sources, and included provisions for stratospheric ozone protection.

The CAA required all regions in the United States to develop ***Prevention of Significant Deterioration (PSD)*** standards to maintain air quality if in compliance

FIGURE 9.2 The regulatory pyramid illustrates that facilities must comply with all local, state, and federal regulations for a facility.

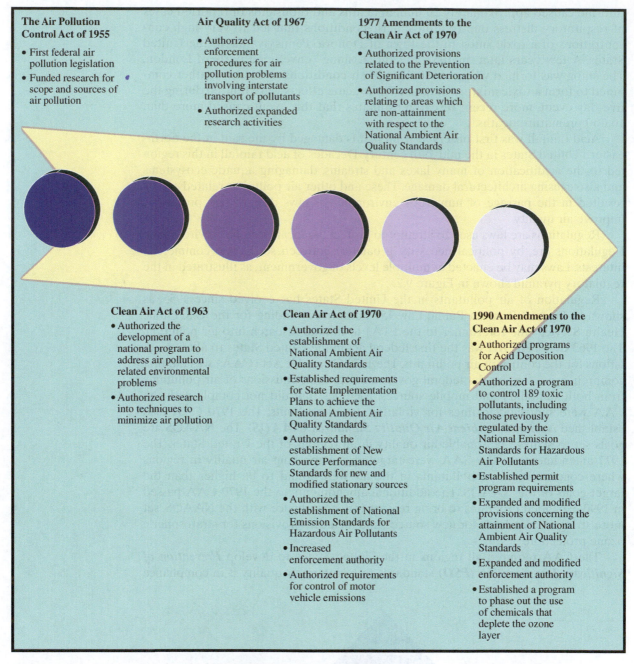

The Air Pollution Control Act of 1955

- First federal air pollution legislation
- Funded research for scope and sources of air pollution

Air Quality Act of 1967

- Authorized enforcement procedures for air pollution problems involving interstate transport of pollutants
- Authorized expanded research activities

1977 Amendments to the Clean Air Act of 1970

- Authorized provisions related to the Prevention of Significant Deterioration
- Authorized provisions relating to areas which are non-attainment with respect to the National Ambient Air Quality Standards

Clean Air Act of 1963

- Authorized the development of a national program to address air pollution related environmental problems
- Authorized research into techniques to minimize air pollution

Clean Air Act of 1970

- Authorized the establishment of National Ambient Air Quality Standards
- Established requirements for State Implementation Plans to achieve the National Ambient Air Quality Standards
- Authorized the establishment of New Source Performance Standards for new and modified stationary sources
- Authorized the establishment of National Emission Standards for Hazardous Air Pollutants
- Increased enforcement authority
- Authorized requirements for control of motor vehicle emissions

1990 Amendments to the Clean Air Act of 1970

- Authorized programs for Acid Deposition Control
- Authorized a program to control 189 toxic pollutants, including those previously regulated by the National Emission Standards for Hazardous Air Pollutants
- Established permit program requirements
- Expanded and modified provisions concerning the attainment of National Ambient Air Quality Standards
- Expanded and modified enforcement authority
- Established a program to phase out the use of chemicals that deplete the ozone layer

FIGURE 9.3 Timeline of major air quality legislation in the United States.

with the NAAQS or to develop *State Implementation Plans (SIPs)* to improve air quality if the NAAQS goals were exceeded in the region. The NAAQS are defined for the six primary pollutants shown in Table 9.1 that had demonstrable negative human health impacts. Secondary standards are required for air pollutants that may have potentially negative environmental effects, including plant damage, decreased visibility, and architectural damage. The actions required by the state regulatory agency for compliance with NAAQS goals are summarized in Table 9.2. The CAA

TABLE 9.1 U.S. EPA primary and secondary NAAQS standards for primary pollutants

POLLUTANT [FINAL RULE CITE]		PRIMARY/ SECONDARY	AVERAGING TIME	LEVEL	FORM
Carbon monoxide [76 FR 54294, Aug. 31, 2011]		Primary	8 hours	9 ppm	Not to be exceeded more than once per year
			1 hour	35 ppm	
Lead [73 FR 66964, Nov. 12, 2008]		Primary and secondary	Rolling 3-month average	0.15 $\mu g/m^{3(a)}$	Not to be exceeded
Nitrogen dioxide [75 FR 6474, Feb. 9, 2010] [61 FR 52852, Oct. 8, 1996]		Primary	1 hour	100 ppb	98th percentile averaged over 3 years
		Primary and secondary	Annual	53 ppb	Annual mean
Ozone [73 FR 16436, Mar. 27, 2008]		Primary and secondary	8 hours	0.070 ppm	Annual fourth highest daily maximum 8-hour concentration, averaged over 3 years
Particle pollution [71 FR 61144, Oct. 17, 2006]	$PM_{2.5}$	Primary	Annual	12 $\mu g/m^3$	Annual mean, averaged over 3 years
		Secondary	Annual	12 $\mu g/m^3$	Annual mean, averaged over 3 years
		Primary and secondary	24 hours	35 $\mu g/m^3$	98th percentile, averaged over 3 years
	PM_{10}	Primary and secondary	24 hours	150 $\mu g/m^3$	99th percentile of 1-hour daily maximum concentrations, averaged over 3 years
Sulfur dioxide [75 FR 35520, June 22, 2010]		Primary	1 hour	75 ppb	99th percentile of 1-hour daily maximum concentrations, averaged over 3 years
[38 FR 25678, Sept. 14, 1973]		Secondary	3 hours	0.5 ppm	Not to be exceeded more than once per year

[a] Final rule signed October 1, 2008. The 1978 lead standard (1.5 $\mu g/m^3$ as a quarterly average) remains in effect until one year after an area is designated for the 2008 standard, except that in areas designated non-attainment for the 1978 standard remains in effect until implementation plans to attain or maintain the 2008 standard are approved.

TABLE 9.2 Definition of attainment regions for NAAQS

ATTAINMENT REGION	UNCLASSIFIABLE REGION	NONATTAINMENT REGION
Air quality is in compliance with NAAQS	Insufficient data to determine compliance with NAAQS	Air quality violates NAAQS
PSD applies	PSD applies	New source review applies

resulted in significant reductions of the primary pollutants shown in Figure 9.4. Despite improvements in air quality in some regions and regulations of new sources, nearly 124 million people lived in counties that did not comply with one or more of the NAAQS criteria, as shown in Figure 9.5.

The Environmental Protection Agency has developed an *Air Quality Index (AQI)* to help communicate air quality trends to the public and to provide warnings to the public on days that a region is not in compliance with the NAAQS. The AQI relates daily air pollution concentrations for ozone, particle pollution, NO_2, CO, and SO_2 to health concerns for the general public. Each compound is given a numerical value, based on a scale, where 100 is equivalent to the NAAQS standard.

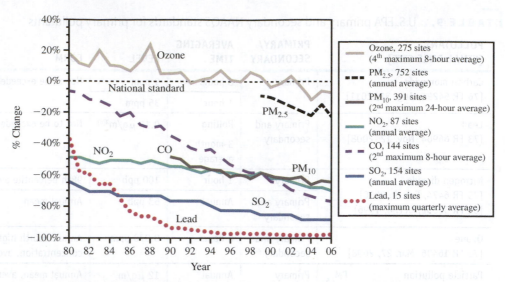

FIGURE 9.4 Comparison of national levels of the six principal priority pollutants to national ambient air quality standards, 1980–2006. National levels are averages across all sites with complete data for the time period.

Source: U.S. EPA (2007a). A Plain English Guide to the Clean Air Act. Office of Air Quality, Planning and Standards, U.S. Environmental Protection Agency. Research Triangle Park, NC. EPA-456/K-07-001.

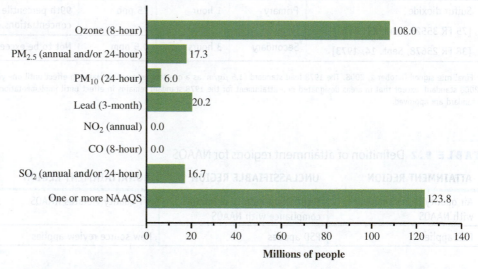

FIGURE 9.5 Number of people (in millions) living in counties with air quality concentrations above the level of the primary (health-based) National Ambient Air Quality Standards (NAAQS) in 2010.

Source: U.S. EPA (2011a). *The Benefits and Costs of the Clean Air Act from 1990 to 2020: Summary Report*. Office of Air and Radiation, U.S. Environmental Protection Agency. Research Triangle Park, NC. EPA-454/R-12-001.

An AQI score below 100 is considered acceptable, as shown in Figure 9.6, and a value greater than 100 is considered unhealthy, especially for sensitive populations. Figure 9.7 shows the number of days on which the AQI exceeded 100 for 35 metropolitan cities in the United States.

$$AQI = \frac{C_{\text{environment}}}{C_{\text{NAAQS}}} \times 100 \tag{9.1}$$

Air Quality Index (AQI) Values	Levels of Health Concern
0–50	Good
51–100	Moderate
101–150	Unhealthy for sensitive groups
151–200	Unhealthy
201–300	Very unhealthy
301–500	Hazardous

FIGURE 9.6 U.S. EPA Air Quality Index.

Source: U.S. EPA (2009b). *Ozone and Your Health*. Office of Air and Radiation, U.S. Environmental Protection Agency. Washington, DC.

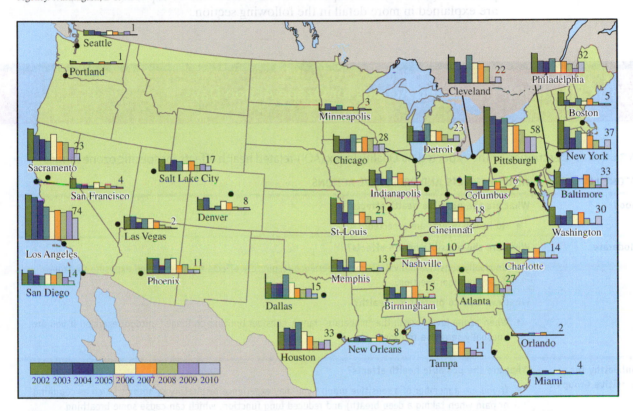

FIGURE 9.7 Number of days on which AQI values were greater than 100 during 2002–2010 in selected cities.

Source: U.S. EPA (2011a). *The Benefits and Costs of the Clean Air Act from 1990 to 2020: Summary Report*. Office of Air and Radiation, U.S. Environmental Protection Agency. Research Triangle Park, NC.

EXAMPLE 9.1   Air Quality Index Values

The ozone level in a city was measured at 86 ppb at 8:50 a.m. What AQI value should be reported on the 9:00 a.m. morning news broadcast for the city?

$$AQI = \frac{C_{environment}}{C_{NAAQS}} \times 100 = \frac{86\ ppb_v\ \dfrac{1\ ppm_v}{1{,}000\ ppb_v}}{0.075\ ppm_v} \times 100 = 115$$

The AQI is level orange, between 101 and 150, and is unhealthy for sensitive groups.

The EPA performed a cost-benefit analysis of implementation of the 1990 CAA, evaluating the expected benefits of the CAA between 1990 and 2020. The findings show a significant reduction of over 230,000 early deaths associated with decreased exposure to air pollutants (Table 9.4). It is estimated that the direct benefits of the CAA will reach almost $2 trillion (U.S.) by the year 2020. Due to the high uncertainty associated with the benefits projection, a wide range in expected benefits is projected with a low estimate of benefits equal to three times the cost, the expected benefits equal to 30 times the cost, and the high estimate equal to 90 times the cost. All benefits should exceed the $65 billion (U.S.) cost of implementing the CAA. Cleaner air is expected to lead to better health and higher productivity for American workers as well as savings on medical expenses due to health-related problems (Box 9.1). The expected health effects due to exposure to air pollutants are explained in more detail in the following section.

BOX 9.1 Health Effects and Actions for Ozone Action Days

TABLE 9.3 Description of the expected Air Quality Index (AQI)-related health effects for specific ozone ranges

OZONE LEVEL	HEALTH EFFECTS AND PROTECTIVE ACTIONS
Good	**What are the possible health effects?** • No health effects are expected.
Moderate	**What are the possible health effects?** • Unusually sensitive individuals may experience respiratory effects from prolonged exposure to ozone during outdoor exertion. **What can I do to protect my health?** • When ozone levels are in the "moderate" range, consider limiting prolonged outdoor exertion if you are unusually sensitive to ozone.
Unhealthy for Sensitive Groups	**What are the possible health effects?** • If you are a member of a sensitive group,[a] you may experience respiratory symptoms (such as coughing or pain when taking a deep breath) and reduced lung function, which can cause some breathing discomfort. **What can I do to protect my health?** • If you are a member of a sensitive group,[a] limit prolonged outdoor exertion. In general, you can protect your health by reducing how long or how strenuously you exert yourself outdoors and by planning outdoor activities when ozone levels are lower (usually in the early morning or evening). • You can check with your state air agency to find out about current or predicted ozone levels in your location. This information on ozone levels is available on the Internet at *www.airnow.gov*.

Unhealthy	**What are the possible health effects?**

- If you are a member of a sensitive group,[a] you have a higher chance of experiencing respiratory symptoms (such as aggravated cough or pain when taking a deep breath) and reduced lung function, which can cause some breathing difficulty.
- At this level, anyone could experience respiratory effects.

What can I do to protect my health?

- If you are a member of a sensitive group,[a] avoid prolonged outdoor exertion. Everyone else—especially children—should limit prolonged outdoor exertion.
- Plan outdoor activities when ozone levels are lower (usually in the early morning or evening).
- You can check with your state air agency to find out about current or predicted ozone levels in your location. This information on ozone levels is available on the Internet at *www.airnow.gov*.

Very Unhealthy	**What are the possible health effects?**

- Members of sensitive groups[a] will likely experience increasingly severe respiratory symptoms and impaired breathing.
- Many healthy people in the general population engaged in moderate exertion will experience some kind of effect. According to EPA estimates, approximately:
  - Half will experience moderately reduced lung function.
  - One-fifth will experience severely reduced lung function.
  - 10 to 15% will experience moderate to severe respiratory symptoms (such as aggravated cough and pain when taking a deep breath).
- People with asthma or other respiratory conditions will be more severely affected, leading some to increase medication usage and to seek medical attention at an emergency room or clinic.

What can I do to protect my health?

- If you are a member of a sensitive group,[a] avoid outdoor activity altogether. Everyone else—especially children—should limit outdoor exertion and avoid heavy exertion altogether.
- Check with your state air agency to find out about current or predicted ozone levels in your location. This information on ozone levels is available on the Internet at *www.airnow.gov*.

[a]Members of sensitive groups include children who are active outdoors; adults involved in moderate or strenuous outdoor activities; individuals with respiratory disease, such as asthma; and individuals with unusual susceptibility to ozone.

Source: U.S. EPA (1999). *Smog – Who Does It Hurt?: What You Need to Know About Ozone and Your Health*. Office of Air and Radiation, U.S. Environmental Protection Agency. Washington, DC. EPA 452/K-99-001.

TABLE 9.4 Differences in key health effect outcomes associated with fine particles (PM$_{2.5}$) and ozone between the With-CAA90 and Without-CAA90 scenarios for the 2010 and 2020 study target years (in number of cases avoided, rounded to two significant digits). The table shows the reductions in risk of various air pollution-related health effects achieved by the 1990 Clean Air Act Amendments programs with each risk change expressed as the equivalent number of incidences avoided across the exposed population

HEALTH EFFECT REDUCTIONS (PM$_{2.5}$ & OZONE ONLY)	POLLUTANT(s)	YEAR	
		2010	2020
PM$_{2.5}$ adult mortality	PM	160,000	230,000
PM$_{2.5}$ infant mortality	PM	230	280
Ozone mortality	Ozone	4,300	7,100
Chronic bronchitis	PM	54,000	75,000
Acute bronchitis	PM	130,000	180,000
Acute myocardial infarction	PM	130,000	200,000
Asthma exacerbation	PM	1,700,000	2,400,000
Hospital admissions	PM, Ozone	86,000	135,000
Emergency room visits	PM, Ozone	86,000	120,000
Restricted activity days	PM, Ozone	84,000,000	110,000,000
School loss days	Ozone	3,200,000	5,400,000
Lost work days	PM	13,000,000	17,000,000

Source: U.S. EPA (2011a). *The Benefits and Costs of the Clean Air Act from 1990 to 2020: Summary Report*. Office of Air and Radiation, U.S. Environmental Protection Agency. Research Triangle Park, NC.

9.2 Health Effects of Air Pollutants

The acute and chronic human health effects of air pollutants are often difficult to predict from direct exposure. Air pollutants enter the body through the respiratory system (see Figure 9.8). The adverse health effects of most air pollutants are related to the cardiopulmonary system. *Acute* or short-term symptoms are often much more severe in individuals predisposed to adverse respiratory events. Common short-term effects include irritated mucous membranes, inflammation of the bronchial tubes, and possible airway restriction. *Chronic* or long-term effects of exposure to air pollutants may include chronic inflammation of the bronchial tubes, sustained airway restriction, or pulmonary emphysema. For example, pulmonary emphysema, which results in shortness of breath due to destruction of the alveoli membranes in the lungs, has been linked to exposure to airborne asbestos dust.

The human respiratory system consists of three distinct regions, each affected differently by different constituents in the air. Air enters the body through the nasopharyngeal region or exothoracic airway. The *nasopharyngeal region* includes the nose, mouth, and larynx. Water-soluble compounds in the air may be absorbed by the mucous membranes in the nasal cavity. Airflow restrictions in the larynx result in the deposit of large airborne particles in the mucous lining of the larynx, increasing the risk of malignant tumors in the larynx.

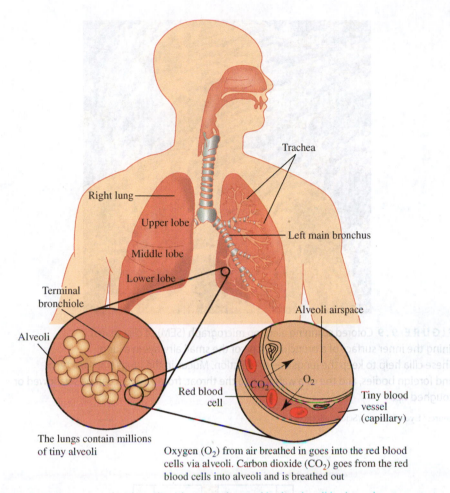

The lungs contain millions of tiny alveoli

Oxygen (O_2) from air breathed in goes into the red blood cells via alveoli. Carbon dioxide (CO_2) goes from the red blood cells into alveoli and is breathed out

FIGURE 9.8 Oxygen and carbon dioxide are exchanged in the alveoli in the pulmonary region of the respiratory system.

Source: Based on Patient.co.uk.

The ***tracheobronchial region*** of the respiratory system is characterized by the branching of the airway bronchi through the lungs, conveying the air to the pulmonary region. The tracheobronchial region includes the cilia, trachea, primary and secondary bronchi, and terminal bronchioles. The cilia are fine, hair-like particles (Figure 9.9) that line the entire region and help trap small to medium-size particles that make it past the larynx. The particles become covered in the mucous lining of the bronchi, and the cilia use a whip-like motion to propel the particles out of the respiratory system and into the throat. The bronchi are the conduits that move the air into and from the lungs. There are between 20 and 25 branching steps, or generations of branching, from the trachea (called the first-generation respiratory passage) to the alveoli (see Figure 9.10).

Carbon dioxide is released from the blood into the lungs, and oxygen is moved from the air to the bloodstream in the pulmonary region of the lungs (see Figure 9.8). The pulmonary region consists of 300 to 500 million very small, interconnected groups of porous membranes called alveoli. The total air exchange area of the alveoli region is about 50 square meters in the average adult. White blood cells,

FIGURE 9.9 Colored scanning electron micrograph (SEM) of cilia (hairs) lining the inner surface of a bronchiole, one of the small airways in a lung. These cilia help to keep the lungs free of infection. Mucus in the lungs traps bacteria and foreign bodies, and the cilia waft it up to the throat, from where it can be swallowed or coughed up.

Source: Eye of Science/Science Source.

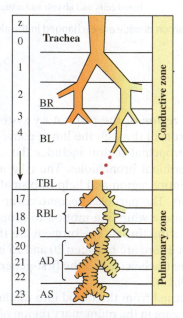

FIGURE 9.10 Idealization of the first 16 generations of branching (z) of the human airway.

Source: Based on Weibel, E.R. (1962). *Morphology of the Human Lung*. Berlin: Springer-Verlag

TABLE 9.5 Oxygen concentrations in the air and symptoms associated with oxygen deficiency

% O_2 IN AIR	DESCRIPTION OF SYMPTOMS
20.9	Normal O_2 concentration in air
17	Hypoxia occurs with deteriorating night vision, increased heart beat, accelerated breathing
14–16	Very poor muscular coordination, rapid fatigue, intermittent respiration
6–10	Nausea, vomiting, inability to perform, unconsciousness
<6	Spasmodic breathing, convulsive movements, death within minutes

Source: Based on Heinsohn, R.J. (1991). *Industrial Ventilation: Engineering Principles.* New York: John Wiley & Sons.

called macrophage, destroy most infectious bacteria and fungi, with some notable exceptions such as tuberculosis.

Respiratory distress symptoms are related to an upset in the process of air exchange in the lungs. **Hypoxia** is caused by an inadequate supply of oxygen to the body. The onset of hypoxia in a healthy adult can occur if oxygen levels in an enclosed space become too low, as shown in Table 9.5. Cyanosis, a symptom of hypoxia, is caused when there is not enough oxygen in the blood and the blood turns blue. **Hypercapnia** is a condition that occurs when too much carbon dioxide builds up in the bloodstream, causing distress.

Other respiratory diseases may lead to irritation of the respiratory system where air passages become irritated and are physiologically constricted. Constricted airways may lead to edema and secondary infections in the lungs. If cell tissue in the lungs is damaged, the airways may not be able to transfer oxygen and carbon dioxide from the blood, leading to necrosis and edema. Excess proteins in the blood or infectious agents may cause fibrosis or stiffening of the lung tissue that may restrict airflow and lung function. Foreign particles or chemical exposure may cause oncogenesis, or tumor formation, in the lungs that could become malignant. Pneumonia is an inflammation of lung tissue usually caused by a bacterial or virus infection of the lungs. Pneumonia is not normally fatal in high-income countries, but it is still a leading cause of death, particularly for children and seniors, in low-income countries.

9.2.1 Carbon Monoxide

Carbon monoxide (CO) is a colorless and odorless gas emitted from combustion processes and subject to NAAQS standards, as shown in Table 9.6. Carbon monoxide is similar in size and molecular mass to diatomic oxygen (O_2) in the air. Carbon monoxide will replace oxygen in the bloodstream if it is present in air at high concentrations, as illustrated in Figure 9.11. The carbon monoxide replaces oxygen at the active hemoglobin site, thereby reducing oxygen transport to the brain, organs, and muscles. The severity of symptoms described in Table 9.6 associated with carbon monoxide depends on the extent of carbon monoxide that bonds with hemoglobin in the blood. As carbon monoxide levels increase in the bloodstream, exposed individuals experience headaches and reduced hand–eye coordination. If oxygen deprivation becomes severe due to elevated levels of carbon monoxide, symptoms may include fainting, coma, and death, as shown

TABLE 9.6 Sources and effects of NAAQS criteria air pollutants

POLLUTANT	SOURCES	HEALTH EFFECTS
Ozone (O_3)	Secondary pollutant typically formed by chemical reaction of volatile organic compounds (VOCs) and NO_x in the presence of sunlight.	Decreases lung function and causes respiratory symptoms, such as coughing and shortness of breath; aggravates asthma and other lung diseases leading to increased medication use, hospital admissions, emergency department (ED) visits, and premature mortality.
Particulate matter (PM)	Emitted or formed through chemical reactions; fuel combustion (e.g., burning coal, wood, diesel); industrial processes; agriculture (plowing, field burning); and unpaved roads.	Short-term exposures can aggravate heart or lung diseases, leading to respiratory symptoms, increased medication use, hospital admissions, ED visits, and premature mortality; long-term exposures can lead to the development of heart or lung disease and premature mortality.
Lead	Smelters (metal refineries) and other metal industries; combustion of leaded gasoline in piston engine aircraft; waste incinerators; and battery manufacturing.	Damages the developing nervous system, resulting in IQ loss and impacts on learning, memory, and behavior in children. Cardiovascular and renal effects in adults and early effects related to anemia.
Oxides of nitrogen (NO_x)	Fuel combustion (e.g., electric utilities, industrial boilers, and vehicles) and wood burning.	Aggravates lung diseases, leading to respiratory symptoms, hospital admissions, and ED visits; increased susceptibility to respiratory infection.
Carbon monoxide (CO)	Fuel combustion (especially vehicles).	Reduces the amount of oxygen reaching the body's organs and tissues; aggravates heart disease, resulting in chest pain and other symptoms leading to hospital admissions and ED visits.
Sulfur dioxide (SO_2)	Fuel combustion (especially high-sulfur coal); electric utilities and industrial processes; and natural sources such as volcanoes.	Aggravates asthma and increases respiratory symptoms. Contributes to particle formation with associated health effects.

Source: U.S. EPA (2012b). *Our Nation's Air: Status and Trends through 2010.* Office of Air Quality, Planning and Standards, U.S. Environmental Protection Agency. Research Triangle Park, NC.

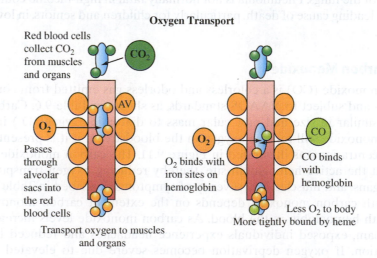

Oxygen Transport

Red blood cells collect CO_2 from muscles and organs

O_2
Passes through alveolar sacs into the red blood cells
Transport oxygen to muscles and organs

O_2 binds with iron sites on hemoglobin

O_2
CO binds with hemoglobin
Less O_2 to body
More tightly bound by heme

FIGURE 9.11 Oxygen transport of blood and carbon monoxide interference in oxygen transfer.

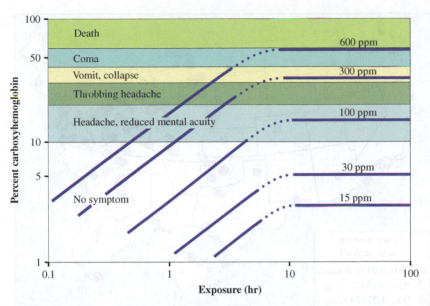

FIGURE 9.12 Response to carbon monoxide as a function of concentration (ppm$_v$) and time. The OSHA 8-hr permissible exposure limit (PEL) is 35 ppm$_v$, and the U.S. EPA primary air quality standard is 9 ppm$_v$.

Source: Based on Seinfeld, J.H. (1986). *Atmospheric Chemistry and Physics of Air Pollution*. New York: John Wiley & Sons.

in Figure 9.12. Minor symptoms usually are alleviated within 2 to 4 hours if the exposure to carbon monoxide ceases. Individuals with cardiovascular disease may experience myocardial ischemia (reduced oxygen to the heart), often accompanied by chest pain (angina), when exercising or under increased stress. Even short-term CO exposure can lead to severe symptoms for people with heart disease. The majority of carbon monoxide emissions are from mobile sources, especially in urban areas. Urban automobile exhaust is a major source of carbon monoxide in the atmosphere. Carbon monoxide concentrations of 5 to 50 ppm$_v$ have been measured in some urban areas, and concentrations above 100 ppm$_v$ have been measured along some heavily congested highways.

9.2.2 Lead

Lead is found in both manufactured products and the natural environment, and it has been used as an additive in gasoline to improve engine performance. Lead is subject to NAAQS standards (Table 9.1). Lead emissions in the United States decreased by 95% from 1980 to 1999, as lead was phased out as a fuel additive. Consequently, the concentration of lead found in the air in the United States decreased by 94% from 1980 to 1999.

Major emission sources of lead in the United States today include lead smelters, ore and metal processing facilities, and aviation fuel. Lead concentrations measured in ambient air in the United States in 2010 are shown in Figure 9.13. Individuals may be exposed to lead in drinking water, lead-contaminated food, or soil. Ingestion of dust from lead-based paint in older homes is one of the primary routes of lead exposure. Once ingested, lead accumulates

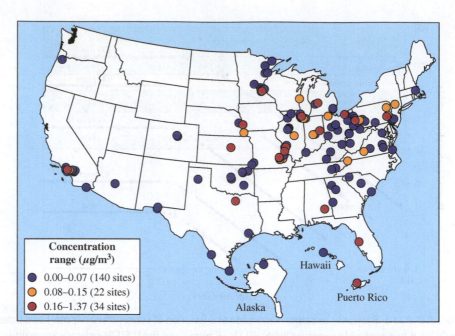

FIGURE 9.13 Lead concentrations reported in $\mu g/m^3$ measured in the United States in 2010. All 34 sites that exceeded the NAAQS standard are located near stationary lead sources.

Source: U.S. EPA (2012b). *Our Nation's Air: Status and Trends through 2010.* Office of Air Quality, Planning and Standards, U.S. Environmental Protection Agency. Research Triangle Park, NC.

in the bones. Medical symptoms are related to the amount of lead to which an individual is exposed. Lead negatively affects the nervous system, kidney function, the immune system, the reproductive and development system, the cardiovascular system, and the oxygen-carrying capacity of the blood. Infants and children under the age of 5 are especially vulnerable to the neurological effects of lead exposure, which has been documented to cause behavioral problems, learning disorders, and lowered IQ.

9.2.3 Nitrogen Oxides

Nitrogen oxides (NO_x) are a variety of reactive gases that include nitrous acid and nitrogen dioxide (NO_2). NO_2 is formed from the emissions of combustion processes from both mobile and stationary sources (see Figure 9.14). Nitrogen dioxide concentrations measured in ambient air in the United States in 2010 are shown in Figure 9.15. NO_x's contribute to the formation of ground-level ozone and are also linked to direct adverse human health effects. NO_2 is a known irritant to the alveoli after exposures of 30 minutes to 24 hours and reportedly causes emphysema in animals. NO_x in the air increases an individual's susceptibility to pulmonary infections. Studies show a connection between short-term NO_2 exposure and hospital visits for respiratory symptoms. Individuals are generally exposed to the highest level of NO_2 concentrations in automobiles and within 300 feet of a major highway, railroad, or airport. NO_x reacts with ammonia, moisture, and other compounds to form small aerosols. These small particles may be transported deep into the respiratory system,

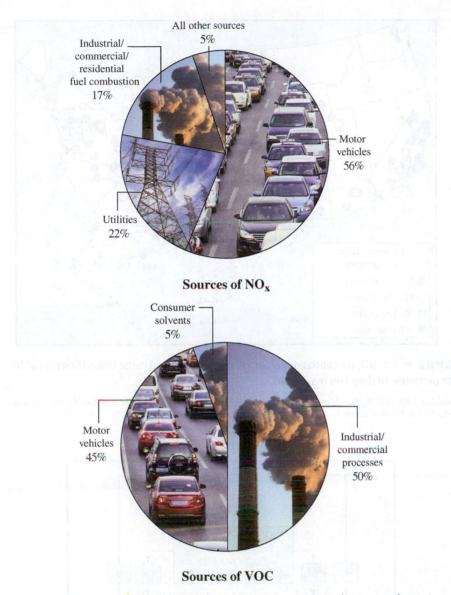

Sources of NO$_x$

Sources of VOC

FIGURE 9.14 Sources of NO$_x$ and VOCs that are precursor compounds to ozone formation.

Source: U.S. EPA (2003b). *Ozone—Good Up High, Bad Nearby*. Office of Air and Radiation, U.S. Environmental Protection Agency. Washington, DC. EPA 451/K-03-001; TonyV3112/Shutterstock.com; Shutter_M/Shutterstock.com; M. Shcherbyna/Shutterstock.com; M. Shcherbyna/Shutterstock.com.

causing or worsening respiratory symptoms. Regulations that require catalytic converters for motor vehicles have led to a 40% decrease in NO$_2$ concentrations in the United States since 1980. NO$_x$ also reacts with volatile organic compounds (VOCs) to form ozone in the presence of heat and sunlight, as illustrated by the chemical pathways shown in Figure 9.16. Ozone forms in the troposphere from reactions with NO$_x$ and organic pollutants. (It is discussed in more detail in the next section.)

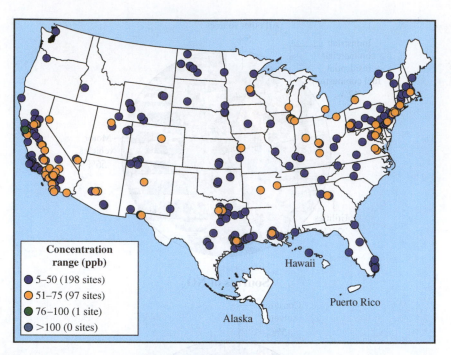

FIGURE 9.15 NO$_2$ concentrations reported in ppb$_v$ measured in the United States in 2010 (98th percentile of daily 1-hr maximum).

Source: U.S. EPA (2012b). *Our Nation's Air: Status and Trends through 2010*. Office of Air Quality, Planning and Standards, U.S. Environmental Protection Agency. Research Triangle Park, NC.

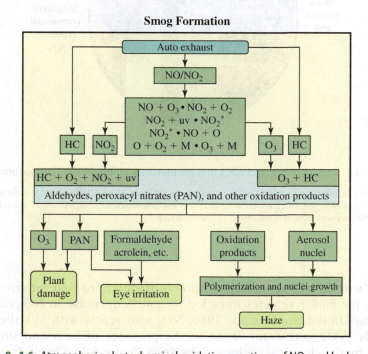

FIGURE 9.16 Atmospheric photochemical oxidation reactions of NO$_x$ and hydrocarbon-based VOCs.

Source: Based on Heinsohn, R.J., and Kabel, R.L. (1999). *Sources and Control of Air Pollution*. Upper Saddle River, NJ: Prentice Hall.

9.2.4 Ozone and Photochemical Smog

Ozone (O_3) is a permanent gas and a beneficial compound in the Earth's *stratosphere*. The ozone in the stratosphere absorbs harmful ultraviolet irradiation from the sun. The *troposphere* is the part of the atmosphere that extends from the surface of the Earth to approximately 17 kilometers above the surface. We live and breathe air in the troposphere. However, ozone in the troposphere is a highly reactive, short-lived gas. *Tropospheric ozone* is formed by the chemical reactions between nitric oxides (NO_x), volatile organic compounds (VOCs), and sunlight, which are the precursors of ozone formation.

An important approach to the classification of air pollutants is to distinguish primary and secondary pollutants. A *primary pollutant* is emitted directly to the atmosphere, and the original chemical form has negative impacts on people or the environment. *Secondary pollutants*, such as tropospheric ozone, are produced in the atmosphere via chemical reactions. Emissions from industrial facilities, combustion processes, motor vehicle exhaust, gasoline vapors, and chemical solvents are major sources of ozone precursor chemicals.

Ozone in the troposphere forms on warmer days when the chemical precursors to ozone are prevalent. Ozone is particularly problematic in urban areas that have heavy traffic. Gasoline and diesel engines emit VOCs and NO_x from automobiles. Ozone concentrations produced from the VOCs, NO_x, and sunlight increase during the daytime. When the light intensity decreases in the evening hours, ozone production slows and eventually ceases, decreasing ozone concentrations during evening hours (see Figure 9.17). Since NO_2 reacts and is consumed in the formation of ozone during the daylight hours, NO_2 concentrations increase and build back up in the evening when ozone formation ceases.

While ozone may be formed in urban areas and along congested highways, it may also be transported long distances by wind, which worsens the air quality in rural areas as well. Low levels of ozone may cause adverse health effects, and for this reason, it is regulated as a NAAQS priority pollutant (see Table 9.1). Ozone concentrations as low as 1 ppm_v constrict and irritate the airways in the lungs

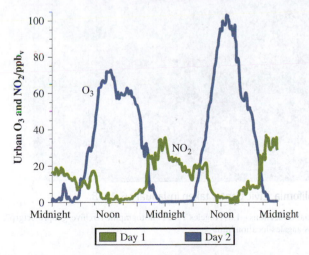

FIGURE 9.17 Diurnal variation of tropospheric ozone and NO_2 concentration during two typical summer days near London.

Source: Based on Palmer, R. (2008). "Quantifying sources and sinks of trace gases using space-borne measurements: current and future science." *Phil. Trans. R. Soc.* A 366. 4509–4528, rsta.royalsocietypublishing.org/content/366/1885/4509.full.pdf+html.

BOX 9.2 Discovering the Cause of Photochemical Smog

In the 1940s, as there became fewer and fewer clear days in Southern California, concern was expressed about what was causing this reduced visibility. During World War II, little was done to investigate this issue, and the situation became progressively worse. A butadiene plant built in downtown Los Angeles experienced severe upsets and produced noxious fumes that caused office buildings to be evacuated. Although we now know that the butadiene plant had little to do with the poor visibility shown in Figure 9.18, the event caused a public outcry and a demand to do something about the air pollution. As early as 1945, residents referred to the city's pall as "smog," in spite of the fact that the combination of smoke and fog—which gave

rise to the term—was not very prevalent in Los Angeles. In fact, when the Los Angeles County Air Pollution Control District (APCD) was created, no one knew exactly what was in smog or what caused it—nor, therefore, how to control it.

Experts were sent for, and they came with their instruments for measuring SO_2 and smoke. These soon proved worthless, however, because the levels of sulfur oxides were very low. In 1947, the County of Los Angeles received the power to issue citations and address air pollution, and the LA County Air Pollution District was formed. Early efforts to reduce air pollution in Los Angeles all focused on reducing SO_2 emissions from such sources as backyard burning, but they ultimately proved

FIGURE 9.18 Los Angeles, California, Civic Center taken in January 6, 1948.

Source: Courtesy of UCLA Library Special Collections – Los Angeles Times Photographic Archive. geoprojectgrp7 .blogspot.com/2015/03/air-pollution-in-los-angeles-location.html.

fruitless. Finally, it was decided that some research was necessary.

Arie Haagen-Smit (Figure 9.19), a biologist working at CalTech on the fumes emitted by pineapples, decided to distill the contents of the air in Los Angeles. He discovered peroxy-organic substances, which were no doubt the cause of eye irritation. The source, if this was true, had to have been the gasoline-powered automobile, and the publication of this research resulted in a firestorm of protest by the auto industry. Scientists at the Stanford Research Institute, which had been doing smog research on behalf of the transportation industry, presented a paper at CalTech accusing Haagen-Smit of bad science. In retaliation, he abandoned his work on pineapples and began to work full time on the smog problem.

FIGURE 9.19 Arie Haagen-Smit.

Source: Courtesy of the Archives, California Institute of Technology.

Some of Haagen-Smit's earlier research had been on the damage to plants caused by ozone, and he discovered that the effect of automobile exhaust on plants produced a similar injury, suggesting that the exhaust contained ozone. This raised the question of where the ozone came from, as it was not emitted by the automobiles directly. He finally hit on the idea of mixing automobile exhaust with hydrocarbons in a large air chamber and subjecting the mixture to strong light—in effect modeling the atmosphere over Los Angeles. With this experiment, he was able to demonstrate that ozone, while not emitted directly from any source, was formed by reactions in the atmosphere, and that the ozone in the smog was formed by reactions that began with the oxides of nitrogen in automobile exhaust.

The powerful automobile and gasoline industries denied that this could be the cause of the smog and suggested instead that the ozone had to be in the smog itself. Throughout the 1950s and 1960s, Southern California air quality officials began to regulate and reduce air pollutants from petroleum-based solvents containing hydrocarbons, landfills emitting toxic gases, power plants emitting nitrogen oxides, and rendering plants that processed animal wastes. These air quality regulations, among the first in the country, significantly reduced emissions; however, smog and ozone levels remained more than four times higher than the NAAQS levels. It quickly became clear that changes were required in automobile emissions to further reduce smog and ozone in Los Angeles. Haagen-Smit's courageous work, taking on the most powerful political forces in California, paved the way for an eventual reduction in smog in Los Angeles and elsewhere.

(Continued)

BOX 9.2 Discovering the Cause of Photochemical Smog *(Continued)*

California required passenger cars to have catalytic converters in the 1975 model year, despite strong resistance from automobile makers. According to Jim Boyd, executive officer of the California Air Resources Board from 1981 to 1996, "in the beginning, they said it could not be done. They said the technology was impossible. That it was incredibly expensive." The catalytic converter requirement was one of the country's first "technology-forcing" regulations, compelling industry to reduce emissions. The downward trend of ozone and smog in the basin area beginning in 1976 is shown in Figure 9.20. Although the smog in the Los Angeles area is still greater than the standard on some days, the condition of California's air pollution has improved. Figure 9.21 shows a comparison of downtown Los Angeles in the years 1968 and 2005; visibility and air quality are much improved.

For more information visit: The Southland's War on Smog: Fifty Years of Progress Toward Clean Air (through May 1997) at www.aqmd.gov/home/research/publications/50-years-of-progress.

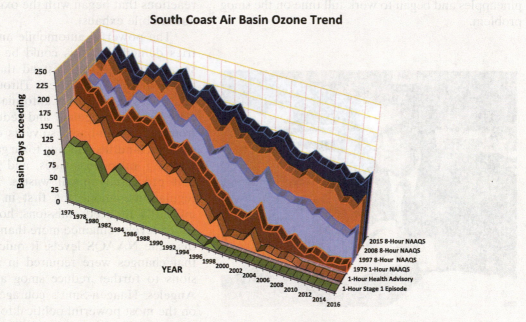

South Coast Air Basin Ozone Trend

2015 8-Hour NAAQS
2008 8-Hour NAAQS
1997 8-Hour NAAQS
1979 1-Hour NAAQS
1-Hour Health Advisory
1-Hour Stage 1 Episode

FIGURE 9.20 Trend in the number of days that exceed the air quality standard for ozone. As shown, the 8-hour ozone standard decreased in 1997, 2008, and again in 2012, decreases due to a change in the standard.

Source: South Coast Air Quality Management District, www.aqmd.gov/home/air-quality/historical-air-quality-data/historic-ozone-air-quality-trend.

FIGURE 9.21 The impact of air pollution control systems on smog formation in Los Angeles.

Source: Downtown Los Angeles smog photographs by the Herald-Examiner Collection (1968, left) and Gary Leonard (2005, right) courtesy of the Los Angeles Public Library, www.lapl.org/#photo-collection, geoprojectgrp7.blogspot.com/2015/03/air-pollution-in-los-angeles-location.html.

Based on: "Biographical Memoir of Arie Jan Haagen-Smit" by James Bonner (1989). *Biographical Memoirs of the National Academy of Sciences*, Vol. 58. Washington, DC: National Academies Press.

(see Figure 9.22). This airway constriction may lead to discomfort and hospitalization for sensitive groups. Long-term effects are unknown for typical urban concentrations of ozone (0.05 to 0.2 ppm) illustrated in Figure 9.23, but ozone may potentially accelerate the aging of lung tissue. Individuals with preexisting respiratory illness, children, and older adults are particularly susceptible to ozone. Children are most susceptible to ozone because of their increased likelihood of having asthma and because they spend more time outdoors exposed to ozone. Ozone also negatively affects ecosystems by damaging sensitive vegetation.

Respiratory symptoms associated with exposure to ozone include:

- Difficulty breathing
- Shortness of breath and pain when breathing deeply
- Coughing and throat irritation

FIGURE 9.22 Illustration of a healthy lung airway (right) and an inflamed lung airway (left). Ozone can inflame the lung's lining, and repeated episodes of inflammation may cause permanent lung damage.

Source: Stocktrek Images/Getty Images

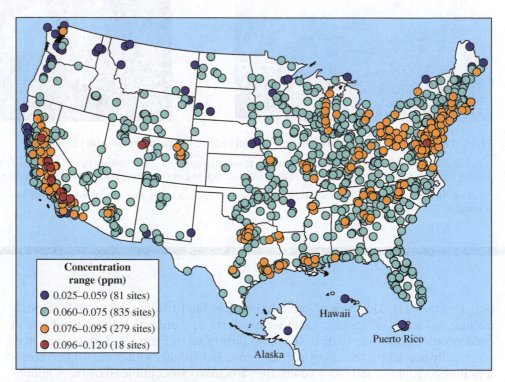

FIGURE 9.23 Ozone concentrations in ppm$_v$ (fourth highest daily maximum 8-hour concentration) measured in the United States in 2010.

Source: U.S. EPA (2012b). *Our Nation's Air: Status and Trends through 2010.* Office of Air Quality, Planning and Standards, U.S. Environmental Protection Agency. Research Triangle Park, NC.

- Inflamed and damaged bronchi
- Worsened existing respiratory diseases such as asthma, emphysema, and chronic bronchitis
- Increased frequency of asthma attacks
- Increased susceptibility to respiratory infections
- Possibly increased risk of premature death from heart or lung disease
- Potentially damaged lungs without obvious symptoms

ACTIVE LEARNING EXERCISE 9.1 Outdoor Air Quality Data

The U.S. EPA monitors the ambient air throughout the United States. The air quality monitors and the collected data are available online at the EPA's *Interactive Map of Air Quality Monitors* website (*www.epa.gov /outdoor-air-quality-data/interactive-map-air-quality-monitors*). Find the air quality monitor closest to your home by clicking on the link to the *Air Data Map App* or use the Air.now website (*www.airnow.gov/*). Choose the Ozone (active) layer from the priority pollutant list and find the monitoring location closest to you. Click on the pinned monitoring station closest to your home to find the data for the most recent year. Download the data and open the file in a spreadsheet program. How many primary and secondary exceedances of the NAAQS standards occurred in the most recent year? What was the maximum (First Maximum Value) ozone concentration that occurred during the most recent year? Calculate the AQI for the maximum ozone concentration and note the action level for the day on which the maximum ozone level occurred, as well as the possible health impacts from ozone at that concentration.

9.2.5 Particulate Matter (PM)

Particulate matter is any very small airborne solid or liquid mixture that may contain acids, organic chemicals, soil, or dust (see Table 9.7 and Figure 9.24). The size of the particle is directly related to the particle's potential to cause negative health effects. Particles larger than 10 micrometers (or microns) in size are typically removed in the nasopharyngeal region of the respiratory system and are not

TABLE 9.7 Constituents of atmospheric particles and their major sources

Aerosol species	PRIMARY (PM $< 2.5\ \mu$m)		PRIMARY (PM $> 2.5\ \mu$m)		SECONDARY PM PRECURSORS (PM $< 2.5\ \mu$m)	
	Natural	Anthropogenic	Natural	Anthropogenic	Natural	Anthropogenic
Sulfate (SO_4^{2-})	Sea spray	Fossil fuel combustion	Sea spray	—	Oxidation of reduced sulfur gases emitted by the oceans and wetlands and SO_2 and H_2S emitted by volcanism and forest fires	Oxidation of SO_2 emitted from fossil fuel combustion
Nitrate (NO_3^-)	—	Mobile source exhaust	—	—	Oxidation of NO_x produced by soils, forest fires, and lightning	Oxidation of NO_x emitted from fossil fuel combustion and in motor vehicle exhaust

(Continued)

TABLE 9.7 *(Continued)*

Aerosol species	PRIMARY (PM < 2.5 μm)		PRIMARY (PM > 2.5 μm)		SECONDARY PM PRECURSORS (PM < 2.5 μm)	
	Natural	Anthropogenic	Natural	Anthropogenic	Natural	Anthropogenic
Minerals	Erosion and re-entrainment	Fugitive dust from paved and un-paved roads, agriculture, forestry, construction, and demolition	Erosion and re-entrainment	Fugitive dust from paved and unpaved roads, agriculture, forestry, construction, and demolition	–	–
Ammonium (NH_4^+)	–	Mobile source exhaust	–	–	Emissions of NH_3 from wild animals and undisturbed soil	Emissions of NH_3 from motor vehicles, animal husbandry, sewage, and fertilized land
Organic carbon (OC)	Wildfires	Prescribed burning, wood burning, mobile source exhaust, cooking, tire wear, and industrial processes	Soil humic matter	Tire and asphalt wear, paved and unpaved road dust	Oxidation of hydrocarbons emitted by vegetation (terpenes, waxes) and wildfires	Oxidation of hydrocarbons emitted by motor vehicles, prescribed burning, wood burning, solvent use, and industrial processes
Elemental carbon (EC)	Wildfires	Mobile source exhaust (mainly diesel), wood biomass burning, and cooking	–	Tire and asphalt wear, paved and unpaved road dust		
Metals	Volcanic activity	Fossil fuel combustion, smelting and other metallurgical processes, and brake wear	Erosion, re-entrainment, and organic debris	–	–	–
Bioaerosols	Viruses and bacteria	–	Plant and insect fragments, pollen, fungal spores, and bacterial agglomerates	–	–	–

Dash (–) indicates either very minor source or no known source of component.

Source: U.S. EPA (2009a). *Policy Assessment for the Review of the Particulate Matter National Ambient Air Quality Standards*. Office of Air Quality Planning and Standards, U.S. Environmental Protection Agency. Research Triangle Park, NC.

classified as particulate matter (PM) by the EPA. The EPA groups PM into two categories illustrated in Figure 9.25 and described as follows:

> ***Inhalable coarse particles*** (PM_{10}), such as those found near roadways and dusty industries, are larger than 2.5 microns and smaller than 10 microns in diameter.

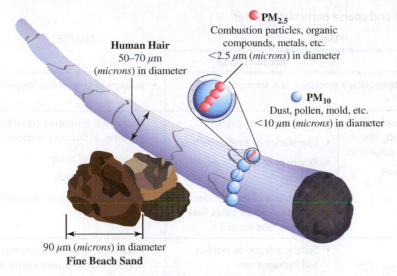

FIGURE 9.24 Illustration of the size of EPA-defined particulate matter.

Source: Based on U.S. EPA, www.epa.gov/airquality/particlepollution/basic.html.

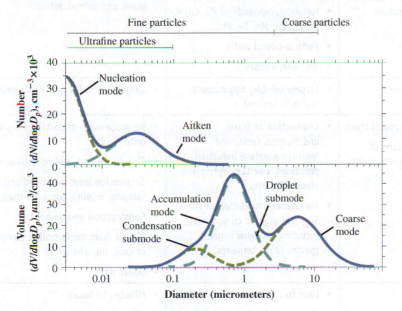

FIGURE 9.25 Particle-size distributions by number and volume.

Source: U.S. EPA (2009a). *Policy Assessment for the Review of the Particulate Matter National Ambient Air Quality Standards.* Office of Air Quality Planning and Standards, U.S. Environmental Protection Agency. Research Triangle Park, NC.

Fine particles (PM$_{2.5}$), such as those found in smoke and haze, are 2.5 micrometers in diameter and smaller. These particles can be directly emitted from sources such as forest fires, or they can form when gases emitted from power plants, industries, and automobiles react in the air (see Table 9.8).

Particulate matter is deposited in different regions of the respiratory system (see Figure 9.26). In the tracheobronchial region, particles are deposited by settling

TABLE 9.8 Characteristics of fine and coarse particulate matter

	FINE		COARSE
	Ultrafine	**Accumulation**	
Formation Processes	• Combustion, high-temperature processes, and atmospheric reactions		• Breakup of large solids/droplets
Formed by	• Nucleation of atmospheric gases including H_2SO_4, NH_3, and some organic compounds • Condensation of gases	• Condensation of gases • Coagulation of smaller particles • Reactions of gases in or on particles • Evaporation of fog and cloud droplets in which gases have dissolved and reacted	• Mechanical disruption (crushing, grinding, abrasion of surfaces) • Evaporation of sprays • Suspension of dusts • Reactions of gases in or on particles
Composed of	• Sulfate • Elemental carbon (EC) • Metal compounds • Organic compounds with very low saturation vapor pressure at ambient temperature	• Sulfate, nitrate, ammonium, and hydrogen ions • EC • Large variety of organic compounds • Metals: compounds of Pb, Cd, V, Ni, Cu, Zn, Mn, Fe, etc. • Particle-bound water • Bacteria, viruses	• Nitrates/chlorides/sulfates from HNO_3/ HCl/SO_2 reactions with coarse particles • Oxides of crustal elements (Si, Al, Ti, Fe), $CaCO_3$, $CaSO_4$, NaCl, sea salt • Bacteria, pollen, mold, fungal spores, plant and animal debris
Solubility	• Not well characterized	• Largely soluble, hygroscopic, and deliquescent	• Largely insoluble and nonhygroscopic
Sources	• High-temperature combustion • Atmospheric reactions of primary, gaseous compounds	• Combustion of fossil and biomass fuels, and high-temperature industrial processes, smelters, refineries, steel mills, etc. • Atmospheric oxidation of NO_2, SO_2, and organic compounds, including biogenic organic species (e.g., terpenes)	• Re-suspension of particles deposited onto roads • Tire, brake pad, and road wear debris • Suspension from disturbed soil (e.g., farming, mining, unpaved roads) • Construction and demolition • Fly ash from uncontrolled combustion of coal, oil, and wood • Ocean spray
Atmospheric half-life	• Minutes to hours	• Days to weeks	• Minutes to hours
Removal processes	• Grows into accumulation mode • Diffuses to raindrops and other surfaces	• Forms cloud droplets and rains out dry deposition	• Dry deposition by fallout • Scavenging by falling raindrops
Travel distance	• <1 to 10s of km	• 100s to 1,000s of km	• <1 to 10s of km (100s to 1,000s of km in dust storms for small coarse particles)

Source: Based on Wilson, W.E., and Suh, H.H. (1997). "Fine particles and coarse particles: Concentration relationships relevant to epidemiologic studies." *Journal of the Air and Waste Management Association* 47:1238–1249.

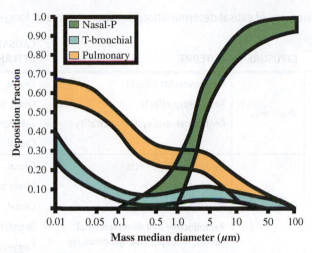

FIGURE 9.26 Predicted regional deposition of particles in the respiratory system for a tidal volumetric flow rate of 21 L/min. The shaded area indicates the variation resulting from two geometric deviations.

Source: Based on Heinsohn, R.J., and Kabel, R.L. (1999). *Sources and Control of Air Pollution.* Upper Saddle River, NJ: Prentice Hall, pp. 4–33, adapted from Stahlhofen, W., Gebhart, J., Heyder, J., and Scheuch, G. (1983) "New Regional Deposition Data of the Human Respiratory Tract." *Journal of Aerosol Science* 14:186–188.

in the low-velocity regions of the bronchioles and alveolar spaces. Particles may also diffuse to the surface of the bronchial bifurcations and alveoli. Once in the alveoli, some particles will be attacked, encapsulated, and passed up the tracheo-bronchial tree through cilia action. If particles pass into the bloodstream, they may be dissolved or removed by the cellular defenses and transferred to the lymphatic drainage system. Some particles may be permanently retained where they may remain benign or in some cases cause lung damage though fibrosis or malignancy.

Exposure to particulate matter has been linked to a variety of adverse health effects, including:

- Increased respiratory symptoms such as airway irritation, coughing, and difficulty breathing
- Decreased lung function
- Aggravation of asthma
- Chronic bronchitis
- Negative cardiovascular symptoms such as irregular heartbeat and heart attacks
- Premature death in people with heart or lung disease

Detailed information regarding the relationship between PM exposure and health effects is presented in the 2009 EPA report entitled "Integrated Science Assessment for Particulate Matter." The EPA found that PM less than 2.5 microns in diameter ($PM_{2.5}$) was likely to cause cardiovascular and respiratory effects and early mortality (Table 9.9).

Particulate matter also contributes to reduced visibility, or haze, as reported in Table 9.10. It may damage the environment by contributing to sedimentation in water bodies, depleting nutrients in soil, damaging sensitive plants, and consequently decreasing ecosystem diversity. PM may also be detrimental to architectural structures, potentially damaging or staining stone features, statues, and monuments.

TABLE 9.9 Summary of causal determinations for short-term and long-term exposure to $PM_{2.5}$ and PM_{10}

SIZE FRACTION	EXPOSURE	OUTCOME	CAUSALITY DETERMINATION
$PM_{2.5}$	Short-term	Cardiovascular effects	Causal
		Respiratory effects	Likely to be causal
		Central nervous system mortality	Inadequate
			Causal
	Long-term	Cardiovascular effects	Causal
		Respiratory effects	Likely to be causal
		Mortality	Causal
		Reproductive and developmental	Suggestive
		cancer, mutagenicity, genotoxicity	Suggestive
$PM_{10-2.5}$	Short-term	Cardiovascular effects	Suggestive
		Respiratory effects	Suggestive
		Central nervous system mortality	Inadequate
			Suggestive
	Long-term	Cardiovascular effects	Inadequate
		Respiratory effects	Inadequate
		Mortality	Inadequate
		Reproductive and developmental	Inadequate
		cancer, mutagenicity, genotoxicity	Inadequate

Source: U.S. EPA (2009a). *Policy Assessment for the Review of the Particulate Matter National Ambient Air Quality Standards*. Office of Air Quality Planning and Standards, U.S. Environmental Protection Agency. Research Triangle Park, NC.

TABLE 9.10 Summary of causality determination for welfare effects for short-term and long-term exposure to $PM_{2.5}$ and PM_{10}

WELFARE EFFECTS	CAUSALITY DETERMINATION
Effects on visibility	Causal
Effects on climate	Causal
Ecological effects	Likely to be causal
Effects on materials	Causal

Source: U.S. EPA (2009a). *Policy Assessment for the Review of the Particulate Matter National Ambient Air Quality Standards*. Office of Air Quality Planning and Standards, U.S. Environmental Protection Agency. Research Triangle Park, NC.

9.2.6 Sulfur Oxides

Sulfur oxides (SO_x) are a priority pollutant and are subject to NAAQS standards as shown in Table 9.1. Fossil fuel combustion accounts for 73% of SO_x emissions. Other industrial sources account for 20% of SO_x emissions. Sulfur oxides are highly soluble and are absorbed in the upper respiratory system, where they irritate and constrict the airways. Studies also show a connection between short-term exposure and increased visits to emergency departments and hospital admissions for respiratory illnesses, particularly in at-risk populations, including children, seniors, and asthmatics. Asthmatic patients may suffer brochioconstriction at concentrations of only 0.25 to 0.5 ppm$_v$. If particulate matter is present in the air along with sulfur oxides, the sulfur oxide sorbs onto the particulate matter, forming an aerosol that can be transported deep into the respiratory system and causes symptoms that are three to four times more severe than would be the case for SO_x alone. The adverse human health effects associated with the exposure dose are illustrated in Figure 9.27.

9.2.7 Hazardous Air Pollutants

The EPA has developed a list of 188 compounds (see www.epa.gov/haps) in addition to the six priority air pollutants that are potentially carcinogenic, mutagenic, or teratogenic. These 188 compounds are defined as **hazardous air pollutants (HAPS)** or, as they are sometimes called, toxic air pollutants. A partial list of suspected and confirmed carcinogenic compounds commonly used in industry is shown in Table 9.11. EPA's 2005 estimates of the increased cancer risk from

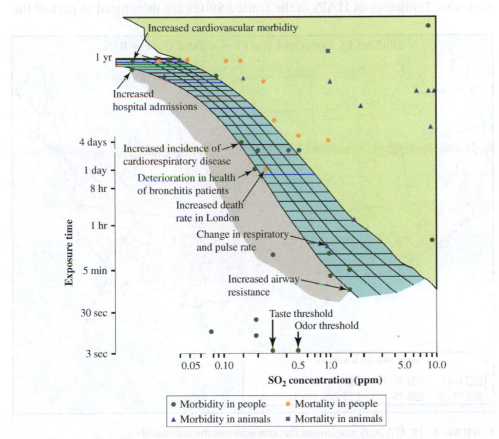

FIGURE 9.27 Health effects due to various exposures to SO_2. The shaded blue area represents the range of exposure where excess deaths have been reported. The gray area represents the range of exposures where health effects are suspected.

Source: Based on Seinfeld, J.H. (1986). *Atmospheric Chemistry and Physics of Air Pollution*. New York: John Wiley & Sons.

TABLE 9.11 List of commonly used industrial chemicals that are suspected or known carcinogens

Arsenic	Carbon tetrachloride	1,2-Dibromoethane	Nitrosamine
Asbestos	Chloroform	1,2-Dichloroethane	Perchloroethylene
Benzene	Chromium	Inorganic lead	Polycyclic aromatic hydrocarbons
Cadmium	1,4–Dioxane	Nickel	Vinyl chloride

air toxic emissions in the United States are shown in Figure 9.28. Formaldehyde and benzene emissions, which contribute to 60% of the national increase in risk, had the greatest nationwide cancer risk impact in the United States based on the 2005 analysis. Anthropogenic activities are responsible for most HAP emissions, although natural sources such as volcanic eruptions and forest fires may also release HAPs. The EPA has developed Health Effects Fact Sheets for HAPs (www.epa.gov/haps/health-effects-notebook-hazardous-air-pollutants) that provide a summary of information on the human health effects of these air toxins. Emissions of HAPs in the United States are monitored as part of the

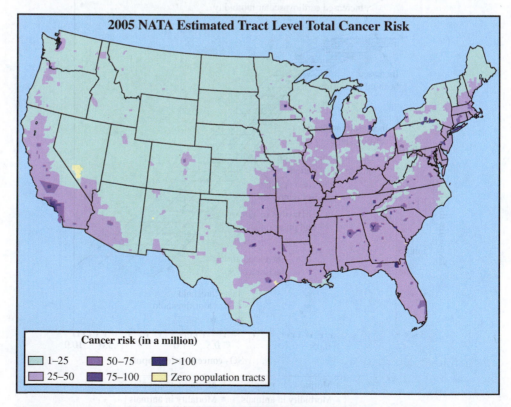

FIGURE 9.28 EPA 2005 assessment that characterizes the nationwide chronic lifetime cancer risk estimates and noncancerous hazards from inhaling air toxics.

Source: U.S. EPA (2011b). *An Overview of Methods for EPA's National-Scale Air Toxics Assessment*. Office of Air Quality Planning and Standards, U.S. Environmental Protection Agency. Research Triangle Park, NC. Cancer Map 1. www.epa.gov/ttn/atw/natamain/.

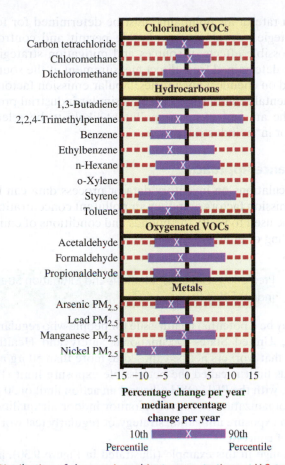

FIGURE 9.29 Distribution of changes in ambient concentrations at U.S. toxic air pollutant monitoring sites, 2003–2010 (percent change in annual average concentration).

Source: U.S. EPA (2012b). *Our Nation's Air: Status and Trends through 2010*. Office of Air Quality, Planning and Standards, U.S. Environmental Protection Agency. Research Triangle Park, NC. www.epa.gov/airtrends/2011/dl _graph.html.

EPA's Toxic Release Inventory (TRI) program (*www.epa.gov/tri*). The TRI provides information on economic analysis, risk, and pollution prevention information in an effort to help industry decrease annual emissions of toxic compounds. Figure 9.29 shows the change in ambient air concentrations of several HAPs from 2003 to 2010 at monitoring sites in the United States.

9.3 Estimating Emissions of Air Pollutants

Air pollutants may adversely affect human health, plant and animal health, architectural and historically significant structures, visibility, and ecosystem diversity. Some of these effects are well understood and have been the basis of environmental legislation and regulations, while other effects such as mixtures of compounds and endocrine disruptors are still being studied. Reducing the risk and negative impacts associated with air pollutants requires either reducing their emissions at the source or adding pollution control equipment to prevent emissions from escaping into the environment. Regardless, the ability to estimate and measure air pollutant emissions is a prerequisite to controlling emissions and reducing their negative impacts.

The emission rate of air pollutants must be determined for formulating emission control strategies, developing applicable permit and control programs, and identifying the possible effects of sources and mitigation strategies. Air pollutant emissions may be determined from direct measurement at the source, mass balance calculations based on chemical inventories, tabular emission factors for similar processes, or fundamental relationships based on similar industrial processes. *Emission factors* express the amount of an air pollutant likely to be released based on a determinant factor in an industrial process.

9.3.1 Mass Balance Approach

Mass balance calculations on inventory data or process data can be useful to estimate pollutant emission factors or indoor air pollutant concentrations. Mass balance models can also be used to verify the bounds and conditions of emissions calculated from direct sampling or emission factors.

EXAMPLE 9.2	Predicting Lead Concentrations and Mitigation Strategies in an Indoor Firing Range

Lead exposure may be a potential health issue for people who regularly work in indoor firing ranges. The United States Occupational Safety and Health Administration (OSHA) requires that workers not be exposed to more than 50 $\mu g/m^3$ of lead over an 8-hour period. This limit is called the permissible exposure limit (PEL). In order to ensure compliance with the PEL, OSHA has set an action limit of 30 $\mu g/m^3$. The action limit requires an organization to actively monitor indoor air quality, install an alarm system, develop an exposure mitigation strategy, or regularly test workers' blood levels for the pollutant of concern.

In the firing range in this example (illustrated in Figure 9.30), large-caliber rifles are used that contain 48 μg of lead per round of ammunition. The rifles fire at a maximum rate of 650 rounds per minute. The average firing rate in the range is closer to 150 rounds per minute. The National Institute for Occupational Safety and Health (NIOSH) requires that the room have a minimum ventilation velocity of 75 feet per minute. Lead has built up in the building from years of use, so the background lead concentration in the building is 15 $\mu g/m^3$, even when no one is firing a rifle in the range. Exhaust from the building passes through three stages of fabric high-efficiency

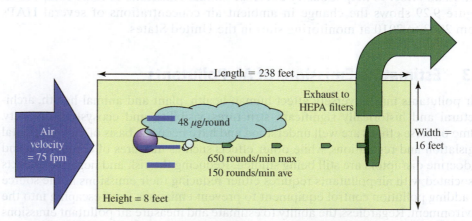

FIGURE 9.30 Schematic of lead emissions in an indoor firing range.

particulate arrestor (HEPA) filters that remove 95% of overall lead emissions prior to being emitted to the atmosphere.

Determine the following information about individual potential lead exposure in the firing range:

a. The average lead concentration from firing at the maximum firing rate, C_{max}, neglecting background lead levels

b. The total lead concentration averaged over an 8-hour period, C_{tot}, when firing at the average firing rate, including background lead levels, C_b

c. The expected average emission rate from the range with the HEPA filters

a. We can calculate the volumetric flow rate of air in the range by multiplying the airflow velocity, v_{NIOSH}, by the cross-section area, A_{cs}, perpendicular to the airflow:

$$v_{NIOSH} = 75 \text{ fpm}$$

$$A_{cs} = 16 \text{ ft} \times 8 \text{ ft} = 128 \text{ ft}$$

$$Q = v_{NIOSH} \times A_{cs} = 75 \text{ (fpm)} \times 128 \text{ (ft}^2\text{)} = 9,600 \text{ cfm}$$

Converting to m³/min gives us

$$Q = \frac{9,600}{35.29} = 272 \frac{m^3}{min}$$

The emission rate from firing the rifle at the maximum rate of 650 rounds per minute is

$$M = \left(48 \, \mu g \, \frac{Pb}{round}\right)\left(650 \, \frac{rounds}{min}\right) = 31,200 \, \mu g \, \frac{Pb}{min}$$

Rearranging the equation for the relationship between mass emission rate, volumetric flow rate, and concentration yields

$$C_{max}\left(\frac{\mu g}{m^3}\right) = \frac{M}{Q} = \frac{\left(31,200 \, \mu g \, \dfrac{Pb}{min}\right)}{\left(272 \, \dfrac{m^3}{min}\right)} = 115 \, \frac{\mu g}{m^3}$$

The lead concentrations at the maximum firing rate are expected to be well above the PEL, at least for short periods of time when firing at the maximum rate, even neglecting background concentrations.

Similarly, if we use an 8-hour average firing rate of 150 rounds per minute, then

$$M = \left(48 \, \mu g \, \frac{Pb}{round}\right)\left(150 \, \frac{rounds}{min}\right) = 7,200 \, \mu g \, \frac{Pb}{min}$$

Rearranging the equation for the relationship between mass emission rate, volumetric flow rate, and concentration yields

$$C_{ave}\left(\frac{\mu g}{m^3}\right) = \frac{M}{Q} = \frac{\left(7,200 \, \mu g \, \dfrac{Pb}{min}\right)}{\left(272 \, \dfrac{m^3}{min}\right)} = 26 \, \frac{\mu g}{m^3}$$

b. The concentration from firing at the average flow rate is still near the action level, and the background concentration has not yet been included. The total expected concentration in the firing range, C_{tot}, averaged over an 8-hour period is

$$C_{tot} = C_{ave} + C_b = 26 \frac{\mu g}{m^3} + 15 \frac{\mu g}{m^3} = 41 \frac{\mu g}{m^3} > \text{action limit of } 30 \frac{\mu g}{m^3}$$

Therefore, further action must be taken within the firing range to reduce exposure to those working in the facility.

c. The average emissions to the environment from the facility may be determined from the average mass emission rate and subtracting the amount removed by the HEPA filters:

$$EM_{ave} = C_{tot}Q\left(1 - \frac{EF}{100}\right) = \left(41 \frac{\mu g}{m^3}\right)\left(272 \frac{m^3}{min}\right)\left(1 - \frac{95\%}{100}\right)\left(\frac{60 \, min}{hr}\right)\left(\frac{mg}{1,000 \, \mu g}\right) = 33 \frac{mg}{hr}$$

9.3.2 Emission Factors

Direct analysis and mass balance approaches to determining emissions may not always be possible. Sampling can be very costly and may be highly variable due to short-term changes in process operations or expected major modifications to an industrial process. A mass balance approach may be limited by a lack of available data from similar processes and changes in inventory. Both direct measurement methods and mass balance approaches may not accurately reflect the variability of emissions over time. Incorporating safety factors when predicting emissions and modeling possible exposure and environmental impacts should account for this variability. Proposed new facilities may not have useful inventory data on which to base emissions and cannot be sampled directly. In these cases, the EPA has developed a comprehensive set of emission factors that attempt to relate the quantity of pollutants released to the atmosphere based on an activity associated with the release of that pollutant. These factors are called AP-42 emission factors, and they are averages of available data associated within a specific industrial category. A summary of the approaches to estimate emissions is provided in Table 9.12.

TABLE 9.12 Summary of methods for estimating source emissions

APPROACH TO DETERMINING EMISSIONS	COMMENTS
Direct sampling	Taken from actual or similar source
	May be expensive to determine change in emissions over time
	Continuous emission monitoring may be required for some processes
Mass balance	Must accurately define the feed stock and process conversion
	Useful for providing boundary conditions to check direct sampling and emission factor estimates
Emission factors	Average value from typical industry performance
	Found in EPA AP-42 tables
	Used for predicting emissions, impacts on ambient air quality, and as an input in dispersion models for new sources or major modifications to existing sources

The general equation for estimating emissions is

$$EM = AC \times EF \times \left(1 - \frac{ER}{100}\right) \tag{9.2}$$

where

EM = pollutant emission rate (mass/time)
AC = activity or production rate (units are varied)
EF = AP-42 emission factor (units vary)
ER = percent overall emissions reduction efficiency, including removal and capture efficiency

EXAMPLE 9.3 Estimating Emissions from Coffee Roasting Using U.S. EPA AP-42 Emission Factors

Professor Coffy buys his coffee from the Coffee Mountain Roasting Company, which roasted 26 million pounds of coffee beans in 2010. Determine the following emission if all 26 million pounds was continuously roasted. Create a table that shows the emissions with and without a thermal oxidizer.

The AP-42 emission factors for coffee roasting are found in Chapter 9 of AP42, Volume I, 5th edition (*www.epa.gov/air-emissions-factors-and-quantification/ap-42-compilation-air-emissions-factors*) under Section 9.13, Miscellaneous Food & Kindred Products: Coffee Roasting. The emission factors for uncontrolled filterable particulate matter, VOCs, methane, CO, and CO_2 are provided for various types of coffee roasting, with and without thermal oxidation VOC control devices in the U.S. EPA tables reproduced in Tables 9.13 and 9.14. The emission factor for filterable particulate matter, EF_{PM}, is given in Table 9.13 as 0.66 pound of PM/ton of coffee roasted. The activity of coffee roasting in tons can be determined from the given information:

$$AC_{coffee} = 26 \times 10^6 \frac{lb}{yr}\left(\frac{ton}{2,000\ lb}\right) = 13,000 \frac{tons}{year}$$

TABLE 9.13 Emission factors for particulate matter for coffee-roasting operations[a]

SOURCE	SOURCE CLASSIFICATION CODE	FILTERABLE PM (lb/ton)	CONDENSIBLE PM (lb/ton)
Batch roaster with thermal oxidizer	SCC 3-02-002-20	0.12	ND
Continuous cooler with cyclone	SCC 3-02-002-28	0.028	ND
Continuous roaster	SCC 3-02-002-21	0.66	ND
Continuous roaster with thermal oxidizer	SCC 3-02-002-21	0.092	0.10
Green coffee bean screening, handling, and storage system with fabric filter[b]	SCC 3-02-002-08	0.059	ND

[a] Emission factors are based on green coffee bean feed. Factors represent uncontrolled emissions unless noted. SCC = Source Classification Code. ND = no data. D-rated and E-rated emission factors are based on limited test data; these factors may not be representative of the industry.

[b] Emission Factor Rating: E

Source: Based on U.S. EPA (2011d). *Compilation of Air Pollutant Emission Factors—Update 2011*. U.S. Environmental Protection Agency. Washington, DC. www.epa.gov/air-emissions-factors-and-quantification/ap-42-compilation-air-emissions-factors.

TABLE 9.14 Emission factors for carbon-containing compounds for coffee roasting operations.[a] Emission Factor Rating: D

SOURCE	SOURCE CLASSIFICATION CODE	VOC[b] (lb/ton)	METHANE (lb/ton)	CO (lb/ton)	CO$_2$ (lb/ton)
Batch roaster	SCC 3-02-002-20	0.86	ND	ND	180
Batch roaster with thermal oxidizer	SCC 3-02-002-20	0.047	ND	0.55	530
Continuous roaster	SCC 3-02-002-21	1.4	0.26[d]	1.5	120[c]
Continuous roaster with thermal oxidizer	SCC 3-02-002-21	0.16	0.15[d]	0.098	200

[a] Emission factors are based on green coffee bean feed. Factors represent uncontrolled emissions unless noted. SCC = Source Classification Code. ND = no data. D-rated and E-rated emission factors are based on limited test data; these factors may not be representative of the industry.

[b] Volatile organic compounds as methane. Measured using GC/FID.

[c] Emission Factor Rating: C

[d] Emission Factor Rating: E

Source: Based on U.S. EPA (2011d). *Compilation of Air Pollutant Emission Factors—Update 2011*. U.S. Environmental Protection Agency. Washington, DC. www.epa.gov/air-emissions-factors-and-quantification/ap-42-compilation-air-emissions-factors.

Substituting the values into Equation (9.2) yields

$$EM_{PM} = AC \times EF \times \left(1 - \frac{ER}{100}\right) = \left(13,000 \; \frac{\text{tons coffee}}{\text{yr}}\right)\left(0.66 \; \frac{\text{lb of PM}}{\text{coffee}}\right)$$

$$= 8,580 \; \frac{\text{lb of PM}}{\text{yr}}$$

Similar calculations may be performed for each major category of emissions, the results of which are shown in Table 9.14 and 9.15.

Notice that for the coffee roasting company, filterable particulate matter, VOCs, methane, and carbon monoxide emissions are reduced by a thermal oxidizer air pollution control system. The thermal oxidizer converts the VOCs, and possibly some of the particulate matter, along with any supplemental fuel needed for the oxidation

TABLE 9.15 Emission estimates from a coffee roasting company with and without a thermal oxidizer for VOC control

EMISSION TYPE	EMISSION FACTORS (lb of pollutant/ton of coffee)		ESTIMATE EMISSION (lb/yr)	
	Without a Thermal Oxidizer	With a Thermal Catalytic Oxidizer	Without a Thermal Oxidizer	With a Thermal Catalytic Oxidizer
PM	0.66	0.092	8,580	1,196
VOC	1.4	0.16	18,200	2,080
Methane	0.26	0.15	3,380	1,950
CO	1.5	0.098	19,500	1,274
CO$_2$	120	200	1,560,000	2,600,000

Source: Based on U.S. EPA (2011d). *Compilation of Air Pollutant Emission Factors—Update 2011*. U.S. Environmental Protection Agency. Washington, DC.

process into carbon dioxide, so that carbon dioxide emissions from the process with the thermal oxidizer would likely be greater than from the same process without a thermal oxidizer. Furthermore, the mass emission rate of carbon dioxide is two to three times greater than the other pollutants. While carbon dioxide is not a direct health threat and is currently unregulated in the United States, it is a prominent greenhouse gas.

The EPA has classified the reliability of the emission factors into the ratings shown in Table 9.16. Emission factors are given a rating of A to E, where A is based on the most reliable data and the data used for E-rated emissions factors are the least reliable. The ratings are subjective in nature and do not imply statistical error bounds or confidence intervals. The emission factors are appropriate for estimating source-specific emission impacts on area-wide pollutant inventories. Emission factors may also be used in dispersion models to help estimate the potential impacts of new sources or process modifications on individual exposure and ambient air quality. Regulatory authorities may use emission factors for screening sources that require compliance permits. Emissions factors may also be used to determine appropriate control and mitigation strategies.

The AP-42 emission factors are published and updated in the EPA's two-volume set entitled *Compilation of Air Pollutant Emission Factors* (*www.epa.gov/air-emissions-factors-and-quantification/ap-42-compilation-air-emissions-factors*). It should be noted that the absence of an emission factor for a process does not indicate that there are no emissions for the process. There may be more than one emission factor for a general category of processes, so table footnotes and details should be read carefully. The data are useful for estimating emissions, but are not accurate enough to indicate how process variations or changes might affect emissions. The emission factors do not imply or represent emissions limits or regulations for a facility or process.

TABLE 9.16 Description of ratings factors for U.S. EPA AP-42 emissions factors

RATING	DESCRIPTION
A	**Excellent.** Factor is developed from A- and B-rated source test data taken from many randomly chosen facilities in the industry population. The source population is sufficiently specific to minimize variability.
B	**Above average.** Factor is developed from A- or B-rated test data from a "reasonable number" of facilities. The facilities tested may not represent a random sample. The population is sufficiently specific to minimize variability.
C	**Average.** Factor is developed from A-, B-, and/or C-rated test data from a reasonable number of facilities. The facilities tested may not represent a random sample. The source population is sufficiently specific to minimize variability.
D	**Below average.** Factor is developed from A-, B-, and/or C-rated data from a small number of facilities. The facilities may not represent a random sample. There is evidence of variability.
E	**Poor.** Factor is developed from C- and D-rated test data. The facilities may not represent a random sample, and there is evidence of variability within the population.

Source: U.S. EPA (1995). *Compilation of Air Pollutant Emission Factors—Volume I: Stationary Point and Area Sources.* U.S. Environmental Protection Agency. Washington, DC. AP-42 Fifth Edition.

Direct sampling should be conducted to confirm emission rates and concentrations if expected values are within an order of magnitude of regulatory thresholds for indoor air limits, ambient air quality compliance, or air toxic requirements. Emission factors are very useful for a first step in estimating the order of magnitude of expected emissions from a process. Emissions factors are also useful in developing airshed regional plans for ambient air quality control.

ACTIVE LEARNING EXERCISE 9.2 Fuel Switching for Energy Development

Using natural gas has become a cost-effective way to produce power. Natural gas producers claim that natural gas is cheaper and less harmful to the environment than coal-burning power facilities. A local company is considering switching from a 50-MMBtu coal-powered boiler to an 85-MMBtu natural gas process. The company claims it can increase its power output and decrease harmful emissions. Find the volume, chapter, and section numbers for U.S. EPA emissions factors for an uncontrolled packaged 85-MMBtu/hr natural gas combustion process.

If the combustion process operates 350 days per year, 24 hours per day, calculate the emission per year for the following pollutants in pounds and kilograms:

- NO_x and N_2O
- CO and CO_2
- Lead
- PM (total)
- SO_2
- Methane
- Benzene
- Cadmium

The company is planning to purchase the 85-MMBtu natural gas unit and says it can "go green" by using the natural gas combustion system to replace a 50-MMBtu medium-volatile uncontrolled bituminous coal spreader stoker boiler. If the coal has a 6% sulfur content, is this statement true? Why or why not?

9.4   Dispersion of Air Pollutants

Air emissions may be transported, dispersed, or concentrated by meteorological conditions. Emissions may be modeled to determine the effects an emission source may have on NAAQS. Various types and complexities of models may be used to represent the movement of air pollutants.

Dispersion occurs owing to the mean air motion that moves pollutants downwind, turbulent fluctuations that disperse pollutants in multiple directions, and mass diffusion of the pollutants that causes pollutant migrations from areas of high to low concentration. The density, shape, and size of the pollutant also influence movement.

A simple three-dimensional box model, illustrated in Figure 9.31, may sometimes represent topographic and meteorological conditions. The box model is useful for estimating worst-case scenarios for exposure or for developing long-term average concentrations in a topographically bound air shed. The model is only a

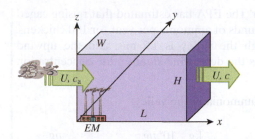

FIGURE 9.31 Box model for estimating air pollutant movement.

long-term average estimate of pollutants, since it cannot account for short-term variations or turbulence.

The box model is most accurate when topographic boundaries are available, such as a city or town that lies in a valley between two mountain ranges. In this case, the width of the model box, W, is the distance between the two mountain ranges. The height of the box is the height of the meteorological boundary layer. It is common for air to become stratified due to changes in density and temperature in a process similar to that discussed in Chapter 7 in regard to lake stratification. An inversion is a common meteorological process, especially in narrow valleys, that occurs and prevents mixing with air at higher elevations. The height of the box model, H, would be the height at which the atmospheric inversion occurs. A mass balance upon the emissions, EM (mass/time), can be developed to calculate the concentration, C, at some distance, L, downwind:

$$V \frac{dC}{dt} = EM + C_a Q_{in} - C Q_{out} \tag{9.3}$$

where

dC/dt = change in concentration with time
$V = H \times W \times L$ = volume of the box
C_a = ambient concentration of the pollutant in the air entering the box upwind of the emission source
C = concentration of the pollutant downwind of the emission source
Q = volumetric airflow rate

$$Q_{in} = Q_{out} = Q = U \times H \times W \tag{9.4}$$

where U is the mean velocity of the airflow through the box. Substituting the dimensions of the box and Equation (9.4) into Equation (9.3) yields

$$(H \times W \times L) \frac{dC}{dt} = EM + C_a(U)(H \times W) - C(U)(H \times W) \tag{9.5}$$

Under steady state conditions, $dC/dt = 0$. Equation (9.5) can be solved for the steady state concentration downwind, C_{ss}:

$$C_{ss} = C_a + \frac{EM}{U \times H \times W} \tag{9.6}$$

EXAMPLE 9.4   Ammonia Concentration in Perdu Valley

Perdu Valley is a 4-kilometer-wide agricultural valley where more than 5 million chickens are raised each year. The farms in the valley are at an elevation of 500 meters. The valley is bordered on two sides by mountains that are 1,200 meters high, which is the same height where inversions typically occur. The EPA has estimated that raising caged chickens in a flush house may result in 30 pounds of ammonia per year for 100 chickens. The average mean airflow velocity through the valley is 1.5 m/s, and the upwind ammonia concentration is 1.0 μg/m^3. What is the downwind steady state concentration of ammonia in the valley in μg/m^3?

We can determine the emission rate of ammonia in the valley:

$$EM = \frac{30 \text{ lb NH}_3}{\text{yr} - 100 \text{ chickens}} \times (5 \times 10^6) \text{ chickens} \frac{\text{kg}}{2.2 \text{ lb}} \frac{10^9 \mu g}{\text{kg}} = 6.82 \times 10^{14} \frac{\mu g}{\text{yr}}$$

Determine the steady state concentration in the valley using Equation (9.6):

$$C_{ss} = C_a + \frac{EM}{U \times H \times W} = 1.0 \frac{\mu g}{\text{m}^3} + \frac{6.82 \times 10^{14} \dfrac{\mu g}{\text{yr}}}{\left(1.5 \dfrac{\text{m}}{\text{s}}\right)\left(31.536 \times 10^6 \dfrac{\text{s}}{\text{yr}}\right)(1,200 \text{ [m]})(4,000 \text{ [m]})}$$

$$= 6.1 \frac{\mu g}{\text{m}^3}$$

The box model may also be used for variable emissions, where $dC/dt \neq 0$. Separating the variables and integrating with respect to time yields

$$\int_{C_o}^{C_t} \frac{dC}{(EM + U(H \times W)C_a) - U(H \times W)C} = \frac{1}{H \times W \times S} \int_0^t dt \qquad (9.7)$$

Rearranging Equation (9.5) yields

$$U(H \times W)(C_{ss}) = U(H \times W)(C_a) + EM \qquad (9.8)$$

Substituting the steady state concentration for the ambient concentration in Equation (9.8) and multiplying both sides of the integrated equation by $U(HW)$ yields

$$\frac{C - C_0}{C_{ss} - C_0} = 1 - \exp\left\{\frac{-Ut}{L}\right\} \qquad (9.9)$$

If the initial concentration is very small compared to the final concentration, then Equation (9.9) simplifies to

$$\frac{C}{C_{ss}} = 1 - \exp\left\{\frac{-Ut}{L}\right\} \qquad (9.10)$$

The simplified box model can be used to model emissions in a room or to cross the defined boundaries of a facility. However, the simplified box model does not account for dispersion and variations with time. The box model also does not account for various meteorological conditions and other atmospheric phenomena.

Air motions that influence the movements of pollutants are strongly affected by the height at which the air pollutants are released or are transported in the atmosphere (see Figure 9.32). There are three general classifications of atmospheric transport. *Macroscopic transport* occurs over scales of thousands of kilometers; these movements are closely related to global air circulation patterns. *Microscopic air circulation patterns* are most commonly modeled to determine

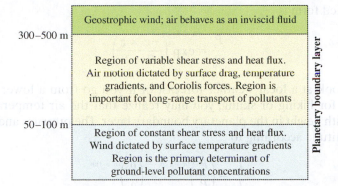

FIGURE 9.32 Earth surface effects on wind speed retardation in the planetary boundary layer.

Source: Based on Seinfeld, J.H. (1986). *Atmospheric Chemistry and Physics of Air Pollution*. New York: John Wiley & Sons.

air pollutant concentrations defined by local phenomena less than 10 kilometers from the source. *Mesoscopic transport* may be affected by both local and large-scale air circulation patterns for pollutants on the order of 100 kilometers from the source.

The troposphere is the region of the atmosphere closest to the Earth. The surface conditions do not typically affect air movements in the geostrophic region, which is about 300 to 500 meters above the Earth's surface. Weather, climate, and long-distance pollutant transport are strongly determined by geostrophic wind patterns. The planetary boundary layer is the layer between the Earth's surface and the upper atmosphere. Terrain and topography variations and temperature gradients strongly affect the friction losses and retardation of wind speed within the boundary layer. General atmospheric conditions must be understood before pollutant movements can be predicted.

Atmospheric conditions may be characterized by a few key parameters. The lapse rate is the negative temperature gradient of the atmosphere, and it is dependent on the weather, pressure, and humidity. The equation that represents the change in air pressure with elevation under isothermal (constant temperature) conditions can be written as

$$\frac{dP}{P} = -\frac{g}{RT_z}dz \tag{9.11}$$

where

P = atmospheric pressure of a parcel of air
g = gravitation constant
R = ideal gas law constant
dz = change in elevation
$T_z = T_0$ = constant temperature at the surface, $z = 0$, and at elevation $= z$

If the atmosphere is at a constant temperature, a relatively rare occurrence, Equation (9.11) may be integrated:

$$\int_{P_0}^{P_z} \frac{dP}{P} = \int_0^z \frac{-g}{RT}dz \tag{9.12}$$

The integrated form is

$$\frac{P_z}{P_0} = \exp\left(\frac{-g \times z}{R \times T}\right) \tag{9.13}$$

If you look at a local weather forecast or if you go from a lower elevation to a mountain for hiking or skiing, you may realize that the air temperature usually decreases with height in the planetary boundary layer. The pressure and density also vary with altitude according to

$$\frac{T_z}{T_0} = \left(\frac{P_z}{P_0}\right)^{(n-1)/n} = \left(\frac{\rho_z}{\rho_0}\right)^{(n-1)/n} \tag{9.14}$$

The change in temperature with respect to the change in elevation can be found by differentiating Equation (9.14):

$$\frac{dT}{dz} = -\frac{g}{R}\left(\frac{n-1}{n}\right) \tag{9.15}$$

where

$n = c_p/c_v$
c_p = constant pressure specific heat
c_v = constant volume specific heat

Under the specific conditions of dry adiabatic air, the ***dry adiabatic lapse rate,*** Γ, is

$$\Gamma = \left(-\frac{dT}{dz}\right)_{\text{dry adiabatic}} = \frac{0.0098°C}{m} = \frac{0.0054°F}{ft} \tag{9.16}$$

The amount of turbulence in the atmosphere is related to the actual lapse rate compared to the dry adiabatic lapse rate (Figure 9.33). If there is a high degree of turbulence in the atmosphere, the atmospheric conditions are classified as unstable. In an unstable atmosphere, if a pocket or parcel of air is disturbed, it is unlikely to approach a new equilibrium and will increase turbulence. A pocket of air that is disturbed in a stable atmosphere will be forced into a position of equilibrium by the surrounding air, and turbulence will be decreased. A stable atmosphere resists vertical mixing. The potential temperature gradient may be used to compare the dry adiabatic lapse rate and the actual or environmental lapse rate and to determine

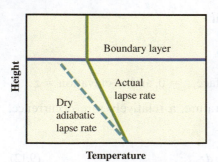

FIGURE 9.33 The actual lapse rate can be compared to the dry adiabatic lapse rate to determine the degree of stability.

TABLE 9.17 Stability classification of atmospheric conditions

$d\theta/dz$	STABILITY CLASSIFICATION	CATEGORY
< 0	Unstable	A, B, C
= 0	Neutral	D
> 0	Stable	E, F, isothermal or inversion

Source: Based on Heinsohn, R.J., and Kabel, R.L. (1999). *Sources and Control of Air Pollution*. Upper Saddle River, NJ: Prentice Hall.

the stability classification shown in Table 9.17. The ***potential temperature gradient*** is defined as the actual lapse rate plus the dry adiabatic lapse rate:

$$\frac{\Delta\theta}{\Delta z} \cong \left(\frac{\Delta T}{\Delta z}\right)_{actual} + \Gamma \tag{9.17}$$

Unstable atmospheric conditions and strong vertical mixing characterize super-adiabatic conditions. The environmental lapse rate is greater than the dry adiabatic lapse rate for stable atmospheres (Figure 9.34). A neutral atmosphere is typically a transitional condition where the environmental lapse rate is equal to the dry adiabatic lapse rate. Stable atmospheric conditions tend to reduce vertical mixing, and the environmental lapse rate is less than the dry adiabatic lapse rate. Isothermal conditions, when there is no temperature change with elevation, create very strongly stable atmospheric conditions. Inversions, which occur when the temperature increases with increasing altitude, are the most stable, and little mixing occurs during an inversion. The stability classification influences wind characteristics and wind speed (Table 9.18). Observations of atmospheric conditions may be used with the Pasquill stability classification system shown in Table 9.19 to estimate stability classifications.

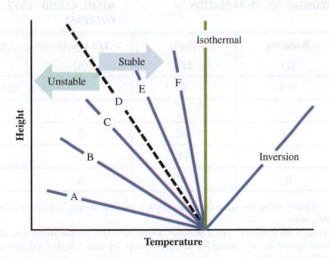

FIGURE 9.34 Stability classifications for the atmosphere.

Source: Based on Heinsohn, R.J., and Kabel, R.L. (1999). *Sources and Control of Air Pollution*. Upper Saddle River, NJ: Prentice Hall.

TABLE 9.18 Beaufort scale of wind speed equivalents at 10-meter elevation

DESCRIPTION	SPECIFICATIONS	U_{10} (m/s)
Calm	Smoke rises vertically	< 0.5
	Smoke drifts but does not move wind vane	0.5–1.5
Light	Wind felt, leaves rustle, ordinary vanes moves	1.5–3
Gentle	Leaves and twigs in constant motion, extends light flag	4–5
Moderate	Raises dust and loose paper, small branches are moved	6–8
Fresh	Small trees may sway, crested wavelets form on lakes	8–10
	Large branches move, whistling heard in wires, umbrellas used with difficulty	11–14
Strong	Whole trees in motion, difficulty walking into wind	15–17
	Breaks twigs off trees, generally impedes progress	18–20
Gale	Slight structural damage occurs (chimneys and roof)	21–24
	Trees uprooted, considerable structural damage occurs	25–28
Whole gale	Rarely experienced, widespread damage	29–33
Hurricane	> 75 miles per hour	> 34

Source: Based on Heinsohn, R.J., and Kabel, R.L. (1999). *Sources and Control of Air Pollution*. Upper Saddle River, NJ: Prentice Hall.

TABLE 9.19 Meteorological conditions for Pasquill stability categories[a]

WIND SPEED @ 10 m	DAYTIME INCOMING SOLAR RADIATION			NIGHT, CLOUD COVER, OR THICKLY OVERCAST	
Class[b]:	Strong (1)	Moderate (2)	Slight (3)	> 1/2 Low Clouds (4)	< 3/8 Clouds (5)
< 2 m/s	A	A–B	B	Strongly stable	
2–3 m/s	A–B	B	C	E	F
3–5 m/s	B	B–C	B–C	D	E
5–6 m/s	C	C–D	C–D	D	D
> 6 m/s	C	D	D	D	D

[a] The neutral category, D, should be assumed for overcast conditions during the day or night. Category A is the most unstable and category F is the most stable, with category B moderately unstable and category E slightly stable.

[b] Class 1: clear skies, solar altitude greater than 60 degrees above the horizontal, typical of a sunny afternoon, very convective atmosphere; class 2: summer day with a few broken clouds; class 3: typical of a fall afternoon, summer day with broken low clouds, or summer day with clear skies and solar altitude from only 15 to 35 degrees above the horizon, class 4: can be used for a winter day.

Source: Based on Hanna, S.R., Briggs, G.A., and Hosker, R.P. Jr. (1982). *Handbook on Atmospheric Diffusion*. Technical Information Center, U.S. Department of Energy.

EXAMPLE 9.5 Stability Classification Based on Temperature Measurements

Determine the environmental lapse rate and potential temperature for the early morning, at 6 a.m., 8 a.m., and 2 p.m. for the data presented in Table 9.20.

We have

$$\frac{dT}{dz} \cong \frac{\Delta T}{\Delta z} = \frac{(14 - 12)}{(100 - 0)} = 0.02°C/m$$

TABLE 9.20 Temperature measurements at a meteorological field station for Example 9.5

TIME	SURFACE TEMPERATURE (°C)	TEMPERATURE AT 100 m (°C)
6:00 a.m.	12	14
8:00 a.m.	15	14
2:00 p.m.	22	18

The potential temperature from Equation (9.17) is

$$\frac{\Delta \theta}{\Delta z} \cong \left(\frac{\Delta T}{\Delta z}\right)_{actual} + \Gamma = 0.02°C/m + 0.0098°C/m = 0.03°C/m$$

From Table 9.17 and Figure 9.34, we see that a positive lapse rate is a stable atmosphere since the positive environmental lapse rate indicates that an inversion condition exists early in the morning. This is a very stable condition with very little vertical mixing.

EXAMPLE 9.6 Stability Classification Based on Meteorological Observations

The following conditions were observed near a location being considered for an industrial site (stability conditions are needed to model predicted emissions). Records show that typical daytime conditions are moderate sunlight with broken clouds and wind can be felt that rustles leaves. Determine the likely average Pasquill stability category.

From Table 9.18, the wind description is indicative of light winds of 1.5 to 3 m/s. From Table 9.17, with moderate incoming solar irradiation, the stability category is most likely B, or possibly A, both of which are unstable conditions.

The stability conditions affect the shape and movement of air released from a smokestack. The amount of dispersion and downwind concentration of pollutants released from a stack is related to the exit velocity and temperature of air emitted from a smokestack, the mean velocity of the air moving past the stack, and the stability classification of the atmosphere. There are six characteristic plume shapes: fanning, looping, coning, fumigation, lofting, and trapping plumes (as illustrated in Figure 9.35).

The movement of air pollutants released from a smokestack depends on how high the smokestack is, particularly in relationship to the altitude at which an

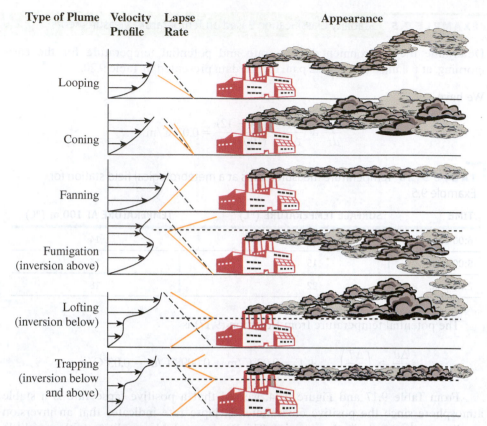

FIGURE 9.35 Types of plumes, typical mean air velocity profiles, environmental lapse rate (in yellow) compared to the dry adiabatic lapse rate (dashed line in gray), and the two-dimensional appearance of the plume type.

Source: Based on Slade, D.H. (ed.) (1968). *Meteorology and Atomic Energy.* U.S. Atomic Energy Commission, Air Resources Laboratories, Environmental Sciences Services Administration, U.S. Department of Commerce, Washington, DC.

inversion may occur. Pollutants trapped in a fumigation plume below an inversion layer have caused incredibly devastating air pollution events, such as the 1948 smog in Donora, Pennsylvania. Modern smokestacks are designed so that the combined height of the stack and the upward momentum of the air jet leaving the smokestack propel pollutants above the inversion layer.

Gases that exit a smokestack with a **specific height**, H_s, have a momentum and buoyancy that are characteristic of the gases emitted and the atmosphere they are being emitted into (see Figure 9.36). As the plume rises, it also travels downwind and mixes vertically and laterally. The height of the **plume rise**, dh, is important in modeling the movement of pollutants and estimating the downwind exposure. The **effective stack height**, H, is the sum of the physical stack height, H_s, and the plume rise:

$$H = H_s + \delta h \qquad (9.18)$$

The buoyancy of the gas emitted from an exhaust stack is accounted for by the Briggs buoyancy flux parameter, F:

$$F \frac{\text{m}^4}{\text{s}^3} = \frac{g v_s D_s^2 (T_s - T_a)}{4 T_s} \qquad (9.19)$$

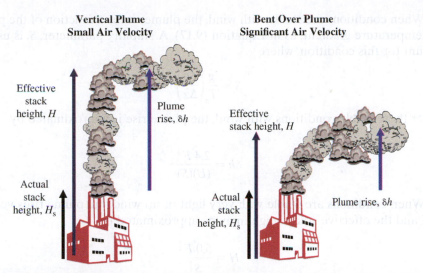

FIGURE 9.36 Plume rise from a smokestack into an atmosphere with a small air velocity and a large air velocity that cause the plume to bend.

where

v_s = exit velocity of gas from the smokestack (m/s)

D_s = inside diameter of the smokestack (m)

g = gravitational acceleration constant (m/s^2)

T_s = temperature of the gas exiting the smokestack (K)

T_a = ambient air temperature at the height of the smokestack, H_s (K)

There are numerous correlations for estimating the plume rise based on meteorological conditions and atmospheric stability. If the atmosphere is unstable or neutral, the plume rise can be estimated as a function of the Bouyancy Flux term, F. The proper equation is based upon the downwind distance from the smokestack, x, and the distance at which atmospheric turbulence begins to dominate parcel movement x^*.

x = downwind distance from the stack source (in m) for a stable atmosphere or for

x^* = the distance at which atmospheric turbulence begins to dominate parcel and pollutant movement and is estimated from

$$x^* = 14F^{\frac{5}{8}} \quad \text{if} \quad F < 55 \, \tfrac{m^4}{s^3} \quad \text{and} \quad x^* = 34F^{\frac{2}{5}} \quad \text{if} \quad F \ge 55 \, \tfrac{m^4}{s^3} \quad (9.20)$$

For unstable or neutral conditions, where $x < 3.5x^*$, the plume rise in meters is given by

$$\delta h = \frac{1.6 \, (3.5x^*)^{\frac{2}{3}} \, F^{\frac{1}{3}}}{U} \quad (9.21)$$

where U is the wind velocity in the x-direction (in m/s).

The Briggs plume rise formula is a simplified model for use when the temperature of the gas plume is much higher than the ambient temperature, such as in combustion-based power generation facilities. If $x > 10 \times H_s$, Briggs found that the following equation more accurately describes the plume rise:

$$\delta h = \frac{1.6 \, (10H_s)^{\frac{2}{3}} \, F^{\frac{1}{3}}}{U} \quad (9.22)$$

When conditions are stable with wind, the plume rise is a function of the potential temperature gradient, as in Equation (9.17). A stability parameter, S, is used to account for this condition, where

$$S = \frac{g}{T_a}\left(\frac{\Delta\theta}{\Delta z}\right) \tag{9.23}$$

Under these stable conditions with wind, the plume rise is approximated by

$$\delta h = \frac{2.4\,F^{\frac{1}{3}}}{(U)(S)} \tag{9.24}$$

When conditions are stable with very light or no wind, the plume rises vertically, and the effective stack height may be approximated by

$$H = \frac{5.0\,F^{\frac{1}{4}}}{S^{\frac{3}{8}}} \tag{9.25}$$

The plume rise can be affected by the geometry of the building and the ratio of the exit gas to the mean air velocity. If the exit velocity is less than 1.5 times the mean air velocity, low-pressure wake regions can be created behind the smokestack or building. The low-pressure wake region may cause the pollutants to become trapped in the low-pressure eddy created by air passing by the smokestack or building. Wake effects may increase exposure near the sources and affect building ventilation. In this case, more detailed calculations are required. Proper stack placement and building ventilation should take into consideration topography, meteorology, and potential building wake effects.

Downwind air pollutant concentrations and subsequent exposure are affected by the stack characteristics and turbulent conditions in the atmosphere. Temperature gradients, convective air currents, and wind shear over surface features naturally cause turbulence. However, turbulence is very difficult to accurately model mathematically. Instead, statistical methods are used to estimate vertical and horizontal dispersion parameters based on the likely magnitude of turbulence associated with the aforementioned atmospheric stability classes. The stability class of the atmosphere governs the degree of vertical mixing. Eddy diffusion governs the degree of horizontal dispersion. There are four types of air pollutant dispersion models: physical, numerical, statistical, and Gaussian plume models. Only the Gaussian plume model will be described in this text, but the many resources listed in the References section of this chapter may be consulted for exploration of other air dispersion models.

A mass balance can be written that includes dispersion of air pollutants with respect to time and location:

$$\frac{\delta c}{\delta t} = -\frac{\delta(cU)}{\delta x} + EM + \frac{\delta\left[\frac{\delta(cD_x)}{\delta x}\right]}{\delta x} + \frac{\delta\left[\frac{\delta(cD_y)}{\delta y}\right]}{\delta y} + \frac{\delta\left[\frac{\delta(cD_z)}{\delta z}\right]}{\delta z} \tag{9.26}$$

where

c = concentration of the pollutant emitted from a smokestack
U = wind velocity in the x-direction
EM = source generation (or decay) term

D_x, D_y, and D_z are the effective dispersion coefficients along each axis that incorporate both eddy and molecular diffusion

Several simplifying assumptions may be applicable to many air pollution sources. If the source is constant and the gases are nonreactive, a steady state assumption can be made, where $dC/dt = 0$. We will also assume that bulk motion due to the mean air velocity governs air pollutant transport in the x-direction, so that dispersion in the x-direction is negligible. The mean air velocity, U, and the dispersion coefficients in the y- and z-directions will be assumed to be constant. Equation (9.26) can then be simplified to

$$U \frac{\delta c}{\delta x} = D_y \frac{\delta^2 c}{\delta y^2} + D_z \frac{\delta^2 c}{\delta z^2} \tag{9.27}$$

The following mathematical boundary conditions are required to solve the mass balance equation: The concentration at the initial point approaches infinity as the size of the release point approaches zero. The concentration approaches zero as the distance from the source approaches infinity. There is no diffusion at ground level. The simplified mass balance using the Gaussian solution to solve Equation (9.27) is visualized in Figure 9.37. The solution to Equation (9.27), based on the Gaussian function used to represent dispersion, is

$$c_i(x, y, z) = \frac{EM_{i,s}}{2\pi x (D_y D_z)^{1/2}} \exp\left\{ -\frac{U}{4x} \left(\frac{y^2}{D_y} + \frac{z^2}{D_z} \right) \right\} \tag{9.28}$$

The theoretical dispersion parameters, D_x and D_y, are in practice replaced with empirically based dispersion coefficients in the horizontal direction, σ_y, and the vertical direction, σ_z, where

$$\sigma_y^2 = \frac{2xD_y}{U} \tag{9.29}$$

$$\sigma_z^2 = \frac{2xD_z}{U} \tag{9.30}$$

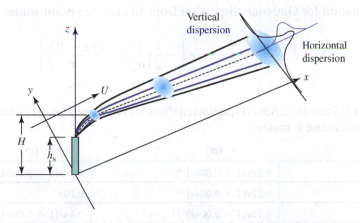

FIGURE 9.37 Visualization of the Gaussian solution to the pollutant dispersion mass balance equation for a constant mean air velocity, U, oriented along the x-direction. The blue lines represent horizontal dispersion in the y-direction. The gray lines represent vertical dispersion in the z-direction.

TABLE 9.21 Approximate curve fit constants for calculating dispersion coefficients as a function of atmospheric stability classification and downwind distance from the air pollution source in a rural environment

STABILITY CLASS	a	b	x < 1 km c	x < 1 km d	x < 1 km f	x > 1 km c	x > 1 km d	x > 1 km f
A	213	0.894	440.8	1.941	9.27	459.7	2.094	9.6
B	156	0.894	106.6	1.149	3.3	108.2	1.098	2.0
C	104	0.894	61.0	0.911	0	61.0	0.911	0
D	68	0.894	33.2	0.725	−1.7	44.5	0.516	−13.0
E	50.5	0.894	22.8	0.678	−1.3	55.4	0.305	−34.0
F	34	0.894	14.35	0.740	−0.35	62.6	0.180	−48.6

Source: Based on Martin, D.O. (1976). "The Change of Concentration Standard Deviation with Distance." *Journal Air Pollution Control Association* 26(2):145–147.

Then Equation (9.28) can be rearranged so

$$c_i(x, y, z) = \frac{EM_{i,s}}{2\pi U \sigma_y \sigma_z} \exp\left\{ -\frac{1}{2}\left(\frac{y}{\sigma_y}\right)^2 - \frac{1}{2}\left(\frac{z}{\sigma_z}\right)^2 \right\} \tag{9.31}$$

The dispersion coefficients are based on the previously discussed stability categories and are referred to as the Pasquill–Gifford coefficients. Different empirical correlations have been developed. General equations for the Pasquill–Gifford dispersion coefficients for rural areas are provided in Equations (9.32) and (9.33), and the coefficients for the empirical equations for rural dispersion in flat terrain are provided in Table 9.21. The equations suitable for urban areas are presented in Table 9.22, where the distance x has the units of kilometers:

$$\sigma_y = a(x)^b \tag{9.32}$$

$$\sigma_z = c(x)^d + f \tag{9.33}$$

For air pollutants emitted from a smokestack, the x-axis is aligned with the prevailing wind direction and the effective stack height must be considered. Thus, the general equation for Gaussian dispersion from an elevated point source becomes

$$c_i(x, y, z) = \frac{EM_{i,s}}{2\pi U \sigma_y \sigma_z} \exp\left\{ -\frac{1}{2}\left(\frac{y}{\sigma_y}\right)^2 - \frac{1}{2}\left(\frac{z - H}{\sigma_z}\right)^2 \right\} \tag{9.34}$$

TABLE 9.22 Pasquill–Gifford dispersion coefficients for urban sites, with downwind distance, x, measured in meters

STABILITY	σ_y (m)	σ_z (m)
A–B	$0.32x(1 + 0.004x)^{-0.5}$	$0.24x(1 + 0.0001x)^{0.5}$
C	$0.22x(1 + 0.004x)^{-0.5}$	$0.20x$
D	$0.16x(1 + 0.004x)^{-0.5}$	$0.14x(1 + 0.0003x)^{-0.5}$
E–F	$0.11x(1 + 0.004x)^{-0.5}$	$0.08x(1 + 0.0015x)^{-0.5}$

Source: Based on Griffiths, R.F. (1994). "Errors in the Use of the Briggs Parameterization for Atmospheric Dispersion Coefficients." *Atmospheric Environment* 28(17): 2861–2865.

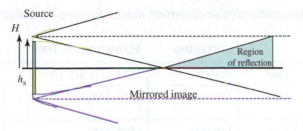

FIGURE 9.38 Maximum downwind concentration estimates may assume that the ground reflects pollutants.

The maximum ground-level concentration can be determined for estimating compliance with NAAQS and for estimating the maximum exposure to an individual downwind. The most conservative estimate assumes that pollutants cannot be dispersed into the ground but are instead reflected by the ground (see Figure 9.38). For gaseous pollutants that are reflected by the ground at the centerline of the plume, Equation (9.34) can be modified to account for the reflected particles:

$$c_i(x, 0, 0) = \frac{EM_{i,s}}{\pi U \sigma_y \sigma_z} \exp\left\{ -\frac{1}{2}\left(\frac{H}{\sigma_z}\right)^2 \right\} \tag{9.35}$$

Particles with a particle diameter less than 20 μm typically have a very low settling velocity and move with the mean air stream. Particles less than 20 μm in diameter may be modeled with Equation (9.34), but the particles may adhere to the ground and are reflected to a lesser degree than gaseous particles. Therefore, Equation (9.34) is unreliable for estimating the maximum concentration of particulate matter. Particles greater than about 20 μm in diameter may have a significant settling velocity that should be considered and require modifications to the general Gaussian equation for accurate estimates of downwind concentrations.

Atmospheric dispersion models may be used for developing state implementation plans, new source review, environmental audits, environmental assessments, liability cases, and risk assessments. The information required for the application of most models is similar and is listed in Table 9.23. Various air dispersion models have been developed and adapted to a particular use, depending on topographic conditions, the type of pollutants, and the expected application of model results. A few of the common models are listed in Table 9.24.

TABLE 9.23 Information required and available from air dispersion modeling

SOURCE INFORMATION	METEOROLOGICAL INFORMATION	RECEPTOR INFORMATION
Coordinates of source	Pasquill stability class	Coordinates
Stack height	Wind direction	Elevation of ground level at receptor
Stack top ID	Wind speed	Height of receptor above ground level
Stack gas exit velocity	Temperature	
Stack gas exit temp	Mixing height (urban)	
Directionally dependent building width	Mixing height (rural)	
Directionally dependent building height		
Elevation of stack base		

TABLE 9.24 Recommended air dispersion models for specific application

APPLICATION		SCREENING	REFINED SHORT TERM	REFINED LONG TERM
Flat or simple terrain	Rural	SCREEN	ISCST, OCD, BLP, FDM	ISCST or ISCLT
	Urban	SCREEN	ISCST, RAM, FDM	ISCST or ISCLT
Complex terrain	Rural	CTSCREEN VALLEY COMPLEX-I RTDM	CTDM PLUS	
	Urban	CTSCREEN VALLEY SHORTZ LONGZ		

ACTIVE LEARNING EXERCISE 9.3 Air Dispersion Modeling

Construction of a waste-to-energy plant in the Latah Creek Valley is under consideration. You have been asked to evaluate the proposed facility in terms of its potential impact on the air quality in the valley. You are given the following information about the proposed facility and site:

Solid waste loading rate: 800 tons (English) per day
Generating capacity: 22,000 kW
Furnace temperature: 2,500°C
Type: Municipal multichamber refuse incinerator
Stack exit flow rate: 3.18 m³/s

- What will be the uncontrolled emission rate in kg/s of each pollutant?
  - Particulates
  - SO_2
  - CO
- Use the following information to determine the atmospheric stability on a bright sunny afternoon:
  - The temperature at ground level is 15°C.
  - The temperature at 10 m is 14.7°C.
  - The wind speed at the exit of the smokestack is 6 m/s and the ambient temperature around the stack is 14°C.
- Determine the plume rise using the following information:
  - The inside diameter of a 30-m-tall smokestack is 2 m.
  - The flow rate of air through the smokestack is 10 m³/s at 25°C.
  - The proposed facility is in a rural area.
- What will the concentration of each pollutant be at an elementary school that is 680 meters directly downwind from the proposed facility?

EXAMPLE 9.7   Modeling Emissions from an Asphalt Processing Facility

A company proposes to build a drum mix/hot mix asphalt plant that uses a No. 2 fuel oil-fired dryer to produce 1.5×10^6 Mg of asphalt per year. The plant is being promoted as an opportunity to bring new jobs to the community. The company plans to operate the plant around the clock, 365 days a year, 24 hours a day in this rural community. The exit to the smokestack on the proposed plant is 10 meters above ground and will have a diameter of 0.91 m. Waste gas will exit the smokestack at a flow rate of 5.6 m^3/s (actual) and 675 K. The air pressure within the stack will be essentially equal to the ambient air pressure of 1 atm.

The average ambient temperature measured at the stack height is 300 K. The average temperature measured at 110 meters on a nearby radio tower is 299.02 K. The prevailing wind speed is 6 m/s directly toward a mobile home park.

The community wants to know what impact the plant will have on the air quality at the mobile home park that is 400 m directly downwind from the plant. Assume for the purposes of ambient air analysis that all the nitric oxide compounds are in the form of NO_2. The maximum reported NO_2 concentration in the ambient air upwind of the plant is 2 ppb. We can assume the pollutants are not reflected by the ground.

  a. Determine the mass emission rate of NO_x in mg/s.

  b. What is the effective height of the smokestack in m?

  c. Calculate what the concentration of NO_x will be in the mobile home park at ground level directly downwind from the asphalt plant.

  a. In Chapter 11 of the U.S. EPA AP-42, the EPA lists the emission factor for nitrogen oxide emissions from an asphalt plant. According to the AP-42 list, nitrogen oxide emissions will be 0.055 lb per ton. The emission rate is calculated from

$$EM = \left(1.5 \times 10^6 \, \frac{Mg}{yr}\right)\left(0.055 \, \frac{lb}{ton}\right)\left(0.5 \, \frac{\frac{kg}{Mg}}{\frac{lb}{ton}}\right)\left(10^6 \, \frac{mg}{kg}\right)\left(\frac{yr}{31{,}536{,}000 \, s}\right)$$

$$= 1{,}300 \, \frac{mg}{s}$$

  b. The stability class must be determined and the buoyancy flux calculated.

The velocity of the gas exiting the smokestack is

$$v_s = \frac{Q}{A} = \frac{4Q}{\pi D_s^2} = \frac{4(5.6)}{\pi(0.91)^2} = 8.61 \text{ m/s}$$

The buoyancy flux may be determined from Equation (9.19):

$$F = \frac{g v_s D_s^2 (T_s - T_a)}{4T_s} = \frac{(9.81)(8.61)(0.91)^2(675 - 300)}{4(675)} = 9.71 \frac{m^4}{s^3}$$

The potential temperature gradient is determined from Equation (9.17):

$$\frac{\Delta\theta}{\Delta z} \cong \left(\frac{\Delta T}{\Delta z}\right)_{actual} + \Gamma = \left(\frac{299.02 - 300}{110 - 10}\right) + 0.0098 = 0 \, \frac{°C}{m}$$

Since $\Delta\theta/\Delta z$ is equal to zero, the atmospheric stability is neutral.

The distance to the point of possible exposure, x, is 400 m, and the stack height, H_s, is 10 m, so that

$$10H_s = 10(10) \text{ m} = 100 \text{ m and therefore } x = 400 \text{ m} > 100 \text{ m}$$

The exhaust is from a combustion source with the temperature exiting the smokestack much higher than the ambient temperature, and the downwind point of concern is greater than ten times the actual stack height. Under these conditions, Equation (9.22) can be used to estimate the plume rise:

$$\delta h = \frac{1.6 \, (10H_s)^{\frac{2}{3}} F^{\frac{1}{3}}}{U} = \frac{1.6 \, [(10)(10)]^{\frac{2}{3}} \, (9.71)^{\frac{1}{3}}}{6} = 12.3 \text{ m}$$

The effective stack height is determined from Equation (9.18):

$$H = H_s + \delta h = 10 + 12.3 = 22.3 \text{ m}$$

c. The general dispersion equation is used to estimate the downwind dispersion. From the potential temperature gradient, the atmosphere is characterized by the neutral, D, stability category.

The dispersion coefficients for a rural atmosphere are estimated from the values in Table 9.21 and Equations (9.32) and (9.33):

$$\sigma_y = a(x)^b = 68 \left(\frac{400 \text{ m}}{1{,}000 \, \frac{\text{m}}{\text{km}}} \right)^{0.894} = 30.0 \text{ m}$$

$$\sigma_z = c(x)^d + f = 33.2 \left(\frac{400 \text{ m}}{1{,}000 \, \frac{\text{m}}{\text{km}}} \right)^{0.725} + (-1.7) = 15.4 \text{ m}$$

The general dispersion Equation (9.34) is used to predict the downwind concentration of the nitric oxide at ground level at the mobile home park:

$$C_{NO_x(400,0,0)} = \frac{EM}{2\pi U \sigma_y \sigma_z} \exp\left\{ -\frac{1}{2}\left(\frac{y}{\sigma_y}\right)^2 - \frac{1}{2}\left(\frac{z-H}{\sigma_z}\right)^2 \right\}$$

$$= \frac{1{,}300}{2\pi(6.0)(30.0)(15.4)} \exp\left\{ -\frac{1}{2}\left(\frac{0}{30.0}\right)^2 - \frac{1}{2}\left(\frac{0-22.3}{15.4}\right)^2 \right\}$$

$$= 0.0264 \, \frac{\text{mg}}{\text{m}^3} \text{ of NO}_x$$

Now we can convert the concentration at the downwind location to ppb, assuming all the nitric oxides are of the form NO_2:

$$C_{NO_2(400,0,0)}[\text{ppb}] = \frac{24.5 \left(0.0264 \, \frac{\text{mg}}{\text{m}^3} \right)}{46 \, \frac{\text{g}}{\text{mol}}} \times 1{,}000 = 14 \text{ ppb NO}_2$$

Since the maximum reported upwind NO_2 concentration is 2 ppb, the downwind concentration increases to 16 ppb. The asphalt plant in this location with these atmospheric conditions will increase the exposure to NO_2 substantially but will still be below the NAAQS standard for NO_2 in Table 9.1.

9.5 Air Pollutants from Combustion Processes

The combustion of fossil fuels adds carbon dioxide, water vapor, and heat to the atmosphere. Combustion is a major source of carbon monoxide, nitrogen oxides, particulate matter, and sulfur oxides—primary pollutants measured as part of the NAAQS. Approximately three-quarters of global sulfur dioxide emissions are from combustion sources. Air pollutants from combustion sources may be controlled, but these control systems generally increase the cost of energy from the combustion process.

Combustion is the chemical process that occurs when oxygen reacts with various compounds; it results in the release of heat. Major stationary combustion sources include coal, oil, and gas used to make steam; cement production; thermal incineration of waste materials; and dryers and ovens used to make bricks, food, and surface coatings. Combustion cycles also power mobile vehicles and aircraft, and supply power for emergency and portable generators.

The maximum theoretical thermal cycle efficiency to generate energy from a combustion process is approximately 70%. Heat loss is the largest source of wasted energy. In 1948, the conversion of fossil fuels to energy through combustion processes was only 22% efficient. By 1968, the conversion efficiency had increased to 34%, and energy efficiency from combustion processes typically is still about 33% to 35% for coal- and oil-fired power plants.

The emissions associated with combustion sources depend on the mixture of the gases in the combustion zone. Incomplete combustion of the fossil fuels occurs when the amount of fuel exceeds the amount of oxygen required for combustion, which is known as a fuel-rich mixture. Fuel-rich mixtures may lead to the inefficient use of fuel and excess emissions of air pollutants. If the amount of oxygen in the combustion zone exceeds that which is required for the combustion of a given amount of fuel, it is known as a lean fuel mixture. Lean fuel mixtures increase nitrogen oxide emissions from the combustion zone and increase heat losses.

The stoichiometry of the complete theoretical combustion for pure hydrocarbons is described by this equation:

$$C_xH_y + (b)O_2 + 3.76(b)N_2 \rightarrow xCO_2 + \frac{y}{2}H_2O + 3.76(b)N_2 \tag{9.36}$$

where

C_xH_y = general hydrocarbon formula

$b = x + y/4$ = the stoichiometric number of moles of O_2 required for complete combustion. 3.76 is the number of moles of N_2 present in air for each mole of O_2

Carbon monoxide is formed by reactions between hydrocarbon radicals and the hydroxyl radical (OH*). It is a precursor to the formation of carbon dioxide, as shown in Equations (9.37) through (9.44). Carbon dioxide is formed by the reaction between carbon monoxide and the hydroxyl radical. If the combustion zone lacks sufficient oxygen, there will be insufficient hydroxyl radical production, leading to increased carbon monoxide emissions.

$$HCHO + OH^* \rightarrow HCO^* + H_2O \tag{9.37}$$

$$HCO^* + OH^* \rightarrow CO + H_2O \tag{9.38}$$

$$HCO^* + O_2 \rightarrow HO_2^* + CO \tag{9.39}$$

$$CO + OH^* \rightarrow CO_2 + H^* \tag{9.40}$$

$$H^* + CO_2 \rightarrow CO + H^* + O^* \tag{9.41}$$

$$CO + O_2 \rightarrow CO_2 + O^* \tag{9.42}$$

$$CO + O^* + M \rightarrow CO_2 + M \qquad (9.43)$$

$$CO + HO_2^* \rightarrow CO_2 + OH^* \qquad (9.44)$$

Oxygen is most often added to the combustion zone through the addition of air. Air contains both nitrogen and oxygen, so nitrogen is also introduced into the combustion zone when air is added to ensure complete combustion of the fuel and to minimize carbon monoxide production. Excess nitrogen in the combustion zone is reactive with oxygen radicals in the combustion zone, as illustrated by the reactions shown in Equations (9.45) through (9.48), and nitrogen oxides are formed.

$$N^* + O_2 \leftrightarrow NO + O^* \qquad (9.45)$$

$$O^* + N_2 \leftrightarrow NO + N^* \qquad (9.46)$$

$$O_2 + M \leftrightarrow O^* + O^* + M \qquad (9.47)$$

$$N^* + OH^* \leftrightarrow NO + H^* \qquad (9.48)$$

The formation of nitrogen oxides is proportional to the temperature of the flame. High flame temperatures (3,000° to 3,600°F) produce 6,000 to 10,000 ppm_v of NO_x. As the gas cools to 300° from 600°F, the NO_x concentration may drop to less than 1 ppm_v. Typical NO_x concentrations exiting coal-fired power plants range from 300 to 1,200 ppm_v. McKinnon (1974) developed an empirical relationship to predict the nitrogen oxide concentration exiting a combustion zone:

$$C_{NO} = 5.2 \times 10^{17} \left\{ \exp\left(-\frac{72,300}{T}\right) \right\} y_{N_2} y_{O_2}^{1/2} t \qquad (9.49)$$

where

C_{NO} = NO concentration (ppm_v)
y_{N_2} = the molar fraction of the nitrogen component
y_{O_2} = the molar fraction of the oxygen component
T = absolute temperature (K)
t = residence time of the gas in the combustion chamber (s)

The U.S. EPA recommends that combustion zone conditions and chemistry be tightly controlled to limit the formation of nitrogen oxides. Control of the combustion process may include reducing the peak temperature of the flame zone, reducing the residence time of gases in the flame, or reducing the oxygen concentration in the flame zone through control of the *air-to-fuel ratio*, as illustrated in Figure 9.39. Controlling nitrogen oxides must be a compromise with limiting carbon monoxide emissions, as the

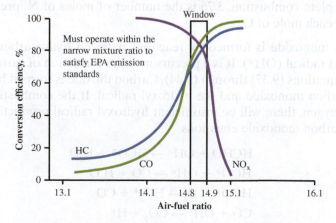

FIGURE 9.39 Air-to-fuel ratio that minimizes both carbon monoxide and nitrogen oxide emissions.

Source: Cooper, C.D., and Alley, F.C. (1986). *Air Pollution Control: A Design Approach*, Fig. 15.8, p. 475. Prospect Heights, IL: Waveland Press, Inc.

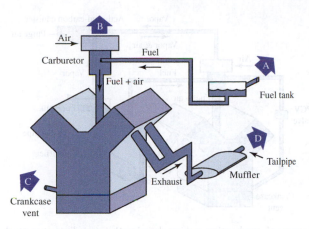

FIGURE 9.40 Internal combustion engine showing the sources of air pollutants.

conditions that limit nitrogen oxide formation favor carbon monoxide formation, and vice versa. The exhaust gas may be treated to remove nitrogen oxides; post-combustion treatment is common in the European Union (EU) and in Japan.

Although many of the above control techniques can apply to moving sources as well as stationary ones, one very special moving source—the automobile—deserves special attention. The automobile has many potential sources of pollution, but only a few important ones require control (see Figure 9.40):

- Evaporation of hydrocarbons (HC) from the fuel tank
- Evaporation of HC from the carburetor
- Emissions of unburned gasoline and partially oxidized HC from the crankcase
- NO_x, HC, and CO from the exhaust

The evaporative losses from gas tanks have been reduced by improving gas tank caps, which prevent the vapor from escaping. Losses from carburetors have been reduced by using activated carbon canisters, which store the vapors emitted when the engine is turned off. The hot gasoline in the carburetor vaporizes. When the car is restarted, the vapors can be purged by air and burned in the engine (see Figure 9.41).

The third source of pollution, the crankcase vent, is eliminated by closing off the vent to the atmosphere and recycling the blowby gases into the intake manifold. The positive crankcase ventilation (PCV) valve is a small check valve that prevents the buildup of pressure in the crankcase. The most difficult problem to control is the exhaust, which accounts for about 60% of the HC and almost all the NO_x and CO.

Air pollutants from mobile emissions are difficult to control completely due to the transient nature of parameters within the combustion chamber. Emissions from mobile combustion processes, such as the internal combustion engine (ICE), include carbon monoxide, nitrogen oxides, sulfur oxides (if there is sulfur in the fuel), and organic compounds that are not completely oxidized, often called ***products of incomplete combustion (PICs)***. Typical hydrocarbon, carbon monoxide, and nitrogen oxide emissions for an ICE at various air-to-fuel ratios are shown in Figure 9.42.

One immediate problem is how to measure these emissions. It is not as simple as sticking a sampler up the tailpipe because the quantity of pollutants emitted changes with the mode of operation. The effect of the operation on emissions is shown in Table 9.25. Note that, when the car is accelerating, the combustion is efficient (low CO and HC) and the high compression produces a lot of NO_x. In contrast, decelerating results in low NO_x and very high HC due to partially burned fuel.

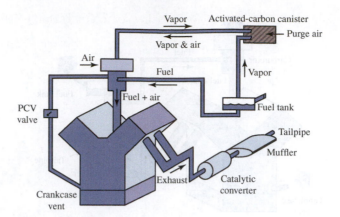

FIGURE 9.41 Internal combustion engine showing the air pollutant control devices associated with an automotive vehicle.

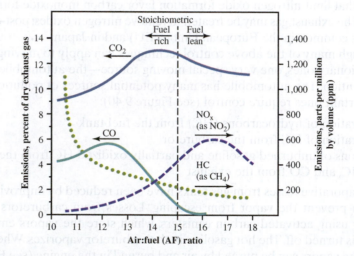

FIGURE 9.42 Carbon monoxide, hydrocarbon, and nitrogen oxide emissions as a function of air-to-fuel ratios.

Source: Watson, A.Y., Bates, R.R., and Kennedy, D. (eds.) (1988). *Air Pollution, the Automobile, and Public Health*, Fig. 2, p. 42. Washington, DC: National Academy Press. ISBN: 0-309-56826-9.

Because of these difficulties, the EPA has instituted a standard test for measuring emissions. This test procedure includes a cold start, acceleration, and cruising on a dynamometer to simulate a load on the wheels and a hot start.

Carbon monoxide and nitrogen oxide may be reduced by introducing oxygen in the molecular formula of the fuel rather than by adding excess air. Ethanol and MTBE have been added to fuels to decrease VOCs and air toxic emissions in the U.S. EPA standard for reformulated gasoline and gasoline standards used in the state of California. The oxygenated fuel standards are given in Table 9.25.

In addition to maintaining the air-to-fuel ratio between 14.8 and 14.9 and adding oxygen to the fuel, the control of emissions from mobile sources is also done through the catalytic conversion of air emissions. Typically, a three-way catalytic converter is used to perform both chemical oxidation and reduction in one unit, as shown in Figure 9.43. Carbon monoxide and PICs are oxidized to CO_2, and NO is

TABLE 9.25 Composition of reformulated gasoline for internal combustion engines in the United States and California

CONSTITUENT % IN THE FUEL	BASELINE GASOLINE	REFORMULATED FUEL	CALIFORNIA FORMULATION
Oxygen	0	≥ 2.0	≥ 2.0
Benzene	1.53	≤ 1.0	≤ 1.0
Lead		0.0	0.0
Aromatics	32	≤ 25	≤ 25
Olefin			≤ 6
Sulfur, ppm	339		≤ 40
NO_x		No increase	No increase
Estimated % reductions in emissions			
VOCs/ozone		25	25
Toxics		≥ 20	≥ 20

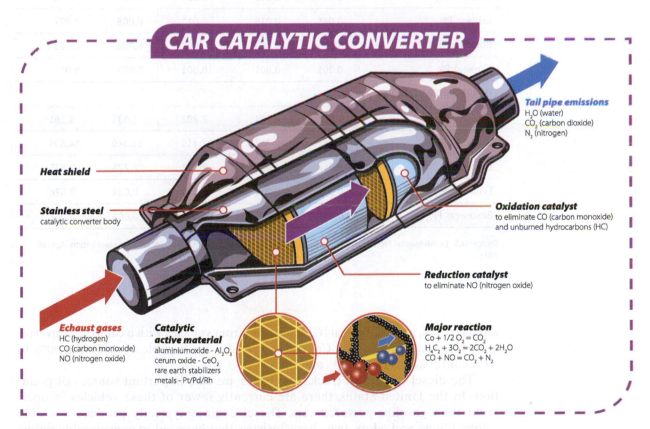

CAR CATALYTIC CONVERTER

Tail pipe emissions
H_2O (water)
CO_2 (carbon dioxide)
N_2 (nitrogen)

Heat shield

Stainless steel catalytic converter body

Oxidation catalyst to eliminate CO (carbon monoxide) and unburned hydrocarbons (HC)

Reduction catalyst to eliminate NO (nitrogen oxide)

Echaust gases
HC (hydrogen)
CO (carbon monoxide)
NO (nitrogen oxide)

Catalytic active material
aluminiumoxide - Al_2O_3
cerum oxide - CeO_2
rare earth stabilizers
metals - Pt/Pd/Rh

Major reaction
$Co + 1/2 O_2 = CO_2$
$H_4C_2 + 3O_2 = 2CO_2 + 2H_2O$
$CO + NO = CO_2 + N_2$

FIGURE 9.43 Typical configuration for a catalytic converter.

Source: EreborMountain/Shutterstock.com.

TABLE 9.26 Estimated U.S. National Average Vehicle Emissions Rates per Vehicle by Vehicle Type using Gasoline (Grams per mile)

	2000	2005	2010	2015	2020
LIGHT-DUTY VEHICLES					
Total HC	1.953	1.297	0.851	0.433	0.280
Exhaust CO	20.360	13.456	9.280	5.763	4.152
Exhaust NO_x	2.173	1.475	0.951	0.401	0.192
Exhaust $PM_{2.5}$	0.022	0.016	0.011	0.006	0.004
Brakewear $PM_{2.5}$	0.003	0.003	0.003	0.003	0.003
Tirewear $PM_{2.5}$	0.001	0.001	0.001	0.001	0.001
LIGHT-DUTY TRUCKS					
Total HC	2.454	1.297	0.804	0.574	0.339
Exhaust CO	30.622	17.119	11.088	8.663	5.422
Exhaust NO_x	3.304	2.002	1.333	0.817	0.376
Exhaust $PM_{2.5}$	0.028	0.016	0.011	0.008	0.007
Brakewear $PM_{2.5}$	0.003	0.003	0.003	0.003	0.003
Tirewear $PM_{2.5}$	0.001	0.001	0.001	0.001	0.001
HEAVY-DUTY VEHICLES					
Total HC	3.947	2.933	2.202	1.631	1.161
Exhaust CO	54.487	44.217	32.115	23.140	14.894
Exhaust NO_x	6.042	4.628	3.743	2.228	0.875
Exhaust $PM_{2.5}$	0.097	0.064	0.048	0.034	0.026
Brakewear $PM_{2.5}$	0.006	0.005	0.005	0.005	0.006

Source: U.S. Environmental Protection Agency, Office of Transportation and Air Quality, personal communication, Apr. 30, 2021.

reduced to N_2 and O_2. Typical ICE emissions from vehicles with a catalytic converter are presented in Table 9.26. Catalytic converters are capable of achieving 90% to 99% conversions of CO, NO_x, and PICs.

The diesel engines in trucks and buses are also important sources of pollution. In the United States, there are currently fewer of these vehicles in operation than gasoline-powered cars. Diesel engines typically produce a visible smoke plume and odors, two characteristics that have led to considerable public irritation with diesel-powered vehicles and concerns for public health. Diesel engines are typically more fuel efficient than gasoline-powered engines, however, and therefore emit less CO_2 per mile. Design improvements have resulted

TABLE 9.27 Estimated U.S. National Average Vehicle Emissions Rates per Vehicle by Vehicle Type using Diesel (Grams per mile)

	2000	2005	2010	2015	2020
LIGHT-DUTY VEHICLES					
Total HC	2.283	1.688	1.190	0.266	0.143
Exhaust CO	39.892	27.357	19.401	5.028	3.640
Exhaust NO_x	2.863	1.926	1.223	0.321	0.129
Exhaust $PM_{2.5}$	0.053	0.051	0.036	0.005	0.002
Brakewear $PM_{2.5}$	0.003	0.003	0.003	0.003	0.003
Tirewear $PM_{2.5}$	0.001	0.001	0.001	0.001	0.001
LIGHT-DUTY TRUCKS					
Total HC	1.417	1.087	0.954	0.688	0.308
Exhaust CO	18.362	10.287	7.361	5.795	2.458
Exhaust NO_x	6.282	5.939	4.930	3.633	1.804
Exhaust $PM_{2.5}$	0.326	0.322	0.280	0.184	0.078
Brakewear $PM_{2.5}$	0.003	0.003	0.003	0.003	0.003
Tirewear $PM_{2.5}$	0.002	0.002	0.002	0.002	0.002
HEAVY-DUTY VEHICLES					
Total HC	0.935	0.969	0.796	0.464	0.269
Exhaust CO	4.599	4.661	3.606	2.599	2.000
Exhaust NO_x	24.929	18.397	12.409	6.923	4.169
Exhaust $PM_{2.5}$	1.047	0.835	0.561	0.255	0.106
Brakewear $PM_{2.5}$	0.010	0.009	0.009	0.009	0.009

Source: U.S. Environmental Protection Agency, Office of Transportation and Air Quality, personal communication, Apr. 30, 2021.

in cleaner emissions, and U.S. EPA restrictions on the sulfur content of fuel will further improve emissions. Typical emissions from diesel engines are listed in Table 9.27.

The greatest advance in engine development, however, has been the complete redesign of engines to produce fewer emissions. For example, the geometric configuration of the cylinder in which the combustion occurs is important because complete combustion requires that all the gasoline ignite at the same time and burn as a steady flame during the very short time required. Cylinders that have nooks and crannies in which the air-gasoline mixture can hide will produce partially combusted emissions, such as hydrocarbons and carbon monoxide. In addition, fuel injection accurately measures the exact amount of gasoline needed by the engine, preventing excess fuel from entering the carburetor and subsequently emitting incompletely burned fuel through the exhaust.

Even with extensive investment in engineering, however, it will be difficult to manufacture a totally clean internal combustion engine. Electric cars are clean but can store only limited power; thus, their range is limited. In addition, the electricity used to power such vehicles must be generated in power plants, which create more pollution. Fuel cells are being explored to replace the internal combustion engine.

9.6 Air Pollution Control Technologies

This section provides a broad understanding of the control of toxic air emissions. The control technologies and physical chemical properties that make control systems effective are quite complex. However, each type of control technology is generally applicable to a wide variety of pollutants. Therefore, the technologies can be grouped by the characteristics of an exhaust air stream that they control effectively. In order to properly select and identify applicable control technology for a given source, one must first characterize the waste stream and regulated pollutants that are associated with the source.

Air pollution control devices are conveniently divided into those used for controlling particulates and those used for controlling gaseous pollutants. The reason, of course, is the difference in the size of the pollutants. Gas molecules have diameters of about 0.0001 μm; particulates range up from 0.1 μm.

Any air pollution control device can be treated as a black box separation device. The removal of the pollutant is calculated by using the principles of separation covered in Chapter 4. If a gas (such as air) contains an unwanted constituent (such as dust), the pollution control device is designed to remove the dust. The required *removal efficiency* of any separating device is expressed by

$$R_1 = \frac{x_1}{x_0} \times 100 \tag{9.50}$$

where

R_1 = removal efficiency (%)
x_1 = amount of pollutant collected by the treatment device per unit time (kg/s)
x_0 = amount of pollutant entering the device per unit time (kg/s)

Difficulties may arise in defining the x_0 term. Some dust, for example, may be composed of particles so small that the treatment device cannot be expected to remove this fraction. These very small particles should not, in all fairness, be included in x_0. Yet, it is common practice to simply let x_0 equal the contaminants to be removed, such as total particulates.

EXAMPLE 9.8 Determining the Required Removal Efficiency of Air Pollution Control Equipment

An air pollution control device is to remove a particulate that is being emitted at a concentration of 125,000 μg/m³ at an airflow rate of 180 m³/s. The device removes 0.48 metric ton per day. What are the emission concentration and the required removal efficiency of the air pollution control equipment?

We will start with the general mass balance equation. Consider the mass of the particulate matter entering and exiting the air pollution control system.

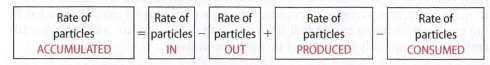

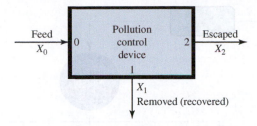

FIGURE 9.44 An air pollution control device as a black box for Example 9.8.

Since this is in steady state and no particulates are produced or consumed, the equation reduces to

Rate of particles IN	=	Rate of particles OUT

Figure 9.44 shows the black box. The particulates IN is the feed and is calculated as the flow rate times the concentration, which yields the mass flow rate:

$$(180 \text{ m}^3/\text{s})(125{,}000 \text{ } \mu\text{g/m}^3)(10^{-6} \text{ g}/\mu\text{g}) = 22.5 \text{ g/s}$$

The particulates OUT consists of the particles that escape and those that are collected. The latter is calculated as

$$\frac{(0.48 \text{ metric ton/day}) (10^6 \text{ g/metric ton})}{(86{,}400 \text{ s/day})} = 5.6 \text{ g/s}$$

The material balance yields

$$\text{Escaped particulates} = 16.9 \text{ g/s} \cong 17 \text{ g/s}$$

$$\text{Emission concentration} = \frac{(16.9 \text{ g/s}) (10^6 \text{ } \mu\text{g/g})}{(180 \text{ m}^3/\text{s})}$$

$$\cong 94{,}000 \text{ } \mu\text{g/m}^3$$

The percentage of particles required to be removed by the air pollution control system is

$$R_1 = \frac{5.6 \text{ g/s}}{22.5 \text{ g/s}} (100) = 25\%$$

Both mobile sources and stationary sources may require controls under the Clean Air Act. Mobile sources are most commonly regulated by the type of fuel required or the control systems sold as part of the vehicle, such as the catalytic control system on most automobiles as illustrated in Figure 9.45. Stationary sources may be required to have additional pollution abatement installed.

Stationary sources may be considered major or minor sources of a particular priority pollutant. Determining whether a source is major or minor depends in part on the mass emission rate of a particular pollutant and the geographic location of the source. The geographic location is important because sources are regulated more strictly if they are in a non-attainment region, which is defined as a region that does not meet the current goals for the NAAQS listed in Table 9.1. In addition to the

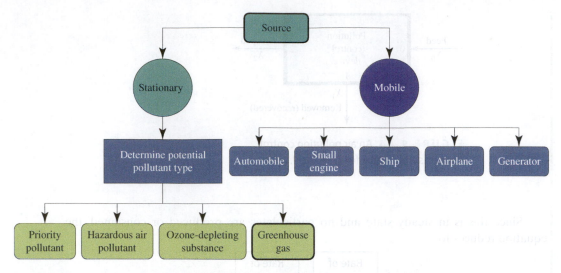

FIGURE 9.45 Criteria for determining stationary and mobile sources.

priority pollutants listed above, the U.S. EPA regulates 187 potentially hazardous air pollutants under the *National Emission Standards for Hazardous Air Pollutants* (NESHAP) rule. Any new or reconstructed facilities that emit 100 tons/year or more of a listed HAP or 25 tons/year or more of any combination of HAPs must comply with the control standards specified in the NESHAP as shown in Figures 9.46 and 9.47. In non-attainment areas where ambient pollution levels exceed the NAAQS, major and minor sources are determined by the specific type and amount of a given constituent emitted. Alternative sites must be considered for major sources of emissions. If no alternative site is chosen, the pollution control technology that can achieve the *lowest achievable emission rate* (LAER) must be implemented as illustrated in Figure 9.48.

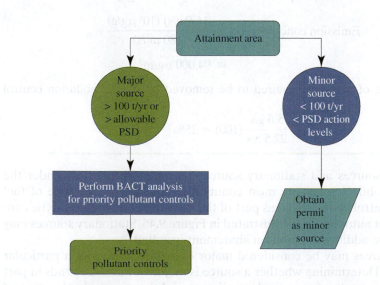

FIGURE 9.46 Emissions estimates and regional ambient air quality information are used to determine if the pollutants are regulated as major or minor sources under an attainment or non-attainment regulatory framework.

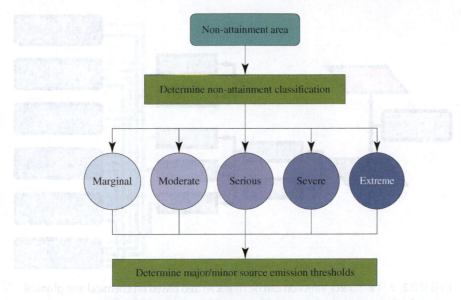

FIGURE 9.47 Criteria for determining classification of sources in NAAQS non-attainment areas.

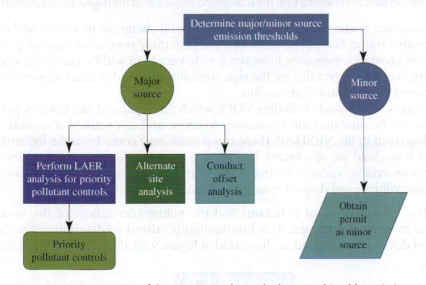

FIGURE 9.48 Determination of the criteria to achieve the lowest achievable emission rate (LAER).

The various control technologies can be related to the chemistry of the regulated compounds. These regulated compounds can be broadly grouped into three categories:

1. Inorganic compounds including sulfur oxides (SO_x), nitrogen oxides (NO_x), and carbon monoxide (CO), which are all priority pollutants. These compounds are most commonly associated with combustion sources such as boilers and incinerators. Inorganic compound controls are largely based on pollution avoidance within the combustion zone discussed in the previous section or additional acid gas scrubbers.
2. Particulate matter (PM) including lead particulates, which are also considered priority pollutants. Particulate matter can be generated from industrial

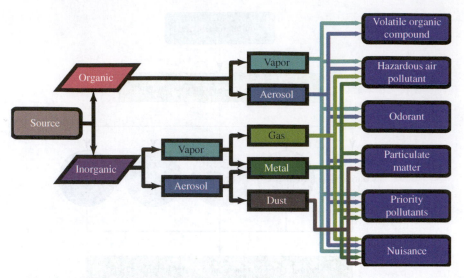

FIGURE 9.49 Source emission can be characterized based on chemical and physical properties; however, the regulations that govern the control of these emissions are not always directly correlated with the sources and properties of the regulated constituents.

processes, loading and unloading areas of solid chemicals or wastes, and automotive traffic. Particulate emissions are preferably controlled through pollution avoidance measures; however, a wide variety of add-on control devices are available depending on the size, concentration, and desired removal efficiency of the polluted air stream.

3. Organic compounds including VOCs, which are regulated like priority pollutants because they are a precursor to ozone and compounds of concern described in the NESHAP. These compounds may come from the industrial or household use of solvents, fuels, pesticides, and cleaners. A wide variety of commercially available control devices can be used based on the concentration, volume, and desired removal efficiency of the polluted air stream.

The concentration of pollutants and the volumetric airflow of the waste air stream must be determined. This information is critical to determine if pollutant control devices are required, as illustrated in Figure 9.50. If pollution control devices

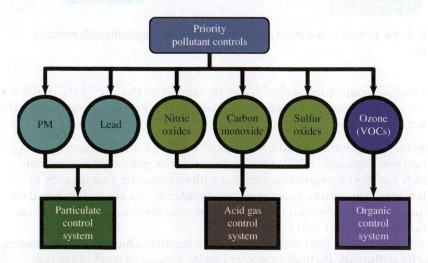

FIGURE 9.50 Characterization of air streams and source determination.

are required, then the characteristics of the waste air stream are used in selecting the appropriate control technology.

9.6.1 Control Devices for Particulate Matter

Particulate matter is generated from a wide variety of natural and anthropogenic sources. The choice of a particular type of control system is dependent on the particulate size, composition, and the desired removal efficiency. In general, most PM control devices fall into one of the following categories:

- Settling chambers
- Cyclonic collectors
- Cascade impactors
- Spray chambers and packed towers
- Venturi scrubbers
- Filtration systems, including baghouses
- Electrostatic precipitators

The following considerations and characteristics of the exhaust air stream are important in the choice of a PM control device:

- What is the size of the particulate matter to be controlled?
- What is the desired removal efficiency for the PM?
- Are the particles "tacky" or likely to adhere to a surface?
- What footprint size or area is required by the control system?
- For the control of very small particles (<100 μm), are they electrically conductive or electrically resistive?

The particle size for which the desired control efficiency must be achieved is the paramount factor in selecting particulate control systems (see Figure 9.51).

For large particles (>100 μm), gravitational or momentum-based control devices may be the most economical control strategies. These devices rely on the mass momentum of the particle overcoming the velocity streamlines of air through the control device to impact and collect PM on a surface. A *settling chamber* is the simplest device of this type and the easiest to visualize. Settling chambers use a simple horizontal flow chamber to reduce the air velocity within the chamber, causing the gravitational acceleration of the particles to overcome the horizontal acceleration due to

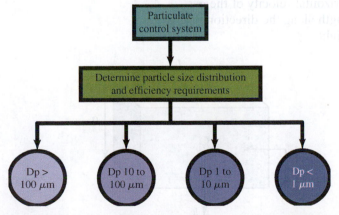

FIGURE 9.51 Determining the PM size as a parameter in selecting particulate control technologies.

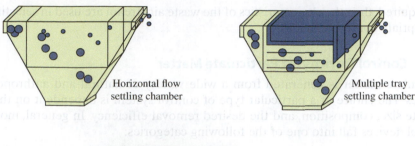
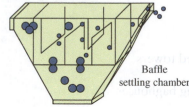

FIGURE 9.52 Particle settling chambers that remove large particles from exhaust air streams.

the moving air stream (Figure 9.52). When this occurs, the particles with large enough size and mass fall downward or settle in the bottom of the chamber. Settling chambers can also take advantage of momentum shifts due to sudden changes in the direction of the airflow, as illustrated in the baffle settling chamber and the multiple tray settling chamber. Settling chambers have a low capital cost and low associated pressure drop. Unfortunately, they can have low removal efficiencies and require a large footprint.

The size of the chamber and the settling velocity of the particle determine the efficiency of a settling chamber. The objective of the settling chamber is to achieve a large enough space to allow quiescent, low-velocity airflow. If the airflow rate is sufficiently low, then Stokes' settling law, Equation (7.46), may be applied to calculate the settling velocity of a particle in air. Notice that the density of the fluid and the fluid viscosity are the values used for air (not water as in earlier chapters). The removal efficiency for a particle of size i, E_i, of a horizontal flow settling chamber illustrated in Figures 9.52 and 9.53 is

$$E_i = \frac{v_t L}{v_x H} \tag{9.51}$$

where

v_t = vertical terminal settling velocity
v_x = horizontal velocity of the gas
L = length along the direction of flow
H = height

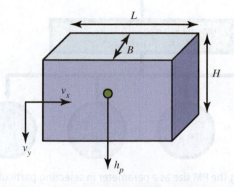

FIGURE 9.53 Simplified schematic of a single ($N_c = 1$) horizontal flow settling chamber.

Combining the theoretical efficiency equation and Stokes' settling law yields the following equation to estimate particle removal efficiency in a horizontal settling chamber:

$$E_i = f(d_p)^2 \left(\frac{g \times \rho_p \times B \times L \times N_c}{18\mu \times Q} \right) \qquad (9.52)$$

where

N_c = number of parallel chambers
f = empirical factor relating actual and theoretical conditions, typically 0.5
B = width of chamber

The overall collection efficiency, E_{tot}, is the sum of the individual particle size removal efficiencies multiplied by the weight fraction of the particles of size i, W_i:

$$E_{tot} = \sum E_i W_i \qquad (9.53)$$

In order to obtain higher removal efficiency and also reduce the size of control devices, cyclonic separators are used. Cyclonic devices are the most prevalent type of mechanical or momentum-based PM control devices. **Cyclonic separators** illustrated in Figure 9.54 remove particles by significantly accelerating the air stream and forcing the streamlines of the airflow to change direction. Particles larger than

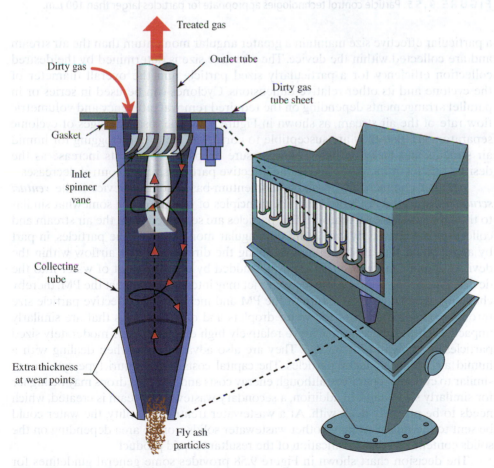

Treated gas
Dirty gas
Outlet tube
Dirty gas tube sheet
Gasket
Inlet spinner vane
Collecting tube
Extra thickness at wear points
Fly ash particles

FIGURE 9.54 Cyclonic dust separator.

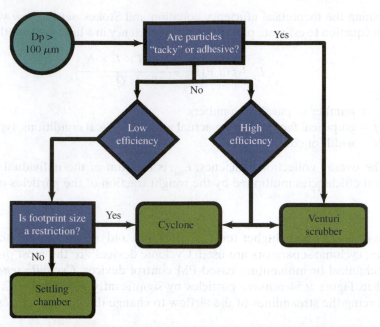

FIGURE 9.55 Particle control technologies appropriate for particles larger than 100 μm.

a particular effective size maintain a greater angular momentum than the air stream and are collected within the device. The effective size is determined by the desired collection efficiency for a particularly sized particle and the overall diameter of the cyclone and its other relative dimensions. Cyclones can be used in series or in parallel arrangements depending on the required removal efficiency and volumetric flow rate of the air stream, as shown in Figure 9.55. The disadvantages of cyclonic separators are that they are susceptible to corrosion, wear, and clogging for humid air streams and tacky particles. The pressure drops and fan costs increase as the desired efficiency increases and as the effective particle size for control decreases.

Another commercially available momentum-based control device is the *venturi scrubber* illustrated in Figure 9.56. The principles of operation are somewhat similar to those of a cyclonic separator in that particles are separated from the air stream and collection is achieved by increasing the angular momentum of the particles, in part by accelerating the air stream and changing the direction of the airflow within the device. Another mechanism of removal is added by injecting a jet of water into the device. Fine droplets created by the water jet may intercept or impact the PM, thereby changing the angular momentum of the PM and increasing its effective particle size through conglomeration of the water droplets and other particles that are similarly impacted. These devices can achieve relatively high efficiencies for moderately sized particles, as shown in Figure 9.57. They are also advantageous when dealing with a humid air stream or tacky particles. The capital costs of a venturi type system are similar to cyclonic separators, although energy costs and pressure drops may be higher for similarly sized units. In addition, a secondary wastewater stream is created, which needs to be properly dealt with. At a wastewater treatment facility, the water could be sent to the final clarifier or other wastewater solids storage area depending on the solids content and the classification of the resultant liquid product.

The decision chart shown in Figure 9.58 provides some general guidelines for determining the applicable control technologies depending on the characteristics of the exhaust air stream and the facilities control requirements.

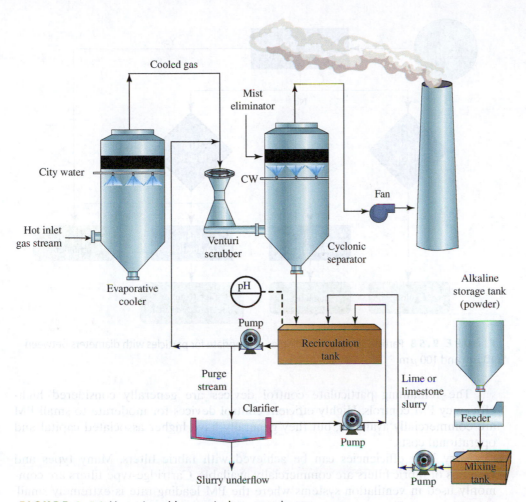

FIGURE 9.56 Venturi scrubber particulate control system.

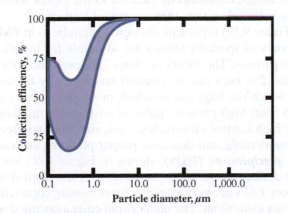

FIGURE 9.57 Typical efficiencies for several types of particulate wet scrubbers.

Source: EPA, www.epa.gov/apti/bces/module6/matter/control/control.htm#collect.

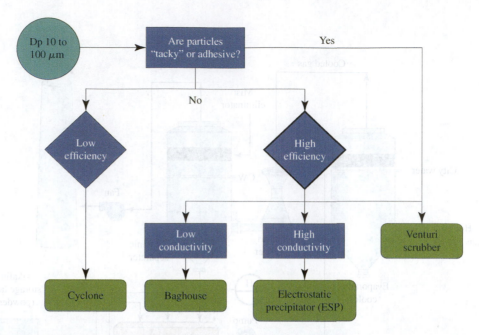

FIGURE 9.58 Particle control technologies appropriate for particles with diameters between 10 μm and 100 μm.

The remaining particulate control devices are generally considered high-efficiency PM controls. Highly efficient control devices for moderate to small PM are commercially available, but they generally have higher associated capital and operational costs.

Very high efficiencies can be achieved with fabric filters. Many types and varieties of fabric filters are commercially available. Cartridge-type filters are commonly used in ventilation systems where the PM loading rate is extremely small. Cartridge-type filters are defined by their efficiency, as shown in Figure 9.58. They generally take up little space, have low capital costs, and require little maintenance except for periodic replacement. Cartridge-type filters can be expensive, however, if PM loading rates are high, and they must be replaced frequently. In addition, disposal of the spent cartridges results in a solid waste problem. In order to deal with these shortcomings, continuously cleaned fabric filters were designed. When incorporated into a control system, they are called baghouses.

Baghouses (Figure 9.59) represent the newest standards in PM control technologies. A wide variety of specialty fabrics are available for installation in commercially available baghouses. The filters or "bags" are periodically cleaned following a regular schedule. The bags can be cleaned mechanically by shaking or rapping the frames onto which the bags are installed, or by periodically reversing the airflow or sending a short high-pressure pulse of air through the bags. The baghouses can achieve very high control efficiencies. Cost, size, and maintenance issues with baghouses are considerable, and therefore proper planning and care are required.

Electrostatic precipitators (ESPs), shown in Figure 9.60, are commonly used pollution control devices associated most often with the control of emissions from combustion sources. ESPs are very effective for removing conductive particles from large volumes of dry exhaust air. The high capital costs associated with ESPs may be offset by lower fan and maintenance costs for some exhaust air streams. For moderately sized particles (approximately 1 to 100 μm), a wide variety of control systems

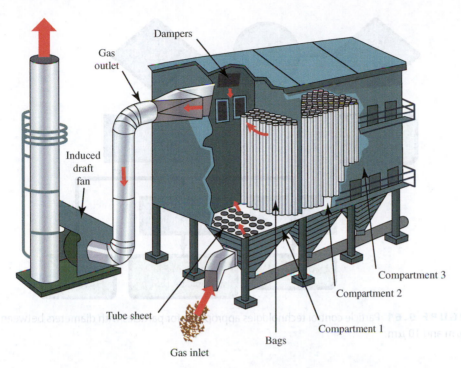

FIGURE 9.59 Reverse air baghouse particulate pollution control system.

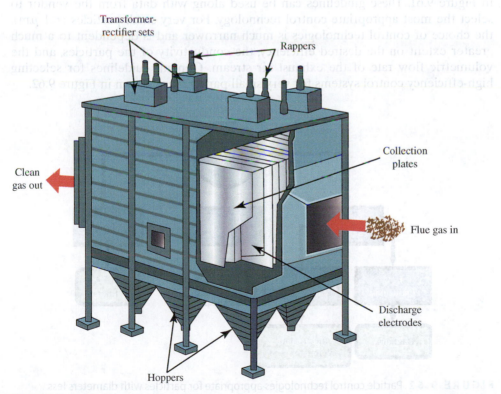

FIGURE 9.60 Electrostatic precipitator.

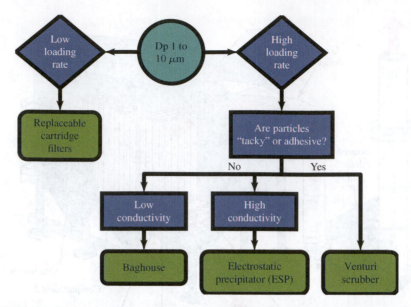

FIGURE 9.61 Particle control technologies appropriate for particles with diameters between 1 μm and 10 μm.

may be used, depending on the desired removal efficiency and other characteristics of the exhaust air stream.

General guidelines for narrowing down the types of control systems are shown in Figure 9.61. These guidelines can be used along with data from the vendor to select the most appropriate control technology. For very small particles (<1 μm), the choice of control technologies is much narrower and is dependent to a much greater extent on the desired efficiency, the conductivity of the particles, and the volumetric flow rate of the exhaust air stream. General guidelines for selecting high-efficiency control systems for very small particles are shown in Figure 9.62.

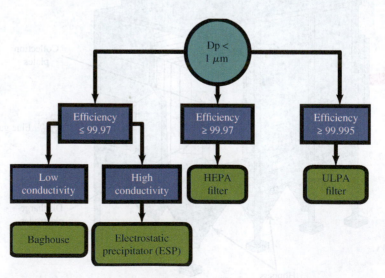

FIGURE 9.62 Particle control technologies appropriate for particles with diameters less than 1 μm.

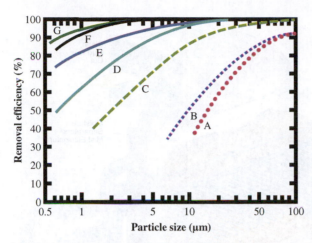

FIGURE 9.63 Comparison of approximate removal efficiencies. A = settling chamber, B = simple cyclone, C = high-efficiency cyclone, D = electrostatic precipitator, E = spray tower wet scrubber, F = venturi scrubber, G = bag filter.

Source: Based on Lappe, C.E. (1951). "Processes Use Many Collection Types." *Chemical Engineering* 58:145.

The efficiencies of the various control devices vary widely with the particle size of the pollutants. Figure 9.63 shows approximate collection efficiency curves, as a function of particle size for the various devices discussed.

9.6.2 Control Devices for Inorganic Air Toxics

Sulfur oxides, nitrogen oxides, and carbon monoxide are inorganic air pollutants generated by combustion processes. The combustion processes are designed to minimize the generation of all three types of compounds by regulating the temperature and the amounts of oxygen and fuel entering the combustion chamber. Carbon monoxide is controlled almost exclusively at the source and is therefore not commonly removed in air pollution control systems. A brief description of technologies appropriate for the control of major sources of SO_x or NO_x follows; they are identified in Figure 9.64.

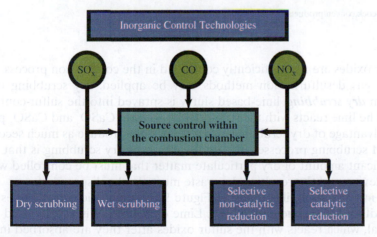

FIGURE 9.64 Hierarchy of inorganic control technologies.

Source: WEF (1995a). *Odor Control in Wastewater Treatment Plants.* Water Environmental Federation, Alexandria, VA.

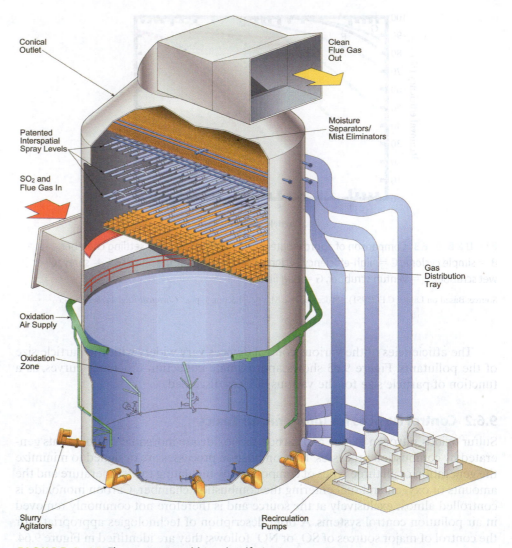

FIGURE 9.65 Flue gas wet scrubbing desulfurization process.

Source: www.babcock.com/en/products/wet-scrubbers-fgd.

If sulfur oxides are not sufficiently controlled in the combustion process, either of two flue gas desulfurization methods may be applied: dry scrubbing or wet scrubbing. In *dry scrubbing*, lime-based slurry is sprayed into the sulfur-containing air stream. The lime reacts with the sulfur oxides to form $CaSO_3$ and $CaSO_4$ precipitates. An advantage of dry scrubbing is that it does not produce as much secondary waste as wet scrubbing processes. The disadvantage of dry scrubbing is that it creates a significant amount of dry particulate matter that must be controlled with an additional device, and the resulting dry waste must also be disposed of.

During *wet scrubbing* illustrated in Figure 9.65, the acidic sulfur oxides come in contact with a basic aqueous solution. Lime or limestone is usually used as the base material, which reacts with the sulfur oxides after they are absorbed into the liquid. Gypsum ($CaSO_4 \cdot 2H_2O$) is formed and must be settled out from the aqueous solution as sludge and disposed of. This process is relatively safe due to the creation of liquid slurry instead of a dry, hot waste stream that could potentially create flammable conditions in the downstream control device. However, the costs of disposing of the resulting sludge are much higher than the disposal costs associated with the more concentrated dry solid waste formed during the dry scrubbing process.

Thermal reduction processes use ammonia-based compounds to reduce NO_x to N_2. There are two main types of thermal reduction processes: selective catalytic reduction (SCR) processes and selective non-catalytic reduction (SNR) processes.

Selective catalytic reduction processes use noble metals, base metals, or a zeolite as the catalytic surface. These catalysts are arranged in various geometric configurations to provide adequate surface area and contact time with the contaminated air stream. A large volume of catalyst is required, and the air stream must be operated within a narrow temperature band for effective NO_x removal and to prevent degradation of the catalyst. In addition, these catalysts are subject to fouling or "poisoning" by particulate matter and sulfur oxides. SCR devices are most often used for treating relatively clean exhaust gases with chemicals that quickly degrade in contact with the catalysts and that do not contain PM or sulfuric compounds.

Temperature-selective non-catalytic reduction of NO_x is used for air streams that contain PM or SO_x. During this process, ammonia-based compounds are injected into the gas stream to reduce the NO_x. This process is very temperature sensitive; the air temperature must remain above 1,070 K for adequate reactions to occur.

9.6.3 Control Devices for Organic Air Pollutants

Organic pollutants originate from a wide variety of potential sources, including solvents, cleaners, fuels, pesticides, and naturally occurring degradation products. Organic vapors may be regulated as VOCs, HAPs, or ozone-depleting substances (ODSs). The control technologies for organic emissions generally fall into one or a combination of the following types:

- Thermal oxidation
- Condensation
- Adsorption
- Absorption
- Advanced oxidation
- Biofiltration

It is extremely important to properly characterize the exhaust air stream and the requirements or expectations of the control technology. The following questions should be considered when choosing an organic pollution control technology:

- What is the total concentration of organics in the air stream?
- What is the concentration of each chemical constituent in the air stream?
- Do the constituents of the air stream have a low boiling point or a high boiling point?
- Do the constituents contain chlorinated compounds or other potential acidic formation products?
- Is the operation continuous or intermittent?
- Are the constituents water soluble?
- Are the constituents biodegradable?

The first and perhaps the most important step in choosing an organic pollutant control technology is to determine the total concentration of pollutants in the exhaust air stream. Figure 9.66 shows five general classifications based on the

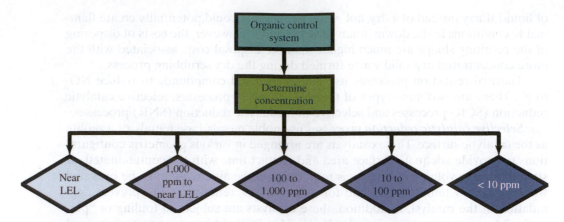

FIGURE 9.66 The concentration of organic emissions in the exhaust air stream is the most important consideration in choosing an appropriate air pollution control technology.

concentration of organics in the exhaust air. Most industrial applications of air pollution treatment involve the control of very large volumetric airflow rates with low or very low concentrations of pollutants (< 100 ppm).

The ***lower explosive limit (LEL)*** is the concentration of an organic constituent in air that can sustain combustion without any additional fuel once a spark is present. If organic emissions are present in the air at concentrations near the LEL, the organic material may be recoverable through condensation or a sorption process and recycled, as illustrated in Figures 9.67 and 9.68. If the recycled value of the material is low and the exhaust temperature of the exhaust air stream is high, then the air stream may be used as a supplemental fuel and heat source for the facility.

The task of selecting a cost-effective organic control device for low-concentration/high-volume air streams is a difficult one. The combination of very low concentrations and high flow rates makes most systems somewhat large. Therefore, the available area for a control system may be an important factor in choosing the technology that best suits a facility's needs. The costs and benefits of various air

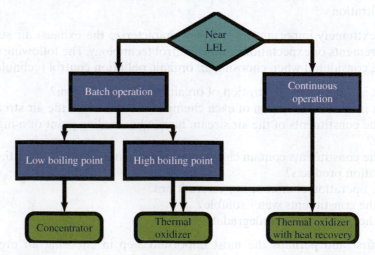

FIGURE 9.67 Organic air emission control technologies appropriate for exhaust streams with organic concentrations near the constituent's LEL.

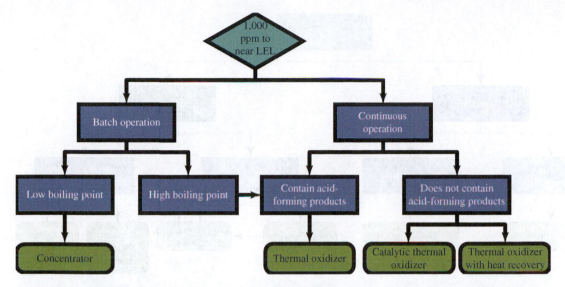

FIGURE 9.68 Organic air emission control technologies appropriate for exhaust streams with organic concentrations between 1,000 ppm$_v$ and the constituent's LEL.

pollution control processes in mid-range (100 to 1,000 ppm$_v$) concentration air streams are largely dependent on whether the industrial process is operated intermittently as a batch operation or as a continuous operation, as shown in Figures 9.69 and 9.70.

Absorption systems involve the selective transfer of the pollutant from a gas to a contacting liquid. The pollutant of concern must have preferential solubility in the liquid. The soluble pollutants diffuse from the gas through a gas-liquid interface via molecular diffusion or turbulent (eddy) diffusion, and the pollutants are

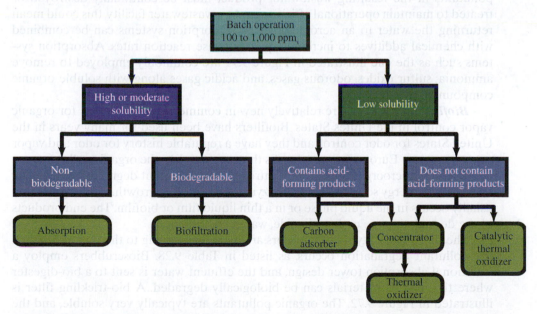

FIGURE 9.69 Organic air emission control technologies appropriate for batch process exhaust streams with organic concentrations between 100 ppm$_v$ and 1,000 ppm$_v$.

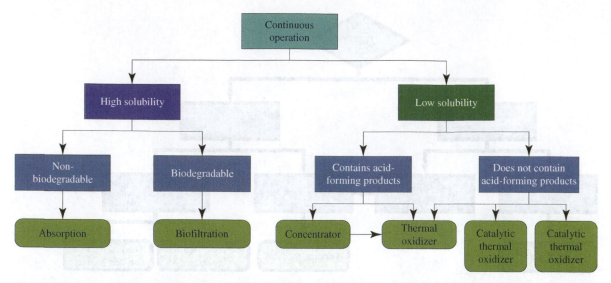

FIGURE 9.70 Organic air emission control technologies appropriate for continuous process exhaust streams with organic concentrations between 100 ppm$_v$ and 1,000 ppm$_v$.

subsequently dispersed in the liquid. Absorption systems may be applicable when pollutant concentrations are higher than 10 ppm for very water-soluble compounds. The efficiency of absorption control systems normally ranges from 80% to greater than 98%.

Absorption systems are ideal for controlling water-soluble organic pollutants such as alcohols and organic acids. They do not use fossil fuels and thus do not produce carbon dioxide or NO$_x$ emissions as byproducts. Exhaust temperatures should be lower than 100°C in order to prevent significant evaporative losses. The soluble pollutants in the resulting wastewater product must be continually destroyed or treated to maintain operational efficiencies. At a wastewater facility, this could mean returning the water to an aerobic digester. Absorption systems can be combined with chemical additives to increase aqueous phase reaction rates. Absorption systems such as the one illustrated in Figure 9.71 are commonly employed to remove ammonia, sulfur oxides, odorous gases, and acidic gases along with soluble organic compounds.

Biofiltration systems are relatively new in commercial applications for organic vapor control in the Unites States. Biofilters have been used for many years in the United States for odor control, and they have a reputable history for odor and vapor control in many European countries. As the name implies, the organic pollutants are degraded by microorganisms in the control system. Pollutant degradation normally serves as an energy source or added enzyme for microbial growth. This degradation usually occurs in the liquid phase or in a thin liquid film or biofilm. The end products of the degradation are carbon dioxide, water, and biomass.

The three typical types of biofilters are named according to the region in which the pollutant degradation occurs, as listed in Table 9.28. Bioscrubbers employ a traditional absorption tower design, and the effluent water is sent to a bio-digester where the organic materials can be biologically degraded. A bio-trickling filter is illustrated in Figure 9.72. The organic pollutants are typically very soluble, and the majority of the degradation occurs within the filter media. Because bio-trickling filters have the ability to "store" the contaminants as a food source, they are often

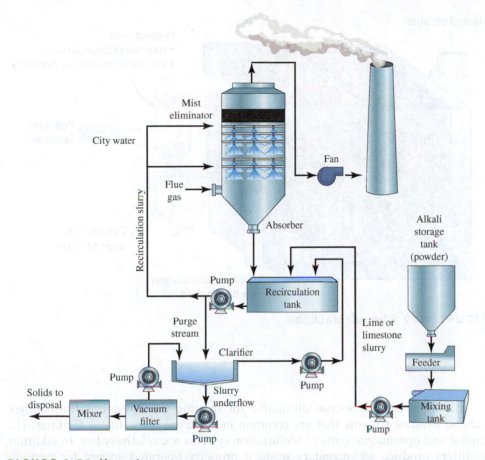

FIGURE 9.71 Absorption system.

TABLE 9.28 Types of biofilters

CONTROL TYPE	MICROORGANISM LOCATION	WATER ADDITION
Biofilter	Fixed in biofilm	Stationary liquid phase; water added intermittently
Bio-trickling filter	Fixed in biofilm	Continuous flowing liquid phase; water added continuously
Bioscrubber	Suspended in digester	Continuous flowing liquid phase; water added continuously

applied in situations that have variable influent loading rates. Biofilters may be open to the atmosphere or completely enclosed. Closed systems are more expensive but are more reliable and less likely to be affected by ambient temperature fluctuations and oversaturation due to precipitation. The degradation of the organic pollutants occurs within the biofilm layer in the biofilter, and water and nutrients (nitrogen and phosphorus) are added periodically to maintain an environment that is conducive to microbial growth.

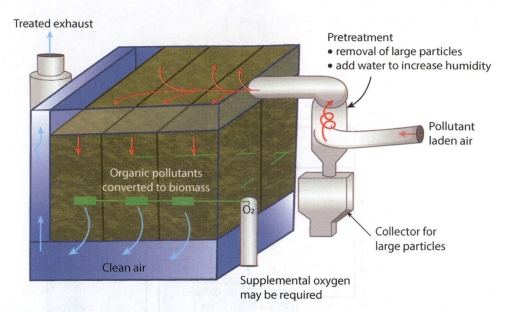

FIGURE 9.72 Schematic of a biofilter.

Biofilters are a low-cost alternative for treating the low-concentration/high-volume exhaust streams that are common in wastewater treatment facilities. The capital and operational costs of biofiltration systems are relatively low. In addition, biofilters produce no secondary waste if properly operated, except for periodic media replacement. Biofilters do not consume fossil fuels to destroy pollutants and therefore do not produce secondary waste products associated with combustion. Biofilters are limited in their widespread application due to the variable removal efficiencies of current systems caused in part by their intolerance to fluctuations in airflow rates, concentrations, and temperatures. Like absorption systems, which depend on water, biofilters are similarly limited to exhaust streams with temperatures between 10°C and 90°C. In addition, biofilters typically require a very large area compared to other types of organic control technologies. Compounds that are particularly well suited for biodegradation are listed in Table 9.29.

TABLE 9.29 Pollutants and their potential for control by biofiltration

WELL SUITED FOR BIOFILTRATION	MODERATELY SUITED FOR BIOFILTRATION	POORLY SUITED FOR BIOFILTRATION
Alcohols	Aromatic compounds	Aliphatic hydrocarbons
Aldehydes	Sulfur-containing compounds	Ethers
Ammonia		Halogenated compounds
Amines		Nitriles
Carbonic acids		Nitrogen oxides
Hydrogen sulfide		
Ketones		

Carbon adsorption systems utilize the attractive van der Waals forces on porous granulated activated carbon (GAC) surfaces to capture and contain organic pollutants. Materials such as GAC to which pollutants adhere and are removed from passing air streams are called adsorbents. The pollutants that are removed by the GAC are called the adsorbate. GAC adsorbent systems are commonly applied in a wide variety of organic control situations due to their simplicity of design, low capital costs, and minimal maintenance requirements. If the systems are properly monitored, removal efficiencies typically range from 90% to greater than 98%. Carbon adsorption systems are ideal for small applications constrained by space and low capital investment.

The activated carbon acts as a capture and control device and does not destroy the organic pollutants. The organic compounds adhering to the surface of the GAC eventually consume all available surface sites in a given layer. This creates three zones within the carbon bed. The mass transfer zone (MTZ) where pollutants are being removed gradually moves from the inlet side of the carbon bed to the outlet side. As time passes and more pollutants are adsorbed, the amount of exhausted or spent carbon increases. Breakthrough and reduced removal efficiency occur when the MTZ reaches the exit of the carbon bed and there is no longer an excess of active carbon sites available. When the carbon adsorption bed efficiency decreases, as the bed nears saturation, the carbon must be replaced or regenerated. Passing steam through the bed, which causes the pollutants to desorb from the surface, can regenerate the carbon. The pollutants are then normally destroyed by thermal oxidation subsequent to regeneration of the carbon. The vendor of the material most commonly carries out the regeneration phase off site. The operational costs of replacement and regeneration of the carbon may be considerable.

Monitoring the carbon adsorption system is critical to prevent and detect breakthrough of the pollutants as the bed nears the end of its useful life span. Both the inlet and outlet concentrations should be measured to determine pollutant removal and reclamation efficiency. Regular monitoring of the system allows accurate measurement of cycle times, and service contracts should be set up to minimize episodes of poor efficiency.

Thermal oxidation systems reliably maintain removal efficiencies of 98% to greater than 99.99%. The temperature requirements for destruction range from typical design temperatures of 870°C (1,600°F) to 980°C for most organic compounds and up to 1,200°C (1,800° to 2,200°F) for halogenated and extremely toxic pollutants. Thermal oxidation systems destroy organic compounds and can maintain high destruction efficiencies even with wide fluctuations in concentration. However, thermal oxidation systems do not tolerate wide flow rate fluctuations well. Thermal oxidation systems also consume large quantities of fossil fuels and, as a result, they are sources of nitrogen oxides, carbon dioxide, and possibly acid gases for halogen- and sulfur-containing waste streams. Furthermore, thermal oxidation systems are typically run 24 hours a day due to the long start-up and shut down times required for operation. A thermal oxidation system is shown schematically in Figure 9.73.

The volumetric flow rates of the fuel and air streams along with the size of the reaction chamber are important factors in ensuring adequate retention time and destruction efficiency. For low-concentration air streams, the burner airflow rate or additional make-up air is not necessary if there is sufficient oxygen in the polluted air stream to maintain combustion. A mass balance and enthalpy balance should be performed to estimate the required fuel flow rate to maintain

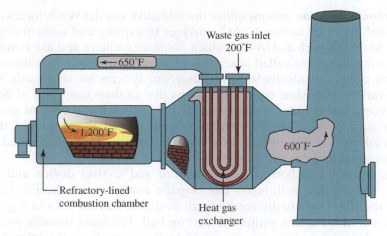

FIGURE 9.73 Schematic of a thermal oxidizer with a recuperative heat exchanger.

the desired operational temperature (see Figure 9.74). The mass and enthalpy balances should consider the following:

F_i, the flow rate of the polluted gas stream

F_f, the flow rate of the fuel (typically methane or propane)

F_a, the burner airflow rate

F_e, the exhaust flow rate

Fuel costs make up the majority of operational costs and must be considered in selecting the most cost-effective treatment systems. Adequate design and operation of thermal oxidation systems are dependent on the temperature, residence time of the gas, and turbulence or mixing within the reaction chamber. These variables are dependent on one another. The kinetic rate constants increase exponentially with temperature. Reaction times on the order of 0.1 to 0.5 second are usually sufficient to allow the reactants to reach the desired degree of chemical destruction. Turbulence within the reaction chamber ensures sufficient mixing. Therefore, a higher reaction temperature results in a shorter residence time, a smaller combustion chamber, and lower capital costs. A longer residence time lowers the operating temperature, resulting in less fuel usage and higher capital costs. The ratio of operational costs to capital investment should be considered during the selection and design of thermal oxidation systems. Factors that should be considered when determining the capital costs include the construction materials, instrumentation, costs of heat exchangers, engineering fees, and construction fees.

Catalytic oxidation systems use thermally activated catalysts to oxidize and destroy organic pollutants but at a lower temperature than traditional thermal

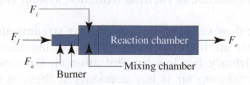

FIGURE 9.74 Schematic of a mass balance approach to the design of a thermal oxidation system.

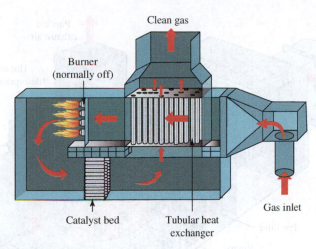

Clean gas

Burner
(normally off)

Catalyst bed

Tubular heat
exchanger

Gas inlet

FIGURE 9.75 Schematic of a catalytic oxidizer.

oxidation systems (see Figure 9.75). The catalyst is a substance that increases the speed of the chemical reactions in any chemical change to the catalytic material itself. The reaction chamber of a catalytic oxidizer is filled with a structured medium on which the catalyst is deposited. The catalytic material usually consists of a noble metal coated onto a ceramic support material that has a very high surface area-to-volume ratio. As a result of contact with the catalyst, temperatures of 320°C to 540°C (600°F to 1,000°F) are sufficient to destroy the organic pollutants.

The lower operating temperature greatly reduces the fuel requirements for the control systems, which in turn significantly reduce operating costs. In addition to fuel savings, catalytic systems require less insulating material and reduce the fire hazards that may be associated with higher temperature oxidation. Destruction efficiencies for catalytic control devices may be somewhat lower than for traditional thermal oxidation process, but they are still extremely good—typically ranging from 95% to greater than 99%. Fuel savings for catalytic systems are somewhat offset by their higher initial costs and the costs due to periodic catalyst replacement. Catalysts that are consumed through thermal failure or poisoning cannot be regenerated. The spent catalysts must be disposed of and replaced with a new catalytic medium. The catalyst can also be inactivated or poisoned by some contaminants and is therefore not suited for all waste streams. Polluted air streams that contain high levels of particulates and acid gases will poison the catalyst and should be dealt with using an alternative control strategy.

Hybrid control systems and concentrators combine one or more of the functional advantages of one technology synergistically with a second technology. The most common hybrid or combined system uses a carbon- or zeolite-based adsorption technology to collect organic pollutants on a surface. While one section of the adsorbent is removing pollutants from the exhaust air stream, another section is being regenerated by heat or steam as illustrated in Figure 9.76. The resultant air stream flowing through the regenerative section typically has a higher pollutant concentration and a lower total volumetric flow rate. Thus, this process is often referred to as a concentration step and the adsorption unit as a "concentrator." A destructive technology such as biofiltration or some form of thermal oxidation is then used to destroy the pollutants in the air passing through the regenerative section of the concentrator.

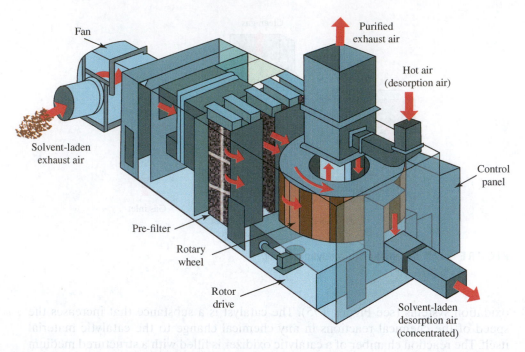

FIGURE 9.76 Pre-concentrator type adsorption system.

The overall removal efficiency is typically limited by the removal efficiency of the concentrator. Removal efficiencies may range from 85% to 95%. Concentrators may be greatly advantageous when an exhaust stream has a variable flow rate, has a variable concentration, or is intermittent in nature. The benefits of using a concentrator with the various organic control technologies are described in Table 9.30.

Besides the lower removal efficiency, the other major drawback to using a concentrator is the higher initial capital investment required. Maintenance issues associated with the airflow systems, pumps, fans, valves, and monitoring equipment are also more complex for combined systems.

A facility's requirements for a pollution control technology are dependent on four major factors:

Concentration of the pollutants in the air stream
Flow rate of the exhaust air stream controlled
Chemistry of the polluted air stream
Costs associated with the control technologies, both capital and operational costs

TABLE 9.30 Benefits of using a concentrator

TECHNOLOGY	BENEFITS
Biofiltration	Reduces fluctuations in concentration; reduces size and footprint
Carbon adsorption	Regenerates carbon onsite or "in situ"; reduces carbon replacement costs
Thermal oxidation	Provides higher concentration and higher temperature source; reduces fuel requirements
Catalytic oxidation	Reduces size and cost of catalyst bed; provides higher temperature source; reduces fuel consumption

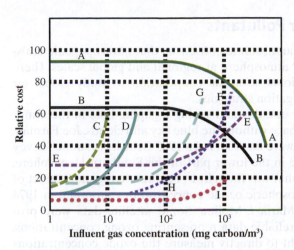

A. Direct combustion with heat recovery
B. Catalytic combustion with heat recovery
C. Non-regenerative adsorption
D. Regenerative adsorption
E. Adsorptive concentration with direct combustion of the concentrated gas
F. Water scrubbing
G. Chemical scrubbing
H. Bioscrubbing
I. Biofiltration

FIGURE 9.77 Cost effectiveness of various air pollution control technologies as a function of the organic carbon concentration in the exhaust gas.

Source: Chou, M.S., and Lu, S.L. (1998). "Treatment for 1,3-Butadiene in an Air Stream by Biotrickling Filter and a Biofilter." *Journal of Air and Waste Management Association* 48(8):711–720, Fig. 1, p. 711.

Each application by its very nature is unique and requires an original approach to control. However, guidelines for determining which technology to consider are useful for narrowing the list of vendors. The characteristics of the air stream need to be identified by answering the questions discussed in this chapter. If the air stream is properly characterized, adequate control technologies, which are commercially available, can be found.

The variations of treatment costs with different control technologies are illustrated in Figure 9.77. The graph shows that thermal oxidation-based technologies are generally most cost effective for organic concentrations on the order of or higher than 1,000 mg of carbon molecules per cubic meter of air (mg C/m³). A wide variety of pollution control processes may be applicable depending on the type of process and the characteristics of the organic waste for exhaust streams with organic waste concentrations between 10 and 1,000 mg C/m³. Absorptive and biological air pollution control technologies are very cost effective when organic waste concentrations in the exhaust are lower than 10 mg C/m³.

The selection of particulate control equipment requires that cost and effectiveness be compared against the physical and chemical composition of the gas stream. The selection of organic control equipment must match available resources with the proper system for controlling the major constituents of the polluted gas stream. In some cases, more than one technology may meet all the requirements of the facility. In other situations, no single technology in and of itself may be ideal. Control technologies should be selected and designed in coordination with regulatory authorities. In some cases, a regulatory authority may define control requirements, particularly when regulations call for the Maximum Available Control Technology (MACT). Experience and judgment are inherent and necessary parts of designing an effective air pollution control system.

9.7 Global Impacts of Air Pollutants

Scientific evidence has developed an absolute consensus that anthropogenic emissions have significantly altered our atmosphere at regional and global scales. There has been unprecedented international scientific cooperation to study global air pollutant effects and potential mitigation strategies.

Imagine looking up at the sky one day and seeing a gaping hole in the atmosphere that allows you to see into space without the blue sky and clouds. Joe Farman, Brian Gardiner, and John Shankin were members of a 1985 British Antarctic Survey (BAS) that studied the atmosphere in the lower part of the Southern Hemisphere. These scientists were looking for what was thought to be a relatively small effect of chlorinated chemicals on the stratospheric ozone concentrations proposed in 1974 by Frank Sherwood Roland and Maria J. Molina. Some satellite data were producing what was thought to be unreliable data representing ozone concentrations. The British Antarctic Survey sought to directly measure the ozone concentrations in the atmosphere over Antarctica. The British research team "looked" into the sky with their equipment to measure ozone, and they could not believe what they "saw." Their first presumption was that their instrumentation was not working correctly. Only after rechecking their instruments and data did they trust their own observations. The direct ozone measurements and satellite data showed a 40% decrease in ozone concentrations over Antarctica by 1984. By 1999, ozone concentrations in the Antarctic had decreased by 57% in what had become known as the "Antarctic ozone hole," illustrated by the National Aeronautics and Space Agency (NASA) for the year 2000 in Figure 9.78. Drs. Rowland and Molina received a Nobel Prize for their work in atmospheric chemistry that explained how parts-per-billion concentrations of chlorine in atmosphere could react to destroy ozone.

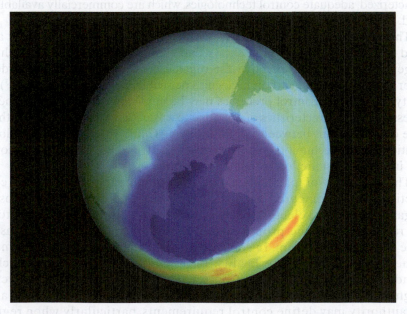

FIGURE 9.78 NASA visualization of the Antarctic ozone hole from satellite data collected on September 10, 2000.

Source: NASA/Goddard Space Flight Center, Scientific Visualization Studio.

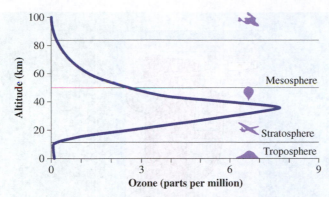

FIGURE 9.79 The concentration of ozone varies with altitude. Peak concentrations, an average of 8 molecules of ozone per million molecules in the atmosphere, occur between an altitude of 30 and 35 kilometers.

Source: NASA, ozonewatch.gsfc.nasa.gov/facts/SH.html.

Approximately 90% of the ozone (O_3) in the Earth's atmosphere is found in the stratosphere (see Figure 9.79). The ozone in the atmosphere absorbs nearly all of the UVc ultraviolet radiation from the sun, most of the UVb irradiation, and about 50% of the UVa irradiation (Figure 9.80).

The "hole" in the stratospheric ozone layer has been mainly attributed to the widespread use and emissions of chlorofluorocarbons (some of which are illustrated in Figure 9.81). *Chlorofluorocarbons (CFCs)* were invented in the early 1930s and used for a wide variety of commercial and industrial applications, including aerosol spray cans, coolants, refrigerants, and fire retardants (Figure 9.81). The simplest CFCs contain one or two carbon atoms and chlorine or fluorine, such as CFC-12 (CF_2Cl_2). The CFCs are relatively inert, insoluble, and transparent, and they do not degrade with time. They persist in the atmosphere until they migrate to the upper atmosphere, where cold Antarctic stratospheric clouds convert inert chlorine (Cl) molecules to reactive chlorine forms that react with ozone to form diatomic oxygen (O_2).

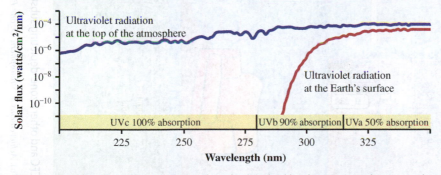

FIGURE 9.80 Solar ultraviolet radiation is largely absorbed by the ozone in the atmosphere—especially the harmful, high-energy UVa and UVb. The graph shows the flux (amount of energy flowing through an area) of solar ultraviolet radiation at the top of the atmosphere (top line) and at the Earth's surface (lower line). The flux is shown on a logarithmic scale, so each tick mark on the y-axis indicates 10 times more energy.

Source: NASA, ozonewatch.gsfc.nasa.gov/facts/SH.html.

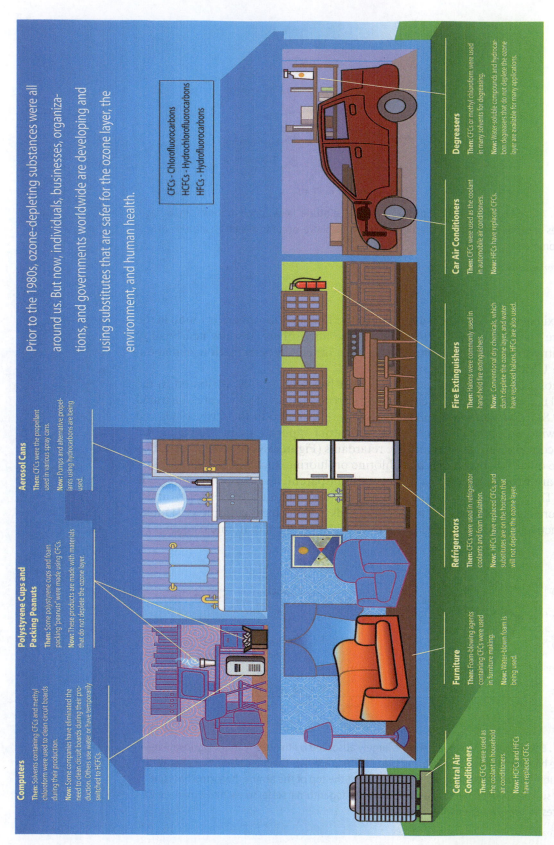

Prior to the 1980s, ozone-depleting substances were all around us. But now, individuals, businesses, organizations, and governments worldwide are developing and using substitutes that are safer for the ozone layer, the environment, and human health.

CFCs - Chlorofluorocarbons
HCFCs - Hydrochlorofluorocarbons
HFGs - Hydrofluorocarbons

Computers
Then: Solvents containing CFCs and methyl chloroform were used to clean circuit boards during their production.
Now: Some companies have eliminated the need to clean circuit boards during their production. Others use water or have temporarily switched to HCFGs.

Polystyrene Cups and Packing Peanuts
Then: Some polystyrene cups and foam packing 'peanuts' were made using CFCs.
Now: These products are made with materials that do not deplete the ozone layer.

Aerosol Cans
Then: CFCs were the propellant used in various spray cans.
Now: Pumps and alternative propellants using hydrocarbons are being used.

Central Air Conditioners
Then: CFCs were used as the coolant in household air conditioners.
Now: HCFCs and HFGs have replaced CFCs.

Furniture
Then: Foam-blowing agents containing CFCs were used in furniture making.
Now: Water-blown foam is being used.

Refrigerators
Then: CFCs were used in refrigerator coolants and foam insulation.
Now: HFGs have replaced CFCs, and substitutes are on the horizon that will not deplete the ozone layer.

Fire Extinguishers
Then: Halons were commonly used in hand-held fire extinguishers.
Now: Conventional dry chemicals, which don't deplete the ozone layer, and water have replaced halons. HFGs are also used.

Car Air Conditioners
Then: CFCs were used as the coolant in automobile air conditioners.
Now: HFGs have replaced CFCs.

Degreasers
Then: CFCs or methyl chloroform were used in many solvents for degreasing.
Now: Water-soluble compounds and hydrocarbon degreasers that do not deplete the ozone layer are available for many applications.

FIGURE 9.81 CFC and other ozone-depleting substances used in the home prior to ratification of the 1987 Montreal Protocol.

Source: U.S. EPA (2007b). *Achievements in Stratospheric Ozone Protection: Progress Report.* Office of Air and Radiation. U.S. Environmental Protection Agency. Washington, DC.

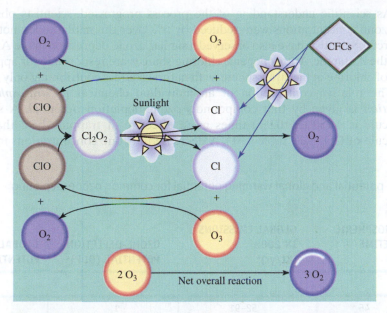

FIGURE 9.82 Degradation reactions of ozone, catalyzed by chlorine in the atmosphere.

Source: Based on Stolarski, R.S. (1985). *A Century of Nature: Twenty-One Discoveries That Changed Science and the World*. Chapter 18: A hole in the Earth's shield. Chicago: University of Chicago Press.

The production of reactive chlorine requires sunlight (Figure 9.82), which initiates the following reactions:

$$2Cl + 2O_3 \rightarrow 2ClO + 2O_2 \tag{9.54}$$

$$ClO + ClO + M \rightarrow Cl_2O_2 + M \tag{9.55}$$

$$Cl_2O_2 \xrightarrow{uv} Cl + ClO_2 \tag{9.56}$$

$$ClO_2 + M \rightarrow Cl + O_2 + M \tag{9.57}$$

The net reaction for chlorine in the stratosphere is

$$2O_3 \xrightarrow{Cl, uv, \& M} 3O_2 \tag{9.58}$$

Bromine, another active stratospheric halogen, goes through a similar process:

$$ClO + CBrO \rightarrow Br + Cl + O_2 \tag{9.59}$$

$$Cl + O_3 \rightarrow ClO + O_2 \tag{9.60}$$

$$Br + O_3 \rightarrow BrO + O_2 \tag{9.61}$$

Chlorine monoxide (ClO) forms when the South Pole begins to receive solar radiation in the spring. Chlorine monoxide (ClO) atoms combine to form a dimer (Cl_2O_2) that reacts with the sunlight to release diatomic oxygen and form more reactive chlorine compounds that begin the chain reaction again. It has been estimated that one chlorine molecule may destroy over 100,000 ozone molecules before the chlorine is removed from the atmosphere. The chlorine atom acts as a catalyst in the chain reaction proposed by Rowland and Molina that ultimately converts ozone into the diatomic oxygen (O_2), which absorbs much less UV irradiation.

As scientific understanding of the process progressed, scientists observed that the ozone concentrations were reduced by 5% over the midlatitudes, not just in the Antarctic. Scientists have also observed a similar ozone depression in the Arctic as well as in the Antarctic. Anthropogenic chlorine sources are responsible for approximately 84% of chlorine in the stratosphere, with natural sources contributing only about 16%.

The relative contribution of an individual chemical to **stratospheric ozone depletion** is given by that compound's ozone-depletion potential. A single CFC molecule is 4,750 to 10,900 times more effective at trapping heat than a single molecule of CO_2 (see Table 9.31). A decrease in ozone concentrations in the upper

TABLE 9.31 Ozone-depletion potential and global warming potential for common ozone-depleting substances and their alternatives

HALOGEN SOURCE GASES	ATMOSPHERIC LIFETIME (yr)	GLOBAL EMISSIONS IN 2008 (Kt/yr)[a]	OZONE-DEPLETION POTENTIAL (ODP)[c]	GLOBAL WARMING POTENTIAL (GWP)[c]
Chlorine gases				
CFC-11	45	52–91	1	4,750
CFC-12	100	41–99	0.82	10,900
CFC-113	85	3–8	0.85	6,130
Carbon tetrachloride (CCl_4)	26	40–80	0.82	1,400
HCFCs	1–17	385–481	0.01–0.12	77–2,220
Methyl chloroform (CH_3CCl_3)	5	~20	0.16	146
Methyl chloride (CH_3Cl)	1	3,600–4,600	0.02	13
Bromine gases				
Halon-1301	65	1–3	15.9	7,140
Halon-1211	16	4–7	7.9	1,890
Methyl bromide (CH_3Br)	0.8	110–150	0.66	5
Very short-lived gases (e.g., $CHBr_3$)	Less than 0.5	[b]	[b]Very low	[b]Very low
Hydrofluorocarbons (HFCs)				
HFC-134a	13.4	149 ± 27	0	1,370
HFC-23	222	12	0	14,200
HFC-143a	47.1	17	0	4,180
HFC-125	28.2	22	0	3,420
HFC-152a	1.5	50	0	133
HFC-32	5.2	8.9	0	716

[a] Includes both human activities (production and banks) and natural sources. Emissions are in units of kilotons per year (1 kiloton = 1,000 metric tons = 1 gigagram = 10^9 grams).

[b] Estimates are very uncertain for most species.

[c] 100-year GWPs. Values are calculated for emission of an equal mass of each gas.

Source: Based on Fahey, D.W., and Hegglin, M.I. (2010). *Twenty Questions and Answers About the Ozone Layer: 2010 Update*. United Nations Environment Programme.

atmosphere from reactions with stratospheric chlorine results in more UV radiation reaching the Earth's surface. Increased exposure to UV radiation may increase the risk of skin cancer, eye damage, and suppression of the immune system. Increased UV radiation can also damage marine phytoplankton and crops, such as soybeans, and reduce the crop yield. Furthermore, CFCs and other ***ozone-depleting substances (ODSs)*** have a disproportionate effect on the greenhouse effect by absorbing heat re-radiated from the Earth's surface. The global warming potential (GWP), which will be discussed in more detail in subsequent chapters, compares the CFC's effectiveness as a greenhouse gas to carbon dioxide.

As a result of these unprecedented observations of anthropogenic impacts on the Earth's atmosphere, the Montreal Protocol was adopted in September 1987. The ***Montreal Protocol*** was an international treaty, ultimately signed by 191 countries. The treaty phased out and banned the use and production of CFCs and ODSs. The protocol was amended in 1990 to end CFC and other ODS use by 2000 in economically developed countries, and funds were allocated to help developing countries take similar steps. Emissions of ODSs have been declining for some time, and the total inorganic chlorine concentration in the stratosphere appears to have peaked in 1997 and 1998. As a result of the international cooperation, the ozone layer has not significantly thinned since 1998 and appears to be recovering. Some estimates suggest that Antarctic ozone levels could recover and return to pre-1980 levels by the year 2075 (see Figure 9.83).

Many measureable benefits have been realized from the collaborative effort to implement the Montreal Protocol and subsequent ODS policies (see Box 9.3). Highlights of the Montreal Protocol achievements include:

1. Elimination of production of most CFCs, methyl chloroform, and halons
2. Increased use of existing hydrochlorofluorocarbons (HCFCs)
3. New production of a wide range of industrial fluorine-containing chemicals, including new types of HCFCs, hydrofluorocarbons (HFCs), and perfluorocarbons (PFCs)

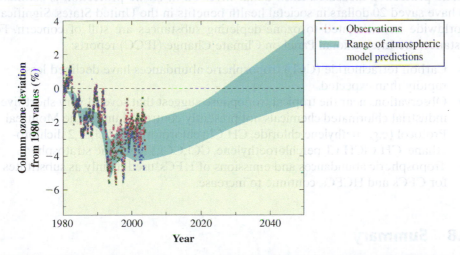

FIGURE 9.83 Range of atmospheric ozone depletion from 1980 and projected to 2050.

Source: Based on IPCC/TEAP, Metz, B., et al. (eds.) (2005). *Safeguarding the Ozone Layer and the Global Climate System: Issues Related to Hydrofluorocarbons and Perfluorocarbons Summary for Policymakers and Technical Summary*. IPCC, Geneva, Switzerland.

BOX 9.3 Changes in Mass Emissions, Ozone-Depletion Potential (ODP), and Global Warming Potential (GWP) as a Result of Implementing the Montreal Protocol and Other International Treaties

CFCs are inexpensive and effective coolants that have been widely used and manufactured. As a result of increasing industrialization, CFC and chlorinated organic compound manufacturing was expected to increase in an exponential fashion had the effects of the CFCs on the ozone layer not been identified and quick action taken to change those trends. The Montreal Protocol was only the first of many sequential actions taken to reduce the levels of reactive chlorine in the stratosphere. Substitute chemicals have been developed to replace CFCs and other ODSs, but environmental impacts have been found to be associated with the replacement compounds (see Table 9.31).

It is important to note that while the mass of total emissions from the industrial use of hydrofluorocarbon (HFC) coolants and ODS replacements is expected to increase in the future, the environmental impacts of the replacement chemicals are likely to be much smaller. HFCs have no appreciable reaction in the stratosphere with ozone, so that the emissions of ozone-depleting substances are expected to continue to decrease even as overall industrial emissions increase. Unfortunately, the global warming potential of HFCs ranges from 133 to 14,200 times the GWP of CO_2, so the continued increase in these chemicals are expected to contribute to the anthropogenically enhanced greenhouse gas effect and resultant climate change.

4. Use of nonhalogenated chemical substitutes such as hydrocarbons, carbon dioxide, and ammonia
5. Development of not-in-kind alternative methods such as water-based cleaning procedures

The reduction in ODS emissions has resulted in a decreased warming potential equivalent to 8,900 million metric tons of carbon. Research estimates suggest that over 6.3 million lives were saved in the United States alone by the year 2016 due to ozone protection measures. Each dollar invested in ozone protection is estimated to have saved 20 dollars in societal health benefits in the United States. Significant worldwide issues related to ozone-depleting substances are still of concern. For instance, the International Panel on Climate Change (IPCC) reports:

- Carbon tetrachloride (CCl_4) tropospheric abundances have declined less rapidly than expected.
- Observations near the tropical tropopause suggest that several very short-lived industrial chlorinated chemicals not presently controlled under the Montreal Protocol (e.g., methylene chloride, CH_2Cl_2; chloroform, $CHCl_3$; 1,2 dichloro-ethane, CH_2ClCH_2Cl; perchloroethylene, CCl_2CCl_2) reach the stratosphere.
- Tropospheric abundances and emissions of HFCs, used mainly as substitutes for CFCs and HCFCs, continue to increase.

9.8 Summary

The control and mitigation of pollutants in the air protect ecosystem diversity and human health, which in turn lead to economic and social impacts from the local to the global scale. National regulations and international treaties are credited with saving millions of lives and providing significant economic and environmental benefits. Airborne pollutants in the atmosphere may undergo a variety of chemical transformations. Some air pollutants have direct environmental and health impacts, while others, like the precursors to photochemical smog and ozone-depleting substances, undergo reactions in the environment that yield adverse environmental or health effects. The mass balance principles and emission rates that have been discussed in prior chapters of this book can also be applied to calculate or estimate air pollutant emission rates. Physical and chemical transport models are useful tools that allow engineers and scientists to estimate the potential effects of air pollution and air pollution control technologies on local, regional, and global scales. Depending on the chemical and industrial process, air pollution control technologies may involve treating emissions from an industrial process or reducing or replacing the chemicals used in a process or product. Regulatory air pollution control policies have maintained or improved air quality in many regions of the world. For example, international regulations and the ban on CFCs have helped reduce the magnitude of the ozone hole. However, international agreements on the future control of greenhouse gases have been much more difficult to agree upon and implement.

References

CDC. (2008). "Smoking-attributable mortality, years of potential life lost, and productivity losses—United States, 2000–2004." *Morbidity and Mortality Weekly Report* 57(45):1226–1228. Centers for Disease Control and Prevention.

Chou, M. S., and Lu, S. L. (1998). "Treatment for 1,3-butadiene in an air stream by biotrickling filter and a biofilter." *Journal of Air and Waste Management Association* 48(8):711–720.

Cooper, C. D., and Alley, F. C. (1986). *Air Pollution Control: A Design Approach*. Prospect Heights, IL: Waveland Press.

Fahey, D. W., and Hegglin, M. I. (2010). Twenty questions and answers about the ozone layer: 2010 Update. United Nations Environment Programme.

Griffiths, R. F. (1994). "Errors in the use of the Briggs parameterization for atmospheric dispersion coefficients." *Atmospheric Environment* 28(17):2861–2865.

Hanna, S. R., Briggs, G. A., and Hosker, R. P., Jr. (1982). Handbook on atmospheric diffusion. Technical Information Center, U.S. Department of Energy.

Heinsohn, R. J. (1991). *Industrial Ventilation: Engineering Principles*. New York: John Wiley & Sons.

Heinsohn, R. J., and Kabel, R. L. (1999). *Sources and Control of Air Pollution*. Upper Saddle River, NJ: Prentice Hall.

IPCC/TEAP, Metz, B., et al. (Eds.). (2005). Safeguarding the ozone layer and the global climate system: Issues related to hydrofluorocarbons and perfluorocarbons. Summary for policymakers and technical summary. IPCC, Geneva, Switzerland.

Lappe, C. E. 1951. "Processes use many collection types." *Chemical Engineering* 58:145.

McKinnon, D. J. 1974. Nitric oxide formation in premixed hydrocarbon flames. *Journal of Air Pollution Control Association*. 24(3):237–239.

Palmer, P. (2008). "Quantifying sources and sinks of trace gases using space-borne measurements: Current and future science." *Philosophical Transactions of the Royal Society* A. 366:4509–4528.

Pasquill, F. (1961). "The estimation of the dispersion of windborne material." *The Meteorological Magazine* 99(1063):33–49.

Seinfeld, J. H. (1986). *Atmospheric Chemistry and Physics of Air Pollution*. New York: John Wiley & Sons.

Slade, D. H. (Ed.). (1968). Meteorology and atomic energy. U.S. Atomic Energy Commission, Air Resources Laboratories, Environmental Sciences Services Administration, U.S. Department of Commerce, Washington, DC.

Stolarski, R. S. (1985). *A Century of Nature: Twenty-One Discoveries That Changed Science and the World.* Chapter 18: A hole in the Earth's shield. Chicago: University of Chicago Press.

U.S. EPA. (1995). Compilation of air pollutant emission factors—Volume I: Stationary point and area sources. AP-42, 5th ed. U.S. Environmental Protection Agency, Washington, DC.

U.S. EPA. (1999). Smog—Who does it hurt? What you need to know about ozone and your health. Office of Air and Radiation, U.S. Environmental Protection Agency, Washington, DC.

U.S. EPA. (2003a). Latest findings on national air quality: 2002 status and trends. Office of Air Quality, Planning and Standards, U.S. Environmental Protection Agency, Research Triangle Park, NC.

U.S. EPA. (2003b). Ozone—Good up high, bad nearby. Office of Air and Radiation, U.S. Environmental Protection Agency, Washington, DC.

U.S. EPA. (2005). Guidelines for carcinogen risk assessment, Final Report. Risk Assessment Forum, U.S. Environmental Protection Agency, Washington, DC. cfpub.epa.gov/ncea/cfm/

U.S. EPA. (2007a). A plain English guide to the Clean Air Act. Office of Air Quality, Planning and Standards, U.S. Environmental Protection Agency, Research Triangle Park, NC.

U.S. EPA. (2007b). Achievements in stratospheric ozone protection: Progress report. Office of Air and Radiation. U.S. Environmental Protection Agency, Washington, DC.

U.S. EPA. (2009a). Policy assessment for the review of the particulate matter National Ambient Air Quality Standards. Office of Air Quality Planning and Standards, U.S. Environmental Protection Agency, Research Triangle Park, NC.

U.S. EPA. (2009b). Ozone and your health. Office of Air and Radiation. U.S. Environmental Protection Agency, Washington, DC.

U.S. EPA. (2011a). The benefits and costs of the Clean Air Act from 1990 to 2020: Summary report. Office of Air and Radiation, U.S. Environmental Protection Agency, Research Triangle Park, NC.

U.S. EPA. (2011b). An overview of methods for EPA's National-Scale Air Toxics Assessment. Office of Air Quality Planning and Standards, U.S. Environmental Protection Agency, Research Triangle Park, NC.

U.S. EPA. (2011c). Policy assessment for the review of the particulate matter National Ambient Air Quality Standards. Office of Air Quality Planning and Standards, U.S. Environmental Protection Agency, Research Triangle Park, NC.

U.S. EPA. (2011d). Compilation of air pollutant emission factors—Update 2011. U.S. Environmental Protection Agency, Washington, DC. www.epa.gov/ttn/chief/ap42/index.html.

U.S. EPA. (2012a). National Ambient Air Quality Standards (NAAQS). Air and Radiation, U.S. Environmental Protection Agency. www.epa.gov/air/criteria.html.

U.S. EPA. (2012b). Our nation's air: Status and trends through 2010. Office of Air Quality, Planning and Standards, U.S. Environmental Protection Agency, Research Triangle Park, NC.

Watson, A. Y., Bates, R. R., and Kennedy, D. (Eds.). (1988). *Air Pollution, the Automobile, and Public Health.* Washington, DC: National Academy Press.

WEF. (1995). *Odor Control in Wastewater Treatment Plants*. Water Environmental Federation, Alexandria, VA.

Wilson, W. E., and Suh, H. H. (1997). "Fine particles and coarse particles: Concentration relationships relevant to epidemiologic studies." *Journal of the Air and Waste Management Association* 47:1238–1249.

Key Concepts

Clean Air Act

National Ambient Air Quality Standards (NAAQS)

Prevention of Significant Deterioration (PSD)

State Implementation Plans (SIPs)

Air Quality Index (AQI)

Acute

Chronic

Nasopharyngeal region

Tracheobronchial region

Hypoxia

Hypercapnia

Troposphere

Tropospheric ozone

Primary pollutant

Secondary pollutant

Inhalable coarse particle

Fine particle

Hazardous air pollutants (HAPs)

Emission factors

Dispersion

Dry adiabatic lapse rate

Potential temperature gradient

Specific height

Plume rise

Effective stack height

Combustion

Air-to-fuel ratio

Products of incomplete combustion (PICs)

Removal efficiency

Settling chamber

Cyclonic separators

Venturi scrubber

Baghouses

Electrostatic precipitators (ESPs)

Dry scrubbing

Wet scrubbing

Selective catalytic reduction

Lower explosive limit (LEL)

Absorption

Biofiltration

Carbon adsorption

Thermal oxidation

Catalytic oxidation

Hybrid control systems and concentrators

Chlorofluorocarbons (CFCs)

Stratospheric ozone depletion

Ozone-depleting substances

Montreal Protocol

Problems

9-1 If a typical adult inhales 3.5 L of air per breath and breathes 15 times per minute, how many grams of oxygen will be inhaled in 24 hours if the density of air is 1.225 kg/m^3?

9-2 What is the average molecular weight of "clean air"?

9-3 Research one of the following incidents and list the air pollutants involved. Describe the possible health effects caused by exposure, and describe the pathway for exposure to the pollutants.

   a. Cyanide gas exposure from the Union Carbide plant in Bhopal, India

   b. Chernobyl nuclear reactor in Ukraine

   c. Asbestos and vermiculite mining in Libby, Montana, United States

   d. Fukushima Daiichi nuclear power plant in Okuma, Fukushima, Japan

9-4 Is breathing secondhand smoke a voluntary or involuntary risk?

9-5 Exactly 500.0 mL of CO is mixed with 999,500 mL of air. Calculate the resulting concentration of CO in ppm and in mg/m^3. Would you expect any health impacts from this air? Explain your answer.

9-6 What is the National Ambient Air Quality Standard for the following compounds expressed in $\mu g/m^3$?
  a. Carbon monoxide, 1-hr standard
  b. Carbon monoxide, 8-hr standard
  c. Nitrogen dioxide, 1-hr standard
  d. Nitrogen dioxide, 8-hr standard
  e. Ozone, 8-hr standard
  f. Sulfur dioxide, 1-hr standard
  g. Sulfur dioxide, 3-hr standard

9-7 Calculate the AQI for a city that has a measured 8-hr carbon monoxide concentration of 4 mg/m^3. What AQI-based level of health concerns should be reported?

9-8 Calculate the AQI for a city that has a measured 1-hr nitrogen dioxide concentration of 0.267 mg/m^3. What AQI-based level of health concerns should be reported?

9-9 Calculate the AQI for a city that has a measured 8-hr ozone concentration of 0.110 mg/m^3. What AQI-based level of health concerns should be reported?

9-10 Calculate the AQI for a city that has a measured 24-hr $PM_{2.5}$ concentration of 0.023 mg/m^3. What AQI-based level of health concerns should be reported? What are the likely health effects?

9-11 Describe the potential health effects from sulfur dioxide exposure to:
  a. 0.1 ppm_v SO_2 for 8 hr
  b. 0.5 ppm_v SO_2 for 8 hr
  c. 5.0 ppm_v SO_2 for 8 hr

9-12 Describe the potential health effects from carbon monoxide exposure to:
  a. 30 ppm_v CO for 12 hr
  b. 100 ppm_v CO for 12 hr
  c. 500 ppm_v CO for 12 hr

9-13 Identify a city or metropolitan area in Figure 9.15 that is associated with high concentrations of nitrogen dioxide.
  a. What is the population of the city or metropolitan area?
  b. What is the population density per km^2 of the city or metropolitan area?
  c. What is the population density of automobiles in the city or metropolitan area?

9-14 Identify a city or metropolitan area in Figure 9.23 that is associated with high concentrations of ozone.
  a. What is the population of the city or metropolitan area?
  b. What is the population density per km^2 of the city or metropolitan area?
  c. What is the population density of automobiles in the city or metropolitan area?

9-15 Identify a city or metropolitan area in Figure 9.28 that is associated with increased total cancer risk.
  a. What is the population of the city or metropolitan area?
  b. What is the population density per km^2 of the city or metropolitan area?
  c. What is the population density of automobiles in the city or metropolitan area?

9-16 What are the major natural and anthropogenic sources of primary PM smaller than 2.5 μm and larger than 2.5 μm for the following aerosol species?
  a. Sulfates
  b. Nitrates

c. Minerals

d. Ammonium

e. Organic carbon

f. Elemental carbon (EC)

g. Metals

h. Bioaerosols

9-17 What are the estimated atmospheric half-life and the removal process for these particles?

a. Ultrafine particles

b. Fine (accumulation) particles

c. Coarse particles

9-18 What size particle (in μm) is most likely to be deposited in the pulmonary region of the lungs?

9-19 Describe in what region of the respiratory system the following particles are likely to be deposited and how each particle could be removed from the respiratory system.

a. A grain of dirt with a diameter of 500 μm

b. A grain of sand with a diameter of 90 μm

c. A particle of dust with a diameter of 25 μm

d. A pollen particle with a diameter of 5 μm

e. An aerosol with a diameter of 1 μm

9-20 A person smokes one cigarette in 60 seconds in a waiting room. The concentration of smoke particles in the inhaled smoke is 10^{15} particles/m^3. The smoke particles have a diameter of 0.2 μm and a density of 0.8 g/mL. If the smoker breathes in air at a rate of 5 L per minute, what is the inhalation rate of smoke particles in mg/s during the time they are smoking?

9-21 A miner has a documented diagnosis involving respiratory effects, cancer, and central nervous system impairments.

a. Atmospheric conditions in the work environment could have what degree of causation for the miner's health diagnosis?

b. What, if any, significance would there be in linking workplace exposure to causation of the symptoms if this individual smoked two packs of cigarettes a day, every day, for 30 years?

9-22 What gas analysis method(s) would be appropriate for analyzing hydrocarbons in the smokestacks at a petroleum refinery?

9-23 What gas analysis method(s) would be appropriate for analyzing carbon in the smokestacks from a cement kiln?

9-24 There are 40 customers who are heavy smokers in a small pub. Each person smokes one cigarette every 10 minutes. Each cigarette emits 10^{15} particles of smoke. The smoke particles have a diameter of 0.1 μm and a density of 0.8 g/m^3.

a. What is the mass emission rate of smoke particles into the room in mg/s?

b. If the room is 5 m wide, 2.5 m high, and 8 m long, and the ventilation rate is 10 m^3/hr, what is the steady state smoke concentration in the room?

c. If the background PM$_{2.5}$ concentration is 0.025 mg/m^3, what total steady state smoke concentration is the nonsmoking bartender exposed to during an 8-hr shift?

d. Smoke of this size has been determined to cause what potential health outcomes?

e. Where are these particles deposited in the respiratory system?

9-25 Find the estimated emissions listed below for a spreader stoker cogeneration facility that burns 10,000 tons of bituminous coal per year with a 3% by weight sulfur content. Use the EPA's AP-42 (Volume 1, 5th ed.) Chapter 1: External Combustion Sources Emission Factors for Bituminous and Sub-bituminous Coal Combustion.

a. Tons of SO_x emitted per year
b. Tons of NO_x emitted per year
c. Tons of CO emitted per year

9-26 Find the estimated emissions listed below for an uncontrolled (post-NSPS) large wall-fired boiler natural gas cogeneration facility that burns 500 million standard cubic feet (scf) of natural gas per year. Use the EPA's AP-42 (Volume 1, 5th ed.) Chapter 1: External Combustion Sources Emission Factors for Natural Gas Combustion.

a. Tons of NO_x emitted per year
b. Tons of CO emitted per year

9-27 The following data are from the U.S. EPA AP-42 emission factors for malt beverages. A closed fermenter vents CO_2 = 2,100, VOC (ethanol) = 2.0, and hydrogen sulfide, H_2S = 0.015. The units for the emission factors in the EPA tables are pounds of pollutant per 1,000 barrels (bbl). A brewery produces about 7 million barrels of product per year. Estimate these emissions:

a. kg of carbon dioxide per year from fermenter venting
b. kg of VOCs (ethanol) per year from fermenter venting
c. kg of hydrogen sulfide per year from fermenter venting

9-28 The following data are from the U.S. EPA AP-42 emission factors for malt beverages. For a steam-heated brewers grain dryer, CO = 0.22, CO_2 = 53, and VOC (ethanol) = 0.73. The units for the emission factors in the EPA tables are pounds of pollutant per 1,000 barrels (bbl). A brewery produces about 7 million barrels of product per year. Estimate these emissions:

a. kg of carbon monoxide per year from fermenter venting
b. kg of carbon dioxide per year from fermenter venting
c. kg of VOCs (ethanol) per year from fermenter venting

9-29 How many pounds of the following pollutants are emitted from burning 1 ton of bituminous coal in a domestic hand-fired furnace?

a. PM
b. Sulfur oxides
c. Carbon monoxide
d. Hydrocarbons
e. Nitric oxides
f. HCl

9-30 Assuming each mole of sulfur in fuel oil (#2) forms 1 mole of SO_2, does the energy released from burning 1 ton of oil result in more or less sulfur emission than that released from burning coal with the best available emissions controls for the same amount of energy released?

9-31 Why can't a gasoline engine be tuned so that it produces the minimum amounts of all three pollutantss—CO, HC, and NO_x—simultaneously?

9-32 If a car is designed to burn ethanol (C_2H_5OH), what is the stoichiometric air-to-fuel ratio?

9-33 How much air is needed to burn 2 tons per year of methane generated at a municipal landfill?

9-34 How much air is needed to completely combust a mixture of 4,000 ppm_v butane and 8,000 ppm_v propane?

9-35 An exhaust air stream contains 4,000 ppm methane, 1,000 ppm ethane, and 5% oxygen from a landfill.

a. How many moles of oxygen are required for combustion?

b. Does additional air need to be added to the exhaust gas to reach the required amount of oxygen?

c. If so, what percentage of make-up air must be added?

9-36 An accident causes traffic to come to a halt in an underwater tunnel. The people in their vehicles allow their engines to idle. There are 100 vehicles stopped in the tunnel, and assume their engines emit on average 0.269 g/min of VOC (aldehyde), 3.82 g/min of CO, and 0.079 g/min of NO_x. The tunnel is long and narrow with a height of 10 m, a width of 25 m, and a length of 0.5 km. The ambient air concentrations of CO, VOCs, and NO_x are low. The mean air velocity through the tunnel, when traffic stops, is limited to 0.2 m/s.

a. If traffic remains halted long enough for the concentrations of carbon monoxide, nitric oxides, and hydrocarbons in the tunnel to reach a steady state, what would the expected concentrations of carbon monoxide, nitric oxides, and hydrocarbons be in ppm_v in the tunnel?

b. Do the CO and NO_x concentrations exceed NAAQS?

c. Does the CO concentration exceed OSHA's permissible exposure limit (PEL)?

9-37 The following data are recorded at meteorological field stations. Determine the environmental lapse rate and the stability classification of the atmosphere in each case.

a. Temperature at ground level is 15.0°C. Temperature measured at 10 m is 15.1°C.

b. Temperature at ground level is 15.0°C. Temperature measured at 10 m is 14.7°C.

c. Temperature at ground level is 15.0°C. Temperature measured at 10 m is 15.5°C.

d. Temperature at ground level is 15.0°C. Temperature measured at 10 m is 14.1°C.

9-38 Consider a prevailing lapse rate that has these temperatures: ground = 21°C, 500 m = 20°C, 600 m = 19°C, 1,000 m = 20°C. If a parcel of air is released at 500 m and 20°C, will it tend to sink, rise, or remain where it is? If a stack is 500 m tall, what type of plume do you expect to see?

9-39 Determine the Pasquill stability category for the following scenarios and the type of plume that will be seen from a smokestack.

a. An overcast day with leaves rustling

b. A sunny afternoon with the solar altitude 75 degrees above the horizon and loose paper blowing on the street

c. A sunny afternoon with the solar altitude 75 degrees above the horizon and smoke from a match rising vertically

d. A later summer day with broken low clouds and smoke from a match rising vertically

e. Nighttime when chimney smoke drifts but does not move wind vanes

f. A winter day when chimney smoke drifts but does not move wind vanes

9-40 The inside diameter of a building exhaust smokestack is 2.0 m. The flow rate of ventilation air emitted through the stack is 10 m³/s at 25°C. The exhaust exits the stack into an atmosphere with a 3 m/s average wind speed and an ambient temperature of 21°C. Determine the effective stack height for a 30-m-high smokestack under the stability conditions calculated in Problem 9-37 as assigned by your instructor.

9-41 The inside diameter of an incinerator's smokestack is 0.5 m. The flow rate of ventilation air emitted through the stack is 0.1 m³/s at 800°C. The exhaust exits the stack into an atmosphere with a 3 m/s average wind speed and an ambient

temperature of 23°C at the stack exit. Determine the effective stack height for a 10-m-high smokestack given the following ground-level temperatures:
a. $T = 23.2°C$
b. $T = 23.0°C$
c. $T = 22.0°C$

9-42 A coal-burning power plant emits 300 kg/hr of SO_2. The surrounding atmosphere was classified as neutral with a wind speed of 2 m/s. The σ_y and σ_z values are assumed to be 100 m and 30 m, respectively.
a. What is the ground-level concentration of SO_2 in $\mu g/m^3$ if the town that is 2 km directly downwind of the plant has a stack that produces an effective stack height of 20 m?
b. What will be the ground-level concentration if the power plant installs a 150-m stack?

9-43 What is the maximum ground-level pollutant concentration in mg/m^3 from an urban area elevated source with effective stack height $H = 20$ m, cyanide (HCN) emissions = 0.5 g/s, and average wind velocity = 3 m/s with a stability class C for the atmosphere? Assume $\sigma_z = 100$ m and $\sigma_y = 100$ m.

9-44 A 1,530-MW bituminous coal-fired wet-bottom plant emits mercury into the atmosphere in a rural area. The EPA AP-42 emission factor for the power plant is 16 lb Hg per 10^{12} Btu. The plant operates around the clock, 365 days a year, 24 hours per day. The 20 m^3/s (actual) of exhaust from the plant exits the 2-m-diameter stack at 700°C. The stack is 50 m high. The air pressure within the stack is essentially equal to the ambient air pressure of 1 atm. The average ambient temperate measured at the stack height is 20°C. The average temperature measured at 110 m on a nearby radio tower is 18.5°C. The prevailing wind speed is 3 m/s. Determine:
a. The mass emission rate of Hg in mg/s
b. The plume rise from the stack in m
c. The maximum ground-level pollutant concentration in mg/m^3 from an elevated source in ppm
d. The concentration of mercury 2 km directly downwind from the power plant. Calculate the concentrations at ground level of gaseous pollutants that are not absorbed by the ground.

9-45 An 1,880-MW wet-bottom, wall-fired sub-bituminous coal power plant emits sulfur oxides, nitrogen oxides, and carbon monoxide into the atmosphere in a rural area. The plant uses coal with a 4% sulfur content. The plant operates around the clock, 365 days a year, 24 hours per day. The 25 m^3/s (actual) of exhaust from the plant exits the 1-m-diameter stack at 800°C. The stack is 25 m high. The air pressure within the stack is essentially equal to the ambient air pressure of 1 atm. The average ambient temperate measured at the stack height is 20°C. The average temperature measured at 110 m on a nearby radio tower is 20.0°C. The prevailing wind speed is 2 m/s. Determine:
a. The AP-42 emission factor for the sulfur oxide, nitrogen oxides, or carbon monoxide (as assigned by your instructor)
b. The mass emission rate of the sulfur oxide, nitrogen oxides, or carbon monoxide (as assigned by your instructor) in mg/s
c. The plume rise from the stack in m

9-46 List advantages and disadvantages of the following organic waste control technologies:
a. Settling chamber
b. Cyclone

c. Venturi scrubber

d. Baghouse

e. Electrostatic precipitator

9-47 An air treatment system is to be installed in a heavily populated urban area. Space is at a premium in the facility, so the system must be small. Also, the particles should be collected as a dry solid waste. What type of pretreatment system do you recommend to remove particles with a mean size of 40 μm prior to a baghouse?

9-48 What type of particulate control system do you recommend for a cement manufacturing facility? The permit is written for 95% removal of $PM_{2.5}$. The particle loading rate is very high, and the particles cannot carry an electrical charge.

9-49 A clean room requires that loading rate particulate concentrations in the room be extremely low. A removal efficiency of 99.995% is required for particles larger than 0.2 μm. What type of control system do you recommend?

9-50 What size particle with a density of 1,000 kg/m³ will be collected at 50% efficiency in a standard cyclone? The cyclone is 1 m in diameter, has six effective turns, and treats 2 m³/s of air at 25°C.

9-51 Taking into account cost, ease of operation, and ultimate disposal of residuals, what type of control device do you suggest for the following emissions?

a. A dust with a particle range of 5–10 μm

b. A gas containing 20% SO_2 and 80% N_2

c. A gas containing 90% HC and 10% O_2

9-52 List advantages and disadvantages of the following organic waste control technologies:

a. Condensation

b. Adsorption

c. Absorption

d. Biofiltration

e. Advanced oxidation

9-53 Pollution control systems are required for the air streams listed in Table 9.32. For which type of control or controls would you conduct a size and cost study? Why?

TABLE 9.32 Scenarios for air pollution control equipment

POLLUTANT	CONC.	FLOW RATE (scfm)	EXHAUST TEMPERATURE (°C)	OPERATION	POLARITY
Carbon disulfide	5%	2,000	40	Intermittent batch	Nonpolar
Isopropyl alcohol	200 ppm	15,000	25	8 hr/day	Polar
TCE	15 ppm	5,000	25	8 hr/day	Nonpolar
TCE	15 ppm	50,000	25	24 hr/day	Nonpolar
Styrene	50 ppm	50,000	30	24 hr/day	Nonpolar
Styrene	5,000 ppm	50,000	30	24 hr/day	Nonpolar
Toluene	50 ppm	50,000	45	Intermittent batch	Nonpolar
Methyl ethyl ketone	100 ppm	25,000	50	8 hr/day	Semi-polar
Phosgene	0.2 ppm	10,000	10	6 hr/day	Nonpolar

9-54 A sewage sludge incinerator has a scrubber to remove the particulate material from its emissions, but the scrubber has been acting up. A small pilot bag filter is being tested for treating the particulate emissions. The experimental setup splits the emissions from the incinerator so that the 200 m³/s total flow is divided, with 97% going to the scrubber and only 3% going to the pilot bag filter. The particulate concentration of the untreated emission is 125 mg/m³. The solids collected at the baghouse are collected hourly and average out to 2.6 kg/hr. The water does not have any particulates in it when it enters the scrubber at a flow rate of 2,000 L/min. There is negligible evaporation in the scrubber, and the scrubber water is found to carry solids at a rate of 52 kg/hr.

   a. What is the efficiency of each method of air pollution control?

   b. Regardless of the outcome in the efficiency calculations, is there any reason the scrubber would still be superior as a pollution control device?

9-55 List the chemicals and/or materials that have changed in order to comply with the Montreal Protocol.

Team-Based Problem Solving

9-56 The owners of Wood Spec, Inc., a low-cost furniture manufacturer, are planning to significantly expand their manufacturing facility in a rural area of the southeastern United States. The state regulatory authority has asked them to estimate the effects of plant expansion on the local air quality. The plant expansion will process 3 million m² of particle board (std, UF) daily. You are a consulting engineer hired to determine the air quality ramifications of building the facility. The following data from a similar facility located in the Midwest have been provided.

- The process line operates 5 days a week for 6 hours per day.
- The concentration of formaldehyde emitted from the furniture manufacturing plant is 20 ppm.
- The exhaust gas exits the stack at 145 m/s and its temperature is 300 K.
- The physical height of the exhaust gas stack is 10 meters.
- The data in Table 9.33 were collected from meteorological stack testing.

TABLE 9.33 Emissions from the furniture manufacturer

PM TYPE	PARTICLE SIZE (μm)	EMISSION FACTOR (mg/m²-day)
Chips	> 100	46
Sawdust	1–100	18
Fine PM	0.01–1	4
Total	0.01–100	68

These typical meteorological conditions were obtained from a local weather station:

- Strong solar radiation
- Ambient temperature = 299 K

- Wind speed at 10 m = 2 m/s
- Actual lapse rate = −0.015 K/m

Comments from a meeting with the owners of Wood Spec, Inc., and representatives from the state regulatory authority included the following information:

- Both formaldehyde and particulates are a potential concern.
- The state will assume that 1 mole of formaldehyde reacts in the atmosphere to form 1 mole of ozone (the worst-case reasonable assumption).
- The pollutants should be modeled as if they were reflected by the ground.
- The existing ambient concentration of PM is 10 $\mu g/m^3$ and ozone is 45 $\mu g/m^3$.

The following contractual obligations must be completed before you are paid for your work:

a. Discuss the potential health impacts of the air pollutants, including but not limited to the following:
   i) A description of the region of the respiratory system most likely to be affected by PM consisting of chips, sawdust, and fine PM
   ii) Identification of the region of the respiratory system least likely to be affected by chips, sawdust, and fine PM
b. Estimate the concentration of formaldehyde in ppm in the exhaust stack prior to being emitted into the atmosphere.
c. Report the U.S. EPA's AP-42 emission factors for formaldehyde and particulate matter.
d. Calculate the emission rate of formaldehyde and total particulate matter (PM) in mg/s.
e. Estimate the plume rise from the stack and the effective stack height. Report your findings in meters.
f. Apply the Gaussian dispersion equation to model the dispersion of these air pollutants from the facility, and define the variables used in the equation.
g. Use a spreadsheet or mathematical software to create a graphical illustration of the ground-level concentration of formaldehyde and PM directly downwind from the proposed facility.
h. Report the concentration and distance from the facility of the maximum formaldehyde and PM concentrations.
i. Will this area downwind of the proposed manufacturing facility remain in compliance with the U.S. NAAQS PM and ozone standards? Consider the preexisting ambient concentration of these criteria pollutants in your report.
j. What type of air pollution control system do you recommend for the PM?
k. What type of air pollution control system do you recommend for the formaldehyde?

## CHAPTER 10

The Carbon Cycle and Energy Balances

FIGURE 10.1 One of many cartoons that illustrates the perceived complicated science of climate change.

Source: Tote/Shutterstock.com.

According to a new U.N. report, the global warming outlook is much worse than originally predicted. Which is pretty bad when they originally predicted it would destroy the planet.

—JAY LENO

GOALS

THE EDUCATIONAL GOALS OF THIS CHAPTER are to use basic principles of chemistry, physics, and math to define and solve engineering problems—in particular, mass and energy balance problems related to the Earth's carbon flux. This chapter will examine the modern problems associated with fossil fuel use, carbon dioxide concentrations in the atmosphere, and atmospheric climate change. The global context of climate change will be explored, in particular the relationships between climate change and resources.

OBJECTIVES

At the conclusion of this chapter, you should be able to:

10.1 Describe the scientists who conducted early research on the climate, and discuss how climate science evolved.

10.2 Calculate carbon emissions from various fossil fuels.

10.3 Define and discuss the major repositories and fluxes for carbon.

10.4 Describe and calculate anthropogenic changes to the global energy balance.

10.5 Utilize thermodynamic principles to solve a simplified steady state blackbody radiation model of the Earth.

10.6 Define and discuss the impacts of greenhouse gases and radiative forcing.

10.7 Describe how global climate models predict future climate conditions using carbon dioxide emission scenarios and greenhouse gas concentration estimates.

10.8 Describe the potential environmental and societal impacts of a warming climate, and discuss mitigation, capture, and sequestration technologies to minimize the negative impacts of a changing climate.

Introduction

Climate change has been a hit in the media for at least the past two decades. But when did scientists begin to realize the important link between atmospheric gas concentrations and our climate on Earth? Many people believe that the science related to climate change is a relatively new field. In this chapter, we will examine how scientists over 100 years ago developed the fundamental principles that predict how and why the climate changes. We will explore the fundamental equations and data that show how climate change is tied to anthropogenic emissions. We'll also build on the tools developed in Chapters 1 and 2 to create mathematical models that can be used to predict changes in any system, regardless of whether this is a small laboratory reactor or the entire planet's atmosphere.

Simple but powerful mathematical models allow engineers and scientists to define problems and their solutions. The chemical transformation of fossil fuels through combustion will be evaluated. Mathematical models and fossil fuel usage data will be used to estimate the amount of carbon dioxide added to the atmosphere from societies' energy consumption. The impact of these global changes will be investigated by developing simple physical thermodynamic models and solving an energy balance or radiation flow from the sun through the atmosphere and to the Earth. The basic chemical, physical, and mathematical models developed in this chapter form the basis for complex climate change models that are used to predict the impact of climate change on the environment, economy, and society.

10.1   Climate Science History

In 1827, Jean-Baptist Fourier recognized that the chemical composition of the atmosphere absorbed energy radiated from the surface of the planet and trapped heat similar to a greenhouse. The heat trapped by the gases in the atmosphere results in temperatures on the surface of the planet that are much more conducive to life as we know it than the temperature of the planet without this insulating atmospheric blanket.

John Tyndall measured the effects of carbon dioxide and water vapor on the absorption of energy radiated from the surface of the planet in 1860 and identified the importance of the concentration of these gases on the average surface temperature of the Earth. It wasn't until much later, in 1896, that Svante Arrhenius estimated the effects that increasing carbon dioxide concentrations would have on the Earth's average surface temperature. By balancing the knowledge of the atmosphere's composition and the energy into and out of the atmosphere, these early scientists modeled and predicted the impact of various components of the atmosphere on the greenhouse gas effect and, subsequently, the average surface temperature of the Earth.

Guy Stewart Callendar (1898–1964) was a noted steam and power engineer who was curious about how the climate worked. Callendar (Figure 10.2) examined the work of early scientific studies on the effect of carbon dioxide in the atmosphere and linked increasing carbon dioxide concentrations in the atmosphere to the emissions from fossil fuel use that occurred as a result of the Industrial Revolution. In 1938, Callendar published his paper "The Artificial Production of Carbon Dioxide and Its Influence on Temperature." Callendar's

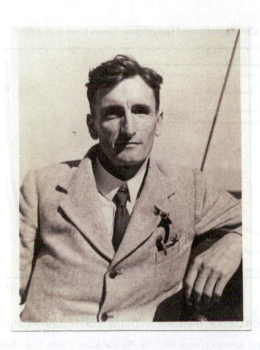

FIGURE 10.2 Photo of Guy Stewart Callendar, one of the "fathers" of climate change science.
Source: G.S. Callendar Archive, University of East Anglia.

paper linked anthropogenic emissions of carbon dioxide and predicted those emissions may impact the Earth's climate. In 1957, Hans Seuss and Roger Revelle referred to the "Callendar effect" as the change in climate brought about by increases in the concentration of atmospheric carbon dioxide, primarily through the processes of combustion. Since that time, scientists around the world have worked with the International Panel on Climate Change and other organizations to examine the fate and transport of carbon dioxide in the atmosphere. These scientists and engineers work to create more accurate models to predict how the changing carbon dioxide concentration of the Earth's atmosphere will influence the climate.

The ***International Panel on Climate Change (IPCC)*** is comprised of a group of scientists from many nations on the planet that are intimately involved in better understanding changes to the Earth's climate. The IPCC (2007a) statement on the current status of this research was summarized:

As climate science and the Earth's climate have continued to evolve over recent decades, increasing evidence of anthropogenic influences on climate change has been found. Correspondingly, the IPCC has made increasingly more definitive statements about human impacts on climate. Debate has stimulated a wide variety of climate change research. The results of this research have refined but not significantly redirected the main scientific conclusions from the sequence of IPCC assessments.

Figure 10.3 illustrates how much the concentration of carbon dioxide, methane, and nitrous oxide has changed over the last two thousand years.

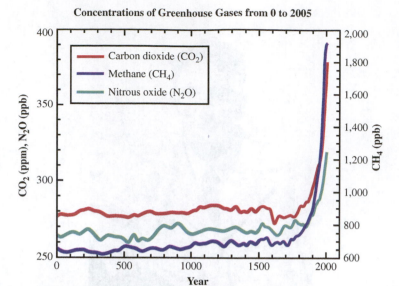

Concentrations of Greenhouse Gases from 0 to 2005

FIGURE 10.3 Carbon dioxide, methane, and nitrous oxide concentrations in the atmosphere.

Source: Based on IPCC (2007). *Climate Change 2007: The Physical Science Basis*. Contributions of Working Group I to the Fourth Assessment Report of the Intergovernmental Panel on Climate Change [S. Solomon, D. Qin, M. Manning, Z. Chen, M. Marquis, K.B. Averyt, M. Tignor, and H.L. Miller (eds.)] Cambridge University Press, Cambridge, United Kingdom, and New York, NY. 996 pp. FAQ 2.1, Fig. l, p. 135.

EXAMPLE 10.1 How Much Has the Atmosphere Changed?

The volume of the atmosphere at standard temperature and pressure (STP) is approximately 3.94×10^{18} m^3. For over a thousand years prior to the Industrial Revolution, the concentration of CO_2 in the atmosphere was approximately 280 ppm. The concentration of CO_2 in the atmosphere in 2011 was approximately 390 ppm. How much (mass) CO_2 has been added to the atmosphere since the Industrial Revolution?

First, we can convert units of ppm$_v$ to mg/m^3 in air for the concentration of carbon dioxide prior to the Industrial Revolution:

$$c_{280}\left(\frac{mg}{m^3}\right) = \frac{280\,(ppm_v) \times 44.01\left(\dfrac{mg}{mmol}\right)}{24.5\left(\dfrac{m^3 - ppm_v}{mmol}\right)} = 503\,\frac{mg}{m^3}$$

Similarly, the approximate concentration of carbon dioxide in mg/m^3 is

$$c_{390}\left(\frac{mg}{m^3}\right) = \frac{390\,(ppm_v) \times 44.01\left(\dfrac{mg}{mmol}\right)}{24.5\left(\dfrac{m^3 - ppm_v}{mmol}\right)} = 701\,\frac{mg}{m^3}$$

Then we subtract the pre-Industrial Revolution concentration from the present-day concentration of CO_2 to determine the increase of carbon dioxide in each cubic meter of air:

$$\Delta CO_2 = 701\ \frac{mg}{m^3} - 503\ \frac{mg}{m^3} = +198\ \frac{mg}{m^3}$$

The total mass of the atmosphere is estimated to be 5.14×10^{18} kg. The density of air at STP = 1.293 kg/m³. If the atmosphere were all compressed and assumed to be at conditions equivalent to STP, then

$$m_{CO_2\ added} = 198\ \frac{mg\ of\ CO_2}{m^3} \times \frac{kg\ of\ CO_2}{10^6\ mg\ of\ CO_2} \times \frac{1\ m^3\ of\ air}{1.293\ kg\ of\ air} \times 5.14 \times 10^{18}\ kg$$

of air in the atmosphere, and

$$m_{CO_2\ added} = 7.87 \times 10^{14}\ kg\ of\ CO_2$$

Approximately 7.87×10^{14} kg of CO_2 have been added to the atmosphere since the Industrial Revolution.

10.2 Carbon Sources and Emissions

Since mass is neither created nor destroyed (except in nuclear reactions, and we will neglect those for the time being), the carbon added to the atmosphere must have come from somewhere. We have already discussed biogeochemical cycles. Just as we have followed water molecules in the hydrologic cycle, so too can we trace the movement of carbon atoms from one repository to another.

Carbon concentrations in the atmosphere have increased significantly since the start of the Industrial Revolution. We can examine the possible sources and sinks of carbon in the Earth's atmosphere.

The total U.S. energy use is shown in Figure 10.4. Information about energy use and fossil fuel consumption can be used to calculate the carbon emissions associated

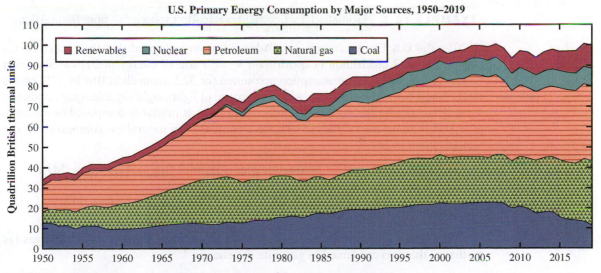

FIGURE 10.4 Historic energy sources and consumption in the United States.

Source: U.S. EIA (2020). *Monthly Energy Review*, April 2020. U.S. Energy Information Administration, Office of Energy Statistics, U.S. Department of Energy, Washington, DC.

with the conversion of fossil fuels to energy. In general, the combustion of fossil fuels can be described by

$$\text{Fuel} + \text{oxygen} \rightarrow \text{energy} + \text{carbon dioxide} + \text{water}$$

Depending on the chemical composition of the fossil fuel, differing amounts of energy and carbon dioxide are produced. Table 10.1 shows the differences in typical net heating values for common fuels.

TABLE 10.1 Typical heating values and default carbon dioxide emission factors for common fuels

FUEL	NET HEATING VALUE (MJ/kg)	DEFAULT CARBON DIOXIDE EMISSION FACTORS (kg CO$_2$/TJ)
Peat	10.4	106,000
Wood, oak	13.3–19.3	95,000–132,000
Wood, pine	14.9–22.3	95,000–132,000
Coal, anthracite	25.8	98,300
Charcoal	26.3	112,000
Coal, bituminous	28.5	94,600
Fuel oil, no 6 (bunker C)	42.5	73,300
Fuel oil, no 2 (home heating oil)	45.5	77,400
Gasoline (84 octane)	48.1	69,300
Natural gas (density = 0.756 kg/m³)	53.0	56,100
Reference:	Davis and Masten (2009)	IPCC (2006)

EXAMPLE 10.2   Estimating U.S. Contributions to Atmospheric CO$_2$ from Natural Gas

In 2019, the U.S. Energy Information Administration estimated that the United States consumed 100.2 quadrillion (1 quadrillion = 10^{15}) Btu of energy in 2019 as shown in Figure 10.5. Natural gas consumption accounted for 32.1 quadrillion Btu, or 32% of the total energy. Natural gas is composed of a mixture of lightweight organic compounds, but we may assume for this estimate that the natural gas is primarily composed of methane (CH$_4$). How many metric tons of CO$_2$ were emitted from natural gas combustion in the United States in 2019?

We begin by converting the English energy units of Btu to megajoules (MJ):

$$32.1 \times 10^{15}\,\text{Btu} \times \frac{1,055\,\text{J}}{\text{Btu}} \times \frac{1\,\text{MJ}}{10^6\,\text{J}} = 3.39 \times 10^{13}\,\text{MJ}$$

We use the heating values from Table 10.1 to calculate the mass of natural gas (as methane) that was consumed to generate the energy:

$$3.39 \times 10^{13}\,\text{MJ} \times \frac{1\,\text{kg}}{53.0\,\text{MJ}} = 6.39 \times 10^{11}\,\text{kg of methane}$$

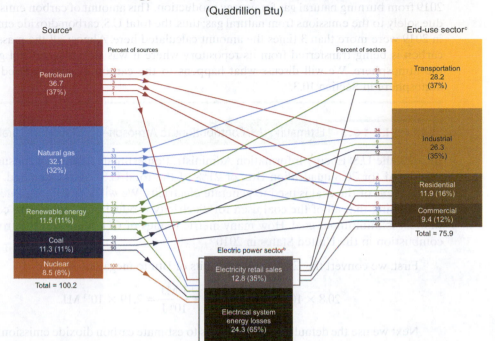

U.S. energy consumption by source and sector, 2019
(Quadrillion Btu)

ᵃ Primary energy consumption. Each energy source is measured in different physical units and converted to common British thermal units (Btu). See U.S. Energy Information Administration (EIA), *Monthly Energy Review*, Appendix A. Noncombustible renewable energy sources are converted to Btu using the "Fossil Fuel Equivalency Approach", see EIA's *Monthly Energy Review*, Appendix E.
ᵇ The electric power sector includes electricity-only and combined-heat-and-power (CHP) plants whose primary business is to sell electricity, or electricity and heat, to the public. Energy consumed by these plants reflects the approximate heat rates for electricity in EIA's *Monthly Energy Review*, Appendix A. The total includes the heat content of electricity net imports, not shown separately. Electrical system energy losses are calculated as the primary energy consumed by the electric power sector minus the heat content of electricity retail sales. See Note 1, "Electrical System Energy Losses," at the end of EIA's *Monthly Energy Review*, Section 2.
ᶜ End-use sector consumption of primary energy and electricity retail sales, excluding electrical system energy losses from electricity retail sales. Industrial and commercial sectors consumption includes primary energy consumption by combined-heat-and-power (CHP) and electricity-only plants contained within the sector.
Note: Sum of components may not equal total due to independent rounding. All source and end-use sector consumption data include other energy losses from energy use, transformation, and distribution not separately identified. See "Extended Chart Notes" on next page.
Sources: EIA, *Monthly Energy Review* (April 2020), Tables 1.3 and 2.1-2.6.

FIGURE 10.5 Primary energy consumption by source and sector in the United States in 2019.

Source: U.S. EIA (2020). *Monthly Energy Review*, April 2020. U.S. Energy Information Administration, Office of Energy Statistics, U.S. Department of Energy, Washington, DC.

We must balance the chemical equation for the combustion of methane to determine the molar ratio of CO_2 to methane:

$$CH_4 + 2O_2 \rightarrow CO_2 + 2H_2O$$

Then the molecular weight of methane and carbon dioxide is

$$MW_{methane} = 1 \times 12 + 4 \times 1 = 16 \frac{g}{mol}$$

$$MW_{carbon\ dioxide} = 1 \times 12 + 2 \times 16 = 44 \frac{g}{mol}$$

Converting the mass of methane consumed to the mass of carbon dioxide produced, we find

$$6.39 \times 10^{11} \text{ kg of methane as } CH_4 \times \frac{1{,}000 \text{ g}}{1 \text{ kg}} \times \frac{1 \text{ mol } CH_4}{16 \text{ g of } CH_4} \times \frac{1 \text{ mol } CO_2}{1 \text{ mol } CH_4}$$

$$\times \frac{44 \text{ g } CO_2}{1 \text{ mol } CO_2} \times \frac{\text{ton}}{10^6 \text{ g}} = 1.76 \times 10^9 \text{ metric tons of } CO_2$$

Over 1 billion metric tons of carbon dioxide were emitted by the United States in 2019 from burning natural gas for energy production. This amount of carbon emissions is due solely to the emissions from natural gas; thus, the total U.S. carbon dioxide emissions in 2019 were more than 3 times the amount calculated here. Almost all the mass of the carbon is being transferred from its repository where it was stored as natural gas into the atmosphere. We will discuss what happens to the carbon dioxide emitted to the atmosphere in Section 10.3.

EXAMPLE 10.3   Estimating U.S. Contributions to Atmospheric CO_2 from Natural Gas

In 2010, the U.S. Energy Information Administration estimated that coal consumption accounted for 20.8 quadrillion Btu, or 21% of the total U.S. energy demand. Ninety two percent of the coal is used to generate electricity. We will simplify our calculation by assuming that all of the coal used for electricity production is anthracite coal and the rest is bituminous coal. How many metric tonnes of CO_2 were emitted from coal combustion in the United States in 2010?

First, we convert the English energy units of Btu to megajoules (MJ):

$$20.8 \times 10^{15}\,\text{Btu} \times \frac{1{,}055\,\text{J}}{\text{Btu}} \times \frac{1\,\text{MJ}}{10^6\,\text{J}} = 2.19 \times 10^{13}\,\text{MJ}$$

Next we use the default emission factors to estimate carbon dioxide emission factors from Table 10.1 for the two sources of coal:

$$1 \text{ terajoule (TJ)} = 10^6 \text{ megajoule (MJ)} = 10^{12} \text{ joule (J)}$$

$$2.19 \times 10^{13}\,\text{MJ} \times 0.92 \times \frac{\text{TJ}}{10^6\,\text{MJ}} \times \frac{98.3\,\text{tonnes }CO_2}{\text{TJ anthracite coal}}$$

$$= 1.98 \times 10^9 \text{ tonnes of carbon dioxide from anthracite}$$

$$2.19 \times 10^{13}\,\text{MJ} \times 0.08 \times \frac{\text{TJ}}{10^6\,\text{MJ}} \times \frac{94.6\,\text{tonnes }CO_2}{\text{TJ bituminous coal}}$$

$$= 1.66 \times 10^8 \text{ tonnes of carbon dioxide from bituminous}$$

Total CO_2 emissions from both anthracite and bituminous coal = 2.15×10^9 tonnes.

Over 2 billion tonnes of carbon dioxide were emitted by the United States in 2010 from burning coal, primarily for electricity production. While slightly less coal energy is used, the carbon dioxide emissions associated with coal use are nearly twice the total carbon dioxide emissions from natural gas. This is due to a higher carbon intensity value for coal than for natural gas.

EXAMPLE 10.4   Comparing Carbon Intensities for Fuels

Due to the chemical structure and energy efficiency associated with the chemical energy stored in different fuels, some fuels emit less carbon dioxide per unit energy than others. In the previous example, we compared the total mass of carbon dioxide emitted from natural gas and coal in the United States. Calculate the greater carbon intensity from burning coal and charcoal (fuels more commonly used in low-income countries) by comparing their carbon dioxide emissions to natural gas.

Carbon intensity of coal compared to natural gas is

$$\frac{98,300 \ \frac{kg}{TJ} \ \text{anthracite coal}}{56,100 \ \frac{kg}{TJ} \ \text{natural gas}} = 1.75$$

So burning coal will emit 175% of the carbon dioxide if the same amount of energy is derived from natural gas.

Carbon intensity of charcoal compared to natural gas is

$$\frac{112,000 \ \frac{kg}{TJ} \ \text{charcoal}}{56,100 \ \frac{kg}{TJ} \ \text{natural gas}} = 2.00$$

Burning charcoal, a practice in many low-income countries with high rates of population growth, emits twice the amount of carbon dioxide compared to if natural gas had been used to generate the energy.

ACTIVE LEARNING EXERCISE 10.1 Draw Your Personal Carbon Cycle

Carbon atoms are essential building blocks for all life forms on Earth. The carbon atom shares electrons with other atoms. The ability to share electrons between carbon molecules and other elements makes the carbon atom unique. Carbon can form bonds with hydrogen to form methane (CH_4). Carbon atoms may share atoms with oxygen to form CO_2, CO in the gas phase, and CO_3^{2-} in aqueous solutions. Organic chemistry is the field that studies carbon–carbon bonded elements that are naturally occurring and can be synthesized (or can form) in the laboratory and for industrial use.

The following steps should be followed in order to develop a carbon mass balance equation.

1. Draw a detailed schematic of the major carbon sources and sink for the Earth's atmosphere. You may use the conceptual carbon cycle model shown in the accompanying figure to start your analysis.
2. Show the system boundaries and define the control volume. Note that the mass of the atmosphere is 5.14×10^{18} kg. For the purposes of our simple atmospheric model, determine the volume of the atmosphere in m³.
3. Define the variables and list the known data and assumption(s) needed to solve a mass balance equation for carbon in the atmosphere.
4. Create a table that lists each variable in units of
   a) Volume = m³ AND liters
   b) Mass = kg AND g
   c) Energy = Btu AND joules
   d) Concentrations in air = ppm_v AND mg/m³
   e) Concentrations in water = ppm_m AND mg/L

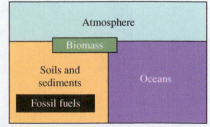

Carbon dioxide and other gases that increase the energy absorbed by the atmosphere, called *greenhouse gases*, are emitted from anthropogenic and natural sources. Approximately 73.2% of greenhouse gas emissions in 2016 were due to energy uses, such as electricity production, heating, and cooling (see Figure 10.6). Agriculture and land use emissions were the second highest source of greenhouse gas emissions in 2016 and accounted for 18.4% of emissions. Carbon dioxide is removed from plant growth, particularly in densely forested areas; the deforestation due to changes in land use for agriculture or the built environment accounted for 2.2% of the increase in greenhouse gas emissions in 2016. Energy use in built infrastructure (17.5%) and transportation (16.2%) produces about one-third of all greenhouse gases. Iron and steel production (7.2%), chemical and petrochemical production (3.6%), and cement production (3%) are the largest industrial contributors to greenhouse gases in the atmosphere. Figure 10.6 illustrates that most greenhouse gases come from our infrastructure and the production of goods and services, with only 3.2% of emissions associated with the treatment of wastewater and landfilling waste.

The sum total of worldwide emissions of carbon dioxide from fossil fuel use has been compared to changes in atmospheric carbon dioxide levels, and as shown in Figure 10.7, there is a strong correlation between these two values. However, not all of the carbon dioxide emitted stays in the atmosphere, as illustrated in Figure 10.8. Where does the carbon emitted from burning fossil fuels go? In order to answer this question, we have to look more closely at the biogeochemical cycle for carbon.

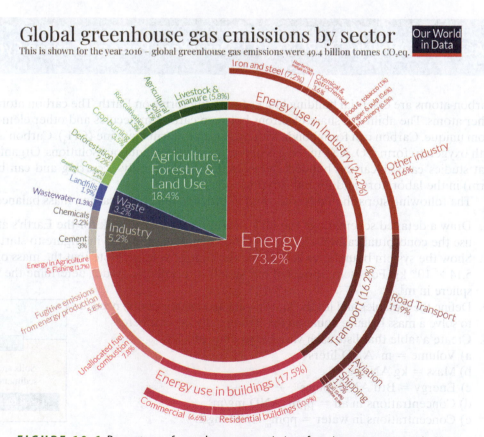

FIGURE 10.6 Percentage of greenhouse gas emissions from important sectors in 2016.

Source: Climate Watch, The World Resources Institute (2020). OurWorldinData.org: www.visualcapitalist.com/a-global-breakdown-of-greenhouse-gas-emissions-by-sector/.

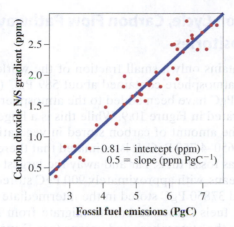

FIGURE 10.7 The difference between CO_2 concentration in the Northern Hemisphere and Southern Hemisphere (y-axis), computed as the difference between annual mean concentrations (ppm) at Mauna Loa and the South Pole (Keeling and Whorf, 2005, updated), compared with annual fossil fuel emissions (x-axis; PgC; Marland et al., 2006), with a line showing the best fit. The observations show that the north–south difference in CO_2 increases proportionally with fossil fuel use, verifying the global impact of human-caused emissions.

Source: Based on IPCC (2007a). *Climate Change 2007: The Physical Science Basis*. Contributions of Working Group I to the Fourth Assessment Report of the Intergovernmental Panel on Climate Change [S. Solomon, D. Qin, M. Manning, Z. Chen, M. Marquis, K.B. Averyt, M. Tignor, and H.L. Miller (eds.)] Cambridge University Press, Cambridge, United Kingdom, and New York, NY. pp. 996. Figure 7.5.

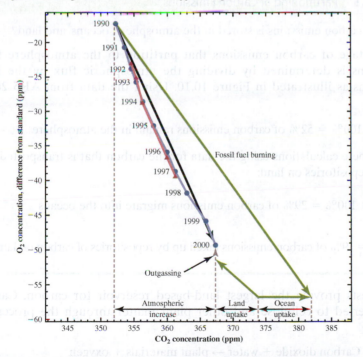

FIGURE 10.8 Partitioning of fossil fuel CO_2 captured by the atmosphere based on the relatively well known $O_2 : CO_2$ stoichiometric relationship of the different fuel types. Uptake by land and ocean is constrained by the known $O_2 : CO_2$ stoichiometric ratio of these processes, defining the slopes of the respective arrows.

Source: Based on Prentice, I.C., Farquhar, G.D., Fasham, M.J.R., Goulden, M.L., Heimann, M., Jaramillo, V.J., Kheshgi, H.F., LeQuere, C., Scholes, R.J., and Wallace, D.W.R. (2001). "The Carbon Cycle and Atmospheric Carbon Dioxide." In J.T. Houghton et al. (eds.), *Climate Change 2001: The Scientific Basis*, pp. 183–237. Cambridge University Press.

10.3 The Carbon Cycle, Carbon Flow Pathways, and Repositories

The atmosphere contains only a small fraction of the world's carbon. Until relatively recently, the atmosphere contained about 589 PgC (1 PgC = 10^{15} grams) of carbon. Over 240 PgC have been added to the atmosphere since the Industrial Revolution, as illustrated in Figure 10.9. While this is a huge mass of carbon, it is small compared to the amount of carbon stored in vegetation, soil, and detritus that contain about 3,650–4,750 PgC. It is estimated that there are also 1,002–1,940 Pg of carbon stored as fossil fuels. Far and away the largest carbon reservoirs on the planet are the oceans, with approximately 900 PgC stored in the surface water and marine biota and 37,850 PgC stored in the intermediate and deep ocean. The combustion of fossil fuels causes carbon to migrate from fossilized deposits in geologic reservoirs to the atmosphere and oceans (see Figure 10.10).

The IPCC Fifth Assessment Report (AR5) has reported the ongoing efforts of several climate models to calculate the net mass flow rate, or flux, from one carbon repository to another. Model results from 1750 to 2011 are shown in Table 10.2. The most recent model, AR5 2002–2011, shows that the net worldwide emissions of carbon from fossil fuel combustion and cement production are estimated to be 8.3 ± 0.7 PgC per year. The atmospheric increase of carbon is only estimated to be 4.3 ± 0.2 PgC per year.

EXAMPLE 10.5   Partitioning of Carbon Emissions

What percent of carbon emissions is stored in the atmosphere, oceans, and land?

The percentage of carbon emissions that partition to the atmosphere from carbon emissions is determined by dividing the atmospheric flux by the total carbon emissions as illustrated in Figure 10.10. Using the data from AR5 2002–2011 yields

$$\frac{4.3}{8.3} \times 100\% = 52\% \text{ of carbon emissions remain in the atmosphere}$$

We repeat these calculations using the data for the carbon that is transported into the oceans and repositories on land:

$$\frac{2.4}{8.3} \times 100\% = 29\% \text{ of carbon emissions migrate into the oceans}$$

$$\frac{1.6}{8.3} \times 100\% = 19\% \text{ of carbon emissions taken up by repositories of carbon on land}$$

Global forests provide the largest land-based reservoir for carbon. Carbon dioxide is converted to cellulose and other plant matter through the process of photosynthesis:

$$\text{Carbon dioxide + water} \rightarrow \text{plant materials + oxygen}$$

A simple chemical process, whereby carbon dioxide can form glucose, which is a common organic energy-containing compound (Figure 10.11), is represented by the chemical reaction

$$6CO_2 + 6H_2O \xrightarrow{\text{uv}} C_6H_{12}O_6 + 6O_2 \tag{10.1}$$

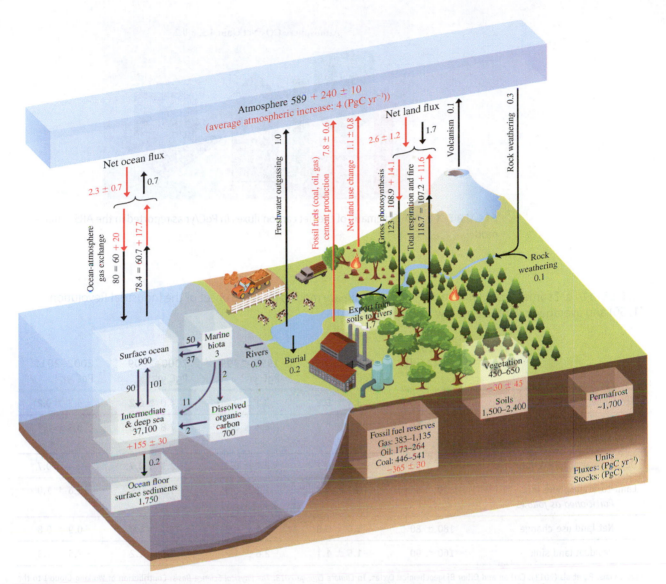

FIGURE 10.9 A simplified schematic of the global carbon flux and biogeochemical pathways. Numbers represent reservoir mass, also called "carbon stocks," in PgC (1 PgC = 10^{15} g C) and annual carbon exchange fluxes (in PgC yr^{-1}). Black numbers and arrows indicate reservoir mass and exchange fluxes estimated for the time prior to the Industrial Revolution in about 1750. Red arrows and numbers indicate annual anthropogenic fluxes averaged over the 2000–2009 time period. Fluxes from volcanic eruptions, rock weathering (silicates and carbonates weathering reactions resulting in a small uptake of atmospheric CO_2), export of carbon from soils to rivers, burial of carbon in freshwater lakes and reservoirs, and transport of carbon by rivers to the ocean are all assumed to be pre-industrial fluxes—that is, unchanged during 1750–2011. The atmospheric inventories have been calculated using a conversion factor of 2.12 PgC per ppm.

Source: Based on Ciais, P., et al. (2013). Carbon and Other Biogeochemical Cycles. In *Climate Change 2013: The Physical Science Basis*. Contribution of Working Group I to the Fifth Assessment Report of the Intergovernmental Panel on Climate Change. Cambridge University Press, Cambridge, United Kingdom, and New York, NY.

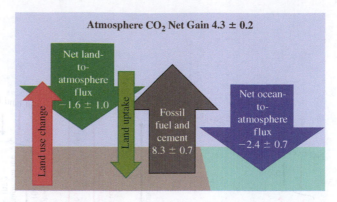

FIGURE 10.10 Schematic of the net carbon fluxes in PgC/yr as reported in the AR5 climate model.

TABLE 10.2 Estimates of global carbon fluxes by climate models from the onset of the Industrial Revolution (1750) and averaged over the past few decades up to 2011

	1750–2011 CUMULATIVE PgC	1980–1989 PgC yr^{-1}	1990–1999 PgC yr^{-1}	2000–2009 PgC yr^{-1}	2002–2011 PgC yr^{-1}
Atmospheric increase	240 ± 10	3.4 ± 0.2	3.1 ± 0.2	4.0 ± 0.2	4.3 ± 0.2
Fossil fuel combustion and cement production	375 ± 30	5.5 ± 0.4	6.4 ± 0.5	7.8 ± 0.6	8.3 ± 0.7
Ocean-to-atmosphere flux	−155 ± 30	−2.0 ± 0.7	−2.2 ± 0.7	−2.3 ± 0.7	−2.4 ± 0.7
Land-to-atmosphere flux *Partitioned as follows*	30 ± 45	−0.1 ± 0.8	−1.1 ± 0.9	−1.5 ± 0.9	−1.6 ± 1.0
Net land use change	180 ± 80	1.4 ± 0.8	1.5 ± 0.8	1.1 ± 0.8	0.9 ± 0.8
Residual land sink	−160 ± 90	−1.5 ± 1.1	−2.6 ± 1.2	−2.6 ± 1.2	−2.5 ± 1.3

Source: Ciais, P., et al. (2013). Carbon and Other Biogeochemical Cycles. In *Climate Change 2013: The Physical Science Basis*. Contribution of Working Group I to the Fifth Assessment Report of the Intergovernmental Panel on Climate Change. Cambridge University Press, Cambridge, United Kingdom, and New York, NY.

Photosynthesis changes atmospheric carbon dioxide to plant tissue material, such as cellulose (see Figure 10.12). Photosynthesis and uptake of carbon by the land-based process historically removed about 120 PgC per year, and since the Industrial Revolution, an additional 2.6 PgC per year has been removed from the atmosphere. However, land-use changes, mainly deforestation associated with agriculture and cooking fuel harvesting of trees, result in an additional 1.6 PgC returned to the atmosphere. Respiration processes from simple organisms as well as more complex organisms account for the addition of 119.6 PgC per year to the atmosphere. The respiration process is the opposite process of photosynthesis and strongly resembles the combustion process of extracting energy from carbon sources. Animals use sugars, like glucose (see Figure 10.11),

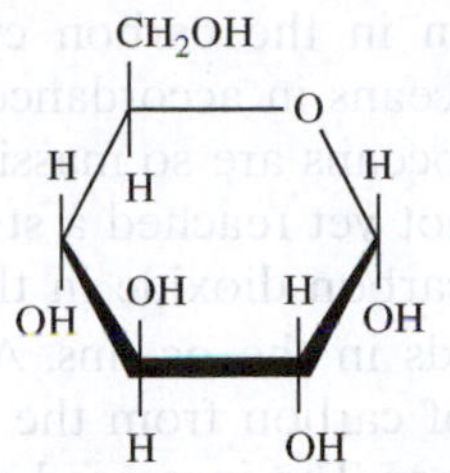

FIGURE 10.11 Chemical structure for glucose.

carbohydrates, and proteins in combination with oxygen to produce energy, cell growth, water, and carbon dioxide via the representative reaction:

$$\text{Glucose} + \text{oxygen} \rightarrow \text{water} + \text{carbon dioxide}$$

$$C_6H_{12}O_6 + O_2 \xrightarrow{\text{energy}} \text{cell mass} + CO_2 + H_2O \qquad (10.2)$$

An important aspect of the movement of carbon from one repository to another is how long the carbon will remain in that repository. The residence time, t_r, is the term used to describe how long a substance remains as a defined volume, such as the atmosphere, a lake, or a small reactor in a laboratory. The residence time can be determined by dividing the system volume, V, by the volumetric flow rate, Q, of the material, or alternatively, by dividing the mass of material, m, in the control volume by the mass flow rate of material, M, flowing through the volume:

$$t_r = \frac{V}{Q} = \frac{m}{M} \qquad (10.3)$$

EXAMPLE 10.6 Calculating Residence Time

The carbon flux to the atmosphere is approximately 55.3×10^{12} kg/yr. The approximate steady state mass of carbon in the atmosphere is 7.35×10^{12} kg. How long, on average, is carbon likely to remain in the atmosphere before cycling through another repository?

$$t_r = \frac{V}{Q} = \frac{m}{M} = \frac{7.35 \times 10^{12}\,\text{kg}}{55.3 \times 10^{12}\,\text{kg/yr}} = 0.133\,\text{years} \times \frac{365\,\text{days}}{\text{year}} = 48.5\,\text{days}$$

Thus, the average time carbon molecules spend in the atmosphere is about 48.5 days.

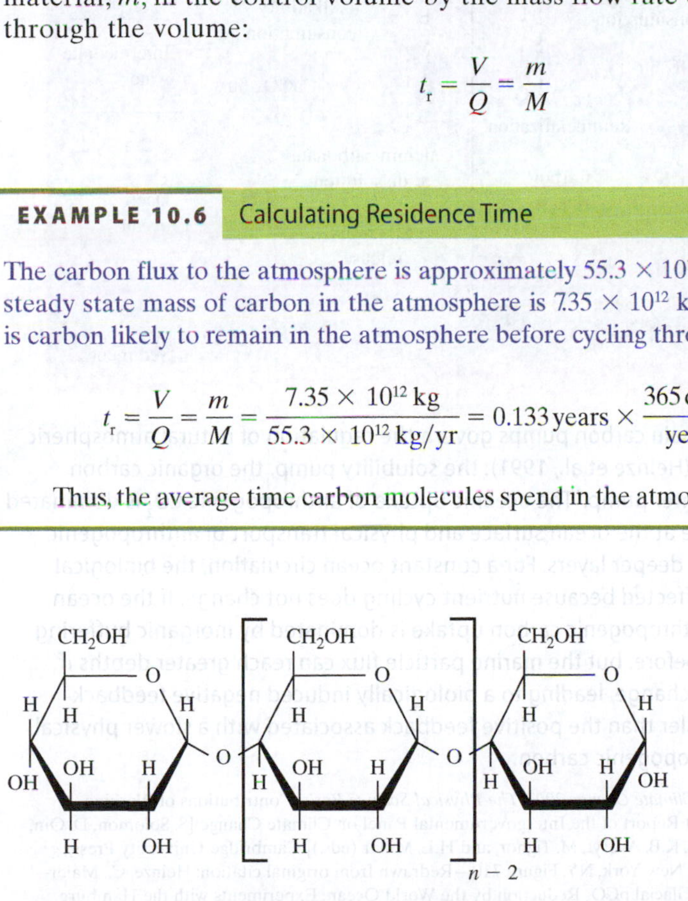

FIGURE 10.12 Chemical structure for cellulose.

By far the largest repository of carbon in the carbon cycle is the Earth's oceans. Carbon dioxide dissolves into the oceans in accordance with Henry's law, as illustrated in Figure 10.13. However, the oceans are so massive that the current atmospheric levels of carbon dioxide have not yet reached a steady state, or equilibrium condition, between the amount of carbon dioxide in the atmosphere and the amount of carbon-containing compounds in the oceans. As the atmospheric levels of carbon dioxide increase, transfer of carbon from the atmosphere to the oceans has also increased by 22.2 PgC per year. The increased ocean temperatures have also increased the amount of carbon dioxide transferred from the oceans to

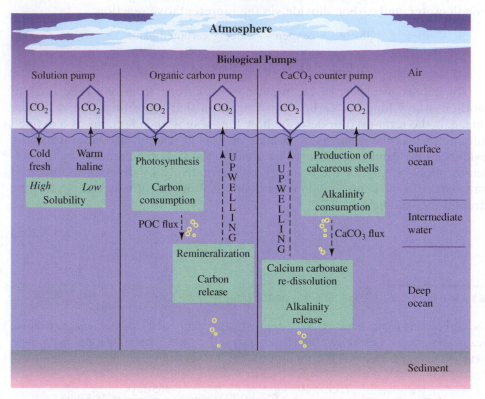

FIGURE 10.13 Three main carbon pumps govern the regulation of natural atmospheric CO_2 changes by the ocean (Heinze et al., 1991): the solubility pump, the organic carbon pump, and the $CaCO_3$ "counter pump." The oceanic uptake of anthropogenic CO_2 is dominated by inorganic carbon uptake at the ocean surface and physical transport of anthropogenic carbon from the surface to deeper layers. For a constant ocean circulation, the biological carbon pumps remain unaffected because nutrient cycling does not change. If the ocean circulation slows down, anthropogenic carbon uptake is dominated by inorganic buffering and physical transport as before, but the marine particle flux can reach greater depths if its sinking speed does not change, leading to a biologically induced negative feedback that is expected to be smaller than the positive feedback associated with a slower physical downward mixing of anthropogenic carbon.

Source: Based on IPCC (2007a). *Climate Change 2007: The Physical Science Basis*. Contributions of Working Group I to the Fourth Assessment Report of the Intergovernmental Panel on Climate Change [S. Solomon, D. Qin, M. Manning, Z. Chen, M. Marquis, K.B. Averyt, M. Tignor, and H.L. Miller (eds.)] Cambridge University Press, Cambridge, United Kingdom, and New York, NY. Figure 7.10—Redrawn from original citation: Heinze, C., Maier-Reimer, E., and Winn, K. (1991) "Glacial pCO₂ Reduction by the World Ocean: Experiments with the Hamburg Carbon Cycle Model." *Paleoceanography* 6(4):395–430.

the atmosphere by 20 PgC per year. The net result of this changing, nonequilibrium situation is a net increase in carbon removed by the atmosphere, so that 2.4 PgC per year, or 29% of the carbon emitted, is transferred to the planet's oceans. This increase in the carbon content of the oceans has already had a profound effect by increasing the acidity of the oceans.

In order to understand the environmental impacts of this change to the carbon cycle, we must evaluate the energy balances used to describe the average surface temperature of the planet. The energy balances calculated in the next section of this chapter will allow us to predict a range of potential impacts of our changing climate on our environment.

10.4 Global Energy Balance

The IPCC has summarized some of the key questions people and policymakers have about our changing climate in Figure 10.14. We have already discussed present and past conditions related to carbon dioxide concentrations in the atmosphere, but we have not yet determined how carbon dioxide concentrations are related to climate. To make this connection, we must review the thermodynamic energy balance Callendar studied to link carbon dioxide concentrations in the atmosphere to the changes occurring in our climate.

In order to analyze the effect of atmospheric gas concentrations on the Earth's climate, we must be able to perform an energy balance on the planet. The law of conservation of energy states that energy cannot be created or destroyed. The energy balance within a given control volume is similar in form to that of the general mass balance equation, where

$$\text{Change in energy} = \text{energy inputs} - \text{energy outputs} \tag{10.4}$$

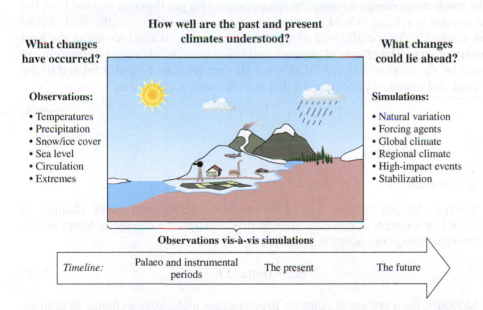

FIGURE 10.14 Key questions about the climate system and its relationship to humankind.

Source: Based on IPCC (2001). *Climate Change 2001: Synthesis Report.* A Contribution of Working Groups I, II, and III to the Third Assessment Report of the Intergovernmental Panel on Climate Change [R.T. Watson and the Core Writing Team (eds.)]. Cambridge University Press, Cambridge, United Kingdom, and New York, NY.

This is the general mass balance equation presented in Chapter 4. The first law of thermodynamics states that energy cannot be created or destroyed, but we can change the form of the energy. (Once again we'll simplify reality slightly by neglecting the processes of nuclear fission and fusion, which are reactions that typically take place only in nuclear reactors.) Energy is defined as the capacity to do work. Work is defined as the force acting on a body through a distance. The SI unit of a joule is defined as the constant force of 1 newton that is applied to a body that is moved one meter in the direction of that force. The first law of thermodynamics relates work and energy by

$$Q_H = U_2 - U_1 + W \tag{10.5}$$

where

Q_H = heat absorbed
U_1 = energy of the system at state 1
U_2 = energy of the system at state 2
W = work

Energy has many different forms, and a complete analysis of general thermodynamics is beyond the scope of this text. However, some basic terms and concepts must be introduced. Energy stored as thermal energy is measured in the system by a calorie. One *calorie* is defined as the amount of energy required to raise 1 gram of water 1 degree Celsius from 14.5°C to 15.5°C. The ***British thermal unit (Btu)*** is defined as the amount of energy required to raise 1 pound of water 1 degree Fahrenheit.

The ***specific heat*** of a substance is the quantity of heat required to increase a unit mass of the substance by 1 degree with units of energy/(mass-temperature). For most solids and liquids, the specific heat varies only slightly with temperature. However, gases may expand when heated, doing work on the environment. Thus it takes much more energy to raise the temperature of a gas that can expand than the heat needed if volume is held constant. For this reason, either specific heat at constant volume, c_v, or specific heat at constant pressure, c_p, is used to define the heat capacity of a gas. The effects of pressure and volume on the energy of the system are related by the term *enthalpy*. Enthalpy is a thermodynamic property related to the material and chemical properties of that material and is defined as

$$H = U + PV \tag{10.6}$$

where

U = internal energy
P = pressure
V = volume

Energy changes can be related to internal energy changes or changes in enthalpy. For a system at constant volume that undergoes a change in temperature, the internal change in energy is described by

$$\Delta U = (m)(c_v)\Delta T \tag{10.7}$$

Similarly, for a system at constant pressure that undergoes a change in temperature, the change in enthalpy is given by

$$\Delta H = (m)(c_p)\Delta T \tag{10.8}$$

For most liquids and solids in the natural environment that do not change phase or change substantially in pressure or volume, $c_v = c_p = c$ and $\Delta U = \Delta H$. The change in the energy stored within the system due to a change in temperature, if c is assumed constant over the temperature range, is given by

$$\text{Change in stored energy} = (m)(c)\Delta T \qquad (10.9)$$

When enthalpy changes are considered, the energy balance equation can be written as

$$\frac{\Delta dH}{dt} = \frac{d(H)_{in}}{dt} + \frac{d(H)_{out}}{dt} \qquad (10.10)$$

The energy accumulation within a system may be a result of the transfer of energy to the system by conduction, convection, or radiation. Conduction is the direct physical transfer of heat via molecular diffusion from a substance of greater heat (or energy) coming into direct contact with a substance of less heat (or energy). Convective heat transfer occurs via fluid motion—for example, by air currents or ocean currents, such as those that move heat from the warmer equator toward the cooler polar regions. Radiation is the heat emitted by an object that is absorbed by the surrounding objects and is due to the emission of electromagnetic radiation.

Heat loss through the walls of a home occurs because of a combination of conductive, convective, and radiant energy exchange. The overall heat transfer process can be estimated by

$$q = \frac{A(T_i - T_o)}{R} \qquad (10.11)$$

where
  q = heat transfer through a surface (energy/time)
  A = area of the surface through which the heat transfers
  T_i = air temperature within the surface
  T_o = air temperature outside the heat transfer surface
  R = resistance or R-value [area-degree/energy (typically m²-°C/W or hr-ft²-°F/Btu)]

Radiative heat transfer is related to the frequency and wavelength of the radiation. Incoming solar radiation is divided into several different wavelengths as shown in Figure 10.15. The wavelength and frequency of the radiation are related by the speed of light, which remains constant:

$$\text{Speed of light} = c_o = 2.998 \times 10^8 \text{ m/s} = (\lambda)(v) \qquad (10.12)$$

where
  λ = wavelength (m)
  v = frequency (1/s or hertz)

The energy associated with the radiation at a given frequency is related by Planck's law:

$$E = (h)(v) \qquad (10.13)$$

where h = Planck's constant = 6.6256×10^{-34} J-s.

As shown in Figure 10.15, energy may be radiated from a body, such as the sun or Earth, over a range of wavelengths. A simplifying assumption in thermodynamic analysis is often to model an object as a blackbody—that is, a plain object of simple geometry. The **blackbody temperature** is defined as the maximum amount of radiation an object or body can emit at a given temperature. The amount of energy at any wavelength is called the spectral emissive power, $E_{b-\lambda}$, and is given by the Planck distribution law:

$$E_{b-\lambda} = \frac{2\pi hc_o^2}{\lambda^5}\left[\exp\left(\frac{hc_o}{\lambda kT}\right) - 1\right]^{-1} \tag{10.14}$$

where

$E_{b-\lambda}$ = spectral emissive power (W/m²-μm)
λ = wavelength of the radiant energy (μm)
k = Boltzmann constant = 1.3805×10^{-23} (J/K)
T = absolute temperature of the radiating body (K)

Integrating over the spectrum of wavelengths for the blackbody yields the Stefan–Boltzmann law of radiation:

$$E_{bb} = \sigma A T^4 \tag{10.15}$$

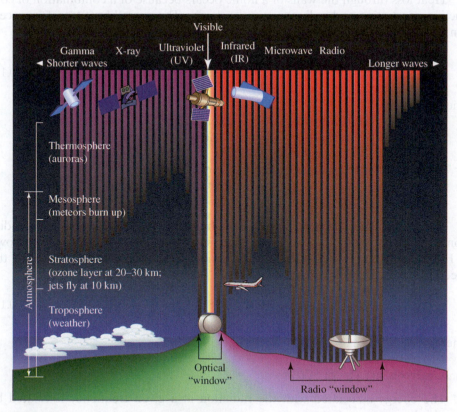

FIGURE 10.15 Types of incoming solar radiation and their penetration into the layers of the Earth's atmosphere.

Source: STScI/JHU/NASA.

where

E_{bb} = total blackbody emission rate (W)
σ = Stefan–Boltzmann constant = 5.67×10^{-8} W/(m²-K⁴)
T = absolute temperature (K)
A = surface area of the object (m²)

10.5 Global Energy Balance and Surface Temperature Model

The Earth can be modeled as a simple blackbody to gain an understanding of the principles of how the sun heats the Earth and to estimate the Earth's average surface temperature. The incoming solar radiation is quite constant, with only small variations noted since AD 1600, as shown in Figure 10.16.

Solar irradiation strikes the Earth at a nearly constant energy of 1,365 W/m². Of this radiation, the atmosphere reflects, on average, 30% of the energy back into space. The percent reflected is called the **Earth's albedo**, and the albedo is a function of the

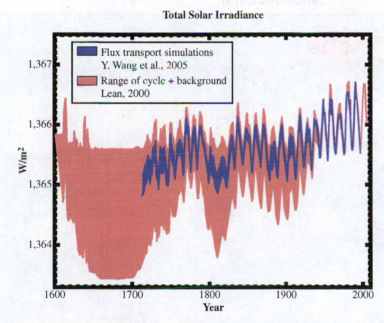

Total Solar Irradiance

FIGURE 10.16 Reconstruction of total solar irradiance time series starting as early as 1600. The upper envelope of the shaded regions shows irradiance variations arising from the 11-year activity cycle. The lower envelope is the total irradiance reconstructed by Lean (2000), in which the long-term trend was inferred from brightness changes in sun-like stars. In comparison, the recent reconstruction of Y. Wang et al. (2005) is based on solar considerations alone, using a flux transport model.

Source: Based on Forster, P., Ramaswamy, V., Artaxo, P., Berntsen, T., Betts, R., Fahey, D.W., Haywood, J., Lean, J., Lowe, D.C., Myhre, G., Nganga, J., Prinn, R., Raga, G., Schulz, M., and Van Dorland, R. (2007). Changes in Atmospheric Constituents and in Radiative Forcing. In: *Climate Change 2007: The Physical Science Basis*. Contribution of Working Group I to the Fourth Assessment Report of the Intergovernmental Panel on Climate Change [S. Solomon, D. Qin, M. Manning, Z. Chen, M. Marquis, K.B. Averyt, M.Tignor, and H.L. Miller (eds.)]. Cambridge University Press, Cambridge, United Kingdom, and New York, NY.

light reflected or deflected. For instance, ice coverage in the polar regions reflects more than 70% of the sun's energy, while the albedo near the equator is less than 20% (see Figure 10.17).

Assuming the Earth is a blackbody that absorbs the solar radiation as shown in Figure 10.18, we develop the following equations:

$$\text{Blackbody radiation absorbed by the Earth} = E_{ebb} = S_o \pi r_e^2 \tag{10.16}$$

$$\text{Energy absorbed by the Earth} = E_{abs} = S_o \pi r_e^2 (1 - \alpha) \tag{10.17}$$

The Stefan–Boltzmann law can be applied to the Earth, assuming the average temperature of the Earth is nearly constant. Balancing the energy absorbed by the Earth using Equation (10.17) with the Earth's blackbody temperature from Equation (10.15) yields

$$S_o \pi r_e^2 (1 - \alpha) = \sigma (4\pi r_e^2) T_e^4 \tag{10.18}$$

Suomi NPP VIIRS Global Land Surface Albedo
20150701-20150731

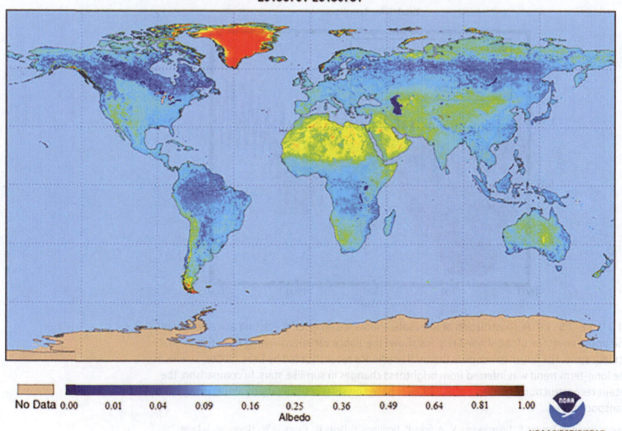

FIGURE 10.17 Surface reflectivity of the Earth based on NASA Suomi NPP VIIRS Global Land Surface Albedo Measurement 2015. Cells with missing data are colored tan.

Source: www.nesdis.noaa.gov/content/reflecting-reflection-suomi-npp-instruments-aid-monitoring-earth%E2%80%99s-albedo.

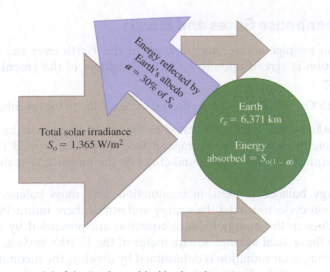

FIGURE 10.18 Model of the Earth as a blackbody radiator in space.

Source: Based on Masters, G.M., and Wendell, E. (2007). *Introduction to Environmental Engineering and Science.* Upper Saddle River, NJ: Pearson Education.

where T_e = average surface temperature of the Earth (K).

Solving for T_e yields

$$T_e = \left[\frac{S_o(1 - \alpha)}{4\sigma}\right]^{\frac{1}{4}} = \left[\frac{1{,}365(1 - 0.30)}{4(5.67 \times 10^{-8})}\right]^{\frac{1}{4}} = 254 \text{ K} = -18°C \qquad (10.19)$$

It is obvious that the average surface temperature of the Earth is not −18°C, which would mean a long-lasting ice age. Therefore, we know our model is not valid. What we have neglected in this model is the insulating capacity of the atmosphere. The Earth's atmosphere provides an insulating effect that keeps the Earth much warmer than the predicted −18°C. This insulation performs the same thermodynamic function as the insulation in a home, a coffee cup, or a greenhouse. The magnitude of the insulation effect is a function of the chemical properties that make up the atmosphere.

EXAMPLE 10.7   An Ice Planet Without an Atmosphere

If the Earth were completely covered with ice, so that the planet's albedo was 80%, what would be the Earth's blackbody radiation temperature?

Using Equation (10.19) and an albedo $\alpha = 0.80$ yields

$$T_e = \left[\frac{S_o(1 - \alpha)}{4\sigma}\right]^{\frac{1}{4}} = \left[\frac{1{,}365(1 - 0.80)}{4(5.67 \times 10^{-8})}\right]^{\frac{1}{4}} = 186 \text{ K} = -87°C$$

This would be a very cold planet indeed without the gases that make up and insulate our atmosphere.

10.6   Greenhouse Gases and Effects

The long-term average surface temperature of the Earth over the last epoch of human evolution is agreed upon to be 15°C. The effect of the greenhouse gases is approximately

$$15°C − T_e = 15°C − (−18°C) = 33°C \text{ effective increase due to the greenhouse gas effect}$$

The **_greenhouse gas effect_**, illustrated in Figure 10.19, makes the planet much more hospitable by increasing its average surface temperature by 33°C. The composition of the atmosphere has a profound effect on the magnitude of the greenhouse effect.

The energy balance is used, in conjunction with mass balance information from the carbon cycle, to model the energy and atmosphere interactions. Typically, numerical values in the energy balance equation are presented by averaging the effects over a theoretical average square meter of the Earth's surface. The effective average incoming solar radiation is determined by dividing the incoming irradiation by the surface area of the Earth:

$$\frac{S_o \pi r_e^2}{4 \pi r_e^2} = \frac{1{,}365\ \frac{W}{m^2}}{4} = 341\ \frac{W}{m^2} \text{ of incoming solar radiation}$$

Approximately 30% of the incoming solar radiation, or 102 W/m², is reflected back into space by the Earth's albedo. In order to maintain a constant temperature

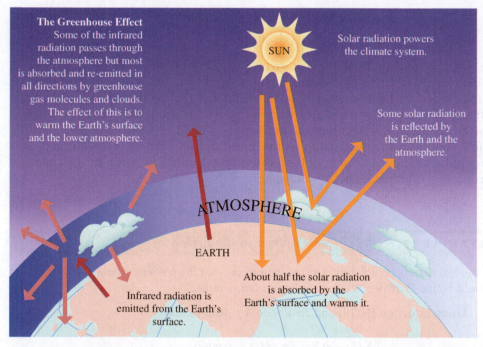

FIGURE 10.19 Schematic of an idealized model illustrating the natural greenhouse effect.

Source: Based on *Climate Change 2007: The Physical Science Basis*. Contributions of Working Group I to the Fourth Assessment Report of the Intergovernmental Panel on Climate Change [S. Solomon, D. Qin, M. Manning, Z. Chen, M. Marquis, K.B. Averyt, M. Tignor, and H.L. Miller (eds.)] Cambridge University Press, Cambridge, United Kingdom, and New York, NY.

Global Energy Flows W/m²

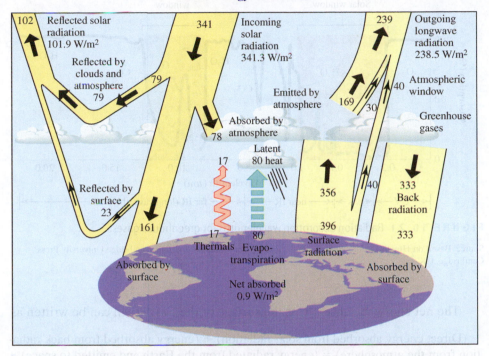

FIGURE 10.20 An energy balance on the Earth that includes atmospheric insulation effects.

Source: Based on Trenberth, K. E., Fasullo, J.R., and Kiehl, J. (2009). "Earth's Global Energy Budget." *Bull. Amer. Meteor. Soc.* 90(3):311–324.

on the Earth, 239 W/m² must be radiated back into space, as shown in Figure 10.20. The atmosphere absorbs about 33%, or 78 W/m², of the incoming irradiation that is not reflected, and the remaining 67%, or 161 W/m², is absorbed by the surface of the Earth.

The Earth absorbs the incoming solar radiation and, acting as a blackbody, radiates heat back into the atmosphere. Not all components of the atmosphere are capable of trapping the heat from the sun and Earth. The major components of the atmosphere, nitrogen as N_2 and oxygen as O_2, absorb very little of the radiation from the sun and Earth. The atmosphere's major energy-absorbing compounds are shown in Figure 10.21, and these greenhouse gases absorb various amounts of radiation, depending on the wavelength of the electromagnetic waves. Water vapor (H_2O) in the atmosphere absorbs radiation with wavelengths near 7 and 20 μm, and it is the most important greenhouse gas due to its fairly high concentration and the wide range of wavelengths over which it absorbs energy. Methane (CH_4) and nitrous oxide (N_2O) absorb radiation with wavelengths near 8 μm. Ozone (O_3) absorbs radiation with wavelengths near 10 μm. Carbon dioxide (CO_2) absorbs radiation with wavelengths near 15 μm. Radiation with wavelengths near 9 μm and from 10 to 13 μm largely passes through the atmosphere, and approximately 40 W/m² is radiated directly from the Earth back into space. The greenhouse gases in the atmosphere retain and radiate a tremendous amount of heat, 333 W/m², of radiation back to Earth.

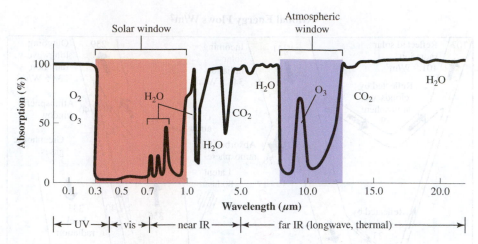

FIGURE 10.21 Radiation absorption wavelengths for greenhouse gases.

Source: Based on Houghton, J.T. (2009). *Global Warming: The Complete Briefing*. Cambridge University Press. Cambridge, UK.

The net energy balance across the surface of the Earth then can be written as

(Direct energy absorbed from solar radiation) + (energy absorbed from back radiation from the atmosphere) = (energy radiated from the Earth and emitted to space) + (energy radiation from the Earth to the atmosphere) + (thermal convective energy) + (energy lost to evapotranspiration)   (10.20)

Substituting the values from Figure 10.20 into Equation (10.20) yields

$$161 \frac{W}{m^2} + 333 \frac{W}{m^2} = 40 \frac{W}{m^2} + 356 \frac{W}{m^2} + 17 \frac{W}{m^2} + 80 \frac{W}{m^2}$$

Greenhouse gases in the Earth's atmosphere effectively act as a temperature regulator for the planet and control the Earth's climate. The ***radiative forcing*** for each greenhouse gas can be computed by calculating the amount of energy each gas absorbs in the atmosphere based on the combination of the concentration of the gas in the atmosphere and the ability of the gas to absorb energy over various wavelengths. Radiative forcings for the atmospheric greenhouse gases were calculated for the period between 1860 and 2000. The concentrations of the major greenhouse gases in 2005 and 2011 and their associated influence on the radiative forcing of the atmospheric greenhouse effect are shown in Table 10.3.

Since many of the concentrations of these gases changed during the period of industrialization from 1860 to 2000, scientists have also calculated the net change in the radiative forcing over this time period (see Figure 10.22). Carbon dioxide has had the largest net impact on increasing radiative forcing, and we can also see that there is very little uncertainty associated with this forcing. Simply put, we have added more insulation to our atmosphere, which will in turn make the planet warmer as carbon dioxide levels and temperatures approach a new equilibrium condition. It is important to notice that calculating the exact rise in the average surface temperature of the planet is made more complicated by additional positive and negative forcings from changes in other greenhouse gases and potential changes in the Earth's albedo due to increased cloud cover and aerosol effects.

TABLE 10.3 Average observed concentrations of greenhouse gases in the atmosphere in parts-per-trillion by volume and the radiative forcings associated with the gases at these concentrations (from 2013 IPCC report)

SPECIES	CONCENTRATIONS (ppt)		RADIATIVE FORCING[a] (W m^{-2})	
	2011	2005	2011	2005
CO_2 (ppm)	391 ± 0.2	379	1.82 ± 0.19	1.66
CH_4 (ppb)	1,803 ± 2	1,774	0.48 ± 0.05	0.47[e]
N_2O (ppb)	324 ± 0.1	319	0.17 ± 0.03	0.16
CFC-11	238 ± 0.8	251	0.062	0.065
CFC-12	528 ± 1	542	0.17	0.17
CFC-13	2.7		0.0007	
CFC-113	74.3 ± 0.1	78.6	0.022	0.024
CFC-115	8.37	8.36	0.0017	0.0017
HCFC-22	213 ± 0.1	169	0.0447	0.0355
HCFC-141b	21.4 ± 0.1	17.7	0.0034	0.0028
HCFC-142b	21.2 ± 0.2	15.5	0.0040	0.0029
HFC-23	24.0 ± 0.3	18.8	0.0043	0.0034
HFC-32	4.92	1.15	0.0005	0.0001
HFC-125	9.58 ± 0.04	3.69	0.0022	0.0008
HFC-134a	62.7 ± 0.3	34.3	0.0100	0.0055
HFC-143a	12.0 ± 0.1	5.6	0.0019	0.0009
HFC-152a	6.4 ± 0.1	3.4	0.0006	0.0003
SF_6	7.28 ± 0.03	5.64	0.0041	0.0032
SO_2F_2	1.71	1.35	0.0003	0.0003
NF_3	0.9	0.4	0.0002	0.0001
CF_4	79.0 ± 0.1	75.0	0.0040	0.0036
C_2F_6	4.16 ± 0.02	3.66	0.0010	0.0009
CH_3CCl_3	6.32 ± 0.07	18.32	0.0004	0.0013
CCl_4	85.8 ± 0.8	93.1	0.0146	0.0158
CFCs			0.263 ± 0.026[b]	0.273[c]
HCFCs			0.052 ± 0.005	0.041
Montreal gases[d]			0.330 ± 0.033	0.331
Total halogens			0.360 ± 0.036	0.351[f]
Total			2.83 ± 0.029	2.64

Notes:

[a] Pre-industrial values are zero except for CO_2 (278 ppm), CH_4 (722 ppb), N_2O (270 ppb), and CF_4 (35 ppt).

[b] Total includes 0.007 W m^{-2} to account for CFC-114, Halon-1211, and Halon-1301.

[c] Total includes 0.009 W m^{-2} forcing (as in AR4) to account for CFC-13, CFC-114, CFC-115, Halon-1211, and Halon-1301.

[d] Defined here as CFCs + HCFCs + CH_3CCl_3 + CCl_4.

[e] The value for the 1750 methane concentrations has been updated from AR4 in this report, thus the 2005 methane RF is slightly lower than reported in AR4.

[f] Estimates for halocarbons given in the table may have changed from estimates reported in AR4 owing to updates in radiative efficiencies and concentrations.

Source: Ciais, P., et al. (2013). Carbon and Other Biogeochemical Cycles. In *Climate Change 2013 The Physical Science Basis*. Contribution of Working Group I to the Fifth Assessment Report of the Intergovernmental Panel on Climate Change. Cambridge University Press, Cambridge, United Kingdom, and New York, NY.

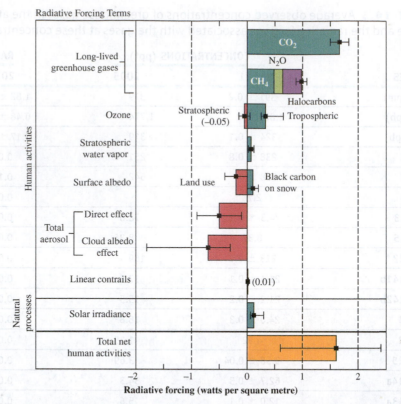

Radiative Forcing Terms

FIGURE 10.22 Summary of the principal components of the radiative forcing of climate change. All these radiative forcings result from one or more factors that affect climate and are associated with human activities or natural processes. The values represent the forcings relative to the start of the industrial era (about 1750). The thin black lines attached to each colored bar represent the range of uncertainty for the respective value.

Source: Based on IPCC (2007). *Climate Change 2007: The Physical Science Basis*. Contributions of Working Group I to the Fourth Assessment Report of the Intergovernmental Panel on Climate Change [S. Solomon, D. Qin, M. Manning, Z. Chen, M. Marquis, K.B. Averyt, M. Tignor, and H.L. Miller (eds.)] Cambridge University Press, Cambridge, United Kingdom and New York, NY, USA.

10.7 Climate Change Projections and Impacts

In this chapter, we have shown how fossil fuel use has resulted in increased carbon dioxide emissions. The basic principle of the law of conservation of mass and the mass balance equation show that the carbon dioxide emitted from fossil fuel use is closely correlated to carbon dioxide concentrations in the atmosphere. The law of conservation of energy and the energy balance equations were used to show that the greenhouse gases in the atmosphere regulate the Earth's climate. The changes in the carbon dioxide concentration in the atmosphere have the greatest impact on our climate, but other factors influence climate change in both positive and negative ways. Basic mass and energy balances form the foundation of sophisticated climate models that are used to predict the type of climate and climatic effects that will occur on Earth as we approach a new equilibrium condition at some point in the future. Figure 10.23 illustrates how emission

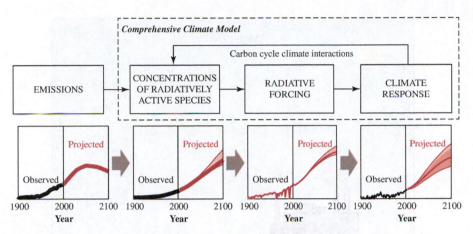

FIGURE 10.23 A schematic illustrating the observed and projected possible scenarios of climate change and how mass and energy fundamentals may be used to predict how the Earth's climate will respond to changing conditions in the atmosphere. Several steps from emissions to climate response contribute to the overall uncertainty of climate model projection. These uncertainties can be quantified through a combined effort of observation, process understanding, a hierarchy of climate models, and ensemble simulations. In comprehensive climate models, physical and chemical representations of processes permit a consistent quantification of uncertainty. Note that the uncertainty associated with the future emissions path is of an entirely different nature.

Source: Based on Ciais, P., et al. (2013). Carbon and Other Biogeochemical Cycles. In *Climate Change 2013: The Physical Science Basis*. Contribution of Working Group I to the Fifth Assessment Report of the Intergovernmental Panel on Climate Change. Cambridge University Press, Cambridge, United Kingdom, and New York, NY.

projections can be used to model and predict the change in concentration of atmospheric greenhouse gases under different assumptions about the future use of fossil fuels. The change in concentration of the future greenhouse gases will create a change in the energy balance due to a change in the radiative forcing that will influence how the Earth's climate will respond over the next century and farther into the future.

10.7.1 Climate Modeling

The climate models used to make current predictions started out with many of the same calculations that have been illustrated in this chapter. The models we have examined are quite simple and take into account only limited variables. Development of a full climate model is well beyond the scope of this textbook and chapter. However, Figures 10.24–10.30 illustrate how climate models have evolved by continuing to include more variables in the analysis, and uncertainties are refined to gain a clearer picture of how our climate will respond to the carbon emissions associated with fossil fuel use.

There is a broad consensus across the scientific community that our climate is changing as a result of fossil fuel emissions from energy-intensive industrialized societies. The IPCC states that "warming of the climate system is unequivocal, as is now evident from observations of increases in global average temperatures, widespread melting of snow and ice and rising global average sea level."

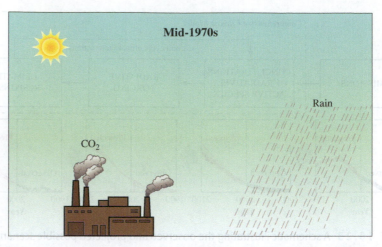

FIGURE 10.24 The complexity and level of detail in models have improved greatly since the early climate models were developed. In the 1970s, carbon dioxide was modeled in conjunction with models geared toward gaining a greater understanding of the effects of acid rain.

Source: Based on IPCC (2007a). *Climate Change 2007: The Physical Science Basis*. Contributions of Working Group I to the Fourth Assessment Report of the Intergovernmental Panel on Climate Change [S. Solomon, D. Qin, M. Manning, Z. Chen, M. Marquis, K.B. Averyt, M. Tignor, and H.L. Miller (eds.)] Cambridge University Press, Cambridge, United Kingdom, and New York, NY.

FIGURE 10.25 In the 1980s, carbon dioxide concentrations in the atmosphere were modeled and changes in the Earth's albedo were evaluated.

Source: Based on IPCC (2007a). *Climate Change 2007: The Physical Science Basis*. Contributions of Working Group I to the Fourth Assessment Report of the Intergovernmental Panel on Climate Change [S. Solomon, D. Qin, M. Manning, Z. Chen, M. Marquis, K.B. Averyt, M. Tignor, and H.L. Miller (eds.)] Cambridge University Press, Cambridge, United Kingdom, and New York, NY.

The IPCC has called for sustainable strategies for development and climate change adaptation and mitigation. The interrelationships among climate, societies' activities, adaptation, and mitigation of climate change are illustrated in Figure 10.32 (see page 730).

The 2013 IPCC climate model report (AR5) further concludes that:

> *Human influence on the climate system is clear, and recent anthropogenic emissions of greenhouse gases are the highest in history. Recent climate changes have had widespread impacts on human and natural systems.*

Anthropogenic greenhouse gas emissions have increased since the pre-industrial era, driven largely by economic and population growth, and are now higher than ever. This has led to atmospheric concentrations of carbon dioxide, methane, and nitrous oxide that are unprecedented in at least the last 800,000 years. Their effects, together with those of other anthropogenic drivers, have been detected throughout the climate system and are extremely likely to have been the dominant cause of the observed warming since the mid-20th century.

FIGURE 10.26 The FAR model included ocean absorption of carbon dioxide.

Source: Based on IPCC (2007a). *Climate Change 2007: The Physical Science Basis.* Contributions of Working Group I to the Fourth Assessment Report of the Intergovernmental Panel on Climate Change [S. Solomon, D. Qin, M. Manning, Z. Chen, M. Marquis, K.B. Averyt, M. Tignor, and H.L. Miller (eds.)] Cambridge University Press, Cambridge, United Kingdom, and New York, NY.

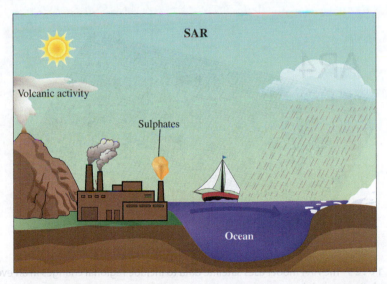

FIGURE 10.27 The SAR model began to investigate the effects of aerosols and their ability to act as a negative radiative forcing.

Source: Based on IPCC (2007a). *Climate Change 2007: The Physical Science Basis.* Contributions of Working Group I to the Fourth Assessment Report of the Intergovernmental Panel on Climate Change [S. Solomon, D. Qin, M. Manning, Z. Chen, M. Marquis, K.B. Averyt, M. Tignor, and H.L. Miller (eds.)] Cambridge University Press, Cambridge, United Kingdom, and New York, NY.

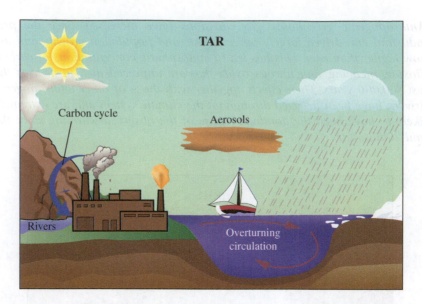

FIGURE 10.28 The TAR model began to combine the previous work evaluated in the FAR and SAR models. This increasingly complex model included most major sources and sinks of carbon and began to evaluate other radiative forcing factors. However, the TAR grid was very large and had difficulty evaluating changes in land use and associated forcings that would occur on small regional scales.

Source: Based on IPCC (2007a). *Climate Change 2007: The Physical Science Basis*. Contributions of Working Group I to the Fourth Assessment Report of the Intergovernmental Panel on Climate Change [S. Solomon, D. Qin, M. Manning, Z. Chen, M. Marquis, K.B. Averyt, M. Tignor, and H.L. Miller (eds.)] Cambridge University Press, Cambridge, United Kingdom, and New York, NY.

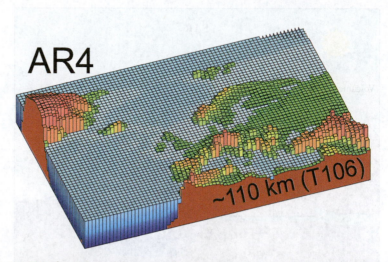

FIGURE 10.29 The AR4 model uses satellite data to include regional changes in an ever more sophisticated climate model.

Source: IPCC (2007a). *Climate Change 2007: The Physical Science Basis*. Contributions of Working Group I to the Fourth Assessment Report of the Intergovernmental Panel on Climate Change [S. Solomon, D. Qin, M. Manning, Z. Chen, M. Marquis, K.B. Averyt, M. Tignor, and H.L. Miller (eds.)] Cambridge University Press, Cambridge, United Kingdom, and New York, NY. 996 pp. Fig. 1.4, p. 113.

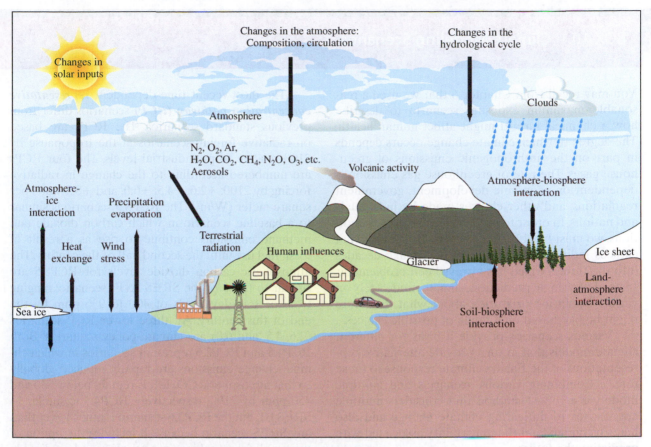

FIGURE 10.30 Current models use high-resolution data to include changes in land use and the most recent understanding of radiative forcing mechanisms to model past, present, and future climates. By replicating climates in the past, these models are calibrated and validated and are able to enhance our understanding of our current climate conditions and our ability to predict how the climate will respond to future changes in societies' use of fossil fuels and alternative energy technologies, with a low degree of uncertainty.

Source: Based on IPCC (2007a). *Climate Change 2007: The Physical Science Basis.* Contributions of Working Group I to the Fourth Assessment Report of the Intergovernmental Panel on Climate Change [S. Solomon, D. Qin, M. Manning, Z. Chen, M. Marquis, K.B. Averyt, M. Tignor, and H.L. Miller (eds.)] Cambridge University Press, Cambridge, United Kingdom, and New York, NY.

> *Continued emission of greenhouse gases will cause further warming and long-lasting changes in all components of the climate system, increasing the likelihood of severe, pervasive, and irreversible impacts for people and ecosystems. Limiting climate change would require substantial and sustained reductions in greenhouse gas emissions which, together with adaptation, can limit climate change risks.*

The climate models described in the IPCC report predict the Earth's average surface temperature will increase by 3.1°C to 5.5°C by the year 2100 due to current levels of GHG emissions.

10.7.2 Climate Model Projections

The magnitude of the increase in the Earth's average surface temperature will depend on the choices people make about their use of materials and energy.

BOX 10.1 Climate Modeling Scenarios

You may recall from Chapter 6 that we used a reasonable maximum exposure scenario to describe how a chemical hazard might affect human health. The degree to which climate change occurs depends in part on the anthropogenic emissions of greenhouse gases. The rate of greenhouse gas emission is dependent on economic development, government regulations, and other choices made by individuals and nations. In order to predict future climate conditions, we must be able to also correctly predict what choices individuals and nations will make decades into the future, which has always been problematic. Instead of a single prediction, climate models rely on various scenarios, or stories, about what the future might be like based on a range of possible choices.

Various scenarios provide a basis for different climate models that in turn provide alternative representations of the Earth's climate response to those forces. Combining various scenarios and multiple models is a robust method that considers multiple approaches to mitigating climate change and also helps us understand the uncertainty associated with predicting the future.

Standard scenarios used by the climate modeling community as input to global climate model simulations and climate projections are presented in IPCC assessment reports and U.S. National Climate Assessments (NCAs). Several iterations of scenarios have been modeled over time; the more recent include those from the Special Report on Emissions Scenarios (SRES) in 2000 and the Representative Concentration Pathways (RCPs) in 2010.

The *Special Report on Emissions Scenarios (SRES)* in 2000 (see Figure 10.31) was based on estimated population projections, laying out a consistent picture of demographics; international trade; flow of information and technology; and other plausible social, technological, and economic characteristics. *Integrated assessment models (IAMs)* based on socioeconomic scenarios were used to derive future emissions. A few benchmark scenarios were modeled, and the results were reported to indicate what might occur under high emissions A1FI (fossil-intensive), mid-high A2, mid-low B2, and low-emission, high-regulation B1 storylines.

The most recent time-dependent *representative concentration pathway (RCP)* scenarios differ from previous standard scenarios. The RCPs are based on radiative forcing scenarios at the tropopause by 2100 relative to preindustrial levels. The four RCPs are numbered according to the change in radiative forcing by 2100: +2.6, +4.5, +6.0, and +8.5 watts per square meter (W/m^2). The RCP8.5 scenario is similar to a baseline scenario in which carbon dioxide and methane emissions continue to rise as a result of fossil fuel use until the second half of the century. The atmospheric carbon dioxide levels for RCP8.5 are similar to those of the SRES A1FI scenario, ranging from present-day levels of 400 to 936 ppm by the end of this century. All three lower RCP scenarios (2.6, 4.5, and 6.0) are climate-policy scenarios. Both RCP4.5 and RCP2.6 represent scenarios in which climate change emissions are rapidly reduced globally so that atmospheric CO_2 levels remain below 550 and 450 ppm by 2100, respectively. RCP4.5 is similar to SRES B1, but the RCP2.6 scenario is much lower than any SRES scenario because it includes the option of using policies to achieve net negative carbon dioxide emissions before the end of the century.

Since the RCP scenarios are based on radiative forcings, the emission patterns and socioeconomic means to meet the emissions targets must be evaluated using IAMs for plausible policies and technological strategies that will prevent the radiative forcing value from being exceeded. The socioeconomic scenarios were modeled independently from the RCPs to create five *shared socioeconomic pathways (SSPs)* with climate forcing that matches the RCP values. The five SSPs are SSP1 ("Sustainability," low challenges to mitigation and adaptation), SSP2 ("Middle of the Road," moderate challenges to mitigation and adaptation), SSP3 ("Regional Rivalry," high challenges to mitigation and adaptation), SSP4 ("Inequality," low challenges to mitigation, high challenges to adaptation), and SSP5 ("Fossil-Fueled Development," high challenges to mitigation, low challenges to adaptation). The outcomes from different forcing scenarios provide policymakers with alternatives and a range of possible futures to consider.

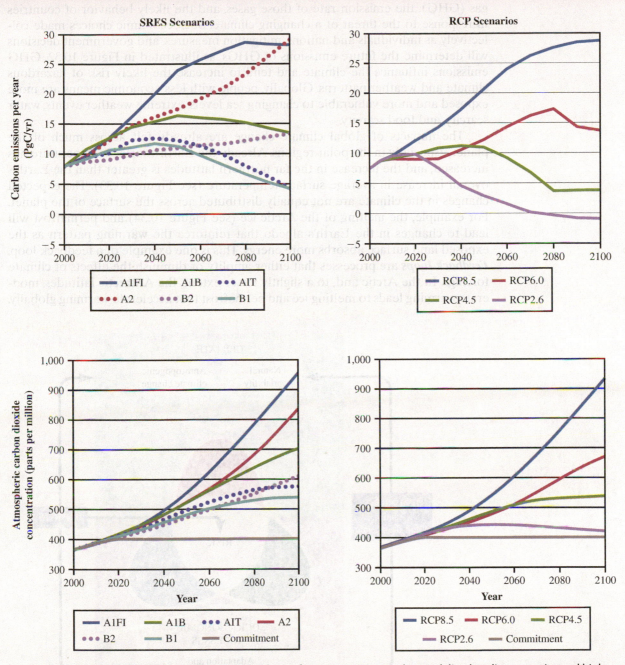

FIGURE 10.31 Socioeconomic scenarios and radiative forcing scenarios used in modeling baseline scenarios and high sustainability scenarios.

Source: Based on Hayhoe, K., Edmonds, J., Kopp, R.E., LeGrande, A.N., Sanderson, B.M., Wehner, M.F., and Wuebbles, D.J. (2017). Climate Models, Scenarios, and Projections. In D.J. Wuebbles et al. (eds.), *Climate Science Special Report: Fourth National Climate Assessment*, Volume I, pp. 133–160. U.S. Global Change Research Program, Washington, DC. doi: 10.7930/J0WH2N54.

In Chapter 6, we studied the risks of hazardous materials based on their concentration in the environment, the effects of specific materials on human health, and the exposure to those hazards based on an individual's behavior. Similarly, the IPCC evaluates the risk of various climate scenarios based on the type of greenhouse gas (GHG), the emission rate of those gases, and the likely behavior of countries in response to the threat of a changing climate. Socioeconomic choices made collectively as individuals and nations, mitigation measures, and government decisions will determine the future emissions of GHGs, as illustrated in Figure 10.32. GHG emissions influence the climate and tend to increase the likely risk of hazardous climate and weather patterns. Globally, people with less economic means are more exposed and more vulnerable to changing sea levels, extreme weather events, water scarcity, and food scarcity.

The impacts of global climate change are already felt across much of the planet, especially in the polar regions. Already the Earth's surface temperature has increased, and the increase in the far northern latitudes is greater than the Earth's overall increase in average surface temperature (see Figure 10.33). The expected changes in the climate are not equally distributed across the surface of the planet. For example, the melting of the Arctic ice (see Figure 10.34) and permafrost will lead to changes in the Earth's albedo that reinforce the warming pattern as the exposed land surface absorbs more energy. This is one example of a feedback loop. *Feedback loops* are processes that either amplify or diminish the effects of climate forcings. In the Arctic and, to a slightly lesser extent, the Antarctic latitudes, moderate warming leads to melting ice and permafrost that accelerate warming globally.

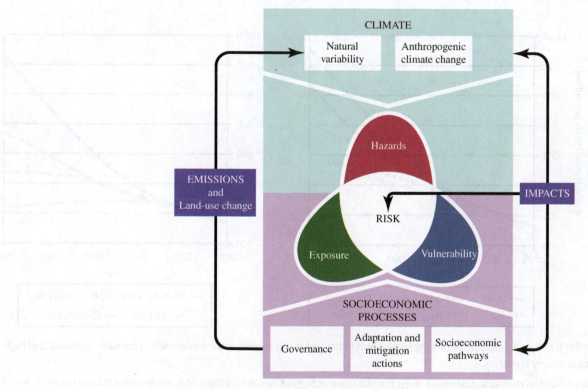

FIGURE 10.32 The relationships among socioeconomic choices, greenhouse gas emissions, climate-related hazards, and risk of exposure and vulnerability to human and natural systems due to the changing climate.

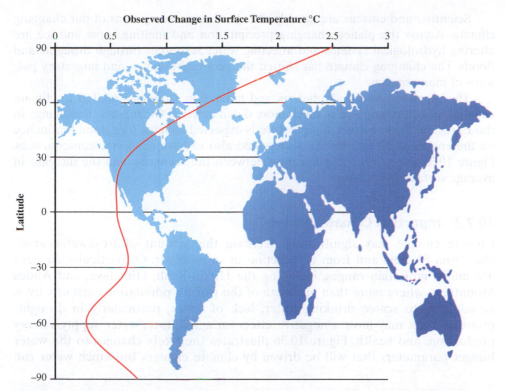

FIGURE 10.33 The change in the Earth's average surface temperature is not uniform across all latitudes. The increase in surface temperature is most dramatic over the far northern latitudes.

Source: Leo Blanchette/Shutterstock.com. Data from UNEP (2011). *Keeping Track of Our Changing Environment: From Rio to Rio +20 (1992–2012)*. Division of Early Warning and Assessment (DEWA), United Nations Environment Programme (UNEP), Nairobi.

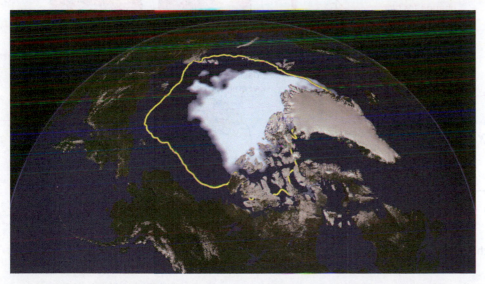

FIGURE 10.34 An image of the Arctic sea ice on September 16, 2012, the day that the National Snow and Ice Data Center announced the minimum reached in 2012. The yellow outline shows the average sea ice minimum from 1979 through 2010. The sea ice is shown with a blue tint.

Source: NASA/Goddard Space Flight Center Scientific Visualization Studio. The Blue Marble data is courtesy of Reto Stockli (NASA/GSFC).

Scientists and citizens are already able to observe many signs of our changing climate. Across the planet, changing precipitation and melting snow and ice are altering hydrological systems and affecting water resources through droughts and floods. The changing climate has shifted the geographic range and migratory patterns of many species.

The possible effects, adaptations, and mitigation strategies related to climate change can be investigated in much more detail in the IPCC reports. The change in the Earth's average surface temperature is expected to have a profound influence on the environment and human society and also to have economic consequences. Figure 10.35 illustrates the connection between the scenarios and the increase in average surface temperature.

10.7.3 Impacts of Climate Change

Climate change may significantly decrease the amount of freshwater available from glaciers and from a reduction in snow cover. Of particular concern are major mountain ranges, including the Hindu-Kush, Himalaya, and Andes Mountains, where more than one-sixth of the world's population currently lives. In addition to scarce drinking water, lack of water, particularly in drought-prone regions, may have a negative effect on agriculture, water supply, energy production, and health. Figure 10.36 illustrates the likely changes to the water budget parameters that will be driven by climate change. Too much water can

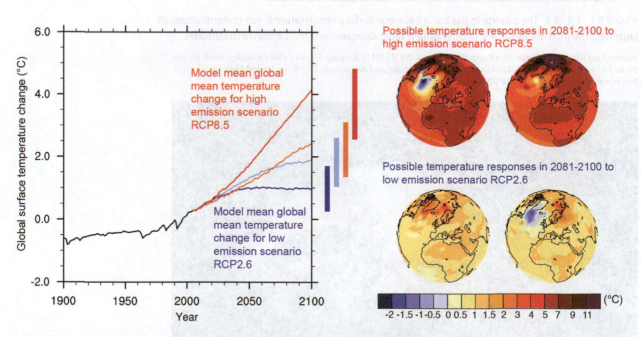

FIGURE 10.35 Projected changes in annual average surface temperature under baseline high emissions and ambitious mitigation scenarios.

Source: IPCC (2014a). Summary for policymakers. In *Climate Change 2014: Impacts, Adaptation, and Vulnerability*. Part A: Global and Sectoral Aspects. Contribution of Working Group II to the Fifth Assessment Report of the Intergovernmental Panel on Climate Change [C.B. Field, V.R. Barros, D.J. Dokken, K.J. Mach, M.D. Mastrandrea, T.E. Bilir, M. Chatterjee, K.L. Ebi, Y.O. Estrada, R.C. Genova, B. Girma, E.S. Kissel, A.N. Levy, S. MacCracken, P.R. Mastrandrea, and L.L.White (eds.)]. Cambridge University Press, Cambridge, United Kingdom, and New York, NY, pp. 1–32.

ACTIVE LEARNING EXERCISE 10.2 Climate Scenarios and Simulation

Global negotiations to reduce greenhouse gas (GHG) emissions have so far failed to produce worldwide agreement. Even if climate negotiations succeed, a binding treaty may be difficult to implement in many nations due to inadequate public support for emissions reductions. The scientific consensus on the reality and risks of anthropogenic climate change has never been stronger, yet public support for action in many nations remains weak. Policymakers, educators, the media, civic and business leaders, and citizens need tools to understand the dynamics and geopolitical implications of climate change. In this exercise, you will use the C-Roads Climate Change Policy Simulator developed by Climate Interactive, MIT, Ventana Systems, and UML Climate Change Initiative. You will evaluate various model scenarios to maintain a global average temperature increase of less than 2 degrees.

- Access the World Climate Summit Simulation website (www.climateinteractive.org/tools/c-roads/).
- Watch the short video on the World Climate Exercise.
- Choose (or you may be assigned) a country for which you would like to be a delegate for climate negotiations and create a brief statement that describes the climate policy position for the country you represent.

  ○ Please remember that in this role, you are a delegate representing the current political interests of that country. For example, if you choose the United States, you will be representing and arguing (using science) for the policies of the current administration.
  ○ Do an online web search to find out if there is additional or more recent information about "your" country's position on a global climate treaty.
  ○ Before using C-Roads, create a 2-minute video or presentation on your country's position. Be sure to include your nation state's name.

- Use the C-roads simulator to develop a short, well-reasoned argument for your nation state's proposal to limit climate change to 1.5°C (or other values consistent with your country's political position on climate change). Include a screen shot of the C-Roads simulation's results of your proposal. Include in your proposal the year each country will have peak GHG emissions, the year reductions will begin, the annual rate of GHG reductions, and plans to reduce deforestation or begin afforestation for the country you represent. Summarize your proposal in a 2-minute video or presentation on your country's strategy to reduce climate change.

also be detrimental, as the increased risk of flooding is likely to pose a significant threat to physical infrastructure and water quality. There is a high degree of confidence that the negative consequences of climate change for water resources outweigh the positive changes. Risks and vulnerabilities vary across different regions of the planet; for example, water-scarce areas of the world face the likelihood of less rainfall and higher evaporation rates, which exacerbate problems with access to clean water. Areas such as the southwestern United States could see a drying period even greater in scope and duration than those that may have resulted in the changes associated with the society of Chaco Canyon (see Chapter 3).

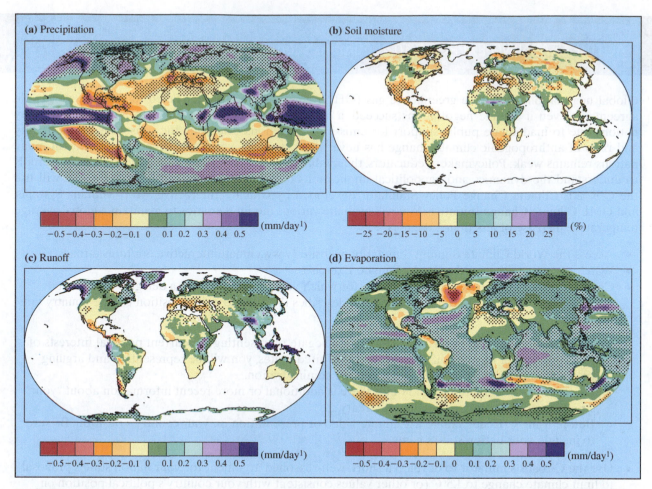

FIGURE 10.36 Multimodel mean changes in (a) precipitation (mm/day), (b) soil moisture (%), (c) runoff (mm/day), and (d) evaporation (mm/day). To indicate consistency in the signs of change, regions are stippled where at least 80% of models agree on the sign of the mean change. Changes are annual means for the SRES A18 scenario for the period 2080–2099 relative to 1980–1999. Soil moisture and runoff changes are shown at land points with valid data from at least 10 models. Details of the method and results for individual models can be found in the IPCC reports.

Source: IPCC (2007a). *Climate Change 2007: The Physical Science Basis*. Contributions of Working Group I to the Fourth Assessment Report of the Intergovernmental Panel on Climate Change [S. Solomon, D. Qin, M. Manning, Z. Chen, M. Marquis, K.B. Averyt, M. Tignor, and H.L. Miller (eds.)] Cambridge University Press, Cambridge, United Kingdom, and New York, NY. 996 pages. Fig. 10.12.

People with less economic means are more exposed and more vulnerable to the risks associated with rapid climate change. The impacts of climate change on society reported by the IPCC (2014a) include the following:

- Increases in climate-related extremes, such as heat waves, droughts, floods, cyclones, and wildfires, threaten ecosystems and many human systems (very high confidence).
- Climate-related hazards exacerbate other stressors, often with negative outcomes for livelihoods, especially for people living in poverty (high confidence).
- Due to rising sea levels projected throughout the 21st century and beyond, coastal systems and low-lying areas will increasingly experience adverse

BOX 10.2 Terminology of Climate Change Adaptation

Exposure: The presence of people; livelihoods; species or ecosystems; environmental functions; services and resources; infrastructure; or economic, social, or cultural assets in places that could be adversely affected by rapid climate change.

Vulnerability: The propensity or predisposition to be adversely affected. Vulnerability encompasses a variety of concepts and elements, including sensitivity or susceptibility to harm and lack of capacity to cope and adapt.

Adaptation: The process of adjustment to actual or expected climate and its effects. In human systems, adaptation seeks to moderate or avoid harm or exploit beneficial opportunities. In some natural systems, human intervention may facilitate adjustment to expected climate and its effects.

Mitigation: The human intervention to reduce the sources or enhance the sinks of greenhouse gases. Mitigation can mean using new technologies and renewable energies, making older equipment more energy efficient, or changing management practices or consumer behavior.

It can be as complex as a plan for a new city, or as simple as improvements to a cook stove design.

Transformation: A change in the fundamental attributes of natural and human systems. Transformation could reflect strengthened, altered, or aligned paradigms, goals, or values toward promoting adaptation for sustainable development, including reducing poverty.

Resilience: The capacity of social, economic, and environmental systems to cope with a hazardous event or trend or disturbance, responding or reorganizing in ways that maintain their essential function, identity, and structure, while also maintaining the capacity for adaptation, learning, and transformation.

Source: IPCC (2014a), Summary for Policymakers. In *Climate Change 2014: Impacts, Adaptation, and Vulnerability*. Part A: Global and Sectoral Aspects. Contribution of Working Group II to the Fifth Assessment Report of the Intergovernmental Panel on Climate Change [C.B. Field, V.R. Barros, D.J. Dokken, K.J. Mach, M.D. Mastrandrea, T.E. Bilir, M. Chatterjee, K.L. Ebi, Y.O. Estrada, R.C. Genova, B. Girma, E.S. Kissel, A.N. Levy, S. MacCracken, P.R. Mastrandrea, and L.L. White (eds.)]. Cambridge University Press, Cambridge, United Kingdom, and New York, NY, pp. 1–32.

impacts such as submergence, coastal flooding, and coastal erosion (very high confidence).

- Many global risks of climate change are concentrated in urban areas (medium confidence). Heat stress, extreme precipitation, inland and coastal flooding, landslides, air pollution, drought, and water scarcity pose risks in urban areas for people, assets, economies, and ecosystems (very high confidence).
- Throughout the 21st century, the impacts of climate change are projected to slow down economic growth, make poverty reduction more difficult, decrease food security, and increase poverty, particularly in urban areas and emerging hotspots of hunger (medium confidence).

For increases in the global average temperature that exceed 1.5°C to 2.5°C, major changes are projected in ecosystem structure and function, species' ecological interactions, and shifts in species' geographical ranges, with predominantly negative consequences for biodiversity and ecosystem goods and services (i.e., water and food supply) (IPCC, 2007b). A large fraction of both terrestrial and freshwater species face an increased risk of extinction due to climate change. This risk increases in part due to other climate factors such as habitat modification, over-exploitation of resources, pollution, and invasive species (high confidence). The IPCC has reported potential ecosystem impacts of climate change, including the following:

- Ecosystems are severely threatened by an unprecedented combination of climate change and disturbances that include flooding, drought, wildfire, insect infestations, and ocean acidification. (IPCC, 2007b)
- For scenarios RCP4.5, 6.0, and 8.5, ocean acidification poses substantial risks to marine ecosystems, especially polar ecosystems and coral reefs, associated with impacts on the physiology, behavior, and population dynamics of individual species from phytoplankton to animals (medium to high confidence). (IPCC, 2014a)

The changing climate may also affect human food supplies, health, and the economy. Throughout the 21st century, climate change is expected to lead to increases in poor health in many regions. Some of the changes in specific regions may include increases in the yearly growing season, with a potential increase in crop yields in those regions. All aspects of food security are potentially affected by climate change, including food access, utilization, and price stability (high confidence). The IPCC (2014a) reports that food production will be affected by climate change in the following ways:

- For the major crops (wheat, rice, and maize) in tropical and temperate regions, climate change without adaptation is projected to negatively impact production for local temperature increases of 2°C or more above late-20th-century levels, although individual locations may benefit (medium confidence).
- Rural communities will see threats to water availability and supply, food security, and agricultural incomes (high confidence). Economically disadvantaged families and female-headed households with limited access to land, modern agricultural inputs, infrastructure, and education are most vulnerable to climate threats.
- Global economic impacts of climate change are difficult to estimate. Estimates completed over the past 20 years vary in their coverage of subsets of economic sectors and depend on a large number of assumptions, many of which are disputable, and many estimates do not account for catastrophic changes, tipping points, and other factors. With these recognized limitations, the incomplete estimates of global annual economic losses for additional temperature increases of ~2°C are between 0.2% and 2.0% of income (±1 standard deviation around the mean) (medium evidence, medium agreement). Estimates of the incremental economic impact of emitting carbon dioxide lie between a few dollars and several hundreds of dollars per ton of carbon (robust evidence, medium agreement).
- The health risks from climate change in developing countries with low incomes are especially serious. There are increased risks of under-nutrition resulting from diminished food production in poor regions (high confidence), lost work capacity and reduced labor productivity in vulnerable populations, and poor health from food- and water-borne diseases (very high confidence) and vector-borne diseases (medium confidence). Positive effects are expected to include modest reductions in cold-related mortality

ACTIVE LEARNING EXERCISE 10.3 Regional Impacts of Climate Change

The IPCC states that "impacts of climate change will vary regionally. Aggregated and discounted to the present, they are very likely to impose net annual costs, which will increase over time as global temperatures increase." Based on the regional differences in climate impacts, estimate whether the current climate change models will likely cause home values in your area to increase or decrease. Consider the influence of the following parameters in making your decision:

- Water quantity
- Water quality
- Landscaping
- Home heating/cooling costs
- Local food production and prices

The following resources may be useful in considering regional impacts:

The National Climate Assessment developed by the U.S. Global Change Research Program (USGCRP), a federal program mandated by the U.S. Congress to coordinate federal research and investments in understanding the forces shaping the global environment, both human and natural, and their impacts on society (www.globalchange.gov/)

The National Oceanic and Atmospheric Administration – Climate Change Impacts (www.noaa.gov/education/resource-collections/climate/climate-change-impacts)

The International Panel on Climate Change (IPCC) Assessment Report (AR) Climate Change: Impacts, Adaptation and Vulnerability, Part B Regional Aspects (www.ipcc.ch/report/ar5/wg2/)

and morbidity in some areas due to fewer cold extremes (low confidence), geographical shifts in food production (medium confidence), and reduced capacity of vectors to transmit some diseases. But globally over the 21st century, the magnitude and severity of negative impacts are projected to increasingly outweigh the positive impacts (high confidence). The most effective measures to reduce vulnerability for health in the near term are programs that implement and improve basic public health measures, such as providing clean water and sanitation, securing essential health care including vaccination and child health services, increasing the capacity for disaster preparedness and response, and alleviating poverty (very high confidence).

10.8 Carbon Dioxide Mitigation, Capture, and Storage

Economic and population growth continue to be the most important drivers of increased CO_2 emissions. Baseline scenarios without additional mitigation measures are predicted to result in global mean surface temperature increases of 3.7°C to 4.8°C compared to pre-industrial levels. Large-scale changes in energy use and development and land use are needed to keep CO_2 levels below 450 ppm.

International cooperation is required to mitigate climate change, since GHGs accumulate over time and mix globally. Social, economic, and ethical considerations are important in considering approaches and agreements to mitigate climate change. Past and future contributions of GHGs to the atmosphere differ between countries and international agreements; these differences are shown in Figure 10.37. Climate change adaptation and mitigation strategies should consider issues of equity, justice, and fairness. Social, economic, and ethical analyses may be used to inform value judgments and may account for values of various sorts, including human well-being, cultural values, and nonhuman values. In some countries, tax-based policies directed toward transportation fuels have helped weaken the link between GHG emissions and subsidized fossil fuels. Regulatory approaches such as energy efficiency standards and labeling programs have been effective in reducing energy use. The **United Nations Framework Convention on Climate Change (UNFCCC)** is the main multilateral forum that focuses on addressing climate change. Other existing and proposed international climate change cooperative agreements may consist of multilateral agreements, harmonized national policies, and regionally coordinated policies.

Behavior, lifestyle, and culture have a considerable influence on energy use and associated emissions. Changing consumption patterns (e.g., mobility demand and mode, energy use in households, choice of longer-lasting products), dietary change, and reduction in food wastes can complement technology-based mitigation strategies. However, individual behavior changes are vastly insufficient for mitigating most climate impacts. Modeling scenarios indicate that rapid increases in energy efficiency and a tripling or quadrupling of renewable energy adoption rates and **carbon dioxide capture and storage (CCS)** are needed by the year 2050 to keep CO_2 levels below 450 ppm. The global annual budget of GHG emission is between

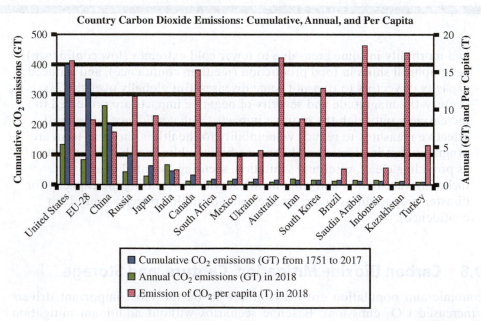

FIGURE 10.37 Numerous considerations are required for cooperative climate change mitigation strategies, and emissions can be interpreted by cumulative emissions, annual emissions, or per capita emissions.

1,320 $PgCO_2eq$ and 1,690 $PgCO_2eq$ to prevent the average surface temperature from rising more than 2°C relative to pre-industrial levels. It is expected that mitigation strategies will produce other positive results, such as reducing air pollution and providing more robust and resilient energy systems.

Decarbonizing (i.e., reducing the carbon intensity of) electricity generation is a key component of cost-effective mitigation strategies. Renewable energy (RE), nuclear, and CCS need to increase from the current share of approximately 30% to more than 80% of energy production by 2050, and fossil fuel power generation without CCS must be phased out almost entirely by 2100 (IPCC, 2014b). Replacing traditional coal-fired power plants with more efficient natural gas combined-cycle power plants or combined heat and power plants significantly reduces GHG emissions associated with energy production. CCS technologies could reduce the life cycle GHG emissions of fossil fuel power plants during the transition to renewable energy. CCS is a three-step process that includes carbon dioxide capture, transport of the captured carbon dioxide, and storage of the carbon dioxide for extended periods of time in biogeochemical repositories other than the atmosphere. The first step in reducing carbon emissions is to separate and capture the carbon either from a carbon-emitting source or from carbon stored in the atmosphere.

Carbon dioxide emitted from combustion processes is at much higher concentrations than carbon dioxide in the atmosphere. The higher the concentration of the carbon dioxide, the easier it is to separate the carbon dioxide from other compounds and capture the carbon. Therefore, most carbon dioxide mitigation, capture, and sequestration methods are targeted toward capture and sequestration of carbon from fossil fuel-burning power plants or other carbon-intensive industries. Several alternative carbon capture processes are illustrated in Figure 10.38, including capturing carbon dioxide after standard combustion, capturing even higher concentration carbon dioxide from combustion with oxygen (Oxyfuel), and separating and capturing carbon from the hydrogen in the fossil fuel.

Pipelines are expected to become the principal means of transporting captured carbon dioxide to geologic storage sinks. A review of the 500 largest sources of carbon dioxide emissions in the United States shows that 95% are within 50 miles of a possible storage site (Dooley, 2006). However, significant work and resources must be allocated to develop the infrastructure required to transport large quantities of captured carbon dioxide.

Once captured and transported, the carbon must be stored or sequestered in such a fashion that it cannot reach the atmosphere for long time periods (hundreds to thousands of years). The carbon may be injected into deep geologic formations, injected into the deep oceans, mineralized (combined with minerals to form an inactive solid), or utilized in other industries. The relationships between the fossil fuels source, fossil fuel use, transport of captured CO_2, and possible sequestration processes are illustrated in Figure 10.39.

Carbon storage includes at least three phases of operation. (1) A pre-injection phase includes extensive geologic site characterization, evaluation of site suitability, and modeling to predict carbon dioxide transport and movement in the subsurface. (2) The operational phase includes injecting the carbon dioxide into a well, monitoring groundwater chemistry, and tracking the subsurface carbon dioxide plume. (3) The final phase of operation should include continuous monitoring of water quality and the carbon dioxide plume near the injection site.

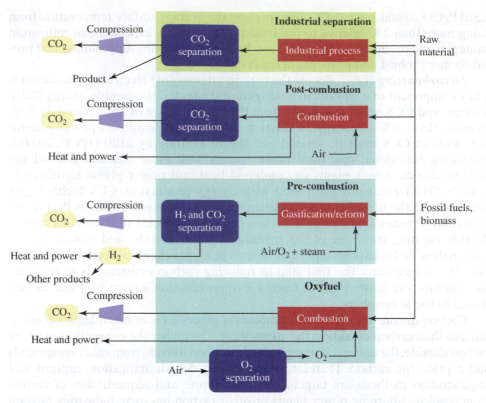

FIGURE 10.38 Schematic of alternative carbon dioxide capture systems.

Source: Based on IPCC (2005). *IPCC Special Report on Carbon Dioxide Capture and Storage.* Prepared by Working Group III of the Intergovernmental Panel on Climate Change [B. Metz, O. Davidson, H.C. de Coninck, M. Loas, and L.A. Meyer (eds.)] Cambridge University Press, Cambridge, United Kingdom, and New York, NY.

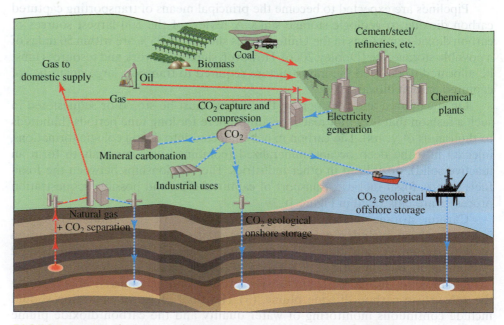

FIGURE 10.39 Alternative carbon capture, storage, and reuse technologies.

Source: Based on The Cooperative Research Centre for Greenhouse Gas Technologies (CO₂ CRC).

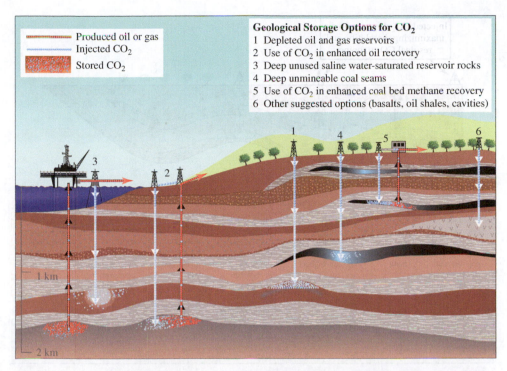

FIGURE 10.40 Schematic of alternative carbon dioxide geological capture and sequestration options.

Source: Based on The Cooperative Research Centre for Greenhouse Gas Technologies (CO₂ CRC).

Subsurface geological sequestration, illustrated in Figure 10.40, may include storage in (1) depleted oil and gas reserves, (2) enhanced oil and gas recovery (EOR and EGR) wells, (3a) offshore and (3b) onshore deep saline formations, and (4) use in enhanced coal bed methane recovery (ECBMR). Carbon dioxide has been used in EOR operations since the 1970s, so in some ways the theory of deep-well carbon dioxide injection is not new. However, the current number of EOR projects is insufficient to meet the expected demands of future carbon dioxide storage. Nonetheless, the U.S. Department of Energy and the International Energy Agency (IEA) estimate that there is enough potential geologic storage in the United States alone to store more than a thousand years of carbon emissions from current coal-fired electricity production in the United States. Potential geologic sequestration capacity worldwide is estimated to be about 2,000 PgCO$_2$ (IPCC, 2005). The majority of large point sources of carbon dioxide, worldwide, are located within 300 kilometers of potential geologic storage sites. Although the capacity for geologic sequestration is available, there are several possible mechanisms for carbon dioxide escape or leakage from geologic repositories (illustrated in Figure 10.41). The environmental and human health risks in general and for each individual site must be thoroughly characterized before carbon injection.

Ocean storage of carbon dioxide (see Figure 10.42) through deep sea injection of carbon produces dense, pressurized liquid carbon dioxide that forms carbon "lakes" that can delay dissolution of the CO$_2$ into the surrounding ocean environment. A smaller portion of large point sources of carbon dioxide is

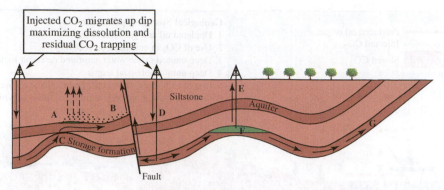

Injected CO_2 migrates up dip maximizing dissolution and residual CO_2 trapping

Potential Escape Mechanisms

A. CO_2 gas pressure exceeds capillary pressure and passes through siltstone	B. Free CO_2 leaks from A into upper aquifer up fault	C. CO_2 escapes through 'gap' in cap rock into higher aquifer	D. Injected CO_2 migrates up dip, increases reservoir pressure and permeability of fault	E. CO_2 escapes via poorly plugged old abandoned well	F. Natural flow dissolves CO_2 at CO_2/water interface and transports it out of closure	G. Dissolved CO_2 escapes to atmosphere or ocean

Remedial Measures

A. *Extract and purify groundwater*	B. *Extract and purify groundwater*	C. *Remove CO_2 and reinject elsewhere*	D. *Lower injection rates or pressures*	E. *Re-plug well with cement*	F. *Intercept and reinject CO_2*	G. *Intercept and reinject CO_2*

FIGURE 10.41 Potential escape mechanisms and remedial measures for geologic sequestration of carbon dioxide.

Source: Based on IPCC (2005). *IPCC Special Report on Carbon Dioxide Capture and Storage.* Prepared by Working Group III of the Intergovernmental Panel on Climate Change [B. Metz, O. Davidson, H.C. de Coninck, M. Loas, and L.A. Meyer (eds.)] Cambridge University Press, Cambridge, United Kingdom, and New York, NY.

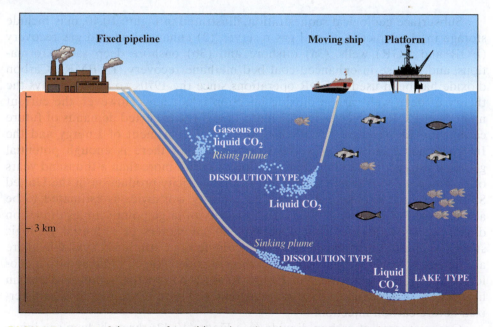

FIGURE 10.42 Schematic of possible carbon dioxide sequestration methods in the oceans.

Source: Based on IPCC (2005). *IPCC Special Report on Carbon Dioxide Capture and Storage.* Prepared by Working Group III of the Intergovernmental Panel on Climate Change [B. Metz, O. Davidson, H.C. de Coninck, M. Loas, and L.A. Meyer (eds.)] Cambridge University Press, Cambridge, United Kingdom, and New York, NY.

located in close proximity to feasible ocean storage sites. However, alternatives to geologic sequestration may be needed. The effects of carbon storage on ocean pH and deep sea ecosystems are under investigation in several regions.

Current cost estimates for CCS technology range widely. The U.S. Department of Energy (DOE) estimates that CCS technology would cost 60 to 114 U.S. dollars (DOE, 2010a) per metric ton of carbon dioxide, with 70% to 90% of that cost associated with capture and compression processes. Based on 2002 electricity generation costs, this would represent an increase of 0.01 to 0.05 U.S. dollars per kilowatt hour ($/kWh), depending on the fuel, the CSS technology, and the location of the facility (IPCC, 2005). Capturing carbon dioxide requires additional energy, so the total energy produced and fuel use from a CCS-equipped power generation plant are estimated to be 10% to 40% more than a comparable non-CCS plant. However, carbon dioxide emissions would be 80% to 90% lower as illustrated qualitatively in Figure 10.43. Depending on the type of carbon capture system implemented at existing coal-powered plants in the United States, the comparable cost of electricity (COE) delivered to the consumer may increase from 40% to 90% (DOEb, 2010), as noted in Table 10.4.

CCS technology may be most cost-effective when integrated into an energy and resource recovery system (see Figure 10.44) based on the principles of industrial ecology that will be discussed in more detail in Chapter 15. The most cost-effective CCS systems based on energy and economic modeling occur in association with energy production. The IPCC estimates that 220 to 2,200 $PgCO_2$ from CSS would be needed to stabilize atmospheric greenhouse gas concentrations between 450 and 750 ppm_v until the year 2100.

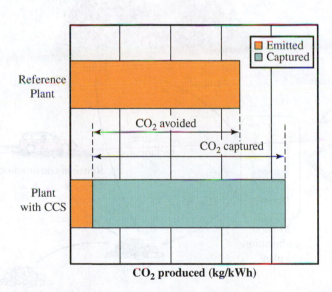

FIGURE 10.43 Comparison of carbon dioxide emissions from reference facilities with and without carbon dioxide capture and storage technology.

Source: Based on IPCC (2005). *IPCC Special Report on Carbon Dioxide Capture and Storage.* Prepared by Working Group III of the Intergovernmental Panel on Climate Change [B. Metz, O. Davidson, H.C. de Coninck, M. Loas, and L.A. Meyer (eds.)] Cambridge University Press, Cambridge, United Kingdom, and New York, NY.

TABLE 10.4 2002 Energy generation costs and carbon dioxide avoidance costs associated with CCS systems. Gas prices are assumed to be 2.8 to 4.4 U.S. dollars/GJ, and coal prices 1 to 1.5 U.S. dollars/GJ

CCS SYSTEM COMPONENT	COST RANGE	REMARKS
Capture from a coal- or gas-fired power plant	15–75 $/ton CO_2 net captured	Net cost of captured CO_2 compared to the same plant without capture
Capture from hydrogen and ammonia production or gas processing	5–55 $/ton CO_2 net captured	Applies to high-purity sources requiring simple drying and compression
Capture from other industrial sources	25–115 $/ton CO_2 net captured	Range reflects use of a number of different technologies and fuels
Transportation	1–8 $/ton CO_2 net transported	Per 250 km pipeline or shipping for mass flow rates of 5 (high end) to 40 (low end) Mt CO_2/yr
Geological storage[a]	0.5–8 $/ton CO_2 net injected	Excluding potential revenues
Geological storage: monitoring and verification	0.1–0.3 $/ton CO_2 net injected	This covers pre-injection, injection, and post-injection monitoring, and depends on the regulatory requirements
Ocean storage	5–30 $/ton CO_2 net injected	Including offshore transportation of 100–500 km, excluding monitoring and verification
Mineral carbonization	50–100 $/ton CO_2 net mineralized	Range for the best case studied. Includes additional energy use for carbonation

[a] Over the long term, there may be additional costs for remediation and liabilities.

Source: Based on IPCC (2005). *IPCC Special Report on Carbon Dioxide Capture and Storage.* Prepared by Working Group III of the Intergovernmental Panel on Climate Change [B. Metz, O. Davidson, H.C. de Coninck, M. Loas, and L.A. Meyer (eds.)] Cambridge University Press, Cambridge, United Kingdom, and New York, NY.

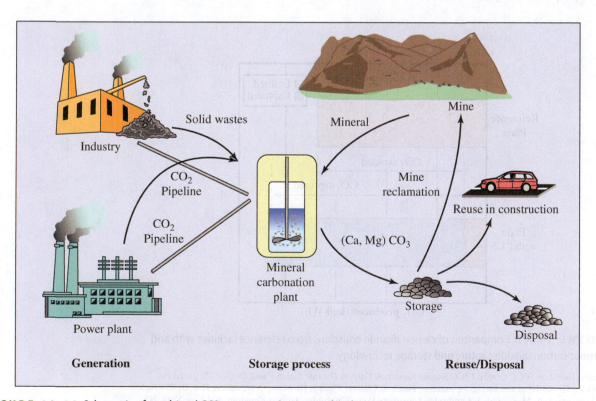

FIGURE 10.44 Schematic of combined CCS energy production and carbon resource recovery options.

Source: Based on IPCC (2005). *IPCC Special Report on Carbon Dioxide Capture and Storage.* Prepared by Working Group III of the Intergovernmental Panel on Climate Change [B. Metz, O. Davidson, H.C. de Coninck, M. Loas, and L.A. Meyer (eds.)] Cambridge University Press, Cambridge, United Kingdom, and New York, NY.

10.9 Summary

Climate change has far-reaching consequences for the environment, society, and economy. As the climate changes, societies must weigh the cost and risks associated with both action and inaction in curbing current levels of carbon dioxide emissions. In this chapter we have examined the fundamental science and theories that explain climate change. The IPCC (2007b, 2007c) suggests a strategy of weighing the costs and the threats in order to deal with the changing climate:

There is high confidence that neither adaptation nor mitigation alone can avoid all climate change impacts. Adaptation is necessary both in the short term and longer term to address impacts resulting from the warming that would occur even for the lowest stabilization scenarios assessed. There are barriers, limits, and costs that are not fully understood. Adaptation and mitigation can complement each other and together can significantly reduce the risks of climate change.

Sustainable development can reduce vulnerability to climate change, and climate change can impede nations' abilities to achieve sustainable development pathways. Making development more sustainable can enhance mitigative and adaptive capacities, reduce emissions, and reduce vulnerability, but there may be barriers to implementation.

References

Ciais, P., Sabine, C., Bala, G., Bopp, L., Brovkin, V., Canadell, J., Chhabra, A., DeFries, R., Galloway, J., Heimann, M., Jones, C., Le Quéré, C., Myneni, R. B., Piao, S., and Thornton, P. (2013). Carbon and Other Biogeochemical Cycles. In *Climate Change 2013: The Physical Science Basis*. Contribution of Working Group I to the Fifth Assessment Report of the Intergovernmental Panel on Climate Change [T. F. Stocker, D. Qin, G.-K. Plattner, M. Tignor, S. K. Allen, J. Boschung, A. Nauels, Y. Xia, V. Bex, and P. M. Midgley (eds.)]. Cambridge University Press, Cambridge, United Kingdom, and New York, NY.

Davis, M. L., and Masten, S. J. (2009). *Principles of Environmental Engineering and Science, 2nd ed*. New York: McGraw-Hill Higher Education.

DOE. (2010a). Cost and Performance Baseline for Fossil Energy Plants. Volume 1: Bituminous Coal and Natural Gas to Electricity. U.S. Department of Energy, National Energy Technology Laboratory.

DOE. (2010b). Industrial Carbon Capture and Storage. U.S. Department of Energy, National Energy Technology Laboratory.

Dooley, J. J. (2006). Carbon Dioxide Capture and Geologic Storage: A Core Element of a Global Energy Technology Strategy to Address Climate Change. College Park, MD, Global Energy Technology Strategy Program.

Field, C. B., Barros, V. R., Mach, K. J., Mastrandrea, M. D., van Aalst, M., Adger, W. N., Arent, D. J., Barnett, J., Betts, R., Bilir, T. E., Birkmann, J., Carmin, J., Chadee, D. D., Challinor, A. J., Chatterjee, M., Cramer, W., Davidson, D. J., Estrada, Y. O., Gattuso, J.-P., Hijioka, Y., Hoegh-Guldberg, O., Huang, H. Q., Insarov, G. E., Jones, R. N., Kovats, R. S., Romero-Lankao, P., Larsen, J. N., Losada, I. J., Marengo, J. A., McLean, R. F., Mearns, L. O., Mechler, R., Morton, J. F., Niang, I., Oki, T., Olwoch, J. M., Opondo, M., Poloczanska, E. S., Pörtner, H.-O., Redsteer, M. H., Reisinger, A., Revi, A., Schmidt, D. N., Shaw, M. R., Solecki, W., Stone, D. A., Stone, J. M. R., Strzepek, K. M., Suarez, A. G., Tschakert, P., Valentini, R., Vicuña, S., Villamizar, A., Vincent, K. E., Warren, R., White, L. L., Wilbanks, T. J., Wong, P. P., and Yohe,

G. W. (2014). Technical Summary. In *Climate Change 2014: Impacts, Adaptation, and Vulnerability*. Part A: Global and Sectoral Aspects. Contribution of Working Group II to the Fifth Assessment Report of the Intergovernmental Panel on Climate Change [C. B. Field, V. R. Barros, D. J. Dokken, K. J. Mach, M. D. Mastrandrea, T. E. Bilir, M. Chatterjee, K. L. Ebi, Y. O. Estrada, R. C. Genova, B. Girma, E. S. Kissel, A. N. Levy, S. MacCracken, P. R. Mastrandrea, and L. L.White (eds.)]. Cambridge University Press, Cambridge, United Kingdom, and New York, NY, pp. 35–94.

Forster, P., Ramaswamy, V., Artaxo, P., Berntsen, T., Betts, R., Fahey, D. W., et al. (2007). Changes in Atmospheric Constituents and in Radiative Forcing. In *Climate Change 2007: The Physical Science Basis*. Contribution of Working Group I to the Fourth Assessment Report of the Intergovernmental Panel on Climate Change. [S. Solomon, D. Qin, M. Manning, Z. Chen, M. Marquis, K. B. Averyt, M. Tignor, and H. L. Miller (eds.)]. Cambridge, UK: Cambridge University Press.

Hayhoe, K., Edmonds, J., Kopp, R. E., LeGrande, A. N., Sanderson, B. M., Wehner, M. F., and Wuebbles, D. J. (2017). Climate models, scenarios, and projections. In *Climate Science Special Report: Fourth National Climate Assessment, Volume I* [D. J. Wuebbles, D. W. Fahey, K. A. Hibbard, D. J. Dokken, B. C. Stewart, and T. K. Maycock (eds.)]. U.S. Global Change Research Program, Washington, DC, pp. 133–160, doi: 10.7930/J0WH2N54.

Heinze, C., Maier-Reimer, E., and Winn, K. (1991). "Glacial pCO_2 Reduction by the World Ocean: Experiments with the Hamburg carbon cycle model." *Paleoceanography* 6(4):395–430.

Houghton, J. T. (2009). *Global Warming: The Complete Briefing*. Cambridge, UK: Cambridge University Press.

Houghton, J. T., Jenkins, G. T., and Ephraums, J. J. (1990). Climate Change, the IPCC Scientific Assessment. Cambridge, UK: Cambridge University Press.

IPCC. (2001). Climate Change 2001: Synthesis Report. "A Contribution of Working Groups I, II, and III to the Third Assessment Report of the Intergovernmental Panel on Climate Change." R. T. Watson and the Core Writing Team (eds.). Cambridge, UK: Cambridge University Press.

IPCC. (2005). IPCC Special Report on Carbon Dioxide Capture and Storage. Prepared by Working Group III of the Intergovernmental Panel on Climate Change. [B. Metz, O. Davidson, H. C. de Coninck, M. Loas, and L. A. Meyer (eds.)] Cambridge, UK: Cambridge University Press.

IPCC. (2006). IPCC Guidelines for National Greenhouse Gas Inventories. Prepared by the National Greenhouse Gas Inventories Programme. [H. S. Eggleston, L. Buendia, K. Miwa, T. Ngara, and K. Tanabe, K. (eds.)] Institute for Global Environmental Studies, Kanagawa, Japan.

IPCC. (2007a). Climate Change 2007: The Physical Science Basis. Contributions of Working Group I to the Fourth Assessment Report of the Intergovernmental Panel on Climate Change. [S. Solomon, D. Qin, M. Manning, Z. Chen, M. Marquis, K. B. Averyt, and H. L. Miller (eds.)] Cambridge, UK: Cambridge University Press.

IPCC. (2007b). Climate Change 2007: Impacts, Adaptations and Vulnerability. Contributions of Working Group II to the Fourth Assessment Report of the Intergovernmental Panel on Climate Change. [M. L. Parry, O. F. Canziani, J. P. Palutikof, P. J. van der Linden, and C. E. Hanon (eds.)]. Cambridge, UK: Cambridge University Press.

IPCC. (2007c). Climate Change 2007: Mitigation. Contributions of Working Group III to the Fourth Assessment Report of the Intergovernmental Panel on Climate Change. [B. Metz, O. R. Davidson, P. R., Bosch, R. Dave, and L. A. Meyer (eds.)]. Cambridge, UK: Cambridge University Press.

IPCC. (2014a). Summary for Policymakers. In Climate Change 2014: Impacts, Adaptation, and Vulnerability. Part A: Global and Sectoral Aspects. Contribution of Working Group II to the Fifth Assessment Report of the Intergovernmental Panel on Climate Change [C. B. Field, V. R. Barros, D. J. Dokken, K. J. Mach, M. D. Mastrandrea, T. E. Bilir, M. Chatterjee, K. L. Ebi, Y. O. Estrada, R. C. Genova, B. Girma, E. S. Kissel, A. N. Levy, S. MacCracken, P. R. Mastrandrea, and L. L.White (eds.)]. Cambridge University Press, Cambridge, United Kingdom, and New York, NY, pp. 1–32.

IPCC. (2014b). Summary for Policymakers. In Climate Change 2014: Mitigation of Climate Change. Contribution of Working Group III to the Fifth Assessment Report of the Intergovernmental Panel on Climate Change [O. Edenhofer, R. Pichs-Madruga, Y. Sokona, E. Farahani, S. Kadner, K. Seyboth, A. Adler, I. Baum, S. Brunner, P. Eickemeier, B. Kriemann, J. Savolainen, S. Schlömer, C. von Stechow, T. Zwickel, and J. C. Minx (eds.)]. Cambridge University Press, Cambridge, United Kingdom, and New York, NY.

Lean, J. (2000). Evolution of the Sun's Spectral Irradiance Since the Maunder Minimum. Geophysical Research Letters, Vol. 27, No. 16, pp. 2425–2428, Aug. 15, 2000.

Myhre, G., Shindell, D., Bréon, F.-M., Collins, W., Fuglestvedt, J., Huang, J., Koch, D., Lamarque, J.-F., Lee, D., Mendoza, B., Nakajima, T., Robock, A., Stephens, G., Takemura, T., and Zhang, H. (2013). Anthropogenic and Natural Radiative Forcing. In *Climate Change 2013: The Physical Science Basis*. Contribution of Working Group I to the Fifth Assessment Report of the Intergovernmental Panel on Climate Change [T. F. Stocker, D. Qin, G.-K. Plattner, M. Tignor, S. K. Allen, J. Boschung, A. Nauels, Y. Xia, V. Bex, and P.M. Midgley (eds.)]. Cambridge University Press, Cambridge, United Kingdom, and New York, NY.

Prentice, I. C., Farquhar, G. D., Fasham, M. J. R., Goulden, M. L., Heimann, M., Jaramillo, V. J., Kheshgi, H. S., Le Quéré, C., Scholes, R. J., and Wallace, D. W. R. (2001). The carbon cycle and atmospheric carbon dioxide. In J. T. Houghton, Y. Ding, D. J. Griggs, M. Noguer, P. J. V. D. Linden, X. Dai, K. Maskell, and C. A. Johnson (eds.), *Climate Change 2001: The Scientific Basis*, pp. 183–237. Cambridge University Press.

Trenberth, K. E., Fasullo, J. T., and Kiehl, J. (2009). "Earth's global energy budget." *Bull. Amer. Meteor. Soc.* 90(3):311–324.

U.S. EIA. (2020). Monthly Energy Review April, 2020. U.S. Energy Information Administration, Office of Energy Statistics, U.S. Department of Energy, Washington, DC.

Wang, Y. M., Lean, J. L. and Sheeley, N. R. 2005. Modeling the sun's magnetic field and irradiance since 1713. *Astrophys. J.*, **625**, 522–538.

Key Concepts

International Panel on Climate Change (IPCC)
Greenhouse gases
Calorie
British thermal unit (BTU)
Specific heat
Blackbody temperature
Earth's albedo
Greenhouse gas effect
Radiative forcing

Special Report on Emissions Scenarios (SRES)
Integrated assessment models (IAMs)
Representative concentration pathways (RCPs)
Shared socioeconomic pathways (SSPs)
Feedback loops
Exposure
Vulnerability
Adaptation

Mitigation
Transformation
Resilience
United Nations Framework Convention
  on Climate Change (UNFCCC)

Carbon dioxide capture and storage
  (CCS)
Decarbonizing

Problems

10-1 When did scientists begin to realize the important link between atmospheric gas concentrations and our climate on Earth?

10-2 What changes have occurred in our atmosphere over the past 2,000 years?

10-3 How well are past and present climates understood?

10-4 What are the possible effects of a changing climate on economic systems, ecosystems, social structures, and technological challenges that may lie in the future?

10-5 Describe the accomplishments and cite the year the accomplishments were achieved by the following scientists who studied the way our atmosphere works:
  a. Jean-Baptist Fourier
  b. John Tyndall
  c. Svante Arrhenius
  d. Guy Stewart Callendar
  e. Hans Seuss and Roger Revelle
  f. The International Panel on Climate Change (IPCC)

10-6 In general, the combustion of fossil fuels can be described by what chemical equation?

10-7 Burning coal will emit how much carbon dioxide compared to the same amount of energy derived from natural gas?

10-8 Complete the following table by listing the petagrams (Pg) of CO_2 stored in the repository or transferred in each transformation pathway listed.

REPOSITORY OR PATHWAY	PETAGRAMS OF CO_2	REPOSITORY OR PATHWAY	PETAGRAMS OF CO_2
Atmosphere		Fossil fuels	
Weathering (atmosphere)		Rivers	
Respiration		Surface ocean	
Gross Photosynthesis		Marine biota	
Land sink		Intermediate and deep ocean	
Land-use changes		Surface sediment	
Vegetation, soil, and detritus		Weathering (geologic)	

10-9 What percent of carbon emissions is stored in the atmosphere, oceans, and land?

10-10 Carbon dioxide is converted to cellulose and other plant matter through the process of photosynthesis. Write the chemical equation in words for the formation of glucose.

10-11 In your own words and using typical chemical equations, describe the respiration process for animals using sugars (like glucose), carbohydrates, and proteins in combination with oxygen to produce energy, cell growth, water, and carbon dioxide.

10-12 Write the equation for the residence time as a function of the system volume, V, and the volumetric flow rate, Q, of the material or, alternatively, by dividing the mass of material, m, in the control volume by the mass flow rate of material, M, flowing through the volume.

10-13 Define the following terms:
  a. Blackbody temperature
  b. Calorie
  c. Planck's constant
  d. Specific heat
  e. Speed of light
  f. Stefan–Boltzmann constant

10-14 The energy balance within a given control volume can be expressed in words as _____.

10-15 List the variable and the equation that relate the first law of thermodynamics to work and energy.

10-16 List the variable and the equation that define the enthalpy of a system.

10-17 List the variable and the equation that describe heat loss through the boundaries of a system.

10-18 Show the equation and variables associated with the integrated form of the Stefan–Boltzmann law of radiation.

10-19 What are the average and range of incoming solar radiation since AD 1600?

10-20 What are the average and range of the Earth's albedo?

10-21 Write the equation that describes the blackbody radiation absorbed by the Earth.

10-22 Write the equation that describes the energy absorbed by the Earth.

10-23 Write the equation that describes the average surface temperature of the Earth as described by a blackbody.

10-24 What is the effect of the greenhouse gases in degrees Celsius on the Earth's average surface temperature?

10-25 What is the effective average incoming solar radiation determined by dividing the incoming irradiation by the surface area of the Earth?

10-26 The net energy balance across the Earth's surface can be written as _____.

10-27 What is the radiative forcing parameter for the following atmospheric gases?
  a. Carbon dioxide
  b. Methane
  c. Nitrogen oxide (N_2O)
  d. Halocarbons
  e. Ozone (stratosphere)
  f. Ozone (troposphere)

10-28 Describe how atmospheric models have evolved and what variables have been added in each evolutionary step since 1970.

10-29 Describe in 500 words or less the evidence that is used to support the following statement: *"Warming of the climate system is unequivocal, as is now evident from observations of increases in global average temperatures, widespread melting of snow and ice, and rising global average sea level."*

10-30 What are measurable changes and what magnitude of those changes is expected in the environment due to the enhanced greenhouse effect?

10-31 What potential economic and social changes may result from the environmental changes associated with the enhanced greenhouse effect?

10-32 What engineering and scientific technologies are currently on the horizon that may mitigate the likely effects of the enhanced greenhouse effect? What "science fiction" technologies could you imagine that are not yet understood by today's body of knowledge, but could be imagined to mitigate or reverse the enhanced greenhouse effect?

10-33 World energy production by source is shown in Figure 10.45. Using Excel, plot the data for crude oil, coal, and natural gas for each decade: 1970, 1980, 1990, and 2000. Establish a linear trend line through your graph for each fuel source. Also add a data set for the summation of each source to approximate world energy use of fossil fuels.

a. Forecast and extend the *x*-axis time and the trend line backwards to 1850 and forward to 2050. Show this trend line on a graph.

b. Show on a sheet of paper an example calculation for CO_2 emissions for the year 1990 for:

i) Oil

ii) Coal

iii) Natural gas

iv) All CO_2 emissions for 1990 from fossil fuel use

10-34 Plot on a graph the CO_2 emissions for each decade from 1850 to 2050 based on your fossil fuel forecasts of Problem 10-33. Show:

a. Oil

b. Coal

c. Natural gas

d. All CO_2 emissions for 1990 from fossil fuel use

10-35 The sum of all CO_2 emissions from 1850 to 2050, from fossil fuel emissions, is equal to the total area under the graph from Problem 10-33.

a. Based on this method, how much CO_2 was emitted from fossil fuels between 1850 and 2010?

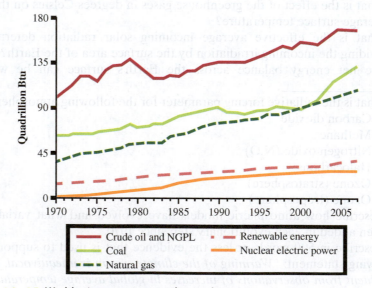

FIGURE 10.45 World primary energy production by source.

Source: U.S. EIA (2012). *Annual Energy Review 2011*. U.S. Energy Information Administration. Office of Energy Statistics. U.S. Department of Energy. Washington, DC. DOE/EIA-0384 (2011).

b. How does this compare to the atmospheric increase in CO_2 between 1850 and 2010, based on measurements of CO_2 in ppm in the atmosphere between 1850 and 2010?

10-36 Based on the summary statistics of sources for the electricity portfolio within the United States (see Table 10.5), electricity generation results in 0.68956 kg of CO_2 per kWh.

a. Calculate the metric tons of CO_2 emitted from all energy electricity generation for each year from 2000 to 2010 in Table 10.5.

b. Plot the net generation from all energy sources and the CO_2 emissions from energy generation for the years 2000 to 2010. Be sure to properly label your graph axes and units.

c. Fit a curve to the graph and use the curve to show predicted energy generation and CO_2 emissions in the year 2050.

10-37 In 2007, the weighted average combined fuel efficiency of cars and light trucks in the United States was 20.4 miles per gallon. The average vehicle traveled 11,720 miles in 2007. The amount of CO_2 emitted per gallon of gasoline burned was 8.92 kg. Assume for the following calculations that the 2007 data are applicable to all registered vehicles. In 2010, there were nearly 32 million registered vehicles in California.

a. Calculate the emissions of CO_2 from California in 2010.

b. It is estimated that there were approximately 250 million registered vehicles in the United States in 2010. Calculate the emissions of CO_2 from all automobiles in the United States in 2010.

c. By the year 2050, it is estimated that there will be perhaps 300 million automobiles in the United States. If this happens, and we assume driving conditions stay approximately the same, what would the CO_2 emissions from automobiles be in 2050 in the United States?

d. In 2012, India was estimated to have about 15 million automobiles. While driving conditions and automobile types in India are much different from those in the United States, what would the emissions of CO_2 be from automobiles in India if the gas mileage and driving patterns were the same as those in the United States?

e. By the year 2050, India is expected to be the largest automobile market in the world. Goldman Sachs has estimated that there will be more than 610 million automobiles in India by 2050. If this happens, and we assume driving conditions stay approximately the same, what would India's CO_2 emissions from automobiles be in 2050?

f. Create a bar graph illustrating the number of vehicles and CO_2 emitted from scenarios in parts (a) through (e).

10-38 An average home in the United States consumed about 12,773 kWh of electricity in 2005. The national average carbon dioxide emission rate for electricity generated in 2007 was 1,293 lb of CO_2 per megawatt-hour. Approximately 7% of the electricity generated is wasted due to losses in the transmission and distribution system.

a. How much CO_2 was generated per year for an average U.S. home due to electricity use?

b. How much CO_2 was emitted from household use of electricity if there were 111.1 million homes in the United States in 2005?

c. How much CO_2 will be emitted from household use of electricity if there are 166 million homes in the United States in 2050?

TABLE 10.5 Summary statistics of energy consumption for the United States, 1999–2012

DESCRIPTION	2010	2009	2008	2007	2006	2005	2004	2003	2002	2001	2000	1999
	Net Generation (thousand megawatt hours)											
Coal (1)	1,847,290	1,755,904	1,985,801	2,016,456	1,990,511	2,012,873	1,978,301	1,973,737	1,933,130	1,903,956	1,966,265	1,881,087
Petroleum (2)	37,061	38,937	46,243	65,739	64,166	122,225	121,145	119,406	94,567	124,880	111,221	118,061
Natural gas (3)	987,697	920,979	882,981	896,590	816,441	760,960	710,100	649,908	691,006	639,129	601,038	556,396
Other gases (4)	11,313	10,632	11,707	13,453	14,177	13,464	15,252	15,600	11,463	9,039	13,955	14,126
Nuclear	806,968	798,855	806,208	806,425	787,219	781,986	788,528	763,733	780,464	768,826	753,893	728,254
Hydroelectric conventional (5)	260,203	273,445	254,831	247,510	289,246	270,321	268,417	275,806	264,329	216,961	275,573	319,536
Other renewables (6)	167,173	144,279	126,101 (R)	1,105,238	96,525	87,329	83,067	79,487	79,109	70,769	80,906	79,423
Wind	94,652	73,886	55,363	34,450	26,589	17,811	14,144	11,187	10,354	6,737	5,593	4,488
Solar thermal and photovoltaic	1,212	891	864	612	508	550	575	534	555	543	493	495
Wood and wood-derived fuels (7)	37,172	36,050	37,300	39,014	38,856	38,856	38,117	37,529	38,665	35,200	37,595	37,041
Geothermal	15,219	15,009	14,840[R]	14,637	14,568	14,692	14,811	14,424	14,491	13,741	14,093	14,827
Other biomass (8)	18,917	18,443	17,734	16,525	16,099	15,420	15,421	15,812	15,044	14,548	23,131	22,572
Pumped storage (9)	-5,501	-4,627	-6,288	-6,896	-6,558	-6,558	-8,488	-8,535	-8,743	-8,823	-5,539	-6,097
Other (10)	12,855	11,928	11,804[R]	12,231	12,964	12,821	14,232	14,045	13,527	11,906	4,794	4,024
All energy sources	4,125,060	3,950,331	4,119,388	4,156,745	4,064,702	4,055,423	3,970,555	3,883,185	3,858,452	3,736,644	3,802,105	3,694,810

Source: U.S. EIA (2012). *Annual Energy Review 2011*. U.S. Energy Information Administration. Office of Energy Statistics. U.S. Department of Energy. Washington, DC.

10-39 An average home in the United States consumed about 47,453 cubic feet of natural gas, 59.1 gallons of liquid petroleum gas, 58.0 gallons of diesel fuel oil, and 0.85 gallons of kerosene.
  a. How much CO_2 was generated per year for an average U.S. home due to natural gas usage?
  b. Assume that 0.0545 kg of CO_2 are emitted per cubic foot of natural gas burned. How much CO_2 would be emitted from household natural gas usage if there were 111.1 million homes in the United States in 2005?
  c. How much CO_2 would be emitted from household natural gas usage if there were 166 million homes in the United States in 2050?

10-40 The default carbon dioxide emission factor for diesel fuel is 74,100 kg/TJ. Which common transportation fuel (gasoline or diesel) has the greater carbon intensity per unit of energy? What are their respective carbon intensities compared to natural gas?

10-41 The default carbon dioxide emission factor for biodiesel fuel is 70,800 kg/TJ, and the emission factor for diesel fuel is 74,100 kg/TJ.
  a. Which diesel fuel, biodiesel or "regular" diesel fuel, has the greater carbon intensity per unit of energy?
  b. What are their respective carbon intensities compared to natural gas?
  c. If a city were to invest in climate-friendly buses for public transportation, should it choose a diesel-powered bus, a biodiesel-powered bus, or a natural gas-powered bus?

10-42 According to the 1980 TAR report from the IPCC, what percent of carbon emissions is stored in the atmosphere, oceans, and land?

10-43 Cellulose can be represented by the chemical formula $(C_6H_{10}O_5)_n$, where n represents a long number of groups of this formula that can be combined. How many molecules of carbon dioxide would be needed to form one molecule of the base structure of cellulose, $C_6H_{10}O_5$?

10-44 The carbon flux to the oceans is approximately 30.3×10^{12} kg/yr. The approximate steady state mass of carbon in the oceans is 40×10^{15} kg. How long, on average, is carbon likely to remain in the atmosphere before cycling through another repository? How much longer or shorter is this than the residence time of carbon in the atmosphere?

10-45 The water flux to the atmosphere is approximately 5.18×10^{17} kg/yr. The approximate steady state mass of water in the atmosphere is 1.3×10^{16} kg. How long, on average, is water likely to remain in the atmosphere before cycling through another repository?

10-46 The water flux to the oceans is approximately 1.54×10^{14} kg/yr. The approximate steady state mass of water in the oceans is 1.35×10^{18} kg. How long, on average, is water likely to remain in the oceans before cycling through another repository?

10-47 If the Earth were completely covered with water so that the planet's albedo was 20%, what would be the Earth's blackbody radiation temperature?

10-48 A medium-growth coniferous tree planted in an urban or suburban setting after being raised in a nursery for 1 year can sequester 23.2 pounds of CO_2 over a 10-year period of growth. If you lived in an average home in the United States, how many trees would you need to plant to remove the CO_2 you would produce from driving your car, using electricity in your home, and heating your home with natural gas for a period of 10 years?

Team-Based Problem Solving

10-49 Estimate the carbon emissions that result from building a cement wall 1,954 miles long. Estimate the volume of concrete based on these specifications:

- Foundation: 6 feet deep, 18 inch radius
- Column: 4 square feet area by 30 feet tall
- Wall panels: 25 feet tall by 10 feet long by 8 inches thick

Concrete, of course, requires reinforcing steel (or rebar). A reasonable estimate for the amount of rebar is about 3% of the total wall volume.

a. Calculate the emissions of CO_2 from manufacturing the concrete needed to build the wall using AP-42 Tables.

b. Calculate the emissions of CO_2 from manufacturing the steel needed to build the wall using AP-42 Tables.

c. Calculate the estimated cost for concrete and steel required just to fabricate the wall.

d. Compare the emissions from constructing the cement wall to the yearly emissions of the United States and China, and report the values for each.

e. How do these emissions compare to the annual emissions of CO_2 from global fossil fuel consumption?

f. This simple analysis of building a cement wall covers only the carbon dioxide emissions from two sources of materials: cement and steel. Life Cycle Analysis (LCA) tries to capture all possible impacts from the cradle (material extraction) to the grave (disposal of all materials).

  i) List some of the other potential sources of emissions of carbon dioxide associated with building the wall.

  ii) What chemical compounds other than CO_2 might be emitted in building the wall?

  iii) What other potential environmental and social impacts might be associated with the construction of the wall?

Energy Conservation, Development, and Decarbonization

FIGURE 11.1 The Earth at night. This satellite image not only clearly illustrates population density and energy consumption, but it also more subtly highlights regions of the world that experience widespread energy poverty such as in sub-Saharan Africa and Asia. Notice the low levels of lighting in these regions. 75% of the world's population that lack access to electricity live in sub-Saharan Africa.

(Source: NASA Earth Observatory/NOAA NGDC)

> While poverty itself is the measure of well-being and enlightenment, the approach of power used is a reliable indicator of the degree of safety, comfort, and convenience, without which the human race would be able to remaining suffering and want and civilization might perish.
>
> —Nikola Tesla

Energy Conservation, Development, and Decarbonization

FIGURE 11.1 The Earth at night. This satellite image not only clearly illustrates population density and energy consumption, but it also more subtly highlights regions of the world that experience widespread energy poverty, such as in sub-Saharan Africa and Asia. Notice the low levels of lighting in these regions; 75% of the world's population that lack access to electricity live in sub-Saharan Africa.

Source: NASA Earth Observatory/NOAA NGDC.

While not exactly a true measure of well being and enlightenment, the amount of power used is a reliable indication of the degree of safety, comfort, and convenience, without which the human race would be subject to increasing suffering and want and civilization might perish.

—Nikola Tesla

GOALS

THIS CHAPTER HIGHLIGHTS THE CHALLENGE OF deep decarbonization facing the world today and enables you to understand how energy use affects the environment. The chapter explores major strategies for reducing energy-related carbon emissions, including conservation, efficiency, heat recovery, renewable energy, electrification, and distributed generation. The interdependence of water, energy, and food sustainability is introduced.

OBJECTIVES

At the conclusion of this chapter, you should be able to:

11.1 Summarize challenges to deep decarbonization of the global energy system and calculate the annual rates of change needed to achieve emission goals.

11.2 Compare and contrast finite and renewable energy resources and calculate the available energy from a variety of resources.

11.3 Explain the role of carbon footprinting and calculate a carbon footprint, avoided emissions, and the benefits of a portfolio of carbon offsets.

11.4 Compare and contrast strategies for energy conservation and efficiency and estimate energy savings from conservation and efficiency improvements.

11.5 Discuss options for renewable energy and calculate the reductions in carbon emissions from using these options.

11.6 Review the role of electrification in decarbonization and size an electric power load.

11.7 Explain how water, energy, and food sustainability are interconnected.

Introduction

As Nikola Tesla indicated, energy is a basic human need, without which necessities such as lighting, cooking, industrial processes, transportation, long-distance communication, refrigeration, and space heating would not be possible. Access to modern energy services does indeed provide "safety, comfort, and convenience," and the universal goal for energy is to have a safe, affordable, abundant, and reliable supply for everyone. Yet about 800 million people worldwide live without electricity, and more than 2.5 billion use cooking facilities that pollute their homes and threaten their health (International Energy Agency, 2020).

Energy use today is characterized by a duality of energy poverty and wealth in the global community. Residents of affluent nations benefit from comprehensive access to electricity, clean cooking and heating fuels instead of traditional biomass (dung, grass, wood, and peat), and reliable energy supplies. Homes, hospitals, schools, offices, businesses, and factories in these societies are rarely threatened by a serious disruption of energy production due to shortages of fuels, costly spikes in the cost of energy, or an inadequate energy infrastructure. In contrast, low-income countries experience a widespread lack of access to modern energy services at the household level, significant inequality in rural and urban access to energy supplies, and an inability to provide an adequate amount of energy to their growing populations.

The deep and urgent challenge facing us today is how to fulfill the world's unmet—and still growing—energy needs while deeply decarbonizing the energy sector. About three-quarters of worldwide greenhouse gases (GHGs) are created by fossil fuels. Shifting to low- and no-carbon energy resources is essential to mitigate climate change (see Chapter 10).

Meeting our energy needs also requires ***energy security***. The concept of energy security generally refers to the ability of a nation to protect itself from the economic, political, and social disruptions caused by an interrupted supply of a critical energy resource, the failure of energy infrastructure, or rapid and steep changes in energy prices. Although such "shocks" do not occur frequently, the results can be devastating when they do happen. As you will learn in this chapter, there are many opportunities to achieve a significant energy transformation and improve energy security. Efficiency and conservation, cleaner energy resources, and electrification are key pathways to a low-carbon energy future and resilient energy systems.

11.1 The Challenge of Decarbonization

Until 2015, there was no *political* consensus about how much to limit global warming. For decades, the scientific community identified an increase of 2°C as the upper limit of manageable risk to the Earth's ecosystems, climate dynamics, and human society. The 2°C limit could be achieved if atmospheric concentrations of carbon dioxide stayed below 450 ppm; scenarios of greenhouse gas emissions consistently showed that a concentration of 450 ppm would occur about 2050 if the world continued on its current emissions trajectory.

A new temperature goal of 1.5°C has emerged. Both the scientific and international communities realized that 2°C would create more severe impacts than initially predicted. Countries that participated in the 2015 Paris Climate Agreement reached political consensus and pledged to keep warming below a temperature target for the first time. Hundreds of nations agreed that keeping the average global temperature increase well below 2°C was a necessity and to try and limit it to 1.5°C. A special report by the Intergovernmental Panel on Climate Change (IPCC) detailed the more severe impacts of a 2°C change (IPCC, 2019) and stressed that global emissions must be cut in half by 2030 to limit warming to 1.5°C and achieve net zero emissions by 2050 (Figure 11.2). To stay within the 2°C limit, global emissions must be cut by 25% by 2030 to achieve net zero emissions by 2070 (see Example 11.1).

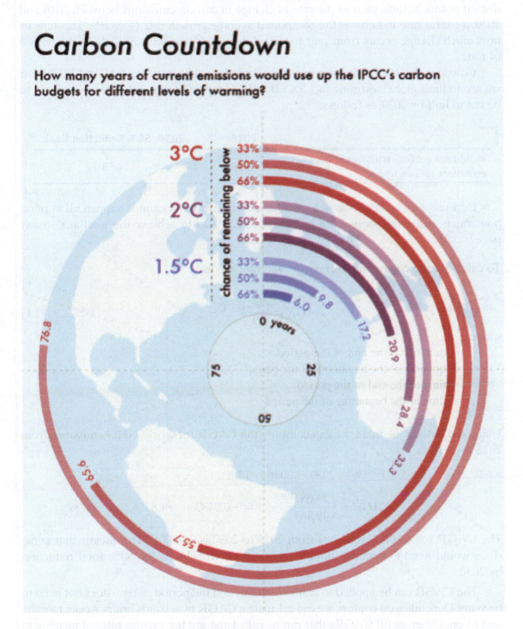

Carbon Countdown

How many years of current emissions would use up the IPCC's carbon budgets for different levels of warming?

FIGURE 11.2 Global carbon budgets. This diagram illustrates the probabilities of limiting global warming to three different temperatures. It explains how long we can continue emitting GHGs at current amounts before the corresponding concentration of atmospheric CO_2 is reached for the specified temperature. Net zero carbon emissions must begin after the indicated number of years to avoid further warming.

Source: *The Carbon Brief*, www.carbonbrief.org/six-years-worth-of-current-emissions-would-blow-the-carbon-budget-for-1-5-degrees.

EXAMPLE 11.1 Calculating Compound Average Growth Rates

Many real-world phenomena fluctuate from year to year and do not increase or decrease at a steady rate. Some examples are population, economic indicators, energy use, and carbon emissions. When we want to analyze growth trends and patterns between two distant points in time, such as the rate of change in carbon emissions between 2016 and 2030, a useful rate to know is the compound average growth rate (CAGR). This tells us how much change occurs from year to year if the rate is constant throughout the period of time.

Consider the total reduction in carbon emissions that must be achieved by 2030 if we are to limit global warming to 1.5°C. The IPCC reported that 2016 emissions need to be cut in half by 2030, as follows:

	2016	2030, 50% Reduction Goal
World total net GHG emissions in CO_2 equivalent (billion tonnes)	49.36	24.68

Because a 50% reduction in emissions is significant and cannot happen all at once, how much would emissions need to decrease each year to achieve this goal at a steady pace?

To calculate the compound average growth rate, we use

$$CAGR = \left(\frac{Q_n}{Q_0}\right)^{\frac{1}{t_n - t_0}} - 1 \tag{11.1}$$

where
Q_n = quantity at the end of the period
Q_0 = quantity at the beginning of the period
t_n = time at the end of the period
t_0 = time at the beginning of the period

Using the data in the table, we can compute the CAGR for world GHG emissions from 2016 to 2030 as

$$t_n - t_0 = 2030 - 2016 = 14$$

$$CAGR = \left(\frac{24.68}{49.36}\right)^{1/14} - 1 = -0.0483 = -4.83\%$$

The CAGR for carbon emissions from 2016 to 2030 is −4.83%. This means that emissions would need to decline about 5% every year to achieve the 50% total reduction by 2030.

The CAGR can be applied to many situations and the period of time does not need to be years. Depending on context, we can calculate a CAGR by seconds, hours, weeks, months, and so on. Other useful CAGRs that can be calculated are the various rates of increase in renewable energy that must be installed to substitute for existing fossil fuel energy.

Energy production and use account for almost 75% of total global greenhouse gas emissions. Other major contributors include agriculture, landfill decomposition, wastewater treatment, and land-use and land-cover change (Figure 11.3). Within the energy sector, the major contributor to global GHGs is electric power production,

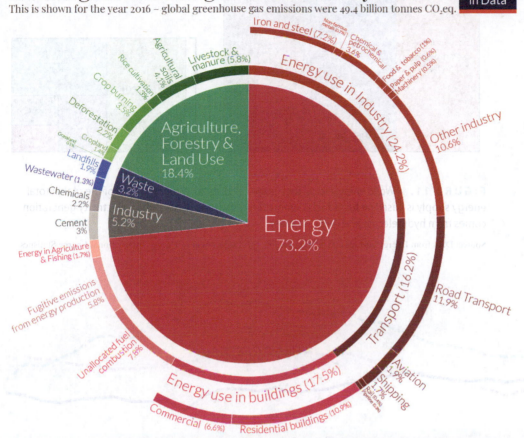

Global greenhouse gas emissions by sector

Our World in Data

This is shown for the year 2016 – global greenhouse gas emissions were 49.4 billion tonnes CO_2eq.

FIGURE 11.3 Global greenhouse gas emissions. Worldwide, energy use accounts for the majority of all greenhouse gas emissions. Virtually all of the emissions from the transport, energy supply, industry, and buildings sectors result from energy consumption and production in those sectors.

Source: Climate Watch, The World Resources Institute (2020). OurWorldinData.org: www.visualcapitalist.com/a-global-breakdown-of-greenhouse-gas-emissions-by-sector/.

which is expanding significantly in developing countries as the preferred way to meet social energy needs. The global energy system is extremely dependent on fossil fuels. Figure 11.4 illustrates that renewable resources account for less than one-fifth of world energy use.

In the United States, fossil fuel combustion accounts for 92% of all direct CO_2 emissions (U.S. EPA, 2021), but the impact of the electric power sector is declining, and transportation is now the largest contributor of GHGs (Figure 11.5). Total U.S. emissions have been on the decline since 2005, an accomplishment given the growth in the U.S. population and economy. Emissions per person and per dollar of GDP output have been declining for over 20 years due to greater energy efficiency and a shift toward services in the economy (Figure 11.6). Nonetheless, this improvement is not enough to mitigate climate change.

Achieving the 1.5°C or 2°C warming limit will require a deep ***decarbonization*** of the world's energy systems. Decarbonization refers to cutting the GHG emissions created by the energy sector and its use of the high-carbon fossil fuels on which it so

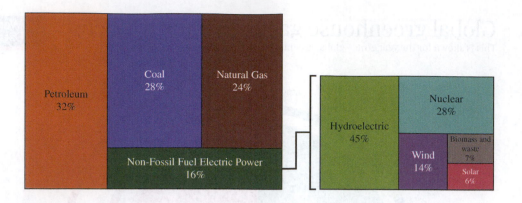

FIGURE 11.4 World energy fossil fuel usage (2018 data). More that 80% of the global total energy supply is based on fossil fuels. Almost three-quarters of no-carbon electricity generation comes from hydroelectric and nuclear power.

Source: Data from Energy Information Administration, U.S. Department of Energy, International Energy Statistics.

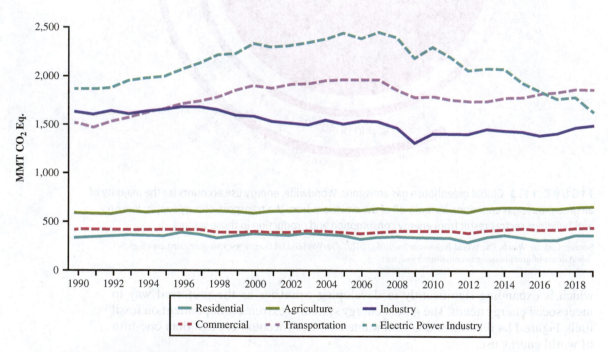

FIGURE 11.5 U.S. direct GHG emissions by source. The electric power industry has been the major source of U.S. GHGs since 1990. Fuel switching to natural gas and the growing use of renewables have lowered the impact of this sector, and transportation now accounts for the largest proportion of U.S. emissions.

Source: Based on U.S. Environmental Protection Agency, *Inventory of U.S. Greenhouse Gas Emissions and Sinks, 1990–2019* (Washington, DC: 2021). www.epa.gov/ghgemissions/inventory-us-greenhouse-gas-emissions-and-sinks-1990-2019.

overwhelmingly depends. As we will discuss later, the scope and scale of decarbonization presents an unprecedented challenge. This goal requires a deeply embedded energy system that has developed over hundreds of years to shift in only about two decades.

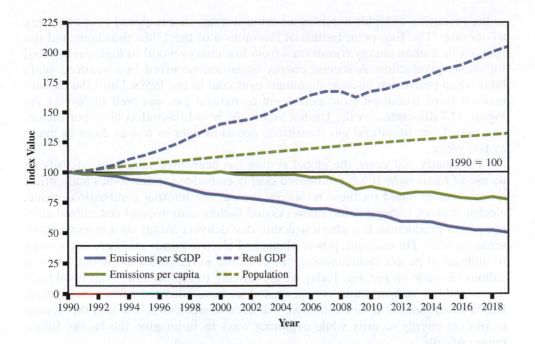

FIGURE 11.6 U.S. greenhouse gas intensity. The U.S. economy is becoming less greenhouse gas intensive. The number of GHGs emitted per dollar of GDP has been falling steadily since 1990.

Source: Based on U.S. Environmental Protection Agency, *Inventory of U.S. Greenhouse Gas Emissions and Sinks, 1990–2019* (Washington, DC: 2021). www.epa.gov/ghgemissions/inventory-us-greenhouse-gas-emissions-and-sinks-1990-2019.

In addition, decarbonization must occur while the global demand for energy continues to increase, primarily because of economic development in low- and middle-income countries. Energy use intensifies with development and rising incomes, and world energy consumption will continue to grow for decades and likely more than a century; more than 3 billion people have either unmet basic energy needs or desire a quality of life that energy use will help them achieve. And the global population is still growing.

11.1.1 Energy Transitions in Historical Context

The shift to a less carbon-intensive energy system is the latest in a series of historical energy transitions. During an ***energy transition***, the dominant fuel used by a society changes; the change can result from resource scarcity, new types of resources, or technological innovation.

Before the Industrial Revolution began in Europe in the 1700s, humankind met its energy needs in the same basic ways for thousands of years. Low-carbon ***traditional biomass***—wood, charcoal, peat, dung, and grass—was the basic fuel for fire and heat. Most physical energy was powered by the labor of people and beasts of burden; they pulled carts, rowed ships, turned grinding stones, plowed land, and lifted heavy objects. Simple machinery leveraged these efforts with pulleys, wheels, gears, levers, and so on.

The Industrial Revolution was a watershed because of the steam engine, which could replace people and animals to power simple machinery. It also led to the invention of highly complex machine systems that required a lot of energy to operate. The European Industrial Revolution of the 1700s thus launched the first major human energy transition—from low-energy wood to high-energy coal for steam production. A second energy transition occurred two hundred years later when petroleum became dominant over coal in the 1950s. Until the climate crisis, a third transition from petroleum to natural gas was well underway. As Figure 11.7 illustrates for the United States, the wood-to-coal, coal-to-petroleum, and petroleum-to-natural gas transitions occurred over as few as 20 or as many as 100 years.

For nearly 300 years, the global system has been expanding and intensifying its use of fossil fuels. It has established deeply embedded technologies and critical infrastructures based on these resources, such as the internal combustion engine, electric motors, transportation networks and fueling stations, and centralized electric power production. It is also a structure that delivers energy on a massive, time-sensitive scale. For example, power plants and electric power grids provide energy to millions of people *simultaneously*. In 2016, the world consumed about 4 billion gallons of crude oil *per day*. Today we are striving to rapidly shift from fossil fuels to no- and low-carbon sources of energy, but we have inherited a fossil fuel system. Rapid decarbonization means that we will need to work largely within this system to protect energy security while exploring ways to re-imagine the energy future more radically.

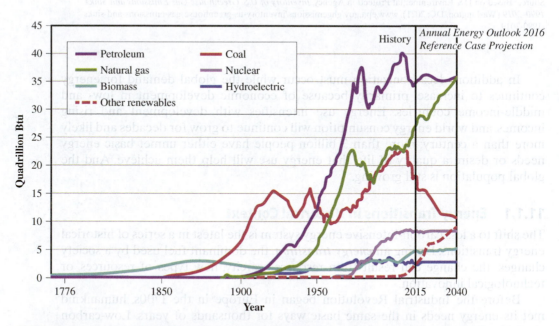

FIGURE 11.7 U.S. energy transitions. The United States has gone through two significant energy transitions. The first was the shift to coal as a dominant fuel about 1880. The second was a shift to petroleum as the dominant fuel about 1950. The projections after 2015 show the United States on a "business as usual" pathway and without a major commitment to net zero emissions by 2050. Note that the U.S. Industrial Revolution began in the 1850s, about a hundred years after it started in Europe.

Source: Based on U.S. DOE, *Today in Energy: Fossil fuels still dominate U.S. energy consumption despite recent market share decline.* July 2016. www.eia.gov/todayinenergy/detail.php?id=26912#.

11.1.2 Energy and Development

Energy is fundamental to human and economic development, but not all societies use energy in equal amounts or with equal energy intensity. The differences can be explained by why and how people meet their energy needs.

Our demand for energy derives from our personal and economic needs for energy services. **Energy services** include lighting, transportation, space heating, refrigeration, telecommunications, agricultural activities, industrial manufacturing, materials fabrication, cooking, and so on.

Any given energy service can potentially be done in a variety of ways. For example, we can walk, ride an animal, row a boat, or pedal a bicycle to get around, and none of these modes of transportation requires the direct consumption of an energy resource other than food for people and animals. Alternatively, we can take a train, drive a car, or fly in a plane, all of which require the supply of an energy fuel for power.

These choices represent different degrees of energy intensity, and the technology on which our choice relies determines the demand for specific energy resources. This concept is known as **derived demand**. For instance, commercial aircraft fly only on jet fuel; the demand for jet fuel is therefore derived from our society's demand for airline services. Train locomotives, on the other hand, can use coal if they are powered by steam or diesel fuel if they have an internal combustion engine. A society's demand for trains therefore results in derived demand for coal and diesel fuel.

The variations that we see in energy use and intensity between countries are due to the kinds of energy services people need, how (and if) those needs are met, and the derived demand for resources that result. In particular, low- and some middle-income countries are not heavily industrialized, which means they have less derived demand for manufacturing energy. Their citizens have lower incomes, and many may not be able to afford electricity or gasoline, let alone the household appliances and cars that this energy will power.

These underlying international differences matter for decarbonization because we are trying to shift the energy system while also expanding energy production and affordable access to the global community. As we will discuss later, many of the world's countries—and much of the world's population—need economic growth and a better quality of life, which in turn will significantly increase their energy use and energy intensity.

Energy Intensity and Economic Growth

The world's most economically developed and wealthy nations account for a disproportionate amount of global greenhouse gas emissions. For example, member nations of the Organization of Economic Cooperation and Development (OECD) represent 38 of the world's most industrialized countries. In 2018, the OECD nations emitted 8.9 tonnes of CO_2 on average per person. This amount was a little more than double the 4.3-tonne average per person in the rest of the world (OECD, 2020).

The higher GHG impacts of wealthy, industrialized countries can be explained by their energy intensity. **Energy intensity** captures the dynamics of a nation's economic system as it industrializes, expands, and deepens its use of energy-consuming technologies. Intuitively, we know that an agrarian economy based on subsistence, labor-intensive agriculture requires less energy for economic production than an economy that makes steel, chemicals, and manufactured goods. Energy intensity is calculated as the amount of energy consumed for each monetary unit of gross domestic product (GDP) created; the U.S. unit of measure for energy intensity is Btu/\$1 GDP.

When societies industrialize, economic growth increases household income. Rising incomes stimulate not just the demand for more consumer products but also the amount of energy consumed at home for appliances, heating, cooling, and so on. Economic development therefore intensifies energy use in two ways: by expanding industrial output to meet social demand and by the growing wealth and consumption of households. Energy intensity does not increase indefinitely with economic growth, however. It rises to a point and then begins to stabilize and possibly decrease. We saw this in Figure 11.6 for the United States, where per capita energy consumption and energy intensity of the economy have been declining even though the population and GDP continue to grow.

Figure 11.8 illustrates the general relationship between energy consumption and affluence. This graph shows the strong association between per capita national income and energy consumption as well as the dominance of industrialized economies at the upper end of the income range. Figure 11.8 also illustrates the large number of people with unmet energy needs. The nine countries at the lower end of the income range have a population of about 3.6 billion people. Most of the future growth in world energy demand will come from these and similar nations as they pursue their economic development and social well-being. Table 11.1 provides key statistics on energy use, population, and gross domestic product (GDP) for a variety of countries. Example 11.2 demonstrates how to calculate per capita energy consumption and economic energy intensity.

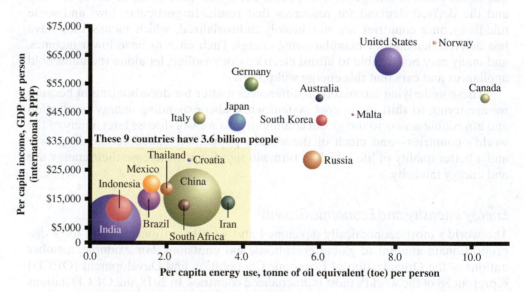

FIGURE 11.8 The relationship between prosperity and energy consumption, 2018. This graph illustrates the dramatically higher levels of per capita energy consumption in wealthy industrialized countries compared to low- and middle-income nations. The increasing future global demand for energy will come from the less industrialized nations as their economies develop and citizens get wealthier. The great challenge is providing for the growing energy needs of these societies in a sustainable manner. The nine nations in the bottom left corner of this graph account for 3.6 billion people, which is half of the world's total population. (The population of each country is represented proportionally in this chart by the size of the "bubble.")

Source: Based on U.S. Department of Energy, Energy Information Administration, *International Energy Statistics;* The World Bank, *Databank: World Development Indicators.*

TABLE 11.1 Energy and economic indicators for selected countries, 2018

COUNTRY	TOTAL POPULATION	ELECTRICITY CONSUMPTION, BILLION KILOWATT-HOUR (billion kWh)	GDP, PURCHASING POWER PARITY EQUIVALENT CURRENT INTERNATIONAL DOLLARS ($)	ELECTRICITY USE PER CAPITA, KILOWATT-HOUR (kWh/person)	TOTAL ENERGY USE, KILOTONNE OIL EQUIVALENT (ktoe)
Australia	24,982,688	234.28	1,255,451,034,245	9,378	149,856
Bolivia	11,353,140	8.35	100,626,844,046	735	8,070
Brazil	209,469,320	528.62	3,146,320,980,860	2,524	321,781
Canada	37,065,178	558.82	1,862,154,210,814	15,077	383,423
China	1,392,730,000	6,453.17	21,739,076,706,358	4,633	3,721,565
Croatia	4,087,843	16.77	116,725,859,609	4,103	10,579
Germany	82,905,782	533.18	4,556,074,544,346	6,431	349,379
Greece	10,732,882	51.36	318,562,046,357	4,785	29,479
Guatemala	16,346,950	10.57	141,651,998,408	647	8,097
Haiti	11,123,183	0.36	34,083,159,241	32	1,255
Iceland	352,721	18.88	20,536,241,304	53,520	5,785
India	1,352,642,283	1,277.17	9,029,375,940,714	944	790,068
Indonesia	267,670,549	248.94	3,116,958,911,116	930	201,559
Iran	81,800,204	254.72	1,128,459,783,071	3,114	294,703
Italy	60,421,760	301.62	2,605,604,157,045	4,992	172,109
Jamaica	2,934,853	3.02	29,256,359,453	1,031	2,984
Japan	126,529,100	939.79	5,275,711,543,015	7,427	485,367
Kenya	51,392,570	8.72	221,233,277,620	170	8,767
Malta	484,630	2.33	21,557,427,818	4,816	3,331
Mexico	126,190,782	270.74	2,556,350,916,361	2,145	201,685
Morocco	36,029,089	29.68	279,075,124,084	824	21,448
Nicaragua	6,465,502	3.74	37,934,761,584	578	2,420
Norway	5,311,916	125.62	370,296,238,032	23,649	47,842
Philippines	106,651,394	85.51	930,064,933,515	802	48,434
Russia	144,477,859	929.25	4,211,362,904,038	6,432	839,905
South Africa	57,792,520	206.52	747,322,170,987	3,574	142,040
South Korea	51,606,633	534.62	2,192,613,180,992	10,359	312,453
Thailand	69,428,454	185.85	1,286,523,956,730	2,677	138,825
United Arab Emirates	9,630,966	119.46	660,454,236,989	12,403	116,771
United States	326,838,199	4,032.83	20,611,860,934,000	12,339	2,551,203
Vietnam	95,545,959	216.99	742,208,673,972	2,271	93,748

Source: U.S. Energy Information Administration, *International Energy Statistics;* The World Bank, *Databank: World Development Indicators.*

EXAMPLE 11.2   Calculating Per Capita Energy Use and Energy Intensity

What is the per capita energy use and the energy intensity of the United States?

$$\text{Per capita energy use} = \frac{\text{total energy use}}{\text{total population}} \qquad (11.2)$$

Using values for the United States in Table 11.1, we have

$$\text{Per capita energy use} = \frac{2{,}551{,}203 \text{ ktoe}}{326{,}838{,}199} = 0.0078 \text{ ktoe per person per year}$$

$$\text{Energy intensity} = \frac{\text{total energy use}}{\text{total GDP}} \qquad (11.3)$$

Using values for the United States in Table 11.1, we have

$$\text{Energy intensity} = \frac{2{,}551{,}203 \text{ ktoe}}{\$20{,}611{,}860{,}934{,}000} = 1.24 \times 10^{-7} \text{ ktoe per \$1 GDP per year}$$

Energy intensity is often misinterpreted by students. Correctly, it reflects the amount of energy used to generate one dollar of GDP. We would therefore say that the United States consumed 1.24×10^{-7} ktoe of energy for every dollar of GDP that it generated. (Students often incorrectly state this in the reverse: that it costs one dollar to generate 1.24×10^{-7} ktoe of energy).

Accounting for Derived Demand

Measures of total or per capita energy use in a country do not show the important roles that different users and their needs play in energy consumption and derived demand for particular energy resources. To more deeply understand energy needs and demand, we analyze the four standard energy *end-use sectors*: the residential, commercial, industrial, and transportation sectors. The residential sector represents households. The commercial sector includes governments, nonprofit organizations, and businesses engaged in service industries such as insurance, real estate, sales, tourism, and health care. The industrial sector represents mining, construction, electric power production, agriculture, and manufacturing. The transportation sector involves the movement of passengers and freight. Figure 11.9 illustrates the shares of total energy in the United States accounted for by the end-use sectors.

The end-use sectors have distinctive patterns of fuel needs and uses. Figure 11.10 illustrates the flow of energy resources to the end-use sectors, where these resources are used to meet sector-specific energy needs. Notable patterns are evident. For example, 92% of all transportation energy comes from petroleum, and 70% of all petroleum used in the United States flows to the transportation sector. This suggests that not only is transportation the "driver" of American oil use, but also transportation systems are very vulnerable to disruptions in the supply or price of petroleum. In addition, the data in Figure 11.10 call attention to the role of electricity in modern society. We see that 38% of all U.S. primary energy production is used to generate electricity, which is then sold as a secondary form of energy to the end-use sectors. All nuclear energy is used for electricity, and electric power accounts for 91% of U.S. coal consumption.

By analyzing the flow of fuels from source to sector, we can begin to develop decarbonization strategies that meet the specific needs that different sectors have for energy services while also providing safe, abundant, affordable, and reliable energy.

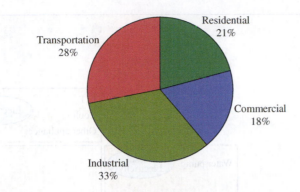

FIGURE 11.9 End-use sector energy consumption in the United States, 2019. The industrial and transportation sectors account for the greatest energy use in the United States, and the vast majority of energy used in the transportation sector is accounted for by passenger cars.

Source: Data from U.S. Energy Information Administration, Annual Energy Review 2020.

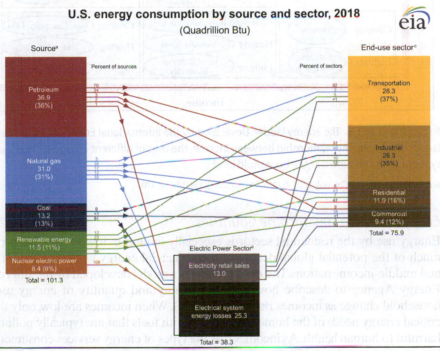

FIGURE 11.10 U.S. primary energy use by source and sector, 2018. This image is one of the most powerful graphical representations of U.S. energy use. The percentages for the bars in each stacked column show the total share of that item. The outflow arrows on the left indicate the percentage distribution of the resource to the end-use sectors; the inflow lines on the right indicate what percentage of the energy used by each sector comes from a specific source. The energy units in this infographic are quadrillion Btu (QBtu).

Source: U.S. Department of Energy, Energy Information Administration; www.eia.gov/todayinenergy/detail.php?id=41093.

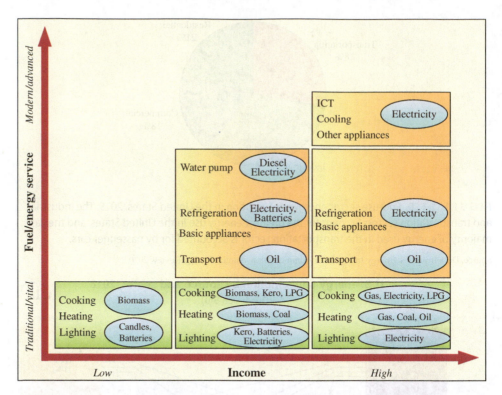

FIGURE 11.11 The energy ladder. Developed by the International Energy Agency, the energy ladder represents the relationship between income, the uses of different energy services, and the types of fuels consumed at the household level.

Source: World Energy Outlook 2002 © OECD/IEA, 2002, Fig. 13.1, p. 370.

Energy Issues in Low-Income Countries

Energy use by the residential sector is especially limited in low-income countries, and much of the potential global demand for commercial energy will come from the low- and middle-income nations. The **energy ladder** is a model developed by the International Energy Agency to describe how both the quality and quantity of energy used in a household change as incomes rise (Figure 11.11). When incomes are low, only the most critical energy needs of the home are met and with fuels that are typically polluting and harmful to human health. As incomes rise, the types of energy services consumed by the household expand to fulfill unmet needs and higher-quality fuels are used. Energy services also become more modern (derived from electricity and other types of commercial energy) and technologically complex. When households "climb" the energy ladder, they benefit from a wider variety of energy services, they transition from low- to high-carbon density fuels, and they increase their overall energy consumption.

Large numbers of households in low-income countries live at the bottom of the energy ladder, a condition of **energy poverty**. Such families are unable to access commercially provided energy. They rely instead on biomass that they gather themselves, like fuelwood, or on limited purchases of kerosene, charcoal, and batteries. The private and social costs of energy poverty are high. Private costs are the negative effects of a condition on the people who directly experience it; the private costs of energy poverty are obvious. **Social costs** are the negative impacts of the condition on society as a whole. The social costs of energy poverty are reflected in low human productivity, health impacts, and lost social capital. For example, gathering fuelwood is women's work in many countries, and it may consume several hours a day. The time that women spend gathering wood is time lost for more productive activities such as running a small business, farming, or attending to other household needs.

Traditional biomass is also harmful to human health, especially when burned indoors. Respiratory infections and asthma are debilitating and disproportionately affect children and women in the home. Other health consequences of energy poverty include the inability to purify water or pump potable groundwater, to refrigerate food and protect it from spoiling, and to sterilize medical equipment in community health clinics. Lost social capital includes the inability of children to do their homework or study at night because of a lack of lighting (Figure 11.12).

Low-income countries are also affected by limited access to electricity. Because of its exceptional versatility, electric power can meet a wide variety of energy needs and can be distributed over long distances. It is therefore the fastest growing type of energy worldwide. Yet in 2018, an estimated 800 million people globally lived without access to electricity. There is widespread lack of access in low-income countries and to a lesser extent in middle-income nations (Figure 11.13). Rural electrification is generally slower than in urban areas, but cities may experience frequent brownouts or blackouts. Low-income nations may consequently have both an *insufficient amount* of electricity and a *lack of reliability*. The result is compromised economic development, in which the scope, scale, and productivity of human ability, business, and industry are lessened.

FIGURE 11.12 An example of a homemade tin kerosene lamp common in many low-income countries. Such lamps provide light at night in rural areas and during the day in urban dwellings that have no direct access to sunlight. Kerosene is a costly and precious commodity to those low on the energy ladder, and lack of lighting carries a high social cost. Children cannot study at night, adults cannot extend their work into evening hours, and villages lose the ability to have community meetings and deliberations when the day's work is done.

Source: Photo by Bradley Striebig.

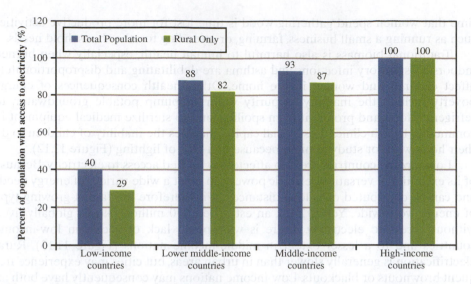

FIGURE 11.13 Global access to electricity, 2018. Low-income and lower middle-income countries lack universal access to electricity. Increases in world electricity demand will come from their economic growth and development. Rural electrification is a challenge in these regions.

Source: Data from The World Bank, Databank: World Development Indicators.

11.2 Energy and Natural Resources

The Earth's energy resources are usually categorized as either finite or renewable. *Finite resources*—in particular, the fossil fuels—are those that took millions of years to form and can be regenerated only on a geological timescale, if at all (Box 11.1). *Renewable resources* are those that replenish within a human lifetime; they include resources such as biomass and surface waters. Infinitely renewable resources are inexhaustible in their supply and result from the physics of our planet and solar system. These include solar, wind, tidal, and geothermal energy.

BOX 11.1 How Long Did It Take for Fossil Fuels to Form?

The history of the Earth represents deep time and is measured on a geological timescale. Key changes in our planet's climate, atmospheric content, living species, and landscape are demarcated by chronological eons and eras that last millions of years. The three fossil fuels (coal, oil, and natural gas) were created over millions of years by the transformation of highly concentrated amounts of organic matter. Coal formations are about 300 million years old, and the median age of oil fields is about 65 million years, although large oil fields in Siberia have been dated at 650 million years. The conditions required to replicate large-scale fossil fuel formation are not generally present on Earth today.

Fossil fuels originated during the Carboniferous Era, which did not have the biological decomposition of the carbon cycle (see Chapter 3) of our current geological era. As a consequence, organic matter today is cycled before dense amounts of carbon can accumulate to be transformed by fossilization into a fuel.

If we converted the planet's 4.5 billion year history to a 24-hour clock, then biologically modern humans have existed on Earth less than 3 seconds of this 24-hour period. In turn, the Industrial Revolution represents only 6/1,000 of a second of that time. As a species, we have consumed, in a fraction of a second, finite resources that took "hours" to form.

11.2.1 Finite Resources

The most significant finite energy resources are the fossil fuels, and discussions about them can be confusing. Resource data are presented as energy density or as volume and with many kinds of units. In addition, non-SI and non-metric units are frequently used to analyze American energy data. Examples include a barrel of oil, a therm of natural gas, a quad of energy, a short ton of coal, and so on.

Table 11.2 provides a comparison of the energy density of common biomass and fossil fuels, and Table 11.3 lists the most common units of energy and their energy content or unit equivalencies. Calculating fuel density equivalencies is demonstrated in Example 11.3. *We cannot caution you enough* to take extra care in understanding

TABLE 11.2 Energy density of common fuels

FUEL	ENERGY DENSITY (MJ/kg)
Peat	10.4
Wood, oak	13.3–19.3
Wood, pine	14.9–22.3
Coal, anthracite	25.8
Charcoal	26.3
Coal, bituminous	28.5
Fuel oil, no. 6 (bunker C)	42.5
Fuel oil, no. 2 (home heating oil)	45.5
Gasoline (84 octane)	48.1
Kerosene	45.6
Natural gas (density = 0.756 kg/m³)	53.0

Source: Based on Davis, M.L., and Masten, S.J. (2004). *Principles of Environmental Engineering and Science*. McGraw-Hill, New York.

TABLE 11.3 Energy conversion units

COMMONLY USED QUANTITIES OF ENERGY AND ENERGY CONTENT

1 British thermal unit (Btu)	The amount of energy required to heat one pound mass of water by one degree Fahrenheit; 1 Btu = 1,055 joules (J)
1 megajoule (MJ)	948 Btu
1 kilowatt-hour (kWh)	3,412 Btu
1 horsepower (hp)	746 watts
1 barrel of crude oil (bbl)	42 U.S. gallons = 5,100,000 Btu
1 ton of oil equivalent (toe)	41.868 GJ
1 therm	100 ft³ natural gas (ccf) = 103,100 Btu
1 gallon of gasoline	125,000 Btu
1 gallon of #2 fuel oil	138,690 Btu
1 gallon of LP gas	95,000 Btu
1 ton of coal	25,000,000 Btu
1 ft³ natural gas	1,031 Btu

(Continued)

TABLE 11.3 *(Continued)*

CONVENTIONAL ENGLISH UNITS FOR PRESENTING ENERGY DATA, ESPECIALLY IN THE UNITED STATES

MMBtu	1 million Btu (Note that in the SI system, M = 1,000,000)
1 Quad	10^{15} Btu (a quadrillion Btu)
ccf	100 ft³
mcf	1,000 ft³
ton	2,000 lb (also known as a short ton)
long ton	2,240 lb

the quantities and units of energy data that are used outside of the strict scientific context. Scientific analyses reliably use SI and metric units. Government, industry, and nongovernmental organizations may, however, use a variety of SI, metric, and alternative but conventional units.

EXAMPLE 11.3 | Calculating Fuel Equivalencies

The significance of fossil fuels as an energy source can best be understood by comparing them to the fuel that they replaced—wood. While we are concerned about the climate impacts of CO_2 emissions today, it is important to remember that the fossil fuels replaced wood as our primary energy source and that energy crises due to the lack of wood occurred throughout human history. Widespread deforestation and other environmental degradation occurred in many parts of the world because wood was the sole source of heat energy for local societies.

1. How much red oak is required to provide the equivalent amount of energy as a ton of bituminous coal if 1 kg of red oak has an energy density of 14.9 MJ?

2. In addition to the amount of the resource, it would be interesting to know how much space is required to store our fuel. How much space would 1,737 kg of red oak take up compared to 1 ton of coal?

1. Using the equivalencies in Tables 11.2 and 11.3, we find

$$1 \text{ ton of bituminous coal} = 2,000 \text{ lb} \times \frac{0.454 \text{ kg}}{1 \text{ lb}} = 908 \text{ kg} \times \frac{28.5 \text{ MJ}}{1 \text{ kg}} = 25,878 \text{ MJ}$$

The energy density of red oak is 14.9 MJ/kg. To continue using the given equivalencies,

$$25,878 \text{ MJ of coal} = \frac{1 \text{ kg red oak}}{14.9 \text{ MJ}} = 1,736.8 \text{ kg red oak}$$

From these calculations, we see that it takes 1,737 kg of red oak to provide the same amount of energy as 1 ton (908 kg) of bituminous coal; we need almost double by mass of oak.

2. A ton of coal by volume is about 38.5 ft³. (This is a cube a little over 1 meter in length per side.) By volume, 20 kg of red oak is equivalent to 1 ft³. To calculate,

$$1,737 \text{ kg red oak} \times \frac{1 \text{ ft}^3}{20 \text{ kg}} = 86.85 \text{ ft}^3$$

By volume, red oak thus requires more than double the space as its equivalent amount of coal, which as given is 38.5 ft³.

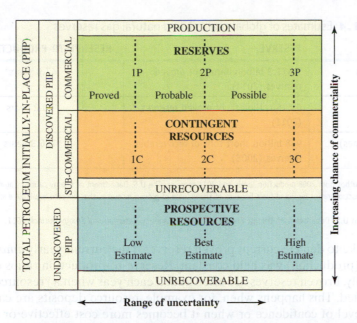

FIGURE 11.14 Classification system for petroleum reserves. The Society of Petroleum Engineers developed the classification system for estimating petroleum reserves, which are classified as either proved, probable, or possible. Note that undiscovered deposits are referred to as "prospective *resources*." Even though estimates are made of the amounts of these resources, they are not counted in estimates of the available supply of petroleum. Proved reserves are those that are technically and economically recoverable and form the basis of petroleum supply and depletion rates.

Source: Based on Society of Petroleum Engineers, Petroleum Resource Management System 2007. www.spe.org /industry/docs/Petroleum_Resources_Management_System_2007.pdf#redirected_from=/industry/reserves/prms.php.

The availability of finite energy resources is discussed in terms of *occurrences*, *resources*, and *reserves*. An occurrence is the geological presence (or deposit) of a substance. We do not know the total occurrences of the fossil fuels on Earth; this quantity is unknowable. Resources are a subset of occurrences and are estimates of the amounts of deposits that we think exist. Resources include deposits that are known to exist as well as those that have not yet been discovered. Quantitative estimates about the amounts of these resources include probabilities about their likely existence. **Reserves** are a subset of resources and are the most accurate calculation of how much of a resource is available. Reserve estimates are based on higher degrees of certainty about (1) the locations and sizes of deposits, (2) the ability to recover deposits using existing technologies, and (3) deposits that are economically profitable to recover (Figure 11.14).

No standardized classification system exists for all finite energy resources. Each substance—coal, uranium, natural gas, and petroleum—has evolved its own system for categorizing resources and reserves. This is due to the very different geophysical characteristics of each substance. These resources share the concept of a *proven* or *recoverable* reserve, however, which means that the geological deposits must be both technically recoverable and economically cost effective to extract.

Table 11.4 provides estimates of global coal, oil, and natural gas reserves. Table 11.4 also presents the **reserve-to-production ratio**, which is the amount of time

TABLE 11.4 Estimates of global coal, oil, and natural gas reserves

RESOURCE	RESERVE	RESERVE-TO-PRODUCTION RATIO
Liquid fuels	1,471.2 billion barrels of proved reserves (2011)	47.0 years*
Natural gas	6,675 trillion ft³ of proven reserves (2011)	60.2 years
Coal (all types)	948 billion short tons of recoverable reserves (2008)	126.3 years

*Estimated by author from 2008 production levels and 2011 reserves. The U.S. Department of Energy cautions against calculating reserve-to-production ratios for liquid fuels because of the volume of proven resources that become available each year.

Source: Based on U.S. Department of Energy, Energy Information Agency, *International Energy Outlook 2011*.

it would take to deplete current proven reserves if the current yearly *amount* of consumption (production) was held constant. Reserve-to-production ratios can change dramatically. Proven reserves are often added each year when a "resource" estimate is reclassified. This happens when, for example, resource deposits are calculated at a higher level of confidence or when it becomes more cost effective or technically possible to extract resources.

The **Hubbert curve** (see Figure 11.15) is used to estimate the point in time at which the geological availability and rate of extraction of a natural resource can no longer meet the rate of demand for it. This point is known as the "peak" because it is the time at which production of the resource is historically greatest. Peak oil is the most intensively discussed Hubbert curve. The Hubbert curve is often mistakenly interpreted as a model of total resource reserves, and the peak is often misinterpreted as the time at which half of all the resources have been consumed. However, the Hubbert model estimates the *rate* of extraction of a resource, which cannot grow forever because the resource is finite. The rate of increase in production must ultimately slow down, stabilize, and then decrease.

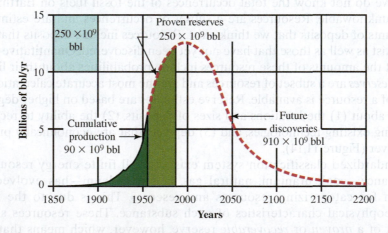

FIGURE 11.15 The Hubbert curve. This image of the original Hubbert curve shows the time at which the *rate of increase* in the production (or supply) of a resource reaches its peak. This is not the same thing as the rate of depletion or absolute depletion of the resource. The peak of the curve is a signal that our ability to supply the resource at a pace that can keep up with growing demand has reached its maximum limit.

Source: Based on Hubbert (1956).

The timing of the peak, which is the point at which the maximum rate of increase in productive output occurs, matters. After that point in time, supply cannot keep up with demand, prices will increase, and demand will change. Peak oil is not a prediction that we are running out of oil, but an indication that major economic, social, and technological changes in the supply and demand for oil and its energy services are imminent. In the United States, we see the political stress of peak oil dynamics in the controversy about whether or not to drill for oil in the Arctic National Wildlife Refuge (see Box 11.2).

BOX 11.2 To Drill or Not to Drill?

The U.S. Arctic National Wildlife Refuge (ANWR) is managed by the U.S. Fish and Wildlife Service and is a federally protected conservation area in northeastern Alaska. Comprising 19 million acres, ANWR is a pristine wilderness that lies north of the Arctic Circle (Figure 11.16); its landscape and habitats represent all six of the ecological regions of the circumpolar north. Biodiversity within the Reserve is considerable, and it hosts hundreds of bird species, major mammals (including the charismatic polar bears, arctic fox, and caribou), and fragile ecosystems. ANWR is valued aesthetically as an undisturbed wilderness and for its role in supporting unique ecological diversity.

At issue in ANWR are 1.5 million acres known as Section 1002. This tract of land has been exempted by Congress from full wilderness protection because of its potential for oil development. On the North Slope of the Brooks Range on the Beaufort Sea,

Section 1002 is near the proven (and rich) commercial Alaskan oil fields. Its potential for oil and its extraordinary beauty have made ANWR a political hotspot, reflecting competing interests in the environment, energy security, and economic development. When the price of oil and gasoline go up, the focus on ANWR is renewed. At present, no drilling is allowed in Section 1002, but that could change with an act of Congress. The environmental impacts of different drilling strategies are highly contested.

Oil in ANWR is a prospective resource and not a proven reserve, and estimated amounts are based on a U.S. Geological Survey (USGS) study in 1998. The USGS estimates that there are 5.7 to 16 billion barrels of technically recoverable oil in ANWR, which represents less than 15% of all technically recoverable, undiscovered, and prospective resources of the United States.

FIGURE 11.16 The U.S. Arctic National Wildlife Refuge. Located in northeast Alaska, ANWR is a source of political controversy because of its ecological significance and the presence of oil.

Source: U.S. Fish and Wildlife Service; Steven Chase, U.S. Fish and Wildlife Service.

As an energy resource, uranium is used exclusively for electricity generation by nuclear reactors. Uranium is a metal found in ore deposits and must undergo significant physical and chemical enrichment to become a usable fuel. Because spent nuclear fuel can be reprocessed and used again, many regard uranium as a renewable resource. There are limits to the number of times the fuel can be reprocessed economically, however, so it is best to regard uranium as a finite but recyclable natural resource.

11.2.2 Renewable Resources

In contrast to finite energy resources that have a fixed stock, renewable energy resources replenish themselves. The most important renewable energy resources are infinitely renewable—our use of these "fuels" does not reduce their physical abundance in any way. Infinitely renewable energy results from nature's own geophysical forces, including the potential energy of the sun, wind, and tides. Example 11.4 shows how to estimate how much solar electric power can be generated at a particular place. Geothermal energy is another type of infinite resource because it exploits the natural temperature gradients of the Earth for heating, cooling, and electricity. Despite the global pandemic, installed capacity for renewable electric power grew by a record-breaking 45% from 2019 to 2020. Wind power continues to be the dominant type of renewable electricity generation (Figure 11.17).

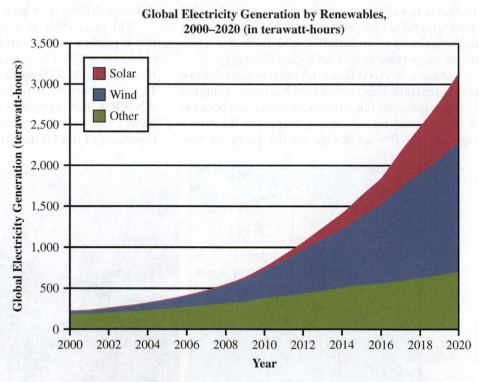

FIGURE 11.17 Electricity from renewables. Major growth in the amount of electricity generated from renewable resources has been driven by wind power. However, growth in solar electricity is starting to escalate as costs decline.

Source: Based on British Petroleum, "Renewable Energy," *Statistical Review of World Energy 2021*, www.bp.com/en /global/corporate/energy-economics/statistical-review-of-world-energy/renewable-energy.html.

EXAMPLE 11.4   How Much Usable Energy Can We Get from the Sun?

The amount of usable energy that we can get from a renewable resource varies by site, climate, and technology. In order to reliably estimate the productive energy that we can get for electricity, heat, or other work, we have to conduct what is called a "resource characterization." Solar energy is one of the easier renewable resources to evaluate because solar radiance is a relatively predictable quantity and has been mapped for all parts of the globe. The challenge is in accounting for the intermittency of solar energy—it varies by *diurnal phase* because of day and night, by season due to the angle of the Earth, and by climate because of variations in precipitation and cloudy days. So, how much electricity can we get from solar energy in San Antonio, Texas?

The electric power output of a solar photovoltaic panel can be calculated using the concept of a peak sun-hour (or just "sun-hour") to account for intermittency. Throughout the world, solar energy is at its most intense at solar noon in the summertime, when the sun is perpendicular to a surface. A peak sun-hour is therefore equivalent to the amount of solar energy over a period of 1 hour at solar noon in the summer, where this energy is striking a 1-square-meter area. This amount of energy is equivalent to 1,000 watt-hours. By adding up all of the available solar radiation over the course of a day and throughout the year, we can determine the average annual peak sun-hours for every location around the world. This average basically tells us how many hours of maximally productive sunlight are available to a solar panel each day at a specific location.

Solar PV panels are rated by their theoretically maximum output capacity. A 2-kW PV panel would produce 2 kW of electricity at its maximum output, which would be during peak solar radiance. A 2-kW PV panel exposed to peak solar radiance for one hour would therefore generate 2 kWh of electricity.

If you know the average number of peak sun-hours in a location, you can then readily estimate the average output of a solar panel for a year:

$$\text{Solar PV output} = \text{rated capacity of the panel} \times \text{average daily peak sun-hours} \times 365 \tag{11.4}$$

A 500-W PV panel in San Antonio, Texas, where the average daily peak sun-hours are 5.3, would produce 967.25 kWh per year, as

$$\text{PV output} = 500 \text{ W} \times \frac{1 \text{ kW}}{1,000 \text{ W}} \times \frac{5.3 \text{ hours}}{\text{day}} \times \frac{365 \text{ days}}{\text{year}} = 967.25 \text{ kWh}$$

Hydrogen could technically be considered an infinitely renewable energy resource because of its abundance as an element. It is used for hydrogen fuel cells, which generate electricity and have no carbon emissions. Fuel cells are used predominantly to power cars and buses, but they are also used as on-site electricity generators for commercial buildings. Hydrogen fuel cells do not emit any pollutants; their emissions are simply water vapor and warm air. Fuel cells are a proven technology but are not in widespread global use at present, in part because of the difficulty in economically mass-producing hydrogen and the lack of an energy infrastructure for hydrogen as a fuel. Hydrogen is also currently synthesized almost exclusively from natural gas, so its industrial process does have a significant carbon footprint. Nevertheless, hydrogen is a very promising fuel for the low-carbon transition. It can be produced through electrolysis using renewable power such as wind and solar.

Other important renewable energy resources are not infinite but are renewable because they regenerate and replenish themselves generally within a human lifetime. Surface water for hydropower is such a regenerative renewable resource. Although surface water used to be considered an infinite renewable resource for hydropower production, this is no longer the case. Localized climate change as well as human overwithdrawals of water from rivers have changed hydrological cycles as well as watershed flows (see Chapter 3). Other renewable resources include many kinds of commercial biomass such as cultivated grasses, bagasse (the pulp from sugarcane), and forest product residues. Algae is being explored as a biomass for biodiesel because of its high oil content. Any fuel that is derived from biomass is a *biofuel*; these include ethanol, biodiesel, and methane created by landfill decomposition (biogas). Renewable energy resources will be discussed again later in this chapter in terms of their role in decarbonization and electrification of energy systems.

11.2.3 Energy and Environmental Degradation

Every kind of energy use has some kind of environmental consequence. Even "clean" renewable energy cannot be harnessed without environmental impacts. For example, hydroelectric dams have had a significant impact on the ecosystems of the U.S. Pacific Northwest because they interfere with salmon spawning, some solar panels contain heavy metals, and biomass creates air contaminants when burned.

The impacts of energy on the environment therefore include not only the fuels that are used but also energy technologies and their operations. As natural resources, the fossil fuels and uranium must be extracted, purified, and processed into fuels. Converting fuels into energy services like transportation or electricity requires a technology like a car or a power plant. Although solar and wind energy are not fuels, we still need photovoltaic solar panels and wind turbines to convert the renewable energy to electricity. As such, we must consider the product life cycles of the equipment used for converting energy as well as the natural resources themselves. As explained in Chapters 5 and 14, the life cycle of a material or product involves all aspects of its production, including natural resource extraction, material processing, fabrication, shipment, and disposal.

Hazardous wastes and air pollutants are created during the materials cycles of the fossil fuels and uranium. The extraction and refining of coal and uranium create wastes that can acidify water supplies and degrade soils, landscapes, and habitats. Oil spills contaminate marine ecosystems; fuel refining industries emit noxious pollutants and generate hazardous wastes. *Fracking*, or hydraulic fracturing, is of particular concern. Hydraulic fracturing is a process in which highly pressurized water blended with chemicals is injected deep into wells to fracture bedrock that is trapping oil and natural gas. This fracturing opens space for the resources to flow more easily and has significantly lowered the cost of natural gas. Fracking is associated with several critical issues, however, including the health effects of water contamination and toxicity, methane leakage during extraction, and the activation of earthquakes.

Combustion of fossil fuels results in a wide variety of air pollutants, including sulfur dioxide (a health threat as well as a precursor to acid rain), nitrogen oxides (the precursors to smog and associated respiratory diseases), particulate matter (a contributor to heart disease), mercury (a heavy metal with neurological effects), and carbon monoxide (a poisonous gas).

As we mentioned, even renewable energy has consequences. With respect to solar photovoltaic panels, much of the concern is over the manufacturing of the panels as well as their end-of-life disposal. Issues associated with wind energy and hydroelectric power focus intensively on wildlife impacts, including loss of habitat and wildlife killed by the operation of these technologies. The sustainability of some biofuels is also questioned, especially for their impacts on agriculture and forestry. Ethanol requires large-scale crop production and has been criticized for perpetuating the environmental impacts of farming monoculture. The widespread cultivation of palm oil for biodiesel is a contributing factor to tropical deforestation and destruction of peat bogs, which affect their role as carbon sinks.

11.3 Carbon Footprinting and Embodied Energy

Fossil fuels do not contribute equally to GHGs because of differences in their chemistry and because the combustion equipment that burns them has different levels of fuel efficiency. Table 11.5 lists the commonly burned fuels, their energy density, and the amount of CO_2 equivalent emitted per million Btu of fuel burned in the

TABLE 11.5 U.S. default factors for calculating CO_2 emissions from fossil fuel and biomass combustion

FUEL TYPE	HEAT CONTENT	CARBON CONTENT (PER UNIT ENERGY)	CO_2 EMISSION FACTOR (PER UNIT ENERGY)	CO_2 EMISSION FACTOR (PER UNIT MASS OR VOLUME)
Coal and Coke	MMBtu/short ton	kg C/MMBtu	kg CO_2/MMBtu	kg CO_2/short ton
Anthracite	25.09	28.24	103.54	2,597.82
Bituminous	24.93	25.47	93.40	2,328.46
Subbituminous	17.25	26.46	97.02	1,673.60
Lignite	14.21	26.28	96.36	1,369.28
Coke	24.80	27.83	102.04	2,530.59
Mixed electric utility/electric power	19.73	25.74	94.38	1,862.12
Mixed industrial sector	22.35	25.61	93.91	2,098.89
Natural Gas	Btu/scf	kg C/MMBtu	kg CO_2/MMBtu	kg CO_2/scf
U.S. Weighted Average	1,028	14.46	53.02	0.0545
Petroleum Products	MMBtu/gallon	kg C/MMBtu	kg CO_2/MMBtu	kg CO_2/gallon
Distillate fuel oil no. 1	0.139	19.98	73.25	10.18
Distillate fuel oil no. 2	0.138	20.17	73.96	10.21
Residual fuel oil no. 6	0.150	20.48	75.10	11.27
Kerosene	0.135	20.51	75.20	10.15
LPG	0.092	17.18	62.98	5.79
Propane (liquid)	0.091	16.76	61.46	5.59
Motor gasoline	0.125	19.15	70.22	8.78

(Continued)

TABLE 11.5 *(Continued)*

FUEL TYPE	HEAT CONTENT	CARBON CONTENT (PER UNIT ENERGY)	CO_2 EMISSION FACTOR (PER UNIT ENERGY)	CO_2 EMISSION FACTOR (PER UNIT MASS OR VOLUME)
Aviation gasoline	0.120	18.89	69.25	8.31
Kerosene type jet fuel	0.135	19.70	72.22	9.75
Asphalt and road oil	0.158	20.55	75.36	11.91
Crude oil	0.138	20.32	74.49	10.28
Biomass Fuels—Gaseous	**MMBtu/scf**	**kg C/MMBtu**	**kg CO_2/MMBtu**	**kg CO_2/scf**
Biogas (captured methane)	0.000841	14.20	52.07	0.0438
Landfill gas (50% CH_4/50% CO_2)	0.0005025	14.20	52.07	0.0262
Wastewater treatment biogas	Varies	14.20	52.07	Varies
Biomass Fuels—Liquid	**MMBtu/gallon**	**kg C/MMBtu**	**kg CO_2/MMBtu**	**kg CO_2/gallon**
Ethanol (100%)	0.084	18.67	68.44	5.75
Biodiesel (100%)	0.128	20.14	73.84	9.45

Note: scf denotes "standard cubic feet." scf measures quantity of gas, not volume. It refers to the amount of gas at standard temperature and pressure. One standard cubic foot of natural gas is equal to 0.0027 cubic feet of natural gas.

Source: Based on 2012 Climate Registry, www.theclimateregistry.org/downloads/2012/01/2012-Climate-Registry-Default-Emissions-Factors.pdf.

United States. These CO_2 emissions factors are not the same throughout the world; they are calculated by country because nations differ in their fuel mix and types of combustion technologies. Decarbonization strategies are based on an understanding that energy fuels have different amounts of carbon in them, which affects their emissions and global warming potential (see Chapter 10). Carbon footprints are the primary accounting method for tracking and analyzing GHG emissions from particular sources or activities, and they enable us to evaluate decarbonization efforts. The concept of embodied energy allows us to look at the overall impact of specific products on energy use and GHGs.

11.3.1 Carbon Footprints

Carbon footprint techniques using mass balance equations (see Chapter 10) are used to calculate GHG emissions for different energy activities as well as the emissions that are *avoided* by implementing carbon mitigation efforts. Institutions and policymakers can also use the carbon footprint to track changes in their GHG emissions by calculating their carbon footprint for a baseline year. The potential climate benefits of investments in resource conservation and green technologies can then be monitored over time. For instance, companies that adopt a renewable energy portfolio can calculate the potential impact of options such as investing in forest protection, planting trees to increase carbon sequestration, or investing in renewable energy. Box 11.3 and Examples 11.5, 11.6, and 11.7 illustrate various techniques for calculating carbon footprints and offsets.

BOX 11.3 Calculating Carbon Footprints

The carbon footprint is the sum of all greenhouse gas (GHG) emissions associated with an individual, organization, or product expressed in CO_2 equivalence. Greenhouse gas emissions other than carbon dioxide are expressed in terms of their CO_2 equivalence based on their 100-year Global Warming Potential.

GHG emissions are calculated using standards determined from approved inventories such as the Global Greenhouse Gas Protocol (World Business Council for Sustainable Development and World Resources Institute, 2004) and the U.S. EPA Environmental Footprint Methodology. The carbon footprint should include direct emissions (scope 1), indirect emissions related to electricity consumption (scope 2), and indirect emissions associated with other products and materials (scope 3), including emissions associated with transportation, waste disposal, and consumer goods as illustrated in Figure 11.18. Data for global GHG emissions and production of goods can be found from various sources, including the Food and Agriculture Organization of the UN (FAOSTAT), the UN Commodity Trade Statistics Database (UN Comtrade), the International Energy Agency Statistics (IEA), and the Ecological Footprint Atlas. Table 11.6 provides examples of several types of industrial production and other facilities for which specific tools are used to calculate GHG emissions.

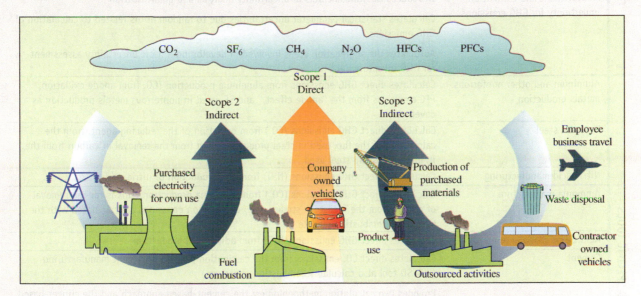

FIGURE 11.18 Direct and indirect GHG emissions for a value chain. This illustration shows how GHG emissions occur from the direct and indirect activities related to producing goods and services.

Source: Based on World Business Council for Sustainable Development and World Resources Institute (2004). *The Greenhouse Gas Protocol: A Corporate Accounting and Reporting Standard, Revised Edition.* (Washington, DC: World Resources Institute), Figure 3, p. 26.

(Continued)

BOX 11.3 Calculating Carbon Footprints (Continued)

TABLE 11.6 Greenhouse gas (GHG) emission calculation tools available online for calculating standardized emission rates for GHGs from various sectors

	CALCULATION TOOLS	MAIN FEATURES
CROSS SECTOR TOOLS	Stationary combustion	Calculates direct and indirect CO_2 emissions from fuel combustion stationary equipment
		Provides two options for allocating GHG emissions from a co-generation facility
		Provides default fuel and national average electricity emission factors
	Mobile combustion	Calculates direct and indirect CO_2 emissions from fuel combustion in mobile sources
		Provides calculations and emission factors for road, air, water, and rail transport
	Hydrofluorocarbons (HFCs) from air conditioning and refrigeration use	Calculates direct HFC emissions during manufacture, use, and disposal of refrigeration and air-conditioning equipment in commercial applications
		Provides three calculation methodologies: a sales-based approach; a life cycle-, stage-based approach; and an emission factor-based approach
	Measurement and estimation uncertainty for GHG emissions	Introduces the fundamentals of uncertainty analysis and quantification
		Calculates statistical parameter uncertainties due to random errors related to calculation of GHG emissions
		Automates the aggregation steps involved in developing a basic uncertainty assessment for GHG inventory data
SECTOR SPECIFIC TOOLS	Aluminum and other nonferrous metals production	Calculates direct GHG emissions from aluminum production (CO_2 from anode oxidation, PFC emissions from the "anode effect," and SF_6 used in nonferrous metals production as a cover gas)
	Iron and steel	Calculates direct GHG emissions (CO_2) from oxidation of the reducing agent, from the calcination of the flux used in steel production, and from the removal of carbon from the iron ore and scrap steel used
	Nitric acid manufacturing	Calculates direct GHG emissions (N_2O) from the production of nitric acid
	Ammonia manufacture	Calculates direct GHG emissions (CO_2) from ammonia production. This is for the removal of carbon from the feedstock stream only; combustion emissions are calculated with the stationary combustion module
	Acidic acid manufacture	Calculates direct GHG emissions (N_2O) from adipic acid production
	Cement	Calculates direct CO_2 emissions from the calcination process in cement manufacturing (WBCSD tool also calculates combustion emissions)
		Provides two calculation methodologies: the cement-based approach and the clinker-based approach
	Lime	Calculates direct GHG emissions from lime manufacturing (CO_2 from the calcination process)
	HFC-23 from HCFC-22 production	Calculates direct HFC-23 emissions from production of HCFC-22
	Pulp and paper	Calculates direct CO_2, CH_4, and N_2O emissions from production of pulp and paper. This includes calculation of direct and indirect CO_2 emissions from combustion of fossil fuels, biofuels, and waste products in stationary equipment
	Semiconductor wafer production	Calculates PFC emission from the production of semiconductor wafers
	Guide for small office-based organizations	Calculates direct CO_2 emissions from fuel use, indirect CO_2 emissions from electricity consumption, and other indirect CO_2 emissions from business travel and commuting

Source: Based on World Business Council for Sustainable Development and World Resources Institute (2004). *The Greenhouse Gas Protocol: A Corporate Accounting and Reporting Standard, Revised Edition*. (Washington, DC: World Resources Institute).

EXAMPLE 11.5   Calculating the Carbon Footprint of an American Household

The U.S. EPA (2013) reported that, on average, an American home consumed the following energy resources in 2010. What is the simplified carbon footprint for average home energy use in the United States?

ENERGY SOURCE	AMOUNT CONSUMED
Delivered electricity	11,319 kWh
Natural gas	66,000 ft³
Gasoline	464 gallons
Fuel oil	551 gallons
Kerosene	108 gallons
Auto transportation (gasoline)	2 vehicles at 11,493 miles each
Auto fuel economy	21.5 mpg

The EPA uses a simplified method of estimating an individual's greenhouse gas emissions in their Household Carbon Footprint Calculator (*www3.epa.gov/carbon-footprint-calculator/*). The carbon dioxide equivalent emission rates are estimated from the conversion factors in Table 11.7.

TABLE 11.7 U.S. EPA carbon dioxide emission factors for household resource consumption

SOURCE	SOURCE UNITS	CO_2 CONVERSION FACTOR	CONVERSION FACTOR UNITS
Electricity	kWh	7.0555×10^{-4}	Tonnes CO_2/kWh
Coal	Railcar	232.74	Tonnes CO_2/90.89 tonnes railcar
Natural gas	ft³	5.44×10^{-5}	Tonnes CO_2/ft³
Oil	Barrel	0.43	Tonnes CO_2/barrel
Gasoline	Barrel	0.2913	Tonnes CO_2/barrel
Kerosene	Barrel	0.42631	Tonnes CO_2/barrel
Propane cylinders used for home barbecues	Cylinder	0.024	Tonnes CO_2/cylinder
Uptake by trees	Urban tree planted	0.039	Tonnes CO_2/tree
U.S. forests storing carbon for one year	1 acre of average U.S. forest	1.22	Tonnes CO_2/acre-year

Source: Based on U.S. EPA. Clean Energy Calculations and References. Accessed May 15, 2013. www.epa.gov/cleanenergy/energy-resources/refs.html.

To answer this question, the individual GHG emissions for each category are calculated and then summed to estimate the total carbon footprint for an average U.S. household.

Step 1: Calculate CO_2 emissions for each energy source using the emissions factors in Table 11.7.

Electricity usage:

$$11{,}319 \text{ kWh} \times 7.0555 \times 10^{-4} \text{ tonnes } \frac{CO_2}{\text{kWh}} = 7.99 \text{ tonnes } CO_2$$

Natural gas usage:

$$66{,}000 \text{ ft}^3 \times 5.44 \times 10^{-5} \text{ tonnes } \frac{CO_2}{\text{ft}^3} = 3.59 \text{ tonnes } CO_2$$

Home gasoline use:

$$464 \text{ gallons} \times \frac{\text{barrel}}{42 \text{ gallons}} \times 0.2913 \text{ tonnes } \frac{CO_2}{\text{barrel}} = 3.22 \text{ tonnes } CO_2$$

Home fuel oil use:

$$551 \text{ gallons} \times \frac{\text{barrel}}{42 \text{ gallons}} \times 0.43 \text{ tonnes } \frac{CO_2}{\text{barrel}} = 5.64 \text{ tonnes } CO_2$$

Home kerosene use:

$$108 \text{ gallons} \times \frac{\text{barrel}}{42 \text{ gallons}} \times 0.42631 \text{ tonnes } \frac{CO_2}{\text{barrel}} = 1.10 \text{ tonnes } CO_2$$

Gasoline for automobile transportation:

$$2 \text{ cars} \times \frac{11{,}493 \text{ miles}}{\text{car}} \times \frac{\text{gallon}}{21.5 \text{ miles}} \times \frac{\text{barrel}}{42 \text{ gallons}} \times 0.2913 \text{ tonnes } \frac{CO_2}{\text{barrel}} = 7.42 \text{ tonnes } CO_2$$

Step 2: Sum the emissions from all energy sources.

The simplified carbon footprint for the average U.S. home in 2010 is

$$(7.99 + 3.59 + 3.22 + 5.64 + 1.10 + 7.42) \text{ tonnes } CO_2 = 28.96 \text{ tonnes } CO_2$$

EXAMPLE 11.6   Estimating Avoided CO_2 Emissions

We can calculate the CO_2 emissions avoided by using cleaner energy resources. The solar PV panel in Example 11.4 generates 967.25 kWh of electricity per year, carbon-free. How much CO_2 do we avoid by using the solar panel?

Step 1: Convert energy equivalences.

To estimate the avoided CO_2, we refer to the emissions factors in Table 11.5. From the available data, we see that we need to convert our kilowatt hour output into million Btu (MMBtu). Using the energy equivalencies provided in Table 11.3, we find

$$\text{Output of PV panel} = 967.25 \text{ kWh} \times \frac{3{,}412 \text{ Btu}}{1 \text{ kWh}} \times \frac{1 \text{ MMBtu}}{1{,}000{,}000 \text{ Btu}} = 3.3 \text{ MMBtu}$$

Step 2: Account for source energy losses.

Electric power plants, as source energy, are not 100% efficient. There are conversion losses in the generation and transmission of electricity. To get electricity to the end user as site energy, we will have to produce more than 100% of the site energy at the power plant. In the United States, conventional thermoelectric power plants powered by fossil fuels are only 34% efficient; that is, 66% of the heat content of the fuel is lost in the process of burning it, converting it to electricity, and distributing it to the end users. We therefore have to multiply our site energy by a factor of 3 to arrive at the source energy:

$$\text{Source energy} = 3.3 \text{ MMBtu} \times 3 = 9.9 \text{ MMBtu}$$

Step 3: Calculate the avoided emissions.

Referring to Table 11.5, we see that CO_2 emissions for a coal-fired electric power plant are 94.38 kg per MMBtu burned.

$$\text{Avoided } CO_2 = 9.9 \text{ MMBtu} \times \frac{94.38 \text{ kg}}{1 \text{ MMBtu}} = 934.4 \text{ kg } CO_2$$

Our solar panel therefore avoids 934.4 kg per year of CO_2.

EXAMPLE 11.7   Estimating a Portfolio of Carbon Offsets

By building on Examples 11.5 and 11.6, we can now estimate the reduction in GHGs from a portfolio of carbon mitigation initiatives. If electricity for the average U.S. home shifted to 50% fossil energy, 15% solar PV, 15% wind, and 20% nuclear energy and we also invested in 1 acre of newly planted trees, what are the yearly offsets of carbon emissions?

Step 1: Calculate the emissions reductions from lower fossil fuel electricity.

In Example 11.5 we see that the total CO_2 emission from electricity is 7.99 tonnes of CO_2 per year, which we assume to be entirely based on fossil energy. A 50% reduction in this emission rate would be 7.99/2, or approximately 4.00 tonnes of CO_2.

Step 2: Calculate GHG emissions for no-fossil energy resources.

Renewable and nuclear energy resources do emit some GHGs as part of their electric power generation process or as a consequence of their embodied energy. We can calculate the rate at which GHGs are emitted from nonfossil energy resources by drawing on their life cycle GHGs averaged over their lifetime power output. This estimate yields GHG emissions produced per kilowatt-hour generated (Table 11.8).

TABLE 11.8 Estimated life cycle GHG emissions per kilowatt-hour by energy source

ENERGY SOURCE	g CO_2 EQUIVALENTS/kWh	ENERGY SOURCE	g CO_2 EQUIVALENTS/kWh
Biopower	18	Wind energy	12
Solar (photovoltaic)	46	Nuclear energy	16
Solar (concentrating solar power)	22	Natural gas	469
Geothermal energy	45	Oil	840
Hydropower	4	Coal	1,001
Ocean energy	8		

Source: Based on Moomaw, W., et al. (2011). Table A.II.4.

Using data from Table 11.8:

Solar (photovoltaic) emissions:

$$11,319 \text{ kWh} \times 0.15 \times 46 \text{ g} \frac{CO_2}{\text{kWh}} \times \frac{1 \text{ tonne}}{10^6 \text{ grams}} = 0.0781 \text{ tonnes } CO_2$$

Wind energy emissions:

$$11{,}319 \text{ kWh} \times 0.15 \times 12 \text{ g} \frac{CO_2}{\text{kWh}} \times \frac{1 \text{ tonne}}{10^6 \text{ grams}} = 0.0204 \text{ tonnes } CO_2$$

Nuclear energy carbon equivalent emissions:

$$11{,}319 \text{ kWh} \times 0.20 \times 16 \text{ g} \frac{CO_2}{\text{kWh}} \times \frac{1 \text{ tonne}}{10^6 \text{ grams}} = 0.0362 \text{ tonnes } CO_2$$

Annual carbon emissions from the nonfossil electricity generation are

$$C_{nonfossil} = (0.0781 + 0.0204 + 0.0362) \text{ tonnes } CO_2 = 0.1347 \text{ tonnes } CO_2$$

Step 3: Calculate total offsets.

The total carbon offset, including the offsets associated with tree planting, is

$$C_{offset} = \text{original GHG emissions} - (\text{fossil emissions} + \text{nonfossil emissions}) + \text{other offsets}$$

Given that 1 acre of newly planted trees sequesters 1.22 tonnes of CO_2 per year,

$$C_{offset} = 7.99 - (4.00 + 0.1347) + 1.22 = 5.075 \text{ tonnes } CO_2 \text{ per year}$$

Each year, just over 5 tonnes of CO_2 would be avoided compared to previous emission profiles, or a 64% reduction in electricity-related GHGs.

11.3.2 Direct and Embodied Energy

We can analyze and account for energy use as direct energy or as embodied energy; each approach gives us important insights into GHGs and how to mitigate them. *Direct energy* represents activities that involve the conversion of an energy fuel or resource into usable energy. When we use gasoline to power a car, that is direct energy use. Other examples of direct energy use are converting sunlight with photovoltaic cells into electricity and burning natural gas to produce steam for an industrial process.

Basically, if we are burning a fuel or generating electricity, we are engaged in direct energy use. Decarbonization of the energy sector focuses heavily on direct energy because this is where GHG emissions occur. Figure 11.10 provided an understanding of direct energy use and the flow of fuels by end-use sectors. Figure 11.19 gives us a more detailed understanding of where and why energy-related GHG emissions are directly emitted; except for some agricultural emissions, all of these are from direct energy use. In the United States, transportation accounts for slightly more GHGs than electricity generation, but together these two uses represent 56% of total U.S. emissions. The next largest source is industry at 22%.

Embodied energy, also referred to as indirect energy, is a different kind of energy accounting. It represents the total amount of energy required to make and transport a finished product or material. Embodied energy includes the energy needed to extract and process natural resources into usable material feedstocks, to manufacture the product, and to ship items from place to place as the product moves through its life cycle (see Chapter 5). Embodied energy represents the sum of all the direct energy that is used to create and transport a specific product.

Embodied energy is accounted for using life cycle analysis, as introduced in Chapter 14. This sort of analysis allows us to think more holistically about energy sustainability. By considering where and how products consume energy throughout

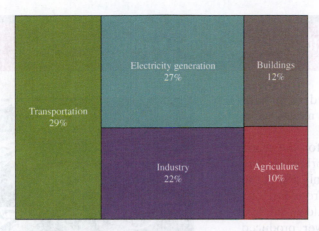

FIGURE 11.19 U.S. GHG emissions by source, 2018. This graph provides more detail about where U.S. GHGs are emitted directly. With the exception of some agricultural emissions, all direct GHGs result from energy use. Understanding these sector patterns can help us target decarbonization strategies more effectively.

Source: Data from National Academies of Sciences, Engineering, and Medicine (2021). *Accelerating Decarbonization of the U.S. Energy System: Technology, Policy, and Societal Dimensions.*

their life cycles, we can target the most energy-intensive stages for decarbonization efforts. Embodied energy also allows us to think about the role of recycling in energy conservation as well as our ability to recover the energy used for our manufactured products. Recycling avoids the energy required to extract and process a natural resource for first use (see Box 11.4). Embodied energy is recovered and put to use when, for example, waste-to-energy facilities burn trash to make process

BOX 11.4 The Value of Recycling Aluminum

Commercial aluminum is refined from aluminum oxide found in bauxite ore, and it is an extraordinarily energy-intensive process. Although a known substance since 1800, aluminum could not be mass produced because of both technological and energy-based barriers to the extraction of aluminum oxide from bauxite and to refining the metal itself. The engineering and economic breakthroughs arrived first with the invention of the electric arc furnace and second with the development of inexpensive hydroelectric power. Commercial aluminum manufacturing began in 1886 with hydroelectricity powered by Niagara Falls.

Metals are important from a sustainability perspective for two reasons. First, as natural resources, all metals are theoretically recoverable in terms of principles of physics and chemistry. As a conse-

quence, it is indeed possible (theoretically) to have a closed-loop materials cycle for metals and metal products where no metallic waste is ever generated. Indeed, nonferrous metals are regarded as being infinitely recyclable without any serious degradation of materials quality or properties.

Second, metal manufacturing is one of the most energy-intensive industrial processes because most metals must be extracted from low-grade ores and also refined by sustained, intense heat. They should therefore be a prime candidate for energy conservation. Regrettably, there are both technical and economic barriers to fully recovering all metals used in human societies and closing the materials loop for these substances. Where technically possible and economically feasible, metal recycling is critically

(Continued)

BOX 11.4   The Value of Recycling Aluminum *(Continued)*

important. Aluminum, copper, tin, and steel have strong recycling markets, as do precious metals (gold, silver, platinum).

Aluminum is a fascinating metal for the scope of materials recovery that occurs worldwide, and because recycling aluminum requires only 5% of the energy needed to process aluminum from raw ore, a 95% energy savings. According to Alcoa, "almost 75% of all the primary aluminum ever produced since 1888 is still in productive use." And, in such industries as automobiles, construction, and electric power systems, the aluminum recycling rates are near 90%, and the U.S. aluminum recycling industry contributes billions of dollars per year to the U.S. economy.

But the real story is in aluminum beverage cans (Figure 11.20). Beverage cans are an infinitely recyclable packaging material; in the United States, the recycled content of a can is 73%. Beverage cans are the most highly recycled container and are "the only packaging material that covers the cost of its own collection and reprocessing." The life of

FIGURE 11.20

Source: Mikael Damkier/Shutterstock.com.

a recycled can is a mere 60 days from the time it is collected as scrap to the time it is back on the shelf as a beverage. Can recycling in a number of countries such as Brazil, Japan, and Switzerland exceeds 90%; Germany and Finland recycle cans at a rate over 95%. There is room for improvement in the United States, where we recycle only about half of our aluminum cans.

steam or electricity. Because processing natural resources into usable material feedstocks for manufacturing is so energy intensive (see Chapter 5), significant decarbonization of embodied energy will happen mostly through dematerialization of our products and through recycling.

Laws of physics govern the potential contributions of embodied energy to sustainability. Newton's first law of thermodynamics states that all energy is conserved in a closed system. However, because of entropy, the second law tells us that not all energy can be recovered because some processes are irreversible. For example, the energy initially contained in a fossil fuel cannot be completely recovered and used again after it has been burned.

Physical embodied energy is commonly accounted for using life cycle assessment, but it is reflected somewhat in the economic concept of *levelized cost of energy (LCOE)*. LCOE allows us to directly compare very different technologies like wind turbines, solar photovoltaics, and coal-fired power plants. LCOE adds up all of the capital, fixed, and operating costs of a power generating technology over its lifetime and divides those costs by the total energy the system is expected to produce during its lifetime. LCOE tells us the average cost per unit of energy created by the system. As seen in Figure 11.21, global LCOE estimates indicate that wind and solar technologies can now generate a megawatt hour of electricity more cheaply than coal in many parts of the world, even without financial subsidies.

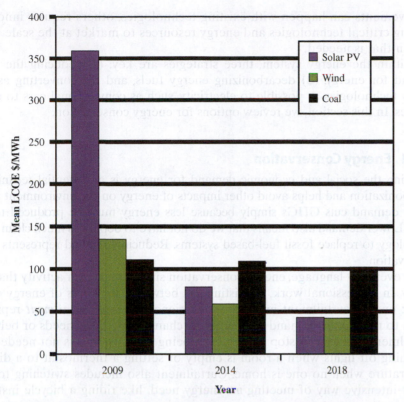

FIGURE 11.21 Levelized cost of energy for key electric power systems. The average levelized cost of energy globally for both wind and solar power (unsubsidized and excluding storage) is now better than coal, making renewable energy very cost competitive for new electric power generation facilities in many parts of the world.

Source: Data from Lazard, Levelized Cost of Energy Analysis Version 12.0 (November 2018). www.lazard.com /perspective/levelized-cost-of-energy-and-levelized-cost-of-storage-2018/.

11.4 Decarbonization Through Energy Conservation

Limiting global warming to either 1.5°C or 2°C relies on major changes to the energy system. The task is daunting, and a swift transition to net zero global carbon emissions requires that we work largely within the existing system and in anticipation of growing energy needs in many parts of the world. As we have reviewed so far, these considerations include the technologies and infrastructures designed around fossil fuels, the impacts of economic development on energy consumption, the different types of energy services and derived demand created by the end-use sectors, the special needs of low-income countries, and the availability of energy alternatives to fossil fuels.

Blueprints for deep decarbonization are provided by the International Energy Agency in *Net Zero by 2050: A Roadmap for the Global Energy Sector* (2021) and the National Academies of Sciences, Engineering, and Medicine in *Accelerating Decarbonization of the U.S. Energy System: Technology, Policy, and Societal Dimensions* (2021). These two significant reports indicate that a rapid transition is possible but will require major governmental policy efforts and societal change; market dynamics alone cannot create the shift. Although many improvements can happen with existing technologies, others require innovation to bring critical technologies and energy resources to market at the scale of production that is needed.

Within the energy system, three strategies are key: (1) reducing the overall demand for energy, (2) decarbonizing energy fuels, and (3) converting as many energy technologies as possible to electricity, such as conventional cars to electric vehicles. In this section, we review options for energy conservation.

11.4.1 Energy Conservation

Reducing the social and economic demand for energy is an essential technique of decarbonization and helps avoid other impacts of energy on the environment as well. Lower demand cuts GHGs simply because less energy must be produced to meet needs. Lower demand also means that we do not have to deploy as much clean energy technology to replace fossil fuel-based systems. Reducing demand represents energy conservation.

In everyday language, energy conservation simply means an activity that saves energy. In professional work, we distinguish between two kinds of energy conservation: energy curtailment and energy efficiency. *Energy curtailment* represents efforts to reduce the demand for energy by changing people's needs or behaviors. Curtailment actions can stop energy from being used when it is not needed, such as turning off lights when a room is empty or setting a thermostat to a different temperature when no one is home. Curtailment also includes switching to a less energy-intensive way of meeting an energy need, like riding a bicycle instead of driving a car.

Curtailment is different from *energy efficiency*. Efficiency refers to the technical performance of a piece of equipment that converts energy from one state to another. It is important to keep these two concepts distinct from both an engineering design and an energy management perspective because they lead to very different opportunities and costs for decarbonization. Example 11.8 shows how to compare the energy conserved by driving less (curtailment) with the energy conserved by buying a new, more fuel-efficient car. In general, when households, businesses, or industry substitute more efficient technology, this is considered a curtailment measure. Therefore, replacing incandescent light bulbs with LED bulbs in a home is a conservation *behavior*—the household has reduced its demand for the energy service of lighting by switching to a more efficient technology.

EXAMPLE 11.8 Estimating Energy Conservation

It can be relatively straightforward to estimate energy conservation for a variety of energy systems and to calculate how much curtailment (behavioral change) is required to equal an improvement in energy efficiency. For example, suppose you are thinking about buying a new car. Your current automobile has a fuel economy of 20 miles per gallon (mpg), and you are thinking about a new vehicle that has a rating of 32 mpg. If you kept your existing car, how many fewer miles would you have to drive per year to achieve the same level of energy conservation as the new car? Assume you currently drive 36,000 miles per year.

Step 1: Calculate your current annual fuel usage.

$$\text{Total fuel usage} = \frac{36{,}000 \text{ miles}}{\text{year}} \times \frac{1 \text{ gallon}}{20 \text{ miles}} = 1{,}800 \text{ gallons per year}$$

Step 2: Calculate your annual fuel usage with the new car.

$$\text{Total fuel usage} = \frac{36{,}000 \text{ miles}}{\text{year}} \times \frac{1 \text{ gallon}}{32 \text{ miles}} = 1{,}125 \text{ gallons per year}$$

Step 3: Calculate the equivalent miles driven on the fuel savings for your current automobile.

$$\text{Equivalent miles driven} = \frac{(1{,}800 - 1{,}125) \text{ gallons}}{\text{year}} \times \frac{20 \text{ miles}}{\text{gallon}} = 13{,}500 \text{ miles per year}$$

Based on these calculations, we see that buying a new car would save 675 gallons of fuel per year, the equivalent of driving 13,500 miles less in your existing vehicle. To achieve the equivalent fuel economy of buying a new car, you would need to reduce your miles traveled by 37.5%.

Behavioral changes are attractive because they can be accomplished with little or no cost. In the residential sector, energy conservation can be achieved in many ways that involve nothing more than changes in personal habits, such as washing laundry in cold water, driving less, turning off lights, lowering thermostats when homes are unoccupied, and so on. However, these changes can be remarkably difficult to achieve because they require that the average person be aware of ways to conserve energy, have the motivation to do so, and then act on it. Unless energy prices are extremely high or people have strong values about energy conservation or climate change, it is often hard to motivate people to change their energy habits.

In the commercial and industrial sectors, conserving energy and saving money are often synonymous. Because of the scale of buildings, technologies, and industrial operations, energy conservation in these sectors is often achieved by using automation that adjusts energy usage. One example is controls that regulate the amount of electric lighting in a room relative to the amount of daylight in it. Another is carbon dioxide sensors that regulate air-handling equipment based on a room's occupancy rather than a fixed rate of fresh air replacement every hour.

Energy conservation is not just for wealthy industrialized countries. A family in a low-income country that switches from open-fire cooking to a cookstove is practicing energy conservation because the family reduces its need for fuelwood (see Box 11.5).

In the United States, there are few public policies that actively and directly promote energy conservation. In 1974, after an energy crisis, the federal government implemented an interstate highway speed limit of 55 miles per hour to conserve oil; this law was repealed in 1995. Today, we see conservation policies most extensively in building codes that mandate specific types of conservation in the design and construction of buildings, such as insulation requirements (see Chapter 16).

BOX 11.5 The Global Clean Cookstoves Movement

Nearly one-third of the world's population, about 2.5 billion people, uses traditional biomass to cook and heat their homes. This unsafe form of energy accounts for the deaths of nearly 2 million people per year (mostly women and children). Emphysema, asthma, pneumonia, cancer, and low birthweights are significant health consequences of indoor fires that burn wood, charcoal, dung, and crop residues; the need for biomass by the world's energy poor is also a notable cause of deforestation and soil erosion. Not only is there a high social cost to the time spent gathering biomass fuels in rural areas, but in urban areas low-income families can also spend one-fifth or more of their scant incomes on charcoal and wood. Traditional cookstoves and open fires are not only costly in terms of health, the environment, and household economics, but they are also extremely inefficient from an engineering standpoint. In the most extreme case of an open fire for cooking, 90% of the heat is lost to the air, with only 10% going to a cooking pot.

The global clean cookstoves movement represents a decades-long initiative to develop more efficient cooking facilities for residents of low-income countries. Engineering an improved cookstove is not trivial because of the wide varieties of fuels, cultural cooking practices, and prepared foods worldwide. In all instances, improved cookstoves are designed to drastically reduce indoor air pollutants, conserve scarce biomass resources, and protect the environment. As a consequence, several dominant technological designs have emerged to meet

FIGURE 11.22 Pakistani woman adding fuelwood to a cookstove. Cookstoves are better than open fire cooking because they dramatically improve indoor air quality and reduce the amount of wood needed for cooking.

Source: Steve Davey Photography/Alamy Stock Photo.

the cooking needs of families in Latin America, Africa, Asia, and eastern Europe. These include advanced biomass cookstoves, alcohol and biogas stoves, electric cookers, planchas (flat griddles used in Latin America), propane gas stoves, solar ovens, and "rocket" cookstoves with specialized combustion chambers. However, new technologies are not enough. The affordability of the stoves and fuel is still a challenge. In addition, there is resistance to some types of cookstoves in many communities because of cultural factors regarding the appearance of food and their effects on cooking practices as well.

Occasionally, state or federal policies may promote conservation through tax credits or other financial incentives; these are designed to offset the costs of adopting specific types of conservation products. In the private sector, electric power and natural gas utilities also sponsor rebate programs for conservation equipment from time to time.

11.4.2 Energy Efficiency

Energy efficiency refers to the productivity of technologies that convert an energy input into a more useful or usable form, called "work." Usable energy can take many forms, including heat, mechanical work, electric power, light, and cooling. Energy efficiency therefore measures how much of the input energy becomes useful and is not lost in the transformation. It is represented by the Greek letter eta (η) and the equation

$$\eta = \frac{\text{useful energy out}}{\text{energy in}}$$

Energy efficiency is therefore the ratio of the amount of energy produced by a technology to the amount of energy that was put into the system. Efficiency can be improved in one of two ways: (1) by creating more work with the same amount of energy or (2) by obtaining the same amount of work with less energy.

Many types of products, machines, and equipment convert energy from one form to another. These include combustion engines, motors, boilers, and lighting as well as anything that generates electricity—from steam turbines to wind turbines and photovoltaic cells. Appliances and many consumer goods are also energy conversion devices, including refrigerators, light bulbs, air conditioners, and hot water heaters. In addition to specific pieces of equipment, we can think about the energy efficiency of large systems. For example, a conventional electric power plant is able to convert only about 60% to 70% of its fossil fuel input into usable electricity. Losses occur mostly in the power plant itself because of the thermodynamics of the process; smaller amounts are lost in the transmission of electric power (Figure 11.23).

Energy efficiency is a significant way to reduce energy consumption. For example, the energy efficiency of American refrigerators has improved dramatically, largely due to government regulation. As shown in Figure 11.24, the average energy

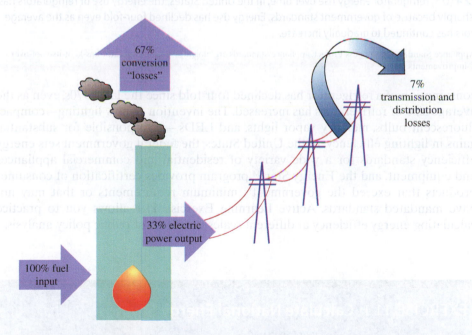

FIGURE 11.23 Efficiency of a conventional thermoelectric power plant. A typical electric power plant burns a fossil fuel (or uses nuclear energy) to create steam for conversion into electricity. The thermodynamic and electromechanical conversion "losses" are high; only about one-third of the fuel energy ultimately becomes electricity. There are also slight transmission losses when electricity is moved through power lines.

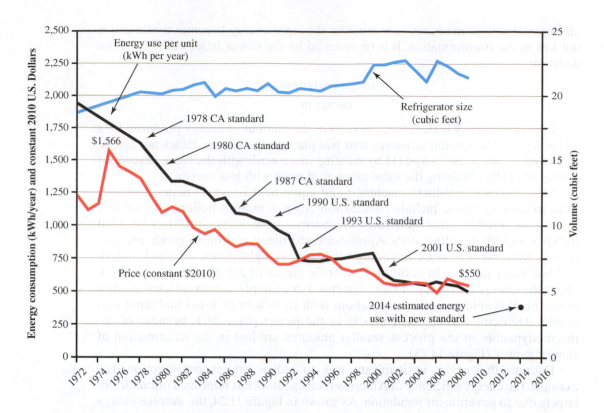

FIGURE 11.24 U.S. refrigerator energy use over time. In the United States, the energy use of refrigerators has been declining sharply because of government standards. Energy use has declined four-fold even as the average size of refrigerators has continued to gradually increase.

Source: Based on Appliance Standards Awareness Project, appliance-standards.org/blog/how-your-refrigerator-has-kept-its-cool-over -40-years-efficiency-improvements.

consumption of a refrigerator has declined four-fold since the late 1970s, even as the average size of refrigerators has increased. The invention of new lighting—compact fluorescent bulbs, mercury vapor lights, and LEDs—is responsible for substantial gains in lighting efficiency. In the United States, the federal government sets energy efficiency standards for a wide variety of residential and commercial appliances and equipment, and the Energy Star™ program provides certification of consumer products that exceed the government's minimum requirements or that may not have mandated standards. Active Learning Exercise 11.1 allows you to practice calculating energy efficiency at different scales as a form of public policy analysis.

ACTIVE LEARNING EXERCISE 11.1 Calculate National Energy Savings

This activity requires you to perform a number of computational steps to come up with the most compelling statistics, but it is readily doable. You will practice the full sequence of conversions, equivalences, and estimates of energy output and consumption that are commonly done in a professional setting.

Assume that you work as an energy analyst for the U.S. Secretary of Energy. The Secretary wants to provide a compelling fact about energy conservation to really grab an audience's attention in a speech. The point to stress is that small actions conducted by many people can add up to significant energy savings and carbon mitigation. The Secretary wants to tell people how much energy the United States would save if every home replaced one incandescent light bulb with a CFL.

Use the following "givens," as well as other information provided in this chapter, to calculate four different statistics that reflect national energy savings and CO_2 mitigation. At least one of these statistics must be the number of baseload electric power plants that can be shut down by switching bulbs.

- There are 120 million households in the United States.
- A "high-use" bulb is one that burns for at least 4 hours per day.
- A 13-watt CFL is equivalent to a 60-watt incandescent bulb.
- The average baseload power plant has a rated output capacity of 800 MW.
- The typical baseload electric power plant in the United States uses a mixture of different kinds of coal.

A major engineering challenge today is presented by the physical limits to efficiency for heat engines. A heat engine is any device or system that converts thermal energy (heat) to mechanical work. Heat engines include the internal combustion engine in cars as well as traditional steam engines. The *Carnot limit* tells us that for any heat engine, there is a maximum efficiency that can be attained. This physical upper limit to efficiency results from the difference between the temperature of the source heat in a system and the cooler temperature at which the heat is discharged. Once the Carnot limit is reached, the only way to improve overall system performance is to capture and utilize the waste heat in some way. The thermal efficiency of a typical automobile engine is less than 20%, compared to the theoretical Carnot limit of 37%.

The federal government and many states encourage energy efficiency and conservation with a wide variety of policies and tools. The most important are mandatory *performance standards* that dictate the minimum energy efficiency of a wide variety of residential, commercial, and industrial equipment, such as home appliances, lighting, industrial boilers, commercial heating and cooling equipment, and vending machines. More than 30 appliance and equipment classes are regulated by the U.S. Department of Energy. The United States and the state of California also regulate the fuel efficiency of automobiles. As with conservation, government tax credits and private sector rebates are used to encourage the purchase of energy-efficient equipment by homeowners, businesses, and industry.

Public and private awareness programs provide information about product energy use to empower consumers when they shop for new products (Figure 11.25). The U.S. Energy Star™ program is a voluntary certification initiative in which private businesses apply to the federal government to get their products certified as being more energy efficient than the regulatory standards require. The government also funds research and development programs to enhance the performance of energy technologies, from batteries to coal-fired power plants.

Energy recovery contributes to energy efficiency by capturing and using energy that is otherwise lost as waste heat. The exploitation of usable waste heat is known

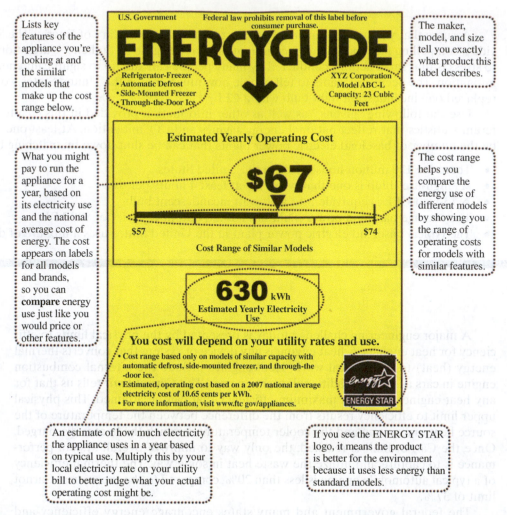

FIGURE 11.25 U.S. energy guide label. The striking black and yellow energy guide label is required by law and is regulated by the U.S. Federal Trade Commission. Its purpose is to promote energy efficiency by allowing consumers to compare the energy costs of products as part of their purchasing decision.

Source: U.S. Federal Trade Commission.

as **combined heat and power (CHP)** (see Box 11.6). CHP, also known as cogeneration, most commonly uses the waste steam or hot gases from electric power production for hot water, heating, and cooling in buildings. Less commonly, it uses the waste heat from high-temperature industrial processes to generate electricity. Integrated gasification combined cycle power plants use gasified coal and CHP heat recovery to raise the efficiency of electric power plants to about 60%; these facilities also have lower carbon and other emissions, which is why they are often referred to as a "clean coal" technology. The United States does have public policies and private incentives from time to time that promote CHP.

BOX 11.6 Combined Heat and Power

Combined heat and power (CHP) systems, also commonly referred to as *cogeneration*, are unsung heroes in the world of sustainable energy and deserve more popular attention. Rather than directly consuming an energy fuel to provide energy, CHP systems exploit waste heat to provide cooling and heating or to generate electricity.

There are two basic types of CHP systems. *Topping cycle plants* use waste steam and heat from electric power production to create steam or hot water that is then used for heating and cooling a building. Because electricity is generated by both steam and gas turbines, the design of CHP systems will vary depending on the generating system

(Figure 11.26). The capacity of topping cycle CHP is extraordinary. For example, Consolidated Edison of New York operates the largest district steam system in the United States and supplies 1,800 major customers (such as the United Nations and the Empire State Building) with hot water, heating, and cooling via a network of 105 miles of lines and pipes. Amazingly, 45% of Con Edison's district steam is CHP from its own electric power plants and those of the Brooklyn Navy Yard. By installing CHP in electric power plants, the overall energy efficiency of these plants increases from 45% to 80% because waste energy is put to productive use.

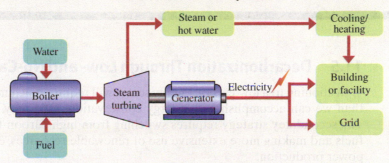

Boiler-based CHP system

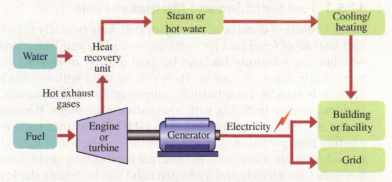

Turbine-based CHP system

FIGURE 11.26 Combined heat and power plants. CHP systems use waste heat for productive purposes. These diagrams illustrate two different types of topping cycle plants. The boiler-based system is a conventional steam turbine in which a fuel is combusted to heat water into steam. Turbine-based systems burn a fuel, and it is the combustion exhaust gas that turns the electric generator, not steam.

Source: U.S. Environmental Protection Agency, www.epa.gov/chp/basic/.

(Continued)

BOX 11.6 Combined Heat and Power *(Continued)*

The other type of CHP is less common and is referred to as a *bottoming cycle plant*. Here, the waste heat from industrial furnaces and other high-temperature manufacturing processes (metals, glass-making) is captured to generate electricity that is typically used within the facility itself. These types of cogeneration systems are much less common than topping cycle plants.

There is room for more combined heat and waste heat recovery. The Combined Heat and Power Alliance reports that 9% of electricity in the United States was generated by heat recovery in 2016, with the potential to increase to 20% of total U.S. electricity generation. If viable projects with a 10-year payback period were built, carbon emissions would decrease from 2016 to 2030 by an amount equivalent to more than eight conventional power plants (Combined Heat and Power Alliance, 2017).

The U.S. military is a leading adopter of CHP systems. In 2012, the Marine Corps Air Ground Combat Center (in California) and the 82nd Airborne at Fort Bragg (North Carolina) won the EPA's CHP Energy Star™ award. At the Air Ground Combat Center, the new CHP system improves plant efficiency to 64%, saves nearly $6 million per year in energy costs, and annually avoids CO_2 emissions that would be equivalent to the electricity used by 2,400 homes. At Fort Bragg, the CHP system saves $1 million each year and avoids CO_2 emissions equivalent to the electricity used by 1,500 residences.

11.5 Decarbonization Through Low- and No-Carbon Resources

Transitioning to a low-carbon energy sector will require much more decarbonization than we can accomplish with energy conservation, efficiency, and recovery alone. The second key strategy requires switching from high-carbon to low- or no-carbon fuels and making more extensive use of renewable resources, especially for electric power production.

11.5.1 Fuel Switching and Alternative Fuels

In the context of decarbonization, *fuel switching* typically refers to burning a cleaner fuel instead of fossil fuel for combustion energy and heat. A major consideration is whether the substitute fuel can be used directly in existing machinery and equipment. Only some alternative fuels can be used without much trouble in existing engines, boilers, and combustion equipment. In other instances, the ability to switch fuels requires retrofitting with alternative equipment. Because the transportation sector is such a large contributor to GHGs, substitute fuels must work within the existing transportation system and infrastructure for at least the next two decades. Liquid fuels are a necessity for internal combustion engines until new vehicle technologies (like electric and hydrogen cars) can be widely deployed.

A common type of fuel switching involves the substitution of natural gas for coal. We see this most extensively in electric power plants and industrial processes. Motor vehicles that operate on compressed natural gas are also a form of fuel switching. Equipment that uses natural gas operates at higher efficiencies and the fuel also burns more cleanly than coal; it has lower carbon emissions and other air pollutants for each unit of equivalent energy that is created. Fuel switching to natural gas in electric power plants and industry is why coal use in the United States has declined significantly since 2007–2008. This trend has also been driven by the lower cost of natural gas due to fracking.

Fuel switching includes biofuels. Biofuels can be liquid or gas, and both are generally considered sustainable resources. Biofuels are renewable resources and have lower carbon emissions when burned for direct energy. However, the life cycle carbon emissions of *first-generation* biofuels can be quite high and offset their value as a low-carbon fuel because they are agricultural crops. First-generation biofuels are derived from food crops such as corn and are usually converted to ethanol or biodiesel as transportation fuels. *Second-generation* biofuels do not derive from edible foods. They are produced from the cellulose of agricultural or forest product waste or from vegetation like grasses. Second-generation biofuels are considered carbon neutral because their life cycle carbon emissions are negligible.

Biogas is a type of biofuel and is a renewable resource. When organic matter breaks down, it produces methane as a byproduct of anaerobic decomposition. Three major sources of fuel-quality biogas are landfills, wastewater treatment plants, and animal manure. Landfill gas can be tapped and piped. Biogas from wastewater and cow manure is obtained from anaerobic digesters. Biogas can be substituted for natural gas in many types of applications, especially hot water production. The carbon benefits of biogas are significant. First, it avoids the GHG emissions that would have been created by allowing the uncaptured gas to enter the atmosphere. Second, it avoids the GHGs that would have been emitted by the extraction, processing, and combustion of the natural gas that the biogas replaces.

Renewable biomass is also used as a solid fuel for electricity generation, but these power plants tend to be smaller than those operated on fossil fuels. Bagasse (sugarcane residue), wood chips, and commercially farmed shrub willow are common biomass feedstocks for electric power production worldwide. Trash is also a renewable fuel. **Waste-to-energy** facilities burn municipal garbage to generate steam or electricity; not only do these energy plants use our trash as a fuel, but they also produce lower carbon emissions and have the added benefit of conserving landfill space. When waste is burned, we are recovering part of the embodied energy in the materials of our waste stream that would otherwise be buried in a landfill.

Fuel switching is attractive because many of these alternative fuels are recovered from waste, such as some biodiesel, biogas, municipal garbage, and biomass residues from agriculture and industry. Although it is an important strategy to adopt, fuel switching has two limitations for deep decarbonization. The first and most significant challenge is the *scale* of the global transportation sector, which consumes about 4 billion gallons of crude oil per day and accounts for about one-quarter of energy-related global GHGs. Because biofuels for transportation are derived from food crops and other commercial biomass like palm oil, biofuels compete with agricultural land for food production. First-generation biofuels simply are not sustainable at the scale required to displace petroleum. This is one reason why there is so much emphasis on electric vehicles. Second, many industries require sustained, extremely high temperature heat that can only be provided by a fossil fuel. Examples of these needs are superheated steam, smelting metals, and firing some ceramics. Many of these industries produce essential materials for the economy, such as steel, chemicals, and cement.

11.5.2 Other Renewable Energy Applications

Many renewable energy resources are used as fuels for combustion. Except for municipal trash, these renewables all come from organic matter, living things that *regenerate* themselves. Biomass for electricity and biofuels are derived from plants and woody vegetation. Biogas is captured from decomposing organic waste. When burned, these are all low-carbon fuels compared to fossil fuels.

The renewable energy resources that come from "planetary" forces—solar, wind, water, geothermal, and tides—and their energy applications create no or limited direct carbon emissions and air pollutants. Wind, tidal, and hydropower are used exclusively for electricity generation. Solar and geothermal energy are used primarily for electricity but also for heat.

Solar thermal collectors are the technology used for solar heating, which is mostly for hot water production and space heating in buildings (see Chapter 16). However, **concentrated solar power** uses solar thermal energy to produce electricity from steam. These systems concentrate solar energy with mirrors and lenses (much like a magnifying glass) onto a massive collector; they have the potential to be important power plants in places with high levels of solar radiation. Solar photovoltaic (PV) panels convert sunlight directly into electricity. Geothermal energy is exploited in two ways. First, because of the natural and stable temperature gradients of the Earth, heating and cooling systems can be preheated or prechilled by circulating their heat-exchanging fluids underground. **Ground source heat pumps** (Figure 11.27 and Example 11.9) are an example of this technology. Second, geothermal energy can be used to generate electricity, known as **geothermal power**. This requires intense heat or steam, and such plants are typically found near active volcanoes, tectonic plates, and natural geysers. In 2020, there were about 16 GW of installed geothermal electric power capacity in the world; Iceland benefits the most from its geothermal resources, which accounts for about one-third of its electricity.

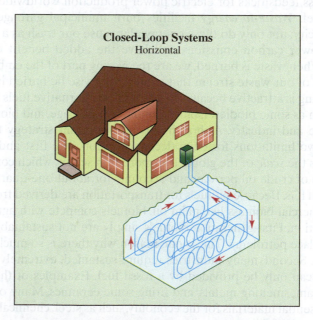

FIGURE 11.27 Ground source heat pump. A closed-loop ground source heat pump cycles compressed fluid through buried piping. During the summer, the heat pump takes advantage of geothermal energy by precooling compressor fluids for air conditioning; in the winter, the compressor fluid is prewarmed by the ground, which is warmer than air temperatures.

Source: U.S. Department of Energy, www.energy.gov/sites/prod/files/styles/large/public/closed_loop_system _horiz.gif?itok=A8hCygmB.

EXAMPLE 11.9 Estimating the Energy Savings of Ground Source Heat Pumps

The energy performance of heating and cooling equipment is measured in two ways. For cooling systems, the rating is called an *energy efficiency ratio (EER)* and is calculated as

$$EER = \frac{\text{Btu of cooling output}}{\text{watt-hour of electric input}} \qquad (11.5)$$

EER units are expressed in Btu per watt-hour.

Heating energy efficiency is represented as the coefficient of performance (COP) and is calculated as

$$COP = \frac{\text{power output in watts}}{\text{power input in watts}} = \frac{EER}{\dfrac{3.412 \text{ Btu}}{\text{watt-hour}}}$$

The COP is a dimensionless ratio and can be understood as a measure of instantaneous work output. Note that to convert the COP to an EER, 3.412 Btu/Wh is applied as a straight conversion factor. (Recall from Table 11.3 that there are 3,412 Btu in a kWh.)

Both the EER and the COP represent the same ratio, which is the amount of work output relative to one unit of electricity input (power). A COP of 3.8, for example, means that a heating system produces 3.8 times more energy in heat than it consumes in electricity. An EER of 11 means that a watt-hour of electricity produces 11 Btu of cooling output.

Suppose that the COP for a conventional heat pump is 2.4 and 3.5 for a ground source heat pump. If a building requires 43.3 million Btu per year for heat, how much electricity does the geothermal system save?

Step 1: Convert the annual heating load to kilowatt-hours.

This question requires us to express the energy savings in units of electricity, which is commonly expressed in kilowatt-hours (kWh). Using the equivalencies from Table 11.3:

$$\text{Annual heating load} = 43{,}300{,}000 \text{ Btu} \times \frac{1 \text{ kWh}}{3{,}412 \text{ Btu}} = 12{,}690.5 \text{ kWh}$$

Step 2: Calculate the input energy required for each system.

Because the COP is dimensionless, it represents a straightforward input/output ratio. Manipulating Equation 11.5 above,

$$\text{Power input} = \frac{\text{power output}}{COP}$$

We can thus calculate the energy input requirements for each system as

$$\text{Power input, conventional heat pump} = \frac{12{,}690.5 \text{ kWh}}{2.4} = 5{,}287.7 \text{ kWh}$$

$$\text{Power input, geothermal heat pump} = \frac{12{,}690.5 \text{ kWh}}{3.5} = 3{,}625.9 \text{ kWh}$$

Step 3: Compute the net energy savings.

Net energy savings = 5,287.7 kWh − 3,625.9 kWh = 1,661.8 kWh

The geothermal heat pump saves us 1,661.8 kWh per year for this particular heating load. This is equivalent to 31% less input energy than the conventional heat pump. We can therefore say that the ground source heat pump is almost one-third more efficient than the conventional system.

Hydrogen fuel cells are often considered a renewable energy technology because they use hydrogen as their fuel (Figure 11.28). Fuel cells create electricity through a chemical process involving hydrogen ions, and they have efficiencies as high as 60%. In addition, if the waste heat is recovered with CHP, the overall efficiency of a fuel cell system can exceed 90%. Fuel cells emit no air pollutants or greenhouse gases, just water and heat. Although fuel cells are now in commercial use—especially in the transportation sector—there are many obstacles to the widespread development and adoption of this technology. Costs are high, and we do not have an effective way of producing mass quantities of hydrogen as a fuel from sustainable resources. Almost all hydrogen fuel today is derived from natural gas. Hydrogen can be obtained by splitting it from water through electrolysis. Intensive research is needed on ways to mass-produce hydrogen economically by electrolysis, which would allow it to be manufactured with electricity and without using a fossil fuel.

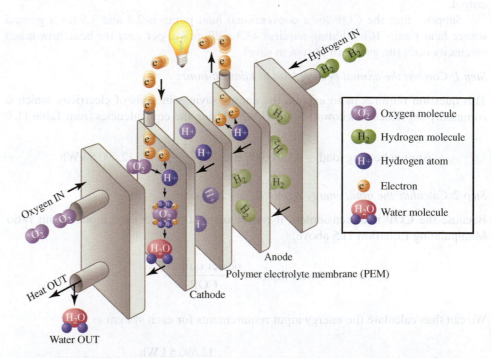

FIGURE 11.28 How fuel cells work. Fuel cells convert the chemical energy in hydrogen molecules to electricity by separating the protons and electrons in hydrogen, then capturing the free electrons. Fuel cells emit no pollutants or greenhouse gases, only water and heat.

Source: Based on U.S. Department of Energy (2010, November). "Energy Efficiency and Renewable Energy, Fuel Cell Technologies Program," *Fuel Cells*.

Electricity generation from wind, solar, water, geothermal, and hydrogen resources are proven low- and no-carbon commercial technologies. However, there are challenges to scaling up electricity generation from renewable energy resources; these are reviewed in the next section. The third critical strategy for deep decarbonization of the world's energy system is to electrify as many energy needs as possible and then provide that electricity with low-carbon power.

11.6   Decarbonization Through Electrification

Getting to net zero carbon emissions requires eliminating the use of fossil fuels except for those energy needs that have no viable clean substitutes. The third major strategy for rapid decarbonization therefore involves converting energy technologies based on combustion heat to electricity and then generating electricity with clean power. Examples of electrification are electric vehicles that run on motors instead of an internal combustion engine, space heating using heat pumps instead of hot water boilers, and steel production with electric arc furnaces instead of a blast furnace.

Not all combustion heat can or must be electrified, but it becomes easier to supply heat with cleaner resources after electrification has been maximized. For example, solar thermal can provide hot water in buildings instead of electric hot water systems, and biogas can be substituted for natural gas in some industrial applications. Figure 11.29 illustrates how heat needs in industry, buildings, and the transportation sector can be met with no- or low-carbon resources by 2050. The most pressing challenge is to transition worldwide electricity generation from about 60% fossil fuels in 2020 to 25% fossil fuels by 2030 (Figure 11.29).

11.6.1 Electricity Generation

Globally, electricity generation accounts for a little more than one-third of all energy-related GHGs. Achieving net zero emissions by 2050 will create major growth in the demand for electricity because of electrification and because of an expected shift from natural gas to electrolysis for the production of hydrogen fuel. At the same time, demand by developing economies will also increase. World consumption of electricity is predicted to increase by 2.5 times between 2020 and 2050 (International Energy Agency, 2021). This growth in demand will be devastating to GHG emissions if renewables do not displace fossil fuels in the electric power sector.

The transformation will involve more than conventional wind and solar energy, although these are the renewables that have the biggest role in the first decade or more of the transition. Hydrogen fuel cells are expected to be deployed more widely later in the transition. In addition, fossil fuels do not disappear entirely from the portfolio of electricity resources. Carbon capture, utilization, and storage can significantly reduce the carbon footprint of fossil fuel electricity generation as the technology develops further and matures. In short, a wide variety of resources and technologies—both "off the shelf" and in development—will be needed to rapidly convert the electric power sector.

A few critical issues affect the widespread use of solar and wind power from 2020 to 2030. Cost is not one of them. Even without government subsidies, both types of electricity generation are cost competitive with coal-fired plants in many parts of the world (Salzman, 2019). Key issues for wind and solar are their intermittency and the rate at which they must be adopted. In its pathways to net zero

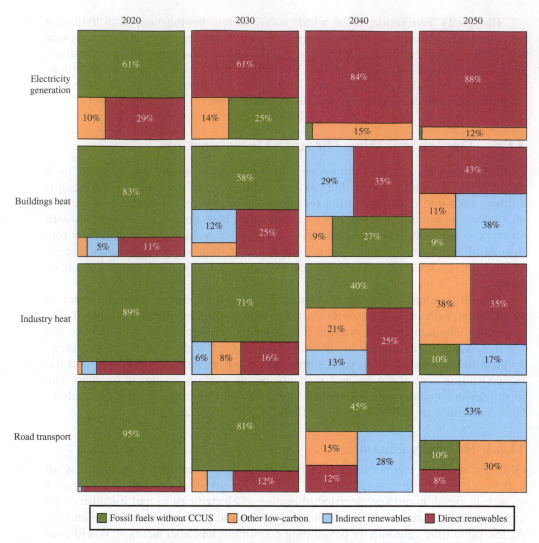

FIGURE 11.29 Pathways to decarbonization. This graphic by the International Energy Agency illustrates how direct energy use in each sector can transition to low- or no-carbon fuels by 2050. Meeting the heat needs of industry, buildings, and transportation is more easily accomplished in tandem with electrification. Fossil fuel use that includes carbon capture, utilization, and storage (CCUS) is included in the "other low carbon" category. Note the dramatic increase in renewables for electricity generation that must occur between 2020 and 2030.

Source: Data from International Energy Agency (2021). *Net Zero by 2050: A Roadmap for the Global Energy Sector*, p. 73.

by 2050, the International Energy Agency (2021) requires that wind and solar grow from 9% to 40% of total global electricity generation between 2020 and 2030.

Electricity produced by wind and solar is *intermittent*—it is not produced in continuous amounts throughout the day. For example, cloud cover affects solar PV electric output during the day and the sun doesn't shine at night, so no power will be generated. Wind speeds vary due to daily weather patterns and usually fall at night. Sunshine and wind are also seasonal; the availability of these resources varies throughout the year. The intermittency and seasonality of these resources mean that, on their own, they cannot provide a steady amount of electricity or quickly adapt to changes in load. Load is the amount of electricity that is being used at any point in time on an electric circuit (see Active Learning Exercise 11.2).

ACTIVE LEARNING EXERCISE 11.2 Conduct a Personal Energy Audit

Designing any type of energy system requires that you size the system to meet the energy "load," or amount of energy required. This exercise will give you practice estimating your personal weekly electricity load. You can do this for yourself regardless of where you live—whether a dormitory room, an apartment, a shared house, or your family home.

- Begin by listing all the devices and equipment that you personally use that require electricity. You should include, for example, lights, computers, game systems, washing machines, and so on. If you cook on an electric stove, you should include that too. If someone else prepares your meals, do not count it. Set up a table or form like the sample below to record your device data and energy calculations.
- Some challenging items to estimate are shared or communal equipment, such as refrigerators, hot water systems, and space heating and cooling machinery. What are various ways of estimating your portion of these loads?
- Determine how much energy is used by an item when it is on. This will be a power rating in watts or kilowatts. The power rating is often on a label attached to the device, and it may be specified in watts, amps, volts, or volt-amps. If you cannot find it, a list of power ratings for a wide variety of consumer and household appliances can be found quickly in an Internet search. Fill in this information on your table.
- Estimate how much time you use each item over the course of a week and complete the table accordingly. When you are done itemizing, add the total watt-hours of electricity you use per week and convert the result to kilowatt-hours.

Sample

ELECTRICAL APPLIANCE, DEVICE, OR PIECE OF EQUIPMENT	POWER RATING, IN WATTS (watts = volts × amps)	TIME USED PER WEEK, IN HOURS OR FRACTIONS OF	WATT-HOURS USED PER WEEK (power × hours used)
Desk lamp	75 watts	28 hours	2,100
Hair dryer	1,100 watts	1 hour	1,100
Laptop computer	40 watts	42 hours	1,680
Item.........			
Item.........			
Total, in kWh			Divide total by 1,000 to convert to kWh

After you have completed your audit, answer these questions:

1. What surprised you about your energy use?
2. What did you find most difficult to estimate? Why?
3. What are some opportunities for energy conservation that you can incorporate in your lifestyle right now?

Utility-scale batteries are therefore an important piece of the infrastructure for an electric grid that runs on renewables (Figure 11.30). To support expanding wind and solar power, the International Energy Agency indicates that utility-scale battery storage must increase from 18 GW to 590 GW between 2020 and 2030, and public

FIGURE 11.30 Utility scale battery storage. Utility scale battery storage is essential for the transition to low carbon electricity. This artistic computer rendering gives a sense of the size and number of batteries that must be deployed at a small-scale storage site.

Source: iStockPhoto.com/Petmal.

electric vehicle charging stations need to expand from 46 GW to 1,780 GW. Both utility and electric vehicle batteries raise a separate set of sustainability issues. At present, we do not have a clear pathway to the safe recycling or end-of-life disposal of these lithium-ion batteries.

Many of the strategies for renewables focus on integrating large-scale solar PV and wind farms into existing electric power grids. These approaches leverage the mature infrastructure for electricity, which has been in place for decades and can serve millions of people simultaneously. A centralized approach alone is not sufficient, however. A complete transformation of electricity production with renewables must also rely on decentralized power production at the site on which it is used or at the community scale. This is known as ***distributed generation***.

11.6.2 Distributed Generation

Centralized electric power plants produce electricity on a mega scale and then transmit and distribute it to locations sometimes hundreds of miles from the source. These plants are commonly known as baseload electric power plants, and they typically generate 1,000–1,500 MW of power with fossil fuels or nuclear energy. Baseload plants operate continuously, and their output can be readily increased or decreased to meet load requirements. There are no substitute renewable fuels or technologies that can produce electricity continuously, with this degree of flexibility, at this volume, everywhere. The world's largest solar and wind farms do come close to (or exceed) these capacities, but facilities of this size cannot be put just anywhere. Typical "large" solar or wind farms have a capacity that ranges from as little as 20 MW to about 500 MW. These systems are connected to the power grid for long-distance electricity transmission and distribution. Hundreds of thousands of these farms—with supporting battery storage—will need to be built worldwide.

A supplement to large-scale, grid-integrated renewables is distributed generation. In a distributed energy system, a small amount of electricity is generated at its point of use (like a building) or at multiple sites throughout a connected system, typically a community or a few city blocks. Distributed generation is integrated with the central power grid. Distributed energy can leverage grid-integrated renewables by reducing the amount of electricity the centralized network needs to provide.

Technical and economic challenges confront our ability to implement distributed generation and effectively manage it in tandem with centralized electricity systems. First, we need a "smart grid" that can dynamically regulate and control the electricity flowing from hundreds of diffused sources (Figure 11.31). A smart grid enables more than distributed generation from renewable energy; it can also enable integrated, highly automated demand management between the central grid and the distributed generation network. Demand management lowers electricity loads to match the amount of electricity available or to avoid unnecessary energy use.

Examples of automated demand management include "smart house" networks; smart appliances; building automation; and the ability of utilities to turn off equipment in homes, businesses, and industry from a remote location. Demand management has the added benefit of strengthening energy security by preventing power outages caused by system instability. It will be critical to managing the electricity generated by intermittent renewables. Major research and development efforts are underway to develop the electronics and communication technologies required for a smart grid.

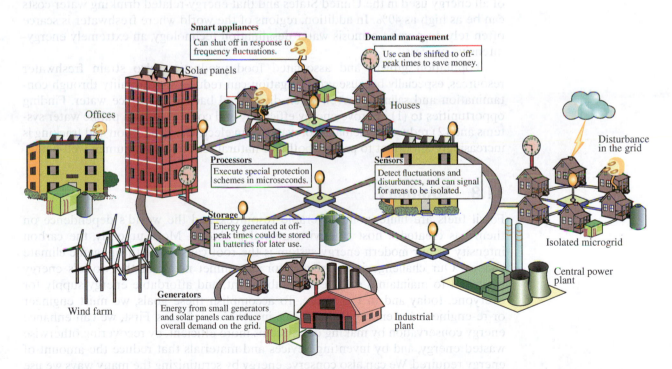

FIGURE 11.31 Smart grid connectivity. Electric power systems of the future will integrate centralized power production with small-scale distributed generation throughout the system. In addition to being able to regulate diffuse sources of energy, a smart grid also has storage capacity, the ability to regulate appliances and energy-consuming equipment at the source, and is more resistant to disturbances in the grid itself.

Source: Based on Smart Grid 2030™ Research Associates, www.smartgrid2030.com/wp-content/uploads/2009/10/SG-Nature.jpg.

A second challenge in the United States and elsewhere is the cost of distributed generation for homes and businesses, which is primarily solar PV. Even when solar is subsidized and pays for itself in energy savings, the up-front cost is still thousands of dollars, and the payback period can be several years. Consequently, there is no widespread consumer or business demand for the installation of distributed generation.

11.7 The Water–Energy–Food Nexus

A critical sustainability issue is the interdependence of water availability, energy consumption, and food production. This relationship is called the *water–energy–food nexus*, and it is of concern because these key societal needs compete for increasingly scarce freshwater resources. In the United States, as much freshwater is withdrawn from the environment for thermoelectric power production as is withdrawn for agricultural irrigation and livestock (U.S. DOE, 2014).

Regarding energy, water is required for both hydroelectric and thermoelectric power generation. Because of climate change, less rainfall and snowpack are responsible for reduced water resources for both types of power production. Even though thermoelectric power plants discharge water back into the environment, evaporative losses do occur. Hydrofracking has also raised concerns about the depletion and contamination of aquifers and other groundwater resources. Regarding water, water supply and treatment systems are energy intensive at all stages. The U.S. EPA (2014) reports that drinking water and wastewater systems account for 3% to 4% of all energy used in the United States and that energy-related drinking water costs can be as high as 40%. In addition, regions of the world where freshwater is scarce often rely on reverse osmosis water desalination technology, an extremely energy-intensive process.

Population growth and associated food production also strain freshwater resources, especially because crop irrigation can reduce water quality through contamination and salination even when discharged back into surface water. Finding opportunities to (1) improve energy efficiency and conservation in public water systems and (2) reduce the demand for water from electricity generation and fracking is increasingly important to protect both our natural resources and human well-being.

11.8 Summary

Fossil fuels are finite, nonrecoverable resources, and the world's dependence on them has created a host of environmental problems. More urgently, the carbon intensity of our modern energy system is the root cause of anthropogenic climate change. Our challenge is to provide for the unmet needs of the world's energy poor and to maintain a safe, reliable, abundant, and affordable energy supply for everyone, today and in the future. To accomplish these goals, we must engineer or re-engineer our energy technologies in a variety of ways. First, we can enhance energy conservation by making our systems more efficient, by recovering otherwise wasted energy, and by inventing devices and materials that reduce the amount of energy required. We can also conserve energy by scrutinizing the many ways we use energy and curtail consumption by making behavioral and lifestyle changes. Second, we can diversify our fuels by using low- or no-carbon substitutes. Finally, we must electrify as many energy applications as possible, and then generate electricity with resources that do not emit GHGs.

References

Combined Heat and Power Alliance. (2017). Combined Heat and Power (CHP) and Waste Heat to Power (WHP) National Overview.

Hubbert, M. K. (1956, March). Nuclear Energy and the Fossil Fuels. Paper presented at the spring meeting of the Southern District Division of Production, American Petroleum Institute, San Antonio, TX.

International Energy Agency. (2020). SDG7: Data and Projections.

International Energy Agency. (2021). *Net Zero by 2050: A Roadmap for the Global Energy Sector.*

IPCC. (2019). Global warming of 1.5°C.

Moomaw, W., et al. (2011). "Annex II: Methodology." In O. Edenhofer et al., *IPCC Special Report on Renewable Energy Sources and Climate Change Mitigation.* Cambridge, UK: Cambridge University Press.

National Academies of Sciences, Engineering, and Medicine. (2021). *Accelerating Decarbonization of the U.S. Energy System: Technology, Policy, and Societal Dimensions.*

OECD. (2020). "Climate change," Environment at a Glance Indicators, www.oecd-ilibrary .org/sites/5584ad47-en/index.html?itemId=/content/component/5584ad47-en& _ga=2.159252862.576286423.1625765149-206341139.1625262482.

Salzman, A. (2019). "Solar and wind power will cost less than coal by 2030, according to one analyst's math." Barron's Online. 20 November. www.barrons.com/articles /retirement-challenges-help-51625849597.

U.S. DOE. (2014, June). The Water-Energy Nexus: Challenges and Opportunities.

U.S. EPA. (2013). Clean Energy Calculations and References. Accessed May 15, 2013, at www.epa.gov/cleanenergy/energy-resources/refs.html.

U.S. EPA. (2014). *Water: Sustainable Infrastructure.* Energy Efficiency for Water and Wastewater Facilities. Accessed March 17, 2014, at www.water.epa.gov/infrastructure /sustain/energy efficiency.cfm.

U.S. EPA. (2021). Inventory of U.S. Greenhouse Gas Emissions 1990–2019: Data Highlights, www.epa.gov/ghgemissions/inventory-us-greenhouse-gas-emissions -and-sinks-fast-facts-and-data-highlights.

World Business Council for Sustainable Development and World Resources Institute. (2004). *The Greenhouse Gas Protocol: A Corporate Accounting and Reporting Standard, Revised Edition.* Washington, DC: World Resources Institute.

Key Concepts

Energy security	Renewable resources
Decarbonization	Reserve
Energy transition	Reserve-to-production ratio
Traditional biomass	Hubbert curve
Energy services	Biofuel
Derived demand	Fracking
Energy intensity	Direct energy
End-use sectors	Embodied energy
Energy ladder	Levelized cost of energy
Energy poverty	Energy curtailment
Social costs	Energy efficiency
Finite resources	Carnot limit

Performance standards

Energy recovery

Combined heat and power (CHP)

Fuel switching

Biogas

Waste-to-energy

Concentrated solar power

Ground source heat pumps

Geothermal power

Distributed generation

Water–energy–food nexus

Problems

11-1 True or false: Petroleum is the fastest growing form of energy worldwide.

11-2 True or false: In the United States, the transportation sector generates more greenhouse gas than the electric power sector.

11-3 True or false: Globally, deforestation contributes more to net greenhouse gas emissions than do the residential and commercial building sectors.

11-4 Order the following sectors in terms of the amount of global greenhouse gases they generate, from highest to lowest: landfills, energy used in industry, energy in agriculture and fishing, aviation, residential buildings.

11-5 List three different kinds of energy services, and identify at least two ways that each of these energy needs can be met.

11-6 Explain how the technologies that we design to satisfy energy needs affect derived demand (the type and amount of energy resources that we use).

11-7 Calculate per capita energy use and energy intensity for the following countries:
   a. China
   b. Australia
   c. Mexico
   d. Kenya
   e. Germany

11-8 For the five countries in Problem 11-7, plot per capita energy use on the x-axis and energy intensity on the y-axis. Do the positions of the country data points make sense? Explain why or why not.

11-9 What are the differences among the residential, commercial, and industrial end-use sectors?

11-10 What is the energy ladder?

11-11 What is energy poverty, and why is it a problem?

11-12 What is the average annual rate of depletion of the global reserves of liquid fuels, natural gas, and coal assuming no new reserves and no increase in annual production/consumption levels?

11-13 Consider the data presented in Figure 11.10.
   a. Which end-use sectors do you think are most vulnerable to an energy shock? Why?
   b. Which fuel has the most diversified applications? Why?

11-14 Calculate the energy content of the following in both MJ and Btu:
   a. 10 tonnes of coal
   b. 2 barrels of oil
   c. 4,000 kWh
   d. A quad of natural gas

11-15 Which has more energy in MJ: 1 toe or 1 bbl?

11-16 Why does a gallon of gasoline have slightly more energy than a gallon of crude oil?

11-17 Consider an electric power plant that generates 5,256,000 MWh of electricity per year. How much of the following resources are required to provide this much power? Report your results in tonnes.

  a. Fuel oil #6

  b. Natural gas

  c. A coal blend composed of 60% bituminous coal and 40% anthracite coal

11-18 What is the distinction between a reserve and a resource?

11-19 Refer to the data in Table 11.4 for the following problems.

  a. Calculate the amounts of liquid fuels, natural gas, and coal consumed globally in the base year given.

  b. Assume that coal consumption is not flat but growing at a rate of 4% a year. What would the reserve-to-production ratio be for this growth rate?

  c. What would the annual rate of increase in natural gas reserves need to be in order to have a reserve-to-production ratio of 125 years? (Assume consumption does not increase.)

  d. Assume that the consumption of liquid fuels is increasing by 6.3% a year. What would the annual rate of increase in reserves need to be in order to have a reserve-to-production ratio of 75 years?

11-20 Why is the Hubbert curve for peak oil not a model of the rate of oil depletion?

11-21 Make the argument that uranium is (a) a finite resource, (b) a renewable resource, and (c) an infinitely renewable resource.

11-22 The table below lists some major U.S. cities and their available sun-hours.

  a. How much electricity does a 4-kW solar PV system in Chicago generate in a year?

  b. How large would a PV system need to be in Fairbanks, New York City, and Miami to generate the same amount of electricity in a year as a 2-kW system in Phoenix?

  c. Miami is farther south than Phoenix, so in principle it should have more sun-hours per year than Phoenix. What are some reasons it does not?

CITY	SUN-HOURS	CITY	SUN-HOURS
Fairbanks, Alaska	3.99	Portland, Oregon	4.51
Phoenix, Arizona	6.58	Chicago, Illinois	3.14
Los Angeles, California	5.62	Miami, Florida	5.62
New York, New York	4.08	San Antonio, Texas	5.30

11-23 Countries in certain kinds of climates do not have the four seasons of winter, spring, summer, and fall. They have relatively little temperature variation, and "wet seasons" and "dry seasons" that can last for several weeks each. What are the implications of this climate pattern for the design and installation of:

  a. A solar hot water heating system?

  b. A solar PV system that is independent of the electric power grid (e.g., it stores all of its excess electricity in batteries)?

11-24 What is EROI, and why is it a helpful energy metric?

11-25 Consider the energy content of the different fossil fuels and their EROI. Discuss why it will be difficult to shift the global energy system away from these resources.

11-26 What are the energy conservation gains from the following actions?
 a. Switching from a single 60-watt incandescent bulb that burns for 6 hours per day to a 13-watt CFL
 b. Using pure biodiesel instead of motor gasoline for an automobile that travels 60,000 miles per year with a fuel economy of 28 mpg
 c. Replacing a heat pump with a COP of 3.2 with a heat pump with a COP of 2.6 if the heating load is 43.3 million Btu per year
 d. Replacing an air conditioner with an EER of 11 with a unit that has an EER of 14 if the cooling load is 22 million Btu per year
 e. Upgrading a natural gas furnace from one with 88% efficiency to one with 92% efficiency if the heating load is 55 million Btu per year

11-27 Estimate the avoided yearly CO_2 emissions from the following.
 a. Switching from a single 60-watt incandescent bulb that burns for 6 hours per day to a 13-watt CFL
 b. Using pure biodiesel instead of motor gasoline for an automobile that travels 60,000 miles per year with a fuel economy of 28 mpg
 c. Replacing an air conditioner with an EER of 11 with a unit that has an EER of 14 if the cooling load is 22 million Btu per year
 d. Upgrading a natural gas furnace from one with 88% efficiency to one with 92% efficiency if the heating load is 55 million Btu per year

11-28 What are the two different kinds of CHP systems?

11-29 Define and discuss the features of distributed generation.

11-30 Which aspects of the water–energy–food nexus are most important where you live?

Team-Based Problem Solving

Design a solar PV system (with battery storage) for a rural village in the Andes Mountains of Bolivia. The community would like to have five days of backup power. This village needs power to meet the following energy needs:

11-31 A community center that can burn three lights for 6 hours each evening.

11-32 A ¼-HP water pump that can supply the village with 100 gallons of freshwater daily.

11-33 A radio that will be turned on for 3 hours per day.

11-34 A small refrigerator to store vaccines.

## PART 3

Designing Resilient and Sustainable Systems

CONTENTS

## CHAPTER 12

Designing for Sustainability

FIGURE 12.1 To be sustainable by design, we need to think green and innovate.

Source: iStockPhoto.com/Animaflora.

You don't filter smokestacks or water. Instead, you put the filter in your head and design the problem out of existence.

—WILLIAM McDONOUGH

GOALS

IN THIS CHAPTER, WE HIGHLIGHT HOW sustainable design can be used in practice as part of technical design. We identify common philosophies of designing for sustainability and introduce the green principles that inform and guide design decisions. We also explain strategies used in technical design to achieve more sustainable buildings, materials, products, processes, systems, and end-of-life value recovery. Because of the importance of participatory design, we suggest people-centered tools that enable engineers to engage individuals and communities in the design process.

OBJECTIVES

At the conclusion of this chapter, you should be able to:

12.1 Describe how design affects the human–environment interface, the activities of traditional technical design, and what it means to design for sustainability.

12.2 Discuss the context in which philosophies on sustainable design have emerged and the unifying philosophies that they share.

12.3 Explain how ecological design as a philosophy is used in design practice and give examples.

12.4 Summarize chemistry, carbon, and circularity as a design philosophy.

12.5 Compare and contrast the 12 principles of green engineering and green chemistry and explain how they can be applied to sustainable design at different stages of the product life cycle.

12.6 Explain and give examples of specific design techniques that make products more sustainable and make simple calculations of the benefits from dematerialization.

12.7 Identify and summarize how Design for X strategies achieve designing for value recovery and analyze the obstacles to closing material loops through recycling and extended producer responsibility.

12.8 Identify the design characteristics of more sustainable processes and systems. Compare and contrast industrial ecology and lean manufacturing as analytical approaches to process and system improvement.

12.9 List and discuss three ways that people and communities can be effectively involved in engineering design.

Introduction

The purpose of this chapter is to explicitly introduce the concept of design and demonstrate its role in promoting more sustainable engineering and human development. This text has presented many problems and challenges that people create for the environment and the ability of Earth to support life. It has also provided the fundamental biological, chemical, and physical principles through which Earth regulates itself, energy, and matter. We now need to provide you with insights and tools that can help you begin to remedy the problems of the past for a more sustainable future. The solutions are rooted in design thinking. Our problems today have been created by the design choices of yesterday; to navigate our way out of them, we need to re-think how we design. As the architect William McDonough tells us, "You don't filter smokestacks or water. Instead, you put the filter in your head and design the problem out of existence."

This chapter provides an orientation to design as both a human and an engineering activity. It then focuses specifically on approaches to sustainable design that you can use in practice. Although many approaches to sustainable design exist, you will see that they are based on just a few unifying principles: that we limit our use of finite resources, protect the natural systems that provide ecosystem services, think holistically in terms of the life cycle of material flows, and avoid using or creating hazardous substances. We will introduce sustainable design philosophies that incorporate these unifying approaches and then move progressively to design principles and methods that use them. We conclude with people-centered design, which reminds us that the purpose of technical design is to meet a human need. At the end of the day, we are designing for people.

12.1   Sustainable Design in Context

As you have learned throughout this text, the Earth's natural systems interact dynamically with one another. Matter and energy flow in a continuous process of mass balancing through the biogeochemical cycles and solar energy. Living organisms adapt to each other, their environments, and the nutrients that cycle in their ecosystems. Waste is not intrinsic to any of these systems, nor is toxicity. These planetary dynamics are the support system for all life on Earth. Even though this is an idealized representation of ecology and equilibrium—which are far more complex, brutal, and dynamic in real life—it does reflect the key point that there are laws of biology, chemistry, and physics that fundamentally determine the capacity of Earth to support life. There is also the dimension of *time*. Some processes happen almost instantaneously. Others occur over centuries or longer. **Sustainability science** is the emerging discipline that investigates how human activity affects Earth's mass balancing, energy flows, and ecosystems (Abraham, 2006).

12.1.1 Design, the Environment, and Human Nature

Human society and communities are not like those of other living organisms. The intersection of people and the natural world is called the **human–environment interface** (Figure 12.2). It is in this interface that social, technological, and ecological systems interact. The interface is where we extract natural resources, artificially create energy, discharge waste and pollution, and reshape habitats and landscapes. Unlike other living things, the relationship of people to the environment is mediated by three defining and unique human activities: our material

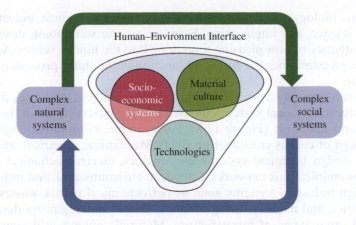

FIGURE 12.2 The human–environment interface. Unlike other living things, our interactions with the natural world are structured by our material culture, our socioeconomic systems, and our technologies. These determine what natural resources we extract, how we create energy, the characteristics of our pollution and waste, and how we reshape landscapes and habitats. These interactions are mostly by human design, so we can redesign them to become more sustainable.

culture, our socioeconomic systems, and technology. What all three of these activities have in common is that they are *design* activities. **Design** is the process of human problem solving that results in an "artifact," something that persists over time as either a tangible thing or a structured pattern of behavior. Material culture, socioeconomic systems, and technology are all forms of human problem solving and they are all artifacts. They are intentional responses to meeting needs and they persist over time.

For example, a fishing pole is a tangible object of material culture. It exists in many forms in many cultures throughout human history. But it is also an object designed to meet the need of catching fish for food. Democracy and capitalism are intentionally designed political and economic systems that address the problems of governing people and allocating our use of resources. You cannot touch these systems, as they are not tangible. However, they are highly structured patterns of social behavior with rules and norms that persist over time. The internal combustion engine is a technology designed to overcome the limitations of people and animals as sources of physical work by providing extraordinary mechanical power.

The design choices that we make for things like fishing poles, our governance, or combustion engines have environmental consequences. All human design choices do—and they always will—because our lives depend on the resources of the natural world, and we are part of that world. There is always a human–environment interface, but this interface can be more sustainable or less sustainable. As engineers, you are directly involved in the design of material objects, technologies, systems, and the processes for making them. However, design as a disciplinary activity has not traditionally included sustainability.

12.1.2 The Traditional Requirements of Technical Design

Design is a creative and intentional process in which we imagine and specify how materials, objects, products, technological systems, and structures are made. Many disciplines and professions are involved in technical design—principally architecture, biology, chemistry, physics, environmental science, industrial design, computer science, and engineering. From an artistic standpoint, design thinking reflects aesthetics, or how pleasing something is to the human senses. As a practical matter, design establishes the form and function of the thing, process, or system we are creating.

Engineers typically design five kinds of things and the processes that relate to them: products, technical systems, industrial systems, structures in the built environment, and materials (Figure 12.3). For example, automotive engineers work on the design of cars as consumer products. Mechanical, electrical, and computer engineers design technical systems like engines, electromechanical equipment, and telecommunications networks. Chemical, environmental, and industrial engineers design industrial systems, such as petrochemical plants, wastewater treatment facilities, and manufacturing assembly lines. Civil engineers design bridges, roads, and other types of infrastructure. Materials engineers design substances that exhibit specific chemical and physical properties so that these materials can be used to make things.

It is easy to see from these examples that engineers are involved with designing, making, and building at very different physical and technological scales and different levels of complexity. Materials are designed at the atomic and molecular level. Products range from something as small and simple as a pencil to as large and complex as a passenger aircraft. Industrial systems can be modest factories with basic machinery or acres of robotics operated by highly sophisticated control systems. Technologies include miniature semiconductors as well as jumbo hydropower turbines. Products, technical systems, industrial systems, and built structures share three basic elements of design even though they represent different scales and complexities: form, function, and materials selection.

FIGURE 12.3 Engineers design a variety of things, including industrial systems, structures in the built environment, technologies, products, and materials.

Source: iStockPhoto.com/SUMITH NUNKHAM; iStockPhoto.com/Nastasic; iStockPhoto.com/Dmytro Skrypnykov; iStockPhoto.com/StockImages_AT; iStockPhoto.com/Nikola Ilic.

Functional design determines the use that something has and how it physically works to achieve its purpose. Functional design is the core feature of design because it establishes how we meet a need. For example, a hair dryer meets the need of drying hair faster than natural evaporation. This consumer product achieves its purpose through its functional design: An electric motor powers a fan that blows hot air created by an electric heating element. If the hair dryer has adjustable speeds or temperature settings, these are also part of its functional design. Performance requirements are intrinsic to and inform functional design choices. How fast the fan turns and how hot the heating coil gets are part of functional design. The product's overall reliability (crudely, the probability that it will not fail under conditions of normal use) is also part of functional design.

Form design reflects the overall shape, size, and appearance of objects being designed, including any connected parts they might have (Figure 12.4). Form design must accommodate the detailed geometries, dimensions, tolerances, fit, and joinery of components, which is the domain of functional design. Because of this, function affects form. Likewise, constraints on form—shape, size, and appearance—affect functional design. Form and function are often interdependent design decisions, not stand-alone choices.

Technical design also includes the materials from which something is made. Materials selection affects an object's form, its physical properties, and how it can be manufactured. Form, function, and materials selection are highly interactive design choices and do not occur in isolation from one another. The combined result of all

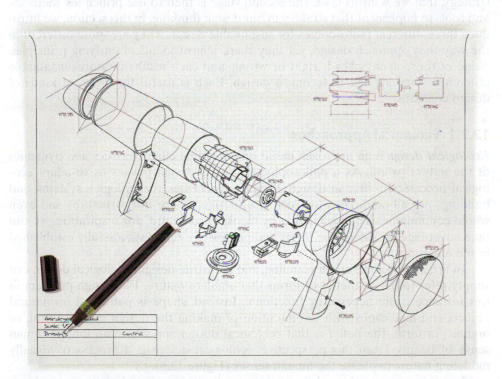

FIGURE 12.4 This CAD drawing of a hair dryer illustrates not only form design, but also how the assembly and fit of parts are integral to a product's functional design.

Source: iStockPhoto.com/anilyanik.

design elements and choices is a material, product, process, technical system, industrial system, or structure that fulfills an intended purpose at some specified level of quality, reliability, durability, and safety.

Designing for sustainability means that we are actively incorporating principles of sustainability as we make technical design decisions about form, function, and materials. As we will see in discussions throughout this chapter, designing for sustainability requires that we make design choices with the beginning and the end of a designed object's life in mind, as well as all of the stages in between.

12.2   Sustainable Design Philosophies

As you can see from discussions throughout this text, sustainability in an engineering context encompasses a wide range of challenges that require us to manage pollution, reduce toxic and hazardous wastes, conserve natural resources, and analyze material and energy flows. We are also required to be mindful of economic costs, to minimize risks to people and the environment, and to avoid destabilizing the Earth's climate. This is a very large set of expectations.

We need useful approaches to design that let us include the urgent needs for sustainability in our work. For example, as a practical matter, how might we go about improving (redesigning) the hair dryer shown in Figure 12.4 to make it a more sustainable product? Getting to the hands-on aspects of designing for sustainability is a two-stage process. The first stage is deciding on the design philosophy, or basic strategy, that we want to take. The second stage is then to use principles, methods, and tools to implement that strategy in our design thinking. In this section, we introduce three different philosophies for sustainable design. They are quite different in the way they approach design, yet they share almost identical unifying principles. None of these approaches is right or wrong, and each results in more sustainable outcomes when applied to technical design. Each is useful for different kinds of design contexts.

12.2.1 Ecological Approaches

Ecological design is an approach based on emulating characteristics and dynamics of the natural world. As a philosophy, ecological design pushes us to adapt ecological processes to human design. Products, processes, technologies, systems, and buildings are all candidates for ecological design, as are infrastructures and even whole communities. The idea is that by "looking to nature" for inspiration, we can take advantage of sustainable forms and processes already successfully established in the environment.

In art, architecture, urban planning, and industrial design, ecological design can simply refer to the aesthetics of forms that emulate nature. The design problem in this instance is not necessarily functional. Instead, shape is patterned on natural objects and landscapes with the intention of making the design more natural, or organic, in form. The belief is that ecological design appeals to an innate human sense of beauty. From this perspective, ecological design is a process of visually modeling nature to please the human senses (Figure 12.5).

In engineering, ecological design best applies to functional processes, and it is practiced in architecture, urban planning, and industrial design as well. The heating and cooling system of Eastgate Centre in Harare, Zimbabwe, is an example of ecological design that integrates form and function to achieve the desired energy performance of the building (see Box 12.1).

FIGURE 12.5 Fallingwater. Architect Frank Lloyd Wright pioneered what is referred to as "organic architecture," a style of design in which buildings and their surrounding landscape are perceived as an integrated composition. Fallingwater is a home in Pennsylvania and one of Wright's iconic designs.

Source: Education Images/Contributor/UIG via Getty Images.

BOX 12.1 What Can We Learn from Termites?

Architect Mick Pearce's Eastgate Centre in Harare, Zimbabwe (Figure 12.6), is one of the most celebrated examples of ecological design in architecture. The ventilation, heating, and cooling system in this building is inspired by the natural convection processes of *Odontotermes transvaalensis* termite mounds; their structural "chimney" effect draws and moves cool, warm, and fresh air throughout the mound. The process is commonly referred to as passive heating and cooling.

Eastgate Centre incorporates both passive and active energy in its ventilation and cooling systems.

Electric fans move and exhaust air throughout the building. Natural convection in the chimneys, driven by internal and external air temperature differentials, enhances fan-powered ventilation. Active and passive energy are used to move large volumes of evening air through the building and cool its thermal mass overnight. Eastgate starts its morning cooled down and able to draw on thermally "stored" night air for daytime cooling. This ecological design has lowered the energy needed for cooling to about 10% of that needed for conventional air conditioning.

(Continued)

BOX 12.1 **What Can We Learn from Termites?** *(Continued)*

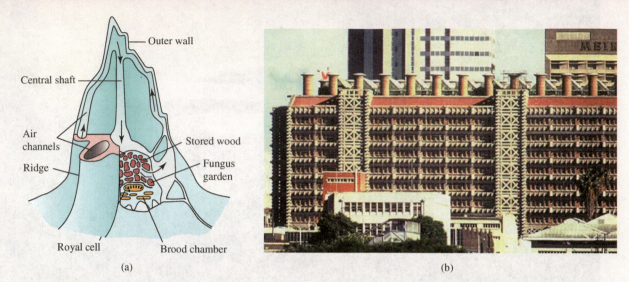

(a)

(b)

FIGURE 12.6 The Eastgate Centre in Zimbabwe's capital city of Harare is an exemplar of ecological design. Patterning the natural dynamics of ventilation in (a) termite mounds, architect Mick Pearce constructed (b) a building of considerable beauty and low environmental impact.

Source: (a) Based on what-when-how, Nervous System (Insects), what-when-how.com/insects/nervous-system-insects; (b) Courtesy of Mick Pearce.

The elegance of the design inspiration is deceptively simple. Substantial computer simulation and modeling were required by the engineering firm OveArup in order to achieve the architect's vision and structural features. Parameters for direct sunlight, window operability, and interior light control had to be established for the passive system to work properly. In addition, the feasibility of this design is tied to its geographic location—Zimbabwe's climate is especially conducive to large-scale passive cooling. Engineering knowledge and skill are thus essential elements in translating nature's designs into human use.

A critical goal of ecological design is to eliminate waste from human systems. Very simply, ecosystems don't create unmanageable trash, garbage, or toxic byproducts. Organisms take only what they need from their environment, and their excrement is cycled back into minerals, elements, and harmless nutrients for other living things in the ecosystem. Dead organisms decompose the same way. Mass balance is preserved in ecosystems through closed-loop matter cycling.

William McDonough and Michael Braungart popularized this aspect of ecological design in their seminal work *Cradle-to-Cradle: Remaking the Way We Make Things* (2002). The cradle-to-cradle approach reimagines the traditional linear materials economy as an idealized closed-loop materials cycle (see Chapter 5). In this industrial ecosystem, products at the end of their life are processed into biological or **technical nutrients** to be used over and over again (Figure 12.7).

Engineering methods, tools, and models that apply ecological principles include waste minimization (Chapter 5), industrial ecology (Chapter 13), and life

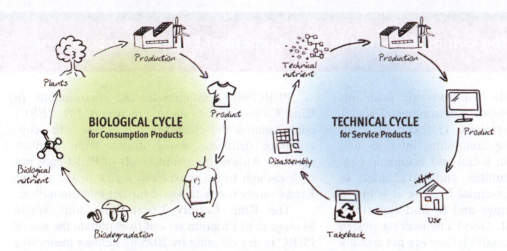

FIGURE 12.7 Cradle-to-cradle design represents a closed-loop flow of materials in production systems, creating an industrial ecosystem. Plant-based resources biodegrade at their end of life into biological nutrients. Engineered materials and product components are recovered, processed, and cycled as "technical nutrients."

Source: EPEA GmbH, epea.com/en/about-us/cradle-to-cradle.

cycle analysis (Chapter 14). These approaches help us shift away from the linear model of resources to a model in which resources, materials, and "waste" are dynamically cycled and reused in industrial production systems. Our technological processes consequently become more "ecological" in their operation. In industrial ecology, we call this activity *industrial metabolism*.

12.2.2 Green Engineering

In the spring of 2003, 65 professional chemists and engineers met at a conference with the title *Green Engineering: Defining the Principles*. Over several days, they synthesized a set of holistic principles from existing thinking about sustainable design. Their efforts are known as the Sandestin Declaration (Abraham and Nguyen, 2003). This declaration is an overarching philosophy on designing for sustainability from an engineering perspective, referred to simply as ***green engineering***.

The nine principles that constitute the Sandestin Declaration on green engineering can be streamlined into five essential points:

1. *Be holistic and use life cycle thinking.* Approach product and process design holistically, and always use life cycle thinking. Incorporate tools for systems analysis and environmental impact assessment.
2. *Conserve natural resources.* Protect not just human health and well-being but ecosystems as well. Limit depletion and degradation of Earth's resources, which also includes preventing waste.
3. *Avoid hazardous substances.* Materials and energy use should be as benign and safe as possible.
4. *Design with people in mind.* Use engineering design that is sensitive to local contexts and cultures and that actively involves community members and stakeholders. (See Box 12.2.)
5. *Innovate.* To achieve greater sustainability, design inventive solutions that improve current technologies and know-how.

BOX 12.2 Engaging Communities in the Development of Engineering Solutions

Developing sustainable solutions to problems requires integrating technical, environmental, social, and economic considerations. This can be challenging because of the competing interests and needs often reflected in social and economic concerns. Involving communities and stakeholders in the design process is essential because it helps to identify barriers to change and to find opportunities for mutual benefit. Good engineering design options will not be successful if they are not socially accepted.

The King County Local Hazardous Waste Management Program in Washington state works to protect the county community and environment. One area of its focus is dry cleaning. Studies have shown that people chronically exposed to perchloroethylene (PERC), a chemical commonly used in dry cleaning, have an increased risk for several cancers and other serious health effects (Figure 12.8).

PERC also contaminates the environment. In King County alone, there are about 189 PERC-contaminated dry-cleaning sites. These sites contaminate drinking water drawn from shallow aquifers. Although current levels of PERC are not high enough to be considered a risk to the public, King County wants to avoid further contamination.

The King County Local Hazardous Waste Management Program set out to eliminate the use of PERC in dry cleaning by 2025 by helping businesses switch to safer alternatives. They recognized that most dry-cleaning owners in King County are first-generation Korean immigrants whose shops are family owned and operated. Employees are usually of Latin American origin or descent. Because owner-operated businesses are not typically subject to state occupational safety regulations, many workers did not have the information they needed to protect themselves from hazardous PERC exposures. In addition, most

FIGURE 12.8 Dry cleaners have traditionally used PERC, a toxic solvent, in their operations.

Source: iStockPhoto.com/Deepblue4you.

of the information about chemical hazards was available only in English, which was not the first language of most local workers in this industry.

King County staff set out to inform, engage, and incentivize the dry cleaners. They created educational materials written in Korean and Spanish and engaged ambassadors for the program to better know the proprietors and build trust. The program offered owners one-time $20,000 grants to switch from PERC to safer wet-cleaning methods. Wet cleaning uses water and detergent rather than a carcinogenic organic solvent and is effective for all "dry clean only" fabrics.

One of the first business owners to receive the grant reported that he no longer suffered from health problems working in the shop and also had lower utility bills. Because wet-cleaning technology performs well, is cost effective, and is safer, King County decided to provide the grants until every PERC dry cleaner makes the switch.

King County's work with dry cleaners provides an excellent example of how to "design with people in mind." King County partnered with wet-cleaning manufacturers, other government agencies, the King County Board of Health, and the local community to facilitate a win-win transition to a more sustainable technology.

Source: Based on *Public Health Insider* (2018, October 24). Helping dry cleaners switch to safer alternatives: Towards a PERC-free King County by 2025, publichealthinsider.com/2018/10/24/helping-dry-cleaners-switch-to-safer-alternatives-towards-a-perc-free-king-county-by-2025/.

The core themes of the Sandestin Declaration were echoed by the Organisation for Economic Cooperation and Development (OECD). The OECD distilled the essential requirements for sustainable materials management into four policy principles for its member countries (OECD, 2012):

1. Preserve natural capital (see Chapter 5).
2. Design and manage materials, products, and processes for safety and sustainability from a life cycle perspective.
3. Use the full diversity of policy instruments to stimulate and reinforce sustainable economic, environmental, and social outcomes.
4. Engage all parts of society to take active, ethical responsibility for achieving sustainable outcomes.

As with green engineering, the OECD emphasizes life cycle thinking, protecting natural resources and ecosystems, and avoiding hazards as core philosophies for sustainable design.

The OECD principles for sustainable materials management also highlight the role of policy ethics in sustainability problem solving. Policy principles form the basis for government legislation and regulations that promote the common good, and they often capture specific ethical obligations to society. One common obligation is *transparency*, which requires governments to be open about their decision-making processes and to disclose risks to the public. For example, the European REACH regulation (discussed in Chapter 6) requires that chemicals have publicly available toxicological data before they can potentially be sold commercially. The "no data, no market" directive of REACH restricts businesses that do not meet this transparency obligation from selling their chemical products. Transparency also underlies the principle of *right to know*, in which workers have a legal right to know about any hazardous chemicals and materials in their workplace.

Another example of policy ethics is the ***precautionary principle***. This principle states that when the potential harm (risk) of an innovation or action is unknown, highly uncertain scientifically, or potentially irreversible, it should not be approved for public use. The precautionary principle does not have a common definition and is used as a legal standard in varying degrees around the world. An important feature of the principle is that the burden of proof is on innovators to show that their innovations or activities will not cause harm. Both the REACH legislation in Europe and the revised Toxic Substances Control Act in the United States represent a more precautionary approach and reflect the value that chemicals should be proven safe before being placed on the market.

12.2.3 Chemistry, Carbon, and Circularity

Mikhail Davies of Interface, a global commercial flooring company, proposed "three lenses" through which we can evaluate the impact of products and materials on the well-being of people, the environment, and the climate. These lenses are green chemistry, circular economy, and embodied carbon (Interface, Inc., undated). ***Green chemistry*** involves the development of materials and processes that lower or avoid the use of hazardous materials (Chapter 6). The circular economy represents the recovery and reuse of products and materials at their end of life (Chapter 5). Embodied carbon reflects the greenhouse gas impacts of the energy required to make a product—from materials extraction and processing to manufacturing the final product (Chapters 10, 11, 13, and 14). By using these three lenses, we can explore the sustainability implications of any particular product, material, process, industrial system, or design. Heine and Whittaker (2021) simplified the three lenses into the catchphrase *chemistry, carbon, and circularity*. They also expanded the philosophy into a more concrete approach to materials design and manufacturing.

Chemistry, carbon, and circularity is explicitly a life cycle approach. It requires designers to consider all stages of a material or product's production and use, then incorporate these life cycle insights into design choices. The goals of more sustainable design here are to fundamentally limit or avoid toxicity and hazard in the chemicals and materials used to make a product, to reduce the carbon footprint of that product, and to improve or ensure the product's ability to be cycled as a technical or biological nutrient.

The three philosophies reviewed in this section rely on simple directives, overlap one another, and stress complementary guidance about how to approach designing for sustainability from an engineering perspective. As philosophies, these fundamentals do not give us much in the way of methods or decision tools for sustainable design. However, they do change the way that we think about technical design and what we should try to accomplish in the design of products, technologies, processes, industrial systems, the built environment, and materials. So, now that we have established these philosophies, how do we apply them in practice? The rest of this chapter will explore how designing for sustainability can be implemented.

12.3 Ecological Approaches to Design in Practice

Ecological approaches to design are most commonly used in the built environment, product design, and materials development. Each of these contexts is summarized here. Although nutrient and waste cycling is intrinsic to ecological design, it is also common to all sustainable design philosophies. As a consequence, we will review strategies for "closing loops" later in this chapter.

12.3.1 The Built Environment

Ecological principles are applied to the built environment in two basic ways. One is in the design of building systems, especially for energy and water use. The other way is to manage the impacts of a structure on its site and surrounding landscape.

Ecological design of buildings and their systems has a number of established approaches and practices in architecture and engineering. Many criteria are now codified in green building design standards, such as the Leadership in Energy and Environmental Design (LEED) rating system, the Living Building Challenge, and Passivhaus, which are all discussed in greater detail in Chapter 16. Passivhaus is a particularly good example because its performance standards require that almost all building heat be provided by passive energy—from the sun and the body heat of building occupants. This is hardcore ecological design, as passive solar energy is the activation energy in photosynthesis and the body heat of occupants is metabolic heat. Other examples of ecological design in action include the convective heating and cooling system at Eastgate Centre (see Box 12.1) and greywater recycling and reuse in commercial buildings.

Ecological design is used to mitigate the impact of structures on the landscape. Stormwater management is a good example of this application. Stormwater runoff from streets, roadways, and parking lots is a significant nonpoint source of water pollution. It is a problem created specifically by our built environment, which causes water to run over hard surfaces like asphalt, concrete, and roof tops, collecting contaminants as it flows. In order to prevent flooding in streets and neighborhoods, stormwater is usually channeled through buried stormwater drains or open concrete aqueducts. Accumulated debris, garbage, automotive oils, chemical pollutants, animal waste, and other contaminants move rapidly to the nearest stream, river, or lake.

To avoid such water degradation, stormwater is now frequently managed by containing it locally. Detention ponds halt the flow of contaminants to natural bodies of water by retaining stormwater and letting it evaporate. Ponds are sized to accommodate a defined volume of water from a storm surge (Figure 12.9). Detention ponds are not, however, ecological systems. The water in them is usually so contaminated with chemicals or nutrients that they are not healthy for wildlife or aquatic organisms.

Alternatively, low-impact development (see Chapter 16) is an environmental engineering model in which runoff is managed at its source by imitating the hydrology of natural systems. Several techniques are used for low-impact development, but one common practice is rain gardens. Water filtration zones are planted in shallow

FIGURE 12.9 Stormwater detention pond. A typical stormwater detention pond is designed to retain stormwater and prevent the flow of contaminants into natural bodies of water.

Source: Photo by Bradley Striebig.

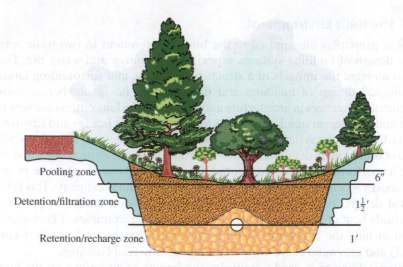

Pooling zone

Detention/filtration zone

Retention/recharge zone

6″

1½′

1′

FIGURE 12.10 Rain garden schematic. Rain gardens are modeled after the Earth's own natural water regulation and purification system in which vegetation, topography, and geology all dynamically interact to control the flow of water and clean it of contaminants. Rain gardens are a key design feature of low-impact development that manages stormwater runoff at its source, rather than channeling it to natural bodies of water and polluting them.

Source: Based on City of Des Moines, www.dmgov.org/Departments/Parks/PublishingImages/rain_garden.gif.

depressions, and stormwater accumulates in these zones to be naturally purified as it percolates through the soil (Figure 12.10). Rain gardens have the added benefit of being aesthetically beautiful gardens that can become miniature habitats. Rain garden principles can be scaled up to handle larger volumes of water through bioretention ponds and constructed wetlands, which are themselves complex engineered ecosystems (Figure 12.11). Low-impact development imitates the ecosystem services

FIGURE 12.11 Constructed wetlands. This constructed wetland is artificial and is basically a scaled-up rain garden. However, at this larger scale, it begins to function as a defined ecosystem with a broader range of dynamic cycles and species.

Source: Photo by Bradley Striebig.

of water regulation and purification, while at the same time creating small-scale eco-systems and habitats for wildlife, pollinators, and other organisms.

12.3.2 Functional Design and Biomimicry

Ecological design goes beyond adapting ecosystem dynamics and biogeochemical cycles to engineered systems. It also has methods for replicating nature to achieve specific functions in product design. ***Biomimicry*** is an emerging design discipline based on the understanding that living organisms are themselves masterful engineers, having evolved over millions of years through processes of natural selection. What living systems literally embody are successful adaptations to their environment through a vast array of forms and functions.

As a disciplinary approach to design, biomimicry establishes methods to analyze functional design in the natural world. The underlying designs of nature can be the basis for sustainable innovation in energy, transportation, nontoxic materials, waste processing, buildings, and so on. For example, the Vitalis water bottle is a biomimetic design based on pine trees that have spiraled, or helical, fibrous structures (Figure 12.12). This allows the trees to withstand heavy load and force vertically, horizontally, and diagonally. Vitalis incorporated this ecological design as a grooved geometry in their bottles, making them ultralight yet strong. Both the water bottle and Sharklet technology are examples of how form design can achieve function (see Box 12.3). Biomimicry can be applied to all design scales, including nano-, micro-, and molecular.

FIGURE 12.12 Vitalis water bottle. The Portuguese brand Vitalis has an elegantly designed, ultralight PET (a type of plastic) water bottle. The company was able to reduce materials yet integrate strength by analyzing the spiral growth structures of whitebark pine tree fibers. In 2010, Vitalis reduced its use of raw materials by 250 tons through the improved design.

Source: Unicer.

BOX 12.3 Sustainable Design Thinking in Action: "A Shark Tale"

Sharklet Technologies, Inc. invented the first antibacterial material that does not require any chemical, antibiotic, or antiseptic agents. Its antibacterial properties are based exclusively on its material patterning and texture, which inhibits bacterial growth and is biomimetic of sharkskin. Sharklet's use on surfaces and medical devices could inhibit the development of antibiotic-resistant bacteria because it does not interfere with the biological processes of the organisms. Below is an excerpt from the company's website that describes the design thinking and process that led to this award-winning innovation. As can be seen from Dr. Brennan's story, good powers of observation and a persistent curiosity can lead to fundamentally new engineering ideas.

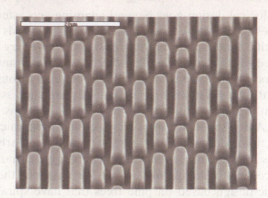

FIGURE 12.13 Sharklet™ technology.

Source: Courtesy of Sharklet Technologies, Inc.

Dr. Anthony Brennan, a materials science and engineering professor at the University of Florida, was visiting the U.S. Naval base at Pearl Harbor in Oahu as part of Navy-sponsored research. The U.S. Office of Naval Research solicited Dr. Brennan to find new antifouling strategies to reduce use of toxic antifouling paints and trim costs associated with dry dock and drag.

Dr. Brennan was convinced that using an engineered topography could be a key to new antifouling technologies. Clarity struck as he and several colleagues watched an algae-coated nuclear submarine return to port. Dr. Brennan remarked that the submarine looked like a whale lumbering into the harbor. In turn, he asked which slow moving marine animals don't foul. The only one? The shark.

Dr. Brennan was inspired to take an actual impression of shark skin, or more specifically, its dermal denticles. Examining the impression with scanning electron microscopy, Dr. Brennan confirmed his theory. Shark skin denticles are arranged in a distinct diamond pattern with tiny riblets (Figure 12.13). Dr. Brennan measured the ribs' width-to-height ratios which corresponded to his mathematical model for roughness—one that would discourage microorganisms from settling. The first test of Sharklet™ yielded impressive results. Sharklet reduced green algae settlement by 85% compared to smooth surfaces.

While the U.S. Office of Naval Research continued to fund Dr. Brennan's work for antifouling strategies, new applications for the pattern emerged. Brennan evaluated Sharklet's ability to inhibit the growth of other microorganisms. Sharklet proved to be a mighty defense against bacteria.

Similar to algae, bacteria take root singly or in small groups with the intent to establish large colonies, or biofilms.

Similar to other organisms, bacteria seek the path of least energy resistance. Research results suggest that Sharklet keeps biofilms from forming because the pattern requires too much energy for bacteria to colonize. The consequence is that organisms find another place to grow or simply die from inability to signal to other bacteria.

Dr. Brennan's and Sharklet Technologies' research has demonstrated Sharklet's success in inhibiting the growth *of Staph a.*, *Pseudomonas aeruginosa*, VRE, *E. coli*, MRSA, and other bacterial species that cause illness and even death.

Source: Sharklet Technologies, Inc. "Inspired by Nature." Accessed February 2, 2014 at sharklet.com/technology.

Biomimicry as a design discipline has the potential to transform many aspects of engineering. It has developed a taxonomic classification of eight fundamental functions found in nature from which 30 subgroups and 162 specific functions have been derived (Figure 12.14). The challenge over the coming years is to translate

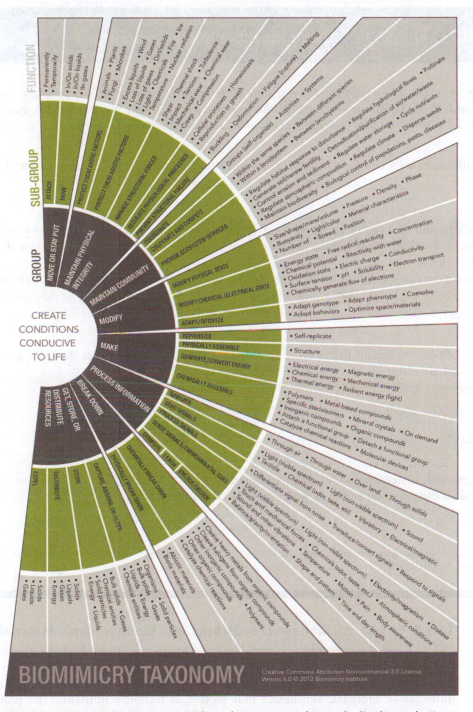

FIGURE 12.14 Biomimicry taxonomy. Biological processes can be standardized into a basic functional typology. These categories guide designers who are looking for ways in which nature has already developed solutions to particular kinds of design "problems."

and adapt our biological knowledge to the language and processes of engineering. Engineers must also adapt their problem-solving approaches to gain insights from nature.

ACTIVE LEARNING EXERCISE 12.1 Using the Biomimicry Taxonomy

Biomimetic design requires us to problem-solve a little differently, because we have to align our thoughts with the underlying functions of nature. In this exercise, you will take a challenge presented by the Biomimicry 3.8 Institute so that you can better understand how the process works.

Design Problem

You are designing a building in an area of low rainfall. You want your building to collect rainwater and store it for future use.

Use the Taxonomy

1. Refer to the taxonomy, and find the verbs. Move away from any predetermined ideas of what you want to design, and think more about what you want your design to do. Try to pull out single functional words in the form of verbs.

2. Go to the online taxonomy at *asknature.org/biological-strategies/*. The questions to pose through the Search or Browse options might be "How would nature…"

    * Capture rainwater?
    * Store water?

3. Try a different angle. Some organisms live in areas that don't experience any rain, yet they still get all of the water they need. So other questions to pose might be: "How would nature…"

    * Capture water?
    * Capture fog?
    * Absorb water?
    * Manage humidity?
    * Move water?

4. Turn the question around. Instead of asking how nature stores water, you might think about how nature protects against excess water or keeps water out: "How would nature…"

    * Remove water?
    * Stay dry?

Source: The Biomimicry 3.8 Institute, AskNature.com, "Biomimicry Taxonomy: Biology Organized by Challenge," www.asknature.org/article/view/biomimicry_taxonomy. [Accessed 5 October 2012].

12.3.3 Circularity, Biobased Feedstocks, and Biodegradable Materials

Cradle-to-cradle, or circular economy design, is fundamentally an ecological approach to product design. As Figure 12.7 illustrates, the flow of materials that make up consumer products can be renewable plant-based resources that "recycle" as biological nutrients or inorganic substances that recycle as technical nutrients. The practical implication of the biological cycle in Figure 12.7 is to select biobased and biodegradable materials for product design. Although biobased feedstocks and biodegradable components are desirable, we must be sensitive to the subtleties of these materials.

Biobased feedstocks are usually derived from living organisms and represent starting materials at the beginning of a product's life cycle. Biobased feedstocks are considered to be renewable natural resources because they regenerate themselves and are not finite. However, it is possible to deplete a renewable resource by harvesting it faster than it can replenish itself. The sustainability of any given biobased resource must be considered within a system of competing rates—the rate of regeneration versus the rate of extraction. There are also social and economic trade-offs to consider when dedicating land and other resources for growing biobased feedstocks, such as the trade-off between using a crop for food or for a biofuel (see Chapter 11). The bioeconomy value pyramid illustrates the relationship between the commercial value added of different biobased feedstocks and the production volume of the feedstock (Figure 12.15).

Biodegradability refers to the potential environmental fate of a material at the end of its useful life. In a circular economy, biodegradable materials serve as nutrients to generate biobased resources. It is important to recognize that biobased and biodegradable are not inherently connected material properties, however. A biobased feedstock is not necessarily biodegradable, and biodegradable materials are not necessarily biobased. For example, the plastic PET can be made from sugar cane. However, the polymerization process that transforms sugar cane into PET creates a plastic that is not biodegradable.

In general, environmental fate refers to what happens to a molecule once it is released into the environment. Chemicals and materials break down in the environment through both biotic and abiotic processes as represented in the biogeochemical cycles. Depending on a substance's chemical structure and properties, it may persist for a long time in the environment or quickly become harmless. Materials that break down quickly by biological processes are referred to as ***biodegradable***.

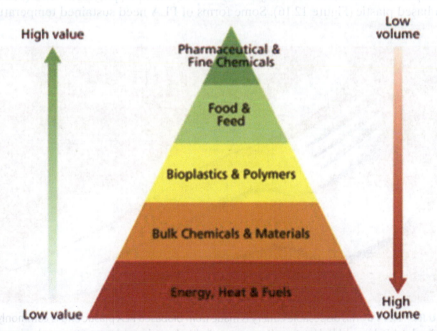

FIGURE 12.15 The bioeconomy value pyramid shows the relationship between value added and volume of biobased materials.

Source: Stegmann, P., et al. (2020, May). "The Circular Bioeconomy: Its elements and role in European bioeconomy clusters." *Resources, Conservation & Recycling: X*. Volume 6, doi.org/10.1016/j.rcrx.2019.100029.

Biodegradation happens in soil, sediment, and water when microorganisms like bacteria and fungi "digest" a substance and convert it into harmless metabolic waste.

Material can also break down abiotically through chemical and physical processes. Some molecules are degraded by ***photolysis***, which occurs when the molecules break down into smaller units by absorbing light. ***Hydrolysis*** refers to the breakdown of substances due to reaction with water. Abiotic mechanisms that can cause solid materials to physically fragment into increasingly smaller pieces include oxidation, weathering (freezing and thawing), and wave action.

From a sustainability standpoint, the three key considerations in the biodegradability of a material at its end of life are (1) the extent of degradation, (2) how long the process takes, and (3) the conditions in which it occurs. With respect to the extent of biodegradation, ***ultimate biodegradation*** (or mineralization) occurs when a substance fully degrades to become carbon dioxide, water, mineral salts, and new biomass for the organisms that consume it. If it only partially degrades and is no longer recognizable as the original or parent substance—but it is not fully degraded either—this is ***primary biodegradation***. With respect to how long the process takes, claims of biodegradability can be applied only to those materials that degrade within weeks or a few months. So, even though a substance may eventually degrade completely through an organic process, if it does not happen fast enough it isn't considered biodegradable.

If a substance does biodegrade, the process generally occurs only when particular environmental conditions are met. Temperature matters. So does oxygen, which determines whether the process is aerobic or anaerobic. Some materials biodegrade in freshwater but not saltwater. The medium (soil, sediment, water) matters, too. Each of these environmental conditions can foster or hinder biodegradation. However, the ability of a substance to biodegrade does not necessarily mean that decomposition will occur naturally outdoors. For example, polylactic acid (PLA) is a biobased plastic (Figure 12.16). Some forms of PLA need sustained temperatures

FIGURE 12.16 This disposable cutlery is made from biobased PLA plastic and is commonly described as biodegradable. However, it cannot easily biodegrade outdoors or in most backyard composting systems. It must be broken down in commercial composting facilities that achieve the right environmental conditions.

Source: iStockPhoto.com/Ajcespedes.

over 130°F to biodegrade, which are not natural surface temperatures. Such PLA requires commercial composting facilities to achieve the appropriate environmental conditions that will break it down; this isn't going to happen in backyard bins.

Because of the variable conditions under which different types of materials might biodegrade, test methods are designed to measure the extent of biodegradation of a material (primary or ultimate) under defined conditions and within a specified time frame. Scientists and engineers should look closely at any claims about biobased materials and biodegradability to make sure they do not reflect greenwashing (see Box 12.4).

BOX 12.4 Greenwashing

The term *whitewashing* comes from the practice of covering surfaces, such as walls, bricks, or fences, with a simple lime and water mixture to improve their appearance. It is a metaphor for the act of covering up or glossing over what would normally be seen as troublesome behavior or facts.

The term **greenwashing** is a play on words that describes a deliberate attempt to make an organization or its products appear more environmentally friendly than they really are. Savvy people can be optimistic about sustainability while still knowing how to detect greenwashing. TerraChoice Environmental Marketing, Inc., helps us recognize greenwashing with a certain amount of tongue-in-cheek humor. TerraChoice studied product marketing claims using the *Guides for the Use of Environmental Marketing Claims* (Green Guides) published by the U.S. Federal Trade Commission (FTC). The FTC is a federal agency that regulates truth in advertising and unfair competition. The Green Guides were developed to clarify acceptable and unacceptable product claims about the environment. In their study, TerraChoice found "seven sins of greenwashing":

1. The hidden trade-off. These are claims that are very narrow and that ignore other significant impacts. For example, a plastic container might be praised for having a small carbon footprint but may still contain toxic chemicals and not be recyclable. The FTC suggests that marketers analyze and communicate any trade-offs from beneficial attributes they are claiming.
2. No proof. These are environmental claims that cannot be easily fact-checked. An example is a

product that claims a certain percentage of its content is from recycled materials but has no data to back up the claim. It could be true, but there is no evidence. The FTC states that claims must be readily substantiated.
3. Vagueness. These are claims that are so poorly defined or broad—like using the word *green*—they are likely to be misunderstood by the consumer and can't be substantiated. The FTC indicates that marketing should not make broad, unqualified general claims about environmental benefit.
4. Irrelevance. Such claims may be truthful but are unimportant or unhelpful to consumers seeking environmentally preferable products. A product that boasts it doesn't harm the ozone layer with CFCs is irrelevant because these chemicals have been banned since 1978. The FTC argues that marketing should not feature useless claims.
5. Lesser of two evils. Some claims may be true, but they do not disclose a significant health or environmental impact of the product. For example, claiming that benzene (a toxic chemical) has been synthesized from plants instead of petroleum ignores the fact that benzene is harmful. As with trade-offs and irrelevant benefits, marketing should reveal environmental complications resulting from the benefit it claims.
6. Fibbing. These claims are simply false and beyond greenwashing. The FTC legally requires that advertisers have a reasonable basis for all expressly stated and implied marketing claims.

(Continued)

BOX 12.4 Greenwashing *(Continued)*

7. False labels. These are misleading claims about third-party endorsements. An example is using a fake certification-style graphic or green jargon like "eco-preferred." The FTC legally requires that advertisers use only environmental certifications that clearly and prominently convey the basis for certification and specific environmental benefits.

The FTC Green Guides are not mandatory regulations for advertisers. Rather, they describe the types of environmental claims the FTC may find deceptive. The FTC can and has taken enforcement action against deceptive advertising, including ordering companies to cease deceptive practices and fining those that violate these orders.

Sources:

Green Washing: Do You Know What You Are Buying? Richard Dahl Environmental Health Perspectives, June 1, 2010, doi.org/10.1289 /ehp.118-a246.

U.S. Federal Trade Commission, Environmental Claims: Summary of the Green Guides, www.ftc.gov/tips-advice/business-center/guidance /environmental-claims-summary-green-guides.

U.S. Federal Trade Commission (2012, October 1). U.S. Issues Revised "Green Guides," www.ftc.gov/news-events/press-releases/2012/10/ftc -issues-revised-green-guides.

To summarize, ecological design in practice results in a wide variety of innovations that apply ecosystem principles to different scales and types of design problems, from materials to products to buildings.

12.4 Chemistry, Carbon, and Circularity in Practice

The chemistry, carbon, and circularity approach can be applied to product and process design as well as materials development. It asks us to make design decisions that lower (or eliminate) the toxicity and hazards of chemicals, materials, and processes used for a product over its life cycle; to reduce the amount of embodied carbon-based energy; and to enable a product to be recovered and used as technical or biological nutrients at its end of life. In practice, this requires that engineers understand the individual stages of the product life cycle and the activities relevant to design choices at each stage (see Chapters 5 and 14). It also requires strategies for green chemistry, decarbonization, and designing for value recovery.

The principles of green chemistry are similar to the principles of green engineering; these approaches will be discussed together in the next section. Decarbonization can be accomplished in several ways that are discussed elsewhere in this text. Chapter 13 on industrial ecology enables you to understand the factors that contribute to embodied energy and tells how to calculate this for material flows. Chapter 14 on life cycle analysis introduces you to the software tools and techniques for quantifying the total energy and carbon footprint of a product and for using these tools in engineering design. Chapter 11 on energy conservation and decarbonization provides an overview of practices for decarbonizing direct energy use; this is the source of embodied energy in material flows. Chapter 16 on the built environment provides techniques for reducing energy use in buildings.

Designing for value recovery represents the body of practices and techniques that enable a closed-loop, cradle-to-cradle production system. Because value recovery is endemic to all sustainable design philosophies, it will be explored as a separate topic later in this chapter.

12.5   Green Engineering and Green Chemistry in Practice

The Sandestin Declaration established a guiding philosophy for green engineering that requires holistic life cycle thinking, conservation of natural resources, avoiding hazardous substances, design with people in mind, and innovation. This broad approach was informed by (among other resources) the 12 principles of green chemistry (Anastas and Warner, 1998) and the 12 principles of green engineering (Anastas and Zimmerman, 2003). These two sets of principles are very similar and in some cases identical; both are intended to give scientists and engineers practical guidance for applied design thinking. Table 12.1 organizes the principles of green chemistry and green engineering by grouping them according to their life cycle stage and design purpose.

The unifying mandate for both green engineering and green chemistry is to avoid the use of hazardous materials at all stages of the life cycle (Table 12.1). Substances should be safe because of *inherency*. This means that the material or chemical is intrinsically harmless. Inherency is a very different design goal than that of traditional risk management, where safety is *circumstantial*. This means that people and the environment are protected from the risk of a hazardous substance by controlling their *exposure* to it (see Chapter 6). However, exposure controls can and

TABLE 12.1 A Comparison of the Principles of Green Engineering and Green Chemistry

LIFE CYCLE STAGE	DESIGN GOAL	12 PRINCIPLES OF GREEN ENGINEERING	12 PRINCIPLES OF GREEN CHEMISTRY
ALL STAGES	Avoid hazardous substances	*Inherent Rather Than Circumstantial* • Designers need to strive to ensure that all materials and energy inputs and outputs are as inherently nonhazardous as possible.	*Less Hazardous Chemical Syntheses* • Wherever practicable, synthetic methods should be designed to use and generate substances that possess little or no toxicity to human health and the environment. *Inherently Safer Chemistry for Accident Prevention* • Substances and the forms of substances used in a chemical process should be chosen to minimize the potential for chemical accidents, including releases, explosions, and fires. *Design Safer Chemicals* • Chemical products should be designed to effect their desired function while minimizing their toxicity.
RESOURCE EXTRACTION	Use renewable resources	*Renewable Rather Than Depleting* • Material and energy inputs should be renewable rather than depleting.	*Use of Renewable Feedstocks* • A raw material or feedstock should be renewable rather than depleting, whenever technically and economically practicable.

(Continued)

TABLE 12.1 (Continued)

LIFE CYCLE STAGE	DESIGN GOAL	12 PRINCIPLES OF GREEN ENGINEERING	12 PRINCIPLES OF GREEN CHEMISTRY
MATERIAL PROCESSING	Prevent process waste	**Prevention Instead of Treatment** • It is better to prevent waste than to treat or clean up waste after it is formed.	**Prevention** • It is better to prevent waste than to treat or clean up waste after it has been created. **Real-Time Analysis for Pollution Prevention** • Analytical methodologies need to be further developed to allow for real-time, in-process monitoring and control prior to the formation of hazardous substances.
MANUFACTURING	Resource efficiency	**Maximize Efficiency** • Products, processes, and systems should be designed to maximize mass, energy, space, and time efficiency. **Output-Pulled Versus Input-Pushed** • Products, processes, and systems should be "output pulled" rather than "input pushed" through the use of energy and materials. **Integrate Local Material and Energy Flows** • Design of products, processes, and systems must include integration and interconnectivity with available energy and materials flows.	**Design for Energy Efficiency** • The energy requirements of chemical processes should be recognized for their environmental and economic impacts and should be minimized. If possible, synthetic methods should be conducted at ambient temperature and pressure. **Atom Economy** • Synthetic methods should be designed to maximize the incorporation of all materials used in the process into the final product.
	Conservation of materials and resources	**Design for Separation** • Separation and purification operations should be designed to minimize energy consumption and materials use.	**Catalysis** • Catalytic reagents (as selective as possible) are superior to stoichiometric reagents. **Reduce Derivatives** • Unnecessary derivatization (use of blocking groups, protection/deprotection, temporary modification of physical/chemical processes) should be minimized or avoided, if possible, because such steps require additional reagents and can generate waste. **Safer Solvents and Auxiliaries** • The use of auxiliary substances (e.g., solvents, separation agents) should be made unnecessary wherever possible and innocuous when used.
CONSUMPTION AND USE	Avoid over-designed products	**Meet Need, Minimize Excess** • Design for unnecessary capacity or capability (e.g., "one size fits all") solutions should be considered a design flaw. **Durability Rather Than Immortality** • Targeted durability, not immortality, should be a design goal.	

LIFE CYCLE STAGE	DESIGN GOAL	12 PRINCIPLES OF GREEN ENGINEERING	12 PRINCIPLES OF GREEN CHEMISTRY
END OF LIFE	Material cycling	**Minimize Material Diversity** • Material diversity in multicomponent products should be minimized to promote disassembly and value retention. **Design for Commercial "Afterlife"** • Products, processes, and systems should be designed for performance in a commercial "afterlife." **Conserve Complexity** • Embedded entropy and complexity must be viewed as an investment when making design choices on recycle, reuse, or beneficial disposition.	**Design for Degradation** • Chemical products should be designed so that at the end of their function they break down into innocuous degradation products and do not persist in the environment.

Sources: Adapted from Anastas and Warner (1998), Anastas and Zimmerman (2003).

do fail. In some instances, it is because of day-to-day handling of the material in the workplace, such as the use of PERC in dry cleaning (see Box 12.2). Failure to control exposure can also be abrupt and catastrophic, such as in Bhopal, India, where a release of methyl isocyanate killed thousands of people in 1984. By improving the inherent safety of a material, we can reduce or eliminate risk and the costs of preventing exposure. As the U.S. EPA's *Safer Choice* motto explains, "If it's not in your product, you don't have to worry about it" (see Box 12.5).

BOX 12.5 "If It's Not in Your Product, You Don't Have to Worry About It"

This motto of the U.S. EPA's Safer Choice Program implies that it is wise to design products without using hazardous chemicals, even if exposure is low (Figure 12.17). The Safer Choice program is a voluntary EPA product-labeling program like Energy Star; the EPA certifies only cleaning products that contain low-hazard ingredients.

Focusing on a material or product's inherent hazard rather than exposure is especially important to protect children, who often interact with products in ways that were not intended. For example, a 4-year-old boy died tragically in 2006 after he swallowed a silver-toned charm from a bracelet that was given away with the purchase of children's shoes. The charm was made almost entirely of lead. Lead in toys created a great deal of public alarm and has been found in many products, including paints on dolls and trucks.

Before the outcry about lead, banning substances in children's playthings seemed to be impossible. But in the first important reform of the Consumer Product Safety Commission, not only did such an action pass, but it also did so with a veto-proof majority. The Consumer Product Safety Improvement Act of 2008 ensured that toys are virtually free of not only neurotoxic lead but also six ortho-phthalates used as plasticizers known to disrupt the hormonal development of children. Phthalates are a class of chemicals used to make plastics flexible, but they can be absorbed when children put toys in their mouths. Three of the targeted phthalates had long been prohibited

(Continued)

FIGURE 12.17 Products and materials should be inherently nontoxic.

Source: iStockPhoto.com/Monkeybusinessimages.

in the European Union: di-(2-ethylhexyl) phthalate (DEHP), dibutyl phthalate (DBP), and benzyl butyl phthalate (BBP). There are safer alternatives, and many are produced by the same companies that make ortho-phthalates.

The Act also greatly improved toy safety in general. Typically, toys are not tested for chemical hazards in advance and are recalled only when problems are discovered after they hit the market. The Act turns this process around and requires that samples of all toys be independently tested for lead and other hazards before they can be sold.

Avoiding hazardous chemicals in products, whether they are added intentionally or unintentionally, requires a level of stewardship that can be challenging. Many retailers and manufacturers have little to no knowledge of the chemical composition of the products they sell because this information is often proprietary to protect trade secrets. A growing number of laws now require greater transparency about chemical and material constituents in products both to the public and throughout the supply chain.

12.5.1 Green Engineering

When viewed through a life cycle lens, the 12 principles of green engineering provide guidance about the selection and use of materials, the design of industrial processes, and the functional and form design of products. The principles can inform design decisions at multiple scales, from molecular to factory.

Most of the principles of green engineering in Table 12.1 are readily understandable, but a few require more explanation. The principle of *output-pulled versus input-pushed* derives from the Le Châtelier principle in chemistry. This insight explains that a system uses less material and energy when its output is drawn off in small amounts as needed (output-pull) instead of being driven by large amounts of input (input-push). This principle is fundamental to chemical processes, but it can also be applied to other systems. Just-in-time manufacturing and just-in-time delivery are management strategies that apply the output-pulled principle at the scale of technological systems and supply chains. Output-pulled can be widely applied in industrial engineering, where it is relevant to product lines as well as machine-scale unit operations. Wherever there are material or energy flows, this is potentially a useful design approach.

The directive to *integrate local material and energy flows* is complementary to *output-pulled versus input-pushed* and likewise focuses on process and systems design. This principle improves efficiency by using the materials and energy available within a system, which could be at the molecular, machine, production line, factory, or community scale. By applying it to process or systems design, we can close waste and energy loops during the production process, not just at the end stage of a product's life. For example, waste steam can be captured to make hot water or electricity. Iron ore waste at a steel foundry can be sintered onsite and put back into the foundry's blast furnace. Waste fryer oil from local restaurants can be the feedstock for community biodiesel manufacturing. Industrial ecology provides methods and tools for evaluating production operations to meet this design goal.

The principle that *complexity should be preserved* at the end of the life cycle is an important design concept. When materials are recovered at the end of a product's life, this principle tells us that we do not necessarily want to reduce all products to their base materials. Complex products have high value added because they incorporate many parts and components, may have intricate technical functions, are technologically intensive to process and manufacture, may have high embodied energy, and could use expensive elements such as noble metals and rare earths.

Complex products embody considerable amounts of resources, time, investment, intellectual property, and technological sophistication that make them costly to manufacture from scratch. Rather than break these items down into their separate base materials, it would be better to maintain and reuse the item's complexity, as with inkjet cartridges (Figure 12.18).

FIGURE 12.18 Inkjet cartridges are a simple example of the principle of preserving complexity. The integrity of these products can be completely preserved by refurbishing and re-inking them, a process that is more cost effective than manufacturing new cartridges. Some manufacturers make cartridges that can be productively re-inked and reused up to 100 times.

Source: Photo by Bradley Striebig.

12.5.2 Green Chemistry

Chemicals are essential substances in our engineered world. They are used as ingredients in food, cosmetics, and pharmaceuticals; they are feedstocks or additives for materials from which products are made; they serve innumerable roles as solvents, surfactants, lubricants, coatings, surface treatments, and reagents in industrial operations and manufacturing; and they are end products in their own right, such as cleaners, personal hygiene products, disinfectants, fertilizers, and pesticides. Three potential characteristics of chemicals present threats to human health and the environment: their long-term persistence in the environment, the degree to which they might bioaccumulate in organisms, and their toxicity.

The purpose of green chemistry is to reduce or eliminate the creation and use of hazardous (or potentially hazardous) substances by designing new chemicals (Anastas and Warner, 1998). The 12 principles of green chemistry provide design guidance on how to achieve this goal. Although most of the principles of green chemistry in Table 12.1 are self-explanatory, the principle of *catalysis* to conserve resources requires a bit of explanation. In chemical reactions, catalysis is desirable because catalytic reagents lower the activation energy of the reaction and make it more efficient. In addition, many catalysts are not consumed by the reaction. Therefore, less of them are required, and they can be recovered and used repeatedly. In contrast, stoichiometric reagents are consumed by the reaction and must be replaced each time.

When we reduce the presence of hazardous materials in products or industrial operations, we call it **detoxification**. Detoxification can be accomplished in two main ways. The first is to substitute existing, less hazardous materials for those that are problematic. A classic example of detoxification through chemical substitution is the replacement of ozone-depleting chlorofluorocarbons with alternative hydrofluoroalkanes. Another example is the substitution of rapidly biodegradable surfactants for surfactants that degrade into toxic byproducts.

The other major way to accomplish detoxification is to design new chemicals and materials that are benign with respect to human health and the environment. This is where green chemistry is especially valuable in changing the performance, toxicity, and fate of chemicals used in industrial processes and manufactured products. Green chemistry demonstrates that it is possible to intentionally design new molecules, compounds, and chemical processes to reduce hazards and improve sustainability. Innovations that have resulted from applying the principles of green chemistry include eliminating hazardous reagents, using catalysts to eliminate steps in a chemical synthesis, creating inherently benign or biodegradable chemicals as replacements for chemicals of concern, and using biobased- instead of petroleum-based feedstocks. Box 12.6 summarizes three award-winning examples of green chemistry that demonstrate several of the 12 principles of green chemistry.

From an engineering design perspective, detoxification can be a challenging strategy. It requires knowing the chemical content and hazard potential of materials selected for a product or used for an industrial process. This information is not necessarily easily available. In addition, substitutes need to be explored or new chemicals designed. Practical and detailed guidance on where to look for chemical and material information is given in Heine and Whittaker (2021). A useful approach to detoxification and making design decisions is alternatives assessment, which is explored in Chapter 15.

The principles of green engineering and green chemistry give us specific and actionable guidance about how to improve sustainability through technical design.

BOX 12.6 Preventing Hazardous Waste Through Green Chemistry and Innovation

Green chemistry can be applied in just about any industrial sector. Here are three examples derived from winners of the U.S. EPA Green Chemistry Challenge Awards program. Many other impressive examples are summarized on the EPA website.

In 1998, Rohm and Haas Company won the award for designing the marine antifoulant called Sea-Nine. Fouling is the unwanted growth of plants and animals on a ship's surface; it costs the shipping industry approximately $3 billion a year. A significant portion of this cost is due to the increased fuel consumption needed to overcome hydrodynamic drag. Historically, organotins such as tributyltin oxide (TBTO) were used to provide antifouling properties to boat paints. While they were effective at preventing fouling, they are toxic and persistent in the environment. The negative environmental effects associated with TBTO include acute toxicity, bioaccumulation, decreased reproductive viability, and increased shell thickness in shellfish. Rohm and Haas developed a chemical that protected ships from fouling by a wide variety of marine organisms without causing harm to nontarget organisms. Company scientists selected an isothiazolone derivative, 4,5-dichloro-2-n-octyl-4-isothiazolin-3-one. This compound was found to be an effective antifoulant and to degrade very rapidly in seawater and even faster (1 hour) in sediment. Sea-Nine is acutely toxic to selected marine organisms, which is a desirable characteristic of an antifoulant. But Sea-Nine degrades rapidly and is not chronically toxic, nor does it bioaccumulate. The chemical does its job but does not cause harm beyond its immediate function. TBTO not only persists in the environment but also bioaccumulates, creating ongoing toxic effects in marine food webs. Chapter 15 contains a discussion of alternative approaches to controlling biofouling that may not need biocides at all.

EDEN Bioscience Corporation received an award in 2001 for the product Messenger, which stimulates plants' natural defenses against pests and disease without altering plant DNA. Messenger is based on a class of nontoxic, naturally occurring proteins called "harpins." When applied to crops, harpins support plant health, resulting in increased plant biomass, photosynthesis, nutrient uptake, and root development. Ultimately, such increases lead to greater crop yield and quality of product. Estimated yearly crop losses from pests are estimated at $300 billion worldwide. Growers have typically pursued two strategies to limit economic losses and to increase yields: (1) using traditional chemical pesticides or (2) growing crops genetically engineered for pest resistance. Each of these approaches has negative consequences, particularly on nontarget organisms and in terms of public perception. Harpins have been shown to have virtually no adverse effect on any of the organisms tested, including mammals, birds, honeybees, plants, fish, aquatic invertebrates, and algae. Because Messenger is based on fragile molecules that degrade rapidly by ultraviolet light and microorganisms, it has no potential to bioaccumulate, and it will not contaminate surface or groundwater resources. Unlike most agricultural chemicals, harpin-based products are made in a water-based fermentation system that uses no harsh solvents or reagents, requires only modest energy inputs, and generates no hazardous chemical wastes.

In 2020, Johns Manville was recognized for developing a biobased, formaldehyde-free fiberglass mat used in the carpet industry to stabilize carpet tile and other types of flooring. Fiberglass is a widely used composite material made of glass fiber bound by a polymer matrix, usually a thermoset resin. Formaldehyde is often used in thermoset binders and is considered a probable carcinogen. The award-winning fiberglass binder uses dextrose, a renewable carbohydrate that is nonhazardous and is produced from corn, potato, or wheat starch. The new biobased binder is made using a biodegradable acid catalyst at temperatures about 40°C lower than conventional polymer binders. The dextrose-based binder has a shelf life of about 12 months compared to less than 1 month for traditional binders,

(Continued)

thus reducing waste generated from expired unused product. Mats made with the Johns Manville binder are more stable to heat and humidity and have better delamination strength than the former technology, adding a performance advantage on top of improved safety and environmental features. The company estimates that the new product reduces water usage by as much as 17,000 gallons per day and energy usage by the equivalent of 138,000 cubic feet of natural gas per day.

Some green chemistry innovations represent incremental improvements, while others are more disruptive to the marketplace, changing existing ways of doing things.

However, we cannot design products, processes, systems, and technologies that will optimize all principles at once. Trade-offs will emerge in practice. A very crude example is the use of nuclear energy to generate electricity. It is carbon-free electricity, but the fuel produces a large amount of low- and high-level hazardous waste across its materials cycle. Weighing the risks of climate change against the risks of nuclear power is a sobering task. In Chapter 15, we introduce alternatives assessment as a decision support tool to help us select from competing design options that present different types of environmental benefits and risks. In the following sections, we will explore specific methods and techniques that convert green design principles into more sustainable products, life cycles, processes, and systems.

12.6   Product Design Strategies

Technical product design is grounded in the need to develop products that provide a specific function at a defined level of quality, performance, durability, reliability, and safety. Designing for sustainability cannot compromise a product's functional integrity (this is also true for systems, processes, materials, and buildings). There are, nonetheless, several opportunities to incorporate sustainability in product design, including selecting materials and dematerialization, avoiding the "overdesign" of products, and designing for value recovery.

12.6.1   Materials Selection and Dematerialization

Three types of design decisions about materials can enhance sustainability. First, we can substitute more sustainable materials for an existing product of a given design. Second, we can design a product's form and functionality so that more sustainable materials can be used. Third, we can reduce the amount of material (mass) required for a product and its components. All three design decisions require very careful engineering. The statics and mechanics of different material options relative to the performance, quality, durability, reliability, and safety specifications of the product must be considered. Using a biobased, recyclable plastic instead of a metal is terrific as an abstraction, but not if the material will easily fail. As discussed in Chapter 5 with LEGO toy bricks, finding more sustainable alternatives is not necessarily easy.

Using more sustainable materials in product design means selecting parts, components, and engineered material feedstocks that are ideally created from renewable resources, are nonhazardous, and are rematerialized. ***Rematerialized feedstocks*** are substances that have been recycled at the end of life and reprocessed into usable inputs. In general, more sustainable materials selection also results in lower embodied energy and smaller carbon footprints. Practical strategies for improving sustainable materials selection include using biobased natural resources, detoxification, and biomimicry (Table 12.2). New biobased feedstocks are becoming viable material alternatives, especially for petroleum-based plastics. Detoxification can involve material substitution or innovation through green chemistry. Biomimicry may inspire alternative materials for a variety of design needs (see Box 12.7).

TABLE 12.2 Design strategies, methods, and tools for sustainability

LIFE CYCLE STAGE	DESIGN GOALS	GREEN DESIGN PRINCIPLE OR PHILOSOPHY	DESIGN STRATEGIES, METHODS, AND TOOLS
ALL STAGES	Avoid hazardous substances	Inherent rather than circumstantial	Detoxification Life cycle analysis Biomimicry
	People-centered design	Design with people in mind	Participatory design Use renewable energy
RESOURCE EXTRACTION	Use renewable resources	Renewable rather than depleting	Biobased feedstocks Rematerialized feedstocks Remanufacturing Design for recycling
MATERIAL PROCESSING	Prevent process waste	Prevention instead of treatment	Rematerialized feedstocks Remanufacturing Dematerialization
	Resource efficiency	Maximize efficiency Output-pulled versus input-pushed Integrate local material and energy flows	Life cycle analysis Industrial energy auditing Lean manufacturing Materials flow analysis Biomimicry
MANUFACTURING	Conservation of materials and resources	Design for separation	Use renewable energy
CONSUMPTION AND USE	Avoid overdesigned products	Meet need, minimize excess Durability rather than immortality	Biomimicry Design for repair Design for remanufacturing (DfRem)
END OF LIFE	Material cycling and value recovery	Minimize material diversity Design for commercial "afterlife" Conserve complexity Design for degradation	Minimize material diversity (dematerialization) Design for disassembly (DfD) Design for recycling (DfR) Design for remanufacturing (DfRem) Biodegradable materials selection

BOX 12.7 Butterfly Effects

The *butterfly effect* is a popularized social metaphor that helps us understand how small changes in complex systems can have profound impacts in unexpected places. But the butterfly (Figure 12.19) is also a powerful source of engineering innovation through biomimicry.

Nature's beauties are inspiring a host of pragmatic innovations that promise a more sustainable future. One of the most significant biological features

of many species of butterfly is that their color *is not* generated by a pigment but by cellular structures that are arranged as shingled plates at the nanoscale. So what can we learn from this? That color can be created by optical interference. This property was adapted by Qualcomm, an electronics company that invented a display screen using electromechanical manipulation of thin films. The display requires only ambient light; it has no internal light source of its own. The result is highly readable color displays even in bright sunlight, using only 10% of the power required by conventional LCD screens.

Morphotex®, a nanotechnology-based synthetic fiber no longer in production, used the structural chromogenic properties of butterfly color to create color without pigments, potentially reducing water consumption, dyes, and chemical waste. Similarly, ChromaFlair™ is a color-shifting pigment based on structural chromogenics as well. On automobiles, this paint is a rainbow of iridescent color.

Butterfly biomimicry is not all about color aesthetics. Chinese scientists are exploring how to develop more efficient solar energy technology by studying the deep black in butterfly wings, which functions as a heat trap. Parts of butterfly wings are also self-cleaning, inspiring research that will help us develop materials that do not require detergents, surfactants, and large amounts of water to stay clean.

FIGURE 12.19

Source: Photo by Bradley Striebig.

Life cycle analysis tools can help evaluate total energy use, carbon footprints, air emissions, and water contamination for different materials options.

Designing for more sustainable materials use also involves technical design that enables end-of-life value recovery, which "closes the loop" of product and material life cycles. End-of-life value recovery requires multiple design strategies and will be discussed later in this chapter.

Dematerialization is a longstanding practice in product design and has traditionally been done to lower product costs. Dematerialization can reduce the amount of material needed for a product, the diversity of materials used in a product, or both. Dematerialization enhances sustainability by reducing the demand for natural resources and lowering the environmental impacts of materials processing (see Table 12.2). In product design, dematerialization is often achieved by ***product simplification*** (reducing the number of necessary parts and components), material substitution, or making the product smaller. Dematerialization can also include reducing or eliminating the use of packaging materials, which is often called "light-weighting."

Dematerialization is also a subtle component of design in the service sector (see Example 12.1). Taken to its extreme, dematerialization can eliminate tangible products altogether, such as replacing hard copy books with virtual e-books.

EXAMPLE 12.1   Dematerialize Your Fizzy Water

Dematerialization is a design strategy that may be used in service industries. An example is bringing and using our own cups in coffee shops. Dematerialization can be applied to our lifestyles as well. When we take reusable bags to the grocery store, use cloth instead of disposal diapers, or replace paper towels with Swedish cloths, we are dematerializing. We can estimate the sustainability benefits of our personal dematerialization with some simple calculations.

Many people enjoy drinking carbonated water, also known as club soda, sparkling water, seltzer water, and fizzy water. We can buy fizzy water as a packaged beverage in single-use plastic bottles and aluminum cans. Or we can avoid creating a beverage container waste stream by making fizzy water at home with a sparkling water machine. In doing so, we significantly "dematerialize" our club soda consumption and improve the life cycle sustainability of our carbonated water use.

The Soda Club sparkling water machine is composed of a reusable 1-liter bottle, a refillable CO_2 cartridge, and the housing that holds the CO_2 cartridge and carbonates the water. Each CO_2 cartridge can make 110 liters of soda. To make club soda, you fill the bottle with your own chilled tap or filtered water, screw the bottle into the machine housing, and press a button to release CO_2 into the bottle. After a few moments, a loud noise created by pressure in the bottle indicates that the water is carbonated and ready to drink.

Assume that the Soda Club machine costs $79; it includes a CO_2 cartridge and the reusable bottle. The bottle is indefinitely refillable unless it gets damaged. When the cartridge is empty, you can return it to the store to be refilled for $15. Assume that your family currently drinks 6 liters of club soda each week at $1.50 per bottle. You purchase these by the case (12 bottles), with one trip to the grocery store per case. Compare the cost, transportation footprint, and waste generated from purchasing individual bottles of club soda to using the Soda Club system.

First, we can determine the situation for purchasing individual bottles:

$$\text{Number of individual bottles} = \frac{(6 \text{ L/week})(52 \text{ weeks/year})}{1 \text{ L/bottle}} = 312 \text{ bottles}$$

$$\text{Cost} = 312 \text{ bottles/year} \times \$1.50/\text{bottle} = \$468/\text{year}$$

$$\text{Trips} = \frac{(312 \text{ bottles/year})(1 \text{ trip/case})}{12 \text{ bottles/case}} = 26 \text{ trips}$$

Next, we determine the situation for purchasing the Soda Club system:

$$\text{Number of cartridges} = \frac{(6 \text{ L/week})(52 \text{ weeks/year})}{110 \text{ L/cartridge}} = 2.8 = 3 \text{ cartridges}$$

$$\text{First year's cost} = \$79 + \$15(2) = \$109$$

$$\text{Subsequent years' costs} = \$15(3) = \$45$$

The data are summarized in the following table.

OPTION	COST ($ PER YEAR)	NUMBER OF BOTTLES	NUMBER OF TRIPS TO STORE
Purchase bottles by the case	468	312	26
Soda Club system (first year)	109	1	3
Soda Club system (following year)	45	0	3

The results make it clear that the Soda Club system will dematerialize your family's consumption of club soda by using far fewer bottles, lower your costs of fizzy water, and reduce your transportation footprint. Your family's use of the Soda Club machine also avoids all the life cycle impacts of single-use plastic and aluminum containers, even if these materials were recycled at their end of life.

12.6.2 Avoid "Overdesign" of Products

Many sustainability impacts of technical design occur at the consumer stage of the life cycle, when the product is being used. Two principles of green engineering address a particular concern about product end use, which is that products can be overdesigned. *Overdesign* means that products are designed with an abundance of features or capacity that provides more functionality than any individual user might need. In colloquial language, we say that a product has too many "bells and whistles." Overdesign represents excess capacity and unnecessary value added, which means too much cost, energy, material, and time are embodied in the good. The green design principle is to *meet need and minimize excess*. Biomimicry is a design approach that can potentially find examples of functional design in the natural world that meet needs with the greatest efficiency (see Table 12.2).

Overdesign is also a concern with regard to product durability. Every product has a *design lifetime*, which is how long it can physically withstand normal wear and tear (including minor repair) and still function or operate. Eventually all things will reach the end of their material, mechanical, and electrical life. Durability is the combined result of form, functional design, materials, and the reliability of the working parts in the product. The physical design lifetime of a product is different from its *commercial lifetime*, which is how long a typical user will have the product before replacing it or simply no longer wanting it. The commercial lifetime also reflects how long parts and components may be available to repair and maintain a product, either by the original owner or in secondary markets.

The green engineering principle is *targeted durability, not immortality*. This design goal aims to bring the physical lifetime of a product closer to its commercial lifetime. We do not want to overinvest in materials, components, and manufacturing to make products whose quality and cost far exceed what is needed, thus wasting money, energy, time, materials, intellectual property, and so on.

An example of potential product overdesign in terms of features and durability is tapered roller bearing assemblies (Figure 12.20). This product is used in wheel hubs for cars and boat trailers, among other uses. Many diverse applications of these bearings can use the exact same size of assembly. In a car, however, the unit must last for tens of thousands of miles and failure can be catastrophic. Bearing assemblies for cars are made of high-performance steel and lubricants designed to withstand specific conditions of speed, stress, vibration, temperature, load, driving distance, and so on. These same bearings could be used in a boat trailer, but they

FIGURE 12.20 This high-performance steel tapered roller bearing is an example of a product that is overdesigned if used in some applications.

Source: iStockPhoto.com/Aleksandr Yurkevich.

would be overdesigned (and very expensive) in that application. The driving conditions, total lifetime distance, and commercial life of a boat trailer are very different from those of a car. As a consequence, trailer bearings can be made of low-carbon steel—with very different material properties—yet still be suitable for the need. This is an interesting example of how form design can truly be identical, but how material selection drastically affects the ability of product functionality to meet user needs.

Durability is a perplexing issue in sustainable product design. Targeted durability is an attractive sustainability goal, but products often have design lifetimes that exceed their commercial lifetimes. One reason for this is safety and liability concerns, which require products with high durability and reliability. Another is the fact that many engineered materials have an intrinsic level of durability simply because of their chemical and physical nature. The sometimes-unavoidable mismatch between durability and commercial life is one more reason it is important to close the material loops in a product life cycle. "Junkyards" for cars have exploited the mismatch for decades by supplying salvaged auto parts that have many years of life remaining in them.

In addition, the commercial lifetime of a product can be surprisingly short because of *planned obsolescence*. Planned obsolescence is a deliberate business strategy that forces customers to buy product upgrades or replacements so the company can maintain or grow its future sales. Strategies for planned obsolescence include making frequent design changes so that the product goes "out of fashion" quickly, intentionally designing for limited durability or product failure, not making spare parts available, and tightly controlling the ability to repair an item. Consumer

FIGURE 12.21 The boombox is a classic example of planned obsolescence. Consumer electronics are intentionally designed to have a very short commercial life, typically about 3 years.

Source: iStockPhoto.com/EuToch.

electronics have been designed with planned obsolescence since their beginning because of **Moore's law** (Figure 12.21). Moore's law is the observation that the number of integrated circuits on microchips doubles every two years, thus exponentially increasing the processing power of those chips, which are the technological foundation of modern digital computing. There appears to be a fundamental physical limit to Moore's law because integrated circuits are now so densely packed on microchips that their heat cannot be effectively dissipated.

The tension between physical design life, commercial life, and planned obsolescence affects waste minimization as a sustainability strategy because it limits the ability of products to be reused and repaired. Reuse and repair can extend the commercial and design life of a product, which avoids all of the economic costs, environmental burdens, and waste stream impacts of replacing products more frequently.

As a consequence, a strong "right to repair" consumer movement has emerged in the United States, especially with respect to electronics and "tech" products. In 2021, the Federal Trade Commission voted unanimously to affirm the right to repair and to regulate companies that unduly restrict independent repairs to their products or the manufacture of replacement parts.

To sum up, trying to avoid overdesigning products is challenging because many technical, social, and economic dynamics encourage a long design lifetime but a short commercial life. The way to bring these competing forces into a sustainable balance is to avoid overdesign when possible, improve the ability to lengthen commercial life, and create closed-loop product and material flows. As with materials selection and dematerialization, life cycle analysis is a tool that can help quantify the environmental benefits of helping design lifetime and commercial lifetime to converge.

ACTIVE LEARNING EXERCISE 12.2 How Long Can It Last?

During World War II, resources were scarce and common items like gasoline, sugar, and butter were rationed to support the war effort. Every scrap of material was collected for recycling and flowed into wartime production. People were discouraged from buying anything new, and the U.S. War Advertising Council promoted the slogan "Use it up, wear it out, make it do, or do without." Every consumer product was pushed to the absolute limit of its physical design life to extend its commercial usefulness. People became very skilled at patching, maintaining, and repairing the products of their daily lives. When an item was truly beyond saving, they dismantled complex products into separate waste streams for recycling.

The purpose of this exercise is to help you visualize and think critically about the sustainability implications of design life, commercial life, and planned obsolescence in products. It also has you reflect on the requirements for closed-loop material flows.

Gather up about five products you use in your daily life; they should represent a range of product types like your phone, an appliance, an article of clothing, and so on. For each item, answer these questions:

1. How long do you think this product can physically last and still be functional? What do you think about its durability and reliability? Why? Do you think it was purposefully designed to become commercially obsolete? How long do you think its intended commercial life is, and why?
2. How long do you think you will keep and use this product? How long would you *like* to keep and use it? (This is its commercial life for you.)
3. How do its physical life and commercial life compare? Is there a mismatch? Why?
4. What will you do with this product if it is fine but you don't want it anymore? Will you throw it away, donate or sell it for secondhand use, or try to recycle it? What influences your choice?
5. What will you do with this product if it shows wear or breaks? Will you refurbish or repair it (why or why not), throw it in the trash, or try to recycle it? Can you refurbish/repair it yourself, or will you have to pay someone else to do that?

Compare your insights about each product. Are there meaningful differences in the physical and commercial lifetimes of these items? Why? If any products are overdesigned, how could you make them more sustainable? What limits your ability to extend the commercial life of an item? What does this exercise suggest about the right to repair and our ability to reuse products as sustainability strategies?

12.7 Designing for Value Recovery

Value recovery happens at the end of life for a material or product. This is the point in time at which the user will dispose of them. Items may become unwanted, such as when they are replaced with a new product, someone gets rid of possessions, or a company goes out of business. Items might also be broken beyond repair, or they may have exhausted the durability of their material, mechanical, and electrical life. This is also the point in time at which we want to avoid end-of-life waste and the permanent disposal of materials and products.

Designing for value recovery involves choosing products and materials that "decompose" into technical nutrients that can be cycled back into an industrial ecosystem. Not all materials are inherently recyclable (see Chapter 5). When recycled material is reprocessed, sometimes it can be rematerialized into its source substance and recycled relatively indefinitely, like aluminum and glass beverage containers.

In other instances, the material might be downcycled (used in a lesser value-added application) or upcycled (used for higher value-added products). This depends on what happens to the material's properties as a result of reprocessing. The green engineering principle of *conserving complexity* tells us that technical nutrients don't always have to be material feedstocks. Instead, we can potentially recycle intact manufactured parts, components, and products back into production systems.

Technical design has direct impacts on the ability to recover and reuse products and materials at their end of life. Designing for value recovery means that we are intentionally designing the form and function of an object, and selecting its materials, in anticipation of giving it a meaningful next use. Although we typically think of end-of-life products as consumer goods that individuals buy, they can also be *capital goods*. Capital goods create the products and services that companies sell. Examples include ATMs in a bank, medical equipment in a hospital, robotics in a factory, and photocopiers in an office.

Some solutions to waste stream diversion and value recovery are social. Household products that are still useful but no longer wanted can be donated to thrift stores, sold as secondhand items, or advertised in freecycle social media. Creating this kind of cultural shift for unwanted items is not design for value recovery. Instead, it is engineering that intentionally creates and captures technical nutrients when products and materials are no longer wanted, when they are irredeemably broken, or when they are just worn out.

Two conditions are necessary for value recovery at a product's end of life. One condition is that the product itself is capable of being dismantled to extract its value for reuse. We will introduce Design for X strategies as approaches to sustainable product design for end-of-life fate (see Table 12.2). The second condition is that a closed-loop material flow actually exists in the system. If there is no loop to logistically reprocess materials and products, then no recycling can occur. We will explore how recycling markets work and other methods of product recovery to better understand how loops are closed.

12.7.1 Design for X

Few products are composed of just a single component or material. A good example is a simple drinking cup, which has no parts at all and is made of just one substance such as glass or plastic. Most products have many connected parts and components that enable them to function. A car is an extreme example and is commonly estimated to have about 30,000 unique and discrete parts; each part is made of a material that can potentially be recovered and recycled.

Design for Excellence, or *Design for X* (DfX), is a family of specialized design guidelines, rules, and metrics. The X represents a specific design dimension, which can be one of dozens of dimensions like assembly, manufacturability, risk, inspection, safety, and so on. Design for environment (DfE) falls under the umbrella of DfX and specifically addresses the impact of design choices on human health and the environment. The subspecialties within DfX and DfE can overlap because sustainability has become of greater interest. Three particular strategies within DfX are relevant to end-of-life value recovery: design for disassembly, design for remanufacturing, and design for recycling. The environmental impacts of all three design strategies can be assessed with life cycle analysis tools, and their design guidelines may have evaluation metrics as well.

Disassembly is a critical precondition for value recovery. Complex products composed of many parts must be disassembled into their components to recover them for recycling. End-of-life products can re-enter the industrial ecosystem not only as recycled feedstocks, however, but also as items for remanufacturing. ***Remanufacturing*** is the process of refurbishing, repairing, and restoring used parts, components, and products to like-new conditions for resale. Disassembly is likewise needed for remanufacturing and to repair products before their end of life.

Design for disassembly (DfD) involves a variety of design techniques that make product disassembly easier and more cost effective. One strategy is simplification, which reduces the number of separate parts in a product; product simplification is also a technique of dematerialization. Another is changing the sequence in which parts and components are put together and/or the ways in which they are attached. Changing the assembly sequence and method (solder, adhesive, fasters, etc.) can make it easier to take products apart by hand or even to automate the disassembly process. DfD also focuses on designing the logistics for taking things apart. Logistical considerations affect the cost effectiveness of DfD and include how, where, and in what volume products will be disassembled. DfD is also used as a design method for buildings so that they can be dismantled and recycled at the end of their commercial or structural life.

Design for remanufacturing (DfRem) fulfills the green engineering principles to *preserve complexity* and to *design for a commercial afterlife* beyond simple material recycling. Remanufacturing involves the disassembly of products, after which parts and components are sorted, inspected, cleaned, repaired, refurbished, and restored to "like new" condition. These components can then be used to rebuild a product like the one they were taken from or for updated models. For some companies, DfRem is therefore a strategy in which future generations of a product are designed to use the same parts as past models. DfRem can work especially well for capital goods where durability and quality may—by intrinsic material properties and by intentional design—exceed the commercial life of products. The classic example of a company that uses this approach to remanufacturing is Xerox Corp., which has been doing so since the 1980s. Many other examples of remanufacturing can also be found.

ACTIVE LEARNING EXERCISE 12.3 Leaders in Design for Remanufacturing

Design for remanufacturing is a way to conserve the value-added complexity of materials, costs, technologies, and intellectual property in products. It also avoids the environmental costs of making parts and products from scratch with first-use raw materials. Below is a list of companies with leading examples of design for remanufacturing. Select one and research its remanufacturing strategy. What are the design features of the products that allow remanufacture? How is the product a capital good? Explore how the company acquires its products to remanufacture—what are the logistics of the take back or return programs? Finally, how is remanufacturing good for customers and the competitive advantage of the business?

- ABB: industrial robots
- Applied Materials: semiconductor, solar, and display materials
- Caterpillar: construction equipment
- Cisco: computer networking equipment
- Colborne Foodbotics: food manufacturing automation
- GE Healthcare: medical imaging technology
- Xerox: office document and photocopier equipment

Design for recycling (DfR) improves the recycling and reuse of a product's material content. While the environmental value of recycling is often associated with conserving finite natural resources and reducing the needs for landfill space, rematerialized feedstocks can dramatically reduce the overall ecological footprint of a product. The environmental impacts of materials are usually most intense at the front end of the product life cycle. Converting natural resources into cleaned, purified, synthesized, and uniform material feedstocks is usually heat, chemical, and water-intensive; recycling avoids these material processing impacts. For example, recycling aluminum avoids more than 90% of the energy required to make it from raw minerals; recycling a ton of paper conserves 7,000 gallons of water.

DfR design choices include things like using more recyclable materials, replacing hazardous materials, reducing the diversity of materials in a product, making recycling more cost effective, making it easier to recycle a product (which overlaps with DfD), and using more rematerialized feedstocks in manufacturing (which overlaps with materials selection decisions in design). DfR isn't just about using existing materials, however. It can also involve innovating new chemicals and materials that are inherently recyclable to replace those that are not.

All three Design for X strategies (disassembly, remanufacturing, and recycling) require a way to capture products at the end of life and reprocess them. One of the biggest barriers to a circular economy is that there are not enough linkages to close material loops in the system. It is not necessarily cost effective to recycle materials or for companies to take back their products. This creates a dilemma for designers, who may be trying to design for a material or product recovery that does not exist.

12.7.2 Extended Producer Responsibility and Recycling Markets

To achieve a circular economy, technical and biological nutrients must cycle. In an industrial ecosystem, this is accomplished in two basic ways. One is through ***extended producer responsibility*** programs. These are programs in which companies are responsible for the fate of their products at end of life. The second way is through recycling programs and markets.

Extended Producer Responsibility

Extended producer responsibility (EPR) can be voluntary or mandated by government policy. Remanufacturing is a type of EPR, and it is practiced in the United States and Europe by companies that do it as part of their competitive strategy. As explained, remanufacturing refurbishes used parts, components, and products and restores them to as-new conditions for resale. No data are systematically collected about how much remanufacturing takes place. A study conducted by the U.S. International Trade Commission found that remanufacturing generated at least $43 billion in sales and represented at least 180,000 full-time jobs in the United States in 2011 (U.S. ITC, 2012).

This activity is clearly not trivial. In 2011, however, total U.S. manufacturing amounted to $5.6 trillion and full-time equivalent employment was 12 million, so remanufacturing represents a small fraction of the American industrial system. Remanufacturing is most widely practiced in the aerospace, heavy-duty and off-road equipment, auto parts, machinery, information technology, medical devices, and retreaded tire industries. The U.S. Remanufacturing Industries Council suggests that lack of awareness about remanufacturing among industry, consumers, and policymakers is one of the biggest barriers to creating more closed-loop production.

Extended producer responsibility happens in other ways, too. Companies create ***take back programs*** when their products become post-consumer waste and the companies take responsibility for them as waste. Products are then recycled to the greatest

BOX 12.8 Take Back Programs and Extended Producer Responsibility

Many manufacturers around the world now participate in voluntary and mandatory take back programs. A take back program is one in which a product (usually postconsumer) can be returned to the manufacturer at the end of its life. The producer then becomes physically responsible for the product and may reprocess it back into a feedstock for production; may refurbish components and modules for remanufacture or reuse; may recycle and reclaim raw materials; or may arrange for final end-of-life disposal.

Take back programs result from environmental policies and regulations that create the principle of extended producer responsibility. Extended producer responsibility makes a manufacturer responsible for the environmental consequences of its product over a greater range of the product's life cycle; it is intended to capture the cost of environmental protection in producer costs and consumer prices.

Extended producer responsibility is a waste management principle that first emerged in Sweden, then Germany, and is now practiced throughout Europe. Scarce landfill space, the presence of potentially hazardous substances in product componentry, and the growing cost to government of waste disposal are strong motives for this type of management innovation. The Green Dot® logo represented in Figure 12.22 is used within the EU to designate products for which there is a take back packaging program supported by the manufacturer. The Green Dot program is not mandatory; it is one of several strategies that producers can select from to comply with EU regulations.

By shifting the financial and environmental burden of waste disposal back onto manufacturers and consumers, economic markets can then work to encourage producers to adopt less environmentally harmful materials and to develop products that can be more readily disassembled. High embodied costs of product recovery and disposal will also signal consumers to buy products that have a smaller environmental footprint. Extended producer responsibility and take back programs are also intended to stimulate innovation in materials recovery and recycling technology.

There are no federal mandatory take back programs or requirements in the United States, although more than 20 states have some sort of take back law with respect to electronics waste, commonly referred to as E-waste. Many large consumer electronics and office machinery companies have well-known take back initiatives, including Xerox, Dell, Hewlett Packard, IBM, and Sony. Other firms are active, too, in such industries as power tools, automobiles, and cameras.

Through the Carpet America Recovery Effort, the U.S. carpet industry has set a goal of 40% waste diversion for discarded carpeting. Carpet constitutes a significant portion of the U.S. waste stream: About 3.2 million tons were landfilled in 2010, which is a little less than 1.5% of all municipal waste by weight. For large organizations such as schools and office buildings, the cost of carpet disposal in landfills can be significant. In response to growing pressures for carpet waste diversion to both conserve landfill space and reduce costs to consumers, the U.S. carpet industry has initiated a substantial take back initiative. More than 50% of all carpeting manufactured in the United States is made of nylon 6 or nylon 6,6; these materials can be economically reused as new feedstock for carpet (as is the case for nylon 6) or downcycled for automobile components and other plastics (nylon 6,6). Most major carpet manufacturers now sponsor take back programs. However, according to the U.S. Environmental Protection Agency, as of 2010 only 9% of discarded carpeting was being successfully diverted from the waste stream.

Der Grüne Punkt –
Duales System Deutschland GmbH

FIGURE 12.22 The Green dot logo indicates a product that has a take back packaging program in the EU.

Source: The Green Dot, the financing symbol of Duales System Deutschland GmbH.

extent possible or otherwise disposed of in an environmentally responsible manner. One example is the Carpet America Recovery Program (see Box 12.8). Another is Hewlett-Packard's take back of inkjet printer cartridges. Take back programs are becoming increasingly popular for consumer products such as medicine, electronics, and clothing. And, in 2021, Maine became the first state to enact EPR for consumer packaging. Companies must pay for the municipal costs of recycling their consumer packaging material, which should act as an incentive to reduce packaging overall. Legal mandates for EPR like the one in Maine create a market dynamic for dematerialization that did not previously exist.

Recycling Markets

To close the loop at the end of a product or material's life, we can divert the waste stream though recycling. Recycled products and materials are subject to the economic laws of supply and demand, which creates much of the variability we see in recycling today.

Nonhazardous waste and recycling markets can be different for consumers and industry. As consumers, what we see and are most familiar with is recycling as waste management in our communities. However, industries do produce waste, called **scrap**, as part of their manufacturing processes. Ideally, scrap can be reused onsite; if it cannot be used but has value, it may be sold to a scrap dealer for recycling. In addition, entire industries exist to recycle waste from certain kinds of products, like cars. Cars are the most recycled consumer products in the world: Each year, nearly 95% of the cars that reach their end of life are recycled. It is a multistage process that involves removing and processing hazardous as well as salvageable parts from the auto. Because 75% of a car by weight is metal, cars are then crushed, shredded, and put through a process of metals separation. Similar industries exist for refrigerators and freezers.

Nonhazardous waste in the United States generally falls into two categories: municipal solid waste (MSW) and construction and demolition debris. The amount of construction waste is more than twice the amount of municipal solid waste each year, and about 75% of construction and demolition debris is diverted from landfills for recycling and reuse in some manner. About half of all MSW is landfilled.

Municipal recycling programs in the United States are generally not designed to recover materials in complex assembled products. Key exceptions are major appliances and electronic waste. Municipal recycling focuses primarily on items composed of only a single material. The most commonly recycled materials are glass, paper, cardboard, metals, and plastics. Food and yard waste may also be diverted from the waste stream and composted into fertilizer or mulch.

It is challenging to talk about MSW because all waste material is not the same. By weight, cardboard and paper account for 66% of all recycled MSW in the United States, compared to about 5% for plastic. But cardboard and paper simply have more mass than plastic. A better indicator of how "closed" a material loop might be is the **recycling rate**, which is the amount of a material recovered for recycling as a proportion of the total amount of that material generated as waste. In the United States, the recycling rate for cardboard and paper is 66%; for corrugated cardboard, it is an astonishing 97% (U.S. EPA, 2020). The situation for plastic is less impressive. Only 9% of plastic waste is recycled; PET (#1 plastic) and HDPE (#2 plastic) have recycling rates of 29% each (Figure 12.23).

Figure 12.23 shows the U.S. municipal solid waste recycling rates for a variety of products and materials. Lead-acid batteries have almost 100% recycling rates because they are classified as a hazardous waste and must be recovered by federal law. As a consequence, there are many options for collecting them that are easy for consumers. High-quality technical nutrients—like aluminum beverage cans and glass containers—are not at all close to being universally recycled, even though they can be recovered and reprocessed almost indefinitely as feedstocks.

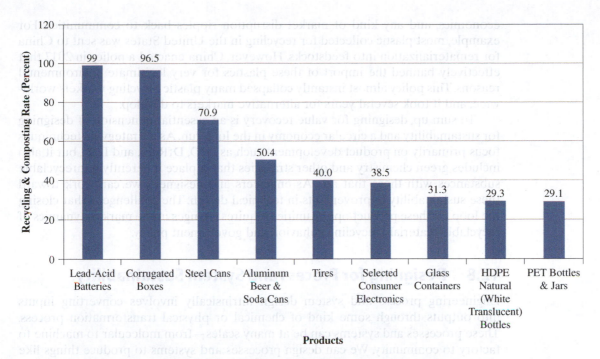

FIGURE 12.23 U.S. recycling rates for various municipal solid waste products, 2018. Recycling rates do not include material that is burned for energy recovery, only items that have been collected for rematerializing.

Source: U.S. EPA (2020).

There are complicated reasons for low recycling rates. One is that recycling may not be easily available in some communities. It can cost more for local governments to handle recyclables than they can get from selling this material as scrap. To avoid losing money through a recycling program, local governments often limit the recycling options available to households. There might be no community recycling at all. Alternatively, there could be recycling, but people are responsible for separating their own materials and taking them to a collection site. There could be curbside recycling, but only for a limited number of materials that must be separated by the household first and then collected on different days. Or there could be recycling, but people have to pay a fee to participate. All of these options present barriers to household participation in recycling.

Recycling has become more cost effective as a public service because the technologies for **single-stream recycling** have become more widely available. In single-stream recycling, materials do not need to be separated and collected by type. Instead, all recyclable items can be combined into a single waste stream and separated at an industrial facility, which then sells the material to scrap markets. These sorting facilities use a wide variety of technologies to screen the waste stream, including lasers, optical scanners, air jets, and magnets.

Nonetheless, local recycling is very sensitive to the cost effectiveness of regional and global markets for recycled material. It costs money to collect, sort, and bundle waste material to sell to scrap dealers. Waste is heavy, so transportation costs must be considered. Scrap dealers expect to make money reselling this material to businesses that will rematerialize it as a feedstock. Those feedstocks are then sold to manufacturers, who expect it to be less expensive than feedstocks made from first-use raw materials. Each type of material has its own value chain and cost economics, and any kind of market disruption ripples back to communities. For example, most plastic collected for recycling in the United States was sent to China for rematerialization into feedstocks. However, China enacted a policy in 2017 that effectively banned the import of these plastics for very legitimate environmental reasons. This policy almost instantly collapsed many plastic recycling markets worldwide, and it took several years for alternative markets to develop.

To sum up, designing for value recovery is an essential dimension of designing for sustainability and a circular economy in the long run. As a strategy, its techniques focus primarily on product development such as DfD, DfRem, and DfR, but it also includes green chemistry and other strategies that replace inherently nonrecyclable substances with those that are. As engineers and designers, we can work toward these sustainability improvements in technical design. The challenge is that closing the loop on these product opportunities requires changes in the market dynamics of recyclable materials, recycling behavior, and government policy.

12.8 Designing for Process and System Sustainability

Engineering process and system design intrinsically involves converting inputs into outputs through some kind of chemical or physical transformation process. These processes and systems can be at many scales—from molecular to machine to factory to community. We can design processes and systems to produce things like nanosilver, chocolate bars, chemicals, and cars. The design principles, methods, and tools for process and system sustainability are shown in Table 12.2 in the materials processing and manufacturing stages of the product life cycle. These approaches are not limited to these two stages of the life cycle, however; they easily apply to other contexts for process and system design.

The engineering expertise required for this kind of design is highly specialized. Nevertheless, there are basic principles for designing more sustainable systems and processes: conserve energy and resources, avoid waste, and be as efficient as possible. Because avoiding hazardous substances is a sustainability directive for all types of design, this principle also applies to process and systems design. The collective insights of green engineering and green chemistry offer guidance for our design decisions.

Two types of process and systems analysis can help us apply green design principles to industrial production. The first is industrial ecology (see Chapter 13). A number of concepts and methods in this approach allow us to identify opportunities and leverage points within a system to make it more sustainable. These include industrial metabolism, type II systems, material flow methods, and embodied energy.

In practice, industrial ecology focuses on ways we can lower resource requirements by reducing waste (the green engineering principle to maximize mass, energy, space, and time efficiency), obtaining energy and materials locally, and converting waste streams into closed loops. An example is for a company to reuse its own scrap waste onsite for an industrial process. Installing heat recovery systems to capture waste heat closes an energy loop. ***Industrial energy auditing*** is a practice that can be used to identify opportunities for saving energy at different scales, such as process, product line, and factory.

Another example of applied industrial ecology is the way electronic components are cleaned during their manufacturing process. Conventional approaches often use solvents that evaporate into the air. A more sustainable approach is to close the solvent loop by capturing and distilling solvent emissions to reuse the fluid, which also improves air quality. The company that makes Tencel™, a textile fiber, has made closed-loop production an art form (see Box 12.9).

BOX 12.9 Closing Process Loops in Textile Manufacturing

Reports in the press on the sustainability of the global textile industry have been consistently negative since Elizabeth Cline's *Overdressed: The Shockingly High Cost of Fast Fashion* was published in 2012. Cline's exposé, as well as the widely publicized Bangladesh Rana Plaza disaster in 2013, illustrated a suite of horrors—toxic chemicals, widespread violations of environmental regulations, and flagrant disregard for human life.

The economic role of textiles and apparel is significant. Global textile mill revenues are over $800 billion a year; downstream retail sales of clothing alone exceed $1 trillion. Projected global growth in apparel is eye-popping. Between 2015 and 2030, total garment consumption is expected to increase by the equivalent of half a trillion T-shirts. However, only 20% of post-consumer clothing is reused or recycled in some manner worldwide. Rates of waste stream diversion in the EU and United States are lower than 20%. The production of textiles—particularly clothing and household linens—can represent hundreds of discrete industrial processes and dozens of stops along a globally diffuse supply chain. Almost anything that can be done to close loops in these systems improves sustainability.

The Lenzing Group in Austria manufactures the textile fiber *lyocell*, which consumers know by the brand name Tencel™. Lyocell is unusual because it is a synthetic fiber made from plant cellulose; it isn't a natural fiber like cotton or wool. It isn't a purely synthetic polymer either, because it is biobased and not petroleum-based. Lyocell is almost identical to cotton as fiber, as a finished textile, and in terms of its physical properties (Figure 12.24). Lenzing has a closed-loop process for manufacturing lyocell and an integrated supply chain for cellulose. The result is a fiber with dramatically improved sustainability not just in its manufacture but also over its life cycle. One independent life cycle assessment of Tencel™ found that it had a significantly lower environmental impact than cotton and performed better than polyester, the two fibers that together account for 80% of all textiles today (Shen et al., 2010).

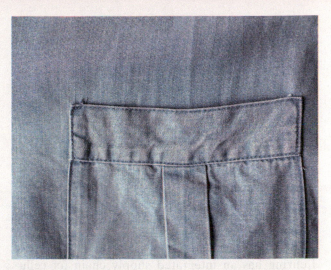

FIGURE 12.24 Lyocell is a biobased synthetic fiber often used to make denim. It has the same look and feel as cotton.

Source: iStockPhoto.com/Fascinadora.

Because it is synthesized from plant cellulose, lyocell is classified as a viscose fiber. It is best compared to the process for making rayon, the original viscose fiber that is still manufactured today. To convert cellulose into usable viscose fiber, it must be chemically dissolved to form a liquid and then regenerated as a solid, continuous filament. This is done through precipitation, wet extrusion, and spinning processes. The finished filaments are spun into yarn, which is then woven into fabric.

The process for making rayon viscose is nasty. It uses highly toxic carbon disulfide (CS_2) as a solvent as well as caustic sodium hydroxide (NaOH) and sulfuric acid (H_2SO_4) in various baths for converting and regenerating the cellulose. The dissolved cellulose is a chemical derivative, not the same compound as the original plant material. The regenerated fibers must be desulfurized and bleached in a water-intensive process. Liquids from the viscose process *can* be recycled as either CS_2 or H_2SO_4 and reused again within the process. But about half of any CS_2 that is not recovered volatilizes into the atmosphere.

(Continued)

BOX 12.9 Closing Process Loops in Textile Manufacturing *(Continued)*

The lyocell process for making viscose rayon is a radical innovation. Cellulose is dissolved directly—no derivative is created—and respun into a filament without caustic or acidic chemicals. The amino oxide NMMO (*N*-methylmorpholine *N*-oxide) is used in all stages of the viscose process. It is nontoxic and nonhazardous, and 99% of this liquid is recovered and reused in a closed-loop process. The 1% that is lost as waste is essentially benign. The use of NMMO reduces the number of processing steps and the energy intensity that is required to produce viscose. It also eliminates the need for desulfurizing and bleaching the regenerated filament. As a textile fiber, lyocell fully biodegrades and can be recycled.

In addition to these chemical process changes, Lenzing has an integrated supply chain for cellulose from sustainably managed forests to its lyocell factory gate in Austria. Wood pulping (to make the

feedstock for lyocell) is done at the same industrial complex that makes the fiber. This integration has a variety of life cycle benefits, including significantly lower energy and carbon footprints. Residual wood waste is used onsite for biomass energy production, and there is extensive heat recovery and reuse. Energy at the Lenzing lyocell plant is therefore 100% renewable and almost carbon free.

Tencel™ has received numerous third-party sustainability certifications and awards, including the U.S. Department of Agriculture's certified biobased product, the EU Ecolabel, TÜV Austria Belgium NV certification for biodegradability and compostability, and the European Award for the Environment (for resource efficiency and low ecological impacts). Lenzing's lyocell also meets Standard 100 by Oeko-Tex, which certifies that the fiber is harmless to human health.

The second type of process and systems analysis that informs process and system design is lean manufacturing. ***Lean manufacturing*** is a business operations strategy that focuses on continuous quality improvement of production systems. The purpose of lean manufacturing is to maximize the efficiency of production, and it directly embodies the green engineering principle to maximize mass, energy, space, and time efficiency. Lean manufacturing methods and techniques are based on identifying the "eight wastes of lean" (Figure 12.25) and reducing those wastes.

1. DEFECTS
Waste from a product that does not meet specifications

2. OVERPRODUCTION
Waste from making more products than can be sold

3. WAITING
Waste from idle time between processing steps

4. UNUSED TALENT
Waste from not fully benefiting from people's skills, talent, and knowledge

5. TRANSPORTATION
Wasted time, cost, and resources from moving products/materials unnecessarily

6. INVENTORY
Waste from products and materials waiting for their value-added to be used

7. MOTION
Wasted time and effort from unneeded movements by people

8. EXTRA PROCESSING
Waste related to overdesigning products or unneeded processing operations

FIGURE 12.25 The eight wastes of lean manufacturing.

This is accomplished by detailed process flow mapping, statistical analysis of various system processes and operations, and appropriate redesign of processes and operations. An interesting technique of lean manufacturing management is the *kanban* method of production scheduling; it relies on the green engineering principle of output-pulled instead of input-pushed.

12.9 People-Centered Design

The philosophy of green engineering and other sustainable design approaches calls for engaging people, stakeholders, and communities in the design and problem-solving processes. "People-centered design" adds layers of complexity to engineering. As a consequence, tools are available to help us consider both technical and nontechnical design factors. These tools address the selection, design, and implementation of engineering design solutions. For example, *capacity factor analysis* is a tool that assists with the development of sustainable water, sanitation, and household energy solutions by evaluating a community's capacity to manage its own technology (Henriques and Louis, 2011).

Because of the need to integrate social considerations into both the *product* and the *process* of developing solutions, engineers and designers are increasingly turning to participatory and community-based design tools. *Participatory design* focuses on directly involving individuals who use the technology or information as co-designers, and on enabling those affected by a design to drive the collaborative design process (Simonsen and Robertson, 2012). *Community-based design* can be thought of as participatory design at the community scale (DiSalvo, Clement, and Pipek, 2012). In the context of international development, *participatory development* is a bottom-up, people-centered approach aimed at cultivating the full potential of people, especially the underprivileged, at the grassroots level (Forsyth, 2005).

IDEO, a design and innovation consulting firm, uses *design thinking* as a tool for problem solving and people-centered design. In its guide for educators, IDEO (2011a) defines design thinking as being human-centered and focused on understanding the needs and motivations of key stakeholders. It is a collaborative methodology that requires the participation of multiple individuals in order to gain diverse perspectives about a particular problem and the possible solution space. It is experimental because the process encourages exploring and trying out new ideas or approaches. As seen in Table 12.3, IDEO structures its *design process*—which

TABLE 12.3 IDEO's Five Design Process Phases

IDEO PHASE	EXAMPLE METHOD OR APPROACH
1. Discovery	Defining and researching (including through fieldwork) the problem and gathering inspiration (e.g., contextual immersion, or learning from individuals, groups, experts, peers, etc.)
2. Interpretation	Telling stories that capture learning, finding themes, and framing opportunities
3. Ideation	Idea generation and refinement
4. Experimentation	Prototyping and testing
5. Evolution	Integrating feedback, defining success

Source: Based on IDEO (2011a, b).

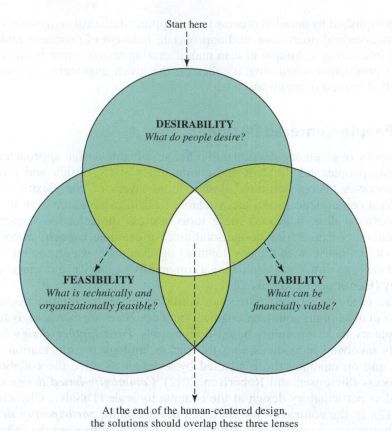

FIGURE 12.26 IDEO's three core values related to sustainable design. The IDEO design process requires that we create solutions that meet people's desires, that are technically and organizationally feasible, and that are economically viable.

Source: Based on IDEO (2011a, b).

operationalizes design thinking—into five key phases: discovery, interpretation, ideation, experimentation, and evolution. IDEO holds three key values related to sustainable design: human desirability, technical feasibility, and technical viability (Figure 12.26).

Although the IDEO process is similar to other engineering design and system analysis processes, it is absolutely distinctive for its focus on creating collaborative environments that facilitate creativity and innovation. Given the essential role that social factors play in engineering design for development, the approach adapts well to the contexts of low-income communities. IDEO partnered with the Bill & Melinda Gates Foundation to adapt its methodology, which has been used successfully by large corporations, to develop a toolkit for nonprofit organizations, social enterprises, and individuals who create solutions for those living on less than $2 per day.

The methodology of the toolkit begins by first defining a specific design challenge and then following through with an exploration of three themes known as Hear, Create, and Deliver (Table 12.4). For each theme, the toolkit describes a number of techniques that can be used to accomplish the goals of the theme. The toolkit is a powerful resource for engineers who want to incorporate significant social considerations into their design.

TABLE 12.4 IDEO Toolkit Design Method

THEME	DESCRIPTION	STEPS
Hear	During the Hear phase, your design team will collect stories and inspiration from people. You will prepare for and conduct field research.	1. Identify a design challenge 2. Recognize existing knowledge 3. Identify people to speak with 4. Choose research methods (e.g., expert interview, in context immersion, etc.) 5. Scenario-based questions (or Sacrificial Concepts) 6. Develop your mindset
Create	In the Create phase, you will work together in a workshop format to translate what you heard from people into frameworks, opportunities, solutions, and prototypes. During this phase, you will move together from concrete to more abstract thinking in identifying themes and opportunities, and then back to the concrete with solutions and prototypes.	1. Develop the approach (e.g., participatory co-design, emphatic design, etc.) 2. Share stories 3. Identify patterns (e.g., create frameworks) 4. Create opportunity areas 5. Brainstorm new solutions 6. Make ideas real (i.e., prototyping) 7. Gather feedback
Deliver	In the Deliver phase, you will begin to realize your solutions through rapid revenue and cost modeling, capability assessment, and implementation planning. This will help you launch new solutions into the world.	1. Develop a sustainable revenue model 2. Identify capabilities required for delivering solutions 3. Plan a pipeline of solutions 4. Create an implementation timeline 5. Plan mini-pilots and iterations 6. Create a learning plan (e.g., track indicators, evaluate outcomes, etc.)

Source: Based on IDEO (2011a, b).

12.10 Summary

When defined at the level of sustainable development, sustainable design is the design of products, processes, or systems that enable people to enjoy long, healthy, and creative lives while protecting and preserving the natural resources that allow us to do so now and in the future. We can compare alternative designs and their impacts on sustainable development broadly by using sustainability indexes like the IPAT equation or the ecological footprint (see Chapter 1).

In this chapter, we expanded our understanding of designing for sustainability by exploring how it can be done in practice as technical design. We identified unifying philosophies of applied sustainable design, which include life cycle thinking, avoiding the use of hazardous substances, using renewable resources, and limiting waste. We surveyed the principles of green chemistry and green engineering that serve as guides for design decisions. Methods currently used for more sustainable buildings, materials development, product and process design, and end-of-life value recovery were introduced. Many of these methods include their own evaluation techniques, and life cycle analysis (see Chapter 14) is a tool that can be used to quantify the environmental impacts and benefits of almost all strategies for designing for sustainability. People-centered tools, like participatory design, enable engineers to engage individuals and communities and more fully integrate social and economic considerations into the design process.

References

Abraham, M. A., and Nguyen, N. (2003). "Green engineering: Defining the principles"—Results from the Sandestin Conference. *Environmental Progress* 22:233–236.

Abraham, M. (Ed.). (2006). *Sustainability Science and Engineering: Defining the Principles*. San Diego: Elsevier.

Anastas, P. T., and Warner, J. C. (1998). *Green Chemistry: Theory and Practice*. Oxford, England: Oxford University Press.

Anastas, P., and Zimmerman, J. (2003, March 1). "Design through the twelve principles of green engineering." *Environmental Science and Technology* 37(5):94A–101A.

DiSalvo, C., Clement, A., and Pipek, V. (2012). "Communities: Participatory design for, with, and by communities." In J. Simonsen and T. Robertson (eds.), *Routledge International Handbook of Participatory Design*. New York: Routledge.

Forsyth, T. (2005). *Encyclopedia of International Development*. New York: Routledge.

Heine, L., and Whittaker, M. (2021). "Determining the true value of a sustainable chemical technology." In T. J. Clarke and A. S. Pasternak (eds.), *How to Commercialize Chemical Technologies for a Sustainable Future*, pp. 31–54. New York: Wiley.

Henriques, J. J., and Louis, G. E. (2011). "A decision model for selecting sustainable drinking water supply and greywater reuse systems for developing communities with a case study in Cimahi, Indonesia." *Journal of Environmental Management* 92(1):222.

IDEO. (2011a). Design Thinking for Educators, v. 1.

IDEO. (2011b). Human Centered Design: Toolkit. IDEO and Bill & Melinda Gates Foundation.

Interface, Inc. (undated). Three Lenses of Health and Materials: Green Chemistry, Circular Economy, and Embodied Carbon. www.interface.com/US/en-US/sustainability/product-transparency/three-lenses-of-health-and-materials-en_US.

McDonough, W., and Braungart, M. (2002). *Cradle-to-Cradle: Remaking the Way We Make Things*. New York: North Point Press.

OECD. (2012, October). *Sustainable Materials Management: Making Better Use of Resources*. Paris: Organisation for Economic Cooperation and Development.

Shen, L., et al. (2010). "Environmental impact assessment of man-made cellulose fibres." *Resources, Conservation and Recycling* 55(2):260–274. doi.org/10.1016/j.resconrec.2010.10.001.

Simonsen, J., and Robertson, T. (2012). "Preface." *Routledge International Handbook of Participatory Design*. New York: Routledge.

U.S. EPA. (2020). Advancing Sustainable Materials Management: 2018 Fact Sheet (Washington, DC: U.S. Environmental Protection Agency).

U.S. ITC. (2012). *Remanufactured Goods: An Overview of the U.S. and Global Industries, Markets, and Trade*. Washington, DC: U.S. International Trade Commission.

Key Concepts

Sustainability science	Functional design
Human–environment interface	Form design
Design	Ecological design

Technical nutrients	Planned obsolescence
Green engineering	Moore's law
Transparency	Designing for value recovery
Precautionary principle	Capital goods
Green chemistry	Design for X
Biomimicry	Remanufacturing
Biobased feedstock	Design for disassembly
Biodegradable	Design for remanufacturing
Photolysis	Design for recycling
Hydrolysis	Extended producer responsibility
Ultimate biodegradation	Take back program
Primary biodegradation	Scrap
Greenwashing	Recycling rate
Detoxification	Single-stream recycling
Rematerialized feedstock	Industrial energy auditing
Dematerialization	Lean manufacturing
Product simplification	Capacity factor analysis
Overdesign	Participatory design
Design lifetime	Community-based design
Commercial lifetime	Participatory development

Problems

12-1 Explain form, functional design, and materials selection in the context of a BIC Cristal ballpoint stick pen.

12-2 Suppose you want to design a house. Compare and contrast ecological design, green engineering, and "chemistry, carbon, and circularity" as approaches to the sustainable design of this house. How might the form, functional design, and materials of the house differ for each philosophy?

12-3 From a sustainability perspective, which is better: a biobased feedstock or a biodegradable material? Why?

12-4 What is greenwashing, and why is it bad?

12-5 Review the 12 principles of green engineering. Which three are your favorites? Why?

12-6 Reflect on the principles of green engineering and green chemistry. Do you think it's easy to be green? What are common obstacles to greener engineering and design?

12-7 What strategies can you use to reduce or eliminate the use of hazardous substances through design?

12-8 What is dematerialization, and how can we do it? As a design technique, does it create other advantages besides sustainability?

12-9 Consider Example 12.1 about dematerializing your fizzy water.
   a. How would the results change if your family drank 8 L of club soda a week instead of 6 L?
   b. Use the same assumptions as in Example 12.1. When would your investment in the Soda Club machine break even in terms of cost savings from no longer purchasing cases of fizzy water?

12-10 What is Design for X (DfX)?

12-11 Briefly describe DfD, DfRem, and DfR. What techniques are used for each of these design strategies?

12-12 What kind of recycling is available in your community? What limits recycling in your area?

12-13 The Institute of Scrap Recycling Industries gives awards for innovations in design for recycling. Look up recent award winners. Select one and summarize the DfR techniques or approaches that the winner used.

12-14 Identify the design characteristics of more sustainable processes and systems.

12-15 List and discuss three ways that people and communities can be effectively involved in engineering design, and why we would want to do so.

Team-Based Problem Solving

12-16 You work for a U.S.-based clothing manufacturer, Denim Inc. This company is a profitable niche producer of high-end women's jeans that sell for about $185 per pair. Denim Inc. has become competitive by branding itself as "made in America" and by having great-fitting jeans, but it believes this strategy will no longer be enough to protect its market share. The company has pledged to create a zero waste take back program for all of its jeans sold in the United States. In other words, these jeans would create no post-consumer waste if everyone who buys them returns them to the manufacturer. Denim Inc. sells about 5 million pairs of jeans in the United States each year. Design the logistics of a take back program and develop a strategy for these returned jeans so that they create no waste stream.

12-17 Obtain a broken or cheap semi-complex assembled consumer product from a thrift store, like a toaster, electric drill, hair dryer, or DVD player. Assess its form and functionality. Carefully deconstruct the product to the smallest component that you can using guidance from Active Learning Exercise 5.2 in Chapter 5. Fully document the assembly sequence (which you will be doing in reverse), how connected parts are attached to one another (e.g., solder, glue, screws, rivets, snap fits), the number of each kind of part, the material of each part, and their mass. (A bill of materials and a parts tree or assembly diagram is a useful way to keep track of these details.) Redesign the product to be more sustainable. Fully document your approach and design decision-making process. Estimate the environmental benefits of your redesign.

12-18 Lithium batteries are an essential technology for adopting renewable energy on a widespread basis. These batteries are needed in electric vehicles (EVs) and for large-scale power grid electricity storage. Within a decade, the mass production of lithium batteries for EVs will escalate. Analyze the full material life cycle for lithium from natural resource extraction to its end of life as a vehicle battery. What environmental impacts are associated with lithium at each stage of the life cycle? What issues and challenges are associated with lithium battery recycling? Propose product, process, and value recovery design strategies to make this product more sustainable over time. What environmental trade-offs emerge in your design choices?

12-19 Fleas are amazing insects; relative to their body size, they can leap extraordinary distances both horizontally and vertically. They also have unusual sensing mechanisms. Conduct research on fleas and identify all of their biological processes that could be useful from a design perspective. Then classify each of those processes according to the biomimicry taxonomy. Identify different kinds of products and technologies that could benefit from flea-inspired biomimetic design.

## CHAPTER 13

Industrial Ecology

FIGURE 13.1 The ways we produce, use, and dispose of products all contribute to their environmental impacts.

Source: Andy Singer Cartoons, www.andysinger.com/.

The individual serves the industrial system . . . by consuming its products.

—JOHN KENNETH GALBRAITH

GOALS

THE GOAL OF THIS CHAPTER is to introduce the basic principles of industrial ecology by exploring the relationships between industrial and ecological systems.

OBJECTIVES

At the conclusion of this chapter, you should be able to:

13.1 Define the similarities and differences between natural ecological systems and industrial ecological systems.

13.2 Develop frameworks for conceptualizing complex materials balance problems.

13.3 Perform a simple materials flow analysis.

13.4 Perform a simple embodied energy analysis.

Introduction

Ecology is the scientific field devoted to the study of the relationships that organisms have with each other and the natural environment in which they exist. In previous chapters, we have learned that a significant part of sustainable development efforts is concerned with minimizing the environmental impacts associated with activities. We have also learned about the concept of carrying capacity, which specifies the sustainable rates of resource consumption and waste discharge within an ecosystem. In ecology studies, both resource consumption and waste discharge are generally linked to the population of organisms in a particular ecosystem. This has led some researchers to conclude that the primary driver of environmental impacts leading to unsustainable development is population growth. However, others have identified various other factors that contribute to environmental impacts. Current industrial systems rely on the extraction, processing, manufacturing, transportation, and distribution of natural resources in the form of products. These different processes and stages rely on technologies that generate waste and emissions. Understanding the interrelationships among people, products, technology, and the environment is necessary for developing and implementing sustainable practices.

13.1   Industrial Metabolism

We have explored the role of technology in impacts as well as specific ways to quantify these impacts. However, sustainable engineering requires a change in the way technology is used and, in general, a minimization of impacts from industrial processes.

Industrial ecology proposes the development of technological systems that have minimal impacts on the surrounding "natural system" by mimicking the way nature handles materials and energy transformations. "The concept requires that an industrial system be viewed not in isolation from its surrounding systems but in concert with them" (Graedel and Allenby, 2003).

Since biological organisms tend to adopt metabolic patterns that keep them in balance with their environment, we can use the analogue of biological systems to study and improve the relationship of industrial systems with the environment. To this end, we can assume that the analogue of the biological organism is the industrial process or processes that produce a particular product or family of products.

Both natural and industrial systems have cycles of energy and nutrients or materials. Natural systems utilize a complex system of feedback mechanisms that initiate appropriate responses when certain limits are reached. This is what industrial ecology aims to do for industrial systems. The field of industrial ecology has recently expanded to address consumption activities with proactive decision making.

Industrial systems are characterized by flows of energy and materials. These flows form the basis for design and analysis in industrial ecology. The production and consumption activities of biological systems can also be described in terms of these flows. Three types of ecosystems, as detailed in the following sections, have been identified based on the way materials and energy flow through them.

FIGURE 13.2 A type I system.

Source: Based on Allenby, B.R. (1992, March). "Industrial Ecology: The Materials Scientist in an Environmentally Constrained World," *MRS Bulletin* 17, no. 3: 46–51.

13.1.1 Type I System

This is a linear and open system that relies completely on materials and energy from outside the system (Figure 13.2). Materials and energy enter and leave as either products or wastes. No recycle or reuse occurs within this system. The supply of materials and energy is typically not infinite, and the capacity for the surrounding system to absorb waste is also finite; as a result, the system is ultimately unsustainable.

13.1.2 Type II System

In a type II system (Figure 13.3), a quasi-cyclic materials flow takes place. Some material is recycled or reused within the system. The rest leaves the system as waste, so some resources are still required externally. This represents the current state of industrial systems where some recycle and reuse occur.

13.1.3 Type III System

The type III system refers to a completely integrated closed system in which only some energy crosses its boundary (Figure 13.4). All materials are recycled and reused within the system. This represents an ideal sustainable state and the ultimate goal of industrial ecology where the effluents from one process serve as the raw material for another.

If we think of human beings as existing in an ecosystem with Earth's natural systems, we can infer that a type I system, with its infinite available resources and infinite waste sink, is not representative of our interaction with Earth's system.

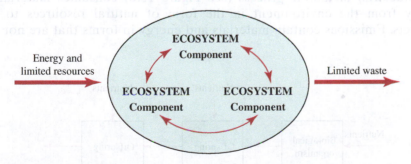

FIGURE 13.3 A type II system.

Source: Based on Allenby, B.R. (1992, March). "Industrial Ecology: The Materials Scientist in an Environmentally Constrained World," *MRS Bulletin* 17, no. 3: 46–51.

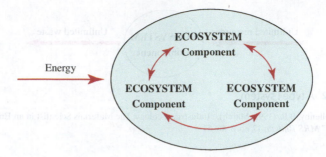

FIGURE 13.4 A type III system.

Source: Based on Allenby, B.R. (1992, March). "Industrial Ecology: The Materials Scientist in an Environmentally Constrained World," *MRS Bulletin* 17, no. 3: 46–51.

A type III system represents an ideal that can be aspired to, but it will likely take a very long time to achieve given the current state of technology. A significant quantity of materials that flow into industrial systems cannot be recycled easily, unlike most naturally occurring materials that may be easily recycled.

A type II system is, however, feasible with changes to our industrial process systems' overall consumption patterns. The objective should be to significantly reduce the extraction of new resources from the Earth's natural system, maximizing the recycling/reuse of resources, and thereby significantly reducing the amount of waste that is ultimately generated. This model can be sustainable if the extraction rates are limited to natural rates of replenishment and the waste discharge rates are limited to the natural rates of waste absorption.

13.1.4 Biological Metabolism

In the metabolic process of biological organisms (Figure 13.5), the organisms take in nutrients from the environment and utilize them to grow and reproduce while generating waste. Ultimately, through the waste generated over the lifetime of the organism and from the eventual death of the organism, the nutrients taken from the environment end up back in the environment.

13.1.5 Industrial Metabolism

The industrial metabolic process (see Figure 13.6) consumes materials and energy from the environment in the form of natural resources to make products. Emissions contain materials and energy in forms that are not easily

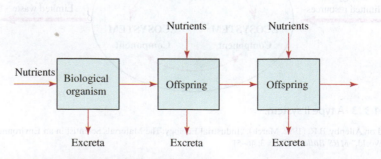

FIGURE 13.5 Metabolism in a biological organism.

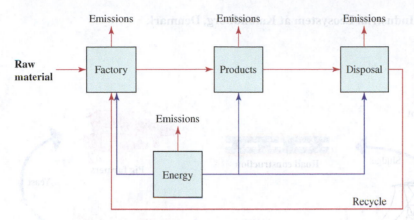

FIGURE 13.6 Metabolism in an industrial organism.

retrievable. The use and disposal of the products also consume energy and generate emissions. A sustainable industrial ecological system will minimize the emissions generated and increase the recovery of materials and energy within the system.

ACTIVE LEARNING EXERCISE 13.1 Ecological System Models in Society

Discuss your view on a society modeled after a type III ecological system. What would this society be like? What advances in science and technology will be required? What changes to current social and economic models will be necessary?

13.2   Eco-Industrial Parks (Industrial Symbiosis)

One of the earliest applications of an ecological approach to the design of industrial systems was the development of industrial ecosystems, also known as eco-industrial parks. This concept is based on the recognition that industrial activities in the modern world rely on the interaction and interdependence of many industries. An *eco-industrial park* is a community of manufacturing and service businesses that collaborate in the management of environmental and resource issues for improved environmental and economic performance. A symbiotic relationship is developed between participants wherein materials, energy, and waste resources are exchanged, thereby minimizing the need for a new influx of materials into the park and minimizing the total waste generated by the park. By adopting principles of industrial ecology and pollution prevention, this model can provide one or more of the following benefits:

• Reduced use of virgin materials
• Reduction in pollution

Industrial Ecosystem at Kalundborg, Denmark

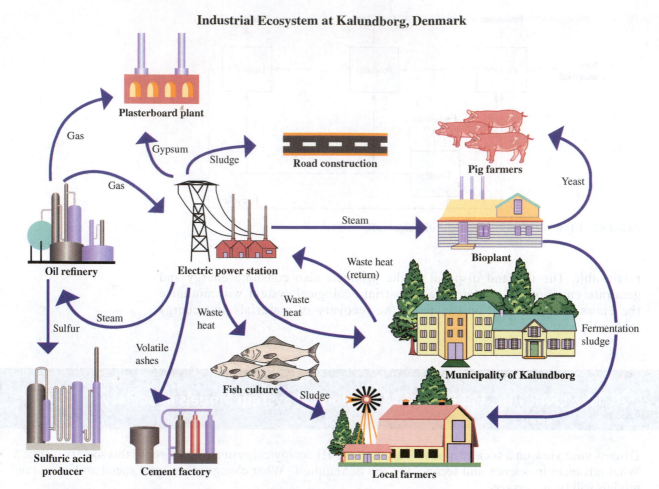

FIGURE 13.7 Industrial symbioses in Kalundborg, Denmark.

Source: Based on Pollution Issues, Industrial Ecology, www.pollutionissues.com/Ho-Li/Industrial-Ecology.html.

- Increased energy efficiency
- Reduced volume of waste products that require disposal
- Increase in the amount and types of process outputs that have market value

A good example is the industrial ecosystem of Kalundborg, Denmark (Figure 13.7). The potential cumulative environmental impact from each industry is minimized by treating the waste products from one operation as an input resource for others.

ACTIVE LEARNING EXERCISE 13.2 Industrial Symbiosis

Can you think of any potential environmentally beneficial industrial symbiotic relationships that can be developed in your community?

BOX 13.1 Case Study: Kalundborg Eco-Industrial Park

One of the best-known examples of industrial symbiosis is the Kalundborg Eco-Industrial Park in Denmark. This park, one of the first demonstrated models for industrial symbiosis, was the result of a collaboration between many industries and businesses. With the objective of minimizing overall waste, industries, local farmers, small businesses, and a municipal government agreed to accept one another's waste materials and energy streams as useful inputs to their processes.

A power company supplies residual steam to a refinery and, in exchange, receives refinery gas that used to be combusted as waste. The power plant burns the refinery gas to generate electricity and steam. It sends excess steam to a fish farm that it operates, to a district heating system serving 3,500 homes, and to the pharmaceuticals bioplant. Sludge from the fish farm and bioplant processes becomes fertilizer for nearby farms. The power plant sends fly ash to a cement company, while gypsum produced by the power plant's desulfurization process goes to a company that produces gypsum wallboard. The refinery removes sulfur from its natural gas and sells it to a sulfuric acid manufacturer.

Some of the reported results include:

- Annual CO_2 emissions are reduced by 240,000 tons.
- 3 million cubic meters of water are saved through recycling and reuse.
- 30,000 tons of straw are converted to 5.4 million liters of ethanol.
- 150,000 tons of yeast replace 70% of soy protein in traditional feed mix for more than 800,000 pigs.
- Recycling of 150,000 tons of gypsum from the desulfurization of flue gas (SO_2) replaces the import of natural gypsum ($CaSO_4$).

13.3   Materials Flow Analysis (MFA)

Industrial ecology is concerned with the development and implementation of solutions based on the analysis of industrial systems. The interconnection of industries with their flows of materials and energy means that an individual component cannot be tracked without the tools for systems analysis. One of these tools is *materials flow analysis*, which is used to evaluate the metabolism of anthropogenic systems. It is a systematic assessment of the flows of materials within the boundaries of a system in space and time. Materials flow analysis relies on the law of conservation of mass and is performed by simple mass balances comparing inputs, outputs, depletions, and accumulations within systems and subsystems. Some relevant terms used in materials flow analysis are:

- **Materials:** Substances as well as goods.
- **Process:** The transport, transformation, or storage of materials.
- **Reservoir:** A system that holds materials. It can be a source, from which materials come; a sink, into which materials go; or both.
- **Stocks:** The quantity of materials held in reservoirs within a system.
- **Flows/Fluxes:** The ratio of the mass per time that passes through a system boundary is termed the *flow*, while the *flux* is the flow per cross section.

Consider the reservoir system shown in Figure 13.8, where

$\dot{m}_{i,1}$ and $\dot{m}_{i,2}$ are mass flows into the reservoir

$\dot{m}_{o,1}$ and $\dot{m}_{o,2}$ are mass flows out of the reservoir

$\dot{m}_s$ is the rate of change of stock in the reservoir

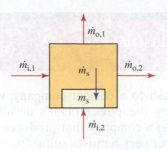

FIGURE 13.8 A materials reservoir system showing input and output flows.

Let us recall the general mass balance equation:

$$\text{Input} - \text{output} = \text{accumulation} \tag{13.1}$$

For a steady state system, accumulation is zero and

$$\text{Input} = \text{output} \tag{13.2}$$

but steady state systems are rare.

We can solve the mass balance for the reservoir:

$$\sum \dot{m}_i - \sum \dot{m}_o = \dot{m}_s \tag{13.3}$$

By definition,

$$\dot{m}_s = \frac{\Delta m_s}{dt} \tag{13.4}$$

where Δm_s is the change in quantity of the stock in the reservoir.

Combining the two equations, we have

$$\Delta m_s = \left(\sum \dot{m}_i - \sum \dot{m}_o \right) dt \tag{13.5}$$

The change in the stock in a reservoir is given by the expression

$$\Delta m_s = \int_{t_1}^{t_2} \left(\sum \dot{m}_i - \sum \dot{m}_o \right) dt \tag{13.6}$$

13.3.1 Efficiencies in Mass Flow Systems

One of the applications of mass flow analysis in industrial systems is the ability to track how much of the original materials going into the process actually end up as part of the desired product stream. Applying the principles of conservation of mass to a system, we can see what portion of the raw materials entering the system becomes the product, as well as what fraction ends up in the waste streams. The efficiency of the process is defined as the ratio of the mass in the desirable product to the total mass entering the system. For a 100% efficient system, no emissions are generated, and all of the raw material is converted to useful product. This is probably not a realistic goal for industrial systems. A more feasible aspiration is for industrial systems to mimic a type II ecosystem in which limited waste is generated across the entire industrial system.

Figure 13.9 illustrates a typical industrial materials flow system of a substance, where

m_E is the mass of the substance in raw materials extracted and processed

m_M is the mass of the substance in materials sent to product manufacturing

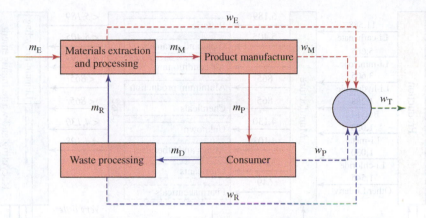

FIGURE 13.9 Materials flow system of a substance in an industrial system.

m_P is the mass of the substance in the consumer product

m_D is the mass of the substance recovered after consumer use

m_R is the mass of the substance in recycled material

w_E is the mass of the substance in the waste from extraction and processing

w_M is the mass of the substance in the waste from product manufacturing

w_P is the mass of the substance of the unrecovered waste after consumer use

w_R is the mass of the substance of the waste from recycling

w_T is the total mass of the substance in all waste streams

The efficiencies of each stage within the system can be defined as

$$\text{Extraction efficiency: } E_{\text{eff}} = \frac{m_M}{m_E + m_R} = \frac{m_M}{m_M + w_E} \qquad (13.7)$$

$$\text{Manufacturing efficiency: } M_{\text{eff}} = \frac{m_P}{m_M} = \frac{m_P}{m_P + w_M} \qquad (13.8)$$

$$\text{Recovery efficiency: } D_{\text{eff}} = \frac{m_D}{m_P} = \frac{m_D}{m_P + w_P} \qquad (13.9)$$

$$\text{Recycling efficiency: } R_{\text{eff}} = \frac{m_R}{m_D} = \frac{m_R}{m_R + w_R} \qquad (13.10)$$

In each stage, reduction of the mass in the waste stream increases the efficiency of that stage.

The system efficiency can be calculated as

$$S_{\text{eff}} = E_{\text{eff}} \times M_{\text{eff}} \times D_{\text{eff}} \times R_{\text{eff}} \qquad (13.11)$$

$$S_{\text{eff}} = \frac{m_M}{m_E + m_R} \times \frac{m_P}{m_M} \times \frac{m_D}{m_P} \times \frac{m_R}{m_D} = \frac{m_R}{m_E + m_R} \qquad (13.12)$$

The most efficient system is the one in which the total waste emitted from the system is minimal.

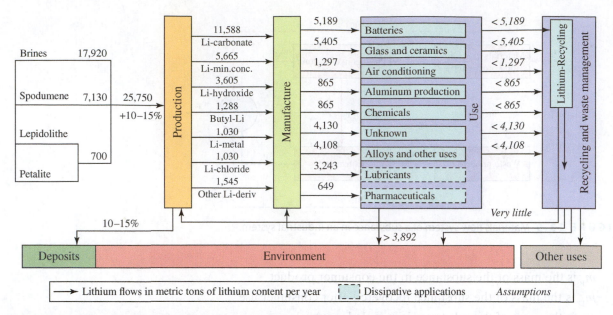

FIGURE 13.10 Material flow model of lithium in 2007.

Source: Reprinted from Ziemann, S., Weil, M., and Schebek, L. (June 2012). "Tracing the Fate of Lithium—The Development of a Material Flow Model." *Resources, Conservation and Recycling* 63:26–34. Copyright 2012, with permission from Elsevier.

The *reuse factor* is defined as the amount of recycled material that can be recovered and used in the process to make a product. This can be calculated as

$$\psi = \frac{m_R}{m_M} \tag{13.13}$$

The global flow of lithium in 2007 is depicted in Figure 13.10. A close inspection indicates that a significant fraction of the lithium is not recycled. Thus the global material flow of lithium is very inefficient. Identifying and quantifying these "losses" provide a basis for problem solving with a goal of reducing impacts on the environment by increasing the efficient use of resources. Wastes that may go unnoticed in conventional monitoring systems can be identified using materials flow analysis. Because of the large values of the masses that are sometimes encountered in materials flow analysis, the unit prefixes listed in Table 13.1 are used to make assessments easier.

TABLE 13.1 Unit prefixes used in materials flow analysis

PREFIX	ABBREVIATION	MULTIPLIER VALUE
kilo	k	10^3
mega	M	10^6
giga	G	10^9
tera	T	10^{12}
peta	P	10^{15}
exa	E	10^{18}
zetta	Z	10^{21}
yotta	Y	10^{24}

13.3.2 Constructing a Materials Flow System

The boundaries of a materials flow system can be drawn across different spatial levels, from a single industrial facility to national and even global levels.

All materials flow analyses entail the construction of a materials budget, which requires:

1. Determining the material(s) to be evaluated.
2. Identifying all relevant reservoirs and flow streams.
3. Quantifying the contents of the reservoir and the magnitudes of the flows. This information is obtained from data if available, taking measurements where possible, and finally making estimations if necessary.

EXAMPLE 13.1   Material Analysis

The following questions refer to the materials flow system shown in Figure 13.11:

1. Calculate the values of the
    i.   Extraction efficiency
    ii.   Manufacturing efficiency
    iii. Recovery efficiency
    iv.   Recycling efficiency
    v.   System efficiency
    vi.   Reuse factor

2. Which subsystem is the least efficient?

1. The best approach is to conduct a mass balance across each subsystem. The mass balance for a steady state system is

$$\text{Input} - \text{output}$$

The materials extraction stage is illustrated in Figure 13.12.

$$5 \text{ Gg} + m_R = 2 \text{ Gg} + m_M$$

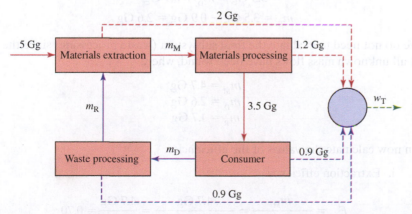

FIGURE 13.11 Life cycle assessment of material extraction and processing.

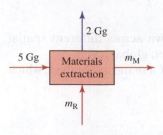

FIGURE 13.12 Materials extraction unit process.

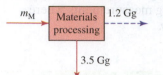

FIGURE 13.13 Materials processing component of the example life cycle assessment.

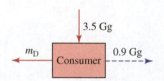

FIGURE 13.14 Consumer use component of the example life cycle assessment.

There are two unknowns in this subsystem that cannot be evaluated until we solve the mass balance for other subsystems. For the materials processing subsystem shown in Figure 13.13,

$$m_M = 3.5 \text{ Gg} + 1.2 \text{ Gg} = 4.7 \text{ Gg}$$

We may now return to the previous equation and substitute for m_M, which yields

$$5 \text{ Gg} + m_R = 2 \text{ Gg} + 4.7 \text{ Gg} = 6.7 \text{ Gg}$$
$$m_R = 6.7 \text{ Gg} - 5 \text{ Gg} = 1.7 \text{ Gg}$$

For the consumer subsystem shown in Figure 13.14,

$$3.5 \text{ Gg} = m_D + 0.9 \text{ Gg}$$
$$m_D = 3.5 \text{ Gg} - 0.9 \text{ Gg} = 2.6 \text{ Gg}$$

We do not need to evaluate the final subsystem (waste processing) since the values of all unknown mass flows have been found, where

$$m_M = 4.7 \text{ Gg}$$
$$m_D = 2.6 \text{ Gg}$$
$$m_R = 1.7 \text{ Gg}$$

We can now calculate the values of the efficiencies:

    i. Extraction efficiency

$$E_{eff} = \frac{m_M}{5 \text{ Gg} + m_R} = \frac{4.7 \text{ Gg}}{5 \text{ Gg} + 1.7 \text{ Gg}} = \frac{4.7 \text{ Gg}}{6.7 \text{ Gg}} = 0.70$$

ii. Manufacturing efficiency

$$M_{eff} = \frac{3.5\ Gg}{m_M} = \frac{3.5\ Gg}{4.7\ Gg} = 0.74$$

iii. Recovery efficiency

$$D_{eff} = \frac{m_D}{3.5\ Gg} = \frac{2.6\ Gg}{3.5\ Gg} = 0.74$$

iv. Recycling efficiency

$$R_{eff} = \frac{m_R}{m_D} = \frac{1.7\ Gg}{2.6\ Gg} = 0.65$$

v. System efficiency

$$S_{eff} = E_{eff} \times M_{eff} \times D_{eff} \times R_{eff}$$

$$S_{eff} = 0.70 \times 0.74 \times 0.74 \times 0.65 = 0.25$$

vi. Reuse factor

$$\psi = \frac{m_R}{m_M} = \frac{1.7\ Gg}{4.7\ Gg} = 0.36$$

2. The *least efficient* subsystem is the waste processing subsystem.

13.4 Embodied Energy

Energy is the ability to do work, and it is an essential need for a functional industrial system. The sustainability objective is to reduce and minimize the energy requirements of product manufacturing. Energy flows in industrial systems are also tracked closely with materials and water flows because they are often intrinsically linked. One-way energy flows are assessed with the concept of energy efficiency.

Due to the second law of thermodynamics, whenever energy from any source is converted into work, some of the energy input is dissipated to the surroundings. The efficiency of this process is defined as the ratio of the useful work obtained to the total energy input:

$$\text{Efficiency} = \frac{\text{useful work output}}{\text{total energy input}} \tag{13.14}$$

For example, approximately 25% of the energy derived from the combustion of gasoline is used to propel a car. The rest is dissipated to the surroundings as heat. Thus, the efficiency of an average automobile is approximately 0.25.

The total amount of energy required in producing a product or service across the product chain is called the embodied energy. A significant amount of energy

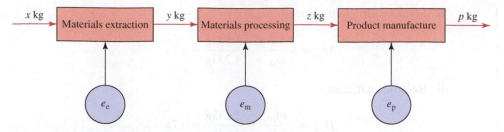

FIGURE 13.15 Example of a materials and energy flow system in product manufacture.

used to manufacture and use products is hidden and not always accounted for by manufacturers and consumers.

Consider the materials processing system shown in Figure 13.15. If e_e, e_m, and e_p are the energies consumed per unit mass (J/kg) during the materials extraction, materials processing, and product manufacture stages, respectively, the energy used in each stage of the process is obtained from the product of the energy consumed per unit mass and the total mass processed in that stage.

The total energy (in joules) used in the manufacture of p kg of product is then given by the sum of the energies consumed across all stages:

$$E_{Total} = (x \times e_e) + (y \times e_m) + (z \times e_p) \qquad (13.15)$$

The energy used in the manufacture of a unit mass of product can then be calculated from

$$\alpha = \frac{E_{Total}}{p} \qquad (13.16)$$

If some of the materials processed in the product manufacture stage are recycled back into the materials processing stage as shown in Figure 13.16, the energy flow balance can be recalculated as

$$E_{Total} = (x \times e_e) + ((y + \mu) \times e_m) + (z \times e_p) \qquad (13.17)$$

Since

$$z = y + \mu \qquad (13.18)$$

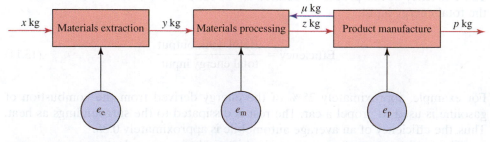

FIGURE 13.16 Example of a materials and energy flow system in product manufacture with a recycle flow stream.

Then

$$E_{Total} = (x \times e_e) + ((y + \mu) \times e_m) + ((y + \mu) \times e_p) \tag{13.19}$$

$$E_{Total} = (x \times e_e) + ((y + \mu) \times (e_m + e_p)) \tag{13.20}$$

EXAMPLE 13.2   Embodied Energy Analysis

Calculate the total energy consumed per kilogram of product manufactured in the system shown in Figure 13.17.

The energy consumed in each stage is the product of the energy per unit mass required and the mass processed in that stage.

Mass processed in materials extraction = 16 kg

Mass processed in materials processing = 12 kg + 2 kg = 14 kg

Mass processed in product manufacture = 10 kg

Note that the mass processed in each stage is equal to the total mass input with no regard to how much ends up in useful product or waste.

$$E_{Total} = \left(16 \text{ kg} \times 1.2 \frac{kJ}{kg}\right) + \left(14 \text{ kg} \times 2.1 \frac{kJ}{kg}\right) + \left(10 \text{ kg} \times 0.8 \frac{kJ}{kg}\right)$$

$$E_{Total} = 19.2 \text{ kJ} + 29.4 \text{ kJ} + 8 \text{ kJ} = 56.6 \text{ kJ}$$

The energy consumed per unit mass of product is

$$\alpha = \frac{56.6 \text{ kJ}}{5 \text{ kg}} = 11.32 \text{ kJ/kg}$$

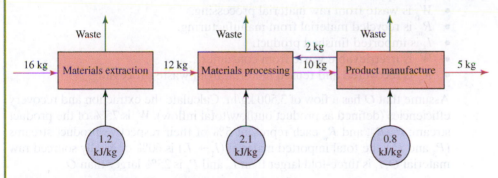

FIGURE 13.17 Flow diagram for Example 13.2.

13.5   Summary

Industrial ecology, as the name implies, is concerned with the development and implementation of solutions at the system level. A key factor in this approach is the notion that the behavior of unit components in a system cannot be completely understood without reference to the system itself and their interactions within it. This system perspective dictates an approach to improving the environmental performance of industrial systems. Much like biological ecological systems, industrial ecological systems rely on the flow of materials and energy. In industrial systems, this is directly related to the extraction of resources and the disposal of wastes to the environment. One way to minimize the overall environmental impact in industrial systems is to develop eco-industrial parks based on industrial symbiotic relationships in which the waste from one industry is useful input to another.

References

Allenby, B. R. (1992, March). "Industrial ecology: The materials scientist in an environmentally constrained world." *MRS Bulletin* 17(3):46–51.

Graedel, T. E., and Allenby, B. R. (2003). *Industrial Ecology*, 2nd ed. Upper Saddle River, NJ: Pearson Education, Prentice Hall.

Key Concepts

Ecology

Industrial ecology

Eco-industrial park

Materials flow analysis (MFA)

Problems

13-1   Explain the concept of embodied energy.

13-2   Figure 13.18 shows the flows of a material through a product life cycle. All symbols on the diagram refer to total mass flows:

- O is locally sourced raw material.
- I_1 is imported raw material.
- W_p is waste from raw material processing.
- R_m is recycled material from manufacturing.
- I_2 is imported finished product.
- W_c is irretrievable waste from consumers.
- R_w is recovered and reused material from waste processing.

Assume that O has a flow of 3,500 kg/hr. Calculate the extraction and recovery efficiencies (defined as product outflow/total inflow). W_c is 7.5% of the product stream; P_5, R_m, and R_w each represent 5% of their respective product streams (P_3 and P_6); the total imported material ($I_1 + I_2$) is 60% of locally sourced raw material O, I_1 is three-fold larger than I_2, and P_6 is 25% larger than O.

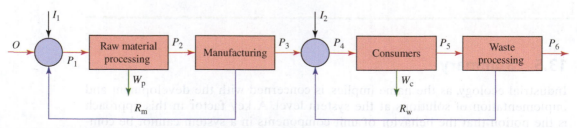

FIGURE 13.18 Illustration for Problem 13-2.

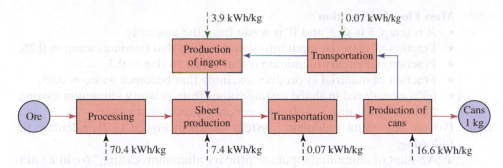

FIGURE 13.19 Illustration for Problems 13-3 and 13-4.

13-3 Calculate the total energy consumed per kg of Al cans produced if 23% of total production is recycled from Figure 13.19.

13-4 If the materials efficiencies processing, sheet production, and production of ingots are 85%, 92%, and 98%, respectively, calculate the new total energy consumed per kg of Al cans for the system shown in Figure 13.19.

13-5 For the aluminum production system shown in Figure 13.20, calculate the following:

a. Amount of aluminum input to "primary aluminum casting" (α, in kg aluminum).

b. Amount of aluminum input into "secondary aluminum casting" (β, in kg aluminum).

c. Total energy consumed.

d. If the waste, W, from the product assembly is sent to shape casting to be reused, recalculate all three values for the previous parts (α, β, and total energy consumed).

Energy Flow Information

- e_p = 2.0 kWh/kg
- e_s = 0.50 kWh/kg
- e_f = 0.75 kWh/kg
- e_m = 0.25 kWh/kg

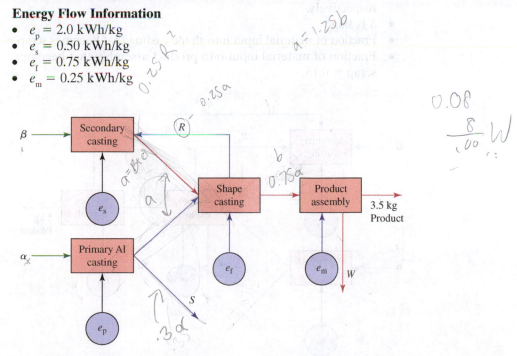

FIGURE 13.20 Illustration for Problem 13-5.

Mass Flow Information

- R is scrap, S is slag, and W is waste from the assembly.
- Fraction of material input into shape casting that becomes scrap = 0.25.
- Fraction of primary aluminum α that is lost to slag = 0.3.
- Fraction of material in product assembly that becomes waste = 0.08.
- 60% of material in shape casting comes from primary aluminum casting and the remaining 40% from secondary aluminum casting.

13-6 For the aluminum production system shown in Figure 13.21, calculate the following:

a. Amount of aluminum input to "primary aluminum casting" (α, in kg aluminum).

b. Amount of aluminum input into "secondary aluminum casting" (β, in kg aluminum).

c. Total energy consumed.

d. If the waste, W, from the product assembly is sent to shape casting to be reused, and the percentage ratios from primary casting and secondary casting still remain 60:40, recalculate all three values for the previous parts (α, β, and total energy consumed).

Energy Flow Information

- e_p = 1.8 kWh/kg
- e_s = 0.75 kWh/kg
- e_f = 1.3 kWh/kg
- e_m = 0.25 kWh/kg

Mass Flow Information

- R_s and R_a are scrap flows from shape casting and product assembly, respectively.
- W_s and W_a are waste flows from shape casting and product assembly, respectively.
- S is slag.
- Fraction of material input into shape casting that becomes scrap = 0.25.
- Fraction of material input into product assembly that becomes scrap = 0.15.

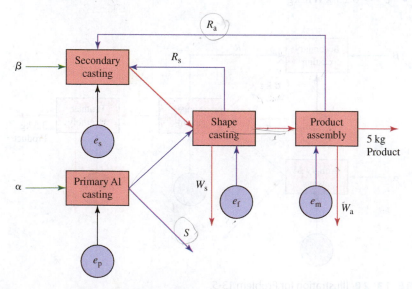

FIGURE 13.21 Illustration for Problem 13-6.

- Fraction of primary aluminum α that is lost to slag = 0.3.
- Fraction of material in product assembly that becomes waste = 0.08.
- Fraction of material in shape casting that becomes waste = 0.03.
- 60% of material in shape casting comes from primary aluminum casting and the remaining 40% from secondary aluminum casting.

13-7 Consider the materials flowchart shown in the Figure 13.22. The wastes from primary production and manufacturing are 17% and 24% of their respective stocks. The mass of material from secondary production is 23% of the mass from primary production. Calculate the masses of materials going into primary and secondary production, Ω and Ψ.

13-8 Figure 13.23 shows the flows of two materials α and β used in the manufacture of product $\alpha\beta$. Product $\alpha\beta$ contains 60% α and 40% β by mass. Calculate the mass values of flow streams A, B, C, D, E, F, G, and H, given $F = C$.

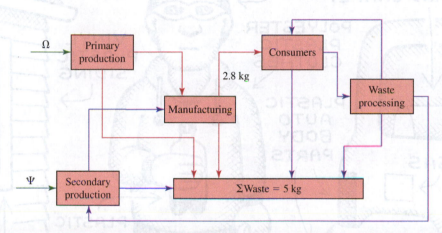

FIGURE 13.22 Illustration for Problem 13-7.

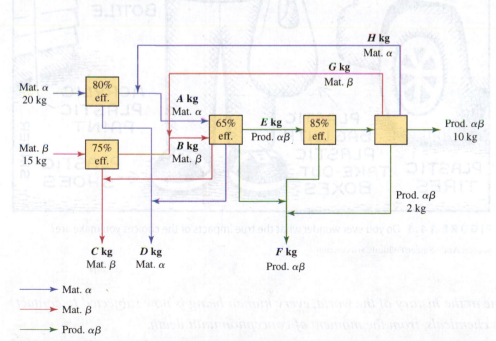

FIGURE 13.23 Illustration for Problem 13-8.

## CHAPTER 14

Life Cycle Analysis

FIGURE 14.1 Do you ever wonder what the true impacts of the choices you make are?

Source: Andy Singer/PoliticalCartoons.com.

For the first time in the history of the world, every human being is now subjected to contact with dangerous chemicals, from the moment of conception until death.

—RACHEL CARSON, 1962

GOALS

THE GOALS OF THIS CHAPTER ARE to introduce the concept of product life cycle thinking and to present elementary principles of life cycle assessment methodologies.

OBJECTIVES

At the conclusion of this chapter, you should be able to:

14.1 Develop a conceptual understanding of the life cycle of a product or process.

14.2 Describe the framework used to conduct a life cycle assessment of a product or process.

14.3 Understand how impact category factors are developed that compare the effects of various chemical substances using standardized metrics.

14.4 Describe the impact categories commonly used to illustrate the results from life cycle assessment models.

14.5 Illustrate the complexity of determining the human and environmental impacts of simple products across their life cycles.

Introduction

In previous chapters, we learned about the relationship between technology and environmental impacts. One of the objectives of sustainable development is to supply society's needs with minimal harm to both the environment and the ability of future generations to meet their needs. The current needs of society are met in a wide variety of ways due to advancements in technology. When we are confronted with multiple options to achieve a particular societal need, it is a major challenge to determine which options are the least harmful and thus the most sustainable.

In the past couple of decades, studies have shown that emissions produced during industrial and technological manufacturing make up just a small percentage of the overall emissions that result from technological products. The extraction of raw materials and the production, transportation, use, and eventual disposal of products all result in emissions and various impacts that must be considered in calculating the full impact of technology on the environment. There is a need for methods that can assess the sustainability of different technologies or options over the entire life of a product. The life cycle of a product is defined as "consecutive and interlinked stages of a product system, from raw material acquisition or generation from natural resources to final disposal" (ISO 14040). This has led to the use of life cycle assessment (LCA), which has been defined as "the compilation and evaluation of the inputs and outputs and the potential environmental impacts of a product system throughout its life cycle" (ISO 14040, ISO 14044). In this context, a product is defined in a broad sense to include physical goods as well as services.

As an example, your textbooks may be printed on paper that was produced from pulp extracted from wood, which itself was obtained from trees using a chemical process. After use, textbooks may be disposed of in various ways, including recycling, incineration, or burial in landfills. In addition, each of these stages requires energy drawn from various sources. Every stage in the life of your textbooks contributes to the overall impact associated with the production and use of textbooks. Life cycle assessment allows us to compare the impacts associated with different products or processes to determine which have the least impact. It may also allow us to determine which stage in the overall life cycle of products contributes the most to the overall impact.

According to the Food and Agriculture Organization (FAO) of the United Nations, the systems that support the world's current dietary patterns account for 21% to 37% of total greenhouse gas emissions. The FAO predicts that global meat consumption will approximately double between 2001 and 2050. Consider this: An average of 18% of global greenhouse gas emissions from all human activities comes from livestock production (FAO, 2010). Transportation, which includes emissions from fuel combustion, accounts for approximately 14%. The equivalent of up to 36.4 kg of CO_2 are emitted in the production of 1 kg of beef without considering the emissions from transportation and managing infrastructure (Ogino et al., 2007). The largest percentage of these emissions is due to deforestation and desertification resulting from animal husbandry (Figure 14.2). The livestock sector emits 37% of all anthropogenic methane and accounts for 65% of anthropogenic nitrous oxide (NO_x), mostly coming from manure. Methane and nitrous oxide have global warming potentials (GWP) of 25 and 298, respectively. Life cycle assessment studies can help us determine which processes in our diet contribute the most toward climate change and other environmental impacts.

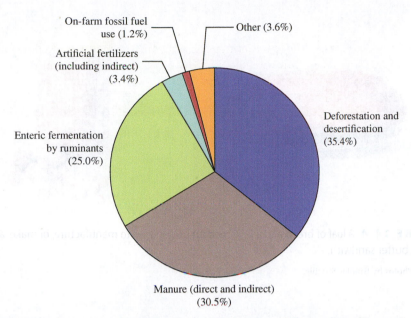

FIGURE 14.2 Proportion of greenhouse gas emissions from different parts of livestock production.

Source: Based on McMichael et al. (2007). Adapted from the Food and Agriculture Organization.

14.1 Life Cycle Thinking

In earlier chapters, we learned about cycles and the paths and processes by which matter and energy move through different systems. Life cycle thinking applies the same principles to understand the complexity behind even simple products. This goal requires comprehensive tracking of the way materials and energy flow through the product system—from the extraction of virgin raw material from the environment through the various stages of processing, refinement, packaging, sales, and use all the way to the ultimate disposal back into the environment.

Let us consider products we use and conveniences that you have become accustomed to. Do you ever think of the various stages and processes that were required to bring you these things? For example, consider what it takes to make a peanut butter sandwich. You need just two slices of bread and some peanut butter, as illustrated in Figure 14.3.

FIGURE 14.3 Ingredients to make a peanut butter sandwich.

Source: Photos by Bradley Striebig.

FIGURE 14.4 A loaf of bread and a jar of peanut butter used to manufacture, or make, a peanut butter sandwich.

Source: Photos by Bradley Striebig.

However, bread usually is not made in individual slices, so your slices probably come from a whole loaf; meanwhile, the peanut butter likely comes from a jar, as shown in Figure 14.4.

Even though sliced bread and peanut butter are readily available in supermarkets and stores, they have to be made first. Bread is made using flour, water, and yeast. Flour is obtained from the milling of cereal grains, typically wheat. Wheat has to be grown and requires land for planting, water, and sources of plant nutrients, as shown in Figure 14.5.

Similarly, peanut butter is made by crushing and grinding roasted peanuts that also have to be grown, which requires land, water, and plant nutrients, as illustrated in Figure 14.6.

FIGURE 14.5 Simplified life cycle of bread.

Sources: Photos by Bradley Striebig, Andy Sacks/The Image Bank/Getty Images.

FIGURE 14.6 Simplified life cycle of peanut butter.

Sources: Photos by Bradley Striebig, Rick Rudnicki/Lonely Planet Images/Getty Images, nanao wagatsuma/Moment /Getty Images, Peeter Viisimaa/E+/Getty Images.

Note that, in this example, we have ignored crucial stages and processes that include packaging and transportation as well as the processes involved in the production of the other ingredients. However, we can appreciate the concept that all products come to us through processes that we often do not think about while we use the products.

Consider the following requirements for manufacturing a peanut butter sandwich (National Peanut Board, 2014):

- It takes about 540 peanuts to make a 12-ounce jar of peanut butter.
- There are enough peanuts in 1 acre of land to make 30,000 peanut butter sandwiches.
- The average American consumes more than 6 pounds of peanuts and peanut butter products each year.
- The average child eats 1,500 peanut butter and jelly sandwiches before they graduate from high school.
- Americans consume on average over 1.5 billion pounds of peanut butter and peanut products each year.
- Peanut butter is consumed in 90% of U.S. households.
- Americans eat enough peanut butter in a year to make more than 10 billion peanut butter and jelly sandwiches.
- The amount of peanut butter eaten in a year could wrap the Earth in a ribbon of 18-ounce peanut butter jars one and one-third times.
- Peanuts account for two-thirds of all snack nuts consumed in the United States.
- Peanut butter is the leading use of peanuts in the United States.

Each of the process steps considered in making peanut butter involves the use of materials and energy and may generate wastes, as illustrated in

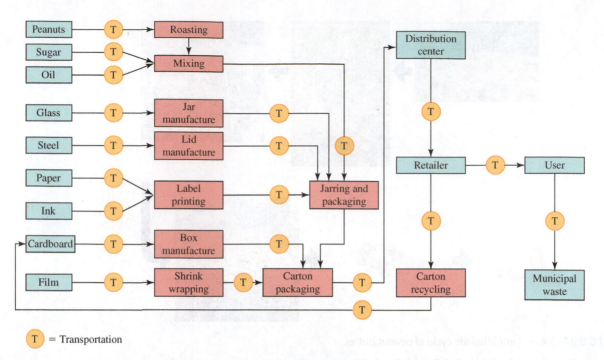

T = Transportation

FIGURE 14.7 The life cycle of packaged peanut butter.

Figure 14.7. We must also consider the potential impacts associated with eating a peanut butter sandwich, possibly relating to the use of fertilizers, herbicides, and pesticides in crop production. Consideration of unintended consequences and impacts associated with products and manufacturing is the subject of the cartoon shown in Figure 14.8, which illustrates the impacts associated with using ethanol as a gasoline additive.

Figure 14.9 shows a simple schematic of the inputs and outputs from a single stage in a product life cycle system. This is the basic building block of life cycle assessments. Figure 14.10 shows the connections among multiple stages in the life cycle of a product.

Electrical energy is generated from other forms of energy in power plants and must be transmitted through high-voltage wires to local utility companies before being transformed into useful electricity for work in our homes, schools, and offices. Carbon-based forms of chemical energy such as coal, petroleum, and natural gas are combusted to generate thermal energy, which in turn generates electrical energy. The combustion of these fossil fuels generates CO_2, which has been linked to global climate change, as discussed in Chapters 10 and 11. This means that when generated from fossil fuels, every unit of electrical energy we use has a direct impact on the global climate. Companies may market "zero emissions" electrical processes, as illustrated by the cartoon in Figure 14.11, but this is deceptive marketing since the production of electricity itself has measurable environmental impacts.

In technological systems, all raw materials are initially extracted from the environment in one form or another. The environment is often called the cradle, the source of all materials. These materials are processed in various industrial stages whose boundaries are called gates before they are ultimately disposed back into

FIGURE 14.8 Cartoon showing the complexity of choices that become apparent from a life cycle perspective.

Source: Andy Singer Cartoons, www.andysinger.com/.

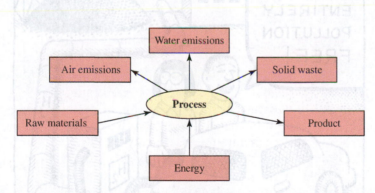

FIGURE 14.9 Simple schematic of a single stage in a product.

the environment. The methods and location of waste disposal are often described as the grave for the waste material. Thus, the environment is both the cradle and the grave. The changes that occur to materials from cradle to grave and the impacts associated with extraction, processing, and disposal can be evaluated using *life cycle assessment (LCA)*.

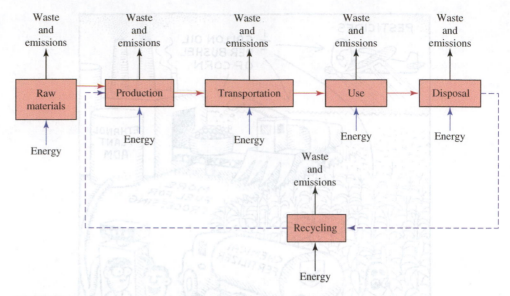

FIGURE 14.10 A simple schematic of the life cycle of a product.

FIGURE 14.11 Cartoon showing how life cycle assessments may reveal unapparent impacts.

Source: Andy Singer Cartoons, www.andysinger.com/.

LCA is an objective process used to quantitatively evaluate the environmental burdens associated with a product, process, or activity throughout its entire life cycle (through the cradle, through all the gates, to the grave). It utilizes various mass and energy balance protocols as well as environmental impact evaluation techniques described in previous chapters to model the associated impacts across every stage of the life of a product. This may include the impacts associated with processing materials as well as the impacts associated with subsidiary actions. For example, in the extraction of natural resources, LCA includes the impacts associated with both the extracted materials and the extraction process.

Consider the cutting of trees to make wood products. The impacts associated with this activity include the environmental impacts from the loss of the trees as well as from the logging technology itself. Depending on the objective for which an LCA is performed, the scope may include the following ranges:

1. Raw materials extraction to the disposal of finished goods (i.e., cradle-to-grave)
2. Raw materials extraction to finished goods (i.e., cradle-to-gate)
3. One processing stage to another (i.e., gate-to-gate)

ACTIVE LEARNING EXERCISE 14.1 Life Cycle Thinking

Pick a typical household product and construct the life cycle of the product from raw material extraction to disposal. How is this life cycle different from a comparable product that fulfills the same function?

14.2 Life Cycle Assessment Framework

An LCA contains three phases: goal and scope definition, inventory analysis, and impact analysis. Each phase is followed by interpretation according to the generally agreed structure shown in Figure 14.12.

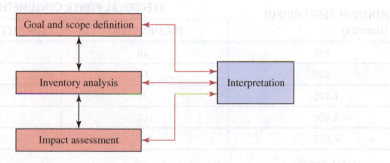

FIGURE 14.12 The LCA framework.

Source: Based on ISO 14040:2006, Environmental management, Life cycle assessment – Principles and framework.

14.2.1 Goal and Scope Definition

The goal specifies the reasons for carrying out the LCA. The goal is important because the parameters to be used in the assessment are usually dependent on what the intended objectives are. The goal also specifies the intended application of the LCA, as well as the intended audience.

The scope helps establish the system boundaries and the limits of the LCA. For example, if an LCA is used to compare products, it is usually not the products themselves that are the basis of the comparison but the functions they provide. A useful term, the *functional unit*, is defined in the scope.

The functional unit is used to establish a basis for comparison of two products by identifying a common function and how each product achieves that function over its life. The functional unit is quantitative and measurable, and a careful and proper identification of the functional unit is crucial, particularly if the LCA is used for comparisons.

For example, if an LCA is to be performed comparing reusable plastic cups and disposable paper cups, it would be incorrect to compare the impacts associated with one paper cup and those associated with one plastic cup since the plastic cup will continue to perform its function after the paper cup is discarded. Rather, it is more appropriate to determine how many times one plastic cup will be used before disposal and to calculate the equivalent number of paper cups needed to fulfill the same function. So, the comparison will be the impacts associated with one plastic cup and the functional equivalent number of paper cups.

Another example involves comparing the life cycle impact of incandescent bulbs and compact fluorescent lamps. One fluorescent lamp uses significantly less energy than an incandescent bulb to produce the same amount of visible light. Since the function of the bulb and the fluorescent lamp is to produce light, it would be incorrect to compare a 40 W incandescent bulb to a 40 W fluorescent lamp. The wattage of a light source refers to how much electrical power it typically consumes under normal operating conditions. A 40 W incandescent bulb and a 40 W fluorescent lamp both consume 40 W of power under normal operating conditions, but they produce significantly different amounts of light. From Table 14.1, we see that a 9 W fluorescent lamp produces approximately the same amount of light as a 40 W

TABLE 14.1 Energy consumption of incandescent bulbs and compact fluorescent lamps

MINIMUM LIGHT OUTPUT (lumens)	ELECTRICAL POWER CONSUMPTION (watts)	
	INCANDESCENT	COMPACT FLUORESCENT
450	40	9–13
800	60	13–15
1,100	75	18–25
1,600	100	23–30
2,600	150	30–52

Source: U.S. EPA (2014). www.energystar.gov/index.cfm?c=cfls.pr_cfls_lumens.

incandescent bulb. The functional unit would be the amount of visible light required, and then LCA would be performed comparing the number of incandescent light bulbs and compact fluorescent lamps required to provide the required amount of light. However, a fluorescent lamp also lasts significantly longer than an incandescent bulb, so one would have to use multiple incandescent bulbs over the lifetime of one fluorescent lamp. Incandescent light bulbs are typically rated to last an average of 1,000 hours, while compact fluorescent lamps are rated to last up to 8,000 hours. This means that, in addition to comparing the two choices based on the amount of light that is needed, the functional unit will also include the length of time that the light will be provided.

Consider that our functional unit is 450 lumens of light for 5,000 hours. For this we would compare the life cycle impact of one 9 W compact fluorescent lamp to five 40 W incandescent light bulbs.

ACTIVE LEARNING EXERCISE 14.2 Determining Appropriate Functional Units

Identify three sets of two or more products on which you would like to perform comparative LCAs. Determine the functional unit of each set.

14.2.2 Inventory Analysis

Inventory analysis involves determining the quantitative values of the materials and the energy inputs and outputs of all process stages within the life cycle. This includes

- Raw materials/energy needs
- Manufacturing processes
- Transportation, storage, and distribution requirements
- Use and reuse
- Recycle and end-of-life scenarios such as incineration and landfilling

Inventory analysis is usually initiated with a flowchart or process tree identifying the relevant stages and their interrelations. Relevant materials and energy data are then collected for each process stage using materials and energy balance protocols to account for unknown values. Standardized units are used to make calculations and comparisons easy. It is also in this phase that the system boundaries necessary to meet the predetermined goal and scope of the LCA are determined. The system boundaries are the limits placed on data collection. For example, if the goal of the LCA is a comparative study of two products for which some life cycle stages are the same with the same materials and energy input–output values, then the system boundary may be drawn to exclude the data related to those stages. Including these would only burden the LCA without providing additional information.

14.2.3 Impact Assessment

Impact assessment entails determining the environmental relevance of all the inputs and outputs of each stage in the life cycle. This includes the environmental impacts associated with the production, use, and disposal of the products. Relevant impact categories are selected, such as degradation of ecological systems, depletion of natural resources, and impacts on human health and welfare. For example, if we conducted the LCA of paper-based textbooks, examples of ecological systems degradation can include the impact of cutting trees as well as the impact of chemicals discharged during pulping on water quality. Trees as natural resources are renewable but not replaceable rapidly enough to immediately offset the impact of logging. The impact of the chemicals used on the health of the populations directly connected to the paper processing industries will also be assessed.

There are many models for impact assessment. Some provide assessments for a narrow range of impact categories; hence, they are mostly used for specific impacts of interest. ISO 14040, the International Standard for performing LCAs, requires that the impact assessment methodology used be explicitly stated. This is because the results of the LCA can depend significantly on the impact assessment method.

14.2.4 Interpretation

LCA results present opportunities to reduce or mitigate identified environmental impacts arising from the manufacture, use, and disposal of products. The product design may be changed, the materials used may be replaced with less impactful materials, and the entire industrial process may even be changed based on the results of an LCA. The LCA represents the most objective tool currently available to inform decisions on the environmental sustainability of products and processes.

A critical step during the interpretation of LCA results is the identification of which of the life cycle stages contributes the largest share of the emissions or impacts of the entire LCA. For example, compared to the impacts associated with manufacturing a vehicle, the impacts associated with the use stage are typically greater and provide the best opportunity for impact reduction.

14.3   Impact Categories

Emissions to the environment and the extraction and use of natural resources are called environmental stresses. What are the consequences of the emissions and the use of resources that are quantified in the inventory analysis? The impact assessment phase translates these into the relevant impact categories. We do this because the actual impact categories are easier to relate to and understand. Impact assessment also makes results more comparable to guide decision making. For example, several gases contribute to the greenhouse effect. Simply knowing the quantities of these gases that may be emitted from a system is not sufficient, but evaluating how these emissions contribute to global temperature increase, climate change, sea level rise, biodiversity changes, and food availability makes it easier to address specific concerns.

Impacts to the environment are typically of three types:

- **Depletion**—The removal of resources from the environment
- **Pollution**—The emission of waste and the dispersion of environmentally harmful materials
- **Disturbances**—Detrimental changes to the environment

Sustainable engineering challenges us to develop alternatives that will substitute current technologies and their associated emissions and impacts. To do this, however, we need to be able to quantify impacts in order to properly evaluate if the alternatives we propose are truly better. Correlating emission sources, emitted species, and the resultant impacts is necessary in quantifying the collective impact due to technology use.

A single emission source can result in multiple impacts, while multiple emission sources can lead to a single impact. For example, a global scale impact like climate change has been linked to the emission of multiple chemical species such as carbon dioxide, methane, nitrous oxide, and chlorofluorocarbons (Table 14.2). These chemicals are emitted in varying quantities and contribute to this impact in varying degrees. Moreover, the amount of time that each of these impact-causing chemicals are active in the environment also varies widely.

Since emissions occur from various sources at different times and since the quantity of each chemical species emitted can vary from source to source and from time to time, common indices called impact potentials were developed to help quantify the total impact anticipated from various sources. This helps us compare these impacts even if the specific chemicals emitted may be different. Next, we will discuss these impact potentials to better understand what they mean and how they can be used.

TABLE 14.2 Emissions from some technology-based activities

ACTIONS	PRIMARY EMISSIONS
Biomass combustion for energy demand Land clearing for agriculture, grazing, construction	CO_2, CO, NO_x, methane, hydrocarbons, particulates
Crop production	CO_2 from soil organic matter, methane from some crops, N_2O, nitrates, and phosphates from fertilizer, herbicides, pesticides
Animal husbandry	Methane, ammonia, hydrogen sulfide
Carbon-based energy production and use Coal Oil Gas	Mining/production: Drilling waste, CO_2, hydrocarbons, methane Combustion: CO_2, CO, hydrocarbons, NO_x, SO_2, particulates
Incineration	CO_2, HCl, metals, dioxins
Landfills	Methane, water contaminants
Industrial manufacturing	Chlorofluorocarbons, CO_2, particulates, hydrocarbons, metals

ACTIVE LEARNING EXERCISE 14.3 Choices We Make and Their Trade-offs

Discuss your thoughts on the accompanying cartoon.

Source: Andy Singer Cartoons, www.andysinger.com/.

14.3.1 Greenhouse Gases and Global Warming Potential (GWP)

Consider this: The greenhouse effect occurs because of the presence of greenhouse gases (GHGs) in the atmosphere. The molecular vibrations of greenhouse gases occur at frequencies that enable them to absorb and emit infrared radiation. The most abundant GHGs in the atmosphere are water, H_2O (yes, water is a greenhouse gas!); carbon dioxide, CO_2; methane, CH_4; nitrous oxide, N_2O; and ozone, O_3, as discussed in Chapter 10. Other relevant GHGs include sulfur hexafluoride and halocarbons like hydrofluorocarbons (HFCs) and perfluorocarbons (PFCs).

The greenhouse gases come from both natural and anthropogenic sources. N_2O is emitted from the combustion of fuels and more importantly from fertilizers. Halocarbons are industrial products, such as chlorofluorocarbons (CFCs) and hydrochlorofluorocarbons (HCFCs), that are used as refrigerants and solvents. Methane is emitted from natural gas releases, coal mining, sewage treatment plants, landfills, cows, rice paddies, and so on. The major source of carbon dioxide is burning fossil fuel. Another major contribution to the buildup of carbon dioxide in the atmosphere is change in land use, due mostly to tropical deforestation.

Greenhouse gases vary in their capacity to trap infrared radiation. For example, one molecule of methane is 26 times more effective at IR absorption than one molecule of CO_2. For this reason, the concept of *radiative forcing* was developed and is used as a quantitative comparison of the strengths of different greenhouse gases in their contribution to global warming. Radiative forcing is a measure of the influence a substance has in altering the balance of incoming and outgoing energy in the Earth's atmosphere. It is the difference between the energy received from the sun and the energy reflected back into space caused by a substance and measured in watts per square meter (W/m^2) of Earth surface. The radiative forcing for many greenhouse gases is shown in Table 14.3. Oftentimes climate scientists are concerned with estimating the impacts occurring due to the changes in the amount of a particular compound in the atmosphere; for example, what would happen if the

TABLE 14.3 Greenhouse gases with significant contributions to radiative forcing

GREENHOUSE GAS	AVERAGE ATMOSPHERIC CONCENTRATION	RADIATIVE FORCING (W/m^2)
CO_2	379 ± 0.65 ppm	1.66
CH_4	1,774 ± 1.8 ppb	0.48
N_2O	319 ± 0.12 ppb	0.16
CFC-11	251 ± 0.36 ppt	0.063
CFC-12	538 ± 0.18 ppt	0.17
CFC-113	79 ± 0.064 ppt	0.024
HCFC-22	169 ± 1.0 ppt	0.033
HCFC-141b	18 ± 0.068 ppt	0.0025
HCFC-142b	15 ± 0.13 ppt	0.0031
CH_3CCl3	19 ± 0.47 ppt	0.0011
CCl_4	93 ± 0.17 ppt	0.012
HFC-125	3.7 ± 0.10 ppt	0.0009
HFC-134a	35 ± 0.73 ppt	0.0055
HFC-152a	3.9 ± 0.11 ppt	0.0004
HFC-23	18 ± 0.12 ppt	0.0033
SF_6	5.6 ± 0.038 ppt	0.0029
CF_4 (PFC-14)	74 ± 1.6 ppt	0.0034
C_2F_6 (PFC-116)	2.9 ± 0.025 ppt	0.0008

Source: Based on IPCC Fourth Assessment Report (2007). Chapter 2, archive.ipcc.ch/publications_and_data/ar4/wg1/en/ch2s2-10-2.html.

concentration of carbon dioxide were doubled. In this case, it is convenient to use the concept of **radiative efficiency**. The radiative efficiency is the radiative forcing that occurs due to a unit increase in the atmospheric abundance of a substance C_i. Typically, the concentration of concern for a given greenhouse gas is 1 part per billion (ppb). Thus, the units of the radiative efficiency for substance C_i that changes by 1 ppb in the atmosphere are $W/(m^2\text{-ppb})$.

The contribution of any greenhouse gas to global warming also depends on its lifetime in the atmosphere. Some compounds are unstable and highly reactive, so they have short lifetimes. Others are stable and unreactive, so they stay in the atmosphere for longer periods of time. Typically, the rate of decay is proportional to the concentration, C, and follows a first-order kinetic relationship:

$$\frac{dC}{dt} = -kC \tag{14.1}$$

$$\frac{C}{C_0} = e^{-kt} \tag{14.2}$$

where

$$k = \frac{1}{\tau} \tag{14.3}$$

and τ is the atmospheric lifetime.

To quantify the contribution of GHGs to global warming, a common index was developed called the **global warming potential (GWP)**. CO_2, being the most significant contributor to anthropogenic radiative forcing, is used as the benchmark. The relative effectiveness of a unit mass of greenhouse gas when compared to an equivalent unit mass of CO_2 is its global warming potential (GWP) and is calculated from

$$GWP_i = \frac{\int_0^{ITH} a_i C_i \, dt}{\int_0^{ITH} a_{CO_2} C_{CO_2} \, dt} \tag{14.4}$$

where a_i and a_{CO_2} are the radiative efficiencies of substance i and CO_2, respectively, and ITH is the integrated time horizon.

As we have previously discussed, the concentrations of substances in the environment change over time at different rates. Unstable and reactive substances decay at faster rates and thus have shorter residence times in the environment regardless of the value of their radiative forcing efficiencies. Stable substances, on the other hand, remain in the environment for longer periods, thus causing impacts for longer periods.

For example, methane remains in the atmosphere for approximately a decade, while CO_2 has a variable lifetime, with about 50% being removed from the atmosphere within a century and about 20% remaining for thousands of years. Consequently, in the near term, a given mass of methane makes a greater contribution to global warming than the equivalent mass of CO_2 released at the same time; however, after 100 years, the methane will have long been removed from the atmosphere and the CO_2 will still be present and contributing to global warming. For this reason, the impact of greenhouse gases in the atmosphere is also dependent on the time that has passed since the gasses were emitted. The **integrated time horizon (ITH)** allows us to calculate the relative impact of substances over different time periods. The integrated time horizon is typically chosen as 20, 100, and 500 years to model short-term, medium-term, and long-term global warming potential values.

From Equation (14.4), we can see that the GWP of CO_2 is equal to 1. Carbon dioxide (CO_2), methane (CH_4), and nitrous oxide (N_2O) are released in large volumes but have relatively low global warming potentials. Other gases, such as perfluorocarbons, are released in much smaller volumes but have much higher global warming potentials.

The product of the global warming potential for a greenhouse gas and the emitted mass gives the equivalent mass of CO_2 that would result in the same radiative forcing effect:

$$\text{Equivalent mass } CO_2 = GWP_i \times m_i \qquad (14.5)$$

where m_i is the emitted mass of the greenhouse gas.

For emissions containing multiple greenhouse gases, the equivalent CO_2 values of each GHG may be summed to obtain the **total equivalent CO_2**:

$$\text{Total equivalent } CO_2 = \sum_i (GWP_i \times m_i) \qquad (14.6)$$

The total equivalent CO_2 allows us to compare the contributions to global warming by two or more systems that may or may not have identical emitted species in different quantities.

EXAMPLE 14.1 Calculating Equivalent CO_2 Using Global Warming Potentials

Calculate the equivalent CO_2 for 40 kg of methane.

We can use Equation (14.5):

$$\text{Equivalent mass } CO_2 = GWP_i \times m_i$$

The GWP value for methane obtained from Table 14.4 using the 100-year time horizon is 25, so we have

$$\text{Equivalent mass } CO_2 = 25 \times 40 \text{ kg} = 1,000 \text{ kg equivalent } CO_2$$

TABLE 14.4 Global warming potential for selected greenhouse gases

INDUSTRIAL DESIGNATION OR COMMON NAME	CHEMICAL FORMULA	APPROX. LIFETIME (years)	GLOBAL WARMING POTENTIAL FOR GIVEN TIME HORIZON		
			20-yr	100-yr	500-yr
Carbon dioxide	CO_2	Variable	1	1	1
Methane	CH_4	12	72	25	7.6
Nitrous oxide	N_2O	114	289	298	153
CFC-11	CCl_3F	45	6,730	4,750	1,620
CFC-12	CCl_2F_2	100	11,000	10,900	5,200

(Continued)

TABLE 14.4 (Continued)

INDUSTRIAL DESIGNATION OR COMMON NAME	CHEMICAL FORMULA	APPROX. LIFETIME (years)	GLOBAL WARMING POTENTIAL FOR GIVEN TIME HORIZON		
			20-yr	100-yr	500-yr
CFC-13	$CClF_3$	640	10,800	14,400	16,400
CFC-113	CCl_2FCClF_2	85	6,540	6,130	2,700
CFC-114	$CClF_2CClF_2$	300	8,040	10,000	8,730
CFC-115	$CClF_2CF_3$	1,700	5,310	7,370	9,990
Halon-1301	$CBrF_3$	65	8,480	7,140	2,760
Halon-1211	$CBrClF_2$	16	4,750	1,890	575
Halon-2402	$CBrF_2CBrF_2$	20	3,680	1,640	503
Carbon tetrachloride	CCl_4	26	2,700	1,400	435
Methyl bromide	CH_3Br	0.7	17	5	1
Methyl chloroform	CH_3CCl_3	5	506	146	45
HCFC-22	$CHClF_2$	12	5,160	1,810	549
HCFC-123	$CHCl_2CF_3$	1.3	273	77	24
HCFC-124	$CHClFCF_3$	5.8	2,070	609	185
HCFC-141b	CH_3CCl_2F	9.3	2,250	725	220
HCFC-142b	CH_3CClF_2	17.9	5,490	2,310	705
HCFC-225ca	$CHCl_2CF_2CF_3$	1.9	429	122	37
HCFC-225cb	$CHClFCF_2CClF_2$	5.8	2,030	595	181
HFC-23	CHF_3	270	12,000	14,800	12,200
HFC-32	CH_2F_2	4.9	2,330	675	205
HFC-125	CHF_2CF_3	29	6,350	3,500	1,100
HFC-134a	CH_2FCF_3	14	3,830	1,430	435
HFC-143a	CH_3CF_3	52	5,890	4,470	1,590
HFC-152a	CH_3CHF_2	1.4	437	124	38
HFC-227ea	CF_3CHFCF_3	34.2	5,310	3,220	1,040
HFC-236fa	$CF_3CH_2CF_3$	240	8,100	9,810	7,660
HFC-245fa	$CHF_2CH_2CF_3$	7.6	3,380	1030	314
HFC-365mfc	$CH_3CF_2CH_2CF_3$	8.6	2,520	794	241
HFC-43-10mee	$CF_3CHFCHFCF_2CF_3$	15.9	4,140	1,640	500
Sulfur hexafluoride	SF_6	3,200	16,300	22,800	32,600
Nitrogen trifluoride	NF_3	740	12,300	17,200	20,700
PFC-14	CF_4	50,000	5,210	7,390	11,200
PFC-116	C_2F_6	10,000	8,630	12,200	18,200

TABLE 14.4 (Continued)

INDUSTRIAL DESIGNATION OR COMMON NAME	CHEMICAL FORMULA	APPROX. LIFETIME (years)	GLOBAL WARMING POTENTIAL FOR GIVEN TIME HORIZON		
			20-yr	100-yr	500-yr
PFC-218		2,600	6,310	8,830	12,500
PFC-318		3,200	7,310	10,300	14,700
PFC-3-1-10		2,600	6,330	8,860	12,500
PFC-4-1-12		4,100	6,510	9,160	13,300
PFC-5-1-14		3,200	6,600	9,300	13,300
PFC-9-1-18		>1,000	>5,500	>7,500	>9,500
Trifluoromethyl sulfur pentafluoride		800	13,200	17,700	21,200
HFE-125		136	13,800	14,900	8,490
HFE-134		26	12,200	6,320	1,960
HFE-143a		4.3	2,630	756	230
HCFE-235da2		2.6	1,230	350	106
HFE-245cb2		5.1	2,440	708	215
HFE-245fa2		4.9	2,280	659	200
HFE-254cb2		2.6	1,260	359	109
HFE-347mcc3		5.2	1,980	575	175
HFE-347pcf2		7.1	1,900	580	175
HFE-356pcc3		0.33	386	110	33
HFE-449sl (HFE-7100)		3.8	1,040	297	90
HFE-569sf2 (HFE-7200)		0.77	207	59	18
HFE-43-10pccc124 (H-Galden 1040x)		6.3	6,320	1,870	569
HFE-236ca12 (HG-10)		12.1	8,000	2,800	860
HFE-338pcc13 (HG-01)		6.2	5,100	1,500	460
PFPMIE		800	7,620	10,300	12,400
Dimethyl ether		0.015	1	1	<1
Methylene chloride		0.38	31	8.7	2.7
Methyl chloride		1	45	13	4

Source: Based on IPCC Fourth Assessment Report (2007). Chapter 2, archive.ipcc.ch/publications_and_data/ar4/wg1/en/ch2s2-10-2.html.

EXAMPLE 14.2 Calculating Equivalent CO_2 Using Global Warming Potentials

Consider two process systems with the following emission values:

SPECIES	SYSTEM A	SYSTEM B
CO_2	22 kg	18 kg
CH_4	3 kg	5 kg
N_2O	1.2 kg	1.5 kg

Which system has the greater equivalent CO_2 value?

SPECIES	GWP	SYSTEM A	EQUIVALENT CO_2	SYSTEM B	EQUIVALENT CO_2
CO_2	1	22 kg	22 kg	18 kg	18 kg
CH_4	25	3 kg	75 kg	5 kg	125 kg
N_2O	298	1.2 kg	357.6 kg	1.5 kg	447 kg
			Σ454.6 kg		Σ590 kg

We can use Equation (14.6) and Table 14.4 for the GWP values for the 100-year time horizon.

System B has the greater equivalent CO_2 value and hence makes the greater contribution to global warming.

14.3.2 Ozone Depletion

Ozone is the triatomic molecule of oxygen. It occurs naturally in the atmosphere, where it is unevenly distributed. The presence of ozone (O_3) results in different impacts, depending on which layer of the atmosphere it exists in. Ozone in the troposphere, the layer closest to the Earth's surface, acts as an undesirable pollutant contributing to photochemical smog. However, ozone in the stratosphere, the layer just above the troposphere, is a necessary constituent forming a layer that absorbs harmful ultraviolet radiation. The depletion of the ozone in the stratosphere is a significant global concern. Excessive exposure to ultraviolet radiation is harmful to animals and vegetation. The ozone layer is a naturally formed protective shield for life on Earth that has been affected by human-made ozone-depleting substances (ODS).

Chlorofluorocarbons (CFCs) and halogenated compounds have been identified as contributing significantly to the destruction of stratospheric ozone. CFCs are highly stable chemical structures composed of carbon, chlorine, and fluorine. One important example is trichlorofluoromethane, CCl_3F or CFC-11.

At high altitudes and in the presence of solar radiation, CFCs dissociate to produce chlorine atoms that act as catalysts in the conversion of ozone to oxygen. The reactions can be represented as

$$CFC \xrightarrow{\text{Solar radiation}} Cl + \text{other products}$$

$$Cl + O_3 \rightarrow ClO + O_2$$

$$ClO + O \rightarrow Cl + O_2$$

$$\text{Net:} \quad \overline{O_3 + O \rightarrow 2O_2}$$

From the net reaction above, we can see that the chlorine atom is not used up in the reaction and remains in the atmosphere, catalyzing the conversion of more ozone to oxygen. Other halogenated ozone-depleting substances employ similar mechanisms.

Just as with the greenhouse gases, a common index has been developed to quantify the relative contribution of ozone-depleting substances to the depletion of stratospheric ozone. This index is called the **ozone depletion potential (ODP)** and uses CFC-11 as a benchmark (see Table 14.5). The ODP is the effectiveness in the degradation of stratospheric ozone by the emission of a unit mass of an ozone-depleting substance relative to the depletion caused by CFC-11:

$$\text{Equivalent mass of CFC-11} = ODP_i \times m_i \tag{14.7}$$

The same rule applies to emissions containing multiple ODPs:

$$\text{Total equivalent mass CFC-11} = \sum_i (ODP_i \times m_i) \tag{14.8}$$

TABLE 14.5 Ozone depletion potential of some ozone-depleting substances

CHEMICAL NAME	ODP	CHEMICAL NAME	ODP
CFC-11 (CCl_3F) Trichlorofluoromethane	1	$C_2H_2FBr_3$	0.1–1.1
CFC-12 (CCl_2F_2) Dichlorodifluoromethane	1	$C_2H_2F_2Br_2$	0.2–1.5
CFC-113 ($C_2F_3Cl_3$) 1,1,2-Trichlorotrifluoroethane	0.8	$C_2H_2F_3Br$	0.7–1.6
CFC-114 ($C_2F_4Cl_2$) Dichlorotetrafluoroethane	1	$C_2H_3FBr_2$	0.1–1.7
CFC-115 (C_2F_5Cl) Monochloropentafluoroethane	0.6	$C_2H_3F_2Br$	0.2–1.1
Halon 1211 (CF_2ClBr) Bromochlorodifluoromethane	3	C_2H_4FBr	0.07–0.1
Halon 1301 (CF_3Br) Bromotrifluoromethane	10	C_3HFBr_6	0.3–1.5
Halon 2402 ($C_2F_4Br_2$) Dibromotetrafluoroethane	6	$C_3HF_2Br_5$	0.2–1.9
CFC-13 (CF_3Cl) Chlorotrifluoromethane	1	$C_3HF_3Br_4$	0.3–1.8
CFC-111 (C_2FCl_5) Pentachlorofluoroethane	1	$C_3HF_4Br_3$	0.5–2.2

(Continued)

TABLE 14.5 (Continued)

CHEMICAL NAME	ODP	CHEMICAL NAME	ODP
CFC-112 ($C_2F_2Cl_4$) Tetrachlorodifluoroethane	1	$C_3HF_5Br_2$	0.9−2.0
CFC-211 (C_3FCl_7) Heptachlorofluoropropane	1	C_3HF_6Br	0.7−3.3
CFC-212 ($C_3F_2Cl_6$) Hexachlorodifluoropropane	1	$C_3H_2F_2Br_4$	0.2−2.1
CFC-213 ($C_3F_3Cl_5$) Pentachlorotrifluoropropane	1	$C_3H_2F_3Br_3$	0.2−5.6
CFC-214 ($C_3F_4Cl_4$) Tetrachlorotetrafluoropropane	1	$C_3H_2F_4Br_2$	0.3−7.5
CFC-215 ($C_3F_5Cl_3$) Trichloropentafluoropropane	1	$C_3H_2F_5Br$	0.9−14
CFC-216 ($C_3F_6Cl_2$) Dichlorohexafluoropropane	1	$C_3H_3FBr_4$	0.08−1.9
CFC-217 (C_3F_7Cl) Chloroheptafluoropropane	1	$C_3H_3F_2Br_3$	0.1−3.1
CCl_4 Carbon tetrachloride	1.1	$C_3H_3F_3Br_2$	0.1−2.5
Methyl chloroform ($C_2H_3Cl_3$) 1,1,1-trichloroethane	0.1	$C_3H_3F_4Br$	0.3−4.4
Methyl bromide (CH_3Br)	0.7	$C_3H_4FBr_3$	0.03−0.3
$CHFBr_2$	1	$C_3H_4F_2Br_2$	0.1−1.0
CH_2FBr	0.73	$C_3H_4F_3Br$	0.07−0.8
C_2HFBr_4	0.3−0.8	$C_3H_5FBr_2$	0.04−0.4
$C_2HF_2Br_3$	0.5−1.8	$C_3H_5F_2Br$	0.07−0.8
$C_2HF_3Br_2$	0.4−1.6	C_3H_6FBr	0.02−0.7
C_2HF_4Br	0.7−1.2	CH_2BrCl Chlorobromomethane	0.12

Source: Based on U.S. EPA, Ozone-depleting Substances, www.epa.gov/ozone-layer-protection/ozone-depleting-substances.

14.3.3 Other Impact Categories

Similar equivalent impact factors have been developed for other impact categories.

Photochemical Ozone Creation Potential (POCP)

While ozone in the stratosphere is essential for reducing the amount of ultraviolet radiation that falls on Earth and its depletion is a global-scale concern, tropospheric ozone or ground-level ozone (smog) is a regional-scale concern. It is caused by the degradation of volatile organic compounds (VOCs) in the presence of nitrogen oxides (NO_x)

TABLE 14.6 Photochemical ozone creation potential of selected compounds

CHEMICAL	POCP	CHEMICAL	POCP	CHEMICAL	POCP
Acetylene	0.168	Ethanol	0.268	Propane	0.420
Acetaldehyde	0.527	Ethylene	1.000	Propene	1.030
Acetone	0.178	Formaldehyde	0.421	Toluene	0.563
Benzene	0.189	Methane	0.007	o-Xylene	0.666
Ethane	0.082	Methanol	0.123		

Source: Based on Heijungs, R., et al. (1992). *Environmental Life Cycle Assessment of Products: Guide and Backgrounds*, Ed. CML (Center of Environmental Science), Leiden.

and ultraviolet radiation. Tropospheric ozone causes damage to vegetation at low concentrations and is hazardous to human health at high concentrations.

The ***photochemical ozone creation potential (POCP)*** is an index used to quantify the contribution of chemical species to the creation of tropospheric ozone (Table 14.6). Similar to the global warming potential (GWP), it is calculated using the primary contributing substance as the reference. In this case, ethylene is used and has a POCP value of 1. Other substances that contribute to photochemical ozone formation can then be converted to equivalent ethylene values using

$$\text{Equivalent mass of ethylene} = POCP_i \times m_i \qquad (14.9)$$

Acidification Potential (AP)

Another regional-scale concern is acid precipitation. "Normal rain" is usually acidic with a pH of around 5.6 due to the natural presence of CO_2 in the atmosphere. However, the presence of other acidifying substances in the atmosphere leads to precipitation with lower pH values. Falling acid precipitation causes damage to soils, surface waters, and plants. Acidic rain changes the chemistry of soils and affects the way plants take up nutrients. Low pH values can also affect fish in surface waters.

The acidification potential is the index that was developed to quantify the contribution of acidifying substances to acid precipitation and uses sulfur dioxide (SO_2) as a benchmark (Table 14.7). This is determined by

$$\text{Equivalent mass of } SO_2 = AP_i \times m_i \qquad (14.10)$$

TABLE 14.7 Acidification potential of some compounds

CHEMICAL	ACIDIFICATION POTENTIAL
SO_2	1.00
NO	1.07
N_2O	0.70
NO_x	0.70
NH_3	1.88
HCl	0.88
HF	1.60

Source: Based on Heijungs, R., et al. (1992). *Environmental Life Cycle Assessment of Products: Guide and Backgrounds*, Ed. CML (Center of Environmental Science), Leiden.

TABLE 14.8 Eutrophication potential of some substances

SUBSTANCE	FORMULA	EP
Nitrogen oxide	NO	0.200
Nitrogen dioxide	NO_2	0.130
Nitrogen oxides	NO_x	0.130
Ammonium	NH^{4+}	0.330
Nitrogen	N	0.420
Phosphate	PO_4^{3-}	1.000
Phosphorus	P	3.060
Chemical oxygen demand (COD)		0.022

Source: Based on Heijungs, R., et al. (1992). *Environmental Life Cycle Assessment of Products: Guide and Backgrounds*, Ed. CML (Center of Environmental Science), Leiden.

Eutrophication Potential (EP)

Eutrophication is the enrichment of nutrients in soils and water associated with excessive levels of compounds that contain nitrogen and phosphorus. Current agricultural practices involve the application of fertilizers to soils to boost crop production. Fertilizers add nutrients to soils, but excess nutrients in soil can also get washed into surface water bodies. In water, this leads to accelerated algae and phytoplankton growth that slowly leads to depletion of dissolved oxygen values, ultimately choking life in the water body. Eutrophication can also be caused by the industrial emission of nutrient-rich wastewater.

The eutrophication potential (Table 14.8) was developed to quantify the contribution to eutrophication of water bodies using phosphate (PO_4^{3-}) as a benchmark similar to previously discussed indices:

$$\text{Equivalent mass of } PO_4^{3-} = EP_i \times m_i \tag{14.11}$$

We can use these impact potential values to compare the impacts that may occur even if the actual emitted species are not the same. It also allows us to a have a quantitative basis for decision making.

EXAMPLE 14.3   Using Multiple Impact Potentials to Compare Systems

The table shows the emissions from two systems. Compare systems 1 and 2 on the basis of their contribution to global warming, acidification, and photochemical ozone creation.

CHEMICAL EMISSION	SYSTEM 1 kg	SYSTEM 2 kg
Methane (CH_4)	5.0	21.0
Nitrous oxide (N_2O)	0.2	0.1

To calculate the contribution to global warming, we will use Equation (14.5) and data for a 100-year time horizon from Table 14.4.

CHEMICAL EMISSION	GWP	SYSTEM 1 kg	EQUIVALENT CO$_2$ kg	SYSTEM 2 kg	EQUIVALENT CO$_2$ kg
Methane (CH$_4$)	25	5.0	125.0	21.0	525.0
Nitrous oxide (N$_2$O)	298	0.2	59.6	0.1	29.8
			Σ184.6		Σ554.8

To calculate the contribution to acidification, we will use Equation (14.10) and data from Table 14.7.

CHEMICAL EMISSION	AP	SYSTEM 1 kg	EQUIVALENT SO$_2$ kg	SYSTEM 2 kg	EQUIVALENT SO$_2$ kg
Methane (CH$_4$)	0.007	5.0	0.00	21.0	0.00
Nitrous oxide (N$_2$O)	0.700	0.2	0.14	0.1	0.07
			Σ0.14		Σ0.07

To calculate the contribution to photochemical ozone creation, we will use Equation (14.9) and data from Table 14.6.

CHEMICAL EMISSION	POCP	SYSTEM 1 kg	EQUIVALENT SO$_2$ kg	SYSTEM 2 kg	EQUIVALENT SO$_2$ kg
Methane (CH$_4$)	0.007	5.0	0.035	21.0	0.147
Nitrous oxide (N$_2$O)	0	0.2	0	0.1	0
			Σ0.035		Σ0.147

System 1 contributes more to acidification, while system 2 contributes more to global warming and photochemical ozone creation.

14.4 Impact Assessment

Impact assessment requires three elements and can include four optional ones. These are as follows.

Mandatory Elements

1. *Impact Category Definition*

   Impact categories represent environmental issues of concern, some of which are shown in Table 14.9. Different LCA practitioners use various models to make these assessments, depending on the context and specific impacts considered to be of concern. Defining the impact categories of an LCA involves specifying which impacts are relevant to the goal and scope. Not all impact categories are of relevance in every environment of concern. The Dutch

TABLE 14.9 Commonly used life cycle impact categories

IMPACT CATEGORY	SCALE	RELEVANT LCI DATA (i.e., CLASSIFICATION)	COMMON CHARACTERIZATION FACTOR	DESCRIPTION OF CHARACTERIZATION FACTOR
Global warming	Global	Carbon dioxide (CO_2) Nitrogen dioxide (NO_2) Methane (CH_4) Chlorofluorocarbons (CFCs) Hydrofluorocarbons (HCFCs) Methyl bromide (CH_3Br)	Global warming potential	Converts Life Cycle Inventory (LCI) data to carbon dioxide (CO_2) equivalents Note global warming potentials can be 50-, 100-, or 500-year potentials
Stratospheric ozone depletion	Global	Chlorofluorocarbons (CFCs) Hydrofluorocarbons (HCFCs) Halons Methyl bromide (CH_3Br)	Ozone depletion potential	Converts LCI data to trichlorofluoromethane (CFC-11) equivalents
Acidification	Regional Local	Sulfur oxides (SO_x) Nitrogen oxides (NO_x) Hydrochloric acid (HCl) Hydrofluoric acid (HF) Ammonia (NH_4)	Acidification potential	Converts LCI data to hydrogen (H^+) ion equivalents
Eutrophication	Local	Phosphate (PO_4) Nitrogen oxide (NO) Nitrogen dioxide (NO_2) Nitrates Ammonia (NH_4)	Eutrophication potential	Converts LCI data to phosphate (PO_4) ion equivalents
Photochemical smog	Local	Nonmethane hydrocarbon (NMHC)	Photochemical oxidant creation potential	Converts LCI data to ethane (C_2H_6) ion equivalents
Terrestrial toxicity	Local	Toxic chemicals with a reported lethal concentration to rodents	LC_{50}	Converts LC_{50} data to equivalents
Aquatic toxicity	Local	Toxic chemicals with a reported lethal concentration to fish	LC_{50}	Converts LC_{50} data to equivalents
Human health	Global Regional Local	Total releases to air, water, and soil	LC_{50}	Converts LC_{50} data to equivalents
Resource depletion	Global Regional Local	Quantity of minerals used Quantity of fossil fuels used	Resource depletion potential	Converts LCI data to a ratio of quantity of resource used versus quantity of resource left in a reserve
Land use	Global Regional Local	Quantity disposed of in a landfill	Solid waste	Converts mass of solid waste into volume using an estimated density

Source: Based on *Life Cycle Assessment: Principles and Practice, May 2006,* U.S. Environmental Protection Agency and Science Applications International Corporation.

guide to LCA (Guinée et al., 2002) distinguishes between three sets of impact categories:

- Group A: Baseline impact categories
  These are typically found in almost all LCA studies and include the following:
  - Depletion of abiotic resources
  - Impacts of land use (land competition)
  - Climate change
  - Stratospheric ozone depletion
  - Human toxicity
  - Ecotoxicity (freshwater aquatic ecotoxicity, marine aquatic ecotoxicity, and terrestrial ecotoxicity)
  - Photooxidant formation
  - Acidification
  - Eutrophication

- Group B: Study-specific impact categories
  These may be included depending on the goal and scope of the LCA and on whether data are available.
  - Impacts of land use (loss of biodiversity and life support function)
  - Ecotoxicity (freshwater sediment ecotoxicity, marine sediment ecotoxicity)
  - Impacts of ionizing radiation
  - Odor (air)
  - Noise
  - Waste heat
  - Casualties

- Group C: Other impact categories
  These are impact categories for which research is still ongoing and require additional studies before being included.

  - Depletion of biotic resources
  - Desiccation
  - Odor (water)

2. *Classification*
   Classification is the process of assigning emissions, wastes, and resource use to relevant impact categories. This process relies on established cause-and-effect relationships, also known as stressor–impact relationships. As we have previously discovered, multiple sources can contribute to one type of impact (e.g., CO_2 and CH_4 contribute to global climate change), and a single source can contribute to multiple impacts (e.g., NO_x contributes to acidification as well as eutrophication).

3. *Characterization*
   Characterization is quantitatively determining the impacts arising from the environmental stresses determined in the inventory stage. Different methods are used to characterize different impact categories. Some impacts are calculated using equivalence factors and impact potential values. For example, the global warming contribution of emissions can easily be calculated using equivalent CO_2 values.

Optional Elements

4. *Normalization*

   Some LCAs include a normalization step. This step involves dividing the impact values obtained from characterization by reference values in order to determine the relative magnitude and significance of the impacts caused by the system under study. The reference values are dependent on the context in which the comparison is being made. For example, global reference values will be different from regional reference values.

5. *Grouping*

   This element refers to sorting the impact results from characterization into different defined groups for analysis and presentation. For example, impacts can be sorted under global-, regional-, and local-scale concerns or high-, medium-, and low-priority impacts.

6. *Weighting*

   In this step, the relative importance of one impact is weighted against other impacts. Weighting factors are assigned in a qualitative or quantitative process that is based on perceived or actual relative importance. Weighted factors can be subjective, and there is often no agreement value used, which can sometimes be challenging for engineers who tend to favor objective metrics. It is, however, acceptable to use different values as long as they are appropriately documented.

7. *Data Quality Analysis*

   Life cycle assessments involve the collection and processing of large amounts of data that may come from a wide variety of sources. Sometimes, these data may have been estimated or extracted from literature sources that may not adequately represent the system being studied. As a result, there may be consequent uncertainty in the data and in the results. Data quality analysis is a series of statistical methods used to test the validity of the data and results obtained.

BOX 14.1 Case Study: Life Cycle Assessment of Supermarket Shopping Bags in the UK

A study commissioned by the Environment Agency, which is the leading public body charged with protecting and improving the environment in England and Wales, assessed the life cycle environmental impacts of the production, use, and disposal of different shopping bags for the UK in 2006.

The report considered only the types of shopping bags available from UK supermarkets and considered neither the ability and willingness of consumers to change behavior nor the potential economic impacts of each choice on UK businesses.

The following types of shopping bags were studied and compared.

- A conventional, lightweight carrier made from high-density polyethylene (HDPE)

- A lightweight HDPE carrier with a prodegradant additive designed to break down the plastic into smaller pieces
- A biodegradable carrier made from a starch-polyester (biopolymer) blend
- A paper carrier
- A "bag for life" made from low-density polyethylene (LDPE)
- A heavier, more durable bag, often with stiffening inserts made from nonwoven polypropylene (PP)
- A cotton bag

Functional Unit

Each of these shopping bags is designed for a different number of uses. Those intended for longer, multiple uses need more resources in their production and are therefore likely to produce greater environmental impacts if compared on a bag-for-bag basis. A comparison of life cycle environmental impacts should be based on a comparable function (or "functional unit") to allow a fair comparison of the results. The researchers decided to compare the impacts from the number of bags required for carrying one month's shopping in 2006/2007. The shopping bags studied are, however, also of different volumes, weights, and qualities, so the researchers conducted a survey that found that over a four-week period supermarket shoppers purchased an average of 446 items. They defined their functional unit as

Carrying one month's shopping (483 items) from the supermarket to the home in the UK in 2006/07.

The researchers used the conventional high-density polyethylene bag (HDPE) as the reference. They then calculated how many times each of the other types of bags would have to be used to reduce its contribution to global warming to below that for conventional HDPE bags. They considered different scenarios that included 100% secondary reuse of the HDPE bags and 40% reuse as well as no secondary reuse; the results are shown in Table 14.10.

The bags were also compared for other impacts, including resource depletion, acidification, eutrophication, human toxicity, freshwater aquatic ecotoxicity, marine aquatic ecotoxicity, terrestrial ecotoxicity, and photochemical oxidation (smog formation).

Since different raw materials, manufacturing processes, and disposal methods are applicable to the different bags being compared, it was necessary for the researchers to define an appropriate systems boundary to encompass all these life cycle stages. The system boundary used in the study is shown in Figure 14.13. A cradle-gate model was used that incorporated consideration of impacts of sub-processes such as extraction, transport, packaging, use, reuse, recycling, and disposal as applicable to each type of bag.

The study found the following results.

- The conventional HDPE bag had the lowest environmental impacts of the lightweight bags in eight of nine impact categories. The bag performed well because it was the lightest bag considered.

TABLE 14.10 Number of times primary use required to bring the global warming potential of other reusable bags below that of HDPE bags with and without secondary reuse

TYPE OF BAG	HDPE BAG (NO SECONDARY REUSE)	HDPE BAG (40.3% REUSED AS BIN LINERS)	HDPE BAG (100% REUSED AS BIN LINERS)	HDPE BAG (USED 3 TIMES)
Paper bag	3	4	7	9
LDPE bag	4	5	9	12
Nonwoven PP bag	11	14	26	33
Cotton bag	131	173	327	393

Source: Based on Environment Agency (2011). *Life cycle assessment of supermarket carrier bags: a review of the bags available in 2006*, Report: SC030148, www.environment-agency.gov.uk.

(Continued)

BOX 14.1 Case Study: Life Cycle Assessment of Supermarket Shopping Bags in the UK *(Continued)*

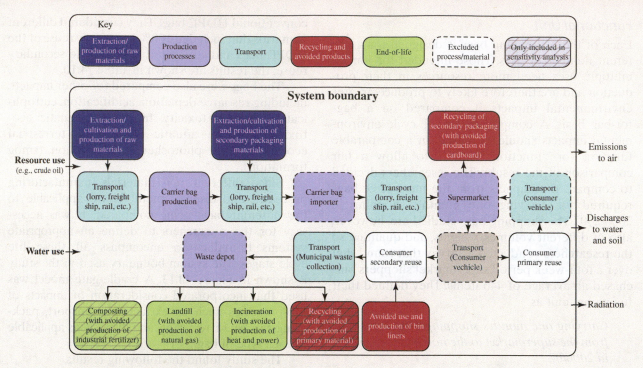

FIGURE 14.13 System boundary applied in the study.

Source: Based on Environment Agency (2011). *Life cycle assessment of supermarket carrier bags: a review of the bags available in 2006,* Report: SC030148, www.environment-agency.gov.uk.

- The starch-polyester (biopolymer) bag had the highest impact in seven of the nine impact categories considered. This was partially due to it having approximately twice the weight of the conventional HDPE bags and partially due to the high impacts of raw material production, transport, and the generation of methane from landfill at disposal.
- The environmental impact of all types of shopping bags is dominated by resource use and production stages. Transport, secondary packaging, and end-of-life management generally had a minimal influence on their performance.

- Whatever type of bag is used, the key to reducing the impacts is to reuse it as many times as possible. Where reuse for shopping is not practicable, other reuse (e.g., to replace bin-liners) is beneficial.
- The reuse of conventional HDPE and other lightweight carrier bags for shopping and/or as bin-liners is pivotal to their environmental performance, and reuse as bin-liners produced greater benefits than recycling bags.
- Starch-polyester (biopolymer) blend bags have a higher global warming potential and abiotic depletion than conventional polymer

bags, due both to the increased weight of material in a bag and higher material production impacts.

- The paper, LDPE, nonwoven PP, and cotton bags should be reused at least 3, 4, 11, and 131 times, respectively, to ensure that they have lower global warming potential than conventional HDPE carrier bags that are not reused. The numbers of times each would have to be reused when different proportions of conventional (HDPE) carrier bags are reused are shown in Table 14.10.
- Recycling or composting generally produces only a small reduction in global warming potential and abiotic depletion.

Figure 14.14 shows the comparison of all the bags based on their contribution to global warming (measured as equivalent CO_2 emitted over the life cycle). The results show that the polypropylene bag has the largest impact in this category. It also shows that a significant portion of this impact comes from the extraction and production of raw materials for all bags.

While the HDPE bag has the lowest impact across all impact categories, Figure 14.15 shows that this impact can be significantly reduced with increased secondary reuse. Secondary reuse is, however, a function of consumer habits and choices, which need to be considered in impact reduction decisions.

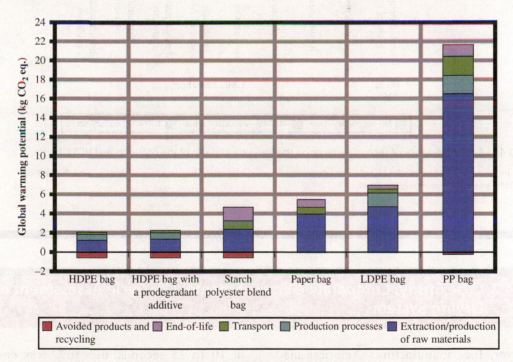

FIGURE 14.14 The life cycle contribution of each shopping bag to global warming.

Source: Based on Environment Agency (2011). *Life cycle assessment of supermarket carrier bags: a review of the bags available in 2006,* Report: SC030148, www.environment-agency.gov.uk.

(Continued)

BOX 14.1 Case Study: Life Cycle Assessment of Supermarket Shopping Bags in the UK (Continued)

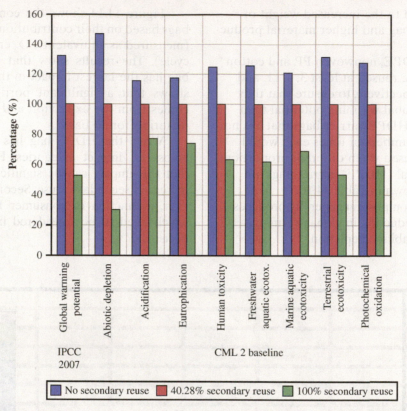

FIGURE 14.15 The influence of secondary reuse on the life cycle impacts of the conventional HDPE bag.

Source: Based on Environment Agency (2011). *Life cycle assessment of supermarket carrier bags: a review of the bags available in 2006*, Report: SC030148, www.environment-agency.gov.uk.

BOX 14.2 Case Study: Comparative Environmental Life Cycle Assessment of Hand Drying Systems

Excel Dryer, Inc., manufactures American-made hand dryers for schools, hospitals, restaurants, and many other commercial facilities. Their hand dryers, brand named XLERATOR®, feature patented technology that is marketed to dry hands in 10 to 15 seconds, use 80% less energy than conventional dryers, and offer 95% cost savings over paper towels. The company believed that, while a comparison to other hot-air hand dryers in the use phase of the product life cycle was easy, it was also necessary to know how their product compared with conventional hand dryers in other stages of the life cycle considering a complete set of environmental performance metrics. The company commissioned Quantis, a consulting firm specializing in life cycle assessments, to compare the life cycle impacts of several systems for drying hands in public restrooms: the XLERATOR dryer, a conventional electric dryer, and paper towels containing between 0% and 100% recycled content.

Functional Unit

The primary function of the products studied is to dry hands after washing in a public restroom. The functional unit provides a basis for comparing all life cycle components on a common basis: namely, the amount of that component required to fulfill the described function. The researchers assumed that the lifetime of a single installation of an electric dryer is 10 years, and on the basis of approximately 500 uses per week, they defined their functional unit as

Drying 260,000 pairs of hands

The system boundaries for all three choices included raw materials extraction and processing, production and assembly of subcomponents, consumer use, and disposal scenarios. Figure 14.16 shows the system boundary used for the XLERATOR system, Figure 14.17 shows the system boundary

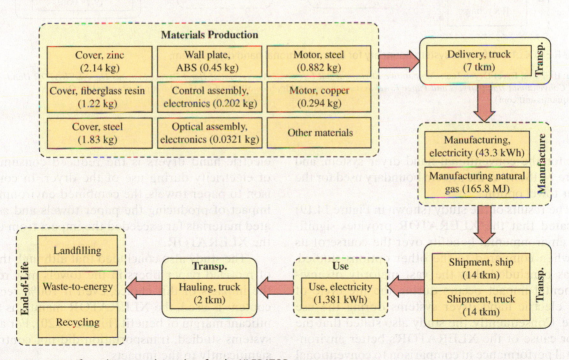

FIGURE 14.16 Life cycle system boundary for the XLERATOR system.

Source: Based on Excel Dryer, Inc., *Comparative Environmental Life Cycle Assessment of Hand Drying Systems: The XLERATOR Hand Dryer, Conventional Hand Dryers and Paper Towel Systems*, www.exceldryer.com/PDFs/LCAFinal9-091.pdf, prepared by Quantis (www.quantis-intl.com).

(Continued)

BOX 14.2 Case Study: Comparative Environmental Life Cycle Assessment of Hand Drying Systems *(Continued)*

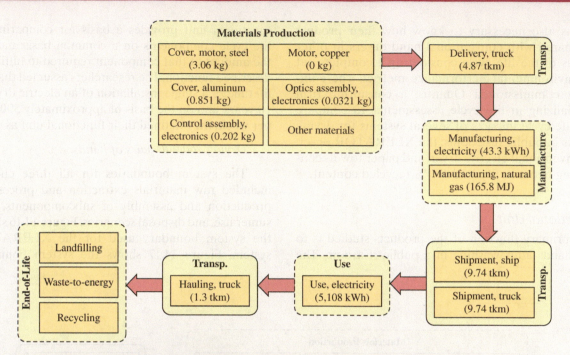

FIGURE 14.17 Life cycle system boundary for the conventional hand dryer system.

Source: Based on Excel Dryer, Inc., *Comparative Environmental Life Cycle Assessment of Hand Drying Systems: The XLERATOR Hand Dryer, Conventional Hand Dryers and Paper Towel Systems*, www.exceldryer.com/PDFs/LCAFinal9-091.pdf, prepared by Quantis (www.quantis-intl.com).

used for the conventional hand dryer system, and Figure 14.18 shows the system boundary used for the paper towel option.

The results of the study (shown in Figure 14.19) indicated that the XLERATOR provides significant environmental benefits over the course of its life when compared to the other options studied. It was concluded that the vast majority of environmental impact occurs during the life cycle of both electric hand dryer systems during the use phase. Consequently, the study also stated that the major cause of the XLERATOR's better environmental performance in comparison to conventional

electric hand dryers is the reduced consumption of electricity during use of the dryer. In comparison to paper towels, the combined environmental impact of producing the paper towels and associated materials far exceeded the impact from use of the XLERATOR.

The study also concluded that although the use of recycled paper fibers in the towels may reduce the impacts of this choice, even if 100% recycled content is used, the XLERATOR maintains a significant margin of benefit (Figure 14.20). For all the systems studied, transportation did not contribute significantly to the impacts.

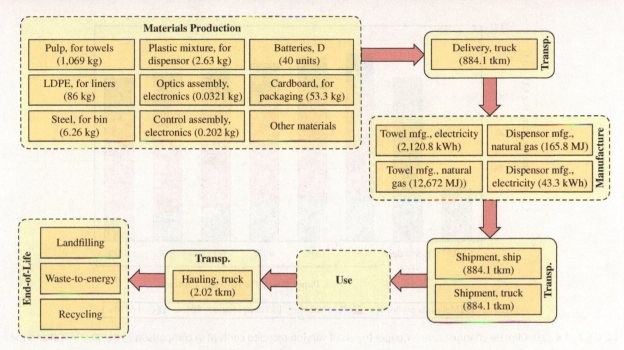

FIGURE 14.18 Life cycle system boundary for the paper towels system.

Source: Based on Excel Dryer, Inc., *Comparative Environmental Life Cycle Assessment of Hand Drying Systems: The XLERATOR Hand Dryer, Conventional Hand Dryers and Paper Towel Systems*, www.exceldryer.com/PDFs/LCAFinal9-091.pdf, prepared by Quantis (www.quantis-intl.com).

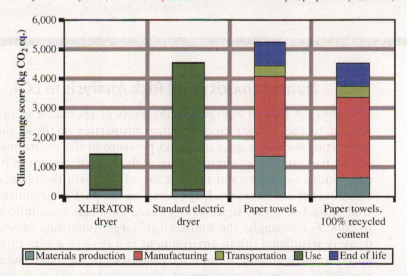

FIGURE 14.19 Total life cycle climate change score for each of the systems.

Source: Based on Excel Dryer, Inc., *Comparative Environmental Life Cycle Assessment of Hand Drying Systems: The XLERATOR Hand Dryer, Conventional Hand Dryers and Paper Towel Systems*, www.exceldryer.com/PDFs/LCAFinal9-091.pdf, prepared by Quantis (www.quantis-intl.com).

(Continued)

BOX 14.2 Case Study: Comparative Environmental Life Cycle Assessment of Hand Drying Systems *(Continued)*

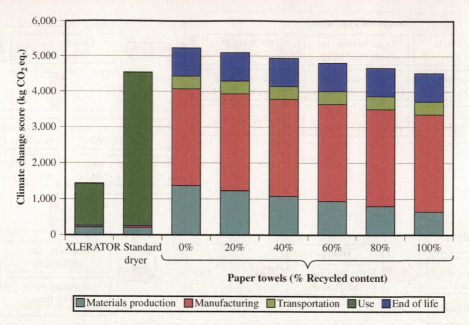

FIGURE 14.20 Climate change score for paper towels of varying recycled content in comparison to the XLERATOR and the conventional electric hand dryer.

Source: Based on Excel Dryer, Inc., *Comparative Environmental Life Cycle Assessment of Hand Drying Systems: The XLERATOR Hand Dryer, Conventional Hand Dryers and Paper Towel Systems*, www.exceldryer.com/PDFs/LCAFinal9-091.pdf, prepared by Quantis (www.quantis-intl.com).

14.5   Human Toxicity and Risk Analysis in LCA

The physical flow of energy and the mass of chemicals in the environment are related to the physical and chemical properties of those chemicals and other properties such as temperature and pressure in the environment. An evaluation of the true impacts of emissions in a given environment is insufficient without an attempt to understand the risk associated with the changing levels of emissions in that environment. Depending on the particular emission of concern, the true impact may vary significantly based on social, economic, and environmental factors. For example, the emission of large quantities of some chemicals in a densely populated urban environment can lead to a very different impact compared to the emission of the same quantities of the same chemicals in a sparsely populated or desert environment. Unfortunately, life cycle analysis is limited in how well it helps us determine the true impacts. For this, we use other tools for risk analysis coupled with life cycle analysis. Many measures of risk may affect

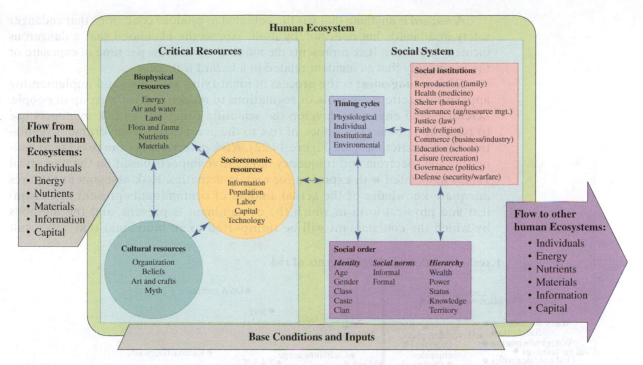

FIGURE 14.21 The human ecosystem model, demonstrating considerations for risk analysis when evaluating and comparing the potential risks between industrialized processes.

Source: Based on Machlis, G.E., Force, J.E., and Burch, W.R. Jr. (1997, July 1). "The human ecosystem Part I: The human ecosystem as an organizing concept in ecosystem management." *Society & Natural Resources: An International Journal* 10(4):347–367. Copyright © 1997 Routledge, reprinted by permission of the publisher (Taylor & Francis Ltd, www.tandf.co.uk/journals).

human health, behavior, and communities, as illustrated in the diagram shown in Figure 14.21.

The ability to calculate this risk is subject to many interpretations in toxicity and exposure data. Certain assumptions must also be made to determine the appropriate risk. For example, an environmentally related risk calculation might assume that a person lives in the same home location for 70 years and is exposed to a given environmental substance present in the home or local environment 24 hours per day, 365 days per year for 70 years. Exposure in the workforce might be calculated based on a 30-year career, with exposure 8 hours per day for 250 days per year. Risk factors need to be estimated, and these estimates may be used to make decisions about publicly acceptable levels of risk compared to publicly accepted levels of regulation. Typically, a human health risk assessment includes four steps:

- Hazard identification
- Dose–response assessment
- Exposure assessment
- Risk characterization

A *hazard* is anything that has the potential to produce conditions that endanger safety and health, but does not, by itself, express the likelihood that a dangerous incident will occur. Risk represents the number of incidents per time of exposure or the probability that an incident related to a hazard will occur.

Risk management is the process of identifying, selecting, and implementing appropriate actions, controls, or regulations to reduce risk for a group of people. Scientists and engineers develop the scientific comparisons risk managers use to relate the public acceptance of risk to the acceptance of government regulations and restrictions (see Figure 14.22). *Risk assessment* is the process of estimating the spectrum and frequency of negative impacts based on the numerical values associated with exposure for certain activities. Risk assessment requires adequate knowledge of the actual amount of contaminants present, the chemical and physical form in which the contaminant is present, and the methods by which the contaminant will be transported to an individual that will result

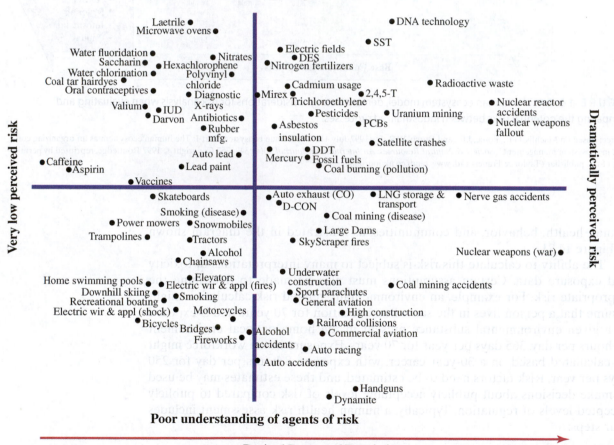

FIGURE 14.22 The public's perception of risk may differ dramatically from that of someone trained in risk assessment. Public perception may be related to the degree of the perceived risk and the understanding of the perceived risk.

Source: Reprinted with permission from Stern, P.C., and Fineberg, H.V. (Eds.) (1996). *Understanding Risk: Informing Decisions in a Democratic Society.* Committee on Risk Characterization, Commission on Behavioral and Social Sciences and Education, Figure 2.4, page 62, National Research Council, 1996, by the National Academy of Sciences, Courtesy of the National Academies Press, Washington, DC.

in an exposure pathway. Toxicity tests have been developed to enhance the understanding of how organisms and ecosystems respond to contaminants. *Toxicology* is the study of the adverse effects of contaminants on biological systems. Toxicity tests typically involve experiments on animals (often microorganisms), large populations, and extrapolation to broader ecological responses or human effects. Toxicity tests may be used to comply with government water quality standards and to identify the toxicity of specific substances. It has been found that the toxicity of effluent discharges of wastewater correlates well with toxicity measurements for similar contaminants in receiving waters and ecological community responses for many toxicity measurement procedures.

The *dose* of a contaminant is defined as the amount of a chemical received by an individual that can interact with an individual's biological functions and receptors. The dose typically has units measured in mg of contaminant per kg of the subject's body mass (mg/kg). The dose may be any one of the following:

- The amount of a contaminant that is administered to the individual
- The amount administered to a specific location of an individual (for example, the kidneys)
- The amount available for interaction after the contaminant crosses a barrier such as the skin or stomach wall

The response of an organism to a contaminant may be classified as an acute or chronic effect. An acute effect is a response to short-term hazards that result in immediate symptoms. Chronic effects are long-term effects that result from an exposure, and these effects may or may not disappear when the hazard is removed. Some types of contaminants with long-term chronic effects include:

- Carcinogens
- Teratogens
- Neurotoxins
- Mutagens

Epidemiology is the study of the cause of disease and adverse responses. Epidemiological work may be based on historical data and toxicity data. Epidemiological studies are often complicated by a lack of information about the duration and exposure to a contaminant, the change in exposure as individuals and populations change locations, long lag times, and multiple pathways of exposure to multiple contaminants. The lack of available data and data variability create a high level of uncertainty in the results of epidemiological and hazard assessment. However, historical examples of a few specific environmental exposure cases have led to several models that are useful to estimate human risk based on exposure data and toxicological data.

Extrapolation of toxicity data to human response is quite difficult. Hazard identification or assessment involves predicting adverse human responses and the severity of the response to contaminants. Human-based risk assessment for environmentally caused illness is confounded by several factors, including:

- Environmental illness may be indistinguishable from other diseases.
- The relationship between a response (such as cancer) and the environment is not always recognizably different from other populations and environments.
- Chronic effects may be complicated by long latency periods, and the negative response may not occur until long after the exposure has stopped.

- Serious diseases such as respiratory disorders, neurological disorders, and cancer are not generally recognized or classified as potentially work-related by health care providers.

The process of estimating the duration, frequency, and magnitude of contact with a contaminant in a given population is called *exposure assessment*. The total exposure to a study population may be estimated from measured concentrations in the environment, consideration of models of chemical transport and fate, and estimates of human intake over time. The exposure assessment must also consider both the pathway of exposure (how a contaminant moves though the environment to contact the study population) and the route of exposure (how the contaminant enters an individual's body).

In both health assessment and toxicity evaluations for other species, bioaccumulation of contaminants in the environment is an important consideration in risk evaluation. Risk assessment is a complex process with a high level of uncertainty that should be considered as part of the analysis. Data for risk assessments should be collaboratively gathered and reviewed in order to make quantitative decisions about risk.

A full life cycle assessment attempts to characterize and communicate effectively the environmental and human health impacts of all the emission from processes or products. These effects can be negative effects caused by pollutants or positive, for example wastewater treatment processes have a long history of demonstrated positive impacts on human health and well-being. Life cycle assessment models attempt to use quantifiable relationships for how all chemicals emitted in the process steps are:

- Transported in the environment
- React in the environment
- Follow a pathway for human exposure
- Cause human health impacts due to the potential toxicity of the substances being modeled

As we have seen throughout the textbook, each step of this process is dependent upon many variables, such as the specific chemicals involved in the study; the mass of each substance; the characteristics of the local air, water, and soil; the toxicity of the chemicals; and the possible routes of exposure as presented in prior sections of this text. Human health impacts themselves are highly variable based upon age, access to appropriate healthcare, and natural differences in human physiology. Therefore, any life cycle assessment study is limited by the assumptions made both in the limitations of the data available for the specific case being investigated and in the limitations of the life cycle inventory database used in making reasonable assumptions within the model framework. Therefore, the result of LCA studies can lead to important insights and improve our understanding of the steps that might be critical in improving the sustainability of a process or product. However, it is also critical to understand that the assumptions and limitations that make most LCA studies useful for broad system analysis also make LCAs imprecise in evaluating a very specific case or condition.

Risk assessment tools are useful for estimating or predicting environmental and human health risks associated with various processes. These risks contain significant uncertainties that should be communicated to the public policymakers and designers.

ACTIVE LEARNING EXERCISE 14.4 How Would You Quantify Impacts?

Identify the major sources of impact in the life cycle of your school. Which phase contributes most to the impact?

14.6 Summary

Life cycle assessment is a useful tool for evaluating the comprehensive impacts from the manufacture, use, and disposal of products. LCA techniques rely on compiling databases of relevant energy, material inputs, and environmental impacts; evaluating potential impacts associated with process waste; and providing a simplified scale for interpreting modeling results. The results of an LCA can illustrate comparative differences in environmental and health impacts, such as whether product A or product B has a greater likely impact on climate change, as shown in the previous examples.

There are, however, significant limitations to LCA results that the user must recognize and interpret. Comprehensive LCAs require large amounts of resources and time. Accurate data collection is central to a reliable assessment, and the value of an LCA is only as good as the data used. LCAs are usually performed with truncated boundaries to limit the amount of extraneous data, which implies a compromise for practicality. While LCAs offer insights into the environmental performance of products, they do not provide information on cost effectiveness, product efficiency, or any social considerations. LCAs produce good global-scale impact results, though assessments of local and contextual impacts are still challenging. In general, LCAs are considered to be decision support tools that are used in addition to other points of consideration. The simplified LCA modeling results are suitable for relative comparisons of environmental and health impacts, but they do not indicate actual measurable responses or the risk levels associated with the environmental and health risks. For quantifiable estimations of environmental impacts, the environmental models presented in earlier chapters should be used, and detailed risk analysis procedures are required to quantify risk and exposure to specific emissions or scenarios.

The U.S. EPA and ISO standard 14043 recommend three key steps in interpreting the results of an LCA:

1. Identification of the significant issues based on the LCA
2. Evaluations that consider completeness, sensitivity, and consistency checks
3. Conclusions, recommendations, and reporting

Due to the large quantities of data that have to be collected and processed for any LCA, computer software is used to perform most LCAs. Many institutions and companies have developed their own software using data specific to their needs and concerns. There is, however, commercially available software that utilizes comprehensive databases of materials, energy, and emissions inventories as well as multiple impact assessment methods. Some software is used to simply model the *Inventory Analysis* phase of the LCA, while other software is useful for more complete LCAs that include *Impact Assessment* and *Interpretation* of results. Table 14.11 shows examples of some currently available LCA software and their developers.

TABLE 14.11 Some currently available LCA software

TOOL	WEBSITE	DEVELOPERS
EIO-LCA	*eiolca.net*	Carnegie Mellon University Green Design Institute
GaBi	*gabi-software.com*	Sphera Solutions GmbH
GREET	*greet.es.anl.gov*	Argonne National Laboratory
IDEMAT	*idematapp.com*	Delft Univ. of Technology
SimaPro	*pre-sustainability.com*	PRé Consultants

References

Environment Agency. (2011). Evidence: Life Cycle Assessment of supermarket carrier bags: A review of the bags available in 2006. Report: SC030148, www.environment-agency.gov.uk.

Excel Dryer, Inc. Comparative Environmental Life Cycle Assessment of Hand Drying Systems: The XLERATOR Hand Dryer, Conventional Hand Dryers and Paper Towel Systems (www.exceldryer.com/PDFs/LCAFinal9-091.pdf), prepared by Quantis (www.quantis-intl.com).

FAO. (2010). Greenhouse gas emissions from the dairy sector, a Life Cycle Assessment. Food and Agriculture Organization of the United Nations, Rome, Italy.

Guinée, J. B., Gorrée, M., Heijungs, R., Huppes, G., Kleijn, R., Koning, A. de, et al. (2002). *Handbook on Life Cycle Assessment.* Operational guide to the ISO standards. I: LCA in perspective. IIa: Guide. IIb: Operational annex. III: Scientific background. Dordrecht: Kluwer Academic Publishers.

Hoekstra, A. Y. (2010). "The water footprint of animal products". In J. D'Silva and J. Webster (Eds.). *The meat crisis: Developing more sustainable production and consumption. London,* UK: Earthscan, pp. 22–33.

McMichael, A. J., Powles, J. W., Butler, C. D., and Uauy, R. (2007). "Food, livestock production, energy, climate change, and health." *The Lancet* 370: 1253–1263.

National Peanut Board. (2014). Accessed at www.nationalpeanutboard.org/classroom-funfacts.php.

Ogino, A., Orito, H., Shimada, K., and Hirooka, H. (2007). "Evaluating environmental impacts of the Japanese beef cow–calf system by the life cycle assessment method." *Animal Science Journal* 78:424–432.

U.S. Environmental Protection Agency and Science Applications International Corporation. *LCAccess—LCA 101.* 2001. Exhibit 4-1, Retrieved from www.epa.gov/ORD/NRMRL/lcaccess/lca101.htm.

U.S. EPA. Accessed 2014 at www.energystar.gov/index.cfm?c=cfls.pr_cfls_lumens.

Key Concepts

Life cycle assessment (LCA)

Functional unit

Inventory analysis

Impact assessment

Depletion

Pollution

Disturbances

Radiative forcing

Radiative efficiency

Global warming potential (GWP)

Integrated time horizon (ITH)

Total equivalent CO$_2$

Ozone depletion potential (ODP)

Photochemical ozone creation potential (POCP)

Hazard

Risk management   Dose
Risk assessment   Epidemiology
Toxicology   Exposure assessment

Problems

14-1 Consider a simple cotton T-shirt. Develop a life cycle diagram similar to those shown in Figures 14.5 and 14.6 for the cotton shirt. Identify the raw materials, major manufacturing processes, distribution processes, use phase, and end-of-life options for the T-shirt in your diagram.

14-2 Search for a life cycle analysis case study on manufacturing a cotton T-shirt. Review the LCA report and answer the following questions:
  a. Does the case study clearly define/identify the boundaries for the system/product/process being studied?
  b. What are the data sources for the case study?
  c. What, if any, are the indications for data quality? Is there any discussion of error related to the data employed?

14-3 Create a life cycle diagram similar to those shown in Figures 14.5 and 14.6 for the conversion of miscanthus grass to ethanol that may be used in developing a life cycle analysis. The pathway should include, at a minimum, the following components:
  a. Miscanthus farming
  b. Miscanthus transportation
  c. Ethanol plant production
  d. Ethanol transport and distribution
  Identify the inputs required for the completion of each component step in your pathway.

14-4 Describe how you would develop a life cycle analysis of the following grocery store bag options: lightweight plastic bags, mid-weight paper bags, and heavyweight cotton reusable bags.
  a. Describe the goal and scope of your analysis.
  b. List the units of measure that are important in the analysis.
  c. Describe in one paragraph which option you think is likely to be considered most sustainable.
  d. Find a comparative analysis of grocery bag options in the literature and compare the findings of the LCA to your assumptions above. Describe the similarities and differences in a few concise paragraphs. Remember to cite the source of your information.

14-5 The worldwide demand for palm oil has grown over the past few decades at a rate of 7.1% per year. The versatility of palm oil in various applications has made it one of the top 17 oil and fat sources in the world. It not only assists in meeting the demands of edible oil worldwide, but it is also used extensively for oleo-chemicals and biofuel production. Biodiesel is an alternative to fossil diesel and is produced from vegetable oils. The conversion process is a simple chemical reaction that splits the glycerine out of the oil to form an ester. The reaction is carried out by adding methanol or ethanol to the oil, with an additional catalyst, to produce a methyl or ethyl ester, which is what we call biodiesel. Malaysian palm oil can be converted into a methyl ester (PME). In practice, any commercial biodiesel production operation may use a range of feedstock oils to produce biodiesel, rather than a single oil. The main stages of palm oil production are summarized in the table below. You have been asked to create a proposal to conduct an LCA to compare palm-based biodiesel to petroleum-based diesel fuel.

a. Describe the goal and scope of your study.
b. Provide a detailed schematic that illustrates the pathway of material, energy, and waste flows considered in your study.
c. Specify the functional units to be used in your analysis and comparison.
d. Describe the social considerations, environmental considerations, and economic considerations and limitations of an LCA model.

STAGE	PALM OIL METHYL ESTERIFICATION PROCESS
Crop cultivation	Farming inputs to growing crops
Crop processing	Production of palm oil on site
Feedstock transport	Transport of raw palm oil to a refinery
Feedstock processing	Refining palm oil
Conversion	Conversion of palm oil to PME and byproduct glycerine production
Biodiesel transport	Importation of PME to your location from Malaysia
Use	Use in tractor-trailer truck transportation

14-6 U.S. retail sales of candles are estimated at approximately $3.2 billion (2017 USD) annually, excluding sales of candle accessories. You are in a team competition to design the "greenest," most "environmentally friendly" container candle. For this analysis, we will assume the candle wax will be held within a container of the material of your choice, and a container candle is a non-flammable container filled with wax and a wick. The design will be judged for intended use and sustainability. The chosen design elements will be incorporated into a candle that you can build for yourself (if you so choose).

a. Describe the weight of the candle.
b. Describe the diameter of the candle.
c. Create a list of materials you need to make the candle.
d. Describe how you will judge whether your candle is more sustainable (greener, more environmentally friendly) than the candles of other teams. Create a list of metrics related to this.
e. Give a unit of measure for each of the metrics you listed in part d.
f. Briefly describe how each *material* in the candle affects the properties of the design related to making a properly burning candle. Online resources may be useful to help you develop your ideas.
   i) Briefly describe the burning process and the *role of the wax* in that process.
   ii) Briefly describe the *role of the wick* in the burning process, including a description of what type of wick might be most appropriate for your intended use.
g. For each of the materials chosen in part f, describe the *raw* materials associated with them. Find a possible geographic location and repository for the raw materials.
h. Sketch the material flow process from cradle to grave (or cradle to cradle) for each material choice.

i. Describe the *purpose* of your group's life cycle assessment of a candle.
j. What types of output do you envision for your expected product from the LCA? Describe categories of impacts and their possible relevant units.
k. Determine the total mass of material needed to produce a year's worth of candles. Determine the distance from the source of each material, through your process from raw materials to manufacturing, to the location of your manufacturing facility. Make any appropriate assumptions that are necessary. Estimate the carbon dioxide emissions involved in moving this amount of material if diesel fuel is used.
l. Note the expected *limitations* required to begin the LCA.

## CHAPTER 15

Assessing Alternatives

FIGURE 15.1 Every decision made about a product or process design has a ripple effect—not unlike what you might observe when you throw a pebble into a pond. The impacts that are identified depend on your perspective, the scope of your analysis, and the tools you use to assess them.

Source: YJ.K/Shutterstock.com.

Do the best you can until you know better. Then when you know better, do better.

—MAYA ANGELOU

GOALS

THE EDUCATIONAL GOALS OF THIS CHAPTER are to introduce alternatives assessment as a structured decision support framework and to show how it can be applied to advance safer and more sustainable chemicals, materials, products, and processes through governmental regulations and private industry. This chapter builds on earlier chapters that introduced hazard assessment, risk assessment, exposure assessment, life cycle assessment, and more. It describes how results from these other assessment tools and frameworks can be incorporated into decisions made using the broader umbrella of alternatives assessment. Readers should also recognize how alternatives assessment can bring together scientific information and social values in a transparent way to support decision making.

OBJECTIVES

At the conclusion of this chapter, you should be able to:

15.1 Define alternatives assessment.

15.2 Describe the elements of alternatives assessment.

15.3 Cite examples of alternatives assessment used in practice.

15.4 Describe how alternatives assessments are being used by governments.

15.5 Describe how alternatives assessments are being used by businesses.

15.6 Identify important resources that support alternatives assessment.

Introduction

Throughout this text, we have sought to strengthen and clarify the linkages between traditional engineering practices and sustainability. We argue that the best way to do so is to treat sustainability as a proactive design imperative rather than as a reactive response intended to mitigate the negative impacts of waste, toxins, and overconsumption. Earlier chapters focused on traditional treatment technologies as well as sustainable design philosophies, principles of green and sustainable chemistry and engineering, and sustainable materials management. We also reviewed tools such as life cycle assessment, hazard assessment, risk assessment, and industrial ecology, all of which are designed to generate data in support of design goals. However, what has been missing until this point is a discussion of how to use the information derived from disparate assessment tools to inform decision making that drives continual improvement toward sustainable design objectives. As William McDonough wrote, "You don't filter smokestacks or water. Instead, you put the filter in your head and design the problem out of existence." To do so requires systems thinking based on a shared understanding of the systems, tools, and metrics that allow scientists and engineers to measure progress. Every decision made about a product or process design has a ripple effect—not unlike what you might observe when you throw a pebble into a pond. The impacts that are identified depend on your perspective, the scope of your analysis, and the tools you use to assess them. No one tool or framework can assess all potential impacts. *Alternatives assessment (AA)* is a way of gathering and integrating information from different assessment tools to inform decision making and to avoid regrettable substitutions. AA is intended to help people avoid going from the proverbial frying pan into the fire when it comes to moving away from unsustainable materials and products and toward those that are more sustainable. The poet and activist Maya Angelou proffered that we should do the best we can with what we know. And that when we know better, we should do better. Alternatives assessment is intended to help with the journey of continual improvement.

15.1   Alternatives Assessment

Alternatives assessment is an emerging science and policy approach to assessing *aspects* and *impacts* associated with materials, products, technical systems, industrial systems, or structures. The philosopher David Hume wrote about the gap between what "is" and what "ought" to be. The scientific method is about what is; it does not tell us what to do about it. Policy is intended to help drive what society perceives as what ought to be. When we deal with science and policy approaches, it is important not to jump from what is to what ought to be.

The terms *aspects* and *impacts* come from their use in environmental management systems. An aspect is an element of an organization's activities, products, or services that has or may result in an impact, positive or negative, on the environment or human health. For example, a discharge of cleaning agents in wastewater is an aspect that may or may not result in the negative impact of water pollution.

In AA, the relevant aspects and impacts are identified and assessed individually; then a decision analytic framework is applied to allow for a holistic comparison of alternatives with consideration of all the relevant aspects and impacts. Examples of aspects and impacts addressed in AA include hazard, exposure, cost and availability, performance, energy quality and quantity, waste generation and recovery, and stakeholder engagement. AA is intended to drive *continual improvement* toward sustainable design objectives, including the use of safer chemicals, reduced carbon footprints, and greater material circularity that helps to preserve natural capital (chemistry, carbon, and circularity). This is consistent with the sustainable design objectives described in Chapter 12.

Chemical alternatives assessment (CAA) is a more focused form of AA. It is a process for identifying and comparing potential chemical and nonchemical alternatives that can be used as substitutes to replace chemicals of high concern.

Several developments led to the emergence of alternatives assessment. In the late 1990s, a shift in thinking began among environmental policymakers and regulators. They went from focusing primarily on identifying and regulating highly toxic or problematic substances to also identifying inherently safer ones. The term "whack-a-mole" comes from an arcade game that originated in Japan as "Mogura Taliji." Players use a mallet to pound down small plastic moles that pop up from holes. As soon as one mole is pounded back into its hole, another pops up from another hole. This phrase has also been applied to the experience of regulating toxic substances. Too often, regulating one substance has led to societal use of replacements that are also problematic. Replacing a toxic regulated substance with another toxic substance is called a ***regrettable substitution***. An example of a regrettable substitution occurred when the use of methylene chloride, a suspected carcinogen and endocrine disruptor used in brake cleaners, was phased out through regulations. The regulations did not specify options to use instead. Many brake shops switched to the use of *n*-hexane, which was subsequently found to be neurotoxic. Society is not best served by playing whack-a-mole with such critical matters. A few key paradigms that emerged at that time are cradle-to-cradle design, green chemistry and engineering, and design for the environment. These design paradigms are still very relevant today and seek to proactively promote products and processes that are safe and sustainable by design.

Implementing sustainable design paradigms required the development of new tools, including the ability to assess and consistently compare chemicals for their inherent hazard. European REACH legislation was implemented in 2007 based in part on the principle "no data, no market." In other words, chemicals should have sufficient toxicological data for review before they are allowed on the market in consumer products. The Globally Harmonized System of Classification and Labelling of Chemicals (GHS) was launched in 2003 and is updated every two years (United Nations, 2021). Additional chemical hazard assessment methods were developed that built on GHS and added additional hazard endpoint criteria as needed for inherent chemical attributes, such as persistence in the environment, bioaccumulation potential, and endocrine disruption. As chemical hazard assessments proved their utility in supporting decision making, the need emerged to centralize and increase chemical hazard assessment information in databases in order to scale access to it and to share information along the supply chain.

With the shift from focusing on chemicals of concern to focusing on safer alternatives, and with improved ability to assess and compare chemicals for hazard, it became apparent that a safer alternative may not be a viable alternative if the only consideration is hazard. Alternatives must also be acceptable with respect to exposure, performance, cost and availability, life cycle impacts, and social impacts. In her book *Making Better Environmental Decisions* (2000), Mary O'Brien made the case for AA. She wrote that "one of the most essential, and powerful steps to change is understanding that there are alternatives."

AA as a methodology was further developed in frameworks and guidance documents produced by the National Academy of Sciences and the Interstate Chemicals

Clearinghouse (IC2), a coalition of state governments in the United States. The frameworks and guidance outline how to use the various components of AA in an iterative way and how to apply decision analysis. Many of the tools discussed in earlier chapters, including hazard assessment, risk assessment, life cycle assessment, economic analysis, and so on, help to answer specific questions that are important but not sufficient for making decisions. For example, life cycle assessment helps to answer questions about comparative impacts primarily related to energy consumption and material inputs and outputs. Economic analysis helps to answer questions about costs that may be incurred over the life cycle of a product. Alternatives assessment, meanwhile, helps to guide decision making in a transparent way when there are multiple attributes to consider.

When faced with the challenge of an unsustainable chemical, material, product, or process, perhaps the first and most important question to ask is "Is it necessary?" Some things are overengineered with respect to performance. If the product is deemed necessary, then the alternatives assessment process begins in order to see if the need can be met with the use of other chemicals, materials, product designs, and so on. People typically look first to see if there might be a simple replacement that would not involve major changes to the current system. This is called a ***drop-in replacement***. For example, if a paint stripper contains methylene chloride, which is toxic, why not simply replace it with water, which is not? One reason is that water doesn't work as a paint stripper. Sometimes the properties of the chemical of concern that make it toxic are related to the properties that make it functional. In other situations, a single chemical substitute *can* work, but it may mean having to reconfigure

BOX 15.1 But There Is a Bridge!

Imagine a traveler standing by an icy mountain river which they intend to cross over to the other side. A team of four risk assessors review this situation. The first risk assessor is trained in toxicology and recommends that the traveler wade across the river because the risks are low; it is not toxic, only cold. The second risk assessor is trained in cardiology and recommends that the traveler wade across the river because the traveler seems young and is not yet chilled, and therefore the risk of cardiac arrest is low. The third risk assessor has a specialty in hydrology and recommends that the traveler wade across the river because, based on the study of other rivers, this one is no more than 4 feet deep and probably has no whirlpools and therefore the risk of drowning is low. Finally, the fourth risk assessor is trained in ecological risk assessment. That person recommends that the traveler wade across the river because, compared to global climate change, ozone depletion, and loss of species diversity, the risks to ecosystems from

the traveler crossing the river are trivial. Each of the assessors argues that the risks of great harm from wading across the river are low.

Despite all of this, the traveler refuses to wade across the river. When asked why, the traveler points and states: "Because there is a bridge over there."

The risk assessors in this story are evaluating the risks of only one option: wading across an icy river. The traveler is evaluating the alternatives, one of which involves crossing the river on a nearby bridge. The traveler does not really care whether getting wet in the icy stream will kill them or not, because it doesn't make sense to them to even become chilled, or even wet, in light of their options. This is an illustration of how alternatives assessment helps to answer questions that may not be addressed by tools with narrower scope, such as risk assessment.

Source: Based on O'Brien, M. (2000). Making Better Environmental Decisions: An Alternative to Risk Assessment. Cambridge, MA: The MIT Press.

the formulation. Imagine wanting to substitute almond flour for wheat flour to make a chocolate cake. It can be done, but you cannot just assume that almond flour is a drop-in replacement that can be added in the same way and amount as wheat flour and cooked for exactly the same time. If you do, the result will likely be a (tasty) flop. You will need to adjust the amount of flour, liquid, and fat.

When there are no chemicals that can be readily substituted, designers need to think more broadly. Perhaps the solution is not a chemical substitute at all. Can paint be stripped from furniture with heat, sandblasting, or light? If so, what are the trade-offs? How do these other options compare with respect to performance and cost? Why does the paint need to be stripped in the first place? Perhaps the long-term solution is an alternative to paint that does not require stripping. Can products be decorated in ways that do not require paint stripping, or even paint for that matter? Maybe an alternative material can be used that already contains colorant and that can be recycled at the end of its useful life. Like the ripples from dropping a pebble into a pond, ideas for solutions may start very close to current practices but then extend out further and further as creative minds think of solutions at different scales. Remember, the point is to consider the service that is being provided and to think of effective and desirable ways to provide the same service. The point is not necessarily to simply find a safer paint stripper that will work with conventional practices. While it is human nature to resist being forced to change, alternatives assessment can be a driver for innovation and continual improvement.

15.2   Elements of AA

15.2.1 NRC Framework

The U.S. National Academies of Sciences National Research Council (NRC) convened a committee to develop a framework for the design and evaluation of safer chemical substitutions that could be used to inform government and industry decisions. The framework outlines 13 steps that are structured to support decision making about alternatives to chemicals of concern (see Figure 15.2). The framework is designed to be flexible enough so that an assessor can complete certain steps sequentially, in parallel, or iteratively, allowing the assessor to customize the approach to fit the scenario. Some steps are considered essential, while others are considered optional.

> Step 1: Identifying the ***chemical of concern*** is the first step. A chemical of concern is a chemical used in products that has toxicity or other effects that are deemed to create undesirable risks. The chemical of concern may be identified through regulations, but it can also be driven by corporate policies or public concern. For example, some manufacturers of water bottles worked to eliminate bisphenol A (BPA) from plastics used for water bottles in advance of any regulatory requirements.

> Step 2: Next is scoping and problem formulation. While scoping may sound rather perfunctory, it is hard to underestimate the importance of this step. During the scoping step, the assessor defines and documents the goals, principles, and decision rules that will guide the rest of the assessment and thus the outcome of the assessment. Many decisions in an AA are not purely technical, but rather are value-driven or context-dependent. Step 2 is the planning phase of the assessment because it involves determining which health effects, exposure pathways, life cycle segments, and performance attributes will be

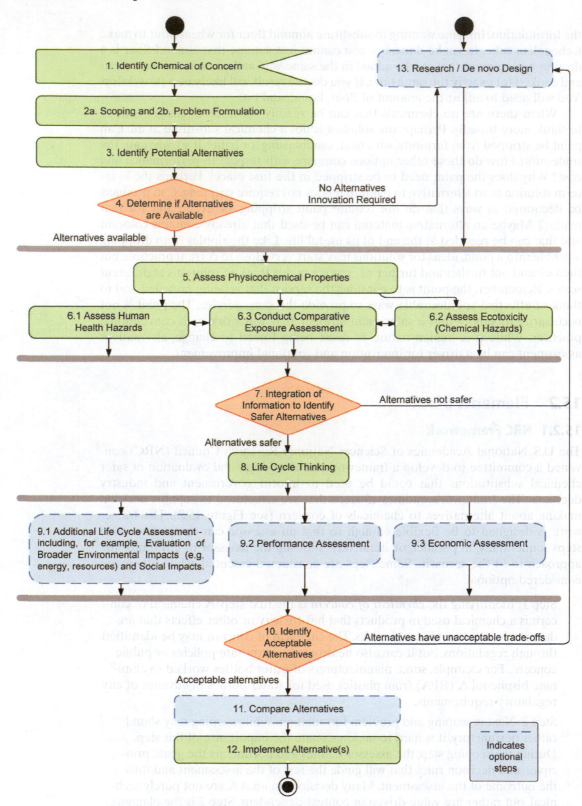

FIGURE 15.2 The NRC 13-step framework.

Source: National Research Council (2014). *A Framework to Guide Selection of Chemical Alternatives*. Washington, DC: The National Academies Press.

considered. In this step, the preferences of the decision maker are made explicit in the form of decision rules or algorithms to be applied to resolve trade-offs among different attribute domains (e.g., toxicity, material and energy use, cost) and to address uncertainty. It is in step 2 that decision rules are set that will be used to define a "safer" alternative. Are there specific hazards of interest? Are there function and performance requirements that are essential for alternatives to be considered viable candidates? The scoping step also defines the role of stakeholders. Who is affected by the decisions, and whose perspectives should be considered in the decision analysis? Stakeholders may be engaged in the assessment process, and their input can help determine what goals, principles, and decision rules will guide the process. This is necessary because decisions will need to be made when there are trade-offs. And there are always trade-offs.

Step 3: Potential alternatives are identified. The scope of alternatives selected for evaluation has major impacts on the outcome of the AA. If the scope is very narrow, then the available options will be limited and probably quite similar to the chemical of concern. If the scope is very broad, then alternatives may be considered that are very different from current practices. For example, if the chemical of concern is a per- or polyfluorinated substance (PFAS) added to disposable paper-based food packaging, then a narrow scope may include other paper additives like waxes or plastic coatings. If the scope is broad, then alternatives may include non-paper materials like aluminum or glass and business models that enable the packaging products to be reused. The scope of alternatives may be dictated by regulations and is influenced by the person who is doing the AA. If the AA is being done by a manufacturer of paper-based packaging, the focus is unlikely to be on non-paper solutions. Initial scoping can also eliminate options early on that do not meet minimum requirements.

Step 4: Now a decision needs to be made. If no alternatives are found, then it may make sense to initiate research to develop new alternatives or improve existing ones as indicated in step 13 until alternatives are available.

Step 5: Once alternatives are identified, it is time to dive into the analysis. It is useful to start by collecting data on the physicochemical properties of alternatives. Physicochemical properties include properties such as water solubility, boiling point, vapor pressure, biodegradability (persistence), bioaccumulation potential, and so on. The physical properties of a chemical are closely related to the exposure potential of the chemical. For example, if a chemical has a very high vapor pressure, then it will evaporate quickly into the atmosphere and increase inhalation exposure to those around it if it is used in a closed environment. Physicochemical data can be obtained quickly and inexpensively, and therefore it is practical to gather them early.

Step 6: This step is divided into three parts:

- Step 6.1: An assessment of hazards to human health
- Step 6.2: An assessment of ecotoxicity hazards
- Step 6.3: An assessment of comparative exposure

These assessment steps can be completed concurrently because the findings are interrelated, and assessments or conclusions from one step may affect the conclusions from other steps. You may want to refer to Chapter 6 for a more extensive discussion of assessing exposure as well as hazards to humans and environmental species.

Step 7: The assessor can begin to integrate the information gathered up to this point. It is hoped that at least a few alternatives appear to be safer than the chemical of concern. Step 7 is a decision point like step 4. If no alternatives are found, then it may make sense to initiate research to develop new alternatives or improve existing ones as indicated in step 13. If alternatives have been found, then it is time to expand the analysis to consider impacts across the product life cycle.

Step 8: The assessor is guided to apply life cycle thinking. This is different from life cycle assessment, which is a structured methodology described in Chapter 14. Life cycle thinking means thinking about whether there are likely to be significant impacts to human health, the environment, or society at points in the product or process life cycle. If those impacts are expected to differ in significant ways between the chemical of concern and proposed alternatives, then a deeper analysis may be necessary. The point of life cycle thinking is to avoid regrettable substitutions that may result from shifting the burden from one life cycle stage to another. For example, if an alternative is inherently safer as used in a consumer product but the manufacturing process used to make it is terribly polluting, then further analysis should be done to ensure that the negative impacts are not simply shifted to a different receptor (i.e., people who live near the manufacturing site) at a different stage of the life cycle. Life cycle thinking is applied early in the process with existing information, and it can help the assessor decide if a full life cycle assessment is necessary. While it would be nice to have perfect information, decisions must be made with the best information currently available, so life cycle thinking can help determine if the differences are likely to be great enough to merit the additional time and cost of data gathering and analysis.

Step 9: At a minimum, according to the NRC framework, steps 1–8 should be considered in every assessment. The framework presents the following elements of step 9 as optional:

- Step 9.1: Additional life cycle assessment
- Step 9.2: Performance assessment
- Step 9.3: Economic assessment

The presentation of step 9 as optional is a little misleading. There is obviously no point in suggesting that a safer alternative exists if it is not effective or if it is prohibitively expensive. Ideally, the initial scoping of alternatives has selected a set that are within reasonable parameters for cost and performance. What step 9 is emphasizing is that a deep dive into performance and cost may not be necessary until it is time to actually adopt the alternative and make it work. Unlike physiochemical and hazard properties, cost and performance can change depending on market forces and product design.

Step 10: The assessor identifies acceptable alternatives. As with steps 4 and 7, this is a decision point. If acceptable alternatives are not found, then it may make sense to initiate research to develop new alternatives or improve existing ones as indicated in step 13.

Step 11: If acceptable alternatives do exist, then it is in this step that the assessor recommends one or more alternatives to be further evaluated.

Step 12: The information is passed to those who can actually implement the option(s). Work will need to be done to substitute the alternative(s), to troubleshoot performance issues, and to manage trade-offs.

Step 13: While this step is numbered last, it is referred to throughout the process. Ongoing research, design, and development are needed for continual improvement and innovation.

ACTIVE LEARNING EXERCISE 15.1 Evaluating Alternative Cakes

This exercise is about chocolate cake (or another flavor if you prefer). Break into three teams, where each team will prepare one chocolate cake. Find and compare three recipes for a rich chocolate cake. One is to be based on wheat flour, which contains gluten; one is to be based on almond flour, which is gluten free; and the third is to be flourless. Each team should make one cake to share with the other groups.

1. Prepare a spreadsheet with the following information. You may add other parameters if you believe they are important.
   a. Identify the ingredients needed for each cake. What are the quantities of each ingredient, and what is the total estimated cost per cake based on the ingredients?
   b. What is the total time to make each cake? Separate out preparation time from cooking time. Time may translate into labor costs.
   c. Rate the cakes based on taste. You will need to establish a scoring method to do this.

2. Compare the cakes using the following two scenarios.
   a. You are about to open a bakery on a busy main street in a medium-sized city. Which is your pre-ferred cake, and why?
   b. You are preparing a cake for a gathering at which several of the participants have celiac disease. Consider gluten to be a "hazard" in this scenario. (People who have celiac disease certainly consider it to be a hazard for them). Which is your preferred cake, and why?

15.2.2 IC2 Guide

The most comprehensive, step-by-step guidance available for AA is the Interstate Chemicals Clearinghouse *Alternatives Assessment Guide* v1.1 (IC2 AA Guide). The first version of the IC2 guide was written in 2014 prior to the NRC framework; it was revised in 2017. Many elements of the two resources are similar, including the scope, identification of alternatives, screening for physiochemical properties, and assessment of individual parameters that are referred to in the guide as modules. There are modules for the following assessments:

- Chemical hazards
- Exposure
- Cost and availability
- Performance
- Life cycle thinking/life cycle assessment
- Materials management (waste management)
- Stakeholder engagement
- Social impacts

Two features are particularly useful. First, the guide provides extensive guidance on how to generate data for each parameter and guides the assessor to do so in an iterative way. Each parameter module defines progressive levels (see Figure 15.3). Basic or initial levels allow for simple comparisons. The higher the level, the more resource and data intensive are the requirements. For example, a Level 1 performance assessment may simply require evidence

Level 1	*Basic Exposure Evaluation:* Identifies potential exposures concerns along with how the concerns may be addressed. Decisions in this level are based upon a qualitative assessment using readily available data.
Level 2	*Expanded Exposure Evaluation:* Builds on the previous level by increasing the quality and quantity of information. More detailed quantitative data is required to evaluate the importance of exposure in the AA process.
Level 3	*Detailed Exposure Evaluation:* This level builds on previous levels and requires detailed scientific studies as the basis for decisions. If these studies are not available, they are conducted and the data used to determine the importance of exposure in the AA process.
Advanced	*Full Exposure Assessment:* This level requires a complete and detailed exposure assessment as defined in the Risk Assessment Process by the National Academy of Sciences.

(a)

Level 1	*Basic Performance Evaluation:* Identifies a few, very basic questions about whether the alternative performs the required function in the product. This level uses qualitative information readily available from manufacturers and other sources to evaluate alternatives.
Level 2	*Extended Performance Evaluation:* Builds upon the information obtained in Level 1 to determine whether the alternative performs the required function in the product. It uses quantitative information of existing data reviewed by technical experts in the field to evaluate alternatives.
Level 3	*Detailed Performance Evaluation:* Expands upon the previous levels. It uses quantitative information to evaluate alternatives based upon results of specified tests reviewed and validated by technical experts.

(b)

FIGURE 15.3 The IC2 guide takes an iterative approach with increasingly data-intensive requirements at each level for each module. (a) Exposure assessment levels. (b) Performance levels.

Source: theic2.org/article/download-pdf/file_name/IC2_AA_Guide_Version_1.1.pdf.

that an alternative is available in the marketplace and that it is being used. A higher-level performance assessment may entail testing against established industry performance standards. It is not always necessary to go beyond lower-level assessments to make a decision; this is especially true when the differences between alternatives using one module overwhelm the differences between alternatives for another. For example, if one chemical used in fragrances is carcinogenic, persistent, and bioaccumulating while another works well, is slightly less volatile, and is of inherently low hazard for all the hazard endpoints, then there is no point in analyzing the nuances of exposure differences. They will not affect the outcome.

The second feature of the IC2 guide that is particularly useful is its focus on decision analysis. Decision analysis provides a structured approach to making decisions when there are multiple variables. The decision framework used to compare alternatives can affect which alternatives are considered acceptable. The IC2 guide presents three decision frameworks (see Figure 15.4). The first is the linear or sequential decision framework, the second is the simultaneous decision framework, and the third is a hybrid decision framework. In the ***linear or sequential framework***, each alternative is compared against the criteria for each parameter (module) in a sequential way. For example, if the alternative meets the requirements for hazard, then it is accepted and analyzed against the requirements for the next module. If not, then the alternative is kicked out of the pool of options, or at least held in reserve in case the criteria are revised. With the sequential framework, users may reject options along the way, winnowing down the pool of options until only those that meet all requirements remain.

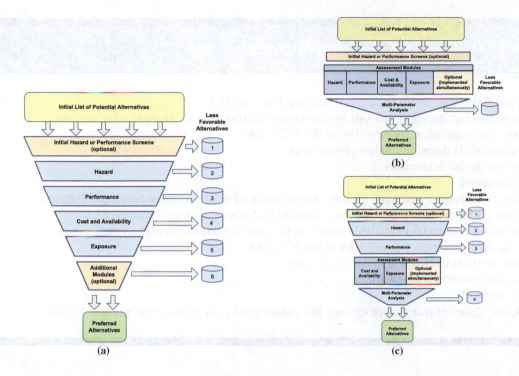

FIGURE 15.4 Decision frameworks from the IC2 guide. (a) Sequential framework.
(b) Simultaneous framework. (c) Hybrid framework.

Source: theic2.org/article/download-pdf/file_name/IC2_AA_Guide_Version_1.1.pdf.

In the *simultaneous or multiparameter framework*, no options are initially ruled out. Rather, options are scored or rated according to criteria within each module. For example, the assessor may assign hazard ratings from 1 to 5, with 1 being most hazardous. Each option is then assigned a hazard score ranging from 1 to 5. The same is true for performance, cost and availability, exposure, and any other modules as dictated in the problem formulation. Once each alternative is rated for each parameter, the assessor applies an algorithm. The alternative that receives the most favorable score is then the most acceptable alternative. Some people are uncomfortable with using algorithms to make decisions. However, the algorithms are created by the assessor with input from stakeholders, so the algorithm should reflect the values that are important to them. Algorithms may include weightings. For example, if performance is critical, then the assessor may assign a multiplier that amplifies the results of the performance scoring. Multiparameter decision analysis is a tool to aid people in better understanding their own values in decision making; it is not intended to force people to make decisions they don't want to make. What is essential is that the assessor is transparent about how the algorithm is developed and the weightings are assigned.

The *hybrid framework* is a blend of the sequential and simultaneous decision frameworks. The hybrid approach allows for the early elimination of options that do not meet essential requirements. For example, if an alternative does not meet the minimum criteria for hazard or performance, then it is ruled out. Those alternatives that make it through the first cut are then scored for each remaining parameter and compared using the simultaneous approach. The IC2 guide allows for flexibility as long as the rules underlying the decisions are transparent.

ACTIVE LEARNING EXERCISE 15.2 Decision Analysis

1. Use the data from the bakery scenario in Active Learning Exercise 15.1.
   a. Show how you would use decision analysis based on the existing data to analyze your options using each of the three decision frameworks in the IC2 guide:
      i) Linear (sequential) decision analysis framework
      ii) Simultaneous decision framework
      iii) Hybrid framework
   b. Does the choice of decision framework change the outcome of your recommendations? Explain.
2. Use the data from the gluten-sensitive participants scenario in Active Learning Exercise 15.1.
   a. Show how you would use decision analysis based on the existing data to analyze your options using each of the three decision frameworks in the IC2 guide:
      i) Linear (sequential) decision analysis framework
      ii) Simultaneous decision framework
      iii) Hybrid framework
   b. Does the choice of decision framework change the outcome of your recommendations? Explain.

15.3 Example: Assessing Alternatives to Antifouling Boat Paints in Washington

Biofouling on boats can harm boat structures and increases drag, resulting in slower and less efficient boats that burn more fuel (see Figure 15.5). In 2011, the Washington State Legislature passed the Recreational Water Vessels—Antifouling Paints Law, Revised Code of Washington (RCW) Chapter 70.300 to address the impacts of copper-based antifouling paints on small recreational vehicles, with restrictions starting in 2018. The state of Washington commissioned an alternatives assessment to examine alternatives to antifouling paints that use copper in their formulations to determine how the alternatives affect marine organisms and water quality. Copper is the most commonly used biocide in modern antifouling paints, replacing tributyltin (TBT), which was banned globally in 2008 following the discovery that it causes imposex in gastropods. (Imposex is a disorder in which female snails develop male sex organs in response to marine pollutants.) Copper can also negatively affect aquatic life, especially young salmon. Even low levels of copper can interfere with a salmon's development, reproduction, and ability to avoid predators (Northwest Green Chemistry and TechLaw, 2017). The driver for the AA was that copper was found in Washington marinas in excess of safe limits. No one disputes the need to take action to prevent biofouling, but given the aquatic toxicity associated with currently available biocides, there was interest in finding and adopting safer alternatives.

By design, **biocidal coatings** kill aquatic organisms that attempt to bind to the boat surface; biocidal antifouling paints are designed to slowly release biocide. The biocide may leach from the paint, or the paint that contains biocide may slowly slough off the boat. Both approaches prevent fouling but contaminate the water.

FIGURE 15.5 Biofouling on a boat hull affects a boat's performance, hull integrity, and fuel efficiency.

Source: Svetlana Yudina/Shutterstock.com.

The AA work was performed by Northwest Green Chemistry and TechLaw. The scoping and problem formulation step of the AA required that the assessors use the IC2 guide and apply the modules for hazard, exposure, cost and availability, and performance. Stakeholders were engaged throughout the AA process. The other modules were considered optional, and the choice to use them depended on whether or not they were perceived as necessary to discriminate between options. The state of Washington limited consideration of alternatives to those that were available in Washington at the time, thus limiting the scope of alternatives that could be considered. The AA took about a year to complete, in part due to regular progress updates and periodic public calls and meetings to gather feedback along the way.

The assessors considered options from a range of technology types presented in Figure 15.6 under biocidal and nonbiocidal technologies. The assessors looked at alternative biocides, alternative biocidal coating formulations, nonbiocidal coating formulations, noncoating solutions, and marina-wide solutions. After defining technology types, the assessors looked for available product options within each technology type. This involved researching specific trade name products sold at marinas and via the Internet for use in Washington along with pricing and whatever performance data could be gleaned from the manufacturer and public literature.

The assessors first compared alternative biocides primarily for hazard. Table 15.1 lists the biocides that are registered for use in antifouling paints in Washington state.

Options may also include paint formulations that contain one or more of the registered biocides. Just because a paint contains a specific biocide does not mean it will perform the same way as other paints that contain that same biocide. Some paints are formulated more effectively than others. As a result, they may last longer, require more or less paint to be applied to the boat, or work well at different concentrations of biocide.

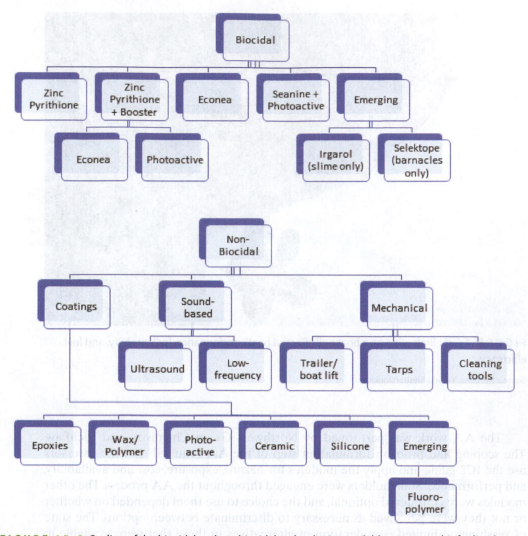

FIGURE 15.6 Outline of the biocidal and nonbiocidal technologies available to prevent biofouling boats.

Source: Northwest Green Chemistry and TechLaw (2017). *Washington State Antifouling Boat Paint Alternatives Assessment Report. Final Report*. static1.squarespace.com/static/5841d4bf2994cab7bda01dca/t/59d40515c534a598eeb6c18a/1507067168544 /Washington+CuBPAA_Final_2017.pdf.

Nonbiocidal coatings were considered. Another name for them is ***foul release coatings***. Foul release coatings work by making it difficult for fouling organisms to attach to the boat. When coupled with the shear from the boat moving through water, foul release coatings can be effective in preventing fouling. Hard foul release coatings are gaining more favor with the use of nanoscale (nano) chemistry. In the nano form, materials like silica can be added to coatings that make surfaces extremely hard and from which organisms such as barnacles can be easily removed. Soft foul release technologies are based on silicones and fluoropolymers that make the surface slippery and subsequently hard for marine life to stick to. Like hard foul release coatings, they work best when the boat is used; movement helps to remove organisms from the surface. Ironically, there are anecdotes that slippery boat bottoms can be hard to work with when the boat is out of the water and held in a boat cradle.

Noncoating technologies were also identified. Small vessels may be lifted out of the water when not being used, or a waterproof material may be used as a barrier between the hull and the surrounding organisms when the boat is moored. There

TABLE 15.1 Biocides registered for use in antifouling paints in Washington state

BIOCIDE ACTIVE INGREDIENT	PC CODE	TARGET ORGANISMS
Copper (flakes or powder)	22501	Broad effect, some algae resistant
Copper pyrithione (Copper Omadine)	88001	Algae, marine plants
Copper thiocyanate	25602	Broad effect, some algae resistant
Cupric Oxide, Copper(II) Oxide	42401	Broad effect, some algae resistant
Cuprous Oxide, Copper(I) Oxide	25601	Broad effect, some algae resistant
Tralopyril (Econea)	119093	Molluscs
Cybutryne (Irgarol 1051)	128996	Algae, marine plants
DCOIT (Sea-Nine 211)	128101	Broad effect
Silver	72501	Slime (microorganisms)
Zinc	129015	Algae, marine plants
Zinc pyrithione (Zinc Omadine)	88002	Algae, marine plants

Source: Department of Ecology, State of Washington (2019). *Antifouling Paints in Washington State Report and Recommendations. Report to the Legislature Pursuant to SHB 2634 (2018)*. Publication 19-04-020. apps.ecology.wa .gov/publications/documents/1904020.pdf.

are also ultrasonic antifouling systems that work by transmitting ultrasound to the protected surfaces (see Figure 15.7). Increasing and decreasing pressure where the water touches the protected surface creates ultrasound-induced cavitation. This microscopic movement of water disrupts the first stages of the food-chain and stops the development of a biofilm. The effect does not damage the surface and prevents the attachment of larger organisms like barnacles and mussels. Ultrasonic anti-fouling works well with materials of uniform density like steel, aluminum, GRP, and FRP; it is not suitable for wooden boats, as wood dampens ultrasound transmission.

FIGURE 15.7 Ultrasonic Antifouling systems include a control unit and ultrasound transducers that are mounted to the hull or other protected equipment.

Source: Courtesy of Sonihull, NRG Marine Ltd.

FIGURE 15.8 The drive-in boat wash developed in Sweden is now being adopted at marinas in the United States.

Source: Motorboat hull cleaning, Drive-in Boatwash; driveinboatwash.com/discover-machines/.

Finally, the AA noted the emergence of marina-wide solutions. Boats may be scrubbed regularly by divers or cleaned in a drive-in boat wash, similar to a drive-in car wash (Figure 15.8). A drive-in boat wash is not dependent on an antifouling or foul release coatings at all. The drive-in boat wash was developed in Sweden and was not available in Washington at the time of the AA. Since that time, however, multiple drive-in boat washes have been installed in North America.

15.3.1 Results

Given the analysis of available options, how can one decide on the best option? There are many variables, and the definition of a best option depends on whether you are a regulator charged with environmental protection or boat owners who want an economical but effective way to protect their boats. There is also variability in preferences among boat owners. For example, if the boat is constantly on the move, then foul release coatings may be effective because of the constant shear. Perhaps the owner has a racing boat and wants the bottom to be as fast as possible, or perhaps the boat rarely sails but is used primarily while sitting in a harbor, which makes it an easy target for biofouling. Each use case may have different performance requirements.

Rather than pointing to a single option as a possibility, the assessors prepared a selection guide. The selection guide presents antifouling options, with ratings for each parameter that was evaluated: hazard, exposure, cost and availability, and performance. The selection guide was intended to enable users to make informed decisions that align with their own preferences. It laid out each antifouling option and rated each parameter based on extensive supporting data. A simplified version of the selection guide is presented in Table 15.2. Ratings are identified as High (H), Moderate (M), Low (L), or Data Gap (DG) for the parameters most relevant to the AA. Users have the option to find the safest options and then to explore if they are a good fit for their personal situation. The H/M/L/DG ratings could be converted to numerical ratings for comparison using the simultaneous decision framework.

Several challenges were associated with the AA. One of the biggest challenges was the absence of data for key parameters, especially performance. For recreational boats, there were no authoritative and objective sources to rate performance. The only data available were based on panel tests designed to compare panels painted with different biocidal coatings. The panels are submerged in water

TABLE 15.2 A simplified selection guide to help users select their preferred alternative

TECHNOLOGY TYPE	PRODUCT	HUMAN HEALTH HAZARDS (HIGH IS WORST)	ENVIRONMENTAL HEALTH HAZARDS (HIGH IS WORST)	INITIAL COST ($)	COST OVER 5 YEARS ($) (INCLUDES HAUL OUT AND RECOATING COSTS)	PERFORMANCE RATINGS (HIGH IS BEST)	EXPOSURE TO TOXIC BIOCIDES OR OTHER IMPACTS TO AQUATIC LIFE (HIGH IS WORST)
Biocidal coating	Ecopaint ABC	M	M	225	6,891	M-H	M-H
Biocidal coating	NextBX	M-H	H	270	8,672	H	H
Foul release coating	Newsa	M	L-M	500	4,000	DG	L
Ultrasound	US LTD	L	L-DG	5,259	5,259	DG	L-DG

and evaluated and compared for the extent of biofouling over time. Panel tests are not relevant to other antifouling technologies such as ultrasound or foul release coatings. There was no credible way to compare the performance of the incumbent biocidal coatings against the performance of nonbiocidal coatings and noncoating technologies other than by reviewing marketing materials and customer comments, which are not considered to be highly reliable. Performance data are often not as difficult to obtain for government or military specifications because performance testing is typically a prerequisite for fair and transparent purchasing decisions; however, that is not the case for consumer products. Only one product was used for both military and recreational purposes, and its ability to meet military specifications served as evidence for good performance. One of the weaknesses of AA is that there are usually more data gaps for new products than for the incumbent products. The incumbent products tend to serve as the benchmark, which can limit recognition of other benefits of disruptive technologies.

15.3.2 Adoption of Alternatives

The AA provided information on options currently available in Washington, answering some questions and raising new ones and resulting in a somewhat disappointing delay in implementing the legislation until 2026. For one thing, there were no viable copper-free antifouling paints for wooden boats, so wooden boats were given an exemption. For nonwooden boats, the other biocides were not considered much safer than copper. Stakeholders were reluctant to move to a biocidal paint that they considered to perform worse if it was not decidedly safer, and they were very hesitant to adopt foul release coatings and noncoating technologies that had limited to no data on performance. The drive-in boat wash was considered extremely promising for marinas in Washington and resulted in a visit by the Swedish manufacturer to the region. There are now several boat washes in the United States, and case studies show that in some marinas in Europe, over 90% of boats moored in a harbor with a boat wash can successfully remain free of biofouling using no antifouling coatings at all. This is what is meant as a ***disruptive technology***, a product or technology that creates a new market and displaces established markets. Perhaps the delay should not be too disappointing if it results in a shift to the adoption of boat washes rather than alternative biocidal coatings that are at best only incrementally better for water quality and aquatic life. But this is still to be determined. New foul release coatings are currently in development. The Netherlands government has recommended the development of methods to test and compare a range of technology types.

While the problem of antifouling boat paints may not affect most people, especially landlubbers, it provides a nice example of a service that can be met with a variety of technologies and a range of parameters that illustrate how AA can be applied in support of effective regulations. Exposing organisms in the aquatic environment to toxic biocides needs to be weighed against performance, cost and availability, and social willingness to adopt alternatives. Just because there are alternatives does not guarantee that they will be adopted in the marketplace. Sometimes old technologies become entrenched, and it takes a great deal of effort to replace them. In other cases, newer technologies are adopted rapidly, leading to disruptive shifts in the marketplace. Entrepreneurs create products hoping for rapid adoption and financial success.

15.4   Governmental Uses of AA

There is growing use of AA by governments in support of legislation, particularly in the states of California and Washington and in the European Union. In general, governments use AA to make sure that they do not ban a chemical of concern that has important uses to society if there are no alternatives. Sometimes a chemical of concern has many uses and all but one or two of those uses can be substituted. In such cases, governments authorize ongoing use of the chemical for a limited set of applications, for a limited period of time (i.e., five years). They will then revisit the decision to see if the chemical of concern is still considered essential and if viable alternatives have emerged. It takes time to transition away from chemicals that are entrenched in the supply chain and to ensure that the alternatives meet performance requirements.

In practice, governmental jurisdictions differ with respect to which agency oversees the AA, the chemicals and products to which the AA applies, who is responsible for doing the AA, and the potential regulatory response options. Typically, AA is applied to specific chemical/product combinations. A chemical of concern is identified, and those products in which it is used are targeted for AA work and, hopefully, substitution. The selection of product types is prioritized based on those that pose the greatest risks to human health or the environment. For example, some U.S. states are restricting per- and polyfluorinated alkyl substances (PFASs) for some but not all applications. They start with applications such as food packaging or firefighting foam, where there is likely to be exposure that results in risk. AAs are specific; the alternatives that can replace PFASs to provide grease and water repellency in food packaging are not likely to be the same as the alternatives that suffocate fires.

15.4.1 Washington

The state of Washington has specified AAs in three recent laws and regulations. The approach continues to evolve with experience. The first AA was driven by the law prohibiting copper antifouling paint in recreational boats as discussed above and specified use of the IC2 guide. Consultants were hired to perform the AA, and results from the AA were used to determine if and when a ban on copper antifouling paints for recreational boats would be implemented. The second was a law (RCW 70A.222.070) prohibiting the manufacture, sale, and distribution of food packaging in Washington to which per- and polyfluoroalkyl substances (PFASs) have been added in any amount. This law also specifies use of the IC2 guide and requires demonstrating that there are alternatives that are inherently safer from the chemical toxicity perspective and that are also readily available and cost competitive. Food packaging includes many product types ranging from pastry papers to pizza boxes and takeout containers. As viable alternatives are found for each product type, the state of Washington can then prohibit the sale of products containing PFASs for that specific product type. So, while it may be easy to replace PFAS-coated pastry paper with safer wax-coated pastry paper, it may not be as simple to replace other product types. The AA work will continue until alternatives for all product applications are found. For the PFAS AA, a consultant was hired for the first round and the report was reviewed by an ad hoc committee from the Washington National Academy of Sciences. The third law is Safer Products for Washington (2019, Substitute Senate Bill 5135). Safer Products for Washington requires the state to select five priority chemicals of concern and to prioritize consumer products in which they are found. The AA work is done by the Washington Department of Ecology with extensive stakeholder input. The Washington DOE is not required to use the IC2 guide, but it must demonstrate that alternatives meet minimum requirements established for determining if the alternatives are safer, feasible, and available. Depending on the outcome of the AA, Washington state has the option to restrict or prohibit the products, to require labeling or other forms of notification that the products contain the chemical, or to take no action. This work is to occur in four-year cycles, with the next five chemical/product combinations of concern selected for AA after year four. (see Figure 15.9).

15.4.2 California

In 2008, the California legislature passed the Green Chemistry Law authorizing the Department of Toxic Substances Control (DTSC) to adopt regulations to identify and prioritize chemicals of concern in consumer products and to evaluate alternatives.

Comparing CA, WA, and EU REACH approaches to AA

Agency/Legislation	Applies to	Who does the AA	Regulatory Response
CA Dept. of Toxic Substances Control: Safer Consumer Products (2013)	Selected combinations of chemicals on the Candidate Chemicals List and Priority Products	Responsible entities such as manufacturers	Options include: 1. Provide information to consumers 2. Restrictions on chemical and consumer products that use it 3. Prohibition of product sales 4. Requirements to engineer safety measures 5. Requirements for end of life management 6. Investment in green chemistry or engineering
WA Department of Ecology: Safer Products for Washington (2019)	Priority chemical classes (5 at a time); consumer products containing priority chemicals	WA Department of Ecology	Options include: 1. Require notice 2. Restrict/prohibit 3. No action
European Chemicals Agency (ECHA): REACH (2007)	Substances of Very High Concern (SVHC) added to 1. Candidate List and 2. Authorization List	ECHA with Member State and EC consultation (steps 1 and 2); Companies who want to use the substance after the sunset date (step 3) must do the alternatives assessment	Authorization or Restriction steps include: 1. Recommend SVHC for Candidate List 2. Recommend SVHC for Authorization List with sunset date for use in the marketplace 3. Application for Authorization after sunset date for specific applications (Companies)

FIGURE 15.9 Examples of jurisdictions with legislation that requires the use of AA: some key similarities and differences.

These are known as the California Safer Consumer Products regulations. The four steps in the process are shown in Figure 15.10. DTSC first established an extensive list of chemicals of concern called the Candidate Chemicals List. They then established a list of Priority Products that contain those chemicals to help focus their efforts. DTSC then spent several years developing the tools and guidance for performing AA and implementing the regulation, engaging stakeholders and a Green Ribbon Science Panel to advise along the way. Unlike Washington, California requires that the manufacturers and providers of specified priority products containing the chemical of concern perform the AA and that they submit the AA work for review by DTSC. In step four, based on the results of the AA, California DTSC is authorized to take the following actions:

- Require that manufacturers provide information to consumers.
- Restrict the use of the chemical (e.g., limit concentrations).
- Prohibit sales of the product containing the chemical.

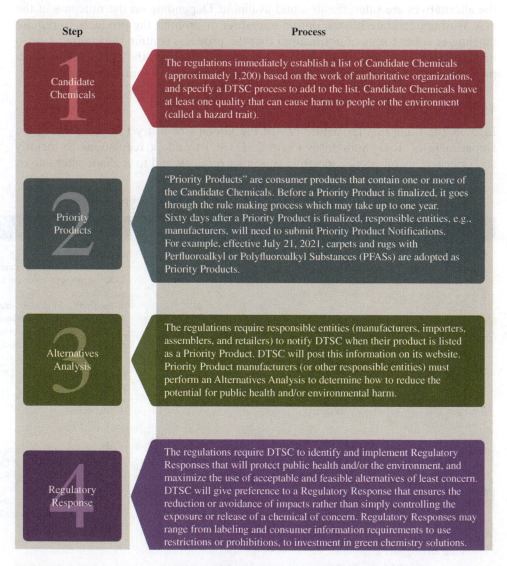

Step	Process
1 Candidate Chemicals	The regulations immediately establish a list of Candidate Chemicals (approximately 1,200) based on the work of authoritative organizations, and specify a DTSC process to add to the list. Candidate Chemicals have at least one quality that can cause harm to people or the environment (called a hazard trait).
2 Priority Products	"Priority Products" are consumer products that contain one or more of the Candidate Chemicals. Before a Priority Product is finalized, it goes through the rule making process which may take up to one year. Sixty days after a Priority Product is finalized, responsible entities, e.g., manufacturers, will need to submit Priority Product Notifications. For example, effective July 21, 2021, carpets and rugs with Perfluoroalkyl or Polyfluoroalkyl Substances (PFASs) are adopted as Priority Products.
3 Alternatives Analysis	The regulations require responsible entities (manufacturers, importers, assemblers, and retailers) to notify DTSC when their product is listed as a Priority Product. DTSC will post this information on its website. Priority Product manufacturers (or other responsible entities) must perform an Alternatives Analysis to determine how to reduce the potential for public health and/or environmental harm.
4 Regulatory Response	The regulations require DTSC to identify and implement Regulatory Responses that will protect public health and/or the environment, and maximize the use of acceptable and feasible alternatives of least concern. DTSC will give preference to a Regulatory Response that ensures the reduction or avoidance of impacts rather than simply controlling the exposure or release of a chemical of concern. Regulatory Responses may range from labeling and consumer information requirements to use restrictions or prohibitions, to investment in green chemistry solutions.

FIGURE 15.10 An overview of the California Safer Consumer Products program.

Source: Based on dtsc.ca.gov/what-are-the-safer-consumer-products-regulations/.

- Require the manufacturer to engineer safety measures.
- Require end-of-life management practices.
- Invest in green chemistry research and development.

15.4.3 Europe

In Europe, REACH legislation implemented in 2007 (discussed in Chapter 6) calls for the use of AA in support of restricting and authorizing the ongoing use of substances of very high concern (SVHCs). SVHCs are chemicals of concern subject to regulatory obligations. Chemicals that meet the criteria for SVHCs are placed on the Candidate List. If a chemical is identified as a SVHC, then it undergoes pressure to be substituted, either directly or indirectly. Direct substitution pressure comes through regulatory restrictions and limited authorization of its use. Indirect pressure for substitution comes from the requirements to make information about hazards freely and publicly available on the European Chemical Agency (ECHA) website and to notify the public about the presence of hazardous substances in products. Such transparency is intended to help deselect the use of SVHCs in the marketplace.

Alternatives assessment has the same objective in Europe as it does in Washington and California: to ensure that substances of very high concern are progressively replaced by safer alternatives. However, the process is somewhat different. Under REACH, selected SVHCs from the Candidate List are added to the Authorization List, which allows companies to apply for authorization for continued (or a new) use. AA is required in support of petitions for authorization. The SVHC is headed for restriction unless it is authorized for specific uses and for a limited time. Manufacturers must apply for authorization to use chemicals on the Authorization List and must submit an analysis of alternatives as part of that application. These analyses are publicly available and are reviewed by ECHA. They consider not only the technical feasibility of alternatives but also the social and economic impacts.

AA is envisioned as a tool that can help to (1) authorize limited ongoing uses of a chemical of concern when viable alternatives are not available, (2) drive the adoption of safer alternatives when they exist, and (3) drive the innovation and the development of safer and more sustainable options. These are very different outcomes, and it is still to be determined how best to use AA to achieve them. Europe is currently seeking to augment the effectiveness of REACH legislation by creating additional policies intended to promote a toxic-free environment. The European Commission published Chemicals Strategy for Sustainability on October 14, 2020 (*ec.europa.eu/environment/pdf/chemicals/2020/10/Strategy.pdf*). It is part of the EU's commitment to zero pollution, which is a key goal of the European Green Deal.

Rather than focusing all efforts on eliminating and remediating toxic chemicals or even minimizing and controlling them, the EU is looking for policy drivers that will promote safe and sustainable chemicals. The investment and promotion of safe and sustainable chemicals are viewed as a target and not as a hoped-for outcome of controlling toxic substances, as illustrated in Figure 15.11.

The key actions recommended as part of the Chemicals Strategy (European Commission) include:

1. Banning the most harmful chemicals in consumer products and allowing their ongoing use only when deemed essential
2. Accounting for the cocktail effect of chemicals when used in mixtures and when exposure to them comes from multiple sources
3. Phasing out PFASs except for essential uses

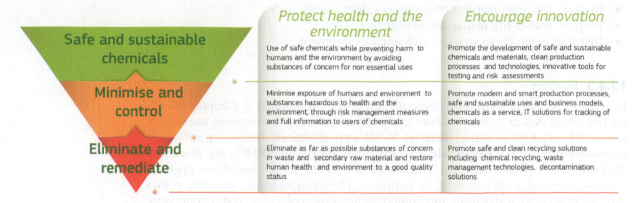

	Protect health and the environment	Encourage innovation
Safe and sustainable chemicals	Use of safe chemicals while preventing harm to humans and the environment by avoiding substances of concern for non essential uses	Promote the development of safe and sustainable chemicals and materials, clean production processes and technologies, innovative tools for testing and risk assessments
Minimise and control	Minimise exposure of humans and environment to substances hazardous to health and the environment, through risk management measures and full information to users of chemicals	Promote modern and smart production processes, safe and sustainable uses and business models, chemicals as a service, IT solutions for tracking of chemicals
Eliminate and remediate	Eliminate as far as possible substances of concern in waste and secondary raw material and restore human health and environment to a good quality status	Promote safe and clean recycling solutions including chemical recycling, waste management technologies, decontamination solutions

FIGURE 15.11 The EU toxic-free hierarchy—a new hierarchy in chemicals management.

Source: ec.europa.eu/environment/pdf/chemicals/2020/10/Strategy.pdf.

4. Boosting investment and innovation for chemicals that are safe and sustainable by design throughout their life cycle
5. Promoting resilience of the EU's supply chain, including critical chemicals
6. Establishing a "one substance one assessment" process for hazard and risk assessment
7. Being a global leader by championing and promoting high standards and not exporting chemicals banned in the EU

15.4.4 Safe and Sustainable by Design

Safe and sustainable by design (SSbD) is a key theme that runs throughout the European Green Deal. It takes the agenda of the Circular Economy Action Plan, which is designed to improve the durability, reusability, upgradability, and reparability of materials and products, and combines it with the Chemicals Strategy for Sustainability, which aims to catalyze the shift toward chemicals, materials, and products that are inherently safe and sustainable, from production to end of life. The European Commission has committed to developing criteria for safe and sustainable by design by 2022. Their strategy proposes a clear roadmap and timeline for the transformation of industry with the aim of attracting investment in safe and sustainable products and production methods. Perhaps this initiative is the kind of vehicle necessary to support disruptive technologies and transformations!

In addition to establishing safe and sustainable by design criteria for chemicals, the European Commission intends to establish an EU-wide SSbD support network to promote cooperation and sharing of information across sectors and the value chain and to provide technical expertise on alternatives. They intend not only to advocate for this transformation but also to fund it. The initiative will ensure the development, commercialization, deployment, and uptake of SSbD substances, materials, and products through financial support, in particular to small and medium-sized enterprises through relevant EU funding and investment instruments and public–private partnerships. They also plan to map and address competence gaps to implement SSbD by ensuring the development of adequate skills at all levels—including in vocational and university education, research, industry, and among regulators. Finally, they will establish metrics and indicators in cooperation with stakeholders to measure progress in the industrial transition toward the production of safe and sustainable chemicals.

15.5 Business Uses of AA

In order for products to become safe and sustainable by design, there must be cooperation between government and industry. Some companies have pioneered their own approaches rather than waiting for government to lead them. These leaders, or "frontrunners" have learned a lot about what it means to be safe and sustainable by design and should be part of the discussion to define the criteria that will then be applied to them. Collaboration between government and industry and alignment on SSbD criteria help to create a stable economic and regulatory environment for businesses and help with *future-proofing* their products and operations. Future-proofing refers to taking actions that pre-emptively address future regulations and anticipated roadblocks. It can also help drive innovation. For example, if an electronics manufacturer used the flame retardant deca-brominated diphenyl ether, it would have been subjected to regulatory restrictions and purchasing restrictions based on public concern. If it jumped to a similar chemical, the regrettable substitution deca-brominated diphenyl ethane, then it may have bought a little time, but not much. The ethane was also targeted for restriction as per the whack-a-mole approach discussed earlier in this chapter. This may not sound like a big problem until one realizes what is involved in ensuring that the chemical of concern is not in one's global supply chain. One company estimated that it cost $6 million and a lot of effort to ensure that a restricted flame retardant was removed from its supply chain and subsequently from all of its products. One would not want to make the same mistake twice and have to go through the same exercise again. By screening chemical additives ahead of time and by selecting inherently benign chemicals for products, including flame retardants, the company can future-proof its products and avoid unnecessary costs and challenges.

AA can be used to advance safe and sustainable design. Frontrunners have indicated some of what they see as necessary to support the transition to an economy that is safe and sustainable by design, including progress toward safer chemistry, reduced carbon footprints, and circular materials. Two scientists from the chemical company BASF advocate for sustainability and green chemistry programs with the caveat that they must be based on data to ensure that a company's products are not promoted using superficial marketing tools (Harmon and Otter, 2018). For example, some products are marketed as "free of" some commonly used chemicals found to be toxic, yet the marketing materials may not indicate what chemicals have been used instead and if those chemicals are in fact safer. The product could contain a regrettable substitution. The authors argue for the use of chemical hazard assessment tools to determine if enough hazard information is available for the chemicals and if those data indicate sufficiently low hazard to support their long-term use. They advocate for the use of tools and indicators that provide evidence of positive improvements in the life cycle and safety of materials and products. Governments can reinforce the use of credible metrics behind marketing claims.

Some businesses use a few elements of AA for evaluating their products (e.g., hazard and exposure), while others use a broader and more comprehensive set of parameters (hazard, exposure, carbon footprint, recyclability, alignment with UN Sustainable Development goals, and more).

The SAFR AA framework reviewed in Box 15.2 is based on the parameter's hazard and exposure. Other companies have developed AA frameworks to apply

BOX 15.2 Systematic Assessment for Flame Retardants (SAFR)

ICL Industrial Products is the industrial chemicals division of ICL Group Ltd., a company that mines and processes minerals. Some of these minerals are used to make halogen and phosphorus-based flame retardants. The company recognizes market trends moving away from the use of halogenated organic flame retardants and the importance of transitioning to safe and sustainable chemicals. ICL created an AA framework that it applies internally to its own products called the Systematic Assessment for Flame Retardants (SAFR). SAFR considers both hazard and exposure. Hazard assessment is based on one or more of the GHS-based chemical hazard assessment methods (i.e., GHS only, GreenScreen, cradle-to-cradle). Exposure assessment is based on (1) the frequency of exposure and (2) the inherent properties of the flame retardant in combination with its ability to leach or migrate from the plastic or other material to which it is added (see Figure 15.12). Frequency of exposure is based on the likelihood that the user will

OUR ASSESSMENT OF FLAME RETARDANTS IN THEIR USES

HAZARD / EXPOSURE	LOW	MEDIUM	HIGH	UNACCEPTABLE
LOW POTENTIAL	RECOMMENDED	RECOMMENDED	ACCEPTABLE	
MEDIUM POTENTIAL	RECOMMENDED	ACCEPTABLE	NOT RECOMMENDED	TO BE PHASED OUT
HIGH POTENTIAL	ACCEPTABLE	NOT RECOMMENDED	NOT RECOMMENDED	

HOW WE ASSESS EXPOSURE

Our exposure assessment has a two-tiered approach. We consider both:

1. The frequency of contact during the intended use (eg. TV, computer, car seats, insulation boards);
2. The potential emissions of the FR used due to either migration to surface (blooming), leaching or volatilization.

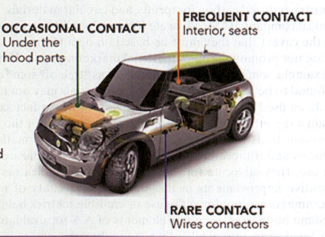

OCCASIONAL CONTACT
Under the hood parts

FREQUENT CONTACT
Interior, seats

RARE CONTACT
Wires connectors

FIGURE 15.12 Systematic Assessment for Flame Retardants (SAFR).

Source: www.icl-ip.com/safr/.

have direct contact with the material that contains the flame retardant. For example, flame retardants used in automobiles may be used in a part of the engine that would likely result in minimal direct contact, or they may be used in materials in the interior of the car that would result in frequent contact. Some flame retardants are reacted into polymers and plastics (reactive flame retardants), while others are mixed with the polymer but not reacted in (additive flame retardants). Reactive flame retardants are not as mobile as additive flame retardants. The amount of leaching and migration of the flame retardant from the plastic is related to whether or not it is reactive or additive and the nature of the polymer or plastic itself. Some polymers hold on more tightly than others to chemical additives.

Once the hazard and exposure data are gathered, the chemicals are subjected to the SAFR analysis. Chemicals with high inherent hazard that meet the criteria for chemicals of concern are rated as unacceptable and targeted for phaseout regardless of their potential for exposure. Chemicals with high hazard (but not meeting the criteria for a chemical of concern) are acceptable only under conditions of low potential for exposure. Under conditions of medium or high potential for exposure, they are considered "not recommended." Chemicals with low inherent hazard and low potential for exposure are "recommended," while chemicals with low inherent hazard and high potential for exposure are deemed "acceptable." Even low hazard chemicals can result in some risk, however, if the exposure is high enough, and that is why a chemical with low hazard but potentially high exposure is considered acceptable but not recommended.

The public commitment by ICL Industrial Products to using AA in prioritizing products to bring to the market shows leadership. More and more companies are recognizing the benefits of this kind of initiative to help future-proof their product lines. It is impressive that a manufacturer pre-emptively phases out its own products ahead of regulations and preferentially markets those with better hazard and exposure profiles.

to their product lines that are more comprehensive with respect to the number of parameters used in the assessment. For example, the Swiss company Clariant makes a variety of chemical additives, including flame retardants, with a focus on nonhalogenated flame retardants. The company developed an initiative called the Product Value Portfolio program. Every product is evaluated against 36 criteria derived from a wide range of sustainability topics. The topics cover the entire product life cycle and are grouped according to whether they primarily impact people, the planet, or product performance. The program was developed by internal company experts in collaboration with an independent nonprofit organization called the Centre on Sustainable Consumption and Production. (see Figure 15.13).

Products that meet minimum sustainability criteria for all parameters and that excel in some parameters are eligible for labeling as ECOTAIN products. See the website *(www.clariant.com/en/Sustainability/Discover-Ecotain/Our-Initiative)*. Products that score poorly are prioritized for replacement with alternatives. While all products have sustainability trade-offs, what is nice about the ECOTAIN criteria is that none of the trade-offs are extreme. Knowing that all ECOTAIN products meet minimum criteria for all sustainability parameters, and excel in others, creates a roadmap toward the ongoing development of safe and sustainable products.

Another example of a comprehensive AA framework used by a company to develop and market household cleaning products is MethodHome and the Compass of Clean (see Box 15.3).

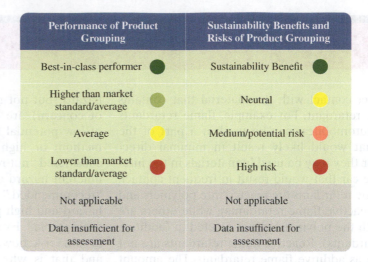

Performance of Product Grouping		Sustainability Benefits and Risks of Product Grouping	
Best-in-class performer	🟢	Sustainability Benefit	🟢
Higher than market standard/average	🟢	Neutral	🟡
Average	🟡	Medium/potential risk	🟠
Lower than market standard/average	🔴	High risk	🔴
Not applicable		Not applicable	
Data insufficient for assessment		Data insufficient for assessment	

FIGURE 15.13 The Clariant Portfolio Value Program includes tools to manage product portfolios for sustainability.

Source: Based on Coles, N., Nicolau, M., and Brüggemann, N. (2015). Developing Tools for Sustainable Product Portfolio Management. CSCP and Clariant.

BOX 15.3 Method and the Compass of Clean

MethodHome (Method) was founded by two innovators who wanted to shake up an industry that they believed had shown little innovation for a number of years. Adam Lowry and Eric Ryan decided to modernize consumer cleaning products and to integrate sustainability into the DNA of product development without marketing their products as "green" products. The company was the first to come up with original and appealing fragrances (cucumber, rose, lemon, lavender, etc.) and produced highly concentrated laundry soaps that allowed for many more washes per package. They used packaging materials that had attractive shapes and that could be readily recycled. In some cases, they used packaging to educate consumers about environmental issues by making some of their plastic packaging from plastics taken from beach and ocean cleanup efforts. Method formulators selected only chemical ingredients that are inherently benign and biodegradable, and feedstocks that are renewable or that are made using green chemistry principles. They also prioritized continual improvement toward a low carbon footprint. Like other companies, they converged on safe chemistry, reduced carbon footprints, and circularity as appropriate for their product type. They also developed metrics to assess the internal and external community-related benefits associated with their products. Figure 15.14 illustrates the full range of aspects and impacts associated with Method products, including those that are community-related, health-related, and environment-related. Every product and its packaging is benchmarked against these criteria and used to measure its current status and to drive ongoing improvement toward the ideal. As the old business adage goes, "You only manage what you measure." By having internal programs that are sophisticated in evaluating and producing products that are safe and sustainable by design, companies can guarantee that they are on a path that will help them continue to innovate and to future-proof their products, staying ahead of regulations that could stymie their business.

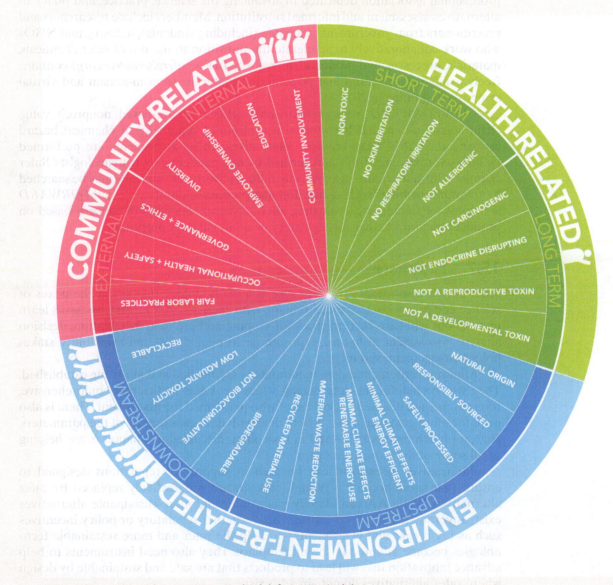

FIGURE 15.14 Greenskeeping: How Method defines sustainability.

Source: Methodhome.com.

15.6   Resources

A growing number of tools and resources are available to support alternatives assessment. The OECD Substitution and Alternatives Assessment Toolbox *(www.oecdsaatoolbox.org)* is a compilation of resources relevant to chemical substitution. There are four resource areas:

- AA Tool Selector
- AA Frameworks
- Case Studies
- Regulations and Restrictions

The Association for the Advancement of Alternatives Assessment (A4) is a professional association dedicated to advancing the science, practice, and policy of alternatives assessment and informed substitution. Members include researchers and practitioners from government, academia (including students), industry, and NGOs who work collaboratively to accelerate the transition to the use of safer chemicals, materials, processes, and products. The website *(www.saferalternatives.org)* contains freely available recordings and presentation materials from in-person and virtual symposia beginning in 2015.

Chem*FORWARD (www.chemforward.org)* is a science-based, nonprofit, value chain collaboration that has created a centralized repository of chemical hazard assessments with a focus on safer alternatives. These assessments are performed by licensed toxicology firms and verified by independent expert toxicologists. Safer chemical alternatives are organized into industry "verticals" and they may be searched by relevant attributes such as function and product applications. Chem*FORWARD* is helpful for finding and evaluating alternatives to chemicals of concern based on product function whether for regulatory or nonregulatory purposes.

15.7   Summary

Alternatives assessment is here to stay as a methodology that sits at the nexus of science and policy, but challenges remain. With each new AA project, assessors learn more about what are best practices for scoping and problem formulation, decision analysis, assessment of hazard, exposure performance, cost and availability, stakeholder engagement, and more.

Minimum criteria and data requirements are needed and are being established. They will be helpful for determining when an AA is sufficiently comprehensive, transparent, and detailed, regardless of who performed the assessment. There is also need for more guidance about how to deal with data gaps for any of the parameters. New approach methods (NAMs) in toxicology, discussed in Chapter 6, are helping to fill some of the hazard data gaps.

Alternatives assessment can be used in support of regulations designed to ensure that substances of very high concern are progressively replaced by safer alternatives. While AA can identify when safer and more sustainable alternatives exist, governments still need instruments—whether regulatory or policy incentives such as tax breaks and investments—to help the safer and more sustainable technologies become adopted in the marketplace. They also need instruments to help advance innovation that will lead to products that are safe and sustainable by design where safer alternatives do not already exist.

AA is used by private companies to support product design and development. Safe and sustainable by design appears to be emerging as an area of common interest between industry and government. Companies benefit from their use of AA to support SSbD efforts because they can develop products that appeal to customers who are concerned about sustainability and they can future-proof their product lines with respect to regulations. Creating an AA framework that is customized for their product lines can also help drive innovation and continual improvement. With common goals and criteria for safe and sustainable by design, and with common tools and frameworks for alternatives assessment, we may be at a new juncture, where government can spend less time policing polluting industries and more time collaborating to develop effective policies that advance sustainability for the common good.

References

Department of Ecology, State of Washington. (2019). Antifouling Paints in Washington State Report and Recommendations. Report to the Legislature Pursuant to SHB 2634 (2018). Publication 19-04-020. apps.ecology.wa.gov/publications/documents/1904020.pdf.

European Commission. (2020). Chemicals Strategy for Sustainability Towards a Toxic-Free Environment. Fact Sheet. ec.europa.eu/environment/pdf/chemicals/2020/10/chemicals-strategy-factsheet.pdf.

European Commission. (2020). Communication from the Commission to the European Parliament, the Council, the European Economic and Social Committee and the Committee of the Regions. COM (2020) 667 final. Brussels. ec.europa.eu/environment/pdf/chemicals/2020/10/Strategy.pdf.

Harmon, J. P., and Otter, R. (2018). "Green chemistry and the search for new plasticizers." *ACS Sustainable Chemistry & Engineering* 2:2078–2085.

Interstate Chemicals Clearinghouse. (2017). Alternatives Assessment Guide v1.1. theic2.org/article/download-pdf/file_name/IC2_AA_Guide_Version_1.1.pdf.

National Research Council. (2014). *A Framework to Guide Selection of Chemical Alternatives.* Washington, DC: The National Academies Press.

Northwest Green Chemistry and TechLaw. (2017). Washington State Antifouling Boat Paint Alternatives Assessment Report. Final Report. static1.squarespace.com/static/5841d4bf2994cab7bda01dca/t/59d40515c534a598eeb6c18a/1507067168544/Washington+CuBPAA_Final_2017.pdf.

O'Brien, M. (2000). *Making Better Environmental Decisions: An Alternative to Risk Assessment.* Cambridge, MA: The MIT Press.

United Nations. (2021). Globally Harmonized System of Classification and Labeling of Chemicals. Rev.9. unece.org/transport/standards/transport/dangerous-goods/ghs-rev9-2021. (Note that GHS is revised every two years and the most current version should succeed the older version.)

Key Concepts

Alternatives assessment (AA)
Aspects
Impacts
Continual improvement
Chemical alternatives assessment
Regrettable substitution
Drop-in replacement
Chemical of concern
Linear or sequential decision framework

Simultaneous or multiparameter decision framework
Hybrid framework
Biocidal coatings
Foul release coatings
Disruptive technology
Safe and sustainable by design
Future-proofing

Problems

15-1 All of the biocides presented in the hazard summary table in Figure 15.15 are classified as GHS Category 1 for acute aquatic toxicity.

a. What are the criteria for defining a substance as GHS Category 1 for acute aquatic toxicity for fish? The values are typically given in mg/L for a 96-hour test period in the GHS system. Look up the values in the environmental criteria in the most current version of the GHS (*unece.org /about-ghs*).

b. What are the criteria for defining a substance as GHS Category 1 for chronic aquatic toxicity for a substance that is readily biodegradable? How do the criteria differ for a substance that is not readily biodegradable?

c. If you had to choose one of the biocides in this table based on the hazard summary table alone, which one would you choose and why?

15-2 The biocide 4,5-dichloro-2-octyl-2H-isothiazol-3-one is also known by the trade name Sea-Nine 211 (CAS# 64359-81-5). It has an LC_{50} of 0.34 ppm to Balanus amphitrite (barnacles) (taken from *www.bencide.co.kr/data/SEA -NINE%20211N.pdf*), which indicates high aquatic toxicity. However, it is an organic substance that is rapidly biodegradable. Metals are not biodegradable. What are the pros and cons of using a substance like Sea-Nine instead of copper, zinc, or tributyl tin in biocidal paint formulations used on boats? Is it possible that a biodegradable substance can still build up in the environment? If so, under what conditions?

Key: vL = very low; L = low; M = moderate; H = high; vH = very high; DG = data gap; Italics = lower confidence; Bold = higher confidences.

CAS #	Name	GreenScreen Benchmark	Carcinogenicity	Mutagenicity	Reproductive Toxicity	Developmental Toxicity	Endocrine Activity	Acute Toxicity	Systemic Toxicity	Systemic Toxicity, Repeated dose*	Neurotoxicity	Neurotoxicity, Repeated dose*	Skin Sensitization*	Respiratory Sensitization*	Skin Irritation	Eye Irritation	Acute Aquatic Toxicity	Chronic Aquatic Toxicity	Persistence	Bioaccumulation	Reactivity	Flammability
1317-39-1	Cu₂O	1	L	L	L	M	DG	M	DG	M	DG	DG	L	DG	L	M	vH	vH	vH	M	L	L
13463-41-7	ZnPy	1TP	L	L	L	M	M	vH	vH	H	M	H	L	H	L	vH	vH	vH	H	vL	L	L
1314-13-2	ZnO	1	L	M	L	L	DG	L	L	H	DG	DG	L	H	L	L	vH	vH	vH	DG	L	L
64359-81-5	Seanine	2	L	L	L	L	M	vH	M	L		L	H	DG	vH	vH	vH	vH	L	vL	L	L
122454-29-9	Econea	2	L	L	L	L	L	vH	DG	H	DG	H	L	DG	M	M	vH	vH	H	vL	L	L
28159-98-0	Irgarol	2	M	L	M	L	M	L	M	M	M	DG	M	DG	L	L	vH	vH	H	L	L	L
86347-14-0	Medeto-midine	1	L	L	L	L	L	vH	DG	DG	M	DG	L	DG	L	L	vH	vH	vH	vL	L	L

FIGURE 15.15 Comparing the toxicological profiles of the biocides evaluated as part of the Washington state boat paint alternatives assessment.

Source: *Washington State Antifouling Boat Paint Alternatives Assessment Report*, Northwest Green Chemistry, static1.squarespace.com/static /5841d4bf2994cab7bda01dca/t/59d40515c534a598eeb6c18a/1507067168544/Washington+CuBPAA_Final_2017.pdf.

15-3 Look up both acute and chronic saltwater critical concentration levels for copper, zinc, and tributyl tin *(www.epa.gov/wqc/national-recommended-water-quality-criteria-aquatic-life-criteria-table#table)*.

a. What is the difference between the Criterion Maximum Concentration (CMC) and the Criterion Continuous Concentration (CCC)? Which is associated with acute aquatic toxicity and which is associated with chronic aquatic toxicity?

b. Put the substances in order of toxicity (high to low) for both CMC and CCC values.

15-4 You are a manufacturer of biocidal antifouling coatings for boats. What other factors would you consider besides the biocide when selecting ingredients for your product formulation?

15-5 The following table provides the cost per gallon of several different antifouling boat coatings (not all of them are biocidal). Some of them require thicker coatings than others and so, in addition to costs per gallon, the table includes the cost of painting a boat with a surface area of 100 ft². Some of them require recoating every year, while others require less frequent recoating based on the manufacturer's recommendations. Assume that the average cost of hauling a 35-foot boat out of the water to paint it is $900. The cost of the actual painting (labor only) is estimated at $1,645 for a 35-foot boat with a 100-ft² bottom. Assume that all boats have 100-ft² bottoms and are 35 feet in length.

COMPANY NAME	PRODUCT NAME	MANUFACTURER CLAIM FOR LONGEVITY (YEARS)	AVERAGE COST PER GALLON ($)	AVERAGE COST PER 100 ft² COVERAGE ($)
Ripple	e-Cover	1	145	78
Swell	Go fast	2	223	85
Waves	Ahoy	3	225	290
Float	Hard coat	5	510	165
Skim	Foul-off	3	270	103
Sail	Oceana	1	170	165
Race	Solution	2	250	150

a. Calculate the cost of antifouling protection if you own the boat for 1 year (just one coating).

b. Compare the cost of each product over 5 years based on the manufacturer's recommended frequency of recoating.

c. List the products in order of cost (highest to lowest) considering cost per gallon.

d. List the products in order of cost (highest to lowest) considering the cost of ownership for 5 years.

15-6 You are a manufacturer of cleaning ingredients for electronic products. You need a solvent and are considering methanol, ethanol, and/or isopropyl alcohol. Comprehensive toxicology information is available at echa.europa.eu.

SOLVENT NAME	CAS NUMBER	ACUTE MAMMALIAN TOXICITY (ORAL, LD_{50}) IN mg/kg BODY WEIGHT	GHS H STATEMENT CODES ACCORDING TO ECHA HARMONIZED CLASSIFICATIONS
Methanol	67-56-1	1,187	H225, H331, H370, H301, and H311
Ethanol	64-17-5	8,300	H225
Isopropyl alcohol	67-63-0	5,840	H225, H319, and H336

a. Which substance is most toxic based on acute mammalian toxicity LD_{50} values?

b. Define each of the hazard statement codes associated with each of the solvents.

c. Estimate the lethal dose of ethanol for your body weight using the LD_{50} value provided above. (Note that LD_{50} values for ethanol in the scientific literature vary depending in part on how much water is present in the ethanol tested.)

d. What non-acute human health hazards are associated with each solvent?

e. Look up and compare the vapor pressures for each of the solvents. Make sure that the values are in the same units and that they are measured at equivalent temperatures (at least within 5°C). In an enclosed environment, which solvent would result in the greatest inhalation exposure?

f. Assuming that the performance and cost for the solvents are equivalent, which solvent would you choose? Why?

Sustainability and the Built Environment

FIGURE 16.1 The Philip Merrill Environmental Center. This facility houses the headquarters of the Chesapeake Bay Foundation in Annapolis, Maryland. The foundation promotes conservation and restoration of the highly threatened Chesapeake Bay, the largest marine estuary in North America (and the second largest in the world). As a consequence, this NGO practices what it preaches and designed a green building that is LEED Platinum certified. Among other features, the Philip Merrill Center uses over 90% less water than a comparable office building, and more than half of its construction materials came from within 300 miles of the site.

Source: Photo by Freddie Shanks

We had the design and build the most wonderful place in the world. But it takes people to make a dream a reality.

—WALT DISNEY

Sustainability and the Built Environment

FIGURE 16.1 The Philip Merrill Environmental Center. This facility houses the headquarters of the Chesapeake Bay Foundation in Annapolis, Maryland. The foundation promotes conservation and restoration of the highly threatened Chesapeake Bay, the largest marine estuary in North America (and the second largest in the world). As a consequence, this NGO practices what it preaches, and designed a green building that is LEED Platinum certified. Among other features, the Philip Merrill Center uses over 90% less water than a comparable office building, and more than half of its construction materials came from within 300 miles of the site.

Source: Photo by Bradley Striebig.

You can design and create, and build the most wonderful place in the world. But it takes people to make a dream a reality.

—WALT DISNEY

GOALS

THIS CHAPTER INTRODUCES YOU TO THE concepts of the built environment, low-impact development, conservation design, and green building. We begin with a short overview of land-use and land-cover change and the relationship between urban planning and sustainable development. Sustainable land-use planning principles are introduced through low-impact development and conservation design strategies, which are then linked to green building rating systems and the interactions between structures and their sites. The chapter concludes with a discussion of energy conservation in buildings. Understanding the impact of design choices on green buildings and rating systems will enhance your ability to use the techniques, skills, and modern engineering tools necessary to communicate principles of sustainable design for the built environment.

OBJECTIVES

At the conclusion of this chapter, you should be able to:

16.1 Explain land-use and land-cover change and how development of the built environment affects it.

16.2 Identify the basic techniques of land-use planning, and summarize how smart growth principles contribute to more sustainable cities and land-use practices.

16.3 Compare and contrast the goals and techniques of environmentally sensitive design, low-impact development, and conservation design as approaches to reducing the impacts of buildings on their sites and in the landscape.

16.4 Explain what makes a building "green," and discuss (with examples) how "green" labeling and green building codes encourage more sustainable buildings.

16.5 Identify, explain, and give examples of the ways in which we can reduce energy consumption in buildings. Include a consideration of building design, construction, building energy systems, energy building codes, and beyond code certification programs.

Introduction

Some of the most significant impacts that humankind has on the natural world can be attributed to our ***built environment***. The built environment represents all human-made structures, engineered alterations to land forms, and constructions of every sort. The built environment includes not only our homes and buildings, but also roads, dams, parks, bridges, recreational spaces, flood control systems, and the full landscape of our cities.

The built environment involves different types of land use, and we refer to such human activity and its impact on the natural world as ***land-use and land-cover change***. Over time, because of land-use and land-cover change, we have altered local climates through deforestation, reduced the ability of watersheds to filter and purify water, and degraded ecosystems. Indeed, the greatest threat to biodiversity and species loss in the world has been—and continues to be—the loss of natural habitats resulting from the built environment.

The sustainability of human structures is intensely important because of the scale of urbanization in the world today. More than half the population in more developed regions already live in urban areas; between 2018 and 2050, almost all population growth will be absorbed by urban regions worldwide. The United Nations estimates that the world's cities will grow by a total of 2.5 billion people during this time (United Nations, 2019). Urban communities provide spaces for people to live, work, and play, and require extensive construction of buildings, roads, energy infrastructures, communication networks, water purification and distribution systems, sewage treatment plants, and sanitation and waste management facilities. Very simply, the spatial expansion of cities puts tremendous pressure on environmental resources.

In this chapter, we will focus on the basics of sustainability and the built environment by exploring land-use planning generally and environmentally sensitive development and green building specifically. We will start at the largest spatial scales by reviewing the nature of land-use and land-cover change. Then we will explore how urban planning, low-impact development, conservation design, and green building are changing the way we think about, design, and build our communities.

16.1   Land-Use and Land-Cover Change

Scientists classify the Earth's surface into basic categories of land cover that represent vegetation, ecosystems, and types of human development. Figure 16.2 illustrates one typology commonly used for land-cover analysis.

Because the Earth's surface provides many important environmental functions—including ecosystem services for humans, habitats that support biodiversity, and biogeochemical cycling of matter—the loss of critical land cover can affect these environmental functions in ways that increase economic costs and threats to human life. Land-use and land-cover changes are now monitored intensely, especially with new techniques in satellite imagery and remote sensing.

Of particular interest to us is the scope of land change due to physical ***development***. In the context of land-cover change and land-use planning, the term *development* refers to tracts of land that have been newly built on or transformed in some other way for human use. Development represents the conversion of open land and natural areas to residential neighborhoods, airports, industrial parks, hydroelectric dams, shopping centers, parks, and so on. Development reflects the inherent urbanization of communities and is driven most substantially by population growth and economic industrialization.

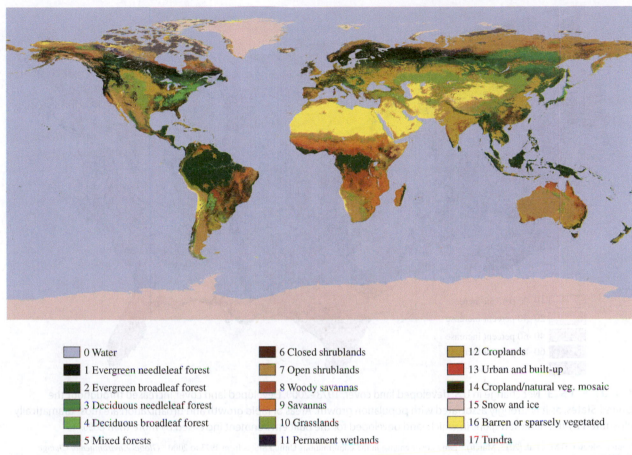

0 Water	6 Closed shrublands	12 Croplands
1 Evergreen needleleaf forest	7 Open shrublands	13 Urban and built-up
2 Evergreen broadleaf forest	8 Woody savannas	14 Cropland/natural veg. mosaic
3 Deciduous needleleaf forest	9 Savannas	15 Snow and ice
4 Deciduous broadleaf forest	10 Grasslands	16 Barren or sparsely vegetated
5 Mixed forests	11 Permanent wetlands	17 Tundra

FIGURE 16.2 Global land-cover patterns. This map illustrates the ways in which we classify Earth's surface into types of land cover. Common categories include surface water (water, snow, ice), types of vegetation (savannah, grassland), ecological systems (forests, permanent wetlands), and human development (cropland, urban development). At the global scale of this map, it is hard to see the scope of the human built environment.

Source: NASA, "Satellites and the City," www.nasa.gov/images/content/121557main_landCover.jpg.

Figure 16.3 illustrates the scope of development in the United States from 1973 to 2000. As you can see from this map, the rate of development is not evenly distributed; it varies considerably by region. In some places, developed land cover increased by more than 80%, notably in the Midwest (Indiana and Ohio), in the Pacific Northwest (northern Washington and Idaho), and in a vast area of the southwestern United States covering portions of California, Nevada, Utah, Colorado, Arizona, and New Mexico.

Urbanization and land-cover change are of particular concern for two reasons. First, they can create unsustainable environmental change by altering local climates, stimulating the loss of economically useful natural resources, and affecting a myriad of ecosystem services. Deforestation, the loss of wetlands, and the destruction of critical habitats are especially problematic. Second, the built environment can also put people in harm's way. Destructive forest fires, mudslides, and flooding are examples of hazards to human life and property that can occur when development occurs in areas naturally prone to fire, unstable geology, or floods.

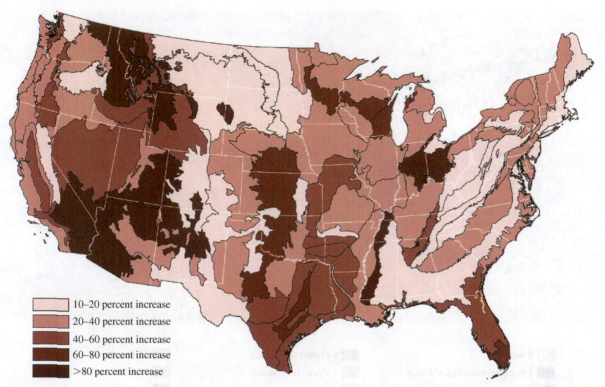

10–20 percent increase
20–40 percent increase
40–60 percent increase
60–80 percent increase
>80 percent increase

FIGURE 16.3 Net change in U.S. developed land cover, 1973–2000. Developed land cover increased throughout the United States, as it is strongly associated with population growth. Areas of rapid growth and urbanization are most dramatically illustrated by the darkest regions in which land developed for the built environment increased by more than 80%.

Source: Sleeter, B.M., et al. (2013, March). "Land-cover change in the conterminous United States from 1973 to 2000," *Global Environmental Change* 23:733–748, p. 744.

16.2 Land-Use Planning and Its Role in Sustainable Development

Land-use planning is a form of public policy that governs *how* land is used by community residents within a defined geographical area, such as a state, province, county, district, or city. In theory, land-use planning attempts to manage human activities through laws and regulations designed to make the most efficient use of land, natural resources, and water. In practice, this is not always the case. As we will see, many land-use regulations result in an inefficient use of space and degradation of the environment.

Land-use planning is often referred to as urban planning when it focuses on cities and large, complex metropolitan areas. It is widely practiced in high-income nations; middle- and low-income countries face many challenges in designing and enforcing land-use and urban planning laws. Land use in these countries tends to be governed by national efforts to protect natural resources, such as forest policies or the establishment of parks and wildlife preserves. Because of budget constraints, low-income countries often struggle to successfully implement and enforce conservation laws as well.

In the United States, settlement and land-use patterns over the past 75 years have resulted in an increasingly unsustainable form of development known as ***urban sprawl***. After World War II, the United States began to suburbanize, particularly because of the "baby boom" and the rise of automobile culture. As land-use patterns, sprawl and suburbia are characterized by low-density housing that spreads along major streets and arterial roadways. Cities lost their ***mixed-use*** urban centers in which homes, shops, businesses, and recreational spaces all coexisted with relatively high levels of ***population density*** (the number of people living in a square mile or square kilometer). Figure 16.4 illustrates the contrast between urban mixed-use space and suburbs in the United States.

FIGURE 16.4 Mixed-use urban spaces and suburban neighborhoods. U.S. cities began to spread in a new pattern of development after World War II. Historically characterized by mixed-use spaces in which residential neighborhoods, offices, and shops were commingled with one another, urban areas spread out and transformed into highly segregated land-use patterns. Residential neighborhoods were isolated from businesses and shops and became less dense in terms of housing. This overall pattern of spread is generally referred to as sprawl.

Source: trekandshoot/Shutterstock.com; Courtesy of Turner Construction Company and Robert Benson Photography.

The benefits associated with urban sprawl are typically related to less expensive and larger single-family housing and smaller, more manageable political jurisdictions. However, sprawl results in longer work commutes, greater conversion of agricultural land and open spaces, worsening water quality, and higher levels of material usage and cost. Sprawl also resulted in the hollowing out of many American cities, making them much less livable spaces. As people migrated from mixed-use urban neighborhoods to the suburbs, amenities declined. Urban centers became shopping districts at best—and sites of decline and decay at worst.

16.2.1 Zoning and Land-Use Planning

Zoning is the primary tool of land-use planning, and its principal purpose is to define the legally permitted uses of land. In the United States, zoning is often practiced as *single-use zoning*. Community activities are highly segregated from one another and only one type of land use is allowed in a designated zone. Residential neighborhoods are usually isolated from farms and commercial districts, for example, and are often allocated to very large tracts of land. Indeed, the widespread practice of single-use zoning is one of the reasons sprawl emerged as such a basic pattern of the American urban landscape. The most commonly used zoning designations are residential, commercial, industrial, agricultural, mixed use, and recreational spaces (Figure 16.5).

Legend:
- Agricultural District
- Light Industrial Park
- General Business
- Parks and Recreation
- Industrial Zone
- Flood Plain District
- Mixed-Use Redevelopment Zone
- Planned Commercial Center
- Residential Multifamily
- Residential Single Family

FIGURE 16.5 Single-use zoning patterns. In the United States, single-use zoning is a common practice in which different types of land uses are segregated from one another. This zoning map clearly shows how residential, business, and town centers are distinctly separated. Single-use zoning is a common policy in suburban communities and has been a contributing factor to sprawl.

Source: Serdar Tibet/Shutterstock.com.

With respect to residential areas, zoning laws also define **housing density**, which specifies the number and type of dwellings allowed, typically per acre of proposed residential space. Residential housing density is commonly classified as low-density residential (LDR), medium-density residential (MDR), and high-density residential (HDR). LDR housing typically ranges from 0.2 to 6 single- or two-family housing units per acre, while MDR ranges from 5 to 15 single- or multifamily units per acre. HDR usually reflects large, multifamily structures like apartment complexes, and usually has a density of more than 15 units per acre. A variety of zoning requirements for Spokane County, Washington, are shown in Table 16.1.

Zoning ordinances codify the rules and regulations that govern the allowed uses for buildings or dwellings in an area. Ordinances are often extremely detailed, and they can specify building placement, structure heights, parking requirements, signage, landscaping requirements, and other performance standards that affect building engineering and design.

Zoning provisions can affect the sustainability of the built environment by influencing the land area required for construction and the amount of building material needed to comply with zoning regulations. For example, the zoning provisions discussed previously—housing density, the requirements for open space, and single-use zoning—all contribute to less compact and more spread out developments. Other zoning provisions that can affect sustainable land-use practices include requirements for the building envelope, setback, the angle of bulk plane, and floor-to-area ratios.

TABLE 16.1 Density definitions for Spokane County, Washington

	LOW-DENSITY RESIDENTIAL	LOW-DENSITY RESIDENTIAL PLUS	MEDIUM-DENSITY RESIDENTIAL	HIGH-DENSITY RESIDENTIAL
Density (units/acre)	1–6	1	6–15	>15
Max building coverage (%)	55	55	65	70
Max height (ft)	35	35	40	50
Minimum lot area (ft²)				
Permitted uses	6,000	6,000	6,000	6,000
Single family	5,000	43,560	4,200	1,600
Duplex	10,000	n/a	8,400	3,200
Min. frontage (ft)				
Permitted uses	50	60	60	60
Single family	50	90	50	20
Duplex	50	n/a	50	40
Min. setback (ft)				15 (20)
Front (garage)		15 (20)	5 + 1 for each additional foot	
Side		5	of structure height over 25 feet	
			to a maximum of 15 feet.	
Rear setback	5 + 1 for each additional foot of structure height over 25 feet to a maximum of 15 feet.			

Source: Based on Spokane County Public Works Department (2004). *Zoning Code*, Spokane County, Washington. 1026 West Broadway Ave., Spokane, WA.

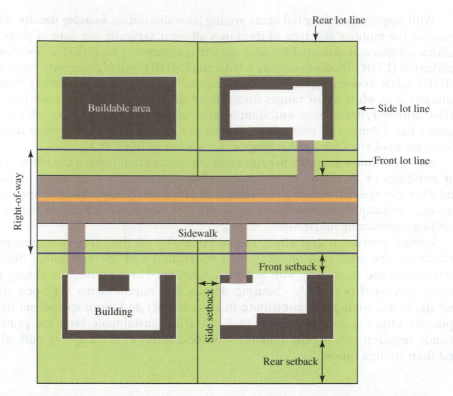

FIGURE 16.6 Buildable area and setbacks for residential structures. Zoning laws usually specify the area of a parcel of land that can be built upon and the minimum distances from property lines (setbacks) required for construction. While these requirements are often for safety and to minimize nuisances to neighbors, setbacks and buildable area requirements can also contribute to sprawl by requiring more space than is necessary.

The buildable area, or ***building envelope***, is the region of a parcel of land that may have a structure legally built upon it, as shown in Figure 16.6. Four square lots are illustrated in this figure, with two lots on each side of the street. In addition to the lots, there is a ***right-of-way*** composed of the street itself and a narrow strip of land (commonly called an ***easement***) adjacent to each side of the street; the easement is often preserved for utilities and other types of infrastructure. No permanent building structure is allowed within the right-of-way. Zoning laws regarding the size of roads and easement requirements can aggravate sprawl by requiring larger land areas to achieve desired lot sizes and roadways.

It is also typical in the United States to have a required distance—called a ***setback***—between a structure and an adjacent feature such as a property line, street, another structure, or a stream. Figure 16.6 also illustrates setback requirements for a residential dwelling. As seen in the image, the building structure must be located completely within the buildable area (building envelope) established by the setback distances and size of the right of way. As with single-use zoning, U.S. residential setback requirements have also been criticized for contributing to urban sprawl because they encourage larger lot sizes and lower density development.

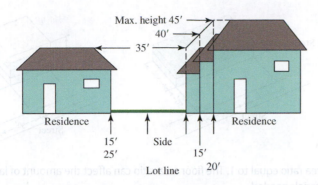

FIGURE 16.7 Maximum building height for residential structures. Zoning ordinances usually restrict the heights of structures. This diagram illustrates how even small variations in height can affect the bulk and visual appearance of a building.

Zoning ordinances can also specify a maximum building height (Figure 16.7) and may define the ***angle of bulk plane*** in an attempt to prevent buildings from literally overshadowing smaller, neighboring structures. The angle of bulk plane is typically determined by calculating the angle from the edge of the lot to be developed to the steepest angle of the proposed building, as illustrated in Figure 16.8. Depending on permitted building heights and angle of bulk plane requirements, zoning laws can also result in lower-density residential areas.

Zoning codes may also specify other key building parameters such as the floor-to-area ratio. The ***floor area ratio (FAR)*** is the ratio of the building's footprint (floor area) to the area of the lot on which the building is to be constructed. Buildings cannot exceed the specified FAR, although multiple building

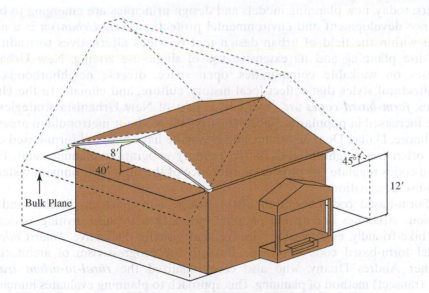

FIGURE 16.8 Angle of bulk plane. This illustration provides an example of a structure encroaching upon the angle of bulk plane.

Source: Based on *Boulder Revised Code Home*, Chapter 9-7: Form and Bulk Standards.

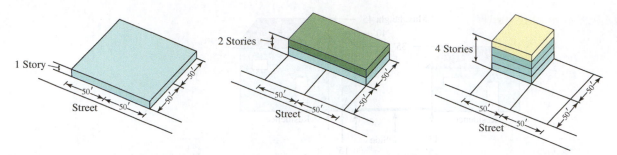

FIGURE 16.9 Configurations for a floor area ratio equal to 1. The floor area ratio can affect the amount of land used for a structure as well as the amount of building materials needed.

Source: Based on City of Boulder. *Understanding Density and Floor Area Ratio: Examples of Residential, Mixed Use, Commercial and Industrial Developments in Boulder.*

configurations are possible. An example of three methods to comply with a FAR equal to one is illustrated in Figure 16.9. Note that requirements for the floor area ratio will have an impact on the amount of building material required to construct a building.

In summary, land-use zoning designations and ordinances define the basic manner in which land can be used for the built environment, and their provisions can affect sustainable land use in important ways. Permitted uses, housing density, building envelopes, setbacks, widths of right of ways, open space requirements, floor area ratios, angle of bulk plane, and many other specifications are *necessary regulations* for safety, convenience, and pleasing spaces. But individually and collectively, these factors affect the location and arrangement of our structures, which ultimately affects land cover and its environmental benefits.

Some zoning is more conducive to a sustainable built environment than others; today, new planning models and design principles are emerging to better balance development and environmental protection. ***New Urbanism*** is a movement within the field of urban design that provides alternatives to traditional land-use planning and its extensive use of single-use zoning. New Urbanism focuses on walkable communities, open space, diverse neighborhoods, and architectural styles that reflect local history, culture, and climate. In the United States, ***form-based codes*** are emerging as part of New Urbanism strategies and have increased in popularity since the early 2000s. Major metropolitan areas like Baltimore, Dallas, Denver, Miami, and Nashville have adopted form-based codes that often use neighborhoods as the primary geographic planning unit. Form-based codes regulate the physical appearance of the built environment instead of land-use designations.

Form-based codes result in urban spaces with more predictable and harmonious structures and spatial patterns, spaces that are also inviting, pedestrian and bike-friendly, ecologically sensitive, and visually distinctive. *SmartCode* is a model form-based code developed through the progressivism of architect and planner Andrés Duany, who also conceptualized the ***rural-to-urban transect*** (the Transect) method of planning. This approach to planning evaluates human settlements and the physical environment along a continuum of rural-to-urban spaces and establishes appropriate form-based regulations in each of the transect zones. New Urbanism and form-based codes are also part of a broader design shift in land-use planning known as *smart growth*.

16.2.2 Smart Growth

Smart growth is an alternative strategy to traditional urban and regional development, and it is practiced within the larger framework of land-use planning. **Smart growth** represents land-use development and urban design principles intended to curb sprawl, conserve natural resources, and create more livable cities. (In Europe, the smart growth concept goes by the term *Compact City*.) It advocates **urban intensification**—higher levels of housing density in combination with mixed-use zoning, the greater availability of green space and natural areas, and enhanced mass transit. The Smart Growth Network, a coalition of organizations concerned about sustainable development, distilled 10 basic principles to manage growth in the built environment:

1. Mix land uses.
2. Take advantage of compact building design.
3. Create a range of housing opportunities and choices.
4. Create walkable neighborhoods.
5. Foster distinctive, attractive communities with a strong sense of place.
6. Preserve open space, farmland, natural beauty, and critical environmental areas.
7. Strengthen and direct development toward existing communities.
8. Provide a variety of transportation choices.
9. Make development decisions predictable, fair, and cost effective.
10. Encourage community and stakeholder collaboration in development decisions.

Smart growth requires a wide variety of urban planning, landscape design, and community development strategies. Three prominent land-use techniques are urban infill, brownfield redevelopment, and cluster developments.

Urban infill is a strategy in which abandoned, derelict, unused, or underutilized space in urban areas is redeveloped. The infill land use may be purely residential or commercial, but ideally it has a mix of housing options as well as parks, commercial, and business space. The cost economics of urban infill is often problematic, as it can be much more expensive to redevelop existing urban spaces than to simply build on the suburban periphery (another contributing factor to sprawl). Urban infill is most widely practiced in cities that have an **urban growth boundary** (a zoning designation that legally limits the geographic spread of growth and new development) or physical barriers to expansion such as mountains or existing dense urban development. Portland, Oregon, is a city renowned for its urban growth boundary and successful infill projects; however, escalating housing prices and unfavorable patterns of land values have become growing concerns in the region. Most recently, **urban agriculture** has emerged as a type of infill land use, where small farms and community gardens are interspersed in a city's neighborhoods.

Brownfield redevelopment also involves the redevelopment of existing space. In the United States, a **brownfield** has traditionally represented an inactive industrial site with some degree of real or perceived environmental/chemical contamination. Environmental remediation may be required to effectively reuse the site for some types of land use; brownfields are nonetheless potentially attractive sites for growth because they can involve large acreages in desirable locations. Atlantic Station in Atlanta, Georgia, is an award-winning, mixed-use brownfield redevelopment project that successfully repurposed 138 acres in midtown Atlanta (Figure 16.10).

FIGURE 16.10 Atlantic Station in Atlanta, Georgia, is an example of a substantial urban brownfield and infill redevelopment initiative, the largest in the United States. It provides mixed use, green spaces, and other amenities to make the neighborhood a livable community for its residents.

Source: ESB Professional/Shutterstock.com.

Cluster developments are subdivisions or building tracts that are designed to concentrate housing and other structures in one area, protecting much of the landscape from development. In a cluster development, the overall zoned density of land does not change, but the building envelope, setback requirements, and housing density are adjusted to create an alternative arrangement of buildings (Figure 16.11). Ideally, cluster developments contain outdoor recreational areas and different types of housing options to accommodate a range of resident ages, incomes, and lifestyles. They should also be highly walkable with sidewalks, low-traffic areas, and pedestrian ways. What makes cluster developments special from a planning perspective is that they usually violate several aspects of local zoning laws and require special regulations to permit their use.

Smart growth policies and principles are fundamentally designed to plan for growth in ways that reduce traffic congestion, provide a wider variety of housing options at reasonable costs, and increase convenience and quality of life for local residents. Smart growth also purposefully protects the natural environment. Burchell and Mukherji (2003) reported that managed growth may decrease land consumption by 21% compared to conventional methods. Significant savings can be realized for infrastructure and utility costs as well (Table 16.2) because higher density growth reduces overall demand on public services and infrastructure.

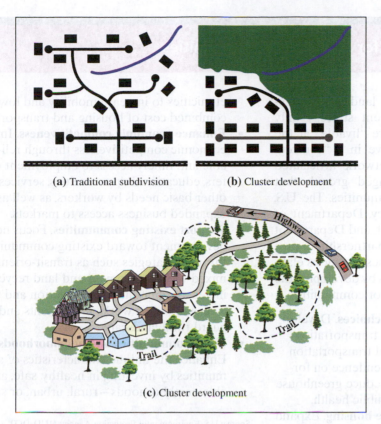

(a) Traditional subdivision **(b)** Cluster development

(c) Cluster development

FIGURE 16.11 Arrangement of cluster developments. Cluster developments differ from traditional suburban neighborhoods and subdivisions in the arrangement of their homes. Parts (a) and (b) both have the exact same number of housing units, but the cluster development in part (b) arranges the dwellings more compactly, opening up a green space. As seen in part (c), this green space can be used for community recreation.

Source: Based on Triangle J Council of Governments, *Green Growth for Clean Water* (Research Triangle Park, NC, 2012), pp. 17, 29.

TABLE 16.2 Projected water and sewer infrastructure under conventional development and managed growth scenarios in the United States, 2000–2025

REGION	TOTAL WATER AND SEWER DEMAND, gal/day (millions)			TOTAL WATER AND SEWER LATERALS (thousands)			TOTAL INFRASTRUCTURE COSTS, $ (millions)		
	Conventional Development	Managed Growth	Demand Savings	Conventional Development	Managed Growth	Lateral Savings	Conventional Development	Managed Growth	Cost Savings
Northeast	1,451	1,444	7	3,406	3,068	338	16,015	14,751	1,264
Midwest	2,935	2,915	21	7,110	6,604	505	30,393	28,839	1,556
South	7,942	7,870	72	21,243	19,116	2,126	84,573	79,026	5,547
West	5,794	5,737	56	14,108	12,456	1,652	58,786	54,544	4,242
Total	18,121	17,965	156	45,867	41,245	4,621	189,767	177,160	12,609

Source: Based on Burchell, R.W., and Mukherji, S. (2003). "Conventional Development Versus Managed Growth: The Costs for Sprawl." *American Journal of Public Health* 93(9):1534–1540.

BOX 16.1 Design Values for Sustainable Communities

Smart growth as an approach to land use is effective as a sustainable development strategy only if our communities become more "livable" at the same time—no one wants to live in a "concrete jungle." The Smart Growth Network articulated 10 design principles for managed growth that reflect the need for livable communities. The U.S. Environmental Protection Agency, Department of Housing and Urban Development, and Department of Transportation also created a partnership to promote sustainable communities. These agencies elaborated the smart growth principles by developing the following "Livability Principles" for communities:

- **Provide more transportation choices**. Develop safe, reliable, and economical transportation choices to decrease household transportation costs, reduce our nation's dependence on foreign oil, improve air quality, reduce greenhouse gas emissions, and promote public health.
- **Promote equitable, affordable housing**. Expand location- and energy-efficient housing choices for people of all ages, incomes, races, and ethnicities to increase mobility and lower the combined cost of housing and transportation.
- **Enhance economic competitiveness**. Improve economic competitiveness through reliable and timely access to employment centers, educational opportunities, services, and other basic needs by workers, as well as expanded business access to markets.
- **Support existing communities**. Focus new development toward existing communities—through strategies such as transit-oriented, mixed-use development and land recycling—to increase community revitalization and the efficiency of public works investments and to safeguard rural landscapes.
- **Value communities and neighborhoods**. Enhance the unique characteristics of all communities by investing in healthy, safe, and walkable neighborhoods—rural, urban, or suburban.

Source: U.S. Environmental Protection Agency, HUD-DOT-EPA Partnership for Sustainable Communities, "Livability Principles."

To sum up, smart growth strategies involve urban intensification by redeveloping unused (or underused) land for productive purposes and by creating more compact, higher-density neighborhoods at the urban/suburban interface. When combined with other design techniques for walkability, green spaces, mixed use, and so on, smart growth can result in cities with pleasing amenities, greater conveniences, and more local employment opportunities. Smart growth does, however, require a major rethinking of zoning laws, and it reflects an intensive effort at progressive urban design (see Box 16.1).

Higher densities of buildings and people tend to minimize urban sprawl and require less transportation, lower energy costs in building operations, and fewer materials for infrastructure. Nevertheless, high-density development does come with challenges. It may stress water, sanitation and waste management systems. It can also be more costly to implement, and as seen in some areas, increase the cost of housing. Urban intensification is also not necessarily easy to "retrofit" into existing communities because large, multistory buildings may negatively affect their surroundings in important ways.

16.3 Environmentally Sensitive Design

Environmentally sensitive design (also known as environmentally sensitive development) is a term that generically describes a broad range of landscape design strategies and on-the-ground techniques to integrate both development and environmental conservation. Environmentally sensitive design is very different from a conventional *environmental impact assessment*, which is an analysis that catalogues the expected environmental impacts a project will have at a site and indicates how those impacts might be mitigated. Environmental impact assessments *are not* design strategies. The Maryland Office of Planning (1995) stated the foundational design principle of environmentally sensitive development quite succinctly, which is to "let the land shape your plan." The goal of environmentally sensitive design is to site and construct the built environment in a manner that protects sensitive areas and minimizes the degradation associated with development.

No matter where we live, certain types of "sensitive" landscape features consistently require conservation and protection because of their profound role in ecosystem services and biogeochemical cycling. These include:

- Forest cover
- Floodplains
- Streams and their riparian buffers
- Wetlands
- Steep slopes
- Habitats of threatened, rare, and endangered species

Nonsensitive features of the landscape have important environmental roles as well. For example, a meadowland provides habitat for native pollinators and small mammals, contributing to the overall biological complexity of a region. Large expanses of undeveloped or agricultural land create wildlife corridors, regulate the local climate, support local food systems, and offer recreational opportunities. Environmentally sensitive design therefore focuses on protecting sensitive areas, conserving open space, and maintaining the integrity of the environmental and social functions supported by the landscape.

In addition to the loss of sensitive landscape features, open space, and habitats, the most significant land-cover impacts of the built environment are associated with the disturbance of soil and the disruption of a location's natural hydrology. Site development commonly involves excavating and re-grading land, laying down large areas of impervious surfaces, redirecting or eliminating natural bodies of water, and constructing stormwater systems to manage rainwater runoff. (*Impervious surfaces* are those areas in the urban landscape that do not absorb water or allow water to infiltrate into the ground, roads, parking areas, and rooftops.)

As a consequence, wetlands have been eliminated at an astounding rate in the United States, and it is likewise common to bury small surface streams or channelize them into engineered drainage systems. Resculpted land, vast areas of impervious material, and the disturbance of surface waters result in the loss of productive soil, sedimentation of streams and rivers, accelerated stormwater runoff, increased non-point-source water pollution, declining biodiversity, and the destruction of critical water habitats. Figure 16.12 illustrates some of these problems by showing how development can affect the hydrologic dynamics of a forested site.

The social and environmental problems associated with stormwater runoff are particularly significant in urban and suburban areas. Rainfall rapidly moves across impervious roadways, parking lots, and rooftops, gathering contaminants as it goes; it also tends to flow in sheets across the sloped lawns and grassy areas common in suburbia. Without stormwater management systems, localized flooding can result, which is clearly a socially undesirable condition. However, traditional stormwater systems direct runoff to local water bodies in a manner and at rates that would not occur naturally. Water contamination, downstream flooding, and degraded stream

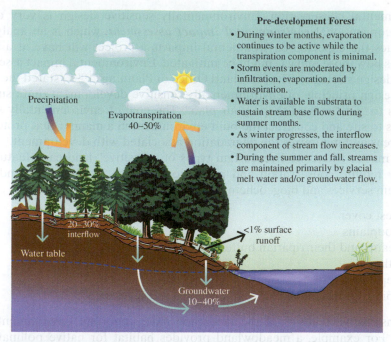

Pre-development Forest

- During winter months, evaporation continues to be active while the transpiration component is minimal.
- Storm events are moderated by infiltration, evaporation, and transpiration.
- Water is available in substrata to sustain stream base flows during summer months.
- As winter progresses, the interflow component of stream flow increases.
- During the summer and fall, streams are maintained primarily by glacial melt water and/or groundwater flow.

Precipitation

Evapotranspiration 40–50%

<1% surface runoff

20–30% interflow

Water table

Groundwater 10–40%

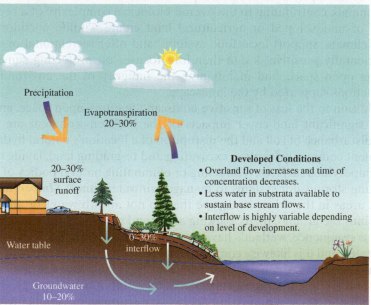

Precipitation

Evapotranspiration 20–30%

Developed Conditions

- Overland flow increases and time of concentration decreases.
- Less water in substrata available to sustain base stream flows.
- Interflow is highly variable depending on level of development.

20–30% surface runoff

0–30% interflow

Water table

Groundwater 10–20%

FIGURE 16.12 Disruption of site hydrology from development. Development can have significant consequences for hydrologic cycles. The image at the top shows how this site's natural topography and vegetation regulate the level of the water table, the volume of groundwater, the rate of interflow, the amount of surface water runoff, and the degree of evapotranspiration. After development, the reduced forest cover diminishes evapotranspiration, and the accelerated runoff reduces groundwater recharge. Over time, the environmental dynamics of this site will be significantly changed in a manner that will alter its microclimate, degrade its water flow and quality, and diminish its ability to support complex vegetation.

Source: Based on Hinman, C. (2005). *Low Impact Development: Technical Guidance Manual for Puget Sound*. Puget Sound Action Team, Olympia, WA.

ecology are the unintended consequences of these systems (Box 16.2). In some places, runoff may not be directed to a structured drainage system and erodes soil in problematic ways. Mudslides can occur, for example, when sloped land has not been adequately protected from erosion due to runoff from the built environment.

BOX 16.2 Impervious Surfaces and the Ecological Integrity of Streams

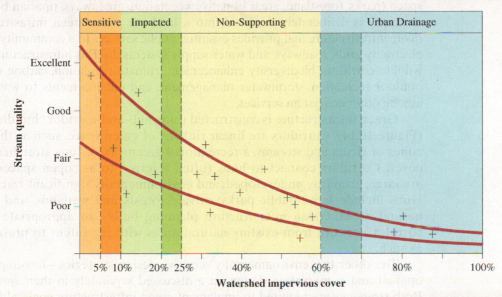

FIGURE 16.13 The relationship between impervious cover and stream quality. This chart represents the degree of impervious land cover in a watershed as a cone of acceptable conditions that is wider at lower impervious cover and progressively narrows at higher impervious cover.

Source: Based on Schueler, T., and Fraley-McNeal, L. (2008). *The Impervious Cover Model revisited: A review of recent ICM research.* Presented at the Symposium on Urbanization and Stream Ecology, May 23 and 24, 2008. chesapeakestormwater.net/2009/11/the-reformulated-impervious-cover-model/.

Streams in a watershed with less than 10% impervious surface can maintain natural levels of biodiversity and experience little harm to their normal stream channel and habitats. Streams become significantly affected by development when impervious areas represent 11% to 25% of the watershed, as illustrated in Figure 16.13. When an impervious surface exceeds about 25% of the land area, streams, waterways, and their surrounding riparian habitat become irreparably damaged. As a consequence, they cannot function as a healthy ecosystem or as an effective hydrologic channel in the watershed. Streambank erosion, high water velocity during storm events, increased sedimentation, low dissolved

oxygen levels, and nutrient loading are some of the consequences of extensive impervious surfaces that *are not* mitigated by environmentally sensitive design. This has two important results. First, it substantially diminishes a stream's biodiversity and ecology, which has a key role in the surrounding ecosystem and for maintaining water quality. Second, after repeated storm events, streambank erosion and sedimentation tend to "straighten" a stream, which increases water velocity. The result is downstream flooding during heavy rain and the rapid movement of contaminants into larger bodies of water within and outside of the local watershed. Over time, vital streams effectively become ecologically sterile drainage pipes.

By "letting the land shape the plan," environmentally sensitive design does several things. First, it requires that *landscape features* be the primary determinant of how a site or region is developed. Second, it protects critical environmental features by literally designing around them and disturbing them minimally, if at all. Third, it mitigates stormwater flow and erosion by containing and processing runoff as close to its source as possible and by imitating natural processes.

Three complementary environmentally sensitive design paradigms have emerged, and all of them may be used in conjunction with one another. These are green infrastructure, low-impact development, and conservation design. *Green infrastructure* refers to a landscape pattern in which a connective network of green space (parks, forestland, areas of native vegetation, greenways, riparian buffers, and so forth) is deliberately designed into a land-use plan. Green infrastructure is a *living* infrastructure that provides essential public services to a community much like electricity grids, roadways, and water supply systems do. This infrastructure provides wildlife corridors, biodiversity enhancement, climate regulation, carbon mitigation, outdoor recreation, stormwater management, and enhancements to water quality, among other ecosystem services.

Green infrastructure is constructed using hub-and-corridor "building blocks" (Figure 16.14). Corridors are linear ribbons of green space, such as the riparian zones of rivers and streams, a recreational greenway, or long stretches of forest cover. Corridors connect to hubs (often referred to as "open spaces"), which are large, primarily undeveloped, and environmentally significant tracts of land. Hubs include large public parks, private forests and wetlands, and protected natural areas. Green infrastructure planning builds an appropriate hub-and-corridor network from existing natural areas with the intent to preserve them as such.

The other two environmentally sensitive design strategies—low-impact development and conservation design—are discussed separately in their own sections. Both strategies are required to implement green infrastructure goals, although it is common to have low-impact development and conservation design *without* a green infrastructure land-use plan. Each of these also represents site-specific techniques and methods that you should be familiar with because they are rapidly becoming a suite of best management practices in a number of professions.

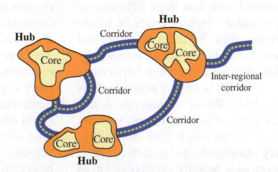

FIGURE 16.14 Interconnected hubs and corridors are the basic design elements for green infrastructure. Hubs are open spaces and usually environmentally important; corridors provide pathways for wildlife and may also protect rivers and streams.

Source: U.S. EPA, "Going Green" in the Mid-Atlantic, *Natural Infrastructure,* www.epa.gov/region03/green /infrastructure.html.

16.3.1 Low-Impact Development

Because of stormwater management issues in particular, low-impact development (LID) has emerged as a strategy to *avoid* runoff when possible and to *minimize* the disruption to natural hydrologic processes from development. The goal of LID is to mimic the natural hydrologic characteristics of a site so that it is soundly within the environmentally sensitive design philosophy of letting the land shape the plan. Although LID is often used to casually describe a variety of sophisticated design strategies, it is most properly understood as a set of ***best management practices (BMPs)*** to manage runoff at a site. BMPs are techniques based on demonstrated, successful engineering practice; they are measures that consistently achieve the desired results in a cost-effective manner. LID BMPs are now mandatory in the land-use, transportation, and water quality laws of many states and localities.

LID uses water management systems that retain, treat, and filter runoff as close to its source as possible. This enhances water quality and also provides the benefit of soil conservation. LID BMPs are designed to maintain predevelopment water infiltration rates and to minimize downstream sediment loading, streambank erosion, and flooding. LID works in part by diverting and slowing stormwater runoff to natural bodies of water. A large number of techniques and practices are commonly used in LID depending on the nature and location of a project; several examples can serve to illustrate the scale and scope of LID efforts.

At the residential scale, gutters and downspouts traditionally move rainwater off of a rooftop and away from the foundation of a house, where water can be structurally damaging. However, downspouts are commonly channeled to the street, where rain then moves rapidly as stormwater runoff. Discharging downspouts to a lawn is better but may still result in stormwater flow; LID promotes the use of rain gardens (as discussed in Chapter 12) as a simple way of capturing roof water and allowing it to infiltrate naturally on site. Impervious surfaces can be reduced or avoided at the residential scale by using pervious pavers instead of concrete or asphalt for hardscapes.

At commercial scales or in the urban environment, runoff from surfaces (as well as rooftops) can be managed through a number of LID techniques, many of which involve bioretention and filtration. For example, surface runoff and downspouts can be directed to ***vegetated swales***, which are basically a bigger, engineered version of residential rain gardens. Vegetated swales are most commonly used as bioretention areas to hold and filter runoff for groundwater recharge (Figure 16.15). However, they can also be used as drainage *channels*, filtering runoff while also

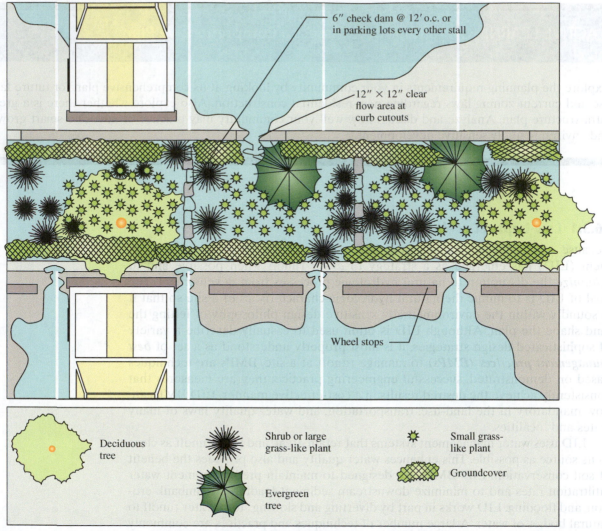

6" check dam @ 12' o.c. or
in parking lots every other stall

12" × 12" clear
flow area at
curb cutouts

Wheel stops

Deciduous
tree

Shrub or large
grass-like plant

Small grass-
like plant

Groundcover

Evergreen
tree

Note: At least 50% of the facility shall be planted with grasses or grass-like plants,
primarily in the flow path. Large grass-like plants can be considered as shrubs.

Swale area = Approx. 400 sq. ft.
(not to scale)

FIGURE 16.15 Bioretention systems for stormwater management. Bioretention areas adjacent to parking lots and roadways are commonly referred to as vegetated swales. These engineered biological systems are shallow depressions in the ground constructed to collect and filter stormwater, both cleaning it and allowing it to either recharge naturally as groundwater or slowly make its way to a body of surface water. This diagram shows a bioretention zone in a parking lot.

Source: Based on City of Portland. (2004). *Stormwater Management Manual: Revision #3*. Environmental Services, Clean River Works, City of Portland, OR.

slowing and directing it to a body of surface water (Figure 16.16). Biorentention and filtration can be implemented at very small scales, such as with the use of tree boxes in stormwater drains.

Bioretention systems preserve stream health by cleaning, reducing, and slowing runoff. Stream health is further protected through a number of additional LID techniques. One is simply not to bury or otherwise channel a stream itself, something that was done with surprising frequency in conventional development. (Restoring

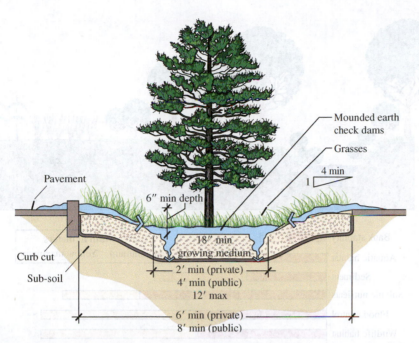

Pavement

6″ min depth

Mounded earth
check dams

Grasses

4 min
1

Curb cut

Sub-soil

18″ min
growing medium

2′ min (private)
4′ min (public)
12′ max

6′ min (private)
8′ min (public)

FIGURE 16.16 Cross section of a vegetated swale channel. Vegetated swales can also be used to *move* or *redirect* the flow of stormwater. This profile view illustrates how they are designed and constructed.

Source: Based on City of Portland. (2004). *Stormwater Management Manual: Revision #3*. Environmental Services, Clean River Works, City of Portland, OR.

a buried stream in an urban area to natural conditions is referred to as **stream daylighting**.) Another technique is to use a **riparian buffer zone**. Although riparian buffer zones are commonly discussed in the context of agriculture, they are also used in LID. These zones offer significant protection of surface streams by first keeping built structures at an extended distance from the water. Riparian buffer zones characteristically grow a mixed variety of grasses, plants, shrubs, and trees adjacent to the stream channel (Figure 16.17); the wider the buffer, the greater the variety of environmental benefits that result, including flood control and wildlife habitat. Streams that have lost their natural riparian buffers can be restored by planting appropriate new vegetation.

Although low-impact development is most commonly discussed in terms of stormwater prevention, management, and control, it focuses on water conservation as well. LID advocates reducing the demands for freshwater created by new development. Consequently, LID water conservation strategies include rainwater catchment methods that allow harvested rainwater to be used for landscape and garden irrigation and potentially for indoor toilet flushing. Cisterns are commonly used to capture rainwater in the large volumes required for these applications. LID water conservation can also involve onsite water recycling, such as with **gray water** use for irrigation and indoor toilet flushing. Gray water is the wastewater from activities such as bathing, dishwashing, and laundry; because it is not contaminated with human body waste (known as sewage), it can be safely reused/recycled for nonpotable needs. However, zoning laws throughout the United States often prohibit the

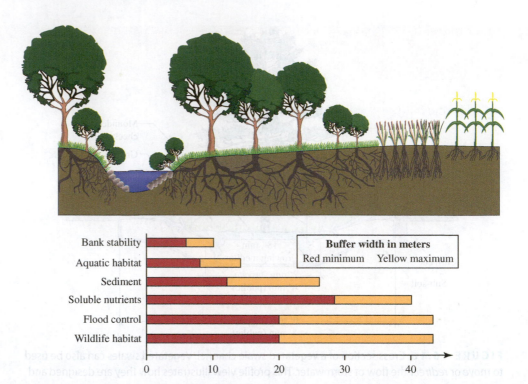

FIGURE 16.17 Benefits of different riparian buffer widths. Riparian buffers are used to protect and restore stream health in agricultural settings (as seen in this image) as well as in low-impact development. As buffers get wider, more environmental benefits result.

Source: Based on Brinson, M.M. et al. (eds.) 2002. *Riparian Areas: Functions and Strategies for Management.* Committee on Riparian Zone Functioning and Strategies for Management. Water Science and Technology Board. Board on Environmental Studies and Toxicology. Division on Earth and Life Sciences. National Research Council. Washington, DC: National Academy Press.

use of graywater systems. Low-impact development BMPs are most commonly used in and applied to the landscape, but they can also involve the design of buildings and building water systems.

16.3.2 Erosion Challenges

Productive topsoil is a nonrenewable resource that can take hundreds of years to form. It is critical not just to provide our food, but also to sustain the plants and photosynthesis that form the basis of life on Earth. In the United States, soils are estimated to be eroding at a rate 17 times faster than they are produced; in Asia, Africa, and South America, the rate of loss is estimated at 30 times the rate of production. A good deal of this soil loss occurs because of agriculture and its associated practices; nonetheless, it is important that development of the built environment not worsen these conditions.

Fluvial (water) erosion is caused when runoff carries away soil particles, and it takes place in four typical forms: (1) sheet, which occurs evenly over the slope of a land surface; (2) rill, when the rainfall collects into very small but well-defined temporary channels; (3) gully, when rills enlarge; and (4) streambank, when the stream channel widens or collapses. Fluvial erosion of all types may be caused by the increased runoff from development.

BOX 16.3 Universal Soil Loss Equation

The *universal soil loss equation* is one method that may be used to estimate soil loss through erosion:

$$A = R \times K \times L \times S \times C \times P \qquad (16.1)$$

where

A = annual soil loss due to erosion (mg/ha-yr)

R = rainfall factor, which is the measure of energy expected from raindrops hitting the soil during a 30-minute storm

K = soil erodibility factor based on the ratio of silt, sand, and organic matter

L = length of the slope

S = gradient of the slope

C = soil cover factor

P = a "support practice" factor based on the type of agricultural practice. For construction practices, this is assumed to be equal to one.

Typical values for the soil cover factor, C, are provided in Table 16.3. The remaining terms of the universal soil loss equation can be determined from site-specific maps and in soil maps from organizations like the U.S. Department of Agriculture, the U.S. Geological Survey, and local soil conservation agencies.

TABLE 16.3 Typical soil cover factors used in the universal soil loss equation to estimate erosion

COVER TYPE	NO COVER	STRAW OR HAY TIE DOWN	CRUSHED STONE COVER	WOOD CHIPS	SOD OR LAWN
Cover factor	1.0	0.02–0.06	0.02–0.08	0.02–0.08	0.01

Eolian (wind) erosion occurs when high winds remove soil, and it is worsened when soils are dry or not covered by vegetation. Development can trigger wind erosion temporarily because of the earth moving involved in the construction process and by creating areas of uncovered topsoil. Development can also exacerbate wind erosion permanently by changing the topography of the land and by removing natural wind breaks, such as dense vegetation.

Development most commonly contributes to eroded soil because of disturbances to the natural hydrologic conditions of a location, including the loss of natural vegetative land cover. LID best management practices for stormwater management consequently go far in conserving soil lost through water erosion. With respect to wind erosion, hedgerows, tree rows, and fencerows may be used as windbreaks to avoid conditions that create a permanent risk of soil erosion. A design with diverse plantings of various heights and densities will not only buffer winds but also provide habitat.

Environmentally sensitive design is especially concerned with steep slopes, one of several sensitive landscape features that in principle either should not be developed or should be aggressively protected from erosion and instability. Steep slopes may be geologically stabilized through a variety of retaining wall systems (Figure 16.18). Soil erosion may be prevented with fabric-based methods that use

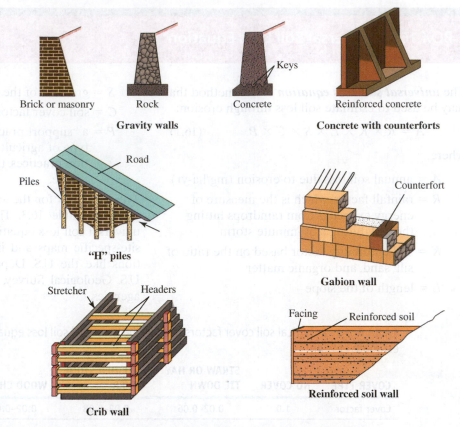

Gravity walls

Brick or masonry   Rock   Concrete

Keys

Concrete with counterforts

Reinforced concrete

Road

Piles

"H" piles

Counterfort

Gabion wall

Stretcher   Headers

Crib wall

Facing   Reinforced soil

Reinforced soil wall

(a) Common Types of Retaining Structures

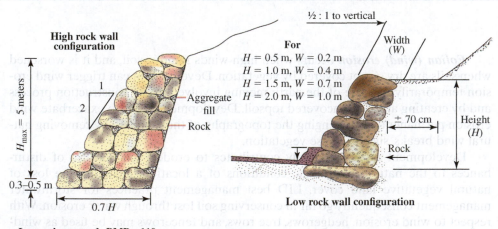

High rock wall configuration

½ : 1 to vertical

Width (W)

H_{max} = 5 meters

1
2

Aggregate fill

Rock

For
H = 0.5 m, W = 0.2 m
H = 1.0 m, W = 0.4 m
H = 1.5 m, W = 0.7 m
H = 2.0 m, W = 1.0 m

± 70 cm

Height (H)

Rock

0.3–0.5 m

0.7 H

Low rock wall configuration

Low-volume roads BMPs: 110

(b) Typical Rock Wall Construction

FIGURE 16.18 Common retaining systems for steep slopes. Development that occurs on or near steep slopes requires that the earth itself be held in place by a substantial retaining structure.

Source: Keller, G., and Sherar, J. (2003). *Low Volume Roads Engineering: Best Management Practices Field Guide.* USDA Forest Service/USAID, Washington, DC. 182p. Figure 11.4.

BOX 16.4 Stokes' Settling Law

The design and maintenance of sediment control systems for runoff management require knowledge of the sedimentation loss rate, runoff velocity, and estimated particle size distribution and weight. Sediments may be removed from runoff when the downward velocity of the particle is greater than the runoff velocity. **Stokes' settling law** [see Equations (7.41) through (7.44)] may be used to estimate particle removal as a function of the runoff flow velocity and expected residence time in the sedimentation removal device. The settling velocity for a single particle may be estimated by

$$V_s = \frac{2g\left(\dfrac{\rho_s}{\rho_f} - 1\right)r^2}{9\eta_f} \qquad (16.2)$$

where

V_s = settling velocity (m/s)
g = gravitational constant (m/s^2)
ρ_s = density of the particle (kg/m^3)
ρ_f = density of the fluid (kg/m^3)
r = particle radius (m)
η_f = kinematic viscosity of the fluid (m^2/s)

The settling time, t_s, of a particle is given by the relationship between the depth, h, and settling velocity:

$$t_s = \frac{h}{V_s} \qquad (16.3)$$

plastic netting, wood mesh, nylon mesh, or paper twine to hold soils, seeds, and new vegetation in place. Synthetic **geotextile** and turf reinforcement materials can be used on very steep slopes and will lock soil in place but allow water to permeate. Geotextiles are synthetic materials purposefully designed to protect soil and manage water flow in a variety of ways; they are straightforward to install but may be unsightly and expensive to implement and repair.

Some development may nonetheless permanently generate soil and sediments. Long-term controls will be required for these sites; BMPs include settlement, infiltration, and wetland basins and must be sized relative to the amount of estimated sedimentation (see Box 16.4). Permanent controls are designed to last 50 years or more and require little maintenance. In practice, permanent basins are rarely effective at controlling sedimentation that long. They also consume valuable land space and are expensive to construct.

In short, low-impact development and BMP erosion control measures implement the goals of environmentally sensitive design through on-the-ground techniques. These strategies primarily target the water- and soil-based environmental damage created by stormwater runoff and other development impacts.

16.3.3 Green Infrastructure and Conservation Design

Today, one of the most sophisticated, site-specific approaches to environmentally sensitive development is known as **conservation design** (also referred to as conservation development). It integrates principles of cluster development, low-impact development, and green infrastructure to achieve communities that are aesthetically beautiful, highly functional, and ecologically sound. In conservation design, the open spaces created by cluster development are used for some combination of wildlife corridors, land preservation, historic preservation, recreation, and water resources protection. It is a holistic approach to environmentally sensitive design that is most commonly used to develop large tracts of land into residential neighborhoods or mixed-use communities. For localities that have formally adopted a green infrastructure plan, conservation design is an essential tool for implementing that plan in new developments.

Conservation design provides many environmental advantages compared to traditionally developed properties. In addition to avoiding the development of sensitive landscape features, this approach can also protect upland buffers, preserve habitat and biodiversity, reduce erosion and runoff, and improve groundwater infiltration. Research has also shown that homes on smaller lots designed using conservation principles not only sell more quickly, but also appreciate in value faster than comparable homes in traditionally designed neighborhoods.

In practice, conservation design literally starts with letting the land shape the plan. Landscape architect Randall Arendt developed a conservation design methodology that is supported by the Natural Lands Trust, the American Planning Association, and the American Society of Landscape Architects. It consists of four steps:

1. Identify conservation areas from resource maps.
2. Locate the housing sites.
3. Design street alignments and trails.
4. Draw lot lines.

After the first step, *identify conservation areas from resource maps*, the design process becomes progressively more detailed by establishing (a) areas of the landscape for conservation, (b) the buildable areas of the site, and (c) the methods for water conservation. Figure 16.19 illustrates how, by overlaying land use and resource maps, areas of highest conservation value can be identified. The buildable area will be sited in a manner that does not disturb the conservation zones.

Figure 16.19 also illustrates the contrast between conventional site development and conservation design in terms of step 2, *locate the housing sites*. In part (b), only the wetland (a primary area of conservation) was protected. The rest of the site is fully apportioned to building lots, a strategy that is typical of conventional developments. Secondary conservation features, such as the natural drainage of the property, are built over. The conservation approach in part (c) contains the exact same number of housing units as in part (b), making it a ***density-neutral*** design. But the homes are clustered into more compact arrangements through smaller lot sizes and narrower streets, enabling protection of both the wetland and the natural drainage. The conservation approach also opens spaces for low-impact development stormwater management strategies—vegetated swales are distributed throughout the development, contributing to the green infrastructure of the site. These conservation areas can reduce erosion and sediment control costs for the developer.

Figure 16.20 illustrates all four steps of the conservation design method. Part (a) represents the conservation areas identified through resource mapping. Highly sensitive features of the landscape, such as steep slopes, wetlands, and hydric soils, are designated as primary conservation areas. For step 1, part (b) diagrams secondary features of the landscape to be incorporated into the plan, such as attractive views, a majestic old tree, woodlands, and a wildflower meadow.

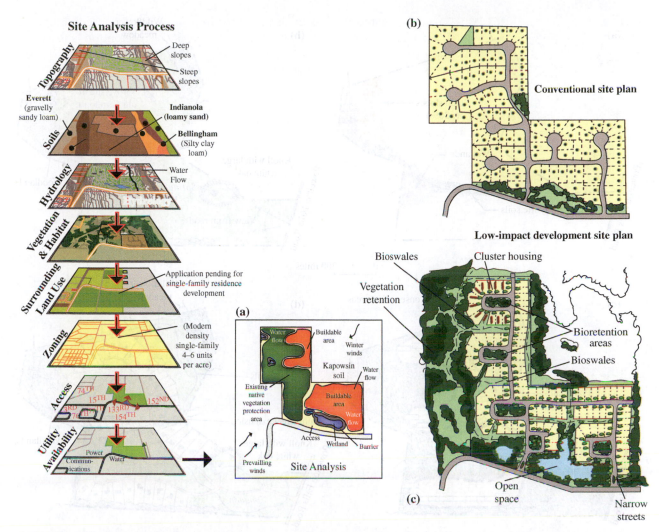

FIGURE 16.19 Resource mapping and conservation design. In conservation design, resource mapping is the critical first step. By evaluating landscape features, geology, and land cover, natural drainage flows can be identified as well as areas for conservation. (a) Resource mapping results in a preliminary site analysis; the buildable area is shown in brown, the conservation area in green, and a wetland in blue. Part (b) demonstrates how this site would be developed using conventional siting approaches; part (c) represents a conservation design strategy. The total housing density of both (b) and (c) is the same, but the conservation approach incorporates cluster development, LID techniques, and densely vegetated greenspace.

Source: Based on Hinman, C. (2005). *Low Impact Development: Technical Guidance Manual for Puget Sound*. Olympia, WA: Puget Sound Action Team, 2005, PSAT 05-03. pp. 24, 47.

Part (c) represents step 2, locating the housing sites and buildable area of the property. In part (c), the buildable area stays clear of all of the primary and secondary areas of conservation, places the housing at a distance from the nuisance of the road, and orients the homes toward attractive views. Finally, part (d) completes the plan with steps 3 and 4: *designing the street alignment and trails* and *drawing the lot lines.*

Because density-neutral conservation design strategies usually do not comply with local zoning laws—they often have smaller lot sizes and setbacks, larger

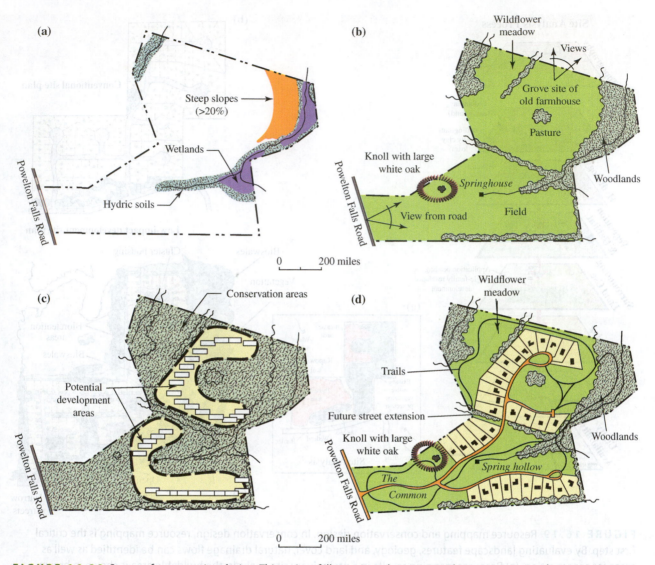

(a)

Steep slopes (>20%)

Wetlands

Hydric soils

Powelton Falls Road

(b)

Wildflower meadow

Views

Grove site of old farmhouse

Pasture

Knoll with large white oak

Springhouse

Woodlands

View from road

Field

Powelton Falls Road

0   200 miles

(c)

Conservation areas

Potential development areas

Powelton Falls Road

(d)

Wildflower meadow

Trails

Future street extension

Knoll with large white oak

Woodlands

Spring hollow

The Common

Powelton Falls Road

0   200 miles

FIGURE 16.20 Stages of conservation design. This series of illustrations demonstrates how conservation design progresses from landscape resource mapping to complete site plans.

Source: Based on Arendt, R. G. (1996). *Conservation Design for Subdivisions: A Practical Guide to Creating Open Space Networks*. Island Press.

building envelopes, and narrower streets, for example—they are simply not possible in most locations without a special review and permitting process from local planning authorities. The case study of Jackson Meadow in Minnesota (Box 16.5) shows how one community negotiated a conservation development in a highly sensitive landscape.

Environmentally sensitive design moves land-use planning and sustainable development a large step forward in its use of green infrastructure, low-impact development, and conservation design. These practices are far from standard, however, and it will take years of effort to change zoning laws, professional practice, and public perception to make them so.

BOX 16.5 Conservation Design at Jackson Meadow in Marine on St. Croix, Minnesota

Jackson Meadow, a residential development in Marine on St. Croix, Minnesota, provides one of the most complete examples of conservation design in Minnesota.

Located to the north and east of the Twin Cities, Jackson Meadow is nestled in the rolling farmland and wooded bluffs directly west of the historic village center of Marine on St. Croix, Minnesota. The site's great natural beauty and unique cultural setting prompted the design team to take a different approach to developing the site. The design team first identified the resources they hoped to preserve, and then laid out the 64 clustered home sites in a pattern influenced by the city of Marine and adjacent St. Croix River. The unique home sites are surrounded by 275 acres of protected woodlands, restored prairies, and farmland.

The preserved land within Jackson Meadow is part of a river city with historic character; it is adjacent to William O'Brien State Park; and it is located along the bluffs of the St. Croix River, which is a designated Wild and Scenic River. With this context in mind, the project team had the following goals for the development:

- Respect the sense of place of Marine on St. Croix by using its historic architecture and form to influence the project's housing styles and layout.
- Minimally impact the bluffs and forest, restore native vegetation, and maintain an agricultural buffer around the development.
- Provide an extensive pedestrian trail system and other amenities.
- Use innovative stormwater and wastewater treatment techniques to minimize water resource impacts.

The Jackson Meadow design team worked diligently with citizen groups, the Marine Planning Commission, and the City Council to design the most appropriate development for this unique site and historic community. Through this process, the city—historically resistant to new development proposals—was encouraged to rewrite its ordinances to achieve greater performance from future developments.

This process, however, required a great deal of effort by all parties. More than 40 meetings were held before the plat received final approval. And, as with all developments, certain compromises were made.

As in many new developments in the urban fringes, the cost of housing is beyond the means of many homebuyers. In addition, though sensitive to the site, the project still adds more traffic to Marine's streets and furthers Marine's transformation to a commuter community. However, with Jackson Meadow, the city of Marine struck an important balance between adding additional housing units to the city and simultaneously preserving its small-town identity.

One of Jackson Meadow's greatest lessons is that the design process can often combine many different strategies to preserve open space and achieve the desired housing densities. For example, the design team clustered homes to save open space within the development, acquired adjacent sensitive woodlands with the assistance of the Department of Natural Resources, and employed a unique variation of Transfer of Development Rights (TDR) to preserve adjacent farmland.

Source: Based on Minnesota Land Trust, "Jackson Meadow," *Conservation Design Portfolio Case Study 1.*

16.4 Green Building

Sustainability of the built environment can be improved by changing the way we develop land. However, we also have to consider the environmental impacts of the structures themselves. Green building has emerged as a concept that can inform a more sustainable approach to building design, construction, and operation.

Green building is defined by the U.S. Office of the Federal Environmental Executive (2003) as "the practice of (1) increasing the efficiency with which buildings and their sites use energy, water, and materials, and (2) reducing building impacts through better siting, design, construction, operation, maintenance, and removal—the complete building life cycle." Similarly, the U.S. Environmental Protection Agency (2016) defines green building as "the practice of creating structures and using processes that are environmentally responsible and resource-efficient throughout a building's life-cycle from siting to design, construction, operation, maintenance, renovation, and deconstruction. The practice expands and complements the classical building design concerns of economy, utility, durability, and comfort."

Green building consequently embodies three concepts that have been previously explored in our text—environmental footprints, life cycles, and industrial ecology. First, green building reduces the environmental footprint of a building on its site by protecting as many of the natural features and functions of the location as possible. Second, we can think of a building as a product with its own life cycle. As with the life cycle principles that you saw in Chapters 5 and 12, the goal of green building is to conserve natural resources and minimize waste over the material life of the building. Third, the building itself is a production system. Buildings dynamically provide services for their occupants such as water, sewage disposal, light, fresh air, heat, cooling, and electricity. In this sense, the principles of industrial ecology can be applied to optimize the "production processes" of the building. It is not inconceivable that buildings could become nearly closed-loop systems in the future.

Green building is not the same thing as the ecological design discussed in Chapter 12, although ecological design can significantly enhance the green properties of a building. As we saw with the termite biomimicry used in Zimbabwe, ecological principles can dramatically reduce the energy required by a building. Ecological design is also concerned with aesthetics in a way that is not represented in the practicalities of designing and building structures with improved environmental outcomes.

The fundamental challenge with green building is to meaningfully differentiate a building that is "green" from a conventional structure. Establishing a set of green characteristics and performance standards that can be applied to the actual practice of designing and constructing buildings is no easy task. The fundamental question is, *what actually makes a green building objectively more sustainable than one that is not?*

There is no uniform law or government-mandated "code" that defines what a green building is or requires that new construction be built as such. What has evolved over time is *endorsement labeling*, a concept that is discussed in more detail in Chapter 17. Endorsement labeling is a type of voluntary certification in which an independent, not-for-profit third party verifies that a product meets certain criteria in its construction, material content, performance, and operation. An example of such a third party is Underwriters Laboratories, a product safety certification organization more than 100 years old with operations in over 40 countries.

Two voluntary green building endorsement/certification systems are most commonly used in the architecture and construction industries today, and both were first introduced in the 1990s. The *Leadership in Energy and Environmental Design (LEED)* program is used for commercial buildings, residential homes, and even entire neighborhoods. LEED was developed by the U.S. Green Building Council and has been adapted for use in a number of countries. In Europe, the *Building Research Establishment Environmental Assessment Method (BREEAM)* is well established in the UK and commonly used in Germany, the Netherlands, Norway, Spain, and Sweden. BREEAM similarly certifies a wide variety of buildings and communities. Lesser known green building rating systems include EarthCraft in the southeastern United States, Green Globes in North America,

Green Star in Australia and New Zealand, and the Comprehensive Assessment System for Built Environment Efficiency (CASBEE) in Japan.

These programs are based on recommended best practices, and buildings typically earn credits toward certification for the practices or materials they adopt. The sustainability of the building is expected to result from its conformance to the design criteria for which it received credits. In sharp contrast, the ***Living Building Challenge*** is the only holistic, performance-based green building certification system in the world. Unlike credit-based methods, the Living Building Challenge is based on actual building performance. A building must be in continuous operation for at least a year, at which time it is audited for compliance with the standards and requirements. Unlike other common green building certifications, the Living Building Challenge prohibits the use of particular materials and requires projects to have "net zero energy" (which we discuss later in this chapter).

16.4.1 The LEED Rating and Certification System

The LEED rating system is sponsored by the U.S. Green Building Council. Different certification systems are available for commercial buildings (offices, libraries, churches, hotels), retail spaces (shops, banks, restaurants, big box stores), K-12 schools, health care facilities, residential homes, and neighborhood communities. The rating systems undergo periodic revision to reflect emerging best practices and to incorporate new techniques in the certification process. LEED recognizes that sustainable building criteria may vary regionally because of different environmental opportunities and challenges; regional councils consequently designate regionally specific green criteria that projects can earn bonus points for achieving.

All LEED certifications use the same approach. All projects must meet a number of prerequisites with regard to sites, water use, energy, and other criteria. These prerequisites do not earn points but are mandatory building elements. Each rating system has 100 base points. A project is credited for different green features it incorporates from the ratings list, and each feature earns a specific number of points, or "credits." To be LEED-certified, a building must earn a minimum of 40 points. Buildings that exceed the minimum standard can be rated as silver, gold, or platinum, where each successive tier requires more points and has increasingly more favorable environmental benefits (Table 16.4). Box 16.6 summarizes key green features incorporated in a LEED Platinum university residence hall.

All of the rating systems also share five fundamental categories through which certification credits can be earned:

1. *Sustainable sites* criteria minimize a building's impact on ecosystems and water resources.
2. *Water efficiency* criteria enhance water conservation within the building and for its outdoor landscapes.

TABLE 16.4 LEED Building Certification Levels

LEED RATING	EARNED POINTS
Certified	40–49
Silver	50–59
Gold	60–79
Platinum	80 points and over

Source: Based on U.S. Green Building Council.

BOX 16.6   A LEED Platinum Residence Hall Renovation

FIGURE 16.21 Wayland Hall at James Madison University. The back side of Wayland Hall hosts native plantings and a bicycle shelter, both of which are seen in this photo. There is also a 10,000-gallon, below-ground cistern that harvests rainwater for use in the building's toilet system. All of these features helped the building earn a LEED Platinum rating.

Source: Photo by Scott Smith.

Wayland Hall (Figure 16.21) is a dormitory on the campus of James Madison University in Harrisonburg, Virginia. At the time it was renovated, Wayland was one of only four full-scale LEED platinum residence halls in the country, and the first completely renovated residence hall to achieve platinum status under the USGBC's new construction and major renovation guidelines. The $11.6-million renovation transformed the 41,000-square-foot space into an innovative new living–learning community dedicated to the visual and performing arts. The renovated building includes a gallery, music practice rooms, an art studio, and a performance and exhibition room. All aspects of the renovation, including an ambitious reconfiguration of the bedroom spaces, are designed to encourage interaction, promote sustainable living, and expose students to the discipline and joy of the arts.

A variety of rigorous design strategies contribute to an expected 38% reduction in energy consumption and savings of over 1.3 million gallons of water each year. Energy conservation features include a heat recovery unit that captures heat from exhaust air, window sensors that cut off heating and cooling when windows are open, drain water heat recovery, and a ground source heat pump. Notable contributions to water conservation include systems that collect, filter, and cycle rainwater for toilet flushing.

Site improvements along and behind the residence hall replaced a parking lot with a series of landscaped terraces. The new design reduces impervious site cover, improves pedestrian connectivity, and provides new opportunities for residence life to extend outside. In all, the site design creates a stormwater management strategy that is as environmentally friendly as it is beautiful.

Source: Based on VMDO Architects, "James Madison University's Wayland Hall Renovation," www.vmdo.com/project.php?ID=26.

3. *Energy and atmosphere* criteria reduce energy consumption and promote the use of renewable energy.

4. *Materials and resources* criteria foster the use of sustainable building materials and reduced waste.

5. *Indoor environmental quality* criteria enhance the comfort of occupants with access to natural daylight, outdoor views, and better indoor air quality.

The BREEAM rating system and certification operates in a similar manner to LEED, although the specifics are different. BREEAM likewise has mandatory requirements, and buildings earn credits in a number of categories, such as management, health and well-being, energy, transport, water, materials, waste, land use, innovation, and pollution. BREEAM is also a 100-point system, but with five rating categories instead of four.

16.4.2 Green Building and Land-Use Planning

Green building rating systems have the capacity to substantially enhance more sustainable land-use and urban planning. Both BREEAM and LEED certify neighborhoods and communities using a breadth of sustainable development criteria that promote conservation design and low-impact development. The Living Future Institute also has a Living Community Challenge similar to the Living Building Challenge.

In addition, the U.S. Environmental Protection Agency provides a variety of resources related to green communities. For example, the *Sustainable Design and Green Building Toolkit for Local Governments* offers guidance to local governments about how to evaluate and revise existing codes and ordinances to support environmentally, economically, and socially sustainable communities. The EPA also offers evaluation tools that enable communities to evaluate their green infrastructure and its effects on water quality. The EPA *Water Quality Scorecard: Incorporating Green Infrastructure Practices at the Municipal, Neighborhood, and Site Scale* helps local governments identify opportunities and remove barriers that will enhance green infrastructure and water quality.

Although our discussion of green building rating systems has focused literally on *buildings*, rating systems are available for other types of constructions. The **Envision™ rating system** of the American Society for Civil Engineers provides an assessment tool for evaluating the sustainability of all civil infrastructure, including roads, pipelines, power lines, telecommunication towers, landfills, and water treatment plants. Several transportation-specific rating systems exist; Box 16.7 summarizes New York State's efforts to promote a sustainable transport infrastructure through its GreenLITES program.

ACTIVE LEARNING EXERCISE 16.2 Compare and Contrast LEED

Compare and contrast LEED green building criteria with the Living Building Challenge and the Envision™ sustainable infrastructure rating system. How are these similar and different? Does one system do better than another in capturing what we mean by "sustainability" in their rating criteria? Explain your answer.

BOX 16.7 GreenLITES and Transportation Sustainability

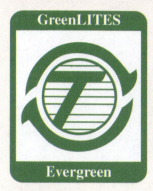

FIGURE 16.22

Source: N.Y. Department of Transportation, www.dot.ny.gov /programs/greenlites/repository /Evergreen.jpg.

Transportation infrastructure is a significant form of land use and has major impacts on air quality, water quality, and natural resource consumption. Transportation agencies recognize these problems and have responded with the green transportation rating systems modeled after LEED. The New York State Department of Transportation designed GreenLITES (Leadership in Transportation and Environmental Sustainability), a self-evaluation certification that is the first U.S. rating system for transportation projects.

GreenLITES helps New York measure performance, recognize good practices, and identify areas for improvement. The tool reflects a points system in five categories—sustainable sites, water quality, materials and resources, energy and atmosphere, and innovation/unlisted. Because GreenLITES extends beyond just road projects, there are a large number of credit opportunities (more than 200), but not all credits are necessarily relevant or available for a particular project. The rating system classifies successful projects as Certified, Silver, Gold, or Evergreen depending on the total number of points earned.

The first GreenLITES awards honored four Evergreen projects: three highway projects and one greenway/multiuse trail. A notable award was for the reconstruction of three miles of New York State Route 30. This highway segment is located along the Adirondack Trail Scenic Byway and is bordered by environmentally sensitive wetlands, a forest preserve, and Tupper Lake. The project included multiple environmentally and socially sustainable design elements including:

- A wider shoulder along the side of the road that balances the needs of the traveling public and of environmentally sensitive Adirondack ecosystems.
- Fencing installed along wetland borders to protect turtles from the trafficway.
- Relocated utility lines underground to enhance scenic views along the highway and at four designated overlooks.
- A closed storm drainage system to capture sediment and reduce flow of pollutants into the lake.

Green transportation rating systems modeled after GreenLITES include the Illinois Livable and Sustainable Transportation (I-LAST) Rating System and the adaptation by Pennsylvania Turnpike Commission.

Two additional green transportation rating systems are of interest because of their unique features. The BE²ST rating system, created in partnership with the University of Wisconsin, uses qualitative measures to first screen road projects and then rate projects with quantitative measures. BE²ST is also unique because it incorporates both environmental life cycle assessment and economic life cycle cost analysis. STEED (Sustainable Transportation Environmental Engineering and Design) was created by the private consulting firm Lochner Engineering and is a checklist for roadway projects that can track how the sustainability of a project changes throughout the engineering design cycle.

16.4.3 Green Building and Construction Codes

Building codes are government regulations that establish legal requirements and standards for the safety, occupancy, siting, and construction of buildings and structures. Building codes address a wide range of issues, such as fire resistance, energy efficiency, the ability to withstand extreme weather events, building statics and strains, acceptable materials, ventilation, and sewage, among many others. Depending on a country's government, building codes can be enacted at the national, regional, or local level, or some combination of all three. Codes are almost always enforced at the local level.

Building codes are highly technical and writing them requires expertise. Another challenge is to keep them current as new technologies and systems are developed or as broader environmental regulations regarding energy and sustainability are adopted. Consequently, many government jurisdictions voluntarily adopt **model codes** that are written by third-party, professional standards organizations. An example is the model National Electrical Code in the United States, a set of wiring and electrical equipment standards developed by the National Fire Protection Association.

Although similar to the voluntary green building rating and certification systems, green building codes are legally mandatory minimum requirements for all new construction. They are highly specific requirements for the actual design and construction of buildings, not broad-based goals. In the United States, dozens of localities and the State of California have implemented a variety of green building and construction codes. Boulder, Colorado, is a city that has aggressively incorporated mandatory green building requirements into its zoning law (see Box 16.8). CalGreen is California's Green Building Standards Code; as of 2021, it is the only state-level mandatory green building code in the United States.

Green building and construction codes are especially challenging to develop because of the need for a holistic approach and the incredibly diverse range of

BOX 16.8 Boulder, Colorado: Rocky Mountain Green

The city of Boulder, Colorado, is nestled in the foothills of the Rocky Mountains. It has a long-standing reputation as an environmentally progressive community. Boulder was an early pioneer of green building rating systems and is notably one of the few places in the world with rigorous, *legally mandated* green construction requirements. Boulder's green building initiatives started in the early 1980s with water and energy conservation provisions. In 1996, Boulder introduced Green Points, a rating system that scores green building attributes. Green Point performance criteria are now written into Boulder's building code. Put simply, the city will not issue a construction building permit for a project unless it earns the required number of points in the Green Point system, many of which are physically verified by building inspectors. Projects can earn points for energy efficiency, practices like xeriscaping (landscaping to save water and grow native plants), managing stormwater runoff, reusing existing buildings, and effective waste management

FIGURE 16.23

Source: © Steve Krull City Town & Street Scene Images / Alamy.

during the construction phase. The Green Points program is mandatory for all new residential construction and remodeling/addition activities over 500 ft².

buildings with unique code requirements (such as hospitals, hotels, schools, and multistory high rises). Introduced in 2010, the ***International Green Construction Code (IGCC)*** developed by the International Code Council is the preeminent model code for green building. The 2018 version covers all commercial buildings, mixed-use, multifamily residential structures higher than three stories, and some industrial buildings. The sustainability scope of this code is substantial. It addresses the building site, energy and water efficiency, materials and resource use, indoor environmental quality, and building emissions. The IGCC integrates a highly complementary model code, ASHRAE 189.1, *Standard for the Design of High-Performance Green Buildings Except Low-Rise Residential Buildings*. ***ASHRAE*** is the American Society of Heating, Refrigeration, and Air Conditioning Engineers. The U.S. Green Building Council endorses both the IGCC and ASHRAE 189.1.

16.5   Energy Use and Buildings

Green building rating systems like LEED and BREEAM are holistic approaches to the built environment, incorporating the sustainability of a building on its site, the surrounding landscape, water usage, material inputs, energy consumption, and interior comfort. However, few projects make the effort to become certified as a green building relative to the total amount of new construction that takes place each year, and mandatory green construction codes have not yet been widely adopted.

Because of the extraordinary environmental, social, and economic issues associated with climate change, considerable effort is made to reduce the impacts of building energy use as a specific component of green building. In the United States, residential and commercial buildings consume 40% of the nation's primary energy; more than 76% of all electricity is used by residential, commercial, and industrial buildings (U.S. Department of Energy, 2015). This figure *excludes* electricity used for industrial processes and manufacturing.

Residential, commercial, and office buildings all use energy for the same basic energy services, called *loads* when referring to the amount of energy each of these needs consume. Common energy needs include space heating, cooling, hot water production, ventilation, lighting, refrigeration, and electrical appliances and devices. What varies among types of buildings is how much energy is accounted for by different loads. For example, about half of the energy used by commercial buildings in the United States is for heating, cooling, and lighting. In contrast, these three loads account for almost two-thirds of residential energy use in the United States (Figure 16.24). Reducing building energy use requires that we understand how different kinds of structures use energy so that we can implement effective strategies for energy efficiency and conservation.

16.5.1 Strategies for Building Energy Conservation

Improved energy use in buildings is largely accomplished in four ways. First, we can conserve energy by virtue of a building's architectural design. For example, Eastgate House in Zimbabwe was designed for convection like that of termite mounds, significantly reducing its energy needs for cooling. Similarly, the use of skylights and careful placement of windows provide natural sunlight during the day, reducing the need for artificial light. This design strategy is known as ***daylighting*** and uses a variety of products and architectural features. ***Passive solar design*** is likewise a form of architecture that takes advantage of solar energy to help heat homes. Design

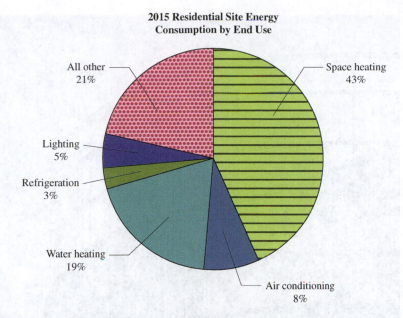

**2015 Residential Site Energy
Consumption by End Use**

All other
21%

Space heating
43%

Lighting
5%

Refrigeration
3%

Water heating
19%

Air conditioning
8%

FIGURE 16.24 Energy loads in the U.S. residential sector. U.S. residential energy consumption is heavily driven by the need for space heating and cooling, which accounts for just over half of home energy use nationally. Water heating is significant as well, and these three energy services combined represent almost three-fourths of household energy use.

Source: Based on U.S. Energy Information Administration, *Energy Explained: Energy Use in Homes,* www.eia.gov /energyexplained/use-of-energy/homes.php.

strategies can be simple or extraordinarily innovative and complex; Figure 16.25 shows 1 Bligh Street, a skyscraper in Australia that has a double-skin glass facade that provides insulation, ventilation, and daylighting for the building.

Second, energy conservation can be achieved by our choice of construction materials and the structure of a building's exterior envelope. Because space heating and cooling represent such significant energy loads for buildings, constructing a tight thermal layer will moderate heat loss and gain between the indoors and outdoors. Double- and triple-paned windows, careful weatherstripping, and substantial insulation in walls, attics, and foundations are examples of ways in which we conserve building energy through the envelope. Structural insulated panels (SIPs) are advanced building products that significantly lower energy use in homes and small commercial buildings. SIPs are used instead of framing lumber and plywood for the sheathing of a house and provide higher insulating values compared to conventional construction. Although SIPs cost more, they shorten construction time considerably and thus save on labor costs.

Third, energy savings can be realized by selecting highly energy-efficient equipment, appliances, and devices in the building. It is especially critical to optimize heating and cooling given their building energy loads, as well as lighting in commercial structures. Systems requiring motors for fans or pumps—such as air handling and water supply—also need to be considered. Motors with large horsepower and long *duty cycles* can consume a tremendous amount of energy. (Duty cycle refers to the length of time a device is operating over a defined period.) The efficiency of equipment can vary notably; engineers and designers must carefully select the systems that create the major energy loads in a building.

Efficiency in buildings is enhanced by dynamic controls. For example, occupancy and photo sensors adjust lighting by turning lights off when rooms are vacant

FIGURE 16.25 1 Bligh Street, Sydney, Australia. This skyscraper in Sydney has a double-skin glass facade. Two layers of glass about 2 feet apart make up the exterior envelope of the building, providing natural daylight and insulation for the interior spaces. An innovative cooling system is incorporated into the glass skin as well. 1 Bligh Street reportedly has 42% less CO_2 emissions than an office complex of similar size.

Source: © Martin Cameron/Alamy.

and by actively adjusting lighting levels during the daytime relative to the amount of available daylight in a room. CO_2 sensors allow air-handling equipment to ventilate a room relative to the actual air quality, instead of replacing large volumes of air in rooms that may only be minimally occupied. Heat recovery systems are also ways to achieve greater system efficiencies within a building; heat recovery techniques are used for water heating, space heating and cooling, and combined heat and power

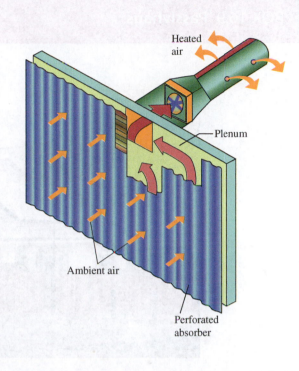

Heated air

Plenum

Ambient air

Perforated absorber

FIGURE 16.26 Solar air heating systems, such as the plenum on this apartment building, use passive solar heating and active ducting to provide space conditioning inside buildings.

Sources: Solar wall: Hadrian/Shutterstock.com; Solar Air Heating Systems: Based on www.ecobuildtrends.com/2011/02/solar-air-heating.html.

systems. Microturbine technology in particular is significantly increasing the opportunity for small, onsite combined heat and power systems, which is an asset for large office buildings and multifamily housing in particular.

Finally, the carbon footprint of building energy can be reduced by using renewable energy. Photovoltaic panels can actively provide onsite electricity, as can wind turbines if the wind resource is substantial enough. Solar hot water heating and solar air heating systems (Figure 16.26) are both highly cost-effective solar technologies. When renewable energy technologies are used to substitute, in part, for a building's conventional materials, we refer to it as a ***building integrated system***. The solar air system in Figure 16.26 is an example of a building-integrated heating system.

Combinations of these four energy conservation strategies can achieve dramatic reductions. The Passivhaus concept (Box 16.9) is a good case to demonstrate this. These homes rely on passive solar gain, the body heat of their occupants, and the heat generated from equipment and appliances to provide virtually all of the space heating that a Passivhaus structure requires. However, energy conservation measures *do* interact with one another, and their interactive effects represent an instance in which "one plus one does not equal two." If we implement three different energy-efficient technologies in a building, each of which individually might contribute a 10% energy savings, the net savings from all three combined will not be 30%. For example, we can use daylighting techniques to reduce the demands for electricity for lighting during the daytime, but more (or larger) windows will also generate more heat loss during cooler seasons, thus requiring more energy for heating. As a consequence, the total benefit from multiple energy efficiency and conservation efforts

BOX 16.9 **Passivhaus**

FIGURE 16.27 Although Passivhaus buildings are often strikingly modern in design, they can also be constructed in traditional architectural styles, such as this row of terraced houses in Houghton-le-Spring, United Kingdom.

Source: © Ashley Cooper pics/Alamy.

Passivhaus construction is a voluntary, certified low-energy building that has been implemented most widely in Germany and Scandinavia. More than 20,000 Passivhaus structures—including homes, schools, and offices—had been built in these countries by about 2010. A Passivhaus building (Figure 16.27) has the rather startling characteristic of needing no major active heating system—passive solar gain, as well as the heat generated by a building's occupants and appliances, are all that a structure constructed to Passivhaus standards usually requires for comfortable space heating during most of its heating season. High insulation values, passive solar design, very low air infiltration, and other properties help a Passivhaus achieve this level of heating performance. Because of other energy efficiency measures incorporated into the building, Passivhaus buildings use 80% to 90% less energy than their conventional equivalents.

must be carefully modeled to understand how much energy will be saved overall. Many commercial software products are available to make such calculations in the design stage for the building. Robust, but preliminary, estimates can be made with free software such as DOE-2, OpenStudio, EnergyPlus, eQuest, and RETScreen.

16.5.2 The Role of Energy Building Codes

The minimum energy efficiency performance of residential and commercial buildings can be legally mandated by building codes. These codes set the baseline energy efficiency and consumption that buildings must achieve in order to comply with local construction laws. Different requirements will apply for commercial and residential structures. Codes affect the building envelope, energy-consuming equipment, and the energy consumption of a building's overall systems, such as ventilation.

The International Code Council sponsors the model *International Energy Conservation Code (IECC)* that covers both residential and commercial structures. Most states and/or communities in the United States use the IECC model code in their own ordinances. The IECC is updated about every three years through a public process to account for improvements in technology, best practice construction techniques, and climate change. The other model building energy standard is the ASHRAE Standard 90.1; *Energy Standard for Buildings except Low-Rise Residential Buildings.* Like the IECC, ASHRAE 90.1 is revised about every three years. LEED-certified commercial buildings must, at a minimum, comply with ASHRAE 90.1 standards.

Building energy codes vary to allow for differences in climate and the cost effectiveness of different energy conservation measures. *Heating degree days* is the metric used by the IECC to distinguish broad and narrow geographic variations in climate; many aspects of the energy code vary based on the designated climate zone (Figure 16.28). Degree days represent an estimate of the amount of energy required to heat or cool a building; heating degree days are calculated by summing the extent to which the average outdoor temperature deviates by 1 degree from a baseline temperature, below which it is assumed a building will need heat. In the United States, this baseline temperature is 65°F (18°C). To illustrate, a winter day with an average daily temperature of 28°F will have 37 heating degree days for that specific day. Climate zones are calculated by adding up the total number of heating degree days for the entire year, which can be a few hundred days in a warm climate or thousands in a cold area.

16.5.3 "Beyond Code" Energy Rating Systems and Models

Building energy codes simply set the *minimum* energy conservation characteristics of a building. As with green building rating systems, it is also possible to certify that

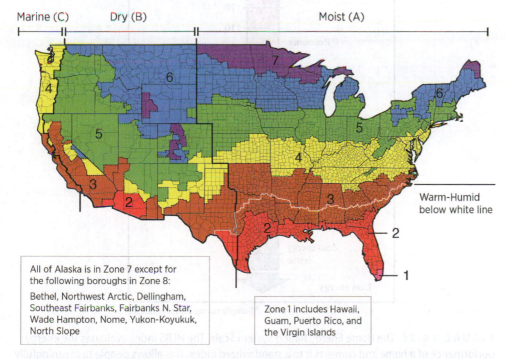

FIGURE 16.28 IECC climate zones are based largely on heating degree days to establish regionally specific energy efficiency building code standards.

Source: Pacific Northwest National Laboratory, basc.pnnl.gov/images/iecc-climate-zone-map.

a building exhibits a level of energy efficiency that is significantly greater than that required by building energy codes. These building energy rating systems are *not* the same as LEED or BREEAM, because they evaluate only the energy characteristics of buildings, not their entire potential environmental impacts. Such rating systems are often referred to as ***beyond code programs***. There are several third-party building energy efficiency rating systems; the most widely known program is that of the Energy Star building certification system sponsored by the U.S. Department of Energy. It is a surprise to many people that the Energy Star endorsement labeling program certifies not only appliances and equipment, but also residential and commercial buildings.

The Residential Energy Services Network (RESNET) in the United States has also developed a standardized scoring tool to rate a home's overall energy consumption, known as the Home Energy Rating System (HERS). RESNET conducts highly detailed analyses of a residence's energy conservation and standardizes the resulting evaluation to a 100-point rating index (Figure 16.29). The HERS score allows consumers to compare the energy properties of different homes, and may also enable them to qualify for a more favorable loan known as an energy-efficient mortgage.

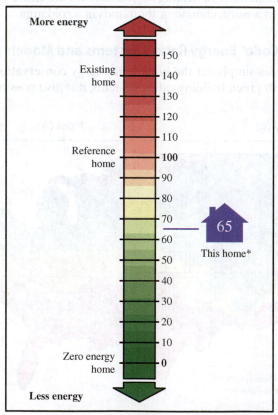

*Sample rating representation.

FIGURE 16.29 The Home Energy Rating System Scale. The HERS index evaluates the energy performance of a home and converts it to a standardized index. This allows people to meaningfully compare housing choices.

Source: Based on RESNET, Home Energy Rating, www.resnet.us.

Internationally, the Passivhaus (see Box 16.9) is a formally certified structure with respect to energy conservation; the Passive House Institute sets the standards that are used worldwide for both residential and nonresidential structures. Passivhaus standards govern the requirements for heating systems within a building and limit the building's total energy use. Passivhaus structures fall into a category known as *ultra-low energy buildings* or *nearly zero energy buildings*. An actual **net zero building** exhibits no *net* energy consumption and therefore also has a net zero carbon footprint. Also known as zero energy buildings, these structures typically integrate passive solar design, solar thermal hot water heating, highly energy efficient technologies, and active solar PV or wind energy systems that generate clean electricity.

The net zero building principle is that over the course of a year, a building will generate as much carbon-free energy as it uses. As a consequence, the building in theory has no net impact on energy resources or CO_2 levels. However, there is ambiguity in the net zero building concept, because a building can be net zero at its site or with respect to its source emissions. Net zero site energy represents a building that generates as much energy as it consumes at its site. Zero net *source energy*, however, means that a building offsets the carbon emissions created when energy is produced elsewhere and is delivered to the building. An example is electricity generated at a power plant. The United States now requires that all new federal buildings that began their design process in 2020 or later must be net zero buildings. The European Union has similarly directed its member states to require that all new buildings after 2021 must be nearly zero energy buildings.

ACTIVE LEARNING EXERCISE 16.3 Design Beyond Code

Assume that you are planning to build an 1,800-ft^2 home where you are currently located and that you would like this home to be "beyond code." Develop a set of design and construction criteria that you will use for your new house.

16.6 Summary

Engineers and architects designing buildings and communities in the future must understand the contemporary principles and best management practices for more environmentally benign land-use planning and construction. This includes smart growth, environmentally sensitive design, low-impact development, and conservation design. Professionals must also be aware of building codes and third-party rating systems that allow structures and communities to be certified at the highest levels of energy and environmental conservation, including LEED, BREEAM, HERS, Passivhaus, and net zero buildings.

In this chapter, we have explored several approaches to land development and buildings that attempt to create more sustainable infrastructure. Traditional infrastructure development patterns have produced low-efficiency patterns of development that lead to high rates of erosion, urban sprawl, inefficient energy consumption, and degraded water resources. Development patterns that utilize smart growth planning, conservation design principles, best management practices for erosion and stormwater management, and green building techniques minimize unsustainable water resource impacts, conserve energy, and mitigate the associated emissions from energy production and use. Industrial ecology principles and life cycle assessment tools can not only help us assess the low-impact development techniques described in this chapter, but they can also help us rethink the ways buildings use energy.

Reducing the land-, resource- and energy-intensive nature of new urban development may have significant economic, environmental, and social benefits for a planet already witnessing a significant shift in climate and weather patterns. Future infrastructure investments will need to be more sustainable and more resilient to meet the demands of growing populations and limited available resources. Green building standards, conservation design principles, and energy-efficient construction all involve good engineering practices that reflect the following:

- Identifying and providing buffers and setbacks associated with environmentally sensitive areas
- Limiting disturbances and development in floodplains
- Clustering development to minimize impervious surfaces
- Incorporating open spaces and low-impact development into site design
- Achieving the highest levels of energy conservation in new construction

References

Burchell, R. W., and Mukherji, S. (2003). "Conventional development versus managed growth: The costs for sprawl." *American Journal of Public Health* 93(9):1534–1540.

Maryland Office of Planning. (1995). Achieving Environmentally Sensitive Design in Growth Areas Through Flexible and Innovative Regulations. Baltimore: Maryland Office of Planning.

United Nations. (2019). World Urbanization Prospects 2018 Highlights. Department of Economics and Social Affairs, Population Division. ST/ESA/SER.A/421.

U.S. Department of Energy. (2015, September). Quadrennial Technology Review, "Chapter 5: Increasing Efficiency of Building Systems and Technologies." Washington, DC.

U.S. Environmental Protection Agency. (2016). "Green Building." archive.epa.gov /greenbuilding/web/html/about.html.

U.S. Office of the Federal Environmental Executive. (2003, September). "The Federal Commitment to Green Building: Experiences and Expectations." Washington, DC.

Key Concepts

Built environment	Floor area ratio (FAR)
Land-use and land-cover change	New Urbanism
Development	Form-based codes
Urban sprawl	Rural-to-urban transect
Mixed use	Smart growth
Population density	Urban intensification
Single-use zoning	Urban infill
Housing density	Urban growth boundary
Zoning ordinances	Urban agriculture
Building envelope	Brownfield
Right-of-way	Cluster development
Easement	Environmentally sensitive design
Setback	Environmental impact assessment
Angle of bulk plane	Impervious surfaces

Green infrastructure
Best management practices (BMPs)
Vegetated swales
Stream daylighting
Riparian buffer zone
Gray water
Fluvial (water) erosion
Universal soil loss equation
Eolian (wind) erosion
Stokes' settling law
Geotextile
Conservation design
Density neutral
Green building
Endorsement labeling
LEED
BREEAM

Living Building Challenge
Envision rating system
Building codes
Model codes
International Green Construction Code
   (IGCC)
ASHRAE
Daylighting
Passive solar design
Duty cycle
Building integrated system
International Energy Conservation
   Code (IECC)
Heating degree day
Beyond code program
Net zero building

Problems

16-1 Identify and explain three to four factors that contribute to sprawl.

16-2 Identify and explain three to four strategies intended to counteract sprawl.

16-3 Why are sprawl and other forms of low-density development unsustainable?

16-4 How do form-based codes differ from traditional zoning codes?

16-5 Walk through a nearby residential area. Identify and estimate the front setback distance and the maximum building height.

16-6 List the 10 smart growth principles.

16-7 Why might communities be resistant to implementing smart growth land-use practices?

16-8 Which of the following statements about cluster developments is FALSE?
   a. They provide for more open green space.
   b. They have a lower housing density than a conventional subdivision.
   c. They don't necessarily implement the features of low-impact development.
   d. They often require a change in local zoning codes.

16-9 What does "let the land shape your plan" mean?

16-10 What are the benefits of large expanses of undeveloped or agricultural land?

16-11 Summarize the environmental impacts and consequences of the built environment for land cover and water quality.

16-12 A watershed with 40% impervious cover would probably have what kind of stream quality?

16-13 A stormwater settling basin collects particles from a parking lot. The temperature of the runoff water is 22°C. The water has a density of 997.8 kg/m^3 and a kinematic viscosity of 0.9565 mm^2/s. The average radius of the silt particle is 0.01 mm, and the silt has a density of 2,690 kg/m^3. The average radius of the sand particle is 0.10 mm, and the sand has a density of 2,660 kg/m^3. The average radius of the gravel is 2.0 mm, and it has a density of 2,590 kg/m^3. Compare the settling times for the silt, sand, and gravel particles from the parking lot if the settling height is 0.8 m.

16-14 Explain the two basic design features of green infrastructure.

16-15 Identify and explain five different BMPs for low-impact development.

16-16 How is low-impact development different from conservation design?

16-17 What are the rating criteria for a LEED-certified green building? How many LEED points are required for LEED certification, LEED Silver certification, LEED Gold certification, and LEED Platinum certification?

16-18 The Seattle Library was an innovative green building project launched in 1998. The building design included considerations for sustainable design as described in the report entitled "Design for Social Sustainability at Seattle's Central Library," available online (*doi.org/10.3992/jgb.2.1.1*). Describe how the design of the Seattle Central Library incorporates sustainable design elements in each of the following areas:
   a. Size and placement
   b. Building/material choices
   c. Landscape and ecology
   d. Energy
   e. Transport
   f. Water

16-19 You are designing a building on 0.4 acre. It will be three stories high, and each story will have about 12,000 ft². Determine the development density descriptor from Table 16.1 in Spokane County for the proposed building.

16-20 Explain what type of development—one house per acre with 20% impervious cover or eight houses per acre with 65% impervious cover—generates more stormwater and why.

16-21 Research the local development planning policies and codes for your city or county.
   a. Do the planning policies encourage open space preservation in concept? Explain.
   b. Do the planning policies encourage open space preservation in practice? Explain.
   c. Do the planning policies follow conservation design theories? Which procedures are utilized and which are not in the growth management act?

16-22 Draw a building envelope for low-density residential zoning described in Table 16.1 with six homes per acre.
   a. What is the minimum lot size?
   b. What is the maximum square footage of the home?
   c. For a home with a three-car garage, what would be the remaining square footage of the first floor?

16-23 How do conventional buildings and green buildings differ?

16-24 What are the LEED rating categories and possible points? Which rating categories have greater weights (e.g., more credits are possible)? Why might this be so? Do you think the categories are unbalanced? Why or why not?

16-25 Refer to the Fossil Ridge High School LEED summary online (www.calmac.com/stuff/contentmgr/files/0/da4335bf91f3ba5e8f9a22f04e27f88a/files/calmac_fossil_ridge_case_study.pdf).
   a. Estimate and assign LEED points for the Fossil Ridge High School design.
   b. In what areas of design did "green building" increase costs the most?
   c. In what areas of design did "green building" result in the greatest potential savings?

16-26 Use Google Earth, Google Maps, or other aerial imagery to select a potential residential development site in your area. Draw a yield plan if the property is divided into
   a. 0.25-acre lots
   b. 1.0-acre lots
   c. 5.0-acre estates

16-27 From the Google Earth, Google Maps, or other aerial imagery developed in Problem 16-26, complete the following:
   a. Identify primary conservation areas and briefly label and explain them.
   b. Identify secondary conservation areas and briefly label and explain them.
   c. Identify potential development areas and briefly label and explain them.
   d. Locate potential housing sites.
   e. Locate transportation routes.
   f. Identify lots using conservation design principles and maintaining 30% open space.
   g. Does your conservation design look different from the yield plan in Problem 16-26?
   h. Compare and contrast the advantages and disadvantages of the yield plan [choose part (a), (b), or (c) from Problem 16-26] and the conservation base design.

16-28 What are the four basic ways we can improve energy use in a building?

16-29 Refer to Figure 16.24. Based on this graph, a 10% reduction in which category of energy end use would yield the greatest energy savings?

16-30 Refer to Figure 16.24. Discuss how you might reduce energy consumption for refrigeration, lighting, cooling, heating, and water heating through various green building strategies.

16-31 What is a heating degree day, and how is it relevant to building codes?

16-32 Discuss how the function of a building and the behavior of the occupants might influence the appropriate exterior design temperature of the following:
   a. Office building
   b. Gymnasium
   c. Hospital

16-33 Lighting can account for up to what percentage of a building's electricity usage?

16-34 How much energy could be saved by replacing five 60-W incandescent bulbs each of which burns 4 hours per day with 8-W LED (light-emitting diode) light bulbs?

Team-Based Problem Solving

16-35 Consider a home that uses the equivalent of 16,000 kWh per year for the following energy loads:
   • Domestic hot water: 3,500 kWh
   • Refrigeration: 1,300 kWh
   • Lighting: 1,500 kWh
   • Heating: 4,800 kWh
   • Cooling: 2,400 kWh
   Make this house a net zero home.

16-36 The site shown in Figures 16.30, 16.31, and 16.32 is in Washington State, and we will assume that the area is slated for development. The area for development is 400 acres. Suppose the Sierra Club, local home owners, and other interested parties have told the land owner they would sue and seek a court injunction to prevent development if the zoning were changed from agricultural to mixed-use residential. Pretend that at least 10-acre lots are required

FIGURE 16.30 Photo for Problem 16-36.

Source: USDA – Farm Service Agency.

FIGURE 16.31 Photo for Problem 16-36.

Source: USDA – Farm Service Agency.

FIGURE 16.32 Photo for Problem 16-36.

Source: USDA – Farm Service Agency.

under the current zoning code. They are interested mainly in protecting the (real) high-value green corridor and wetlands that make up 20% of the property. These wetlands are used by a wide variety of migratory birds, including potentially endangered species. Your company has been selected over competitors to develop this site based on your preliminary submittal. You have been requested to provide more development details in order to ensure the environmental impacts due to development are minimized. You may incorporate previous suggestions into your final plans along with the technical details listed below. As part of your presentation, be sure to address the following:

a. Provide yield under existing zoning codes and LDR-P.
b. Explain conservation areas.
c. Cite potential differences in a "green design" approach.
d. Explain how you would incorporate LEED practices into your development plan and construction methods. Estimate the categories and points associated with the LEED certification that you will try to attain.
e. Estimate your lot size and discuss how it relates to the current zoning codes in your proposal.
f. On a separate diagram, complete the following for the roadways designed for the development:
   i)   Label the roadway type.
   ii) Determine average daily traffic volume (ADT) on each road.

        iii) Provide a schematic detail of each cross section.

        iv) Discuss how your transportation plans fits into the LEED analysis.

  g. Provide an erosion and sedimentation control plan.

        i) Label the phases of development.

        ii) Determine temporary and permanent structures.

        iii) Provide a schematic detail of each type of control method.

        iv) Provide a maintenance schedule and identify who is responsible for maintenance.

        v) Discuss how your erosion and sedimentation plan fits into the LEED analysis.

  h. Provide a water (water, wastewater, and stormwater) management plan.

        i) Determine temporary and permanent structures.

        ii) Provide a schematic detail of each type of control method.

        iii) Provide a maintenance schedule and identify who is responsible for maintenance.

        iv) Discuss how your water management plan fits into the LEED analysis.

  i. Estimate the cost(s) of finished homes (property and built home) on the proposed property.

  j. Estimate the amount your company would be willing to pay per acre.

Challenges and Opportunities for Sustainability in Practice

FIGURE 17.1 Wangari Muta Maathai was a Kenyan social and environmental activist most widely known for founding the Green Belt Movement, a nongovernmental environmental organization focused on reforestation, environmental conservation, and women's rights. In 2004, she was the first African woman to be awarded the Nobel Peace Prize for her contribution to sustainable development, democracy, and peace. She believed that "you cannot protect the environment unless you empower people, you inform them, and you help them understand that these resources are their own, that they must protect them."

Source: John Moore/AP/Shutterstock.com

> Change sometimes builds a new model that makes the existing model obsolete.
> — BUCKMINSTER FULLER

Challenges and Opportunities for Sustainability in Practice

FIGURE 17.1 Wangarĩ Muta Maathai was a Kenyan social and environmental activist most widely known for founding the Green Belt Movement, a nongovernmental environmental organization focused on reforestation, environmental conservation, and women's rights. In 2004, she was the first African woman to be awarded the Nobel Peace Prize for her "contribution to sustainable development, democracy and peace." She believed that "you cannot protect the environment unless you empower people, you inform them, and you help them understand that these resources are their own, that they must protect them."

Source: John Mcconnico/AP/Shutterstock.com.

To change something, build a new model that makes the existing model obsolete.

—BUCKMINSTER FULLER

GOALS

THE PURPOSE OF CHAPTER 17 is to provide you with a greater context for the social and economic factors that shape the widespread acceptance and use of more sustainably designed products, processes, techniques, and behaviors. In this chapter, we explain the dynamics of the adoption and diffusion of innovations, the economic concepts that help us understand why achieving greater environmental sustainability is a challenge, and the role of governmental policymaking in overcoming those obstacles.

OBJECTIVES

By the end of this chapter, you should be able to:

17.1 Summarize and analyze the social, cultural, technical, and economic factors that affect the potential adoption and diffusion of an innovation.

17.2 Conduct a simple benefit–cost analysis comparing two investment options.

17.3 Identify and give examples of government policy tools that help to overcome barriers to the adoption and diffusion of innovations.

17.4 Summarize the issues associated with social justice and sustainability in wealthy industrialized nations.

Introduction

To achieve a more sustainable world, people must change their ideas, their behavior, and their vision of the future. It isn't enough to wish or hope for things to be different; action is required. Although news reports often paint the state of the environment as dire, there is reason to be encouraged. As we have seen from the past several chapters, sustainable development is *actionable. Workable solutions* exist to prevent us from repeating the environmental harms of the past and enable us to live more sustainably on the planet. The human community has a surprising array of models, concepts, tools, land-use approaches, technologies, conservation behaviors, construction techniques, and analytical methods to put the principles of sustainable development into practice. So long as we use them more extensively and continue to build our repertoire of strategies, we can work to ensure that conditions will improve in the future.

People are often resistant to change, however, and some features of our social and economic systems make implementing sustainable development difficult. To create compelling change, Buckminster Fuller advocated for designing in a manner that makes the old way of doing things obsolete. His approach to design science captured the principles of sustainable development decades before sustainability concerns became commonplace. Meeting basic human needs, cooperating and collaborating to solve problems, doing more with less, and conserving natural resources are central to Fuller's approach to successful change through design.

To conclude our text, we consequently want to explore some of the issues that affect how people, communities, and societies respond to products and techniques engineered to be more sustainable. Social sustainability concepts are closely related to human psychology and sociology. These are beyond the scope of our text, however; we will instead examine factors that you can meaningfully address as engineers in the design process. In particular, we will focus on dynamics that affect the diffusion and adoption of innovations, the nature of economic decision making, and the role of government in offsetting social and economic constraints. We will also address issues of social justice in industrialized countries.

17.1 The Diffusion and Adoption of Innovations

Chapters 7 through 16 highlighted a wide array of approaches that can result in a more sustainable society generally and a more sustainable engineering practice specifically. Ecological design, industrial ecology, life cycle analysis, technologies for energy conservation, green building, conservation design, low-impact development, and smart growth represent proven opportunities to reduce human impacts on the environment. These approaches are not widely practiced even though most are relatively well known. We have to ask ourselves why this is.

Environmental sustainability and sustainable development represent radical new ways of thinking both socially and economically. Acting on these goals means changing our values, our lifestyles, our methods of economic accounting and valuation, and the criteria we use for making decisions about what we buy, use, build, and make. Sustainability has achieved a great deal of influence today as a driving force for social change through a process that is referred to as a ***social movement***. Social movements occur when new ways of thinking become institutionalized in a society and are broadly advocated by groups that have political and social power. There is no doubt that a social movement is occurring, but sustainability has not yet become a cultural standard against which we weigh decisions about what we do. It will take time for it to become thoroughly established and internalized as a broad societal goal and common habit; understanding these psychological and sociological processes is unfortunately beyond the scope of our text.

However, it *is* possible to discuss why people and organizations may or may not use the sustainability practices discussed in Chapters 12 through 16. The term used to describe when, why, and how we do things differently—whether in our behavior or in the types of products and technologies that we buy—is ***adoption and diffusion of innovations***. Home owners who buy a CFL light bulb for the first time are innovating—they are adopting a new technology for energy conservation in their house. Companies that begin to practice life cycle assessment are focusing on innovation to reduce their global footprint associated with manufacturing and energy use. Local governments that implement green building ordinances are innovating—they have adopted new practices. When many individuals and organizations adopt a specific innovation (such as CFL lighting), the innovation begins to diffuse—or spread—throughout society. By anticipating obstacles to the adoption and diffusion of an innovation, *we can proactively address them in engineering design*.

The degree and rate at which new innovations are adopted and diffused throughout a society can be modeled by a logistics S-curve. For any given innovation, there will be some defined group of adopters (people or organizations) who could potentially use it; this is represented by the *y*-axis in Figure 17.2. To illustrate, if three-fourths of all U.S. households purchased energy-saving CFL light bulbs, then we would say that CFLs as an innovation have diffused to 75% of all potential

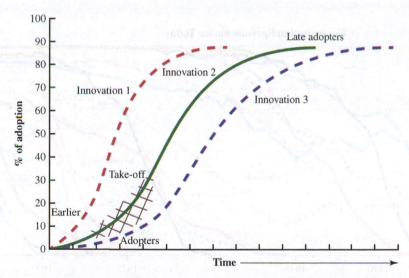

FIGURE 17.2 Logistics curves for the adoption of innovations. The *y*-axis represents the percentage of total potential adopters that ultimately use the innovation. The *x*-axis represents the speed at which change takes place (the rate of adoption over time).

Source: Based on Rogers, E.M. (1963). *Diffusion of Innovations, 3rd ed.* New York: Free Press.

adopters. A critical issue is *how many* potential adopters have actually decided to use an innovation, whether it is a practice, product, or process. Very simply, it does not matter how terrific or wonderful an innovation might be *if nobody uses it.* Without adoption and diffusion, society simply cannot benefit from the existence of something new.

The rate of innovation matters as well, and the logistics curve also models how fast an innovation might be accepted. As you can see from the *x*-axis in Figure 17.2, in the early stages of an innovation, only a few people or organizations use it (the earlier adopters); then the rate of adoption accelerates as more people become aware of the innovation and its benefits (the take-off period). The rate of adoption slows down over time because there are fewer and fewer adopters who could potentially use the innovation. Finally, the rate flattens out and plateaus with the "late adopters." Figure 17.3 shows the diffusion curves for a variety of consumer electronics and home appliances in the United States as a way of illustrating these concepts with real innovations.

Broadly speaking, the widespread social diffusion of innovations requires that four conditions be met: that potential users are *aware* of an innovation, that they have the *ability* to acquire and use it, that they *accept* that it will improve their lives in some way, and that they then choose to *adopt* the innovation in practice. These conditions are relevant in all societies, including low-income communities where appropriate technologies are essential. The story of the Jaipur foot (see Box 17.1) illustrates how a remarkable innovation languished until a dedicated government official created a way to mass produce and distribute it free of charge to those most in need. In doing so, he eliminated the diffusion barrier to people's ability to acquire the innovation. Another award-winning appropriate technology, the Chotukool, is confronting adoption challenges, even though its inventor has carefully worked out both its manufacture and distribution. The Chotukool is a low-cost source of refrigeration for rural Indian families without access to electricity

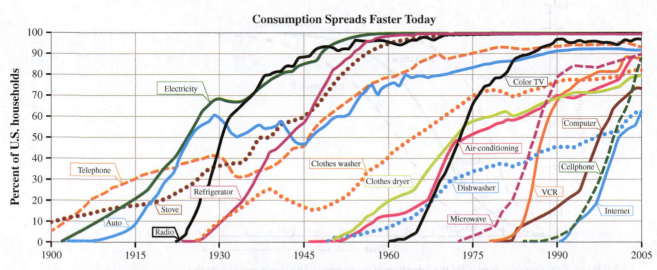

FIGURE 17.3 The diffusion of consumer electronics and appliances in the United States. As seen here, each innovation achieves different levels of total adoption (compare radios to the Internet, for example). In addition, the speed at which the diffusion occurs is also variable; VCRs achieved 70% adoption within a few years compared to decades for washing machines.

Source: Based on Felton, N., *The New York Times*, 10 February 2008.

BOX 17.1 Diffusing the Jaipur Foot

The Jaipur foot is a prosthetic limb celebrated as a model of design for its extreme affordability. The prosthesis was invented in Jaipur, India, in 1968 as a result of the collaboration between orthopedic surgeon Dr. Pramod Karan Sethi and master craftsman Pandit Ram Chandra Sharma. The foot is an extraordinary biomechanical invention because unlike traditional prosthetics, it allows people to walk, climb, squat, and sit cross-legged; it more closely mimics the human foot than any other device. In addition, it can be custom made and fit to its owner in a few hours, and only costs about U.S. $45 to make. The cost of a Western prosthesis would be well over $10,000.

Although the Jaipur foot is extremely inexpensive compared to other prosthetics, only about 60 of these prosthetics were fitted between 1968 and 1975. Significant barriers to its diffusion existed: not only was its cost still out of reach to India's impoverished—those most in need of the device—but there was also no ability to readily make, fit, and distribute it to the tens of thousands of amputees who wanted it.

This excerpt from an article by D. V. Sridharan in the online magazine *Good News India* explains how the Jaipur foot overcame obstacles to its diffusion through the efforts of Devendra Raj Mehta. Within 6 years, Mehta's nonprofit organization, the Bhagwan Mahaveer Viklang Sahayata Samiti (BMVSS), was fitting almost 3,000 Jaipur limbs per year; within 10 years, that number had risen to 6,000. Today BMVSS has fitted more than a half a million of these prosthetics free of charge, and currently fits about 20,000 artificial limbs using the Jaipur foot technology a year, including in countries with significant human limb loss because of war and land mines. The most current and detailed information on Jaipur foot technology, manufacturing, and distribution may be found at the Bhagwan Mahaveer Viklang Sahayata Samiti website (www.jaipurfoot.org).

After a severe traffic accident in 1969, Mehta received physical therapy in Jaipur, the capital of Rajasthan in Northern India. The Jaipur foot had just been invented around that time, and Mehta was deeply affected by the large number of maimed and impoverished people coming to Sawai Man Singh (SMS) Hospital trying to get a foot.

"He recalled that experience years later. He was back in government service and Principal Secretary to the Chief Minister of Rajasthan. 1975 was the 2,500th birth year of Mahavira, the founder of Jainism, which the Indian government wanted all states to celebrate fittingly. Mehta suggested rescuing the Jaipur foot from its neglect and delivering it to the needy.

Technology by itself cannot bring about change, however elegant it may be. The Jaipur foot was no doubt a technical winner, but what made it languish, unavailable to the needy? *There were no processes to manage its delivery in the thousands.*

"The first thing I wanted changed was the approach to visitors," says Mehta. "It had to become human." It was common for the maimed to arrive at SMS Hospital and be made to wait for days for mere registration. When BMVSS began operations, the first practice put in place was that registration must be done on arrival, around the clock. Then the patient is given food and a bed. They and a caretaker are hosted until their limb is custom fitted. And they walk out upright in dignity, with return fare in hand. The whole service is free.

There is an ordered assembly line approach to fabrication. The amputee's stump is covered with a knitted sock and a plaster mold is made. From this socket, a plug is made which is an exact replica of the limb. High Density Polyethylene Pipe [HDPE] is warmed and stretched over the plug. A vulcanized rubber foot is attached and suitable straps are provided to fasten the limb to the body. Most of the time, fitment is on the same day, and comfort with using it is achieved in hours.

The Jaipur foot costs about Rs.1,300 to make. It can be made by a team of five with average skills, which can produce 15 limbs a day. It is often regarded as a limb suited for Indian use: wearing it,

(Continued)

BOX 17.1 Diffusing the Jaipur Foot *(Continued)*

one can work in the fields, run, pedal a bicycle, squat on the floor, or even climb a tree. In contrast, a prosthesis from the West is mostly cosmetic.

Pandit Ram Chandra Sharma's original Jaipur foot was made from wood and then aluminum. Later, while visiting a pipe factory in Hyderabad, he was struck by a narration of high density polyethylene's virtues: light weight, low cost, moldability, and strength. Today, body-colored HDPE pipe is manufactured specially for BMVSS use."

Source: Based on Sridharan, D.V., "Jaipur foot: the real story," *Good News India* (undated, circa 2005).

(about 400 million people according to the World Bank). In the first two years of its market introduction, only about 15,000 coolers had been sold relative to millions of potential adopters. It remains to be seen whether the Chotukool will experience a diffusion take-off phase.

In industrialized countries, an extraordinary number of factors can influence the awareness, ability to use, acceptance, and ultimate adoption of an innovation (Figure 17.4). The slow acceptance of compact fluorescent light bulbs (CFLs) readily illustrates problems that new innovations confront. Although CFLs are very energy efficient—for example, a CFL requires only 13 watts of electricity to produce the same amount of light as a 60-watt incandescent—they are not widely seen in the United States (Figure 17.5). One problem is their price; early on, CFLs could cost several dollars compared to an average of 25 to 50 cents for a conventional light bulb. "Sticker shock" was not the only problem that CFLs faced with consumers, however. Poor light rendering—colors in the lighting spectrum that are visually harsh—was a frequent complaint, as was the slow illumination of these bulbs. Because CFLs are fluorescent, they do not achieve full brightness immediately. Older bulbs could take several minutes to completely light a space, a condition that presents a safety hazard in some locations and is not culturally acceptable to people accustomed to instant light at the flip of a switch. The spiral shape of CFLs also prevented people from using them with existing lamps, because clip-on lamp shades could not be readily attached to the bulb. CFL technology was slow to be adopted due to several challenges:

- They seemed prohibitively expensive compared to the common light bulb, and consumers could not easily calculate the energy savings from these bulbs relative to their up-front cost.
- Lighting quality was poor both in terms of the color spectrum and the length of time required to achieve full brightness.
- Bulbs were incompatible with existing lighting equipment, which would require that consumers buy new lamps and lampshades.

Although CFL technology has improved over the past decade and achieved stronger market acceptance, new innovations have surpassed CFL light bulbs. As seen in Figure 17.6, LED lighting has come to dominate the U.S. market. LED bulb manufacturers learned from the issues that plagued CFL bulbs and have avoided many of the same pitfalls. LED bulbs are available in the traditional shape and size of older incandescent bulbs, are lower in price, and often feature variable brightness and color.

Adopter's Human Resources	Adopter's Organizational Structure	Adopter's Organizational Culture and Decision Process
■ Motivation ■ Skills ■ Technical knowledge resources ■ Specialization and professionalism ■ Commitment ■ Managerial attitudes and support	■ Size and resources ■ Centralization ■ Complexity ■ Communication/ administrative intensity ■ Formalization ■ Flexibility	■ Cooperation and openness ■ Organizational support for innovation ■ Technology champions ■ Innovation proneness ■ Orientation (outward v. inward) ■ Organizational position and role of decision maker
Adopter's Market Context	**Industry Characteristics**	**Communication Channels and Social Networks**
■ Unionization ■ Competitive strategy ■ Location ■ Growth strategy ■ Knowledge of competitors' behavior ■ Market scope	■ Heterogeneity ■ Concentration ■ Regionalization ■ Government regulation ■ Wage rates ■ Growth rate ■ Inter-firm competitiveness	■ Boundary spanners ■ Word-of-mouth ■ Opinion leaders ■ Professional and trade associations ■ Word-of-mouth ■ Informal and indirect links
Technical Attributes of the Innovation	**Economic Attributes of the Innovation**	**Supplier/Vendor Characteristics**
■ Complexity-crudeness ■ Communicability ■ Divisibility ■ Type of innovation (process or product) ■ Complementarities required ■ Relative improvements in old technologies ■ Compatibility (values and practice) ■ Learning by doing ■ Relation to innovator product class schemas ■ Radical v. incremental ■ High, medium, and low tech	■ Continuing cost ■ Initial cost ■ Expectations about future prices ■ Expectations about future tech trajectory of innovation ■ Labor saving v. materials saving ■ Scale neutral v. lumpy ■ Uncertainty/risk ■ Profitability ■ Start-up investment ■ Time savings ■ Rate of recovery of cost	■ Technical capabilities and support ■ Public relations ■ Expertise in monitoring deployment ■ Communications skills

FIGURE 17.4 Factors affecting the adoption of an innovation. For businesses, a large number of factors affect their awareness of, ability to use, acceptance of, and ultimate adoption of an innovation. Drivers range from the communication and social networks in which they are engaged to industrial structure and their competitive environment, to the technical characteristics of the innovation.

Source: Based on Koebel, C. T., Papadakis, M., Hudson, M., and Cavell, M. (2003). *The Diffusion of Innovation in the Residential Building Industry*. Blacksburg, VA: Virginia Polytechnic Institute and State University.

Although many of these diffusion factors are beyond your ability to influence as an engineer, you do have the ability to address several important considerations, including quality, user needs, cost factors, and compatibility with existing equipment and technologies (among others). You can engage in people-centered and participatory design for all types of problems (not just those of the underprivileged), which will likely increase the potential for your innovations to become widely adopted. This is because people-centered and participatory design are (1) grounded in user-defined needs assessment, (2) sensitive to the values and preferences of users, (3) shaped by an awareness of the economic constraints of potential adopters, and (4) informed by the requirements for technical compatibility of the innovation.

FIGURE 17.5 Market presence of CFL light bulbs. These graphs show the sales trends of CFL light bulbs compared to incandescent lighting. Both the sales index in the top image and the blue market penetration bars on the bottom graph are representations of the diffusion logistics curve for CFLs. Market use of this innovation plateaued at about 25% after four years.

Source: Based on Bickel, S., Swope, T., and Lauf, D. (2010). Energy Star CFL Market Profile: Data Trends and Market Insights. U.S. Department of Energy. Silver Spring, MD.

The relatively slow adoption and diffusion of solar thermal technologies (see Box 17.2) illustrates the many challenges that our global society faces in encouraging a rapid uptake of more environmentally sustainable technologies.

In sum, there are many reasons why an innovation that can enhance environmental sustainability may not be widely accepted or used. Awareness, costs, cultural

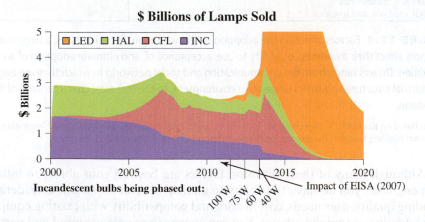

FIGURE 17.6 Diffusion of different lighting technologies. After setting the standard for 100 years of electric lighting, incandescent bulbs (in blue, above) are being replaced by alternative lighting choices. The Energy Independence and Security Act (EISA) of 2007 made it hard for incandescent bulbs to achieve the mandated levels of efficiency; CFL and halogen technology function as transitional innovations as we move to LEDs.

Source: Based on Knott, J. (2012, October 15). "LEDs Will Eliminate All Other Lamp Sources in 6 Years," *CEPro*.

factors, values, access, and usability all shape people's willingness to try something new. Sustainable development requires that we understand the deeply rooted social, organizational, cultural, and economic contexts that frame the suitability of innovations for individuals, communities, and businesses and their ultimate willingness and ability to change. However, there is no doubt that cost considerations are foremost among many considerations and that cost differences are distorted because we do not account for the environmental harm created by the traditional way of doing things. Nonpoint-source water pollution, climate change, and the impacts of land-use and land-cover change all have associated social costs that are not reflected in the market prices of goods that embody these environmental problems.

BOX 17.2 Barriers to the Diffusion of Solar Thermal Technologies

Although most people think of photovoltaic electricity when they think of solar energy, the sun provides an extraordinary amount of energy needed for *heat*. A variety of solar thermal technologies and systems are available for hot water production and space heating and cooling. Passive solar architecture, solar hot air collectors, and thermal storage systems are proven systems that still are not widely used. Industrial operations that can be supplemented by solar thermal energy include hot water production, drying methods, and a variety of chemical processes, among others. Barriers to the development and widespread adoption of these technologies include technical, economic, legal, cultural, and behavioral obstacles.

For example, technical problems with residential solar hot water systems occurred early in their product life and tended to involve their inability to store hot water for long periods of time or have backup sources of water heating. Consumers were regularly disappointed by the technology, which created a poor reputation in the market for what is now a very reliable form of hot water production and storage. Early technical problems were the result of engineering design that did not account for system complexity and poor installation of even relatively simple systems. In Europe, continued challenges include the inability of some domestic appliances, such as dishwashers and washing machines, to use water from multiple water lines for hot and cold water. Industrial solar thermal technologies

confront a wide array of technical challenges that can only be met with further research and development (Philibert, 2006).

Almost all solar thermal technologies face economic challenges. The competitiveness of solar thermal products is a function of the available sunshine in a region, the local cost of conventional energy, and the cost of conventional heating alternatives. Both installation costs and the energy costs of alternative equipment must be considered. The OECD reports that in Greece, with a sunny Mediterranean climate, a solar hot water system for a family would cost about 700 euros (€) (Philibert, 2006). In Germany, with less sunshine and freezing temperatures, an appropriate comparable system would cost €4,500. Although the hot water needs of the German family could be met with a solar heating system, the operating cost is equivalent to as much as €0.16 to €0.42 per kilowatt hour. (The average cost of electricity in Germany is about €0.25.)

Legal, cultural, and behavioral barriers also exist. Passive solar architecture does not cost appreciably more than a conventional building and results in significant energy savings; solar hot air collectors are likewise highly cost effective. Lack of awareness may be one reason these technologies are overlooked in new construction. Better explanations include a general lack of willingness to innovate in the construction industry, the inability to create standardized building designs using these systems because of the need for careful site orientation, and a local workforce

(Continued)

BOX 17.2 Barriers to the Diffusion of Solar Thermal Technologies (Continued)

without the requisite trade skills. Landlords will have little incentive to adopt innovative residential solar thermal technologies if they cannot recoup their investments through higher rents. In some U.S. neighborhoods, solar hot water systems cannot be installed because they are unsightly, a practice that is known as "restrictive covenant."

Cultural cooking norms are an important reason solar cookers and ovens have been met with widespread disinterest around the world, in spite of the fact that these resources require no fuel, emit no smoke, and may enable people to spend less

time attending to meal preparation. Because solar cookers generally cannot brown food, they result in visually unappetizing meals for cultures that rely on frying (such as in Asia). People who prepare meals in a communal cooking environment, like some cultures in Africa, are mystified by a cookstove that a person would use by themselves. In industrialized countries, solar cookers are not useful in the evening when most people arrive home from work. Families that live in dense urban areas with multistory dwellings simply do not have access to the amount of sunshine needed for solar cooking.

ACTIVE LEARNING EXERCISE 17.1 The Diffusion and Adoption of a Packing Peanut Innovation

FIGURE 17.7

Source: Mike Flippo/Shutterstock.com.

Packing peanuts, those ubiquitous little Styrofoam™ nuggets that protect products during shipment and storage, are actually made from a variety of materials and come in a range of colors that signify their material properties. According to the radio program *Living on Earth* (2012), Americans dispose of 9.5 million tons (19 *billion* pounds!) of packing peanuts each year. Because this material is largely inert and takes centuries to break down in a landfill, the latest innovation is a completely biodegradable peanut that dissolves in water.

For this exercise, you should first conduct Internet research on the different kinds of packing peanuts on the market today and their product attributes (appearance, performance characteristics, material content, cost, and so on). If you are a manufacturer of peanuts, what are you most likely to consider in the design and manufacture of this product? Would this change if you conducted a life cycle analysis or if you considered the fundamental principles of green engineering? (*Hint*: You should research the term *bioplastics* to get at this quickly.)

If you are a large-scale shipper of products that require peanuts for safe transport and to protect your goods from damage, what factors are you likely to consider as a bulk purchaser of peanuts? If you are a large-scale retailer of products packaged in peanuts that you store in inventory, what are you likely to prioritize when choosing packing material for the products that you buy?

After reflecting on and analyzing these different perspectives, what factors do you think constitute the major opportunities and obstacles to the widespread diffusion of the biodegradable peanut as an innovation? Who is the actual "adopter" in this context? Do you think it likely that biodegradable materials will come to dominate the peanut market? Be able to discuss why or why not. If you think not, what circumstances might create the conditions necessary for high rates of market adoption of this innovation?

17.2   The Economics of Sustainability

Sustainability presents many challenges to us from an economic perspective. In particular, there are four types of economic issues that we see routinely in environmental problems that affect our ability to develop more sustainable solutions:

1. The fundamental affordability of greener products, technologies, and systems
2. The opportunity cost of money and the way individuals and businesses make decisions about how to invest and spend their money
3. The problem of externalities and the fact that the burden of environmental degradation is not generally reflected in the prices of the goods and services that we buy
4. The difficulty in establishing a monetary value for a clean, healthy environment and sustainable business practices

Engineers are often frustrated by these economic factors, especially when they have designed a product or process that reflects a high degree of technical efficiency or reduces impacts on the environment. By better understanding the economic logic at work in human decision making, you will be better able to anticipate how the successful adoption of your design might be affected.

17.2.1   The Fundamental Affordability of Greener Goods and Services

Affordability is an obvious issue when we are talking about the ability of the world's underprivileged—especially those living in $1- or $2-dollar-a-day poverty—to purchase simple life improvements that also reduce their impact on the environment. As seen in Figure 17.8, in 2020 nearly 10% of the global population lived at a poverty level equivalent to $1.90 per day (that's almost *a billion people*). When scarce household cash needs to be protected for medicine, food, or other critical necessities, many families simply cannot afford to invest in more sustainable life improvements. Most of us have no trouble comprehending the basic fact that if you *have no money*, you cannot afford to buy something even if it is of great benefit to you.

Paradoxically, our altruistic impulses to simply give low-income people what they need—subsidized cookstoves, solar-powered electricity, water sanitation systems, and other goods—have been shown time and again not to work. The failure of such charitable initiatives was touched on previously in this chapter, and it deserves emphasis again here. People are most likely to accept, adopt, and use innovations when their needs are truly being met, when they have made a financial investment by paying something toward the cost of the innovation, and when there is a local capacity for operation, maintenance, and repair. In economic terms, people *value what they are willing to pay for*. As a consequence, human-centered engineering design must be accompanied by an economic program that will enable people to contribute in some manner to the purchase of goods and services that will make them better off. This is why microcredit is often critical to the success of sustainability initiatives, such as those provided by Grameen Shakti for solar PV panels (see Chapter 1) and Agua del Pueblo for community water systems (Box 17.3). Not only does it increase the likelihood that an innovation will be used, but it also helps build the long-term financial and institutional capacity of the community. Enabling

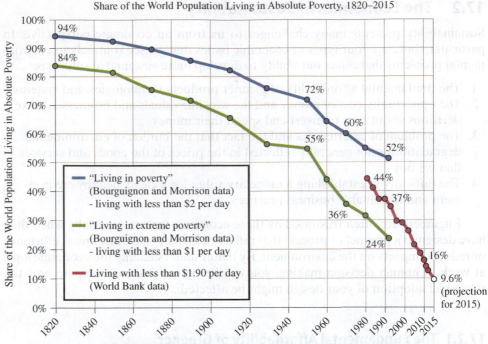

FIGURE 17.8 Percentage of the global population at different poverty levels. These definitions from 1820 to 1992 correspond to poverty lines equal to consumption per capita of $2 and $1 a day expressed in 1985 purchasing-power parity (PPP), and from 1980 onward, the share of people living below the international poverty line, which is $1.90 at 2011 PPP.

Source: Based on Roser, M., and Ortiz-Ospina, E. (2013). "Global Extreme Poverty." Published online at OurWorldInData.org. Retrieved from ourworldindata.org/extreme-poverty.

people to pay at least some token amount toward these goods and services also protects their dignity and self-esteem and enhances their sense of being able to provide for their families.

Affordability in the context of the world's wealthier nations is a bit more perplexing, since our purchasing power is significantly greater than that of people in developing nations. Nonetheless, one's ability to pay is also a consideration because more sustainable products, goods, and services often carry a **green premium**. That is, they are more costly than "normal" products and services. For example, consider an energy-efficient refrigerator that costs $250 more than its less efficient counterpart, or a new automobile with better pollution control technology that costs thousands of dollars more than an older, more polluting, used car. If people cannot afford the marginally higher cost of the green premium, they will not make the more environmentally sustainable purchase. We cannot escape the fact that people must live within their means.

BOX 17.3 Agua del Pueblo (*The People's Water*)

Agua del Pueblo (AdP, "the people's water") is an NGO that helps rural Guatemalan communities design, build, finance, and operate their own community water supply systems. Founded in 1972, the organization has enabled hundreds of villages to develop a secure supply of potable water. In the process of doing so, these communities have improved their health, expanded economic opportunities, enhanced the status of women, improved the social capital of their villages, and uplifted the technical skill sets of virtually all of the adults living in the village.

In addition to causing the diseases and illnesses described in Chapter 7, the lack of potable water has high *social costs* in Guatemala. The majority of rural families in Guatemala have no water in their homes, and in this culture, women have the role of providing water. They must make numerous daily trips to a river, lake, well, or mudhole. They go to wash clothing, to obtain water for household use, and to bathe. Women must carry a 2- or 3-gallon jug home from the water source several times a day and must make several trips a week to this source to wash clothing. In one community served by Agua del Pueblo, women spent as many as 45 working-days per year in transit just to obtain water. This is an incredible loss of time that could be spent more productively (the social cost), even within the scope of traditional gender roles in this culture. With more time available to them, women could pay more attention to performing household sanitary measures (improving family health), preparing better meals (improving family nutrition), or producing such salable items as weaving or pottery (improving family income).

In this context, a community water supply system is an innovation that faces many obstacles and barriers to its adoption and long-term viability. The cost of the system and the ability to repair and maintain it over time are the two biggest challenges to potable water supply in rural Guatemala. The subsistence communities and villages of Guatemala are small, people have many demands on their time,

FIGURE 17.9 Community water systems reduce the amount of physical labor and time spent collecting potable water by women and girls in rural Guatemala.

Source: Milosz_M/Shutterstock.com.

and few (if any) adults have the required technical skills to design or manage such systems.

An early pioneer of reproducible methods for the adoption of appropriate technology, AdP addresses these challenges in a manner that can be easily copied from place to place in Guatemala, enabling the "uptake" of water supply systems to occur much more easily. The AdP strategy is anchored in community-based decision making as well as the training of paraprofessional local water supply technicians.

(Continued)

BOX 17.3 Agua del Pueblo (*The People's Water*) (Continued)

In order to receive technical support and financial assistance from AdP, a village must elect a committee to oversee community input into the project, because participation is required for all stages of the project—planning, financing, design, construction, education, and maintenance. The village is responsible for deciding, with technical guidance, the nature of the water supply system to be constructed. Choices include, for example, whether to have water taps in individual homes or in central strategic locations throughout the community. Local residents build the water system (using volunteer labor and labor-intensive construction techniques) and supply locally available construction materials. This not only motivates a strong sense of community ownership of the system, but also lowers the costs of outside inputs. AdP educates a cadre of regional and local rural water supply technicians to design and supervise the construction of the systems, as well as train villagers to perform the basic system maintenance and repair. Villagers (both men and women) commonly rotate responsibility for monitoring and fixing problems with the water supply.

The local community must also contribute to the cost of the system, even if it is very modest. A small cash down payment is made, and AdP provides some matching funds as well as low-cost loans.

By being required to pay for a valuable service, the community not only demonstrates its commitment to the system, but also develops what the World Bank refers to as a "habit of payment for other worthwhile goods" (World Bank, 1975).

The results of AdP's efforts for sustainable development have been striking. Hundreds of communities have chosen to adopt water systems (involving thousands of miles of ditches and small pipelines) in spite of the costs and demanding physical labor. In addition to the health benefits of clean water, the projects have dramatically increased the social capacities of their communities. Elected village committees for the water projects evolved into robust political systems for many other community development initiatives. These include the establishment of schools, the construction of better homes, the implementation of adult education programs, and empowerment in negotiating with plantation land owners and district politicians. Employment opportunities increased because of the enhanced technical skills of the populace. The status of women increased with their greater involvement in both the technical management of the water system and community decision making. The microfinancing models were extended by villages to other projects, such as schools and community centers, as well as buying materials to build better homes.

This relative trade-off in what we get for our money is understood in economic terms as *opportunity cost*. If we need a car but can only spend so much money on it because we also need to pay rent and buy food, then there are opportunity costs between rent, food, and a more fuel-efficient car. The more income we have, the more we can afford a wide variety of goods and services, and the opportunity cost of money becomes a more subtle and complex decision that is not based on the absolute affordability of a product or service, but on our needs, wants, and desires. Should I buy the more energy-efficient refrigerator, use the extra $250 to go on a weekend holiday, or save the money for other needs at a later date? This relative weighing of opportunity cost is what drives essentially all economic decisions for consumers, businesses, industry, and governments.

17.2.2 The Opportunity Cost of Money

No one has an infinite amount of money. We must make choices about what to buy, how to save, and alternative investments. Opportunity cost is an economic concept that represents what we gain (or lose) by choosing one option over another. For example, by choosing to go to college, you gave up the opportunity to get a full-time job and begin earning an income right away. The opportunity cost of this decision is the amount of money you could have earned by working for four years; however, you willingly incurred the cost of a university education on the assumption that your lifetime earnings with an advanced degree would exceed your income from working immediately after high school graduation or working part time and attending college part time.

Virtually all economic decision making has inherent opportunity costs associated with it. As consumers, we aren't even consciously aware of how often we consider these trade-offs. Whenever you quickly choose the cheapest box of cereal on the shelf, you have made a decision about the opportunity cost of your money: You would rather have the benefit of the money you did not spend instead of the additional tastiness or quality that a more expensive cereal might offer.

Opportunity cost is an important consideration in our analyses of why more sustainable products, services, buildings, and industrial practices do not rapidly take hold. Compared to conventional choices, the green premium of environmentally friendly goods and services acts as an adoption barrier for people and organizations for which the opportunity cost of the "greener" purchase is high. The implication for engineering design is rather straightforward: try to make more sustainable products and processes no more costly than their conventional alternatives; better still, make them less expensive.

If a green premium is unavoidable, one of several things must happen in order to motivate people to spend more money up-front and change their perception of opportunity cost. One option is to appeal to environmental values, such as through green marketing in which products may be labeled as "eco-friendly." Another example is to get a product formally certified by an independent third party as meeting specific sustainability criteria, a process known as *endorsement labeling* (see Box 17.4). LEED buildings (Chapter 16) and Energy Star products (Chapter 11) are examples of endorsement labeling that protects consumers from *greenwashing*, which is a form of deceptive environmental advertising, marketing, and public relations. It also enables businesses and industry to more easily comply with their own environmental management goals.

Although sustainable products and systems may have higher initial costs, they can save money in the long run. A second way in which initial opportunity costs may be overcome is therefore to demonstrate the long-term benefits of the more sustainable choice. Known broadly as benefit–cost analysis, there are many techniques for comparing the long-term advantages and disadvantages of different economic choices. Commonly used methods are life cycle analysis and return-on-investment ratios. Example 17.1 demonstrates the technique of life cycle costing and return-on-investment ratios with respect to household light bulbs. It shows that the more expensive energy-saving bulb not only pays for itself quickly in energy savings (a concept known as *payback period*) but also generates long-term financial savings. Benefit–cost analysis works well for analyzing many kinds of sustainability choices, in particular for energy and water conservation efforts, waste management, and improved materials selection. Box 17.5 illustrates how benefit–cost techniques were used to compare traditional stormwater management controls with low-impact development best management practices (BMPs).

BOX 17.4 The Benefits of Endorsement Labeling

Endorsement labeling, commonly known as green certification, has emerged as a way to help consumers and businesses make informed choices and buy more sustainable products. Endorsement labeling requires that an independent third party develop standards and protocols for determining whether or not a product or process has achieved the required sustainability characteristics.

Green certifications are not required by government. Rather, manufacturers and producers voluntarily submit their products for certification (endorsement) as a way of distinguishing themselves in the marketplace. The certification may be done by a third-party institution itself (such as Energy Star and Green Seal) or by those trained and licensed to do so independently (such as Passivhaus and the HERS home energy rating programs).

You have seen examples of endorsement labeling previously. Energy Star products certify that equipment and appliances have a level of energy efficiency well above that required by law. Chapter 16 reviewed a number of green certification systems for green building, including LEED, BREEAM, Earthcraft, Passivhaus, Energy Star, and net zero building. Other green certification programs exist, including Green Seal (cleaning products in the United States) and Green Dot (packaging in the European Union). The Forest Stewardship Council is an international NGO that certifies that timber and lumber products are harvested from sustainable managed forests. Endorsement labeling programs can therefore operate effectively at national, regional, and global scales.

Endorsement labeling/green certification has many benefits. First, it provides an independent standard about what makes a product or service more sustainable, and it avoids many problems associated with corporate greenwashing. Second, certified products are the only ones legally allowed to exhibit the endorser's logo, such as the Green Seal

FIGURE 17.10 The Green Seal endorsement logo. Green Seal is a U.S. nonprofit that has been providing third-party certifications of environmentally safe cleaning products since 1989.

Source: www.greenseal.org.

emblem shown in Figure 17.10. This enables products to legitimately "brand" themselves as green, giving manufacturers a competitive edge in the marketplace.

Third, it allows consumers to overcome information asymmetries. Individuals and small businesses often do not have the time or ability to evaluate the environmental implications of their purchasing decisions, and a third-party endorsement enables people to more easily make choices based on their environmental values. Over time, the expectation is that strong consumer preferences for endorsed products will "pull" sustainability into the marketplace by creating competitive pressures on all competitors to be more green.

Fourth, companies that want their products to be green can use third-party certifications as a way of guiding product attributes and performance specifications. Finally, because endorsement labeling is voluntary, it reduces the burden on government to make and enforce regulations.

BOX 17.5 Cost–Benefit Analysis for an Urban Stormwater Control Project

The city of Caldwell, Idaho, incorporated low-impact development techniques into a daylighting project for its local Indian Creek and a downtown redevelopment project. This is a brief case summary of how planners incorporated cost–benefit analysis to compare conventional stormwater management options with low-impact development (LID) best management practices.

Indian Creek is an excellent focus stream because it is a high-priority subwatershed to the lower Boise River and is contaminated by sediments, bacteria, and phosphorus. The project recognized that the existing collection systems could be upgraded to mitigate the discharge of untreated stormwater into a highly visible, restored stream. The LIFE™ model was used to determine a cost–benefit ratio. LIFE is a physically based, continuous simulation tool that represents the stormwater and runoff processes that occur within bioretention facilities, vegetated swales, green roofs, infiltration devices, and other LID controls.

Indian Creek meanders through farmland, residential, and industrial areas before it enters downtown Caldwell, Idaho, where it joins the lower Boise River. As Caldwell grew, it buried Indian Creek beneath asphalt and concrete. The daylighting project provides important habitat improvement along Indian Creek and is one component of a larger effort aimed at creating an attractive core area for the revitalization of downtown Caldwell. This work builds on existing efforts in the Indian Creek watershed, including a 2002 redevelopment design charette and an existing "Urban Ecology Design Manual for the Lower Boise River." This project recognized that the existing collection systems could be upgraded to mitigate the discharge of untreated stormwater into a highly visible, restored stream.

The final consideration for selected sites was a favorable cost–benefit ratio as predicted by the LIFE model. LIFE also accounts for runoff generated from all categories of land cover, including roadways, landscaping, and buildings over a variety of land uses and soil types, for new development and redevelopment. The model was used to estimate improvements in stormwater runoff quality resulting from application of LID technologies versus traditional stormwater control methods. Basins that drain to critical pinch points in the existing creek were delineated based on historic storm sewer mapping and "ground-truthed" (validated by on-site visual inspection) with recent field visits. In addition, 70.4 acres in the downtown core were delineated by land use. Results of this land-use characterization are

- Impervious –75%
  - Sidewalk –26%
  - Parking –22%
  - Roof –16%
  - Road –11%
- Pervious –25%

This information, as well as localized data on rainfall, runoff coefficients, and soil properties, was input into the model. Results of the modeling predicted the following runoff improvements:

- If all sidewalks were converted to permeable pavers, annual stormwater flows would be reduced by 13.2%.
- If half of all parking areas were retrofitted to include bioretention cells, annual stormwater flows would be reduced by 31.6%.
- If all roofs were retrofitted to discharge rainwater into underground swales, annual stormwater runoff would be reduced by 16.3%. If all roofs were also converted to green roofs, then an additional reduction of 0.7% of annual stormwater flows could be achieved.

These reductions in flows would correspond to a reduction in sediment discharge of 6,110 pounds a year (34% reduction versus traditional stormwater controls). These reductions in flows would correspond to a reduction in phosphorus discharge of 34 pounds a year (33% reduction versus traditional stormwater controls).

(Continued)

BOX 17.5 Cost–Benefit Analysis for an Urban Stormwater Control Project (Continued)

TABLE 17.1 Summary of Load Removal Predictions at Demonstration Site

	SEDIMENT	PHOSPHORUS
Traditional upstream technologies	586 lb/year 5%	1.9 lb/year 5%
LID technologies	2,400 lb/year 32%	14 lb/year 30%

Source: Excerpted from Doran, S., and Cannon, D. (2006, October). "Cost-Benefit Analysis of Urban Storm-water Retrofits and Stream Daylighting Using Low Impact Development Technologies." Water Environment Federation Technical Exhibit and Conference. ©Water Environment Federation.

On a site-specific basis, LIFE predicted increased improvements due to the LID stormwater design as compared to traditional upstream technologies (Table 17.1). From a cost–benefit perspective, traditional stormwater controls have a cost of $8,500, with a removal efficiency of 5% (the resulting cost–benefit ratio is $1,700 per % of load removal).

In contrast, the LID technologies cost more ($20,648), but the removal efficiency is much higher (32%) for a resulting cost–benefit ratio of $645 per % of load removal. On a cost–benefit basis, the improvements in LID certainly suggest a better return on investment.

Yet even though it is a valuable tool, benefit–cost analysis still may not result in a decision to purchase (or invest in) the more sustainable option, for several reasons. First, consumers and companies may truly not be able to afford the more costly product up front. Second, they may not have the time or skill to evaluate the long-term economic benefits of their purchase, which is a problem known as *information asymmetries* that is especially characteristic of households. Not many people would be able to carefully evaluate the long-term energy savings from a more efficient refrigerator, for example, while they are standing in a store trying to decide which model to buy. Small businesses likewise suffer from information asymmetries for many of the same reasons as households and individuals. Third, businesses and industry are under pressure to compete and make profits. Consequently, the payback periods for more sustainable options may be too long, or alternative and less sustainable choices may generate larger returns on investment.

In sum, the nature of opportunity costs drives economic decision making, and rational economic choices may not necessarily be wise environmental decisions. Options for making our choices more sustainable—by achieving both environmental goals *and* economic goals—include designing "greener" products and processes in order to be cost-competitive with conventional choices, designing products and processes in a manner that may qualify for an endorsement label, and making appropriate use of benefit–cost analysis to demonstrate the long-term advantages of a sustainable product or technique.

EXAMPLE 17.1 How Much Is That Light Bulb?!

As discussed earlier in this chapter, the up-front cost of CFL light bulbs is a disincentive to many consumers. However, these bulbs pay for themselves many times over in electricity savings. Life cycle costing (LCC) is one type of analysis that can be used to compare the economic costs of different purchasing decisions. The choice between a CFL and an incandescent bulb readily illustrates the basic technique.

Life cycle costing sums all expenses associated with owning and operating a device over its lifetime. These costs include the initial purchase and installation price, money spent for maintenance and repair, replacement of components, costs of operation, and any income received for salvage or sale of used equipment at the end of its life cycle.

FIGURE 17.11

Source: Somchai Som/Shutterstock.com.

In this example, we will assume that our choice is between a 60-W incandescent light bulb and a 13-W CFL. These bulbs provide equivalent light output in lumens, so the CFL provides the same amount of lighting but uses 78% less energy. The cost of the CFL is $2.25, while the incandescent is $0.50. A CFL bulb will last for 10,000 hours of operation, but the incandescent will last only 1,000 hours. The electricity rate is $0.13/kWh.

The following table is a simple life cycle cost table of this problem. A life cycle cost analysis must specify the time period, known as the life cycle period, over which we will conduct our analysis. Because the CFL is a longer-lasting bulb, we will choose 10,000 hours as our life cycle period.

	INCANDESCENT	CFL
Initial purchase cost	$0.50	$2.25
Maintenance and repair	—	—
Replacement costs	$4.50	—
Energy use	$78.00	$16.90
Salvage or resale	—	—
Total life cycle cost	**$83.00**	**$19.15**

This table illustrates several basic life cycle costing elements. We see the initial purchase cost of the bulbs, and we note that there is no maintenance or repair required for light bulbs over their life cycle period (if a bulb breaks, we cannot fix it!). However, because the incandescent bulb burns out after 1,000 hours, we will have to replace it nine times to achieve the equivalent amount of lighting that our single, 10,000-hour CFL provides. Replacement costs are $4.50 (9 bulbs × $0.50 per bulb).

Energy use is where our CFL bulb really starts to shine. The wattage difference adds up to dramatically higher operating costs for the incandescent bulb:

$$60 \text{ W} \times 10,000 \text{ hours} \times \frac{\$0.13}{\text{kWh}} \times \frac{1 \text{ kW}}{1,000 \text{ W}} = \$78.00 \qquad (17.1)$$

If we repeat the same calculation for the CFL, we see that total energy costs for the CFL are only $19.15. Finally, neither type of light bulb has a salvage or resale value because once they are burned out they have no useful life left and no material value as scrap.

When we add up the initial purchase cost, replacement expenses, and energy costs over the full 10,000-hour life cycle of the project, we see that the total life cycle cost of using an incandescent bulb is $83.00, whereas it is only $19.15 for the CFL. The CFL bulb creates a *net savings* (or net benefit) of $63.85 (the difference between $83.00 and $19.15).

We can use the same information to calculate a return-on-investment (ROI) ratio. ROI is generally calculated as

$$\text{ROI} = \frac{(\text{benefit from the investment} - \text{cost of the investment})}{\text{cost of the investment}} \qquad (17.2)$$

Using Equation (17.2) and the net savings already calculated, the ROI for our CFL is

$$\text{ROI} = \frac{(\$63.85 - \$2.25)}{\$2.25} = \$27.38$$

This result means that every $1 invested in our CFL yields a return of $27.38.

Unfortunately, we are still confronted by a major challenge when it comes to more effectively demonstrating—in economic terms—the value of sustainable design. That limitation lies in the difficulty of capturing the real environmental costs of pollution and environmental degradation in the prices of the goods and services that we use. If this were to happen, environmentally harmful products and services would ultimately be far more expensive than their more benign counterparts. This difficulty is associated with an economic concept known as externalities.

17.2.3 The Problem of Externalities

As you learned early on in this text, pollution and environmental degradation are the result of industrial activity, transportation, energy production, and so on. In economic terms, these harmful environmental impacts represent a ***negative externality***—a cost (or harm) experienced by someone other than the buyers and sellers of the good or service that resulted in the externality. For example, sulfur dioxide is a major emission from electric power production, and it contributes to respiratory disease and acid rain. The people experiencing the health effects and consequences of acid rain are often different from those buying electricity.

The price of electricity to the consumer simply reflects the costs to the utility of making the energy (including any pollution control measures) as well as a modest rate of profit. However, the social burden of externalities from electric power production is not captured in electricity rates: This price does not reflect the health expenses for people made sick by SO_2 or the economic losses to fisheries and forests because of acid rain. Negative externalities are a challenging problem because they are literally *outside of the market*. This means that they cannot be represented intrinsically through the dynamics of supply and demand that create prices in market transactions. If the social costs of negative externalities could be reflected in market prices, higher prices would be a disincentive for pollution. This, in turn, would create opportunities for correcting the problem. To continue our example, if externalities could be effectively captured in price, renewable energy would be much more cost-competitive with fossil fuel electricity because the price of electricity generated by fossil fuels would be considerably higher if social health costs and economic losses to others were included.

Because economic markets cannot fix the problem of negative externalities, they are one of several types of problems referred to as a ***market failure***. Government regulations are often used to protect society from negative externalities and other kinds of market failures. Environmental protection laws that limit pollutants are a way of reducing contaminants to less risky levels. For example, in the 1970s, the U.S. Clean Air Act limited SO_2 emissions from new electric power plants to 1.2 pounds of SO_2 per one million Btu of fuel burned. This type of emission standard is referred to as ***command and control regulation*** because it represents laws that control industrial behavior through statements about what is permissible activity and what is illegal. Command and control regulation is accomplished through a variety of policy tools and mechanisms, including permits, licenses, fees, emission standards, technology standards, and penalties for noncompliance. Today, most laws governing industrial air and water pollution are command and control regulations.

Market-based incentives are an alternative to command and control regulation and are a relatively new economic mechanism for mitigating environmental and public health externalities. As just mentioned, U.S. command and control regulations for sulfur dioxide required new power plants to limit their emissions to 1.2 pounds of CO_2 per one million Btu of fuel burned. In 1990, the Clean Air Act replaced this regulation with a ***cap-and-trade*** market mechanism. The U.S. Acid Rain Program is now widely regarded as the most successful market-based pollution control mechanism in the world (see Box 17.6).

Cap-and-trade systems can be successful only under particular circumstances—they are not necessarily effective for all pollutants. These systems must have an actual cap that is either permanent or tightened over time; allowance trading mechanisms must be in place and easily used with clear rules; and measurement and verification methods must be robust and strictly enforced. The concept has been extended to other pollutants. The European Union inaugurated the European Union Emissions Trading System in 2005 with a carbon cap-and-trade system designed to reduce greenhouse gases by 20% in 15 years. The United States does not have a governmental cap-and-trade system for carbon, and the one voluntary scheme initiated by the private sector, the Chicago Climate Exchange, failed in 2010.

Instead, markets for ***renewable energy credits (RECs)*** are emerging in the United States as a viable alternative for carbon mitigation. In these markets, 1 REC represents 1 megawatt of electric power generated by a clean energy source. RECs are used by utilities to demonstrate the amount of electricity sourced from renewable energy, and they are also used by consumers who want to purchase green electric power. As with cap-and-trade systems, REC markets help provide a price signal regarding the desirability of clean energy and enable it to literally be bought and sold.

In sum, environmental externalities reflect the indirect social costs and harms created by economic activity, and are outside the ability of private markets to correct through the classic price dynamics of supply and demand. As a consequence, government policy is required to protect society and the environment, and this

BOX 17.6 The U.S. Sulfur Dioxide Cap-and-Trade Program

The U.S. Acid Rain Program uses an innovative market-based system to reduce pollution in a manner that is as economically efficient as possible. The cap-and-trade process works by establishing a permanent, maximum yearly limit on total SO_2 emissions for the United States, referred to as a "cap." SO_2 polluters, mainly electric power plants, are then given a number of "allowances," where one allowance is equal to 1 ton of emitted SO_2. The total number of allowances is equivalent to the national cap on SO_2. At the end of the year, the power plants must produce an allowance for each ton of SO_2 that they emit. If a power plant emits more SO_2 than it has allowances for, it is penalized with a hefty fine.

What makes this a market-based incentive is that the allowances are *tradeable*—they can be bought and sold. If a polluter emits less SO_2 than it has allowances for, it can sell its extra allowances or "bank" them for the future (Figure 17.12). Polluters that discharge more SO_2 than they have allowances for must purchase allowances to make up the difference. Over time, it will be more cost effective for a polluter to begin reducing emissions than to pay for needed allowances. Today, U.S. SO_2 emissions are 40% lower than they were in 1980.

Capping and Trading Emissions: The Concept

Before the Program

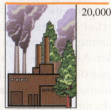 

With no reductions required, Unit 1 and Unit 2 each emit 20,000 tons a year.

The "CAP"

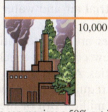 

The cap requires a 50% cut in emissions—e.g., from 20,000 to 10,000 tons.

Emissions trading under the CAP

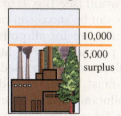 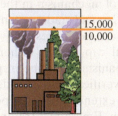

If Unit 1 can efficiently reduce 15,000 tons of emissions and Unit 2 can only efficiently reduce 5,000 tons, trading allows each unit to act optimally while ensuring achievement of the overall environmental goal. Unit 1 can hold on to (and "bank") its excess allowances or can sell them to Unit 2, whereas Unit 2 must acquire allowances from Unit 1 or from another source in the program.

FIGURE 17.12 Cap-and-trade basics. This graphic shows how the U.S. SO_2 cap-and-trade system works to reduce emissions through market mechanisms. U.S. SO_2 emissions declined by 64% after the onset of this program; by 2009, actual SO_2 emission levels for all sources regulated under the Acid Rain Program were well below the national limit set by the EPA.

Source: Based on U.S. EPA (2002, May). *Clearing the Air: Facts About Capping and Trading Emissions.*

protection usually takes the form of command and control regulation. New efforts to develop market-based incentives through cap-and-trade programs have been shown to be successful and are useful because they allow businesses to mitigate pollution in a manner that is most economically effective for themselves. When rigorous caps are set and the rules of cap-and-trade are enforced, the desired level of environmental quality is achieved through the most efficient economic means. Many environmental regulations dealing with externalities have profound implications for engineering design, because the technological systems and processes resulting in the externalities must comply with government standards or target limits. As businesses place more emphasis on pollution prevention (rather than control), greater opportunities arise to practice the methods of industrial ecology and life cycle analysis described in Chapters 13 and 14.

17.2.4 The Difficulty of Environmental Valuation

Environmental sustainability challenges traditional economic approaches in yet another way, which is how we monetarily value and account for the impacts of our actions on natural resources and a clean environment. We see this with the emergence of the ecosystem services concept (see Chapter 5) as well as sustainable development's "three pillars" of society–environment–economy.

Several important changes in economic culture are taking place throughout the world today in order to accommodate the growing emphasis on *social and environmental accounting* by business enterprises. Social and environmental accounting is a formal assessment of how business activity benefits people and the environment. In business, two terms are used somewhat synonymously to reflect social and environmental accounting: *corporate social responsibility* and *triple bottom line*. The British visionary John Elkington has been an advocate of corporate social responsibility for over three decades; he also coined the term *triple bottom line*. Both notions stress the idea that business accountability to investors and society involves more than just the traditional financial reporting of profits and losses. Firms also build (or destroy) social and environmental capital, which they can and should account for as standard practice through their corporate reporting systems. In this manner, businesses can more realistically reflect how their operations holistically affect social, environmental, and economic conditions. The triple bottom line thus represents business outcomes in which "people, planet, and profit" benefit simultaneously.

Social and environmental accounting requires new methods, performance measures, and metrics. The ecological footprint, which we discussed in Chapter 1, is one such indicator, as is the United Nations' System of Integrated Environmental and Economic Accounting for a country's macroeconomy. Most initiatives to create social and environmental accounting techniques are global in nature because of the scope of international trade and commerce and concerns about social justice. Businesses around the world—especially transnational corporations— should ideally use common terminology and methods for assessing the social and environmental outcomes of their business practices. Examples include the Global Reporting Initiative standardized sustainability measurement reporting system and the ISO 14000 and ISO 26000 standards from the International Organization for Standardization (ISO). As yet, however, there are no standard, universally adopted protocols.

As with social and environmental accounting, valuation of ecosystem services challenges traditional economic/financial concepts and methods. Because most ecosystem services are not directly bought and sold, there is no market and no price for them. Assigning economic value becomes a complex effort and is based on methods of *nonmarket valuation* (assigning monetary value that is not based on the price of actually buying and selling the good or service in question). With individual services such as purified water, valuation is relatively straightforward: It is estimated from the cost of operating water treatment and supply facilities and/or the supply of potable water. The process becomes more difficult once we allow for the complexity of ecological systems, such as the role of healthy forests in preventing soil erosion or water contamination.

ACTIVE LEARNING EXERCISE 17.2 Who's Green, and Why?

A number of media organizations regularly issue lists of green companies. *Newsweek* magazine in particular provides an annual ranking of American and global companies that achieve high ratings with respect to their environmental impacts and social responsibility. Because the ability to be green varies by type of industry, *Newsweek* provides ratings for different industrial and service sectors.

Following is a list of *Newsweek's* most highly ranked green businesses in each of these 19 sectors for 2020. Select one of these companies and investigate why it is considered exceptional within its class. The company's annual reports and corporate website, online articles, and the *Newsweek* ratings (see *www.newsweek.com/americas-most-responsible -companies-2020*) should get you all of the information you need. Is your company ISO 14000 compliant? Has it won awards? What does it actually *do* to deserve its rating? Once you have completed your work, prepare a brief digital presentation summarizing your findings and analysis.

- Hewlett Packard
- Cisco

- Dell
- Intel
- Microsoft
- NVIDIA
- Citigroup
- General Mills
- Comerica
- Jones Lang LaSalle
- Lam Research
- American Express
- Amgen
- Proctor & Gamble
- VMware
- PVH
- Cummins
- Micron Technology
- MetLife
- Merck & Co
- Target
- Emerson Electric
- Huntington Bancshares
- Kellogg's
- Regeneron Pharmaceuticals

17.3 The Role of Government

Governments have a significant role to play in creating opportunities for sustainable development or removing the obstacles and barriers that exist. However, the role of government in fostering sustainability varies worldwide, simply because each nation defines the proper and legitimate role of government based on its own society, culture, and needs. As a consequence, the scope and range of governmental behavior are considerable as we reflect on the breadth of global well-being and circumstance.

In low-income nations, governmental policies and actions (often in partnership with nongovernmental organizations) may focus on achieving several of the sustainable development goals of the United Nations. For nations that have many people struggling to meet basic needs, sustainable development is a matter of meeting these needs while preserving and protecting the environment. As discussed earlier in this chapter, the role of appropriate technology on appropriate scales is critical in providing for human material quality of life. Pollution prevention and control regulations often are not as critical as those for education, public health, water supplies, and economic growth. Natural resource conservation policies related to forestry and agriculture are also areas of policymaking common in low-income nations, and they are usually designed to prevent environmental degradation.

In high-income nations, public policy has historically focused on environmental protection and natural resource conservation. Clean air and water laws are well over 60 years old in the United States and parts of Europe; sewage and sanitation regulations date back 100 years or more. Forestry and agriculture policies are complex and vary considerably by nation. Even though Europe and the United States share a common history in terms of their environmental movements of the 20th century, after the 1973 oil embargo and oil shock of the 1970s, their policies began to diverge significantly. Europe's policy evolution—as reflected by many individual nations as well as the EU—has moved decidedly toward laws and guidance that promote sustainable development. The EU is therefore considerably more aggressive about carbon and climate change mitigation; many European nations also regulate land use and urban growth more proactively than the United States.

In almost all instances, government policies are intended to correct market failures of one kind or another. Market failures contribute not just to negative externalities, but also to underinvestment in a range of socially and economically desirable activities such as R&D, technological innovation, and more energy-efficient products. Governments therefore have a role to play not just for traditional environmental protection, but also for stimulating both innovation and the social adoption of more sustainable products, processes, and behaviors. The tools and techniques that government laws and policies commonly use to stimulate both innovation and adoption include the following:

- *Tax incentives*—Tax credits can be used to offset the higher cost of more sustainable technologies, such as solar and wind energy systems for homes and business. R&D tax credits facilitate innovation in the corporate sector.
- *Standards*—Performance standards are widely used to achieve higher levels of energy efficiency in products, machinery, and equipment. Examples include fuel economy and carbon emission standards for automobiles, and energy efficiency standards for heating and cooling equipment. We also include building energy codes in this category.
- *Mandatory best practice requirements*—Government regulations can mandate that best practices be used for pollution prevention and control, as well as other forms of environmental conservation. Examples include state and local governments that mandate low-impact development BMPs for stormwater management.
- *Research and development*—Government can itself invest in and conduct R&D through national research laboratories, or it can give grants to universities and private companies to develop innovative new products and processes.

In general, governments in Western, capitalist societies are reluctant to force people to behave in certain ways or to directly manipulate market prices. As a consequence, many public policies work indirectly to provide incentives for changing the social and economic conditions that can contribute to sustainable development.

17.4 Social Justice and Sustainability in Wealthy Countries

The concept of sustainable development explicitly stresses equity, quality of life, and social justice as part of our global transformation toward environmental stewardship. Much of our textbook has focused on and given examples for low-income countries—for obvious reasons.

Wealthy industrialized nations confront challenges to social justice and sustainability as well. In the United States particularly, the underprivileged and minority communities may live at the margins of the "sustainability movement" in a variety of ways. An affordable, "green" quality of life is beyond the reach of many Americans, 13.7% of whom lived in poverty in 2021 (Giannarelli, Wheaton, and Shantz, 2021). Low-income families must often live in substandard housing that potentially presents many environmental hazards, including lead paint, poor indoor air quality, insect and rodent infestations, and contaminated water from aging plumbing systems. Children are especially vulnerable to the environmental conditions of poverty; the incidence of asthma among impoverished children is notably higher than among the general population, as is asthma among low-income African Americans (National Center for Health Statistics, 2012). These higher rates of disease reflect a portfolio of causes, including cockroaches, inner city or industrial neighborhood air quality, mold, mice, and poor air ventilation. The two U.S. zipcodes with the highest number of children living in poverty had asthma incidence rates over 40% (NBC News, 2014).

High household energy costs due to old appliances, inadequate heating equipment, and poorly insulated buildings are also common. Higher proportions of disposable income may consequently go to paying utility bills; the federal government provided nearly $3 billion in assistance through the Low Income Home Energy Assistance Program in 2013 to help families meet their basic energy needs. Very simply, low-income families in the United States are commonly subject to some of the oldest or most poorly maintained housing stock in America and have little influence over their landlords. The green building innovations explored in Chapter 16 and the energy-saving measures noted in Chapter 11 are far beyond the means of many Americans. In addition, developers are unlikely to build more sustainable low-income housing simply because they cannot recover the higher construction costs through rent.

Environmental justice is also a compelling issue in the United States, and it reflects a type of *spatial discrimination*. Many low-income, minority, and tribal communities face environmental health risks such as contaminated drinking water and exposure to toxic chemicals, or they confront environmental nuisances such as noxious odors from landfills and factories. In Appalachia, mountaintop removal—a technique used in coal mining—subjects communities to air pollution, degraded water resources, and constant dynamite explosions. The nature of such discrimination is complex. These communities often lack the political power to resist siting unwanted land uses or to compel environmental clean-up. In other instances, people must live in areas with environmental degradation and contamination simply because it is more affordable.

U.S. laws help only to a limited extent. The National Environmental Policy Act (NEPA) requires the federal government to address environmental justice considerations for the actions of its own agencies. NEPA does not apply to private

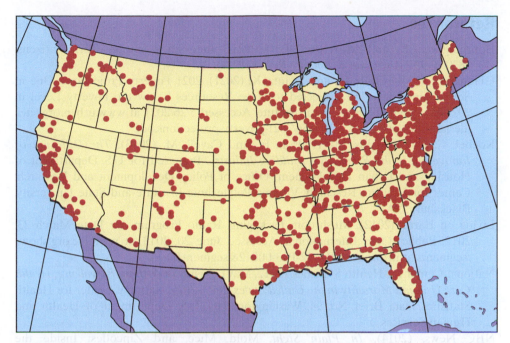

FIGURE 17.13 Location of National Priorities List sites for U.S. Superfund remediation. This map shows the location of hazardous waste sites that qualify for the U.S. Superfund. It can take years for remediation to begin, and many more communities technically qualify for Superfund than are actually on the National Priorities List.

Source: Based on National Atlas, nationalatlas.gov.

organizations, however, and it is often criticized for being weakly enforced at the federal level. The U.S. Comprehensive Environmental Response, Compensation, and Liability Act of 1980, commonly known as Superfund, provides funding to clean up the most contaminated sites in the United States; there are over 1,000 sites on the National Priorities List (Figure 17.13). The Superfund list is not exhaustive, however; many other communities technically qualify but are not put on the priority list for remediation. With respect to low-income housing, state and local public housing authorities are often large landlords, but struggle to upgrade the quality of housing because of budgetary constraints. Overall, weak enforcement of existing laws as well as limited policy protection in general characterize the politics of environmental justice. Lack of affordability and access to legal representation likewise prohibits communities from seeking and receiving help through the legal system.

17.5 Summary

Achieving sustainable development requires that we more widely adopt and diffuse innovations that enhance sustainability. Yet there are many obstacles to doing so. By understanding the dynamics of the adoption and diffusion process, we can engage in better engineering design to minimize them as much as possible. In addition, a variety of economic conditions act as barriers to the adoption of green innovations. These conditions can be addressed with a variety of techniques, including endorsement labeling, cost–benefit analysis, implementation of corporate social responsibility and environmental accounting systems, and selective use of governmental policy tools.

References

Bourguignon, F., and Morrisson, C. (2002). "Inequality among world citizens: 1820–1992." *American Economic Review* 92(4): 727–744.

Giannarelli, L., Wheaton, L., and Shantz, K. (2021). 2021 Poverty Projections: One in Seven Americans Are Projected to Have Resources below the Poverty Level in 2021. Washington, DC: Urban Institute. Accessed 5/26/2021 at www.urban.org/sites /default/files/publication/103656/2021-poverty-projections.pdf.

Koebel, C. T., Papadakis, M., Hudson, E., and Cavell, M. (2003). *The Diffusion of Innovation in the Residential Building Industry*. Prepared for U.S. Department of Housing and Urban Development, Office of Policy Development and Research; Center for Housing Research, Virginia Polytechnic Institute and State University, Blacksburg, VA.

Living on Earth. (2012). Mushroom Packaging. Radio Program Air Date March 23, 2012, Public Radio International. Accessed June 20, 2013, at www.loe.org/shows /segments.html?programID=12-P13-00012&segmentID=4.

National Center for Health Statistics. (2012, May). *Trends in Asthma Prevalence, Health Care Use, and Mortality in the United States, 2001–2010*. National Center for Health Statistics Data Brief No. 94. Washington, DC: U.S. Department of Health and Human Services.

NBC News. (2014). *In Plain Sight*. Mold, Mice, and Zipcodes: Inside the Childhood Asthma Epidemic. Accessed April 6, 2014, at inplainsight.nbcnews .com/_news/2014/01/03/22149240-mold-mice-and-zip-codes-inside-the-childhood -asthma-epidemic?lite.

Philibert, C. (2006). *Barriers to Technology Diffusion: The Case of Solar Thermal Technologies*. Paris: OECD.

Rogers, E. M. (1963). *Diffusion of Innovations*, 3rd ed. New York: Free Press.

World Bank. (1975). Issues in Village Water Supply. Public Utilities Department, Report No. 793.

World Bank. (2021). PovcalNet: An online analysis tool for global poverty monitoring. Updated March 16, 2021. Accessed June 9, 2021 at iresearch.worldbank.org /PovcalNet/.

Key Concepts

Social movement	Market failure
Adoption and diffusion of innovations	Command and control regulation
Green premium	Market-based incentives
Social costs	Cap-and-trade program
Opportunity cost	Renewable energy credits
Endorsement labeling	Social and environmental accounting
Greenwashing	Corporate social responsibility
Payback period	Triple bottom line
Information asymmetries	Nonmarket valuation
Negative externality	Environmental justice

Problems

17-1 Identify three different, *real* organizations that have made sustainability an explicit consideration in their operations.

17-2 The author Thomas Friedman is credited with saying that "pessimists are usually right and optimists are usually wrong but all the great changes have been accomplished by optimists." Do you consider yourself an optimist or a pessimist about society's ability to accomplish change that leads to more sustainable development? Why?

17-3 Explain the concept of an opportunity cost and how that affects the decision by individuals, organizations, and businesses to buy (or invest in) more sustainable products and practices.

17-4 What are the four ways that we can offset opportunity costs for "green" decision making?

17-5 List five benefits of endorsement labeling/green certification.

17-6 Identify and explain three reasons people living in poverty don't necessarily use more sustainable products that are charitably given to them.

17-7 Why are environmental externalities so difficult to remedy using traditional market dynamics and financial criteria?

17-8 Give an example of a specific, real government policy for each of the following that promotes the adoption and diffusion of more sustainable products and techniques:
   a. Tax incentives
   b. Standards
   c. Best management practices
   d. Research and development

17-9 Explain three reasons a company might not select an investment that would improve its environmental performance and save it money in the long run.

17-10 Voluntary compliance with third-party standards is an increasingly common way that companies are creating social and environmental accountability. Why are these standards usually developed for management systems and not specific products?

17-11 Refer to Chapter 7 and the discussion of ceramic water filters in Benin (see Tables 7.4 to 7.6). Develop two different cost-benefit analyses for a household as follows:
   a. Construct a life cycle cost table comparing the ceramic water filter to purchased Pur™ filtered water. What is the total life cycle cost of each option, and what is the return-on-investment ratio for the ceramic filter?
   b. What is the cost benefit of each system expressed as the cost per % reduction in the coliform bacteria load?

Team-Based Problem Solving

17-12 Below is a list of technical innovations (products, methods, or techniques) that enhance sustainability. Select one to research and analyze with respect to (a) how it might balance the three pillars of sustainable development and (b) the barriers and obstacles that it confronts with respect to its adoption and diffusion.
   • Hydrogen fuel cells
   • Biomimetic design
   • Solar ovens
   • Combined heat and power systems
   • Soy-based building insulation
   • Biodegradable plastic

- Green roofs
- Ceramic water filters for low-income countries
- Solar walls
- Wind turbines rated at 5 kW or less
- Biodiesel

17-13 Many efforts have been made to apply the success of the U.S. cap-and-trade program for sulfur dioxide to other environmental problems. Below are four efforts that have been introduced as an actual cap-and-trade system or as a credit trading scheme. Select one, analyze its social, economic, and environmental contexts, and critique the basic principles of the trading or credit scheme designed to mitigate this problem. If you do not think a trading program will be successful for the activity you selected, explain why and defend your position.

- Nutrient trading (in watersheds)
- Carbon emissions
- Mitigation banking (of wetlands)
- Regional NO_x trading

17-14 You work for a company that wants to use the Global Reporting Initiative's G4 Standard to develop its first Sustainability Report. It has assigned you to a team that will summarize the criteria and the process, and develop a 12-month project timeline in which the report can be completed. Prepare a 15-minute presentation of your team's summary and recommended project plan.

17-15 Consumer electronics and electronic waste are emerging as an extraordinary problem in industrialized countries, especially the United States. Reflect on concepts related to the adoption and diffusion of innovation from this chapter. Develop a solution to the e-waste problem in your state that incorporates all of these insights as well as government policies to overcome market failures.

17-16 Identify the active Superfund sites in your state. Explore and analyze the original source of contamination of these sites, and explain the nature of the contamination, hazard, and risk confronting communities located there. Conduct a socioeconomic and demographic analysis of these sites across the state. What are the environmental justice issues associated with them?

Conversion Factors

CONVERSIONS BETWEEN U.S. CUSTOMARY UNITS AND SI UNITS

U.S. Customary unit		Times conversion factor		Equals SI unit	
		Accurate	Practical		
Acceleration (linear)					
foot per second squared	ft/s^2	0.3048*	0.305	meter per second squared	m/s^2
inch per second squared	$in./s^2$	0.0254*	0.0254	meter per second squared	m/s^2
Area					
circular mil	cmil	0.0005067	0.0005	square millimeter	mm^2
square foot	ft^2	0.09290304*	0.0929	square meter	m^2
square inch	$in.^2$	645.16*	645	square millimeter	mm^2
Density (mass)					
slug per cubic foot	$slug/ft^3$	515.379	515	kilogram per cubic meter	kg/m^3
Density (weight)					
pound per cubic foot	lb/ft^3	157.087	157	newton per cubic meter	N/m^3
pound per cubic inch	$lb/in.^3$	271.447	271	kilonewton per cubic meter	kN/m^3
Energy; work					
foot-pound	ft-lb	1.35582	1.36	joule (N·m)	J
inch-pound	in.-lb	0.112985	0.113	joule	J
kilowatt-hour	kWh	3.6*	3.6	megajoule	MJ
British thermal unit	Btu	1,055.06	1,055	joule	J
Force					
pound	lb	4.44822	4.45	newton ($kg·m/s^2$)	N
kip (1,000 pounds)	k	4.44822	4.45	kilonewton	kN
Force per unit length					
pound per foot	lb/ft	14.5939	14.6	newton per meter	N/m
pound per inch	lb/in.	175.127	175	newton per meter	N/m
kip per foot	k/ft	14.5939	14.6	kilonewton per meter	kN/m
kip per inch	k/in.	175.127	175	kilonewton per meter	kN/m
Length					
foot	ft	0.3048*	0.305	meter	m
inch	in.	25.4*	25.4	millimeter	mm
mile	mi	1.609344*	1.61	kilometer	km
Mass					
slug	$lb-s^2/ft$	14.5939	14.6	kilogram	kg
Moment of a force; torque					
pound-foot	lb-ft	1.35582	1.36	newton meter	N·m
pound-inch	lb-in.	0.112985	0.113	newton meter	N·m
kip-foot	k-ft	1.35582	1.36	kilonewton meter	kN·m
kip-inch	k-in.	0.112985	0.113	kilonewton meter	kN·m

CONVERSIONS BETWEEN U.S. CUSTOMARY UNITS AND SI UNITS (Continued)

U.S. Customary unit		Times conversion factor		Equals SI unit	
		Accurate	**Practical**		
Moment of inertia (area)					
inch to fourth power	in.^{4}	416,231	416,000	millimeter to fourth power	mm^4
inch to fourth power	in.^{4}	0.416231×10^{-6}	0.416×10^{-6}	meter to fourth power	m^4
Moment of inertia (mass)					
slug foot squared	slug-ft^2	1.35582	1.36	kilogram meter squared	kg·m^2
Power					
foot-pound per second	ft-lb/s	1.35582	1.36	watt (J/s or N·m/s)	W
foot-pound per minute	ft-lb/min	0.0225970	0.0226	watt	W
horsepower (550 ft-lb/s)	hp	745.701	746	watt	W
Pressure; stress					
pound per square foot	psf	47.8803	47.9	pascal (N/m^2)	Pa
pound per square inch	psi	6,894.76	6,890	pascal	Pa
kip per square foot	ksf	47.8803	47.9	kilopascal	kPa
kip per square inch	ksi	6.89476	6.89	megapascal	MPa
Section modulus					
inch to third power	in.^{3}	16,387.1	16,400	millimeter to third power	mm^3
inch to third power	in.^{3}	16.3871×10^{-6}	16.4×10^{-6}	meter to third power	m^3
Velocity (linear)					
foot per second	ft/s	0.3048*	0.305	meter per second	m/s
inch per second	in./s	0.0254*	0.0254	meter per second	m/s
mile per hour	mph	0.44704*	0.447	meter per second	m/s
mile per hour	mph	1.609344*	1.61	kilometer per hour	km/h
Volume					
cubic foot	ft^3	0.0283168	0.0283	cubic meter	m^3
cubic inch	in.^{3}	16.3871×10^{-6}	16.4×10^{-6}	cubic meter	m^3
cubic inch	in.^{3}	16.3871	16.4	cubic centimeter (cc)	cm^3
gallon (231 in.^{3})	gal.	3.78541	3.79	liter	L
gallon (231 in.^{3})	gal.	0.00378541	0.00379	cubic meter	m^3

*An asterisk denotes an *exact* conversion factor

Note: To convert from SI units to USCS units, *divide* by the conversion factor

Temperature Conversion Formulas

$$T(°C) = \frac{5}{9}[T(°F) - 32] = T(K) - 273.15$$

$$T(K) = \frac{5}{9}[T(°F) - 32] + 273.15 = T(°C) + 273.15$$

$$T(°F) = \frac{9}{5}T(°C) + 32 = \frac{9}{5}T(K) - 459.67$$

PRINCIPAL UNITS USED IN MECHANICS

Quantity	International System (SI)			U.S. Customary System (USCS)		
	Unit	Symbol	Formula	Unit	Symbol	Formula
Acceleration (angular)	radian per second squared		rad/s^2	radian per second squared		rad/s^2
Acceleration (linear)	meter per second squared		m/s^2	foot per second squared		ft/s^2
Area	square meter		m^2	square foot		ft^2
Density (mass) (Specific mass)	kilogram per cubic meter		kg/m^3	slug per cubic foot		$slug/ft^3$
Density (weight) (Specific weight)	newton per cubic meter		N/m^3	pound per cubic foot	pcf	lb/ft^3
Energy; work	joule	J	$N \cdot m$	foot-pound		ft-lb
Force	newton	N	$kg \cdot m/s^2$	pound	lb	(base unit)
Force per unit length (Intensity of force)	newton per meter		N/m	pound per foot		lb/ft
Frequency	hertz	Hz	s^{-1}	hertz	Hz	s^{-1}
Length	meter	m	(base unit)	foot	ft	(base unit)
Mass	kilogram	kg	(base unit)	slug		$lb\text{-}s^2/ft$
Moment of a force; torque	newton meter		$N \cdot m$	pound-foot		lb-ft
Moment of inertia (area)	meter to fourth power		m^4	inch to fourth power		$in.^4$
Moment of inertia (mass)	kilogram meter squared		$kg \cdot m^2$	slug foot squared		$slug\text{-}ft^2$
Power	watt	W	J/s $(N \cdot m/s)$	foot-pound per second		ft-lb/s
Pressure	pascal	Pa	N/m^2	pound per square foot	psf	lb/ft^2
Section modulus	meter to third power		m^3	inch to third power		$in.^3$
Stress	pascal	Pa	N/m^2	pound per square inch	psi	$lb/in.^2$
Time	second	s	(base unit)	second	s	(base unit)
Velocity (angular)	radian per second		rad/s	radian per second		rad/s
Velocity (linear)	meter per second		m/s	foot per second	fps	ft/s
Volume (liquids)	liter	L	$10^{-3} m^3$	gallon	gal.	$231 in.^3$
Volume (solids)	cubic meter		m^3	cubic foot	cf	ft^3

SELECTED PHYSICAL PROPERTIES

Property	SI	USCS
Water (fresh)		
weight density	9.81 kN/m³	62.4 lb/ft³
mass density	1,000 kg/m³	1.94 slugs/ft³
Sea water		
weight density	10.0 kN/m³	63.8 lb/ft³
mass density	1,020 kg/m³	1.98 slugs/ft³
Aluminum (structural alloys)		
weight density	28 kN/m³	175 lb/ft³
mass density	2,800 kg/m³	5.4 slugs/ft³
Steel		
weight density	77.0 kN/m³	490 lb/ft³
mass density	7,850 kg/m³	15.2 slugs/ft³
Reinforced concrete		
weight density	24 kN/m³	150 lb/ft³
mass density	2,400 kg/m³	4.7 slugs/ft³
Atmospheric pressure (sea level)		
Recommended value	101 kPa	14.7 psi
Standard international value	101.325 kPa	14.6959 psi
Acceleration of gravity (sea level, approx. 45° latitude)		
Recommended value	9.81 m/s²	32.2 ft/s²
Standard international value	9.80665 m/s²	32.1740 ft/s²

SI PREFIXES

Prefix	Symbol	Multiplication factor		
tera	T	10^{12}	=	1 000 000 000 000
giga	G	10^{9}	=	1 000 000 000
mega	M	10^{6}	=	1 000 000
kilo	k	10^{3}	=	1 000
hecto	h	10^{2}	=	100
deka	da	10^{1}	=	10
deci	d	10^{-1}	=	0.1
centi	c	10^{-2}	=	0.01
milli	m	10^{-3}	=	0.001
micro	μ	10^{-6}	=	0.000 001
nano	n	10^{-9}	=	0.000 000 001
pico	p	10^{-12}	=	0.000 000 000 001

Note: The use of the prefixes hecto, deka, deci, and centi is not recommended in SI.

Earth and Environmental Physical and Chemical Data

Mass = 5.9726×10^{24} kg
Volume = 108.321×10^{10} km^3
Equatorial radius = 6,378.1 km
Polar radius = 6,356.8 km
Volumetric mean radius = 6,371.0 km
Mean density = 5,514 kg/m^3
Surface gravity = 9.798 m/s^2
Bond albedo = 0.306
Visual geometric albedo = 0.367
Solar irradiance = 1,367.6 W/m^2
Black-body temperature = 254.3 K

Terrestrial Atmosphere

Surface atmospheric pressure = 1,014 mb
Surface atmospheric density = 1.217 kg/m^3
Atmospheric scale height = 8.5 km
Mass of atmosphere = 5.1×10^{18} kg
Mass of hydrosphere (oceans) = 1.4×10^{21} kg
Mass of stratosphere = 0.5×10^{18} kg
Mass of water in the atmosphere = 1.3×10^{16} kg
Average temperature = 288 K (15°C)
Diurnal temperature range = 283 K to 293 K (10 to 20°C)
Wind speeds = 0 to 100 m/s
Mean molecular weight of the dry atmosphere = 28.97 g/mole

Source: NASA.gov: Earth Fact Sheet

WATER SOURCE	WATER VOLUME, IN CUBIC MILES	WATER VOLUME, IN CUBIC KILOMETERS	PERCENT OF FRESH WATER	PERCENT OF TOTAL WATER
Oceans, seas, & bays	321,000,000	1,338,000,000	–	96.5
Ice caps, glaciers, & permanent snow	5,773,000	24,064,000	68.7	1.74
Groundwater	5,614,000	23,400,000	–	1.69
Fresh	2,526,000	10,530,000	30.1	0.76
Saline	3,088,000	12,870,000	–	0.93
Soil moisture	3,959	16,500	0.05	0.001
Ground ice & permafrost	71,970	300,000	0.86	0.022

(Continued)

WATER SOURCE	WATER VOLUME, IN CUBIC MILES	WATER VOLUME, IN CUBIC KILOMETERS	PERCENT OF FRESH WATER	PERCENT OF TOTAL WATER
Lakes	42,320	176,400	–	0.013
Fresh	21,830	91,000	0.26	0.007
Saline	20,490	85,400	–	0.006
Atmosphere	3,095	12,900	0.04	0.001
Swamp water	2,752	11,470	0.03	0.0008
Rivers	509	2,120	0.006	0.0002
Biological water	269	1,120	0.003	0.0001

Source: Based on Igor Shiklomanov's chapter "World fresh water resources" in Peter H. Gleick (editor), 1993, *Water in Crisis: A Guide to the World's Fresh Water Resources* (Oxford University Press, New York).

PHYSICAL CONSTANTS FOR DRY AIR AT STP*

CONSTANT	VALUE
Average molecular weight	28.96
Specific heat	
At constant pressure	1,004.2 J/kg °C
At constant volume	719.6 J/kg °C
Density	1.293 kg/m^3
Viscosity	1.72 x 10^{-4} poise
Coefficient of heat conductivity	0.0209 W/m °C
Speed of sound in air	331.4 m/sec

*Standard temperature and pressure, denoted STP, is a temperature of 0°C and a pressure of 1 atm.

AREA, BIOMASS, AND PRODUCTIVITY OF ECOSYSTEM TYPES

ECOSYSTEM TYPE*	AREA (10^{12} m^2)	MEAN PLANT BIOMASS [kg(C)/m^2]	AVERAGE NET PRIMARY PRODUCTIVITY [kg(C)/m^2/yr]
Tropical forests	24.5	18.8	0.83
Temperate forests	12.0	14.6	0.56
Boreal forests	12.0	9.0	0.36
Woodland and scrubland	8.0	2.7	0.27
Savanna	15.0	1.8	0.32
Grassland	9.0	0.7	0.23
Tundra and alpine meadow	8.0	0.3	0.065
Desert scrub	18.0	0.3	0.032
Rock, ice, and sand	24.0	0.01	0.015

(Continued)

ECOSYSTEM TYPE*	AREA (10^{12} m²)	MEAN PLANT BIOMASS [kg(C)/m²]	AVERAGE NET PRIMARY PRODUCTIVITY [kg(C)/m²/yr]
Cultivated land	14.0	0.5	0.29
Swamp and marsh	2.0	6.8	1.13
Lake and stream	2.5	0.01	0.23
Open ocean	332.0	0.0014	0.057
Upwelling zones	0.4	0.01	0.23
Continental shelf	26.6	0.005	0.16
Algae bed and reef	0.6	0.9	0.90
Estuaries	1.4	0.45	0.81

*For a description of each of the major types of ecosystems (deserts, boreal forests, estuaries, etc.) see Whittaker (1970), Whittaker and Likens (1973), and Ehrlich et al. (1977).

Carbon Sources and Equivalence

TABLE C.1 Physical properties of CO_2

PROPERTY	VALUE
Molecular weight	44.01
Critical temperature	31.3°C
Critical pressure	73.9 bar
Critical density	467 kg/m^{-3}
Triple point temperature	−56.5°C
Triple point pressure	5.18 bar
Boiling (sublimation) point (1.013 bar)	−78.5°C
Gas Phase	
Gas density (1.013 bar at boiling point)	2.814 kg/m^{-3}
Gas density (@ STP)*	1.976 kg/m^{-3}
Specific volume (@ STP)*	0.506 m^3/kg^{-1}
Cp (@ STP)*	0.0364 kJ (mol^{-1} K^{-1})
Cv (@ STP)*	0.0278 kJ (mol^{-1} K^{-1})
Cp/Cv (@ STP)*	1.308
Viscosity (@ STP)	13.72 μN·s m^{-2} (or μPa·s)
Thermal conductivity (@ STP)*	14.65 mW (mK^{-1})
Solubility in water (@ STP)*	1.716 vol vol^{-1}
Enthalpy (@ STP)*	21.34 kJ mol^{-1}
Entropy (@ STP)*	117.2 J mol K^{-1}
Entropy of formation	213.8 J mol K^{-1}
Liquid Phase	
Vapor pressure (at 20°C)	58.8 bar
Liquid density (at −20°C and 19.7 bar)	1,032 kg/m^{-3}
Viscosity (@ STP)*	99 μN·s m^{-2} (or μPa·s)
Solid Phase	
Density of carbon dioxide snow at freezing point	1,562 kg/m^{-3}
Latent heat of vaporization (1.013 bar at sublimation point)	571.1 kJ/kg^{-1}

*Standard Temperature and Pressure, which is 0°C and 1.013 bar.

Source: Based on Air Liquid gas table; Kirk-Othmer (1985); NIST (2003).

TABLE C.2 Approximate equivalents and other definitions

TO CONVERT	INTO THE FOLLOWING UNITS	MULTIPLY BY
1 t C	t CO_2	3.667
1 t CO_2	m^3 CO_2 (at 1.013 bar and 15°C)	534
1 t crude oil	Bbl	7.33
1 t crude oil	m^3	1.165

TABLE C.3 Characterization of coals by rank (according to ASTM D388-92A)

CLASS GROUP	FIXED CARBON LIMITS (dmmf basis)[a] % Equal to or greater than	Less than	VOLATILE MATTER LIMITS (dmmf basis)[a] % Greater than	Equal to or less than	GROSS CALORIFIC VALUE LIMITS (mmmf basis)[b] MJ kg^{-1} Equal to or greater than	Less than	Agglomerating Character
Anthracite							Non-agglomerating
Meta-anthracite	98	–	–	2	–	–	
Anthracite	92	98	2	8	–	–	
Semi-anthracite	86	92	8	14	–	–	
Bituminous coal							Commonly agglomerating
Low volatile	78	86	14	22	–	–	
Medium volatile	69	78	22	31	–	–	
High volatile A	–	69	31	–	32.6[d]	–	
High volatile B	–	–	–	–	30.2[d]	32.6	
High volatile C	–	–	–	–	26.7	30.2	
					24.4	26.7	Agglomerating
Sub-bituminous coal							Non-agglomerating
A	–	–	–	–	24.4	26.7	
B	–	–	–	–	22.1	24.4	
C	–	–	–	–	19.3	22.1	
Lignite							
A	–	–	–	–	14.7	19.3	
B	–	–	–	–	–	14.7	

[a] Indicates dry-mineral-matter-free basis (dmmf).

[b] mmmf indicates moist mineral-matter-free basis; moist refers to coal containing its natural inherent moisture but not including visible water on the surface of the coal.

[c] If agglomerating, classified in the low volatile group of the bituminous class.

[d] Coals having 69% or more fixed carbon (dmmf) are classified according to fixed carbon, regardless of gross calorific value.

Source: Based on ASTMD388-92A.

TABLE C.4 Typical ultimate analysis of petroleum-based heating fuels

COMPOSITION %	NO. 1 FUEL OIL (41.5°API)[a]	NO. 2 FUEL OIL (33°API)[a]	NO. 4 FUEL OIL (23.3°API)[a]	LOW SULFUR, NO. 6 FUEL OIL (33°API)[a]	HIGH SULFUR, NO. 6 FUEL OIL (15.5°API)[a]	PETROLEUM COKE[b]
Carbon	86.4	87.3	86.47	87.26	84.67	98.5
Hydrogen	13.6	12.6	11.65	10.49	11.02	3.08
Oxygen	0.01	0.04	0.27	0.64	0.38	1.11
Nitrogen	0.003	0.006	0.24	0.28	0.18	1.71
Sulfur	0.09	0.22	1.35	0.84	3.97	4.00
Ash	<0.01	<0.01	0.02	0.04	0.02	0.50
C/H Ratio	6.35	6.93	7.42	8.31	7.62	29.05

[a] Degree API = $(141.5/s) - 131.5$; where s is the specific density at 15°C

[b] Reference; Kaantee et al. (2003)

TABLE C.5 Typical natural gas composition

COMPONENT	PIPELINE COMPOSITION USED IN ANALYSIS Mol % (dry)	TYPICAL RANGE OF WELLHEAD COMPONENTS (Mol %) LOW VALUE	HIGH VALUE
Carbon dioxide, CO_2	0.5	0	10
Nitrogen, N_2	1.1	0	15
Methane, CH_4	94.4	75	99
Ethane, C_2H_6	3.1	1	15
Propane	0.5	1	10
Isobutane	0.1	0	1
N-butane	0.1	0	2
Pentanes	0.2	0	1
Hydrogen sulphide	0.0004	0	30
Helium	0.0	0	5
Heat of combustion (LHV)	48.252 MJ·kg^{-1}	–	–
Heat of combustion (HHV)	53.463 MJ·kg^{-1}	–	–

TABLE C.6 Chemical analysis and properties of some biomass fuels

	PEAT	WOOD (SAW DUST)	CROP RESIDUES (SUGAR CANE BIOGAS)	MUNICIPAL SOLID WASTE	ENERGY CROPS (EUCALYPTUS)
Proximate Analysis					
Moisture	70–90	7.3	–	16–38	–
Ash	–	2.6	11.3	11–20	0.52
Volatile matter	45–75	76.2	–	67–78	–
Fixed carbon	–	13.9	14.9	6–12	16.9
Ultimate Analysis					
C	45–60	46.9	44.8	–	48.3
H	3.5–6.8	5.2	5.4	–	5.9
O	20–45	37.8	39.5	–	45.1
N	0.75–3	0.1	0.4	–	0.2
S	–	0.04	0.01	–	0.01
Heating Value MJ kg^{-1} (HHV)	**17–22**	**18.1**	**17.3**	**15.9–17.5**	**19.3**

Source: Based on Sami et al., Hower 2003.

TABLE C.7 Direct emissions of non-greenhouse gases from two examples of coal and natural gas plants based on best available control technology, burning specific fuels

EMISSIONS	COAL (SUPERCRITICAL PC WITH BEST AVAILABLE EMISSION CONTROLS)	NATURAL GAS (NGCC WITH SCR)
NOx, g GJ^{-1}	4–5	5
SOx, g GJ^{-1}	4.5–5	0.7
Particulates, g GJ^{-1}	2.4–2.8	2
Mercury, mg GJ^{-1}	0.3–0.5	N/A

Source: Based on Cameron, 2002.

TABLE C.8 Direct CO_2 emission factors for some examples of carbonaceous fuels

CARBONACEOUS FUEL	HEAT CONTENT (HHV) (MJ kg^{-1})	EMISSION FACTOR (GCO_2 MJ^{-1})
Coal		
Anthracite	26.2	96.8
Bituminous	27.8	87.3
Sub-bituminous	19.9	90.3
Lignite	14.9	91.6
Biofuel		
Wood (dry)	20.0	78.4
Natural Gas	(kJ m^{-3})	
	37.3	50
Petroleum Fuel	(MJ m^{-3})	
Distillate Fuel Oil (#1, 2, & 4)	38,650	68.6
Residual Fuel Oil (#5 & 6)	41,716	73.9
Kerosene	37,622	67.8
LPG (average for fuel use)	25,220	59.1
Motor Gasoline	–	69.3

IPCC (2005). *IPCC Special Report on Carbon Dioxide Capture and Storage.* Prepared by Working Group III of the Intergovernmental Panel on Climate Change [Metz, B., Davidson, O., de Coninck, H.C., Loas, M., and Meyer, L.A. (eds.)] Cambridge University Press, Cambridge, United Kingdom and New York, NY, USA. 442pp.

Source: Based on NIES, 2003.

APPENDIX D

Exposure Factors for Risk Assessments

Source: Based on U.S. EPA (2011). *Exposure Factors Handbook: 2011 Edition*. National Center for Environmental Assessment, Office of Research and Development, U.S. Environmental Protection Agency. Washington, D.C. U.S.A.

TABLE D.1 Summary of exposure factor recommendations

	PER CAPITA INGESTION OF DRINKING WATER				CONSUMERS-ONLY INGESTION OF DRINKING WATER			
	Mean		95th Percentile		Mean		95th Percentile	
	mL/day	mL/kg-day	mL/day	mL/kg-day	mL/day	mL/kg-day	mL/day	mL/kg-day
Children								
Birth to 1 month	184	52	839[a]	232[a]	470[a]	137[a]	858[a]	238[a]
1 to <3 months	227[a]	48	896[a]	205[a]	552	119	1,053[a]	285[a]
3 to <6 months	362[a]	52	1,056	159	556	80	1,171[a]	173[a]
6 to <12 months	360	41	1,055	126	467	53	1,147	129
1 to <2 years	271	23	837	71	308	27	893	75
2 to <3 years	317	23	877	60	356	26	912	62
3 to <6 years	327	18	959	51	382	21	999	52
6 to <11 years	414	14	1,316	43	511	17	1,404	47
11 to <16 years	520	10	1,821	32	637	12	1,976	35
16 to <18 years	573	9	1,783	28	702	10	1,883	30
18 to <21 years	681	9	2,368	35	816	11	2,818	36
Adults								
>21 years	1,043	13	2,958	40	1,227	16	3,092	42
>65 years	1,046	14	2,730	40	1,288	18	2,960	43
Pregnant women	819[a]	13[a]	2,503[a]	43[a]	872[a]	14[a]	2,589[a]	43[a]
Lactating women	1,379[a]	21[a]	3,434[a]	55[a]	1,665[a]	26[a]	3,588[a]	55[a]

[a] Estimates are less statistically reliable based on guidance published in the *Joint Policy on Variance Estimation and Statistical Reporting Standards on NHANES III and CSFII Reports: NHIS/NCHS Analytical Working Group Recommendations* (NCHS, 1993).

TABLE D.2

	INGESTION OF WATER WHILE SWIMMING			
	Mean		Upper Percentile	
	mL/event[a]	mL/hour	mL/event	mL/hour
Children	37	49	90[b]	120[b]
Adults	16	21	53[c]	71[c]

[a] Participants swam for 45 minutes.
[b] 97th percentile
[c] Based on maximum value.

TABLE D.3

MOUTHING FREQUENCY AND DURATION

	Hand-to-Mouth				Object-to-Mouth			
	Indoor Frequency		Outdoor Frequency		Indoor Frequency		Outdoor Frequency	
	Mean contacts/ hour	95th Percentile contacts/ hour	Mean contacts/ hour	95th Percentile contacts/ hour	Mean contacts/ hour	95th Percentile contacts/ hour	Mean contacts/ hour	95th Percentile contacts/ hour
Birth to 1 month	–	–	–	–	–	–	–	–
1 to <3 months	–	–	–	–	–	–	–	–
3 to <6 months	28	65	–	–	11	32	–	–
6 to <12 months	19	52	15	47	20	38	–	–
1 to <2 years	20	63	14	42	14	34	8.8	21
2 to <3 years	13	37	5	20	9.9	24	8.1	40
3 to <6 years	15	54	9	36	10	39	8.3	30
6 to <11 years	7	21	3	12	1.1	3.2	1.9	9.1
11 to <16 years	–	–	–	–	–	–	–	–
16 to <21 years	–	–	–	–	–	–	–	–

Object-to-Mouth

Duration

	Mean minute/hour	95th Percentile minute/hour
Birth to 1 month	–	–
1 to <3 months	–	–
3 to <6 months	11	26
6 to <12 months	9	19
1 to <2 years	7	22
2 to <3 years	10	11
3 to <6 years	–	–
6 to <11 years	–	–
11 to <16 years	–	–
16 to <21 years	–	–

- No data

TABLE D.4

SOIL AND DUST INGESTION

	Soil				Dust		Soil + Dust	
	General Population Central Tendency mg/day	High End			Central Tendency mg/day	General Population Upper Percentile mg/day	General Population Central Tendency mg/day	General Population Upper Percentile mg/day
		General Population Upper Percentile mg/day	Soil-Pica mg/day	Geophagy mg/day				
6 weeks to <1 year	30	–	–	–	30	–	60	–
1 to <6 years	50	–	1,000	50,000	60	–	100	–
3 to <6 years	–	200	–	–	–	100	–	200
6 to <21 years	50	–	1,000	50,000	60	–	100	–
Adult	20	–	–	50,000	30	–	50	–

- No data.

TABLE D.5

INHALATION

Long-Term Inhalation Rates		
	Mean m³/day	95th Percentile m³/day
Birth to 1 month	3.6	7.1
1 to <3 months	3.5	5.8
3 to <6 months	4.1	6.1
6 to <12 months	5.4	8.0
1 to <2 years	5.4	9.2
Birth to <1 year	8.0	12.8
2 to <3 years	8.9	13.7
3 to <6 years	10.1	13.8
6 to <11 years	12.0	16.6
11 to <16 years	15.2	21.9
16 to <21 years	16.3	24.6
21 to <31 years	15.7	21.3
31 to <41 years	16.0	21.4
41 to <51 years	16.0	21.2
51 to <61 years	15.7	21.3
61 to <71 years	14.2	18.1
71 to <81 years	12.9	16.6
≥81 years	12.2	15.7

TABLE D.6

Short-Term Inhalation Rates, by Activity Level

	Sleep or Nap		Sedentary/Passive		Light Intensity		Moderate Intensity		High Intensity	
	Mean m³/minute	95th m³/minute	Mean m³/minute	95th m³/minute	Mean m³/minute	95th m³/minute	Mean m³/minute	95th m³/minute	Mean m³/minute	95th m³/minute
Birth to <1 year	3.0E-03	4.6E-03	3.1E-03	4.7E-03	7.6E-03	1.1E-02	1.4E-02	2.2E-02	2.6E-02	4.1E-02
1 to <2 years	4.5E-03	6.4E-03	4.7E-03	6.5E-03	1.2E-02	1.6E-02	2.1E-02	2.9E-02	3.8E-02	5.2E-02
2 to <3 years	4.6E-03	6.4E-03	4.8E-03	6.5E-03	1.2E-02	1.6E-02	2.1E-02	2.9E-02	3.9E-02	5.3E-02
3 to <6 years	4.3E-03	5.8E-03	4.5E-03	5.8E-03	1.1E-02	1.4E-02	2.1E-02	2.7E-02	3.7E-02	4.8E-02
6 to <11 years	4.5E-03	6.3E-03	4.8E-03	6.4E-03	1.1E-02	1.5E-02	2.2E-02	2.9E-02	4.2E-02	5.9E-02
11 to <16 years	5.0E-03	7.4E-03	5.4E-03	7.5E-03	1.3E-02	1.7E-02	2.5E-02	3.4E-02	4.9E-02	7.0E-02
16 to <21 years	4.9E-03	7.1E-03	5.3E-03	7.2E-03	1.2E-02	1.6E-02	2.6E-02	3.7E-02	4.9E-02	7.3E-02
21 to <31 years	4.3E-03	6.5E-03	4.2E-03	6.5E-03	1.2E-02	1.6E-02	2.6E-02	3.8E-02	5.0E-02	7.6E-02
31 to <41 years	4.6E-03	6.6E-03	4.3E-03	6.6E-03	1.2E-02	1.6E-02	2.7E-02	3.7E-02	4.9E-02	7.2E-02
41 to <51 years	5.0E-03	7.1E-03	4.8E-03	7.0E-03	1.3E-02	1.6E-02	2.8E-02	3.9E-02	5.2E-02	7.6E-02
51 to <61 years	5.2E-03	7.5E-03	5.0E-03	7.3E-03	1.3E-02	1.7E-02	2.9E-02	4.0E-02	5.3E-02	7.8E-02
61 to <71 years	5.2E-03	7.2E-03	4.9E-03	7.3E-03	1.2E-02	1.6E-02	2.6E-02	3.4E-02	4.7E-02	6.6E-02
71 to <81 years	5.3E-03	7.2E-03	5.0E-03	7.2E-03	1.2E-02	1.5E-02	2.5E-02	3.2E-02	4.7E-02	6.5E-02
≥81 years	5.2E-03	7.0E-03	4.9E-03	7.0E-03	1.2E-02	1.5E-02	2.5E-02	3.1E-02	4.8E-02	6.8E-02

TABLE D.7

	SURFACE AREA	
	Total Surface Area	
	Mean m²	95th Percentile m²
Birth to 1 month	0.29	0.34
1 to <3 months	0.33	0.38
3 to <6 months	0.38	0.44
6 to <12 months	0.45	0.51
1 to <2 years	0.53	0.61
2 to <3 years	0.61	0.70
3 to <6 years	0.76	0.95
6 to <11 years	1.08	1.48
11 to <16 years	1.59	2.06
16 to <21 years	1.84	2.33
Adult Males		
21 to <30 years	2.05	2.52
30 to <40 years	2.10	2.50
40 to <50 years	2.15	2.56
50 to <60 years	2.11	2.55
60 to <70 years	2.08	2.46
70 to <80 years	2.05	2.45
≥80 years	1.92	2.22

TABLE D.7 (Continued)

Adult Females

21 to <30 years	1.81	2.25
30 to <40 years	1.85	2.31
40 to <50 years	1.88	2.36
50 to <60 years	1.89	2.38
60 to <70 years	1.88	2.34
70 to <80 years	1.77	2.13
≥80 years	1.69	1.98

Percent Surface Area of Body Parts

	Head	Trunk	Arms	Hands	Legs	Feet
			Mean Percent of Total Surface Area			
Birth to 1 month	18.2	35.7	13.7	5.3	20.6	6.5
1 to <3 months	18.2	35.7	13.7	5.3	20.6	6.5
3 to <6 months	18.2	35.7	13.7	5.3	20.6	6.5
6 to <12 months	18.2	35.7	13.7	5.3	20.6	6.5
1 to <2 years	16.5	35.5	13.0	5.7	23.1	6.3
2 to <3 years	8.4	41.0	14.4	4.7	25.3	6.3
3 to <6 years	8.0	41.2	14.0	4.9	25.7	6.4
6 to <11 years	6.1	39.6	14.0	4.7	28.8	6.8
11 to <16 years	4.6	39.6	14.3	4.5	30.4	6.6
16 to <21 years	4.1	41.2	14.6	4.5	29.5	6.1
Adult Males ≥21	6.6	40.1	15.2	5.2	33.1	6.7
Adult Females ≥21	6.2	35.4	12.8	4.8	32.3	6.6

Surface Area of Body Parts

	Head Mean m²	Head 95th m²	Trunk Mean m²	Trunk 95th m²	Arms Mean m²	Arms 95th m²	Hands Mean m²	Hands 95th m²	Legs Mean m²	Legs 95th m²	Feet Mean m²	Feet 95th m²
Birth to 1 month	0.053	0.062	0.104	0.121	0.040	0.047	0.015	0.018	0.060	0.070	0.019	0.022
1 to <3 months	0.060	0.069	0.118	0.136	0.045	0.052	0.017	0.020	0.068	0.078	0.021	0.025
3 to <6 months	0.069	0.080	0.136	0.157	0.052	0.060	0.020	0.023	0.078	0.091	0.025	0.029
6 to <12 months	0.082	0.093	0.161	0.182	0.062	0.070	0.024	0.027	0.093	0.105	0.029	0.033
1 to <2 years	0.087	0.101	0.188	0.217	0.069	0.079	0.030	0.035	0.122	0.141	0.033	0.038
2 to <3 years	0.051	0.059	0.250	0.287	0.088	0.101	0.028	0.033	0.154	0.177	0.038	0.044
3 to <6 years	0.060	0.076	0.313	0.391	0.106	0.133	0.037	0.046	0.195	0.244	0.049	0.061
6 to <11 years	0.066	0.090	0.428	0.586	0.151	0.207	0.051	0.070	0.311	0.426	0.073	0.100
11 to <16 years	0.073	0.095	0.630	0.816	0.227	0.295	0.072	0.093	0.483	0.626	0.105	0.136
16 to <21 years	0.076	0.096	0.759	0.961	0.269	0.340	0.083	0.105	0.543	0.687	0.112	0.142
Adult Males ≥21	0.136	0.154	0.827	1.10	0.314	0.399	0.107	0.131	0.682	0.847	0.137	0.161
Adult Females ≥21	0.114	0.121	0.654	0.850	0.237	0.266	0.089	0.106	0.598	0.764	0.122	0.146

TABLE D.8

MEAN SOLID ADEHERENCE TO SKIN (mg/cm²)

	Face	Arms	Hands	Legs	Feet
Children					
Residential (indoors)[a]	–	0.0041	0.0011	0.0035	0.010
Daycare (indoors and outdoors)[b]	–	0.024	0.099	0.020	0.071
Outdoor sports[c]	0.012	0.011	0.11	0.031	–
Indoor sports[d]	–	0.0019	0.0063	0.0020	0.0022
Activities with soil[e]	0.054	0.046	0.17	0.051	0.20
Playing in mud[f]	–	11	47	23	15
Playing in sediment[g]	0.040	0.17	0.49	0.70	21
Adults					
Outdoor sports[i]	0.0314	0.0872	0.1336	0.1223	–
Activities with soil[h]	0.0240	0.0379	0.1595	0.0189	0.1393
Construction activities[j]	0.0982	0.1859	0.2763	0.0660	–
Clamming[k]	0.02	0.12	0.88	0.16	0.58

[a] Based on weighted average of geometric mean soil loadings for 2 groups of children (ages 3 to 13 years; $N = 10$) playing indoors.

[b] Based on weighted average of geometric mean soil loadings for 4 groups of daycare children (ages 1 to 6.5 years; $N = 21$) playing both indoors and outdoors.

[c] Based on geometric mean soil loadings of 8 children (ages 13 to 15 years) playing soccer.

[d] Based on geometric mean soil loadings of 6 children (ages ≥ 8 years) and 1 adult engaging in Tae Kwon Do.

[e] Based on weighted average of geometric mean soil loadings for gardeners and archeologists (ages 16 to 35 years).

[f] Based on weighted average of geometric mean soil loadings of 2 groups of children (age 9 to 14 years; $N = 12$) playing in mud.

[g] Based on geometric mean soil loadings of 9 children (ages 7 to 12 years) playing in tidal flats.

[h] Based on weighted average of geometric mean soil loadings of 3 groups of adults (ages 23 to 33 years) playing rugby and 2 groups of adults (ages 24 to 34) playing soccer.

[i] Based on weighted average of geometric mean soil loadings for 69 gardeners, farmers, groundskeepers, landscapers, and archaeologists (ages 16 to 64 years) for faces, arms and hands; 65 gardeners, farmers, groundskeepers, and archaeologists (ages 16 to 64 years) for legs; and 36 gardeners, groundskeepers, and archaeologists (ages 16 to 62) for feet.

[j] Based on weighted average of geometric mean soil loadings for 27 construction workers, utility workers, and equipment operators (ages 21 to 54) for faces, arms, and hands; and based on geometric mean soil loadings for 8 construction workers (ages 21 to 30 years) for legs.

[k] Based on geometric mean soil loadings of 18 adults (ages 33 to 63 years) clamming in tidal flats.

- No data.

TABLE D.9

BODY WEIGHT

	Mean kg
Birth to 1 month	4.8
1 to <3 months	5.9
3 to <6 months	7.4
6 to <12 months	9.2
1 to <2 years	11.4
2 to <3 years	13.8
3 to <6 years	18.6
6 to <11 years	31.8
11 to <16 years	56.8
16 to <21 years	71.6
Adults	80.0

TABLE D.10

FRUIT AND VEGETABLE INTAKE

	Per Capita		Consumers-Only	
	Mean g/kg-day	95th Percentile g/kg-day	Mean g/kg-day	95th Percentile g/kg-day
	Total Fruits			
Birth to 1 year	6.2	23.0[a]	10.1	25.8[a]
1 to <2 years	7.8	21.3[a]	8.1	21.4[a]
2 to <3 years	7.8	21.3[a]	8.1	21.4[a]
3 to <6 years	4.6	14.9	4.7	15.1
6 to <11 years	2.3	8.7	2.5	9.2
11 to <16 years	0.9	3.5	1.1	3.8
16 to <21 years	0.9	3.5	1.1	3.8
21 to <50 years	0.9	3.7	1.1	3.8
≥50 years	1.4	4.4	1.5	4.6
	Total Vegetables			
Birth to 1 year	5.0	16.2[a]	6.8	18.1[a]
1 to <2 years	6.7	15.6[a]	6.7	15.6[a]
2 to <3 years	6.7	15.6[a]	6.7	15.6[a]
3 to <6 years	5.4	13.4	5.4	13.4
6 to <11 years	3.7	10.4	3.7	10.4
11 to <16 years	2.3	5.5	2.3	5.5
16 to <21 years	2.3	5.5	2.3	5.5
21 to <50 years	2.5	5.9	2.5	5.9
≥50 years	2.6	6.1	2.6	6.1

[a] Estimates are less statistically reliable based on guidance published in the *Joint Policy on Variance Estimation and Statistical Reporting Standards on NHANES III and CSFII Reports: NHIS/NCHS Analytical Working Group Recommendations* (NCHS, 1993).

TABLE D.11

FISH INTAKE

	Per Capita		Consumers-Only	
	Mean g/kg-day	95th Percentile g/kg-day	Mean g/kg-day	95th Percentile g/kg-day
	General Population—Finfish			
All	0.16	1.1	0.73	2.2
Birth to 1 year	0.03	0.0[a]	1.3	2.9[a]
1 to <2 years	0.22	1.2[a]	1.6	4.9[a]
2 to <3 years	0.22	1.2[a]	1.6	4.9[a]
3 to <6 years	0.19	1.4	1.3	3.6[a]
6 to <11 years	0.16	1.1	1.1	2.9[a]
11 to <16 years	0.10	0.7	0.66	1.7
16 to <21 years	0.10	0.7	0.66	1.7
21 to <50 years	0.15	1.0	0.65	2.1
Females 13 to 49 years	0.14	0.9	0.62	1.8
≥50 years	0.20	1.2	0.68	2.0

General Population—Shellfish

All	0.06	0.4	0.57	1.9
Birth to 1 year	0.00	0.0[a]	0.42	2.3[a]
1 to <2 years	0.04	0.0[a]	0.94	3.5[a]
2 to <3 years	0.04	0.0[a]	0.94	3.5[a]
3 to <6 years	0.05	0.0	1.0	2.9[a]
6 to <11 years	0.05	0.2	0.72	2.0[a]
11 to <16 years	0.03	0.0	0.61	1.9
16 to <21 years	0.03	0.0	0.61	1.9
21 to <50 years	0.08	0.5	0.63	2.2
Females 13 to 49 years	0.06	0.3	0.53	1.8
≥50 years	0.05	0.4	0.41	1.2

General Population—Total Finfish and Shellfish

All	0.22	1.3	0.78	2.4
Birth to 1 year	0.04	0.0[a]	1.2	2.9[a]
1 to <2 years	0.26	1.6[a]	1.5	5.9[a]
2 to <3 years	0.26	1.6[a]	1.5	5.9[a]
3 to <6 years	0.24	1.6[a]	1.3	3.6[a]
6 to <11 years	0.21	1.4	0.99	2.7[a]
11 to <16 years	0.13	1.0	0.69	1.8
16 to <21 years	0.13	1.0	0.69	1.8
21 to <50 years	0.23	1.3	0.76	2.5
Females 13 to 49 years	0.19	1.2	0.68	1.9
≥50 years	0.25	1.4	0.71	2.1

[a] Estimates are less statistically reliable based on guidance published in the *Joint Policy on Variance Estimation and Statistical Reporting Standards on NHANES III and CSFII Reports: NHIS/NCHS Analytical Working Group Recommendations* (NCHS, 1993).

Recreational Population—Marine Fish—Atlantic

	Mean g/day	95th Percentile g/day
3 to <6 years	2.5	8.8
6 to <11 years	2.5	8.6
11 to <16 years	3.4	13
16 to <18 years	2.8	6.6
>18 years	5.6	18

TABLE D.11 (Continued)

	Recreational Population—Marine Fish—Gulf	
3 to <6 years	3.2	13
6 to <11 years	3.3	12
11 to <16 years	4.4	18
16 to <18 years	3.5	9.5
>18 years	7.2	26
	Recreational Population—Marine Fish—Pacific	
3 to <6 years	0.9	3.3
6 to <11 years	0.9	3.2
11 to <16 years	1.2	4.8
16 to <18 years	1.0	2.5
>18 years	2.0	6.8

TABLE D.12

MEATS, DAIRY PRODUCTS, AND FAT INTAKE

	Per Capita		Consumers-Only	
	Mean g/kg-day	95th Percentile g/kg-day	Mean g/kg-day	95th Percentile g/kg-day
	Total Meats			
Birth to 1 year	1.2	5.4[a]	2.7	8.1[a]
1 to <2 years	4.0	10.0[a]	4.1	10.1[a]
2 to <3 years	4.0	10.0[a]	4.1	10.1[a]
3 to <6 years	3.9	8.5	3.9	8.6
6 to <11 years	2.8	6.4	2.8	6.4
11 to <16 years	2.0	4.7	2.0	4.7
16 to <21 years	2.0	4.7	2.0	4.7
21 to <50 years	1.8	4.1	1.8	4.1
≥50 years	1.4	3.1	1.4	3.1
	Total Dairy Products			
Birth to 1 year	10.1	43.2[a]	11.7	44.7[a]
1 to <2 years	43.2	94.7[a]	43.2	94.7[a]
2 to <3 years	43.2	94.7[a]	43.2	94.7[a]
3 to <6 years	24.0	51.1	24.0	51.1
6 to <11 years	12.9	31.8	12.9	31.8
11 to <16 years	5.5	16.4	5.5	16.4
16 to <21 years	5.5	16.4	5.5	16.4
21 to <50 years	3.5	10.3	3.5	10.3
≥50 years	3.3	9.6	3.3	9.6

Total Fats

Birth to 1 month	5.2	16	7.8	16
1 to <3 months	4.5	12	6.0	12
3 to <6 months	4.1	8.2	4.4	8.3
6 to <12 months	3.7	7.0	3.7	7.0
1 to <2 years	4.0	7.1	4.0	7.1
2 to <3 years	3.6	6.4	3.6	6.4
3 to <6 years	3.4	5.8	3.4	5.8
6 to <11 years	2.6	4.2	2.6	4.2
11 to <16 years	1.6	3.0	1.6	3.0
16 to <21 years	1.3	2.7	1.3	2.7
21 to <31 years	1.2	2.3	1.2	2.3
31 to <41 years	1.1	2.1	1.1	2.1
41 to <51 years	1.0	1.9	1.0	1.9
51 to <61 years	0.9	1.7	0.9	1.7
61 to <71 years	0.9	1.7	0.9	1.7
71 to <81 years	0.8	1.5	0.8	1.5
≥81 years	0.9	1.5	0.9	1.5

[a] Estimates are less statistically reliable based on guidance published in the *Joint Policy on Variance Estimation and Statistical Reporting Standards on NHANES III and CSFII Reports: NHIS/NCHS Analytical Working Group Recommendations* (NCHS, 1993).

TABLE D.13

GRAINS INTAKE

	Per Capita		Consumers-Only	
	Mean g/kg-day	95th Percentile g/kg-day	Mean g/kg-day	95th Percentile g/kg-day
Birth to 1 year	3.1	9.5[a]	4.1	10.3[a]
1 to <2 years	6.4	12.4[a]	6.4	12.4[a]
2 to <3 years	6.4	12.4[a]	6.4	12.4[a]
3 to <6 years	6.2	11.1	6.2	11.1
6 to <11 years	4.4	8.2	4.4	8.2
11 to <16 years	2.4	5.0	2.4	5.0
16 to <21 years	2.4	5.0	2.4	5.0
21 to <50 years	2.2	4.6	2.2	4.6
≥50 years	1.7	3.5	1.7	3.5

[a] Estimates are less statistically reliable based on guidance published in the *Joint Policy on Variance Estimation and Statistical Reporting Standards on NHANES III and CSFII Reports: NHIS/NCHS Analytical Working Group Recommendations* (NCHS, 1993).

TABLE D.14

	Mean g/kg-day	95th Percentile g/kg-day
HOME-PRODUCED FOOD INTAKE		
Consumer-Only Home-Produced Fruits, Unadjusted[a]		
1 to 2 years	8.7	60.6
3 to 5 years	4.1	8.9
6 to 11 years	3.6	15.8
12 to 19 years	1.9	8.3
20 to 39 years	2.0	6.8
40 to 69 years	2.7	13.0
≥70 years	2.3	8.7
Consumer-Only Home-Produced Vegetables, Unadjusted[a]		
1 to 2 years	5.2	19.6
3 to 5 years	2.5	7.7
6 to 11 years	2.0	6.2
12 to 19 years	1.5	6.0
20 to 39 years	1.5	4.9
40 to 69 years	2.1	6.9
≥70 years	2.5	8.2
Consumer-Only Home-Produced Meats, Unadjusted[a]		
1 to 2 years	3.7	10.0
3 to 5 years	3.6	9.1
6 to 11 years	3.7	14.0
12 to 19 years	1.7	4.3
20 to 39 years	1.8	6.2
40 to 69 years	1.7	5.2
≥70 years	1.4	3.5

Consumer-Only Home-Caught Fish, Unadjusted[a]

1 to 2 years	–	–
3 to 5 years	–	–
6 to 11 years	2.8	7.1
12 to 19 years	1.5	4.7
20 to 39 years	1.9	4.5
40 to 69 years	1.8	4.4
≥70 years	1.2	3.7

Per Capita for Populations that Garden or Farm

	Home-Produced Fruits[b]		Home-Produced Vegetables[b]	
	Mean g/kg-day	95th Percentile g/kg-day	Mean g/kg-day	95th Percentile g/kg-day
1 to <2 years	1.0 (1.4)	4.8 (9.1)	1.3 (2.7)	7.1 (14)
2 to <3 years	1.0 (1.4)	4.8 (9.1)	1.3 (2.7)	7.1 (14)
3 to <6 years	0.78 (1.0)	3.6 (6.8)	1.1 (2.3)	6.1 (12)
6 to <11 years	0.40 (0.52)	1.9 (3.5)	0.80 (1.6)	4.2 (8.1)
11 to <16 years	0.13 (0.17)	0.62 (1.2)	0.56 (1.1)	3.0 (5.7)
16 to <21 years	0.13 (0.17)	0.62 (1.2)	0.56 (1.1)	3.0 (5.7)
21 to <50 years	0.15 (0.20)	0.70 (1.3)	0.56 (1.1)	3.0 (5.7)
>50 years	0.24 (0.31)	1.1 (2.1)	0.60 (1.2)	3.2 (6.1)

Per Capita for Populations that Farm or Raise Animals

	Home-Produced Meats[b]		Home-Produced Dairy	
	Mean g/kg-day	95th Percentile g/kg-day	Mean g/kg-day	95th Percentile g/kg-day
1 to <2 years	1.4 (1.4)	5.8 (6.0)	11 (13)	76 (92)
2 to <3 years	1.4 (1.4)	5.8 (6.0)	11 (13)	76 (92)
3 to <6 years	1.4 (1.4)	5.8 (6.0)	6.7 (8.3)	48 (58)
6 to <11 years	1.0 (1.0)	4.1 (4.2)	3.9 (4.8)	28 (34)
11 to <16 years	0.71 (0.73)	3.0 (3.1)	1.6 (2.0)	12 (14)
16 to <21 years	0.71 (0.73)	3.0 (3.1)	1.6 (2.0)	12 (14)
21 to <50 years	0.65 (0.66)	2.7 (2.8)	0.95 (1.2)	6.9 (8.3)
>50 years	0.51 (0.52)	2.1 (2.2)	0.92 (1.1)	6.7 (8.0)

[a] Not adjusted to account for preparation and post cooking losses.

[b] Adjusted for preparation and post cooking losses.

- No data

TABLE D.15

TOTAL PER CAPITA FOOD INTAKE

	Mean g/kg-day	95th Percentile g/kg-day
Birth to 1 year	91	208[a]
1 to <3 years	113	185[a]
3 to <6 years	79	137
6 to <11 years	47	92
11 to <16 years	28	56
16 to <21 years	28	56
21 to <50 years	29	63
≥50 years	29	59

[a] Estimates are less statistically reliable based on guidance published in the *Joint Policy on Variance Estimation and Statistical Reporting Standards on NHANES III and CSFII Reports: NHIS/NCHS Analytical Working Group Recommendations* (NCHS, 1993).

TABLE D.16

HUMAN MILK AND LIPID INTAKE

	Mean		Upper Percentile	
	mL/day	mL/kg-day	mL/day	mL/kg-day
Human Milk Intake				
Birth to 1 month	510	150	950	220
1 to <3 months	690	140	980	190
3 to <6 months	770	110	1,000	150
6 to <12 months	620	83	1,000	130
Lipid Intake				
Birth to 1 month	20	6	38	8.7
1 to <3 months	27	5.5	40	8.0
3 to <6 months	30	4.2	42	6.1
6 to <12 months	25	3.3	42	5.2

TABLE D.17

ACTIVITY FACTORS

	Time Indoors (total) minutes/day		Time Outdoors (total) minutes/day		Time Indoors (at residence) minutes/day	
	Mean	95th Percentile	Mean	95th Percentile	Mean	95th Percentile
Birth to <1 month	1,440	-	0	-	-	-
1 to <3 months	1,432	-	8	-	-	-
3 to <6 months	1,414	-	26	-	-	-
6 to <12 months	1,301	-	139	-	-	-
Birth to <1 year	-	-	-	-	1,108	1,440
1 to <2 years	1,353	-	36	-	1,065	1,440
2 to <3 years	1,316	-	76	-	979	1,296
3 to <6 years	1,278	-	107	-	957	1,355
6 to <11 years	1,244	-	132	-	893	1,275
11 to <16 years	1,260	-	100	-	889	1,315
16 to <21 years	1,248	-	102	-	833	1,288
18 to <64 years	1,159	-	281	-	948	1,428
>64 years	1,142	-	298	-	1,175	1,440

	Showering minutes/day		Bathing minutes/day		Bathing/Showering minutes/day	
	Mean	95th Percentile	Mean	95th Percentile	Mean	95th Percentile
Birth to <1 year	15	-	19	30	-	-
1 to <2 years	20	-	23	32	-	-
2 to <3 years	22	44	23	45	-	-
3 to <6 years	17	34	24	60	-	-
6 to <11 years	18	41	24	46	-	-
11 to <16 years	18	40	25	43	-	-
16 to <21 years	20	45	33	60	-	-
18 to <64 years	-	-	-	-	17	-
>64 years	-	-	-	-	17	-

TABLE D.17 *(Continued)*

	Playing on Sand/Gravel minutes/day		Playing on Grass minutes/day		Playing on Dirt minutes/day	
	Mean	95th Percentile	Mean	95th Percentile	Mean	95th Percentile
Birth to <1 year	18	-	52	-	33	
1 to <2 years	43	121	68	121	56	121
2 to <3 years	53	121	62	121	47	121
3 to <6 years	60	121	79	121	63	121
6 to <11 years	67	121	73	121	63	121
11 to <16 years	67	121	75	121	49	120
16 to <21 years	83	-	60	-	30	-
18 to <64 years	0 (median)	121	60 (median)	121	0 (median)	120
>64 years	0 (median)	-	121 (median)	-	0 (median)	-

	Swimming minutes/month	
	Mean	95th Percentile
Birth to <1 year	96	-
1 to <2 years	105	-
2 to <3 years	116	181
3 to <6 years	137	181
6 to <11 years	151	181
11 to <16 years	139	181
16 to <21 years	145	181
18 to <64 years	45 (median)	181
>64 years	40 (median)	181

TABLE D.18

	Occupational Mobility	
	Median Tenure (years) Men	Median Tenure (years) Women
All ages ≤16 years	7.9	5.4
16 to 24 years	2.0	1.9
25 to 29 years	4.6	4.1
30 to 34 years	7.6	6.0
35 to 39 years	10.4	7.0
40 to 44 years	13.8	8.0
45 to 49 years	17.5	10.0
50 to 54 years	20.0	10.8
55 to 59 years	21.9	12.4
60 to 64 years	23.9	14.5
65 to 69 years	26.9	15.6
≥70 years	30.5	18.8

	Population Mobility			
	Residential Occupancy Period (years)		Current Residence Time (years)	
	Mean	95th Percentile	Mean	95th Percentile
All	12	33	13	46
- No data.				

TABLE D.19

	LIFE EXPECTANCY
	Years
Total	78
Males	75
Females	80

BUILDING CHARACTERISTICS

	Residential Buildings	
	Mean	10th Percentile
Volume of Residence (m³)	492	154
Air Exchange Rate (air changes/hour)	0.45	0.18
	Non-Residential Buildings	
	Mean (Standard Deviation)	10th Percentile
Volume of Non-residential Buildings (m³)		408
Vacant	4,789	510
Office	5,036	2,039
Laboratory	24,681	1,019
Non-refrigerated warehouse	9,298	476
Food sales	1,889	816
Public order and safety	5,253	680
Outpatient healthcare	3,537	1,133
Refrigerated warehouse	19,716	612
Religious worship	3,443	595
Public assembly	4,839	527
Education	8,694	442
Food service	1,889	17,330
Inpatient healthcare	82,034	1,546
Nursing	15,522	527
Lodging	11,559	1,359
Strip shopping mall	7,891	35,679
Enclosed mall	287,978	510
Retail other than mall	3,310	459
Service	2,213	425
Other	5,236	527
All Buildings	5,575	
Air Exchange Rate (air changes/hour)	1.5 (0.87) Range 0.3-4.1	0.60

TABLE D.21 Age Dependent Adjustment Factor (ADAF)

AGE (years)	EXPOSURE FACTORS	EXPOSURE DURATION (years)	ADAF
0 to <2	Child	2	10
2 to <6	Child	4	3
6 to <16	Adult	10	3
>16 to <30	Adult		1

Source: Donohue, J.M., and Simic, M.J. (2011). *Age Dependent Adjustment Factor (ADAF) Application*. United States Environmental Protection Agency. Office of Water (4304T).

Glossary

A

abiotic resources: Nonliving natural resources and raw materials.

absorption: The selective transfer of a substance from a gas to a contacting liquid.

accelerated eutrophication: increase in the rate of nutrient enrichment due to the introduction of large quantities of nutrients resulting from anthropogenic activities.

acid: Any substance that can donate a hydrogen ion, H^+ (or proton).

activated sludge: The portion of the recycle stream containing "active" microorganisms that is returned to the activated sludge reactor.

activated sludge system: A wastewater treatment system that utilizes the recycling of microorganisms to maintain an "active" microbe population in the reactor.

activity coefficient: A numerical factor used to relate the standard chemical activity and the conditional chemical reactivity.

acute health effect: Short-term symptoms in response to a pollutant or pathogen.

adaptation: The process of adjustment to the actual or expected climate and its effects.

adoption and diffusion of innovations: The processes through which people and organizations begin to use new or different products or begin to change their own behaviors.

advanced materials: Novel, high-value-added substances with enhanced properties that are often created with highly specialized, sophisticated manufacturing techniques.

aerobic organisms: Life forms that require molecular oxygen (O_2) for respiration.

affluence: A measure of the quality of life of individual members of a society.

Air Quality Index (AQI): A numerical scale that communicates air quality trends to the public and provides warnings to the public on days when a region is not in compliance with the NAAQS.

air-to-fuel ratio: The ratio of air to carbon fuel in a combustion process.

algae: Chlorophyll-containing eukaryotic organisms that produce oxygen.

algae blooms: Conditions when algae reproduction rates are extraordinarily high.

alloy: A metal made by combining metals or elements—for example, steel, bronze, and brass.

alternatives assessment (AA): The process of gathering and integrating information from assessment tools to use for decision making.

ampholyte: A substance that can either donate or receive a proton.

anaerobic organisms: Life forms that do not require molecular oxygen; they may obtain oxygen from inorganic ions, such as nitrates, sulfates, or proteins.

angle of bulk plane: The angle from the edge of the lot to be developed to the steepest angle of the proposed building.

appropriate technology: Engineering design that takes into consideration the key local social, economic, environmental, and technical factors that influence the success or failure of a design solution.

aquifer: A geologic stratum that contains a substantial amount of groundwater.

arbitrary-flow reactor: A nonideal reactor that neither completely plugs flow nor completely mixes flow.

artesian well: Water trapped in a confined aquifer that is pressurized enough to rise to the surface.

ASHRAE: American Society of Heating, Refrigeration, and Air-Conditioning Engineers.

aspects: Elements of an organization's activities, products, or services that have resulted or may result in an impact, positive or negative, on the environment or on human health.

atmosphere: A layer of gases and suspended particulates that surround the Earth and are held in place by gravity.

B

background consumption: The normal biological functioning of all organisms to meet physical and/or psychological needs in order to survive and reproduce.

bacteria: Unicellular organisms that do not have a nuclear membrane.

baghouses: Bag filter-based particulate removal systems.

base: Any substance that can accept an H^+ ion (or proton).

best management practices (BMPs): Techniques based on successful engineering practice that consistently achieve results in a cost-effective manner.

beyond code program: A method for acknowledging that a building exhibits a level of energy efficiency that is significantly higher than that required by building energy codes.

bioaccumulation potential: The ability of a material to be retained in animal tissue to the extent that organisms higher up the trophic level will retain increasingly higher concentrations of this chemical.

biobased feedstock: Materials derived from living organisms that are starting materials at the beginning of a product's life cycle.

biocapacity: The ecosystem's capacity to produce biological materials used by people and to absorb waste material generated by humans, under current management schemes and extraction technologies.

biocentric outlook: A viewpoint that values each organism as a center of life pursuing its own good in its own way.

biochemical oxygen demand (BOD): The amount of oxygen consumed by microorganisms in water in the process of turning a substrate into cell mass, energy, and carbon dioxide.

biocidal coatings: Material coatings or paint used to kill aquatic organisms to prevent fouling of a boat's hull.

bioconcentration: The uptake and accumulation of an aquatic substance in life forms directly from the environment bioconcentration factor (BCF).

Bioconcentration factor (BCF): A factor used to estimate the equilibrium concentration in fish (in mg/kg) from a given concentration of a contaminant in water (in mg/L).

biodegradable: Materials that may be broken down into smaller molecules by microorganisms.

biofiltration: A filtration process in which organic pollutants are degraded by microorganisms in the control system.

biofuel: Any fuel that is derived from biomass.

biogas: Gas created from anaerobic processes that contains methane gas and other combustible gases.

biogeochemical cycle: The biological and chemical reservoirs, agents of change, and pathways of flow from one reservoir of a chemical on Earth to another reservoir.

biological nitrogen removal (BNR): The two-step nitrification–denitrification process used to remove nitrogen in wastewater treatment.

biological treatment: The use of microorganisms to metabolize waste materials into benign end products.

biomimicry: A design strategy that systematically analyzes biological processes and forms with the intent of using them for engineering solutions.

biotic resources: Nonliving natural resources and raw materials.

blackbody temperature: The maximum temperature emitted by a solid object or body from the adsorption of energy from the sun.

breakpoint dose: The point at which added chlorine results in a residual-free chlorine level.

British thermal unit (Btu): The amount of energy required to raise the temperature of 1 pound of water 1 degree Fahrenheit.

brownfield: An inactive industrial site, usually with some degree of environmental/chemical contamination.

Brundtland definition of sustainability: Development that meets the needs of the present without compromising the ability of future generations to meet their own needs.

building codes: Government regulations that establish legal requirements and standards for the safety, occupancy, siting, and construction of buildings and structures.

building envelope: The region of a parcel of land that may have a structure legally built on it.

building integrated system: A system in which renewable energy technologies are used to substitute for a building's conventional construction materials.

Building Research Establishment Environmental Assessment Method (BREEAM): A European building rating system that certifies a wide variety of buildings and communities.

built environment: All human-made structures, engineered alterations to landforms, and constructions.

C

C-distribution: A plot of concentration or normalized concentration versus time that represents the behavior of a reactor.

calorie: The amount of energy required to raise 1 gram of water 1 degree Celsius, from 14.5°C to 15.5°C.

capacity factor analysis: A tool that assists with the development of sustainable water, sanitation, and household energy solutions by evaluating a community's capacity to manage its own technology.

cap-and-trade program: A market-based incentive program that allows polluters to buy and sell pollution allowances, but in a manner that caps, or limits, the total amount of pollution that can occur in a region.

capital goods: Goods that are used to create the products and services that companies sell.

carbon adsorption: A method that uses the attractive van der Waals forces on porous granulated activated carbon (GAC) surfaces to capture and contain organic pollutants.

carbon capture and sequestration (CCS): A three-step process that includes carbon dioxide capture, transport, and storage for extended periods of time in repositories other than the atmosphere.

carbonaceous oxygen demand (COD): The amount of oxygen consumed to utilize a carbon-based substrate.

carbonate hardness: A measure of the portion of the diprotic ions that can combine with carbonates to form scaling.

carcinogens: Substances that evidence indicates may cause cancer.

Carnot limit: The maximum efficiency that can physically be achieved by a heat engine.

carrying capacity: The maximum rate of resource consumption and waste discharge that can be sustained indefinitely in a given region without progressively impairing the functional integrity and productivity of the relevant ecosystem.

catalytic oxidation: A process that uses thermally activated catalysts to oxidize and destroy organic pollutants, but at a lower temperature than traditional thermal oxidation.

categorical imperative: An unconditional set of rules for making value-laden decisions that one would wish all people obeyed.

ceramics: Compounds of metallic and nonmetallic elements—for example, cement, pottery, brick, porcelain, and tile.

characteristic wastes: Wastes that do not qualify as "listed" but exhibit any of the properties of ignitability, corrosivity, reactivity, or toxicity.

chemical activity: A standardized measure of chemical reactivity within a defined system.

chemical alternatives assessment (CAA): A process for identifying and comparing potential chemical and nonchemical alternatives to replace chemicals of high concern.

chemical coagulants: Chemicals such as aluminum sulfate (alum) or calcium hydroxide (lime) used to enhance the removal of solids, BOD, and phosphorus.

chemical of concern: A chemical used in products that has toxicity or other effects that are deemed to create undesirable risks.

chemical reactivity: A chemical's overall tendency to participate in a reaction.

chemical treatment: The use of chemical processes to reduce the ignitability, corrosivity, reactivity, or toxicity of wastes.

chlorofluorocarbons (CFCs): Halogen-containing molecules used for a wide variety of commercial and industrial applications, including aerosol spray cans, coolants, refrigerants, and fire retardants.

chronic daily intake: An estimate of exposure used in calculating single-exposure pathway risk.

chronic health effects: Long-term effects of exposure; may include chronic inflammation, cancers, or pulmonary emphysema.

clarifiers: Settling tanks used to remove particles.

Clean Air Act: The first federal act in the United States to allow regulations for the control of air pollutants.

cluster development: A design of sub-divisions or building tracts in which housing and other structures are concentrated in one area of a development, leaving an open and community-shared green space.

coagulation: A chemical process that alters the surface charge on particles suspended in the water to form larger particles.

code: A signal to tell a machine how to divide the individual particles in the stream.

combined heat and power generation: The exploitation of usable waste heat in order to heat and cool buildings or to generate electricity.

combined residual: The presence of all the various forms of chlorine that remain after reaction in the system.

combined sewer overflow (CSO): Discharge from sewer systems in which stormwater and sewage are mixed together.

combustion: The chemical process that occurs when oxygen reacts with various compounds, releasing heat and energy.

command and control regulation: A form of governmental regulation in which the government specifies or mandates the exact ways in which companies must comply with environmental protection laws.

commercial lifetime: The length of time a typical user has a product before replacing it or simply no longer wanting it.

community-based design: Participatory design at the community scale.

composites: Materials fabricated from two or more constituents, often by wrapping, weaving, or adhesive bonding.

concentrated solar power: Solar thermal energy used to produce electricity from steam.

condensation: The conversion of water in the gas phase to liquid water by cooling the water molecules.

Conference on Environment and Development: The first world environmental congress to which heads of state were invited as participants; held in Rio de Janeiro, Brazil, in 1992.

confined aquifer: The groundwater trapped between two impervious layers.

conservation design: A sophisticated, site-specific approach to environmentally sensitive design that integrates principles of cluster development, low-impact development, and green infrastructure.

conservative and instantaneous signal: An ideal tracer placed into the flow of material as the material enters the reactor.

constantly mixed flow reactor (CMFR): A reactor in which the contents are equally mixed and equal to the concentrations of the contents exiting the reactor.

constructed wetlands: The use of wetland plants attached to an engineered reactor for stormwater or wastewater treatment.

consumer: An animal that uses a producer's molecules as a source of energy.

consumption: The selection, use, reuse, maintenance, repair, and disposal of goods and services.

contact time: The length of time water is in contact with chlorine in a disinfection process.

containment: A waste stabilization option used if there is no need to remove the offending material and/or if the cost of removal is prohibitive.

continual improvement: The use of safer chemicals, reduced carbon footprints, and greater material circularity to preserve natural capital.

control volume: The boundaries around a specified system.

corporate social responsibility: The idea that businesses are accountable for more than just their financial profits and losses to shareholders; they are also accountable to society for the environmental and social consequences of their business operations.

critical particles: The particles in a separator that have a size at which those with lower settling velocities are not removed and those with higher settling velocities are removed.

critical point: The point at which the lowest oxygen concentration is expected in the stream or river.

cyclonic separators: The most prevalent type of mechanical or momentum-based particulate matter control device.

D

daylighting: Using skylights and the careful placement of windows to reduce the need for artificial light by providing sunlight during the day.

decarbonization and decarbonizing: Reducing or eliminating carbon emissions from electricity generation.

decomposer: An organism that uses detritus as an energy source.

deep ecology: The ethical framework in which humans have no greater importance than any other component of our world.

dematerialization: The redesign of products to minimize their materials content.

demographers: People who study trends in population.

density neutral design: The design of residential developments in a manner that does not affect the total number of housing units that can be built at the development site.

depletion: The removal of resources from the environment.

derived demand: The idea that the social demand for a particular energy fuel or energy-producing technology results from the critical need for a service that the energy provides; for example, the demand for coal derives from the social demand for electricity.

design: Problem solving that results in an "artifact," something that persists over time as either a tangible thing or a structured pattern of behavior.

design for disassembly (DfD): A variety of design techniques that make product disassembly easier and more cost effective.

design for environment: The idea that environmental protection should be designed into products and processes rather than managed as an after-the-fact harm.

design for recycling (DfR): A design technique that improves the recycling and re-use of a product's material content.

design for remanufacturing (DfRem): A design to fulfill the green engineering principles to preserve complexity and to plan for a commercial afterlife beyond simple material recycling.

Design for X: A family of specialized design guidelines, rules, and metrics.

design lifetime: The length of time a product or process can physically withstand normal wear and tear and still function or operate.

designing for value recovery: Choosing products and materials that "decompose" into technical nutrients that can be cycled back into an industrial ecosystem.

detoxification: Reduction in the use of hazardous materials in products or industrial operations.

detritus: Dead organic matter.

development: The conversion of open land and natural areas to the built environment, such as residential neighborhoods, airports, industrial parks, hydroelectric dams, shopping centers, parks, and so on.

direct energy: Energy that is consumed in the act of providing an energy service, such as electricity for lighting; the opposite of direct energy is embodied energy.

dispersion model: A mathematical or statistical model that is used to estimate the transportation and concentration of pollutants.

disruptive technology: A product or technology that creates a new market and displaces established markets.

dissolution of salts: The process of a salt dissolving in a solution.

dissolved inorganic carbon: Dissolved carbon dioxide gas, carbonic acid (H_2CO_3), bicarbonate ion (HCO_3^-), and the carbon ion (CO_3^{2-}).

dissolved oxygen (DO): Diatomic oxygen dissolved in an aqueous solution.

dissolved solids: Salts and minerals that have been dissolved through the natural weathering of soils or anthropogenic processes.

distributed generation: Decentralized power production at the site on which the power is used or at the community scale.

disturbances: Detrimental changes to the environment.

dose: The amount of a chemical received by an individual that can interact with an individual's biological functions and receptors.

dose–response curve: A plot of the measured response versus the dose administered.

downcycling: Using a recycled material or product in an application that is of lower quality or more limited functionality than its original purpose.

drop-in replacement: A simple replacement to the current system.

droughts: Periods when natural or managed water systems do not provide enough water to meet established human and environmental uses.

dry adiabatic lapse rate: The cooling rate with respect to elevation for dry adiabatic air.

dry scrubbing: The process in which lime-based slurry is sprayed into the sulfur-containing airstream to form $CaSO_3$ and $CaSO_4$ precipitates.

duty cycle: The amount of time a device operates in order to complete a functional cycle—for example, the number of hours it takes a pump to move a million liters of water.

E

Earth's albedo: The percent of incoming solar radiation that is reflected by the Earth.

easement: *See* right-of-way.

eco-industrial park: A community of manufacturing and service businesses that collaborate in the management of environmental and resource issues.

ecological design: Design based on the idea that by "looking to nature" for inspiration, we can take advantage of sustainable forms and processes already successfully established in our environment.

ecological footprint: An accounting method used to measure the relationship between consumption and supply of natural resources and their impacts on the environment.

ecology: The study of plants, animals, and their physical environment.

ecosystem services: Benefits to humans provided by a variety of ecological dynamics that support humankind through the generation of tangible and intangible provisioning, regulating, and cultural services.

effective concentration (EC): The concentration of a contaminant that produces a specified response (such as lack of cell reproduction) in a specific time period.

effective stack height: The sum of the physical stack height, H_s, and the plume rise.

effluent: Flow out of a system.

electron acceptor: A molecule that receives or accepts electrons from another molecule during a redox reaction.

electrostatic precipitators (ESPs): Electric corona devices that impart a charge to airborne particles for particle control.

embodied energy: The total quantity of energy directly and indirectly consumed in the production, distribution, and use of a product.

emission factors: Numerical values that express the amount of an air pollutant likely to be released based on a determinant factor in an industrial process.

endogenous respiration: The process in which microorganisms convert their own cell tissue into energy for cell maintenance.

endorsement labeling: A type of voluntary certification in which an independent, not-for-profit, third party verifies that a product meets certain criteria in its construction, material content, performance, or operation.

end-use sectors: The residential, commercial, industrial, and transportation sectors.

energy curtailment: Reduction in the demand for energy caused by changing people's behaviors and lifestyle needs, such as turning off the lights in an unoccupied room.

energy efficiency: The technical productivity for an energy-consuming device or piece of equipment, usually measured as the physical property of work.

energy intensity: Energy consumed per monetary unit of gross domestic product.

energy ladder: A model that describes how both the quality and quantity of energy used in a household increase as household income rises.

energy poverty: The inability to access or afford commercially provided energy; energy poverty is commonly understood as a lack of access to electricity or clean cooking fuels.

energy recovery: The ability to capture waste heat from industrial processes or the energy embodied in solid waste and use it for energy production.

energy return on investment (EROI): A ratio of the amount of usable energy from a source to the amount of energy required to produce the resource; viable resources have EROIs greater than 1.

energy security: The ability of a nation to protect itself from the economic, political, and social disruptions caused by an interrupted supply of a critical energy resource, the failure of an important energy infrastructure, or rapid and steep changes in energy prices.

energy services: The specific functions that energy provides for people, such as lighting, heating, cooling, refrigeration, and industrial production.

energy transition: A change in the dominant fuel used by a society.

environmental ethic: The recognition that we are, at least at the present time, unable to explain rationally our attitude toward the environment and that these attitudes are deeply felt, not unlike the feeling of spirituality.

environmental impact assessment: An analysis that catalogues the expected environmental impacts a project will have at a site and indicates how those impacts might be mitigated.

environmental justice: An ethical principle that tries to eliminate the spatial discrimination that results in low-income, minority, and tribal communities being unfairly affected by the environmental actions of others.

environmentally sensitive design: A broad range of design strategies and on-the-ground techniques to integrate both development and environmental conservation.

envision rating system: An assessment tool for evaluating the sustainability of all civil infrastructure.

eolian (wind) erosion: Erosion that occurs when wind removes soil particles.

epidemiology: The study of the determinants and distributions of diseases and adverse responses in specified populations.

epilimnion: The uppermost layer of a water body, characterized by warmer, less dense water.

ethics: Ideas that provide a framework for making difficult choices when we face a problem involving moral conflict.

eukaryotes: Microorganisms that have a nucleus or nuclear envelope and an endoplasmic reticulum, or interconnected organelles; include protozoa, fungi, and green algae.

eutrophication: The nutrient enrichment of aquatic ecosystems; a natural process that can be accelerated by excessive anthropogenic emissions of nitrogen and phosphorus.

evaporation: The process of converting liquid water from surface water sources to gaseous water that resides in the atmosphere.

exceptional quality sludge: Treated wastewater solids that meet pollutant concentration limits, pathogen reduction standards, and vector attraction reduction criteria.

exponential growth: Growth in which the rate of change is proportional to the instantaneous value of the variable that is changing.

exposure: The presence of people; livelihoods; species or ecosystems; environmental functions; services and resources; infrastructure; or economic, social, or cultural assets in places that could be adversely affected.

exposure assessment: The process of estimating the duration, frequency, and magnitude of contact with a contaminant in a given population.

ex-situ: Off-site.

extended producer responsibility: The idea that a manufacturer is responsible for the environmental consequences of its product over the full length of the product's life cycle.

extraction and treatment: The pumping of contaminated groundwater to the surface for either disposal or treatment, or the excavation of contaminated soil for disposal or treatment.

F

F-distribution: A plot of the fraction of the signal that has left the reactor with respect to time, t.

facultative microorganisms: Small life forms that use oxygen when it is available as an energy source and can also use anaerobic reactions if oxygen is not available.

feedback loops: Processes that either amplify or diminish the effects of climate forcing.

feedstocks: Raw material used as an input in the manufacturing process.

fertilize: Provide nutrients to allow more plants and animals to grow and develop.

fertilizer: "A substance that makes fertile" because of its nutrient content.

fine particles: Particles that are 2.5 micrometers in diameter and smaller.

finite resources: Resources that formed over a period of millions of years and can be regenerated only on a geological timescale.

first-order reaction: A reaction in which the reaction rate is directly proportional to the amount of material.

flocculation: A process in which mixing wastewater and coagulants increases the overall size of the particles.

floor area ratio (FAR): The ratio of the building's footprint (floor area) to the area of the lot on which the building is to be constructed.

fluvial (water) erosion: Erosion that occurs when rainwater runoff carries away soil particles.

food chain: A series of processes in which energy flows through an ecosystem.

food-to-microorganism ratio (F/M): The amount of BOD divided by the microorganism growth rate.

food web: Interactions among the various organisms in a food chain.

form-based codes: Land-use regulations that use neighborhoods as the primary geographic planning unit rather than larger zoning districts.

form design: The overall shape, size, and appearance of objects being designed, including any connected parts they might have.

fossil water: Groundwater that has been isolated in an aquifer for thousands or millions of years.

foul release coatings: Coatings that make it difficult for organisms to attach to the hull of a boat.

fracking: A process in which highly pressurized water blended with chemicals is injected deep into wells to fracture bedrock that is trapping oil and natural gas.

free chlorine: HOCl and OCl⁻ molecules that are available to react with organisms in a disinfection process.

frequency analysis: An estimate of streamflow and demand to determine reservoir capacity.

fuel switching: Burning a cleaner fuel instead of a fossil fuel for combustion energy and heat.

functional unit: The basis for comparison of two or more products; used to identify a common function and study how each product achieves that function during its life.

future-proofing: Taking actions that preemptively address future regulations and anticipated roadblocks.

G

geotextiles: Synthetic materials purposefully designed to protect soil and manage water flow.

geothermal power: Electricity generated from geothermal energy.

Globally Harmonized System of Classification and Labelling of Chemicals (GHS): A worldwide initiative to promote guidelines for ensuring the safe production, transport, handling, use, and disposal of hazardous chemicals.

global warming potential (GWP): A common index used to quantify the contribution of greenhouse gases to global warming, using CO_2 as a benchmark.

gray: The quantity of ionizing radiation that results in the absorption of 1 joule of energy per 1 kilogram of absorbing material.

graywater: Wastewater from activities such as bathing, dishwashing, and laundry that can be safely reused/recycled for non-potable water needs.

green building: The practice of reducing environmental impacts in the construction, operation, and demolition of buildings.

green chemistry: An area of chemistry that focuses on the design of materials and processes to decrease or avoid the use of hazardous substances.

green engineering: A practice similar to sustainable engineering that provides design rules and principles for more sustainable products and processes.

green infrastructure: A landscape pattern that uses a connective network of green space to provide essential public services to a community.

green premium: The higher cost of using a more sustainable product rather than products that are less sustainable.

greenhouse gas effect: The warming of the planet's surface due to the trapping of atmospheric heat by greenhouse gases.

greenhouse gases: Gas components of the atmosphere that absorb and store energy.

greenwashing: A form of deceptive environmental advertising, marketing, and public relations that leads consumers to believe a product is more sustainable than it truly is.

ground source heat pump: A technology used in buildings that takes advantage of the natural and stable temperature gradients of the Earth by circulating the heat-exchanging fluids used in heating and cooling systems underground to preheat or pre-chill them.

groundwater: All water that occupies the zone of saturation in the ground.

H

half-life: The time required for a quantity to decrease by one-half.

hazard: Anything that has the potential to produce conditions that endanger safety and health.

hazard classification: Categories that indicate the types of chemical hazards and their severity.

hazard endpoint: An adverse outcome that occurs from exposure to chemicals that have certain inherent hazardous properties.

hazard index (HI): Sum of the hazard quotients when there is suspected exposure to more than one hazardous material.

hazard quotient (HQ): A number that indicates the probability of a noncarcinogenic toxicity risk.

hazardous air pollutants (HAPs): The U.S. EPA list of air pollutants (also called air toxics) that pose potentially significant health risks and may be carcinogenic, mutagenic, or teratogenic.

hazardous waste: Defined in the United States by the EPA under the RCRA as any waste that is dangerous or potentially harmful to our health or the environment.

heating degree day: An estimate of the amount of energy required to heat a building; it is calculated by summing the extent to which the average outdoor temperature deviates from a baseline temperature.

helminths: Parasitic worms that are transmitted by wastewater and wastewater solids.

Henry's law constant: The constant that describes the slope of the linear relationship between the liquid phase and gas phase concentrations.

high-level waste: Highly radioactive materials produced as byproducts from reactions that occur inside nuclear reactors.

housing density: The number and type of dwellings allowed, typically per acre of proposed residential space.

Hubbert curve: A tool used to analyze the rates of growth in demand for a fossil fuel and the rate at which new reserves become available; commonly associated with peak oil discussions.

Human Development Index (HDI): A numerical index developed by the United Nations Development Program (UNDP) to compare the state of human development in countries based on three indicators: life expectancy, education, and income.

human–environment interface: The intersection of people and the natural world.

hybrid control systems and concentrators: A control system that combines one or more of the functional advantages of one technology synergistically with a second technology.

hybrid framework: A blend of the sequential and simultaneous decision frameworks.

hydrologic cycle: The sequence in which water passes from one biogeochemical cycle to another.

hydrology: The study of the waters of the Earth; their occurrence, circulation, and distribution; their chemical and physical properties; and their reaction with the environment, including the relationships to living things.

hydrolysis: The breakdown of substances due to reaction with water.

hydrosphere: The portion of the Earth that accounts for most of the water storage; consists of oceans, lakes, streams, and shallow groundwater bodies.

hypercapnia: A condition that occurs when too much carbon dioxide builds up in the bloodstream, causing distress.

hypolimnion: The lower layer of a water body characterized by cooler, denser water that usually has a lower DO concentration.

hypoxia: An inadequate supply of oxygen to the body.

I

impact assessment: Determination of the environmental relevance of all the inputs and outputs at each stage in the life cycle.

impacts: Factors that contribute to diminishing the sustainable rate of resource consumption and waste discharge.

impervious surfaces: Areas in the urban landscape that do not absorb water or allow water to infiltrate into the ground, such as roads and parking areas.

indicator: A measurement or metric based on verifiable data that can be used to communicate important information to decision makers and the public about processes related to sustainable design or development.

indicator organisms: Life forms in water that may or may not cause disease but indicate the possible presence of pathogens.

industrial ecology: The study of technological systems that mimic the way nature handles materials and energy transformations.

industrial energy auditing: A practice that can be used to identify opportunities for saving energy at different scales, such as process, product line, and factory.

infiltration: The process of precipitation seeping into the ground.

inflow and infiltration (I/I): Flow into a sewer system through appurtenances, like manholes covers, cracks, and leaks in the system.

influent: Flow into a system.

information asymmetries: Situations in which different parties possess unequal knowledge, time, or skill to evaluate the long-term consequences of a decision.

inhalable coarse particles: Particles larger than 2.5 microns and smaller than 10 microns in diameter.

in-situ: Description of treating waste while leaving the hazardous constituents in place.

integrated assessment models (IAMs): Socioeconomic scenarios used to derive future emissions models.

Integrated Risk Information System (IRIS): U.S. EPA human health assessment database.

integrated time horizon (ITH): The value chosen for calculating the global warming potential for various greenhouse gases.

intergenerational equity: Equitable distribution of resources between the current generation and future generations.

intergenerational ethics: The obligation to consider the consequences of our actions and decisions for future generations.

International Energy Conservation Code (IECC): A model building code developed by the International Code Council that covers both residential and commercial structures.

International Green Construction Code (IGCC): A model code for green building developed by the International Code Council.

International Organization for Standardization (ISO): An international NGO with the primary mission of developing voluntary international standards governing most aspects of technology, manufacturing, and business management operations.

International Panel on Climate Change (IPCC): A research group composed of a group of scientists from many nations that are intimately involved in better understanding changes to the Earth's climate.

inventory analysis: Determination of the quantitative values of materials and energy inputs and outputs of all stages in the life cycle.

involuntary risk: A risk imposed on individuals because of circumstances beyond their control.

ionic reactions: Reactions in which there is a change in ion–ion interactions and relationships.

ionic strength: The estimate of the overall concentration of dissolved ions in solution.

ionizing radiation: The energy emitted by radioactive decay that is strong enough to strip electrons and sever chemical bonds.

isotope: One of two or more forms of the same element that have the same atomic number (number of protons) but different mass numbers (number of neutrons and protons).

L

land ethic: A principle that encourages people to extend their thinking about communities to which we should behave ethically to include soil, water, plants and animals, and collectively, the land.

land-use and land-cover change: How changes in human activity affect the use of land and vegetation, topography, and landscape characteristics.

lapse rate: The negative temperature gradient of the atmosphere; dependent on the weather, pressure, and humidity.

law of conservation of mass: Mass cannot be created or destroyed.

law of electroneutrality: The sum of all the positive ions (cations) in solution must equal the sum of all the negative ions (anions) in solution, so that the net charge of all natural waters is equal to zero.

Leadership in Energy and Environmental Design (LEED): A building rating system developed by the U.S. Green Building Council and used for commercial buildings, residential homes, and neighborhoods.

lean manufacturing: A business operations strategy that focuses on continuous quality improvement of production systems.

lethal concentration (LC): The concentration of the contaminant that results in the death of a specific percentage of organisms in a specific time period (e.g., 96-h LC_{50}).

lethal dose (LD): The amount of a chemical that kills half of the study population in a toxicity test.

life cycle assessment (LCA): The compilation and evaluation of the inputs and outputs and the potential environmental impacts of a product system throughout its life cycle.

life cycle models: Models that analyze the material and energy intensity of a product from the sourcing of raw materials through end-of-life disposal.

linear framework: *See* sequential framework.

linear materials economy: An economic model in which natural resources are extracted, processed, and engineered into feedstocks for manufacturing goods. At the end of their commercial or physical life, the goods are disposed of.

listed wastes: Wastes that are included on the F-list and the K-list; more than 50,000 chemicals are currently identified.

lithosphere: The soil crust that lies on the surface of the Earth.

Living Building Challenge: A green building certification standard developed by the Living Building Institute that specifies the actual performance characteristics of the building itself.

lower explosive limit (LEL): The concentration of an organic constituent in air that can sustain combustion without any additional fuel once a spark is present.

lowest observable effect concentration (LOEC): The lowest measured value that produces a result statistically different from the control population.

low-level radioactive waste: Naturally nonradioactive items that have become contaminated with radioactive material or have become radioactive due to neutron radiation.

M

manure: An organic (carbon-containing) nutrient source that can be applied to the land to fertilize crops and improve soil quality.

market-based incentives: An alternative to command-and-control regulation, these incentives allow private companies some degree of freedom to choose how to comply with governmental, environmental, or energy efficiency standards.

market failure: An economic situation in which the general dynamics of supply and demand do not result in the most economically efficient outcome or result in a socially suboptimal outcome.

material balance: The situation when the mass accumulated in a system is equal to the mass flowing into a system minus the mass flowing out of a system plus the mass produced minus the mass consumed.

materials flow analysis (MFA): A method for performing inventory analysis.

maximum allowable toxicant concentration (MATC): The allowable concentration that does not cause any adverse effects to aquatic life.

mean cell residence time (MCRT): *See* solids retention time.

measure: A value that is quantifiable against a standard at a point in time.

metalimnion: The layer of a water body in which there is a rapid change in temperature and density over a very small change in depth.

metals: Elemental substances; the metallic elements of the periodic table.

metric: A standardized set of measurements or data relating to one or more sustainability indicators.

microplastics: Plastic particles 5 mm or less in length, or slightly shorter than 0.25 inch.

Millennium Development Goals: United Nations development goals focused on increasing the standard of living for the world's existing population and also for future generations as the world population continues to grow.

misconsumption: The use of resources by individuals in a way that undermines their own well-being even if there are no aggregate effects on the collective.

mitigation: Human intervention to reduce the sources or enhance the sinks of greenhouse gases.

mixed-batch reactor: A reactor with no flow in or out and a signal that is mixed instantaneously.

mixed liquor: The mixture of the liquid wastewater and the microorganisms undergoing aeration.

mixed liquor suspended solids (MLSS): The particles in an activated sludge reactor.

mixed use: A description of neighborhoods that serve more than one purpose, such as a combined residential and commercial space.

mixed wastes: Radioactive waste mixed with RCRA hazardous waste.

mobility: The ability of a chemical compound to move or be moved; a parameter of increasing regulatory importance, especially when combined with persistence.

model codes: Generic building codes written by a third party following professional standards.

Montreal protocol: An international treaty to phase out and ban the use and production of CFCs and ODSs.

Moore's law: The number of integrated circuits on microchips doubles every two years, thus exponentially increasing the processing power of those chips.

morals: The values people adopt to guide the way they ought to treat one another.

morphology: The study of the structure of a lake or body of water.

multiparameter framework: *See* simultaneous framework.

mutagens: Materials that may cause genetic mutations.

N

nasopharyngeal region: The airway that includes the nose, mouth, and larynx.

National Ambient Air Quality Standards (NAAQS): Goals set to achieve reasonable air quality in all regions of the United States.

natural attenuation: A strategy of monitoring and allowing a material to degrade if there is no threat to life and if the chemical will metabolize into harmless end products.

natural capital: The world's stock of natural resources; viewed as the Earth's wealth that thus should not be overconsumed or degraded.

natural resources management: The field of study concerned with determining how to harvest or extract resources at a rate that is in balance with the rate at which the resource can regenerate or reproduce itself.

negative externality: A cost or harm experienced by someone other than the buyers and sellers of the good or service.

net zero building: A structure that produces as much energy as it consumes, resulting in no net energy consumption and a net zero carbon footprint.

neurotoxins: Substances that may cause impaired brain or nervous system function.

new approach methods (NAMs): Analytical methods that predict human toxicity without using animals.

new urbanism: A movement within the field of urban design that moves away from traditional land-use planning and single-use zoning by creating walkable communities with open space.

nitrification: The conversion of ammonia compounds to nitrate compounds.

nitrogenous oxygen demand (NOD): The amount of oxygen consumed to metabolize a nitrogen-based substrate.

no observed adverse effect level (NOAEL): The level of exposure of an organism at which there is no measurable response.

noncarbonate hardness: The difference between total hardness and carbonate hardness.

nonexceptional quality sludge: Treated wastewater solids that meet regulatory requirements (although less stringent than for EQ sludge) for specific land application purposes.

nonmarket valuation: Assigning a monetary value that is not based on the price of actually buying and selling the good or service in question.

nonrenewable resources: Resources that regenerate either extremely slowly (over hundreds or thousands of years) or not at all.

nutrient: A substance that provides energy for plant growth; nitrogen and phosphorus in water.

O

obligate aerobe: A microorganism that must dissolve oxygen to survive because it uses oxygen as the electron acceptor.

obligate anaerobe: A microorganism that must use anaerobic decomposition processes.

operating curve: The relationship between the flow rate, percent solids recovery, and cake solids concentration.

opportunity cost: An economic concept that represents what is gained or lost by choosing one option over another.

overconsumption: The situation when the level of consumption based on individual and collective choices undermines a species' own life support system.

overdesign: The situation when products are designed with an abundance of features or capacity that provides more functionality than any individual user might need.

overflow rate: The velocity of the critical particle in a clarifier.

oxygen deficit: The difference between the potential saturation oxygen concentration and the actual measured oxygen concentration in a water body.

ozone-depleting substances: Compounds that react to convert ozone (O_3) in the atmosphere to diatomic oxygen (O_2).

ozone depletion potential (ODP): A common index used to quantify the contribution of human-made substances to the depletion of the ozone layer using CFC-11 as a benchmark.

P

partial pressure: The pressure that each part of a gas exerts on its surroundings.

participatory design: Design focused on directly involving individuals who use the technology or information as co-designers and on enabling those affected by a design to drive the collaborative design process.

participatory development: A bottom-up, people-centered approach aimed at cultivating the full potential of people, especially the underprivileged, at the grassroots level.

passive solar design: A form of architecture that takes advantage of solar energy to help heat homes.

payback period: The amount of time it takes for an investment to pay for itself in terms of profit or savings.

performance standards: Specifications of the minimum energy efficiency characteristics of residential, commercial, and industrial equipment.

peri-urban areas: Living areas surrounding urban cities that are characterized by very high population densities; they lack the infrastructure to distribute energy, water, and sanitation services.

persistence: The ability of a substance to break down in various environmental media.

photochemical ozone creation potential (POCP): An index used to quantify the contribution of chemical species to the creation of tropospheric ozone.

photolysis: The decomposition of molecules into smaller units by absorbing light.

photosynthesis: The process in which plants use nutrients and carbon dioxide to convert light energy into chemical energy by building high-energy molecules of starch, sugar, proteins, fats, and vitamins.

planned obsolescence: A deliberate business strategy that forces customers to buy product upgrades or replacements so that the company can maintain or grow its future sales.

plug-flow reactor (PFR): A lengthwise tube or reactor into which a continuous flow is introduced; the flow experiences no longitudinal mixing.

plume rise: The rise of effluent gases due to momentum and buoyancy forces.

pollution: The emission of wastes and the dispersion of environmentally harmful materials.

polymers: Natural and synthetic macromolecules composed of chains of thousands of repeating, smaller molecular units called monomers.

polyphosphate-accumulating organisms (PAOs): Microorganisms in wastewater treatment that retain large amounts of phosphorus in their cell tissue.

population: The number of individual organisms present in a particular ecological system.

population density: The number of people living in a unit of area, such as a square mile or square kilometer.

potable water: Water for drinking and household use that is treated to applicable standards.

potency: A dose–response assessment; the higher the potency, the higher the risk from a chemical substance given the same dose or exposure conditions.

potential temperature gradient: The actual lapse rate plus the dry adiabatic lapse rate.

precautionary principle: When the potential harm (risk) of an innovation or action is unknown, highly uncertain scientifically, or potentially irreversible, it should not be approved for public use.

precipitate: The resulting solid form of a substance created by oversaturated concentrations in solution.

precipitation: The process in which substances (liquids and solids) are removed from a fluid (typically air or water) by gravity.

Prevention of Significant Deterioration (PSD): Standards to maintain air quality if in compliance with the NAAQS.

primary biodegradation: The change that occurs when a substance only partially degrades and is no longer recognizable as the original or parent substance—but it is not fully degraded.

primary pollutant: A substance that is emitted directly into the atmosphere, where the original chemical forms have negative impacts on people or the environment.

primary wastewater treatment processes: Processes that remove wastewater solids; typically flow equalization, coagulation, flocculation, and clarification.

producer: A plant that uses sunlight to manufacture high-energy molecules.

product simplification: A reduction in the number of necessary parts and components.

products of incomplete combustion (PICs): Organic compounds that are not completely oxidized in the combustion zone.

prokaryotes: Organisms that have the simplest cell structure; classified by their lack of a nucleus membrane that contains cellular DNA.

protozoa: Motile, usually single-celled, microscopic eukaryotes.

pseudo-first-order reaction: A second-order or higher reaction by nature, but the system has been altered to make it simulate a first-order reaction.

publicly owned treatment works (POTWs): Systems designed to reduce the concentrations of organic matter, suspended solids, nutrients, and bacteria in municipal wastewater generated by homes and businesses.

R

radiative efficiency: The radiative forcing that occurs due to a unit increase in the atmospheric abundance of a substance.

radiative forcing: The impact a component or condition within the atmosphere has on the Earth's average surface temperature.

radioactive: Emission of energy caused by the decay of isotopes.

radioactivity: The process in which isotopes decay by emitting protons, neutrons, or electromagnetic radiation to carry off energy.

radioisotopes: Isotopes that decay via radiation.

radon: A naturally occurring gas product of the decay of uranium.

rate: A measure of the flow or change of mass or volume per unit time.

raw materials: Naturally occurring resources that are extracted from the environment and processed into food and useful products for humans.

reaction rate constant: A variable that defines how the mass of a component changes with time in a reactor.

reactive nitrogen: Nr includes all forms of nitrogen except di-nitrogen gas (N_2).

reactor: A structure that contains the unit processes used to change the form (destruction) of one material and produce another material form.

reaeration: The dissolution of molecular oxygen into water.

reasonable maximum exposure (RME): The U.S. EPA linear model of a low-dose response for carcinogenic compounds to human response.

recycling: The process of separating recoverable materials from waste to process them back into their raw substances for use as feedstocks.

recycling rate: The amount of a material recovered for recycling as a proportion of the total amount of that material generated as waste.

reduction–oxidation process: A reaction in which the oxidation state of participating atoms changes.

reference concentration (RfC): Inhalation unit risk factor expressed in units of mg/m^3.

regrettable substitution: The situation in which a regulated toxic substance is replaced with another toxic substance.

rem: Roentgen equivalent man, a unit that accounts for the biological effect of absorbed nuclear radiation.

remanufacturing: Recovering the modules or components of a product for refabrication, reassembly, or reuse.

rematerialized feedstock: Substances that have been recycled at the end of life and reprocessed into usable inputs.

remedial action: Steps that may consist of removing waste material or stabilizing the site so that health problems are less likely.

removal action: A legal order that results in a hazardous material being removed and treated or safely disposed of.

removal efficiency: The percent of the pollutant mass that is removed by a control device.

renewable energy credits: Tradable certificates issued as proof that 1 megawatt-hour of electricity is generated by a clean energy source.

renewable resources: Resources that replenish themselves through a natural process, roughly within a human lifetime.

representative concentration pathways (RCPs): Climate-modeling scenarios based on specific radiative forcing changes.

reserve: The total amount of a natural resource that can be physically extracted at economically viable rates; reserves are based on highly reliable statistical estimates.

reserve-to-production ratio: The amount of time it would take to deplete current proved reserves if the current yearly amount of consumption was held constant into the future.

reservoir: A system that holds materials for any quantity of time.

residence time: The length of time that a compound remains in a defined system.

resilience: The capacity of social, economic, and environmental systems to cope with a hazardous event, trend, or disturbance, responding or reorganizing in ways that maintain their essential function, identity, and structure, while also maintaining the capacity for adaptation, learning, and transformation.

resource: A natural substance that can be physically recovered, although it may not necessarily be economically feasible to do so.

Resource Conservation and Recovery Act (RCRA): U.S. EPA regulatory statutes for solid and hazardous wastes.

respiration: The process of using oxygen and high-energy molecules to produce energy for nourishment and growth.

retention times: The lengths of time between when a material enters and exits the reactor.

return activated sludge (RAS): The portion of the sludge that is pumped from the bottom of the secondary clarifier to the aeration tank in an activated sludge system.

right-of-way: A piece of land adjacent to the building envelope that is preserved for the use of utilities and other types of infrastructure; consists of the street itself and a narrow strip of land.

riparian buffer zone: An area along a streambank where a mixed variety of plants grow and in which structures cannot be built close to the water.

risk: A measure of probability that indicates the likelihood of harm from a hazard.

risk assessment: The process of estimating the spectrum and frequency of negative impacts based on the numerical values associated with exposure for certain activities.

risk management: The process of identifying, selecting, and implementing appropriate actions, controls, or regulations to reduce risk for a group of people.

roentgen: A unit of exposure from gamma or X-ray radiation equal to a unit quantity of electrical charge produced in air.

rotating biological contactor (RBC): A biological film attached to the media in the disc that rotates in the air to maintain aerobic conditions for wastewater treatment.

runoff: Collected precipitation that flows overland.

rural-to-urban transect: An approach to urban planning that establishes form-based regulations based on the physical forms of human settlements and the physical environment.

S

safe and sustainable by design (SSbD): A design approach to improve the durability, reusability, upgradability, and reparability of materials and products, to catalyze the shift toward chemicals, materials, and products that are inherently safe and sustainable.

saturated dissolved oxygen (DO$_s$): Oxygen concentrations in equilibrium with the oxygen in the atmosphere.

scrap: Industrial waste.

secondary pollutant: A pollutant formed from the combination of other materials emitted into the environment, for example, smog formed from VOCs and NO$_x$.

secondary treatment processes: Actions designed to remove organic waste by utilizing microorganisms to degrade the organic waste and gravity to remove the microorganisms.

second-order reaction: A reaction in which the rate of reaction is proportional to the square of the amount of material.

selective catalytic reduction: A process that uses noble metals, base metals, or a zeolite to remove organic substances at lower temperatures than thermal oxidizers.

sensitizers: Substances or materials that may cause hypersensitivity of the skin and/or respiratory tract.

septic tank: A partially aerobic and partially anaerobic treatment system often used in decentralized wastewater treatment.

sequential framework: An arrangement in which each alternative is compared against the criteria for each parameter (module) in a sequential way.

setback: A required distance between a structure and an adjacent feature such as a property line, street, another structure, or a stream.

settling chamber: A horizontal flow chamber used to reduce the air velocity causing the gravitational removal of particles.

shared socioeconomic pathways (SSPs): Five plausible pathways to create model conditions that produce the radiative forcing conditions of the RCP values.

sievert: Unit of an absorbed radiation dose that does the same amount of biological damage to tissue as 1 gray of gamma radiation or X-ray; 1 Sv is numerically equal to 100 rem.

simultaneous framework: An algorithmic decision process that includes all possible options.

single-stream recycling: A process in which all recyclable items are combined into a single waste stream and separated at an industrial facility.

single-use zoning: Zoning that allows for only one type of land use in a designated area.

sludge volume index (SVI): A measure of the ability of sludge to settle.

smart growth: A plan that includes a set of land-use development and urban design principles intended to curb sprawl, conserve natural resources, and create more livable cities.

social and environmental accounting: Business accounting techniques that calculate how business activity benefits people and the environment.

social costs: The negative effects of a condition experienced by society as a whole.

social justice: The fair distribution (or sharing) of the advantages and disadvantages (or benefits and burdens) that exist within society.

social movement: An effort to institutionalize new ways of thinking in a society after they are broadly advocated by groups that have political and social power.

solids retention time (SRT): The average time solids stay in a reactor; longer than the hydraulic retention time when the solids are recycled.

solubility: The property of a substance to dissolve into solution in a solvent.

sorbate: The substance that is transferred from one phase to another.

sorbent: The material into or onto which the sorbate is transferred.

source reduction: Measures designed to reduce the amount of natural resources used in the creation of a product such that the resulting scrap and harmful waste products are minimized.

Special Report on Emissions Scenarios (SRES): A report on climate-modeling scenarios based on year 2000 population

projections; demographics; trade; flow of information and technology; and other plausible social, technological, and economic characteristics.

specific heat: The quantity of heat required to increase the unit mass of the substance by 1 degree; with units of energy/(mass-temperature).

specific height: The physical height of an exhaust stack.

specific target organ toxicants: Substances or materials that may cause disorders of the pulmonary, cardiovascular, skeletal, and/or immune systems.

state implementation plans (SIPs): Plans to improve air quality if the NAAQS goals are exceeded in a region.

steady state: A condition that does not change with respect to time.

Stokes' settling law: A mathematical equation used to estimate particle removal as a function of the runoff flow velocity and expected residence time of water in a sedimentation removal device.

stratospheric ozone depletion: Ozone destruction in the stratosphere caused by anthropogenic emissions.

stream daylighting: Restoring a buried stream in an urban area to natural conditions.

substrate: A food or energy source for microorganisms.

substrate loading: The amount of organic matter (food) added relative to the microorganisms available.

substrate utilization rate: The rate of energy conservation from the substrate to cell maintenance and growth.

suspended solids: Materials that are floating or suspended in the water.

sustainability: Everything that we need for our survival and well-being depends, either directly or indirectly, on our natural environment. Sustainability creates and maintains the conditions under which humans and nature can exist in productive harmony and that permit fulfilling the social, economic, and other requirements of present and future generations. Sustainability is important to making sure that we have and will continue to have the water, materials, and resources to protect human health and our environment.

sustainability index: A numerical scale used to compare alternative designs or processes with one another.

sustainability science: An emerging discipline that studies how human activity affects Earth's mass balancing, energy flows, and ecosystems.

sustainable design: The design of products, processes, or systems that balance our beliefs in the sanctity of human life and promote an enabling environment for people to enjoy long, healthy, and creative lives, while protecting and preserving natural resources for both their intrinsic value and the natural world's value to humankind.

sustainable development: Development that reflects the desire to improve the worldwide standard of living while considering the effects of economic development on natural resources.

switch: The portion of a device that separates a material according to the code.

T

take back program: A program in which a product is returned to the manufacturer at the end of its life for reprocessing, refurbishing, recycling, or final disposal.

technical nutrients: Materials that are highly stable and can be reused within a closed-loop manufacturing cycle.

teratogens: Substances that may cause physical or functional defects in an embryo.

theoretical oxygen demand (ThOD): The theoretical amount of oxygen consumed to oxidize a substrate to carbon dioxide.

thermal oxidation: The process of using high temperatures to oxidize organic pollutants.

thermal stratification: The arrangement of layers in a body of water created by water density and temperature differences.

thermocline: The layer in a large body of fluid in which there is a rapid change in temperature and density over a very small change in depth.

thermoplastics: Plastic materials characterized by chemical bonding that enables them to melt when heat and pressure are applied.

thermosets: Plastics that are irreversibly hardened in the polymerization process and do not burn or melt when exposed to heat.

three pillars of sustainable development: The environment, society, and the economy.

total equivalent CO_2: The sum of the global warming potential associated with each molecule in a system of study.

toxicity: The quality of being toxic or poisonous.

toxicity characteristic leaching procedure (TCLP): An extraction procedure in which solidified waste is crushed, mixed with weak acetic acid, and shaken for a number of hours to identify wastes that are likely to leach concentrations of contaminants that may be harmful to human health or the environment.

toxicology: The study of the adverse effects of chemicals, substances, or situations on biological systems.

tracer: A chemical, dye (color), or radioactive element (with a long half-life) used to trace the flow of materials in a system or reactor.

tracheobronchial region: The portion of the respiratory system characterized by the branching of the airway bronchi through the lungs, conveying the air to the pulmonary region.

traditional biomass: Wood, charcoal, peat, dung, and grass—basic fuels for heat.

transformation: A change in the fundamental attributes of natural and human systems.

transparency: A characteristic of agencies being open about their decision-making processes and disclosing risks to the public.

transpiration: The process that occurs when water is conveyed from living plant tissue, especially leaves, to the atmosphere.

trickling filter: A water treatment system in which an active biological film forms on the surface of the media and the biological organisms use the organic material in the wastewater dripping over the media as their food (or energy) source.

triple bottom line: An accounting framework that evaluates business outcomes in terms of "people, planet, and profit" benefiting society simultaneously.

trophic level: The productivity of a body of water determined by the organic matter (usually represented by chlorophyll) or the nutrient concentration in the water.

troposphere: The region of the atmosphere in which humans breathe the air from the Earth's surface; extending to approximately 17 kilometers above the surface of the Earth.

tropospheric ozone: Ozone formed by the chemical reactions between nitric oxides (NO_x), volatile organic compounds (VOCs), and sunlight.

U

ultimate biodegradation: The level of degradation at which a substance fully breaks down to become carbon dioxide, water, mineral salts, and new biomass for the organisms that consume it.

Uncertainty factor (UF): Factor that takes into account variability and uncertainty that are reflected in differences between animal and human toxicity data and variations in human populations.

unconfined aquifer: An aquifer that is underlain by an impervious stratum.

United Nations Framework Convention on Climate Change (UNFCCC): The main multilateral forum that focuses on addressing climate change.

universal soil loss equation: An equation used to estimate soil loss through erosion.

urban agriculture: A type of infill land use in which small farms and community gardens are interspersed in a city's neighborhoods.

urban growth boundary: A zoning designation that legally limits the geographic spread of growth and new development.

urban infill: Abandoned, derelict, unused, or underutilized space in urban areas that is redeveloped for more productive use.

urban intensification: A form of development that seeks to create higher levels of housing density in combination with mixed-use zoning, greater availability of green space and natural areas, and enhanced mass transit.

urban sprawl: Low-density housing that spreads along major streets and arterial roadways.

V

vadose zone: The zone of aeration near the surface of the Earth where the soil pore spaces contain both air and water.

vegetated swales: Areas that collect rainwater runoff planted heavily with vegetation; these areas allow rainwater to percolate naturally through the soil as groundwater.

venturi scrubber: Pollution control equipment that separates particles by combining the increasing angular momentum of the particles and interception by water droplets.

viruses: Intracellular parasites composed of a nucleic acid core with an outer protein coating.

vitrification: A process that turns waste into a solid glass form that is stable and nearly impervious to the environment.

volatile fatty acids (VFAs): Partial degradation products of wastewater constituents.

volatile suspended solids (VSS): A measure determined by the weight of any particles remaining on a filter that evaporate after the filter is heated to 550°C.

volumetric flow rate: The rate of a given volume of fluid flow in a system.

voluntary risk: A risk individuals take of their own free will.

vulnerability: The propensity or predisposition to be adversely affected.

W

waste: A byproduct of a complex system that has no useful purpose and must therefore be disposed of; waste is usually considered a uniquely human concept, as ecological systems do not generate unused or unusable waste.

waste activated sludge (WAS): The portion of the settling solids that are collected for disposal (and not recycled) in an activated sludge system.

waste hierarchy: A ranking of the most preferred to least preferred strategies for waste management.

waste management: The collection, transport, processing, disposal, and monitoring of waste materials in a manner that minimizes the impacts of waste on human health and the environment.

waste-to-energy: Waste-to-energy facilities typically burn solid waste to generate steam or electricity.

water budget: The sum of all the inputs and outputs into a watershed.

water demand management: Practices that consider the total quantity of water abstracted from a source of supply using measures to control waste and excessive consumption.

water hardness: The sum of the concentrations of the divalent cations (species with a charge of 2+) in water.

water–energy nexus: The interdependence between water resource use and energy production; reflects generally the need for water to generate electricity as well as the energy intensity of water supply systems, domestic water use, and wastewater treatment.

Water Poverty Index (WPI): A measure used to understand the relationship between poverty and five water-related criteria: resources, access, capacity, use, and environment.

water table: The surface of the zone of saturation.

watershed: The region that collects rainfall.

wet scrubbing: Removing acid gases with an aqueous base solution.

Z

zero-order reaction: A reaction in which the reaction rate is constant in a system and is not a function of the amount of material.

zero waste: A waste management concept that encourages the elimination of waste through complete recovery and recycling of materials in the waste stream.

zone of aeration: *See* vadose zone.

zone of saturation: The groundwater zone in which the void spaces are filled with water.

zoning ordinances: Rules and regulations that govern the allowed uses for buildings or dwellings in an area.

Index